ISBN 978-1-5277-2235-4
PIBN 10884398

1 MONTH OF
FREE
READING

at

www.ForgottenBooks.com

By purchasing this book you are
eligible for one month membership to
ForgottenBooks.com, giving you
unlimited access to our entire
collection of over 1,000,000 titles via
our web site and mobile apps.

To claim your free month visit:

www.forgottenbooks.com/free884398

English
Français
Deutsche
Italiano
Español
Português

www.forgottenbooks.com

Mythology Photography **Fiction**
Fishing Christianity **Art** Cooking
Essays Buddhism Freemasonry
Medicine **Biology** Music **Ancient**
Egypt Evolution Carpentry Physics
Dance Geology **Mathematics** Fitness
Shakespeare **Folklore** Yoga Marketing
Confidence Immortality Biographies
Poetry **Psychology** Witchcraft
Electronics Chemistry History **Law**
Accounting **Philosophy** Anthropology
Alchemy Drama Quantum Mechanics
Atheism Sexual Health **Ancient History**
Entrepreneurship Languages Sport
Paleontology Needlework Islam
Metaphysics Investment Archaeology
Parenting Statistics Criminology
Motivational

TEXT-BOOK

OF

HUMAN PHYSIOLOGY

INCLUDING

HISTOLOGY AND MICROSCOPICAL ANATOMY

WITH ESPECIAL REFERENCE

TO THE

PRACTICE OF MEDICINE

BY

DR. L. LANDOIS

PROFESSOR OF PHYSIOLOGY AND DIRECTOR OF THE PHYSIOLOGICAL INSTITUTE IN THE
UNIVERSITY OF GREIFSWALD

TENTH REVISED AND ENLARGED EDITION

EDITED BY

ALBERT P. BRUBAKER, M.D.

PROFESSOR OF PHYSIOLOGY AND HYGIENE IN THE JEFFERSON MEDICAL COLLEGE; PROFESSOR OF PHYSIOLOGY
IN THE PENNSYLVANIA COLLEGE OF DENTAL SURGERY; LECTURER ON PHYSIOLOGY AND HYGIENE
IN THE DREXEL INSTITUTE OF ART, SCIENCE AND INDUSTRY, PHILADELPHIA

TRANSLATED BY

AUGUSTUS A. ESHNER, M.D.

PROFESSOR OF CLINICAL MEDICINE IN THE PHILADELPHIA POLYCLINIC; PHYSICIAN TO THE PHILADELPHIA
HOSPITAL; ASSISTANT PHYSICIAN TO THE PHILADELPHIA ORTHOPEDIC HOSPITAL
AND INFIRMARY FOR NERVOUS DISEASES

With 394 Illustrations

PHILADELPHIA

P. BLAKISTON'S SON & CO.

1012 WALNUT STREET

1904

PRESS OF
WM. F. FELL COMPANY
PHILADELPHIA

PUBLISHERS' PREFACE.

The fourth English edition of Professor Stirling's translation of Landois' "Physiology," published in 1891, has been out of print for some years. Since the date of publication of this English edition, the work has passed through three more large editions in Germany. On each occasion Professor Landois still further enhanced its merits by incorporating all those results of physiological investigation which in his judgment would have a permanent value not only for advanced students but for practitioners of medicine as well; and hence there is probably no work which so thoroughly and satisfactorily represents the existing state of physiological science and its relations to pathology and clinical medicine, as that of Professor Landois.

For the reason that pathological processes are but variations of physiological processes in one direction or another, there is appended to almost every section those variations which are regarded by the clinician as pathological. In this way the student is made to realize not only the close interdependence of physiology and pathology, but the necessity for a thorough and accurate knowledge of the former for an intelligent comprehension of the latter. The work of Landois thus becomes a guide which conducts the student from the physiological laboratory to the work of the clinician. That it has been successful in this respect, and that it meets the needs of students and practitioners of medicine, is the only explanation that can be offered for the extraordinary fact that it has passed through ten large editions in Germany in twenty years, and, in addition, has been translated into French, Russian, English, Italian, and Spanish.

The continued success of each successive edition in Germany and the frequent requests for an English edition have convinced the publishers that a new translation would be acceptable to students and practitioners of the present day and decided them to issue the work in its present form. The translation was intrusted to Dr. A. A. Eshner, Professor of Clinical Medicine in the Philadelphia Polyclinic. In this he was ably assisted by Drs. Bernard Kohn, E. A. Shumway, Maurice Ostheimer, R. Max Goepp, Brooke M. Anspach, C. A. Fife, D. J. McCarthy, and W. B. Stanton. The text has been revised and edited by Dr. Albert P. Brubaker, Professor of Physiology in the Jefferson Medical College. The proof has been largely read and the index prepared by Dr. Colin C. Stewart, Assistant Professor of Physiology in Dartmouth College.

The publishers wish to express their appreciation of the conscientious care which has been given by each and all to the preparation of this edition.

PREFACE TO THE TENTH EDITION.

In spite of the short interval that has elapsed between the appearance of the ninth and that of this new edition all sections of the book have been subjected to extensive revision, with the inclusion of the results of the most recent investigations. The number of illustrations has been increased and some have been replaced by better ones.

Since the book has been placed in the hands of students and physicians in ten large editions, as well as in several English editions, a Russian translation in a second edition, an Italian, a French, and a Spanish translation, I have become more firmly established in the conviction that the plan according to which I have labored, both as author of this book and as teacher, is the correct one. Physiology is the foundation of internal medicine, and it should, therefore, be so taught that the physician can continue to build upon it and find support in it. This has been my endeavor; in this sense the book has been revised uniformly throughout in this edition.

L. LANDOIS.

GREIFSWALD.

FOREWORD.

TENDENCY AND PURPOSE OF THE WORK.

In the preparation of the forelying concise Textbook of Physiology the author has been governed by an endeavor to provide for physicians and students a book that should supply the needs of the practising physician in larger measure than is done by the majority of similar works. With this end in view a brief outline of pathological variations is appended in every section to the description of the normal processes. This is done for the purpose of directing the attention of the student from the outset to the field of his future professional activity and of pointing out the extent to which the morbid process represents a derangement of the normal. On the other hand, opportunity is by this means afforded the practising physician to renew acquaintance readily with the theoretical doctrines that as a rule slip away from him all too soon in the pursuit of his vocation. Here he can without effort look back from the morbid phenomena under treatment to the normal processes and in the recognition of these obtain new suggestions for correct interpretation and treatment. From this standpoint the author has described fully all those methods of investigation that may be employed by the practitioner with great advantage and that as a rule are but briefly treated in books on physiology. Reference may be made here to the following sections: Blood-examination; graphic study of the normal and abnormal heart-beat; heart-sounds and heart-murmurs; the pulse; the venous pulse; transfusion; normal and abnormal respiratory murmurs; ventilation; examination of the air in dwellings; the sputum; deviations from the normal digestive processes; diabetes; cholemia; the digestion in febrile patients; thermometry and calorimetry in the febrile state; examination of drinking-water; meat and meat-preparations; excessive deposition of fat and muscle, and the means for its relief; examination of the normal urine and the determination of all pathological constituents, as well as of urinary concretions; uremia, ammoniemia, uric-acid diathesis; morbid disturbances in retention and evacuation of urine; pathological alterations in the sudoriferous and sebaceous secretions; galvanic conductivity through the skin; gymnastics and therapeutic gymnastics; pathological alterations in the motor functions; laryngoscopy and rhinoscopy; pathology of phonation and articulation; physiological principles underlying the therapeutic application of electricity; constant currents and electrical apparatus.

In the consideration of every individual nerve and the different nerve-centers a sketch of the pathological manifestations is added. With relation to the nerve-centers the derangements of the reflexes, those of conduction to the central organs, those of the respiratory center, together with the means for resuscitating asphyxiated persons, and the group of angioneuroses, have received especial consideration. Particular importance has been attached to the physiological topography of the surface of the cerebrum in man with reference to modern investigation into the localization of the functions of the brain. The same principle has been followed also with relation to the physiology of the organs of special sense. Evidence of this will be found in the discussions of abnormalities of ocular refraction, the use of spectacles, ophthalmoscopy, the orthoscope, color-blindness and its practical significance, further investigations into the functions of the other special senses and their principal disorders. The embryological section has given especial consideration to the subject of developmental defects, and to malformations as the most important of these; and also to the means for determining the period of development reached by human embryos.

In description it was the aim of the author to be as concise and comprehensive as possible. Elaborate discussions have been scrupulously avoided. At the same time the typography has been so arranged that the more important and purely physiological matters are presented in conspicuous type. Also, the beginner can without disadvantage pass over the pathologic-physiological sections; the student during the period of clinical instruction will, however, with advantage review the field of normal physiology from the latter.

The author has, further, considered it advisable to add to each physiological section a brief outline of the historical development of the subject in hand, and likewise a summary of the comparative physiology of the animal kingdom. Finally, the histology and microscopic anatomy have been more fully considered in each section than is the case with most textbooks of physiology.

On the basis of the plan thus outlined the appearance of the fore-lying work is I believe justified. That this plan has not been fallacious is indicated by the numerous discussions in the medical journals of North and South Germany, Austria, Switzerland, Hungary, Russia, France, England, Italy, Scandinavia, America, which have received the book with favor, and recognition. The author, however, is particularly gratified that the book has been received with approval by physiologists. In order to dispel any anxiety on the part of those who perhaps may fear that the scientific eminence of our science, of fundamental importance in the entire domain of medicine, may suffer from the attempted association of physiology with the practical department of medicine, I shall quote a few words from a letter written by one of our most illustrious and most versatile physiologists:

"Should anyone publish a handbook like that of yours, of which the first half is before me, he will be entitled to the thanks not only of the students, but also of the teacher and investigator. And as it is my ambition to combine in myself the three qualities indicated, my thanks are tendered you with all my heart. Your pathological descriptions are in their condensed brevity so masterfully clear that I promise myself from your book a most beneficial action and reaction upon the field of clinical medicine." . . . . .

If these words have been realized I should find in this fact a perfect reward for my endeavors. It has always appeared to me in my academic activity as a teacher that my principal aim must lie in the thorough preparation of physicians for physiological thought. And if to this, my aim, there be apposed the statement of prouder sound, "we make physiologists," this would not deflect me from my course as a teacher, of which I believe, in the words of the master Herophilus: ἔστω ταῦτω εἶναι πρῶτα εἰ χαι μή ἐστι πρῶτα.

L. LANDOIS.

TABLE OF CONTENTS.

INTRODUCTION.

PHYSIOLOGY OF THE BLOOD.

THE GASES OF THE BLOOD.

PHYSIOLOGY OF RESPIRATION.

PHYSIOLOGY OF DIGESTION.

PHYSIOLOGY OF ABSORPTION.

PHYSIOLOGY OF ANIMAL HEAT.

PHYSIOLOGY OF METABOLISM.

SYNOPSIS OF THE MOST IMPORTANT SUBSTANCES USED AS FOOD.

PHENOMENA AND LAWS OF METABOLISM.

SUMMARY OF THE CHEMICAL CONSTITUENTS OF THE ORGANISM.

THE SECRETION OF URINE.

THE ORGANIC CONSTITUENTS OF THE URINE.

THE INORGANIC CONSTITUENTS OF THE URINE.

FUNCTIONS OF THE EXTERNAL INTEGUMENT.

PHYSIOLOGY OF THE MOTOR APPARATUS.

SPECIAL MOVEMENTS.

VOICE AND SPEECH.

GENERAL PHYSIOLOGY OF THE NERVOUS SYSTEM AND ELECTRO-PHYSIOLOGY.

ELECTRO-PHYSIOLOGY.

PHYSIOLOGY OF THE PERIPHERAL NERVES.

THE CEREBRAL NERVES.

PHYSIOLOGY OF THE NERVOUS CENTERS.

THE SPINAL CORD.

THE BRAIN.

PHYSIOLOGY OF THE ORGANS OF SPECIAL SENSE.

THE VISUAL APPARATUS.

THE AUDITORY APPARATUS.

THE OLFACTORY APPARATUS.

LIST OF ILLUSTRATIONS.

INTRODUCTION.

THE SCOPE AND AIM OF PHYSIOLOGY AND ITS RELATION TO ALLIED BRANCHES OF PHYSICAL SCIENCE.

Physiology is the science of the vital phenomena of organs, or, briefly, the study of life. In accordance with the classification of organisms the following divisions are made, namely, Animal Physiology, Vegetable Physiology, and the Physiology of the Lowest Forms of Life, which occupy the boundary between animals and plants, the protists, microorganisms or microbes, and the elementary organisms or cells occupying the same plane. It is the aim of physiology to establish these phenomena, to determine their regularity and their causes, and to correlate these with the general fundamental laws of natural science, especially those of physics and chemistry. The relation of physiology to allied branches of natural science is shown in the following scheme:

BIOLOGY,
The science of organized beings or organisms (animals, plants, protists, and elementary organisms).

MORPHOLOGY.
The study of the form of organisms.

General Morphology.	*Special Morphology.*
The study of the formed elementary constituents of organisms (*Histology*):	The study of the parts and organs of organisms (Organology, Anatomy):
(*a*) Histology of plants.	(*a*) *Phytotomy.*
(*b*) Histology of animals.	(*b*) *Zootomy.*

PHYSIOLOGY.
The study of the vital phenomena of organisms.

General Physiology.	*Special Physiology.*
The study of vital phenomena in general:	The study of the functions of individual organs:
(*a*) Of plants.	(*a*) Of plants.
(*b*) Of animals.	(*b*) Of animals.

EMBRYOLOGY.
The study of the generation and development of organisms.

Morphologic division of the study of development, that is, the study of the conformation at different stages of development:	1. Developmental history of the individual being (for instance, man) from its germ, germinal history (*Ontogeny*):	Physiologic division of the study of development, that is, the study of functional activity during development:
(*a*) General.	(*a*) In plants.	(*a*) General.
(*b*) Special.	(*b*) In animals.	(*b*) Special.
	2. Developmental history of entire species of organisms, from the lowest forms of creation upward, family history (*Phylogeny*):	
	(*a*) In plants.	
	(*b*) In animals.	

17

If it be desired to give a special position in the system of organisms to those beings that occupy the lowest plane of development and that, representing to a certain degree the prototype in the family history, have as yet not been differentiated into animal and vegetable, the so-called protists (Haeckel), these likewise would occupy a distinct place in the foregoing arrangement by the side of animals and plants.

Morphology and physiology are coördinate branches of biology. A knowledge of morphology is a prerequisite for the comprehension of physiology, inasmuch as the functions of an organ can be correctly understood only if its external form and its internal structure are previously known. The developmental history occupies an intermediate position between morphology and physiology. It is a department of morphology in so far as it has to do with a description of the parts of the developing organism; it is a physiologic study in so far as it investigates the functions and vital phenomena during the period of development of the organism. In all the branches of biologic science it is necessary to enter upon a consideration of physical and chemical principles.

MATTER.

The entire visible world, including all organisms, consists of matter, that is, of the material or substance that occupies space. A distinction is made between *ponderable* matter (in ordinary language often designated simply matter), which can be weighed upon the scales; and *imponderable* matter, which cannot be weighed upon the scales. The latter is designated *ether* (also luminiferous ether or light-ether). Ponderable matter or bodies possess *form* (or shape), that is, the outline of their limiting surfaces; also *volume*, that is, the amount of space they occupy; and finally an *aggregate condition*, which takes a solid, liquid, or gaseous form.

The ether fills the space of the universe, at any rate, with certainty to the most remote visible stars. This light-ether, notwithstanding its imponderability, possesses quite definite mechanical properties. It is infinitely more attenuated than any other known form of gas, and nevertheless its behavior corresponds rather with that of a solid body than with that of a gas. It more nearly resembles a gelatinous mass than air. It takes part in the vibrations of the atoms of the most distant stars associated with the luminous phenomena of the latter, and it is thus the carrier of light, which through its vibrations it conducts to the visual apparatus with inconceivable rapidity (300,000 kilometers in the second).

Imponderable matter (ether) and ponderable matter (substance) are not sharply delimited from each other; on the contrary, the ether penetrates the interstices present in the smallest particles of ponderable matter.

If ponderable matter be conceived to be divided into gradually smaller and smaller parts, in the process of progressive subdivision parts would eventually be reached whose aggregate condition would still be recognizable. These are designated *particles*. Particles of iron would still be recognized as solid, those of water as fluid, and those of oxygen as gaseous. If it be conceived that the process of division of the particles be carried to a further degree, a point will finally be reached beyond which further division cannot be effected either by mechanical or by physical means. In this way the *molecule* is obtained. A molecule, accordingly, is the smallest portion of a body that is capable of existence in a free state, and that, further, as a unit no longer exhibits the aggregate condition. The molecule is, however, not the ultimate unit of the body. On the contrary, every molecule consists of a collection of the smallest units, which are known as *atoms*. An atom is incapable of occurring alone in a free state, but atoms unite with other atoms of the same or of different character to form atom-complexes, designated molecules. Atoms are unconditionally insusceptible of division; whence the name. Atoms, further, are conceived to be of constant size and solid in themselves. From the chemical standpoint the atom of an elementary body (element) is the smallest amount of an element that is capable of entering into chemical combination. Just as ponderable matter consists in its ultimate parts of ponderable atoms, so also does the ether, imponderable matter, consist of analogous particles of smallest size, namely, ether-atoms.

Within ponderable matter the ponderable atoms are arranged in quite a definite order with relation to the ether-atoms. The ponderable atoms are drawn mutually toward one another (attraction); the ponderable atoms likewise attract the imponderable atoms; but the ether-atoms mutually repel one another. It thus comes about that in the ponderable mass ether-atoms are collected about every ponderable atom. These collections, designated "dynamids" by Redtenbacher, tend, in accordance with the powers of attraction of the ponderable atoms, to approach one another, but only so far as permitted thus to do by the repellent power of the surrounding ether-atoms. Therefore the ponderable atoms can never cohere without interstices, but the entire mass of matter must be considered as loose in texture in consequence of the interposed ether-atoms, which prevent immediate contact between ponderable atoms.

The aggregate condition of the body depends therefore upon the mutual arrangement of the molecules (namely, those small particles of matter that may still occur isolated in a free state).

Within solid bodies, which are characterized by constancy of volume, as well as independence of form, the molecules are arranged in a fixed and unchangeable relation with one another. In fluid bodies, which are characterized by constancy of volume, though by variability of form, the molecules are in constant movement, just as in a mass of moving worms or insects the individual animals are incessantly changing their place with relation to one another. If this movement of the molecules attains such proportions that the individual molecules scatter in all directions (just as the moving collection of insects separates into its constituent parts), the body becomes gaseous, and is characterized in this form both by its inconstancy of form and its variability in volume. The study of molecules and their motor phenomena is the part of physics.

FORCES.

Gravitation; Work of a Force.—All phenomena appertain to matter. They are the appreciable expression of the forces inherent in matter. The forces themselves are not appreciable; they are the causes of the phenomena. The first of the forces to be considered is gravitation. According to the law of gravitation every particle of ponderable matter in the universe attracts every other particle with a certain degree of force. This force diminishes inversely as the square of the distance between the two bodies. The power of attraction is further directly proportional to the quantity of the attracting matter, though without any relation to the quality of the body. The intensity of the force of gravitation can be measured by the extent of the movement that it communicates to a freely falling body previously supported in a vacuum but deprived of its support. This figure is 9.809, because the force of gravity operating for one second upon the freely falling body imparts to this a velocity of 9.809 meters.

The final velocity of the freely falling body at the end of the first second (determined experimentally) is designated thus, g = 9.809 meters. The velocity, v, of the freely falling body is in general proportional to the time, t, occupied in falling.
Therefore $v = gt$ ......(1), that is, at the end of the first second $v = g$, 1 = g = 9.809 meters.

The distance through which the body falls, $s = \dfrac{g}{2}t^2$ ......(2); that is, the distance through which a body falls is as the square of the time occupied in falling.
From (1) and (2) there follows (by eliminating t) $v = \sqrt{2gs}$ ......(3).
The velocity is as the square root of the distance traversed in falling.

$$\text{thus } \frac{v^2}{2g} = s \ \ldots\ldots\ldots\ldots\ldots (4)$$

A freely falling body, and also in general every mass in movement, possesses kinetic energy (actual energy); it is to a certain degree a repository of force. The kinetic energy of a body in movement is always equal to the product of its weight (determinable by scales) and the height to which it would rise from earth if it were raised from the earth with the velocity peculiar to it.

If the kinetic energy of the moving body be designated W and its weight P, then $W = P$, s; then, from (4), $W = P\dfrac{v^2}{2g}$ (5).

The kinetic energy of a body is therefore proportional to the square of its velocity.

If an accelerating force operating on a body (pressure, traction, or tension) drives it for some distance in the direction of its activity, the force thus expends work. This is equal to the product that is obtained if the amount of pressure or traction that propels the body is multiplied by the length of the path traversed.

If K represents the pressure or the traction with which the force operates upon the body and S the path, then the work $A = KS$. In the same way the attraction between the earth and a body raised above it (as, for instance, a ram) is a source of work.

It is customary to express the value of K in kilograms, but, on the other hand, that of S in meters. Accordingly the unit of work is the kilogrammeter (according to some the grammeter), that is, the force that is capable of raising 1 kilo (according to some 1 gram) to the height of 1 meter.

Potential Energy.—*Transformation of Potential Energy into Kinetic Energy, and the Reverse.*—In addition to the kinetic energy referred to, bodies may possess also mechanical potential energy. By this designation is understood an aggregation of forces that are still inhibited in their free evolution, and that, further, are causes of movement, without themselves being movement. The wound clock-spring prevented from unwinding by a catch, the stone resting upon the cornice of a tower, are illustrations of bodies possessing potential energy. Only an impulse is required to evolve actual from potential energy or to convert the potential into kinetic energy. The stone resting upon the cornice of the tower was raised to that place by means of work (A).

A = p, s, p representing the weight and s the height. p = m, g, thus the equivalent of the product of the mass (m) and the force of gravity (g); therefore A = m, g, s.

This is at the same time the expression for the potential energy residing within the stone. This elastic energy may readily be converted into kinetic energy by causing the stone to fall from the edge of the tower by means of a slight push. The actual energy of the stone is equal to the terminal velocity with which it reaches the ground.

$$v = \sqrt{2gs} \text{ (see 3)}.$$
$$v^2 = 2gs$$
$$mv^2 = 2mgs$$
$$\frac{m}{2}v^2 = mgs$$

m, g, s represents the potential energy residing within the stone at rest in its elevated position; $\frac{m}{2}v_2$ is thus the kinetic energy corresponding to this potential energy.

Actual energy and mechanical potential energy can be transformed into each other under most varied conditions; they can also be conveyed from one body to another.

Of the first statement the movement of a pendulum furnishes a striking illustration. The pendulum-bob, located at the highest point of the excursion, and which must be considered to be in a state of absolute rest at this point for a moment, is, exactly as the resting stone in the previous illustration, provided with potential energy. In the free movement that now takes place this potential energy is, converted into kinetic energy, which is greatest when the bob with greatest movement is in the vertical plane. Rising again from this point, the kinetic energy, with diminution in the free movement, is transformed into potential energy, which again attains its maximum at the resting-point at the height of the excursion. In the absence of constantly operating resistances (resistance of the air, friction) this play of the alternate transformation of kinetic energy into potential energy and the reverse taking place in the pendulum would continue uninterruptedly (as in a mathematical pendulum). If it be conceived that the swinging pendulum-bob encounters exactly in the vertical plane a movable body resting at this point, such as a sphere, then (assuming perfect elasticity on the part of the pendulum-bob and the sphere) the kinetic energy of the pendulum-bob would be transmitted directly to the sphere: The pendulum would come to rest, while the sphere would continue in movement with equal kinetic energy (again providing there is no resistance). This is an instance of the transmission of kinetic energy from one body to another. Finally it may be conceived that a coiled clock-spring in unwinding causes another to become coiled. This would be an instance of the transmission of potential energy from one body to another.

From the illustrations given the general proposition may be deduced: If in a system the individual moving masses approach a final condition of equilibrium, the sum of the kinetic energies in the system will be increased; and if the particles are removed from the final condition of equilibrium, then the sum of the potential energies is increased at the expense of the kinetic energies; that is, the kinetic energies diminish.

The pendulum approaching the vertical plane (the position of equilibrium for a 'resting pendulum) from the highest point of its excursion possesses in this position the greatest amount of kinetic energy; and ascending to the highest point of its excursion on the other side it attains, at the expense of the progressively diminishing movement and thereby also the kinetic energy, again gradually the maximum of potential energy.

Heat : Its Relation to Kinetic Energy and to Potential Energy.—If a leaden weight be thrown from the summit of a tower to the earth and there encounter an unyielding surface, its movement in mass will come to rest, but the kinetic energy, which to the eye appears dissipated, is transformed into an actively vibratory movement of the atoms. On striking the ground heat is generated, the amount of which is proportionate to the kinetic energy that is transformed by the impact. At the moment of contact on the part of the falling weight the atoms are set into vibration by the concussion. They impinge upon one another and then rebound in consequence of the potential energy that tends to prevent their immediate apposition; they separate to a maximum degree in so far as the power of attraction of the ponderable atoms permits and they oscillate to and fro in this manner. All atoms oscillate like a pendulum until their movement is transmitted to all the surrounding ether-atoms, that is, until the heat of the heated mass is radiated. Heat is a vibratory movement of the atoms. As the amount of heat generated is proportionate to the kinetic energy that is transformed by the impact, it must be possible to find an adequate measure for both forms of force.

The heat-unit (calory), that is, the energy that raises the temperature of 1 gram of water 1° C., serves as the measure of the amount of heat. This heat-unit corresponds to 425.5 grammeters; that is, the same amount of energy that raises the temperature of 1 gram of water 1° C. is capable of raising a weight of 425.5 grams to a height of 1 meter; or, a weight of 425.5 grams falling from the height of 1 meter would in its impact generate so much heat as would raise the temperature of 1 gram of water 1° C. The mechanical equivalent of the heat-unit is therefore 425.5 grammeters.

It is evident that from the impact of masses in motion an amount of heat of immeasurable degree may be generated. If this statement be applied to the planets, their impact would result in the production of an amount of heat greater than could be generated by any form of earthly combustion. If the earth were suddenly checked in its course and if through the force of gravitation it plunged into the sun [in the course of which it would eventually have acquired a terminal velocity of 630.7 kilometers in a second] an amount of heat would be generated in consequence of the collision equivalent to that produced by the combustion of more than 5000 equally heavy masses of pure carbon. In this manner the demonstration can be made scientifically, that even the sun's heat may have been produced by the impact of cold matter. If the cold matter of the universe were thrown into space, and there left to the attraction of its particles, the impact of these masses would eventually extinguish the light of the stars. In the same way numerous cosmic bodies still collide in space, and innumerable meteors constantly plunge into the sun (from 9400 to 188,000 billions of kilos in each minute). Thus, the action of the force of gravitation is in fact perhaps the exclusive origin of all heat. The following is an instance of the transformation of kinetic energy into heat: The smith makes a piece of iron hot by hammering. The following is an instance of the transformation of heat into kinetic energy: The hot steam of the steam-engine causes the piston to rise. The following is an illustration of the transformation of potential energy into heat: The unwinding of a coiled metallic spring, rubbing upon a rough surface, produces heat by friction. Examples of like character, as well as of other transformations, could be readily given in any number.

Chemical Affinity of Atoms : Relation to Heat.—While the force of gravitation acts upon the particles of matter without reference to the character of the body, still another form of force is found in the realm of atoms, which is effective between the atoms of chemically different bodies, namely, chemical affinity. This is the force by means of which the atoms of chemically different bodies unite in chemical combination. The energy itself is extremely variable between the atoms of different chemical bodies.

A distinction is made between strong chemical affinities (or relations) and weak affinities. Just as it is possible to determine the kinetic energy of a body in motion from the amount of heat that it generates in its impact upon an unyielding surface, so the degree of chemical affinity can be determined from the amount of heat that is produced, as the atoms of chemically different bodies unite in chemical combination; for if a complex body is formed from individual, chemically different atoms heat is, as a rule, generated. If as a result of the force of affinity the atoms of 1 kilo of hydrogen and 8 kilos of oxygen unite to form the chemical combination water, an amount of heat is generated that is equal to that developed by the impact of a weight of 47,000 kilos in falling from a height of 300 meters above the surface of the earth. One gram of hydrogen converted into water by addition of oxygen yields 34,460 heat-units (calories). One gram of carbon converted into carbon dioxid yields 8080 calories.

Whenever in the course of chemical processes considerable affinities are satisfied heat is set free, that is, generated from the force of affinity. The force of affinity is a form of potential energy acting between the various atoms that in the course of the chemical process is transformed into heat. It is thus likewise explicable that in the course of those chemical processes through which strong affinities are dissolved, in which the chemically united atoms are again separated, cooling takes place, or, as is commonly stated, heat becomes latent. That is, the energy of the heat rendered latent is transformed into chemical potential energy, and this in turn, after disintegration of the complex chemical body, appears between its isolated, individual atoms as chemical affinity.

LAW OF THE CONSTANCY OF ENERGY.

Julius Robert v. Mayer (1842) and Hermann Helmholtz (1847) have established the important law that in a system that receives no influence or impression from without the sum of all the contained kinetic energies is always equal. The energies may be transformed one into another, so that the potential energy may be converted into kinetic energy, and the reverse, but never is even the slightest amount of the energy lost. The transformation that takes place in the energies occurs in a definite manner, so that from a definite measure of a given force an equally definite measure of the new-appearing force always results.

The forces occurring in the animal organism appear in the following modifications:

1. *As movement in mass* (generally designated simply movement),

such as the movement of the entire body, of the extremities and many of the viscera; also appreciable even microscopically in cells.

2. *As movement of the atom: in the form of heat.* As is well known, the vibration of atoms results in the production of heat or of light or in chemically active waves in accordance with the number of vibrations in the unit of time. The smallest number of vibrations are those of heat, the highest those that are chemically active, and between the two are the vibrations of light. In the human body only heat-waves have of these three been observed, but some lower forms of life are capable of causing also luminous phenomena.

In the human organism movements in mass are constantly transformed in certain organs into heat, as, for instance, the kinetic energy in the circulatory organs, and which is transformed into heat by the resistance within the vascular apparatus. The measure of these transformations also is the unit of energy = 1 grammeter, and the unit of heat = 425.5 grammeters.

3. In the form of *potential energy* (latent energy) the organism contains many chemical combinations characterized especially by great complexity of constitution and imperfect saturation of the contained affinities, and, therefore, by their great tendency to break down into simpler bodies. The body is capable of generating both heat and kinetic energy from potential energies; kinetic energy, however, is always in combination with heat, while heat may be produced alone. The simplest measure of the potential energies is the amount of heat that can be obtained by the combustion of the chemical bodies in question representing the potential energy. As a secondary matter the number of equivalent units of energy can be determined in turn from the amount of heat generated.

4. It is known that the phenomena of *electricity, magnetism* and *diamagnetism,* may make themselves manifest in two directions, namely, in the form of movement of minutest particles, which may be recognized in the incandescence of a thin wire (the seat of great resistance) traversed by a strong current; and also in the form of movement in mass, as exhibited in the attraction or repulsion of the magnetic needle. In the body electric phenomena appear in the muscles, nerves, and glands; but as compared with other forms of energy they are of subordinate importance. It is not improbable that the electric energy of the body is transformed almost wholly into heat. The endeavor to obtain a measure for electric energy, the unit of electricity, as a means of direct comparison with the heat-unit and the unit of energy, has likewise been attended with definite success.

Luminous phenomena do not occur in the bodies of the most highly developed animals. The significant investigations of Hertz have shown that the phenomena of light exhibit the greatest analogy with those of electricity in the most important connections, so that the relations between the two forms of energy must accordingly be admitted.

It is certain that in the body also the different forms of energy can be transformed one into another in a definite and constantly invariable degree, and that new energy never develops spontaneously in the body, while that present is never destroyed; and thus also the organisms are a theater in which the law of the constancy of energy is in unceasing operation.

The original statement of Julius Robert v. Mayer may be appropriately quoted at this point: "There is but one energy, which operates with unceasing change in dead and in living things, and nowhere in either does any change take place without alteration in the form of energy. Physics has but to investigate the metamorphoses of energy, as chemistry has to investigate the transformations of matter. The generation as well as the destruction of energy is beyond the range of human thought and action: Nothing comes from nothing, nothing can give rise to nothing. If chemistry teaches the immutability of matter, then it is the obligation of physics to demonstrate the quantitative immutability of energy notwithstanding all variability in form. Gravitation, motion, heat, magnetism, electricity, chemical difference, are all but varying modes of manifestation of one and the same natural force that reigns throughout the universe, for any one can under special conditions be converted into another." (Lucretius Carus, born 95 B. C., had already said: "Nullam rem a nihilo gigni, ... neque ad nihilum interimat res.")

ANIMALS AND PLANTS.

Locked up in the constituent elements of the animal body is an aggregation of chemical potential energies (Lavoisier, 1789). The total amount of these in the human body could be measured if the entire cadaver were completely burned in a calorimeter and the number of heat-units generated were noted as a result of its combustion. The chemical combinations in which are bound up the potential energy are characterized by complexity in the arrangement of their atoms, by imperfect saturation of the affinities of the atoms, by a relatively small oxygen-content, and by a great tendency to and readiness of disintegration.

It may be conceived that food is withheld from an individual. The fasting person loses hourly 50 grams of body-weight; the tissues in which his potential energy is bound up are thus consumed. Through the taking up of oxygen combustion continually takes place, and as a result of this process the complex elements of the body are converted into simpler ones, whereby the potential energy forming the connecting link between them is transformed into kinetic energy. It is a matter of indifference whether the process of combustion takes place rapidly or slowly; the same amount of chemical matter always yields the same amount of kinetic energy, as, for instance, heat. After the lapse of a certain time the fasting person becomes conscious of the state of threatened exhaustion of his stored potential energy, and the condition of hunger sets in. The hungry person takes food; all food for the animal kindgom is derived either directly or indirectly from the vegetable world. Even carnivorous animals, which eat the flesh of other animals, consume in the final analysis organized material formed from vegetable food. Thus, the existence of the animal kingdom necessarily implies unconditionally the previous existence of the vegetable kingdom.

Vegetable structures thus contain all of the nutritive materials necessary for the animal body. In addition to water and inorganic matters, vegetables contain, among other organic combinations, especially also the three principal representatives of nutrient bodies, namely, fats, carbohydrates, and proteids. All of these are the seat of abundant potential energy in accordance with the complexity of their chemical constitution.

Fats contain: $\begin{cases} C_n H_{2n-1} O \ (OH) = \text{fatty acids} \\ + \ C_3H_5 \ (OH)_3 = \text{glycerin} \end{cases}$

Animal fats contain: $\begin{cases} C \ 76.5 \\ H \ 12.0 \\ O \ 11.5 \end{cases}$

Carbohydrates contain: $C_6H_{10}O_5$

Proteids contain in percentages: $\begin{array}{l} C_{50-55} \\ H_{6.6-7.3} \\ N_{15-19} \\ O_{19-24} \\ S_{0.3-2.4} \end{array}$

Man, who partakes of a certain amount of these nutrient materials, adds to them through the respiratory process the oxygen of the air, whence there results a process of combustion, in the course of which chemical potential energy is converted into heat. It is evident that the products of this combustion must be bodies of simple constitution, bodies with uniform arrangement of their atoms, with most complete saturation of the affinities of their atoms, of great constancy, partly rich in oxygen and possessing slight or no chemical potential energy. These bodies are carbon dioxid (CO_2), water (H_2O), and, as the most important representative of the nitrogen-containing derivatives, urea ($CO(NH_2)_2$), which, while endowed with a small measure of potential energy is, outside of the body, readily transformed into CO_2 and ammonia (NH_3).

Thus, the animal body is an organism in which, through the intermediation of oxidation-phenomena, the complex nutritive matters of the vegetable world, representing high potential energy, are transformed into simple chemical bodies, in the course of which the potential energy is transformed into an equivalent amount of kinetic energy (heat, work, electric phenomena).

The question naturally arises, How do plants, which, as the first products of creation, found for their nourishment no preexisting materials endowed with potential energy, and still suffer from no lack thereof —how do plants form the complicated nutrient matters mentioned, rich in stored-up potential energy? This potential energy of vegetable life must obviously have been derived from some other form of energy, for it cannot be created out of nothing. This kinetic energy is furnished plants through the light of the sun, whose chemical rays they absorb. Without sunlight there can be no vegetable life. From the air and the earth the vegetable organism obtains CO_2, H_2O, NH_3, and N, of which carbon dioxid, water, and ammonia (from urea) constitute also the excrementitious matters of the animal body. The plant obtains from the rays of the sun the kinetic energy of its light and converts it into potential energy, which, as in all vegetable matter, so also in the nutrient material produced, accumulates in the process of the growth of the plant. This formation of complex chemical combinations takes place in association with elimination of oxygen.

The Papillonaceæ, as, for instance, peas, beans, lupines, acacias, are capable of assimilating the free nitrogen of the air in the tissues of their root-bulbs, through the agency of symbiotic micro-organisms lodged upon these, Rhizobium leguminosarum. Thus, these plants are capable of building up their nitrogen-containing tissues even in soil entirely free from nitrogen. In this way they play an important fertilizing role in agriculture (lupine) and forestry (acacia). Also lower forms of vegetable life, as, for instance, the anaerobic bacterium, Clostridium pasteurianum, is capable of assimilating free nitrogen.

At times plants also exhibit free kinetic energy such as it is customary to encounter in the case of animals. Certain plants, as, for instance, the aroids and others, develop considerable amounts of heat during the flowering-period. It is also to be borne in mind that, in the development of the solid parts of plants, the transformation of formative fluids into solid matter causes heat to be set free. Absorption of oxygen and elimination of carbon dioxid have also been observed in plants. These processes are, however, so insignificant as compared with those described as typical in the vegetable kingdom, that they may be considered as of little or no importance.

Thus, plants are, on the whole, organisms that through the agency of reduction-processes convert simple stable combinations into complex ones, with the transformation of kinetic solar energy into the chemical potential energy of vegetable matter. Animals are living organisms in which through the agency of processes of oxidation the atom-groups of complex construction furnished by plants are split up, the potential energy being transformed into kinetic energy, which makes itself manifest in the animal. Thus a circulation of materials and a constant interchange of energy take place between animals and vegetables. All of the energy of animals is derived from plants and all of the energy of plants is derived from the sun. Therefore, the latter is the cause, the ultimate source of all of the energy of organism, that is, of life as a whole. As the generation of the sun's heat and light can be explained by the gravitation of masses, so it is possible that the force of gravitation is the sole ultimate form of energy for all living things.

"The sun is the constantly bent spring that brings about the activity in the atmosphere, that raises the waters to the clouds, that causes the tides. Light, the most mobile of all forms of force, intercepted by the earth in flight, is transformed by plants into a rigid state, for plants produce upon it a continuous sum of chemical difference, constitute a reservoir in which the fugitive rays of the sun are fixed and, adapted for useful purposes, are deposited. Plants take one form of energy, light, and reproduce another, chemical difference. In the course of the processes of life, but one transformation, both of matter, as well as of energy, takes place, but never is the one or the other produced " (Julius Robert v. Mayer, 1845). ("Omnia mutantur, nihil interit."—Ovid.)

The generation of kinetic energy in the animal body from the potential energy of the plant can be made readily comprehensible by means of a comparison. The atoms of the matter generated in organisms may be conceived to be simple small bodies, spherules or blocks. So long as these lie in a single layer or at least arranged in a few layers upon the ground, a condition of rest and constancy will prevail in consequence of this simple and stable arrangement. If, however, an artificially arranged formation of unstable construction is built up from the small bodies, there will be required (1) the motor force of the constructing agency, which raises and combines the units. As soon, however, as (2) an impulse from without acts upon the completed unstable structure, the atoms collapse and the impact of their fall generates heat (eventually also kinetic energy in the course of other complicated transformations), that is, the energy applied by the constructing agency is transformed into the form of energy last named. In plants the complicated unstable construction of the atom-groups takes place, the sun being the constructing agency. In the animal body, wherein the plant is consumed, the atomic structure is disintegrated into simpler elements, with the generation of kinetic energy.

KINETIC ENERGY AND LIFE.

The forms of kinetic energy that are active in organisms, namely, plants and animals, are precisely the same as those that are recognizable in inanimate matter. A so-called "vital energy," which is supposed to act as a special form of force of peculiar character and cause and control the vital phenomena of living organisms, does not exist. The forces of all matter, both organic and inorganic, are bound up in their smallest particles, the atoms. As, however, the smallest particles of organized matter are generally united in a most complex manner, in contrast to the ordinarily much simpler constitution of inorganic bodies, the forces inherent to the smallest particles of organism will appear in much more complicated phenomena and combinations, and as a result the explanation of the vital phenomena in the organism by the simple principles of physics and chemistry is rendered extremely difficult and in many respects appears impossible.

Metabolism as an Index of Life.—A special form of interchange in matter and energy appears peculiar to the living organisms of the earth. This consists in the ability to adapt themselves to the materials of their environment, and to assimilate them, so that for a time they represent integral parts of the living being, later again to be given off. The complete chain of these phenomena is designated *"metabolism,"* which consists accordingly in ingestion, assimilation, reduction and excretion.

It has already been suggested that metabolism differs in character in animals and in plants. As a matter of fact, this is, as has been shown, actually the case in animals and plants typically and characteristically developed. There is, however, a large group of organisms that in their complete organization exhibit such atypical development that they must be considered as undifferentiated fundamental forms of organisms. They cannot be recognized as either plants or animals, but represent the simplest form of animate matter. These organisms, as the earliest and most primitive forms, have been designated protists. It must be assumed absolutely that these also have a simple metabolism as a condition of life, but with respect to this adequate observations are wanting.

PHYSIOLOGY OF THE BLOOD.

PHYSICAL PROPERTIES OF THE BLOOD.

The *color* of the blood varies from bright scarlet-red in the arteries to the deepest dark bluish-red in the veins. Oxygen, therefore also the air, makes it bright red, while deficiency in oxygen renders it dark. The oxygen-free venous blood is dichroic, that is, it appears dark red in reflected light and green in transmitted light. In thin layers the blood is opaque, as one can readily convince himself, if blood be poured upon a glass plate and be permitted to flow off, by attempting to read printed matter through it. The blood thus behaves as a covering pigment, as its coloring matter is suspended in the plasma in the form of small granules, namely, the red blood-corpuscles.

For this reason the granular coloring matter of the blood can be separated from the blood-plasma by filtration. This, however, is possible only after admixture of the blood with fluids that render the blood-corpuscles rough or viscid. If mammalian blood is mixed with one-seventh of its volume of concentrated sodium sulphate, or if frog's blood is mixed with two per cent. solution of cane sugar, and then filtered, the blood-corpuscles will remain upon the filter.

The *reaction* of blood is alkaline from the presence of disodium phosphate (Na_2HPO_4). The alkalinity rapidly diminishes in intensity after escape from the vessel, and the more rapidly the greater the previous alkalinity. The change depends upon the development of an acid, in which the red blood-corpuscles take part in consequence of a decomposition of as yet undetermined origin. This generation of acid is increased by high temperature and the addition of alkali.

The alkalinity of the blood is diminished (A) by active muscular exercise, in consequence of the development of acid in the muscular tissue. (B) By coagulation. Fresh clot has a more intensely alkaline reaction than blood-serum. (C) After the persistent use of soda the alkalinity of the blood is increased, and after the use of acid it is diminished. (D) Old blood or blood dissolved with water from dry places generally has an acid reaction. The blood of children and women exhibits a lesser degree of alkalinity than that of men, and that of nursing women a lesser degree of alkalinity than that of pregnant women. The alkalinity is less also during digestion than during fasting.

Method of Examination.—As in consequence of the normal color of the blood red litmus-paper cannot be employed directly in testing the reaction, the following plan is pursued: Blood is mixed with an equal volume of concentrated solution of sodium sulphate, and the mixture is placed upon highly porous and sensitive lilac-tinted litmus blotting-paper. The blood-corpuscles remain upon the surface while fluid is taken up by the paper and gives rise to the reaction.

For the quantitative estimation of the alkalinity dilute tartaric acid is added to a volume of blood (7.5 grams of crystalline tartaric acid to 1 liter of water, 1 cu. cm. of which saturates 3.1 mg. of soda) until the blue paper is reddened. One hundred cu. cm. of human blood contains the alkaline equivalent of from 260 to 300 mg. of soda (in guinea-pigs 150 mg., in carnivora 180 mg. of soda).

Landois' method for the quantitative determination of the alkalinity of the blood with only a few drops of blood: Tartaric acid in the concentration already stated is employed to neutralize the alkalinity of the blood. Of this the following mixtures are made by addition of concentrated solution of neutral sodium sulphate: (1) 10 parts of tartaric-acid solution and 100 parts of concentrated sodium-sulphate solution; (2) 20 parts of tartaric-acid solution and 90 parts of sodium-sulphate solution; (3) 30 parts of tartaric-acid solution and 80 parts of sodium-sulphate solution; (4) 40 parts of tartaric-acid solution and 70 parts of sodium-sulphate solution; (5) 50 parts of tartaric-acid solution and 60 parts of sodium-sulphate solution; (6) 60 parts of tartaric-acid solution and 50 of sodium-sulphate solution; (7) 70 parts of tartaric-acid solution and 40 parts of sodium-sulphate solution; (8) 80 parts of tartaric-acid solution and 30 parts of sodium-sulphate solution; (9) 90 parts of tartaric-acid solution and 20 parts of sodium-sulphate solution; (10) 100 parts of tartaric-acid solution and 10 parts of sodium-sulphate solution. To each glass an excess of crystallized sodium sulphate is added to the point of insolubility.

Of the blood to be examined 1 drop is mixed in a graduated tube prepared for the purpose with an equal-sized drop of the acid-sulphate mixture. Into a glass tube with a diameter of 1 mm. and drawn out at one extremity mercury is sucked to a height of about 8 mm. so that the tube is filled to the tip. The upper extremity of the thread of mercury is marked by the scratch of a file. The mercury is now drawn into the tube until its lower border reaches the file-mark. The upper border of the mercury is now marked with another file-scratch. In this way the small measuring apparatus is improvised.

In order now to test the blood, one drop of the tartaric-acid sodium-sulphate mixture is sucked up to the lower mark, and then, after scrupulously drying the tip, the blood is drawn up until the fluid reaches the upper mark. After again cleansing the tip of the tube its contents are blown into a watch-glass, are well stirred and then tested with reagent-paper. Successively the mixtures 2, 3, 4, etc., are treated in the same way. The reagent-paper is cut into strips 3 mm. wide, and these are partially dipped in the blood-specimens in the respective watch-glasses. The blood-corpuscles collect about the immersed extremity of paper, while the fluid is sucked up beyond and indicates the reaction. If the test has been made successively in this manner with the mixtures from 1 to 10 it will be readily seen when the blue tint of the alkaline reaction ceases and the red tint of the acid reaction begins.

In human beings the blood can always be obtained directly from a small needle-puncture. Exact suction into the tube can be effected with certainty and convenience if the upper extremity of the measuring glass is connected by means of a short rubber tube with a hypodermic syringe, the movement of whose piston through a twisting motion facilitates an exact degree of suction. All of the tests must be completed with equal rapidity and at the same temperature.

The degree of alkalinity in the adult will in general be satisfied by mixture 5 or 6, and in the child by mixture 4. If all parts of the blood are uniformly dissolved previously by addition of water this solution, which obviously can no longer be designated blood, exhibits a somewhat higher degree of alkalinity. If blood is tested slowly by the method described the alkalinity will be that of such a solution.

Pathological.—Persistent vomiting and chlorosis are attended with increased alkalinity, while diabetes, as well as cachectic states, rheumatism, uremia, leukemia, profound anemia, high fever, cholera, carbon-monoxid poisoning, and degeneration of the liver are attended with diminished alkalinity. Poisons that cause destruction of red blood-corpuscles likewise bring about reduction in the alkalinity.

Blood has a peculiar *odor.*

This "halitus sanguinis" differs in human beings and in animals, and depends upon the presence of volatile fatty acids. If sulphuric acid be added to blood, and these acids are in consequence set free from their combination with the alkali of the blood, the characteristic odor appears more distinctly.

The blood possesses a *salty taste*, derived from the salts dissolved in the blood-plasma.

The *specific gravity* of the blood is 1058 (from 1046 to 1067) in men, and from 1051 to 1055 in women, while the blood of children has a lower specific gravity. The specific gravity of the red blood-corpuscles is 1105, that of the plasma from 1027 to 1028.3. This fact explains the tendency of the former to sink to the bottom.

Method of Determination.—For clinical investigation the following method (a modification of that described by Roy) can be recommended. In a glass tube, narrow at the bottom and covered with a rubber cap, a fresh drop of blood obtained by puncture with a needle is permitted to enter from below. The tube is at once immersed in a glass vessel filled with a solution of olive-oil in chloroform, and by pressure upon the rubber cap the drop of blood is expelled into the fluid. Various concentrations of the latter with a specific gravity between 1050 and 1070 are prepared, and that solution in which the drop remains suspended indicates the specific gravity of the blood.

The specific gravity is dependent principally upon the hemoglobin-content of the blood, much less upon the number of erythrocytes. It is high in the newborn, namely, 1066. The drinking of water and hunger will reduce the specific gravity temporarily, and it falls also after loss of blood and is lower in the presence of anemia, chlorosis, marasmus, and nephritis (down to 1025). It is increased by thirst, the digestion of solid food, by sweating, acute loss of water through the intestines and the kidneys, as well as cyanotic stasis (down to 1068). The entrance of an increased amount of salts into the blood is shortly followed by dilution, while the salts of the biliary acids, on the other hand, exert a concentrating influence. The specific gravity is increased by vasomotor contraction of the vessels and, conversely, it is diminished by vascular dilatation. The blood-serum of women is heavier than that of men. If blood is made artificially to pass repeatedly through an organ its specific gravity increases in consequence of the taking up of dissolved substances and the giving off of water.

For the determination of the specific gravity of the red blood-corpuscles, these must be isolated by sedimentation. This takes place rapidly in the case of horses' blood. The erythrocytes are said to be somewhat heavier in women and to contain more hemoglobin than those of men.

The *freezing-point* of the blood is about —0.56° C. It increases as the oxygen-content diminishes.

MICROSCOPIC EXAMINATION OF THE BLOOD.

The **red blood-corpuscles** or *erythrocytes* (Fig. 1) were discovered in man by Leeuwenhoeck in 1673 and in the frog by Swammerdam in 1658.

Physical Properties.—Human erythrocytes are coin-shaped discs with biconcave surfaces and rounded margins. The diameter is 7.5 μ, the thickness of the edge 2.5 μ, and the central thickness from 1.8 to 2 μ (Fig. 1).

In health the diameter varies from 6 to 9 μ; the average being from 7.2 to 7.8 μ. The corpuscles are diminished in size by inanition, elevation of the bodily temperature, carbon dioxid and morphin, and increased in size by oxygen, a watery state of the blood, cold, ingestion of alcohol, quinin, hydrocyanic acid. [Pathological conditions are discussed on p. 50.]

The volume of an erythrocyte equals 0.000000077217 cu. mm., the superficies 0.000128 sq. mm. If the total volume of the blood in man be assumed to be 4400 cu.cm., all of the contained blood-corpuscles have a superficies of 2816 square meters, that is, the equivalent of a square with sides of 80 paces. In a second 176 cu.cm. of blood are driven into the lungs and whose blood-corpuscles exhibit a superficies of 81 square meters, that is, a square with sides 13 paces. The volume of all of the erythrocytes can be approximately determined by introducing the blood into a narrow graduated glass tube ("hemokrit" of Hedin), either unmixed or defibrinated or mixed with an equal amount of a preservative fluid capable of preventing coagulation, as, for instance, 2.5 per cent. potassium-bichromate solution or 0.86 per cent. sodium-chlorid solution with some ammonium oxalate, and subjecting it to centrifugation. Treated in this manner healthy human blood is found to contain from 42 to 48 per cent. of corpuscles (anemic blood 30 per cent. and less). The erythrocytes, however, undergo changes in volume, at least after escape of the blood, by the taking up or giving off of fluid material, as exhibited beyond doubt by shrunken and distended forms. Venous blood contains a greater volume of erythrocytes than arterial blood.

The *weight* of an erythrocyte can be determined by multiplying its volume by its specific gravity (1105) = 0.000000085325 mg.

Alexander Schmidt determined the weight of the red blood-corpuscles in 100 parts of blood in the following manner: He ascertained (1) the percentage of dry residue of the blood = T; (2) the percentage of dry residue of the corresponding blood-serum = t; (3) the dry residue of the erythrocytes contained in 100 grams of blood = r; the dry residue of the serum obtained from 100 grams of blood is then T — r, the corresponding amount of serum $\frac{100 \times (T - r)}{t}$; further, the weight of the erythrocytes in 100 parts of blood = $100 - \frac{100 \times (T - r)}{t}$; the latter equals 48 grams in 100 grams of blood from a man and 35 grams in the same amount of blood from a woman.

Number.—In men the number of red blood-corpuscles is more than 5,000,000, while in women it is about 4,000,000 in 1 cubic millimeter, making 25 billions in 5 kilos of blood. The number is in inverse proportion to the amount of the plasma, and from this fact it will be seen

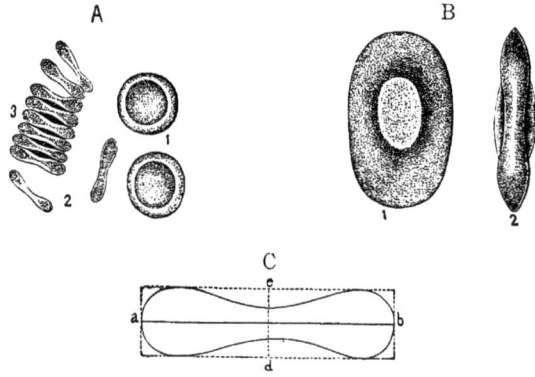

FIG. 1.—A, human colored blood-corpuscles: 1, on the flat; 2, on edge; 3, rouleau of colored corpuscles. B, amphibian colored blood-corpuscles: 1, on the flat; 2, on edge. C, ideal transverse section of a human colored blood-corpuscle magnified 5000 times linear: *ab*, diameter; *cd*, thickness.

that the number must vary in accordance with the state of contraction of the vessels, conditions of pressure and diffusion-currents and the like.

The number of red blood-corpuscles is increased in venous blood (at times in small cutaneous veins and in the presence of stasis), after the ingestion of solid food, after rest at night, after marked loss of water through the skin, the intestine or the kidneys, during inanition (in consequence of the consumption of blood-plasma), in the blood of the newborn, at times after late ligation of the umbilical cord (from the fourth day the number again becomes reduced), in persons of vigorous constitution and in residents of the country. The number is diminished during pregnancy and after copious libations. The capillaries contain relatively few blood-corpuscles. Apparent increase or diminution must also accompany variations in the amount of plasma, and to this fact special attention should be given in investigating the effect of certain influences upon the number of erythrocytes. Thus, for instance, the increased number observed in those residing at a high altitude may depend, wholly or in part, upon a greater or lesser reduction in the plasma. In the earlier stages of fetal life the number is from ½ to 1 million in 1 cu. mm.

Method of Counting Blood-corpuscles.—An exact mixing apparatus for the dilution of the blood is the first requirement. For this purpose the mixer of Potain will answer (Fig. 3). This is a carefully calibrated, pipet-like glass instrument, whose tip is dipped into the blood, which by suction through a rubber tube is drawn into the pipet either to the mark $\frac{1}{2}$ or to the mark 1. The tip carefully dried is then immersed in 3 per cent. sodium-chlorid solution, which is sucked up until it reaches the mark 101. By shaking the mixer a spherule (a) in the bulbous enlargement of the apparatus is moved about so as to effect a homogeneous mixture. If the blood be sucked up to the mark $\frac{1}{2}$ the mixture will be as 1 to 200, and if up to the mark 1 as 1 to 100.

For the enumeration of the cells a small amount of the blood-mixture is introduced into the Abbe-Zeiss counting-chamber (Fig. 2), the first few drops being thrown away. Upon a slide is cemented a glass cell, 0.1 mm. deep, upon whose floor are etched a series of squares and which is surrounded by a groove or depression and is provided with a cover-glass to be placed over it. The space overlying each square has a capacity of $\frac{1}{4000}$ cu. mm. The number of cells in each square is estimated and this multiplied by 4000 gives the number of corpuscles in each

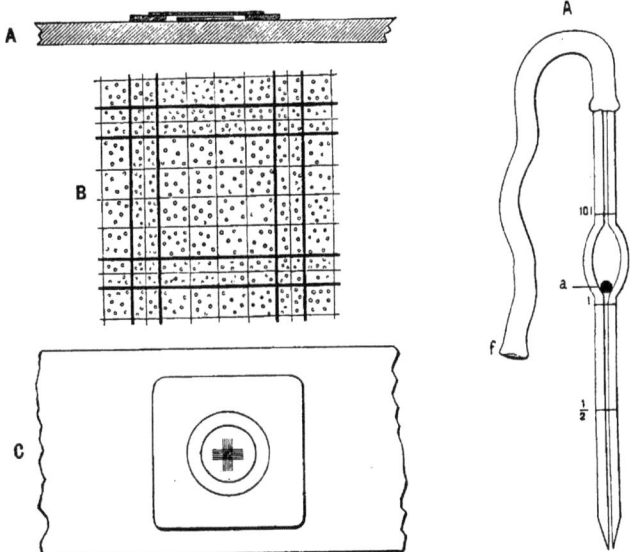

FIG. 2.—Apparatus of Abbé and Zeiss for Counting the Corpuscles: A, in section; C, surface view without cover-glass; B, microscopic appearance with the blood-corpuscles.

FIG. 3.—The *Melangèur*, pipet or mixer.

cu. mm. The result thus obtained must be multiplied by 100 or 200, according as the blood has been diluted 100 or 200 times. To ensure greater accuracy the contents of a large number of squares should be counted and the average taken. Vierordt, Malassèz, Gowers, and others have devised similar forms of apparatus for the same purpose.

To count the white blood-corpuscles alone in the chamber the blood is mixed with 10 parts of a $\frac{1}{3}$ per cent. solution of acetic acid, which dissolves out the red corpuscles. It is advisable to stain the leukocytes in the blood-mixer, and this can be done with some such solution as the following: 50 cu. cm. of a $\frac{3}{4}$ per cent. of solution of sodium chlorid with 5 drops of a 5 per cent. alcoholic solution of gentian-violet or hexamethyl-violet.

The red blood-corpuscles are characterized by their great *elasticity*, *flexibility*, and *softness*.

THE RED BLOOD-CORPUSCLES (ERYTHROCYTES).

Individually the red corpuscles are of a yellowish color with a greenish tint. They are unprovided with either capsule or nucleus, but consist throughout of a homogeneous mass. This consists (1) of a framework of exceedingly pale, soft protoplasm, the stroma or cytoplasm, and (2) of the red blood coloring-matter, the hemoglobin, which impregnates the stroma (like paraplasm), in the same way as a sponge takes up fluid.

INFLUENCES AFFECTING THE VITAL PHENOMENA OF RED BLOOD-CORPUSCLES.

Blood-corpuscles retain in unimpaired degree their vital and functional activities in shed blood and even in defibrinated blood subsequently returned to the circulation. Heat has an influence upon their vitality. If blood be heated to a temperature in the neighborhood of $52°$ C. the vital activity of the erythrocytes is destroyed. This fact is evident from the circumstance that the corpuscles in such blood are soon dissolved when returned to the circulation. Kept in

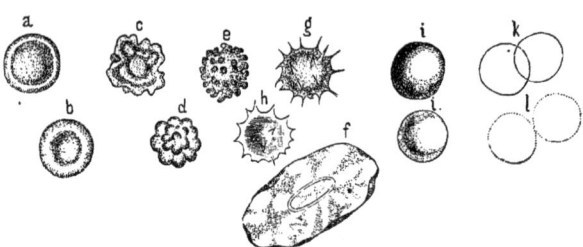

FIG. 4.—Red Blood-corpuscles: a, b, normal human red corpuscles, the central depression more or less in focus; c, d, e, mulberry, and g, h, crenated forms; k, pale corpuscles decolorized by water; l, stroma; f, frog's blood-corpuscle acted on by a strong saline solution.

the cold—in a flask exposed to the influence of ice-water—mammalian blood may retain its functional activity for 4 or 5 days. Removed from the body for a longer period of time and then returned to the circulation the red corpuscles rapidly undergo destruction—an evidence that they have lost their vital activities within this time.

The erythrocytes in blood recently removed from a vessel frequently exhibit changes in form that result in their assuming a mulberry-like appearance. These have been attributed to active contraction on the part of the stroma. Nevertheless, it must as yet be considered doubtful whether this is to be looked upon as an obvious vital phenomenon. It is true, however, that Max Schultze has observed active contractility and motility in the red blood-corpuscles of quite young embryo chickens. In support of the vital activity of the red corpuscles the fact may be cited that certain substances dissolved in the plasma are not capable of diffusing into the red blood-corpuscles, as, for instance, solutions of potassium, of iron, and of manganese, although other substances do enter, as, for instance, sugar and chloroform.

Nucleated erythrocytes are undoubtedly cells, while the non-nucleated erythrocytes cannot properly be so considered. The latter have, therefore, been designated blood-plastids.

INFLUENCES AFFECTING THE PHYSICAL PHENOMENA OF RED BLOOD-CORPUSCLES.

The color of the red corpuscles is changed characteristically by a number of gases. Thus, oxygen, therefore also the air, renders the blood scarlet red, deficiency of oxygen renders it dark bluish red, carbon monoxid renders it cherry red, nitrogen monoxid renders it violet red. All agents that cause marked contraction of the erythrocytes induce a bright scarlet-red color; as, for instance, concentrated solution of sodium sulphate, from the action of which the corpuscles become mulberry-shaped or distorted into the shape of a key, and in a measure attenuated. The color thus produced is brighter than is ever observed in the arteries. Those agents that make the corpuscles globular, as particularly water, cause the color of the blood to become darker.

If a dry preparation of blood be treated with concentrated solution of methylene-blue diluted half with water some of the erythrocytes, particularly degenerated ones, become stained. It is the larger ones that are especially numerous in the presence of anemia and leukemia.

Change in Position and Form.—A phenomenon frequently observed in recently shed blood is the arrangement of the corpuscles like rolls of coin (Fig. 1, A, 3).

The conditions that increase the coagulability of the blood favor this phenomenon, which is to be attributed, in addition to the attraction of the discs, to the formation of a viscid substance. The condition is favored by warming moderately the slide upon which the fresh drop of blood is received. If under such circumstances agents are added to the blood capable of causing the corpuscles to swell, the rolls separate as the individual corpuscles are transformed into globules. The adhesive substance uniting the erythrocytes, and which not rarely is drawn out into filamentous threads, is derived from the peripheral layer of the corpuscles. It consists of the stroma-fibrin, formed on the surface of the corpuscles in consequence of the inception of an injury at the periphery, and which has become viscid.

The changes in shape that the erythrocytes may gradually undergo after leaving the body, up to the point of dissolution, are of especial interest. Some agents bring this series of changes about in rapid succession. If, for instance, blood is exposed to the action of the spark of a Leyden jar, all of the corpuscles become at first mulberry-shaped, that is, the surface becomes rough and soon covered with at times small, at other times large, round nodules (Fig. 4, c d e). If the action be more pronounced the blood-corpuscles become almost globular, with many projecting points, thorn-apple-like (g h); this is probably an indication of the death of the corpuscle. At a further stage the action causes the corpuscles to assume a perfectly globular shape (i i). In this form they appear smaller than normal, as their disc-shaped mass is contracted into a sphere with a lesser diameter. The globules thus formed are viscid, and adjacent corpuscles readily adhere to one another and like fat-globules they may unite to form larger spheres. If the action be continued for a still longer time, the blood coloring matter eventually separates from the stroma (k), and the blood-plasma consequently becomes reddened, while the stroma is recognizable only as a faint shadow (1). The changes in shape described represent the effects also of a number of other injurious agents causing dissolution of the red blood-corpuscles. Thus, for instance, all of the changes in shape can be observed also in putrid fluid.

Influence of Heat.—If a blood-preparation be heated upon a warm stage the corpuscles will be seen to undergo remarkable changes in shape when the temperature reaches 52° C. They become in part globular, in part drawn out into the shape of a biscuit, at times perforated, or larger or smaller drops of the substance of the body are completely constricted off and float about in the surrounding fluid. This is an evidence that considerable degrees of heat destroy the histological individuality of the corpuscles. If the temperature be high and its influence long continued, the erythrocytes are finally entirely dissolved. In the case of burns the blood-corpuscles may undergo the same changes within the vessels.

The addition to blood of a concentrated solution of urea acts in the same way as heat. Blood-corpuscles can be broken into fragments in microscopic preparations by strong pressure. The disintegration of blood-corpuscles into fragments may be designated erythrocytotrypsy, in contradistinction from their dissolution, which is known as erythrocytolysis.

If a finger moistened with blood be passed over a hot glass plate so that the thin layer of fluid is rapidly dried, the most remarkable forms of long drawn-out distorted blood-corpuscles can be seen. This experiment demonstrates in a striking manner their marked softness and elasticity.

If blood be mixed with a concentrated solution of mucilage and if, while being examined under the microscope, concentrated solution of sodium chlorid is added, the corpuscles become drawn out into longitudinal masses (dragon-shaped). The same change is observed if blood be admixed with an equal amount of liquid gelatin at a temperature of 36° C., and sections are made after the gelatinous mass has hardened.

PRESERVATION OF RED BLOOD-CORPUSCLES.

The following are admirable preservative fluids for red blood-corpuscles:

Pacini's Mixture.	*Hayem's Fluid.*
Mercuric chlorid, 2.	Mercuric chlorid, 0.5.
Sodium chlorid, 4.	Sodium sulphate, 5.
Glycerin, 26.	Sodium chlorid, 1.
Distilled water, 226.	Distilled water, 200.
To be diluted with two parts of distilled water before being used.	

In order to avoid all influence of the air in the examination of fresh blood the following procedure is recommended: A drop of Pacini's fluid is placed upon a portion of the skin, which is then punctured with a fine needle through the fluid. In this way the blood rises into the preservative fluid without having at any time come in contact with the air and the form of the corpuscles is at once fixed.

In examining blood for medico-legal purposes the microscope is naturally always employed. Dried spots are carefully softened by means of concentrated or 30 per cent. solution of potassic hydrate, or with some preservative fluid, without friction. By softening them with the aid of concentrated tartaric-acid solution the leukocytes appear with especial distinctness. Often, however, search for the presence of blood-corpuscles will be fruitless. Red, suspicious fluids are examined directly. If the blood-corpuscles in the fluid have possibly already become pale, or if they are present only as stroma, the addition of a wine-yellow aqueous solution of iodin-potassium-iodid to the microscopic preparation will at times render them much more distinct. Saturated solution of picric acid, 20 per cent. solution of pyrogallic acid and 30 per cent. solution of silver nitrate have also been recommended for this purpose.

PERMEABILITY OF ERYTHROCYTES.—ISOTONIA (HYPERISO-TONIA AND HYPISOTONIA).—DEMONSTRATION OF THE STROMA-LAKE COLORATION OF THE BLOOD.

All substances soluble in water attract water with a certain intensity. The energy by means of which this attraction takes place is known as *hygroscopic energy* or *osmotic tension*. The manner in which this behaves with regard to living cells was discovered by de Vries (1884). A vegetable cell consists of a membrane, which is permeable to salts and to water. This membrane is in contact by its inner surface with the adjacent cell-protoplasm, which likewise is permeable to water, but not to salts. If fresh vegetable cells are placed in distilled water, this passes through the cell-membrane and through the cell-protoplasm, and causes the cells to swell. If, however, the cells are placed in a strong saline solution, the cell-contents shrink, because water is abstracted from them. The shrinking of the cellular protoplasm is shown by the fact that the protoplasm contracts upon all sides and becomes detached from the cell-membrane. This detachment of the shrunken cell-body from the cell-wall in consequence of loss of water is designated *plasmolysis* by de Vries.   ·

Plasmolysis is the more pronounced the more concentrated the saline solution surrounding the vegetable cell. The saline concentration that brings about the first signs of plasmolysis can be determined experimentally for every variety of cell. The different salts must be employed in various concentrations, in order to bring about the same degree of plasmolysis. Solutions of different salts that exert the same effects in the process of plasmolysis are designated *isotonic solutions*. The necessary concentrations are to each other as the molecular weights of the different salts. For instance, a 0.58 per cent. solution of sodium chlorid causes the beginning of plasmolysis in the same way as a 1.01 per cent. solution of potassium nitrate, or as a 1.5 per cent. solution of sodium iodid. The molecular weights of the three substances are 58, 101, and 150 respectively. Isotonic solutions have the same freezing-point, which always becomes lower with increasing concentration; and also the same boiling-point, which becomes higher with the degree of concentration.

There is thus for the red blood-corpuscles a given concentration for certain but not all substances in which they neither shrink nor swell. For mammalian erythrocytes this is a 0.9 per cent. solution of sodium chlorid—for the frog 0.6 per cent. If the equally effective degree of centration is determined for other salts, the isotonic solutións will be established. Obviously the blood-plasma likewise is such an isotonic solution, as the erythrocytes retain their form perfectly within it. Those solutions are *hyperisotonic*, that is, of greater concentration, that abstract water from the erythrocytes and therefore cause them to shrink; while those solutions are designated *hypisotonic*, that is, of feebler concentration, that yield up water to the erythrocytes and therefore cause them to swell.

Although the erythrocytes preserve their form in isotonic solutions, nevertheless an interchange may take place between the soluble substances in their interior and those of the surrounding fluid. Thus, chlorids, phosphates, and proteids, for instance, pass from one to the other. Under such circumstances, however, the isotonia is preserved. If, therefore, substances pass from the erythrocytes into the surrounding blood-plasma, other substances must, conversely, pass into them in order to preserve the isotonia. The red corpuscles thus possess the property of maintaining a constant degree of osmotic tension with reference to certain substances. If, for instance, small amounts of an acid, and also carbon dioxid, be added to blood, albumin and phosphates pass from the corpuscles into the plasma, while, conversely, chlorids pass from the latter into the erythrocytes to maintain the isotonia. In consequence, the corpuscles become somewhat globular and their diameter diminishes in size. The blood-corpuscles exhibit the reverse interchange and effect in shape after addition of small amounts of alkali.

Van 't Hoff discovered in 1887 the law that the interchange of substances in solution takes place according to the same laws as those applicable to gases, namely, the osmotic pressure corresponds entirely to the tension of a gas. The laws of gases laid down by Boyle-Mariotte are, therefore, applicable also to substances in solution. Accordingly, and by reason of the diversity of the soluble substances contained within the cells and in the surrounding fluids currents must arise between the two in consequence of the osmotic pressure. If, therefore, erythrocytes, which behave like sacs filled with saline solutions, are placed in another saline solution, phenomena appear entirely analogous to those that occur when a sac filled with gas is introduced into another gas.

The erythrocytes floating in a solution retain their volume only if the fluid is isotonic; that is, if it exerts the same osmotic pressure and if the substances dissolved in the surrounding solution cannot enter the corpuscles. If the osmotic pressure in the surrounding fluid is diminished the corpuscle swells until it becomes completely dissolved in water, whose osmotic pressure is zero. The blood then becomes lake-colored. Exactly the same effect as is produced by distilled water must be produced also by the solution of a substance, quite independently of the degree of its osmotic pressure, if the substance in solution readily penetrates the blood-corpuscles, and therefore can exert no pressure upon its wall. Under such circumstances also the corpuscle will undergo dissolution and the blood become lake-colored.

The phenomenon of the blood becoming lake-colored, which is easily recognizable, indicates, therefore, that the blood-corpuscles are either in a solution of low osmotic pressure or in a solution whose osmotic pressure is not manifest because the wall of the corpuscles is impervious to the substance in solution. Among those solutions in which the blood-corpuscles are dissolved, independently of the degree of osmotic pressure of the solution, urea occupies the first place. The ammonium salts, with the exception of the sulphate, behave in the same manner. Certain exceptions to which the laws of osmotic pressure for the blood-corpuscles do not appear to apply H. Koeppe has been able to explain according to the theory of solutions of van 't Hoff. The circumstance must be taken into consideration, as Ostwald was the first to point out, whether, in accordance with the concentration of their solution, the dissolved substance has or has not completely dissociated itself into its ions.

Many agents separate the coloring-matter from the stroma. In consequence the hemoglobin is dissolved in the blood-plasma, and the blood becomes transparent, as it contains the coloring-matter in the form of a transparent pigment. It is, therefore, designated lake-colored. Lake-colored blood is dark red. In the dissolution of the erythrocytes the change does not affect the aggregate condition, but it consists only in a transposition of the hemoglobin, which leaves the stroma and passes over into the blood-plasma. Therefore, no reduction in temperature takes place.

Method.—For the microscopic demonstration of the stroma it is recommended that a one per cent. solution of tartaric acid blood mixed with an equal volume of concentrated sodium sulphate be carefully added. In order to obtain an abundance of stroma for chemical examination, defibrinated blood is mixed with 10 volumes of a solution of sodium chlorid containing 1 volume of the concentrated solution and from 15 to 20 volumes of water. In this the stromata are precipitated as a whitish sediment.

The following agents effect separation of stroma and hemoglobin:

(a) *Physical agents:* (1) Heating of the blood to a temperature of 60° C. The degree of heat differs, however, in different animals. (2) Repeated freezing and thawing. (3) The static spark, although not after salts have been added to the blood, and the constant and induced currents.

(b) *Chemically active substances generated within the body:* (4) Bile or bile-salts. (5) Serum from other species of animals. Thus, for instance, the serum of dogs' blood and of frogs' blood dissolves the blood-corpuscles of the rabbit in a few minutes. According to Rummo, Maragliano, and Castellino the blood-serum in cases of acute infectious disease and chronic dyscrasias is said to be destructive to the erythrocytes of healthy individuals. (6) Lake-colored blood from a number of other species of animals.

(c) *Other chemical reagents:* (7) Water. (8) Exposure to the vapors of chloroform, ether, amylene; small amounts of alcohol, paraldehyd, thymol, nitrobenzol, ethylic ether, acetone, petroleum ether, and others. (9) Antimony hydrid, hydrogen arsenid, carbon disulphid. (10) Solutions of certain salts may be mixed with blood in a definite concentration without causing change in the red blood-corpuscles. If the saline solution is made either more dilute or more concentrated, dissolution of the corpuscles takes place. This is the case, for instance, with sodium chlorid. Traces of alkali render the erythrocytes more resistant to such solutions, while traces of acid exert an injurious effect. According to Bernstein and Becker salts cause an increase in the resistance to physical solvents, but a reduction to chemical solvents. (11) Addition of boric acid, 1 per cent., to amphibian blood causes the red mass, which at the same time surrounds the nucleus when present and is designated zooid, to escape from the stroma, which is designated ecoid, to withdraw from the periphery to the interior of the corpuscles, and seem entirely to pass out. Brücker, therefore, considers the stroma to a certain degree a repository within which is lodged the remaining substance of the blood-corpuscles especially endowed with vital phenomena. (12) Strong acid solutions dissolve the blood-corpuscles, while weaker solutions cause precipitates in the hemoglobin. This can be distinctly observed in the case of carbolic acid. (13) Alkalies in moderate concentration cause sudden dissolution. Addition of potassic-hydrate solution of about 10 per cent. to the blood from the margin of a cover-glass permits the process of dissolution to be readily observed microscopically. At first the corpuscles abruptly become globular in jerks and thus apparently smaller; later they swell up like soap-bubbles.

The influence of the gaseous content of the red blood-corpuscles upon their solubility is remarkable. The corpuscles in blood containing much carbon dioxid are dissolved most readily; those in blood containing much oxygen are much less readily dissolved; while between the two are the corpuscles containing much carbon monoxid. Total removal of the gases of the blood causes of itself the development of a lake-color.

The erythrocytes possess a certain degree of resistance to the action of solvents.

The following method may be employed to determine this degree readily. A drop of blood is mixed with an equal amount of a 3 per cent. solution of sodium chlorid, and then as much distilled water is added as is required to dissolve all of the red blood-corpuscles. The method is carried out as follows with human blood: With the aid of the blood-mixer of the blood-corpuscle counting-apparatus (Fig. 3) blood is collected from a puncture of the skin up to the mark 1, and is expelled for microscopic examination into a concave glass cell, in which previously an equal amount of a 3 per cent. solution of sodium chlorid had been placed. Well admixed, all of the erythrocytes will be preserved. Now, by means of the same apparatus, distilled water is added, and the changes observed under the microscope until all of the red corpuscles are dissolved. The glass cell is covered after each addition in order to prevent evaporation. The erythrocytes of some persons are more readily dissolved than is normal, being soft and plastic and undergoing striking changes. In addition, reference may be made to the following states: All blood-mixtures that jeopardize the normal condition of the erythrocytes, such as cholemia, intoxications with substances that cause dissolution of the blood-corpuscles and high grades of venosity. Interesting observations may be made further in the presence of blood-diatheses and infectious processes, hemoglobinuria, and burns. The resistance appears diminished in case of anemia and of fever.

FORM, SIZE, AND NUMBER OF ERYTHROCYTES IN DIFFERENT ANIMALS.

All mammals, with the exception of the camel, the llama, the alpaca, and related animals, as well as the cyclostomata among fish, for instance the lamprey, have coin-shaped circular erythrocytes. The mammalia excepted have oval erythrocytes without nuclei, while birds, reptiles, amphibia (1, B) and fish, with the exception of the cyclostomata, have similarly shaped erythrocytes with nuclei.

Coin-shaped Blood-corpuscles.		Size—μ = 0.001 Millimeter. Oval Blood-corpuscles.	
		Short Diameter.	Long Diameter.
Elephant,	9.4 μ	Llama, 4.2 μ	7.5 μ
Man,	7.5 "	Pigeon, 6.5 "	14.7 "
Dog,	7.2 "	Frog, 16.3 "	23.0 "
Rabbit,	7.16 "	Triton, 19.5 "	29.3 "
Cat,	6.2 "	Proteus, 35.6 "	58.2 "
Sheep,	5.0 "	The corpuscles of the amphiuma are about a	
Goat,	4.25 "	third larger than those of proteus.	
Musk-deer,	2.5 "		

Among vertebrates, the blood of the amphioxus is colorless. The large blood-corpuscles of many amphibia can be seen with the naked eye. In those of the frog a nucleolus is demonstrable. It is readily explicable that the larger the blood-corpuscles the smaller must be their number and their total superficies in a given volume of blood. Only in birds is the number relatively larger than in other classes of vertebrates, notwithstanding the greater size of the corpuscles. This probably depends upon the fact that in them metabolism exhibits the greatest energy. Among mammals carnivora have a larger number of blood-corpuscles than herbivora. In goats the blood contains 19,000,000 blood-corpuscles in the cubic millimeter; in the llama, 13,186,000; in the bull finch, 3,600,000; in the lizard, 1,292,000; in the frog, 408,900; in the proteus, 33,600. During the sleep of winter Vierordt observed the number of blood-corpuscles in the marmot diminish from 7,000,000 to 2,000,000 in a cu. mm.

In invertebrates the blood is generally colorless, with colorless cells. In some invertebrates, for instance the earth-worm, the larva of the large gnat, and others, the plasma is red and contains hemoglobin, but the blood-corpuscles are colorless. Red, violet, brownish, greenish, opalescent blood, with colorless corpuscles (ameboid cells), is found in some mussels. In the cephalopods and in certain snails and crabs a bluish, globulin-like coloring-matter is present in the blood, containing copper and combining with oxygen, hemocyanin, which is decolorized by a deficiency of oxygen. Certain round-worms have a green respiratory pigment, chlorocruorin, while other animals have a yellow, red, or brown pigment of similar function.

DEVELOPMENT OF RED BLOOD-CORPUSCLES.

A. The *embryonal development* of the blood-corpuscles begins in the chicken as early as the first day. The corpuscles develop in groups within large globules of protoplasm that detach themselves from the walls of the vascular spaces resulting from the apposition of the formative cells. At first they are globular, rough, nucleated, larger than the permanent cells and unpigmented. At a later period they take up the coloring-matter and attain their definite form, with retention of the nucleus. Only when the vessels enter into communication with the heart, are the corpuscles swept away or isolated in groups, and then become set free in the circulation. Remak demonstrated all stages of their multiplication by division. Cells dividing by mitosis are observed most abundantly between the third and the fifth day of incubation, but no longer after their escape.

Multiplication takes place by division also in the larvæ of amphibia, as well as during fetal life in mammals in the spleen, the bone-marrow and the liver, and in the circulating blood. Neumann, further, found in the liver of the embryo, protoplasmic cells—descendants of the vascular endothelium or of the liver-cells—enclosing red blood-corpuscles. Besides, there were found in the liver cells with large nuclei, in part containing hemoglobin, in part free from hemoglobin, which divided by mitosis and then, with shrinking of the nucleus, became transformed into definitive blood-corpuscles. Foa and Salvioli observed endogenous formation in the lymphatic glands, in addition to the liver and spleen, also within large protoplasmic cells. The spleen also is considered a seat for the formation of the red blood-corpuscles, though only during embryonal life. Here the red corpuscles are believed to be formed of yellow, round, nucleated cells, representing transitional forms.

From the embryonal bodies (erythroblasts), always at first nucleated, there result, in the later stages of embryonal life, the characteristically shaped and at the same time non-nucleated corpuscles; the nucleus, together with a portion of the protoplasm, disappearing. In the human embryo only nucleated corpuscles are present in the fourth week. In the third month they constitute only from one-eighth to one-quarter of all the erythrocytes, while at the end of fetal life nucleated corpuscles are found only with great rarity (**Fig.** 8).

According to some observers, mammalian erythrocytes contain a nucleus-like central body, which Lavdowsky considers as the remains of nuclear substance. According to J. Arnold, the central body sometimes observed consists of a granular-filamentous transformation of the previous nucleus. This body, designated nucleoid, is surrounded by a zone of paraplasm, enclosing hemoglobin and granular and hyaline matter in a filamentous framework. Nucleoid and paraplasm may under certain conditions be extruded from the erythrocytes. Perhaps these contribute to the formation of blood-plates.

B. *Development of Vessels and Blood-corpuscles in the Earliest Post-embryonal Period.*—Following J. Arnold, Golubew believes that the blood-capillaries present in the tail of frog-larvæ form in various situations at first solid buds that grow more and more deeply into the tissues, enter into anastomotic union with adjacent buds and finally become hollow, with disappearance of their protoplasmic contents. The capillaries would thus like an intricate branched network make their way into the tissues and spread like a foreign intruder. Ranvier observed the same process of growth in the omentum of newborn cats.

The development of the capillaries and at the same time of the blood-corpuscles in their interior has been observed in an especially instructive manner in the large omentum of the young rabbit. When a week old, the omentum in these animals exhibits dull-white spots in whose interior lie so-called vessel-forming or vaso-formative cells (**Fig.** 5), that is, strongly refracting cellular elements varying widely in shape, and provided with protoplasmic processes (a). The protoplasm of these cells resembles that of the lymph-cells, particularly with respect to its markedly refracting character. In the interior of these cellular structures can be seen rod-shaped nuclei arranged longitudinally (K K) and red blood-corpuscles (r r), both surrounded by protoplasm.

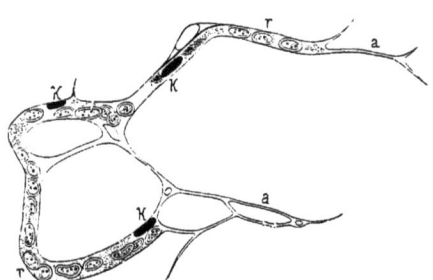

FIG. 5.—Formation of Red Blood-corpuscles within "Vaso-formative Cells," from the Omentum of a Rabbit Seven Days Old: r, r, the formed corpuscles; K, K, nuclei of the vaso-formative cell; a, a, processes which ultimately unite to form capillaries.

From the vessel-forming cells protoplasmic shoots and processes arise, which in part terminate free and in part unite to form a delicate network. In some places nucleated connective-tissue corpuscles arranged longitudinally lie upon the structures. These constitute the beginning of the connective-tissue perivascular sheath.

The vessel-forming cells appear in various shapes, either longitudinally cylindrical, with pointed extremities, or round or oval, rather resembling large lymph-cells or connective-tissue cells. These cells are always the seat of origin of non-nucleated erythrocytes, which thus arise in the protoplasm of the vessel-forming cells, as the chlorophyl-grains or starch-granules arise in the protoplasm of vegetable cells. Only after the blood-corpuscles have thus formed within their interior do these cells unite through their processes with the vascular system. Their tubular arrangement becomes connected with adjacent vessels and the blood-corpuscles are washed away. In rabbits from four to six weeks old these areas contain fewer and fewer corpuscles. If it be borne in mind that Schäfer observed similar formative processes in the subcutaneous connective tissue of young rats, the question must arise whether such blood-forming stations do not exist in many parts of the body and constitute seats for the regeneration of the blood.

For purposes of demonstration it is only necessary to observe omentum of suitable age in a fresh state in peritoneal fluid, evaporation being prevented by applying paraffin to the edges of the cover-glass. Landois saw preparations of this highly interesting developmental process in the laboratory of Ranvier at Paris with such a degree of distinctness as to leave in his mind no doubt as to the accuracy of the observation. Neumann saw analogous formations in the embryonal liver, Wissotzky in the amnion of the rabbit, Nicolaides in the mesentery of the guinea-pig, Klein in the amniotic sac of the chicken's egg, Bayerl in the cartilaginous capsules of ossifying cartilage, Leboucq and Hayem in other situations, all indicative of the fact that the blood-cells develop endogenously in certain cellular structures of considerable size whose protoplasm serves at the same time for the formation of the vessel-wall.

C. *At a later period of life* the red blood-corpuscles develop from special nucleated cells, the *erythroblasts*. It is believed that the latter gradually assume the form and color of perfect erythrocytes. According to Neumann they possess blood coloring-matter from the outset. In caudate amphibia and fish the spleen, and in all other vertebrates, the bone-marrow constitutes the seat for the formation of those juvenile forms that multiply by division. Particularly in the latter all stages of the transformation may be seen, especially pale, contractile cells resembling white blood-corpuscles, and later on red nucleated corpuscles that must be considered as the progenitors of the red corpuscles and that are capable of undergoing multiplication by mitosis.

After copious loss of blood the process of transformation and the entrance into the blood-stream is said to be observed in especially marked degree. J. Arnold found in the protoplasm of the nucleated erythrocytes of bone-marrow granules resembling those of hemoglobin-free cells. In the process of transformation into red blood-corpuscles these granules become invisible through transformation. The products of the mitotic division of the pale cells especially are to be considered as the progenitors of the nucleated erythrocytes. In the red bone-marrow, perhaps also in the spleen, the small veins and most of the capillaries have no definite wall. The formed erythrocytes accordingly can at any time be swept into the circulation from these parts.

The bones of the skull and most of those of the trunk contain red (blood-forming) marrow, while the extremities contain only fatty marrow, or only the upper portions of the femur and the humerus contain red marrow. When active regenerative processes are taking place in the blood the fatty marrow may be transformed into red marrow, and indeed from the upper portion of the bones named downward, even through all the bones of the extremities. Red, blood-corpuscle-forming marrow may develop even in the ossified laryngeal cartilages and in pathological bony tumors.

DESTRUCTION OF RED BLOOD-CORPUSCLES.

As erythrocytes are being constantly formed, it must be assumed that they are being constantly destroyed. Further, the situations are known in which this occurs especially. Among these is first the liver, as the elements of the bile are formed from blood coloring-matter and the blood of the hepatic veins contains a smaller number of red blood-corpuscles. The splenic pulp also contains cells indicative of disintegration of erythrocytes. These are the blood-corpuscle-containing cells described in connection with the spleen. The investigations of Quincke have rendered it probable that the red blood-corpuscles —whose span of life may cover more than three or four weeks—if they are to be eliminated are taken up by the white blood-corpuscles of the liver-capillaries and by perhaps identical cells of the splenic pulp and of the bone-marrow, and preferably deposited in the liver-capillaries, the spleen, and the bone-marrow. The erythrocytes taken up are, without having previously been dissolved, converted in part into yellow and in part into colorless iron-albuminates, *hematosiderin*, which can be demonstrated microchemically in part in granular, in part in soluble form, giving rise to a greenish discoloration on addition of ammonium sulphid. In the spleen and in the bone-marrow, in part perhaps also in the liver, these are again employed for the regeneration of red blood-corpuscles, while another portion of the iron is eliminated through the liver.

Latschenberger has found pigmented and colorless plates in the blood, the latter at times in flakes of fibrin, and these he considers as the terminal products of the disintegration of all morphological blood-elements. The pigmented plates are derived from the erythrocytes and exhibit in part the iron-reaction of hematosiderin, and in part that of biliary coloring-matter. These plates are retained and further transformed in the spleen and in the bone-marrow.

As a sign of the degeneration of the erythrocytes that may precede their death Ehrlich mentions their property of staining violet with eosin-hematoxylin or blue with methylene-blue. The rarity with which cells containing blood-corpuscles are found in the general circulation justifies the conclusion that corpuscles are taken up within the spleen, the liver, and the bone-marrow, being favored by the slowness of the circulation in these parts.

Pathological.—Among pathological conditions there may be quantitative disturbances in the processes of blood-destruction and blood-formation. Accumulation of iron-containing materials from red blood-corpuscles may take place in the spleen, the bone-marrow, and the liver-capillaries: (1) if the destruction of red blood-corpuscles is increased, as, for instance, in cases of anemia; (2) if the formation of new red elements from old material is retarded. If elimination through the liver-cells is interfered with, the iron accumulates in them, and it is then present in the blood-plasma also in increased amount, and it may be eliminated by other glands, although a deposit of iron may take place there (cortex of the kidney, pancreas) within the glandular cells and in the tissue-elements of other organs.

After abundant regeneration of blood in dogs the leukocytes of the liver-capillaries are in the course of four weeks enormously rich in iron-containing granules; likewise the cells of the spleen, of the bone-marrow, of the lymphatic glands, further the liver-cells and the epithelium of the cortex of the kidney. The iron-reaction in the two situations last named takes place also after introduction of hemoglobin or of iron-salts into the blood.

Within thrombi and also in extravasations of blood that diffuse into the surrounding living tissue hematosiderin likewise develops, in addition to hematoidin, which forms when not in contact with the tissues. The stage of iron-reaction of the products of the disintegration of the erythrocytes is, however, not of consequence, as in the progress of time the residuum no longer exhibits this reaction. V. Recklinghausen designates as hemochromatosis a brownish discoloration of the tissues dependent upon abnormal dissolution of erythrocytes or local extravasations of blood, and which is caused by the iron-containing hematosiderin and the iron-free hemofuscin derived from it. Landois observed these conditions after extensive transfusion.

If it be remembered that after repeated copious loss of blood and after every menstruation the blood is regenerated within a relatively short period of time, it is evident that an active process of regeneration must take place. As to the amount of corpuscles destroyed daily the amount of biliary and urinary pigment formed from the blood coloring-matter affords some idea.

THE WHITE BLOOD-CORPUSCLES (LEUKOCYTES), THE BLOOD-PLATES AND ELEMENTARY GRANULES.

Through the lymph-stream colorless cells, designated *white blood-corpuscles* or *leukocytes*, are swept into the blood. In addition to the blood they are found in the lymph, in adenoid tissue, in bone-marrow and as wandering cells in the connective tissues of various parts, as well as between glandular and epithelial cells. They consist of globular masses of viscid, bright or granular, highly refracting, soft, motile, unencapsulated protoplasm (Fig. 6). In the fresh state (A) they exhibit no nucleus, which appears only after addition of water or acetic acid (B), and in consequence of which also the definition becomes more distinct. Water, besides, renders the contents more granular and more turbid, while acetic acid causes them to clear up. The nucleus contains one or more nucleoli. The diameter of the cells varies from 4 to 13 μ. The leukocytes are dissolved by peptone.

In accordance with their form and size leukocytes are differentiated as follows: (1) *Small lymphocytes*, approximating erythrocytes in size, with a large, round, deeply staining nucleus and a thin margin of protoplasm. (2) Large cells, with an extensive oval, feebly staining nucleus and a heavy cortical layer of protoplasm. (3) Cells resembling those last described except that the nucleus is constricted. (4) Somewhat smaller cells, constituting about three-quarters of the total number, with polymorphous, lobulated or variously convoluted nuclei, or nuclei separated into from one to four parts. The last three forms of cells have a genetic connection.

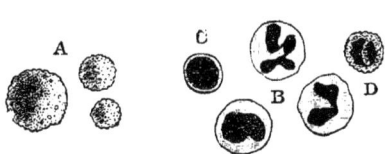

FIG. 6.—A, human white blood-corpuscles, without any reagent; B, after the action of water; C, after acetic acid; D, frog's corpuscles, changes of shape due to ameboid movement.

The leukocytes increase by division, in part by mitosis, in part by amitosis especially in their germ-centers, that is, the lymphatic glands and adenoid tissues. Division has not as yet been observed in the small lymphocytes found in the lymphatic glands (Fig. 8, *o o*). Perhaps these represent juvenile forms. Also sessile cells in the connective tissue may undergo multiplication by division and send their offspring into the blood through the lymph-stream.

The number of leukocytes in a given division of the vascular system may differ widely. At times they may be found increased in one place or another, as, for instance, as a result of chemotaxis, while at other times a large number may be sent into the blood-stream from the lymphatic apparatus. The increase is designated leukocytosis.

The number of leukocytes is considerably less in shed blood than in circulating blood. Immediately after removal from the vessels nine-tenths of all of the leukocytes are destroyed (fibrin-formation).

Local heat diminishes, and cold increases, the number of leukocytes in the vessels of the part of the body treated, as they are restrained in the blood-vessels contracted by cold.

NUMBER OF LEUKOCYTES IN PROPORTION TO THE RED BLOOD-CORPUSCLES IN SHED BLOOD.

UNDER NORMAL CONDITIONS.	IN VARIOUS SITUATIONS.	UNDER VARIOUS CONDITIONS.
1 : 335, Welcker, 1 : 357, Moleschott, 1 : 500–800, v. Jaksch. (In children the number is said to be somewhat greater than in adults.)	Splenic vein, 1 : 60, Splenic artery, 1 : 2260, Hepatic vein, 1 : 170, Portal vein, 1 : 740, The number is in general greater in the veins than in the arteries.	The number is increased by digestion, blood-let- ting, long-continued suppuration, menstru- ation, the puerperium, the death-agony, tonic medicaments (quinin, bitters), ingestion of nuclein. gout. The number is dimin- ished by hunger and impaired nutrition.

The *movement of the leukocytes*—observed by Wharton Jones in 1846 in the ray, and by Davaine in 1850 in man—which has been designated ameboid, because it corresponds entirely with that of the ameba, is due to alternate contraction and relaxation of the protoplasm surrounding the nucleus. It can be recognized especially from the fact that processes are sent out from the surface and withdrawn (Fig. 7) like the pseudopods of the ameba. At the same time the protoplasm has an internal current, which can be seen particularly in the polymorphonuclear cells. Movement has been seen also in the nucleus itself. The movement is attended with two sets of phenomena: (1) *The migration of the cells*, inasmuch as they draw themselves along by means of protrusion and retraction of their viscid processes. In this way they may migrate even through the interstices of intact vessels. Arnold considers the capability of certain wandering cells to develop into epithelioid or giant cells as demonstrated. (2) *The taking up of small granules*, such as fat, pigment, foreign bodies, which at first adhere to the surface and through the internal current are drawn into the interior of the leukocytes and which finally may be again extruded, in the same way as amebæ take up food. Thus they take up fat-globules, peptones and albuminous bodies that have gained entrance into the blood-stream and which they may later deposit elsewhere.

Metschnikoff dwells upon the activity of the leukocytes in retrogressive processes, the parts to be broken down being taken up in the forms of particles and therefore in a measure devoured. He designates the cells with these activities as devouring cells—*phagocytes*. Thus they act as *chondroclasts* and *osteoclasts* in the absorption of cartilage and bone respectively. Cells of similar activity are found in the tails of batrachia, and which take up portions of the tissue, as, for instance, fragments of fibrils, in the disappearance of the tails during the process of metamorphosis. (See also absorption of the deciduous teeth.) Thus, schizomycetes or particles of other substances that have gained entrance into the blood have been found taken up in part by leukocytes. Later, the leukocytes yield up these substances to the endothelial cells of the capillaries of the liver and the lungs, less commonly of the spleen. The motility of the leukocytes is destroyed by quinin.

The leukocytes exhibit still another interesting peculiarity, namely that of *chemotaxis* (chemotropism), which consists in the attraction of freely motile cells—like some lower organisms—by certain substances, and their repulsion by certain others. Especially the metabolic products of pathogenic and non-pathogenic microörganisms exert a strong attractive influence upon the leukocytes.

If. therefore. colonies of staphylococcus (bacteria of suppuration) collect at a given part of the body their metabolic products attract the leukocytes from the neighboring vessels, and in this way inflammatory reaction and suppuration result. The poison is either eliminated with the pus or is destroyed by the phagocytic activity of the leukocytes. The leukocytes also secrete special chemical substances that destroy the injurious microörganisms. These substances are known as alexins.

In warm-blooded animals the leukocytes exhibit movement for a long time upon a warm stage—at a temperature of 40° C. for about two or three hours; a temperature of 47° C. induces rigidity: heat-rigidity and death. The lowest degree of temperature at which ameboid movement is possible is 14° C. In cold-blooded animals, such as the frog the leukocytes can be seen to make their way out of a small coagulated blood-clot in a moist chamber and move about in the expressed serum. v. Recklinghausen observed motile phenomena on the part of leukocytes in a moist chamber for as long as three weeks. Oxygen is necessary for the movement. Induction-currents cause the leukocytes suddenly to become round, like irritated amebæ through retraction of all of

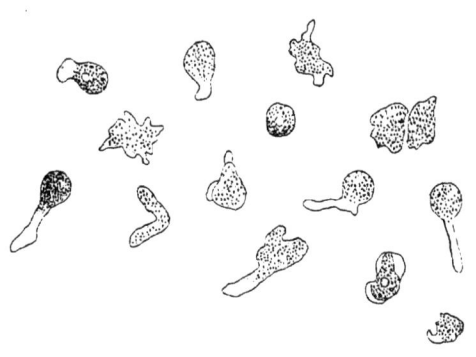

FIG. 7.—Human Leukocytes, Showing Ameboid Movements.

their processes. If the electric current be not too strong, the leukocytes resume their movements in the course of a short time. Strong and long-continued currents destroy them, causing them further to swell and undergo complete disintegration. The dissolution of white blood-corpuscles is known as *leukocytolysis.* It occurs as a normal phenomenon in the circulating lymph and in the blood in limited degree.

With regard to the source and the functional significance of the different varieties of leukocytes complete knowledge is as yet wanting. An attempt has been made to obtain a sharper differentiation of the leukocytes through the property of the smallest granules within the protoplasm of the cells to stain only with acid or with basic or with neutral pigments.

Method.—Recently shed blood is spread in a thin layer upon a cover-slip, dried in the air, then placed in an air-bath at a temperature of 125° C. for two hours. Next it is stained, washed with water, dried in the air and enclosed in Canada balsam.

The granules of the oxyphile or eosinophile cells (Fig. 8. a.b. with unstained nucleus; in c the nucleus is stained violet with hematoxylon) are stained only by acid pigments, such as a saturated solution of eosin in 5 per cent. carbolglycerin. The source of these cells is the bone-marrow. In normal human blood they constitute only about 10 per cent. of all of the leukocytes, but in cases of leukemia they pass in large number from the bone-marrow into the blood-stream—myelogenous leukemia.

The fine granules of the large mononuclear cells of normal blood are stained only by basic pigments, such as a concentrated watery solution of methylene-blue (f. g), as well as those of the majority in lymphemic blood. The cells known as mast-cells contain basophile granules of other size (d, e). These cells are rare in normal blood, but they often occur in large number in leukemic blood. Mast-cells may be found also in the connective tissue of other organs in the vicinity of the epithelial layer, as, for instance, in cutaneous areas the seat of chronic inflammation, and from which they then find their way into the blood.

Fine neutrophile granules are rendered visible by neutral stains, as, for instance, acid fuchsin neutralized with methylene-blue. These cells exhibit peculiarly sharp, polymorphous nuclear figures (h) or apparently several small nuclei. They are encountered in abundance in normal blood and in the presence of leukocytosis (i is such a cell in the fresh state, while in k and l the nucleus alone is stained). The smaller number of neutrophile cells contain a large nucleus, surrounded by a thin layer of protoplasm (m n). They are derived from the spleen and the bone-marrow. Between these two forms (h and m n) there are transitional varieties. The leukocytes h i migrate in the presence of inflammation. In cachectic states the mononuclear cells (m n) preponderate, while both forms are increased in number in association with acute leukocytosis. The lymphocytes o o, with a large reticulated nucleus, are derived from the lymphatic glands.

Neusser found numerous granules of nucleoalbumin in the leukocytes in cases of gout as the forerunners of uric-acid formation. The leukocytes exhibit the reaction for glycogen in the presence of progressive suppuration.

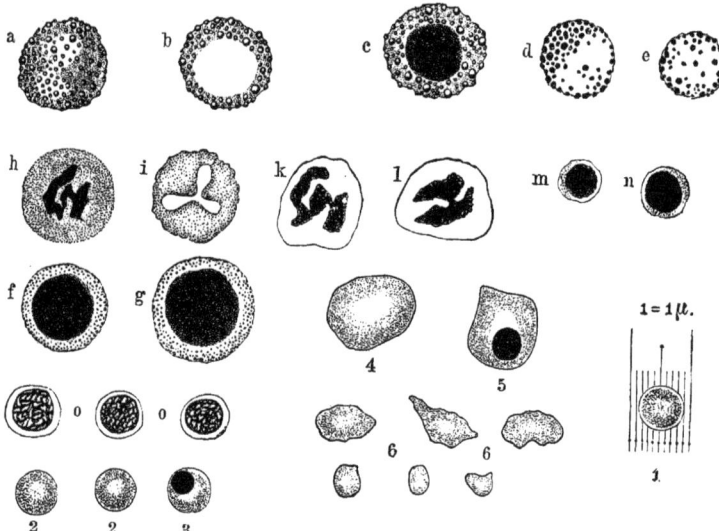

FIG. 8.—Various Forms of Leukocytes and Erythrocytes. × 1000. [All figures are drawn after the same scale: 1, a normal erythrocyte drawn into the scale; 1 = 1·μ.]

The *blood-plates* of Bizzozero (**Fig. 9**) deserve especial consideration as a third morphological constituent of the blood. These are pale, color-less, viscid, biconcave discs of varying size, averaging 3 μ in diameter. One cu. mm. contains 245,000 plates. Bizzozero has observed them in the circulating blood—in the mesentery of the guinea-pig and the wing of the bat. They collect in large numbers upon a thread immersed in fresh blood. They can be obtained from escaping blood after ad-mixture with one per cent. solution of osmic acid or with Hayem's fluid (**Fig. 9, 3**). In shed blood they rapidly undergo transformation into varied shrunken forms (5), disintegrating into small particles and being finally dissolved. Where they are collected together they readily cohere into masses (7), and pass over into aggregations resembling stroma-fibrin, which in coagulated blood may be united with shreds of fibrin (6, 8).

Bizzozero believes that they furnish the material for the fibrin in the process of coagulation, and he, as well as Eberth and Schimmelbusch, attribute the initial formation of white thrombi to them. According to Löwit they are formed from disintegrated leukocytes, and according to Lilienfeld from the nuclein and albumin of the nuclei of these cells. According to Wooldridge they are globulin-precipitates from the plasma. J. Arnold followed their extrusion and detachment from erythrocytes; in smaller measure they are derived from leukocytes. Halla found them increased in pregnant women, Mosen after hemorrhage, Afanassiew in the presence of regenerative states of the blood, Cadet in association with hunger, Hayem after the crisis of certain infectious diseases, and Pusari in cases of afebrile anemia. They are diminished in the presence of fever, as well as of severe infections and blood-stasis, and also after injection of leech-extract. The blood of cold-blooded animals and of birds contains also small spindle-shaped, nucleated cells.

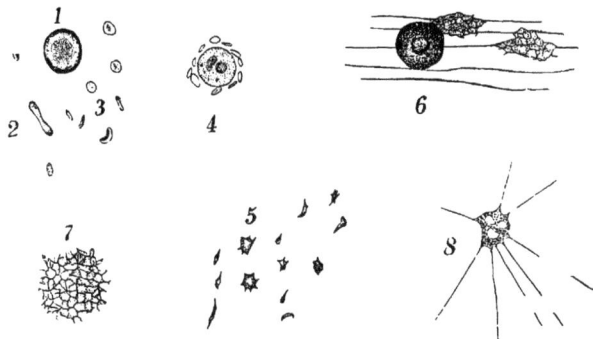

Fig. 9.—"Blood-plates" and Their Derivatives: 1, a red blood-corpuscle on the flat; 2, on the side; 3, unchanged blood-plates; 4, lymph corpuscle, surrounded by blood-plates; 5, altered blood-plates; 6, lymph corpuscle with two heads of fused blood-plates and threads of fibrin; 7, group of fused blood-plates; 8, small group of partially dissolved blood-plates with fibrils of fibrin.

Demonstration in Mass.—If 10 parts of blood are mixed with 1 part of a 0.2 per cent. solution of ammonium oxalate in 0.7 per cent. solution of sodium chlorid, and the mixture is centrifugated, a grayish-red layer principally of leukocytes will form above the erythrocytes, and over this a white layer consisting almost solely of blood-plates, while above all is the clear plasma.

In addition, a few small granules, so-called *elementary granules*, occur in the blood. These are irregular masses of protoplasm derived from disintegrated leukocytes or blood-plates.

According to H. F. Müller there are constantly present also, especially after the ingestion of food, minute, globular, highly refracting granules, which are not fat, and which he designates *blood-dust*, or *hemokonien*.

Coagulated blood contains delicate threads of fibrin (Fig. 9, 6, 8), strung like a spider's web between the corpuscles. They become isolated after dissolution of the corpuscles. Where many such threads occur together a nodular accumulation takes place.

ABNORMAL CHANGES IN THE RED AND WHITE BLOOD-CORPUSCLES.

Loss of blood is always followed by diminution in the *number* of erythrocytes in proportion to the extent of the hemorrhage, and the number may fall to even less than 400,000 in the cu. mm. The loss is soon made good by the absorption of lymph from the tissues. Menstruation furnishes an indication that moderate loss of red blood-corpuscles may be replaced in twenty-eight days. In case of considerable loss of blood, causing a reduction in all of the formative processes, this period may be prolonged to five weeks. In cases of acute febrile disease the elevation of temperature is generally attended with a reduction in the number of red blood-corpuscles, though with an increase in the number of white corpuscles. Chronic diseases diminish the number and often the hemoglobin-content of the erythrocytes in still greater degree. In some individuals, in whom the red blood-corpuscles are deficient in resistance, these undergo dissolution in consequence of the action of profound cold upon peripheral portions of the body, as, for instance, from the application of ice-water, while the blood-plasma becomes reddened and hemoglobinuria may even develop.

Diminished regenerative activity on the part of new erythrocytes will also cause reduction in their number, as blood-corpuscles are constantly undergoing destruction. If with this there be associated direct loss of blood, as, for instance, menstruation, the reduction may become considerable. In the case of chlorosis a congenital deficiency in the development of the blood-forming and blood-propelling apparatus, that is the vascular system, appears to constitute the cause. The heart and the vessels are small, and the absolute number of blood-corpuscles may be reduced even one-half. In the blood-corpuscles themselves, whose relative number may be either maintained or even reduced as much as one-third, the hemoglobin is reduced about one-third. The total volume of erythrocytes has been found diminished. The iron-content of the blood has been reduced, even to one-half. Courses of treatment with iron again increase the amount of hemoglobin and iron in the blood. So-called progressive pernicious anemia, which is characterized by the fact that the progressive impoverishment of the blood may even finally terminate fatally, is probably dependent upon some profound derangement of the blood-forming organs. In the presence of this disease the erythrocytes are reduced in number, while their hemoglobin-content is increased. Involution-forms, disintegrating-products (microcytes and poikilocytes) and earlier developmental stages of erythrocytes (nucleated erythrocytes of normal and of excessive size: normoblasts and megaloblasts) are also present. Numerous chronic intoxications, as with lead, swamp-miasm or syphilis, are likewise attended with reduction in the number of blood-corpuscles.

The *size* of the corpuscles varies in disease between 2.9 and 12.9 μ, with an average size of from 6 to 8 μ. Dwarf blood-corpuscles (6 μ and below, microcytes) have been considered as juvenile forms and are found in abundance in almost all forms of anemia (Fig. 8, 6). Giant corpuscles (megalocytes, 10 μ and above) are found constantly in cases of pernicious anemia, occasionally in cases of leukemia, chlorosis, and cirrhosis of the liver (Fig. 8, 4, 5, represents a nucleated megalocyte as the forerunner of a non-nucleated megalocyte). If the erythrocytes exhibit marked variation in form and size, they are designated poikilocytes (Fig. 8, 6).

Abnormalities in the *form* of the red blood-corpuscles have been observed after severe burns. The corpuscles appear much reduced in size and the thought suggests itself that under the influence of the heat accompanying the burn droplets of the corpuscles have become detached, in the same way as this can be observed in microscopic preparations on application of heat. Disintegration of blood-corpuscles in many such droplets (erythrocytotrypsy) has been observed in connection with various disorders, as, for instance, severe malarial fevers. These particles represent fragments of blood-corpuscles and not independent, intact, small, individual corpuscles. From these fragments there result dark pigment-particles closely related to hematin and which at first float about in the blood (melanemia). This condition can be developed artificially in rabbits by introducing carbon disulphid (7 parts to 90 parts of oil) subcutaneously. The leukocytes take up a number of these particles, which later on are found deposited in various tissues, particularly the spleen, the liver, the brain, and the bone-marrow.

In some cases the red blood-corpuscles exhibit *abnormal softness*, so that they undergo marked changes in form as the result even of slight extraneous influences. With regard to lessened resistance on the part of the erythrocytes, reference may be made to p. 35. The nitrogen-content of the erythrocytes is diminished in cases of secondary anemia, and it is increased in cases of pernicious anemia. In the interior of the erythrocytes of birds, frogs, turtles, etc., low forms of animals develop at times in the form of round pseudovacuoles, and out of which free blood-worms subsequently develop. Also in cases of malarial infection in human beings microbes of varying form (hemameba, Laveran) have been observed within the erythrocytes, and which probably are conveyed by stinging insects (mosquitos)—in the same way as Texas fever is conveyed by ticks. They destroy the red blood-corpuscles and in turn are destroyed by quinin.

The *white blood-corpuscles* are generally increased in all acute diseases in which exudation takes place. They exhibit excessive increase in association with so-called leukemia. In this disease the proportion of red to white blood-corpuscles may be as 2 to 1. In consequence, the blood acquires an appearance as if it were mixed with milk. At the same time the number of erythrocytes is diminished. Leukemia depends upon hyperplasia of the lymphoid tissue or the bone-marrow. These causes are responsible for lymphatic and myelogenous leukemia respectively. Lymphocytes and myelocytes are to be carefully differentiated. The enlargement of the spleen is only secondary; therefore a pure variety of lienal leukemia is not accepted. Myelogenous leukemia belongs probably among the active forms of leukocytosis. An active leukocytosis is one that results through movement or migration of leukocytes into the blood-current. This may involve the polynuclear—neutrophile or eosinophile—or the mixed cells—the latter with involvement of mononuclear elements containing granules: myelemia. The passive form of leukocytosis comprises the various forms of lymphemia.

CHEMICAL CONSTITUENTS OF THE RED BLOOD-CORPUSCLES.

The *blood coloring-matter* hemoglobin—abbreviated Hb—causes the red color of the blood. It is found besides in muscular tissue and in traces, probably only as a contamination through dissolved cells, in the blood-plasma. In the spectroscope it exhibits an absorption-band in the green (Fig. 15, 4). Its percentage-composition according to Hüfner is for the blood of swine, as compared with that for the ox, in parentheses, C, 54.71 (54.66); H, 7.38 (7.25); N 17.43 (17.70); S, 0.479 (0.447); Fe, 0.399 (0.336); O, 19.602 (19.543). For one atom of iron there are two atoms of sulphur in the horse, and three in the dog. According to Hüfner, the formula is $C_{638}H_{1025}N_{164}FeS_3O_{181}$; the molecular weight is 14,129. Hemoglobin is soluble in water; when heated it coagulates only with decomposition, retaining the sulphur in firm union. Although it is a colloidal substance, it nevertheless undergoes crystallization in all classes of vertebrates from which it has thus far been obtained, in figures belonging to the rhombic system, principally in rhombic plates or prisms, and from the guinea-pig in rhombic tetrahedra. The squirrel, however, forms an exception, inasmuch as its crystals appear as hexagonal plates. The crystals simply separate in all classes of vertebrate after slow evaporation of blood rendered lake-colored, though with varying degrees of readiness.

It is to be inferred that the variations in the form of the crystals in different animals are dependent upon slight changes in chemical constitution. The hemoglobin is readily crystallized from the blood of man, the dog, the mouse, the guinea-pig, the rat, the marmot, the cat, the leech, the horse, the rabbit, birds, and fish; and with difficulty from the blood of sheep, oxen, and swine; and not at all from the blood of the frog. Rarely the hemoglobin of a single blood-corpuscle can be seen to form a small crystal with inclusion of the stroma, as Landois also observed in the case of rabbits' blood that had stood for a long time. Within the large blood-corpuscles of fish the small crystal lies at times within the stroma by the side of the nucleus. In this class of vertebrates colorless crystals also have at times been observed.

The crystals of hemoglobin are doubly refracting and pleochromatic, that is, they appear bluish red in transmitted light and scarlet red in reflected light. The crystals, which contain from 3 per cent. to 9 per cent.

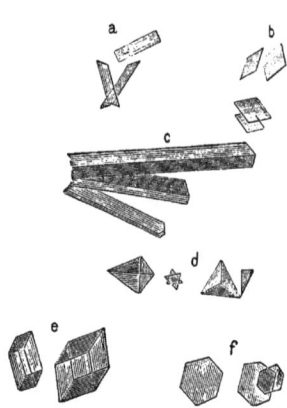

of water of crystallization and therefore become disintegrated from escape of this water on exposure to the air, are always soluble in water, though different varieties dissolve with varying degrees of facility. They are more readily soluble in dilute alkali. The solutions are dichroic, that is, they appear red in reflected light and greenish in transmitted light. They are insoluble in alcohol, ether, chloroform. and fats.

As a result of the process of crystallization the hemoglobin itself appears to undergo an internal change. Previous to crystallization it does not diffuse as a true colloidal body, but it actively decomposes hydrogen dioxid. Dissolved in the form of crystals, however, it is slightly diffusible, and does not decompose hydrogen dioxide, through the action of which it is decolorized. The crystals of hemoglobin collect like an acid at the positive pole of an electric current. As the hemoglobin thus exhibits alterations after its separation from the erythrocytes, Hoppe-

FIG. 10.—Hemoglobin-crystals: a b, from human blood; c, from the cat; d, from the guinea-pig; e, from the marmot; and f, from the squirrel.

Seyler believed that the oxyhemoglobin was united with lecithin within the erythrocytes, and also the hemoglobin. The former combination he designated *arterin* and the latter *phlebin*.

PREPARATION OF HEMOGLOBIN-CRYSTALS.

Method of Rollett.—Defibrinated blood, made lake-colored by freezing and thawing, is poured into a shallow vessel, whose bottom is covered therewith to a height of only 1½ mm. Evaporation is permitted to take place slowly in a cool place and as a result the crystals separate.

Method of Hoppe-Seyler.—Defibrinated blood is mixed with 10 volumes of a solution of sodium chlorid or of sodium sulphate (1 volume of a concentrated solution to 9 volumes of water) and permitted to stand. After the lapse of two days the clear supernatant layer is removed with a pipet, while the thick sediment of blood-corpuscles is washed with water into a glass flask, and shaken with an equal volume of ether until the blood-corpuscles are dissolved. After standing for a short time the supernatant ether is removed, and the lake-colored fluid filtered in the cold; then one-fourth volume of cold (o°) alcohol is added. This mixture is permitted to stand for several days at a temperature of —5° C. The crystals that will thus have formed in abundance can be collected upon a filter and dried by pressure between blotting-paper. Through the gradual action of the alcohol upon the hemoglobin-solution, by introduction into a dialyzer, it is possible to obtain crystals several millimeters long.

Method of Gscheidlen.—Gscheidlen obtained the largest crystals, several centimeters in length, by melting in small glass tubes defibrinated blood that had been exposed to the air for 24 hours, and preserving for several days at a temperature of 37° C. Spread upon a glass plate the crystals readily appear.

QUANTITATIVE ESTIMATION OF THE HEMOGLOBIN.

(a) *From Its Iron-content.*—As in the dry state (100° C.) hemoglobin contains 0.42 per cent. of iron by weight, the amount of hemoglobin can be estimated from the amount of iron in the blood. If m represents in percentage the weight of metallic iron found, the percentage of hemoglobin in the blood will be as 100 m : 0.42. The mode of procedure is as follows: A measured amount of blood is reduced to ash and this is exhausted with hydrochloric acid for the preparation of ferric chlorid. Next the ferric chlorid is converted into ferrous chlorid, and this is titrated with a solution of potassium permanganate.

(b) *Colorimetric Method.*—A dilute watery solution of crystallized hemoglobin is prepared, the exact strength of which is thus known. With this are compared watery dilutions of the blood to be examined, water being added to the latter until the color is the same as that of the hemoglobin-solution. The specimens to

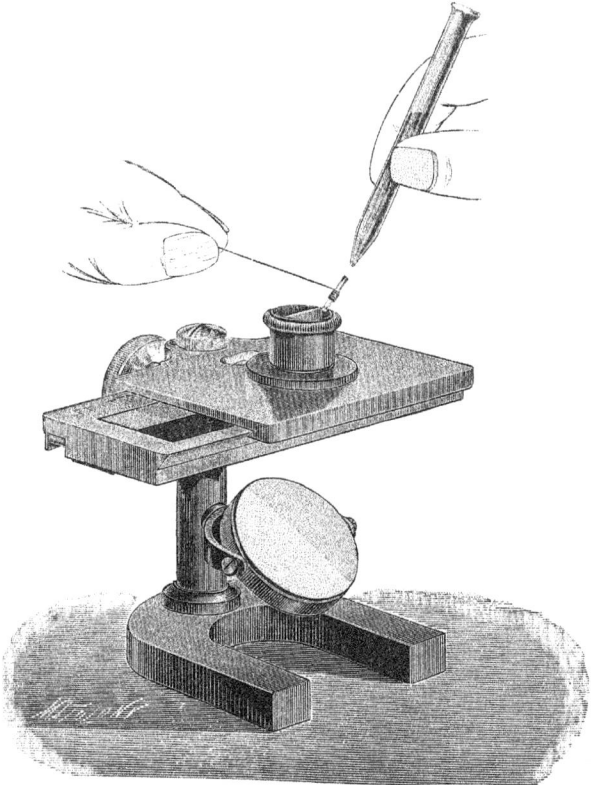

FIG. 11.—V. Fleischl's Hemometer.   To wash out the graduated pipet the larger tube held over it is employed.

be compared are contained in similar vessels of exactly the same thickness (hema-tinometer). Hoppe-Seyler has recently devised a colorimetric double pipet for this purpose. The blood-specimens are saturated with carbon monoxid.

For *clinical purposes* v. Fleischl's hemometer is recommended (Fig. 11). This consists of a cylinder mounted upon a metallic plate and divided into two equal parts, which are closed at one extremity by a disc of glass. Each half is filled with water, and then a measured amount of blood, obtained with a pipet of deter-mined capacity from a punctured wound, is introduced into the one half and dissolved. The color of the red solution thus produced is compared with that of a ruby-red glass wedge viewed through the clear water in the other half of the cylinder and capable of being moved forward and backward by a screw, until the color appears the same in both. The illumination of the blood-solution and the red wedge takes place from below by means of the light of a lamp. The glass wedge is provided with a scale, and when the colors in the two halves of the cylinder are alike the number on the wedge indicates the amount of hemoglobin in terms of percentage of the normal blood; thus. for instance, the figure 80 indicates that the examined blood contains 80 per cent. of the hemoglobin in normal blood.

(c) *With the aid of the spectroscope* Preyer found that a solution of 0.8 per cent. of oxyhemoglobin in water—1 cm. thick—yielded in addition to red and yellow the first band of green in the spectroscope (Fig. 15. 1). Of the blood to be examined about 0.5 cu. cm. is taken and is diluted with water until the identical of effect in the spectroscope is obtained. In addition to having the layers of fluid equal thickness—namely 1 cm.—the width of the slit in the spectroscope and the distance between this and the vessel, as well as the intensity of the source of light (stearin candle), must be the same. If k represents the amount of hemoglobin in percentage that permits the passage of the green color (0.8 per cent.), and b the volume of blood to be examined (about 0.5 cu. cm.), and w the amount of water necessary for dilution. then x equals the amount of hemoglobin in the blood to be examined expressed in percentage, that is $x = k (w-b): b$. It is advantageous to add a trace of potassic hydrate to the blood and to saturate it with carbon monoxid.

The amount of hemoglobin is in men 13.77 per cent. of the total volume of blood, in women 12.59 per cent., in pregnant women—with progressive diminution—from 12 to 9 per cent. According to Lichtenstern and Winternitz the hemoglobin is most abundant in the blood of the newborn, but this is no longer the case after the age of ten weeks. Between six months and five years of age it is smallest in amount and reaches its second maximum between twenty-one and forty-five years, after which it falls again. The hemoglobin in female blood grows less after the tenth year. The ingestion of food is followed by transitory diminution in the amount of hemoglobin in consequence of the dilution of the blood.

The amount of hemoglobin in different animals is as follows: 9.7 per cent. in the dog; 9.9 per cent. in cattle; 10.3 per cent. in sheep; 12.7 per cent. in swine; 13.1 per cent. in the horse, and from 16 to 17 per cent. in birds.

In moist erythrocytes Hoppe-Seyler found the hemoglobin to constitute 40.4 per cent. of all the organic elements, while in the dry corpuscles the amount was 95.5 per cent., the amount being smaller in the nucleated corpuscles of animals.

Pathological.—A reduction in the amount of hemoglobin in the blood takes place during convalescence from febrile diseases, as well as in the presence of pulmonary tuberculosis. carcinoma, ulcer of the stomach, diseases of the heart, chronic disease, chlorosis, leukemia. pernicious anemia. and in conjunction with vigorous mercurial treatment for syphilis. In the presence of hunger the hemoglobin is more resistant than the remaining solid elements of the blood.

EMPLOYMENT OF THE SPECTROSCOPE FOR HEMOGLOBIN EXAMINATION.

The spectroscope (Fig. 12 and Fig. 161) consists (1) of a tube A, having at its peripheral extremity a slit S. which can be made larger and smaller. At the other extremity is a double convex lens C, known as a collimator, so adjusted that the slit is placed exactly at the focus of this lens. Light, from the sun or a lamp, illuminating the slit, passes therefore in parallel lines through C. (2) The prism P, by means of which parallel rays are refracted and broken up into the spectral colors, r–v. An astronomic telescope, inverting the image, is directed toward the spectrum r–v, which appears magnified from 6 to 8 times to the view of the observer B with the aid of the telescope. (3) The tube O contains a delicate scale M etched upon glass, and the image of which when illuminated is thrown upon the surface of the prism, whence it is in turn reflected to the eye of the observer. In this way the observer can see the spectrum and in or over it the scale. To exclude extraneous, disturbing light, the prism and the inner extremities of these tubes are enclosed within a metallic capsule whose interior is colored black.

Absorption-spectra.—If a colored medium, as, for instance, a solution of blood, be placed between the slit of the spectroscope and a source of light, the interposed solution does not permit the passage of all of the rays of white light, but some of these are absorbed. Therefore, that portion of the spectrum whose rays are not permitted to pass appears dark to the observer.

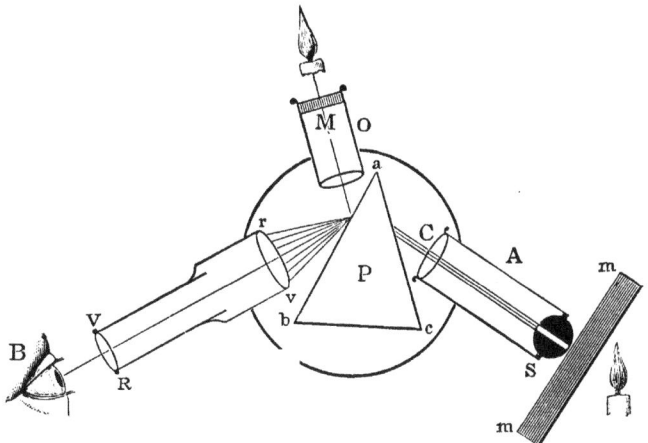

FIG. 12.—Diagrammatic Representation of the Spectroscope for Study of the Absorption-spectra of the Blood.

Flame-spectra.—If combustible substances are permitted to burn before the slit in a non-luminous (gas) flame at the extremity of a platinum wire the elements of the ash yield bands of a special color occupying a definite position. Thus, sodium gives rise to a yellow, potassium to a red and a violet line, which are found on combustion of the ash of almost all organs. If sunlight alone is permitted to pass through the slit the spectrum exhibits a large number of lines (Fraunhofer's lines) occupying definite positions within the colors and according to which different parts of the spectrum can be localized. These are designated A, B, C, D, etc., a, b, c, etc. (Fig. 15).

OXYGEN-COMBINATIONS OF HEMOGLOBIN: OXYHEMOGLOBIN AND METHEMOGLOBIN.

Oxygen-hemoglobin or Oxyhemoglobin—abbreviated to O-Hb—is readily developed when hemoglobin comes in contact with oxygen or with air (details on p. 78). Oxyhemoglobin is somewhat less readily soluble than hemoglobin. On spectroscopic analysis it exhibits two dark absorption-bands in the yellow and the green, whose position and width in an 0.18 per cent. solution are shown in Fig. 15 (2).

Oxyhemoglobin is contained within the erythrocytes in the circulating blood of the arteries and capillaries, as may be demonstrated by spectroscopic examination of the ear of the rabbit and of the thin layers of skin between two fingers placed in apposition. It is an exceedingly unstable chemical combination, yielding its oxygen even through the influence of such agents as release absorbed gases, as, for instance, setting free of gas through the action of an air-pump or the passage of other gases, particularly carbon monoxid, and heating to the boiling-point. Also in the circulating blood the oxygen is readily given up to the tissues of the body, so that in animals dead from suffocation only gas-free — reduced — hemoglobin is found in the veins.   Also constituents of the serum and sugar remove the oxygen.   By addition of reducing substances to a solution of oxyhemoglobin, as, for instance, ammonium sulphid, the two bands of oxyhemoglobin disappear and reduced gas-free hemoglobin results (Fig. 15, 4).   This is recognizable from its wide ill-defined absorption-band.   Agitation with air, however, at once restores both bands through the formation of oxyhemoglobin.   Solutions of oxyhemoglobin are readily distinguished by their scarlet color from the wine-violet-red tint of reduced hemoglobin.

The yellowish-green color of the solar spectrum thrown isolated upon the closed upper eyelid causes a sensation of dark. If the base of two fingers be ligated to the point of interrupting the circulation it will be seen on spectroscopic examination of the intervening red cutaneous seam that the oxyhemo-

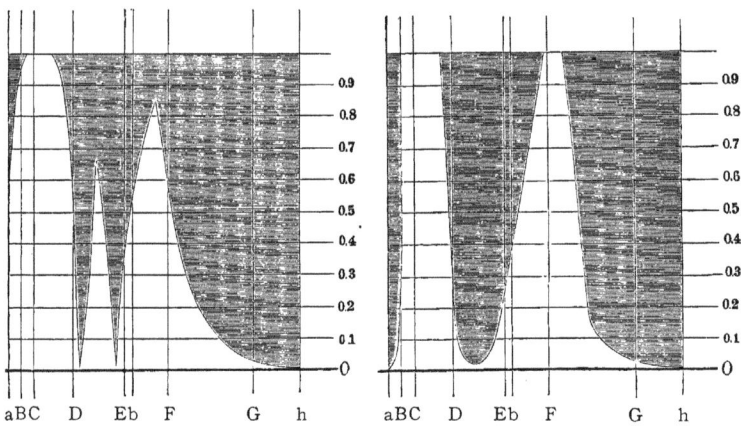

Figs. 13 and 14.—The Absorption-spectra of Oxyhemoglobin (Fig. 13) and of Gas-free Hemoglobin (Fig. 14) with Increasing Concentration. The letters of the lower line indicate the Fraunhofer lines. The figures at the side indicate the percentage-strength of the solutions (after Rollett).

globin is soon transformed into reduced hemoglobin. This reaction is delayed under the influence of cold; it is accelerated in youth, during muscular activity or with suppression of breathing and generally also in the presence of fever. A beating heart also exerts a reducing influence upon oxyhemoglobin. The absorption-spectra naturally vary with the concentration of the solution; in the presence of a greater amount of hemoglobin the bands are wider and may become confluent, and finally the largest part of the spectrum may thus become dark. Figs. 13 and 14 show how the absorption-bands appear in solutions of varying strengths: from a 1 per cent. solution (above) the concentration progressively diminishes downward by gradations of 0.1 per cent., until at O O the fluid is without hemoglobin. The thickness of the layers of fluid is placed at 1 cm.

Spectroscopic examination of small blood-spots, possibly for medico-legal purposes, may be of the greatest importance. Often a minute spot is sufficient. Dissolved with one or two drops of distilled water it may be introduced longitudinally in a thin glass tube before the narrow slit of the spectroscope, and the two bands of oxyhemoglobin appear.

Preserved in alcohol, oxyhemoglobin is transformed into a modification insoluble in water but otherwise identical, namely parahemoglobin.

A second oxygen-containing isomeric, but chemically more stable crystallizable combination is *methemoglobin*, whose molecule contains the same amount of oxygen as oxyhemoglobin, but in different arrangement. Its spectrum closely resembles that of hematin in acid solution (Fig. 15, 5). The band toward the red is the heaviest, while the others are narrow and are in part designated as not characteristic.

Demonstration.—1. By oxidizing substances, such as ozone, potassium iodid, chlorates. nitrates. 2. By reducing substances, such as nascent hydrogen and pyrogallol. 3. By indifferent influences, such as prolonged heating or slow desiccation of the blood. Potassium permanganate, potassium ferrocyanid and ferricyanid exert an intense effect, while nitrites transform the oxyhemoglobin into

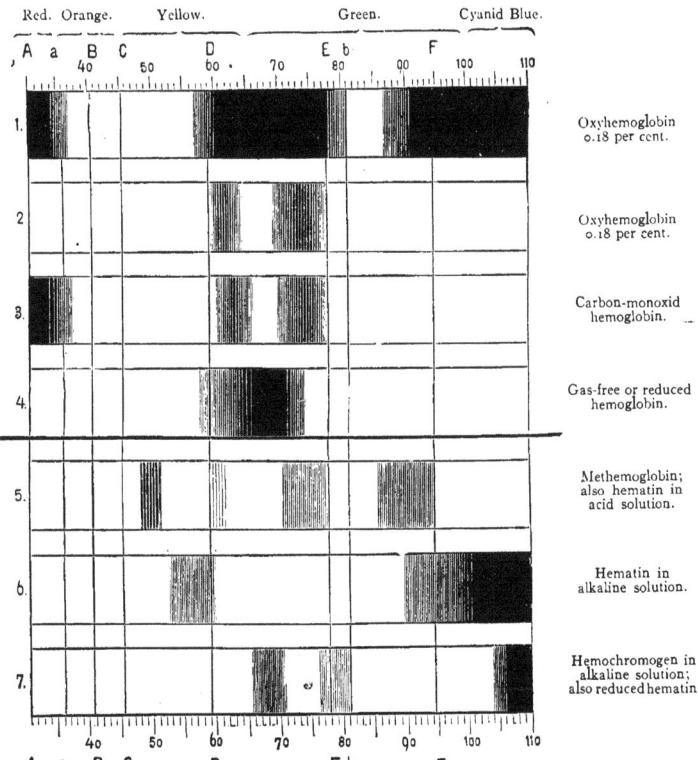

FIG. 15.—The Various Absorption-spectra of Hemoglobin. In all of the spectra the various Fraunhofer lines and a scale in millimeters are drawn.

a mixture of methemoglobin and nitrogen-monoxid hemoglobin. Not alone lake-colored blood, but also the hemoglobin of intact erythrocytes may be transformed into methemoglobin, as, for instance, by potassium chlorate, antifebrin and other substances, and also by intoxication with these substances. Often both conditions are present in combination. The occurrence of methemoglobin in solutions in the blood-plasma of a poisoned individual is designated *methemoplasmia*, and the occurrence of the methemoglobin in the preserved blood-corpuscles *methemacytosis*. Lesser degrees of the latter may recede spontaneously in the body without destruction of the erythrocytes   Profound influences resulting in the production of methemoglobin destroy the blood-corpuscles and require transfusion.

Preparation of Crystals.—To the solution of isolated erythrocytes described on p. 37 is added double its volume of a concentrated solution of ammonium sulphate, and evaporation is permitted to take place in the cold.   There form brownish-red needles, prisms or plates with marked pleochroism.   Methemoglobin develops in part spontaneously in the body, as, for instance, in bloody urine, in the sanguinolent contents of cysts, in old extravasates and in dried blood-crusts.   The addition of a trace of ammonia to a solution of methemoglobin produces an *alkaline solution of methemoglobin*, which exhibits two bands similar to those of oxyhemoglobin, but of which the first is the wider and extends the more toward the red.   If a reducing solution of ammonium sulphid be added to solutions of methemoglobin, reduced hemoglobin develops.

CARBON-MONOXID HEMOGLOBIN AND CARBON-MONOXID POISONING.

Carbon-monoxid hemoglobin is a more stable combination than the preceding and is produced when carbon monoxid is brought into contact with hemoglobin or oxyhemoglobin. It is cherry-red in color, not dichroic, and it exhibits in the spectrum two absorption-bands that closely resemble those of oxyhemoglobin, but are somewhat closer together and more toward the violet (Fig. 15, 3). It can be readily recognized, however, from the fact that reducing substances, which influence the oxyhemoglobin, do not dissolve these bands, that is, do not transform the carbon-monoxid hemoglobin into reduced hemoglobin. A further means of recognition consists in the sodium-test: a 10 per cent. solution of sodium hydroxid added to carbon-monoxid hemoglobin and heated gives rise to a cinnabar-red color. The same solution added to oxyhemoglobin produces a black-brown-greenish mass. The spectrum-analytical examination and the sodium-test permit the recognition of three-tenths carbon-monoxid hemoglobin mixed with seven-tenths oxyhemoglobin.

Carbon-monoxid hemoglobin reactions: Modified sodium-test: The blood is diluted 20 times and an equal amount of sodium hydroxid of a specific gravity of 1.34 is added in a test tube.   Carbon-monoxid blood assumes a beautiful red color after addition of ammonium sulphid—2 grams of sulphur being added to 100 grams of yellow ammonium sulphid—and 30 per cent. acetic acid, while normal blood assumed a greenish-gray coloration.   Both kinds of blood exhibit also differences in color when treated as follows: Dilute potassic hydrate is added, and then a few drops of a watery solution of pyrogallic acid; the mixture is shaken at once and permitted to stand protected from the air. For the purpose of the test, blood made lake-colored with water may be used, as well as blood in which the erythrocytes are preserved by addition of concentrated solution of sodium sulphate.   Three cu. cm. of blood are diluted with ,100 cu. cm. of water; 10 cu. cm. of this are mixed with 2 cu. cm. of 2 per cent. solution of grape-sugar and 2 cu. cm. of saturated solution of barium carbonate or lime-water, and the whole is heated almost to the boiling-point. From 4 to 5 volumes of lead acetate added to the blood cause a distinct difference accordingly as oxygen or carbon-monoxid blood is present.

Oxidizing substances, as, for instance, solutions of potassium permanganate—0.025 per cent.. potassium chlorate—5 per cent., and dilute chlorin-water. render solutions of carbon-monoxid hemoglobin cherry-red, while they render solutions of oxyhemoglobin pale yellow.   Both varieties of hemoglobin thus treated acquire the bands of methemoglobin, the carbon-monoxid hemoglobin considerably later. Subsequent addition of ammonium sulphid transforms the forms of hemoglobin thus altered back again into oxyhemoglobin and carbon-monoxid hemoglobin.

By reason of its greater constancy carbon-monoxid hemoglobin resists putrefaction for a long time, as well as the action of hydrogen sulphid.

If carbon monoxid be inspired it gradually displaces, volume for volume, the oxygen of the hemoglobin, and death finally results; 1000 cu. cm. of carbon monoxid will kill human beings if breathed at once. Small amounts of carbon monoxid in the air ($\frac{1}{4000}$–$\frac{1}{1000}$), however, suffice to generate comparatively large amounts of carbon-monoxid hemoglobin within a short time. As by means of long-continued treatment of carbon-monoxid hemoglobin with other gases, particularly oxygen,—passing them through—the carbon monoxid may be gradually again separated from the hemoglobin, with the re-formation of oxyhemoglobin, so in the body also the carbon monoxid is eliminated through the respiratory process in the course of a few hours, a portion of the carbon monoxid apparently being oxidized into carbon dioxid.

Poisoning with **Carbon Monoxid.**—Carbon monoxid results from incomplete combustion of carbon, as, for instance, through premature closure of stove-valves and badly smoking lamps. It occurs in illuminating gas in a proportion of from 12 to 28 per cent. As carbon monoxid has 200 times as great an affinity for hemoglobin as oxygen, more and more of the latter is displaced from the blood by the breathing of air containing carbon monoxid, and life naturally can continue only so long as sufficient oxygen is conveyed by the blood as is necessary to maintain the processes of oxidization essential to life. Death occurs amid peculiar phenomena, even before all of the oxygen is expelled from the blood; under the most unfavorable circumstances one-fifth of the oxygen will be retained in the blood.

Applied directly to nerve and muscle the gas has no influence whatever. Acting through the blood, however, phenomena appear that are indicative primarily of stimulation, but secondarily of paralysis of the nervous system. Thus, there occur at first severe headache, great restlessness, excitement, increased cardiac and respiratory activity, salivation, tremor, twitching, and spasm. Later, mental confusion, exhaustion, drowsiness, and paralysis set in, and even loss of consciousness, labored stertorous breathing, finally complete loss of sensibility, cessation of breathing and of the heart-beat and death. The temperature at the beginning exhibits an elevation of perhaps a few tenths of a degree C.; then there follows a decline of about 1° C. and more. The pulse-beat at first exhibits increased energy, while later the pulse becomes small and frequent.

Garland-like constrictions of the vessels, followed later by marked dilatation, with hyperemia of the viscera, accompanied by a fall in the blood-pressure, indicate primary stimulation and secondary paralysis of the vasomotor center. The change in temperature mentioned is to be referred to the same cause. This would also explain the appearance of sugar in the urine sometimes observed—in dogs only after abundant feeding of proteid. After the intoxication has terminated the excretion of urea is said to be increased, because the albuminates exhibit a greater tendency to disintegration. In cases of poisoning the great hyperemia of the viscera with fluid cherry-red blood and the dilatation of the vessels are conspicuous. Further, there are friability and softening of the brain, marked catarrh of the respiratory organs and granular degeneration of the muscles. Liver, kidneys, and spleen appear hyperemic, large, flabby, in a state partly of granular and partly of fatty degeneration. All of the muscles and viscera exhibit an exquisite cherry-red color. The spots of postmortem lividity are bright red.

Poisoned persons if still living should be at once brought into the fresh air. High degrees of intoxication demand transfusion. After recovery from the poisoning, sometimes paralysis, rarely anesthesia, trophic disorders and derangement of cerebral activity persist. If mixed with pure oxygen carbon monoxid acts less rapidly.

OTHER HEMOGLOBIN-COMBINATIONS.

Nitric-oxid hemoglobin is formed when nitric oxid enters into combination with hemoglobin.

As this gas in contact with oxygen is at once transformed into nitrous acid, all of the oxygen must first be removed from the blood and the apparatus, possibly through the passage of hydrogen, in the preparation of nitric-oxid hemoglobin. For this reason it cannot be formed within the body. Nitric-oxid hemoglobin is a still more active chemical combination than carbon-monoxid hemoglobin. It is of a bluish-violet color and in the spectrum it exhibits two absorption-bands, pretty much like those of the two other gas-combinations, but less intense, and not dissolved by reducing substances.

The three combinations of hemoglobin with oxygen, carbon monoxid and nitric oxid just considered crystallize like gas-free hemoglobin. They are isomorphous and their solutions are not dichroic. All three gases unite in equal amounts with hemoglobin and they can be expelled in a vacuum.

Hydrocyanic acid also forms readily decomposed combinations with hemoglobin. These develop in cases of hydrocyanic-acid poisoning, and they exhibit two bands that are situated somewhat nearer the violet than those of oxyhemoglobin and are slowly obliterated by reducing substances. This hydrocyanic-acid hemoglobin appears to consist of hydrocyanic acid plus oxyhemoglobin. There is, besides, a further combination of hydrocyanic acid with oxygen-free hemoglobin.

DECOMPOSITION OF HEMOGLOBIN.

Hemoglobin can be decomposed into: (1) iron-containing, pigmented hematin and (2) albuminoid, colorless globin, containing sulphur: (a) by addition of all acids, even feeble carbon dioxid in the presence of much water; (b) by strong alkalies; (c) by all agents that coagulate albumin, as well as by heat at a temperature of from 70° to 80° C.; (d) by ozone.

Hematin.—$C_{32}H_{32}N_4FeO_4$ represents about 4 per cent. of the hemoglobin in the dog. It is of blackish-blue color in reflected light, brown in transmitted light, insoluble in water, alcohol and ether, but soluble in dilute alkalies and acids, as well as in alcohol containing sulphuric acid or ammonia. It does not occur within the body. Hematin thus developed appears in an amorphous form, although it has also been possible to produce it crystallized in needles and rhombic plates.

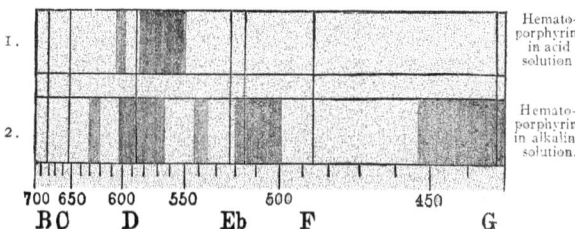

FIG. 16.—The Absorption-spectra of Hematoporphyrin, with the Fraunhofer Lines and a Scale Whose Figures Indicate the Wave-lines of Light in Millionths of a Millimeter.

In the decomposition of hemoglobin containing oxygen hematin at once results, oxygen being bound. On the other hand, oxygen-free hemoglobin yields in a similar process of decomposition, at first a forerunner of hematin deficient in oxygen, namely purple-red hemochromogen ($C_{34}H_{36}NFe_4O_5$). This, however, is transformed into hematin in the presence of oxygen by taking up the latter. Hematin therefore represents an oxidization-stage of hemochromogen. The latter substance is soluble, with exclusion of oxygen, in dilute alkalies, with the formation of a cherry-red color. and exhibits two absorption-bands, namely, one between D and E, and another and narrower between E and b (Fig. 15, 7).

Hemochromogen can be prepared in crystalline form by mixing upon a glass slide one drop of defibrinated blood with one drop of pyridin and covering the whole. The preparation exhibits the absorption-bands and at times also small crystals arranged in the form of stars or sheaves. In the bloody extract of spirit-preparations no longer fresh putrefaction often produces the beautiful red hemochromogen in alkaline solution.

Dilute acids in alcoholic solution withdraw the iron from the hemochromogen and there thus results hematoporphyrin—$C_{16}H_{18}N_2O_3$, which is isomeric with bilirubin, and is permanent in the air. This can also be prepared from hematin by means of strong sulphuric acid. It exhibits in acid solution a small absorption-band in the orange and a wider band in the yellowish-green (Fig. 16, 1). The spectrum of the same substance in alkaline solutions is shown in Fig. 16, 2.

Hematin occurs in solution as—

(A) *Hematin in acid solution.* If acetic acid be added to a solution of hemoglobin the latter becomes mahogany-brown in color, as hematin in acid solution develops and is recognized by four absorption-bands in the yellow and the green (Fig. 15, 5).

(B) If this solution be over-saturated with ammonia *hematin in alkaline solution* develops, exhibiting an absorption-band at the junction between the red and the yellow (Fig. 15, 6).

(C) Addition of reducing agents causes disappearance of this band and produces two wide bands in the yellow, due to the *reduced hematin* thus formed (Fig. 15, 7), and which, according to Hoppe-Seyler, is identical with the hemochromogen in alkaline solution.

Hematin is prepared in substance by precipitation from a solution of hemin in a weak alkali by addition of a dilute acid.

Hemoglobin is transformed into green sulphur-methemoglobin by hydrogen sulphid. This substance also causes the green coloration of putrid portions of the cadaver.

Hematin when reduced in alkaline solution with tin and hydrochloric acid yields *urobilin.* The latter results likewise through the action of hydrogen dioxid on acid.hematin.

Urobilin is occasionally found in cysts, exudates, and transudates. It forms likewise in sterile blood kept at the temperature of the body.

HEMIN (HEMATIN CHLORID); IDENTIFICATION OF BLOOD BY MEANS OF THE HEMIN-TEST.

Teichmann prepared in 1853 from the anhydrid of hematin crystals that Hoppe-Seyler recognized as hematin chlorid—$C_{32}H_{30}N_4O_3FeHCl$. As these may be obtained in characteristic form even from traces of blood they play an important rôle in forensic medicine. The demonstration of their presence depends upon the fact that the hemoglobin dried and heated with an excess of water-free acetic acid—so-called glacial acetic acid, which must burn on a glass rod held in the flame—and addition of sodium chlorid yields hemin-crystals (Figs. 17 and 18). These appear in the form of small rhombic plates, columns, or rods, although they probably belong to the monoclinic system. Not rarely they take the form of hemp-seeds or shuttles or paragraph-signs. At times some lie crossed or in tufts. In crystalline form the hemin-crystals of all varieties of blood examined are identical. They are doubly refracting, appearing yellow and glistening under the polarization-microscope, in contrast with their dark surroundings. with marked absorption of the light parallel with the longitudinal axis of the crystal. They are pleochromatic, that is, bluish-black and glistening like polished steel in reflected light and mahogany-brown in transmitted light.

(1) *Preparation from Dry Blood-stains.*—Several particles of the dry mass are placed upon a glass slide, two or three drops of glacial acetic acid and a minute crystal of sodium chlorid are added. and after the cover-slip has been placed in position heat is carefully applied some distance above a spirit-lamp until a number of small bubbles form. On cooling the crystals will be visible in the preparation (Fig. 18).

(2) *Preparation from stains upon porous bodies*, from which the hemoglobin cannot be scraped. The stained object—fabric, wood—is extracted with a dilute solution of potassic hydrate and then with water. To both filtered solutions a solution of tannic acid is added, and finally acetic acid until an acid reaction is produced. The resulting precipitate is washed upon a filter, then to a portion thereof upon a glass slide a crystal of sodium chlorid is added, and the whole is dried. Finally, the dried object is treated according to the method just described.

(3) *Preparation from Liquid Blood.*—The blood should always have been previously dried slowly and carefully. Then the process is continued as in the first method.

FIG. 17.—Hemin-crystals: 1, from a human being; 2, from a seal; 3, from a calf; 4, from a pig; 5, from a lamb; 6, from a pike; 7, from a rabbit.

FIG. 18.—Hemin-crystals Prepared from Blood-stains.

(4) *Preparation from Dilute Solutions Containing Hemoglobin.*—To the fluid is added ammonia, next tannic acid and then acetic acid until the reaction is acid. A blackish precipitate of hematin tannate forms rapidly. This is washed upon a filter with distilled water, then dried and heated in the same way as according to the first method, except that instead of sodium chlorid a crystal of ammonium chlorid is added.

Not rarely at least small hemin-crystals can be obtained from putrid and lake-colored blood, but under such circumstances the test often fails. Dried with iron-rust, as upon weapons, blood usually no longer yields the reaction. Under such circumstances the matter is, according to Heinrich Rose, scraped away and boiled with dilute potassium-hydrate solution. If blood be present the dissolved hematin forms a fluid that in thin layers presents a bile-green color, but in thick layers a red color.

Hemin-crystals have been demonstrated in all classes of vertebrates, as well as in the blood of the earth-worm. From some kinds of blood, as, for instance, that of cattle and of swine, only irregular masses; scarcely recognizable as having crystalline form, at times develop. Hemochromogen, hematoporphyrin, blood rubbed with sand or animal charcoal, addition of certain salts of iron, lead, mercury, and silver and lime prevent the development of the reaction. The crystals of hemin are insoluble in water, alcohol, ether, and chloroform. They are dissolved by concentrated sulphuric acid, with expulsion of hydrochloric acid and the development of a violet-red color. They are dissolved by dilute alkalies. If a solution of hemin-crystals in ammonia is evaporated, then heated to 130° C., next treated with boiling water, which removes the ammonium chlorid formed, hematoporphyrin results. This is a bluish-black, amorphous powder, becoming brown when rubbed. Its solutions in caustic alkalies are dichroic: that is brownish-red in reflected light, garnet-red in a thick layer with transmitted light and olive-green in a thin layer. The acid solutions are monochromatic—brown.

For the preparation of hemin-crystals in large amount, it is advisable to heat dry horses' blood with 10 parts of formic acid until bubbles form. If the hemin-crystals are suspended in methyl-alcohol, they dissolve after addition of iodin and application of heat, with the development of a purple color, which becomes brown after addition of bromin and green after the passage of chlorin-gas. All of these exhibit a characteristic appearance in the spectroscope. The glacial acetic acid may be replaced by an alcoholic solution of oxalic or tartaric acid, and the sodium chlorid by salts of iodin or bromin. In the latter event bromin-hematin or iodin-hematin is formed.

HEMATOIDIN.

An important derivative of hemoglobin is sorrel-colored hematoidin—$C_{32}H_{36}N_4O_6$ (Fig. 19), which forms in the body from hematin through loss of iron and taking up of water when-ever blood stagnates outside of the circula-tion and undergoes decomposition, as, for instance, in apoplectic extravasations of blood, in coagulated plugs in blood-vessels (thrombi). It develops regularly in every Graafian follicle from the drop of blood poured out at the menstrual rupture of the follicle. It is free from iron, crystallizes in clinorhombic prisms, and is soluble in chloroform and in warm alkalies. Probably it is identical with the biliary coloring-matter, bilirubin.

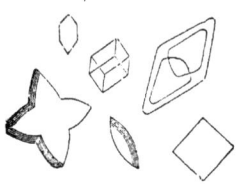

FIG. 19.—Hematoidin-crystals.

Pathological.—After extensive dissolution of blood in the vessels, as, for instance, after transfusion with foreign blood, hematoidin-crystals have been observed in the urine.

THE COLORLESS PROTEID OF HEMOGLOBIN.

This is designated *globin* and is closely related to histon.

Demonstration.—A solution of hemoglobin is made feebly acid with hydro-chloric acid, then one-fifth volume of alcohol is added and the mixture is shaken with ether. The coloring-matter is taken up by the ether and is precipitated by the ammonia. Hydrochloric or nitric acid likewise precipitates the globin, which, however, is redissolved on boiling. Hematin and globin are probably not the sole products of the decomposition of hemoglobin. As hemoglobin-crystals can be decolorized under special conditions, it is most probable that they owe their form to the proteid body. On introducing hemoglobin-crystals with alcohol in a dialyzer surrounded by ether acidulated with sulphuric acid Landois succeeded in decolorizing the crystals.

PROTEID BODIES IN THE STROMA.

These constitute from 5.10 to 12.24 per cent. of the dry red blood-corpuscles of man, including a *globulin* participating in fibrin-formation and possible traces of a *sugar-forming ferment*. Under special conditions it has been observed that the stromata, coherent in masses, form a substance—*stroma-fibrin*—resembling fibrin.

L. Brunton has found in the nuclei of nucleated red blood-corpuscles a *mucin-containing* body, Miescher *nuclein* and Kossel *histon* united with the latter.

THE REMAINING CONSTITUENTS OF THE RED BLOOD-CORPUSCLES.

The red corpuscles contain further: *Lecithin*, 1.867 per cent. in dry erythrocytes; *urea*, equally divided between erythrocytes and serum; *cholesterin*, 0.151 per cent.; no fats; lactic acid, in the dog.

Lecithin and cholesterin can be obtained by agitating considerable amounts of stroma or isolated blood-corpuscles with ether. If the ether is permitted to evaporate the characteristic globular myelin-forms of lecithin and the crystals of cholesterin will be recognized.

Water, 631.63 in the thousand.

After abstraction of considerable quantities of blood the amount of water diminishes and the amount of dry substance, as well as the nitrogen of the erythrocytes, increases. The opposite effect is brought about by infusion of physiologic salt-solution.

Inorganic matters, 7.28 in the thousand, particularly combinations of potassium and phosphoric acid. The phosphoric acid is derived only from consumed lecithin, the sulphuric acid in large part from the hemoglobin consumed in the analysis. Some manganese also is present.

Blood-analysis.—One thousand parts by weight of horses' blood are made up as follows:

344.18 parts blood-corpuscles, with 128 of solids—383 in the dog,
655.82 parts plasma, with 10 per cent. of solids—617 in the dog.

One thousand parts by weight of moist blood-corpuscles are made up as follows:

Solids, ...367.9 (swine), 400.1 (cattle), 435 (horse),
Water, ...632.1 (swine), 599.9 (cattle), 565 (horse).

The solids include:

Hemoglobin, ...	261	(swine)	280.5	(cattle)
Albumin, ...	86.1	"	107	"
Lecithin, cholesterin and other organic matters,..	12.0	"	7.5	"
Inorganic matters,............................	8.9	"	4.8	"
Including potassium, ...	5.543	"	0.747	"
magnesium, ...	0.158	"	0.017	"
chlorin, ...	1.504	"	1.635	"
phosphoric acid,.................	2.067	"	0.703	"
sodium, ...	0	"	2.093	"

CHEMICAL CONSTITUENTS OF THE LEUKOCYTES.

Leukocytes from the plasma of lymphatic glands, as well as pus-corpuscles contain proteids as follows: little albumin, alkali-albuminate and an albuminate resembling myosin and coagulating at 48°, two globulins coagulable at 48.5° and 75° C. respectively, together with serum-globulin, peptone and a coagulating ferment, further considerable nucleins from the nuclei, nucleo-histon, little glycogen, lecithin, cerebrin, cholesterin, fats, protagon, inosite, amidovalerianic acid.

Lymphocytes contain 11.5 per cent. of dry matter. In 100 parts by weight of dry pus there are 0.416 earthy phosphates, 0.143 sodium chlorid, 0.606 sodium phosphate, 0.202 potassium, in part in the form of monopotassium phosphate.

THE BLOOD-PLASMA AND ITS RELATION TO THE SERUM.

The unmodified fluid of the blood is known as *plasma*. In this, how-ever, there separates, generally soon after escape of the blood from the vessels, a fibrillated substance, namely *fibrin*. After this separation, the remaining clear fluid, which no longer undergoes coagulation spon-taneously, is known as *serum*. The plasma is a clear, transparent, somewhat consistent fluid, which in most animals is almost colorless, but in human beings is yellowish and in the horse of citron-yellow color.

DEMONSTRATION OF PLASMA.

(A) Without admixture. As plasma cooled to a temperature of 0° C. does not undergo coagulation, the blood flowing from a vein—particularly of the horse, which is peculiarly suitable on account of the slowness of coagulation and the rapidity with which sedimentation of the blood-corpuscles takes place—is received into a narrow, graduated cylinder standing in a cold mixture. In the blood, which remains fluid, the erythrocytes sink to the bottom within a few hours, and the plasma forms above a clear fluid, which can be removed with a cooled pipet. If this is further passed through a filter upon an ice-cold funnel the plasma will also be freed from leukocytes.

The *amount* can be read from the graduated cylinder, but only approximately, because of the presence of plasma between the sedimented corpuscles. If heated, the plasma, in so far as it contains leukocytes, is transformed, through the forma-tion of fibrin, into a tremulous jelly. If, however, it be whipped with a rod the fibrin will be obtained as a stringy mass. Plasma free from leukocytes is not capable of coagulation.

If the amount of fibrin in a volume of plasma isolated by whipping (varying between 0.7 and 1.0 per cent.) and in the same manner the amount in a volume of blood be determined the two results afford a basis for estimating the amount of plasma in the blood.

(B) With saline admixture. If the blood flowing from a vein into a graduated cylinder be mixed with agitation with ¼ volume of concentrated solution of sodium sulphate or with a 25 per cent. solution of magnesium sulphate (1 volume to 4 volumes of blood), the cells sink to the bottom in a cool place, while the clear supernatant *saline plasma*, which can be measured, is pipetted off. If the salt be removed from the plasma by means of the dialyzer coagulation takes place. The same result is brought about by dilution with water.

FIBRIN: ITS GENERAL PROPERTIES; COAGULATION.

Fibrin is the substance that brings about coagulation in shed blood as well as in plasma and likewise in lymph, and in the chyle, by solidification. If the fluids mentioned are placed at rest and left to themselves the fibrin forms innumerable microscopically delicate (Fig. 9) doubly refracting filaments, which hold the blood-cells together like a spider's web, and with the cells form a mass of gelatinous consistency that is known as *blood-clot (placenta sanguinis)*. At first this is quite diffluent and it is only in the course of from two to fifteen minutes that a number of filaments appear upon the surface that can be removed with a needle, while the interior of the blood-mass is still liquid. In a short time the filaments extend throughout the entire mass. The blood in this stage of coagulation has been designated *cruor*. Later, in the course of from twelve to fifteen hours, the threads of fibrin contract more and more firmly about the corpuscles, and there then results the more solid, gelatinous, tremulous substance, which can be cut with a knife, and which has expressed a clear fluid, known as *blood-serum (serum sanguinis)*. The blood-clot takes the shape of the vessel in which the blood has been received. By solution with water of the blood-corpuscles in the broken-up blood-clot the fibrin can be isolated.

If the blood-corpuscles sink rapidly in the blood, and if the advent of coagulation be delayed, the upper layer of the blood-clot is only stained yellow on account of the absence of enclosed erythrocytes. This is the rule with horses' blood, but it has been observed in the case of human blood, particularly when inflammation was present in some part of the body. Therefore, this layer has also been designated *crusta phlogistica*. Such blood is richer in fibrin and therefore coagulates more slowly.

The crusta forms also under other conditions, but the cause of its formation is not always clear. Thus it occurs when the specific gravity of the blood-corpuscles is increased or that of the plasma is diminished, as in cases of hydremia and chlorosis, in consequence of which the corpuscles sink more rapidly, and during pregnancy. The taller and narrower the vessel, the higher is the crusta.

It can be readily understood why the blood-clot undergoes greater contraction and appears more contracted in the neighborhood of the unpigmented layer free from corpuscles.

If freshly shed blood is whipped with a rod the filaments of fibrin that form collect about the rod, and in this way the fibrin is obtained as a fibrous, grayish-yellow mass from the blood now become defibrinated.

The plasma exhibits analogous phenomena, but it forms only a soft, tremulous jelly, by reason of absence of the resistant blood-corpuscles. The plasma undergoes coagulation only when it contains leukocytes. If these be removed by filtration the plasma is no longer coagulable.

Although the fibrin appears voluminous, it constitutes only from 0.1 to 0.3 per cent. of the mass of the blood. In this connection, it is noteworthy that in two different specimens of the same blood the amount of fibrin may vary considerably.

Fibrin is insoluble in water or ether. Alcohol causes it to shrink by dehydration, while hydrochloric acid causes it to swell and assume a vitreous appearance, with transformation into syntonin. In the fresh state fibrin is tough and elastic. If dried, it becomes horn-like, translucent, brittle, and pulverizable.

Fresh fibrin is capable of actively decomposing hydrogen dioxid into water and oxygen, just as other fresh animal or vegetable tissue is likewise capable of doing. Boiled or preserved in alcohol it loses this power. In the fresh state it is soluble in from 6 to 8 per cent. solutions of sodium nitrate or sodium sulphate, with the formation of globulin; and in dilute alkalies and ammonia, with the formation of alkali-albuminate. These solutions are not coagulated by heat. Also weak solutions of haloid salts (sodium chlorid, ammonium chlorid, potassium iodid, sodium iodid, sodium fluorid, ammonium fluorid) dissolve fibrin at a temperature of 40°, as, for instance, sodium-chlorid solution, from 7 to 20 parts in the thousand, with the production of globulin-bodies and pro-peptone. Fibrin from swine is dissolved by 0.5 per cent. hydrochloric acid and also by malic, oxalic, butyric, acetic, citric, and lactic acids; fibrin from cattle, with greater difficulty. Fibrin exposed to air for a considerable time is not soluble in nitric acid, although it is soluble in neurin. As a result of putrefaction it likewise undergoes solution, with the formation of albumin. Fibrin contains lime, iron, and magnesium.

According to Schmiedeberg the fibrin obtained from plasma has the elementary formula $C_{108}H_{162}N_{30}SO_{34}$, while blood-fibrin has the following composition: $C_{112}H_{168}N_{30}SO_{35} + \frac{1}{2}H_2O$.

GENERAL PHENOMENA ATTENDING COAGULATION.

Blood does not undergo coagulation in immediate contact with the living and unaltered vessel-wall.   Therefore, Brücke was able to preserve uncoagulated for eight days blood cooled to 0° in the still beating heart of dead turtles.   The blood coagulates rapidly within the dead heart or vessels (but not in the capillaries) or within other channels, as, for instance, the urethra.   If blood stagnates in a living vessel, coagulation takes place in the central axis, because it is here not in contact with the living vessel-wall.   Coagulation is of the greatest importance in the control of hemorrhage from injured vessels, which otherwise might terminate fatally. The injured and necrotic tissues of the wound and the vessel-wall lead to the formation of the occluding thrombus by coagulation.

If the vessel-wall is altered by pathological processes, as, for instance, rough or inflamed in consequence of a lesion of the intima, coagulation may take place in such a situation even though the circulation be maintained.

Coagulation of the blood is prevented or retarded:

(a) By addition of alkalies or of ammonia, even in small amounts; further, of concentrated solutions of neutral salts of alkalies and earths —alkaline chlorids, also sulphates, phosphates, nitrates, carbonates ; disodium phosphate in 3 per cent. solution, soluble salts of calcium, strontium and barium dissolved in the blood to the extent of 0.5 per cent.   Simultaneous addition of sodium chlorid inhibits coagulation in still further degree.   Magnesium sulphate—1 volume of a 28 per cent. solution to 3½ volumes of horses' blood—acts most effectively in inhibiting coagulation.

(b) By precipitation of the calcium by means of oxalic acid.

Feeble acids also exert an inhibiting effect.   Thus, coagulation ceases after addition of acetic acid to the point of producing an acid reaction.   The presence of a large amount of carbon dioxid likewise retards coagulation; therefore, venous blood—and also the blood after asphyxiation—coagulates more slowly than arterial blood.

(c) By addition of egg-albumin, sugar-solution, glycerin, soaps or much water.   If uncoagulated blood be brought in contact with a layer of already separated fibrin coagulation is retarded.

(d) Cold (0° C.) retards coagulation for as long as an hour.   If blood be permitted to freeze at once, it will still be liquid on thawing, when it undergoes coagulation.   Coagulation is retarded also when the shed blood is exposed to high pressure; likewise when it is brought in contact with foreign substances to which it does not adhere, as, for instance, anointed substances.

(e) The blood of embryo birds does not coagulate at all before the twelfth or fourteenth day on account of the absence of fibrin-forming cells, and that of the hepatic veins but slightly.   Blood from the dog passed only through the heart and the lungs does not coagulate for a long time.   Blood from the renal vein, also blood cut off from circulation through the liver and intestines, does not coagulate at all.   Fetal blood at the moment of birth coagulates early, but slowly, as the amount of fibrin it contains is small.   Menstrual blood exhibits a slighter tendency to undergo coagulation if admixed with a considerable amount of alkaline mucus from the genital canal.

(f) In cases of *bleeders' disease*—hemophilia—coagulation appears to be want-ing on account of deficiency in the fibrin-generators, in consequence of which wounds of the vessels are not occluded by fibrinous thrombi. The *peptic ferment of the pancreas* dissolved in glycerin and injected into the blood inhibits its coagu-lation, as does also the *diastatic ferment.* Schmidt-Mülheim noted the same result after injection of *pure peptone* into the blood of dogs—0.5 gram to 1 kilo of dog, and 1.5 of rabbit. This is effective, however, only in the presence of the liver. The buccal secretion of the leech, the poison of vipers and the highly toxic substance in the serum of eels' blood likewise inhibit coagulation.

Coagulation is accelerated:

(a) By contact with foreign substances to which the blood adheres, as, for instance, threads and needles introduced into the veins. Also the entrance of air-bubbles into the vessels or the passage of other indifferent gases, as, for instance, nitrogen and hydrogen, exerts an accelerating effect. Removed from the vein, the blood coagulates quickly on the walls of the container, on its surface exposed to the air, on the rod with which it is whipped, etc.

(b) Many products of the *retrogressive metamorphosis of albuminates,* including uric acid, glycin, taurin, leucin, tyrosin, guanin, xanthin, hypoxanthin (not urea), as well as the biliary acids, further lecithin, cholin hydrochlorate, protagon, accelerate coagulation through in-creased ferment-formation. Added in excess, however, they exert an inhibiting effect. Solutions of gelatin injected into the veins cause the blood to coagulate almost instantly after escape from the vessels.

(c) If hemorrhage takes place rapidly the last amounts of blood coagulate earliest. Fresh fibrin, if permitted to remain for a consider-able time in blood, is again dissolved in part.

(d) Heating to a temperature of from 39° to 55° C. accelerates coagulation.

In the shed blood of man coagulation begins in the course of three minutes and forty-five seconds; in that of woman after two minutes and thirty seconds. Hunger exerts an accelerating effect.

Among vertebrates the blood of birds coagulates almost instantly, that of cold-blooded animals distinctly more slowly, while the blood of mammals occupies an intermediate position. The blood of invertebrates, which mostly is colorless, forms a soft, white fibrinous coagulum.

As the process of coagulation involves a change in the aggregate state, heat demonstrable with the thermometer must be set free.

In blood removed from a vein the degree of alkalinity diminishes up to the point of completed coagulation, probably from the formation of acid in the blood as a result of decomposition-processes.

In the process of coagulation a diminution in the amount of oxygen in the blood has been observed, although this takes place also in blood that has not yet undergone coagulation. There is, likewise, elimination of traces of ammonia. Both processes, however, appear not to stand in causal relation with the formation of fibrin.

NATURE OF COAGULATION.

Alexander Schmidt discovered in 1861 that coagulation is a fermen-tative process that consists in the transformation of the soluble albumin of the plasma into the solid substances of the fibrin through the activity of an enzyme that is designated *fibrin-ferment* or *thrombin.* This pro-teid is nothing but *fibrinogen.*

The enzymes or hydrolytic ferments behave in common in the organism in such a manner that they break up the bodies upon which they act into two other substances by taking up water. It, therefore, appears probable that as a result of the action of thrombin decomposition of the fibrinogen into fibrin and a lesser amount of a globulin-body that remains liquid and that Hammarsten has designated *fibrin-globulin*, takes place, with the taking up of water.

Demonstration of Fibrinogen—$C_{112}H_{168}N_{30}SO_{35}$.—Pulverized sodium chlorid is added to lymphatic transudate to the point of saturation. The fluid poured out into the serous sac surrounding the testicle (hydrocele) is especially useful for this purpose. The precipitated fibrinogen is collected upon a filter. This substance is found also in the lymph and in the chyle.

Saline plasma also is capable of precipitating fibrinogen by admixture of equal volumes of plasma and a concentrated solution of sodium chlorid. For purposes of purification it may then be dissolved rapidly and repeatedly in a dilute—8 per cent.—solution of sodium chlorid and again precipitated by a concentrated solution of sodium chlorid. The fibrinogen contained in the sodium-chlorid solution is precipitated by addition of water and is rapidly changed so that it resembles fibrin. Fibrinogen in saline solution coagulates at a temperature of from $52°$ to $55°$ C. Solutions free from salt do not coagulate if quickly brought to the boiling-point.

Fibrinogen behaves like globulin. It is soluble in dilute alkalies and it is precipitated from such solutions by the passage of carbon dioxid. It is further soluble in dilute solution of sodium chlorid, while addition of large amounts of sodium chlorid causes its precipitation as a soft, viscous, tough mass. It is dissolved also by dilute hydrochloric acid, although it is soon transformed into a body resembling syntonin (acid albuminate). In the fresh state it actively decomposes hydrogen dioxid. Its specific rotatory power is $52.2°$.

Demonstration of Fibrin-ferment—Thrombin.—Blood-serum from cattle, which contains a larger amount of ferment than the serum of carnivora, is admixed with twenty times its volume of strong alcohol. The resulting precipitate is collected upon a filter after the lapse of from two to four weeks. It contains the coagulated albumin and the ferment. It is dried over sulphuric acid and reduced to powder. One dram of this powder is stirred for ten minutes in 65 cu. cm. of water. If the mixture is not filtered, the ferment, dissolved in water, alone passes through the filter.

Thrombin is formed from a forerunner, a zymogen, which is present within the leukocytes and is designated prothrombin. Both are soluble with greater difficulty in an excess of acetic acid than globulins. Even small amounts of the ferment may cause coagulation of fluids containing fibrinogen and most readily at a temperature of $40°$ C. Prothrombin is destroyed at a temperature of $65°$, thrombin at a temperature between $70°$ and $75°$. The amount of ferment formed in the blood is the greater the longer the time that has elapsed between the escape and the coagulation of the blood. Blood flowing directly from the vein in alcohol yields no ferment.

Coagulation.—If the separate solutions (1) of the fibrinogenous substance and (2) of the ferment are admixed fibrin-formation takes place at once. The most favorable temperature for this is that of the body. A temperature of $0°$ C. prevents coagulation, while the boiling temperature destroys the ferment. The amount of ferment is a matter of indifference. Larger amounts cause more rapid, but not increased, separation of fibrin. For the formation of fibrin the presence of a certain amount of salt in the fluid is requisite—one per cent. sodium chlorid. Otherwise the process takes place but slowly and is only partial. The presence of a calcium-salt favors coagulation. If the calcium is precipitated by alkali-oxalate this prevents coagulation, although it is true that the presence of a large amount of ferment in the blood is capable of neutralizing the influence of the calcium. Fibrinogen and fibrin contain equal amounts of calcium. Probably the action of the calcium bears some relation to the formation of the fibrinferment, for the plasma contains a substance that exerts a marked coagulative effect after addition of calcium-salts.

According to Kossel and Lilienfeld the leukonuclein contained in the nuclei of the leukocytes, and the nucleinic acid resulting from its decomposition, accelerate coagulation.

If coagulation has taken place in the plasma of the blood, all of the fibrinogenous material in the serum is utilized for the formation of fibrin. On the other hand, fibrin-ferment will still be present in the serum in sufficient amount. Therefore, if blood-serum be added to a fluid containing fibrinogen, as, for instance, hydrocele-fluid, coagulation will at once take place anew.

SOURCE OF THE FIBRINOGENOUS SUBSTANCES.

Alexander Schmidt has found that both fibrin-factors are formed from the destruction of leukocytes. In the circulating blood of man and of mammals, the fibrinogenous substance is already dissolved in the plasma as a soluble product of the physiologic involution-processes of the white cells. The circulating blood, however, contains a much larger number of leukocytes than was previously believed. As soon as the blood is shed, large numbers of white blood-corpuscles are dissolved —according to Alex. Schmidt 71.7 per cent. in the horse. The decomposition-products dissolve in the blood-plasma, and as a result the fibrinferment develops, to a certain extent as a cadaveric product, causing the separation of fibrin. Accordingly the fibrin-ferment does not preëxist within the uninjured corpuscles. Also the so-called transitional forms between colorless cells and erythrocytes in mammalian blood furnish the fibrin-factors as a result of their destruction, which takes place immediately after escape of the blood; likewise perhaps also the bloodplates. The ferment develops with the escape of the blood, and its formation reaches the maximum during the process of coagulation itself.

The influence of adhesion in favoring coagulation depends upon the fact that as a result the blood-corpuscles are caused to give up a portion of their contents—phosphoric acid and alkaline phosphates—to the plasma, to combine with salts of calcium and magnesium present principally in the plasma. If the calcium be precipitated from the blood by means of oxalic acid—1 gram of potassium oxalate to 1 liter of blood—coagulation no longer takes place. If, however, calcium chlorid be again added to this mixture coagulation will result.

In the blood of amphibia and birds it is the red blood-corpuscles that after escape undergo destruction in large numbers and furnish the fibrin-forming materials. In the blood of these animals Alex. Schmidt convinced himself at the same time that also the fibrinogenous substance was originally a constituent of the bloodcorpuscles.

It is thus clear that as soon as the fibrin-factors pass into solution in consequence of dissolution of the blood-corpuscles the separation of fibrin must take place through the combination of the two substances.

If considerable amounts of leukocytes are introduced into the circulation of an animal they are quickly dissolved in large numbers in the blood, so that even death may take place in consequence of widespread coagulation. If the animal survive immediate death by reason of the moderate extent of coagulation, the blood subsequently will be wholly incoagulable in consequence of the absence of leukocytes.

All protoplasmic structures may in combination with plasma set the fibrin-ferment free. The nitrogenous metabolic products of proteids are likewise capable of producing fibrin-ferment in plasma free from cells. These latter active substances can be extracted from the tissues—cells of the liver, the spleen, the lymph-glands, red and white blood-corpuscles, frog-muscle—by means of alcohol. If after alcoholic extraction the residue of such tissues is extracted with water, this watery extract absolutely inhibits coagulation. The substance thus extracted by water is designated by Alex. Schmidt *cytoglobin*, which is the forerunner of fibrinogen and also of serum-globulin.

In accordance with the preponderance in the plasma of either of the substances capable of extraction with alcohol or cytoglobin, coagulation is induced or inhibited respectively. Within the living body the inhibitory action of the cells preponderates, while outside the body the coagulating effect is operative. Those substances, such as the cytoglobin, that inhibit coagulation within the circulation furnish outside of the body the material for the formation of fibrin. As Alex. Schmidt, after addition of cytoglobin to filtered plasma, induced coagulation by addition of extractives in large amount, the amount of fibrin was more than doubled. The blood retains its fluidity in the circulation as long as the amount of cytoglobin exceeds that of the proteid metabolic products of the tissues. The blood may, however, remain fluid also because both of these do not pass over into the plasma.

Pathological.—From the investigations of Alex. Schmidt in collaboration with his pupils Jakowicki and Birk, it has been shown that even healthy functionating blood contains some fibrin-ferment from the destruction of white blood-corpuscles normally undergoing dissolution, and in greater amount in venous than in arterial blood. Nevertheless, it is always more abundant in shed blood. The fact, however, is particularly noteworthy that the amount of fibrin-ferment in the blood in cases of septic fever may increase to such a degree that spontaneous coagulation-thrombosis takes place and even terminates fatally. After injection of putrid matters leukocytes are dissolved in large number, but the ferment is present rather abundantly also in the blood of febrile patients generally. Also injection of peptone, of hemoglobin and in lesser degree of distilled water is followed by dissolution of numerous leukocytes. There are thus true blood-diseases in which the products of the dissolution of the leukocytes accumulate in the blood-plasma. In consequence, spontaneous coagulation naturally occurs within the circulatory organs, and as a result death may even be brought about. At least febrile elevation of temperature usually takes place. At the termination of such conditions the coagulability of the blood is naturally diminished.

Wooldridge showed that a fibrinogen—*tissue-fibrinogen*—occurs in the chyle and in the lymph as a product of the lymphatic glands. In human beings in whom blood-stasis exists in any part of the body, coagulation may take place, with the formation of thrombi, through admixture of lymph, as a certain amount of ferment is already present in the blood. The intestinal mucosa, the skin, and the lungs also appear to produce small amounts of fibrinogen constantly, while the liver and the kidneys constantly destroy it.

RELATIONS OF THE RED BLOOD-CORPUSCLES TO FIBRIN-FORMATION.

After it had been determined by a number of investigators that also the erythrocytes of birds, of the horse, of the frog, may contribute to the production of fibrin, Landois was able in 1874 to follow directly under the microscope the transformation of the stromata of the red blood-corpuscles of mammals into fibrin-fibers. If a drop of defibrinated rabbit's blood be introduced into frog's serum, without agitation, it will be observed that the erythrocytes attach themselves to one another. They become viscous upon the surface, and on pressure on the cover-slip it will be seen the adhesion can be broken up only with a certain amount of force, the adjoining surfaces of the swollen, globular corpuscles often being drawn out into threads. Even after the process has been in operation for a short time, all of the corpuscles are transformed into globules of lesser diameter and those lying nearest the periphery permit their hemoglobin to escape. The decolorization progresses from the periphery of the drop to the center, and finally only a coherent mass of stroma remains. The substance of the stroma exhibits great tenacity. At first the round contours of the individual blood-corpuscles can still be recognized, but as soon as a current is set up in the surrounding fluid by pressure upon or movement of the cover-glass, the stroma-mass becomes agitated to and fro and the stromata lying close together and adherent to one another become drawn out into delicate filaments and bands, with simultaneous disappearance of the previous contour of the cells. In this way the formation of fibrin-filaments from the stromata of the red blood-corpuscles can be followed step by step. Erythrocytes from human beings and from animals undergoing dissolution in the serum of different animals often exhibit the same phenomena.

Stroma-fibrin can be prepared also in the following simple manner: A one per cent. solution of sodium chlorid is shaken in a reagent-glass with ether and a few drops of defibrinated blood. The mixture soon becomes lake-colored. Put aside, the ether, which rises to the top, carries with it the filamentous stroma-fibrin to the surface of the fluid.

Stroma-fibrin and Plasma-fibrin.—Landois has designated *stroma-fibrin* that which arises directly from the stroma of the erythrocytes. On the other hand, the fibrin that is produced through the combination of the fibrin-factors dissolved in the coagulating fluid—plasma—is *plasma-fibrin*, or ordinary fibrin. Both designations are fully justified, if only to indicate the mode of origin of the fibrinous mass.

Substances that cause rapid dissolution of the erythrocytes bring about extensive coagulation, as, for instance, injection of bile or salts of the biliary acids, or of lake-colored blood into the veins. The effective agent under these circumstances is the stroma, through the development of the ferment, and in lesser degree the hemoglobin. As foreign blood after injection often undergoes rapid disintegration in the blood-stream of the recipient, extensive coagulation is often observed under such circumstances, while at the same time the individual smaller vessels are often occluded by plugs of stroma-fibrin.

CHEMICAL CONSTITUTION OF THE BLOOD-PLASMA AND THE SERUM.

The *proteids* constitute about 8 or 10 per cent. of the plasma. Of these only about 0.2 per cent. are bodies producing fibrin. If these be eliminated through the process of coagulation, the plasma is transformed into serum. The specific gravity of human serum is between 1027 and 1029. The blood-plasma contains, besides, the following proteids:

(a) *Serum-albumin*—$C_{75}H_{120}N_{20}SO_{24}$—from 3 to 4 Per Cent.—Its percentage-composition is C 53.1, H 7.1, N 15.9, S 1.9, O 22, Ash 0.22. Its coagulation-temperature is from 51° to 53° C.; its specific rotatory power —61°. In the horse and the rabbit it crystallizes in hexagonal prisms, with a pyramid upon one side. The crystals are doubly refracting, up to 1 cm. in length, and are coagulable by heat.

It is a remarkable fact that serum-albumin is absent from the blood of starving snakes and it makes its appearance only after feeding.

(b) *Serum-globulin*—also known as *fibrinoplastic substance* or *paraglobulin* and also as *serum-casein*—from 2 to 4 per cent. If magnesium sulphate in substance is added to serum to the point of saturation, serum-globulin is precipitated at a temperature of 35° C. It is washed upon a filter with concentrated solution of magnesium sulphate. It is soluble in a 10 per cent. solution of sodium chlorid, and coagulates at a temperature of from 69° to 75° C. Its specific rotatory power is—47.8°, and its formula is $C_{117}H_{174}N_{30}SO_{38}$.

After precipitation of the serum-globulin from the serum by means of magnesium sulphate the *serum-albumin* is precipitated by further saturation with sodium sulphate. Neutral ammonium sulphate, added to the point of saturation, precipitates all of the proteids of the blood-serum, and also those of egg-albumin and of milk ; further, propeptone, but not peptones. Globulin can be precipitated also by dialysis of the serum, as it is insoluble in solutions free from salt.

During hunger the amount of globulin increases, while that of albumin diminishes. After abstraction of blood the amount of globulin in the blood increases. Paraglobulin occurs also in erythrocytes, as well as in the fluids of the connective tissue and the cornea. According to von Jaksch, 100 cu. cm. of blood contain 22.62 grams of albumin, while an equal amount of serum contains more than 8 grams. The latter figure varies under pathological conditions.

Fats—from 0.1 to 0.2 Per Cent.—*Neutral fats*—stearin, palmitin, olein—occur in the form of minute microscopic droplets, whose presence often renders the serum of a milky turbidity after abundant ingestion of fat and also of milk. They are more abundant during hunger and in drunkards. There occur, besides, soaps, lecithin, and its decomposition-product, glycerin-phosphoric acid, and cholesterin. Hürthle found cholesterin oleate and palmitate—0.17 per cent. According to Hanriot a ferment, known as lipase, occurs in blood and which breaks up neutral fat into glycerin and fatty acids. Lipase is found also in the pancreas and in the liver, and traces also in some other parts of the body.

A certain amount of *grape-sugar*—from 0.1 to 0.15 per cent., somewhat more in the blood of the hepatic veins, derived from the liver and the muscles and increased after loss of blood; some *glycogen*—increased in cases of diabetes; a trace of *animal gum*, a reducing substance, insusceptible of fermentation and soluble in ether, *jecorin*, which is a combination of dextrose and lecithin; a dextrose-forming *diastatic ferment*, inactive at a temperature of 65° C. For a discussion of the sugar-destroying power of the blood reference may be made to the section on the liver.

The amount of sugar in the blood is increased by absorption of sugar from the intestinal tract, and in greatest degree in the blood of the portal and hepatic veins. It is increased also in arterial blood, although here it is rapidly changed.

For purposes of demonstration blood is coagulated by boiling after addition of sodium sulphate, and the amount of sugar in the expressed fluid is determined with the aid of Fehling's solution. Pavy digested the blood thrice successively with six times its volume of alcohol, then boiled and expressed the product.   The extract, which is evaporated, contains all of the sugar.

Kreatin, urea—during hunger 0.035 per cent., in the stage of maximum formation 0.153 per cent.; at times *succinic acid, hippuric acid,* and *uric acid* (1 : 6000 in gouty individuals); *guanin* (? carbamic acid); in the blood after death also *sarcolactic acid.*   All of these are present in exceedingly small amount.

Inorganic matters — 0.85 per cent.; principally sodium-combinations.   The amount of salts is increased by a meat-diet, while it is diminished by a vegetable diet.   Ammonium is present in the proportion of 1 mg. to 100 cu. cm., and three or four times as abundantly in the blood of the portal vein.

Human blood-serum contains the following *salts* :

        Sodium chlorid,............................4.92 in 1000.
        Sodium sulphate,.........................0.44   "
        Sodium carbonate,........................0.21   "
        Sodium phosphate, ...0.15
        Calcium phosphate,   ⎫
        Magnesium phosphate, ⎭ 0.73

The alkaline reaction of the serum depends principally upon the sodium carbonate present.   It is only half that of the blood.

The serum of blood containing carbon dioxid in large amount exhibits a more pronounced alkaline reaction and the amount of chlorin contained is diminished. This is dependent upon the fact that hydrochloric acid and water enter the blood-corpuscles, while the alkali remains behind.

If salts in considerable amount are introduced into the blood, the larger amount disappears in the course of a few minutes, diffusing principally into the tissues. Gradually they are eliminated from the body through the kidneys.   The same statement is applicable to sugar and peptone.

Water—about 90 per cent.

Yellowish pigments.

One pigment can be separated by agitation with methyl-alcohol.   It exhibits two absorption-bands of lipochrome, like lutein.   Hydrobilirubin was found by Maly, and choletelin by MacMunn.

Blood, and also blood-serum free from cells, as well as lymph, possess bactericidal properties, which are augmented by increase in the alkalinity, but, on the other hand, disappear on addition of water, on heating to a temperature of 55° C., on exposure to diffuse daylight, and likewise if mineral matters are removed by dialysis.   Egg-albumin and fresh milk exhibit the same properties.   The corpuscle-destroying—globulicidal—action of fresh serum is peculiar to the latter, in conjunction with its bactericidal effect after bacterial invasion.   Both properties are due to certain proteid bodies known as alexins.   The serum of an individual rendered immune by inoculation to any infectious disease exerts an antitoxic effect against the poison of the corresponding infectious agent, and it can therefore be employed against the latter for curative purposes.

Large numbers of microbes may gain entrance into the blood-stream during the death-agony.

The serum of individuals suffering from typhoid fever contains a substance of diagnostic importance, designated agglutinin, which causes agglutination of typhoid bacilli in cultures.

THE GASES OF THE BLOOD.

ABSORPTION OF GASES BY SOLID BODIES AND BY FLUIDS.

Between the particles of solid, porous bodies and gaseous substances there exists a marked attraction of such a character that the gases are attracted by the solid bodies and condensed within their pores; that is, the gases are absorbed by the solid bodies.   Thus, for instance, one volume of boxwood charcoal, at a temperature of 12° C. and a pressure of 760 mm. of mercury, absorbs 35 volumes of carbon dioxid, 9.4 volumes of oxygen, 7.5 volumes of nitrogen, 1.5 volumes of hydrogen. The absorption of the gases is invariably attended with the generation of heat, which is in proportion to the energy with which absorption takes place. Non-porous bodies are in an analogous manner surrounded intimately upon their surface by a layer of condensed gas.

Fluids are in like manner capable of taking up or absorbing gases. In this connection it has been learned that a given amount of fluid at different pressures nevertheless always absorbs an equal volume of gas. Whether the pressure be great or small, the volume of gas absorbed is always the same. It is, however, known, according to the law of Boyle-Mariotte, governing the compression of gases, that with twice, thrice or greater amounts of pressure, twice, thrice or greater amounts of gas by weight are contained within an equal volume of gas. From this there is formulated the law that while at varying pressures the volume of gas absorbed remains the same, the amount of gas by weight contained within the same volume is directly proportional to the amount of pressure. If, therefore, the pressure is zero the amount of the absorbed gas must likewise be zero; whence it follows that fluids under the air-pump in a vacuum may be deprived of their absorbed gases.

The coefficient of absorption represents that volume of gas that is absorbed by 1 volume-unit of a fluid at a given pressure and temperature. From what has been said with regard to the volume of absorbed gases the coefficient of absorption must be wholly independent of the pressure.

The temperature has an important influence upon the coefficient of absorption. When the temperature is low the coefficient is highest, declining at a higher temperature and becoming zero when the fluid boils. From this it follows that absorbed gases can also be expelled from fluids by heating the latter to the boiling-point. The coefficient of absorption increases, however, for various fluids and gases with increasing temperature in a peculiar, and by no means uniform, manner, which must be determined empirically for each. At the temperature of the body the coefficient of absorption of carbon dioxid is 0.5283, of nitrogen 0.0119, of oxygen, at a pressure of 699 mm., 0.0231.

DIFFUSION OF GASES; ABSORPTION OF GASEOUS MIXTURES.

Gases that do not enter into chemical combination with one another are capable of forming a uniform mixture. If, for instance, the necks of two flasks are connected of which the lower contains carbon dioxid and the upper, placed vertically and inverted above the other, contains hydrogen, both gases combine, independently of differences in specific gravity, within each flask so as to form identical mixtures. This phenomenon is known as the *diffusion of gases*. If a porous membrane be previously interposed between the two gases the interchange of gases takes place just the same. Nevertheless different gases pass through the interstices of the membrane with unequal rapidity in the same way as in the case of fluids in the process of endosmosis, so that at first a larger amount of gas will be present upon the one side than upon the other. According to Graham the rapidity with which gases pass through the interstices is inversely as the square root of their specific gravity, but according to Bunsen, not exactly so.

Gases mutually exert no pressure upon one another. Therefore a gas escapes from a space containing another gas as from a vacuum. If, accordingly, the surface of a fluid in which a gas is absorbed be placed in communication with a large amount of another gas, the absorbed gas passes over into the other gas. Therefore, absorbed gases can be removed if the fluids containing them are treated with other gases by agitation or by passing them through.

If two or more gases in mixture lie over a fluid within a closed space the separate gases will be absorbed, and according to weight in proportion to the pressure to which each gas would be exposed if it were alone present in the space. This pressure is known as *partial pressure*. The amount of gas absorbed from mixtures is therefore proportionate to the partial pressure. The partial pressure of a gas in a space partially filled by a fluid is at the same time an expression of the tension of the absorbed gas in this fluid.

The air contains 0.2096 volume of oxygen and 0.7904 volume of nitrogen. If, therefore, one volume of air is present at a pressure P over water, the partial pressure under which oxygen is absorbed is 0.2096 x P, and that for nitrogen equals 0.7904 x P. At a temperature of 0° C. and at 760 mm. pressure 1 volume of water absorbs 0.02477 volume of air, consisting of 0.00862 volume of oxygen and 0.01615 volume of the nitrogen. It accordingly contains 34 per cent. of oxygen and 66 per cent. of nitrogen. Water, therefore, absorbs from the atmospheric air an amount of gas that is by percentage richer in oxygen than the air itself.

SEPARATION OF THE GASES OF THE BLOOD.

The expulsion of the gases of the blood and their collection for chemical analysis are effected by means of the mercurial air-pump. The Pflüger pump for the extraction of gases is illustrated diagrammatically in Fig. 20. It consists of a blood-receptacle (A), a glass flask with a capacity of from 250 to 300 cu. cm., drawn out above and below into tubes, each of which can be closed by means of a stop-cock (a b). The cock b is an ordinary stop-cock, while the cock a has a channel passing through its longitudinal axis and opening at x in such a manner that in accordance with its adjustment it leads either into the receptacle (position x a) or downward through the lower tube (position x′ a′). This receptacle is first completely deprived of air by application to a mercurial air-pump and is then weighed. Next, the extremity x′ is tied in an artery or a vein of an animal and by placing the lower cock in the position x a the blood is permitted to flow into the receptacle. When the desired amount has been collected the lower cock is again placed in the position x′ a′, the exterior is carefully cleaned and the receptacle is weighed in order to determine the weight of the blood collected.

The second portion of the apparatus is the froth-vessel chamber (B), likewise drawn out above and below into tubes, which can be closed by means of the cocks c and d. The purpose of the froth-chamber is to take up the froth formed in consequence of the active escape of the gases from the blood. Below, the froth-chamber is connected with the receptacle by means of a ground-glass tube and above likewise through a well-fitting tube with the drying apparatus (G). This consists of a U-shaped tube expanded below into a glass bulb. The latter is half filled with sulphuric acid, while each arm contains bits of pumice-stone saturated with sulphuric acid. In passing through this apparatus, which likewise may be closed by means of the two stop-cocks e and f, the gases of the blood yield up their watery vapor to the sulphuric acid, so that they may be conveyed through the cock f in a perfectly dry state.

The short tube D is similarly connected with the prolongation from f by means of a properly ground surface, and it is provided with a small manometer from which the degree of vacuum can be read. The tube D communicates with the pump-apparatus proper. This consists of two large glass flasks, E and F, terminating above and below in open tubes, the lower of which, Z and w, are connected by means of a rubber tube G. Both flasks and the tube are filled with mercury to about half the height of the flasks. The flask E is secured, while the flask F can be raised and lowered by means of a pulley-apparatus attached to a stand. When F is raised E becomes filled, and when F is lowered E is emptied. The upper extremity of E divides into two tubes, g and H, of which g is connected with D. The tube h, passing upward, becomes greatly narrowed and further on is so curved that its free extremity, i, dips into a basin containing mercury, v, with its opening below the tube for the reception of the gases. J (eudiometer-tube) completely filled with mercury. At the junction of g and H there is a cock with a double channel, which in the position H connects the flask E with A B G D, and in the position K closes A B G D and connects the flask E with the tube J.

In the first place, B G D is completely exhausted of air by the following steps: The stop-cock is placed in the position K; and F is raised until globules of mercury pass from the free tube i, which is as yet not placed below J, into the basin. Then the stop-cock is placed in the position H, when F is depressed. Next, the cock is placed again in the position K, and so on, until the manometer y indicates that evacuation has taken place. Now, J is placed over i. If the cocks c and b are opened, so that the receptacle A communicates with the remainder of the apparatus, the gases of the blood pass actively into B, with the generation of foam, and through G, dried, to E. The depression of F brings them principally into E. Finally, the cock is placed in the position K, while F is raised, and the gases are conveyed to J above the mercury. Repeated depression and elevation of G with appropriate adjustment of the cock will finally bring all of the gases into J.

The removal of the gases from the blood is materially facilitated by placing the recipient A in a vessel containing water at a temperature of 60° C. It is advisable in the analysis of the gases of the blood to evacuate at once the blood discharged from the vein into the receptacle, because on standing outside of the body the amount of oxygen undergoes a diminution.

Mayow, in 1670, was the first to observe gases arise from the blood in a vacuum, and Priestley demonstrated the presence of oxygen and Davy that of carbon

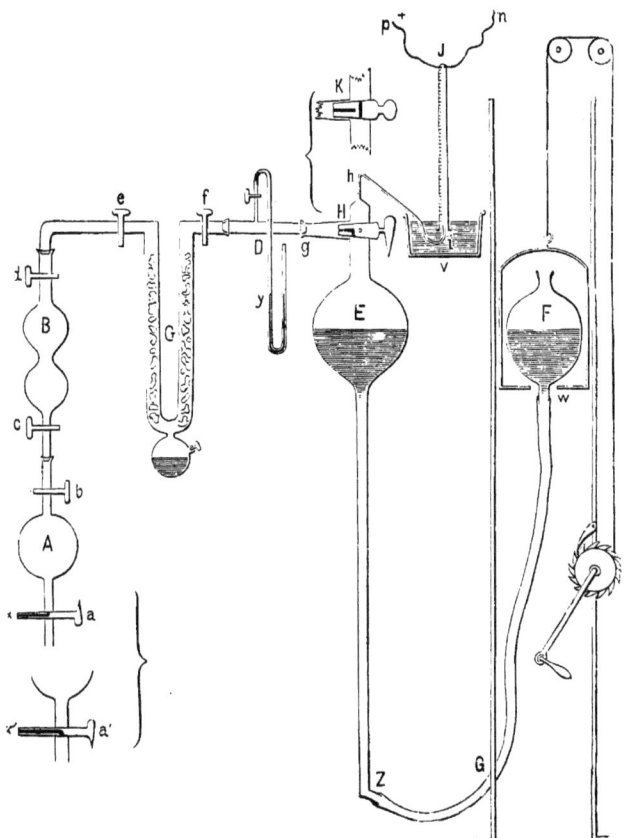

Fig. 20.—Diagrammatic Representation of Pflüger's Pump for the Extraction of the Gases of the Blood.

dioxid. Magnus, in 1857, investigated the percentage-composition of the gases of the blood. The important recent investigations have been made principally by Loth. Meyer, in 1837, and by C. Ludwig and Pflüger and their pupils.

QUANTITATIVE ESTIMATION OF THE GASES OF THE BLOOD.

The evacuated gases consist of oxygen, carbon dioxid, and nitrogen.

The gases of the blood obtained with the aid of the pump will be found in the eudiometer-tube (Fig. 20, J), an accurately graduated glass tube in whose closed upper portion two platinum wires, p n, are soldered. The eudiometer is closed below by mercury.

Estimation of the Carbon Dioxid.—A globule of potassic hydrate fused to a platinum wire and moistened on its surface is brought from below through the mercury into the gaseous mixture. The carbon dioxid unites with the potassium hydrate to form potassium carbonate. After remaining in place for a considerable period of time, the globule is removed in the same way. The diminution in the volume of the gases indicates the volume of the carbon dioxid removed.

Estimation of the Oxygen.—In the same way as in estimating the carbon dioxid a globule of phosphorus is introduced into the eudiometer-tube by means of a platinum wire and which takes up the oxygen for the formation of phosphoric acid; or a dry globule of coke or papier maché saturated with a solution of pyrogallic acid in potassic hydrate, which eagerly takes up oxygen. After removal of the globule the diminution in volume of the gases indicates the amount of oxygen.

The oxygen can be determined most accurately and most rapidly, according to Volta and Bunsen, by explosion in the eudiometer. An abundance of hydrogen, whose volume is carefully determined, is introduced into the eudiometer-tube. Then an electric spark is made to pass through the tube between the wires p and n. The oxygen and the hydrogen combine to form water. In consequence a reduction in the volume takes place in the eudiometer, of which a third represents the oxygen required for the formation of the water.

Estimation of the Nitrogen.—If the carbon dioxid and the oxygen are removed from the gas-container according to the methods described the remainder consists of nitrogen.

SPECIAL FACTS CONCERNING THE GASES OF THE BLOOD.

Oxygen is present in arterial blood from the dog on an average to the amount of 18.3 volumes per cent., at a temperature of 0° C. and 1 meter of mercurial pressure. Arterial blood is saturated, according to Pflüger, to $\frac{9}{10}$, according to Hüfner that of the dog to $\frac{14}{15}$, with oxygen. By means of thorough artificial respiration in animals in the state of apnea or by active agitation of the blood with air the amount of oxygen can be brought up to 23 volumes per cent. veno_us blood contains on the average 8.15 volumes per cent. less of oxygen than arterial blood, although the amount of oxygen varies widely in accordance with the tissues and the circulatory conditions. Sczelkow found 6 volumes per cent. in the blood of resting muscles. Only traces are present in the blood after asphyxiation. In the more highly colored blood of active glands, such as the salivary glands and the kidneys, oxygen is undoubtedly present in larger amount than in ordinary, darker venous blood.

The oxygen occurs in the blood as follows:

(*a*) From 0.1 to 0.2 volume per cent. are in a state of simple absorption in the plasma—thus only a minimal portion, not exceeding that which distilled water at the temperature of the blood and at the partial pressure of oxygen in the air of the lungs would take up.

(*b*) Almost all of the oxygen of the blood is combined chemically, and with the hemoglobin of the erythrocytes, with which it forms oxyhemoglobin ; it is therefore not subject to the laws of absorption. The total amount of blood acts with regard to the chemical absorption of oxygen like a gas-free solution of hemoglobin, except that the absorption of oxygen by the blood takes place more rapidly than by a solution of hemoglobin. At a temperature of 0° and at moderate atmospheric pressure—760 mm. of mercury—1 gram of hemoglobin takes up from 1.6 to 1.8 cu. cm. of oxygen—according to Hüfner 1.592 cu. cm.

The absorption of oxygen on the part of the blood is thus independent of the pressure. This is seen also in shed blood, which, on the one hand, permits more abundant escape of the chemically combined oxygen only when the pressure becomes reduced to about 30 mm. of mercury (at a temperature of 12° C. with increasing temperature at a lower pressure), while, on the other hand, it takes up only little more oxygen even if the air-pressure be enormously high, up to six atmospheres. The same phenomenon is exhibited by the blood in the living body, for both on the highest mountains as well as in the deepest valleys it takes up oxygen in accordance with its requirements. Also, animals breathing in a closed space are capable of abstracting the oxygen from the surrounding air down to the minutest trace.

In spite of the chemical combination existing between the hemoglobin and the oxygen, the total amount of oxygen in the blood can be driven out by those agents that set free absorbed gases: (a) by evacuation ; (b) by boiling ; (c) by the passage of the gases ; because the chemical union of oxyhemoglobin is so feeble that it is broken up by the physical procedures named.

Among chemical agents, reducing substances, such as ammonium sulphid, hydrogen sulphid, solutions of alkaline subsalts, iron filings, etc., extract oxygen from the blood.

The amount of iron present in the blood—0.55 in 1000 parts—is in direct proportion to the amount of hemoglobin, this to the number of erythrocytes and the latter in turn approximately to the specific gravity of the blood. The amount of oxygen taken up by the blood has been shown to be almost proportional to the specific gravity of the blood. It is, therefore, also proportional to the amount of iron in the blood. According to Hoppe-Seyler 1 atom of iron may combine with 2 atoms of oxygen in the blood. According to Bohr the combination is said to be an unstable one. The latter investigator even differentiates several varieties of combination between oxygen and hemoglobin, in accordance with the amount of bound oxygen—namely, 0.4 or 0.75 or 3 cu. cm. of oxygen, at a temperature of 15° C. and an oxygen-pressure of 150 mm.—to 1 gram of hemoglobin. Also carbon monoxid is believed by Bohr to be taken up in varying amounts in an analogous manner.

Immediately after escape of the blood a slight loss of oxygen takes place as a physiological manifestation of tissue-respiration within the living blood. After having been outside the circulation for some time the amount of oxygen is found to undergo progressive diminution, and after a long time and at a high temperature the oxygen may have wholly disappeared from the blood. This latter loss of oxygen is due to decomposition within the shed blood, in consequence of which reducing substances form and these take up the oxygen. Not all varieties of blood act in this connection with equal energy in the destruction of oxygen. The venous blood of active muscles acts most energetically, while the blood of the hepatic veins is scarcely at all active. In place of the oxygen that has disappeared carbon dioxid makes its appearance in the blood, whose color becomes dark. At times the amount of carbon dioxid is even larger than that of the oxygen destroyed.

AS TO THE PRESENCE OF OZONE IN THE BLOOD.

On account of the varied and in part active oxidation-processes that take place through the intermediation of the blood, the question has been raised whether the oxygen in the blood may not be present in the form of ozone (O_3). However, neither in the blood itself nor yet in the gases evacuated from the blood can ozone be found. Nevertheless, the red blood-corpuscles, as well as the hemoglobin, have a definite relation to ozone.

The hemoglobin acts as a conveyer of ozone, that is, it is capable of taking away the ozone from other bodies, and conveying it to other oxidizable substances.

Oil of turpentine that has been exposed to the air for a considerable time always contains ozone. Among reagents for ozone are potassium-iodid paste, which becomes blue, as the ozone releases the combination of iodin and potassium, and the iodin causes the starch-paste to become blue; further, freshly prepared solution of guaiac-resin in alcohol, which also is made blue by ozone. A solution of guaiac is dropped in water, the resin forming a milky precipitate, and oil of turpentine is added. At first no reaction occurs, but if blood or hemoglobin be added, with agitation, a bluish discoloration appears, that is, the blood takes the ozone from the oil of turpentine and conveys it to the guaiac-resin.

It has been stated that hemoglobin acts as an ozone-producer; that is, it is capable of generating ozone from the inactive oxygen of the air with which it comes in contact. For this reason, red blood-corpuscles alone also cause guaiac to become blue. The reaction is most successful if the solution of guaiac is permitted to dry upon blotting-paper and then several drops of blood diluted from 5 to 10 times are added. That under these circumstances the condition is one of stimulation of the surrounding oxygen through the hemoglobin, is shown by the observation that even red blood-corpuscles containing carbon monoxid bring about the blue coloration, naturally not when the extraneous oxygen of the air is excluded. According to Pflüger these reactions take place only with decomposition of the hemoglobin, and for this reason it is believed that the blood-corpuscles as such do not act as producers of ozone.

Also hydrogen sulphid is decomposed by the blood, as by ozone itself, into sulphur and water. Hydrogen dioxid likewise is decomposed by the blood into oxygen and water. This can be prevented by the addition of a small amount of hydrocyanic acid. Crystallized hemoglobin does not bring this result about, and hydrogen dioxid can be cautiously injected into the veins of animals. From this it would appear that unaltered hemoglobin has no ozone-producing effect.

There are three varieties of oxygen: (1) Ordinary or inactive oxygen (O_2), as, for instance, that of atmospheric air. (2) Active or nascent oxygen (O), which can never occur in the free state, but which on its development at once enters into chemical combination as a most powerful oxidizing agent. This is capable of oxidizing water into hydrogen dioxid, the nitrogen of the air into nitrous and nitric acids, and also carbon monoxid into carbon dioxid—which ozone is not capable of doing. This gas certainly plays an important rôle in the organism. (3) Ozone (O_3) forms through the breaking up of certain molecules of ordinary oxygen (O_2) into two atoms each (O), and union of each of these atoms with an undecomposed molecule of oxygen. Ozone is a form of oxygen compressed to two-thirds of its volume.

CARBON DIOXID AND NITROGEN IN THE BLOOD.

Carbon dioxid is present in arterial blood in from 34 to 38 volumes per cent., at a temperature of 0° C. and a pressure of 1 meter; in venous blood on the average in 9.2 volumes per cent. more than in arterial blood, varying greatly in accordance with the situation and the circulatory conditions. The total amount of carbon dioxid in the blood does not equal even one-half of that which the blood would actually be capable of taking up. Thus, the blood after asphyxiation may contain as much as 52.6 volumes per cent. The amount of carbon dioxid in the lymph after asphyxiation is less than that in the blood. The carbon dioxid can be completely pumped out of the total volume of blood without the formation of acids in the process of evacuation—in consequence of decomposition of the constituents of the blood—which might take part in driving out the carbon dioxid.

The Carbon Dioxid of the Plasma or the Serum.

(a) This is absorbed in smallest part simply by the blood-plasma.

(b) The largest part of the carbon dioxid is combined chemically with the blood-plasma, independently of the pressure. This combination may take place in the following manner:

1. A portion of the carbon dioxid is loosely combined with sodium carbonate, forming sodium bicarbonate, one equivalent of carbon dioxid being taken up by

the simple carbonate: $CO_3Na_2 + CO_2 + H_2O = 2CO_3NaH$. In this way considerable amounts of carbon dioxid may be bound. As the sodium bicarbonate releases the carbon dioxid but slowly in a vacuum, while blood releases it with violence, it must be borne in mind that perhaps sodium combined with a proteid (serum-globulin alkali) contains the carbon dioxid in a complex combination, from which it readily separates in a vacuum.

2. A minimal portion of the carbon dioxid of the plasma might be combined chemically with neutral sodium phosphate: One equivalent of this salt may combine with one equivalent of carbon dioxid, so that acid sodium phosphate and acid sodium carbonate result: $PO_4Na_2H + CO_2 + H_2O = PO_4NaH_2 + CO_3NaH$. In the process of evacuation the carbon dioxid escapes, with the formation of neutral sodium phosphate. As, however, the sodium phosphate formed in blood-ash has resulted almost wholly from the combustion of lecithin and nuclein, only the small amount of this salt already present in the plasma can be taken into consideration.

The Carbon Dioxid in the Blood-corpuscles.

The erythrocytes also contain carbon dioxid in loose chemical combination. In defibrinated human blood 31.2 volumes per cent. of carbon dioxid have been found in the serum, and only 4.5 in the blood-corpuscles. The combination of the carbon dioxid is effected in part through the hemoglobin, therefore through the formation of carbohemoglobin, in part from the globulin-alkali combinations of the erythrocytes. The leukocytes also combine with carbon dioxid in accordance with the character of the constituents of the serum, and in about the proportion of from $\frac{1}{12}$ to $\frac{1}{8}$ of the absorptive power of the serum.

According to Bohr there are three varieties of carbon-dioxid combination with hemoglobin, which, while closely resembling one another, take up different amounts of carbon dioxid—namely 1.5, 3 and 6 cu. cm. of carbon dioxid respectively to 1 gram of hemoglobin, at the same partial pressure for the carbon dioxid and at the same temperature. Spectroscopically, carbon-dioxid hemoglobin resembles reduced hemoglobin, except that its absorption-band lies somewhat nearer the violet, and it absorbs more light in the green. Hemoglobin can take up oxygen and carbon dioxid at the same time. and each independently of the other. Therefore it is probable that oxygen and carbon dioxid unite with different constituents of the hemoglobin.

The amount of carbon dioxid in the blood is diminished by alcoholic intoxication, while it is increased by inhalation of ether, which reduces the amount of oxygen. Subcutaneous injection of morphin or chloral diminishes the amount of oxygen. After administration of iodin, mercury, sodium oxalate and nitrate there is a reduction in the amount of carbon dioxid in arterial blood. The same result is brought about in the blood of animals by injection of peptone into the veins, and also in the febrile state on account of the lessened alkalinity of the blood.

Nitrogen is present in the blood in the proportion of from 1.4 to 1.6 volumes per cent. in a state of simple absorption.

For every 100 parts of nitrogen there are 2.1 parts of argon, which, however, is present only in the plasma. The blood contains more nitrogen when the number of erythrocytes is larger than when the number is smaller and when the blood is lake-colored. Jolyet and Sigalas believe, therefore, that the erythrocytes, like solid bodies, absorb nitrogen at their surface. On standing outside the body, the blood yields small amounts of ammonia, particularly with access of oxygen and application of heat, perhaps in consequence of decomposition of an as yet unknown ammonium-salt.

ESTIMATION OF THE INDIVIDUAL CONSTITUENTS OF THE BLOOD.

Estimation of the Water and of All of the Solid Constituents of the Total Blood or of the Serum.—About 5 grams of serum or defibrinated blood are evaporated in a crucible of known weight over a water-bath and dried in a drying chamber

6

at a temperature of 110° C. The loss of weight represents the amount of water that was present. The dry residue is determined by subtracting the weight of the crucible. For clinical purposes Stintzing weighs a few drops of blood in a light, covered glass dish. This he dries for six hours at a temperature of 65° C. and weighs the residue. The amount of water was found to be in men 78.3, in women 79.8. The dry residue corresponds approximately with the amount of proteids contained in the blood and it declines in the presence of anemia.

Estimation of the Fibrin.—A measured volume of blood is whipped with a rod. After complete separation, all of the fibrin is collected upon a satin filter and washed with water; then placed in a dish and again washed with water, alcohol and ether ; next dried in a drying chamber at a temperature of 110° C., and finally weighed. Kossler and Pfeiffer estimate the amount of nitrogen in the serum and in the plasma according to the method of Kjeldahl; the difference represents the amount of nitrogen in the fibrin. The fibrin in 100 cu. cm. of plasma contains 39 mg. of nitrogen (from 30.8 to 45). The fibrin is increased in cases of pneumonia, acute articular rheumatism, erysipelas, scarlet fever, peritonitis (to between 80 and 152 mg.).

Estimation of the Fats (Ethereal Extract) in the Serum or the Total Blood.— About 15 grams of defibrinated blood or serum are dried in a dish at first over a water-bath, then in a drying chamber at a temperature of 120° C., rubbed up, and placed in a flask with ether, which is repeatedly renewed.

The method just described is followed in preparing an *alcoholic extract* from the total blood or the serum.

Estimation of the Inorganic Salts in the Total Blood or Serum.—About 25 grams are dried in a weighed platinum crucible and then reduced to ash over a free flame at red heat. The amount of ash is determined by weighing. If this ash be repeatedly extracted with hot water, and the latter be entirely evaporated in a weighed dish, the weight of the salts soluble in water will be obtained.

Estimation of the Total Proteids in Blood or Serum.—E. Salkowski precipitates all albuminates by means of sodium chlorid and acetic acid. For this purpose he places 20 grams of pulverized serum or 50 cu. cm. of blood in a dry flask and adds 100 cu. cm. of a mixture of 7 volumes of concentrated solution of sodium chlorid and 1 volume of acetic acid, agitating for 20 minutes and filtering. The filter is dried and weighed. V. Jaksch takes 1 gram of blood from a cupping glass, estimates the amount of nitrogen contained by the method of Kjeldahl, and multiplies the result obtained by 6.25.

Estimation of the Proteids of the Blood-corpuscles.—If the proteids contained in one part by weight of the total blood and also of the serum have been determined, and if the amount obtained for the serum be deducted from that obtained for the total blood in the proportion in which red blood-corpuscles and serum are present in the total blood, the result will represent the proteids of the blood-corpuscles, although only approximately.

Estimation of the Red Blood-corpuscles by Weight.—Defibrinated blood is mixed with thrice its volume of a concentrated solution of sodium sulphate and filtered. The blood-corpuscles remaining upon the filter are coagulated by immersing the filter in boiling concentrated solution of sodium sulphate. Then the filter can be washed out with distilled water, after which it is dried and weighed. The increase in the weight of the previously weighed filter is due to the presence of the blood-corpuscles.

ARTERIAL AND VENOUS BLOOD.

Arterial blood contains in solution all those materials that are necessary for the nutrition of the tissues, many that are to be employed in secretion and in addition the larger amount of oxygen. Venous blood need contain less of these matters, while the waste materials of the tissues, the products of retrogressive metamorphosis, will be present in greater amount, including a larger quantity of carbon dioxid. As, however, the interchange through the blood takes place rapidly, no great difference in many of these substances càn bé looked for at a given moment. In many respects analysis fails to furnish conclusive evidence. A little consideration, further, will show that the blood from some veins must be characterized by special peculiarities, such as the blood from the portal vein and the hepatic veins. The essential differences between the two kinds of blood may be summarized as follows:

ARTERIAL BLOOD CONTAINS

More	Less
Oxygen, water, fibrin, extractives, salts, at times chlorids, sugar, fat; and the temperature is on an average 1° C. higher.	Carbon dioxid, blood-corpuscles, proteids, alkali, urea.

The bright red color of arterial blood is due to oxyhemoglobin, to which it is peculiar; while the dark color of venous blood is due to a deficiency in oxyhemoglobin and an abundance of reduced hemoglobin. The larger amount of carbon dioxid in venous blood is not responsible for the dark color, for if equal amounts of oxygen be added to two portions of blood and to the one also carbon dioxid, the latter effects no change in color.

THE AMOUNT OF BLOOD.

The amount of blood in the adult equals $\frac{1}{13}$ of the body-weight, in the newborn $\frac{1}{19}$.

According to A. Schücking the amount of blood in the infant when the umbilical vein is ligated immediately after birth is $\frac{1}{15}$, while that in the infant when ligation is practised later is as much as $\frac{1}{6}$ of the body-weight. Immediate ligation, therefore, causes a reduction of the amount of blood in the newborn child of about 100 grams. Further, the number of red corpuscles is less in the blood of the newborn child after immediate ligation than in that of infants in which ligation is practised later.

For the estimation of the amount of blood, first practised by Valentine in 1838 and by Ed. Weber in 1850 by unreliable methods, the following may be employed:

Welcker's Method.—Blood from the incised carotid of a previously weighed animal, with a cannula tied in the vessel, is received into a weighed flask, in which it is defibrinated by agitation with pebbles. It is then measured. A portion of the defibrinated blood is made cherry-red by the passage of carbon monoxid, because ordinary blood possesses varying coloring power in accordance with the amount of oxygen present. Now a ⊢-shaped cannula is tied in both extremities of the divided carotid and a 0.9 per cent. solution of sodium chlorid is permitted to flow steadily from a pressure-vessel, while the resulting wash-water that escapes from the divided jugular veins and the inferior vena cava is collected until it becomes as clear as water. Then the entire body is minced, and with the exception of the weighed contents of the stomach and intestines, whose weight is deducted from that of the body, the mass is extracted with water and expressed after the lapse of 24 hours. This water and the sodium-chlorid wash-water are mixed and weighed. A portion of this mixture is likewise saturated with carbon monoxid. Of this a specimen is placed in a glass chamber with parallel walls, 1 cm. apart—a so-called hematinometer, while in a second chamber water is added to the undiluted blood from a buret until both fluids exhibit the same shade of color. From the amount of water that is necessary to make the dilution of the blood of the same tint as the wash-water the amount of blood present in the latter can be estimated. In mincing the muscles alone the coloring-matter yielded by them can be considered as muscle-pigment and need not be taken into account. By multiplying the volume of blood by its specific gravity the absolute weight of the blood can be determined. As the differences in the color of the specimens can be estimated most accurately this method is to be commended.

The weight of the blood of mice has been found to be from $\frac{1}{13}$ to $\frac{1}{12}$ of the body-weight, exclusive of gastric and intestinal contents; of guinea-pigs $\frac{1}{19.7}$ (from $\frac{1}{22}$ to $\frac{1}{17}$); of rabbits $\frac{1}{20.1}$ (from $\frac{1}{22}$ to $\frac{1}{15}$); of dogs $\frac{1}{13}$ (from $\frac{1}{18}$ to $\frac{1}{11}$); of cats $\frac{1}{21.5}$; of birds from $\frac{1}{13}$ to $\frac{1}{10}$; of frogs $\frac{1}{20}$ to $\frac{1}{15}$; of fish from $\frac{1}{19}$ to $\frac{1}{14}$.

Vierordt's method, which is based upon the determination of the amount of blood by indirect means, is discussed under circulation time.

The specific gravity also should be determined in a study of the blood. In states of inanition the amount of blood has been observed to be reduced. Obese individuals have relatively less blood. After hemorrhage the blood lost is readily replaced by water, while the blood-corpuscles are only gradually regenerated. After extensive, deplethoric transfusion with defibrinated blood Landois, as well as Panum, observed the amount of blood and its specific gravity to be maintained.

In the living animal Gréhant and Quinquaud permitted a measured amount of carbon dioxid to be inspired, then withdrew a quantity of blood and estimated the amount of carbon monoxid present. From this the amount of blood can be readily determined. A quantity of carbon-monoxid blood could also be transfused and shortly thereafter the proportion of shed blood containing carbon monoxid, and that free from carbon monoxid, be estimated.

The estimation of the amount of blood in individual organs is made after sudden ligation of their veins during life. The organs are cut up into small pieces and the amount of blood contained in the wash-water is determined by comparison with a specimen of blood to be diluted. The estimation after death in a state of freezing is to be rejected.

ABNORMAL INCREASE IN THE AMOUNT OF BLOOD OR ITS INDIVIDUAL PARTS.

An increase in the total mass of blood uniformly in all its parts is known as *polyemia* or *plethora*. It may occur as a morbid manifestation in individuals with excessive nutritive and assimilative activity. A marked bluish-red color of the external integument, with swollen veins and large arteries and a hard and full pulse, injection particularly of the capillaries and smaller vessels of the visible mucous membranes are the readily explicable signs, accompanied by cerebral hyperemia, which may give rise to attacks of vertigo, and hyperemia of the lungs, which may give rise to dyspnea. Also after amputation of large portions of the extremities, with avoidance of loss of blood, a relative increase in the amount of blood has been described (plethora apocoptica).

Polyemia can be induced artificially by injection of blood from the same species. If the normal amount of blood be increased up to 83 per cent. no abnormal condition develops; in particular, the blood-pressure does not become permanently raised. The blood finds its way especially into the greatly distended capillaries, which as a result become stretched beyond their normal elasticity. An increase in the amount of blood, however, up to 150 per cent. jeopardizes life directly, with considerable variations in blood-pressure, and which Landois has observed to terminate fatally in consequence of direct rupture of vessels.

Following upon the injection of blood the formation of lymph rapidly increases. Then the serum is disposed of in the course of one or two days, the water being eliminated principally through the urine, and the proteids in part converted into urea. Therefore, the blood at this time appears to be richer in red blood-corpuscles. The red blood-corpuscles undergo destruction much more slowly and the materials furnished by them are converted in part into urea and in part into the biliary pigment, though not constantly. Nevertheless an excess of red blood-corpuscles may be observed for as long as a month.

That as a matter of fact the blood-corpuscles are slowly destroyed in the process of metabolism is shown from the circumstance that the formation of urea is greater when the animal ingests the same amount of blood than if it receives an equal amount by transfusion. In the latter event a moderate increase in the amount of urea persists often for a number of days as a sign of slow destruction of the red corpuscles. Marked plethora is attended, further, with loss of appetite as well as a tendency to hemorrhages from the mucous membranes.

Serous polyemia is the name given to that condition of the blood in which the amount of serum or plasma is increased. The condition can be produced artificially by injecting into the veins of animals serum from the same species. Under such circumstances the water is soon excreted with the urine, while the albumin is decomposed into urea, without passing over into the urine. An animal forms more urea from a given amount of injected serum than from an equal amount of blood—an indication that the blood-corpuscles are capable of being preserved for a longer time than the serum. If, however, an animal be injected with the serum from another species, in which the blood-corpuscles of the recipient undergo solution, as, for instance, if dogs' serum be injected into a rabbit, the blood-cells of the recipient are dissolved and hemoglobinuria develops, and even death may take place if the dissolution be extensive.

Simple increase in the amount of water in the blood, *aqueous polyemia*, occurs as a transitory phenomenon after copious ingestion of fluid, but increased diuresis soon restores the normal conditions. Disease of the kidneys attended with destruction of the secreting parenchyma of the glands induces, together with aqueous polyemia, often general anasarca through the leakage of water into all of the tissues. Ligation of the ureter likewise gives rise to an increase in the watery elements of the blood. Stintzing and Gumprecht found the dry residue of small amounts of blood after evaporation of the water to be from 19.8 per cent. in women to 21.6 per cent. in men, while in cases of anemia it falls to 8.5 per cent.

An increase of the red blood-corpuscles beyond the normal mean—*polycythemic plethora* or *hyperglobulia*—has been thought to be present in robust individuals when hemorrhages that have regularly taken place cease and in general all of the symptoms of polyemia are present. The cessation of menstrual, hemorrhoidal, and nasal hemorrhages is considered as a cause, as well as the omission of venesection previously employed systematically. Nevertheless, the polycythemia under such circumstances is only inferred and not established by enumeration. On the other hand, a condition of polycythemia has been positively observed. Thus, after transfusion of blood from the same species a portion of the blood-plasma is soon consumed, while the blood-corpuscles are preserved for a longer time. An increase in the number of red blood-corpuscles up to 8,820,000 in a 1 cu. cm. in case of severe heart-disease, with marked stasis, in which more water escapes from the vessels by transudation, is a remarkable fact. The number is for the same reason greater also in cases of hemiparesis upon the paralyzed side presenting phenomena of stasis.

After attacks of diarrhea that cause a reduction in the amount of water in the blood there is likewise an increase in the number of red corpuscles, and it is probable that the same result is brought about by profuse sweating and by polyuria. Agents that influence the caliber of the vessels, such as alcohol, chloral hydrate, amyl nitrite, give rise to an increase in number when they cause contraction of the vessels and to a diminution when they cause relaxation. A transitory increase in the ancestors of the red blood-corpuscles is encountered as a reparative process after profuse hemorrhage or after acute disease. In cachectic states the increase is permanent on account of interference with the transformation into red corpuscles. In the last stages of cachectic states the number progressively diminishes, as at this time the production of the ancestral forms also ceases.

The designation *hyperalbuminous plethora* has been applied to an increase of the albuminates in the plasma such as it may be inferred occurs after abundant absorption from the digestive tract. The same condition may be induced experimentally by injection of serum from the same species of animal, the elimination of urea increasing at the same time. Injection of egg-albumin induces albuminuria.

Melitemia or an excess of sugar in the blood. The sugar of the blood is eliminated in part with the urine, in marked degree up to 1 kilo daily, and the amount of urine may be increased to 25 kilos. To replace this loss an abundance of nourishment and much fluid are necessary, and in this way the amount of urea may at the same time be increased threefold. The marked production of sugar also induces destruction of proteid tissue, so that the amount of urea is increased, even if the supply of albumin be insufficient. The patients emaciate, all of the glands, particularly the testicles, undergo atrophy or degeneration, the skin and the bones become thin, while the nervous system resists the longest. The crystalline lens becomes turbid in consequence of the presence of sugar in the fluids of the eye, which abstract water from the lens. Wounds heal badly on account of the abnormal constitution of the blood. If a drop of blood be spread upon a glass slide, then treated with a solution of Bieberich's scarlet or alkaline methylene-blue and heated for ten minutes at a temperature of 35°, it will not take the stain if derived from a case of diabetes, while normal blood is stained. Instead of grape-sugar excessive accumulation of inosite or of milk-sugar has also been found in the blood and in the urine.

Lipemia.—Increase in the amount of fat in the blood occurs normally after the ingestion of food rich in fat, as, for instance, in nursing kittens, so that the serum itself may acquire a milky turbidity. Pathologically, this is observed in still more marked degree in drunkards and in obese individuals. In conjunction with marked destruction of proteids in the body, therefore, in a large number of wasting diseases, the amount of fat in the blood is increased; likewise after abundant administration of easily digestible carbohydrates, together with much fat, in the food. V. Jaksch found traces of fatty acids in the blood of febrile and leukemic patients. After injuries to bones involving the marrow large numbers of fat-globules often pass from the vessels of the marrow, in part unprovided with walls, into the blood-stream, so that fat may even find its way into the urine, and may give rise to dangerous fat-emboli in the lungs.

The *salts* are usually preserved with great tenacity. If sodium chlorid be withheld, albuminuria results; and if salts in general, paralytic phenomena. Excessive administration of salty food, as in the form of pickled meat, has not rarely been followed by death through fatty degeneration of the tissues, particularly of the glands. Withdrawal of calcium and phosphoric acid brings about softening or atrophy of the bones. In the presence of infectious diseases and of anasarca the amount of salts in the blood has often been found increased, while in the presence of inflammation (sodium chlorid is wanting in the urine in cases of pneumonia) and of cholera the amount is diminished.

The amount of *fibrin* in the blood is increased in the presence of inflammation, particularly of the lungs or the pleura. Therefore venesection under such circumstances is followed by the formation of the so-called buffy coat. The fibrin may be increased also in other diseases attended with blood-destruction. Sigm. Mayer observed an increase likewise after repeated venesection. Blood rich in fibrin usually coagulates more slowly than blood deficient in fibrin, although exceptions to this statement are not wanting.

ABNORMAL DIMINUTION IN THE AMOUNT OF BLOOD OR OF ITS INDIVIDUAL CONSTITUENTS.

Reduction in the mass of the blood as a whole—*true oligemia*—occurs after every direct loss of blood. In the newborn a hemorrhage of even a few cu. cm., in children a year old a hemorrhage of 250 cu. cm., and in adults a loss of one-half of their blood may prove dangerous. Women withstand better than men even considerable loss of the blood. In them the regeneration of the blood appears to take place more readily and more quickly in consequence of the periodic restoration of the blood lost at each menstrual period. Obese persons, as well as the aged and the debilitated, are less tolerant to loss of blood. The hemorrhage is the more dangerous the more rapidly it takes place. General pallor and coldness of the skin, a sense of fear and oppression, relaxation, the appearance of spots before the eyes, roaring in the ears and vertigo, loss of voice and syncopal attacks usually accompany profuse hemorrhage. Dyspnea ("and breathing rapidly he exhales life in a purple stream:" Sophocles' Antigone), cessation of glandular secretion, profound loss of consciousness, then dilatation of the pupils, involuntary discharge of urine and feces, and finally general convulsions are the positive premonitions of rapid death from hemorrhage. In the state of greatest danger life can be saved only by transfusion.

As much as one-quarter of the normal amount of blood can be withdrawn from animals without permanently lowering the blood-pressure in the arteries, because the latter by contraction adapt themselves to the smaller volume of blood in consequence of the anemic irritation of the vasomotor center in the medulla oblongata. Loss of blood up to one-third of the volume of blood causes marked reduction in the blood-pressure. Dogs recover after loss of one-half of the volume of blood. If two-thirds be removed one-half of the animals die, while the remaining half recover spontaneously.

If the hemorrhage does not terminate fatally, the water of the blood, with the dissolved salts, is first replaced through absorption from the tissues, with gradual increase in the blood-pressure; and later the proteids. Considerable time is required for the regeneration of the blood-corpuscles. The blood, therefore, contains for a time an abnormal amount of water—*hydremia;* and finally it exhibits an abnormal deficiency in cells—*oligocythemia, hypoglobulia.* With the increased lymph-stream toward the blood the leukocytes are soon considerably increased above their normal number. Also fewer red blood-corpuscles appear to be consumed during the period of restitution, as, for instance, in the formation of bile.

After moderate venesection in animals Buntzen observed the volume of blood restored in a few hours, and after severe hemorrhage in the course from 24 to 48 hours. The red blood-corpuscles, however, were, after venesection of from 1.1 to 4.4 per cent. of the body-weight, fully restored to the normal only after the lapse of from 7 to 34 days. The commencement of the regenerative process could be recognized in the course of 48 hours. During this period of reorganization the number of the embryonal forms of the blood-corpuscles was increased. The newly formed blood-corpuscles appear at first to contain less hemoglobin than normal. Also in human beings the duration of the period of regeneration appears to be dependent upon the amount of hemorrhage. The reduction in the amount of the hemoglobin of the blood after venesection is approximately proportional to the amount of the blood removed.

Of especial significance is the state of metabolism in the body of an anemic patient. The decomposition of proteids is increased, and as a result the elimination of urea is increased. The combustion of fats in the body is, however, correspondingly diminished, and the amount of carbon dioxid given off is correspondingly reduced. Anemic as well as chlorotic patients therefore readily put on fat. The same significance is to be attached to the lipomatosis of anemic convalescents after acute diseases interfering with blood-formation. The fattening of animals is, accordingly, favored by occasional venesection. The same statement is applicable to intercurrent hunger. Aristotle had already pointed out that swine and birds readily take on considerable fat after days of intercurrent hunger.

Anemia results also from failure on the part of the blood-forming organs. The alarming anemia from the presence of the bothriocephalus, which may pursue a course similar to pernicious anemia is remarkable. It is probably dependent upon a toxic effect induced by the parasite, which impairs the vitality of the blood-corpuscles.

Excessive concentration of the blood through loss of water is designated *dry oligemia.* This condition has been observed in human beings after copious, watery diarrhea, particularly in cases of cholera, and the thick, tarry blood stagnates in the veins. Probably copious loss of water through the skin as a result of diaphoretic treatment, particularly in association with restriction of fluids, may give rise to dry oligemia, even though only in moderate degree.

If the proteids of the blood are diminished in abnormal degree a condition of *hypalbuminous oligemia* is present. The proteids may be diminished more than half. In their place an excessive amount of water usually finds its way into the blood, so that the salts of the plasma are likewise diminished. Loss of proteids from the blood is due directly to albuminuria, which may furnish even 25 grams of proteid daily; to long-continued suppuration, extensive weeping cutaneous surfaces, excessive loss of milk, albuminous diarrhea (dysentery). Frequent and copious hemorrhage, also, induces at first hypalbuminous oligemia, as the loss primarily is principally made good by the taking up of water into the vessels. V. Jaksch found that the amount of proteids failed to decline in correspondence with the reduction in the number of blood-corpuscles.

PHYSIOLOGY OF THE CIRCULATION.

CAUSE, PURPOSE, DIVISION.

The blood maintains itself within the vascular system in an uninterrupted circulating movement that, proceeding from the cardiac ventricles through the largest arterial trunks arising therefrom (the aorta and the pulmonary artery) to the furthermost branches of these vessels, then through a system of capillary vessels, from which it is collected into the venous channels, which progressively increase in size by coalescence, terminates finally in the auricles.

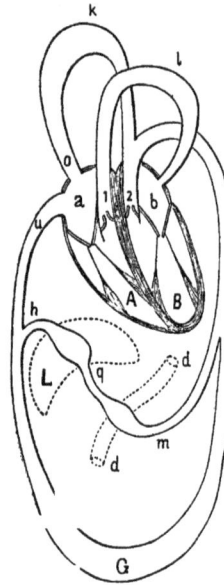

The *cause* of this circulatory movement resides in the difference in pressure to which the blood is exposed in the aorta and the pulmonary artery, on the one hand, and the two venæ cavæ and the four pulmonary veins on the other. The blood naturally flows continuously toward that portion of the closed system of tubes where the pressure is lowest. The greater this difference in pressure the more active will be the movement of the stream. Abolition of this difference in pressure, as after death, will naturally cause a cessation of the flow.

The *purpose* of the circulation is, on the one hand, to carry nourishment through the blood to all the tissues of the body, while on the other, the blood carries away from the tissues to the organs of excretion the waste products of their metabolism.

The circulation of the blood is divided into:

FIG. 21.—Diagrammatic Representation of the Circulation: a, right auricle; A, right ventricle; b, left auricle; B, left ventricle: 1, pulmonary artery; 2, aorta, with semilunar valves; l, lesser circulation; k, greater circulation, including superior vena cava, o; G, greater circulation, including inferior vena cava, u; d d, intestinal tract; m, mesenteric arteries; q, portal vein; L, liver; h, hepatic veins.

1. The *greater circulation*, comprising the pathway from the left auricle and the left ventricle through the aorta and its branches, the capillaries and the veins of the body, to the termination of the two venæ cavæ in the right auricle.

2. The *lesser circulation*, comprising the pathway of the right auricle and the right ventricle, the pulmonary artery, the pulmonary capillaries and the four pulmonary veins arising therefrom up to their point of entrance into the left auricle.

3. The *portal circulation* is occasionally considered as a separate circulatory system, although it is only a second capillary ramification

inserted into a venous pathway. It is composed of the portal vein, which represents the union of the veins of the abdominal viscera—the superior gastric, the superior and inferior mesenteric, and the splenic veins—and which breaks up in the liver into capillaries that again unite to form the hepatic veins, which empty into the inferior vena cava.

Strictly speaking, this differentiation of the portal system into a separate circulation is not justifiable. In many animals similar conditions are found in still other organs, as, for example, the suprarenal of the snake and the kidney of the frog. When an artery breaks up into numerous small branches that shortly reunite, without the intervention of capillaries, to again form an artery, the cluster of branches thus formed is called a "wonderful network," rete mirabile, such as is seen in apes and edentates. Microscopical networks of this character are found in the mesentery of man. The glomerulus of Bowman's capsule in the kidney also is an example of this peculiar arterial division. Analogous formations in the veins are called venous "wonderful networks."

THE HEART.

The mammalian heart-muscle (Fig. 184, 8) is composed of short, closely and finely striated, unicellular elements which are devoid of sarcolemma and, in man, from 50 to 70 μ long and from 15 to 23 μ wide. The ends are rather blunt and generally split, and by these split ends the fibers are joined together anastomotically to form a network. The individual muscle-cells are united by a cement-substance, which is soluble in 33 per cent. potassium-hydrate solution and is stained black by silver nitrate. Each cell at its center contains a nucleus, rarely two smaller nuclei, 14 μ long by 7 μ wide, in its central axis. The transversely striated substance frequently contains molecular granules arranged in rows. The fibrils are placed side by side and are divided by the perimysium into bundles, which, after solution of the connective tissue by boiling, may be isolated. The shape of the bundles on transverse section is rather circular in the auricles, while in the ventricles it is rather flat and laminated; here also several of the smaller bundles may unite to form a thicker band. The interstices between the bundles serve to carry the lymph-vessels.

ARRANGEMENT OF THE MUSCLE-FIBERS OF THE HEART AND THEIR PHYSIOLOGICAL SIGNIFICANCE.

Musculature of the Auricle.—The study of the embryonal heart furnishes the key to the understanding of the complicated arrangement of the muscle-fibers. The simple heart-tube of the embryo exhibits an outer circular and an inner longitudinal layer of muscle-fibers. The septum is formed later, so that it is obvious that both· in the ventricles as well as in the auricles the fibers belong, in part at least, to both halves, as they originally enclose only a single cavity. On the other hand, the fibers of the auricles are generally separated from those of the ventricles by the fibrous ring; nevertheless certain of the muscle-bundles pass from the auricles to the ventricles. In the auricles the embryonal arrangement of the fibers remains fundamentally unchanged. In the ventricles the arrangement is obliterated because during the process of development the fibers here undergo a peculiar bending and looping, as in the stomach, together with a spiral rotation.

The *musculature of the auricles* is in general arranged in two layers: an outer transverse, which is continuous over the two auricles, and an inner longitudinal. The outer fibers can be traced from the entering veins upon the anterior and posterior walls. · The inner fibers are especially prominent where they are attached vertically to the fibrous rings, but in certain parts of the anterior wall in particular they are not arranged continuously. On the septum of the auricles the ring-like muscular layer surrounding the oval fossa, the opening of the oval foramen. in the embryo, is especially prominent. Around the openings of the veins emptying into the auricles are found circular muscle-bundles; these are least well marked around the inferior cava, while around the superior cava they are well developed and extend upward around the vessels for 2.5 cm. (Fig. 22, II). At the entrance of the four pulmonary veins in man and in some mammals, transversely striated muscle-fibers, arranged in an inner circular and an outer longitudinal layer, extend upon the pulmonary veins as far as the hilus of the lung; in other animals (apes, rats) they extend even into the lung itself; indeed, in some mammals (mouse, bat) this muscular layer penetrates the lung so far that in the small veins the entire wall is composed almost wholly of striated muscle-fibers. Muscle-fibers, chiefly circular, are also found at the termination of the great cardiac vein and in the coronary valve of Thebesius. Many elastic fibers are present in the perimysium of the auricles.

From the physiological standpoint the foregoing anatomical data explain the following facts with relation to the contractions of the auricles.

The auricles are able to contract independently of the ventricles; this is particularly manifest in the cessation of the heart's activity, as under such circumstances two or more contractions of the auricle alone are often seen to take place, followed now and then by a single contraction of the ventricle. However, when the action of the heart is

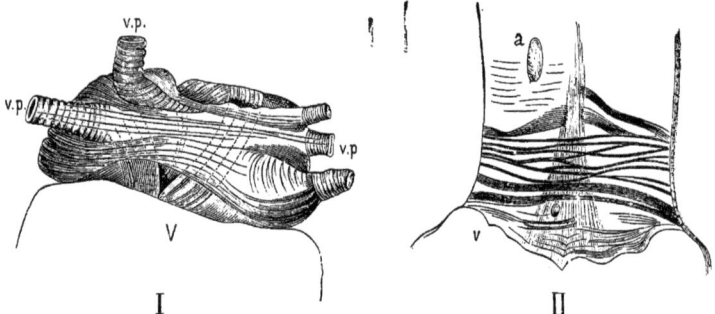

| I | II |

FIG. 22.—I, Course of the Muscle-fibers in the Left Auricle: the outer transverse and the inner longitudinal fibrous layer are visible and in addition the circular fibers of the pulmonary veins, v.p. V, left ventricle (Joh. Reid). II, Distribution of Transversely Striated Muscle-fibers on the Superior Vena Cava (Flischer): a, entrance of the azygos vein; v, auricle.

unimpaired the auricles in their contraction transmit the motor impulse to the ventricles. Whether this stimulation is brought about through nerve-fibers or, as is more probable, through connecting muscle-bundles, has not yet been decided with certainty.

The two chief layers of fibers (transverse and longitudinal), which cross each other, serve to effect uniform contraction of the auricular cavity from all sides, as is the case likewise with most hollow muscular organs.

The circular fibers surrounding the entering venous trunks, through their contraction, which occurs in unison with that of the auricles, cause in part an emptying of blood into the auricle and in part a hindrance to a return of the blood in any considerable measure.

ARRANGEMENT OF THE MUSCULATURE OF THE VENTRICLES.

The Muscle-fibers of the Ventricles.—Beneath the pericardium there is first met an outer longitudinal layer (Fig. 23, A), consisting of only occasional bundles on the right ventricle, while on the left it comprises a compact layer of about one-eighth of the entire thickness of the wall. A second layer of longitudinal fibers lies on the inner surface of the ventricles, being especially well marked at the orifices, as well as inside the perpendicularly placed papillary muscles, while in other situations it is replaced by the irregularly running fibers of the muscular trabeculæ. Between the two longitudinal layers lies the most powerful transverse layer, the fibers of which are separable into individual, leaf-like, ring-shaped bundles. The three layers, however, are not wholly independent and separated from each other, but rather there is a gradual transition between the transverse and the outer and inner longitudinal layers by means of oblique fibers.

The common assumption is that the entire outer longitudinal layer passes gradually into the transverse and this in turn wholly into the inner longitudinal, as is shown diagrammatically in Fig. 23, C. This is not justifiable, and is negatived by the great preponderance in the thickness of the middle layer. In general the outer longitudinal fibers pursue such a course as to intersect the course of the fibers of the inner longitudinal layer at an acute angle. The intervening trans_verse layer constitutes the medium for a gradual transition between these courses.

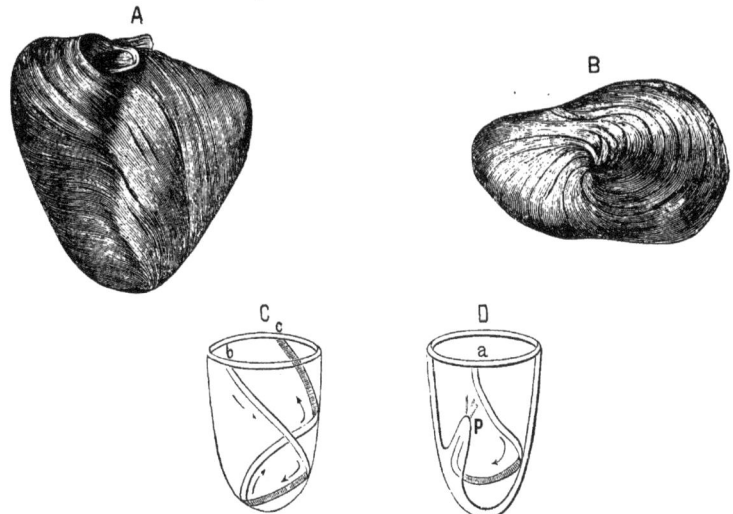

FIG. 23.—Course of the Muscle-fibers in the Ventricles: A, course upon the anterior surface; B, view of the apex with the "whirl" (Henle); C, diagrammatic representation of the course of a muscle-fiber within the wall of the ventricle; D, course of such a fiber into the papillary muscle (C. Ludwig).

At the apex of the left ventricle external longitudinal fibers, uniting in the so-called "whirl" (B), pass in a curved direction inward and upward within the muscle-substance and extend into the papillary muscles (D). Nevertheless it is an error to consider that all of the ascending fibers in the papillary muscles are derived from these vertical muscle-bundles of the outer surface, as many arise independently from the wall of the ventricle. Neither can the origin of these longitudinal fibers on the outer surface of the heart be traced solely to the fibrous rings or to the roots of the arteries. Finally, mention should be made of the special circular layer of fibers that surrounds the left orifice like a sphincter. Numerous lymph-vessels are present in all the interstices between the muscle-fibers and the blood-vessels. These eventually empty into the lymph-vessels and nodes of the mediastinum.

PERICARDIUM; ENDOCARDIUM; VALVES.

The *pericardium*, which includes between its two layers a lymph-space—the pericardial cavity—containing a small amount of lymph, exhibits the structure of a serous membrane; that is, it is composed of connective tissue containing delicate elastic fibers, and is covered on its free surface with a single layer of irregular polygonal, flat, endothelial cells. A rich network of lymph-vessels lies within the pericardium itself, as well as more deeply toward the muscle-mass of the heart. Stomata are wanting in both layers of the pericardium. In the subserous tissue of the pericardium, especially in the sulci for the coronary vessels, are deposits of fat, and lymphatics.

The *endocardium* presents all of the characteristics of a vessel-wall. Facing the cavity of the heart, there is first a single layer of flat, polygonal, nucleated endothelial cells. Then there comes, as the true groundwork of the whole membrane, a layer of delicate elastic fibers (more marked in the auricles, and even forming a fenestrated membrane), in the midst of which but little connective tissue occurs. The latter, much more loosely arranged and intermixed with elastic fibers, is present in larger amount toward the heart-muscle. Scattered bundles of unstriated muscular fibers, usually arranged longitudinally, are found between the elastic elements (in smaller amount in the auricles). These obviously have the task of combating the pressure and the tension exerted on the endocardium during the cardiac contraction; for wherever throughout the body a wall composed of soft parts is exposed to repeated high pressure muscular elements are found, and never elastic tissue alone. The endocardium is non-vascular.

The *valves*—both the arterial (semilunar) and the venous (mitral and tricuspid)—also are a part of the endocardium. The venous and arterial orifices on the right side are separated from each other in the wall of the ventricle, while the two orifices on the left are united into a single large opening. The valves are attached to their basal margins by means of resistant fibrous rings composed of connective-tissue and elastic fibers. They consist of two layers: (1) The fibrous, which is a direct continuation of the fibrous ring, and (2) a layer of elastic elements. The elastic layer of the auriculo-ventricular valves is a direct prolongation of the endocardium of the auricle, and is therefore directed toward that cavity. At their bases the valves are united by their adjacent margins. The tendinous cords are inserted on the free margin and on the under surface of the valves. The semilunar valves possess a thin, elastic layer, thickened at their base and turned toward the arteries.

The auriculo-ventricular valves contain also striated muscle-fibers. Radiating fibers, arising from the auricles, extend into the valves, and it is their function in part to retract the valves toward their bases during the time of auricular systole, and thus to enlarge the passage-way for the flow of blood into the ventricles. Paladino describes still other longitudinal fibers derived from the ventricles. Besides these, there is directed rather toward the ventricular aspect, a concentric muscular layer, following the basal attachment of the valves, which appears to have sphincter-like action—drawing the bases of the valves together during the period of ventricular contraction when the valves are under tension, and thus preventing excessive distention. The larger of the tendinous cords also contain striated muscle-fibers; and the Thebesian and Eustachian valves likewise contain a delicate muscular network.

The name "Purkinje's fibers" has been applied to a grayish network of muscular elements found in mammals and in birds chiefly beneath the endocardium of the ventricle, but occurring also in the muscular mass itself. These appear to represent a stage of embryonal development (on account of the partial striation). They are absent in man and in the lower vertebrates.

Blood-vessels occur in the auriculo-ventricular valves in considerable number only where there are muscle-fibers. In children delicate vessels extend to the free margin of the valve. The semilunar valves are devoid of blood-vessels except under pathological conditions. A network of lymphatics extends from the endocardium to the middle of the valves.

Weight and Size of the Heart.—According to W. Müller, the weight of the heart in children and in older persons having a body-weight up to 40 kilos, is 5 grams for every kilo of body-weight; in individuals having a body-weight of from 50 to 90 kilos, the proportion is 4 grams of heart for each kilo; in individuals having a body-weight of 100 kilos, 3.5 grams of heart for each kilo of body-weight. The auricles become stronger with increasing age. The right ventricle weighs half as much as the left. In man the heart weighs 309 grams; in woman, 274 grams. Blosfeld and Dieberg found the heart in man to weigh 346 grams; in woman, from 310 to 340 grams. The thickness of the left ventricle in man averages 11.4 mm.; in woman, 10.15 mm.; the thickness of the right ventricle, 4.1 and 3.6 mm. respectively.

THE CORONARY VESSELS; AUTOMATIC REGULATION, NUTRITION, AND ISOLATION OF THE HEART.

With reference to the origin of the coronary arteries the question at once arises whether the orifices of these vessels are closed by the elevation of the semilunar valves during systole as a result of the application of the valve-leaflets to the walls of the vessels or whether such occlusion does not take place.

Anatomical.—The two coronary arteries arise from the region of the sinus of Valsalva. The point of origin varies: (1) It is either within the concavity of the sinus; or (2) the mouths of the vessels are not completely within the range of the margin of the valve, and this is frequently the case with the left coronary of man and the ox; or (3) the orifices project beyond the margins of the valve (this is rare). These findings alone make it improbable that closure of the mouths of the coronary arteries by the semilunar valves during ventricular systole is a constant physiological phenomenon.

AUTOMATIC REGULATION OF THE HEART.

According to Brücke the openings of the coronary arteries are covered by the semilunar valves during systole, so that they can be filled only during diastole. The advantage of this arrangement resides in the fact that (a) the diastolic distention of the ventricular vessels stretches the muscular fibers of the ventricular wall and thus correspondingly dilates the ventricle for the reception of the blood that pours in from the auricle during diastole. (b) On the other hand, the systolic distention of the coronary arteries would be useless because the dilatation of the ventricular wall (due to the distention of the arteries already mentioned) would resist the systolic contraction, and because the systolic distention of the coronary arteries and the expulsion of the blood from them would unnecessarily diminish the power of the ventricle. Accordingly, the diastolic distention of the coronary arteries would be most consistent with the mechanical conditions present. This mechanism Brücke has designated the "automatic regulation of the heart."

 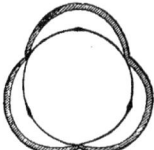

Fig. 24.—Semilunar Valves, Closed.   Semilunar Valves, Opened.

This theory and its underlying principles are untenable, for—

1. The filling under high pressure of the coronary arteries of a dead heart not only is followed by no dilatation, but actually causes a contraction of the cavity of the ventricle.

2. The chief branches of the coronary arteries lie in the sulci of the heart embedded in the loose subpericardial fatty tissue, where their dilatation and contraction could scarcely have any effect upon the size of the cavities of the heart.

3. Brown-Séquard found in animals and v. Ziemssen in a woman with a large deficiency in the wall of the left thorax that the coronary pulse was synchronous with that in the pulmonary artery. Newell-Martin and Sedgwick, by introducing manometers into the coronary and carotid arteries of a large dog, obtained simul--

taneous pulsatory elevations.   In accordance with these observations is the fact that an incised coronary artery spurts continuously, with systolic exacerbations, as do all other arteries.

4. If a strong stream of water is passed intermittently through a sufficiently large tube introduced into the left auricle of a fresh pig's-heart, and it is forced through the auriculo-ventricular orifice on into the aorta; and if the aorta beyond its arch is connected with a large tube directed upward (in order to establish pressure in the aorta), the water will be seen to spurt continuously from the divided coronary artery, with systolic exacerbations.

5. There is constantly present in the sinuses of Valsalva an amount of blood sufficient to fill the arteries in question during systole.

6. The valves when elevated are not applied closely against the wall of the aorta, even with the greatest amount of pressure that can be exerted by the ventricle.   On the contrary, there remains between each valve-leaflet and the aortic wall a semilunar space filled with blood, as is shown in Fig. 24.

7. Undoubted cases of extensive destruction of the semilunar valves that with certainty render closure of the mouths of the coronary arteries impossible are directly opposed to this theory.

8. Observations on muscle have shown that during contraction its small vessels undergo dilatation and the blood-stream through it is accelerated.   It is, therefore, difficult to believe that in the contracted heart-muscle the movement of the blood should cease.

As, during the systole, the small arterial branches lying close to the ventricular cavity are exposed to a pressure greater than that of the aorta a systolic compression of their lumen occurs, with a forcing out of their contents in the direction of the veins.   The ventricular contraction thus aids the flow of the blood in the coronary vessels; marked dilatation of the heart diminishes it.

The capillary vessels of the myocardium are numerous in correspondence with the energetic activity of the heart; they, like the small vessels generally, lie within the muscle-bundles in contact with the muscle-cells.   With their transition into veins several of them coalesce almost at once to form a large vein, from which the extremely easy passage of the blood into the veins is readily understood.   The veins are provided with valves.   As a result it happens that (1) with the systole of the right auricle (therefore during the ventricular diastole) the venous stream is interrupted; (2) with contraction of the ventricle the flow of blood in the cardiac veins is accelerated in the same way as it is in the veins of the muscles.   This systolic acceleration of the venous flow permits of the conclusion that the arterial circulation is not interrupted at this time.

The coronary arteries, between which no anastomoses occur, are characterized by the great thickness of their elastic and connective-tissue intima, and this perhaps explains the frequency of calcification in these vessels.   Many of the lower vertebrates have no vessels in the heart-substance (anangiotic hearts)—for example, the frog; but this statement is disputed.

The motor disturbances and even the complete cessation of action that have been observed in the heart after partial or complete occlusion of the coronary vessels are of importance, particularly as analogous conditions are observed in man in consequence of occlusion or narrowing of the coronary arteries (for example, as a result of calcification).

Method.—In rabbits, under the influence of curare and with artificial respiration, or after previous section of the vagi (in order to exclude the inhibitory influence of this nerve), it is possible to clamp off the coronary arteries close to their origin from the aorta with a spring clamp.   Ligation is less satisfactory, as it cannot be accomplished without wounding the heart.   In dogs it is possible to push a glass rod provided with a button-like extremity from the subclavian artery into the mouth of a coronary artery. Injections of various substances capable of causing occlusion have also been tried.

In 1867, v. Bezold noted in rabbits after clamping off the coronary artery that the heart-beat grew rapidly smaller and smaller; then the contractions occurred in groups, periodically; later on the regular movement of the ventricle ceased entirely, and in its place the muscle-wall exhibited a peculiar fibrillary contraction; finally the heart stood still. As the circulation was reëstablished after removal of the clamp, the phenomena appeared in reverse order until the heart regained its normal beat.

If in a dog the right descending coronary artery and the circumflex artery, together with the artery of the septum, are occluded, the heart soon ceases to beat. The closure of only two of the three arteries caused a cessation of contraction in 9 out of 14 animals; while closure of the septal artery or of the right coronary artery alone had no effect. In almost all instances the auricles likewise cease beating. The heart of a dog that has once ceased to beat recovers only with great difficulty. It appears that the fibrillary contractions are due to irritative injury inflicted during the operation, and not alone to the stasis of the blood.

If in rabbits only the left coronary artery is occluded the beat of the left heart is slowed and weakened, while the right heart pulsates without change. As a result it occurs that the left half of the heart can no longer empty itself completely, so that particularly the left auricle becomes filled to distention with blood, while at the same time the unaffected right heart continues to drive blood into the lungs. In consequence edema of the lungs develops as a result of the high pressure in the lesser circulation which is transmitted from the right heart through the pulmonary vessels into the left auricle. According to Sig. Mayer persistent dyspnea has a similar effect, with earlier weakening of the left than of the right ventricle; the pulmonary edema preceding death can be explained in this manner.

The heart in the higher animals can maintain its activity only when the circulation of blood through its walls is maintained. The heart from which the blood is completely removed rapidly ceases to contract. The coronary circulation must convey the necessary oxygen and nutritive materials to the myocardium, as well as remove the metabolic products from it. The excised "isolated" mammalian heart, which is fed at body-heat through the coronary vessels with bright-red blood, remains active.

Langendorff maintains the circulation in the isolated heart by allowing the coronary arteries to be filled from the aorta. Other fluids, for example, lake-colored blood or serum, are incapable of maintaining the heart's activity. At most, such solutions (as, for example, alkaline salt-solution mixed with egg-albumin—1000 albumin diluted with water, 0.1 sodium chlorid, 0.1 calcium chlorid, 0.075 potassium chlorid), in so far as they exert a slightly irritating effect, are capable of stimulating the heart for a time. If the heart is placed in pure oxygen the pulsation may be maintained for a considerable time by passing serum through the cardiac vessels.

Also the isolated frog's heart can be included in a circulation by means of suitable tubes. To maintain its contractions oxygen and nutritive fluid are necessary to distend its cavities. This object is best fulfilled by arterial blood; indifferent fluids (0.6 per cent. sodium chlorid) quickly bring about a condition of "apparent death," from which, however, the organ can be revived by nutritive fluids.

The frog's heart is less readily exhausted than that of the higher vertebrates.. Serum-albumin, alkaline salt-solutions of blood, or of milk, made slightly viscid with albumin or gum arabic and saturated with oxygen, are capable of main-- taining the activity of the heart for a long time.

Pathological.—In the presence of so-called sclerosis of the coronary arteries in old age there occur acute or chronic attacks of cardiac disability. Weakness of the heart, alterations in rhythm and frequency (to 8 in a minute), constitute, together with dyspnea, syncope, stasis, attacks of pulmonary edema, the most characteristic phenomena; and they may terminate in death from so-called heart-failure. In a case of occlusion of the left coronary artery in a man Hammer saw the pulse fall from 80 to 8, the beats being interrupted by spasmodic vibra-- tion.

THE MOVEMENTS OF THE HEART. VARIATIONS IN TONE.

Method.—In addition to direct observation, the kinematograph may be used' to great advantage for recording and projecting the movements of the heart, par-- ticularly at a slow rate.

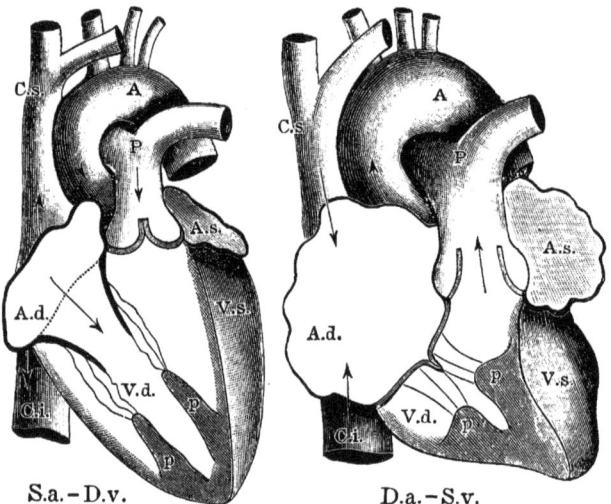

S.a.–D.v.   D.a.–S.v.

Fig. 25.—Diagrammatic Representation of the Auricular Systole with Ventricular Diastole, and of Auricular · Diastole with Ventricular Systole.

The movement of the heart is appreciable as alternate contraction. and relaxation of the heart-walls. The entire motor phenomenon designated the *cardiac cycle* consists of three parts: contraction of the auricles (*auricular systole*); contraction of the ventricles (*ven-- tricular systole*), and the pause, during which the auricles and the ven-tricles are relaxed (*diastole*). During the contraction of the auricles the-- ventricles are at rest, during the contraction of the ventricles the auricles are relaxed. The following phenomena can be noted successively during· a cycle of the heart:

(**A**) *The blood streams into the auricles*, which in consequence are: distended. The cause for this resides in:

1. The pressure of the blood in the venæ cavæ (on the right) and the pulmonary veins (on the left), which is greater than the pressure within the auricles.

2. The elastic traction of the lungs, which after the completed contraction tends to separate the relaxed yielding walls of the auricles lying in contact with each other. The auricular appendages are distended coincidently with the auricles. The appendages serve in a measure as reservoirs for the auricles, to accommodate the large amount of blood flowing in from the veins.

(B) *The Auricles Contract.*—There occur in rapid sequence:

1. The contraction and evacuation of the auricular appendages in the direction of the auricle. Simultaneously, the entering veins are constricted by the contraction of their circular muscular layers, especially the superior vena cava and the site of entrance for the pulmonary veins.

2. The walls of the auricles contract rapidly in a wave-like manner from above downward, particularly toward the auriculo-ventricular orifices, in consequence of which—

3. The blood is forced downward into the relaxed ventricles, which now become considerably dilated. As a result of the auricular contraction there occur:

(a) A slight stasis of the blood in the large venous trunks, such as can be readily observed particularly in rabbits on exposure of the point of junction of the jugular and subclavian veins after division of the muscles of the chest. There is no actual reflux of the blood, but only a slight stasis due to partial interruption of the flow into the auricle, because, as has been stated, the sites of entrance for the veins are narrowed; because, further, the pressure in the superior vena cava and in the pulmonary veins soon counteracts the tendency to regurgitation; and, finally, because in the further ramifications of the inferior and to some extent also of the superior cava and of the cardiac veins, valves prevent the reflux. In the blood thus stagnated in the venæ cavæ the movement of the heart causes a regular pulsating phenomenon that, when abnormally increased, may give rise to the appearance of a venous pulse.

(b) The principal motor effect of the auricular contraction is the distention of the relaxed ventricles, which in small measure are dilated by the elastic traction of the lungs.

Earlier and later investigators have attributed the distention of the ventricles in part to the elasticity of the muscular walls. It has been thought that the strongly contracted ventricular walls, like a compressed rubber bulb, in returning to their resting normal shape, through their own elasticity, aspirate the blood with negative pressure. Such suction-power on the part of the ventricle is, however, effective only in slight degree, if at all.

(c) With the distention of the ventricles the auriculo-ventricular valves at once float upward (Fig. 26), being in part foreed up by the counter-stroke of the blood from the wall of the ventricle; in part they are capable, by reason of their lower specific gravity, to spread out and float horizontally; in part, finally, they are drawn upward by the longitudinal muscular fibers passing from the auricles upon the valves.

(C) *The ventricles now contract* and the auricles relax. In this phase—

1. The muscular walls contract on all sides and reduce the size of the ventricular cavity.

2. At the same time the blood presses against the under surface
of the auriculo-ventricular valves, the inverted margins of which
interdigitate and become hermetically applied to one another (Fig. 26).
The valve-leaflets are prevented from being forced back into the
auricular cavity, because the tendinous cords hold their under surface
and margins firmly like an inflated sail. The approximation of the
edges of adjacent valves is favored further by the circumstance
that the tendinous fibers always pass from one papillary muscle to
the edges of two opposed valves. To the extent that the lower ven-

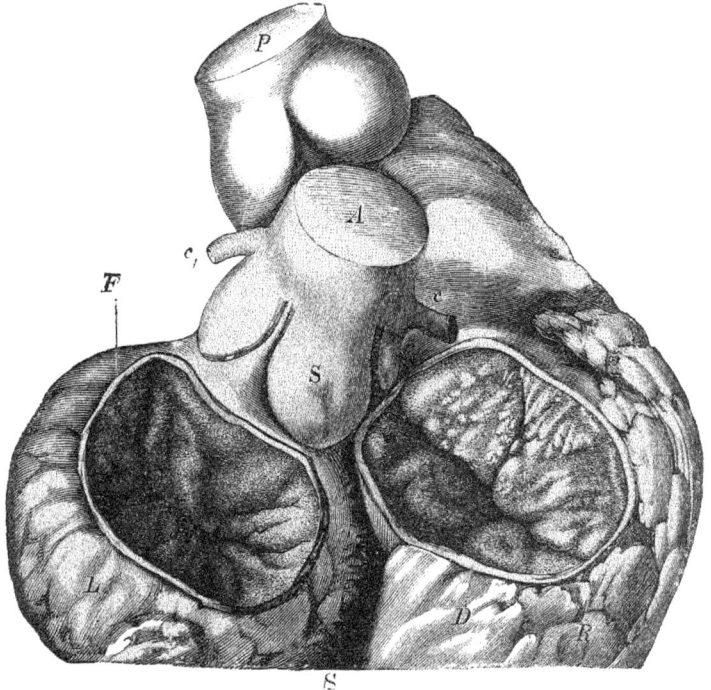

Fig. 26.—Plaster Cast of the Ventricles of the Human Heart, Viewed from Behind and Above. The walls are
removed, only the fibrous rings and the auriculo-ventricular valves being retained: L, left; R, right ventricle;
S, situation of the septum; F, left fibrous ring, with closed mitral valve; D, right fibrous ring, with closed tricuspid valve; A, aorta, with the left (c_1) and the right (c) coronary artery; S, sinus of Valsalva; P, pulmonary artery.

tricular wall approaches the valves during contraction and thus might
render possible a bulging backward of the valves into the auricle, compensation is provided by the shortening of the papillary muscles and
of the large muscle-containing tendinous cords themselves. The valves
when closed present an approximately horizontal surface. There remains, therefore, in the ventricles, even at the height of contraction,
always a remnant of blood, the so-called *residual blood*.

3. When the pressure in the ventricle exceeds that in the arterial

vessels, the semilunar valves are opened, become stretched like tendon above their concave sinuses (Fig. 24), without becoming applied to the arterial wall, and allow the blood to enter.

Goltz and Gaule found, by means of maximal and minimal manometers, a negative pressure in the ventricles during a certain phase of the heart's contraction amounting in the dog to —23.5 mm. of mercury in the left ventricle. They suspected that this phase coincided with the diastolic dilatation and for which they thus assumed a considerable power of aspiration. Moens is of the opinion that this negative pressure prevails in the ventricle shortly before the systole has reached its maximum. He explains the aspiration as being produced by the formation of a vacuum in the ventricle, which must develop as a result of the active movement of the blood, through the aorta and the pulmonary artery, behind the circulating mass of blood, therefore in the ventricle. Gaule and Mink believe that the systolic enlargement of the aorta must at the same time cause a dilatation of the conus arteriosus of the left ventricle.

(D) After the ventricular contraction has attained its height and relaxation has commenced, the semilunar valves close with an audible sound (Fig. 27). The diastole of the ventricle is followed by the pause. Under normal conditions the two halves of the heart contract and relax simultaneously and uniformly.

The heart-muscle exhibits in its activity certain variations in tone, that is, it does not with every systole contract from the same degree of relaxation to the same degree of contraction, but, rather, there follow in rhythmical periods series of contractions that arise from a considerable degree of relaxation of the heart-muscle, alternating with series of contractions that begin in a less complete degree of relaxation. With the latter the degree of contraction is greater than with the former. These variations in tone have been found especially in the auricle of the tortoise-heart. When the arterial blood-pressure is moderately increased, the heart expels a larger amount of blood; if, however, the arterial pressure is greatly increased, the amount of blood expelled at each systole becomes less. Extracts of testicle, suprarenal gland, pituitary gland and spleen in 0.7 per cent. sodium-chlorid solution added to blood exert a tonic effect upon the heart; the extent of the contractions increases and the beats become more regular. Under the influence of alcohol the heart exhibits a marked degree of relaxation and a low degree of contraction. The influence of various poisons is variable. Heat increases the variations in tone.

FIG. 27.—The Closed Pulmonary Semilunar Valves of Man, Viewed From Below.

Whether the relaxation of the heart-muscle is an active dilatation or not has been decided in the affirmative by some investigators. Stimulation of the vagus (likewise digitalis and strychnin) is said to increase the active dilatation; while section of the vagus (likewise atropin) is said to diminish it.

PATHOLOGICAL DISTURBANCE OF THE FUNCTION OF THE HEART.

All obstructions to the blood-flow through the different portions of the heart or of the vessels connecting them give rise to a permanent increase in the work of that portion of the heart especially concerned with relation to the affected section of the circulation, and in consequence to an increase in the thickness of the muscular walls, with dilatation of the cavity. Should the resistance affect not alone one section of the heart, but consecutively other parts further on in the course of the blood-stream, these also will undergo secondary hypertrophy. If, in addition to increasing the muscle-substance of the affected portion of the heart, its cavity is at the same time dilated, as is often the case, the condition is designated excentric hypertrophy, or hypertrophy with dilatation.

The obstructions under consideration in the domain of the vascular channels are: constriction (stenosis) of the arterial or venous orifices and likewise defective closure (insufficiency) of the valves. The latter causes resistance to the blood-flow by permitting regurgitation of a portion of the blood already propelled onward. In this way there results:

1. Hypertrophy of the left ventricle from hindrances to the blood-flow in the territory of the greater circulation, chiefly in the arteries and capillaries, not in the veins. In this category belongs stenosis of the aortic orifice and of the aorta further on; also calcification and loss of elasticity in the large arteries, irregular dilatations of the arterial walls (aneurysm); insufficiency of the aortic valves, as a result of which the left ventricle is continually subject to the aortic pressure; finally, affections of the kidney, in consequence of which a greater arterial pressure is required in order that the urine may be excreted. In the presence of mitral regurgitation also, hypertrophy of the left ventricle is necessary for compensation, and a similar enlargement occurs in the left auricle in consequence of the heightened pressure in the lesser circulation.

2. Hypertrophy of the left auricle results from mitral stenosis and from mitral regurgitation, and also consecutively to aortic regurgitation because the auricle must overcome the uninterrupted aortic pressure that is present in the left ventricle.

3. Hypertrophy of the right ventricle results from (a) hindrances to the blood-flow in the territory of the lesser circulation. These are: (α) atrophy of vascular areas of considerable size in the lungs in consequence of destruction, contraction or compression of the lungs and from loss of numerous capillaries in emphysematous lungs. (β) Overdistention of the lesser circulation with blood in consequence of stenosis of the mitral orifice or of insufficiency of the mitral valve; also consecutively to hypertrophy of the left auricle resulting from aortic regurgitation. (b) Hypertrophy of the right ventricle must occur also in conjunction with insufficiency of the pulmonary valves, which permits the blood to regurgitate into the ventricle, so that the pressure of the pulmonary artery prevails continually in the cavity. This condition is exceedingly rare.

4. Hypertrophy of the right auricle develops consecutively to the condition last mentioned, likewise in association with stenosis of the right auriculo-ventricular orifice, or from insufficiency of the tricuspid valve. This condition is uncommon. When several obstructions in the circulation occur together there is a combination of the resulting phenomena. O. Rosenbach has investigated the manner and method by which the heart maintains its activity after the occurrence of valvular lesions. If the aortic valves were perforated, with or without simultaneous injury to the mitral and tricuspid valves, the heart performed first an increase of work, which counteracted the physical defects, so that the blood-pressure did not fall. The heart, therefore, possesses reserve powers, which are brought into play only when they are required. In consequence of the valvular insufficiency dilatation first develops as a result of the regurgitation of blood into the affected chamber of the heart. Then follows hypertrophy, but until this is completed the compensation must be effected by the reserve power.

Under the conditions that especially render diastole difficult there should yet be mentioned: large effusions into the pericardial sac or pressure on the heart from tumors. The systole is greatly interfered with by adhesions between the heart and the connective tissue of the mediastinum. Under such circumstances the surrounding tissues, even the thoracic wall, must be drawn upon with each contraction of the heart, so that systolic retraction and diastolic projection occur in the situation of the apex-beat.

THE APEX-BEAT. THE CARDIOGRAM.

By the term apex-beat (ictus s. impulsus cordis) is understood the visible and palpable elevation of a circumscribed area of the fifth (less commonly the fourth) left intercostal space, caused by the action of the heart. At times the apex-beat is less distinct, especially when the heart strikes against the fifth rib itself. Changes in the position of the body alter somewhat the situation and the force of the apex-beat. A graphic representation of this movement can be obtained by means of a registering apparatus—the apex-beat tracing or the cardiogram.

Method.—To obtain a tracing of the apex-beat the cardiograph of Marey may be employed. The instrument has been modified by various investigators. The pansphygmograph of Brondgeest is essentially the same as Marey's apparatus, with unimportant changes. Marey's sphygmograph can also be used. In animals the cardiogram can be registered by ligating the tube of the pansphygmograph in the pericardium.

In the normal tracing of the apex-beat of man (A) or of the dog (B) the following details are distinguishable: a b corresponds to the period of the pause and of the contraction of the auricles. As the auricles contract in the direction of the heart's axis from the right and above to the left and downward it is not surprising that the apex of the heart advances toward the intercostal space. In this portion of the tracing there can be seen generally two or even three slight elevations which may be due to the rapidly successive contractions of the venous endings, the auricular appendages, and the auricles themselves.

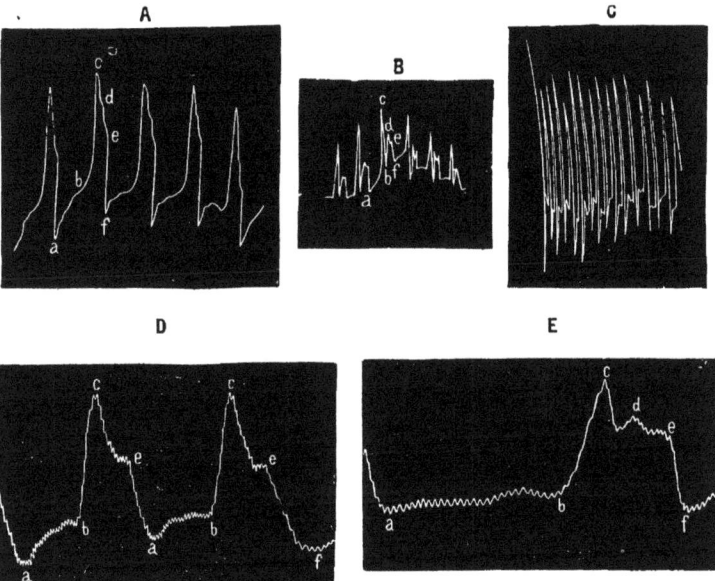

Fig. 28.—A, Normal apex-beat tracing from man. B, from a dog; C, tracing of an accelerated apex-beat from a dog; D and E, normal apex-beat tracings from man recorded upon a vibrating tuning-fork plate. Each serration represents 0.01613 second of time. In all of the tracings a b indicates the auricular contraction, b c, the ventricular contraction; d, the closure of the aortic valves; e, the closure of the pulmonary valves; e f, relaxation of the ventricles.

Naturally the last, occasionally distinct, elevation, occurring shortly before b (corresponding to B v and C v in Fig. 31), will be looked upon as the true auricular contraction; v. Ziemssen and Ter Gregorianz were able to register the elevation of the auricular appendix preceding the auricular contraction in a woman with an exposed heart.

The line b c is caused by the ventricular contraction. It is this alone that is appreciable to the palpating finger as the apex-beat. The first sound of the heart commences with the beginning of the ventricular contraction.

The cause of the ventricular impulse resides in the following factors:

1. The base of the heart (the junction of auricles and ventricles), which in diastole presents the form of a transverse ellipse (Fig. 29, I, *F G*), is contracted to a rather circular figure (*a b*). In this way, the large diameter of the ellipse (*F G*) is naturally diminished and the small diameter (*d c*) is increased, and in consequence the base is brought nearer to the chest-wall (*e*). This alone, however, does not produce the apex-beat, but the base of the heart, thus brought somewhat nearer the chest-wall, and hardened during systole, affords the apex the possibility of making the movement that constitutes the apex-beat.

2. The ventricles, which during the period of relaxation have their apex (Fig. 29, II, *i*) directed obliquely downward in the line of their long

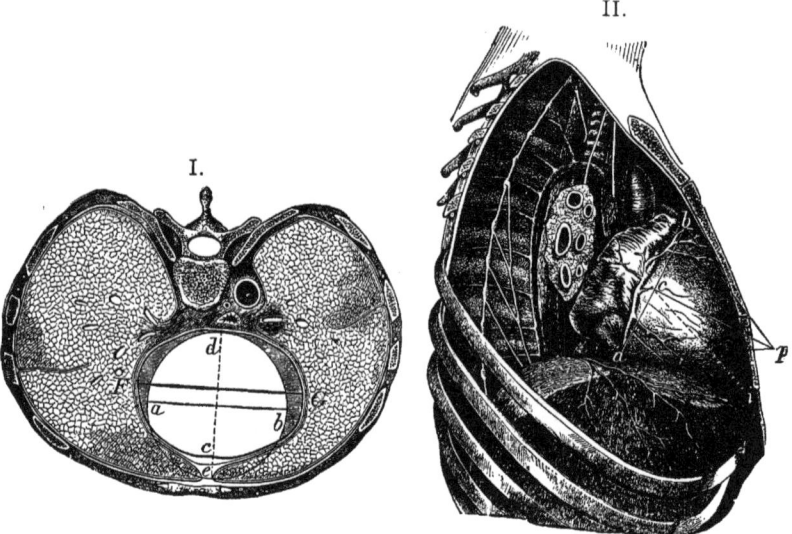

II.

I.

p

FIG. 29.—I. Horizontal Section through the Heart and the Lungs, Together with the Chest-walls, for the Demonstration of the Change in the Shape of the Base of the Heart during the Contraction of the Ventricles: *F G*, transverse diameter of the ventricles during diastole; *c*, position of the anterior ventricular wall; *a b*, transverse diameter of the ventricles during systole with *e*, the position of the anterior ventricular wall during systole. II. Lateral View of the Position of the Heart: *i*, the apex-beat during diastole; *p*, during systole (in part after C. Ludwig and Henke).

diameter, so that the angles (*b c i* and *a c i*) formed by the junction of the ventricular axis with the diameter of the base are unequal, represent a symmetrical cone, with its axis perpendicular to its base. Accordingly, the apex (*i*) must be elevated from below and behind forward and upward (*p*) (W. Harvey: "Cor sese erigere"), and it thus thrusts itself, hardened during systole, into the intercostal space (Fig. 29, II).

3. During the systolic contraction the ventricles of the heart undergo a slight spiral rotation about their long axes ("lateralem inclinationem," W. Harvey), so that the apex is carried from behind slightly forward,

while at the same time a considerable area of the left ventricle is turned forward. This rotation is due to the fact that many of the fibers of the ventricular muscles that arise from the portion of the fibrous ring that is turned toward the chest-wall at the junction of the right auricle and ventricle pass obliquely from above and to the right downward and to the left, in part to the posterior aspect of the left ventricle. Thus, they draw the apex of the heart upward in the direction of their course, and its posterior aspect slightly toward the anterior wall of the thorax. This rotatory movement is favored by the circumstance that the aorta and the pulmonary artery, which are applied to each other in a slightly spiral manner, effect a rotation of the heart in the same direction at the time of systolic tension.

According to an earlier opinion the cardiac impulse was held to be produced or at least increased by:

4. The *recoil* that the ventricles are supposed to experience (like a discharged firearm) at the instant when the column of blood empties itself into the aorta and the pulmonary artery. The apex would, of course, be driven in the opposite direction by this recoil, that is, downward and a little outward. Landois, however, has pointed out that the blood-column is discharged into the vessels 0.08 second after the beginning of the ventricular contraction, while, on the other hand, the apex-beat begins simultaneously with the first sound.

As, however, the apex-beat is observed in bloodless hearts taken from animals after death, and as the apex of the heart is not, as it would be on the theory of the recoil, displaced downward and to the left during systole, but upward and to the right (as has been confirmed by v. Ziemssen in a woman whose heart was exposed), the recoil cannot be regarded as a factor in the problem.

After the ventricles by their systolic movement have traced the greatest part of the apex-beat curve, as far as its apex (c), the curve rapidly descends and the ventricles pass from a state of extreme contraction to one of relaxation. Soon, however, two small elevations appear in the descending limb of the curve at d and e. These are due to the abrupt closure of the semilunar valves, which, being effected with a certain degree of force, is transmitted along the axis of the ventricles as far as the apex, and through the latter even causes concussion of the intercostal space; d corresponds to the closure of the aortic valves, e to that of the pulmonary valves. The valves, therefore, do not close at the same time, there being an interval of about from 0.05 to 0.09 second on the average. Owing to the greater pressure of the blood in the aorta the aortic valves close earlier than those of the pulmonary artery.

While investigators are agreed that the first sound of the heart begins at the point b of the cardiogram, various statements have been made with regard to the point at which the registration of the second sound of the heart takes place. Martius designates the depression between c and d (Fig. 28, E) as the point that corresponds to the second heart-sound; Landois the apices d and e, when the tension of the semilunar valves is increased; Hürthle, Einthoven and Geluk 0.02 second after e; Marey and Fredericq about midway between e and f; and, finally, Edgren at a point immediately in front of f.

Method.—In order to determine the time when the heart-sounds are heard, their vibrations are transmitted to a microphone attached to the thorax. The instrument, which is thrown into vibration by each sound of the heart, opens and closes an electric circuit with each vibration and thus attracts an electromagnet, or sets a capillary electrometer (Fig. 229) in motion. If by means of another contrivance the cardiogram is made to register at the same time, the points on the latter at which the heart-sounds are heard can be seen.

From the point e to the foot of the curve (at f) comprises the time during which complete diastolic relaxation of the ventricles takes place.

THE TIME-RELATIONS OF THE MOVEMENTS OF THE HEART.

Method.—The time-relations of the individual phases of the movements of the heart can be most reliably discerned in the curves of the apex-beat:

When the distance traversed at a uniform rate in a unit of time is known for the registering surface, the time corresponding to each portion of the curve can be ascertained by direct measurement (as in the case of pulse-curves).

Landois determined the time by having the curves traced on a tablet vibrating on the arm of a large tuning-fork (Fig. 60). The curve then contains in all of its segments small undulations due to the vibrations of the tuning-fork. In Fig. 28, D and E represent apex-beat curves of healthy students registered in this way (in D the elevation d is not distinct). A complete vibration of the tuning-fork (from the apex of one undulation to that of the next) corresponds to 0.01613 second; by counting the number of undulations and multiplying by the factor the time is obtained. Although there is a certain regularity in the time of the individual phases of the movement, the readings nevertheless vary between wide limits even in healthy individuals.

The value of a b, which is equivalent to the pause plus the auricular contraction, is subject to the widest variations and depends chiefly on the frequency of the pulse; for, the more rapidly the heart-beats follow one another, naturally, will be the pause, until it finally disappears altogether. Even when the rate of the heart is slow, it is often impossible to distinguish in the curve the portion corresponding to the pause, which, owing to the gradual filling of the heart and the resulting slight bulging of the intercostal space, has a gently ascending form, from that due to the auricular contraction and appearing as a hillock. In one case in which the heart-beats were 55 in a minute, Landois found the pause to be 0.4 second and the auricular contraction 0.177 second. In Fig. 28, A, the pause plus the auricular contraction, when the heart beats 74 times in a minute, is found on measurement to be 0.5 second. In D the corresponding period a b is equivalent to from 19 to 20 vibrations, or 0.32 second; in E the period is equivalent to 26 vibrations, corresponding to 0.42 second.

The ventricular systole is estimated from b, the beginning of the contraction, to e, the completed closure of the semilunar valves of the pulmonary artery. It, therefore, extends from the first to the second heart-sound. This period is also variable, though considerably more constant. When the action of the heart is accelerated, the period becomes less, when the action is slower the period increases; in E it is 0.32 second, in D 0.29 second; when the heart-beats were only 55 Landois found it to be 0.34 second; but when the frequency is exceedingly great it declines to 0.199 second.

Landois was able to ascertain the interesting fact that when the left ventricle is enormously hypertrophied and dilated, the duration of the ventricular contraction does not materially exceed the normal.

That the ventricle contracts more slowly when the action of the heart is weakened is shown when the registering instrument is placed on the ventricle of an animal that has been killed, and the heart-beat is recorded. In Fig. 30, from the ventricle of a rabbit the slow heart-beats (B) are at the same time of longer duration.

This affords an opportunity to determine accurately the length of the period to be allowed for the ventricular systole. Landois thought it wise, in order to avoid misunderstanding, to distinguish the following three separate factors:

1. The interval between the two heart-sounds, that is, from the beginning of the first to the end of the second sound (Fig. 28, b—e).

2. The time occupied by the blood in entering the aorta: This evidently terminates at the depression between c and d (Fig. 28, E); its beginning, however,

does not coincide with b, as from o.o85 to o.o73 to o.o6 second elapses between the beginning of the ventricular contraction and the opening of the semilunar valves of the aorta.   According to this calculation the entrance of the blood into the aorta (aortic inflow) would occupy from o.o8 to o.o9 second.   Landois arrived at this result by the following calculation: The interval between the first sound of the heart and the pulse at the axillary artery is o.137 second.   The propagation of the pulse-wave along the distance from the root of the aorta to the axillary artery, which is equivalent to 30 centimeters, cannot occupy more than o.o52 second of this time (corresponding to the analogous velocity in the distance—50 cm.—from the axillary to the radial artery = o.o87).   Hence, the pulse-wave in the aorta cannot take place earlier than o.137 minus o.o52 = o.o85 second after the beginning of the first sound of the heart.   Landois found in agreement with Hürthle that in some cardiograms the point that marks the beginning of the flow of blood into the arteries, or, what is the same thing, the time of the opening of the semilunar valves, is indicated in the ascending limb by a small interval between b and c.   The current in the pulmonary artery is not interrupted until the point e is reached.

3. Finally, the time occupied by the muscular contraction of the ventricle may be considered.   The contraction begins at b, reaches its greatest degree at c, and is not followed by complete relaxation until f is reached.   The apex of the curve c may, however, be higher or lower, according as the intercostal space yields more or less; the position of c is, therefore, variable.

The time that elapses between d and e, that is, between complete closure of the semilunar valves and of the pulmonary artery, is greater in proportion as the pressure within the aorta exceeds that within the

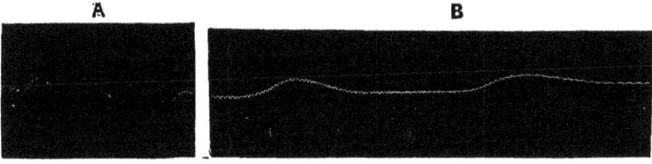

Fig. 30.—Contraction-curves from the Ventricle of a Rabbit Registered on a Plate Attached to a Vibrating Tuning-fork (one vibration = o.o1613 second): A, soon after death; B, taken while the ventricle was in process of dying.

pulmonary artery, as the closure of the valves is effected by the pressure from above.   This interval may vary from o.o5 second to more than twice that length of time; in the latter event the second sound of the heart is also duplicated.   If, however, the tension in the aortic system diminishes and the pressure in the pulmonary artery rises, the interval between d and e may be diminished to such a degree that the two coincide at one point in the curve.

The time occupied by the ventricles in relaxing (e f) after closure of the pulmonary valves is also subject to a certain degree of variation; in healthy adults the average may be given as o.1 second.

When the action of the heart is greatly accelerated, the time occupied by the pause is the first to become shortened, as Donders and Landois have found; then the time occupied by the auricular and ventricular systole also is shortened, in lesser degree, though quite distinctly.   With the highest degree of pulse-frequency the beginning of the auricular systole coincides with the closure of the arterial valves of the preceding heart-beat, a phenomenon that is strikingly illustrated in the tracing from a dog (Fig. 28, C).

As during the registration of apex-beat curves the heart is separated from the registering instrument by the soft parts of the intercostal space, which vary in thickness and in resistance and cannot in every case follow the movements of the heart with entire ease, it cannot be expected that the various portions of the curve shall coincide with mathematical accuracy with the corresponding phases of the heart's movements.

Gibson had the opportunity of taking cardiograms from a case of fissure of the sternum in a man, and obtained the following time-values: Auricular contraction (a b) = 0.115, ventricular contraction (b d) = 0.28, interval between the closure of the valves (d e) = 0.09, ventricular diastole (e f) = 0.11, pause = 0.45 second.

In large mammals (horses) Marey and Chauveau, in 1861, by a most thorough method obtained records of the phases of the movements of the heart in the following manner: Long catheter-like tubes, provided at their lower extremity with a closed and compressible rubber bulb, were connected by means of a flexible piece of tubing attached to the other end with the registering drum of the cardiograph (Fig. 44, KS). It is evident that with every compression of the

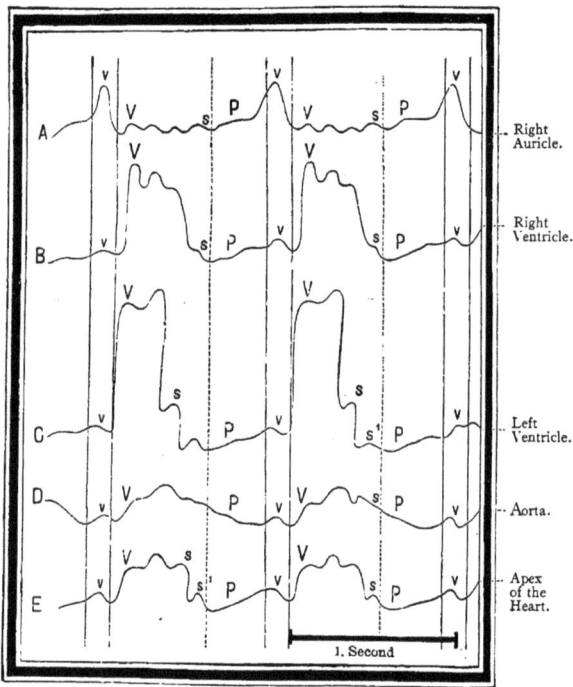

FIG. 31.—Curves Showing the Movements of the Separate Portions of the Heart (Chauveau and Marey).

rubber bulb the stylus connected with the registering drum of the instrument will be elevated.

Fig. 31 shows a number of curves: In making **A** the rubber bulb was in the right auricle, having been introduced through the jugular vein and the superior vena cava; in making B the bulb was introduced into the right ventricle through the tricuspid orifice; in making D it was introduced through the carotid as far as the root of the aorta; in making C, through the semilunar valves of the aorta into the left ventricle; and, finally, in making E the bulb was applied externally to the apex of the heart between this and the inner aspect of the chest-wall. In all of the curves v indicates the auricular contraction, V the ventricular contraction, s the closure of the semilunar valves (which occurred earlier in B than in C), and P the pause. As the recording surface moves at a uniform rate and the scale for the distance covered in each second is given, the individual periods of time can be measured.

It seems probable, however, that the introduction of the tubes into the heart is not without influence on the regular, undisturbed course of its activity.

In order to determine the conditions present coincidently with the pressure in the ventricle and in the aorta in the dog, Hürthle employed his blood-pressure recorders (Fig. 67), which were connected by means of tubes with the interior of the ventricle and of the aorta. A cardiogram was taken at the same time. The vertical lines o, 1, 2, 3 indicate conditions identical in time in the three curves. The point o corresponds with the beginning of the ventricular contraction and the first sound of the heart; while the entrance of the blood into the aorta occurs after an interval, namely at the point 1. The points 2 and 3, according to Hürthle, indicate the closure of the semilunar valves (second sound of the heart). Fredericq obtained similar results by means of other experiments.

One point remains to be cleared up, namely, whether the auricle and the ventricle work in exact alternation, in such a way that the auricle relaxes at the instant when the ventricle begins to contract, or whether the ventricle begins to

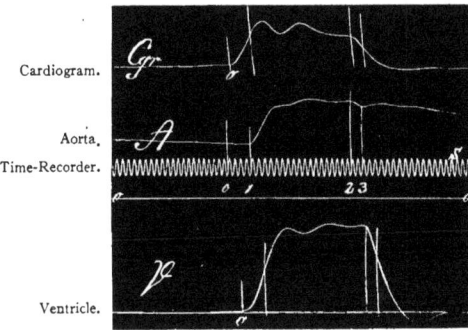

Cardiogram.

Aorta.
Time-Recorder.

Ventricle.

FIG. 32.—Simultaneous Record Showing Cardiogram, the Curve of the Ventricular Pressure and that of the Aortic Pressure, from the Dog. Each division of the time-curve = 0.01 second (K. Hürthle).

contract while the auricle still remains contracted for a short time, so that for a short period of time at least the entire heart is contracted. Heart-beat curves taken from human subjects appear to show that the ventricular contraction begins as the auricular contraction ends; v. Ziemssen and Ter Gregorianz, who made curves directly from the auricle of the exposed heart of a woman, are likewise in accord with the view that the auricular contraction continues for a time while the ventricles are beginning to contract; and also Heigl, on the strength of a similar observation.

A. Fick, who believes that the contractions alternate, considers this alternation as a means for maintaining the pressure in the large venous trunks approximately constant. As the auricle relaxes at the instant when the ventricular systole begins, there is no impediment to the flow of venous blood into the auricle; whereas if the auricular contraction were to persist, the blood would be dammed back. As, further, the auricle contracts at the instant of ventricular relaxation, there will be no abnormal pressure in the veins. In this way the pressure within the auricle may remain more uniform and the blood-stream in the ends of the veins more constant.

PATHOLOGICAL VARIATIONS IN THE HEART-BEAT.

The position of the heart-beat is altered: (1) By the accumulation of fluid (serum, pus or blood) or of gases in one pleural cavity. Copious effusions into the pleural cavity, which at the same time compress the lung and force it upward, may displace the heart as far as the right nipple. Effusions into the right pleura cause displacement of the heart to the left. As the right heart is forced to greater exertion in order to propel the blood through the compressed lung, the apex-beat under such circumstances is usually accentuated. Marked distention of the lungs (emphysema), which depresses the diaphragm, also causes downward and inward displacement of the apex-beat. Conversely, elevation of the diaphragm, as a result of contraction of the lungs or of pressure by the abdominal organs, has the effect of displacing the apex-beat upward—sometimes as far as the third intercostal space—and a little to the left. Thickening of the muscular wall of the heart with dilatation of the cavities (hypertrophy and dilatation), when it affects the left ventricle, causes an increase in the length and breadth of the chamber, and the accentuated apex-beat becomes palpable to the left of the nipple-line, sometimes in the axillary line in the sixth, seventh, or even eighth intercostal space. Hypertrophy and dilatation of the right ventricle cause an increase in the width of the heart: the apex-beat is felt further to the right, sometimes even to the right of the sternum, but at the same time also a certain distance beyond the left nipple-line. In the rare cases of transposition of the viscera, in which the heart is situated in the right half of the thorax, the apex-beat is of course found in exactly the corresponding situation on the right side of the thorax. Landois was the first to take an apex-beat curve from a heart of this kind and found that it presented all of the normal features. When the heart-beat extends to the left beyond the nipple-line or to the right beyond the parasternal line, the area of cardiac impulse is enlarged transversely, a condition that always indicates hypertrophy of the heart. When this transverse enlargement is unusually great, the apex-beat may extend over several intercostal spaces or over both sides of the thorax.

The apex-beat appears abnormally weak in association with atrophy and degeneration of the heart-muscle, or when the innervation of the controlling nerves is impaired. The cardiac impulse may be weakened or even completely obliterated also when the heart is forced away from the chest-wall by an accumulation of fluid or of gas in the pericardium, by a greatly distended left lung, or by an effusion into the left pleural cavity. The same condition results either when the left ventricle is imperfectly filled during contraction (in consequence of marked stenosis of the mitral orifice) or when, owing to extreme narrowing of the aortic orifice, it can empty itself but gradually and slowly.

An increase of the apex-beat is observed in the presence of hypertrophy of the walls of the heart, as well as in association with the most diverse irritative conditions (psychic, inflammatory, febrile, toxic) affecting the heart and its controlling nerves. Extreme hypertrophy of the left ventricle causes a heaving apex-beat, so that a portion of the chest-wall is elevated, with systolic concussion.

In some cases the apex-beat is quite distinct, or even abnormally distinct, while the pulse is quite small. This phenomenon is due to insufficient emptying of the ventricles (spurious contraction of the heart).

Systolic retraction is not infrequently observed on the anterior chest-wall in the third and fourth intercostal spaces on the left side under normal conditions, especially when the action of the heart is accentuated and when there is excentric hypertrophy of the ventricles. As the apex is somewhat displaced with each ventricular contraction and the ventricles at the same time diminish in size, the yielding soft parts of the intercostal space are drawn in to fill the vacuum thus formed. When the heart is adherent to the pericardium and the surrounding connective tissue, movement of the heart during systole becomes impossible and the apex-beat is replaced by systolic retraction of the apical area. Under such circumstances the chest-wall bulges during diastole, in a measure representing a kind of diastolic apex-beat.

The changes in the apex-beat that occur in association with functional disorders of the heart are best studied by tracing apex-beat curves, as has been done by a number of clinicians since Landois first published his method in 1876.

In the curve shown in Fig. 33, P, in reduced size and obtained from a case of marked hypertrophy and dilatation of the left ventricle, the ventricular contraction as a rule is exceedingly large (b c), although the time occupied in contraction by the greatly increased muscular mass of the ventricular wall is not materially longer than under normal conditions. The curves P and Q were obtained from a man with a high grade of excentric hypertrophy of the left ventricle, resulting from insufficiency of the semilunar valves of the aorta. The curve Q was taken purposely at a point near the epigastrium where systolic retraction was present. Although the position of the individual portions of the curve is changed, the individual phases of the heart's action are nevertheless well shown.

Fig. E represents the apex-beat in a case of stenosis of the aortic orifice. The auricular contraction (a b) is quite brief, the ventricular contraction is visibly prolonged and after a short rise (b c) exhibits a series of indentations (c c) caused by the mass of blood forcing its way through the stenotic and roughened entrance to the aorta.

Fig. F represents the apex-beat in a case of insufficiency of the mitral valve; a b is well marked in consequence of the increased activity of the left ventricle; the shock (d) caused by the closure of the aortic valves is slight on account of the diminished tension in the arterial system. On the other hand, the shock of the accentuated pulmonic second sound (e) stands like a huge accent high upon the summit of the curve. In consequence of the tension in the pulmonary artery

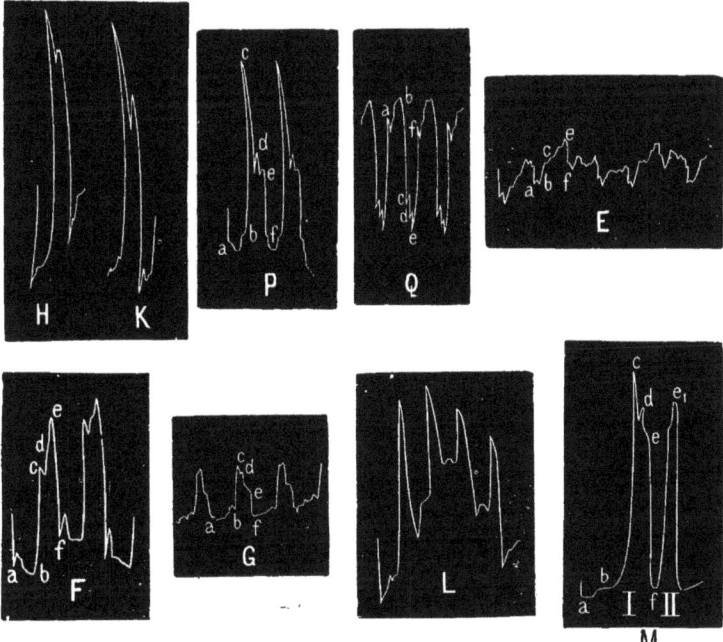

FIG. 33.—Various Forms of Pathological Apex-beat Curves. In all of these curves a b indicates the auricular contraction; b c, the ventricular contraction; d, the close of the aortic semilunar valves; e, that of the pulmonary valves; e f, the time occupied by the relaxation of the ventricles.

the pulmonary second sound may be so accentuated and it may follow so quickly after the second aortic sound (d) that the two almost or quite coincide (H and K).

The curve in a case of stenosis of the left auriculo-ventricular orifice (G) presents first of all a long, irregular, indented auricular contraction (a b), due to the fact that the blood is forced through the narrow orifice with considerable agitation and friction. The ventricular contraction (b c) is feeble on account of the imperfect filling of the left chamber. The closures of the two valves d and e are separated by a comparatively long interval and the ear distinctly hears a duplicated second heart-sound. The aortic valves close rapidly because the aorta receives only a small amount of blood, while the more abundant flow of blood into the pulmonary artery causes retarded closure of the pulmonary valves.

When the heart-beats are rapid and weak and the tension in the aorta and the pulmonary artery is low, the signs of closure of the valves in the latter may be entirely obliterated, as in curve L taken from a girl with exophthalmic goiter who suffered from nervous palpitation of the heart.

In rare cases of mitral insufficiency—a condition in which the right ventricle is greatly overfilled with blood. while the left contains but little, so that the right has to work harder to empty itself than does the left—a peculiar action of the heart has been observed, both ventricles appearing at times to contract together and then again the right ventricle alone (Fig. M after Malbranc). Curve I, which appears in every respect like a normal apex-beat curve, was taken when the entire heart was active; there was present an arterial pulse corresponding to this apex-beat. Curve II, on the other hand, appears to have been recorded by the right heart alone, and it accordingly lacks the closure of the aortic valves (d); nor was there an arterial pulse corresponding to this contraction.

With respect to the cases just considered Landois expressed the opinion as early as 1879 that the phenomenon could not be explained on the mere supposition that the right ventricle alone is active during the phases in question. without any parallel action on the part of the left. He regarded such a condition as impossible, if for no other reason because of the common arrangement of the muscles in the two ventricles and their equally common innervation. The period of apparent rest of the left ventricle is probably no more than a period of exceedingly feeble action, not strong enough to record itself in the apex-beat curve by the closure of the aortic valves and by a pulse in the arteries. This supposition has in fact been confirmed by Riegel and Lachmann, Eger, Eichhorst, Stern, H. E. Hering, and others.

THE HEART-SOUNDS.

On listening over the region of the heart, either directly with the ear applied to the thorax, or with the aid of the stethoscope, or in animals to the exposed heart, two sounds are audible that really do not deserve the name of tones, but which in contradistinction from pathological heart-murmurs are designated *heart-sounds*. As they possess a certain tonal color, it has been possible to determine their musical pitch.

The first sound of the heart is somewhat duller, longer, and lower in pitch by a third or fourth, fluctuating between d sharp and g, not clearly defined, especially at the beginning, and synchronous with the ventricular systole. The second sound of the heart is clearer, more valvular, shorter, and therefore more distinctly marked, varying between f sharp and b flat, clearly defined, and synchronous with the closure of the semilunar valves. The first sound is separated from the second by a short interval, and the second sound from the succeeding first sound by a longer interval. In musical parlance the first sound appears as a rising beat to the second, which is then followed by the pause. The vibration-values and the rhythm may accordingly be expressed as follows:

Bu - túp (lub-dúp)   Bu - túp (lub-dúp)

The first sound is caused by two factors. As it is heard, though faintly, in excised hearts in which the auriculo-ventricular valves are prevented from being stretched and relaxed, and as it is heard also when the movement and closure of the valves are prevented by means of a finger introduced into the auriculo-ventricular orifice, the principal cause of the sound is to be sought in the muscular murmur, produced by the contracting muscular fibers of the ventricles.

The sound is augmented and reinforced by the tension and vibrations of the auriculo-ventricular valves and their tendinous bands at the instant of ventricular contraction.

Wintrich, in 1873, succeeded by the use of suitable resonators in distinguishing one sound from the other; the clearer and shorter valvular sound from the deeper and more protracted muscular tone.

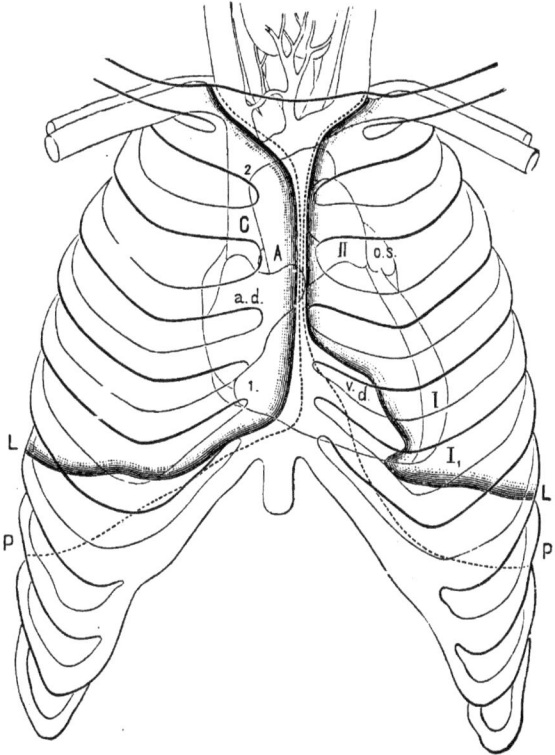

FIG. 34.—Topography of the Thorax and of the Thoracic Viscera: a. d., right auricle; o. s., left auricle; v. d., right ventricle; I, left ventricle with I₁ apex of the heart; A, aorta; II, pulmonary artery; C, superior vena cava; L L, boundaries of the lungs; P P, boundaries of the parietal pleura (v. Luschka and v. Dusch).

Under pathological conditions, such as typhoid fever and fatty heart, in which the heart-muscle is greatly enfeebled, the first sound of the heart may be inaudible. In the presence of insufficiency of the aortic valves, when, owing to the regurgitation of the blood from the aorta into the ventricle, the mitral valve is made tense gradually and before the ventricular systole begins, the first sound of the heart is also not infrequently absent. Both of these pathological instances prove that the coöperation of muscle-tone and valve-tone is required for the production of the first sound of the heart and that when one of these elements is lost the heart-sound may become inaudible. It should further be mentioned that the vibrations of the semilunar valves before or during their closure and the vibrations of the fluid elements of the blood itself have been adduced as contributory factors in the explanation of the first sound of the heart.

The cause of the second sound of the heart, according to the generally accepted view, is the abrupt closure of the semilunar valves. It is, therefore, said to be chiefly a valvular sound. It is, however, in part due also to a sudden concussion of the fluid particles in the large arterial vessels.

Landois has shown from apex-beat curves taken from healthy individuals that the semilunar valves of the aorta and those of the pulmonary artery do not close at the same time. As a rule, however, the difference in time is so slight that the two sets of valves generate only one sound. On the other hand, if, owing to increase of the difference in pressure in the aorta and in the pulmonary artery, this interval becomes greater, a duplication or splitting of the second sound may become quite perceptible. This may occur in perfectly healthy individuals, especially at the end of inspiration or at the beginning of expiration. It is important to remember, however, that although the second sound corresponds with the closure of the semilunar valves, it appears proved that the closure itself gives rise to no sound; it is only an instant later, when the tension of the valves becomes greater, that the second sound becomes audible.

It is generally believed that the points on the chest-wall at which the heart-sounds are heard most distinctly on auscultation correspond to the points in the neighborhood of which they are produced.

The first valvular sound produced at the right auriculo-ventricular orifice is heard most distinctly at the junction of the fifth rib with the sternum on the right side, and is transmitted from that point somewhat inward and obliquely upward along the sternum (Fig. 34, 1). As the left auriculo-ventricular orifice is directed more posteriorly, toward the interior of the thorax, and is covered in front by the arterial orifices, the first mitral valvular sound is heard best at the apex or immediately above it, where a strip of the left auricle is in immediate contact with the chest-wall (I_1, I). As the orifices of the aorta and pulmonary artery are so close together, it is advisable to listen for the aortic second heart-sound in the prolongation of the axis of the aorta, that is, at the right border of the sternum, at the inner extremity of the right costal cartilage (at 2). The pulmonic second heart-sound is heard most distinctly in the second left intercostal space a little to the left and beyond the edge of the sternum (at II). The aortic second sound is clearer, sharper, and shorter, and is heard over a larger area than the pulmonic second sound.

To determine the intensity of the heart-sounds quantitatively H. Vierordt inserts between the chest-wall and the ear a series of solid rubber plugs, which are poor conductors of sound, placed one upon the other in the form of a column.

ABNORMALITIES IN THE HEART-BEAT.

Accentuation of the first sound of the heart in both ventricles indicates a more powerful contraction of the ventricular muscle and a consequent, sudden, and increased tension of the auriculo-ventricular valves.

Accentuation of the second sound is a sign of increased tension in the interior of the corresponding large vessels. Hence accentuation of the pulmonic second sound, which is such an important diagnostic sign, always indicates hyperemia and excessive tension in the lesser circulation.

Feeble heart-sounds are caused by sluggish, weakened heart-action or abnormal ischemia; they are observed particularly in cases of morbid degeneration of the heart-muscle. The cause of weakness of individual heart-sounds can be deduced from the foregoing explanation.

The term *embryocardia* is used when the two sounds of the heart are exactly alike with respect to strength and the intervals between heart-beats, resembling the ticking of a clock; the phenomenon indicates weakening of the heart-muscle.

Irregularities in the structure of individual valves may render the heart-sounds impure by causing irregular vibrations. When pathological cavities filled with air are present in the immediate neighborhood of the heart, they may act as resonators and reinforce the heart-sounds, so that the latter often assume a metallic, ringing character. Both the first and the second heart-sound may be duplicated or split. Duplication of the first sound of the heart is explained by failure of the tricuspid and mitral valves to contract at the same time. Sometimes a sound may be heard that is caused by the contraction of a well-developed auricle and precedes the first sound like a presystolic murmur. As the closure of the aortic valves does not coincide exactly with that of the pulmonary valves, duplication or splitting of the second sound merely represents an exaggeration of physiological conditions. All factors that cause acceleration in the closure of the aortic valves—such as ischemia of the left ventricle—and retardation in the closure of the pulmonary valves—such as the presence of an excessive quantity of blood in the right ventricle, and both factors together when there is stenosis of the left auriculo-ventricular orifice—favor duplication of the second sound.

When the valves of the heart are the seat of irregularities in association with either stenosis or insufficiency, throwing the blood-stream into eddies or oscillations or producing friction, the heart-sounds are replaced by *murmurs*, that is, sounds produced by the fluids and always associated with circulatory disturbances and the valvular changes referred to. It is rare for deposits and new-growths projecting into the ventricle to give rise to murmurs in the absence of valvular lesions or circulatory disturbances. Heart-murmurs are always associated with the systole or diastole. As a rule, systolic murmurs are louder and more accentuated than diastolic. Sometimes they are so loud that even the thorax is thrown into vibration—*purring tremor.*

Diastolic murmurs always depend on structural changes in the mechanism of the heart, such as insufficiency of the arterial valves or stenosis of the venous orifices (usually on the left side only). Systolic murmurs are not always due to disturbances of the cardiac mechanism. In the left heart systolic murmurs may be caused by insufficiency of the mitral valve, stenosis at the aortic orifice and by calcification or abnormal dilatation affecting the ascending aorta. Systolic murmurs in the right heart, which are much more rare, are due to insufficiency of the tricuspid valve or stenosis at the pulmonary orifice.

Systolic murmurs are often present, although never so loud, in cases without any valvular lesion, being caused by abnormal vibration of the valves or of the walls of the arteries. They are heard most frequently at the pulmonary orifice, next at the mitral, and more rarely at the aortic and tricuspid orifices. Anemia and acute febrile affections are the causes of these murmurs.

Heart-murmurs are sometimes produced by the friction of opposed roughened surfaces of the inflamed pericardium (friction-murmurs). The friction-sound may be both audible and palpable.

DURATION OF THE MOVEMENT OF THE HEART.

The excised heart continues to beat independently for a time: in cold-blooded animals for a long period, even for days, in warm-blooded animals for a much shorter time. The last vestige of cardiac action has, however, been observed in the rabbit after $15\frac{1}{2}$ hours, in the mouse after $46\frac{1}{2}$ hours, in the dog after $96\frac{1}{2}$ hours, and in a three-months-old human embryo after 4 hours. The contraction of the excised heart may be reinforced and accelerated by irritation. The contraction of the ventricle first becomes enfeebled, and it is further observed that the contraction of the auricle is not always followed by a ventricular systole, two or more auricular contractions being succeeded by only one feebler ventricular contraction. The contractions of the ventricles, in addition to being more infrequent, require a longer time for their completion, and give the impression of being labored and sluggish (Fig. 30). Later, the ventricles cease to contract altogether and only the auricles continue to beat feebly. Direct irritation of the ventricles, however, as by a prick, is followed by a single contraction. Still later the left auricle ceases while the right auricle continues to beat, and it is the right auricular appendage that continues to beat the longest, being accordingly known to the ancients as "ultimum moriens." The same obser-

8

vation has been made in executed criminals. In the opened heart the papillary muscles fail to contract synchronously with the auricular wall after from two to three minutes. Engelmann made the interesting observation that the muscles of the auricle may lose their power of contracting, in response to irritation of the vagus or as a result of immersion and swelling in water, without losing the power of conducting stimuli. An analogous phenomenon has been observed with respect to the nerves.

After the heart has ceased beating altogether, it can be temporarily roused by direct stimulation, especially by heat; and again the auricles and auricular appendages are the last to react. As a rule, when the heart has been temporarily stimulated to greater activity it ceases to beat the earlier; before the orderly succession of beats ceases altogether tremulous, "undulating" movement of the muscle-bundles usually takes place. In mammals, when the irritability of the heart has ceased, it can be temporarily restored by injecting arterial blood into the coronary vessels. In the frog the heart, which at first becomes rigid, may be revived by filling its cavities with fresh blood. As the heart uses up oxygen and eliminates carbon dioxid, it is quite conceivable that it should beat longer in oxygen than in nitrogen, hydrogen, carbon dioxid, hydrogen sulphid or in a vacuum, even when, to avoid desiccation, aqueous vapor is generated in the vacuum. When the heart, after it has ceased to beat, is returned to a medium containing oxygen, it begins to beat again.

THE CARDIAC NERVES.

The cardiac plexus is formed by: 1. The cardiac branches of the trunk of the vagus nerve; these include cardiac branches from the external branch of the superior laryngeal nerve, the inferior laryngeal nerve, and sometimes the pulmonary branches of the vagus, in larger number on the right than on the left side. 2. The superior, middle, inferior, and lowest cardiac branches from the three cervical ganglia and the first thoracic ganglion of the sympathetic nerve, which frequently vary in number and in size (sometimes one of the branches accompanies the descending branch of the hypoglossus for a part of its course). The branches of the plexus are the deep and the superficial nerves; the latter usually contain a ganglion at the bifurcation of the pulmonary artery beneath the arch of the aorta. The following structures are regarded as belonging to the cardiac plexus:

(a) The right and left coronary plexuses, which convey the vasomotor nerves of the coronary vessels through the vagus portion and the dilators through the sympathetic; and in addition contain sensory fibers derived from the vagus and passing principally to the pericardium. In patients suffering from disease of the heart the presence of sensory nerves is indicated by the occurrence of constant or paroxysmal pain. In the frog, reflex phenomena may be induced from the ventricle in the various portions of the heart, and they probably have their reflex center in the medulla oblongata.

(b) The nerves embedded in the heart-muscle and in the furrows, which are richly supplied with ganglia and which have been designated the automatic motor centers of the heart. The heart contains a circle of nerves richly supplied with ganglia at the edge of the interauricular septum and another at the junction of the auricles and the ventricles. Wherever the two meet they exchange fibers. The ganglia are for the most part found near the pericardium. In mammals the two larger ganglia are situated close to the orifice of the superior vena cava; in birds the largest node of nerve-tissue, containing thousands of ganglia, occupies the posterior point of decussation of the longitudinal and transverse sulci. These nodes of nerve-tissue send smaller branches into the muscular walls of the auricles and ventricles, and these branches in turn are the seat of smaller ganglia.

In the frog a large collection of ganglia, Remak's ganglion, is situated, together with the vagus fibers, within the wall of the sinus of the vena cava (the dilated orifice of the venæ cavæ in the right auricle whose independent movement precedes that of the auricles). From this ganglion the vagus fibers pass as the anterior and posterior septal nerves, each of which is provided with a ganglion at the auriculo-ventricular junction, the ventricular ganglion, or Bidder's ganglion. The nerve-fibers, which are for the most part non-medullated, can be traced further in connection with the ganglia.

The motor fibers terminate with slightly clubbed extremities in each muscle-cell; the sensory, which are derived from medullated fibers, in flat, expanded terminal plexuses, which are quite abundant in the endocardium and the pericardium.

All ganglion-cells are bipolar or multipolar. In the frog most of them are surrounded by a network of fibers; in Bidder's ganglion spindle-shaped cells with two processes, one at each extremity, predominate. In the rabbit and in the frog the ganglion-cells belonging to the sympathetic system have two nuclei, while the vagus ganglia have only one. After division of the vagus branches (in the frog) the spiral process and the pericellular network from which it originates undergo degeneration. The straight process gives off the muscle-nerves. The bulb of the aorta contains numerous nerves for its muscle-fibers; but whether it contains ganglia also is doubtful.

IRRITABILITY OF THE AUTOMATIC MOTOR CENTERS IN THE HEART AND IN THE HEART-MUSCLE.

There are at the present time only two theories with regard to the irritability of the heart and its spontaneous rhythmic action.

1. The older theory teaches that the "automatic centers" that excite the movements and maintain an orderly rhythm are situated within the heart and that this function resides in the ganglia.

2. It is assumed that not one but several such centers are present in the heart and are connected with one another by conducting paths. So long as the heart is intact the various centers are stimulated to rhythmic activity in a definite order from the principal center, the impulse being conveyed through the conducting paths from that center. The forces that excite these regular continuous movements are not known. If, however, diffuse stimuli, of which the simplest is a strong electrical current, are applied to the heart, all of the centers are thrown into action and a spasmodic contraction of the heart takes place without any rhythm of movement. The dominating center is situated in the auricles (in the frog), whence, therefore, the regular progressive movements usually proceed. When its irritability is reduced, as by applying opium to the septum with a cotton pledget, a different set of centers appears to gain control, and the movement may then be propagated from the ventricles to the auricles.

3. The nerve-centers of the auricles are more irritable than those of the ventricles; hence they continue to beat independently for a longer time when the heart is left to itself.

4. All stimuli of moderate strength acting directly on the heart cause primarily an increase in the rhythmic heart-beats; stronger stimuli cause, in a short while, diminution progressing to paralysis, often preceded by spasmodic tremulous "undulation or flickering." Increased activity on the part of the heart exhausts its strength the more rapidly.

5. Individual weak stimuli, such as are insufficient to exert any effect on the heart, may be rendered efficient by repetition, as the heart is capable of summation of the individual stimuli.

6. Even the feeblest stimuli that are at all capable of exciting a contraction always excite an active contraction, that is, "the minimal stimulus has a maximal effect."

7. Each contraction of the heart is followed by a short period during which the heart is less susceptible to subsequent stimuli (Marey's refractory period) and the conducting-power of the muscle-substance is reduced.

8. Stimulation of the heart-centers, apparently reflex, takes place on the inner surface of the heart. Feeble stimuli from this surface are more effective in accelerating and exciting the action of the heart than stimuli from the external surface of the heart. Stronger stimuli, which cause arrest of the heart, also act more readily from the internal than from the external surface of the heart; under such conditions also the ventricular portion is always first to be paralyzed.

9. Portions of the heart that are devoid of ganglia are incapable of independent movement unless a stimulus be applied; they contract only once to a single direct stimulus, or they may beat rhythmically if the stimuli are applied continuously. Such a stimulus may be provided by the continuous pressure of fluid within the cavities of the heart or by means of chemical agents brought in contact with the heart.

10. The pulsations of stimulated portions of the heart devoid of ganglia indicate that the ganglia are not absolutely necessary for the production of rhythmic contractions; but the ganglia are more irritable than the muscle itself. They control also the regular alternating action of the various portions of the heart, so that normal cardiac action must be regarded as under the control of the ganglia.

11. If the heart be cut in such a way that the individual pieces remain in communication, the regular contractions beginning in the auricles and propagated in peristaltic or undulating movements to the ventricles persist for some time. When, however, the heart is completely divided into two pieces, auricle and ventricle, the movements of both continue separately—naturally, no longer in orderly succession, but quite independently.   .

The principal experiments on which the foregoing propositions are based are as follows:

Experimental Division and Ligation of the Heart.—These experiments have been performed chiefly on frogs' hearts. Ligation differs from division in the fact that the physiological connection is destroyed by drawing a ligature tightly around the parts and loosening it again, while the anatomical continuity of the heart-wall and the integrity of the cavities of the heart are maintained.

1. *Stannius' Experiment.*—After separation in a frog's heart of the sinus of the venæ cavæ from the auricle, either by incision or by constriction, the heart is arrested in diastole, while the sinus continues to beat independently. If the heart be again divided at the auriculo-ventricular junction, the ventricle, as a rule, begins at once to beat again, while the auricles continue in diastolic arrest. In accordance with the position of the second line of division the auricles may continue to beat in association with the ventricles, or the auricles alone may contract, while the ventricles remain at rest.

The experiment has been interpreted in the following manner: The sinus of the venæ cavæ contains Remak's ganglion, which is remarkable for its extreme irritability, while Bidder's ganglion, which is situated at the auriculo-ventricular junction, possesses a lesser degree of irritability.. In the normal heart the latter receives its motor impulses from the former. When the sinus of the venæ cavæ is severed, the stimulating Remak's ganglion is without any influence on the heart. The latter becomes arrested for two reasons: because Bidder's ganglion by itself does not possess sufficient power to set the heart in motion, and because the division stimulates the inhibitory nerves of the heart (vagus), which are situated at this point. Pulsation can, however, be induced in a heart that has been arrested in this way by irritation of Bidder's ganglion, as by gently pricking the auriculo-ventricular junction, or by the passage of a moderately strong constant current. In the latter event the ventricular beat sometimes precedes that of the auricles. If, now, the auriculo-ventricular junction be divided, the ventricle begins to pulsate, partly because the procedure stimulates Bidder's ganglion, and partly because the heart is no longer under the influence of the vagus, which had been stimulated by the first division. If the division at the auriculo-ventricular junction is made in such a way as to leave Bidder's ganglion in the auricle, the latter would pulsate and the ventricle remain at rest; if the ganglion is divided into two halves, both the auricles and the ventricles pulsate, because each is stimulated by its own half of the ganglion.

2. When the ventricle alone is divided in the frog's heart by ligature or incision at the auriculo-ventricular furrow, the sinus and the auricles continue to beat undisturbed, while the ventricle is arrested in diastole; the ventricle responds to a local irritant with a single contraction. If the incision is made in such a way as to leave the lower edge of the interauricular septum attached to the ventricle, the latter also continues to pulsate. In the case of the rabbit's heart, also, the ventricles continue to pulsate if a small strip of the auricles is preserved, separated from the auricular nerves.

3. Experiments performed by A. Fick in 1874 first showed that the irritative process in the contractile tissue of the frog's heart is propagated in all directions and that the entire frog's heart acts in a measure like a single continuous muscle-fiber. Thus, for example, a transverse incision, involving the ventricle of the frog's heart, does not prevent the appended flap from taking part in the systolic contraction. This is shown also by the following experiments of Engelmann. If the heart is cut into strips, as by zigzag incisions, in such a manner that the individual pieces remain in connection with one another by means of muscle-substance, the strips pulsate in regular succession, in whatever way they may be connected with one another, as a result of the direction of the incisions. The velocity of propagation, under such circumstances, is from ten to thirty millimeters in the second. These experiments also confirm the observation that the continuous stimulus that propagates the contraction is not conducted by nerve-paths but by the substance of the contractile mass.

4. When the apex of the heart has been separated from the rest of the organ by a ligature it ceases to take part in the contraction of the heart, which continues to pulsate; a direct stimulus, such as a stab of the apex, is followed by only a single contraction. If the heart is filled with saline solution under pressure (both of which act as stimuli), the apex will continue to pulsate. The same thing is observed after poisoning with delphinin or quinin. If a cannula is tied in the ventricle from a point above the auriculo-ventricular junction to the apex, the latter is likewise arrested; if, however, the apical portion is filled through this cannula with oxygenated blood under steady pressure, the apex will pulsate.

The excised apex of the heart resting spontaneously, when stimulated by induction-currents, responds to the weakest efficient stimulation by a maximal contraction; but the application of tetanizing currents is not followed by true tetanus. Closing and opening the constant current applied to the severed apex give rise only to the ordinary closing and opening contractions.

5. When the point of ligation is within the auricles, the pulsations of the heart occur in successive periods (group-formation), and the contractions often increase in strength by regular gradations (stair-case ascent).

6. When the bulb of the aorta, which is devoid of ganglia, is isolated by constriction (frog), it continues to pulsate when the internal pressure is moderate; after it has ceased beating, a single stimulus will give rise to a series of renewed contractions. The number of contractions is increased by raising the temperature to 35° C. and by increasing the internal pressure.

7. The isolated venæ cavæ and their sinuses exhibit normal contractions. If they are still connected with the heart they will control the movements of the heart, that is, contraction of the entire heart may be induced from the position of each of the large veins and the rhythm of the heart may be thus influenced. Conduction takes place only through the muscle-substance and not through the nerves. Porter maintains with regard to the hearts of the dog and the cat that any part of the heart that is excised may continue to pulsate if only it be sufficiently nourished.

In opposition to the doctrine that has just been expounded, namely that the stimulating influence is sent out by the cardiac ganglia, it may be observed that this theory has recently begun to waver. In view of the fact that the embryonal heart, in which it has been impossible to demonstrate the presence of ganglia, pulsates like the heart of certain invertebrates, some recent investigators assert that the automatism of the cardiac action resides in the muscle itself. Similarly, His, Jr., and Romberg, on developmental grounds, teach that the ganglia belong really to the sensory nerves of the heart, and that, therefore, there are no automatic nerve-centers at all. When Krehl and Romberg isolated portions of the rabbit's heart devoid of ganglia by crushing, but in such a way that, so long as the circulation was maintained, they represented anatomical portions of the heart, they found that these pieces continued to pulsate for hours. It is said that even excision of the entire septum of the frog's heart, including Remak's ganglion, has no disturbing effect on the heart-beat.

The propagation of the contraction from the auricles to the ventricles is said to take place through the muscle-fibers that pass from the former to the latter. That the conduction of the stimuli from auricle to ventricle, which does not take place continuously, but periodically in the same rhythm as the heart-beats, is not transmitted through the nerve-paths is proved by the slow rate at which it is effected, the conduction being 300 times slower than in motor nerves.

Engelmann expresses his views upon these questions as follows:

The muscle-cells of the heart itself and not a system of nerve-ganglia constitute the excito-motor central organ; as such they generate the motor stimuli that cause the heart to beat. As those muscle-cells that surround the large veins emptying into the heart are most susceptible to the irritating influence of automatic movement, the systolic contraction occurs first at this point, to spread then in a peristaltic manner successively to the auricles, the ventricles, and the bulb of the aorta. The motor stimulus is propagated directly from muscle-cell to muscle-cell. All of the muscle-cells of the entire heart form together a single physiologically conducting contractile mass. Within each individual portion of the heart—venous trunks, venous sinuses, auricles, ventricles, bulb of the aorta—the motor stimulus is propagated rapidly, in a manner comparable to the contraction of a striated muscle. Those muscle-cells, on the other hand, that form the connecting bridges between the individual portions of the heart conduct slowly, in a manner comparable to unstriated or embryonal muscles. Consequently every individual portion of the heart contracts practically at the same time as a whole; while, on the other hand, the systole of each portion of the heart situated farther on in the course of the blood-stream can take place only after an actual interval, long enough for the blood to be carried from one part of the heart into the next. As the fibers of the heart-muscle, in the act of contraction, temporarily lose their contractility and conducting power, as a sort of fatigue-phenomenon, they contain within themselves the periodicity of contraction and relaxation—systole and diastole. A cycle of the entire heart may be induced from any point in the large veins. When the cardiac stimuli succeed one another slowly, each individual cardiac cycle becomes shorter, but more powerful. The blood is then propelled in larger quantities and with greater force; while if the succession is more rapid, less blood is propelled with a lesser degree of force.

Direct Stimulation of the Heart.—All direct cardiac stimuli act much more vigorously from the internal than from the external surface of the heart. When the stimulation is severe or protracted, the ventricular portion is always paralyzed first.

(a) *Thermic Stimuli.*—Descartes had already observed in 1644 that the eel's heart could be made to pulsate more rapidly by the application of heat. Alex. v. Humboldt explained the acceleration of the pulse that takes place in man in a hot medium in the same way. As the temperature continues to rise, the heart-beats at first often reach a considerable frequency. They then become more infrequent again, and finally cease altogether, and the muscle is found to be contracted. As a rule, the ventricular portion is arrested before the auricles, sometimes after a period of tetanic undulatory spasm. At a temperature of 25° C. and above, the ligated frog's heart immersed in a 0.6 per cent. saline solution, soon becomes arrested, and continues at rest if kept at this temperature. Up to 38° C. Landois has seen it recover if removed quickly. The inner surface of the heart reacts much more readily to all degrees of temperature than the external surface. If the heart, after having been arrested, is removed from the warm bath, it begins to beat rapidly after a pause, which may be interrupted by one or two beats, the frequency gradually diminishing until the normal rate is attained. If the ventricle alone is heated, the frequency of pulsation is not increased. The volume and the extent of the cardiac contractions increase up to a temperature of about 20° C., and beyond that point they begin to diminish again. The functional power increases between 8° and 33° C.; but the frequency increases more than the efficiency of the pulsations. The duration of the contraction at 20° C. is only about one-tenth of what it is at 5° C. The heated heart reacts to rapidly intermittent stimuli, the cold heart only when the intervals are of considerable length. The mammalian heart ceases to beat at from 44.5° to 45° C.

As the heat of the blood diminishes, the heart pulsates more slowly. When a frog's heart is placed on ice between two watch-glasses, its rate diminishes considerably; between 4° C. and 0° C. the pulsations of the frog's heart cease. When a frog's heart is suddenly removed from warm water and placed on ice, the beat is accelerated; conversely, when it is transferred from ice to warm water, the beat is at first slowed and only after a time accelerated.

(b) *Mechanical Stimuli.*—Pressure applied to the outside of the heart causes an acceleration of the cardiac action. In man also light pressure applied to the auriculo-ventricular junction of an exposed heart gave rise to a secondary shorter contraction of both ventricles following each heart-beat. Heavy pressure causes an irregular, undulatory contraction of the muscle, such as may be produced by compressing the excised heart of a warm-blooded animal between the fingers. Increase of the blood-pressure in the interior of the heart effects a similar acceleration, and decrease of the pressure a corresponding diminution in the number of heart-beats. When the intracardiac pressure is excessive, the overstimulation results in irregularity or even slowing of the heart-beat. A resting heart that is still irritable will react by a single contraction to a mechanical impulse (prick).

(c) *Electrical Stimuli.*—A moderately strong constant current passing continuously through the heart produces an increase in its rate. Ziemssen succeeded in accelerating the beat of an exposed heart two-fold or three-fold by passing a strong galvanic current uninterruptedly through the ventricles. Exceedingly strong constant currents, as well as tetanizing faradic currents, produce tetanic undulatory contractions of the heart-muscle, with lowering of the blood-pressure.

If the ventricle of the frog's heart has been permanently relaxed by being clamped at the auriculo-ventricular junction, and one electrode of a constant current is applied to the ventricular wall, and the other to any portion of the trunk, systolic contraction of the ventricle takes place when the current is closed only if the kathode is placed in contact with the ventricle; conversely when the current is opened only if the anode is in contact with the heart-wall. The feeblest faradic currents accelerate the heart-beat; stronger currents produce irregularities, which may go on to fibrillation.

A single induction-impulse applied to the ventricle in systolic contraction has no effect either in the frog or in the mammal. When, however, it is applied to the ventricle in diastolic relaxation, the succeeding systole takes place earlier. The auricles and the apex of the heart, which is devoid of ganglia, but may be excited to activity by suitable stimulation, react in the same way. During their systole an induction-impulse is ineffective, but in diastolic rest the impulse gives rise to a contraction, which is followed by a ventricular contraction. Even strong tetanizing induction-currents applied to the heart are unable to produce tetanus of the entire musculature. There develop between the electrodes localized, white, cylindrical elevations, as in the muscles of the intestines, which may persist for several minutes. After severe and continued tetanization the undulatory contractions outlast the stimulus. Also the isolated apex of warm-blooded animals may exhibit this undulatory contraction only so long as the stimulus lasts. The heart of a previously warmed frog, as well as the isolated apex, reacts to electric stimuli by flickering. The fibrillating or flickering rabbit's heart often returns spontaneously to its normal contractions, the dog's heart with greater difficulty. After the contractions of the frog's heart have become weak and irregular, they can be made regular and isochronous with the rhythm of the stimulus by means of electric stimuli applied in rhythmical succession. The feeblest stimuli that are at all efficient act as well in this connection as the strongest; even with the weakest stimulus the contraction of the heart is the most vigorous possible. Hence, this minimal electrical heart-stimulus is as effective as a maximal stimulus.

V. Ziemssen was unable even with strong induction-currents to cause a variation in the rate of the beat of the exposed human heart. The ventricular diastole alone appeared to be no longer complete, and in addition certain minor irregularities were observed in the contractions. By opening and closing or by reversing a strong constant current applied to the heart of a woman, it was possible to increase the number of heart-beats, and the increased number of pulsations corresponded with the number of the electrical impulses. For example, from a normal of 80 the number of heart-beats was raised to from 120 to 140 to 180 by the application of from 120 to 140 to 180 electrical impulses. Conversely, it was possible also to reduce the normal number of pulsations from 80 to 60 or 50 by applying an equal number of powerful stimuli. In the healthy subject also v. Ziemssen found that he could influence the rhythm and the strength of the heart by applying an electrical current through the chest-wall.

(d) *Chemical Stimuli.*—Many chemical agents, particularly when applied in a state of dilution to the inner surface of the heart, increase the number of pulsations, but when applied in concentrated form or when allowed to act for some time diminish the number or paralyze the heart. Bile and biliary salts diminish the number of heart-beats, as does also absorption of the bile into the blood. In dilute solution, however, both accelerate the action of the heart. The same effect is produced by acetic, tartaric, citric and phosphoric acids. Chloroform and ether when applied to the inner surface of the heart have a distinctly retarding or even paralyzing effect; in small amounts ether accelerates the heart-beats. Opium, strychnin, alcohol, and chloral hydrate have an analogous action. Klug caused blood impregnated with various gases to pass through the frog's heart and found that sulphurous acid, chlorin-gas, nitrous-oxid gas, hydrogen sulphid and carbon monoxid acted as heart-poisons. In the same way, blood saturated with carbon dioxid exhausts the heart, which, however, may recover if the carbon dioxid escapes. A deficiency of oxygen produces a grouped rhythm, in the same way as the phenomena of asphyxiation manifest themselves in the respiratory apparatus in grouped movements.

Rossbach found that local irritation of a circumscribed area of the frog's ventricle by means of mechanical, chemical, or electrical stimuli during contraction causes immediate relaxation in partial diastole of the part to which the stimulus is applied. The immediate after-effect of this form of irritation is a permanent shrinking of the irritated portion of the heart-fibers, and this is likewise strictly confined to the area of irritation. The shrunken portion ceases to functionate and remains permanently robbed of its vital properties. If the same stimuli are applied during diastole, the irritated portion relaxes earlier than the portion that has not been irritated, and the diastole of the irritated portion lasts longer than that of the non-irritated portion. If the weakest stimuli are allowed to act for a considerable length of time on any part of the frog's ventricle, the irritated portion always relaxes earlier than the non-irritated, and the diastole of the irritated portion lasts longer than that of the non-irritated.

Heart-poisons comprise such substances as have a special effect in diminishing or abolishing the movements of the heart. In this respect the neutral salts of potassium are most remarkable. In small doses they accelerate the heart-beat. Yellow potassium ferrocyanid, when injected into a frog's heart, will cause systolic arrest of the ventricles, even when greatly diluted. If blood subsequently enters the ventricle as the result of the contraction of the auricle, the ventricle may again take part in the contraction. Under such conditions, the ventricular muscles sometimes relax in areas after first undergoing reddening. The contraction of the ventricle, which is exceedingly sluggish, later travels from the auriculo-ventricular junction in a peristaltic wave to the apex. The Javanese arrow-poison, antiar, causes systolic arrest of the ventricles, with diastolic arrest of the auricles; muscarin causes diastolic arrest of the heart, which can be overcome by means of atropin. Some of the heart-poisons in small doses cause slowing and in larger doses not infrequently acceleration of the heart-beat: digitalis (and the toxic substances of oleander and the mayflower, which are similar to it), morphin, and nicotin. Others in small doses cause acceleration and in large doses slowing: veratrin, aconitin, camphor.

THE CARDIOPNEUMATIC MOVEMENT.

As the heart during systole occupies a smaller space in the interior of the thorax than during diastole, air must enter the thorax as the heart contracts if the glottis is open. When, however, the heart relaxes in diastole, air must escape through the open glottis as the heart enlarges. A similar influence must be due to differences in the degree of fulness of the intrathoracic vascular trunks. This *cardiopneumatic movement* is, in animals in which during hibernation the respiratory movements are suspended, of the greatest importance for the maintenance of metabolism, which continues in moderate degree. The interchange of carbon and oxygen in the lungs is greatly facilitated by agitation of the pulmonary gases, and this interchange suffices to aërate the blood passing slowly through the lungs.

Method.—The movement may be demonstrated by means of:

1. The manometric flame, the trachea of a curarized animal being opened and connected with a bifurcated tube, one branch of which leads to the gas-tubing and the other to a small gas-flame. As in this manner a free communication is established between the organ of respiration and the gas-supply, the movements of the heart will be transmitted to the gas-flame. In man it is possible, after a little practice, to transmit the movement in an analogous manner to the gas-flame through one nostril after closure of the other nostril and the mouth, or through the mouth after closure of the two nostrils.

2. By acoustic means, namely by introducing an exceedingly sensitive whistle constructed from a hollow sphere, in animals into the trachea divided transversely, in man—especially when the heart's action is stimulated—into the mouth, after closure of the nose, it is possible to demonstrate the cardiopneumatic movement, particularly if the whistle is blown continuously and with extreme softness.

3. By means of the cardiopneumograph (Fig. 35). This consists of a tube, which is held between the lips (D), while respiration is suspended, the glottis is opened and the nostrils are closed. The extremity of the tube, which is bent upward, perforates a small plate (T), over which a delicate membrane consisting of a mixture of collodion and castor-oil is stretched with moderate force. From the center of the membrane a glass thread (H) passes over the free edge of the plate and is provided at its extremity with a delicate hair, which registers the movements of the membrane on a tablet (S) moved by clockwork. Every expiratory movement of air causes depression and every inspiratory movement elevation of the recording point. Attached to the side of the tube is a valve with a sufficiently large opening (K) and which may be opened to allow the individual to breathe freely during a pause. The periodic movements of the respiratory gases propelled by the heart-beat cause associated movements in the delicate collodion membrane, and these are in turn transmitted to the recording lever.

The graphic curve (Fig. 35, A and B) exhibits the following details:

1. The respiratory gases undergo a sudden expiratory movement coincidently with the first sound of the heart because at the instant of the ventricular systole the blood from the ventricles has not yet left the thorax, while venous blood is pouring into the right auricle through the venæ cavæ, and because in the same instant of systole the dilating branches of the pulmonary artery must cause approximately the same quantity of air to escape from the nearest air-passages in the lungs. In fact, the blood contained in the right auricle does not leave the thorax at all; it is only transferred to the lesser circulation. This expiratory movement would often be greater if it were not limited by two factors, namely: (a) because the muscular mass of the ventricle occupies a somewhat smaller volume during contraction, and (b) because the thoracic cavity in the region of the fifth intercostal space is somewhat enlarged outwardly by the apex-beat.

2. There follows immediately a marked inspiratory movement of the respiratory gases, in consequence of which the large ascending limb of the curve is recorded. As soon as the blood-wave has advanced from the root of the aorta to those portions of the large arteries that lie at the boundaries of the thoracic cavity, a much larger quantity of arterial blood begins to leave this cavity, because venous blood is at the same time being poured into it through the venæ cavæ.

This inspiratory movement would also be larger were it not for a slight diminution in the volume of the oral and nasal cavities, attended with an expiratory movement that takes place at the same time on account of the filling of its arteries— oral pulse, nasal pulse.

3. After the second sound of the heart (at 2), which at times causes a slight depression at the apex of the curve, the blood is dammed back in the thorax, in correspondence with the retrograde wave. As a result a second expiratory movement manifests itself in the descending portion of the curve.

4. The subsequent secondary wave-movement of the blood from the heart immediately again causes an inspiratory movement of gases, which produces the recoil elevation in the arteries of the body.

5. More blood now begins to flow into the thorax through the veins with slight fluctuations, and the next heart-beat takes place.

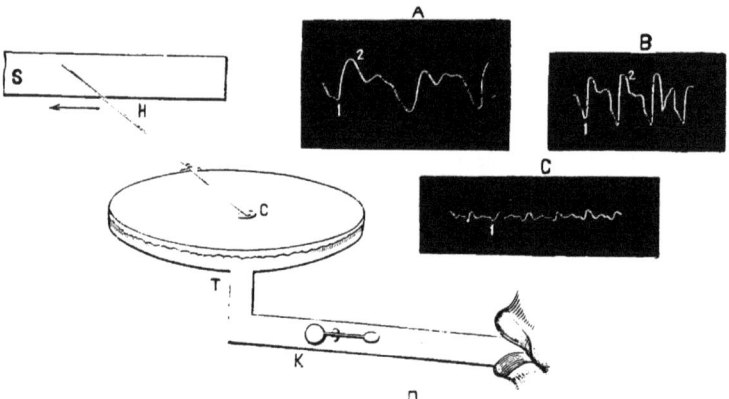

FIG. 35.—Landois' Cardiopneumograph, and Cardiopneumatic Curves Obtained with its Aid. A and B, from man; 1 and 2 correspond to the period of the first and second heart-sounds; C, curves from the dog; D, showing the instrument in use.

Pathological.—In the healthy human subject a crepitating sound is not rarely heard close to the heart, resulting from the movement of the air in the lungs, brought about by the movement of the heart. If there are near the heart abnormally narrow places in the bronchi, through which the respiratory gases are forced, so that they generate a sound or murmur, a fairly loud, sibilant or whistling murmur, known as the *pathological cardiopneumatic murmur*, is heard in rare cases. In the presence of cardiac lesions characterized by considerable fluctuations in the quantity of blood in the vessels of the lesser circulation, the cardiopneumatic movement must be quite marked, as, for example, in cases of insufficiency of the pulmonary and mitral valves.

INFLUENCE OF THE RESPIRATORY PRESSURE ON THE DILATATION AND CONTRACTION OF THE HEART.

The variations in pressure to which all the parts within the thorax are subjected by its inspiratory expansion and expiratory contraction exert a visible influence on the diastole and systole of the heart.

The conditions in various positions of the resting thorax with the glottis open will be considered first. The diastolic dilatation of the cavity of the heart is brought about by the *elastic traction of the lungs*, as well as by the inflow of venous blood and the elastic stretching of the relaxing muscular walls. This traction is greater in proportion as the lungs are more fully expanded (inspiration), and become less effective in proportion as the lungs have already been contracted (expiration). From this it follows:

1. That in the most extreme expiratory position of the thorax, with the greatest possible contraction of the pulmonary tissue, when, there-fore,what is left of the effective elastic traction of the lungs is exceedingly slight, but little blood enters the cavities of the heart; the heart during diastole is small and contains but little blood.   Accordingly, the systolic contractions will be small, that is, a small pulse results.

2. In the most extreme inspiratory position, when the elastic lungs are distended to their utmost, the force of the elastic traction of the lungs is, naturally, greatest, being in fact equivalent to 30 millimeters of mercury.   The effect of this traction may be great enough to counter-act the contractions of the thin-walled auricles and auricular appendages and prevent these structures from emptying their contents completely into the ventricles.   In cases of cardiac weakness it would even appear as if the ventricular activity were impaired by the strong elastic pulmo-nary traction, as the diminution in the strength of the heart-sounds that is sometimes observed attests.   The heart, therefore, is greatly distended in diastole and filled with blood; nevertheless the resulting pulse-waves may be small in consequence of the limitation of auricular activity. Thus, Donders often found the pulse smaller and slower.

3. When the thorax is in the position of moderate rest, a condition in which the elastic traction of the lungs is of moderate strength only, namely, 7.5 millimeters of mercury, the conditions for the action of the heart are most favorable.   On the one hand, diastolic distention of the cavities of the heart is adequate, and, on the other hand, their complete evacuation during systole is not impeded.

A much greater influence on the action of the heart is exerted by the increase or diminution in the intrathoracic pressure produced voluntarily by muscular action.

1. If the thorax is first brought into the position of deepest inspira-tion, then the glottis is closed, and now the space within the chest is greatly reduced with the aid of the expiratory muscles; the cavities of the heart may. be so greatly compressed as to cause momentary sus-pension of the movement of the blood within them.   In this position the elastic traction is greatly diminished, and in addition the pulmonary air, which is under high tension, exerts pressure on the heart and the intrathoracic vessels.   As no venous blood can enter the thoracic cavity from without, the visible veins become enlarged, the blood is driven more rapidly into the left heart, and the latter empties itself into the circulation as quickly as possible.   The lungs are, as a result, anemic and the cavities of the heart empty.   Therefore, there is plethora in the greater circulation, associated with anemia in the lesser and in the heart.   The heart-sounds cease, the pulse disappears.

2. If, conversely, the glottis is closed, while the thorax is in the position of most extreme expiration, and the thoracic cavity is now for-cibly dilated in inspiration, the heart is strongly dilated; for the cavities of the heart are distended not only by the elastic traction of the lungs, but also on account of the extreme rarefaction of the pulmonary air. The contents of the veins are poured copiously into the right heart, and in proportion as the right auricle and the ventricle are capable of

overcoming the outward traction, the blood-vessels of the lungs will be distended with blood. Much less blood will be driven out of the left heart, so that the pulse may even be temporarily arrested. The result is an overdistended, enlarged heart and the presence of an increased amount of blood in the lesser circulation, as compared with the greater.

As, when the breathing is normal, the tension of the pulmonary air is diminished during inspiration and increased during expiration, this normal alternation of pressure tends to assist the circulation: inspiration hastens the venous and lymphatic flow through the venæ cavæ (if the axillary or the jugular vein is opened during an operation, air may be sucked in and cause death) and thus favors complete diastole;

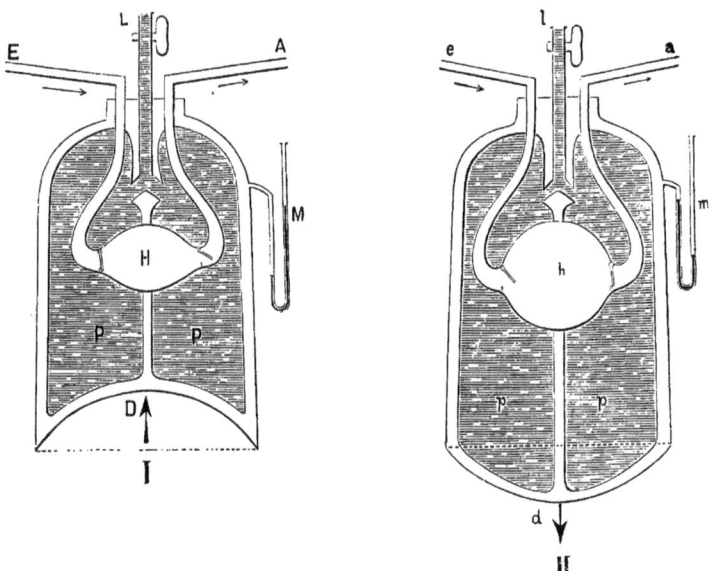

Fig. 36.—Apparatus for the Demonstration of the Influence of Respiratory Expansion (II) and Contraction (I) of the Thorax on the Heart and the Circulation.

expiration hastens the movement of blood into the arterial system and favors systolic emptying of the heart. At the same time the valvular arrangement of the heart secures a constant direction to the accelerated blood-current.

The elastic traction of the lungs also exerts a favorable influence on the lesser circulation, which is contained entirely within the thorax; for the blood within the pulmonary capillaries is under the same pressure as the pulmonary air, while that of the pulmonary veins is under lower pressure, as the elastic traction of the lungs by distending the left auricle necessarily hastens the flow of blood from the pulmonary veins into the left auricle. On the other hand, the elastic traction of the lungs is prevented from interfering to any marked degree with the action of the right ventricle and, therefore, with the movement of blood through the pulmonary artery, because of the sufficient resistance of the blood, right ventricle and the pulmonary artery against the elastic pulmonary traction.

The apparatus illustrated in Fig. 36 shows clearly the influence of inspiratory and expiratory movements on the expansion of the heart and on the current of blood in the large vascular channels leading to and from the heart. The large glass bottle represents the thorax, and its bottom has been replaced at D by an elastic rubber membrane, which represents the diaphragm. P P are the lungs; L the trachea, the entrance to which (glottis) may be closed by means of a stop-cock; H is the heart; E represents the course of the venæ cavæ; and A the aorta. When the tracheal stop-cock is closed and the expiratory position, as shown at I, is established by elevating the membrane D, with diminution in the size of the thoracic cavity, the air in P P is condensed, while at the same time the heart H is compressed; the venous valve closes, while the arterial valve is opened and the fluid is driven out through A. The manometer M, inserted into the flask, shows the increased intrathoracic pressure. Again, when the stop-cock l is closed (in II), and the membrane is strongly depressed, the lungs p p expand, and with them the heart h. The venous valve opens, while the arterial valve closes, and the venous blood enters the heart through e. Thus, inspiration always hastens the venous and inhibits the arterial flow, while expiration inhibits the venous and hastens the arterial flow. If the glottis (L and l) remains open, the air in P P and p p naturally is changed as the thorax passes from the inspiratory to the expiratory position (D and d). Accordingly, the effect on the heart (H and h) and on the blood-vessels is smaller, but even under such conditions it must persist in small measure.

THE MOVEMENT OF THE BLOOD IN THE CIRCULATION.

TORICELLI'S THEOREM ON THE VELOCITY OF ESCAPE OF FLUIDS.

According to Toricelli's law, the velocity (v) with which a fluid escapes, for example, through an opening in the floor of a hollow cylindrical vessel, is equal to the velocity that a freely falling body would attain in falling from the level of the fluid to the level of the opening (the height of the propelling force h).

Hence $v = \sqrt{2 g h}$; in which g = 9.8 meters.

The velocity of outflow increases, as has been shown experimentally, as the height of the propelling force (h) increases, and it preserves the ratio of 1, 2, 3 as the propelling force increases in the ratio of 1, 4, 9; that is, the velocity of outflow is proportionate to the square root of the height of the propelling force. It thus follows that the velocity of outflow depends solely on the distance between the level of the fluid and the opening, and not on the nature of the escaping fluid. Whenever a fluid is found escaping with a definite velocity, the force that causes the flow may be expressed by the height of a column of fluid (h) in a vessel the height of the propelling force.

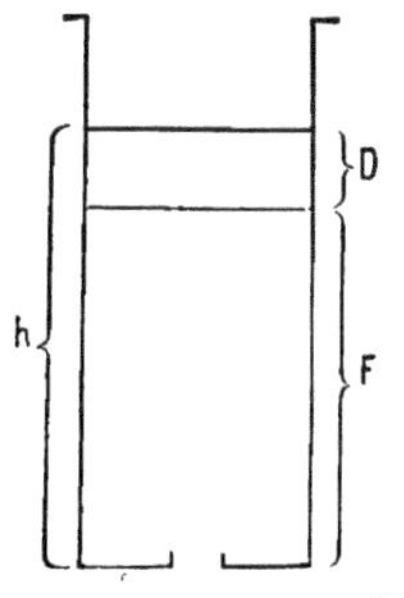

FIG. 37.—Pressure-vessel Filled with Water: h, height of the column of fluid; F, height of the velocity; D, height of the resistance.

Toricelli's law, however, is applicable only when all possible resistance that may be offered to the escape of the fluid is left out of account. As a matter of fact, certain resisting forces are present in any physical experiment of this kind. Hence, the force that is expressed by the height of the propelling force (h) not only causes the escape of the fluid, but also overcomes the sum of all the resistances. These two forces may be expressed by the heights of two columns of water superposed the one upon the other; namely, by the height of the velocity

F (which effects the velocity of escape) and the height of the resistance D (which overcomes any resistance that may be present): hence $h = F + D$.

PROPELLING FORCE, VELOCITY AND LATERAL PRESSURE.

If a fluid passes through a tube (which it completely fills), the first thing to determine is the propelling force h with which the current flows at different points in the tube. The degree of the propelling force depends on two factors:

1. The velocity of the current, v;
2. The pressure (resistance-height) to which the fluid is subjected at different points in the tube, D.

1. The velocity of the current v is determined: (a) from the lumen of the tube l, and (b) from the quantity of fluid q, that passes through the tube in a given unit of time. Then $v = q : l$. Both values, q as well as l, can be determined directly by measurement. The circumference of a circular tube, the diameter of which is d, is $3.14 \times d$. The cross-section (the lumen of the tube) is $l = \frac{3.14}{4} \times d^2$.

After the value of v has been determined in this way, the so-called velocity-height F (of hydraulic engineers) can be estimated from v; that is, the height from which a body would have to fall in a vacuum in order to acquire the velocity of v. This is $F = \frac{v^2}{4g}$ (in which g indicates the distance through which the body falls in one second, or 4.9 meters).

2. The pressure D (resistance-height) is measured directly at various points in the tube by inserting manometer-tubes (Fig. 38).

The propelling force at any selected point in the tube will thus be:

$$h = F + D$$
$$\text{or } h = \frac{v^2}{4g} + D$$

For experimental investigation the large cylindrical pressure-vessel (Fig. 38, A) may be used, within which by a suitable arrangement water can be maintained at a constant level h. The rigid tube a b, passing off from the bottom of the

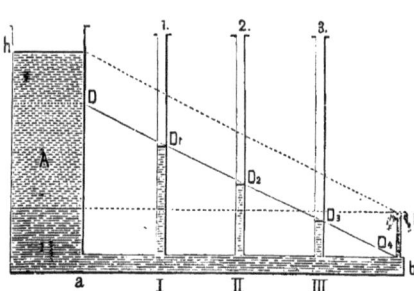

vessel, and of uniform size, is provided with a number of vertical tubes (1, 2, 3) constituting a piëzometer, for the measurement of the pressure; at the extremity b the tube is provided with an opening directed upward. From the latter the water, providing the level at h remains the same, will be thrown to a constant height, and this distance is equivalent to F, the velocity-height. As the pressure D_1, D_2, D_3 in the manometric tubes 1, 2, 3 can be read off directly, it follows that the propelling force of the water at the position of the tubes I, II, III is respectively $h = F + D_1$; $F + D_2$; $F + D_3$.

FIG. 38.—A Pressure-vessel, A, with Outflow Tube, a b, and Manometers, $D_1 D_2 D_3$, Inserted at Different Points.

At the extremity of the tube (at b) where $D_1 = 0$, $h = F + 0$, hence $h = F$. Within the pressure-vessel itself, it is the constant force h that influences the movement of the fluid.

It is, therefore, at once apparent that the propelling force of the water has become progressively smaller from the point where the fluid enters the tube from the pressure-vessel to the end of the tube b. The water in the pressure-vessel falling from h rises at b only to the height F. This diminution in the propelling force is due to the resistances encountered by the current in the tube, which neutralize a part of the kinetic energy (that is, convert it into heat). As, when the water has reached b, the motor power h in the vessel has been reduced to F, the difference having been neutralized by the resistances, the sum of these resistances must be $D = h - F$, from which it follows that $h = F + D$.

METHOD OF ESTIMATING THE RESISTANCES.

When a fluid passes through a tube of uniform caliber throughout its entire length, the propelling force h diminishes progressively in consequence of the resistances that operate uniformly at every point. The sum of all the resistances in the tube is, therefore, directly proportional to its length. In a tube of uniform caliber the fluid passes through each transverse section at a constant velocity; hence v (and, therefore, F) is the same at any point in the tube. The diminution that takes place in the propelling force h can, therefore, be due only to a diminution of the pressure D, as F remains the same everywhere (and h = F + D). The experiment with the pressure-vessel shows, in fact, that the pressure progressively diminishes toward the discharging extremity of the tube. In a tube of uniform width the pressure-height found to prevail in the manometer-tube is the expression of the sum of the resistances that must be overcome by the current in its course from the point examined to the free discharge-opening of the tube.

Forms of Resistance.—The resistances encountered by a stream of fluid reside first of all in the cohesion of the fluid-particles. The outermost parietal layer of the fluid, which is in contact with the tube, remains absolutely quiescent during the passage of the current. All the other layers of the fluid, which may be concerned as a series of concentric cylinders one within the other, move with a progressively increasing velocity from the periphery to the axis of the tube, while the axial thread itself finally represents the most rapidly moving portion of the fluid. In the displacement of these cylindrical layers of fluid at their surfaces of contact, the particles of fluid in juxtaposition must naturally be pulled apart and a portion of the active propelling force will be lost. The degree of resistance depends essentially on the degree of cohesion between the particles of fluid; the more intimate the cohesion between the fluid-particles, the greater will be the resistance; and conversely. It is thus evident that the resistances encountered by the viscous blood in its passage must be greater than those that would be encountered, for example, by water or ether. Four and one-half times as much pressure would be required to drive the same quantity of blood as of water through a tube.

Heat diminishes the cohesion of the particles and it is, therefore, a means for diminishing the resistance encountered by the current. It is also evident that the resistances are only the result of movement, as the forcible separation of the fluid-particles does not begin until the column is set in motion. It is, further, obvious that the greater the velocity of the current—the greater the number of fluid-particles that are torn apart in a unit of time—the greater will be the sum of the resistances. The parietal layer of fluid in contact with the surface of the tube remains, as has been said, in absolute quiescence; it follows, therefore, that the material composing the walls of the tube has no influence on the resistances.

INFLUENCE OF INEQUALITIES IN THE SIZE OF THE TUBE.

When the velocity of the current remains the same, the intensity of the resistances depends on the diameter of the tube; the smaller the diameter the greater the resistance, and the larger the diameter the less the resistance. The resistances, however, increase more rapidly in narrower tubes than the diameter of the tubes increases. This has been proved by experimental investigation.

In tubes that exhibit inequality in size in their course, the velocity of the current varies, being naturally slower in the wide portions and more rapid in the narrower portions. In general the velocity of the current in tubes of unequal caliber is inversely proportional to the transverse section of the different portions of the tube, that is, if the tubes are cylindrical inversely proportional to the square of the diameter of the circular transverse section.

While in tubes of uniform size the propelling force of the moving fluid diminishes uniformly section by section, the diminution is not uniform in tubes of unequal width; for since, as has just been shown, the resistance is greater in a narrow than in a wide tube, the diminution in the propelling force must naturally be greater in the narrow places than in the wide places. At the same time, it has been shown that the pressure in the wider places is greater than the sum of the resistances still to be overcome; while, on the other hand, at the narrower places it is smaller than the sum of these resistances.

Curvature and tortuosity of the vessels give rise to new resistances. In consequence of centrifugal force the fluid-particles cling more closely to the convex side of the arch and thus encounter a greater resistance to their progress than on the concave side.

When the tube divides into two or more branches, the propelling force is also diminished on account of the creation of additional resisting forces. When a current is divided into two smaller currents, some fluid-particles will be retarded, while others will be accelerated on account of the unequal velocity of the various layers of the fluid. Many particles that in the main current, as a part of the axial stream, had the greatest velocity will in the secondary currents when situated in the parietal layers move more slowly; while, conversely, many parietal layers in the main current become more centrally situated in the secondary current with increased velocity. As a result of the resistance thus produced a part of the propelling force is naturally lost. The separation of the fluid-particles as the current divides has a similar effect. If, on the other hand, two tubes join to form a single tube, additional resistance acting in a manner opposite to that described must lessen the propelling force. The sum total of the mean velocity in both branches of the current is independent of the angle formed at the point of division. The opening of a lateral branch that forms part of a tube accelerates the main current to the same degree, irrespective of the size of the angle formed by the lateral branch with the main tube.

MOVEMENT THROUGH CAPILLARY TUBES.

The movement of fluids through capillary tubes is, in accordance with the capillary attraction prevailing in capillary vessels, and in contravention of the laws that have just been developed, governed by certain rules, for the formulation of which credit is due Poiseuille. These rules are as follows:

1. The quantity of fluid that escapes from a capillary tube is proportional to the pressure.

2. The time necessary for the escape of a like quantity of fluid (the pressure, the diameter of the tube, and the temperature remaining the same) is proportional to the length of the tube.

3. The products of the outflow (all other conditions remaining the same) vary with the fourth power of the transverse diameter.

4. The velocity of the current is proportional to the pressure-height and to the square of the diameter, and inversely proportional to the length of the tube.

5. The resistances in the capillary tubes are proportional to the velocities of the current.

CONTINUOUS AND UNDULATORY MOVEMENT IN ELASTIC TUBES.

If an uninterrupted, uniform stream of fluid is permitted to flow through an elastic tube, the movement of this current is subject to the same laws that govern its passage through rigid tubes. If the propelling force increases or diminishes, the elastic tubes are either dilated or constricted, and their relation to the column of fluid is, therefore, simply like that of wider or narrower rigid tubes.

If, however, successive amounts of fluid are introduced at intervals into an elastic tube entirely filled with fluid, the initial portion of the tube will be suddenly distended in accordance with the amount of fluid introduced. The impact imparts to the fluid-particles an oscillatory movement, which rapidly communicates itself to all the fluid-particles from the beginning to the end of the tube; there results a positive wave, which rapidly propagates itself through the entire tube. If the elastic tube be closed at its peripheral extremity, the positive wave will rebound at the point of closure; it becomes a positive recurrent wave and it may even pass backward and forward repeatedly, becoming gradually smaller and smaller, until it finally subsides. Hence, in a closed tube of such character, the sudden periodic impulsion of a mass of fluid produces only a wave-like movement, that is, merely an oscillatory movement or the movement of a form.

3. If, however, additional amounts of fluid are at intervals pumped into the initial portion of an elastic tube entirely filled with fluid already in continuous movement, the continuous movement is combined with the undulatory movement. In such a case the continuous movement of the fluid, that is, the displacement or movement of the fluid in mass through the tube, must be rigidly distinguished from the undulatory or oscillating movement, the movement of the change in form of the column of fluid. The former is a translatory, the latter an oscillatory movement. The continuous movement is slower in elastic tubes, while the undulatory movement is more rapid.

The conditions in the arterial system are the same as those just described. The blood in the arteries is already engaged in continuous motion from the root of the aorta to the capillaries (continuous movement); and the injection at intervals of a mass of blood into the root of the aorta with each systole of the left ventricle produces a positive wave (pulse), which propagates itself with great rapidity to the end of the arterial system, while the constant movement progresses much more slowly.

It is of great importance to compare the movements of fluids in rigid tubes with the movements of fluids in elastic tubes.   When a certain quantity of fluid is forced into a rigid tube under a certain pressure, an equal quantity of fluid will at once escape from the end of the tube, unless such a result is prevented by the development of special resistances.   The conditions are, however, different in the case of an elastic tube.   Immediately after the injection of a definite quantity only a relatively small quantity of fluid escapes at first, the escape of the remainder taking place only after the injecting force has subsided.

If equal quantities of fluid are injected at intervals into a rigid tube, a corresponding amount escapes with each impulse and the discharge continues as long as the impulse, and the pause between each two periods of escape is always equal to the period between two impulses.   In the case of elastic tubes the conditions are different.   As the escape of the fluid continues for some time after the cessation of the impulse, it will always be possible to establish a continuous outflow through elastic tubes by making the interval between two injections shorter than the duration of the outflow that takes place after the impulse has been completed. Thus, the periodic injection of fluid into a rigid tube produces an isochronous, sharply limited outflow of fluid, which can become permanent only when fluid enters the tube in a continuous stream.   In the case of elastic tubes, on the other hand, intermittent introduction of fluid produces under the same conditions a continuous outflow with systolic reinforcement.

Hamel's investigations have shown that elastic tubes permit the passage of more fluid when they are supplied in a rhythmical pulsatory manner than when the fluid enters in an uninterrupted stream under constant pressure.   The advantage of the rhythmical impulse for the propulsion of the circulating fluid, as compared with a uniform pressure, appears to reside in the fact that the alternating movement preserves the elasticity of the arterial walls.

STRUCTURE AND PROPERTIES OF THE BLOOD-VESSELS.

The large blood-vessels in the body are designed solely for the purpose of acting as conducting canals for the mass of blood, while the thin-walled capillary vessels effect the interchange of substances between the blood and the tissues and in the opposite direction.

The Arteries differ from the veins in the possession of thicker walls in consequence of the considerable development of muscular and elastic elements, as well of a greatly developed middle tunic, with a relatively thin adventitial coat.   The walls of the arteries consist of three coats (Fig. 39):

The *intima* is lined on its inner surface by a nucleated endothelium (a) consisting of flat, irregular, oblong cells.   External to the endothelium is a thin, finely granular layer containing more or less distinct fibers and numerous spindle-shaped or stellate protoplasmic cells embedded in a corresponding system of plasma-canals.   To the outer side of this is the inner elastic layer (b), which in the smallest arteries is represented by a structureless or fibrous, elastic membrane and in the medium-sized arteries by a fenestrated membrane; while in the largest it assumes the appearance of a stratified, fibrous or fenestrated, elastic membrane consisting of two or three layers and united by connective tissue.   All of the larger and medium-sized arteries contain longitudinal fibers situated between two elastic plates.   Acting together with the circular fibers they are capable of narrowing the caliber of the vessel; but they possess also the faculty of widening the lumen and maintaining it at a uniform width.   On the other hand, it is improbable that they are capable of independent action or that such independent action is capable of dilating the vessel.

The *middle coat* has for its most characteristic constituent unstriated muscle-fibers (c).   In the smallest arteries this appears to be composed of scattered, transverse, smooth muscle-fibers occupying an intermediate position between the intima and the adventitia.   The connecting material consists of a finely granular tissue traversed by a few delicate elastic fibers.   Passing from the smallest to the

9

smaller arteries, the number of unstriated muscle-fibers increases progressively until they form a strong layer of circular muscle-fibers with almost complete disappearance of the connecting substance. The outer elastic layer forms the boundary between the media and the adventitia. In the large arteries the connecting substance greatly predominates over all other tissues: Separated by layers of delicate fibrous tissue there are numerous (as many as 50) thick, elastic, fibrillated or fenestrated membranes arranged in concentric layers and chiefly in the transverse direction. Scattered here and there between these membranes are occasional smooth muscle-cells arranged transversely, less commonly obliquely, or longitudinally.

The initial portions of the aorta and pulmonary artery, the arteries in bones and the retinal arteries are devoid of muscle-tissue. The descending aorta and the common iliac and popliteal arteries possess oblique and longitudinal muscle-fibers lying among the transverse fibers. The renal, splenic and internal spermatic arteries contain longitudinal bundles at the inner surface of the media; the umbilical arteries, which are exceedingly rich in muscle-tissue, contain longitudinal bundles both on the inner and on the outer surface.

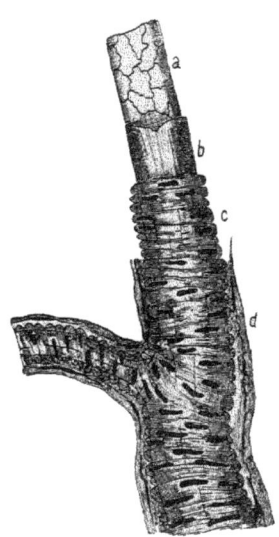

The *external or adventitious coat* in the smaller arteries is a delicate, structureless membrane containing a few protoplasmic cells. In somewhat larger vessels there is an additional layer of elastic tissue of delicate fibers containing strands of fibrillated connective tissue (d). In the medium-sized and largest arteries the greater part of the adventitia consists of bundles of fibrillated connective tissue containing connective-tissue cells, and not infrequently an admixture of fat-cells, running obliquely and crossing each other at numerous points. Among them and chiefly toward the media are found fibrous or fenestrated elastic layers. At the boundary between the adventitia and the media the elastic elements in the smaller and medium-sized arteries fuse to form a more independent elastic membrane (Henle's outer elastic membrane). Longitudinal unstriated muscle-fibers in scattered bundles are found in the adventitia of the arteries of the penis, of the descending aorta, the renal, splenic, internal spermatic, iliac, hypogastric, and superior mesenteric arteries.

Bonnett suggests the following natural division of the layers of the arterial wall: 1. The intima embraces the endothelial tube and the tissues as far as the inner elastic layer. 2. The media contains all those parts that are situated between the inner and the

Fig. 50.—Small Arterial Twig Showing the Individual Layers of the Arterial Wall: a, endothelium; b, elastic inner coat; c, layer of circular muscle-fibers; d, connective-tissue adventitia.

outer elastic layer. 3. The adventitia includes the layers found to the outer side of the elastic membrane.

The Capillaries, which undergo frequent division without suffering diminution in caliber, and in their subsequent course unite again, have diameters varying from 5 to 6 μ (retina, muscles) to from 10 to 20 u (bone-marrow, liver, choroid). The tubes are formed of a single layer of nucleated endothelial cells, with protoplasmic cell-bodies, which in the smaller tubes are spindle-shaped and in the larger vessels are more polygonal (as is the case with the cells of serous cavities); they are connected by numerous intercellular bridges in the depths of the cell-substance (like epithelial cells). The boundaries of the cells are demonstrable as black lines by injection of a solution of silver nitrate. The stained cement-substance exhibits in some places intercalated areas of larger size. Whether these are to be regarded as true openings or stomata, through which it is possible for red and white cells to escape, or merely as denser aggregations of the stained cement-substance is still an undecided question. Delicate anastomosing fibrils derived from non-medullated nerves terminate by small end-plates in the capillary walls. Ganglia in communication with the nerves of capillary vessels are found only in the distribution of the sympathetic nerves. The minute blood-vessels that communicate directly with the capillaries possess, in addition to endothelium, an entirely structureless investing membrane.

The Veins differ from the arteries in the main in the fact that they have a larger caliber than the corresponding arteries and thinner walls on account of the much feebler development of the elastic and muscular elements. Among the latter longitudinal fibers are much more commonly found than transverse. Veins are also distinctly more distensible with the same degree of traction. The adventitia is as a rule relatively the thickest coat. The presence of valves is limited to certain areas of the body.

The *intima* or *internal coat* is provided with short endothelial cells, beneath which, in the smallest veins, is a structureless layer, which in the somewhat larger vessels is composed principally of longitudinal elastic fibers (always thinner than in the arteries). In the large veins this layer may assume the character of a fenestrated membrane, which here and there in the femoral and iliac veins is even duplicated. It is held together by a delicate connective tissue containing spindle-cells. The intima in the femoral and popliteal veins contains a few scattered muscle-fibers.

The *media* or *middle coat* in the larger veins is constituted of alternate layers of elastic and muscular elements, with a fairly abundant fibrillar connective tissue. The media is always thinner, however, than in the corresponding arteries. The number of these alternating layers becomes progressively smaller in the following veins, in the order of their enumeration: popliteal vein, veins of the lower extremity, veins of the upper extremity, superior mesenteric, the remaining veins of the abdominal cavity, the hepatic, pulmonary, and coronary veins. The following veins are altogether devoid of muscle-tissue: the veins of bones, muscles, the central nervous system and its membranes, the retinal veins, the superior cava

FIG. 40.—Capillary Vessels,—the Boundaries of the Cells (Cement-substance between the Endothelial Cells) have been Stained Black with Silver Nitrate and the Nuclei of the Endothelial Cells Made Prominent by Staining.

with the large trunks that empty into it, and the upper portion of the inferior cava. In these vessels the media is much more feebly developed. In the smallest veins the media consists merely of a delicate fibrillar connective tissue in which a few scattered longitudinal and transverse unstriated muscle-cells make their appearance as the center of the circulation is approached.

The *adventitia* or *external coat* of the veins is, generally speaking, thicker than that of the corresponding arteries. It always contains more abundant connective tissue, usually consisting of longitudinal fibers, and on the other hand fewer large-meshed networks of elastic elements. Some veins, however, contain also longitudinal muscle fibers: the renal vein, the portal vein, the inferior cava in the hepatic region, the veins of the lower extremity. The valves consist of finely fibrillated connective tissue in which stellate cells are embedded; the convex surface of the valves is covered with a network of elastic fibers, and both surfaces are invested with endothelium. The valves contain many muscle-fibers.

The *sinuses of the dura mater* are spaces lined with endothelium between duplicatures, or cleft-like invaginations of this membrane.

Cavernous spaces may be regarded as having been produced by numerous divisions and anastomoses of fairly large veins of unequal size, closely following one another. The vessel-wall frequently appears cribriform or like a sponge—the interior traversed by trabeculæ or threads. The surface directed toward the blood is covered with endothelium. The investing wall consists of connective tissue, which is often quite firm and tendinous, as in erectile tissue. It not infrequently contains unstriated muscle-fibers.

An example of an analogous cavernous formation in arteries is found in the coccygeal gland of man. This mysterious structure, which is richly supplied with sympathetic nerve-fibers, consists of nucleated connective tissue and represents a convolution of ampulliform or spindle-shaped dilatations of the median sacral artery, traversed and surrounded by unstriated muscle-fibers.

The *vasa vasorum* do not differ in structure from other vessels of similar caliber.

Intercellular blood-channels devoid of walls are present in the granulation-tissue of wounds. At first nothing but blood-plasma is found between the constituent cells, and it is not until later that blood-cells are driven through the channels by the blood-current. In the incubated egg the primary basis of the blood-vessels is formed in a manner similar to that of the formative cells of the germinal layer. The blood-vessels without walls in the bone-marrow and in the spleen are considered on p. 43.

Among the properties of blood-vessels their contractility should be mentioned first, that is, the ability to contract by virtue of the unstriated muscle-fibers contained in their walls. The intensity and force with which this contraction takes place are proportional to the degree of development of the muscle-tissue.

Heat causes contraction of the blood-vessels (in the mesentery of the frog). Excised arteries contract when filled with dilute alkaline solutions, digitalin, atropin, and antiarin. The isolated apex of the heart also beats more freely in alkaline solutions. When the vessels are filled with a dilute solution of lactic acid they dilate, and the apex of the heart when immersed in such a solution also beats more rapidly. According to Roy, blood-vessels undergo shortening under the influence of heat, if precautions are taken to prevent evaporation and the load remains the same.

If blood containing an admixture of certain substances—such as amyl nitrite, chloral hydrate, morphin, quinin, and atropin—is allowed to flow through the vessels of a recently excised, living organ, dilatation takes place; urea and sodium chlorid have the same effect on the renal vessels; while digitalin and veratrin cause contraction.

The capillaries also possess the power of dilating and contracting, derived from the protoplasmic granules of the cells of which they are composed.

The capillaries have been designated "protoplasm in tubular form," and motor phenomena have been observed in them, especially after irritation in the living animal. Stricker observed this chiefly in the capillaries of young frog-spawn. At a later period of the animal's life the reaction of the capillaries to stimuli is much less distinct. Rouget observed the same phenomena also in new-born mammals. Similar observations have been made by Golubew and Tarchanoff. Accordingly, the shape of individual cells varies with the quantity of blood contained in the vessels. In greatly distended vessels the cells are flat; but when the vessel is collapsed, the cells are more cylindrical and project into the lumen.

Among the physical properties of blood-vessels their elasticity should next be mentioned. The elasticity is slight, that is, the vessels offer little resistance to the distending forces, such as pressure or traction; but it is, at the same time, complete, that is, after the distending force has ceased to act, the vessels regain their previous form.

According to Ed. Weber, Wertheim and A. W. Volkmann, the length of blood-vessels (like that of moist portions of the animal body generally) does not increase in proportion to the weight employed to extend it, but the elongation is considerably less with progressive increase in the weight. Hence the extensibility of the dead artery is greatest when it has been slightly distended by intravascular pressure. After repeated experiments, however, Wundt was led, as a result of experimental observations, to the conclusion that blood-vessels also are subject to the general law of elasticity mentioned. He maintains that it is necessary to take into consideration not only the first distention that occurs after the application of the load, but also the "elastic after-effect" that follows gradually.

This terminal distention, which often proceeds slowly, is so gradual during the last moments that observation with a magnifying lens is necessary to determine when the condition of definitive distention is completed. Deviations from the general law occur; for when a certain load is exceeded, lesser degrees of distention and at the same time permanent changes not infrequently result. A normal vein may be stretched at least 50 per cent. without exceeding the limit of elasticity.

Pathological.—Nutritive disturbances modify the elasticity of the arteries. When death has been preceded by marasmus, the arteries are found relatively more dilated than under normal conditions. Beginning connective-tissue formation in the intima, combined with fatty degeneration, at first increases the distensibility and diminishes the strength of the wall. As the development of the connective tissue progresses in cases of arteriosclerosis, the elasticity and firmness of the arteries are again augmented. Diminished distensibility is found also in connection with atheroma, in cases of nephritis and in the arteries of drunkards.

A property peculiar to the walls of the blood-vessels is their power of cohesion, which enables them to resist rupture, even when the internal tension is considerable. It has been found that the carotid artery does not rupture until the internal pressure has been raised artificially to fourteen times the normal. The resistance of veins to rupture is relatively greater than that of arteries with the same thickness of wall. According to Gréhant and Quinquaud the carotid and iliac arteries in man resist a pressure up to eight atmospheres and the veins more than half of this amount.

Pathological.—Diminished power of cohesion of the blood-vessels, especially the arteries, is not uncommon in old age.

PULSE-MOVEMENT.—TECHNIC OF PULSE-EXAMINATION.

The physicians of antiquity devoted more attention to abnormal excitation of the pulse than to the normal pulse. Thus, Hippocrates (460–377 B. C.) speaks only of the former condition and applies to it the term σφυγμός. Later, Herophilus (300 B. C.) in particular compared the normal pulse (παλμός) with the abnormally excited pulse. He laid especial stress on the time-relations existing between dilatation and contraction of the arterial tube and defined more accurately the properties, volume, fulness (σφυγμός ταχύς) and frequency (σφυγμός πυκνός). His Alexandrian colleague Erasistratus (who died 280 B. C.) was the first to make correct statements in regard to the propagation of the pulse-waves; for he stated distinctly that the pulse appears earlier in the arteries nearer the heart than in the more distant vessels. Erasistratus also felt the pulse below a cannula introduced in the continuity of an artery. Archigenes claims especial interest, particularly with respect to the pathology of the pulse, because he was the first to designate the dicrotic pulse, which he had the opportunity of observing in febrile diseases. Galen (131–202 A. D.) determined more accurately than his predecessors the principles governing expansion and contraction of the arteries during the movement of the pulse. His explanation of the slow pulse was that the time of expansion was prolonged. Galen made also noteworthy observations with regard to the rhythm of the pulse and the effect of temperament, sex, age, season of the year, climate, sleep and waking, emotional influences, and cold and warm baths. Cusanus (1565) was the first to count the pulse-beats with a time-piece.

INSTRUMENTS EMPLOYED IN THE EXAMINATION OF THE PULSE.

It is possible by means of instrumental examination to obtain trustworthy information with regard to the nature of the movement of the pulse. Apart from those instruments by means of which the undulatory movement in the arterial tube can be demonstrated only after this has been opened, the following are worthy of mention:

Poiseuille's Box-sphygmometer.—The exposed artery (Fig. 41, a a) is enclosed for a distance in its continuity in an oblong box (K K), filled with some indifferent fluid. There communicates with the interior of the box a graduated vertical tube (b), filled to a certain point, in which the fluid rises and falls, in accordance with the quantity of blood contained in the artery. The box is constructed like an ordinary box, one half representing the body and the other half the lid. A circular opening is made in each end of the box, one half being contributed by the body and the other half by the lid, in which the artery is hermetically sealed by means of soft fat. Poiseuille found the distention of the carotid during diastole in the horse to be equal to $\frac{1}{13}$, and in the dog to $\frac{1}{12}$ of the entire volume of the arterial segment. The instrument does not record any more minute details in regard to the movement of the artery during the phases of the pulse.

Hérisson's Tubular Sphygmometer (Fig. 42) consists of a glass tube closed at its lower extremity by an elastic membrane and filled to a certain level with mercury. The apparatus is placed vertically on the skin over a pulsating artery, the beats of which set the column of mercury in motion. A similar instrument

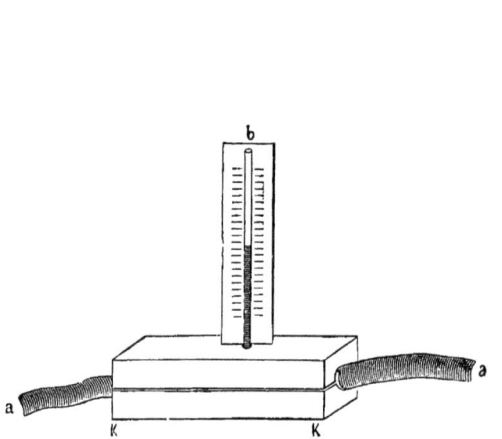

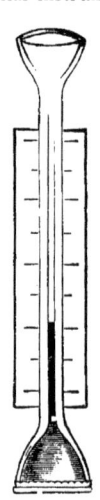

FIG. 41.—Poiseuille's Box-cabinet Sphygmometer: a a, the exposed artery; K K, the surrounding box with the vertical tube and scale b.

FIG. 42.—The Tubular Sphygmometer of Hérisson and Chelius.

was used in 1850 by Chelius, who succeeded with its aid in discovering the double beat of the normal pulse. "After it (the mercury) has been raised by the impact of the blood-wave, it falls again as suddenly to its lowest level, after first making another short pause at some intermediate point."

Marey's Sphygmograph is based on a combination of the lever (which was first employed by Vierordt in 1855 in the construction of his "sphygmograph") with an elastic spring (Fig. 43, A). The latter, which is screwed fast at one extremity (z), while the other extremity is free and provided with a round pad (y), presses against the radial artery with a force equal to that of the spring. To the upper part of the pad is fixed a short vertical ratchet (k), which, when acted upon by a weak spring (e), turns a small cogwheel (t), from the axis of which a light wooden lever (v) extends almost horizontally. This writing lever is provided at its outer extremity with a delicate point (s), which records the movements of the pulse on the smoked surface of a plate (P) made by clockwork (u) to pass in front of the point of the writing lever at a uniform rate. Marey's instrument is trustworthy and is quite extensively used.

Marey's sphygmograph is adapted solely for the radial pulse. It is placed lengthwise on the forearm, where it is steadied by means of two short metallic supports (S) and fastened with a tape, which must not be drawn too tight. The apparatus is also provided with a secondary screw (H), which can be made to act on the spring (A). If the screw is tightened the spring is compressed and rendered shorter, less yielding and movable with greater difficulty; when the pressure is entirely released, the spring (A) has free play, is more yielding and the position of the pad (y) is higher.

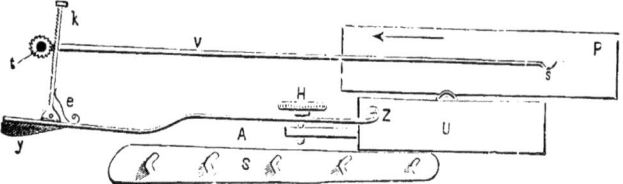

FIG. 43.—Marey's Sphygmograph (Diagrammatic).

Marey's Sphygmograph with Transmission of Air—of which many modifications have been made, for example by Knoll; Fig. 44 illustrates the modification designed by Brondgeest and designated "pansphygmograph"—is constructed on the principle of the pneumatic telegraph. Two pairs of shallow metallic cups—(S S and S′ S′) so-called Upham's capsules—are each pierced from below at their center by a small tube. The ends of these tubes are connected with rubber tubes (K and K′). Over the mouth of each of the four cups a delicate rubber membrane is

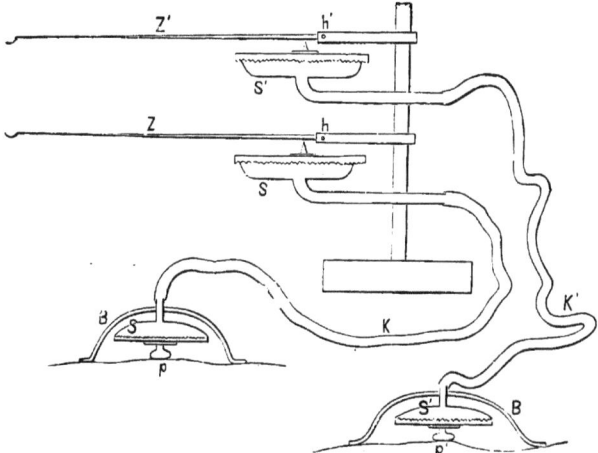

FIG. 44.—Brondgeest's Pansphygmograph Constructed on Upham's and Marey's Principle of the Propagation of Movement through Air-containing Drums Covered with Elastic Membranes. The figure represents also diagrammatically Marey's cardiograph.

stretched and from the middle of each of the two rubber membranes S and S′ there projects a button-shaped process (p and p′), which is applied to the pulsating artery and held in place by metallic arches B B′, the extremities of which rest on the surrounding skin. From the center of each of the other two rubber membranes, which are directed horizontally upward, there projects a knife-edge, which is applied close to the balancing center (h and h′) of the delicate writing levers Z and Z′. It is evident that any pressure applied to the buttons will cause a bulging upward of the membrane of each of the upper cups, the movements of which are propagated to the writing levers.

The instrument sketched in Fig. 44 shows the entire registering apparatus in duplicate. An instrument of this kind may, therefore, be placed with the two pads on two different arteries; for example, when it is desired to demonstrate that the pulse occurs earlier in the arteries near the heart than in more distant vessels.

Although the instruments described are convenient to handle, it has been found by experience that sudden variations in pressure are not accurately recorded in consequence of vibration of the instrument itself; while when the variations in pressure are less sudden, the records may under certain circumstances be fairly accurate. Another disadvantage is that the movement of the writing lever Z is not entirely synchronous with that of the button p. For this reason instruments constructed on this principle are not well adapted for accurate time work. The entire apparatus may also be filled with water, in which event leaden pipes are used instead of the connecting rubber tubes. Thus adapted, the apparatus is more accurate for slower movements, while a pneumatic instrument is better adapted for rapidly varying phases, such as are presented by the movements of the pulse.

Landois' Angiograph.—From one extremity of a plate (Fig. 45, G G) serving as a base, arises a pair of arms, between the upper parts of which the lever (d r) moves freely between two points. The long arm of this lever is provided with a pad (e), directed downward, which is to be applied to the pulse. The short arm of the lever on the other side carries a counter-weight (d), heavy enough to maintain

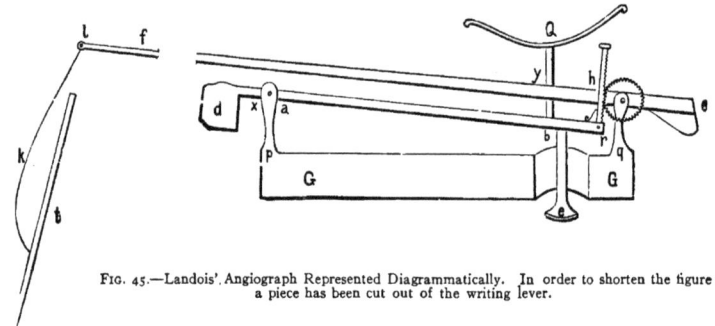

FIG. 45.—Landois'. Angiograph Represented Diagrammatically. In order to shorten the figure a piece has been cut out of the writing lever.

the entire lever in equilibrium. The extremity (r) carries a spring-ratchet, which presses against a cogwheel. The latter is immovably fixed to the axis of the light writing lever c f, which is also suspended between points and is supported by the two uprights q and attached to the opposite end of the base G G. The writing lever also is maintained in perfect equilibrium by means of a small counter-weight. The needle k is suspended from the extremity of the writing lever l, where it is secured by a hinge and is readily movable; it is carried by its own weight toward the tablet (shown in the figure in profile), and as it moves up and down it records the curve with a slight scratching movement on the delicately smoked surface of the tablet.

The lever d r at a point approximately opposite the juncture with the pad e supports on the end of a vertical rod the flat plate q for the reception of weights to increase the load on the pulse. The advantages of the instrument are: (1) The load can be varied at will and can be accurately determined (while in Marey's sphygmograph the pressure of the spring increases as the lever is raised); (2) although the needle is constantly in contact with the smoked surface, it nevertheless records with a minimum degree of friction; (3) the movement of the writing lever is a vertical up-and-down movement and not a curved movement as in Marey's apparatus, thus considerably facilitating an accurate study and measurement of the curves. In the construction of his sphygmograph Sommerbrodt adopted the improvements embodied in Landois' angiograph.

In the choice of a sphygmograph the guiding principle should be that the most complete instrument and the one whose curves most closely correspond with the pressure-variations actually taking place in the artery is that in which the resistance within the apparatus itself is reduced to a minimum, in which those parts that execute the largest movements are as light as possible, but in which the bulk of that portion of the instrument is directly set in motion by the movement of the blood in the artery, is strong enough and heavy enough for its equilibrium to be but slightly disturbed by even considerable force.

Useful sphygmographs have been described by other investigators, as Naumann, Frey, and others. For practical purposes Dudgeon's instrument, which is easily manipulated, may be recommended; the load is applied by the pressure of a spring, or, better, by a weight and beam, and the tablet moves horizontally. A system of lines is recorded together with the curve, making it possible to determine by measurement the size and chronological development of the pulse-beats.

Nomenclature of Pulse-tracings.—In every pulse-tracing (sphygmogram or arteriogram) there are distinguishable the ascending limb, the apex, and the descending limb. Irregular elevations in the course of the descending limb are called catacrotic elevations, while those in the ascending limb are known as anacrotic elevations. The descending limb almost always contains secondary elevations, while the ascending limb almost always appears as a simple rising line. When a recoil-elevation, which will be described more fully later on, occurs once or

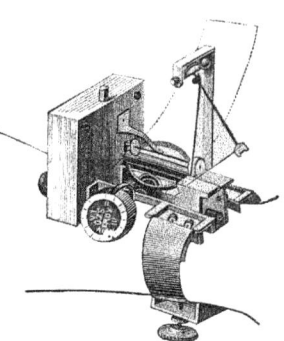

FIG. 46.—Dudgeon's Sphygmograph.

twice in the descending limb, the sphygmographic curve is called dicrotic or tricrotic. When, as happens if the pulse-beats follow one another in rapid succession, the succeeding beat cuts off the recoil-elevation of the preceding curve, the curve is called monocrotic.

Method of Making Sphygmographic Tracings.—The tracings are recorded on smooth glazed paper like that used for visiting cards, which has been covered with a delicate translucent layer of soot by exposure over burning camphor or a smoking lamp. The tracing is fixed by immersing the paper in a solution of shellac and alcohol, after which it is allowed to dry.

Mensuration of Sphygmographic Tracings.—When a tablet is made to move at a uniform rate by means of clockwork, the vertical height and horizontal length of individual portions of the tracing can be measured with a fine rule. The distance traversed by the tablet in a second being known, it is possible by actual measurement to compute the duration of the individual portions of the pulse-movement. Accurate measurements of this kind must be made under the microscope with the aid of an ocular micrometer, a low magnification and direct illumination being employed. The sections to be measured are placed between two lines that, in the case of sphygmographs like Marey's, which make a curved tracing, must be arcs of a circle (of which the writing lever is the radius), and in the case of the angiograph must be vertical.

An especially convenient method consists in recording the curve on a tablet attached to one end of a vibrating tuning-fork (Fig. 60). Another less accurate method consists in recording the vibrations of a tuning-fork on the tablet of the sphygmograph at the same time that a sphygmographic tracing is being recorded, the latter being above the tuning-fork record.

The Gas-sphygmoscope.—To meet the objection that has frequently been urged against instruments for registering the pulse, namely that the secondary elevations observed in the sphygmogram are due to the after-vibrations of the apparatus from inertia, Landois constructed a gas-sphygmoscope, in which the movement of solid bodies is excluded and any after-vibration of inert masses that have been set in motion is, therefore, impossible.

The superficial arteries, whose movement is communicated to the overlying skin, will, naturally, through the movement imparted to this layer of the skin, cause also a movement in the contiguous layers of air. The thin layer of air above the pulsating cutaneous area (Fig. 48) a is excluded by means of a shallow metallic gutter b, which is placed on the skin so that its concavity covers the artery like a small tunnel. The narrow space between the wall of the tunnel and the skin is filled with illuminating gas. To this end one extremity of the metallic tunnel is connected with the gas-tube g, while the other extremity communicates by means of a short rubber connecting piece x q with a small tube t, bent upward at an angle and the point of which is drawn out to a minute opening for the escape of the gas. The gas is allowed to pass through the metallic tunnel. under low pressure, the inflow being regulated so that the flame v is not more than a few millimeters long. It is readily seen that the flame increases in height synchronously with each pulse-beat and that the descent is interrupted by a distinct after-beat. von Kries photographed the image of the flame.

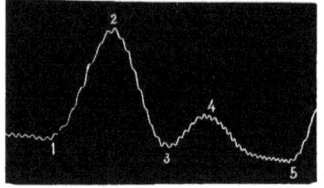

The measurements of the accompanying curve are as follows:

$1-2 = 7.5 = 0.121$ sec.
$1-3 = 16 = 0.258$ "
$1-4 = 22.5 = 0.363$ "
$1-5 = 39.5 = 0.638$ "

FIG. 47.—Sphygmographic Tracing from the Radial Artery Made with Landois' Angiograph Attached to a Vibrating Tuning-fork. Each indentation corresponds to 0.01613 sec.

Hemautography.—If a freely exposed artery be divided in an animal so that the blood-stream spurts forth and is allowed to impinge on a glass plate or a sheet of paper moved vertically at some distance, the resulting tracing will coincide almost perfectly with the normal curve of the artery as recorded by the sphygmograph. In addition to the primary elevation (Fig. 49, P), the recoil-elevation (R) and the elasticity-elevations (e e) are appreciable. This self-registration of the

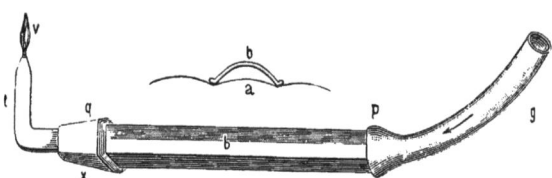

FIG. 48.—Landois' Gas-sphygmoscope.

blood-wave furnishes a convincing proof that the movement is produced in the blood itself and is communicated as an undulatory movement to the arterial wall. By determining the quantity of blood contained in the several portions of the hemautographic tracing it is found that the quantity of blood that escapes from the divided artery during systole is to the quantity that escapes during diastole (that is during contraction and dilatation of the vessel) approximately as 7 : 10. The quantity of blood that escapes during a unit of time while the artery is dilating is equal to a little more than twice the quantity that escapes during a unit of time while the vessel is contracting.

THE PULSE-TRACING, THE RECOIL-ELEVATION AND THE ELASTICITY-ELEVATIONS.

The sphygmogram presents an *ascending limb,* recorded during the distention (diastole) of the artery; the *apex* (Fig. 50, **P**); and the *descending limb,* which corresponds to the contraction (systole) of the artery. The most conspicuous features of the sphygmographic tracing are the two entirely distinct elevations in the descending limb of the curve. The more prominent of the two occupies approximately the center of the descending limb, where it appears as a distinct elevation (R); it is known as the *dicrotic after-beat* or, with reference to its origin, as the *recoil-elevation*.

The sphygmographic tracing reproduces the chronological course of the pressure exerted by the undulatory movement of the blood on the arterial wall, the pad of the sphygmograph, which is supported on a spring, rising and falling with the variations in pressure; the instrument therefore records "pressure-pulse."

ORIGIN AND PROPERTIES OF THE DICROTIC ELEVATION.

The *recoil-elevation* (also designated secondary or dicrotic) is produced in the following manner: After the column of blood propelled into the arterial system by the ventricular systole has generated a positive wave, which, beginning at the aorta, extends rapidly to all of the arteries, even to the minutest arterial branches, in which it disappears, the arteries contract as soon as closure of the semilunar valves prevents the further entrance of blood. The elasticity and the active contraction of the blood-vessels thus exerts a counter pressure on the blood-column. The blood is forced to seek an outlet. In its progress toward the periphery it finds no obstacle in its path, but the portion that escapes toward the center of the circulation recoils from the already closed semilunar valves. The impact of the blood sets up another positive wave, which is again propagated into the arteries and disappears as before in the remotest minute branches. If, however, there is sufficient time for the complete development of the sphygmographic tracing, a second reflected wave is produced in the proximal arteries (especially in the short course of the carotids, but also in the arteries of the upper ex-

FIG. 49.—Hemautographic Tracing from the Posterior Tibial Artery of a Large Dog: P, primary pulse-wave; R, recoil-elevation; e e, elasticity-elevations.

tremities, but not in those of the lower extremities because of their great length) in the same way as the first. Just as the pulse appears somewhat later in the more peripheral arteries than in those nearer the heart, so the secondary wave, produced by the recoil of the blood from the aortic valves, also appears later in the more distant arteries. Both kinds of waves, the primary and the secondary pulse-wave, and possibly also the tertiary recoil-wave, originate at the same point and are propagated in the same way. The longer the distance to be traveled before they reach a given point in the artery, the later will be their arrival at that point.

The following laws with regard to the recoil-elevation have been determined experimentally:

1. The dicrotic elevation appears later in the descending limb of the curve the longer the artery, measured from the heart to the peripheral termination of the artery. (The curves in Figs. 47, 53 and 57 may be measured to confirm this point.)

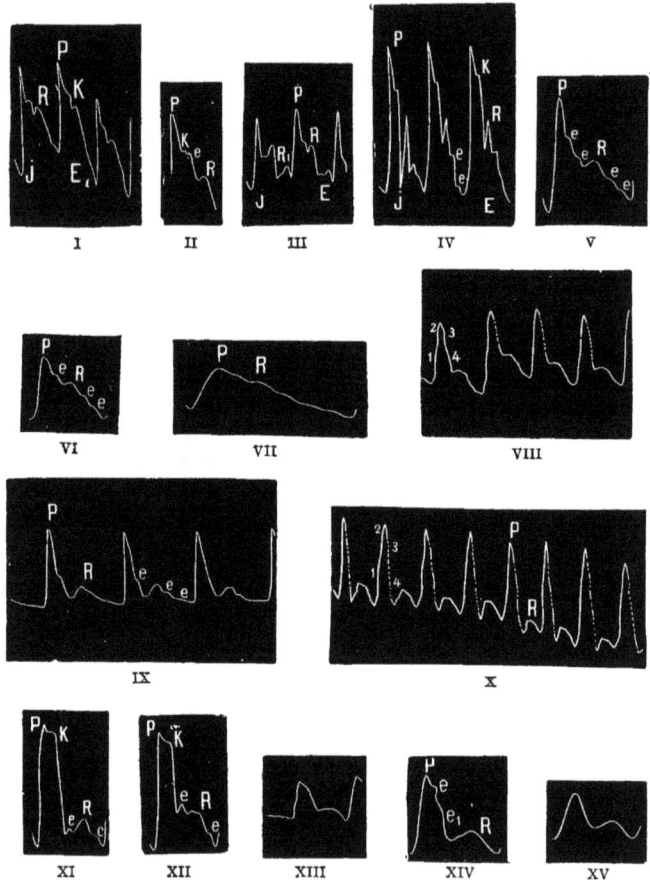

FIG. 50.—I, II, III, Sphygmographic tracings from the carotid artery; IV, from the axillary; V, IX, from the radial; X, bigeminate pulse from the radial; XI, XII, sphygmographic tracings from the femoral; XIII, from the posterior tibial; XIV, XV, from the dorsalis pedis. In all of the tracings P indicates the apex of the curve; R, the dicrotic elevation; e e, the elasticity-elevations; k, the elevation caused by the closure of the aortic semilunar valves.

The shortest accessible arterial course is that of the carotids, where the dicrotic elevation attains its greatest height about 0.35 or 0.37 second after the beginning of the pulse. The next shortest accessible arterial course is that of the upper extremity, where the apex of the dicrotic elevation is traced about 0.36 or [0.38 or 0.40 second after the beginning of the pulse. The longest course is that of the arteries of the lower extremity, in which the apex of the recoil-elevation is formed about 0.45 or 0.52 or 0.59 second after the beginning of the curve, in accordance with the size of the individual.   In children and in small individuals the recoil-elevation occurs accordingly earlier in all of the arteries.   If a rubber tube be connected with the carotid or the femoral artery of a dog, the sphygmographic tracing may be recorded also from this tube. Under such circumstances the interval between the beginning of the ·curve and the dicrotic elevation will naturally be directly proportional to the length of the tube.

2. The dicrotic elevation in the descending limb of the curve will be the lower and the more indistinct the greater the distance of the artery from the heart.   It is not surprising that the secondary wave becomes smaller and more indistinct the further it must travel in the arterial tube.

3. The dicrotic elevation in the pulse will be more distinct the shorter and the more vigorous the primary pulse-wave.   It is, there- fore, relatively largest with a short, powerful systole of the heart.

4. The dicrotic elevation is greater the greater the tension in the arterial tube.

In Fig. 50 IX and X are recorded with low, V and VI with moderate, and VII with high tension of the arterial wall.

Influences Affecting Vascular Tension.—A number of influences are known that affect the tension in the arterial tube.   The tension is diminished by beginning inspiration, vasomotor paralysis, venesection, intermission of the heart's action, heat, and elevation of a part of the body.   The tension is increased by beginning expiration, accelerated heart-action, stimulation of the vasomotor nerves, inter- ference with the flow of blood to the periphery (as by conditions of inflammatory stasis), certain poisons (such as lead), compression of other large arterial trunks, the effect of cold and of electricity on the small vessels of the skin, and inter- ference with the venous flow.   Likewise, exposure of the arterial trunks is followed by increased vascular tension on account of the stimulation caused by the atmos- pheric air coming in contact with the arterial wall.   Increased arterial tension is observed also in association with a variety of morbid conditions.   When the ten- sion is high, the entire sphygmographic tracing is, as a rule, lower.

In conformity with the conditions named, increased tension will be indicated by a lower, more indistinct dicrotic elevation; and diminished tension in the arterial tube, on the other hand, by an enlarged and more distinct dicrotic eleva- tion.   A consideration of the laws governing the dicrotic elevation is of great practical significance in the study of the pulse.   Moens asserts that the interval elapsing between the primary elevation and the dicrotic wave increases directly as the diameter of the vessel, and that the thickness of the wall diminishes as the coefficient of elasticity becomes smaller.

ORIGIN AND PROPERTIES OF THE ELASTICITY-ELEVATION.

In addition to the dicrotic elevation a series of more numerous, though much less distinct, often almost imperceptible, movements are appreciable in the sphygmographic tracing.   These (marked e e in Fig. 50) are produced by the vibrations of the elastic vessel, which behaves like a tense elastic membrane when it is rapidly and vigorously stretched by the pulse-wave, just as a relaxed elastic sheet of rubber undergoes a series of oscillations when it is suddenly and vigorously stretched and made tense.   Similarly, the elastic tube will exhibit oscillatory movements when it passes suddenly from a condition of tension to one of relaxation.   These minor elevations produced in the sphygmographic tracing by the elastic vibrations of the arterial wall are known as *elas- ticity-elevations.*

As the elasticity-elevations are due to the vibrations of the stretched coat of the blood-vessel, the following facts will be readily understood:

1. In the same artery the variations in elasticity increase in number as the tension of the arterial wall increases. Especially high tension has been encountered chiefly during the cold stage of malarial fever (intermittent fever), and precisely in this connection has the most obvious increase in the elevations also been observed.

2. If the tension of the arterial wall is greatly diminished, the elasticity-elevations may disappear. As diminution in the tension favors the development of a dicrotic elevation, the two kinds of elevations have, with respect to their magnitude, an inverse relation to each other.

3. In the presence of diseases of the vessel-wall that diminish or even destroy its elasticity, the elasticity-elevations are either greatly diminished in size or altogether abolished.

4. The greater the distance of the artery from the heart, the greater will be the elasticity-elevations in the descending limb of the curve.

5. When the mean pressure in an artery is heightened on account of interference with the flow of blood in the arteries, the elasticity-elevations are nearer the apex of the curve.

6. The elasticity-elevations vary in number and position in the sphygmographic tracings from the different arteries in the human body.

When the arm is held in the vertical position, relaxation and diminution in the elastic tension appear in the course of five minutes in the arteries of the upper extremity, which at the same time contain less blood.

The elevations that are designated elasticity-elevations are believed by Moens to owe their origin to numerous small waves that appear to be superadded to the dicrotic elevation. Grashey thinks them only in part due to elastic vibrations.

The laws governing the movement of the pulse may be most readily demonstrated by means of investigations in regard to the undulatory movements in elastic rubber tubes, as has been done by Marey, Landois, Moens, Grashey, G. v. Liebig, and others.

THE DICROTIC PULSE.

Under the influence of excessive elevation of temperature the pulse in man is sometimes observed to be composed of two beats (Fig. 50), the first being large and the second small and apparently secondary to the first. A couple of these beats always correspond to a single systole of the heart. By the sense of touch it is quite possible to feel the two unequal beats separately. The study of the pulse with the sphygmograph has taught that the dicrotic pulse is only an exaggeration of the normal pulse. The palpable secondary beat is only a greatly magnified dicrotic elevation, which under normal conditions cannot be recognized by the palpating finger, but which, when increased by some morbid condition, becomes recognizable by the sense of touch. As regards the causes that are responsible for this increase in the size of the dicrotic elevation, Landois' investigations have yielded the following results:

1. The production of a dicrotic pulse is favored by a short primary pulsewave, such as occurs usually in the presence of fever, a condition in which the contractions of the heart are comparatively rapid and unproductive.

2. The dicrotic pulse is favored by reduction of the tension in the arterial system. A short systole combined with diminished arterial tension offers the most favorable condition for the production of the dicrotic pulse. Sometimes the dicrotic pulse is felt only in a certain arterial distribution, while in all the others the pulse-beat is single. This happens especially in the brachial artery on one or other side of the body. Under such circumstances the conditions for the production of dicrotism in the corresponding arterial area must be especially favorable. These conditions will be found in the local diminution of vascular tension in this area in consequence of paralysis of the vasomotor nerves controlling it. If the tension be increased, as can readily be done by compressing adjacent or other arterial trunks of considerable size or the corresponding veins, the dicrotic pulse is converted into a single pulse. In the presence of fever, dicrotism appears to be due to the elevation of temperature (from 39° to 40° C.), which causes greater distention of the artery and shorter and quicker heart-beats.

3. It is absolutely indispensable for the production of the dicrotic pulse that the arterial wall possess its normal elasticity. In old persons with calcified arterial walls dicrotism does not appear.

In Fig. 51, A, B, C illustrate the gradual transition from the normal radial curve (A) to the dicrotic pulse (B, C), in which the recoil-elevation (r) appears as an independent elevation.

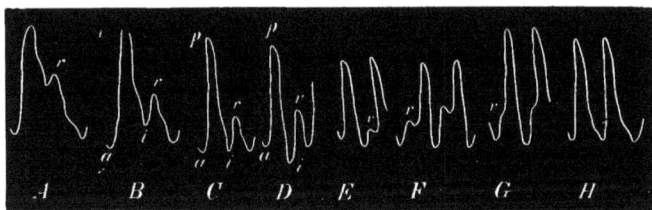

FIG. 51.—Normal Pulse-production of the Dicrotic Pulse. P. caprizans—P. monocrotus.

If in the presence of dicrotism of febrile origin the pulse becomes more and more frequent, the next succeeding pulse-beat may begin before the descending portion of the recoil-elevation is completed (Fig. 51, D, E, F), or it may even begin at the apex (G)—P. caprizans. Finally, if the next succeeding beat begins in the depression (i) between the primary elevation (p) and the recoil-elevation (r), the latter disappears altogether, and the curve (H) assumes the monocrotic form.

DIFFERENCES IN THE TIME-RELATIONS OF THE PULSE.

FREQUENT AND INFREQUENT PULSE.

In accordance with the number of pulse-beats in one minute, the pulse is designated either frequent or infrequent. Under the influence of fever or other agencies the number of pulse-beats may be considerably increased until they reach 120 or more. Reduction of the pulse-beats to about 40 is observed under certain normal conditions (during the puerperium, in states of hunger, and as an idiosyncrasy in some individuals). In rare cases these limits may be exceeded in either direction. In periodic attacks as many as 250 pulse-beats have been counted. Such attacks must be designated pyknocardia (the term tachycardia is incorrect because ταχύς is equivalent to quick). Abnormal infrequency or spanicardia (the term bradycardia is incorrect because βραδύς is equivalent to slow) also occurs; 15, 10, and even 8 beats in the minute have been counted. Under such conditions, disease of the cardiac nerves or of the muscle from over-exertion or disorders in the coronary circulation should be thought of.

Deepening of the respiration without acceleration usually causes some increase in the frequency of the pulse. Accelerated but superficial breathing is without effect, while deep, rapid respirations increase the number of pulse-beats.

QUICK AND SLOW PULSE.

When the development of the pulse-wave is such that the distention of the arterial tube goes on slowly to its maximum and collapse of the distended artery likewise occurs gradually, the slow pulse is produced; while under opposite conditions the quick pulse results. Among the factors that increase the quickness of the pulse are: slowness of cardiac action; greatly diminished resistance of the arterial coats; dilatation of the smallest arteries, diminishing the resistance to the flow of blood; greater proximity to the heart. The curve in a sphygmographic tracing from a quick pulse is high and the apex pointed; a slow pulse yields a low sphygmographic curve, the ascending portion being particularly short, while the apex is broad.

CONDITIONS INFLUENCING THE FREQUENCY OF THE PULSE.

In the normal adult male the number of pulse-beats is 71 or 72 in the minute, in the female about 80.   Other factors that influence the frequency are:

(a) *Age:*

	Beats in the Minute.		Beats in the Minute.
New-born	130–140	10th–15th year	78
1 year	120–130	15th–20th "	70
2 years	105	20th–25th "	70
3 "	100	25th–50th "	70
4 "	97	60th year	74
5 "	94– 90	80th year	70
10 years	about 90	80th–90th year	over 80

(b) The *length of the body* stands in a definite relation to the frequency of the pulse.   Volkmann gives the formula $\frac{P}{P_1} = \frac{L^{\frac{1}{3}}}{L_1^{\frac{1}{3}}}$, in which P and P_1 represent the pulse-frequency and L and L_1 the body-length.   Rameaux suggests the following formula: $N_1 = N\sqrt{\frac{D}{D_1}}$, in which N and N_1 represent the pulse-frequency and D and D_1 the body-length.   By means of this formula the pulse-frequency has been calculated from the body-length in a number of healthy individuals with the following results:

Length of the Body in Units of 10 Cm.	Pulse: Estimated	Observed.
80–90	90	103
90–100	86	91
100–110	81	87
110–120	78	84
120–130	75	78
130–140	72	76
140–150	69	74
150–160	67	68
160–170	65	65
170–180	63	64
Over 180	60	60

As it is possible to determine the pulse-frequency from the body-length, it must also be possible to calculate the body-length from the pulse-frequency.   For this purpose the following is deduced from the foregoing formula:

$$D_1 = \frac{D\,N^2}{N_1^2}$$

These calculations, naturally, have only a theoretical interest, and it is obvious that for purposes of comparison none but perfectly healthy individuals of the same age and sex and living under absolutely identical conditions must be selected.

(c) Of other factors that influence the frequency of the pulse, it has been observed that muscular activity, heightening of the arterial blood-pressure, ingestion of food, elevation of temperature, pain, unpleasant sensations in the alimentary tract, nausea, and psychic or sexual excitement accelerate the pulse.

Further, the pulse is somewhat more frequent in the standing position (also when the body is raised passively) than in the recumbent posture.   Music accelerates the heart-beat in man and in animals and at the same time raises the blood-pressure.   Exposure to increased atmospheric pressure diminishes the pulse-frequency.   In the latter condition the first elasticity-elevation more nearly approaches the summit.

(d) The diurnal *periodicity* of the pulse-frequency is of especial interest.   The variations rarely exceed a few beats and in a general way they correspond with the course of the temperature-curve.   According to Haun the pulse is most frequent with the advent of winter and is least frequent with that of summer.

(e) Frequency of the pulse in various animals: Elephant 28, high-bred stallion about 30 (in mares and work-horses it is a little higher), neat cattle about 50, sheep and swine 75, dog 95, cat 130, rabbit from 120 to 150 in one minute.

VARIATIONS IN THE RHYTHM OF THE PULSE (ALLORRHYTHMIA).

When the finger is applied to the normal artery no special rhythm is observed, the beats apparently succeeding one another at regular intervals, although small differences may be observed in the intervals between the pulse-beats; any more complicated rhythm must be considered an abnormal pulse-movement. Some. times a beat is suddenly dropped from the normal succession—*omission of the pulse*. When this is due simply to weakness of the systole, the pulse is designated *intermittent;* when due to the absence of systole, the pulse is designated *deficient*. The latter occasionally occurs in the obese and has no pathological significance. More rarely a series of pulse-beats is characterized by the successive diminution of individual beats, followed after an interval by a return to the original strength— *P. myurus.* Sometimes a supernumerary pulse-beat appears to be interpolated in the normal series—*intercurrent pulse.* These forms of pulse are not infre. quently produced reflexly through the gastro-intestinal tract, or they are observed in cases of neurasthenia after psychical disturbances, often after intoxi.

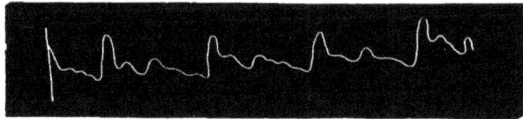

FIG. 52.—Alternating Pulse.

cation with alcohol or tobacco, in the absence of any changes in the heart. Occasionally an intercurrent systole of the auricles takes place in conjunction with the deficient or the intermittent pulse. The regular alternation from a high to a low pulse is known as *alternating pulse.* The peculiarity of the *bigeminate pulse* consists,.according to Traube, in the circumstance that the pulse-beats always occur in pairs, so that the second beat always begins close to the descending limb of the curve of the first. In the same way a *trigeminate* or a *quadri-geminate* pulse may be produced. Knoll found in experiments on animals that these varieties of the pulse occur whenever greater resistances develop in the circulation, increasing the demands on the heart. In man also their occurrence points to a disproportion between the strength of the heart-muscle and the work to be performed. Absolute irregularity of the heart is designated *arrhythmia* or *delirium cordis.*

VARIATIONS IN THE STRENGTH, THE TENSION, AND THE VOLUME OF THE PULSE.

The relative *strength* of the pulse-beat (strong and feeble pulse) may be determined by observing the weight the pulse is capable of raising. For this purpose a weighted sphygmograph may be used, the pad of which is applied to a section of the artery that must be constant in extent. The writing lever naturally ceases to act as soon as the pressure on the artery exceeds the strength of the pulse-beat. The load directly indicates the strength of the pulse. According to G. v. Liebig the pulse in a man with a tendency to pulmonary tuberculosis is readily compressed (feeble) and it has at the same time a tendency to dicrotism.

The pulse appears hard or soft when the artery, in conformity with the mean blood-pressure but independently of the strength of the individual beat, offers a greater or lesser resistance to the palpating finger—*hard* and *soft pulse.* The pulse is said to be *full* when the artery is greatly distended and over-filled, irrespective of the size of the pulse itself, and *empty* when the artery is thin and poorly filled.

In determining the *tension* of an artery and of the pulse, that is, whether the latter is *hard* or *soft*, it should always be noted whether the artery exhibits that quality only during the pulse-wave or also while the vessel is at rest. All arteries are harder during the pulse-beat than in their resting state, but an artery that during the pulse-beat is quite hard may during the pause between the beats appear hard, or under other circumstances soft, as, for example, in cases of aortic insufficiency, in which, after the contraction of the left ventricle, a large quantity of blood flows back into the ventricle through the leaky semilunar valves of the aorta, and the arteries consequently become relatively bloodless. The pulse-tension is lowest in the standing, higher in the sitting, and highest in the recumbent position.

Other things being equal, the volume of the pulse-waves may be directly determined from the size of the sphygmographic tracings. Thus, the following types of pulse are distinguished: the *large* and the *small* pulse; the *unequal* pulse; the extremely weak pulse, which is felt only as a succession of faint tremors (*tremulous* pulse); and the indistinct, scarcely appreciable pulse (*filiform* and *insensible* pulse). A large soft pulse is designated a *dilated* pulse; a small hard pulse a *contracted* pulse; a small pulse of great frequency a *vermicular* pulse; a large, hard, frequent pulse a *serrate* pulse; a large, extremely hard pulse a *vibrant* pulse; and a pulse that is different in two corresponding arteries on opposite sides of the body (due to stenosis, compression or kinking on one side) a *different* pulse.

SPHYGMOGRAPHIC TRACINGS FROM DIFFERENT ARTERIES.

SPHYGMOGRAPHIC CURVE FROM THE CAROTID ARTERY.

(Fig. 50, I, II, III; Fig. 57, C and C_1.)

The ascending limb is exceedingly steep, the apex of the curve (Fig. 50, I, P), traced with a minimum degree of friction, being pointed and prominent. The first elevation below the apex is a small one, the valve-closure elevation (Fig. I, K); this is due to the positive wave, which is produced during the abrupt closure of the semilunar valves at the root of the aorta and is propagated with but little loss of force into the carotid artery. Close to this elevation and visible only in curves traced with a minimum of friction is the highest elasticity-elevation, which is small (Fig. 50, II, e). Further down, but still above the middle of the descending limb, is the dicrotic elevation (R), which is usually larger and is produced by the recoil of the positive wave from the already closed semilunar valves. Relatively, that is, in comparison with the remaining portions of the curve, the dicrotic elevation is slight, in consequence of the high tension prevailing in the carotid artery. After the dicrotic elevation has been formed, the descending limb falls at first abruptly to about the upper third and from this point, in well-traced curves, the writing lever in its downward movement usually traces two more small elevations, the upper of which is an elasticity-elevation, while the lower, which under favorable conditions appears much larger (Fig. 50, III, R_1), represents the second dicrotic elevation. We have here a true tricrotism, which is the more readily recorded in the carotid, because that artery is shorter than the arteries of the extremities.

SPHYGMOGRAPHIC TRACING FROM THE AXILLARY ARTERY.

(Fig. 50, IV.)

The ascending limb of the curve is exceedingly steep. Not far from the apex there is a small valve-closure elevation (K), not unlike that seen in the carotid tracing. Below the middle is found the dicrotic elevation (R), which is fairly high, higher than in the carotid tracing, because in the axillary artery the reduction in arterial tension permits of a greater development of the dicrotic wave. Further down, between the apex of the recoil-elevation and the foot of the curve, two or three smaller elasticity-elevations (e e) are seen.

SPHYGMOGRAPHIC TRACING FROM THE RADIAL ARTERY.

(Fig. 47; Fig. 50, V–X; Fig. 57, R and R'.)

The ascending limb (Fig. 50, V) is of medium height; the ascent is moderately abrupt and suggests the shape of the letter f. The apex (P) is usually well marked. Below the apex there appear, when the tension is considerable, two (V, e e), when the tension is slight, only one elasticity-elevation (VI, IX, e). There then follows at about the middle of the descending limb the recoil-elevation (R), which is usually well marked. This is the more distinct and the better pronounced the larger the number of factors present that favor the development of the secondary wave. It is smallest when the pulse is small and hard, and the artery is greatly distended (Fig. 50, VII, R); larger when the tension is moderate; greatest in the dicrotic pulse.   In the remaining portion of the descending limb, down to the base of the curve, two or three lesser elevations are encountered, the first two being elasticity-elevations (e e) and the lowest appreciable only in rare cases and probably indicating a second recoil-wave.   The sphygmographic curve of the brachial artery at the bend of the elbow is somewhat larger, but does not differ materially from the radial curve.

SPHYGMOGRAPHIC TRACING FROM THE FEMORAL ARTERY.
(Fig. 50, XI, XII.)

The ascending limb is steep and high; on the apex of the curve, which is quite frequently somewhat flat and broad, there is recorded the closure of the semilunar valves (K).   From that point the curve falls in an abrupt manner to about the lower third.   The recoil-elevation (R) appears late after the beginning of the curve, and beyond that point the curve is interrupted in both its ascending and its descending portion by small elasticity-elevations (e e).

SPHYGMOGRAPHIC TRACINGS FROM THE DORSALIS PEDIS ARTERY AND FROM THE POSTERIOR TIBIAL ARTERY.
(Fig. 50, XIV, XV.)   (Fig. 50, XIII, and Fig. 53.)

In the sphygmographic tracing from the dorsalis pedis artery the signs indicating the great distance from the wave-producing apparatus (the heart) are obvious.   Thus, the ascending limb of the curve exhibits a gradual ascent and is low, while the recoil-elevation takes place late.   In the descending limb two elasticity-elevations are found so near the apex (Fig. 50, e e_1) that the upper one usually occupies a point close to the latter.   The elasticity-elevations in the lower portion of the descending limb are, as a rule, poorly developed.   The tracing from the posterior tibial artery in many respects resembles the preceding, especially with regard to the time-relations.

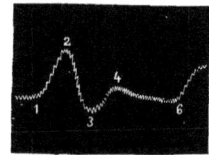

FIG. 53.—Tracing from the Posterior Tibial Artery, Recorded on the Tablet Attached to a Vibrating Tuning-fork by means of Landois' Angiograph.

The tracing shown in Fig. 53 was taken from a medical student, whose height was 180 cm., with the aid of the angiograph, a moderate weight being used and the tracing being recorded on a tablet attached to a vibrating tuning-fork.

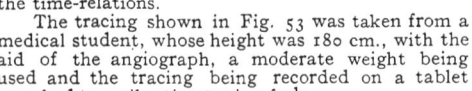

By measurement it is found that
$$\begin{cases} 1-2 \dots\dots 9.5 \\ 1-3 \dots\dots 20 \\ 1-4 \dots\dots 30.5 \\ 1-6 \dots\dots 61 \end{cases}$$
One vibration is equivalent to 0.01613 second
$$\begin{cases} = 0.153 \text{ second} \\ = 0.323 \quad " \\ = 0.492 \quad " \\ = 0.984 \quad " \end{cases}$$

PHENOMENA OF ANACROTISM.

As a rule, the ascending limb in the sphygmographic tracing presents the shape of the letter f, with a rather abrupt rise.   The pulse-beat throws the arterial wall into elastic vibration, as has been explained, the number of vibrations depending largely upon the degree of arterial tension.

In general the distention of the artery, or the tracing of the ascending limb of the curve, which is the same thing, is completed so rapidly that the time is equivalent to a single elastic vibration.   The long-drawn-out f-shaped figure is practically nothing but a long-drawn-out elastic vibration.   When, however, the number of elastic vibrations is small, and the evolution of the ascending limb of the curve is relatively prolonged, two long-drawn-out hump-like curves are sometimes seen in the ascending limb of the tracing.   A condition of this kind, however, is still to be regarded as normal.   (See the elevations in Fig. 50, VIII, at 1 and 2; and at X 1 and 2.)   If, however, a number of closely set elastic vibrations are produced toward the upper portion of the ascending limb of the sphygmographic tracing, so that the apex appears cut off obliquely from the ascending limb and indented, there results the phenomena of anacrotism (Fig. 54, a a), which, like the dicrotic pulse, belong in the domain of pathology.

Anacrotism is observed: 1. When the time occupied by the inflow of blood is longer than the duration of the elastic vibration, for example in cases of dilatation and hypertrophy of the left ventricle. This is illustrated in Fig. 54, A, which represents the radial curve from a patient with contracted kidney. Under such conditions the great mass of blood propelled with each systole requires an abnormally long time to effect distention of the already greatly distended artery.

2. When the distensibility of the arterial tube is diminished, a quantity of blood, which in itself is not increased, will require a longer time to effect distention of the walls. Such a condition is observed in old persons whose arterial walls have acquired great rigidity. As cold tends to contract the arteries, so that they are reduced to a condition of diminished distensibility, it is not difficult to understand that the pulse is likely to assume the characters of anacrotism within an hour after a cool bath (Fig. 54, D). The carotid pulse in the rabbit becomes anacrotic after irritation of the vasomotor nerves.

3. When, owing to blood-stasis as a result of extreme retardation of the blood-current, such as occurs in paralyzed limbs, the quantity of blood injected into the arterial system with each systole is incapable of effecting normal distention of the arterial wall, anacrotic elevations are seen in the sphygmographic tracing (Fig. 54, B).

4. When, after ligation of an artery, the blood can enter the peripheral segment through the relatively small collateral circulation only within a comparatively long time, the distention of the arterial coat will be marked by several elastic vibrations. Wolff succeeded in producing these in tracings from the radial artery not yet possessing distinct anacrotic characters by applying compression above the brachial artery and thus retarding the flow of blood into the radial artery. Also in cases of aortic stenosis, a condition in which the blood can enter the arteries but slowly through the aorta, anacrotism has frequently been observed (Fig. 54, C).

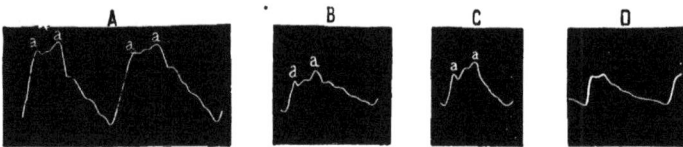

FIG. 54.—Anacrotic Tracings from the Radial Artery: a a, anacrotic notches.

In the same category belongs also the phenomenon of the so-called *recurrent pulse*. When the radial artery is compressed at the wrist, the pulse at once reappears at a point situated peripherally from the site of compression, being transmitted by the arterial palmar arches. The tracing from such a pulse exhibits anacrotism and in addition (as is readily understood) a diminished recoil-elevation, as well as more numerous and more distinct elasticity-elevations.

5. A peculiar form of anacrotism is observed in connection with high grades of aortic insufficiency. The most characteristic sign of this lesion is the permanent patency of the aorta. Hence, not only will waves be propagated in the root of the aorta by the movements of the ventricle, but also the contraction of the hypertrophied left auricle will cause a wave-movement in the ventricular blood that is at once propagated through the patulous orifice of the relatively flaccid aorta and its branches. This is followed by the true pulse-wave, which is produced by the contraction of the ventricle. It is obvious that not only is the wave produced by the contraction of the auricle smaller, but it also precedes the principal wave. The peculiarity of the anacrotism in sphygmographic tracings from large vascular trunks, taken from cases of insufficiency of the aortic valves, is that the auricular wave occurs before the ventricular wave in the ascending limb. This anacrotism manifests itself in curves taken from the larger vascular trunks because the wave, in itself but small, gradually disappears as it advances peripherally toward the smaller vessels.

Fig. 55, I, represents a sphygmographic tracing from the carotid of a man. It exhibits an abrupt ascending limb, caused by the force of the hypertrophied heart. At the apex of the curve there appear quite constantly two sharp indentations, the more anterior of which, having a narrower base, requires less time for its development than the second. The anterior (A) is the anacrotic auricular wave, the second (V) the ventricular wave.

Fig. 55, II, represents a sphygmographic tracing from the subclavian artery of the same individual. It is recognized at once by the peculiarity that the anacrotic notch (a) occupies approximately the junction of the lower and middle thirds of the ascending limb. The recoil-elevation (R) in this curve also is relatively small, for the same reason as in the carotid curve. Below the recoil-elevation are seen feebly developed elasticity-elevations.

Tracings from the femoral artery made with a minimum of friction on the part of the writing stylus exhibit an indentation (Fig. 55, III, a) immediately preceding the ascending limb of the curve, which is blurred in coarse curves. A comparison of this indentation with the anacrotic notch at the lower portion of the ascending limb of the curve from the subclavian artery (Fig. II) will convince the observer that the anacrotic auricular notch must be sought in this well-marked elevation.

It should be mentioned at this point that sphygmographic tracings from cases of aortic insufficiency are characterized further by the following peculiarities:

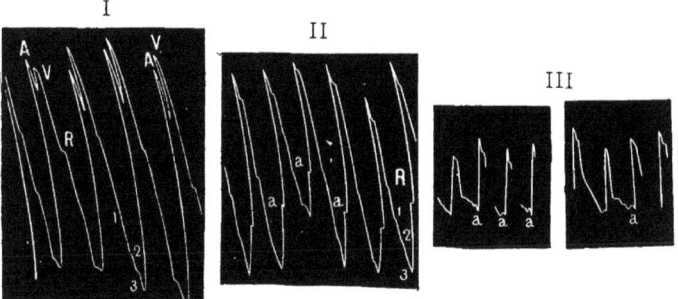

FIG. 55.—I, II, III, Curves Exhibiting Anacrotic Elevation, a, in Association with Insufficiency of the Aortic Valves.

1, the great height of the curve; 2, the rapid fall of the writing lever from the apex. Both of these peculiarities are due to the fact that a large quantity of blood is thrown into the arteries by the enlarged and hypertrophied ventricle, a considerable portion of which flows back into the ventricle after the completion of the systole. In accordance with observations 1 and 2 the pulse is therefore a quick one.   3, A distinct notch is not rarely found at the apex representing an elastic vibration of the greatly distended arterial wall.   4, In tracings taken from cases of aortic insufficiency, as, for example, in that shown in Fig. 55, I, the recoil-elevation (R) is moderate as compared with the size of the curve, because, owing to the lesion of the aortic valves, the pulse-wave in its recoil does not impinge upon a sufficiently large surface.   When the destruction of the semilunar valves is considerable, the recoil-elevation must be produced by the impact of the recurrent wave against the opposite ventricular wall.   Below the recoil-elevation the curve presents two or three faintly marked elasticity-oscillations (1, 2, 3).   The enormous height of the entire curve is sufficiently explained by the massive column of blood injected into the arterial system by the greatly hypertrophied and dilated ventricle.

INFLUENCE OF THE RESPIRATORY MOVEMENTS ON SPHYG-MOGRAPHIC TRACINGS.

The respiratory movements exert a distinct influence on the movements of the pulse by virtue of two different factors: (1) the purely physical diminution of arterial pressure that accompanies each inspiration, and the increase attendant upon each expiration; (2) the variations in blood-pressure, due to excitation of the vasomotor nerve centers, which attend the respiratory movements.

When it is remembered that during inspiration, owing to the dilatation of the thorax, the arterial blood is retained in larger quantities within the chest-cavity, while the venous blood is more actively drawn into the right auricle by aspiration, it is evident that the tension within the arteries must at first diminish during inspiration. The expiratory diminution in the size of the thorax, on the other hand, favors the flow of arterial blood into the vascular trunks and dams the venous blood back toward the venæ cavæ,—two factors that tend to heighten the tension in the arterial system. Furthermore, the expiration that immediately precedes an inspiration allows less blood to enter the heart, so that systolic contractions at the beginning of inspiration throw a somewhat smaller quantity of blood into the aorta; the opposite result attends the inspiration that immediately precedes an expiration.

These variations in tension explain the differences in the size of sphygmographic tracings taken during inspiration and during expiration, as seen in Fig. 56, and in Fig. 50, I, III, IV, in which J indicates the inspiratory, and E the expiratory curve. The differences are as follows: (1) the greater tension in the arteries during expiration causes a general heightening of the level of all curves coinciding with expiration; (2) during expiration the ascending limb is prolonged because the expiratory movement of the thorax tends to increase the force of the wave produced during expiration; (3) the magnitude of the recoil-elevation must be less on account of the increase in pressure during ex-

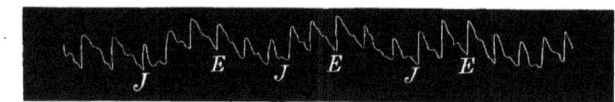

Fig. 56.—Influence of Respiration on the Sphygmographic Tracing (after Riegel).

piration; (4) for the same reason the elasticity-elevations are more distinct and approach more nearly the level of the apex of the curve. During the stage of expiration the pulse is somewhat more frequent than during the stage of inspiration.

This purely mechanical effect of the respiratory movements is modified by the stimulation of the vasomotor center that takes place at the same time. Owing to this nervous influence the arterial pressure—which, it is true, is lowest during inspiration—begins to rise during inspiration and continues to increase until the end of that phase, reaching its maximum at the beginning of expiration. During the remainder of expiration the blood-pressure falls, and again reaches its lowest level at the beginning of inspiration. These influences leave their imprint upon the sphygmographic curves, which, accordingly, present the signs of increasing or diminishing arterial tension, in accordance with the phases of respiration. There is thus to a certain extent a displacement of the pressure-curve to correspond with the respiratory curve.

The statements of different observers vary with regard to the effect of strong expiratory pressure and of forced inspiration on the shape of the pulse-waves. The simplest way of producing strong expiratory pressure is by means of Valsalva's experiment. During this procedure there is at first an increase in the blood-pressure, with the formation of pulse-waves resembling those produced during ordinary expiration—

the recoil-elevation particularly being distinctly less pronounced. If, however, the forced pressure is maintained, the sphygmographic curves begin to exhibit signs of diminished tension. This is due to the influence of the vasomotor center, acting reflexly through the pulmonary nerves. It must be assumed that forced pressure—such as is produced in Valsalva's experiment—when continued, exerts a depressing effect on the vasomotor center. Coughing, singing, and reciting act in a manner similar to Valsalva's experiment; the pulse-frequency being at the same time increased. On the conclusion of Valsalva's experiment the blood-pressure rises until it exceeds the normal by almost as much as it had before been diminished, to return again to the normal after a few minutes.

Conversely, when the circulation is more completely emptied by means of J. Müller's experiment, the sphygmographic curve at first exhibits the characteristic signs of diminished pulse-tension, particularly a higher and more distinct recoil-elevation. After a time, however,

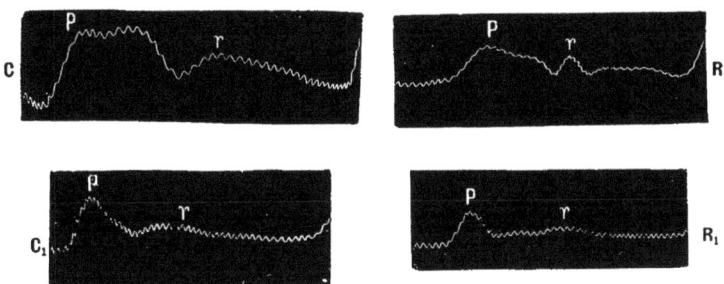

FIG. 57.—The Effect of Marked Expiratory and Inspiratory Pressure on Sphygmographic Curves: C and R, tracings made from the carotid (C) and the radial (R) during Müller's experiment; C_1 and R_1, similar tracings made during Valsalva's experiment. The curves were recorded on a tablet attached to a vibrating tuning-fork.

likewise owing to nervous influences, increased tension may manifest itself. In Fig. 57, C and R represent carotid and radial curves recorded during Müller's experiment, in which the great recoil-elevation clearly shows the diminished tension in the vessels; C_1 and R_1 represent curves taken from the same individual during Valsalva's experiment and clearly show the opposite condition.

Expiration into a vessel like a spirometer (Waldenburg's respiratory apparatus, for example) filled with compressed air has the same effect as Valsalva's experiment, causing after a time a slight lowering of the blood-pressure and a simultaneous increase in the frequency of the pulse. Conversely, inspiration of rarefied air from the same apparatus acts like Müller's experiment, heightening the effect of inspiration, and it may after a time increase the blood-pressure, which, as the experiment is continued, may remain high or fall again.

Inspiration of compressed air lowers the mean blood-pressure, and the after-effect is maintained. The pulse during and after the experiment is increased in frequency. Expiration into rarefied air increases the blood-pressure.

These last-mentioned alterations emanate from the nervous system; they are not produced as readily and are not equally marked in all individuals.

Exposure to compressed air (in the pneumatic chamber) lowers the pulse-curve: the elasticity-oscillations become correspondingly more distinct, as the recoil-elevation diminishes and finally disappears. At the same time the heart's action becomes slower and the blood-pressure is raised.   Exposure to rarefied air has the opposite effect as the sign of diminished tension in the arterial system; but only when as a result the breathing is enfeebled and the pulse is accelerated.

Pathological.—In the presence of adhesions between the heart and the large blood-vessels, on the one hand, and the surrounding structures on the other, the

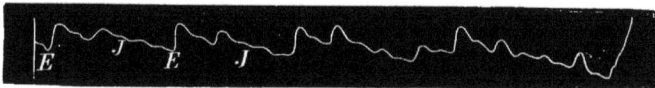

FIG. 58.—Paradoxical Pulse (after Kussmaul).

pulse may be much diminished in size and otherwise altered during inspiration, or it may even disappear altogether.   This phenomenon has been called the paradoxical pulse.   It is due to flattening of the subclavian artery in consequence of elevation of the first rib.   Varieties of this pulse can be produced also in healthy individuals by voluntary alteration of the breathing during inspiration.

THE INFLUENCES OF PRESSURE ON THE SHAPE OF SPHYGMOGRAPHIC TRACINGS.

The changes induced in the movement of the pulse by increasing the pressure upon it affect both the shape of the sphygmographic curves and their time-relations.   Fig. 59 shows at a, b, c, d and e a series of radial curves; a was taken with a minimal pressure and the remainder with a pressure of 100, 200, 250 and 450 grams respectively.   The curves A and B, on the other hand, show the time-relations of curves taken when the pressure was progressively increased.   A study of these curves yields the following results:

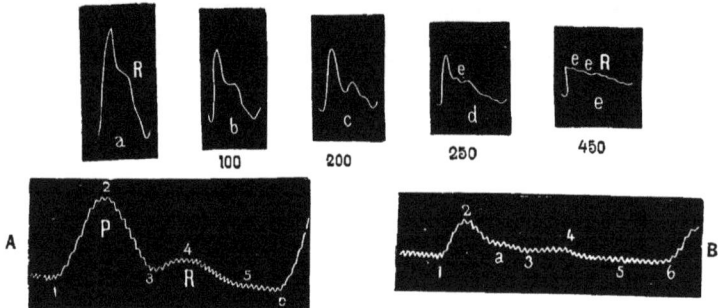

FIG. 59.—Variations in the Shape of Sphygmographic Curves Produced by Increasing the Pressure.

1. With a small load the recoil-elevation is relatively indistinct; the entire curve appears high.

2. With a moderate load, about from 100 to 200 grams, the recoil-elevation is most distinct; the entire curve appears somewhat smaller.

3. As the load is increased, the height of the recoil-elevation diminishes.

4. The smaller elasticity-oscillation immediately preceding the recoil-elevation manifests itself only when the load becomes considerable (from 200 to 300 grams).

5. The quickness of the pulse varies as the load is increased, the time required for the development of the ascending limb being shortened, and that required for the descending limb prolonged.

6. The height of the entire curve diminishes as the load increases.

These points sufficiently emphasize the importance of taking the load of the registering instrument into consideration and the necessity of indicating the actual weight employed, in order to form a correct interpretation of the shape of the pulse-waves.

It appears from an examination of the radial curves A and B, the former of which was taken with a weight of 100 grams, and the latter with a weight of 220 grams, from the same individual and at the same time (1 vibration = 0.01613 second), that changes in the load may produce differences also in the chronological development of the sphygmogram.

When the pressure on an artery is continued for a considerable period of time, the force of the pulse gradually increases. If the greater load is then removed and a smaller one substituted, the sphygmographic curve not infrequently assumes the form of a dicrotic pulse-wave and the recoil-elevation becomes distinctly marked. During the high pressure the blood is forced to make a passage for itself by dilating the collateral vessels. If, then, the main channel is again thrown open, the entire bed of the stream, of course, suddenly becomes much wider. In consequence, there results a greater development of the recoil-elevation. Tracing X in Fig. 50 represents such a dicrotic series, taken after the application of a heavy weight.

VELOCITY OF PROPAGATION OF PULSE-WAVES.

As the pulse-wave passes from the root of the aorta into all the arteries toward the periphery, the pulse is felt earlier in the arteries nearer the heart than in those at a greater distance. This phenomenon was variously confirmed and variously disputed until E. H. Weber determined the movement of rapidity of the pulse-wave from the difference in time of the pulse in the external maxillary artery and in the dorsalis pedis artery and found it to be 9.240 meters in a second. With such great velocity of the pulse-wave, says this investigator, it cannot be regarded as a short wave traveling along the arteries, but so long that a single pulse-wave cannot find room in the entire distance from the beginning of the aorta to the artery of the big toe.

PROPAGATION OF PULSE-WAVES IN RUBBER TUBES.

As it is possible by the intermittent injection of water into rubber tubes to produce waves similar to those produced by the pulse, it is important to learn the results that have been obtained from a study of this undulatory movement.

According to E. H. Weber, the propagation-velocity of these waves is 11.259 meters in one second. Positive and negative waves are propagated with equal velocity and the velocity of the waves is the same whether they have been produced slowly or rapidly.

2. According to Donders, the velocity of the waves is directly proportional to the coefficient of elasticity of the walls of the tubes. It is proportional to the square root of the coefficient of elasticity of the walls of the tubes, with the same lateral pressure.

3. The velocity of the waves increases with the thickness of the walls; it is proportional to the square root of the thickness of the walls, with the same lateral pressure.

4. The velocity is inversely proportional to the square root of the diameter of the tubes, the pressure remaining constant.

5. According to Marey, the velocity diminishes as the specific gravity of the fluid increases. It is inversely proportional to the square root of the specific gravity.

Experiments with Rubber Tubes.—In determining the time-relations Landois employed the following method. He recorded the waves by means of the angiograph on a recording surface attached to a vibrating tuning-fork (Fig. 60). After measuring a certain distance on a long rubber tube, the extremities a and b are placed under the pad of the sphygmograph. B is a compressible bulb, by compression of which a positive wave is thrown into the tube, Q is a portable mercurial manometer, which indicates the pressure in the apparatus. As the pulse-wave first passes through at a and then at b, two elevations, 1 and 2, are recorded. Each small indentation is equivalent to 0.01613 second. The time-relations can be determined by simply counting these indentations.

Propagation-velocity of Water-waves and Mercury-waves within Elastic Tubes.— Landois' experiments, published in 1879, yielded a propagation-velocity of 11.809 meters in 1 second, with an internal pressure of 75 millimeters of mercury.

Landois was unable to find any difference in the propagation-velocity whether the waves were produced rapidly or slowly, or whether they were large or small.

In order to determine whether the material of which the elastic tube is made has any influence on the propagation-velocity of pulse-waves, Landois employed a rather rigid, slightly distensible tube made of gray vulcanized rubber. It was found that the propagation-velocity of the waves in this tube is greater than in a softer and more distensible elastic tube.

This observation is in accord with the fact that the intravascular pressure

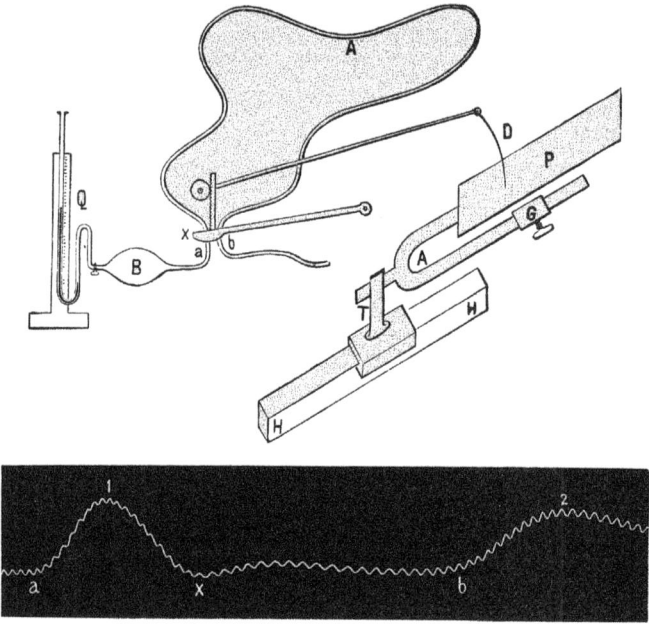

Fig. 60.—Method of Recording the Pulse-curves Obtained from an Elastic Tube on a Tablet Attached to a Vibrating Tuning-fork. Each indentation is equivalent to 0.01613 second.

exerts a demonstrable influence on the propagation-velocity of the pulse-waves; for when the pressure was raised, the waves were propagated with a somewhat diminished velocity. This phenomenon is due to the fact that the distensibility of rubber tubes increases with the pressure, whereas in the arteries the distensibility of the walls diminishes under the same conditions.

The influence exerted by the specific gravity of the fluid was determined by Landois for mercury, the waves of which move with about one-fourth the velocity of waves produced in water.

PROPAGATION-VELOCITY OF THE PULSE-WAVES IN MAN.

Method of Examination.—Landois attached to two different arteries long levers consisting of reeds and so arranged that they both recorded their pulse-curves simultaneously on the same recording surface attached to a vibrating tuning-fork. A quick tap on the fork noted the identical moment on both curves, and by counting the indentations from this point to the beginning of each curve the difference in time was obtained.

In this way Landois developed the following values from a student 174 cm.

in height: The difference between the carotid and the radial was 0.074 second (the distance being estimated as 62 cm.); between the carotid and the femoral, 0.068 second; between the femoral (at the fold of the groin) and the posterior tibial, 0.097 second (the estimated distance being 91 cm.).

Results.—The foregoing observations yield a propagation-velocity for the pulse-waves in the distribution of the arteries of the upper extremities of 8.43 meters in 1 second, and for the arteries of the lower extremities 9.40 meters in 1 second.

It appears that in the less distensible arteries of the lower extremities the propagation-velocity is greater for the same distance than in the arteries of the upper extremities. For the same reason it is less in the peripheral arteries and in the more yielding arteries of the child.

Modifying Influences.—Increase in blood-pressure accelerates, reduction in blood-pressure diminishes, the propagation-velocity of the pulse-wave. Hence, in animals, hemorrhage, slowing of the heart-beat through stimulation of the vagus, division of the spinal cord, dilatation of the vessels (by heat, profound morphin-narcosis or amyl nitrite) cause retardation; while, on the other hand, irritation of the spinal cord causes an acceleration in the movement of the pulse-wave.

The length of the pulse-waves is found by multiplying the time occupied by the entrance of the blood into the aorta, which is from 0.08 to 0.09 second, by the propagation-velocity of the pulse-waves.

A more convenient method is to apply the two tambours of Brondgeest's pan-sphygmograph (Fig. 44) to the two points on the artery to be examined and have one writing-lever record its tracing above that of the other on a plate attached to a tuning-fork. The method may be made quite trustworthy by constructing both apparatus with leaden pipes and filling these with water, in which the propagation of the pulse-wave is quite uniform. A short tap on the tuning-fork (at points indicated by the arrows in Fig. 61) marks the identical instant

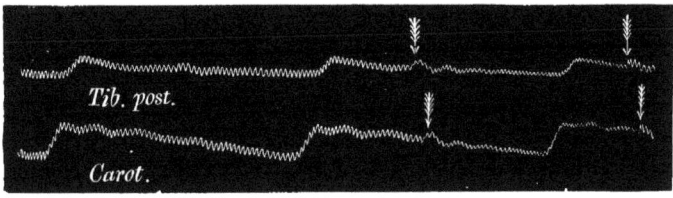

FIG. 61.—Tracings from the Carotid and Posterior Tibial Arteries, Made Simultaneously with Brondgeest's Pan-sphygmograph on a Tablet Attached to a Vibrating Tuning-fork.   The arrows indicate identical moments.

in the two curves. The difference in time is determined by simply counting the vibrations. Fig. 61 shows the curves from the carotid and the posterior tibial taken at the same time from a tall healthy student. The time-difference is 0.137 second.

If the arteries are widely separated or if the observation is made on the heart and on an artery, it is possible to connect the two pads by means of a forked tube with a single writing-lever, and the two pulse-curves, when traced one into the other, can be recognized in the sphygmogram.

In Fig. 62, A is the curve of the ulnar artery, B the same, together with the curve produced by the contraction of the ventricle v H p running through it, and obtained by means of a forked tube. In the curve B, H indicates the apex of the ventricular contraction, P the primary pulse-apex of the ulnar curve; v indicates the beginning of the ventricular contraction, p that of the ulnar pulse. It appears from these curves that the interval between the beginning of the ventricular contraction and the beginning of the pulse in the ulnar artery, in the individual examined, was equivalent to 9 vibrations = 0.15 second.

Grashey applied two sphygmographs to two different arteries and caused the writing-levers to strike sparks into their respective curves from a spark-inductor, so that the sparks marked the identical instant of time in each curve. In this way he determined the propagation-velocity · (from the difference between the radial pulse and that of the dorsalis pedis) to be 8.5 meters in 1 second.

Pathological.—In cases presenting diminished elasticity of the arteries, as, for instance, due to calcification, the propagation of the pulse-wave must be more rapid. Local dilatation of the arteries, such, for example, as has long been known in the form of aneurysms, cause a retardation of the pulse-wave; local stenosis has a similar effect. Relaxation of the vessel-walls during high fever retards the movement of the pulse-wave.

In accordance with what has been said concerning the course of the recoil-wave, its time of appearance must also be affected by the differences mentioned.

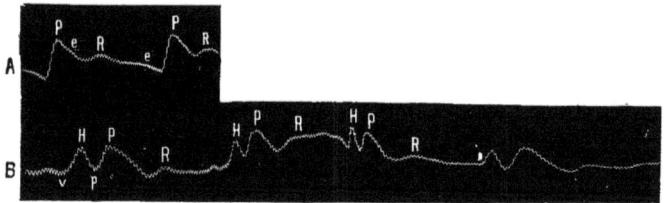

Fig. 62.—Tracing from the Ulnar Artery on a recording surface Attached to a Vibrating Tuning-fork (1 = 0.01613 sec.): P, the apex of the curve; e e, elasticity-vibrations; R, recoil-elevation; B, curves from the same ulnar artery, taken at the same time with v H P = the ventricular contraction of the same individual.

It must appear earlier when the blood-pressure is raised, and also in atheromatous than in healthy arteries; but relatively late in the elastic arteries of the child. The latter point was determined by Landois by mensuration. While in a man, 30 years of age and 172 cm. in height, the apex of the recoil-elevation was reached 0.387 second after the beginning of the radial curve, Landois found that the apex in a girl, 8 years old and 103 cm. in height, occurred at the end of 0.387 second. evidently indicating a relative delay.

OTHER PULSATORY PHENOMENA.

Oral and Nasal Pulse; Tympanic Pulse.—In consequence of the pulsatory movement in the arteries of the soft tissues, the air contained within the oral and nasal cavities is also set into pulsating movement when the glottis is closed, and which can be registered with the aid of the cardiopneumograph. The tracings obtained in this way, and which must closely resemble the sphygmographic tracings from the carotid artery, are of course relatively small, but they can be made larger by increasing the force of the heart. This pulse may be considerably intensified in the presence of pathological enlargement of the heart, dilatation of the left ventricle and thickening of its walls. If a ring containing a soap-bubble be inserted hermetically between the lips, the light-reflex in the bubble (seen in a mirror) reproduces almost perfectly the oscillations of the oral pulse. As a result of the systolic swelling of the vascular soft parts in the tympanic cavity analogous pulsation may be observed in the intact drumhead, or possibly in small bubbles of froth accidentally adherent to openings in a perforated membrane.

If the visual field be darkened, each pulse-beat during violent exertion is often accompanied by a *pulsatory illumination.* Conversely, if the visual field be brightly illuminated, a corresponding obscuration of the field may take place. Pulsation is sometimes observed in the retinal arteries with the ophthalmoscope, especially in cases of aortic insufficiency.

The *orbicularis palpebrarum* muscle under similar conditions contracts synchronously with the pulse. This contraction appears to be due to the fact that the beat of the pulse excites the sensory nerves and reflexly causes a contraction. In this connection attention should be called to an observation made by the brothers Edward and William Weber, which seems to be in accord with this point. They found that, in walking, the pulse and the step not infrequently coincide. Landois believed that this phenomenon may be explained by assuming that the pulse-beat stimulates the muscular mass of the thigh into contraction, to which gradually all the muscles of the thigh accommodate themselves at each step. As the blood-vessels dilate while the muscles are contracting and the movement of the venous blood is accelerated, the coincidence of pulse and step has the additional advantage that the mass of blood to be moved, which is greater during the pulse-beat, is thereby better enabled to pass through the masses of muscle-tissue.

When the *legs are crossed*, the pulse-beat and the recoil-elevation are distinctly recognized in the supported limb.

If with the body at rest in the recumbent position the *lower and upper incisors* are brought gently in contact and kept so, a double beat of the teeth against each other will become audible, as the pulse-wave in the facial arteries elevates the lower jaw. The rapidly succeeding second impact is not due to the recoil-elevation, however, but to the concussion produced by the closure of the semilunar valves.

A pulsatory movement is communicated to the *brain* by the large arteries at its base and in which all the individual features of sphygmographic tracings made from the cerebral arteries are recognized.

Among the *pathological phenomena* of the arterial pulse must be mentioned the systolic *pulsations in the epigastrium*, which are produced in part by the heart in cases of hypertrophy of the right or left ventricle when the diaphragm is depressed, and in part by the forcible pulsation of the abdominal aorta or of the celiac axis, which is usually dilated under such conditions. Abnormal *dilatations (aneurysms) of the arteries* also occasion abnormally strong pulsations in other situations, as, for example, in the trachea in cases of aneurysm of the ascending or transverse position of the aorta.

Hypertrophy and *dilatation of the left ventricle* may cause marked pulsation in the arteries lying nearest the heart. In the presence of similar conditions involving the right ventricle the pulsation of the pulmonary artery in the second left intercostal space is intensified and becomes both visible and palpable (Fig. 34). In cases of *aortic insufficiency* with good compensation in vigorous individuals when the spleen is swollen and palpable (acute infection), this organ also pulsates. Pulsation is visible also in the penis. In cases of *exophthalmic goiter* the spleen may pulsate for months.

VIBRATION OF THE BODY DUE TO THE ACTION OF THE HEART AND THE COURSE OF THE BLOOD-WAVES.

The movement of the heart and of the pulse communicates a vibration to the body as a whole. When a person stands erect on the platform of a spring-scales, the pointer instead of assuming a position of rest plays up and down in accordance with the phases of the heart's action.

In his observations (Fig. 63, I) Landois employed a low box open at the top (K), with a number of rubber bands, close together, stretched across, not far from one of the narrow sides at a b. A quadrangular board (B) was then placed with one extremity resting on the rubber bands and the other on the narrow edge of the box. The subject to be experimented with (A) takes his position on this board and stands erect and steady.

In order to determine the cause of the individual indentations in the curve, the vibration-curve and the curve of the apex-beat were recorded at the same time for the same individual. For this purpose one box (p) of Brondgeest's pansphygmograph (Fig. 44) is applied to the vibrating board, and the pad of the other box to the situation of the apex-beat in the person to be examined. Both writing-levers record their curves on the plate attached to the vibrating tuning-fork: the upper is the vibration-curve, the lower the curve of the apex-beat.

As it is impossible to exclude the marked vibrations in the apparatus itself, the information obtained with regard to the mode of production of the vibrations is only approximately accurate. At the instant of ventricular systole there occurs a short depression, corresponding to the greater pressure of the body on the elastic support; then the body rises suddenly in response to the upward impulse of the blood-wave in the carotid and subclavian arteries. After the closure of the semilunar valves, which is registered by a slight elevation, the blood-wave, as it courses down the body again, causes increased pressure on the platform. The upward movement that now follows may be due to the centripetal wave that precedes the dicrotic wave. The number of inertia-oscillations of the vibrating base that take place until the next heart-beat will depend on the duration of the individual heart-beats.

Pathological.—In cases of insufficiency of the aortic valves the vibration communicated to the body by the action of the heart is marked (Fig. 63, III). The highest apex of the curve, as well as the characteristic drop immediately preceding the ascending limb, corresponds to the ventricular systole. Below the apex of the highest elevation is a small notch, which is produced by a slight vibration communicated to the blood by the partly destroyed semilunar valves in their ineffective effort at closure. The enormous wave of blood that passes through the descending aorta to the iliac artery after the closure of the semilunar valves is the cause of

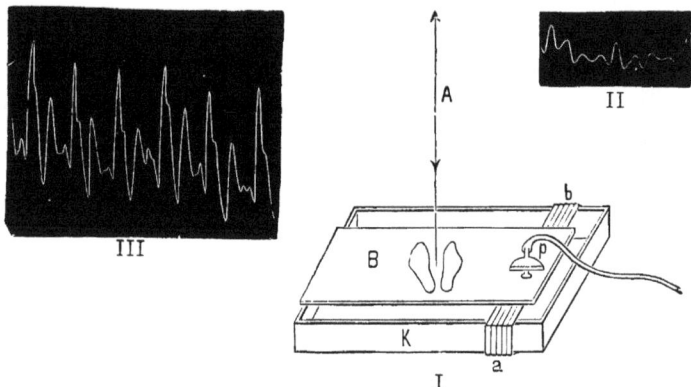

Fig. 63.—I. Elastic Platform for Registering Vibration-curves. II. Vibration-curves Taken from the Body of a Healthy Individual. III. Vibration-curves Taken from a Man Suffering from Aortic Insufficiency and a High Degree of Cardiac Hypertrophy.

the lowest drop of the elastic platform. This is followed by a rise caused by the centripetal movement of the wave. The third rise, which then follows and which is relatively low, appears to correspond with the development of the dicrotic wave in the portion of the arterial system that is directed downward.

THE MOVEMENT OF THE BLOOD.

The closed system of blood-vessels with its many branches, endowed as its walls are with elasticity and contractility, is not only completely filled with blood, but it is in fact overfilled. The volume of the entire mass of blood slightly exceeds the available space within the entire vascular system. It follows, therefore, that the mass of blood everywhere exerts a pressure on the vessel-walls that causes a corresponding distention of the elastic coats. This is true, however, only during life. After death the muscles of the blood-vessels relax and blood-plasma escapes into the tissues, so that the vessels after death are found partially empty.

If the volume of blood be conceived as equally distributed in the entire vascular system, and as everywhere subject to the same pressure, it would be in a condition of passive equilibrium, as is the case shortly before death. If, however, the pressure to which the blood is subjected be heightened at one point of the system of tubes, the blood will escape from this point of increased pressure to some point where the pressure is less; the movement (displacement of the blood-column) is, therefore, the result of the existing difference in pressure. If the venæ cavæ or the aorta in a living animal be suddenly occluded, the blood will continue to flow at a gradually diminishing rate until the differences in pressure in the entire circulation have been equalized.

The velocity of the blood-stream is directly proportional to the difference in pressure and inversely proportional to the resistance en-countered by the blood-current.

The difference in pressure that produces the movement of the blood is created by the heart. In the greater as well as in the lesser circulation the point of highest pressure is at the root of the arterial system, and the point of lowest pressure at the terminal portions of the veins. Hence, the blood constantly flows from the arteries through the capillaries and into the large venous trunks.

The heart maintains the difference in pressure necessary for the circulation of the blood by throwing a certain quantity of blood into the root of the aorta at each systole, after first withdrawing a like quan-tity of blood from the terminations of the venous trunks by means of the diastole of the auricles.

To these laws relating to the causes of the movement of the blood-mass, and which were formulated chiefly by E. H. Weber, must be added an important one by Donders. That investigator demonstrated that the heart, by the work it performs, not only produces the difference in pressure necessary for the movement of the blood, but it also increases the mean pressure existing in the circulatory system. The terminal portions of the large veins that empty into the heart are larger and more elastic than the initial portions of the arteries; and if the heart transfers the same mass of fluid from the veins into the beginnings of the arteries, the arterial pressure must be increased in greater degree than the venous pressure is diminished, and the pressure as a whole must be raised.

The movement of the blood-mass would be jerky or intermittent (1) if the walls of the tube were rigid; for pressure exerted on the fluid contained in rigid tubes is propagated at once throughout the entire length of the tubes, and the movement of the fluid ceases simultaneously with the impact that causes the increase in the pressure. (2) The move-ment would be intermittent also within elastic tubes if the interval between two successive systoles were longer than the duration of the movement of the column necessary to equalize the difference in pressure produced by the systole. If, however, this interval is shorter than is necessary for equalizing the pressure, the current becomes continuous. The more rapidly systole follows upon systole, the greater will be the difference in pressure, the elastic walls of the arterial tubes at the same time undergoing greater distention. In the continuous current thus produced the sudden increase in pressure caused by the systolic injec-tion of a mass of blood corresponding to the size of the ventricular cavity can always be recognized as an intermittent, jerky acceleration of the current (pulse).

This intermittent acceleration of the current is propagated along the arterial pathway with the velocity of the pulse-wave, as both are due to the same cause. Each pulse-beat is therefore attended with a tem-porary, rapidly advancing acceleration of the fluid-particles. Just as the form of the pulse-movement, however, is not simple, so also is this pulsatory acceleration of the current not simple. The latter appears in the complicated form of the current pulse-curve, which likewise exhibits the primary elevation and the recoil-elevation like a (pressure-)sphygmo-graphic curve. Every up-stroke in the limb of the curve corresponds to an acceleration and every down-stroke to a retardation of the moving particles of fluid.

Physical Explanation.—The conditions detailed may be illustrated by means of simple physical experiments. If a rigid tube be connected with the nozzle of a syringe, every movement of the piston will be followed by an intermittent expulsion of water, which will correspond in time exactly to the movement of the piston. The effect of the intermittent injection of fluid into an elastic system of tubes is best exemplified in a fire-hose. Here the air contained in the air-chamber—which is under elastic tension—takes the place of the elasticity of the tubes themselves in the circulatory apparatus. With slow intermittent strokes of the pump, the stream of water is interrupted; but if the movements of the pump are more frequent, the compressed air in the air-chamber effects a continuous outflow, although a distinct acceleration of the stream is seen in correspondence with each stroke of the pump.

Landois was able without difficulty to demonstrate that the particles of water in an elastic tube are set in motion during the passage of the current by every pulsatile wave, in correspondence with the picture presented by the sphygmographic tracing, by introducing in the course of a long elastic tube, in which both a continuous and an undulatory movement could be produced by intermittent pumping, a short glass tube containing a thread passing through an opening in the side and floating to and fro in the stream. Immediately in front of the thread a sphygmograph was connected with the tube. Each pulse-beat caused a synchronous movement of the sphygmograph and of the thread, each upward stroke of the writing lever corresponding to a more marked oscillation of the thread toward the periphery (acceleration), while each downward stroke was marked by a slight diminution in the oscillatory movement (retardation).

In the capillary vessels the pulsatory acceleration of the current ceases with the disappearance of the pulse-wave. The two movements are gradually extinguished by the marked resistance encountered by the blood in the capillary system. It is only when the capillary vessels are greatly dilated and the pressure in the arterial system increases that both pulse and pulsatory acceleration of the current are sometimes communicated to the initial portions of the veins through the capillaries. Such conditions are observed in the vessels of the salivary glands after stimulation of the facial nerve, which dilates the vascular channels. After constriction of the finger with an elastic band, which impedes the return flow of venous blood, and causes an increase in the arterial pressure, with dilatation of the capillaries of the finger, the swollen skin is seen to become intermittently more deeply red isochronously with the well-known throbbing sensation. This is the *capillary pulse.*

Pathological.—The capillary pulse is found sometimes when the action of the left ventricle is greatly increased, for example in cases of aortic insufficiency and of exophthalmic goiter, and often in cases of jaundice.

SCHEMATIC REPRODUCTION OF THE CIRCULATION.

The arrangement of the circulation as described permits a reproduction by physical means, of the most essential conditions, in the so-called *model of the circulation.* Weber's model will be briefly described here. The arterial system and the somewhat larger venous system are represented by portions of animal intestine (Fig. 64).

The system of capillaries between the two is formed by a glass tube of sufficient size, the lumen of which, however, is occupied by a piece of sponge. A short section of intestine into each extremity of which a piece of glass tube is tied represents the heart. The glass tube directed toward the arterial trunk is provided with the necessary valves, which are reproduced by having a piece of small intestine project beyond the edges of the glass tube and securing its free margins with three threads. Through this piece of intestine water can enter only in the direction from the glass tube toward the free intestine, but not in the opposite direction, as the free edges would then come together and close the lumen. From the venous side a similar valve, mounted on the extremity of a separate piece of tube, is inserted into the glass tube directed toward the heart. The two valves open in the same direction. The entire apparatus is moderately distended with water by means of a funnel. By compressing the heart-piece the contents are made to flow through the arterial valve into the arterial portion. When the compression ceases, the contents return from the venous portion through the venous valve into the heart. By means of this apparatus the blood-current becomes continuous when the heart is compressed in rapid succession, and the movement of the pulse can be demon-

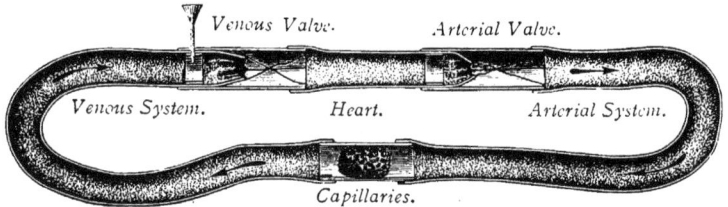

FIG. 64.—Model of the Circulation by Ernst Heinrich Weber.

strated. The latter does not extend beyond the capillary region because the great resistance offered by the many pores of the sponge destroys the force of the pulse-waves.

More complicated models of the circulation, which, however, do not essentially illustrate more than this primitive model by E. H. Weber, have been designed by numerous investigators.

CAPACITY OF THE VENTRICLES.

As the heart creates the difference in pressure necessary for the circulation of the blood by throwing a definite quantity of blood into the roots of the two large arteries every time the ventricles are emptied by systolic contraction, it is desirable to determine this quantity of blood.

As the right and left ventricles must contract simultaneously, and as, in addition, the same quantity of blood must pass through the lesser circulation as through the greater, it follows that the capacity of the right ventricle must be equal to that of the left. It must be remembered, however, that a moderate quantity of blood always remains in the ventricle, as this does not empty itself completely, even at the height of its contraction.

Methods.—1. The capacity of the ventricles is determined *directly* by filling the chambers of the flaccid heart after death with a coagulable material and measuring the coagulated mass. This is an uncertain method, because the pressure in the living ventricles during their diastole, following the contraction of the auricles, is not known.

2. *Indirect Estimation.*—A. W. Volkmann, in 1850, estimated the capacity of the left ventricle in the following manner. The cross-section of the aorta and the velocity of the blood-current in the vessel are determined. From these data the quantity of blood that passes through the aorta in a unit of time is calculated. As the total quantity of blood in the body ($\frac{1}{13}$ of the body-weight) is known, the time required for the passage of this quantity through the aorta can easily be calculated. Finally, if the number of systoles that occur during the time of circulation be known, the quantity of blood for each systole will correspond to the capacity of the ventricle. On the basis of numerous animal experiments Volkmann estimated the ventricular capacity to be equal to $\frac{1}{400}$ of the body-weight; or 187.5 grams for a man weighing 75 kilograms. The accuracy of this method also leaves much to be desired, because the velocity of the current in the aorta, which according to C. Ludwig and Dogiel is subject to considerable

11

fluctuations, can only be determined approximately. Tigerstedt considers Volkmann's figure much too high. He determined the quantity of blood expelled by the left ventricle with each systolic contraction in the rabbit by introducing in the continuity of the aorta an instrument resembling a current-meter. From animal experiments he estimates that in man only 69 cubic centimeters are expelled at each ventricular contraction.

Place calculated as follows: A man uses about 500 liters of oxygen in 24 hours. In order that the venous blood, which contains on the average 7 volumes per cent. less of oxygen than arterial blood, may take up this quantity of oxygen, about 7000 liters of blood must be driven through the lungs in 24 hours. Allowing 100,000 heart-beats for the 24 hours, only 70 cubic centimeters are propelled with each systole.

Other more recent investigators also have calculated that the quantity of blood expelled with each systole is equal only to ⅓ of the capacity of the dead ventricle, or 60 cubic centimeters.

METHODS FOR MEASURING THE BLOOD-PRESSURE.

A. In Animals.—1. *Hales' Tube.*—Stephen Hales, in 1727, first fastened a long glass tube in the lateral wall of a vessel and determined the blood-pressure by measuring the height of the vertical column of blood in the tube.

Hales' tube was fitted at its lower extremity with a short copper tube, bent at a right angle and directed toward the heart; it therefore really represented a so-called Pitot's tube. Pitot, in 1731, used a similar tube to determine the velocity of the current in rivers. The water entering the horizontal portion of the tube, which is directed up-stream, rises in the vertical portion, which projects above the water, to a level proportional to the velocity of the current. This level represents the "velocity-altitude" and it indicates that the water flows with a velocity equivalent to that attained by a body falling freely from a height equal to the velocity-altitude. If a Pitot tube (Fig. 70, II, o p x) be introduced into a closed tube through which flows a fluid under pressure, and an ordinary manometer (x y) be introduced at the same time, the latter will register only the tension of the wall; but in a Pitot tube the fluid will rise to a higher level, for this column of fluid indicates not only the tension of the blood, but also its velocity-altitude. In arteries, however, the latter is extremely small as compared with the former.

2. *Poiseuille's Hematodynamometer.*—Poiseuille, in 1828, used a U-shaped manometer-tube filled with mercury, which he inserted laterally by means of a rigid connecting piece into the wall of the vessel. A ⊢-shaped tube may also be used to connect the blood-vessel with the manometer, the short continuous extremities being inserted into the open vessel (Fig. 65, I, a a) and the vertical limb being connected with the manometer (M) by means of a leaden tube.

3. *Ludwig's Kymograph.*—Carl Ludwig, in 1847, placed a float (Fig. 65, I, d s) on a column of mercury (as James Watt had already done for the manometer of the steam-engine). To the float was attached a vertical wire carrying a writing-contrivance, which records not only the height of the blood-pressure, but also the variations in the pulse-waves on the drum (C), which is made to rotate by clock-work. A. W. Volkmann gave the name of kymograph (wave-tracer) to this instrument. The difference between the levels of the mercurial columns (c d) in the two parts of the tube indicates the pressure within the vessel (the height of the column of mercury multiplied by 13.5 gives the pressure-altitude of the corresponding blood-column). Setschenow added a stopcock at the center of the lower bend of the tube (at b). When this stopcock is turned so as to leave only a narrow orifice of communication, the pulse-waves cease to manifest themselves and the instrument records only the mean pressure. In this form the instrument is the most reliable for this purpose.

The pulsatory variations in pressure are recorded by the kymograph as simple elevations (Fig. 65, III) and, therefore, they do not in the least correspond to the curves obtained with the sphygmograph. After the mercury has once been set in motion by the pulse-beats, it simply undergoes movements up and down by virtue of its own oscillations and all the finer shades of the pulse are completely obliterated. For this reason the kymograph can be used only for recording the blood-pressure, and never for pulse-tracings.

In order to determine the mean pressure from a long blood-pressure tracing presenting numerous elevations and depressions, the planimeter is employed. This instrument is carried over the entire outline of the surface occupied by the curve—

namely the curved line, the abscissa (base) and the initial and terminal ordinates— when the number of square millimeters contained in the entire area can be directly read off on the instrument.   If the paper on which the curve is traced be divided into squares, the size of the area embraced by the curve can be approximately obtained by counting the squares.   A. W. Volkmann cut out the curve-area and weighed it, and then compared with it the rectangle made from the same paper and having the same base-line, so that its altitude naturally represented the mean height of the curve.

4. *A. Fick's Hollow-spring Kymograph*, which was designed in 1864, is constructed on the principle of Bourdon's hollow-spring manometer (Fig. 65, II), which is frequently attached to steam-engines.

A hollow spring bent in the shape of the letter C (F) and filled with alcohol

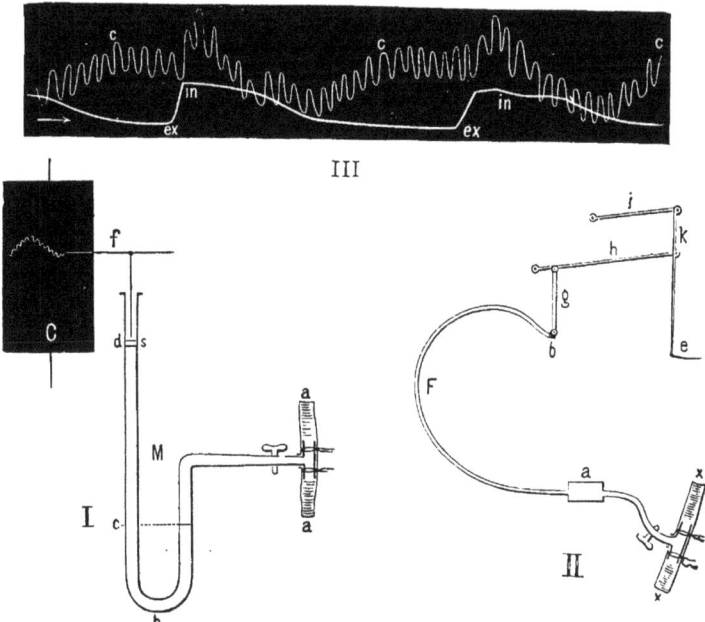

FIG. 65.—I, Carl Ludwig's Kymograph; II, Adolph Fick's hollow-spring kymograph; III, blood-pressure curves (above) and respiratory curves (below), traced at the same time (after C. Ludwig and Einbrodt).

is brought into connection at its lower extremity (a) with the lateral wall of the artery (x x) by means of a suitable cannula, while the other extremity of the spring is closed.   As soon as the internal pressure is increased, the bent spring is straightened out.   The closed extremity (b) is connected with an upright rod (g), which acts on a system of writing-levers (h i k e) composed of delicate pieces of reed, which records the variations in pressure on a moving recording surface.   Both the blood-pressure and the variations in the pulse are recorded: the latter, however, without their characteristic peculiarities.   Hürthle reduced the apparatus to one-fourth of its original size, in which form the results recorded are quite accurate because of the slight displacement of fluid.

5. *A. Fick's Flat-spring Kymograph* (Fig. 66) has been used in preference to any other by its inventor since 1885.   A tube, 1 mm. thick and filled with air (Fig. 66, *a a*), communicates with the blood-vessel by means of a cannula (*c*), and ends in an excavated expansion covered with a rubber membrane, from which a point (*s*) projects downward.   The latter presses upon a tightly stretched hori-

zontal steel spring (*F*), which articulates by means of a connecting piece (*b*) through two joints (*d i*) with a writing-lever (*H*). The parts of the instrument are held in a metallic frame (*R R*). In order to determine the absolute values of variations in pressure the apparatus must first be graduated empirically by comparing it with a mercurial manometer.

6. *Hürthle's Manometer* (Fig. 67) is a similar instrument. A small metallic drum (Fig. 67, d) is intercalated in the course of an artery (c c) by means of tubes. The drum is covered with a thin rubber membrane, from the center of which a process (e) projects. The latter is supported by a spring (F), to which,

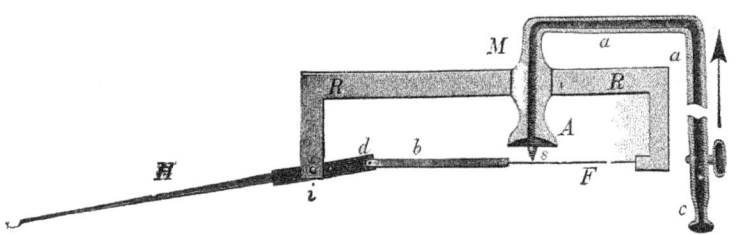

FIG. 66.—Adolph Fick's Flat-spring Kymograph.

at some convenient point that can be varied at will (v), the writing-lever is attached. The whole contrivance is attached to a stationary rod (i i) by means of a carrier (T). This apparatus also, like the preceding one, must first be graduated empirically in order to determine in advance the height to which the point (s) of the writing-lever gradually rises with increasing pressure (from o to 100 mm. of mercury).

Hürthle also constructed a torsion-manometer according to the plan of Roy, the pressure being measured by the torsion of a steel spring.

B. **In man** the blood-pressure within an artery can be measured in the simplest manner by means of a graduated sphygmograph. The weight that just

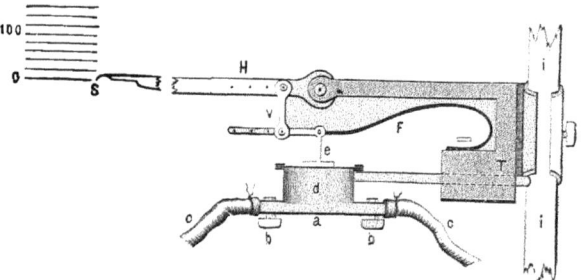

FIG. 67.—Hürthle's Kymograph.

suffices to arrest the movement of the writing-lever corresponds to the tension of the vessel. The radial artery of healthy students examined in this way under Landois' direction and loaded for a distance of 1 cm. exhibited an average blood-pressure of 550 grams.

Manometric Method.—v. Basch determined the blood-pressure by a manometric method, applying his *sphygmomanometer* to the pulsating vessel. The hollow, air-containing cushion applied to the artery communicates with an aneroid barometer, the pointer of which indicates the pressure. As soon as the pressure indicated by the latter slightly exceeds the pressure in the artery, the latter is compressed and pulsation beyond the point of compression is abolished. In the temporal artery the pressure is from 80 to 110 mm. of mercury.

Both of the foregoing methods not only demonstrate the blood-pressure within the arteries, but the pressure exerted by the cushion must exceed the arterial pressure to a degree sufficient to compress the empty artery (which in itself represents a gaping tube). As compared with the blood-pressure, however, the resistance of the artery is extremely slight, being only 4 mm. of mercury, although naturally greater in cases of arteriosclerosis. In the same way the resistance offered by the soft parts superposed upon the artery must also be overcome and in individuals of firm fiber with an abundance of fat this resistance is not inconsiderable. In this way v. Basch found in adults a pressure of from 135 to 165 mm. of mercury in the radial artery; from 80 to 110 mm. in the superficial temporal. Federn thinks it is lower, namely from 80 to 100 mm. of mercury.

In children the blood-pressure increases with age, size, and weight. In the superficial temporal it was found to be 97 mm. between 2 and 3 years of age, and 113 mm. of mercury between 12 and 13 years of age. The blood-pressure rises immediately after exercise; it is higher in the recumbent than in the sitting posture, and in the latter than in the erect posture. After a cold, as well as after a hot, bath the blood-pressure is at first raised and the flow of urine is increased.

Hürthle employs the plethysmograph (Fig. 73) in the following manner for measuring blood-pressure. The glass cylinder communicates with a mercurial manometer. The forearm, first rendered bloodless by firmly bandaging it, is introduced into a cylinder containing water and closed in hermetically. When the blood is allowed to flow freely into the arm, the fluid in the cylinder is displaced and enters the manometer. The blood continues to flow into the arm until the manometric pressure is equivalent to the blood-pressure. The mean pressure in the arm is said to be about 100 mm. of mercury. Sphygmomanometers have been constructed by Marey and Mosso on similar principles.

THE BLOOD-PRESSURE IN THE ARTERIES.

The blood-pressure in the arteries is quite considerable, varying within fairly wide limits. In the larger arteries of large mammals and probably also of man it is between 140 and 160 mm. of mercury.

Examples:

Carotid of the horse, 161 mm. (Poiseuille).
 "   "   " 212–214 mm. (Volkmann).
 "   " dog, 151 mm. (Poiseuille).
 "   "   " 130–190 mm. (Ludwig).
 "   " goat, 118–135 mm. (Volkmann).
 "   " rabbit, 90 mm. (Volkmann).
 "   " chicken, 88–171 mm. (Volkmann).

Aorta of the frog, 22–29 mm. (Volkmann).
Brachial artery of the pike, 35–84 mm. (Volkmann).
Brachial artery in man (after operation) 110–120 mm. (Faivre); perhaps a little too low on account of the traumatism and the disease.

In patients about to be subjected to amputation of the thigh E. Albert, with the aid of a manometer, found the blood-pressure in the anterior tibial artery above the ankle to be between 100 and 160 mm. of mercury. The pulsatory elevation of the column of mercury was from 17 to 20 mm. Coughing caused an increase of between 20 and 30 mm.; firm bandaging of the healthy leg an increase of 15 mm.; passive elevation of the body, in consequence of which the length of the hydrostatic column of blood was augmented, an increase of 40 mm. of mercury.

The pressure in the aorta of large mammals is estimated to be between 200 and 250 mm. of mercury. In general, the blood-pressure is lower in large than in small animals because, on account of the greater length of the blood-channels, a greater resistance is to be overcome. In exceedingly young and exceedingly old animals the pressure is lower than in individuals at the height of their vital activity.

In embryos the arterial pressure is scarcely one-half as great as in the newborn, but the venous pressure is greater. The difference between the arterial and the venous pressure in embryos was found to be scarcely one-half as great as in full-grown animals.

Within the large arteries the blood-pressure undergoes relatively slight diminution toward the periphery, because the differences in the resistance in various sections of the large tubes are inconsiderable. As soon, however, as the arteries undergo frequent division and their caliber accordingly becomes greatly diminished, the blood-pressure rapidly diminishes, because the propulsive power of the blood is weakened by the effort to overcome the increased resistances produced in this way.

The arterial pressure increases directly with the quantity of blood present in the arteries, and conversely. The pressure, therefore,

Increases	Diminishes
1. As the heart's action becomes stronger and more rapid.	1. As the heart's action becomes feebler and slower.
2. In plethoric individuals.	2. In anemic individuals.
3. After considerable increase in the quantity of blood by the direct injection of blood, and also after copious ingestion of food.	3. After profuse hemorrhage or loss from the blood in some other way, as for example, by profuse sweating or copious diarrhea.

The increase and decrease in blood-pressure is not directly proportional to the increase and decrease in the quantity of blood. By virtue of their muscular fibers the blood-vessels possess the faculty of adapting themselves within fairly wide limits to the variable volume of blood. The blood-pressure, therefore, does not rise at once when the quantity of blood is moderately increased. The circumstance that fluid rapidly transudes from the blood into the tissues also assists in maintaining a constant blood-pressure. Moderate venesection, in the dog up to 28 per cent. of the body-weight, is not followed by any noteworthy diminution in the blood-pressure. After slight hemorrhages the pressure may even rise, but the removal of a large quantity of blood is followed by a considerable fall in the blood-pressure, and the loss of from 4 to 6 per cent. of the body-weight reduces it to zero. Increased pressure within the vessels produced by engorgement tends to dilate the cutaneous and muscular vessels, especially those of the extremities, and affects the arteries in the viscera but little. After the pressure has fallen, the visceral blood-vessels return to their original caliber much more promptly than do the cutaneous and muscular blood-vessels.

The arterial pressure rises as the capacity of the arteries is diminished, and conversely. This is accomplished by contraction or relaxation of the unstriated muscle-fibers of the arterial wall.

The pressure within a certain area of the arterial system rises or falls accordingly as the blood-vessels in neighboring areas undergo contraction—or even become impermeable from compression or ligation—or dilatation. The application of heat or cold to a circumscribed portion of the body, also of pressure or diminution of pressure (the latter by introducing an extremity into a closed space containing rarefied air, as, for example, Junod's cupping boot), and the effect of stimulation or paralysis of certain vasomotor areas, furnish striking proofs of the correctness of this statement.

The respiratory movements produce regular variations in the arterial pressure, known as respiratory pressure-variations—the pressure falling with each deep inspiration and rising with each expiration. These variations are readily explained by the fact that at each expiration the blood in the aorta is subjected to the increased pressure of the compressed air in the thorax, while with each inspiration the blood undergoes a diminution in pressure, in consequence of the influence of the rarefaction of the air in the lungs, on the aorta. In addition, the inspiratory expansion of the thorax tends to draw the blood from the venæ cavæ into the heart, while during expiration the blood stagnates, and in this way influences the blood-pressure. The changes are greatest in the arteries nearest the thorax.

The respiratory variations in blood-pressure are in part dependent upon changes in the nervous impulses sent out by the vasomotor center, which coincide with the respiratory movements, and by virtue of which the arteries contract and thus increase the arterial pressure (Traube-Hering's pressure-variations). Fig. 65 III shows a respiratory curve (heavy line) and a blood-pressure curve traced at the same time. This figure shows that at the instant when expiration begins (at ex), the blood-pressure curve rises along with the expiratory pressure, and, conversely, that both curves fall from the instant that inspiration begins (at in); yet the blood-pressure curve begins to rise a little earlier (at c) than expiration itself has begun, that is, during the last part of inspiration. This is due to the contraction of the arteries, which begins a little earlier in obedience to impulses sent out by the vasomotor center. The effect of the arterial contraction is reinforced by the circumstance that during the inspiratory stage the heart is more completely emptied on account of the increased venous flow. The respiratory variations in blood-pressure are observed also during artificial respiration; if this be suddenly interrupted (in curarized animals), the resulting irritation of the medulla oblongata due to the dyspnea causes a considerable rise in the blood-pressure.

In accordance with the depth of the respirations and the corresponding pressure-variations of the air within the thorax, great inequalities are observed in the respiratory fluctuations. This is evident from the fact that in man during quiet inspiration the diminution of pressure in the trachea is equivalent to only 1 mm. of mercury, while during the deepest possible inspiration (with the respiratory canal tightly closed) the diminution is 57 mm. Conversely, quiet expiration in man is attended with an increase in the pressure in the trachea of only 2 or 3 mm., while vigorous contraction of the abdominal muscles causes an increase of 87 mm. of mercury.

Kronecker and Heinricius attribute the variations to mechanical causes, namely to the compression of the heart that accompanies respiration (because, according to them, rhythmical injections of air into the pericardium, which compress the heart, also give rise to analogous variations in blood-pressure). Any interference with the diastole of the heart lowers the blood-pressure; as soon, therefore, as the lung has been distended during inspiration sufficiently to displace the heart, diastole is interfered with and the tension in the aortic system is in consequence lowered. As soon as the air can escape from the lungs and these organs contract, a greater quantity of blood enters the heart, and the arterial pressure rises.

The movements of the pulse cause intermittent variations in the mean arterial pressure, the so-called *pulsatory pressure-variations*. The column of blood injected into the aortic system by the ventricle at each systole, acting in conjunction with the positive wave, produces an increase of pressure in the arterial system corresponding to this positive wave. The increase in pressure finds corresponding expression in the various elevations of the sphygmogram; it also travels along the arteries with the same velocity as the pulse-waves.

In the larger arteries of the horse Volkmann found the pulsatory increase of pressure to be $\frac{1}{13}$, and in the dog $\frac{1}{7}$ of the total pressure. Hürthle, with the aid of his hemodynamometer, found that the pulsatory increase of pressure in the rabbit was equal to almost one-third of the pressure during the interval between pulse-beats.

None of the pressure-recording instruments described shows the form of these pressure-variations with sufficient accuracy; most of them merely record elevations and depressions. Hürthle's kymograph, however, furnishes sufficiently accurate pictures of the pressure-variations in the arteries: these resemble sphygmographic tracings. Hence, the sphygmographic pulse-tracing is at the same time a faithful expression of the pulsatory variations in blood-pressure.

Muscular exertion increases the blood-pressure. At the beginning of a muscular contraction the pressure sometimes undergoes a temporary fall.

When the heart's action is interrupted by continuous stimulation of the vagus or a high positive respiratory pressure, the blood-pressure diminishes enormously in the arteries; while, on the other hand, it increases in the venous trunks because the blood flows from the arteries into the veins in order to equalize the difference in pressure. This experiment shows that when the difference in pressure is (almost) abolished, the resting blood continues to exert some pressure on the blood-vessel walls; that is, in consequence of distention with blood, even in the resting state, a lower pressure is exerted on the walls.

Pathological.—In man it has been found that the blood-pressure, as determined by v. Basch's method, is increased in association with chronic inflammation of the kidneys, arteriosclerosis, lead-colic, after injections of ergotin, and in cases of cardiac hypertrophy with dilatation. It is diminished in the presence of cardiac insufficiency. Digitalis often raises the blood-pressure in cases of cardiac disease; after the injection of morphin the pressure falls. During fever the blood-pressure usually falls, as the shape of the pulse-curves also indicates; in cases of cardiac insufficiency, chlorosis and pulmonary tuberculosis the blood-pressure is also low. If the pressure falls to about 75 mm. in cases of diphtheria (children), the prognosis is grave.

THE BLOOD-PRESSURE IN THE CAPILLARIES.

Method.—Owing to the minute diameter of the capillaries the pressure within these vessels cannot be determined directly. By applying a small glass disc of known dimensions to the vascular substratum and weighting it in a suitable manner until the capillaries become pale, the degree of pressure that just overcomes the pressure within the capillary region is determined approximately. The calculation is made as follows: The pressure (expressed in centimeters of a column of water) is obtained by dividing the number that represents the compressing weight (weight + the weight of the glass disc) by the number of square centimeters contained in the surface pressed upon. In the capillaries of the finger, when the hand is held up, this pressure is 24 mm. of mercury, and with the hand dependent, 62 mm.; in the ear it is 20 mm.; in the gums of the rabbit 32 mm.

Roy and Graham Brown press the vascular area to be examined from below against a rigid glass disc by means of an elastic bladder provided with a manometer; the microscope can then be focused on the glass disc.

The tension of the blood in the capillaries of a circumscribed area is increased by: (1) Dilatation of the small arteries supplying the area. If the latter are dilated, the blood-pressure can be propagated from the large trunks with less loss. (2) Increase of pressure in the small arteries supplying the area. (3) Constriction of the veins draining the capillary area. Occlusion of the veins causes a fourfold increase in the pressure. (4) Increased pressure in the veins, as, for example, by change of position (hydrostatic pressure). Diminution of the blood-pressure in the capillaries is brought about by the opposite conditions.

Also, changes in the diameter of the capillaries must have some influence on the internal pressure. The inherent power of movement (movement of the protoplasm) of the capillary cells, as well as the pressure, swelling, and consistency of the surrounding body-tissues must be considered in this connection. As the resistance to the blood-current is greatest in the small arteries and in the capillary system, the blood—especially in long capillaries—must be subject to different degrees of pressure at the beginning and at the end of such capillaries. In the middle of the capillary system the pressure may not be much less than one-half the pressure prevailing in the main arterial trunks. The capillary pressure exhibits many variations in different parts of the body. Thus, in the erect position, the pressure in the capillaries both of the intestine and of the glomeruli of the kidneys, as well as those of the lower extremities, will be greater than in those of other regions of the body; in the former case on account of the two-fold resistance offered by the duplicate arrangement of the capillaries; in the latter case, from purely hydrostatic influences.

THE BLOOD-PRESSURE IN THE VEINS.

In the large venous trunks near the heart (innominate, subclavian, and common jugular veins) the blood is under a negative pressure, which is on the average equivalent approximately to o.1 mm. of mercury. This enables the lymph-stream to empty itself freely into the large venous trunks.

As the distance from the heart increases, the lateral pressure in the venous trunks gradually increases. In the external facial vein of the sheep it is +o.3 mm., in the brachial 4.1 mm., in branches of the brachial 9 mm.; in the femoral 11.4 mm. The following conditions influence the pressure in the veins:

1. All factors that tend to diminish the difference in pressure existing between the arterial and the venous system, which maintains the circulation of the blood, necessarily increases the pressure in the veins, and conversely.

2. General plethora increases the pressure in the veins, while anemia diminishes it.

3. A special influence on the tension in the large trunks situated near the heart is exerted by the respiration; for during each inspiration the pressure diminishes and the blood rushes toward the thoracic cavity; while with each expiration the pressure increases and the blood stagnates. This effect is intensified in proportion to the depth of the respirations, and when the respiratory passages are closed it must be particularly great.

4. The slight stagnation of the blood in the venæ cavæ that accompanies every contraction of the right auricle has already been discussed in the section devoted to the movements of the heart The respiratory, as well also as the cardiac, fluctuations can sometimes be detected in the common jugular vein of healthy individuals.

5. Changes in the position of the limbs or of the body through hydrostatic influences modify the pressure in the veins in various ways. The highest pressure is found in the veins of the lower extremities, and they are accordingly most abundantly supplied with muscle-tissue. When the muscles and valves in these veins become insufficient, dilatation is likely to develop (varices).

THE BLOOD-PRESSURE IN THE PULMONARY ARTERY.

Method.—Direct estimation of the pressure in the pulmonary artery was made in 1850 by C. Ludwig and Beutner, who opened the left pleural cavity and connected the tube of a manometer directly with the left pulmonary artery, artificial respiration being resorted to. In this way the lesser circulation of the left lung was interrupted completely in cats and rabbits and almost completely in dogs. In addition to this disturbance, the normal flow of the venous blood into the right heart ceases as soon as the thoracic cavity is opened, because the elastic traction of the lungs is abolished and the right heart itself is exposed to the full pressure of the air.

The pressure was found to be in the dog 29.6, in the cat 17.6, and in the rabbit 12 mm. of mercury (in the dog 3 times, rabbit 4 times, and in the cat 5 times less than the pressure in the carotid).

Faivre and Chauveau, in 1856, introduced a catheter into the right ventricle through the jugular vein and connected it with a manometer.

Knoll reached the pulmonary artery through the anterior mediastinum, without opening the pleural cavities, and introduced a cannula laterally into the trunk of the vessel.   By this method he was able to observe the pressure in the artery during spontaneous breathing without restricting the lesser circulation and without displacing the heart.   He thus found a mean pressure of 12.2 mm. of mercury in the rabbit.

Indirect estimation can be made by comparing either the muscular walls of the right with those of the left ventricle, or the thickness of the walls of the pulmonary artery and of the aorta, for it must be assumed that there is a definite relation between the thickness of the walls and the pressure within the vessels.

Beutner and Marey estimate the relation of the pulmonary pressure to the aortic pressure as 1 : 3; Goltz and Gaule, as 2 : 5.   Fick and Badoud, in the dog, found the pressure in the pulmonary artery to be 60 mm., and in the carotid 111 mm. of mercury.   According to Knoll the pulmonary pressure in the rabbit is 6.8 times less than the pressure in the carotid.   In a child the pressure in the pulmonary artery is relatively greater than in the adult.

The pulmonary pressure exhibits certain rhythmical variations due to variations in the tone of the heart's action.   When the air-pressure in the lung falls, the pressure in the lesser circulation also falls, and conversely.

The expansion of the lungs in the thoracic cavity is maintained by the negative pressure on their outer pleural surface.   When the glottis is open, the inner surface of the lungs and the walls of the alveolar capillaries traversing the lungs are exposed to the full pressure of the air.   The heart and the large vascular trunks of the thorax, however, are subject not to the full pressure of the air, but to the pressure of the air minus the pressure corresponding to the elastic traction of the lungs.   The trunks of the pulmonary artery and veins are accordingly subject to the same pressure-conditions.   The elastic traction of the lungs is proportional to the degree of expansion of the lungs.   The blood in the pulmonary capillaries will thus have a tendency to flow from these capillaries into the large vascular trunks.   As the elastic traction of the lungs affects chiefly the more delicate pulmonary veins, and as regurgitation of the blood is prevented by the semilunar valves of the pulmonary artery, as well as by the contraction of the right ventricle, it follows from these pressure-conditions that the capillary blood in the lesser circulation is drained into the pulmonary veins.

Thin-walled tubes embedded within the substance of the walls of an elastic, distensible sac suffer a modification of their lumen, in accordance with the manner in which the sac is distended; for, if the sac is directly inflated so that the air-pressure in its interior increases, the lumen of the tubes is diminished; if, however, the sac is distended by rarefying the air in the closed space surrounding it, the tubes embedded in the wall dilate.   When the distention is brought about in the latter way, namely by the negative pressure of aspiration, the two pulmonary sacs within the thoracic cavity are maintained in a state of distention: therefore the vessels of air-containing lung are more dilated than the vessels of collapsed lung.   Consequently, more blood flows through the lungs when they are distended within the thorax than when they are collapsed.   Inspiratory distention has a similar effect and increases the flow of blood.   The negative pressure prevailing in the lungs during inspiration causes a considerable dilatation particularly of the pulmonary veins, into which vessels, therefore, the pulmonary blood readily flows; whereas the blood of the pulmonary artery, flowing through thick-walled trunks under high pressure, undergoes scarcely any alteration.   The velocity of the blood in the pulmonary vessels is, therefore, increased during inspiration.

The blood-pressure in the lesser circulation is higher also when the lungs are in a state of distention.   Contraction of the vessels, which causes an increase of pressure in the greater circulation, has the same effect in the lesser circulation because more blood flows into the right heart. The vessels of the lesser circulation are exceedingly elastic and their tonicity is slight; hence impermeability even of large pulmonary branches is readily compensated for.

Forcible contraction of the abdominal muscles (straining) causes at first a marked increase in the flow of blood from the pulmonary veins, which, however, gradually ceases, because the blood finds difficulty in entering the pulmonary vessels. When the abdomen is relaxed, the blood again enters the pulmonary vessels in large quantities.

Noteworthy in this connection are the experiments of Severini, who found that the flow of blood through the pulmonary vessels is freer and more rapid when the lungs are filled with air rich in carbon dioxid, than with air containing a larger percentage of oxygen. He believes that these gases affect the vascular ganglia in the lesser circulation that control the size of the vessels.

According to Morel, electrical and mechanical stimulation of the abdominal organs causes a considerable increase of the blood-pressure in the pulmonary artery (dog). According to v. Basch, increase of blood-pressure in the capillaries of the lungs produces greater rigidity and, therefore, diminished elasticity of the alveolar walls.

Pathological.—The pressure in the pulmonary area is increased in man in connection with many morbid disturbances of the circulation and always produces accentuation of the second pulmonic sound, which is such an important pathognomonic sign. It also causes an increase in size and an earlier appearance of the corresponding elevation in the apex-beat curve. But little has been determined with regard to the effect of physiological conditions; temporary suspension of breathing is said always to be followed by an increase in pressure. The influence of the vasomotor nerves on the vessels of the lesser circulation is not so great as that upon those of the greater circulation. Influences that cause a rise or a fall in the blood-pressure in the greater circulation through the agency of the vasomotor or vasodilator nerves have no effect whatever on the pressure in the lesser circulation. Plethora of the pulmonary capillaries is followed by enlargement of the lungs, with more complete distention of the alveoli. The causes may be a diminished flow from the pulmonary veins or disturbances in the left heart. The development of pulmonary edema is discussed on p. 224.

MEASUREMENT OF THE VELOCITY OF THE BLOOD-CURRENT.

The following instruments are used for determining the velocity of the blood-current in the vessels:

1. *Alfred Wilhelm Volkmann's hemodromometer* measures directly the progress of the blood-column through a glass tube in a blood-vessel.

A glass tube shaped like a hairpin, 130 cm. long and 2 or 3 mm. wide and mounted on a scale (Fig. 68, A), is fastened to a metallic basal piece (B) in such a manner that each limb passes to a stopcock perforated all the way through in one direction and halfway through in the other. The basal piece is perforated lengthwise and the two extremities are provided with short cannulæ (c c), which are tied into the two ends of the divided blood-vessel. The entire apparatus is next filled with a 0.6 per cent. sodium-chlorid solution. The stopcocks, which are provided with an arrangement of cogs so that they always turn together, are first placed as shown in Fig. I: the blood then simply flows lengthwise through the basal piece; that is, in the same straight direction as the artery. If at a given moment the stopcocks are turned as shown in Fig. 68, II, the blood is forced to flow through the longer channel represented by the glass tube. The blood will be seen pushing the paler column of water before it and the instant should be noted at which it reaches the extremity of the limb of the tube. The length of the tube being known and the time occupied by the blood in passing through it being determined, the velocity for the unit of time and the unit of length of the course is readily obtained.

Volkmann found the velocity of the current in the carotid of the dog to be between 205 and 357 mm.; in the carotid of the horse, 306; in the facial of the horse, 232; and in the metatarsal artery, 56 mm.

The observation occupies only a few seconds. The tube is narrower than the blood-vessel; nevertheless the blood is said not to flow more rapidly through it than through the larger, uninjured blood-vessel. The intercalation of the tube offers additional resistance to the blood-current, in consequence of which increased retardation must be produced. The apparatus is evidently imperfect; for the larger respiratory and pulsatory variations of pressure in the arterial system do not produce any perceptible changes in pressure.

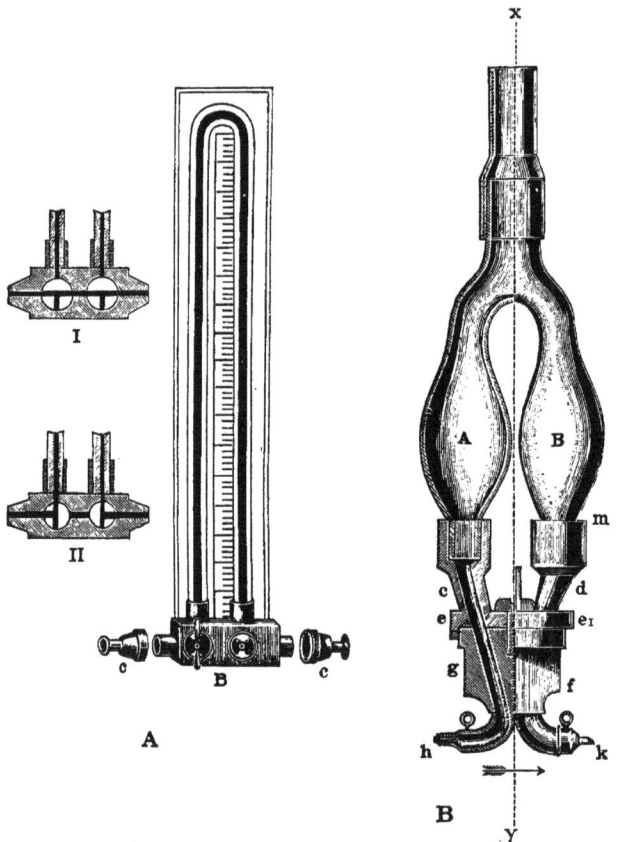

Fig. 68.—**A.** A. W. Volkmann's Hemodromometer. **B.** C. Ludwig's Rheometer.

2. *Carl Ludwig's rheometer* measures the velocity of the blood-stream from the amount of blood that passes from the artery into a communicating graduated glass bulb.

Two communicating glass bulbs (Fig. 68, B, **A** and B), of the same capacity and accurately graduated, are attached by their lower extremities to metallic discs e e_1 by means of tubes c and d. Each disc can be turned about the axis x y in such a way that after it has been turned the tube c communicates with f and the tube d with g; f and g are, in addition, provided with horizontal cannulæ h and k, which are tied into the extremities of the divided artery. When the instrument is in the position shown in the figure, h is tied in the central, and k in the peripheral extremity of the vessel (for example, the carotid). The bulb A is filled with oil and the bulb B with defibrinated blood. At a given moment the blood-current is permitted to enter through h; the oil is displaced by the blood and passes over into B, while the defibrinated blood flows out from B through k into the peripheral portion of the vessel. As soon as the oil reaches m, the time is again noted, and the entire apparatus A B is turned about the axis x y, so that B occupies the place of A. The phenomenon is thus repeated, and the observation may often be continued for some time. By observing the time required by the inpouring blood to fill one of the bulbs the quantity for each unit of time (second) can be calculated.

3. *Carl Vierordt's hemotachometer* measures the velocity of the blood-current by means of a device modeled after Eitelwein's velocity-quadrant, which is constructed on the principle that a pendulum suspended in a moving fluid is deflected by the current in proportion to the velocity.

The apparatus consists of a small metallic box (Fig. 69, I, A) with parallel glass sides and provided at the narrow extremities with two cannulæ (e, a) for the

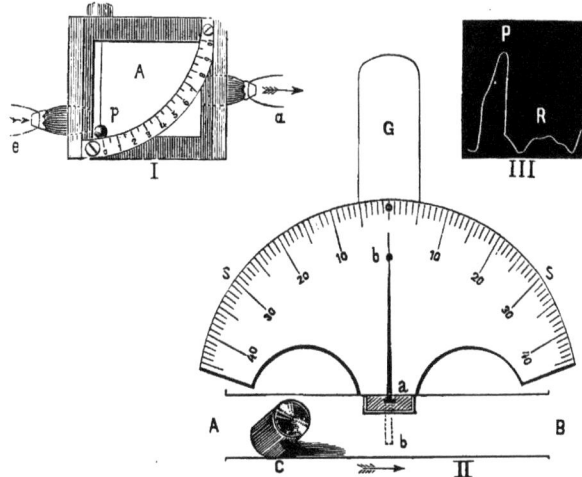

ıɢ. 69.—Vierordt's Hemotachometer: II, Chauveau's and Lortet's dromograph; III, the dromographic curve according to Chauveau.

entrance and exit of the blood. Within the box, opposite the entering blood-current, hangs a small pendulum (p), the oscillations of which are read off on a curvilinear scale and which increase with the velocity of the current. Before making an observation, water is allowed to flow through the instrument for the purpose of determining the velocity of the fluid that corresponds to each degree of deviation of the pendulum.

4. *Chauveau's and Lortet's dromograph* is constructed on the same principle, and is in addition provided with a recording contrivance.

A sufficiently wide tube (Fig. 69, II, A B), provided with a lateral tube C, which can be connected with a manometer, is introduced into the divided artery (carotid of the horse). At a there is a small linear opening closed with a rubber plate through which a light pendulum a b projects into the tube. The pendulum is prolonged upward as a thin indicator (b), which makes excursions proportional to the velocity of the current, and which can be read off on the scale S S. G represents a handle for fixing the instrument. The apparatus is first tested with water to determine the excursions corresponding to the various velocities. As the indicating pendulum is exceedingly light it records the slightest changes in velocity.

The *velocity-curve* (Fig. 69, III) is recorded by permitting smoked paper to pass slowly before the tip of the indicator in the direction of its long axis. The apparatus is of value because it registers the characteristic variations in the velocity of the blood-current that accompany each beat of the pulse. The dromographic curve resembles a pulse-curve, and, like the latter, it possesses a primary (P), as well as a secondary, recoil-elevation (R).

5. *Cybulski's photohemotachometer* is constructed on the principle of Pitot's tube.

When fluid flows through a tube $d\ e$ (Fig. 70, *II*) in the direction indicated by the arrows, the column of fluid stands at a higher level in the manometer p than in the manometer m. While $m\ y$ indicates only the lateral pressure, $p\ x$ indicates the lateral pressure and in addition the velocity-height of the fluid. The velocity of the current in the tube may then be determined from the difference in the two levels. Fluid may be permitted empirically to pass through the tube $II\ d\ e$ with varying velocity and the difference in level between the two tubes $p\ m$ that corresponds to the different degrees of velocity at the current be determined.

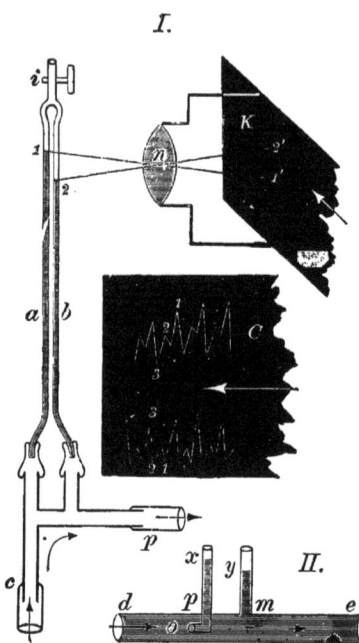

FIG. 70.—I. Diagrammatic Representation of Cybulski's Photohemotachometer: II, Pitot's tube.

The form of Pitot's tube employed by Cybulski is somewhat different, being bent at a right angle ($I,\ c\ p$). The extremity c is tied into the central, and the extremity p into the peripheral, portion of the divided artery. When the blood is allowed to flow freely, the fluid rises to a higher level in the manometer a, which lies in the direction of the current, than in b.

In order to avoid excessive length in the manometers a and b and thus to render the apparatus practically useful, Cybulski connects the manometers a and b by a tube shaped like a hairpin, which is filled with air and can be closed by means of a stopcock (i) applied above the bend. The fluid is allowed to rise to the points 1 and 2. If the stopcock (i) is then closed, the tubes represent an air-manometer in which the difference between the levels 1 and 2 is sharply defined.

As the surfaces of the columns of fluid 1 and 2 continually alter their position with respiration and pulse-beat, that is, as the manometers record the respiratory and pulsatory variations in the velocity of the fluid passing through the tube $c\ p$, the fluctuations of the two levels may be advantageously photographed with a camera provided with a rapidly moving background, K.

Fig. C is a reproduction of the curves obtained from the carotid artery of the dog. During the time represented by the interval between 1_1 and 1 the velocity was 238 mm.; in the phase between 2_1 and 2, 225 mm.; and, finally, between 3_1 and 3, 177 mm. The velocity is greatest at the end of inspiration and at the beginning of expiration. Asphyxia at first increases the velocity. It is increased by paralysis of the sympathetic and becomes smaller when the nerve is stimulated. Division of the vagus increases the velocity, while stimulation of the nerve naturally diminishes it.

THE VELOCITY OF THE CURRENT IN THE ARTERIES, CAPILLARIES, AND VEINS.

In analyzing the results of observation on the velocity of the blood it must be constantly borne in mind that the sectional area of the arterial system beginning with the trunk of the aorta increases progressively by subdivision of the branches, so that in the capillary system the sectional area of the blood-channel is increased 700-fold and more. From this point, owing to the reunion of the venous trunks, the sectional area again diminishes, but it is still greater than at the beginning of the arterial system.

Exceptions are found in the common iliac arteries, which, taken together, are narrower than the trunk of the aorta. The cross-section of the four pulmonary veins, taken together, is also somewhat smaller than that of the pulmonary artery.

An equal quantity of blood must pass through each successive transverse section of both the greater and the lesser circulation. Therefore, the same quantity of blood must flow through the aorta and the pulmonary artery in spite of the great difference between the pressure in the two vessels.

The velocity of the blood-current in the individual transverse sections of the blood-channel must, thus, be inversely proportional to the lumen or their sectional area.

Hence, there is a marked progressive diminution in the velocity from the root of the aorta and pulmonary artery to the capillaries; so that in mammals it is only 0.8 mm. a second (in the frog 0.53 mm.), and in man from 0.6 to 0.9 mm. According to A. W. Volkmann the velocity of the blood in mammals is 500 times less in the capillaries than in the aorta. Therefore, the total cross-section of all the capillaries must be 500 times greater than that of the aorta. In the small afferent arteries Donders found that the velocity was still 10 times greater than in the capillary vessels.

In the venous trunks the velocity again becomes accelerated, being, in the large trunks, from 0.5 to 0.75 times less than in the corresponding arteries.

The velocity of the blood-current does not depend on the height of the mean blood-pressure, and it may accordingly remain the same both in anemic and in plethoric vessels.

On the other hand, the velocity in a given section of the circulation is determined by the difference between the pressure in the cross-section at the beginning and that at the end of the section. It will, therefore, depend on, 1, the vis a tergo (heart's action) and, 2, the amount of resistance at the periphery (dilatation or narrowing of the smaller vessels) to the arterial current.

In accordance with the slight difference in pressure in the arterial and venous systems in the fetus the velocity here is low.

In the arteries every pulse-beat causes an acceleration in the movement of the current (as well as an increase in the blood-pressure) corresponding to the form of the pulse-curve. In large vascular trunks C. Vierordt found the pulsatory increase of velocity to be from $\frac{1}{4}$ to $\frac{1}{2}$ of the velocity during the diastole. These pulsatory variations in the velocity of the current have been recorded by Chauveau by means of his dromograph. Fig. 69 III shows the velocity-curve taken from the carotid of a horse and which corresponds with the pulse-curve in indicating the primary elevation (P), as well as the dicrotic elevation (R). Examination of an extremity with the plethysmograph also discloses this velocity-pulsation or volume-pulsation. In the small arteries an additional pulsatory acceleration is observed, which occurs more rapidly in the first phase than in the later ones. The small trunks themselves are not visibly distended under such circumstances. As the capillary region is approached this phenomenon, like the pulse-movement in general, disappears.

In the arteries the velocity must be retarded by each inspiration and increased by each expiration; but the differences here are exceedingly small.

If what has been said in the foregoing concerning the influence of the respiratory pressure on the dilatation and contraction of the heart, and, therefore, on the movement of the blood, be compared, it will be evident that the respiration must also have an accelerating influence on the blood-current. Likewise, artificial respiration has the same effect: When artificial respiration is suspended in a curarized animal, the blood-current at once becomes slower. If, however, the suspension is continued for some time, the current becomes again accelerated in consequence of the resulting dyspneic irritation of the vasomotor center.

In the veins many derangements in the uniform flow of the blood occur: 1. Regular fluctuations caused by respiration and the movements of the heart at the points where the large trunks empty into the heart. 2. Irregular effects due to pressure, friction in the direction of the current or in the opposite direction, changes in the position either of the body or of the limbs, a pump-like action in the iliac vein due to walking, etc. During extension and outward rotation of the thigh the crural vein relaxes and collapses in the iliac fossa and the internal pressure becomes negative; while when the thigh is flexed and elevated, the vein becomes filled to distention and the pressure rises. By means of this pump-like action the blood (with the aid of the valves) is forced upward. A somewhat similar phenomenon takes place during walking.

ESTIMATION OF THE CAPACITY OF THE VENTRICLES FROM THE CURRENT-VELOCITY BY THE METHOD OF CARL VIERORDT.

There may be considered at this point Vierordt's attempt to estimate the capacity of the ventricles, which is based on the velocity of the blood-current in the innominate artery, in the aorta immediately before the origin of this trunk, as well as in the coronary arteries; although his premises are exceedingly uncertain.

(a) The velocity of the current in the right carotid is 26.1 cm. in a second; the cross-section of the vessel is 0.63 square cm.; hence, the quantity of blood that flows through it is 26.1 × 0.63 = 16.4 cu. cm. (1).

(b) The velocity of the current in the right subclavian artery is 26.1 cm. a second; the cross-section of the vessel is 0.99 square cm.; hence, the quantity of blood that flows through it is 26.1 cm. × 0.99 = 25.8 cu. cm. (2) By adding 1 and 2 the quantity of blood that flows through the innominate artery is obtained: 16.4 + 25.8 = 42.2 cu. cm. The cross-section of this artery is 1.44 square cm.

(c) The cross-section of the aorta immediately before the origin of the innominate artery is 4.39 square cm.; the velocity of the current in the aorta is estimated to be about one-fourth greater than in the innominate, that is, 36.6 cm.; hence, the quantity of blood that flows through it is 161 cu. cm. (3).

(d) The quantity of blood that flows through the two coronary arteries may be assumed to be 4 cu. cm. (4). Hence, the entire quantity of blood that flows through the cross-section of these vessels is $(1 + 2 + 3 + 4)$ 207.2 cu. cm. As the left ventricle must furnish this quantity of blood in a second, and as, in addition, one and one-fifth of the systole corresponds to 1 second, the quantity of blood thrown into the aorta at each systole must be 172 cu. cm., or 180 grams of blood—which is the capacity of the left ventricle.

THE DURATION OF THE CIRCULATION.

The question as to the time required by the blood to make the entire circuit of the circulation was first investigated by Edward Hering, in 1829, in horses by injecting a solution of potassium ferrocyanid into the external jugular vein and noting the time when this substance first appeared in blood withdrawn from the corresponding vein on the opposite side of the neck. Carl Vierordt, in 1858, perfected the technic of these experiments by having a number of cups on a rotating disc pass at uniform intervals beneath the opened vein on the opposite side of the body. The first appearance of the 2 per cent. solution of potassium ferrocyanid is recognized by adding ferric chlorid to the serum separated from the specimen of blood and the development of a Prussian-blue reaction. The duration of the circulation was found to be as follows:

In the horse	31.5 sec.	In the goose	10.89 sec.
" dog	16.7 "	" duck	10.64 "
" rabbit	7.79 "	" buzzard	6.73 "
" hedge-hog	7.61 "	" cock	5.17 "
" cat	6.69 "		

A comparison of these values with the normal pulse-frequency of the same animals yields the following laws:

1. The average duration of the circulation corresponds with 27 contractions of the heart. Applying this figure to man, the duration of the circulation is 22.5 seconds, with 72 pulse-beats in the minute.

If, therefore, the entire quantity of blood passes through the heart in 22.5 seconds, $\frac{1}{22.5}$ of the entire quantity must pass through in 1 second. This quantity is designated the *second-volume of the circulation*. The latter multiplied by 60 gives the *minute-volume*, and as there are 72 heart-beats in the minute, the minute-volume divided by 72 represents the amount of blood propelled at each beat of the heart, that is, the *pulse-volume of the ventricles*. The last calculations, however, are exposed to serious sources of error.

2. In general the mean duration of the circulation in two species of warm-blooded animals is inversely proportional to the pulse-frequency.

Of the influences that affect the duration of the circulation there may be mentioned:

1. A greater length of the vascular channel (for example, from the metatarsal vein of one foot to that of the other) requires a longer time than a shorter channel. This excess in time may be equivalent to about 10 per cent. of the diameter of the circulation.

2. Young animals, with shorter vascular channels and greater pulse-frequency, have a shorter circulation-time than old animals.

3. Rapid and effective contractions of the heart, as during muscular exertion, shorten the time. On the other hand, rapid but ineffective contractions (as after division of both vagi), and slow but correspondingly larger contractions (as with slight irritation of the vagus), appear to have scarcely any effect.

Carl Vierordt has, further, attempted to determine the quantity of blood in man from his investigations in the following manner: In all warm-blooded animals the circulation is completed by 27 contractions of the heart; hence, the entire quantity of blood must be equal to 27 times the ventricular capacity; therefore, in man, 27 times 187.5 grams, or 5062.5 grams. This quantity of blood, estimated as $\frac{1}{13}$ of the body-weight, would correspond to a body-weight of 65.8 kilos.

In 1879 Landois called attention to the fact that potassium ferrocyanid, being a neutral potassium-salt, is a heart-poison, which, in small doses, accelerates, and in large doses paralyzes, the heart. These experiments, in the course of which numerous animals die, thus, of themselves, cause disturbances in the circulation. It was therefore suggested that the experiments be repeated with a substance that truly is chemically indifferent, or perhaps with the microscopic demonstration of particles introduced into the circulation (such as heterogeneous blood-corpuscles, milk-globules or pigment-granules). Accordingly, L. Hermann, in 1884, selected the innocuous sodium ferrocyanid. Wolff thus found the duration of the circulation in the rabbit to be 5.5 seconds, and it is therefore probable that in other animals also the time is shorter than that given by Vierordt. Landois injected mammalian blood-corpuscles into the lateral abdominal vein of frogs and searched for them microscopically on the opposite side. In this way he found the time from 7 to 11 seconds. v. Kries has recently expressed some doubt as to the general applicability of the method even from a physical standpoint. The substances first encountered are carried along only in the axial stream of the blood-vessels, and no conclusion, therefore, can be drawn from their appearance as to the circulation of the entire mass of the blood.

Stewart employed a different method. If the electrical resistance offered by an unopened artery is first determined with a galvanometer, and at a given moment some saline solution is injected into the circulation, the galvanic resistance will be diminished when the saline blood passes through the section in communication with the galvanometer. The instant when this takes place is also noted. In this way Stewart found for the lesser circulation about one-fifth of the entire duration of the circulation (= 10.4 seconds, in the rabbit and in the dog). The duration of the circulation in the kidney was 8 seconds, in the liver 3.8 seconds.

A venous state of the blood increases the duration of the circulation.

Pathological.—In the presence of fever the duration of the circulation appears to be increased.

THE WORK OF THE HEART.

Following the method of Johann Alfons Borelli and Daniel Passavant, Julius Robert v. Mayer estimated the work of the heart according to physical principles. The work performed by a motor is expressed in kilogrammeters, that is, the number of kilos that the motor is capable of raising to the height of 1 meter in the unit of time.

Robert v. Mayer calculated that the left ventricle propels with each systole 0.188 kilo of blood, and, in order to raise it into the aorta, has to overcome the pressure existing in that vessel, corresponding to a column of blood 3.21 meters in length. The work of the ventricle at each systole is, therefore, equivalent to $0.188 \times 3.21 = 0.604$ kilogrammeter. Allowing 75 systoles for each minute, the work of the left ventricle in 24 hours is equal to $0.604 \times 75 \times 60 \times 24 = 65,230$ kilogrammeters. The work of the right ventricle is only about $\frac{1}{3}$ of that of the left, or, in other words, about 21,740 kilogrammeters. The work of the two ventricles taken together is, therefore, 86,970 kilogrammeters. The work performed by a laborer during 8 working-hours equals 300,000 kilogrammeters, thus not quite four times as much as that of the heart. As all of the kinetic energy of the heart is converted by the resistance encountered within the circulation into heat, the work of the heart must result in supplying the body with heat: 425.5 grammeters correspond to 1 unit of heat, that is, the same force that is capable of raising 425.5 grams to a height of 1 meter is also capable of raising the temperature of 1 cu. cm. of water 1° C. The body, therefore, acquires by the conversion of the kinetic energy of the heart about 204,000 units of heat.

As 1 gram of coal yields 8080 units of heat when consumed, the working heart accomplishes as much for the body as if more than 25 grams of coal were burned in it for the production of heat. The values given would be much smaller if the capacity of the ventricles were assumed to be smaller; for example, 60 cubic centimeters; on that basis the work of the heart would be equivalent only to 20,000 kilogrammeters, or $\frac{1}{15}$ of the entire muscular work of the body.

THE MOVEMENT OF THE BLOOD IN THE SMALLEST VESSELS.

In the study of the movement of the blood in the smallest vessels microscopic observation of transparent portions of living animals is the most important method, and it has been repeatedly employed by various investigators since the time of Malpighi, who was the first to observe the circulation of the blood in the pulmonary vessels of the frog.

Method.—Suitable objects for study with transmitted light are the tails of tadpoles and young fishes; the web, the tongue, as well as the mesentery stretched and secured by means of pins on a strip of wax pasted to the object-carrier, or the lung of a curarized frog; in mammals the wing of the bat and the nictitating membrane, drawn out from the orbit and spread out by means of threads over a vertical glass slide; and much less advantageously the mesentery. The following objects can be examined with a low power by reflected light: the blood-vessels of the frog's liver, of the pia mater in the rabbit, of the frog's skin, and of the mucous membrane on the inner aspect of the lip in human beings, as well as of the palpebral and bulbar conjunctivæ.

With respect to the form and arrangement of the capillaries in the various tissues, the following points are worthy of note:

1. The *diameter* of the smallest vessels, which permits the passage of the blood-corpuscles only in single file, may, however, vary from 2 to 5 μ, and in the larger vessels naturally permits the passage of several corpuscles abreast.

2. The *length* is, on the average, about 0.5 millimeter; beyond this limit the vessels either originate by the division of small arteries, or unite to form veins.

3. The *number* of capillaries is extremely variable, being largest in tissues in which metabolism is most active, as the lungs, the liver, and the muscles; and smaller in others, like the sclera and the nerve-trunks.

4. The presence of numerous *anastomoses* is particularly striking, with the formation of plexuses, the shape of which depends principally on the form and structure of the basal tissue. Thus, the capillaries are arranged simply in loops in the papillæ of the skin; as polygonal, retiform meshworks in the serous membranes and on the surface of many glandular acini; as longitudinal tubes running close together between the muscles and the nerve-fibers and between the straight uriniferous tubules; in a radiating manner, converging to a central point in the liver; and in the form of arcade-like loops at the free border of the iris and at the corneo-scleral junction.

With regard to the transition of the smallest arteries into the capillaries, a distinction should be made as to whether the minute arterial twigs are end-arteries—that is, such as do not anastomose with other arterial twigs of the same order, but break up directly into capillaries, and communicate with neighboring arterial twigs only by means of capillaries; or whether before breaking up into capillaries the neighboring arteries communicate by liberal anastomoses, large enough to be called arterial. The presence or absence of arterial anastomoses is important with respect to the nutrition of the region supplied by the vessels.

In observing the blood-current itself it will be seen at once that the red blood-cells progress only along the center of the vessel in the axial stream, while the parietal, transparent layer of plasma remains entirely free from them. The latter, designated *Poiseuille's space*, is recognizable especially in the smallest arteries and veins, in which the axial stream occupies three-fifths, and the light layer of plasma one-fifth, of the entire width of the vessel. It is less distinct in the capillaries. According to Rud. Wagner, Poiseuille's space is wholly absent in the smallest vessels of the lungs and the gills. The red blood-cells pass through the smallest capillaries in single file. In larger vessels they move close together, frequently turning and twisting in their course. On the whole, the rate of progress in the larger vessels is uniform; occasionally, however, as when there is a sharp bend in a vessel, the movement is at times somewhat retarded, at times again accelerated. Wherever the stream divides, a blood-cell occasionally remains attached to the projecting ridge at the point of division, bending at its edges on each side into the bifurcation of the capillary, and appearing somewhat thinned at the center. Often it may adhere in this way for a long while, until, the current becoming accidentally stronger on one side, it is set free, whereupon it rapidly regains its former shape by virtue of its inherent elasticity. When two vessels join to form one, the elasticity of the red blood-cells is again put to proof. Cells at such points are not infrequently heaped up and pushed together in one direction or another. Occasionally, an accumulation of this kind causes a temporary stagnation first in one of the branches and then in the other; the obstruction is then removed, and for some time both capillaries continue to pour their contents into the collecting tube, during which process the corpuscles are shaken up, like dice in a box.

The movement of the white blood-cells is entirely different. They roll along the walls of the blood-vessels, their peripheral zone bathed by the plasma of Poiseuille's space and their inner spherical surface projecting into the procession of red blood-cells. The explanation of this peculiar property on the part of the leukocytes of keeping close to the vessel-wall has been furnished by Schklarewski, who demonstrated by certain physical experiments that in capillary tubes in general (as, for example, glass tubes), containing artificial mixtures of different kinds of granular bodies, those possessing the lowest specific gravity are forced to the wall when a current is set up in the tube, while those having a higher specific gravity move along in the middle of the stream. Thus, when once forced against the wall, the leukocytes must keep on rolling, partly on account of the viscosity of their surface, which causes them to adhere readily to the vessel-wall, and partly because the surface directed toward the axis of the vessel, where the current is swiftest, receives the most effective impulse, often by the direct impact of red corpuscles driven against it. The rolling movement is not rarely intermittent, probably because different parts of the leukocytes adhere with equal tenacity to the vessel-wall. The viscosity of the leukocytes is also in part responsible for their slower movement, which is from ten to twelve times slower than that of the red blood-cells; this is, however, in part also due to the fact that, owing to their parietal position, the larger portion of the body of the leukocyte projects into the peripheral layers of the cylindrical stream, where the current is least rapid.

It is an interesting observation that in the vessels first formed in the incubated egg, as well as in young tadpoles, the movement of the blood from the heart is intermittent.

The velocity of the stream is influenced also by the diameter of the vessels at a given point. The latter is subject to periodical variations, not only in vessels provided with muscular tissue, but also in the capillaries—in the latter in consequence of spontaneous contraction of the protoplasmic cells that form their walls.

In the pulmonary capillaries the blood-stream is more rapid than in those of the greater circulation, whence it may be concluded that the total sectional area of the pulmonary capillaries must be smaller than that of all of the capillaries of the body (of the greater circulation).

THE MIGRATION OF THE BLOOD-CORPUSCLES FROM THE VESSELS; STASIS; DIAPEDESIS.

If the circulation be observed in the mesenteric vessels it is not rarely possible, especially if, after the application of a mild irritant to this vascular tissue (the contact of the air alone is sufficient), an inflammatory process begins to develop, to see the *migration* of leukocytes in varying numbers through the vessel-wall. Instead of rolling along in a jerky manner in the plasmatic zone, the cells gradually move more and more slowly, accumulate in increasing numbers and adhere firmly to the wall; soon they begin to penetrate into the wall and ultimately they make their way completely through it and wander for some distance further into the perivascular tissue.   It is still a matter of doubt whether the corpuscles force their way through interendothelial stomata, supposed to be present, and then enter the lymphatic vascular system, or whether they simply pass through the cement-substance between the endothelial cells.   Several successive steps can be distinguished in this process of migration, which is known as *diapedesis:*—(a) adhesion of the leukocytes to the inner surface of the vessel (after gradual retardation in their progress along the wall up to that point); (b) extension of processes into and through the vessel-wall; (c) withdrawal of the cell-body, which appears constricted at the instant of its passage through the wall of the compression; (d) complete passage through the vessel-wall and the further progress of the leukocyte by virtue of its ameboid movement.

Hering observed that, even under normal conditions, the leukocytes in larger vessels, which are surrounded by lymph-spaces, pass into the lymph-spaces.   This observation explains why cells may be found even in such lymph as has not yet passed through any gland.   The cause of the migration from the vessels resides, in part, in the independent power of movement on the part of the leukocytes; in part it is a physical phenomenon, namely filtration of the colloid mass of the cell-bodies through the force of the blood-pressure, and in the latter connection, therefore, essentially dependent upon the intravascular pressure and the velocity of the blood-current.   Hering regards the migration of leukocytes and even of a few red blood-cells from the small vessels into the lymphatics as a normal process, which he was able to observe in the mesentery of the frog.   The red blood-cells escape from the vessel in the presence of obstruction to the venous flow, which causes, first, escape of blood-plasma through the vessel-wall, and with the plasma the erythrocytes are also forced through, undergoing a marked change of shape on account of the torsion to which they are subjected at the moment when they pass through the vessel-wall, but regaining their shape again after the passage is completed.

The migration of blood-cells had already been described in 1824 by Dutrochet and in 1846 by Waller; the phenomenon was next more carefully studied by Cohnheim.   According to the latter, the migration is a sign of inflammation, and the leukocytes, which accumulate in considerable numbers in the tissue, are to be regarded as true pus-corpuscles, which may later multiply by division.   It should, however, be distinctly stated that, in addition, the connective-tissue cells are also capable, by multiplication, of producing pus-corpuscles, which differ by their greater size from the migrated leukocytes found in pus.

When a vascular part is subjected to severe irritation, hyperemic reddening and swelling of the part are at once observed.   It has been shown by microscopic examination of transparent parts that both the capillaries and the smaller vessels become dilated and engorged with blood-cells; sometimes dilatation is preceded by a temporary contraction of brief duration.   At the same time, a change in the ve-

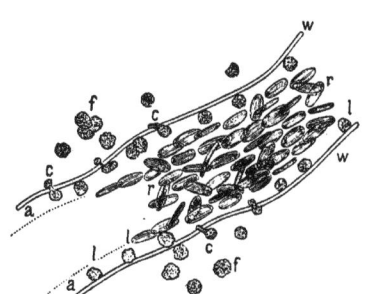

FIG. 71.—Small Mesenteric Vessel from a Frog Showing the Migration of Leukocytes: w w, vessel-wall; a a, Poiseuille's space; r r, red blood-corpuscles; l l, leukocytes moving along the wall, at c c in various stages of migration; f f, migrated cells.

locity of the blood-stream is observed in the vessels.   Rarely, and, as a rule, only for a short time, the blood-stream is accelerated; but generally it is retarded. If the irritation be continued, the retardation soon becomes so great that the current only advances intermittently, and a to-and-fro movement of the blood-column is observed,—a sign that obstruction has already taken place in peripherally situated vascular areas.   Finally, the current in the distended vessels comes to a complete standstill (*stasis*).   Donders points out the greater number of leukocytes in stagnating blood, and believes correctly that this accumulation of leukocytes is a greater obstacle to their progress, as compared with the erythrocytes.   While these processes are going on, the migration of the leukocytes and rarely also of the red cells takes place. Under favorable conditions the stasis may be relieved, generally with a reversal in the order of the phenomena that have attended its development. The escape of blood-corpuscles through the intact wall of the vessel is designated diapedesis. The swelling of inflamed parts is due in part to the dilatation of the vessels, but chiefly to the escape of plasma into the tissues.

THE MOVEMENT OF THE BLOOD IN THE VEINS.

In the smallest veins, which are formed by the union of capillaries, the velocity of the blood-current is greater than in the capillaries, but slower than in the smallest arteries. At the same time, the current is everywhere uniform, and according to hydrodynamic laws the venous current would continue with absolute regularity to the heart, if it were not subject to other disturbances. Such disturbances, however, are operative in various directions. Among special peculiarities of the veins to which interference with the uniformity of the current is attributable the following may be mentioned: 1. The relative relaxation, the great distensibility and compressibility of even the larger trunks; 2, the incomplete distention, which does not increase to any considerable degree the elastic tension of the walls; 3, the numerous and at the same time free anastomoses among neighboring trunks, both in the same tissue-plane and from above downward. By this means it is possible for the blood, when the venous area is partly compressed, to escape through numerous readily distensible channels, and thus the occurrence of actual stasis is prevented; 4, the presence of numerous valves, which permit the blood-current to move only in a centripetal direction. These are wanting in the smallest veins, and they are most numerous in the medium-sized veins. The valves are of great hydrostatic significance, inasmuch as they divide long columns of blood, as, for example, in the crural vein when the body is in the erect position, into sections, thus preventing the entire column from exerting its hydrostatic pressure down to the lowest portions of the vein.

As soon as pressure is exerted on a vein, the nearest valves below the point of pressure close and those next above open, thus leaving a free passage for the blood to the heart. The pressure on the veins may be of varied character: in the first place from without, by contact with various objects. Further, thickened and contracted muscles may compress the veins, especially in the movements of the extremities. That the blood escapes in a stronger stream from an opened vein when the muscles are moved at the same time can be seen whenever venesection is practised. If the muscles are permanently contracted, the venous blood, escaping from the muscles, collects in the parts that are not moved, especially in the cutaneous veins. The pulsatory pressure in the arteries accompanying the veins also tends to accelerate the venous current.

Direct observations have been made as to the velocity of the venous blood-current with the hemodromometer and the rheometer. Thus, Volkmann found a velocity of 225 mm. in a second for the jugular vein; but in view of the low pressure that prevails in the venous system, the employment of instruments for measuring the velocity is necessarily attended with marked deviations from the normal. Reil observed that the quantity of blood escaping from an opening in an artery was two and a half times as great as the quantity of blood escaping from a similar opening in a vein.

As the smaller venous branches unite to form larger ones, the lumen gradually diminishes toward the venæ cavæ: hence the velocity of the current must increase in the same proportion. The velocity in the venæ cavæ may be half as great as that in the aorta.

Borelli estimated the capacity of the venous system as four times as large as that of the arteries. According to A. v. Haller the proportion is as 9 : 4.

As the pulmonary veins are narrower than the pulmonary arteries, the blood moves more rapidly through the former than through the latter.

SOUNDS AND MURMURS IN THE ARTERIES.

The acoustic phenomena observed in the arteries must, from a strictly physical standpoint, be designated as murmurs. Nevertheless it is customary in medical nomenclature, following the example of Skoda, to apply the term *sound* to those acoustic phenomena that are of short duration and sharp definition, like the heart-sounds; while those that are of longer duration and are not distinctly delimited are designated murmurs in the narrower sense. In many cases a sharp distinction between the two is, therefore, impossible.

In the carotid, and more rarely in the subclavian, two distinct sounds are heard in approximately four-fifths of all healthy individuals. These sounds correspond in duration and pitch to the two sounds of the heart and must be interpreted as due to propagation of the sound from the heart by means of the blood as far as the carotid, and they are, accordingly, designated *transmitted heart-sounds*. Sometimes the second sound of the heart alone is heard, as the site of its production is nearer the carotid. The second sound of the pulmonary artery, which is in close contact with the aorta, may also be transmitted to the point mentioned.

Sounds and murmurs occur either spontaneously or only after the application of external pressure, by means of which the lumen of the vessel is narrowed. Accordingly a distinction is made between (1) spontaneous sounds and murmurs and (2) pressure-sounds and pressure-murmurs.

Arterial murmurs are developed most easily by exerting pressure on a circumscribed portion of a large artery, for example, the femoral. The pressure must be so regulated that only a small portion of the lumen remains open for the passage of the blood (*stenotic murmurs*). As a result, a small stream of blood will pass through the stenotic point with great rapidity and force, and enter the wider portion of the artery beyond the site of compression. This so-called pressure-stream throws the fluid-particles into active oscillatory and rotatory movement and thus produces the murmur in the wider, peripheral portion of the vessel. Analogous conditions prevail wherever there is a kink, a sharp bend or a tortuosity in the course of the artery. The phenomenon is, therefore, as a rule a pressure-murmur generated within the fluid. With regard to the question as to the origin of these murmurs, Geigel takes the static that they are due to static transverse vibrations of the vessel-walls. Below the point of compression a thrill is felt in the walls of the large arteries synchronously with the pressure-murmur. In cases of aortic insufficiency, exophthalmic goiter, and circumscribed arteriosclerosis this thrill is much more marked than in normal cases, and it is also appreciable over smaller arteries.

A murmur of like character is that at times heard over the subclavian artery synchronously with the pulse and designated *subclavian murmur*. This is produced by adhesions of the two layers of the pleura at the apices of the lungs, especially in association with tuberculosis and other diseases of the lungs, and in consequence of which the subclavian artery, as a result of torsion and kinking, undergoes local stenosis, which sometimes manifests itself by diminution or absence of the pulse-wave in the radial artery (paradoxical pulse).

Pathological.—It is evident that murmurs will develop in the human body likewise: (a) When, owing to morbid conditions, the arterial tube is dilated at some point where the blood-current is forcibly introduced from a normal portion of the artery. Such dilatations (aneurysms) quite generally give rise to murmurs (bruits). (b) Pressure-murmurs will be generated whenever an organ exerts pressure on an artery, as, for example, by the greatly enlarged uterus during pregnancy, and by a pathological tumor pressing upon a large artery.

In all cases in which there is no external pressure, it is found that the production of spontaneous acoustic phenomena is greatly facilitated if, during the
period of arterial diastole, the arterial wall is as relaxed as possible and, therefore,
becomes suddenly and greatly distended at the time of the pulse-wave, that is,
when the systolic minimum of tension of the arterial wall is rapidly displaced by
the diastolic maximum of tension.   This is particularly the case with aortic insufficiency, a condition in which the arteries are often the seat of widespread
murmurs.   If even during arterial rest the minimum of tension of the arterial
wall is relatively high, the acoustic phenomena are faint and may even disappear
altogether.

The following factors favor the development of arterial murmurs: (1) A sufficient degree of delicacy and elasticity of the vessel-walls; (2) a low peripheral
resistance, that is, accelerated and unobstructed escape of the blood from the end
of the vascular channel; (3) a material difference between the pressure of the fluid
in the stenotic portion and that of the fluid in the peripheral dilatation; (4) large
size of the artery.

Murmurs may be heard also in normal pulsating arteries, especially when
the vessel is the seat of sharp bends or tortuosities.   In almost all cases in which
arterial murmurs are heard, one or several of the foregoing factors can be demonstrated.   It is evident that murmurs of this kind will be most marked when two
or three large arteries are found in close apposition.   Hence the rather loud
murmur generated in the many tortuous and dilated arterial trunks of the gravid
uterus (*uterine* or *placental souffle*) and the much less distinct *funic souffle* in the
two umbilical arteries.   In this category belongs also the so-called *cerebral murmur*
heard in almost one-half of all infants with thin skulls, as well as the murmur
heard over the morbidly enlarged spleen, and the thrill in the thyroid gland in
cases of exophthalmic goiter.

When auscultation is practised over the ulnar artery under the favorable
conditions mentioned, especially in lean individuals, every pulse-beat is found to
be accompanied by two acoustic phenomena, which coincide with the primary
and the dicrotic elevation.   In old persons especially, and in individuals with a
bigeminate pulse, the two sounds are quite distinct.   Friedreich believes the first
sound to be produced by the vessel-wall, that is, the sudden tension of the artery
distended during diastole.   The second murmur naturally is feebler, in correspondence with the lesser degree of distention of the artery by the dicrotic elevation.
Occasionally a third sound is heard between the other two, which corresponds to
the elasticity-oscillations between the apex of the curve and the dicrotic elevation.
In the radial artery and in the dorsalis pedis only a single murmur is, as a rule,
heard synchronously with the pulse-beat.

In cases of aortic insufficiency characteristic acoustic phenomena are present
in the femoral artery.   When the vessel is compressed, there is heard a double
blowing (murmur), the first element of which is due to the fact that a large mass
of blood is driven to the periphery synchronously with the pulse, and the second
to the fact that during the contraction of the artery a large quantity of blood
flows back into the ventricle.   On the other hand, if the artery is not compressed,
two feebler sounds are heard, which are due to the fact that the auricle and the ventricle send a wave of blood into the arterial system in rapid succession (Fig. 55, III).
Gerhardt similarly heard, in cases of insufficiency of the pulmonary valves, two
dull sounds over every portion of the pulmonary surface.   In other cases (when
there is also tricuspid insufficiency) the second sound is produced by the sudden
snapping closure of the valves in the venous veins, caused by the rebound of
the venous blood.   Also, when the arteries are rigid (atheroma) a double sound
is sometimes heard synchronously with the pulse-wave.   This sound is attributed
to the anacrotism of the pulse observed under such conditions.

ACOUSTIC PHENOMENA WITHIN THE VEINS.

The Venous Hum.—Above the clavicle, in the fossa between the origin of the
two heads of the sternocleidomastoid muscle, most commonly on the right side,
there is heard in many individuals (40 per cent.) a sound that may be continuous,
or synchronous with the diastole of the heart, or even with inspiration, and of a
roaring or buzzing, sometimes hissing or singing, character. This sound is generated
within the bulb of the common jugular vein and is called a venous hum.   If
present even when no pressure is exerted with the stethoscope, it is a pathological
symptom.   The phenomenon may be heard in almost any subject if pressure be

exerted and the head is at the same time turned to the opposite side and slightly upward. The pathological venous hum occurs chiefly in young anemic individuals in whom also a thrill is felt over the vessel; it is present also in cases of goiter, at times in youthful individuals, but it becomes less common with advancing age

The cause of the venous hum resides in the whirling entrance of the blood from the relatively narrow portion of the common jugular vein into the dilated bulb situated below. It appears to be generated chiefly when the walls of the thinner portion of the vein are in fairly close apposition, so that the blood-stream is obliged to force its way through. This explains the fact that the occurrence of the phenomenon is favored by pressure and by turning the head to the side and slightly upward. The intensity of the sound depends upon the velocity of the blood as it passes through the narrow portion of the vein, and for this reason the act of inspiration and the diastole of the heart, both factors accelerating the venous flow, intensify the venous hum. The same is true with regard to the favorable influence of the erect posture. In rare cases a sound similar to the venous hum is heard in the subclavian, axillary, thyroid (in cases of goiter), facial and innominate veins, the superior vena cava, the crural vein, and the inferior vena cava at the blunt margin of the liver.

Regurgitant Murmurs.—The expiratory murmur heard at times in the crural vein after sudden efforts at bearing-down is produced by a centrifugal current of blood passing through the vein at the bend of the knee, the valves being incompetent or entirely absent. When the valves in the bulb of the jugular vein are incompetent, a regurgitant murmur may be produced either during expiration (expiratory jugular-valve murmur) or during the systole of the heart (systolic jugular-valve murmur). In the presence of insufficiency of the tricuspid valve a systolic murmur has been heard in the crural vein when its valves were incompetent.

Valvular Sounds in the Veins.—Forced expiration may give rise to valvular sounds in the crural vein, as the valves close with a snap under the pressure of the blood forced back. In the presence of insufficiency of the tricuspid valve a large quantity of blood is thrown back into the venæ cavæ at each ventricular systole. Under such circumstances also the venous valves may close suddenly with the production of a sound. The phenomenon occurs both in the bulb of the jugular vein and in the crural vein at the bend of the knee, but only when the respective valves are competent.

THE VENOUS PULSE. THE PHLEBOGRAM.

Method.—If the movements of a vein are recorded by means of a lightly weighted sphygmograph—a heavy load would compress the vein or at least obliterate the delicate details of the curve—a characteristic form will be observed in a successful venous pulse-curve or phlebogram (Fig. 72).

In the proper interpretation of the details of the phlebogram it is especially important to determine its chronological relations to the phases of the heart's action; hence, it is advisable to record a cardiogram and a phlebogram simultaneously (on a recording surface attached to a vibrating tuning-fork). The beginning of the carotid pulse coincides approximately with the apex of the cardiogram, that is to say, with the descending limb of the phlebogram.

The venous pulse within the common jugular vein is a normal phenomenon. A pulsating movement synchronous with the movements of the heart is frequently observed in the course of this vein. (Compare Fig. 34.) The movement may extend only to the lower portion of the vein, the so-called bulb, or higher up to the trunk of the vein itself. When the valves of the common jugular vein above the bulb are incompetent, a condition that is not at all rare, even in healthy persons, the phenomenon is particularly marked. The undulating movement advances from below upward; as a rule, it is observed only when the subject lies quietly in the horizontal position; it is more common on the right than on the left side, because the course of the right vein is straight and the vessel is nearer the heart than the left vein. The movement is propagated more slowly than the arterial pulse-wave.

The venous pulse possesses the peculiarities of the movement of the heart. The tracing exhibits in a marked degree all of the details of the apex-beat curve, especially in connection with the pathological conditions to be discussed presently, and it therefore closely resembles such a curve, as is shown beyond a doubt by a comparison of the venous pulse-curve (Fig. 72, 1) with the apex-beat curve (Fig. 28, A).

If it be considered that the distended jugular vein, in which the blood is subject only to slight pressure, communicates directly with the auricle, it will be readily understood that a contraction of the auricle will be propagated peripherally into the jugular vein as a positive wave. In Fig. 72, 9 and 10 represent the venous pulse from healthy individuals: the section a b corresponds to the auricular contraction. Landois has occasionally seen this composed of two slight elevations, corresponding to the contraction of the auricular appendage and the auricle. As the blood of the right auricle is subsequently thrown into agitation by the sudden tension of the tricuspid valve, the closure of the latter, which is synchronous with the systole of the right ventricle, sends a positive wave into the jugular vein, and this appears in 9 and 10 as the section b c. Finally, the sudden closure of the pulmonary valves may even be propagated through the blood in the ventricle as far as the auricle and still further up in the jugular vein, and be registered by the production of a small positive wave (e). As the aorta is in immediate contact with the pulmonary artery, a delicate wave may, on sudden closure of the aortic valves (in 9 at d), be generated at this point in a similar manner. During the

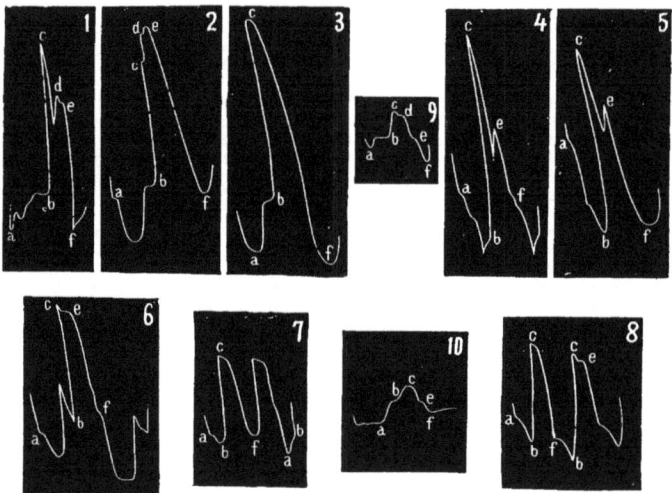

Fig. 72.—Various Forms of Venous Pulse, Chiefly after Friedreich: 1–8, with tricuspid insufficiency; 9 and 10, venous pulse from the jugular vein of a healthy individual. In all of the curves a b indicate contraction of the right auricle; b c, that of the right ventricle; d, closure of the aortic valves; e, closure of the pulmonary valves; e f, diastole of the right auricle.

diastole of the auricle and of the ventricle blood flows freely toward the heart, and in consequence the vein collapses and the writing-lever makes a down-stroke.

According to Knoll the normal jugular pulse is due partly to the positive wave caused by the contraction of the right auricle and partly to the negative wave caused by the dilatation of the ventricle; while the increase in the venous pressure that takes place between these two phases is brought about by interference with the flow of venous blood to the heart during the auricular pause.

In the sinuses of the skull the blood likewise exhibits pulsatory movement, because blood flows freely into the heart during diastolic relaxation. Under favorable conditions this pulsatory movement may be propagated as far as the veins of the retina and thus give rise to the *retinal venous pulse*, which was familiar to the earlier investigators.

Pathological.—The venous pulse may be much larger and much more pronounced in all its characteristic parts in cases of tricuspid insufficiency. A moment's reflection will show that under such circumstances every contraction of the right ventricle must cause regurgitation of a certain quantity of blood into the veins, by which a marked wave may be produced. As a rule, the common jugular vein pulsates quite strongly in cases of tricuspid insufficiency; but when the valves at the bulb of the jugular vein are still competent, the pulse is not propagated into the vein itself. The jugular pulse is, therefore, not a necessary sign of tricuspid insufficiency, but only a sign of insufficiency of the valves of the jugular vein. The ventricular systole, however, is always propagated into the inferior vena cava, which is without valves, and there it produces especially the so-called *liver-pulse*. Each ventricular contraction throws a large quantity of blood as far as the hepatic veins and thus the liver undergoes systolic swelling and distention due to injection.

The figures from 2 to 8 represent tracings from the common jugular vein. In all the curves, a b indicates the auricular contraction; the contracting auricle throws a positive wave into the veins. This portion of the curve appears at times as a simple anacrotic basal elevation (3). Not infrequently (as particularly in 1, representing a curve from one of the thyroid veins) two or three small notches make their appearance at this point, and these may be compared with the analogous elevations in the cardiogram.

In accordance with the tension of the vein, as well as with the freedom of the flow of blood from the vein to the heart, and also with the respiratory position of the thorax, the auricular notch may appear in the descending portion of the foregoing curve, as in 5 and 8; at times alternately as in 3 and 8 (see 7); at other times, a portion of the auricular wave may be in the descending portion of the foregoing curve, while the remainder is found in the ascending portion of the same curve, as in 6, 2 and 4. When the action of the auricle is exceedingly feeble, the auricular wave may even be entirely abortive as in 7 at f.

The ventricular elevation is caused by the large blood-wave thrown back into the vein by the evacuation of the ventricle. The apex of this wave (c) is at times higher, at other times lower, in accordance with the tension in the vein and the pressure of the sphygmograph. It is usually followed by at least one notch (4, 5, 6 e), produced by the sudden closure of the semilunar valves of the pulmonary artery. It is not surprising that the closure of these valves produces an undulatory movement in the ventricle that is propagated through the constantly open tricuspid valve into the auricle and the veins. The adjacent aorta may even produce a small wave next to e by the closure of its valves (as in 1 and 2 d). When the valve-closure becomes feebler in consequence of diminished tension in the large arteries, the aortic-valve wave d is the first to disappear (as in 4 and 5); later also the elevation due to closure of the pulmonary valves e disappears (as in 3 and 7). After the closure of the valves the curve falls, in correspondence with the diastole of the heart, as far as f.

An especially distinct venous pulse may be produced also when the right auricle is greatly overdistended, as in cases of mitral insufficiency or stenosis. In rare instances other veins pulsate in addition to the common jugular, such as the external jugular, some of the facial veins, the anterior jugular vein, the thyroid, the external thoracic, and the veins of the upper and lower extremities. Landois on one occasion saw extensive venous pulsation in a moribund woman without any cardiac lesion, in whom the autopsy revealed an enormous, white, fibrinous clot extending from the right ventricle into the auricle and making closure of the tricuspid valves impossible; even the cutaneous veins on the anterior surface of the thorax could be seen pulsating strongly.

It is evident that pulsations similar to those produced in the veins of the greater circulation in cases of tricuspid insufficiency must also be produced in the pulmonary veins in cases of mitral insufficiency. Such pulsations are, however, not directly visible; although it may be possible to demonstrate their presence by observing the cardiopulmonary movement.

In rare cases the veins on the backs of the hands and the feet are seen to pulsate, because the arterial pulse is propagated to the veins through the capillaries, or possibly through some direct communication between the arterial branches and the veins. This phenomenon may occur even under normal conditions, especially when the peripheral extremities of the arteries are dilated and relaxed, or when the pressure within them becomes high and falls rapidly again, as in cases of aortic insufficiency.

Diastolic collapse of the veins of the neck is observed in association with heart-disease at the instant when the tricuspid valve opens. It is due to deficient contraction of the right auricle. In cases in which the interior of an artery communicates directly with the interior of a vein as a result of traumatism or rupture, the arterial pulse is propagated into the venous channels.

THE DISTRIBUTION OF THE BLOOD.

The *methods* employed for determining the quantity of blood contained in individual organs and members must unfortunately as yet be regarded as inadequate. (1) The quantity of blood contained in the part may be determined after death in frozen cadavers. This method is inaccurate, because after death, particularly through the stimulation of the vasomotor center, the quantity of blood contained in any given part undergoes profound changes in consequence of the fact that different parts of the body die and freeze at different times. (2) A part may be forcibly ligated off from an animal during life, then be at once severed, and the quantity of blood in the tissues be determined while they are still warm. This method is, unfortunately, inapplicable to many internal organs.

J. Ranke determined in this way the distribution of the blood in the living rabbit at rest. He found one-fourth of the entire quantity of blood in (a) the resting muscles, (b) the liver, (c) the circulatory organs (heart and large arterial trunks), (d) the remaining organs taken together; of the last the lungs contained between 7 and 9 per cent.

The amount of blood is influenced by: (1) The anatomical distribution of the vessels in general, that is, the number of vessels in individual parts of the body; (2) especially the size of the vessels, which is dependent upon physiological causes: (a) the blood-pressure within them; (b) the state of irritability of the vasoconstrictor or vasodilator nerves; (c) the condition of the tissues in which the vessels are situated, for example, the intestinal vessels during the absorption of alimentary juices; the muscular vessels during the contraction of the muscles (vessels in inflamed parts).

The most important factor influencing the quantity of blood in an organ is the activity of the latter. In this connection the ancient dictum "ubi irritatio, ibi affluxus" is applicable. Examples are afforded by the salivary glands, the stomach, and the muscles during activity. As, however, under normal conditions of the body, the individual organs in many ways relieve one another, one organ may in the course of a day be found in a condition of greater plethora at one time and another organ at another time. The variations in the distribution of the blood coincide with the alternations in the functional activity of the organs. Thus, while one organ is in a state of increased activity, the remainder often are resting: the process of digestion is attended with muscular lassitude and mental relaxation; severe muscular exertion delays digestion; when the skin is reddened and secreting freely, the action of the kidneys is temporarily in abeyance. Some organs (the heart, the respiratory organs, and certain nerve-centers) appear to maintain a constant level of activity and contain the same quantity of blood at all times.

While an organ is active, the amount of blood present may increase up to 30 per cent. or even to 47 per cent. The organs of locomotion in young and vigorous individuals are likewise relatively more plethoric than those of older individuals with a feebler muscular system.

During mental activity the carotid is dilated, and the dicrotic elevation of the carotid curve is increased, while the radial exhibits reverse conditions, and the pulse is accelerated.

In this condition of greater activity the increased amount of blood usually undergoes more rapid renewal at the same time; for example, after muscular exertion the duration of the circulation is diminished. This circumstance may be affected by a great variety of influences that govern the movement of the blood.

The development of the heart and the large blood-vessels is responsible for certain differences in the distribution of the blood in children and in adults. From childhood to puberty the· heart is relatively small and the vessels are relatively large. After puberty, on the contrary, the heart is large and the arteries are comparatively small. Accordingly, the arterial blood-pressure in the greater circulation must be lower in a child than in an adult. The pulmonary artery is relatively large in childhood, the aorta relatively small; after the onset of puberty both arteries are approximately of the same size. Hence, it follows that the blood-pressure in the pulmonary vessels of the child must be relatively higher than in the adult.

PLETHYSMOGRAPHY.

The plethysmograph is an instrument employed to determine and register the amount of blood in an extremity and its variations. It is a perfected apparatus, modeled after the "box-sphygmometer" described by Chelius in 1850 (Fig. 41). It consists of a long container (G), designed for the reception of an entire extremity. The opening around the introduced part is made air-tight by means of rubber, and the interior of the vessel is filled with water. In the lateral wall of the receptacle is a communicating tube, which also is filled with water to a certain level. As each pulse-beat causes an enlargement of the extremity as a result of the increased flow of arterial blood, the water in the tube will indicate the magnitude of this positive variation in the quantity of blood, which will be transmitted to the drum (T), covered with an elastic membrane, and with which is connected a writing lever moving in a horizontal direction.

The cylinder G may also be filled with air. v. Kries connects the tube with a gas-burner instead of with the registering drum (T), so that the variations in the size of the arm are reproduced in the flame, the flickerings of which may be photographed.

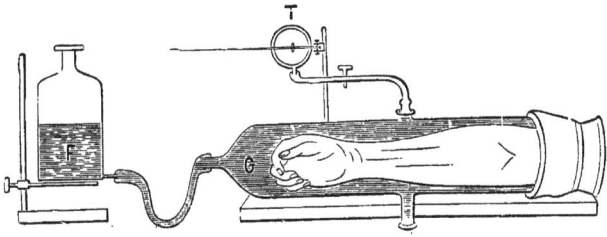

FIG. 73.—Mosso's Plethysmograph: F, communicating flask, by elevation of the level of which the hydrostatic pressure may be increased; T, the inscribing apparatus.

Individual organs (spleen, kidney) may be enclosed in a box-like apparatus in a similar manner for the purpose of observing fluctuations in their size: oncograph.

The fluctuations of the plethysmograph permit recognition of the following phenomena:

1. Pulsatory fluctuations in volume.—As the venous current in ·the resting extremity may be regarded as uniform, any rise in the volume-curve must indicate a greater velocity in the movement of the arterial blood-current toward the periphery, and the reverse. The curves registered by this apparatus represent volume-pulsations and resemble a dromographic curve (Fig. 69, III). A rise in the limb of the curve indicates a greater flow of arterial blood, while a fall indicates a diminution in the flow. If the level of the curve remains the same, it is to be inferred that the arterial inflow of blood is equal to the venous outflow.

At first sight the plethysmographic ·tracing (volume-curve, current-pulse) presents a great similarity to the sphygmographic tracing (pressure-pulse), especially from the fact that both exhibit the dicrotic elevation. More careful examination, however, reveals several differences: In the plethysmographic tracing (current-pulse) the curve descends to a much lower level after the primary apex.

This marked fall, which is not accompanied by a corresponding fall in the pressure, is attributed by v. Kries to a peripheral reflection, that is, one in which a positive wave is reflected as such. The dicrotic elevation (secondary wave) appears, further, somewhat earlier in the plethysmographic curve (current-pulse) than in the sphygmographic curve; although it also has a centrifugal course, as in the sphygmographic curve.

2. The respiratory fluctuations, which correspond to the respiratory fluctuations in blood-pressure. Active breathing and cessation of breathing produce a diminution in volume. Further, the part has been observed to undergo enlargement in consequence of effects at bearing down and coughing, and reduction in size during sobbing. 3. Certain periodic fluctuations, dependent upon periodic-regulatory movements of the vessels, particularly of the smaller arteries. 4. Various fluctuations due to accidental causes that bring about alterations in the blood-pressure, such as change of position producing hydrostatic effects; dilatation or contraction of other large vascular areas. 5. Muscular movements in the extremity introduced into the plethysmograph cause a reduction in volume, because the venous pulse is accelerated, and in addition the musculature itself is somewhat reduced in size, in spite of the fact that the intramuscular vessels are dilated. 6. High (from $33°$ to $36°$ C.) and low (from $4°$ to $8°$ C.) temperature, when applied to the skin of the arm, increase the volume of the member in consequence of paresis of the muscular coat of the blood-vessels caused by the thermic stimuli. 7. Mental exertion diminishes the volume of the extremity; sleep has the same effect. 8. Compression of the afferent artery causes diminution, while constriction of the veins naturally causes an increase in the volume. 9. Irritation of the vasomotor nerves is followed by a decrease, that of the vasodilators by an increase, in volume.

TRANSFUSION OF BLOOD.

Transfusion is the physiological introduction of blood into the vascular system of a living being.

The first mention of direct exchange of blood between two individuals from vessel to vessel takes us back to the time of Cardanus. After the discovery of the circulation of the blood, Potter in England again called attention to the practicability of transfusion. Numerous experiments were made on animals. Attempts were made by the introduction of fresh blood particularly to resuscitate animals that had bled to death. The physicist, Boyle, as well as the anatomist, Lower, took an especially active part in these experiments. The blood of the same or of another species was used. The first transfusion in man was practised by Jean Denis in Paris in 1667 with lamb's blood.

(a) The erythrocytes are the most important constituents to which the resuscitating power of the blood is due. They retain their functions even after the blood has been defibrinated. The changes in the red blood-cells produced by time and by prolonged exposure to high temperatures have been described on p. 36.

(b) With respect to the gases contained in the blood, it is to be remembered that oxygenated blood under no circumstance is injurious. Venous blood can, however, be infused into the blood-vessels of a living being without injury, provided the respiration is sufficient to arterialize the infused blood in its passage through the pulmonary capillaries. Under such circumstances the carbon dioxid contained in the blood is replaced by oxygen in the process of respiration. If the respiration, however, is arrested or if it is not carried on with sufficient activity, the blood, still rich in carbon dioxid, will be conveyed to the left heart and on through the arteries of the medulla oblongata. In consequence there results violent irritation of the centers in that region, followed later by paralysis and even by death.

(c) The fibrin or the substances forming it take no part in the resuscitating activity of the blood. Therefore, defibrinated blood is capable within the body of assuming with equal success all of the functions that belong to non-defibrinated blood,

(d) Investigations, especially by Worm-Müller, have shown that the vascular system (dog) is capable of taking up an excess of foreign blood up to 83 per cent., without injurious consequences. It follows that the vascular system possesses to a certain degree the power of accommodating itself to large quantities of blood, just as it is known to possess the power of adapting itself to a diminished volume of blood, as, for example, after hemorrhage.

Transfusion is practised: 1. In cases of acute anemia, especially after a hemorrhage when it is sufficiently great to threaten the life of the patient. The object under such circumstances is to replace directly with new blood (from 150 to 500 cu. cm.) that which has been lost and is necessary to maintain life.

2. In cases of poisoning in which the blood has been vitiated by the admixture of a toxic substance and has thus become unfit to maintain the vital functions, a large quantity of this vitiated blood may be removed by copious venesection under suitable conditions and normal blood be introduced into the vessels in place of the blood withdrawn (*depletory transfusion*). The chief form of intoxication amenable to this treatment is that with carbon monoxid. Also the admixture of other poisons with the blood, especially those that dissolve the erythrocytes or that cause marked methemoglobinemia, as, for example, potassium chlorate, as well as other toxic substances (ether, chloroform, chloral hydrate, opium, morphin, strychnin, snake-venom), may likewise furnish an indication to replace the poisoned mass of blood with normal blood.

3. Under certain morbid conditions, abnormal states of the blood may develop in the body and threaten its integrity; these may affect both the morphological elements, and the composition of the blood. The morbid alterations in the constitution of the blood include poisoning with urinary constituents (uremia), with biliary constituents (cholemia) and with carbon dioxid. If severe they may cause death. Therefore, in desperate cases of this kind, especially when the cause is a temporary one, the vitiated blood may be in part replaced by normal blood. Whether hydremia, oligocythemia and pernicious anemia are indications for transfusion will depend on the correct interpretation of the underlying disease.

Between a quarter-hour and a half-hour after transfusion, in accordance with the amount of blood introduced, a more or less violent febrile reaction takes place.

The *operative procedure* varies accordingly as defibrinated or non-defibrinated blood is employed. When a defibrination is to be practised, the blood obtained by venesection from a healthy human being is collected in a vessel and beaten with a small rod until the fibrin has been completely removed. The blood is then filtered through an atlas-filter, without pressure, is heated to the temperature of the body by placing the vessel in warm water, and it is conveyed into the opened vessel with the aid of the buret-infuser of Landois or a syringe. The vessel selected may be a vein, as, for example, the basilic at the bend of the elbow, or the long saphenous vein at the internal malleolus. Under such circumstances the blood is injected in the direction toward the heart. The blood may be injected also into an artery (the radial or the posterior tibial), either in the centrifugal or in the centripetal direction. In any event, care must be exercised, especially when the blood is injected into the veins, to guard against the entrance of air, as such an accident might even cause death. Death occurs when the air that has entered the right heart is churned up into froth by the movements of the heart and in this form is pumped into the smaller branches of the lesser circulation, thus arresting the flow of blood through the lungs. After the injection of air into the arterial system a few small bubbles of air may possibly pass through the capillaries of the greater circulation and thus be found everywhere in the vessels. They disappear at once, however, because the oxygen enters into chemical combination and the nitrogen is absorbed.

If defibrinated blood is not to be infused the divided vein of the donor is connected by means of a tube with the vessel of the recipient, so that direct transfusion takes place. The blood may also be taken up with an oiled syringe, to which the blood does not adhere, and transfused at once without defibrination. The latter procedure, however, is attended with the great danger that coagulation may take place during the operation, in consequence of which blood-clots may readily be introduced into the circulation of the recipient. The resulting obstruction and even more so the possible conveyance of coagula to the heart and into the lesser circulation, may even threaten life.

Landois has transfused without injury into animals the non-coagulable blood that has been sucked by leeches after removal from them by stripping. From the cephalic extremity of the leech hardened in alcohol, dried and pulvèrized, a decoction can be prepared by admixture with 0.9 per cent. saline solution (one head is boiled for ten minutes with 6 cu. cm. of a saline solution, and then filtration is practised). This decoction, when mixed in the proportion of 6 cu. cm. to 15 cu. cm. of blood obtained by venesection, suffices to maintain the latter in a fluid state. The mixture will not coagulate for some time and can be used without fear of injury. By this means the dreaded effect of the fibrin-ferment may be avoided.

In Man the Injection of Animal Blood is Unjustifiable under Any Circumstances. —Direct transfusion of blood from the carotid of a lamb into the brachial vein of a man was formerly employed not infrequently for therapeutic purposes. It is to be remembered, however, that the erythrocytes of the sheep are rapidly dissolved in human blood, and in consequence the most efficient constituents of the transfused blood are destroyed. In a general way, it is found that the blood-serum of many mammals has a rapid hemolytic effect upon the blood-cells of other species of mammals. Thus, the serum of dog's blood has a rapid and intense hemolytic action, while that of the horse and of the rabbit is relatively slow in action. The erythrocytes of mammals possess a variable power of resistance to the sera of other species of mammals. Thus, the erythrocytes of the rabbit, when mixed with the blood of another species, are readily dissolved; while the cells of the cat and the dog exhibit much greater resistance. The rapidity with which erythrocytes are destroyed in the blood of another species is proportional to the rapidity with which the blood-cells of the blood of the other species are dissolved in the blood-serum of the recipient. Thus, for instance, rabbit's blood and lamb's blood disintegrate within a few minutes in the circulation of a dog. When there is a difference in the size of the blood-corpuscles of the two species, the hemolysis can readily be observed in small specimens of blood obtained by puncture. As the erythrocytes dissolve, the blood-plasma is stained red by the liberated hemoglobin. A portion of this liberated material may supply the demands of metabolism in the body of the recipient and be utilized for katabolism and anabolism, while part of it is used up in the formation of bile. When, however, the quantity of hemoglobin liberated by the erythrocytes is considerable, hemoglobin is excreted in the urine, and to a less extent in the intestine, in the ramifications of the bronchial tree and into the serous cavities. In the last the hemoglobin may subsequently undergo absorption. Thus, in man hemoglobinuria has been observed after the injection of more than 100 grams of lamb's blood.

When blood from another species is transfused into an animal, the blood-corpuscles of the latter may undergo partial disintegration. This is the case when the erythrocytes of the recipient are readily soluble in the serum of the transfused blood. Upon this fact depends the great danger of transfusing a considerable quantity of heterogeneous blood into the rabbit, whose erythrocytes so readily undergo solution. The same thing would happen if a dog's blood were transfused into the veins of a man. In animals whose erythrocytes readily undergo solution, as, for example, the rabbit, the injection of many kinds of sera, as, for example, that of the dog, of man, of the pig, of sheep, and of the cat, is followed by alarming symptoms, in accordance with the quantity of blood introduced, namely: acceleration of respiratory frequency to the point of dyspnea, convulsions, and even death from asphyxia. Under such circumstances all the stages of hemolysis can be seen in a specimen of blood obtained by puncture. Animals possessing more resistant erythrocytes, such as dogs, tolerate the injection of heterogeneous sera, as, for example, from sheep, neat cattle, horses and pigs, without exhibiting such marked symptoms. The injected foreign serum, being of feeble potency, is disposed of in the circulation of the recipient, before it has time to attack, not to say dissolve, the blood-cells to any great extent.

The process of hemolysis is accompanied by two other phenomena, which render the transfusion of heterogeneous blood especially dangerous: 1. Before the erythrocytes are dissolved, they usually adhere together tenaciously and form small masses, consisting of from 10 to 20 or more blood-cells, which are obviously capable of obstructing large capillary areas. When these masses have been present in the blood for some time they yield up their hemoglobin, leaving only the fused remains of stroma. This forms a viscid, tenacious, s ng mass (stroma-fibrin), which likewise may occlude the smaller vessels. 2. the sudden appearance of large quantities of dissolved hemoglobin in the blood of an animal may cause extensive coagulation, principally in the venous system, but also in the larger vessels throughout a considerable extent. The processes described may produce death either suddenly or after a protracted course. Dissolved hemoglobin causes in the circulation the dissolution of numerous leukocytes, from whose disintegration the fibrin-factors result.   It is curious that hemoglobin exposed to the air gradually loses this property; also fibrin-ferment in contact with hemoglobin is gradually destroyed or rendered inactive.

As numerous small vessels are occluded as a result of the processes described, the signs of impeded circulation and of stasis will be encountered in the different organs of the body.   In man, the injection of lamb's blood is followed by a bluish-red discoloration of the skin.   The obstacles encountered by the blood-current in the lungs cause dyspnea or even laceration of the small vessels in the air-passages and bloody expectoration.   The dyspnea may increase if interference with the free circulation of the blood develops at the respiratory center.   The digestive organs, for the same reason, exhibit increased intestinal peristalsis, diarrhea, evacuation of the bowels, tenesmus, vomiting and abdominal pain.   These phenomena are explained by the fact that any disturbance of the circulation in the abdominal vessels is followed by increased peristaltic movements.   In the kidneys secondary degeneration of the glandular substance takes place in consequence of occlusion of the vessels.   The uriniferous tubules are occluded by casts consisting of coagulated albuminous material.   In the muscles the occlusion of numerous vessels may cause stiffness, or even rigidity from coagulation of myosin, just as in Stenson's experiment, together with increased heat-production.   Also the nervous system, the organs of special sense and the heart may exhibit various disturbances, all of which can be attributed to the occlusion of vessels and the resulting interference with the circulation.   It is interesting to note that the transfusion of foreign blood is followed as a rule within half an hour by the development of active fever.   Finally, it should be mentioned that lacerations of the vessel-walls have also been observed. These explain the obstinate hemorrhages that may occur not only on the free surfaces of mucous and serous membranes, but also in the parenchyma of organs, as well as in surgical wounds.   The blood itself coagulates slowly and imperfectly. By far most of the facts bearing on the transfusion of heterogeneous blood that have been mentioned were discovered through Landois' investigations.

Attempts to inject other substances instead of blood are not to be commended: from 0.75 per cent. to 0.9 per cent. saline solution, while capable of improving the circulatory conditions in a purely mechanical way, and thus exerting a favorable influence, is obviously incapable of supporting life in cases of severe anemia, in which the quantity of blood remaining in the body is insufficient to maintain the vital processes.

THE DUCTLESS GLANDS. INTERNAL SECRETIONS.

Within comparatively recent times there has been attributed to the ductless glands, whose activity is still, for the most part, shrouded in obscurity, a special and important function, namely, the production of substances that enter the circulation and there in some peculiar way excite certain activities, or render innocuous certain poisonous substances generated in the process of metabolism, either by destroying these or by manufacturing an antidote.   In a similar manner it has been asserted of a number of other organs in the body that, in addition to their special function, they exert an important influence on the economy by means of such internal secretion.   Thus, Brown-Séquard and d'Arsonval asserted that the kidneys are in part concerned in rendering innocuous the toxic substances that accumulate in the body after nephrectomy; Tigerstedt and Bergman, that the kidneys produce a substance—renin—that increases the blood-pressure and has a powerful influence on the peripheral nerve-centers.   The substances under consideration can be obtained from the corresponding organs in the form of extracts and their action can then be tested upon the animal body.

The *spleen* is contained in a firm fibrous capsule, which at the hilus gives off an investment for the entering blood-vessels.   From the inner surface of the cap-

13

sule and the surface of the vascular sheaths there pass off numerous intersecting and branching trabeculæ (the trabeculæ of the spleen), which form a rich mesh-work in the interior of the viscus, comparable to the cavities of a sponge. Fibril-lated connective tissue, mixed with elastic and unstriped muscle-fibers, forms the foundation of this portion of the viscus. The interior of the meshes contains a delicate reticulum of adenoid tissue (Fig. 131), which, together with the cellular elements contained in the meshes, is designated the splenic pulp.

The smaller arterial branches, which gradually lose their fibrous sheath, ulti-mately break up into brush-shaped terminal twigs without anastomoses (peni-cils). The points of division of the small arterial branches serve for the lodgment of the whitish Malpighian vesicles, which may attain the size of a pinhead and the structure of which in every respect resembles that of solitary lymph-follicles. The Malpighian bodies are found on examination to be spherical, lymphatic masses that have partially separated from the vascular sheath. In some animals, instead of exhibiting a spherical form, they appear as loose arterial sheaths, in a measure as perivascular lymphatic sheaths, so to speak, which may extend to the smallest arterial twigs. According to Tomsa, lymphatic vessels coming from the Malpigh-ian vesicles are found in the subsequent course of the arterial sheath as far as the hilus of the spleen. Other lymphatics form a network in the capsule.

With regard to the connection between the ends of the arteries and the veins, it is supposed that there is no continuous channel between the smallest capillary ar-terial twigs and the smallest venous branches and that the meshwork of the pulp-reticulum represents an intermediate vascular area devoid of walls. The blood, accordingly, passes through the meshwork of the spleen traversed by the reticu-lum, just as the lymph-stream passes through the spaces of the lymphatic glands. According to another view, there is really a closed vascular channel connecting the ultimate arterial and the corresponding venous capillaries, which, however, con-sists of dilated spaces (like the cavernous spaces in erectile tissues). These inter-mediary spaces are, however, completely surrounded by spindle-shaped endothe-lium.

Within the meshes of the reticulum are found cellular elements of various kinds: (1) White blood-corpuscles of various sizes, some swollen and filled with a granular material; (2) leukoblasts or embryonal forms of leukocytes, which multi-ply by division; (3) erythrocytes; (4) embryonal forms of the latter, also desig-nated erythroblasts, which multiply by mitosis; (5) so-called blood-corpuscle-con-taining cells.

The numerous nerves of the spleen consist of so-called Remak's fibers; they are sensory, motor, and vasomotor.

Of the chemical constituents there should be mentioned globulin and nucleo-albumin, nucleinic acid, leucin, tyrosin, xanthin, hypoxanthin, taurin; further lactic, butyric, acetic, formic, succinic, uric, and glycero-phosphoric (?) acids; as well as fats, cholesterin, a gluten-like body, glycogen, inosite, iron-containing pigments, and even free iron oxid. The pulp becomes black on addition of ammo-nium sulphid. The ash is rich in phosphoric acid and iron, but poor in chlorin-combinations.

With respect to the *function of the spleen*, the following points are note-worthy:

1. The spleen may be removed without injury to the individual, as has been proved both in animals and in man (more than 90 cases, with about 40 recoveries). After removal of the spleen the hematopoietic activity of the bone-marrow appears to be increased. In frogs, extirpation of the spleen has been observed to be fol-lowed by the appearance of brownish-red nodules in the intestine, which have been regarded as vicarious spleens. Tizzoni speaks of splenic neoplasms in the omentum (horse, dog) after obliteration of the parenchyma and blood-vessels of the spleen. In extremely rare cases total absence of the spleen has been observed in man.

2. By virtue of its unstriped muscle-fibers the spleen is capable of undergoing change in volume. Irritation of the spleen or of its nerves (by heat or electricity, by quinin, eucalyptus, ergot, and other agents) causes diminution in the size of the viscus, with anemia and granular change. As the spleen is found to be en-larged a few hours after digestion, at a time when the digestive organs have per-formed their work and contain less blood, the spleen has been regarded as an apparatus for the regulation of the vascularity of the digestive organs.

According to Roy the circulation in the spleen is dependent not alone upon the blood-pressure in the splenic artery, but in marked degree on the contraction of the unstriped muscle-fibers of the capsule and the trabeculæ, and which manifests itself in rhythmical movements lasting one minute.

Paralysis of the splenic nerves, as in connection with certain febrile intoxications (malarial fever, typhoid fever), causes enlargement of the organ. Division of the nerves has the same effect. After extirpation of the small nerve-trunks scattered in the hilus Landois has observed circumscribed enlargement of the organ, with bluish-red discoloration.

3. The spleen has been regarded as a hematopoietic organ. In favor of this view is the fact that after extirpation the erythrocytes are diminished; further, the fact that a splenic infusion (or decoction, also an infusion of bone-marrow), when injected under the skin or into the peritoneal cavity, causes an increase of the erythrocytes. The spleen is also a breeding-place for leukocytes. The blood from the splenic vein always contains numerous leukocytes, many of which are subsequently destroyed in the circulation. Bizzozero and Salvioli discovered that a few days after great loss of blood the spleen became swollen, and the parenchyma was found to be rich in nucleated embryonal erythrocytes.

4. Other investigators regard the spleen as an organ for the destruction of blood-corpuscles, the presence of so-called "blood-corpuscle-containing cells" particularly supporting such a view. These cells are large leukocytes that have taken up red blood-corpuscles after the manner of phagocytes (similar cells are found also in extravasations of blood). The red blood-cells gradually undergo degeneration within the leukocytes and yield as derivatives of hemoglobin iron-containing pigments resembling hematin. The spleen, therefore, contains more iron than can be accounted for by the amount of unaltered blood it contains. If with this fact there be yet compared the occurrence in the spleen of disintegration-products and of higher oxidation-products of the albuminous bodies, the spleen may properly be regarded as an organ for the destruction of erythrocytes. Additional support for this view is found in the appearance of the salts of the red blood-corpuscles in the splenic juice. According to Schiff, extirpation of the spleen has no effect on either the absolute or the relative quantity of the red and white blood-corpuscles.

Even in the normal state the spleen exhibits frequent changes in size in the course of the day, particularly in conformity with varying activity of the digestive organs. In this respect the spleen resembles the arteries. Its vasomotor nerves have their center in the medulla oblongata. Stimulation of that center, especially by asphyxia, causes contraction of the spleen. From the center fibers pass through the spinal cord (which is said to contain between the first and fourth cervical vertebræ ganglionic cells that likewise influence the contraction of the spleen), further through the left splenic nerve and the semilunar ganglion into the splenic plexus. Irritation of the nerves, as well as the direct application of cold to the spleen or even to the splenic region, causes contraction of the viscus. Paralysis of the nerves, by curare or by protracted narcosis, causes enlargement of the spleen. Apparently only the peritoneal investment contains sensory nerves.

Pressure on the splenic vein causes slight enlargement of the spleen. In harmony with this fact is the observation that increased blood-pressure within the splenic vein (in the presence of portal congestion or after the cessation of hemorrhoidal or menstrual bleeding) is frequently attended with splenic enlargement. The injection of splenic extract has an effect opposite to that of injection of suprarenal extract.

The *thymus gland* is relatively well developed during fetal life and continues to grow during the first two years of life; but about the tenth year it becomes stationary in size and later degenerates to form the so-called *thymic fat-body*, the tissues of which still contain the remains of the lymphoid thymus-parenchyma. As long as it persists, the thymus appears to have the function of a lymph-gland; for in the embryo, which possesses no lymph-glands, it is functionally active, and in reptiles and amphibia, which also possess no lymph-glands, it is a permanently functioning organ.

The thymus consists of acini varying in size from 0.5 to 1.5 mm. and possessing the structure of simple lymph-follicles. The lymph-cells lying within the reticulum may exhibit various stages of disintegration. In addition, there are found scattered through the organ peculiar and mysterious *concentric bodies*, especially during the time of involution. Numerous small lymph-vessels in part traverse the interior of the organ and in part spread out upon its surface. Blood-vessels are relatively numerous.

Among the chemical constituents there should be mentioned—in addition to gelatin, albumin, sodium albuminate, sugar and fat—leucin, thymus-nucleinic acid, xanthin, hypoxanthin; formic, acetic, butyric, lactic, and succinic acids.   In the ash, potassium and phosphoric acid preponderate over sodium, calcium, magnesium (ammonium ?), chlorin, and sulphuric acid.

Extirpation of the thymus gland in the frog is fatal.   According to Svehla the infusion of thymus juice causes a fall of blood-pressure and acceleration of pulse, while large doses are fatal.

The *thyroid gland* is an organ provided with vasomotor and secretomotor nerves, and composed of a richly cellular connective-tissue framework, containing closed circular or oval acini (from 0.04 to 0.1 mm. in diameter), which in the embryo and the new-born are lined with a single layer of nucleated, granular, cuboidal cells.   In 50 per cent. of all subjects accessory thyroid glands, up to four, are associated with the main gland; a small detached gland is occasionally found in front of the descending aorta.   In addition, accumulations of epithelial cells are found in the acini and, in embryos, also beneath the common capsule. From birth the cells secrete a colloid substance by a transformation of their protoplasm, at the same time undergoing morphological changes.   Some of the cells are destroyed in this process of colloid degeneration.

The acini of the thyroid gland evacuate their contents in part by rupture, with destruction of the epithelium, in part, in the process of pure colloid-production, by secretion into the intercellular interstices; and in this way the secretion reaches the interfollicular lymph-spaces and then the blood.

Blood-vessels of considerable size and importance enter the organ.   Lymphvessels partly begin in the interior among the acini, and partly form a network in the capsule that surrounds the entire organ.

The constituents of the thyroid gland are colloid, nucleoalbumin, iodothyrin, leucin, xanthin; lactic, succinic, and volatile fatty acids.

According to Schiff, Zesas, J. Wagner and others, extirpation of the thyroid gland is followed by death, with the symptoms of chronic intoxication.   Dysphagia, vomiting and digestive disturbances, acceleration of the breathing; later dyspnea, alteration of the action of the heart, somnolence. slow and hesitating movements with fibrillar twitchings, which may go on to intermittent tonic convulsions (tetany), palsies, alterations in cutaneous sensibility, desquamation of the skin, lowering of the body-temperature and of the blood-pressure, are the symptoms that precede death.   Albuminuria, reduction of the amount of oxygen in the arterial blood and degenerations in the central and peripheral nervous system were observed by Albertoni and Tizzoni, Langhans, Kopp and Capobianco.   In man, also, total extirpation of the thyroid gland (cachexia strumipriva) is a serious matter and often terminates fatally from tetany.

The morbid phenomena may be counteracted, at least temporarily, by the internal administration of fresh or dry thyroid-gland substance, or by the subcutaneous injection of thyroid-gland extract or iodothyrin.   The symptoms may be prevented by grafting a thyroid gland successfully in some other portion of the body, and permitting the organ to form adhesions.   These facts prove that the thyroid gland produces a substance that is indispensable for normal metabolism.   Stated more accurately, the function of the thyroid gland is to neutralize a substance produced in the body, the accumulation of which has a toxic influence · on the nervous system.

The accessory thyroid glands and the hypophysis appear to possess similar functions: they undergo compensatory hypertrophy after extirpation of the thyroid gland.   Other investigators attribute the condition known as myxedema, that is, mucoid infiltration of the subcutaneous tissues of the head and neck, with profound disturbances of the nervous system, to the point of idiocy, to loss of the function of the thyroid.

Especially noteworthy is the enlargement of the thyroid gland, together with the palpitation of the heart and protrusion of the eyeballs, in the condition known as exophthalmic goiter, which appears to be due to simultaneous (toxic?) irritation of the accelerator nerve of the heart, the sympathetic fibers of the unstriated muscles in the orbit and in the eyelids, as well as of the dilator nerves of the vessels of the thyroid gland.   Myxedema and exophthalmic goiter seem to stand in a certain antagonistic relation to each other, the former depending on diminished, the latter on augmented, activity of the thyroid gland (hence extirpation has been recommended in cases of exophthalmic goiter).   Landois observed in dogs that had been fed on thyroid glands a marked increase in the number and force of the cardiac contractions.   The ingestion of thyroid gland causes an increased consumption of oxygen and therefore a more rapid breaking down of the tissues (for which reason it is a familiar therapeutic procedure for reducing weight). According to Schöndorff the body-fat is first transformed, the albumin not being attacked until the fat has been reduced to a certain minimum. The substance (solely?) active in this connection is iodothyrin, a body prepared in 1896 by Baumann, and containing nitrogen, phosphorus, and iodin. In some localities marked enlargement of the thyroid gland (goiter) is quite common, and is not infrequently associated with idiocy and cretinism. In those cases in which the goiter is designated a follicular hyperplasia of the thyroid gland, the condition can be made to disappear by the administration of preparations of the thyroid gland. Fr. Hofmeister found, after extirpation of the thyroid gland in rabbits, degeneration in the cartilages and disturbances in the growth of the bones.

According to Gegenbaur the thyroid gland is an actively functioning organ in some of the remote orders of animals (for example, among the tunicates, in which it appears as a groove and secretes a digestive juice), which in vertebrates has undergone involution.

The *suprarenal bodies* consist of a medullary and a cortical layer, and contain compartments formed by connective tissue and bounded by blood-vessels. In the cortical layer the compartments are oblong and radiate, while in the medullary layer they are rather circular. The former contain (embedded in a reticulum) polyhedral, nucleated, protoplasmic cells without walls, the substance of which contains pigment and fat-granules, and is darker and more resistent than that of the medullary cells. The medullary layer contains also small and multipolar, large sympathetic nerve-cells. Both cortex and medulla are richly supplied with nerve-fibers. The blood-vessels are relatively abundant.

The suprarenal bodies contain the constituents of connective and of nervous tissue, besides leucin, hypoxanthin, benzoic and taurocholic acids, taurin, inosite, fat and pigment-forming bodies. Of inorganic substances potassium and phosphoric acid preponderate.

The function of the suprarenal bodies is practically unknown. After extirpation of one suprarenal body, the other undergoes hypertrophy to double its original size. Bilateral extirpation is followed by death, with the symptoms of poisoning and paralysis. These symptoms, however, do not develop if a small piece is allowed to remain. It appears, therefore, that the suprarenal bodies also are designed to destroy a poisonous substance in the body, which exhibits its injurious effects after extirpation of the glands. The injection of a watery extract of suprarenal body is said to arrest temporarily the toxic symptoms that make their appearance after extirpation.

Injection of the extract obtained from the medullary substance of healthy animals (and which does not contain albumin and is soluble in alcohol) gives rise to marked contraction of the arteries and increase in blood-pressure, slowing of the pulse by central stimulation of the vagus, or even arrest of the auricles. After section of the vagi the heart again becomes more rapid and stronger, owing to the action of the drug on the substance of the heart itself. The extract has the same constricting effect on small blood-vessels and hence raises the blood-pressure. The splanchnic nerve contains vasodilator and secretory fibers for the organ. The breathing is superficial and accelerated. Large doses injected intravenously cause death through enfeeblement of the central nervous system, dyspnea, and cardiac paralysis. In frogs muscular paralysis results.

Brown-Séquard believed that one of the functions of the suprarenal bodies is to inhibit excessive pigment-formation. In agreement with this view, Tizzoni found, after extirpation of the organs (in rabbits), abnormal pigmentations, especially on the lips, and Boinet in the blood and subcutaneous cellular tissues (of rats). In conditions in which erythrocytes are dissolved and converted into pigment the suprarenal bodies are found to be especially rich in pigment. In the medullary layer a substance is formed that becomes brown when exposed to the air or brought in contact with alkaline tissues. In man the skin often presents a bronzed pigmentation (bronzed skin, Addison's disease) when the suprarenal bodies and their capsules have undergone (tuberculous) degeneration. In hemicephalous monsters the organs are atrophic, even when only the anterior halves of the hemispheres are absent.

Hypophysis Cerebri. Coccygeal Gland. Carotid Gland.—But little is known concerning the function of the pituitary body. The posterior portion belongs to the infundibulum, and here the nervous elements are, to a large extent, displaced by connective tissue and blood-vessels; while the anterior portion represents a constricted off and modified part of the invaginated mucous membrane of the pharynx and contains glandular ducts with clear or dark cells. The extract obtained from the pituitary body contains iodin and causes an increase in the blood-pressure, which, however, is less than that caused by an extract of suprarenal gland; the heart-beat becomes slower and more forcible.

The function of the coccygeal gland, which is situated at the extremity of the coccyx, is unknown.

The carotid gland, which occurs in man and mammals, and contains a convoluted plexus consisting of intricately anastomosing capillaries within an epithelioid cellular mass, supported by a reticulum, has been compared by Stilling to the suprarenal bodies. Its function is unknown.

COMPARATIVE.

The heart in fishes (Fig. 74, *I*) and in the gill-bearing larvæ of amphibia is a simple venous organ, consisting of auricle and ventricle. The latter sends the blood to the gills, where it is arterialized, and passing to the aorta it is dis-

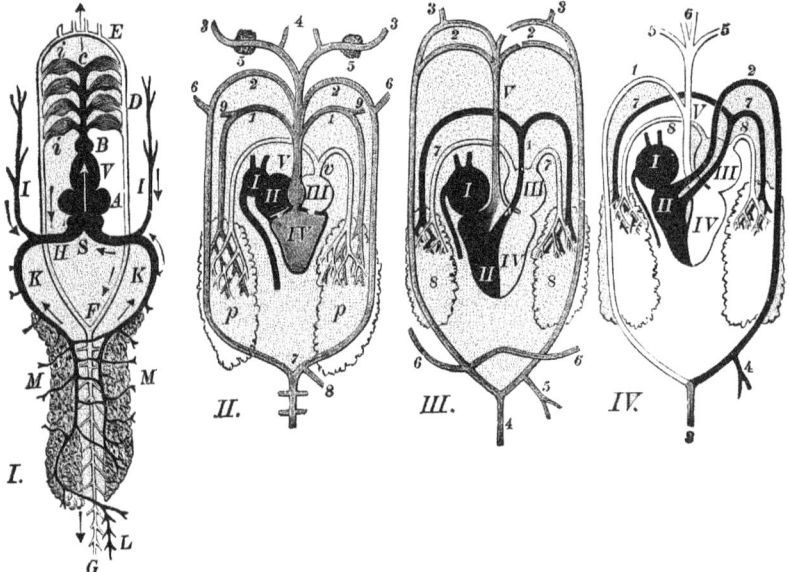

FIG. 74.—Diagrammatic Representation of the Circulation. *I*. In Fish: *A*, auricle with the sinus venosus (*S*); *V*, ventricle; *B*, bulb of the aorta; *c*, branchial arteries; *i i*, branchial vessels; *D*, branchiales veins; *E*, circulus cephalicus aortæ; *F*, common aorta; *G*, caudal artery; *H*, ductus of Cuvier; *I*, anterior cardinal vein; *K*, posterior cardinal vein; *L*, caudal vein; *M M*, kidneys. *II*. In the Frog: *I*, sinus venosus; *II*, right auricle; *III*, left auricle; *IV*, ventricle; *V*, common trunk of the aorta and bulb, giving off the following: 1, pulmonary arteries; 2, arch of the aorta; 3, carotid arteries; 4, lingual arteries (5 carotid gland); 6, axillary arteries; 7, common aorta; 8, celiac artery; 9, cutaneous arteries; *v*, pulmonary veins; *p p*, lungs. *III*. In Saurians: *I*, right auricle with venæ cavæ; *II*, right ventricle; *III*, left auricle; *IV*, left ventricle; *V*, anterior common aorta; 1, pulmonary artery; 2, arch of the aorta; 3, carotid arteries; 4, posterior common aorta; 5, celiac artery; 6, subclavian arteries; 7, pulmonary arteries; 8, lungs. *IV*. In Turtles: *I*, right auricle with venæ cavæ; *II*, right ventricle; *III*, left auricle; *IV*, left ventricle. 1, right aorta; 2, left aorta; 3, posterior common aorta; 4, celiac artery; 5, subclavian arteries; 6, carotid arteries; 7, pulmonary arteries; 8, pulmonary veins.

tributed to all parts of the body, returning finally through the capillaries and the veins to the auricle. The amphibia (frog, *II*) have two auricles and one ventricle. From the latter there arises a single vessel, which, after giving off the pulmonary arteries, becomes the aorta and supplies all the organs of the body. The veins of the greater circulation empty into the right, those of the lesser circulation into the left, auricle. Fishes and amphibia possess a dilated bulbus arteriosus at the beginning of the aorta; and this is partly covered with strong muscular tissue. Among reptiles the saurians (*III*) possess two separate auricles, but the two ventricles are only imperfectly divided. The aorta and pulmonary artery arise separately from the latter. The venous blood of the greater and the lesser circulation, which flows separately into the right and the left auricle, becomes mixed in the cavity of the ventricle. In some reptiles, however, the opening in the ventricular septum appears to be capable of (voluntary or reflex?) closure. The complete separation of the two halves of the heart in turtles is shown in Fig. *IV*. The lower vertebrates possess valves at the orifice of the vena cava, which are rudimentary in birds and in some of the mammals. All birds and mammals, like man, possess two separate auricles and two separate ventricles. In the halicore, a graminivorous marine animal resembling the whale, the ventricular portion of the heart is divided by a deep cleft into two halves. In bats the veins of the wings pulsate. The lowest of all vertebrates, the amphioxus, has no heart at all, but rhythmically contracting vessels.

Of the ductless glands, the thymus and the spleen are found constantly in vertebrates. The latter is wanting only in the amphioxus and in a few fishes.

Among invertebrates closed blood-channels with pulsating movements are only found occasionally, as, for example, in the echinoderms (sea-urchin, star-fish, holothurians) and in the higher worms. Insects possess in the dorsal region a central circulatory organ (the "dorsal vessel"), a contractile, longitudinal duct, capable, by virtue of its muscle-fibers, of dilating, and provided with valves—which propels the blood rhythmically into the interstices of all the organs. Insects have no closed circulation. Shell-fish and snails have a heart and lacunar blood-channels. Cephalopods (sepia, cuttle-fish) have three hearts: an arterial, simple body-heart, and two venous, simple branchial hearts, one at the base of each gill. The circulation in most of these animals is closed. The lowest animals have either (multiple) pulsating vacuoles, which propel the colorless (blood-) juice into the soft body-parenchyma, like the infusoria; or they are totally devoid of any kind of vascular apparatus, the circulation of the juices being effected by the movements of the body (gregarines). In the group of celenterates (polyps, jellyfish) there is a "water-vascular system," which conveys the nutritive juice directly from the digestive cavity, and, at the same time, acts as a respiratory organ, as the water (which contains oxygen) passes through the system of tubes.

HISTORICAL.

The ancients (Empedocles, born 473 B. C.) were familiar with the movement of the blood, but were ignorant of the "circulation." According to Aristotle (384 B. C.) the heart, the acropolis of the body (which is present in every blood-animal), prepares the blood within its cavities and sends it through the arteries as a nutrient fluid to all the different parts of the body, like a system of constantly dividing brooks, irrigating the land and moistening and fertilizing it. The blood however, never flows back to the heart.

Praxagoras (341 B. C.) named the "arteries" (as well as the trachea); he was the first to distinguish arteries from veins. Together with Herophilus and Erasistratus (300 B. C.), the famous physicians of the Alexandrian school, he is responsible for the erroneous view, based on the fact that arteries are empty after death, that the arteries contain air conveyed to them through the respiration (hence the name "artery"). Galen (131–203 A. D.) refuted this error by vivisection. "Whenever," he says, "I injured an artery I saw blood escape. And when I tied a portion of an artery by means of two ligatures at either extremity, I showed that the included portion was full of blood."

Even then the theory of the exclusively centrifugal movement of the blood was maintained; it was erroneously supposed that communicating orifices existed in the septum between the right and the left heart.

Miguel Serveto (a Spanish monk, who was burned as a heretic in Geneva in 1553 at Calvin's instigation) was the first to show that the septum of the heart has no openings. He, therefore, searched for a communication between the right and the left heart and thus succeeded, in 1546, in discovering the lesser circulation: "fit autem communicatio haec non per parietem cordis medium (septum), ut vulgo creditur, sed magno artificio a cordis dextro ventriculo, longo per pulmones ductu, agitatur sanguis subtilis; a pulmonibus praeparatur, flavus efficitur et a vena arteriosa (Arteria pulmonalis) in arteriam venosam (Venæ pulmonales) trans-funditur." Almost a quarter of a century later, in 1589, Caesalpinus traced the course of the greater circulation. He was the first to use the word "circulation." Later, Fabricius ab Aquapendente (Padua, 1574) also recognized and confirmed the centripetal movement of the blood in the veins (which until that time was almost universally believed to be centrifugal, although Vesalius was familiar with the centripetal current in the main trunks) from the position of the valves in the veins, of which he made an accurate study, although they had been men-tioned in the middle of the fifth century after Christ by Theodoretus, Bishop. of Syria, also by Sylvius, by Vesalius (1534) and by Canani (1546). William Harvey, a pupil of Fabricius (until 1604), finally constructed, between the years 1616 and 1619, partly from his own investigations and partly from the results of former observers, the picture of the circulation of the blood, the greatest physiological achievement, which was published in 1628 and marks a new epoch in physiology.

With respect to individual features of the vascular system, the following is yet worthy of mention: According to Hippocrates the heart is a fleshy organ and the root of all the vessels; he was familiar with the large vessels originating from the heart, the valves, the chordæ tendineæ, the auricles, and the closure of the semilunar valves. Aristotle first named the aorta and the venæ cavæ, the school of Erasistratus the carotid; the latter also explained the function of the venous valves. In Cicero mention is made of the distinction between arteries and veins. Celsus, in the fifth century after Christ, pointed out that the veins, when opened below a compressing bandage, bleed. Aretaeus (50 A. D.) knew that arterial blood is bright red and venous blood dark. Pliny (died 79 A. D.) described the pulsating fontanel in man. The presence of a bone in the septum of large mam-mals (ox, stag, elephant) was known to Galen (131–203 A. D.). In his opinion the veins ultimately communicate with the arteries by means of the finest tubes, and this view was later confirmed by de Marchettis (1652) and Blancard (1676) with the aid of injections, and by Malpighi, who made microscopic observations of the circulation of the blood in cold-blooded animals, as well as by William Cowper (1697), who made similar observations on warm-blooded animals. Stenson, who was born in 1638, first demonstrated the muscular nature of the heart, al-though a statement to like effect had already been made by the Hippocratic and Alexandrian schools. Cole demonstrated the progressive increase in the width of the arterial area as the capillary region is approached. Joh. Alfons Borelli (1608–1679) was the first to estimate the power of the heart according to the laws of hydraulics. Craanen, in 1685, described systolic contractions in the pulmonary veins; Leeuwenhoeck (1694) the anatomical arrangement of the heart-muscle fibers among themselves. Chirac, in 1698, ligated a coronary artery of the heart in a dog, without, it is true, producing any result.

According to Aristotle, turtles can live for a short time after the heart has been removed.

Many of the ancients (the Israelites, Empedocles, Kritias, Lucretius) believed that the vital principle of the body, and even the soul (Aristotle and Galen), had its seat in the blood. Aristotle was familiar with the poisonous effects of the vapor of burning charcoal; Porcia voluntarily chose to die by inhaling it. Vene-section was practised by Greek physicians soon after the Trojan war.

The iron in the red blood-corpuscles was discovered by Menghini in 1746.

PHYSIOLOGY OF RESPIRATION.

OBJECTS AND SUBDIVISIONS.

The purpose of respiration is to convey to the body the oxygen necessary for its oxidation-processes, as well as to remove the carbon dioxid resulting from the combustion processes. The activity required for this purpose is most effectively rendered by the lungs. A distinction is made between external and internal respiration. The first embraces the exchange of gases between the outer air and the gases of the blood contained in the respiratory organs (lungs and skin); the second includes the exchange of gases between the capillary blood of the systemic circulation and the body tissues.

STRUCTURE OF THE AIR-PASSAGES AND THE LUNGS.

The lungs are compound tubular (grape-like?) glands that secrete carbon dioxid, and each of which sends its excretory duct (bronchus) to the common air-passage, the trachea.

The *trachea* has for its foundation a number of C-shaped, superposed, hyaline, cartilaginous arches, held together by a rigid fibrous membrane of closely woven elastic network, intermixed with connective tissue, arranged principally in a longitudinal direction. The cartilages serve the function of keeping the lumen of the tube patulous under the varying pressure-relations. They subserve a similar purpose in the bronchi and their branches. They do not occur in air-passages having a diameter of 1 mm. or less; and even in bronchioles of greater size they are less numerous and more irregular, occurring especially at the bifurcations in the form of irregular platelets.

An outer layer of connective and elastic tissue covers the air-passages and branches of the bronchial tree. On the side toward the esophagus this layer is reinforced by additional elastic elements and a few bundles of longitudinal unstriated muscle-fibers. The trachea contains unstriated muscle-fibers, especially arranged transversely, connecting the ends of the cartilaginous arches posteriorly and being inserted into the cartilages by means of elastic tendons. This transverse layer is again covered by longitudinal bundles. The mucous membrane, besides containing connective tissue and leukocytes, is especially rich in longitudinal elastic fibers, which attain their greatest size immediately beneath the epithelial basement membrane. The outer, narrow, scarcely separable submucosa is composed principally of connective tissue, and attaches the mucous membrane to the cartilages with their connecting fibrous membrane. The epithelium of the trachea is a stratified, ciliated epithelium, with the cilia waving toward the glottis, and with many interspersed goblet-cells. Numerous branched, tubular, mucous glands, with larger, brighter cells and smaller, darker ones (Gianuzzi's crescents) are found beneath the muscular layer of the trachea and bronchi. These glands are of a mixed type and have secretory ducts connected with their serous alveoli, but not with the mucous tubules. They secrete the viscid mucus that catches the dust-particles of the inspired air and is then removed from the bronchial tree and larynx by means of the ciliated epithelium. The air-passages are richly supplied with lymph-vessels and lymph-follicles, but are rather poor in nerves and blood-vessels. Ganglia are found on the nerve-trunks.

The direction in which the branches of the bronchi penetrate into their respective lobes corresponds with the inspiratory movement of the chest-wall covering each lobe; for example, the direction of the bronchi in the upper lobe is upward, forward, and outward.

The *small bronchi* are distinguished from the larger ones by a diminution in the amount of cartilage, and by the presence of a complete layer of circular muscle-fibers; mucous glands are wanting, and the epithelium is less developed. Goblet-cells secreting mucus are found as far as the smaller air-passages.

After the small bronchi have by repeated branching become diminished in diameter to from 0.5 to 0.4 mm., they are succeeded by the smallest bronchi, which already bear a few alveoli on their walls. The smallest bronchi still possess ciliated epithelium and unstriated muscle-fibers.

The respiratory bronchioles are the direct continuation of the smallest bronchi. In the bronchioles the cylindrical epithelium is gradually replaced, at first on one side only, by small, squamous cells, and later by a mixed epithelium of large plates and small, squamous cells. At the same time the mural alveoli become more numerous.

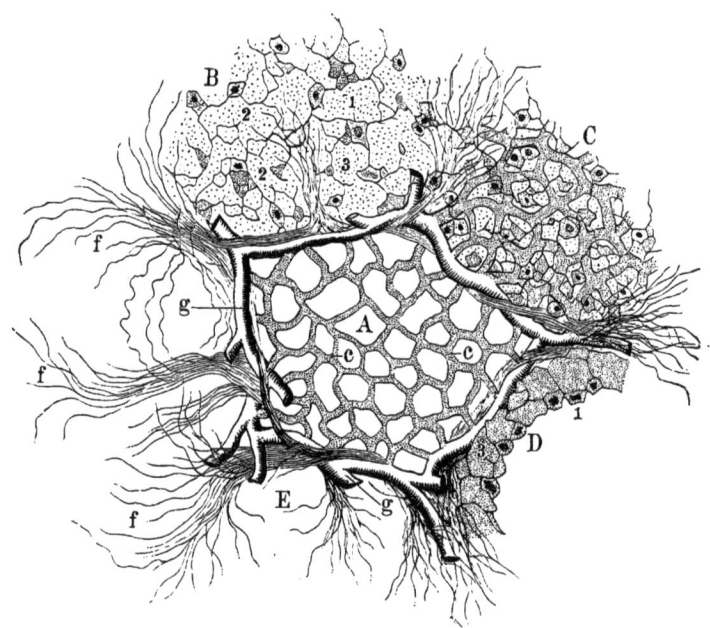

FIG. 75.—Cross-section of Several Pulmonary Alveoli: A, alveolus with the blood-capillaries (c) that arise from larger vessels (g g) bounding the alveoli. B, the epithelium of an alveolus: 1, nucleated cells; 2, non-nucleated platelets; 3, large, fused, non-nucleated plates. C, section of an alveolus with its epithelium and subjacent capillaries. D, alveolus, with its border covered by pulmonary epithelium and plates. E, alveolus whose boundary is indicated only by elastic fibers (f f).

From these respiratory bronchioles there arise, finally, the blind, alveolar ducts, which are completely lined with mixed epithelium, containing the small, squamous cells only in small nests. The alveolar ducts subdivide further, and still contain a few isolated muscle-fibers in their walls. These subdivisions are entirely surrounded by numerous closely packed, hemispherical or spheroidal air-sacs (alveoli).

Concerning *the structure of the alveoli*, the following is to be noted (Fig. 75): (1) The supporting membrane of the sac is structureless, elastic, with enclosed nuclei. Fine pores in the walls of the septa connect neighboring alveoli. (2) Networks of numerous, fine, elastic fibers surround the air-sacs, and give to the pulmonary tissue its great elasticity. As the elastic fibers are characterized by considerable power of resistance, they are often found retaining their characteristic arrangement in the expectoration of patients suffering from pulmonary diseases. This is an infallible sign that the pulmonary tissue is undergoing destruction. (3) The branches of the rich capillary network pass rather toward the lumen of the alveoli. The respiratory epithelium of the alveoli is a single layer of squamous epithelium. In it may be found scattered nucleated, protoplasmic cells (1), which are transformed later into small (from 7 to 15 μ), non-nucleated, bright (2) or dark platelets. Finally, several of the latter unite to form larger (from 22 to 45 μ), non-nucleated plates. (3) Here and there incomplete fissures may be seen in these plates, which indicate previous interspaces between the platelets. The plates have been transformed from original cuboidal cells by the stretching of the lungs during respiration.

See estimates the number of alveoli at 809½ millions, and their respiratory area at 81 square meters (54 times as great as the surface of the body). The alveoli are grouped together by connective tissue into distinct pulmonary lobules. The *blood-vessels of the lungs* belong to two distinct systems:

A. The system of the pulmonary vessels (the lesser circulation). The branches of the pulmonary artery follow those of the air-passages, and are so closely applied to the latter that their pulsations may be communicated to the contained air. The capillaries arising from these branches form a rich network of moderately fine tubules. The pulmonary veins, whose branches likewise accompany the air-passages, are collectively narrower than the pulmonary artery, as a result of the loss of water that the blood undergoes in the lungs.

B. The system of the bronchial vessels conveys the nutrient material for the respiratory organs. The bronchial arteries, following the bronchi, give to them branches, as well as to the lymphatic glands at the hilus of the lungs, the large trunks of the pulmonary vessels (vasa vasorum), and the pulmonary pleura. Numerous anastomoses occur between the branches of the bronchial and pulmonary arteries. Part of the vessels arising from the capillaries communicate with the beginnings of the pulmonary veins; and for this reason any considerable stagnation of blood in the lesser circulation causes a like stagnation in the circulation in the bronchial mucous membrane, with resulting bronchial catarrh. Another part of the bronchial capillaries forms special veins, which, as bronchial veins, traverse the posterior mediastinum, and empty into the trunks of the azygos veins, the intercostal veins, or the superior vena cava. The veins from the smaller bronchi, and even from the bronchi of the fourth class, empty collectively into the pulmonary veins; and the anterior bronchial veins also communicate with the pulmonary vessels.

The interstitial tissue of the lungs is rich in lymphadenoid tissue and is traversed by a network of fine lymph-channels. A coarser, irregular system of lymph-vessels surrounds the pulmonary lobules, larger bronchi, and blood-vessels. These lymph-channels and vessels become injected when animals are made to inhale powdered, soluble dyes. The coloring-matter penetrates the viscid interstitial substance between the epithelium, though according to Klein through small pores that are present.

In the walls of the pulmonary alveoli the finest lymph-tubules form a delicate system of canals lying in the spaces between the blood-capillaries. These canals exhibit enlargements at the points of intersection. Lymph-vessels extend along the bronchi, forming a dense, longitudinally meshed network in the mucosa and submucosa, and finally reaching the lymphatic glands at the roots of the lungs.

The rapidity with which fluids are absorbed in the lungs, even when introduced in considerable quantities, is remarkable. Landois has often seen this after injecting water into the trachea of living animals, and Peiper has demonstrated it for many other substances. Even blood is taken up in like manner, Nothnagel having found blood-corpuscles in the interstitial pulmonary tissue from three to five minutes after injection into the trachea.

In the pulmonary pleura, which is exceedingly rich in elastic fibers, the networks of superficial pulmonary lymph-vessels begin as free stomata. In like manner the lymph-vessels of the parietal pleura communicate by means of stomata in many places (on the diaphragm only in certain localities) with the pleural cavity; according to Klein even with the free surface of the bronchial mucous membrane. The lymph-vessels of the veins of the lesser circulation lie between the media and the adventitia.

The nerves of the lungs, bronchi, trachea, and larynx have ganglia.

It appears that the function of the unstriated muscle-fibers in the trachea and in the entire bronchial tree is to offer resistance within the air-passages to the increased pressure that occurs in all forced expirations, as in speaking, singing, blowing, straining.   According to the testimony of many investigators the vagus is the motor nerve; upon it depends the so-called pulmonary tone when the tension within the air-passages is increased.   Irritation of the vagus, or of the lung directly, does not induce sudden, expiratory movements (as can be seen by fastening a manometer in the trachea).   The only result of irritation of the vagus is an increase in the resistance of the air passing through the small bronchi that have been narrowed by the irritation.   Section of the vagus also is said to increase the volume of the lungs.   Atropin paralyzes, pilocarpin stimulates, the bronchial muscles of the dog, while reflex stimulation takes place through sensory branches of the vagus.   During deepest inspiration the unstriated muscles of the air-passages contract, and during forced expiration they are relaxed.

Pathological.—Irritation of the unstriated muscles, causing spasmodic narrowing of the smaller bronchi, may give rise to asthmatic attacks. If the escape of air from the alveoli is thus made difficult or obstructed, an acute inflation of the lungs—acute emphysema—may result.

According to Sandmann a reflex effect may be produced upon the bronchial muscles from the mucous membrane of the nose and the larynx. This would explain the occurrence of asthma attending nasal affections, such as polypoid growths of the mucous membrane.   In addition to the elements of the connective, elastic, and muscular tissues, and of the mucous membrane, the lungs contain lecithin, inosite, uric acid (taurin and leucin in the ox; guanin (?), xanthin, hypoxanthin in the dog), also sodium, potassium, calcium, magnesium, iron oxid, considerable phosphoric acid, also chlorin, sulphuric acid, silicic acid, and carbon.   In cases of diabetes sugar has been found; in the presence of purulent infiltration glycogen and sugar; in that of renal degeneration urea, oxalic acid, and ammonium-salts; in that of autointoxications leucin and tyrosin.

MECHANISM OF THE RESPIRATORY MOVEMENTS.
ABDOMINAL PRESSURE.

The mechanism of breathing consists in an alternating dilatation and contraction of the thoracic cavity.   The dilatation of the cavity is termed inspiration, and the narrowing expiration. The whole outer surface of both elastic lungs is, by means of its smooth, moist covering of pleura, intimately and hermetically applied to the inner surface of the chest-wall, which in its turn is covered by the parietal pleura.   Hence, it is evident that every expansion of the thorax is accompanied by a corresponding expansion of the lungs, and every contraction compresses those organs.   These movements of the lungs are, therefore, wholly passive, being dependent on the thoracic movements.

By reason of their complete elasticity the lungs are able to follow every change in the capacity of the thorax, without causing the two layers of the pleura ever to separate.   The cavity of the unexpanded thorax is greater than the volume of the collapsed lungs when removed from the body; therefore, the lungs in their natural position within the chest must be stretched, and they are, to a certain degree, in a state of elastic tension.   This tension varies directly with the size of the thoracic cavity.   If the pleural cavity be opened by a perforation from without or by a wound of the lungs from within, the elasticity of the lungs causes them to collapse, and there arises an air-space between the outer surface of the lungs and the inner surface of the thorax (pneumothorax).   The affected lung is incapacitated for respiration.   Double pneumothorax is accordingly fatal.

The degree of the elastic traction of the stretched lung may be measured by introducing a manometer through an intercostal space into the pleural cavity of a dead body.   The elastic tension here is the same as that in the living body during a state of quiet expiration, and is equal to 6 mm. of mercury.   In a patient with perforation of an intercostal space Aron found the elastic tension to be from 4.5 to 6.8 mm. If, however, the thorax is, by force applied from the outside, brought into the expanded position assumed during inspiration, the elastic traction will be increased to 30 mm.

If the glottis be closed during inspiratory dilatation of the thorax, the elastic lungs also will expand, and there will be produced a rarefaction of the air within the lungs, as this air must expand to a greater volume. If the glottis is now suddenly opened, the atmospheric air will enter the lungs, until the density of the air within equals that of the atmosphere. On the other hand, if the chest and the lungs be compressed by expiratory efforts, with a closed glottis, the air in the lungs will become denser, that is, compressed into a smaller volume. If the glottis now be opened, air will escape from the lungs, until the internal and external pressures are equalized. As the glottis is open during ordinary respiration, the adjustment of the diminished or increased air-pressure during inspiration and expiration will occur gradually. It is certain, however, that there exists in the air within the lungs a slight negative pressure during inspiration and a slight positive pressure during expiration. This may be measured in the trachea of persons having wounds of this tube, and equals 1 mm. during inspiration and from 2 to 3 mm. during expiration. According to J. R. Ewald the total figures are only 0.1 mm. and 0.13 mm. respectively.

The so-called abdominal pressure within the abdomen is generally increased during expiration, and declines during inspiration in man and in dogs, while in rabbits it is increased during inspiration. Moderate increase of the abdominal pressure increases somewhat the arterial blood-pressure, as well as the action of the heart; more pronounced increase of abdominal pressure diminishes both.

RESPIRATORY VOLUMES.

The lungs never completely empty themselves of air. Therefore, in filling and emptying the lungs during inspiration and expiration, only a part of the contained air is subjected to change, the amount depending on the depth of the respirations.

Hutchinson in this connection established the following distinctions:

1. *Residual air* is the volume of air that remains in the lungs after complete expiration. This can be estimated approximately after death by collecting over water the air from the lungs after ligating the trachea.

H. Davy and Gréhant estimated the amount during life in the following manner: The subject makes a forcible expiration, and then breathes for a while from and into a spirometer, filled with a measured quantity of hydrogen. If it can be assumed that the residual air has been completely admixed with the hydrogen, the percentage of air in the spirometer after forced expiration will indicate the quantity of residual air. The observers named found the amount to be from 1200 to 1700 cu. cm. Berenstein, by a similar method, estimated the residual air to be equal to from one-fifth to one-fourth of the vital capacity.

The following wholly different method has also been employed to determine the residual air: The amount of an unknown volume of air x can be calculated from the increase in volume that it undergoes when the pressure upon it is lessened, for this increase in volume is directly proportional to the quantity of gas, and to the diminution in the pressure upon it. If P_1 is the original pressure to which the gas is exposed, P_2 the other, lessened pressure, and d the measurable increase in volume of x, then

$$x = (P_2 \times d) : (P_1 - P_2).$$

For carrying out this experiment Pflüger constructed his pneumometer. The subject is placed in a large, hermetically sealed chamber (human cabinet), in which at first the pressure equals that of the atmosphere (P_1). The contained air is then rarefied by means of a pump, until the pressure P_2 is obtained, as indicated by a manometer inserted in the chamber. In this process a part of the residual air (x) will naturally escape during quiet expiration. This is collected and measured (d) by means of a spirometer connected in an air-tight manner with the air-passages. In this way Pflüger found x to be from 400 to 800 cu. cm. Gad, working with different apparatus based on the same principle, estimates the residual air to be half the vital capacity; Schenck gives the proportion of the former to the latter as 1 to 3.7.

2. *Reserve air* is the additional volume of air that can be forced out after a quiet expiration. It measures from 1248 to 1804 cu. cm.

The procedure of H. Davy and Gréhant may also be applied to the estimation of reserve air.

3. *Respiratory or tidal air* is the volume of air that is taken in and given off during quiet respiration. In adults under normal conditions it amounts to about 507 cu. cm.—between 367 and 699 cu. cm., according to Vierordt; in the new-born about one-quarter of this amount.

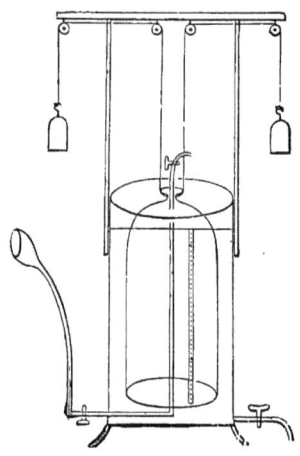

FIG. 76.—Hutchinson's Spirometer.

4. *Complemental air* is the term applied by Hutchinson to the additional volume of air that may be taken in during a forced inspiration immediately succeeding a quiet one.

5. *Vital capacity* indicates the volume of air that escapes from the lungs between the highest phase of inspiration and the lowest phase of expiration. For Germans it amounts to 3222 cu. cm. on an average, and for Englishmen to 3772 cu. cm.

From the foregoing it follows that after a quiet inspiration both lungs contain about from 3000 to 3900 cu. cm. of air (1 + 2 + 3); after a quiet expiration from 2500 to 3400 cu. cm. (1 + 2). From this, as from 3, it follows that during quiet respiration only about one-sixth or one-seventh of the air in the lungs is changed.

If, during a series of quiet respirations, a solitary inhalation of hydrogen be made, and if the expired air be examined to determine how long the hydrogen may be detected in it, it will likewise be found that the air in the lungs completely renews itself (becomes free of hydrogen) in from 6 to 10 respirations.

Donders estimates that the combined bronchial tree and trachea contain about 500 cu. cm. of air.

The vital capacity is determined by means of Hutchinson's spirometer (Fig. 76). The determination is of importance in persons suffering from disease of the thoracic organs. The vital capacity may be influenced by consolidation, destruction, or emphysema of the pulmonary tissue; by the presence of fluids, blood, air, or new-growths in the thoracic cavity; by diminished mobility of the chest; by weakness of the respiratory muscles; by enlargement of the heart or pericardium; or by distention of the abdomen.

By means of a large tube provided with a mouth-piece, the subject (holding his nostrils closed) blows his expiratory air into a graduated gasometer bell-jar that is suspended over water and evenly balanced by a system of weights and pulleys.

After complete expiration the tube is closed, and the increase of air within the jar indicates the vital capacity, provided the water outside and that inside the jar are at the same level. It is also advisable to allow the expired air to cool, until it is of the temperature of the surrounding air.

Of the factors that influence vital capacity the following are known:

1. *Stature.*—Every inch of additional height between 5 and 6 feet is accompanied by about 130 cu. cm. increase in the vital capacity.

2. The *volume of the trunk* is, on the average, seven times that of the vital capacity.

3. *The Body-weight.*—An increase in weight of 7 per cent. above the normal is accompanied by a diminution in the vital capacity of 37 cu. cm. for every additional kilogram.

4. *Age.*—The vital capacity reaches its maximum at thirty-five years; from this up to the sixty-fifth year, and backward to the fifteenth year, 23.4 cu. cm. must be deducted for each year.

5. *Sex.*—Arnold found the average to be 3660 cu. cm. for men, and 2550 cu. cm. for women. For the same stature and chest-measurement, the relation of the vital capacity of men to that of women is as 10 to 7.

6. *Social position* and *occupation* have a decided influence on the physical condition and nutrition, and hence also on the vital capacity. Arnold established three classes, of which each preceding class exceeds the one following by 200 cu. cm. greater vital capacity: (*a*) soldiers and sailors; (*b*) artisans, compositors, police; (*c*) paupers, the nobility, and students.

7. *Miscellaneous.*—The vital capacity is greatest in the standing position, and when the stomach is empty. It is diminished after great effort, and also in debilitated conditions of the body. It is greater in advanced pregnancy than in the puerperium. To a certain extent practice with a spirometer can increase the vital capacity.

THE RATE OF RESPIRATION.

The rate of respiration varies in adults between 12, 16, and 24 in a minute. Four pulse-beats on an average thus occur with every respiration. Many factors influence the rate:

1. *The Position of the Body.*—In adults Guy noted 13 respirations to the minute in the recumbent posture, 19 in the sitting posture, and 23 in the standing posture.

2. *Age.*—In 300 individuals Quetelet found the rate of respiration to be as follows:

Age.	Respirations.	Age.	Respirations.
Up to 1 year	44	Between 20 and 25 years	18.7
At 5 years	26	" 25 and 30 years	16
Between 15 and 20 years	20	" 30 and 50 years	18.1

In the new-born the rate is between 62 and 68.

3. *Activity.*—In children between two and four years old, Gorham counted 32 respirations to the minute in the standing posture, and 24 during sleep. As a result of bodily exertion the rate of respiration increases before that of the heart-beat. The increase in respiratory movements is incited by metabolic products furnished by the muscles engaged in activity. In connection with violent muscular activity the pulse-rate is increased principally by excitation of the center for the cardio-accelerator nerves.

4. *Increase in the surrounding temperature,* also febrile elevation of the bodily temperature, will increase the rate of respiration, which may even assume a dyspneic character.

THE TIME RELATIONS OF RESPIRATORY MOVEMENTS. PNEUMATOGRAPHY.

In order to obtain information with regard to the periodic relations of the various phases of the respiratory movements, it is necessary to trace respiratory curves (pneumatograms) by means of recording instruments.

Method.—The graphic method can be applied in three different ways: 1. The representation of the range of motion in the individual parts of the thorax may be obtained in the following manner:

(a) K. Vierordt and Ludwig arranged an instrument in which the movement of a definite part of the thorax was communicated to a lever, whose longer arm traced the curve on a rotating drum. In like manner Riegel constructed his double stethograph on the principle of the lever. It consisted of two levers on the same support, arranged for use on a patient in such way that one lever was applied to a certain spot on the healthy side of the chest, and the other lever to the corresponding spot on the affected side. A sphygmograph may be employed for recording the respiratory curve, the instrument being placed free outside of the chest upon a stand and applied in such manner that only the pad of the elastic stylus touches the chest-wall at one point. J. Rosenthal constructed a lever to register the movements of the diaphragm in animals (phrenograph); it

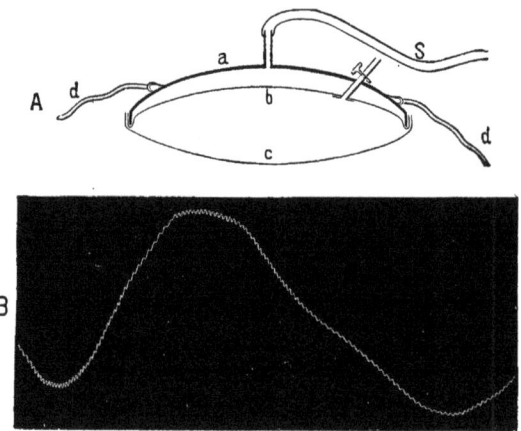

was inserted through an opening in the abdomen, and rested against the diaphragm.

(b) The air-cushion of Brondgeest's pansphygmograph (Fig. 77, A) is constructed on the principle of air-transference. This instrument consists of a saucer-shaped brass vessel (a), over which is stretched a double-layered rubber membrane (b c). Between the layers of this covering there is enough air to make the outer membrane bulge. This cushion is placed on a certain part of the thorax, and fastened with bands (d d) that pass around the chest. Every enlargement of the thorax presses against the membrane c, producing a diminution of the air-space within the capsule. The latter is connected by means of the tube S with the recording chamber that is pictured in Fig. 44.

Instead of this capsular arrangement, Marey, in the construction of his pneumograph, uses a piece of thick, cylindrical, elastic rubber tubing. This is fastened by bands like a girdle around the chest, and is connected by a tube with the recording drum.

2. The variations in the volume of the chest or in the exchanged respiratory gases may be graphically recorded as follows:

For this purpose E. Hering places an animal in an air-tight, closed chamber,

with two openings in its walls.   The trachea of the animal having been previously
cut across, a cannula is fastened in the pulmonary end, and is attached to a tube
passing through one of the openings in the chamber (respiration being conducted
undisturbed through this tube).   Through the other opening passes a manometer-
tube, filled with water, and pro-
vided with a recording float.   The
same experiment may be conducted
with a human subject, provided the
breathing tube be placed in the
mouth and the nose be held closed.
Gad (Fig. 78) has succeeded in recording graphically the variations in
the volume of the respired air by
means of an apparatus: the expired
air lifts a light, balanced box, which
is closed off by water.   In rising,
this box moves a recording lever.
During inspiration the box sinks.

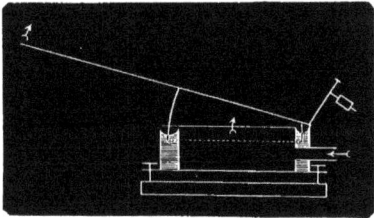

FIG. 78.—Air-volume Recorder (Pneumoplethysmograph)
(after Gad).

3. The variations in the rapidity
with which the respiratory gases are
changed may be recorded as follows:
A tube is fastened in the trachea of an animal, or in the mouth of a human
subject (holding the nostrils closed), in the same way as with the dromograph
(Fig. 69).   The pendulum (made broader for this purpose) will swing to and fro
during inspiration and expiration, and will record the velocity of the currents of
air entering and leaving the lungs.

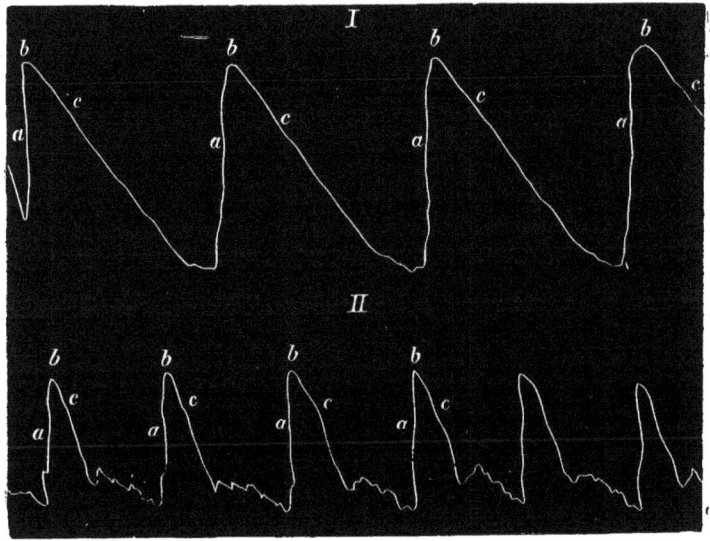

FIG. 79.—Pneumatograms Recorded by Means of Riegel's Stethograph: *I*, normal curve; *II*, curve of a man
with emphysema; *a*, inspiratory limb, *b*, summit, *c*, expiratory limb of the curve. The small elevations are due
to the pulsations of the heart.

The curve in Fig. 77 B was drawn upon a vibrating tuning-fork plate, by
means of the air-cushion of Brondgeest's pansphygmograph, applied to the ensiform process of a healthy man.   The inspiration (ascending limb) begins with
moderate rapidity, is accelerated in the middle, and again becomes slower toward

the end. The expiration begins with moderate rapidity, is then accelerated, and finally becomes much slower in the last part.

Inspiration is somewhat shorter than expiration; in adult males the proportion is 6 : 7, according to Sibson; in women, children, and old persons it is 6 : 8 or 6 : 9. Vierordt found the relation 10 : 14.1 (up to 24.1); J. R. Ewald found it 11 : 12. Cases in which inspiration and expiration are of equal length, or in which the latter is even the shorter, are observed only exceptionally.

Small irregularities may be observed occasionally on various parts of the curve. These are due to the fact that the thoracic movements are at times the result of successive contractions of the respiratory muscles. Now and then powerful heart-beats also cause vibrations of the thoracic wall (Fig. 79).

If respiration proceeds uninterruptedly and quietly, there is usually no real pause, i. e., complete rest of the thorax. Sometimes the lowest flattened part of the expiratory limb is incorrectly taken for the pause. Of course, a pause may voluntarily be made at any phase of the movement.

If the respirations be deep, but slow, an expiratory pause is almost invariably noted; on the other hand, it is always lacking in rapid respiration. An inspiratory pause is never noted under normal conditions, but it may occur under pathological conditions.

TYPES OF RESPIRATORY MOVEMENTS.

Curves recorded from various parts of the thorax throw light upon the so-called type of respiration. Hutchinson was the first to show that women expand the thorax by producing an elevation of the sternum and ribs—costal or thoracic respiration ; while men produce the same effect by depression of the diaphragm—abdominal or diaphragmatic respiration.

If the height of the curves taken in men and women from the manubrium, gladiolus, ensiform process, and epigastrium be compared, it will be seen that the excursion of the sternum is most pronounced in women, while that of the epigastrium (through the diaphragm) predominates in men.

This difference between the sexes, in the type of costal and diaphragmatic breathing, holds good only in quiet respiration. In deep and forced respiration the enlargement of the thoracic cavity is brought about in both sexes principally by a pronounced elevation of the chest and ribs. In this case the epigastrium, even in men, is drawn in rather than forced out. During sleep the type of respiration is thoracic in both sexes, and the inspiratory expansion of the thorax precedes the elevation of the abdominal wall.

It has recently been again pointed out that the costal type arises principally from compression of the lower ribs by corsets or tight belts, especially as a decided abdominal type is encountered in savage women. It is only a conjecture that the costal type may be a natural tendency, the result of pregnancy, during which abdominal respiration may become obstructive and harmful by exerting pressure on the uterus. Some affirm, while others deny, that the difference in type is evident during sleep with the clothing completely removed, and also in young children. Some investigators maintain that the costal type is found in children of both sexes; they attribute this to a greater flexibility of the ribs in children and women, which thus allows the thoracic muscles to exert a more extensive influence on the ribs.

PATHOLOGICAL VARIATIONS IN THE RESPIRATORY MOVE-MENTS.

Changes in the Character of the Movements.—In the presence of affections of the respiratory apparatus the expansion of the thorax may be diminished to the extent of 5 or 6 cu. cm. on one or both sides. When the apices are affected, as occurs so frequently in cases of tuberculosis of the lungs, the subnormal expansion in the upper parts of the thorax is a characteristic feature. Retraction of the intercostal spaces, the ensiform process, and the lower insertions of the ribs occurs during marked inspiratory rarefaction of air in the thorax, such as may take place in the presence of laryngeal stenosis. If this phenomenon be confined principally to the upper parts of the thorax, it shows that the subjacent part of the lung is diseased and capable of little expansion.

In persons suffering from chronic, advanced disease of the respiratory organs, without impairment in the activity of the diaphragm, the insertion of the latter manifests itself on the outer surface of the body by a shallow groove (Harrison's groove), passing horizontally outward from the ensiform cartilage, and due to the marked retraction.

The period of inspiration is lengthened in persons suffering from constriction of the trachea or larynx; that of expiration in those who must call into play all the expiratory muscles, by reason of an emphysematous condition of the lungs (Fig. 72, II). Occasionally, in emphysematous subjects, a short expiratory effort precedes the inspiration.

Changes in the Rhythm of the Movements.—All disturbances of the respiratory apparatus of any degree of magnitude will produce an increase in the frequency or depth of the respirations, or both together. This phenomenon is termed dyspnea. The possible causes of dyspnea are various: 1. Restriction of the respiratory exchange of gases in the blood, as a result of (a) diminution of the respiratory surface (pulmonary diseases), (b) contraction of the air-passages, (c) diminution in the number of red blood-corpuscles, (d) disturbances in the mechanism of respiration (affections of the respiratory muscles and their nerves, painful affections of the thoracic walls), (e) weakness in the circulation, especially the lesser circulation, principally as a result of various cardiac affections.

2. Febrile conditions are a further cause of increase in the frequency of respiration. The febrile blood heats the respiratory center in the medulla oblongata, and thus incites dyspneic respiratory movements up to from 30 to 60 in the minute (heat-dyspnea). If the carotids of animals be placed in hot tubes, the same result is produced. Under the influence of hysterical irritability, a nervous pathological increase in the respiratory rate may be produced in rare cases. Respiratory pauses of considerable duration are uncommon, but they may occur (in one patient with cardiac and renal disease a pause of thirty-seven seconds was observed during sleep).

A remarkable change in the rhythm of respiration is known as Cheyne-Stokes' breathing. This manifests itself as a suffocation-phenomenon in affections that alter the normal supply of blood to the brain, or that change the composition of the blood, for example, cerebral affections, cardiac diseases, or uremic intoxication. Under such circumstances pauses of from one-half to three-fourths of a minute alternate with series of from 20 to 30 respirations, likewise extending over from one-half to three-fourths of a minute. The respirations of a single series are first superficial; they then become deeper and dyspneic, and then again more superficial. After this a pause occurs, and at this time the eyeballs roll, the pupils are contracted and do not react, and the blood-pressure falls. In severe cases complete loss of consciousness, analgesia, abolition of the reflexes, and even inability to swallow, rarely, toward the end of the pause, also muscular twitchings have been observed during the pauses. When the respiratory movements commence again, the pupils become larger and reactive. It has often been observed that consciousness, lost during the pause, has been partially regained whenever the respirations begin.

In agreement with Rosenbach and Luciani the cause of Cheyne-Stokes' breathing is referred to variations in the irritability of the respiratory center, which reaches its lowest point during the pause. Luciani compares the phenomenon with that of the periodically grouped heart-beats. He observed it to set in after injury to the medulla above the respiratory center, after the apnea in animals profoundly poisoned with opium, and finally in the last stage of asphyxia attending respiration in a closed space.

Cheyne-Stokes' respiration is most readily explained by assuming the pause to be a period of asphyxia, and the series of respirations to be agonal. Under the reviving influence of the latter, the respiratory center recovers from the previous state of exhaustion.

During hibernation this form of breathing is normal in the dormouse, the hedge-hog, and the alligator. If frogs are kept immersed in water, or if the aorta be clamped, they lose the power of reaction in a few hours. When taken out of the water, or when the clamp is removed, they recover immediately, and they invariably exhibit the phenomenon of Cheyne-Stokes' respiration. In such frogs the circulation may be interrupted for a time, during which this form of breathing still continues.

Curtailment of the blood-supply in frogs by blood-letting results in periodically grouped respirations. These are followed by a stage of single, infrequent respirations, and finally the breathing stops completely. During the pauses between the periods, each mechanical irritation of the skin will give rise to a series of respirations. Periodic respiration, without variations in the depth of the separate respirations (so-called Biot's respiration), also occurs normally in sleep. While the nervous centers are endeavoring to obtain rest, they forget, to a certain extent, to send out respiratory impulses, and the organism takes no notice of these short pauses. Periodic irregularities in respiration also are frequently of reflex origin. Muscarin, digitalin, curare, chloral, hydrogen sulphid, and the toxins of some infectious diseases (typhoid fever, diphtheria, scarlet fever) are likewise capable of exciting periodic respirations.

SUMMARY OF THE MUSCULAR MECHANISM CONCERNED IN INSPIRATION AND EXPIRATION.

A. INSPIRATION.

I. During quiet inspiration the following muscles are active:

1. The diaphragm (phrenic nerve, from the third and fourth cervical nerves).
2. The external intercostal and intercartilaginous muscles (intercostal nerves).
3. Long and short elevators of the ribs (posterior branches of the dorsal nerves).

In a state of rest the elastic traction of the lungs appears to draw the chest together somewhat with tension on all sides. Accordingly the elastic force thus exerted would act as an aid to the beginning of inspiration. Also Landerer considers the thorax at rest to be an apparatus tending toward the attitude of inspiration, by means of the elasticity in an upward direction of the six upper ribs.

II. During forced inspiration the following muscles are active:

(a) *Trunk-muscles.*

1. The three scalene muscles (muscular branches of the cervical and brachial plexuses).
2. Sterno-cleido-mastoid (external branch of the spinal accessory nerve).
3. Trapezius (external branch of the spinal accessory, and muscular branches of the cervical plexus).
4. Lesser pectoral (anterior thoracic nerves).
5. Posterior superior serratus (dorsal nerve of the scapulæ).
6. Rhomboids (dorsal nerve of the scapulæ).
7. Extensor muscles of the vertebral column (posterior branches of the dorsal nerves).

The assumption that the greater anterior serratus (long thoracic nerve) and the subclavius (brachial plexus) are accessory muscles of inspiration is unwarranted.

(b) *Laryngeal muscles.*

1. Sterno-hyoid (descending branch of the hypoglossus).
2. Sterno-thyroid (descending branch of the hypoglossus).
3. Posterior crico-arytenoid (inferior laryngeal branch of the vagus).
4. Thyro-arytenoid (inferior laryngeal nerve).

(c) *Facial muscles.*

1. Anterior and posterior dilators of the nares (facial nerve).
2. Levator of the ala nasi (facial nerve).
3. The muscles that separate the lips and open the mouth during extreme forced respiration—gasping—(facial nerve).

(d) *Muscles of the palate and pharynx.*

1. Elevator of the veil of the palate (facial nerve).
2. Azygos of the uvula (facial nerve).
3. According to Garland, the pharynx is always narrowed.

B. EXPIRATION.

I. During quiet expiration the size of the thoracic cavity is reduced essentially by the weight of the chest-walls, together with the elasticity of the lungs, costal cartilages, and abdominal muscles.

II. During forced expiration the following muscles are employed:

1. The abdominal muscles (internal or anterior abdominal nerves, branches of the intercostal nerves from the 8th to the 12th).

2. Internal intercostal muscles (the parts lying between the ribs), and the infracostal muscles (intercostal nerves).

3. The triangular muscle of the sternum (intercostal nerves).

4. (?) Posterior inferior serratus (external branches of the dorsal nerves).

5. (?) Quadratus lumborum (muscular branches of the lumbar plexus).

ACTION OF THE INDIVIDUAL RESPIRATORY MUSCLES.

A. Inspiration.—1. The *diaphragm* arises by six processes from the six lower costal cartilages and contiguous osseous parts of the ribs (costal portion), by three processes from the upper four lumbar vertebræ (lumbar portion), and from the ensiform process (sternal portion). It presents a double dome, with the convexity toward the thoracic cavity, and contains the liver in its larger, right-sided concavity, and the stomach and the spleen in its smaller, left-sided cavity. In a state of rest the intra-abdominal pressure and the elasticity of the abdominal wall press these organs against the under surface of the diaphragm, in such a manner that it bulges into the thoracic cavity. This position is aided by the elastic traction of the lungs. The central part of the diaphragm (central tendon) is, to a great extent, fused on its upper surface with the pericardial sac. This part, which supports the heart and is pierced by the inferior vena cava (foramen quadrilaterum), projects downward into the abdominal cavity in a state of rest; and in casts made of the diaphragm it can be recognized as the lowest part of the middle portion (Fig. 80).

During contraction of the diaphragm the two dome-like projections are flattened, and the thoracic cavity is enlarged downward. At the same time the dis-

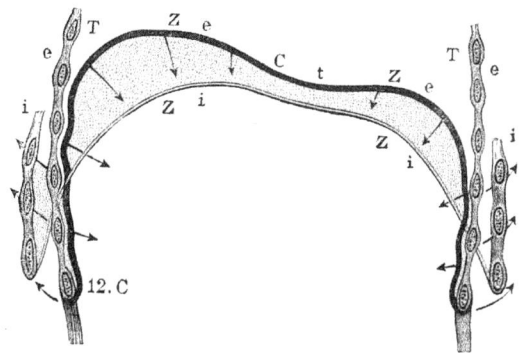

FIG. 80.—Frontal Section of the Thorax at the Extremity of the Twelfth Rib on Each Side (12. C), to Demonstrate the Form of the Diaphragm during Expiration (Z e–Z e) and during Inspiration (Z i–Z i): T e–T e, thoracic wall in a state of expiration; i i, during inspiration; C t, central tendon. The arrows indicate the direction of the movements during inspiration.

tal, arched, muscular parts become flatter, and are drawn away from the chest-wall, to which they are closely applied during expiration. The middle part of the central tendon, upon which the heart rests, takes no considerable part in the movement during quiet inspiration; but during forced inspiration it also is depressed to a certain extent.

In the recumbent posture (especially in men), with full light on the thorax, the contraction of the diaphragm can often be seen directly in the form of a wave-like movement beginning in the sixth intercostal space and running downward through from one to three intercostal spaces in accordance with the depth of inspiration.

The diaphragm undoubtedly plays the most important part in enlarging the thorax. Brücke further maintains that the diaphragm, besides enlarging the thorax in a vertical direction, also expands the lower part in a transverse direction: namely, when it compresses the abdominal organs from above, the latter endeavor to escape laterally, and thus spread out themselves, as well as the adjacent thoracic wall. If the abdominal contents be removed from an animal, the lower ribs are seen to be drawn inward with every contraction of the diaphragm; therefore, the presence of the viscera is necessary for the normal action of the diaphragm.

In order to obtain some idea as to the extent of thoracic enlargement due to the action of the diaphragm, Landois carried out the following experiment: A tracheal cannula was fastened in the body of a well-built, female, newly born child that had died of hemorrhage. The body was completely immersed in water, and the lungs were inflated. The vital capacity was estimated from the amount of water displaced. The abdomen was then opened, the viscera removed, and a wax cast was taken of the under surface of the diaphragm, with uninflated lungs (that is, in the position of expiration). Hereupon, a quantity of air equal to the determined vital capacity was introduced into the lungs, and after the air-passages were closed, a second wax cast was taken of the diaphragm in this last position. The difference in volume between these casts was determined, and it was found that the proportion between the expansion due to the diaphragm and that due to all other causes was $1 : 2\frac{1}{2}$. These figures are, of course, only approximately correct; for, in the first place, the removal of the abdominal viscera permits of unimpeded descent of the diaphragm (which is, to a certain extent, compensated for by the taking of the wax cast); and, secondly, the arch of the actively contracting diaphragm presents a form differing from that produced passively by inflation of the lungs. However, there is no other means at hand for determining the thoracic expansion produced by the diaphragm.

By increasing the intra-abdominal pressure, each diaphragmatic contraction increases the flow of venous blood from the abdominal organs to the inferior vena cava.

The great importance of the diaphragm in the respiratory process can be realized from the fact that bilateral section of the phrenic nerves in young rabbits is followed by death. These nerves contain, as has been shown experimentally, a few sensory fibers for the pleura, the pericardium, and a portion of the peritoneum. In animals, irritation of the lowest five intercostal nerves causes local, inconsiderable contraction of the marginal part of the diaphragm.

The contraction of the diaphragm is not to be regarded as a simple muscular contraction, for its duration is from four to eight times that of the latter. It is, therefore, to be designated as a tetanic movement of short duration.

2. *The Elevators of the Ribs.*—At their vertebral extremity (which lies at a much higher level than the sternal extremity) the ribs are articulated at their heads and tubercles with the bodies and transverse processes of the vertebræ. A horizontal axis passes through both joints, and upon this axis the rib is capable of rotating upward and downward. If the axes of a pair of ribs be prolonged from both sides until they meet in the middle line, an angle is formed that is large (125°) for the upper ribs and smaller (88°) for the lower ones. An imaginary plane may be passed through the arch of each rib, which inclines, in a state of rest, from behind and inward, forward and outward. If the rib turns on its axis, this inclined plane is raised more toward the horizontal. As the axes of the upper ribs pass rather in a frontal direction and those of the lower ribs rather in a sagittal direction, elevation of the upper ribs causes an expansion of the cavity from behind forward, and elevation of the lower ones an enlargement from within outward (as the movements of the ribs inclined downward are perpendicular to their axes). At the same time the costal cartilages undergo slight torsion, which brings their elasticity into play.

All of the inspiratory muscles that act directly on the walls of the thorax produce the desired result by elevating the ribs. In this connection the following points are to be observed: (a) Elevation of the ribs causes a widening of the intercostal spaces. (b) When the upper ribs are elevated, all of the lower ribs and also the sternum are raised at the same time, as all of the ribs are bound together by the soft structures in the intercostal spaces. (c) During inspiration there occurs an elevation of the ribs and a widening of the intercostal spaces. An exception is made of the lowest rib, which does not actually form a part of the thorax. It moves downward, not upward, at least during deep inspiration. (*d*) If, on a preparation of the chest, the ribs be elevated, with widening of the intercostal spaces, as occurs during an inspiratory movement, then all those muscles may be regarded as elevators of the ribs whose origin and insertion approach each other. Hence, only these muscles can be designated as muscles of inspiration. From this point of view the scalene muscles, the long and short elevators of the ribs, and the posterior superior serratus are to be recognized as undoubted inspiratory muscles. They are also to be considered as the muscles having the greatest influence on the ribs during inspiration.

Of the intercostal muscles, according to this experiment, only the external and the intercartilaginous portions of the internal can be designated as inspiratory muscles. The remainder of the internal (the parts covered by the external) are lengthened during elevation of the ribs, and shortened when the ribs are lowered. As a muscle always exhibits its activity by shortening, the internal intercostal muscles have been regarded as depressors of the ribs (that is, as expiratory muscles).

Fig. 81, I, shows that when the rods a and b, representing the depressed ribs, are elevated, the interspace (intercostal space) must become wider: e f > c d. On the left side of the figure it may be seen that when the rods are elevated, the line g h, representing the external intercostal muscles, is shortened (i k < g h), while l m, representing the internal intercostals, is lengthened (l m < o n). Fig. 81, II, shows that the intercartilaginous muscles, designated by g h, and the external intercostal muscles, designated by l k, are shortened by elevation of the ribs. The latter position of these muscular fibers may be represented by the shortened diagonals of the dotted rhomboids.

The controversy over the mechanism of the intercostal muscles dates back to ancient times: Galen (131–203 A. D.) regarded the external intercostal muscles as inspiratory and the internal as expiratory muscles. Hamburger (1727), following Willis' investigations, agreed with this view, and also recognized the intercartilaginous muscles as inspiratory muscles. A. v. Haller, who was Hamburger's direct opponent, considered both internal and external intercostals as muscles of inspiration; while Vesalius (1540) regarded them both as expiratory muscles. Masoin and R. du Bois-Reymond admitted the latter view, but only for forced respiration. Finally, Landerer, who observed that the upper two or three intercostal spaces became narrower during inspiration, believed that both sets were active during both inspiration and expiration. As they hold the ribs together, they have the sole function of transmitting the traction imparted to them simply through the chest-walls. They would, therefore, remain active even when the distance between their points of insertion becomes greater.

After mature consideration of all the conditions, Landois was unable to accept any of these views unconditionally. It is obvious that the external intercostal and intercartilaginous muscles can act together only during inspiration, while the internal can be active only during expiration (the latter statement having been confirmed by Martin and Hartwell in dogs by means of vivisection); but elevation and depression of the ribs are not the chief results attained by the action of these muscles. It was rather Landois' opinion that the chief purpose of the external and intercartilaginous muscles is to counteract the inspiratory widening of the intercostal spaces and the synchronous increase in the elastic traction of the lungs. The function of the internal intercostal muscles is to offer resistance to the expiratory distention that occurs during forced expiratory efforts, as in coughing. Without muscular resistance the intercostal tissues would be so stretched through the uninterrupted traction and pressure that regular respiratory movements would become impossible.

The lesser pectoral (and the greater anterior serratus ?) is capable of assisting in the elevation of the ribs only when the shoulders are held in a fixed position, partly through a firm propping up of the arms, and partly by the rhomboid muscles, as is instinctively done by dyspneic patients.

3. *Muscles Acting upon the Sternum, the Clavicle, and the Spinal Column.*—If the head be held in a fixed position by the muscles of the back of the neck, the sterno-cleido-mastoid can enlarge the thorax in an upward direction by raising the manubrium, together with the sternal extremity of the clavicle, thus assisting the scalene muscles. In like manner, but to a lesser extent, the clavicular insertion of the trapezius may become efficient. A stretching of the dorsal portion of the vertebral column must result in an elevation of the upper ribs and a widening of the intercostal spaces, by means of which the inspiratory capacity is substantially increased. During deep inspiration this stretching is effected involuntarily.

4. In forced respiration every inspiration is accompanied by a descent of the larynx and a widening of the glottis. At the same time the palate is raised, in order to allow the air to enter with the least possible resistance.

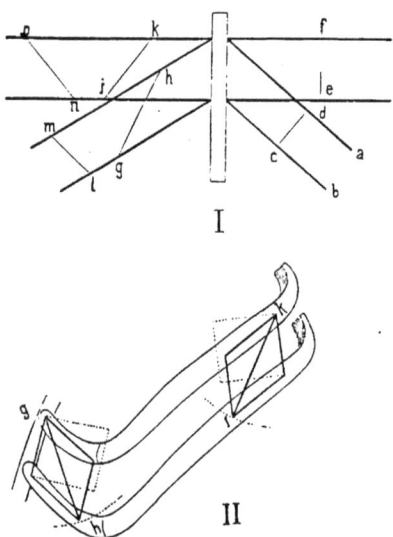

FIG. 81.—I, II. Diagrammatic Representation of the Mechanism of the Intercostal Muscles.

5. Forced respiration is first made evident in the face by an inspiratory dilatation of the nostrils (horse, rabbit). During marked dyspnea the cavity of the mouth is enlarged with each inspiration by a dropping of the jaw (gasping).

B. **Expiration.** — Quiet expiration is accomplished without muscular effort. It is, first of all, dependent principally upon the weight of the thorax, which has a tendency to fall back from its elevated position to the lower expiratory position. This is assisted by the elasticity of the various parts. When the costal cartilages are elevated, their lower borders are slightly rotated from below forward and upward, and their elasticity is thus brought into play. Hence, as soon as the inspiratory forces are relaxed, the cartilages return to their lower and no longer distorted expiratory position. At the same time, the elasticity of the distended lungs draws the thoracic walls, as well as the diaphragm, together on all sides. Finally, the tense, elastic abdominal walls, which become stretched and pushed forward, especially in men, return to their non-distended state of rest when the pressure of the diaphragm from above is released. It is self-evident that when the body is in an inverted position, the effect of the weight of the thorax is removed, and is replaced by the weight of the abdominal viscera pressing upon the diaphragm.

Among the muscles that are brought into action only during forced respiration, the abdominal muscles stand foremost. They diminish the size of the abdominal cavity, and thus press the viscera upward against the diaphragm. The triangular muscle of the sternum draws downward the sternal extremities of the united cartilages and bones of the ribs from the third to the sixth, which have been elevated during inspiration. The posterior inferior serratus depresses the four lowest ribs, the others necessarily following, being assisted by the quadratus lumborum, which is capable of depressing the last rib. According to Henle, however, the posterior inferior serratus fixes the lower ribs so as to withstand the pull of the diaphragm, thus aiding inspiration. Landerer even asserts that in the lower parts of the chest the movements of the ribs enlarge the thoracic cavity downward.

In the erect posture and with a fixed spinal column, deep inspiration and expiration are accompanied by a displacement of the bodily equilibrium. During inspiration the center of gravity is moved slightly forward by the protrusion of the chest and the abdominal walls. In deep inspiration the straightening of the spinal column and the consequent throwing back of the head act as a compensation for the projection of the anterior trunk-wall.

DIMENSIONS AND EXPANSIBILITY OF THE THORAX.

It is of considerable importance for the physician to know the dimensions of the thorax, as well as the extent of its expansion in various directions. With inspiration the thorax is enlarged in all its diameters. The diameters of the thorax are determined by means of calipers; the circumference is measured by means of the centimeter tape-measure.

In well-built men the upper circumference of the chest, close under the arms, measures 88 cm.; in women it is 82 cm. The lower circumference, at the level of the ensiform cartilage, is 82 cm. in men and 78 cm. in women. When the arms are held horizontally the measurement taken during expiration just below the nipples and the angles of the scapulæ equals half the body-length, that is, 82 cm. in men; during deepest inspiration it is 89 cm. At the level of the ensiform cartilage the circumference is about 6 cm. less. In old persons the upper circumference is diminished, being smaller than the lower measurement. Usually the right half of the thorax is some-

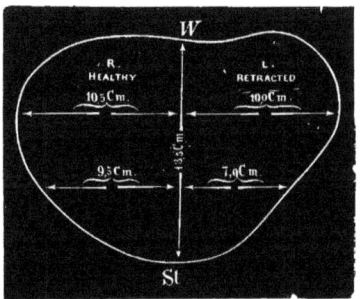

Fig. 82.—Cyrtometer-curve from a Case of Left-sided Retraction of the Thorax in a Twelve-year-old Girl (after Eichhorst).

what larger than the left, on account of the greater muscular development. The longitudinal diameter of the thorax, from the clavicle to the lowest edge of the ribs, varies considerably.

The transverse diameter (distance between the lateral surfaces) is, in men, from 25 to 26 cm., above and below; in women, from 23 to 24 cm. Above the nipples it is about 1 cm. greater. The antero-posterior diameter (measured from the anterior surface of the sternum to the tip of a spinous process) is 17 cm. in the upper part of the thorax, and 19 cm.

in the lower part. Valentin found that during deepest inspiration in men, the circumference of the thorax at the level of the ensiform cartilage increased between $\frac{1}{12}$ and $\frac{1}{7}$; Sibson found this increase to be $\frac{1}{10}$ at the level of the nipples.

Various instruments have been devised to determine the degree of movement (elevation or depression) made by a definite part of the thorax during respiration. The *cyrtometer* of Woillez is quite useful: A measuring chain with stiffly movable

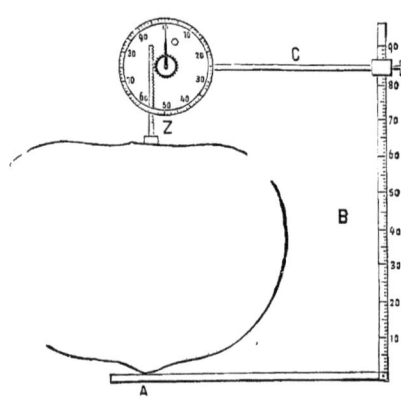

FIG. 83.—Sibson's Thoracometer.

links is applied to the outer surface of the thorax in a definite direction, for example, transversely at the level of the epigastrium or the nipples, or perpendicularly through the mammillary or the axillary line. In two places the links are loosely movable, permitting a removal of the chain without changing its form as a whole. The inner outline of the chain is traced on a sheet of paper, and the form of the thorax is thus obtained (Fig. 82). If the instrument is first applied in the state of expiration, and then during inspiration, there is obtained a diagrammatic representation of the extent of movement in the various parts of the thorax. The same purpose is served by shadow-diagrams or photograms taken at the various periods of respiration. A complicated apparatus has also been constructed of numerous little rods, which rest on the thorax and rise and fall with the respiratory movement and can be fixed in a given position.

The *thoracometer* of Sibson (Fig. 83) measures the elevation of selected parts of the sternum. It consists of two metal rods, joined at right angles, of which one (A) is applied to the spinal column. On B is the movable arm C, which carries at its end the toothed bar (Z) directed perpendicularly downward. The latter is supplied with a spring, and ends below in a ball, which rests upon that part of the sternum to be investigated. The toothed bar, by means of a small cogwheel, moves the indicator (o), which shows the excursions of the sternum on an enlarged scale.

RESPIRATORY EXCURSION OF THE LUNGS.

The boundaries and the size of the lungs in a state of rest on the anterior surface of the thorax are shown in Fig. 34. The shaded boundaries L L indicate the borders of the lungs, while the dotted lines P P show the extent of the parietal pleura (boundaries of the pleural cavity). In the living subject the extent of the lungs can be determined by percussion, that is, by striking the chest-wall (through an interposed thin plate of horn: Piorry's plessimeter) by means of a small cushioned hammer (Wintrich's percussion-hammer). Wherever pulmonary tissue containing air lies in contact with the chest-wall, a sound is obtained like that produced by striking a vessel containing air (resonant percussion-note). Where the underlying parts contain no air, the sound is like that produced by striking the thigh (flat percussion-note). If the parts containing air are thin, or are partly deprived of their air, the note is dull.

Fig. 84 in connection with Fig. 34 shows the boundaries of the lungs on the anterior chest-wall. The apices of the lungs extend above the clavicles anteriorly to a distance of from 3 to 7 cm.; on the posterior surface they extend above the spines of the scapulæ to the level of the seventh spinous process. On the right side the lower border of the lung, in a position of rest, begins at the right edge of the sternum at the insertion of the sixth rib, and extends horizontally outward to about the upper edge of the sixth rib in the mammillary line, and the upper edge of the seventh rib in the axillary line. On the left side (apart from the position of the heart) the lower border of the lung extends downward for the same distance. In Fig. 84 the line at b indicates the lower boundary of the lungs in a state of rest. Posteriorly, both lungs extend to the tenth rib.

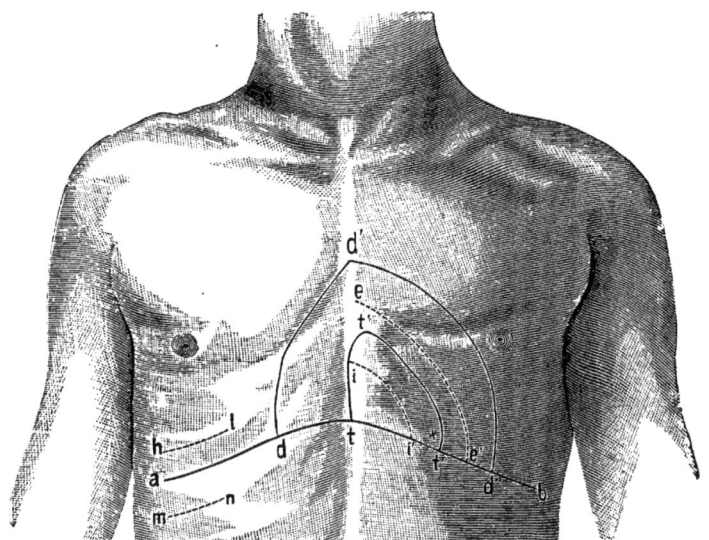

[FIG. 84.—Topography of the Boundaries of the Lungs and the Heart during Inspiration and Expiration (after v. Dusch).

During the deepest possible inspiration the lungs descend anteriorly below the sixth rib as far as the seventh; posteriorly as far as the eleventh rib. At the same time the diaphragm withdraws from the wall of the thorax. During forced expiration the lower borders of the lungs rise almost for the same distance as they sink during inspiration. In Fig. 84 the line m n shows the limit of the border of the right lung during deep inspiration, and h l indicates the same border during complete expiration.

The relation between the border of the left lung and the heart deserves especial attention. In Fig. 34 may be seen an almost triangular space, extending to the left of the sternum from the middle of the insertion of the fourth rib to the sixth rib. This space represents that part of the heart which lies in direct contact with the chest-wall when the thorax is at rest.   Within these limits, represented by the triangle t t' t" in Fig. 84, percussion yields the cardiac dulness; that is, a flat percussion-note is obtained here.

In the larger triangle d d' d" a relatively thin layer of pulmonary tissue separates the heart from the chest-wall, and a dull note is obtained on percussion.   Only outside this triangle is the so-called pulmonary resonance obtained.   On deeper inspiration the inner border of the left lung passes completely over the heart, as far as the mediastinal insertion (Fig. 34), and thus the flat percussion-note is confined to the small triangle t i i'.   On the other hand, during forced expiration the edge of the lung recedes so far that the cardiac dulness embraces the space t e e'.

VARIATIONS FROM THE NORMAL PERCUTORY CONDITIONS IN THE THORAX.

The investigation of the normal percutory conditions and their pathological variations is of the greatest importance for the physician.   Suggestions of percussion (also of the abdomen) are found as far back as Aretaeus (81 A. D.).   The real discoverer, however, is Auenbrugger (d. 1809), whose fundamental work was elaborated especially by Piorry and Skoda; the latter developed the physical theory of percussion (1839).

Over the area of the lungs the otherwise clear, resonant percussion-note is impaired when the lungs have to a greater or lesser extent lost their normal air-content; an airless space of 4 sq. cm. on the outer surface of the lungs will yield a dull note.   The note is impaired also when the lung is compressed from without. The percussion-note is louder or hyperresonant in lean individuals with thin chest-walls, or after deep inspiration, or in the condition of permanent expansion that occurs in emphysematous persons.

It should also be noted whether the percussion-note is of high or of low pitch; this quality being dependent to a certain extent on the degree of tension in the elastic pulmonary tissue, but especially on the tension of the thoracic wall.   As this tension is increased during inspiration, and diminished during expiration, there should be recognized a corresponding difference in the pitch of the note. Deepest inspiration produces a higher pitch, on account of the increased tension of the chest-wall and the lungs; but at the same time the note diminishes in duration and intensity, as the more highly stretched parts possess a diminished amplitude of vibration.   Sometimes in the terminal phase of the deepest possible inspiration there occurs still another change in the percussion-note, in that there is produced a certain restoration of the depth and intensity, falling short, however, of the original volume.   During complete expiration the intensity is lessened and the pitch lowered.

Percussion of the larynx and the trachea yields a clear tympanitic note, whose pitch depends upon the size of the cavity.   The note is highest when the mouth and the nose are open, or when the tongue is protruded, or when straining efforts are made with closed glottis; it becomes lower when the head is extended backward, or during the act of swallowing, as well as during intonation.   It is higher at the end of deep inspiration than during expiration.   Affections of the lungs that lessen the normal tension lower the pitch of the note.

When the percussion-note partakes of a drum-like character, approaching a musical sound, with distinguishable high and low pitch, it is termed tympanitic. If a hollow rubber ball applied to the ear be tapped with the finger, a typical tympanitic sound will result, the pitch of which is higher the smaller the diameter of the ball.   Tapping the trachea in the neck will also yield a tympanitic note. The tympanitic note consists of a primary tone, together with several harmonic overtones, arising from an air-space surrounded by relaxed and movable walls (the non-tympanitic tone consists of the membrane-tone of a tightly stretched wall). The tympanitic note in the chest is always of pathological origin.   It is found in the presence of a cavity within the lung-substance (when the mouth is closed, and especially when the nose is closed at the same time, the note becomes deeper), also in the presence of air in a pleural cavity, as well as in association with diminished tension of the pulmonary tissue.   The tympanitic note is closely allied to metallic tinkling, which arises in large, pathological, pulmonary cavities, as well as when the pleural cavity contains air, when the conditions are suitable for a more uniform reflection of the sound-waves within the cavity. When a percussion-stroke is made over cavities, especially in the upper anterior part of the lung, the air at times escapes with a peculiar ringing and hissing sound—the cracked-pot sound (or coin-sound).

In practising percussion it should be observed by the sense of touch whether the underlying parts offer a feeling of greater or lesser resistance to the stroke; and at the same time the vibratory power may be noted. Under normal conditions small vibratory power is associated with a well-developed bony framework, thick soft parts, and tense muscles. Pathologically, lessened vibration always occurs in connection with an airless condition of the lungs, and is associated with a dull percussion-note. Diminution of the resistance to the percussion-stroke is found normally in slender chests. Pathologically, it occurs when there is a considerable amount of air under the chest-wall, hence in the presence of pneumothorax and of abnormal expansion of the lungs by means of air.

If the handle of a tuning-fork be placed upon the chest-wall, the fork will sound loud over spaces filled with air, and will yield a weak note over spaces containing little or no air (Baas' phonometry).

THE NORMAL RESPIRATORY SOUNDS.

By listening over the chest-wall, either directly or by means of a stethoscope, the *vesicular murmur* can be heard during inspiration, wherever the lungs are in contact with the walls of the thorax. The character of this sound can be imitated if the mouth be placed in the position necessary for the act of sipping, and a sound between f and v be softly emitted. The sound is a sipping, rustling, hissing one. It is due to the sudden expansion of the pulmonary vesicles by the entrance of inspired air (hence the term vesicular) and also to the friction of the air passing through the alveoli. The sound is at times softer, at times louder. It is constantly louder in children under the age of twelve years, as the air-vesicles are one-third narrower than in adults, and cause greater friction with the entering air.

During expiration the air, when leaving the vesicles, gives rise to a weak puffing sound of an uncertain soft character.

The cardiopulmonary murmur heard in the vicinity of the heart when the latter contracts during systole likewise has a vesicular character.

Bronchial breathing may be heard in the larger air-passages during inspiration and expiration, and resembles the sound of a loud, sharp h or sh. Outside of the neck (larynx and trachea) it may be heard between the shoulder-blades at the level of the fourth dorsal vertebra (point of bifurcation), especially during expiration. It is somewhat louder to the right, on account of the larger caliber of the right bronchus. In all other parts of the thorax it is obscured by the vesicular murmur. The bronchial breathing arises entirely in the larynx, from the formation of air-vortices, by reason of the marked constriction of the air-passage at the glottis. This laryngeal stenosis-sound causes a resonance of the tracheo-bronchial air-column, and thus produces the specific character of bronchial breathing, which the listener hears transmitted along the large tubes of the bronchial tree.

It has been maintained that, if the air-filled lungs of an animal be applied to the neck over the larynx or trachea, the bronchial breathing produced there will become vesicular. In that case it must be supposed that vesicular respiration arises from a weakening and acoustic transformation of tubular respiration by its transference through the air-vesicles. Added to this is the fact that it is impossible to produce any sound by forcibly driving air through narrow straws.

During forced respiration rustling sounds often arise at the mouth and nostrils; with these sounds the primary tone of the oral cavity (usually the vowel-sound ah) is often mingled in mouth-breathing.

PATHOLOGICAL RESPIRATORY SOUNDS.

The recognition of the succussion-sound, the friction-sound, and many catarrhal sounds dates back to Hippocrates (460–377 B. C.). The actual foundation of auscultation on a physical basis was laid by Laennec (1816), and its classical development is due to Skoda (1839).

Bronchial breathing arises over the entire area of the lungs, either when the air-vesicles have become airless (through exudation) or when the lungs are compressed from without. In both cases the condensed pulmonary tissue conducts the bronchial respiration to the walls of the thorax. Bronchial breathing will also be heard over pathological cavities of considerable size that communicate with a large bronchus, provided the cavities lie sufficiently near the thoracic wall and have walls of considerable resistance. If there is no movement of air in the cavity, the sounds may be wholly conducted out through the trachea; or during expiration a stenosis-sound (like that at the glottis) may arise in the communicating bronchus, and may be rendered *amphoric* by the resonant cavity.

Amphoric breathing resembles the sound produced by blowing across the mouth of a bottle. It may arise when there occurs in the lungs a cavity at least the size of a fist, through which the air passes in such a manner that there is produced the characteristic sound with a peculiar metallic echo. If the lung is partly expansible and contains air, and the pleural cavity also contains air, the resonance of the latter, together with the exchange of air in the lung, will also produce the amphoric sound.

If the respiratory sounds have no definite character, so that they oscillate between vesicular and bronchial breathing, they are termed *indefinite respiratory sounds*. Frequently a deep respiration or expectoration of mucus will make the character of the sound more evident.

If the air meets with resistance in its passage through the lungs, various phenomena may result: (*a*) At times the air-vesicles are not all filled simultaneously, but intermittently. This occurs (especially at the apices) when partial swelling of the walls of the air-passages obstructs the steady interchange of air; *cogwheel respiration* is the result. Occasionally this is heard in perfectly normal lungs, when the muscles of the chest contract in an intermittent fashion. (*b*) If a bronchus leading to a pulmonary cavity is narrowed in such manner that the air meets with a temporary resistance, the inspiratory sound is at first like that of a loud G, and then goes over during the latter two-thirds of inspiration into a bronchial or amphoric sound. This is termed a *metamorphosing sound*. (*c*) Rales are produced in the larger air-passages when the air causes bubbling of the contained mucus. In the smaller air-spaces rales arise either when the walls of the latter are separated from the fluid contents during inspiration, or when their walls are in contact and are suddenly separated from each other. Rales are distinguished as *moist* (arising in watery contents) or as *dry* (in tough, tenacious contents); further, as *inspiratory* or *expiratory*, or *continuous;* also *coarse, fine,* or *irregular* rales, the high-pitched *crepitant* rales, and finally the *metallic tinkling* rales produced by the resonance of large cavities. (*d*) If the mucous membrane of the bronchi is so swollen or so covered with mucus that the air must force its way through, there arises frequently in the larger passages a deep humming purr—*sonorous rhonchus;* and in the smaller tubes a clear whistling sound—*sibilant rhonchus.* In cases of widespread bronchial catarrh a thrill may often be felt in the chestwall—*bronchial fremitus*—caused by the numerous rales.

When the lung is collapsed and the pleural cavity contains fluid and air, a sound may be heard on shaking the chest, similar to that produced by shaking a large bottle containing water and air—the *succussion-splash* of Hippocrates. Rarely a higher-pitched similar sound may be heard in pulmonary cavities the size of a fist.

When the opposed layers of the pleura are roughened by inflammatory processes, and rub against each other in the act of respiration, a *friction-phenomenon* is produced. This may be partly felt (often by the patient himself) and partly heard. The sound is usually creaking, and may be compared to that produced by bending new leather. Friction-sounds are produced also by the heart's action between the two layers of the diseased roughened pericardium.

During loud speaking or singing the chest-wall vibrates—*vocal fremitus*—as a consequence of the propagation throughout the bronchial tree of the vibrations of the vocal bands. This vibration naturally is most pronounced in the region of the trachea and the large bronchi. If the ear be applied to the chest-wall, the voice can be heard only as an unintelligible hum. If the pleural cavity contains air or a large effusion, or if the bronchi are occluded by large quantities of mucus, the vocal fremitus is weakened or entirely absent. On the other hand, all factors that cause bronchial breathing will increase the vocal fremitus. Hence, the latter will be more marked also in those localities where bronchial breathing is heard, even under normal conditions. The ear under such circumstances will hear the sounds conducted to the chest-wall with increased intensity. This is termed *bronchophony*.

If a pleural effusion or a pulmonary inflammation causes a flattening of the bronchi, the sound of the voice in the chest sometimes assumes a peculiar bleating quality—*egophony*.

Doubtless the gradations of increased or diminished fremitus could be readily demonstrated by means of the sensitive flame (observed in a rotating mirror) or by the use of the microphone. For the former there should be employed an apparatus similar to the gas-sphygmoscope, with the lower part widened in the shape of a funnel.

PRESSURE IN THE AIR-PASSAGES DURING RESPIRATION.

If a manometer be fastened in the trachea of an animal in such a manner that respiration is not interfered with, the instrument will show a negative pressure (—3 mm. of mercury) during inspiration, and a positive pressure during expiration. Donders has modified this experiment for man by introducing a U-shaped manometer-tube through one nostril, and instructing the subject to breathe quietly through the other nostril with the mouth closed. He found that during each quiet inspiration the mercury showed a negative pressure of 1 mm., and during each expiration a positive pressure of 2 or 3 mm. Aron experimented with patients having a tracheal fistula as the result of operation, and found during inspiration a pressure of from —2 to —6.6 mm. of mercury, during expiration from +0.7 to +6.3 mm. of mercury. In speaking, the corresponding fluctuation was from —6 to +7, and when coughing from —6 to +46.1.

As soon as the air is drawn in and expelled with greater force, the fluctuations of pressure become more marked, especially in the acts of speech, singing, and coughing. If forced respiration be practised with the mouth and one nostril closed, so that the respiratory canal communicates only with the manometer, the greatest inspiratory pressure is —57 mm. (between 36 and 74), and the greatest expiratory pressure is +87 (between 82 and 100) mm.

Notwithstanding the higher expiratory pressure, it must not be inferred that the expiratory muscles are stronger than those of inspiration; for during the latter act a series of resisting forces must be overcome, leaving a much diminished supply of force for the aspiration of the mercury. These resisting forces are: (1) The elastic tension of the lungs, which amounts to 6 mm. during complete expiration, but reaches 30 mm. during deepest inspiration. (2) The lifting of the weight of the thorax. (3) The elastic torsion of the costal cartilages. (4) The depression of the abdominal viscera and the elastic distention of the abdominal walls. All these resisting forces aid the expiratory muscles during expiration. With these facts in view, there is no doubt that the combined strength of the inspiratory muscles is greater than that of the expiratory muscles.

As the lungs, by reason of their elasticity, have a tendency to collapse, they naturally exert a negative pressure within the thoracic cavity. In dogs this amounts to from 7.1 to 7.5 mm. of mercury during inspiration, while in expiration it is naturally less, namely only 4 mm. The analogous values obtained by different investigators on the dead body vary; Hutchinson fixes them at 4.5 mm. and 3 mm.

The greatest pressure during inspiration and expiration seems small when compared to the blood-pressure in the large arteries. If, however, the pressure-values obtained for the respired air be estimated for the entire superfices of the thorax, considerable results are obtained.

To measure the muscular respiratory power in case of illness, a U-shaped mercurial manometer may be employed, provided with an attachment suitable for introduction into a nostril or the mouth (Waldenburg's *pneumatometer*). The inspiratory pressure alone may be reduced (in the presence of almost all diseases impairing the expansion of the lungs), or only the expiratory pressure may fall (in cases of emphysema and of asthma), or both may be weakened (as occurs in feeble persons).

If a forced inspiration rarefies the air in the air-passages, the trachea and bronchi become narrowed and shortened; the reverse occurs during expiration.

If a lung be inflated, air will steadily escape through the walls of the alveoli and trachea. The same thing takes place during violent expiratory efforts (cutaneous emphysema attending whooping-cough), so that pneumothorax, entrance of air into the blood-vessels, and even death may result.

If a dog be made to breathe through Müller's valve, by means of which the resistance to respiration may be increased at will, it is found that a pressure of 40 cm. of water is still readily overcome, that a higher pressure can be overcome for a short time, and one of 70 cm. not at all.

Until birth the airless lungs lie collapsed (atelectatic) in the chest-cavity, and fill it, so that pneumothorax is not produced if the thorax be opened in a dead fetus. Even in children that have lived for eight days and have breathed normally, the lungs do not collapse when the pleural cavity is opened, but remain in contact with the chest-wall. It is only after further growth that the thorax becomes so large that the lungs must expand under elastic tension; only then will opening of the thorax cause the lungs to contract into a smaller volume. Hermann calls attention to the fact that a lung containing air cannot be emptied by pressure from without. The reason for this is that the small bronchi will be closed by the pressure before the air can leave the alveoli. The muscles of expiration, therefore, have not the power to compress the lungs until they are airless; but, on the other hand, the inspiratory muscular power is sufficient to expand the lungs beyond the state of elastic equilibrium. Hence, the physical attributes of the lungs limit, to a certain extent, the mechanism of respiration: the muscles of inspiration expand the lungs and at the same time increase their elastic tension, while the expiratory muscles can only diminish the tension, without being able to abolish it altogether.

MOUTH-BREATHING AND NASAL BREATHING.

Quiet respiration is usually performed with the mouth closed, provided the nose be unobstructed. The current of air passes through the naso-pharyngeal cavity, and there undergoes certain changes: (1) Its temperature is increased to the extent of $\frac{5}{6}$ of the difference between its original temperature and that of the body. (2) At this increased temperature it is saturated with aqueous vapor. These changes are made so that the cold, dry air does not irritate the lining of the lungs. (3) Dust-particles may cling to the mucus covering the irregular walls of the air-passages, and are again conveyed outward by the ciliated epithelium. The nasal secretion possesses qualities harmful to many bacteria (for example, anthrax-bacilli), thus demonstrating the salutary effect of nasal breathing when there is danger of contagion. (4) Finally, by means of the sense of smell bad air and air impregnated with injurious admixtures can be recognized. When the mouth is open no current of air passes through the nose during respiration.

Pathological.—Permanent obstruction of the nose, leading to exclusive mouth-breathing, may result in a long series of harmful effects; namely, catarrhal conditions of the pharynx, the air-passages, and the middle ear, abnormal formations in the bones of the mouth and the nose, pains in the facial muscles, changes in speech, disturbances of intellect (difficulty in fixing the attention).

Another important phenomenon is the appearance of edema of the lungs; that is, an exudation of serum from the blood into the pulmonary alveoli. The causes of this condition are: (1) marked obstruction to circulation in the aortic system; for example, after ligation of all of the carotid arteries, or of the arch of the aorta in such a position that only one carotid remains pervious; (2) occlusion of the pulmonary veins; (3) cessation of action in the left ventricle (following mechanical injury), while the right ventricle still continues to beat. All of these causes will produce at the same time anemia of the brain, resulting in anemic irritation of the vasomotor center, and consequent contraction of the small arteries. This will cause an increased amount of blood to enter the veins and the right heart, whose driving power increases the pulmonary edema.

v. Basch believes that an overfilling of the pulmonary capillaries diminishes the elasticity of the alveoli, thus making the latter to a certain extent more rigid. The expansibility, therefore, of the lungs is diminished.

MODIFIED RESPIRATORY ACTS.

There are a number of characteristic, partly involuntary, partly voluntary, variations of the respiratory movements, to which the not altogether suitable term *abnormal respiratory acts* has been applied.

Coughing consists in a sudden violent expiratory effort, usually succeeding a deep inspiration and closure of the glottis, during which effort the glottis is sprung open, and any solid, fluid or gaseous substance that may be irritating the respiratory mucous membrane is expelled. The lips are parted during this act. It may be a voluntary or a reflex act, in the latter case being subject to the will only to a certain degree.

Hawking consists in a rather long expiratory effort through the narrow space between the root of the tongue and the depressed soft palate for the purpose of removing foreign bodies. If the hawking be accomplished in an intermittent fashion, it is accompanied by a springing open of the glottis (mild, voluntary coughing). This act is performed only voluntarily.

Sneezing consists in a sudden expiratory effort through the nose, accompanied by a sudden opening of the naso-pharynx, previously closed by the soft palate. The purpose is to expel mucus or foreign bodies. It is very seldom performed with the mouth open, and is preceded by a single or by repeated spasmodic inspiration. The glottis is always wide open. This act occurs only as a reflex through irritation of the sensory nerves of the nose, or as a result of a bright light suddenly falling upon the retina. The reflex may be to a certain extent inhibited by marked excitation of sensory nerves, such as rubbing the nose, or pressing the hyoid bone forcibly upward. Habitual use of nasal irritants, such as snuff, blunts the sensory nerves against reflex excitation. Coughing and sneezing rarely occur simultaneously.

Snorting and Blowing the Nose ; Snuffing ; Sniffing.—Noisy, forced breathing through the nose is designated snorting. Blowing the nose consists in a strong, noisy, expiratory effort made through nostrils that have been narrowed, either by the fingers or by the muscles of the nose and the upper lip, the object being to remove either foreign bodies or mucus. Snuffing consists of drawing substances up into the nose by a noisy inspiration, the mouth being closed, and the nostrils often being narrowed by the action of the muscles of the nose and the upper lip. Sniffing consists in drawing air up into the nose by a succession of short inspiratory efforts, for the purpose of smelling. The act is frequently accompanied by rustling noises and movements of the nostrils, while the mouth is held closed. All these actions are voluntary.

Snoring results from breathing with the mouth open, the current of air during both inspiration and expiration causing noisy, vibrating movements of the relaxed soft palate. It usually occurs involuntarily during sleep, but it may also be produced voluntarily.

Gargling consists in the noisy slow escape of the expired air in the form of bubbles through a mass of fluid held between the root of the tongue and the soft palate, while the head is thrown back. The act is voluntary.

Crying is called forth by the emotions, and consists in short, deep inspirations, with prolonged expirations, the glottis being narrowed, and the muscles of the face and jaw being relaxed (with contraction of the zygomaticus minor); tears are secreted, and lamenting, inarticulate sounds are often emitted. In conjunction with intense, prolonged crying there often arise sudden, spasmodic; involuntary contractions of the diaphragm, which, when attended with valve-like approximation of the vocal bands, give rise to the inspiratory sound known as **sobbing.** This act is purely involuntary. The sobbing that occurs so frequently during the agonal period may be explained by the electrical influence of the contraction of the heart on the phrenic nerves, which become highly irritable in the act of dying.

Sighing is a prolonged respiratory movement, usually accompanied by a mournful sound, often aroused involuntarily by painful emotions.

Laughing consists in a quick succession of short expirations through vocal

15

bands that are stretched for high notes, and are alternately approximated and separated, while characteristic, inarticulate sounds are emitted from the larynx, with vibrations of the soft palate. The mouth is usually open, and the face is drawn into a characteristic position by the zygomaticus major (not the risorius muscle). Laughing is usually aroused involuntarily by agreeable conceptions, or by feeble, sensory irritation, such as tickling. It may to a certain extent be repressed by the will, as by forcibly closing the mouth and holding the breath; also by painful irritation of sensory nerves, as by biting the tongue or the lips.

 Yawning consists in a prolonged, deep inspiration, with the mouth, the palatal arch and the glottis widely open, successively calling into play numerous inspiratory muscles. Expiration is shorter, and both are often accompanied by a prolonged, characteristic sound. There also occurs frequently a general stretching of the bodily muscles. The act is always involuntary, being usually incited by sleepiness or monotony.

CHEMISTRY OF RESPIRATION.

 The problem here is to estimate qualitatively and quantitatively the gases expelled during respiration. If the results be compared with the gaseous composition of inspired, atmospheric air, a picture may be obtained of the interchange of gases occurring during respiration.

QUANTITATIVE ESTIMATION OF THE CARBON DIOXID, THE OXYGEN, AND THE AQUEOUS VAPOR IN GASEOUS MIXTURES.

Estimation of the Carbon Dioxid.

 The *volume* of carbon dioxid may be estimated by means of Vierordt's *anthracometer* (Fig. 85, II). The gaseous mixture is received and enclosed in a graduated tube r r, previously filled with water, and provided at one end with

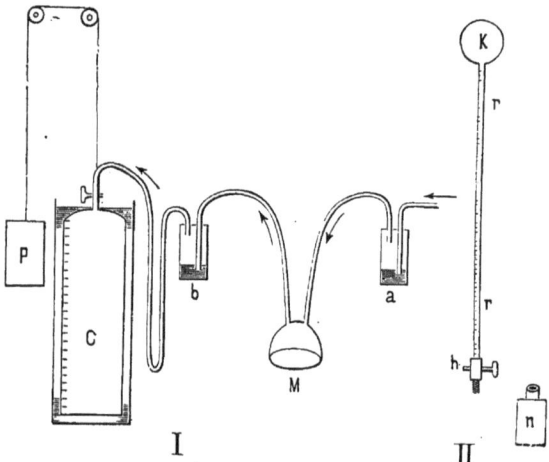

FIG. 85.—I. Apparatus for the Collection of Expired Air (after Andral and Gavarret). II. Carl Vierordt's Anthracometer.

a bulb of known capacity. The bottle n, filled with a solution of potassium hydrate, is then screwed on the end-piece h. The stop-cock is opened, and the potassium-solution is allowed to run up into the tube, the latter being agitated

until all the carbon dioxid is absorbed by the potassium, with the formation of potassium carbonate. Then the solution is allowed to run back into the
bottle, the stop-cock is closed, and the potassium-bottle is removed. The end of
the tube is dipped into water, and the latter is allowed to rise in the tube. The
volume of water thus admitted is equal to the volume of carbon dioxid removed
by the potassium-solution.

Determination by Weight.—A considerable volume of the gaseous mixture
is passed through Liebig's bulbs, filled with a solution of potassium hydrate and
arranged in a combination such as that of Scharling's apparatus (Fig. 86, e, f, g).

Determination by Titration.—A considerable volume of the air to be examined is conducted through a definite quantity of a known solution of barium
hydrate. The carbon dioxid combines to form barium carbonate. The solution
is then neutralized with a titrated solution of oxalic acid. The quantity of oxalic
acid necessary to neutralize the remaining barium hydrate varies inversely with
the amount of barium already combined with the carbon dioxid.

Estimation of the Oxygen.

The volume of oxygen may be determined in two ways: (a) By combining the
gas with potassium pyrogallate.' Vierordt's anthracometer may be employed for
this purpose, substituting a solution of potassium pyrogallate for that of potassium
hydrate. (b) By explosion in an eudiometer.

Estimation of the Aqueous Vapor.

The volume of air to be examined is allowed to pass either through a bulb-
apparatus containing concentrated sulphuric acid, or through a tube filled with
pieces of calcium chlorid. In both cases the water is energetically abstracted,
and the increase in weight will give the amount of water in the air examined.

METHODS OF INVESTIGATION.

Collecting the Expired Air.

If only the gases exhaled from the lungs are to be collected, the bell-jar
of the spirometer (Fig. 76) may be used, suspended in a concentrated solution
of sodium chlorid to limit the gas-absorption. Andral and Gavarret permitted
several successive expirations to be made into a large bell-jar (Fig. 85, I, C).
For this purpose a mouth-piece M was applied in an air-tight manner over the
mouth, the nostrils being closed; the direction of the air-current was regulated by

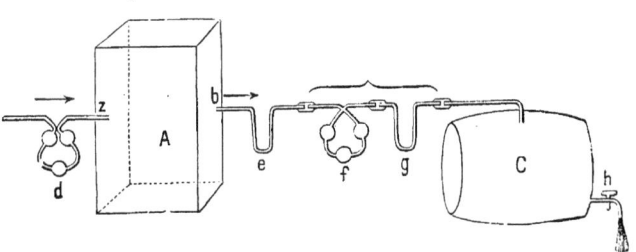

FIG. 86.—Respiration Apparatus of Scharling.

means of two so-called Müller's mercurial valves (a, b), which allowed the air to
pass only in the direction of the arrows.

If the gases given off from the skin during perspiration are to be investigated,
as well as those from the lungs, then the subject must be placed in a closed chamber, from which the gases may be withdrawn for experimental purposes.

The Most Important of the Respiration Apparatus.—(a) The apparatus of Scharling (Fig. 86) consists primarily of a closed chamber A, capable of containing
a human being. The chamber has an afferent opening z, and an efferent opening
b. The latter is connected with an aspirating contrivance C, consisting of a good-
sized barrel filled with water. It is evident that when the water flows out of the

barrel, an uninterrupted stream of fresh air enters the chamber A, and the air in the chamber, mixed with the respired gases, escapes toward the barrel. Connected with the afferent opening z is a set of Liebig's bulbs d, filled with a solution of potassium hydrate through which the entering air passes and is deprived of its carbon dioxid, so that the subject breathes air completely free of carbon dioxid. Upon leaving the efferent opening b the air is first conducted through the tube e, in which the aqueous vapor is absorbed by sulphuric acid, and its amount may be determined by the increase in weight of the tube. Then the air passes through the potassium-bulbs f, where all the carbon dioxid is absorbed. The tube g, filled with .sulphuric acid, is intended for the purpose of absorbing. the aqueous vapor conveyed by the air from f. The increase in weight of f + g represents the weight of the absorbed carbon dioxid. The volume of air interchanged may be estimated from the contents of the barrel.

(b) Regnault and Reiset's apparatus (Fig. 87) consists of a bell-jar R, in which is placed the animal (dog) to be experimented upon. Surrounding this

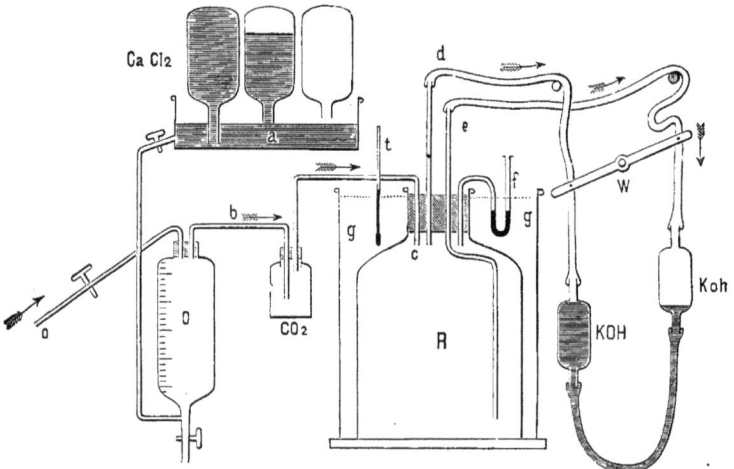

FIG. 87.—Diagrammatic Representation of Regnault and Reiset's Respiration Apparatus.

jar is a cylinder g g, which may be used for calorimetric observations, a thermometer t being introduced for this purpose. The bell-jar has leading into it the tube c, through which is introduced a measured quantity of oxygen (Fig. 87, O), which (Fig. 87, CO_2) has given off to the potassium hydrate any remaining admixture of carbon dioxid. The oxygen in the measuring vessel O is forced toward the bell-jar R by a solution of calcium chlorid, coming from a basin provided with large bottles (Ca Cl_2). From R pass the tubes d and e, connected by rubber tubes with the communicating potash-bottles (K O H, k o h), which may be alternately raised and lowered by means of the scale-beam w. By these means the air is aspirated from R, and the carbon dioxid is absorbed by the solution of potassium hydrate. At the end of the experiment the increase in weight of the bottles represents the quantity of carbon dioxid expired. The amount of oxygen inspired is measured directly in the measuring vessel O. Finally, the manometer f shows whether there is a difference between the air-pressure within the jar and that on the outside.

(c) The most complete apparatus is that of v. Pettenkofer (Fig. 88). A chamber Z, made of metal and provided with a door and a window, has an opening for the entrance of air at a. A double suction-pump P P_1, driven by steam, renews continuously the air in the chamber. This air is first conducted into the vessel b, which is filled with pumice-stone saturated with water. Here the air becomes saturated with aqueous vapor, and then passes through the gasometer c, which indicates the total volume of the interchanged air; the latter is then discharged into the outer atmosphere.

The main tube x, leading from the chamber, carries a mercurial manometer q, for the detection of possible variations in pressure within the room. This tube gives off a branch tube n, through which the air passes for chemical examination. The air in this tube is driven by a suction-apparatus M M_1, constructed on the principle of Müller's mercurial valve, and worked by the same steam-engine as

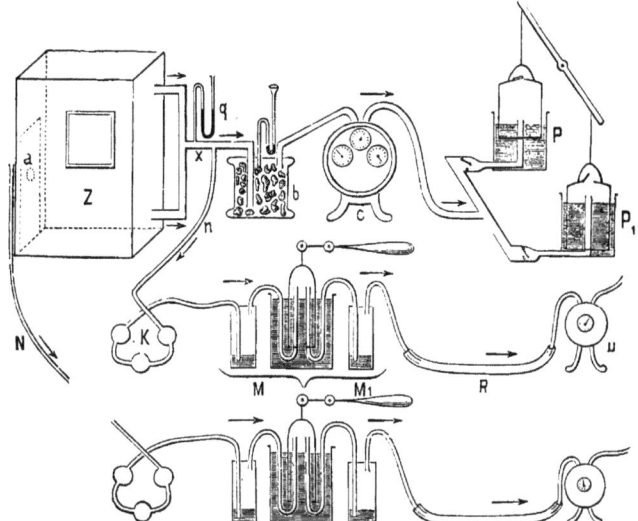

.Fig. 88.—Diagram of v. Pettenkofer's Respiration Apparatus.

the pump P P_1. Before entering the pump the air passes through the sulphuric-acid bulbs, from whose increase in weight the amount of aqueous vapor can be estimated. After leaving the pump the air passes through the tube R, filled with baryta-water, which absorbs the carbon dioxid. The quantity of air passing through the branch tube n is then measured by the gasometer u, after which it finally passes into the atmosphere. The second branch tube N provides for an examination of the air before entering the chamber, by an apparatus identical with that placed on the tube n. The excess of carbon dioxid and water found in n over that in N is due to the respiratory activity of the subject placed in the chamber.

COMPOSITION AND PROPERTIES OF ATMOSPHERIC AIR.

The dry atmosphere contains:

Gas.	Percentage in Weight.	Percentage in Volume.	
O	23.015	20.922	Including 1 per cent. in volume of argon, together with helion, and 1 part of krypton in 20,000 parts of air.
N	76.985	79.02	
CO_2		0.029–0.034	

The air contains likewise xenon, neon, coronium (lighter than hydrogen), and less than one-millionth part of ætherium (which latter possesses a specific gravity only $\frac{1}{10000}$ that of hydrogen, but a power of heat-conduction 100 times as great as that element, and a density of only $\frac{1}{11000}$ that of the air). These elements have not been investigated physiologically. Ætherium is probably a composite substance, and perhaps plays the rôle that has been previously ascribed to luminiferous ether.

Aqueous vapor is always present; its amount usually increases with the height of the temperature. With reference to the humidity of the air there must be distinguished: (*a*) the absolute humidity, that is, the quantity of aqueous vapor contained in a volume of air; (*b*) the relative humidity, that is, the quantity of aqueous vapor contained in a volume of air in relation to its temperature. The latter increases with rising temperature.

The relative humidity is determined either by means of the hygrometer of Klinkerfues or by the psychrometer of August. The latter consists of two accurately graduated thermometers, the bulb of one being kept constantly moistened by means of a wet cloth. As a result of evaporation of the water on the bulb, cooling will take place, and the fall of the thermometer will vary directly with the rapidity of evaporation, that is, with the dryness of the atmosphere. From the difference in the readings of the two thermometers the tension of the aqueous vapor in the air may be calculated according to the formula: $e = e^1 - k \times (t - t^1) \times b$; in which e represents the desired tension of the aqueous vapor in the air at the prevailing temperature, as indicated by the dry thermometer; e^1 the tension of the aqueous vapor that prevails when the air is completely saturated with watery vapor at the temperature of the moist thermometer (to be ascertained from works on physics); b the state of the barometer in millimeters of mercury; t the temperature of the dry thermometer, and t^1 that of the moist thermometer (expressed in degrees Centigrade); and k an empirically obtained constant $= 0.001$. Experience has taught that man breathes best in an air that is not completely saturated with watery vapor in accordance with its temperature, but only to 70 per cent. of that amount. Air that is too dry irritates the mucous membranes of the respiratory organs; while air that is too moist arouses a feeling of uncomfortable oppression, and in warmer air a sensation of oppressive sultriness. At a lower temperature (15° C.) dry air is more comfortable than moist air; at from 24° to 29° C. dry air feels cooler than moist air. With marked dryness of the atmosphere a temperature of 29° C. is well borne; but exceedingly damp air becomes unendurable for any length of time at 24° C. In the living room and in the sick-room attention should, therefore, be paid to the correct degree of atmospheric humidity. (Sprinkling with water, or in winter placing a basin of water on the stove may be resorted to.) Rooms that are too damp, on account of dampness of the walls or the floor, are prejudicial to health.

The following factors are known to influence the absolute quantity of aqueous vapor in the air: (1) At the sea-shore during the day the amount is increased with a rising temperature, and diminished with a falling temperature. (2) In the flat, inland country the humidity rises from sunrise to noon, then diminishes until evening; rises again during the first part of the night, and finally falls again. (3) On high mountains the mid-day decrease in humidity does not occur. (4) Southwestern winds in summer are accompanied by the greatest humidity, while east winds in winter bring the lowest degree of humidity.

With reference to the relative amount of moisture it is to be noted: (1) that it is usually greatest at sunrise, and least toward noon; (2) that it is diminished on high mountains; (3) that it is greater in winter than in summer; (4) that it is usually greater with south and west winds than with north and east winds.

In the course of the year's changes, that air which is found to be the richest in water absolutely is the poorest relatively. For example, the air in summer contains absolutely about three times as much watery vapor as in midwinter, and still the summer air is relatively dryer than that of winter. In the course of the seasons the absolute humidity rises and falls with the mean temperature. The average relative humidity amounts to about 70 per cent. in temperate climates.

With increasing elevation above sea-level the density of the air diminishes.

It likewise diminishes with increase of temperature.

With every increase of about 186 meters in elevation above the surface of the earth, the temperature (irregularly, it is true) falls 1° C.

Above a level of 4000 meters the cold increases in greater proportion; at a level of 7000 meters the most severe degree of cold prevails without variation, being the same at all seasons.

COMPOSITION OF EXPIRED AIR.

The expired air is rich in carbon dioxid, of which it contains on an average 4.38 per cent. by volume (from 3.3 to 5.5 per cent.) during quiet respiration. The amount of carbon dioxid is, therefore, more than 100 times greater than in the atmosphere.

The expired air contains less oxygen (on an average 4.782 per cent. less by volume) than inspired, atmospheric air, namely, only 16.033 per cent. by volume.

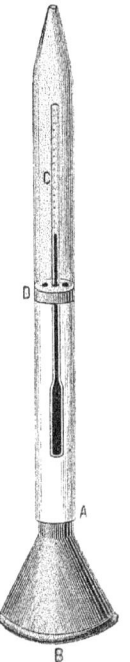

Hence, during respiration there is more oxygen taken into the body from the air than there is carbon dioxid expelled; so that the volume of the expired air is from one-fiftieth to one-fortieth less than the volume of inspired air (under the same conditions of temperature, humidity, and pressure). This relation of the expired carbon dioxid to the inspired oxygen (4.38 : 4.782) is termed the *respiratory quotient:*

$$\frac{CO_2}{O} \left(= \frac{4.38}{4.78} \right) = 0.916.$$

A small excess of nitrogen is admixed with the expired air. It has been found that not all of the nitrogen taken up with the food appears again in the excretions (urine and feces).

The expired air during quiet respiration is saturated with aqueous vapor. It is, therefore, evident that, by reason of the changes in the amount of water contained in the air, a varying quantity of water must be excreted by the body through the lungs. With rapid respirations Moleschott observed the percentage of aqueous vapor to fall. The surrounding temperature also has an influence on the amount of water given off: the minimum occurs at 15° C., while below this point the amount increases moderately, and above the quantity rises rapidly.

The expired air possesses a considerable degree of heat (on an average 36.3° C.), which at moderate external temperatures approaches quite closely that of the body; but even with extreme variations of the surrounding temperature the degree of heat maintains itself within the same limits.

Fig. 89.—Apparatus for Measuring the Temperature of the Expired Air.

Valentin and Brunner employed the instrument represented in Fig. 89 to determine the temperature of the expired air. It consists of a glass tube A **A**, with a mouth-piece B and an inserted, delicate thermometer C. Inspiration is made through the nose, and the air is slowly expelled through the mouth-piece into the tube.

Temperature of the Air.	Temperature of the Expired Air.
− 6.3° C.	+29.8° C.
+17°–19° C.	+36.2–37° C.
+44° C.	+38.5° C.

It would certainly be highly interesting to determine whether the temperature of the expired air undergoes change by reason of inflammations, disturbances of the circulation, or degenerations in the lungs.

Mosso and Rondelli allowed dogs to breathe air at a temperature of from 150° to 160° C., and found that the air in the bronchi was of a higher temperature (39.3° or 37.8° C.) than the rectum.

The diminution in volume of the expired air mentioned already is compensated for by the warming of the inspired air in the air-passages, and by the tension of the contained aqueous vapor, so that the volume of the air expired is even one-ninth greater than that of the air inspired.

Exceedingly small quantities of ammonia are admixed with the expired air, amounting to about 0.0204 gram in twenty-four hours; they are probably evolved from the blood.

Small quantities of hydrogen and marsh-gas (CH_4), both absorbed from the intestines, are likewise exhaled.   Reiset observed that in herbivorous animals the marsh-gas exhaled in twenty-four hours amounted to as much as 30 liters.

The aqueous vapor condensed by low temperature from the expired air of some persons acts as a poison when injected subcutaneously, in consequence of the presence of a volatile base.   These are exceptions, however.

EXTENT OF THE DAILY INTERCHANGE OF GASES.

As more oxygen is normally taken in than is excreted in the carbon dioxid (equal volumes of oxygen and carbon dioxid containing equal quantities of oxygen), a part of the oxygen taken in must be used for other oxidation-purposes.   According to the extent of the latter, there must be considerable variation in the relation of the inspired oxygen to the expired carbon dioxid (the quotient $\frac{CO_2}{O}$, which is given as being on an average 0.916 during normal, quiet respiration).   Within the limits of the normal vital processes, not only may the excretion of carbon dioxid be less than the stated average, but it may even be considerably in excess of the absorption of oxygen.   With such variations it is evident that the estimation of the amount of carbon dioxid alone cannot be a reliable measure of the total interchange of gases.   A complete insight into the gaseous balance can be obtained only by a simultaneous estimation of the oxygen taken in and of the carbon dioxid given off.

SUMMARY OF THE GASEOUS INTERCHANGE.

Absorption in twenty-four hours:
Oxygen, 744 gm. = 516,500 cu. cm.
  (Carl Vierordt).
511–658 gm.(Speck).
(The volumes are determined for 0°
and the mean barometer.)

Excretion in twenty-four hours:
Carbon dioxid, 900 gm. = 455,500 cu.
  cm. (Carl Vierordt), hourly 31.5 –
  33 gm. (J. Ranke);   32.8–33.4
  gm. (v. Liebermeister);   34 gm.
  (Panum);   36 gm. (Scharling).
Water, 640 gm. (Valentin);   330 gm.
  (Carl Vierordt).

FACTORS INFLUENCING THE EXTENT OF THE RESPIRATORY EXCHANGE OF GASES.

The process of carbon-dioxid formation consists probably of two separate stages.   In the first place, through the presence of oxygen in the tissues, there are formed combinations containing carbon dioxid, which are oxidation-products of substances containing carbon. The second step consists in the separation of this carbon dioxid, which can take place even without the absorption of oxygen. Both processes do not always take place uniformly; at times there is a preponderance in the formation of substances destined for decomposition and carbon-dioxid formation, while at other times the liberation of carbon dioxid predominates, with a diminution in the substances mentioned.

The respiratory interchange of gases (also the respiratory quotient) is, within wide limits, independent of the amount of oxygen in the air and the pressure of the atmosphere. According to Schmiedeberg the oxidation in the tissues depends upon a synthesis accompanied by a separation of water, for which purpose the blood supplies the necessary oxygen.

The processes under consideration are affected by:

1. *Age.*—Until the body is fully developed, the output of carbon dioxid increases, while from that point it diminishes with the decline in bodily strength. Hence, in young persons the absorption of oxygen is relatively greater in comparison with the carbon dioxid given off. At all other periods of life both values correspond rather closely. For example:

Age. Years.	CO_2 Excreted in Grams		In Twenty-four Hours. = Carbon.	O Absorbed in Grams.
0.17,	113			
8,	443	grams	= 121 carbon	375 grams.
15,	766	"	= 209 "	652 "
16,	950	"	= 259 "	809 "
18–20,	1003	"	= 274 "	854
20–24,	1074	"	= 293 "	914
40–60,	889	"	= 242 "	757
60–80,	810	"	= 211 "	689

In children the excretion of carbon dioxid is absolutely less, but relatively greater, than in adults; weight for weight, children excrete almost twice as much carbon dioxid as adults. The new-born also consumes relatively more oxygen than adults. In the fetus of the sheep the consumption of oxygen was found to be only one-sixteenth that in the full-grown animal.

2. *Sex.*—Males from the eighth year to advanced age give off about one-third more carbon dioxid than do females. This difference is still more marked at the time of puberty, when it amounts to about one-half. After the cessation of menstruation there is an increase, and in old age again a decrease in the amount of carbon dioxid given off. Pregnancy progressively increases the output, for evident reasons.

Young girls, under otherwise similar conditions, exhale less carbon dioxid than boys; the proportion being 100 : 140. Boys of from nine to twelve years of age exhale from 33 to 34 grams in an hour. In the thirteenth year the excretion of carbon dioxid rises rapidly, and maintains itself at a high level until the nineteenth year (from 42 to 45 grams in an hour). Then it falls between twenty and thirty years to 38 grams; and, finally, between thirty-five and sixty years it is from 34 to 37 grams. Girls between eight and ten years old excrete from 23 to 25 grams; between eleven and thirty years old they exhale from 26 to 32 grams, and at sixty-five years of age 26 grams. Younger and lighter individuals of both sexes, with their greater body surface, give off more carbon dioxid (in proportion to their weight) than older, heavier, and more compact persons.

3. *Constitution.*—As a rule, muscular, active individuals require more oxygen and excrete more carbon dioxid than less muscular

and energetic persons of the same size and weight. The consumption of oxygen and the excretion of carbon dioxid are in inverse proportion to the extent of body surface. In this connection the respiratory gaseous interchange pursues a course parallel with that of heat-production.

4. *Diurnal and Nocturnal Variations.*—In general there is during sleep a diminution in the excretion of carbon dioxid as compared with the waking state (the proportion being 100 : 145, in the most extreme case 100 : 169). This is proportional to the diminution of the general metabolism resulting from the constant heat of the surroundings (the bed), the darkness, the absence of muscular activity, and the abstinence from food (see 5, 9, 6, 7). According to v. Pettenkofer and C. v. Voit and others a slight accumulation of oxygen seems to take place during sleep. After awaking in the morning the respirations become deeper and more rapid, with at first an increase in the excretion of carbon dioxid. In the course of the morning, however, the excretion diminishes again, until the midday meal causes a fresh increase to the maximum. A falling off takes place again in the afternoon, and finally an inconsiderable increase is produced by the evening meal.

During hibernation, in which, together with the taking of food, respiration is entirely discontinued, and the interchange of gases is carried on only by diffusion in the lungs and the cardio-pneumatic movements, the excretion of carbon dioxid falls to $\frac{1}{75}$, and the absorption of oxygen to $\frac{1}{41}$ of the respective amounts during the waking state. Therefore, much less carbon dioxid is given off than there is oxygen absorbed, so that the body weight may even increase in consequence of the excess of oxygen taken up.

5. *Influence of the Surrounding Temperature.*—The bodily temperature of cold-blooded animals is easily raised by an increase in the surrounding temperature. Under such circumstances the animals give off more carbon dioxid than in a cooler state. For example, a frog exposed to a surrounding temperature of 39° C. excreted almost three times as much carbon dioxid as when the temperature was 6° C. Warm-blooded animals behave in a varying manner with changes in the surrounding temperature, accordingly as the bodily temperature remains constant, or is correspondingly raised or lowered. In the latter case, as in cold-blooded animals, a considerable decrease occurs in the excretion of carbon dioxid, when the body is cooled under the influence of cold surroundings. Conversely, elevation of the bodily temperature (also in the presence of fever) gives rise to increase in the excretion of carbon dioxid. The behavior is exactly the reverse when the bodily temperature remains constant on exposure to varying surrounding temperature. With increasing cold of the surrounding medium, the consequent reflex stimulation causes an increase in the oxidation-processes of the body, as well as in the number and depth of the respirations. As a result, more oxygen is taken up and more carbon dioxid is given off. The involuntary muscular movement that occurs when the body is cooled has the most obvious influence on the increase in the gaseous interchange. The season of the year also has an influence on the interchange of gases; in January a man consumed 32.2 grams of oxygen hourly, in July only 31.8 grams. In animals the carbon-dioxid excretion was found to be about one-third higher with a surrounding temperature below 8° C. than with a temperature above 38° C. When the temperature of the air increases (without change in the bodily temperature),

the respiratory activity and the excretion of carbon dioxid diminish, while the pulse remains nearly constant. It has been shown that when there is a sudden change from cold to warm surroundings, the carbon-dioxid output diminishes considerably; and, conversely, when the change is from warm to cold, the excretion increases considerably.

6. *Muscular Exertion* produces a considerable increase in oxygen-consumption and carbon-dioxid elimination, which, for instance, may be three times as great in walking as in a quiet, recumbent position. Every kilogrammeter supplies $3\frac{1}{2}$ milligrams of carbon dioxid; therefore, each additional gram of carbon dioxid formed is the equivalent of 300 kilogrammeters. The establishment of a certain degree of tension in the muscles requires more metabolic change than the maintenance of this tension.

The increase in the interchange of oxygen and carbon dioxid begins almost immediately after the work commences. In a few minutes it attains a constant height of at most from seven to nine times the amount during rest. After the work is finished, the consumption of oxygen falls in from 3 to 15 minutes to the rate during rest. The respiratory quotient remains essentially unchanged during work. During light work there is relatively a little more oxygen consumed than during heavy labor. The production of carbon dioxid is diminished with practice, that is, with a more economically applied exertion of the muscles.

The gaseous exchange is to a certain extent under the influence of the vagus nerve, which in part inhibits and in part accelerates the heart's activity. Irritation of this nerve may produce a diminution in metabolism, characterized by a more pronounced fall in the absorption of oxygen than in the excretion of carbon dioxid; or it may call forth an increase in metabolism, distinguished by a greater rise in the output of carbon dioxid than in the oxygen taken in.

7. *Ingestion of Food* causes a not inconsiderable increase in the carbon-dioxid excretion, which is in general governed by the quantity of food. Hence, the increase is generally most pronounced (about 25 per cent.) from one-half to one hour after the chief meal (dinner). The increase in the consumption of oxygen that follows the introduction of food into the stomach depends in part upon the increased muscular activity of the alimentary canal; nevertheless, the increased exhalation of carbon dioxid cannot be attributed to this alone. It is also, and to a greater extent, dependent on the heat-producing activity of the digestive glands—as in the case of the salivary glands. In addition, some of the carbon dioxid is derived from oxidation, in the course of urea-formation, of a part of the carbon contained in the proteids.

The quality of the food also has some influence. According to Magnus-Levy a proteid diet causes a much greater increase in the consumption of oxygen (about from 70 to 90 per cent.) than does carbohydrate food (which increases the consumption about 39 per cent.), or a fat-diet (which causes an increase of only 15 per cent.), as experiments on dogs show.

A fasting adult weighing 50 kilos inspires in one hour eight liters of air for each kilo; he consumes 0.45 gram of oxygen, and forms 0.5 gram of carbon dioxid. The ingestion of food raises these figures to nine liters of air, 0.5 gram of oxygen and 0.6 gram of carbon dioxid. The deposition of fat following a carbohydrate diet, is attended with an increase in the amount of carbon dioxid given off. This results partly from combustion of the carbohydrates, and partly from their transformation into fat, during which process carbon dioxid is separated. The respiratory quotient is also increased as a result of fat-formation following an abundant carbohydrate diet; the quotient under such conditions may even rise above 1.2.

The absorption of oxygen is uninfluenced by direct injection into the blood either of non-nitrogenous or of nitrogenous substances. The output of carbon dioxid changes to a certain extent in correspondence with the combustion of these substances by means of a constant quantity of oxygen.

Hunger greatly reduces the combustive processes in dogs; but in guinea-pigs it produces at most a small reduction in the consumption of oxygen.

8. *The Number and the Depth of the Respirations* have practically no influence on the formation of carbon dioxid, or on the oxidation-processes in the body, the latter being regulated rather by the tissues themselves through a mechanism as yet unknown. These factors, however, have been observed to exert an evident influence on the removal of the carbon dioxid already formed in the body. An increase in the number of respirations, the depth remaining the same, as well as an increase in their depth, the number remaining the same, results in an absolute increase in the output of carbon dioxid. The quantity seems relatively diminished, however, when viewed with reference to the amount of gases interchanged. Example:

NUMBER OF RESPIRATIONS IN EACH MINUTE	EXCHANGED VOLUME OF AIR.	CONTAINED CO_2	=	PER CENT. CO_2.	DEPTH OF RESPIRATION.	CONTAINED CO_2	=	PER CENT. CO_2.
12	6,000	258 cu. cm.	=	4.3 p.c.	500	21 cu. cm.	=	4.3 p.c.
24	12,000	420 "	=	3.5 "	1000	36 "	=	3.6 "
48	24,000	744 "	=	3.1 "	1500	51 "	=	3.4 "
96	48,000	1392 "	=	2.9 "	2000	64 "	=	3.2 "
					3000	72 "	=	2.4 "

Deep respirations, and also artificial respiratory movements, increase the absorption of oxygen into the blood to the point of saturation. Limitation of the supply of oxygen diminishes its consumption in the body in considerably greater measure than does hunger. Naturally, increased activity of the respiratory muscles causes in itself a greater interchange of gases.

9. *Exposure to Light* causes an increase in the excretion of carbon dioxid in frogs, mammals, and birds, even in frogs deprived of their lungs or of their cerebral hemispheres, or in those in which the spinal cord has been divided high up. At the same time the consumption of oxygen is increased. The same processes occur in individuals without eyes, though to a more limited extent. Rodents and birds show the maximum in red light, toads in violet light. According to Aducco starving pigeons lose weight more quickly in the light than in the dark. Quincke demonstrated that certain tissues, such as leukocytes and parts of fresh tissues, attract more oxygen to themselves under the influence of light than in the dark.

The nitrogenous metabolism of animals remains unchanged during exposure to light. The increased output of carbon dioxid is, therefore, to be attributed to an increased transformation of fat; hence, animals accumulate more fat when kept in the dark.

10. *Blood-letting* produces no diminution in the respiratory exchange of gases, but does cause an increase in the nitrogenous excretion. Profound anemic conditions diminish the interchange of gases.

11. *Changes in the Atmospheric Pressure* produce a slight diminution in the interchange of gases if breathing is made easier; but if

breathing is made more difficult, there is a slight increase. By inspiration of compressed air the absorption of oxygen is increased to an exceedingly small extent. In order to give off one gram of carbon dioxid, a smaller amount of air is needed at a low atmospheric pressure than with a high barometer. There is no diminution in the excretion of carbon dioxid on high mountains. The effects of artificially rarefied air and of the rarefied atmosphere of high altitudes are not the same. A rarefaction of air to 450 mm. of mercury still has no effect, the metabolic changes proceeding unaltered. In the air of high altitudes metabolism is increased, and respiration becomes more frequent and deeper. According to A. and J. Loewy and Zuntz the greater amount of light at high altitudes is the exciting factor.

12. In the presence of artificially induced dyspnea, as by tightly compressing the thorax, the proteid metabolism is increased—the amount of urea being increased—and there is an increase in the excretion of oxalic acid, acetone, ammonia, and sulphur in the urine.

Pathological.—According to the experiments of Gréhant on dogs, it appears that intense inflammation of the bronchial mucous membrane will diminish the output of carbon dioxid, even if there be fever.

In cases of diabetes the body is able to take up the necessary amount of oxygen, but the quantity of carbon dioxid given off is diminished, and the respiratory quotient is low.

Among the poisons, thebain increases the output of carbon dioxid, while morphin, codein, narcein, narcotin, and papaverin diminish it. Curare lowers the metabolism enormously, the absorption of oxygen falling about 35.2 per cent., and the excretion of carbon dioxid about 37.4 per cent. Section of the spinal cord has a similar effect.

DIFFUSION OF GASES WITHIN THE RESPIRATORY ORGANS.

In the pulmonary alveoli the air is richest in carbon dioxid and poorest in oxygen. Further on, from the smallest bronchioles to the larger ones and then onward to the bronchi and the trachea, the respired air becomes, step by step, gradually more like the atmospheric air. Hence it is that if the expired air of a respiration be collected in two halves, the first half (coming from the larger air-passages) contains less carbon dioxid (3.7 volumes per cent.) than the second half (5.4 volumes per cent.). This inequality in the proportion of the gases at various levels of the respiratory organs necessarily causes a continuous diffusion of gases between the various levels, and also, finally, between the gases in the larynx and nasal cavities and the outside atmosphere. The carbon dioxid constantly diffuses from the depths of the air-vesicles toward the outer air, while the oxygen of the latter diffuses toward the gaseous mixture in the pulmonary alveoli. This diffusion is doubtless assisted materially by the constant shaking of the respiratory gases by the cardio-pneumatic movements. During hibernation, and also in cases of apparent death of long duration, this must be the only means for the exchange of gases within the lungs. Ordinarily, however, this mechanism is insufficient for the respiratory process; so that the exchange of air produced by inspiration and expiration must be added to it. By this latter means atmospheric air is introduced into those parts of the lungs lying nearest to the large air-passages, from which and into which the diffusion-currents of oxygen and carbon dioxid pass more readily, on account of the greater differences in the tension of the gases.

If the inspired air contains a diminished quantity of oxygen, the necessary amount of oxygen can still be supplied to a certain extent by more rapid and deeper respirations.

INTERCHANGE OF GASES BETWEEN THE BLOOD IN THE PULMONARY CAPILLARIES AND THE AIR IN THE ALVEOLI.

This interchange of gases is accomplished almost exclusively by chemical processes, independently of the diffusion of gases.

For the determination of the gaseous interchange it is first necessary to ascertain the tension of the oxygen and the carbon dioxid in the venous blood of the pulmonary capillaries. Pflüger and Wolffberg have accomplished this by catheterization of the lungs. An opening is made in the trachea of a dog, and an elastic catheter (Fig. 90, a) is introduced into the bronchus leading to the lower lobe of the left lung. In order to have the bronchus fit closely around the catheter, the latter is made to pierce a rubber sac inflated by means of a communicating rubber-ball pump c. In this way no air from that part of the lung can escape at the side of the catheter. The tube is at first closed at its outlet, and the dog is allowed to breathe independently and as quietly as possible. After four minutes the alveolar air in the closed-off part of the lungs is in complete equilibrium with the blood-gases. By means of a mercurial air-pump the air in the lungs is sucked out of the catheter (at b) and analyzed. The tension of the carbon dioxid and the oxygen in this air will indicate in an indirect way the tension of these two gases in the venous blood of the pulmonary capillaries.

For the direct estimation of the gases in various specimens of blood, these gases are removed by shaking the blood with another kind of gas. The composition of the mixture will indicate the proportions in which the blood-gases have been mixed, and will thus serve to determine their tension. It is desirable to use as much blood as possible with a small quantity of gas; the amount of the latter should be about the same as that supposed to be present in the blood.

In the following table are shown the tension and the percentage of oxygen and carbon dioxid in arterial and venous blood, as well as in the atmosphere and the air of the closed-off alveoli:

I.
Tension of O in arterial blood = 29.6 mm. of mercury; increased by warming; corresponding to a gaseous mixture containing 3.9 per cent. of O.

V.
Tension of O in the alveolar air of the catheterized lung = 27.44 mm. of mercury; corresponding to 3.6 vol. per cent.

II.
Tension of CO_2 in arterial blood = 21 mm. of mercury; corresponding to 2.8 vol. per cent.

VI.
Tension of CO_2 in the alveolar air of the catheterized lung = 27 mm. of mercury; corresponding to 3.56 vol. per cent.

III.
Tension of O in venous blood = 22 mm. of mercury; corresponding to 2.9 vol. per cent.

VII.
Tension of O in the atmosphere = 158 mm. of mercury; corresponding to 20.8 vol. per cent.

IV.
Tension of CO_2 in venous blood = 41 mm. of mercury; corresponding to 5.4 vol. per cent.

VIII.
Tension of CO_2 in the atmosphere = 0.38 mm. of mercury; corresponding to from 0.03 to 0.05 vol. per cent.

If the tension of the oxygen in the atmosphere (VII) be compared with that in venous blood (III) or in the alveoli (V) it will be seen that the absorption of oxygen into the blood during respiration can occur in the form of an equalization of tension. Likewise a comparison of the tension of the carbon dioxid in the atmosphere (VIII) with that in venous blood (IV) or with that in alveolar air (VI) might explain the excretion of that gas in a similar manner. Nevertheless, the respiratory interchange of gases is a chemical process.

According to v. Fleischl the concussion to which the venous blood is subjected on being pumped into the pulmonary arteries provides for a more ready escape of the carbon dioxid, a point that is of the greatest importance with respect to the respiratory process.

The absorption of oxygen from the alveolar air for the purpose of oxidation of the venous blood in the pulmonary capillaries is a chemical process, as the gas-free hemoglobin in the lungs takes up oxygen to form oxyhemoglobin. That this absorption depends, not on diffusion of the gases, but on the atomic combination pertaining to the chemical process, is shown by the fact that the blood does not take up more oxygen when the pure gas is respired than when atmospheric air is respired; further, that animals that are made to breathe in a small, closed space will absorb into their blood all of the oxygen but traces, to the point of suffocation. If the respiratory absorption of oxygen were a diffusion-process, much more oxygen would have to be taken up in the first case in accordance with the partial pressure of the gas; while in the latter case such an extensive absorption could not take place.

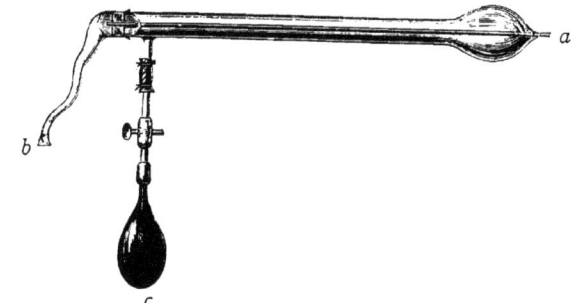

FIG. 90.—Pulmonary Catheter.

Even in highly rarefied air (high balloon-voyages) the absorption of oxygen remains independent of the partial pressure. However, in a space containing rarefied air a longer time and a more vigorous shaking are required for the absorption of oxygen by the blood at the temperature of the body; that is, the absorption of oxygen is not diminished, but is retarded. In this way is explained the death, for example, of the aëronauts Sivel and Croce-Spinelli, during an ascension to a height where the atmospheric pressure is only one-third the normal.

The laws of diffusion come into play in connection with the absorption of oxygen only to the extent that the oxygen, in order to reach the red blood-corpuscles, must, first of all, diffuse into the plasma, where it immediately enters into chemical combination with the erythrocytes.

The excretion of carbon dioxid from the blood into the alveolar air could also be well represented in the form of an equalization of tension (diffusion); but here again chemical processes are operative, although they have not yet been investigated in many details. The absorption of oxygen by the erythrocytes produces at the same time an expulsion of the carbon dioxid. This is proved by the fact that the whole of the carbon dioxid is more easily expelled from the blood if oxygen be at the same time introduced than if all gases are withdrawn. The result is different in the case of the serum, which when subjected to a vacuum will give up only a part of the carbon dioxid, while from 5 to 9 volumes per cent. are still retained; the latter can be released only by the addition of acids. As this carbon dioxid, which exists in firm chemical combination, also escapes on addition of erythrocytes, the corpuscles must contain a substance that acts like an acid in expelling the carbon dioxid.

THE RESPIRATORY GASEOUS EXCHANGE AS A DISSOCIATION PROCESS.

Some forms of gas enter into true chemical combination with other substances, when associated at a certain high degree of partial pressure of the gas in question. This chemical combination, however, is again dissolved as soon as the partial pressure diminishes and reaches a certain low level. Hence, by alternately raising and lowering the partial pressure, a chemical combination of the gas can be formed and again broken up. This process is called *dissociation of gases.* The minimal partial pressure is constant for the various substances and gases in question; but still the temperature, as in the case of the absorption of gases, has a marked influence; namely, increase in temperature diminishes the partial pressure at which dissociation occurs.

Calcium carbonate may be taken as an example to illustrate the dissociation of gases. When this substance is heated in the air to 440° C., carbon dioxid escapes from the chemical combination; but it is gradually taken up again by the calcium, after cooling has taken place.

The chemical combinations containing carbon dioxid, and also those containing oxygen, namely, the oxyhemoglobin and the carbon-dioxid compounds, behave in a similar manner within the blood-stream; these also exhibit the process of dissociation. If these gaseous combinations are placed under conditions in which the partial pressure of these gases is exceedingly low (that is, when they are present in small amounts), the compounds are dissociated; that is, they give off carbon dioxid or oxygen, as the case may be, to the surrounding medium. If, however, they are now again brought into a medium in which, on account of an abundance of these gases, the partial pressure of the oxygen or the carbon dioxid is high, they are again taken up in chemical combination by these gases.

The hemoglobin of the blood in the pulmonary capillaries finds a plentiful supply of oxygen in the alveoli; therefore, it combines with the oxygen, under the high partial pressure of that gas, forming the chemical compound oxyhemoglobin. On its way through the capillaries of the greater circulation, the hemoglobin comes in contact with tissues poor in oxygen; the oxyhemoglobin is dissociated, its oxygen passes to the tissues, and the blood, with gas-free or reduced hemoglobin, returns to the right heart and thence to the lungs, in order to take up oxygen anew.

The carbon dioxid meets the circulating blood in largest amount in the tissues. The high partial pressure of the gas in this situation causes the constituents of the blood to enter into chemical combination with the carbon dioxid. In the lungs, however, the partial pressure for carbon dioxid is low, the gas is dissociated, and it is excreted. It is thus evident that, as concerns the blood, the giving up of oxygen and the absorption of carbon dioxid in the tissues, and, conversely, the absorption of oxygen and the giving up of carbon dioxid in the lungs, are processes that take place simultaneously.

CUTANEOUS RESPIRATION.

Method.—If a human being or an animal is placed in the chamber of a respiration-apparatus (such as Scharling's or v. Pettenkofer's), and the gases passing to and from the lungs are conducted through a respiratory tube, so that none of the gaseous interchange of the lungs enters the chamber, but only the transpiration of the skin, information can thus be obtained concerning the cutaneous respiration. The procedure of leaving the whole head of the subject outside the chamber, the neck being fixed air-tight in its wall, is less correct. The cutaneous respiration of a circumscribed part of the body—for instance, of an extremity—may be studied by enclosing the part in an air-tight cylinder similar to that used for the arm in employing the plethysmograph.

In twenty-four hours a healthy man loses through his skin—which contains the respiratory organ in the moist sweat-glands, richly supplied with blood-vessels—$\frac{1}{67}$ of his entire body-weight, which is greater than the loss through the lungs, since it bears a ratio to the latter of 3 : 2. Of this large loss of weight only from 8 to 10 grams are referable to the carbon dioxid given off. The remainder is comprised in the evaporation of water. Elevation of the surrounding temperature is attended with an increase in the amount of carbon dioxid given off. The excretion at between 29° and 33° C. amounts to 8 grams in twenty-four hours; above 33° C. it is 20 grams (sweating begins at this point); and at 38.4° C. the amount is 27.5 grams. Active muscular exercise likewise produces an increased excretion.

Absorption of oxygen by the skin has also been demonstrated, the amount absorbed being either equal to the volume of carbon dioxid given off, or a little less.

As the excretion of carbon dioxid by the skin is only about $\frac{1}{220}$ of that by the lungs, and as the absorption of oxygen is only about $\frac{1}{140}$ of that by the lungs, it is evident that the respiratory activity of the skin is in any event but slight. ·It is uncertain whether or not the skin gives off gaseous nitrogen or ammonia. According to Funke the skin secretes hourly 0.0824 gram of soluble nitrogen, this quantity being increased in the presence of renal disease.

According to Röhrig, the excretion of carbon dioxid and of water exhibits certain daily variations. It is increased during digestion, after the application of cutaneous irritants, in the presence of obstruction to pulmonary respiration, of hyperemia of the skin, and when the blood contains an increased number of erythrocytes.

In warm-blooded animals. with thick, dry epidermoid structures. the cutaneous interchange of gases is still less than it is in man. In frogs and other amphibia, with a constantly moist skin, cutaneous respiration becomes highly important. The skin here supplies from two-thirds to three-fourths of the total quantity of carbon dioxid excreted, and in hibernating frogs the proportion is still greater. The skin is, therefore, a more important respiratory organ than the lungs. Immersion in oil will, consequently, kill these animals more readily than will ligation of the lungs.

INTERNAL RESPIRATION OR TISSUE-RESPIRATION.

The terms *internal respiration* and *tissue-respiration* are used to designate the interchange of gases between the capillaries of the greater circulation and the tissues. Those organic constituents of the tissues that contain carbon are subjected during their vital activity to a process of gradual oxidation, with the formation of carbon dioxid. Hence, the following inferences may be drawn:

1. The chief seat for the absorption of oxygen and the formation of carbon dioxid is to be found within the tissues themselves. That the oxygen rapidly penetrates from the capillary blood into the tissues is shown by the fact that this blood rapidly becomes richer in carbon dioxid and poorer in oxygen, while oxygenated blood, kept warm outside the body, changes much more slowly and incompletely. Further, if fresh pieces of tissue be placed in defibrinated blood rich in oxygen, the oxygen rapidly diminishes. Also, the circumstance that frogs deprived of their blood exhibit almost as great an interchange of gases as do normal animals indicates that the gaseous interchange takes place in the tissues themselves. Moreover, if the chief seat of oxidation lay, not in the tissues themselves, but in the blood, then, if oxygen were withheld from the blood (during suffocation), those reducing substances that consume the oxygen in the process of oxidation should accumulate in the blood. This is not the case, for even the blood of suffocated animals contains only a trace of reducing substances. The absorption of oxygen into the tissues may occur in the form of a temporary storing of the gas, perhaps with the formation of intermediate lower oxidation-products. This is followed by a period of more rapid separation of carbon dioxid. Thus, the absorption of oxygen and the excretion of carbon dioxid in the tissues do not necessarily proceed on parallel lines and to the same extent.

A clear picture of the development of carbon dioxid in the tissues is furnished by the fact that a larger amount of this gas is found in the cavities of the body and in their gases and fluids than in the blood of the capillaries. Pflüger and Strassburg found the tension of the carbon dioxid (in millimeters of mercury) as follows:

In arterial blood,...........21.28 mm.   In bile,......................50.0 mm.
" the peritoneal cavity, ...58.8   "   " hydrocele-fluid,...........46.5   "
" acid urine,.............68.0   "

The abundance of carbon dioxid in these fluids, as compared with that in the blood, can arise only from the addition to them of the carbon dioxid generated in the tissues.

In the lymph of the thoracic duct the tension of the carbon dioxid (from 33.4 to 37.2 mm. of mercury) is, indeed, greater than in the arterial blood, but it is still considerably less than in the venous blood. This fact does not, however, justify the conclusion that only a small quantity of carbon dioxid is formed in the tissues from which the lymph is collected. It rather permits the inference, either that the lymph possesses less attraction for the carbon dioxid formed in the tissues than does the capillary blood, where chemical forces are active in the production at least of a partial combination of the gas; or that in the course of the slow lymph-current the carbon dioxid is partially given back to the tissues by equalization of tension; or, finally, that carbon dioxid is formed independently in the blood. Furthermore, it is to be pointed out that those muscles that are known to be the principal producers of carbon dioxid furnish this gas abundantly to the blood, their tissues being relatively poor in lymph-vessels.

The amount of uncombined, free carbon dioxid, capable of being pumped out, in the fluids and gases mentioned indicates that the carbon dioxid passes over from the tissues into the blood in an uncombined free state. However, Preyer believes that the gas is carried over into the blood of the veins also in chemical combination.

The interchange of oxygen and carbon dioxid varies considerably in the different tissues. In the first rank belong the muscles, which in a state of activity excrete a large amount of carbon dioxid and consume much oxygen. The interchange of gases in tissues is increased during their activity.   The secreting salivary glands, kidneys, and pancreas are no exception to this rule; for although, in the secreting state, bright red blood flows away from them through the dilated vessels, still this apparently relative diminution in the carbon dioxid of the venous blood is more than compensated for by its absolute increase through the marked increase in volume of the blood passing through these organs.

Active reduction-processes take place in most tissues. If coloring-matters, such as alizarin-blue, indophenol-blue, or methylene-blue, be introduced into the blood of animals, the tissues will soon be stained. Those organs that have an especially strong affinity for oxygen (such as the liver, the cortex of the kidneys, and the lungs), abstract oxygen from these coloring-matters, and change them into colorless reduction-products. The pancreas and the submaxillary gland have almost no reducing power.

2. The blood itself, like all of the tissues, is a seat for the consumption of oxygen and the formation of carbon dioxid.   This is proved by the fact that blood removed from the body quickly becomes poorer in oxygen and richer in carbon dioxid; further, by the circumstance that in the oxygen-free blood of asphyxiated persons and in the blood-corpuscles there are always found small quantities of reducing agents, which become oxidized on the addition of oxygen. At all events, this gaseous interchange is but slight as compared with that occurring in all the other tissues.   It is incontestable that the walls of the blood-vessels, by means of their contained muscular fibers, also consume oxygen and produce carbon dioxid, although this process is so insignificant that the blood undergoes no visible change in color throughout its arterial course.

C. Ludwig and his pupils have proved by specially adapted experiments that transformation into carbon dioxid can actually occur within the blood. If sodium lactate, which is easily oxidized, be mixed with blood, and this mixture be sent through the blood-vessels of a recently excised organ that is still alive (such as the kidney or the lung), a more abundant consumption of oxygen and formation of carbon dioxid will occur in this mixed blood than would occur in pure blood similarly transfused.

3. It may in advance be concluded as probable that the living pulmonary tissue also consumes oxygen and generates carbon dioxid. By passing arterial blood through lungs that have been deprived of air, C. Ludwig and Müller succeeded in demonstrating a diminution in the oxygen and an increase in the carbon dioxid. Bohr and Henriques concluded further from their experiments, in which they restricted to a considerable degree the circulation of blood through the bodily tissues, and found no significant diminution in the excretion of carbon dioxid from the lungs, that the pulmonary tissue is not limited to a mere excretion and absorption of gases, but that it besides possesses the property of forming carbon dioxid from substances that are derived from the other tissues.   In like manner they assumed that oxygen is actively taken up by the lungs; that is, the lungs secrete carbon dioxid and absorb oxygen like a secreting gland.

As the total amount of carbon dioxid and oxygen .in the whole volume of blood at any one time is only about 4 grams, while the amount of carbon dioxid excreted daily is 900 grams, and the amount of oxygen absorbed is 774 grams, it is evident that the interchange of gases proceeds with great rapidity, that the absorbed oxygen must be consumed and the carbon dioxid formed must be excreted quickly.

As a result of an increased introduction of acids into the body there is a diminution in the consumption of oxygen (and in the production of heat), which in a high degree may give rise to an internal asphyxia of the tissues.

RESPIRATION IN A CLOSED SPACE, OR WITH ARTIFICIAL CHANGES IN THE AMOUNTS OF OXYGEN AND CARBON DIOXID IN THE RESPIRED AIR.

Respiration in a closed space results in (1) a gradual diminution of the oxygen, (2) a simultaneous increase of the carbon dioxid, and (3) a diminution in the volume of gas. If the space is only of moderate size, the animal consumes the oxygen almost completely, the blood becomes almost free of oxygen, and death finally results, accompanied by asphyxial convulsions. The absorption of oxygen occurs, therefore, through chemical combination, independently of the laws of absorption.

In larger closed spaces considerable accumulation of carbon dioxid takes place before the oxygen is diminished to such an extent that life is threatened. As the carbon dioxid can be excreted from the body only when its tension is greater in the blood than in the surrounding air, there will be retention of the gas as the amount expired into the enclosed space increases; and, finally, a return of the carbon dioxid into the body may take place. This occurs while the oxygen is still sufficient to support life. Death results, therefore, directly from poisoning by carbon dioxid, with the symptoms of dyspnea of short duration, to which are added stupor and subnormal temperature. This manner of death has been observed in rabbits, after they had reabsorbed some of the carbon dioxid that had been excreted previously by them.

In pure oxygen, or in an atmosphere rich in oxygen, animals breathe in a perfectly normal manner. A little more oxygen is absorbed, but still the amount of carbon dioxid excreted is not increased. In closed spaces filled with oxygen, animals finally die through the reabsorption of their excreted carbon dioxid. Rabbits have thus been observed to die after they had absorbed an amount of carbon dioxid equal to half the volume of their body, although the enclosed air still contained over 50 per cent. of oxygen.

Human beings and animals can still breathe an air-mixture containing only 9 per cent. of oxygen; deepened respirations set in at 10 per cent., and discomfort at 8 per cent. Animals breathe with difficulty and lose consciousness at 7 per cent.; pronounced dyspnea makes its appearance at 4.5 per cent., and quite rapid suffocation at 3 per cent. The air expired by man under normal conditions still contains between 14 and 18 per cent. of oxygen. Mammals placed in a gaseous mixture poor in oxygen consume slightly less oxygen.

The metabolism of animals is unchanged by variations in the amount of oxygen in the respired air between the limits of 10.5 and 87 per cent. If the oxygen falls below 10.5 per cent., there is an increase in the excretion of nitrogen, carbon dioxid, lactic acid, and oxalic acid through the urine.

If the amount of carbon dioxid in the inspired air be increased, the respiratory movements are increased, but the excretion of carbon dioxid and the absorption of oxygen are diminished.

Inspiration is actively stimulated by a deficiency of oxygen, as well as by an excess of carbon dioxid. The dyspnea that is induced under the condition first stated is prolonged and severe, while under the second condition the respiratory activity soon diminishes. A deficiency of oxygen further causes a greater and more prolonged rise in the blood-pressure than does an excess of carbon dioxid. Finally, the consumption of oxygen by the body is less restricted by a diminution of the oxygen in the air than by an excess of carbon dioxid. Death from limitation in the supply of oxygen is preceded by violent irritative phenomena and convulsions, which are absent in case of death from excess of carbon dioxid. Finally, in conjunction with poisoning by carbon dioxid, the excretion of this gas is greatly diminished.

If animals be supplied with a gaseous mixture similar to the atmosphere, but in which the nitrogen is replaced by hydrogen, they breathe quite normally; the hydrogen of the mixture does not undergo any noteworthy change in volume. Increase or diminution in the amount of nitrogen in the air simply causes a greater or lesser absorption of the gas by the fluids of the body.

Cl. Bernard found that if an animal be made to respire in a closed space, it became, up to a certain point, accustomed to the successive deterioration of the air. If he placed a bird under a glass bell-jar, it lived for several hours; but if, before its death, another bird were added from the fresh air, the latter immediately died in convulsions.

It is remarkable that frogs, when placed in air free from oxygen, will for several hours give off just as much carbon dioxid as in air containing oxygen, and this without any obvious disturbances. Hence, the formation of carbon dioxid must be independent of the absorption of oxygen, and the carbon dioxid must be set free in the decomposition of other compounds. Finally, however, complete motor paralysis sets in, while the circulation for a time remains undisturbed.

RESPIRATION OF FOREIGN GASES.

No gas is able to support life without a sufficient admixture of oxygen. Hence, without oxygen, all other gases will quickly cause suffocation (in two or three minutes), even though they be in themselves harmless and indifferent.

Completely indifferent gases are represented by nitrogen, hydrogen, and marsh-gas (CH_4). The blood of an animal breathing any of these gases yields no oxygen to it.

Poisonous Gases.

(a) *Those displacing oxygen:* (1) Carbon monoxid (CO). (2) Hydrocyanic acid (CNH) displaces(?) oxygen from the hemoglobin, with which it forms a more stable compound, and it thus kills with great rapidity. Further, it prevents the formation of ozone from the oxygen in the blood. Blood-corpuscles charged with hydrocyanic acid lose the property of decomposing hydrogen dioxid into water and oxygen.

(b) *Narcotic gases:* (1) Air containing 0.1 per cent. of carbon dioxid has been designated as "bad air"; still, the discomfort experienced in such an atmosphere (for example, in overcrowded rooms) arises rather from offensive exhalations of unknown character than from the carbon dioxid itself. Air containing 1 per cent. of carbon dioxid produces marked discomfort; with 10 per cent. life is endangered, and with a higher percentage death ensues, accompanied by symptoms of coma. (2) When nitrous oxid (N_2O) is respired, mixed with one-fifth its volume of oxygen, it causes in from one and one-half to two minutes a short, evanescent, especially pleasurable state of intoxication (laughing-gas), which is followed by an increased excretion of carbon dioxid. (3) Pure ozonized air produces similar effects; it also causes short, agreeable excitement, then drowsiness and rapidly transient sleep.

(c) *Reducing gases.* (1) Hydrogen sulphid (H_2S) rapidly deprives the erythrocytes of all oxygen, forming sulphur and water by oxidation; death occurs quickly, even before the gas can effect any change in the hemoglobin, with the formation of sulphur-methemoglobin. In addition, hydrogen sulphid forms in the blood sodium sulphid from sodium carbonate, the new compound rapidly causing death.

(2) *Hydrogen phosphid, phosphin* (PH_3), is oxidized in the blood to form phosphoric acid and water, with decomposition of the hemoglobin.

(3) *Hydrogen arsenid, arsin* (AsH_3), and *hydrogen antimonid, stibin* (SbH_3), act like hydrogen phosphid, but in addition they allow the hemoglobin to pass out of the stroma, so that the excreta, as the urine, contain hemoglobin.

(4) *Cyanogen* (C_2N_2) withdraws oxygen and further decomposes the blood.

Irrespirable gases cannot be inspired at all, as they cause reflex spasm of the glottis on entering the larynx. If introduced forcibly into the air-passages, they give rise to violent inflammatory processes, followed by other disturbances and death. Included in this class are hydrochloric acid (HCl), hydrofluoric acid (HFl), sulphurous acid (SO_2), nitrous acid (N_2O_4), nitric acid (N_2O_5), ammonia (NH_3), chlorin, fluorin, iodin, bromin, undiluted ozone, and pure carbon dioxid.

OTHER INJURIOUS SUBSTANCES IN THE INSPIRED AIR.

Particles of dust are among the impurities of the atmosphere that are harmful in large quantities and after long-continued action. Most of these particles are expelled externally by means of the ciliated epithelium of the respiratory organs, whose cilia wave toward the larynx. Some of the dust-particles, however, penetrate the epithelium of the air-vesicles, and thus reach the interstitial pulmonary tissue, from which they frequently pass through the lymph-vessels to the lymphatic glands of the lungs. For this reason coal-dust is found deposited in the lungs of all elderly persons, blackening the alveoli. In moderate amounts these substances are harmless in the tissues; but if the deposits become large, they may cause pulmonary diseases that may finally lead to disintegration of the lungs. The particles penetrate between the alveolar epithelium into the interstitial pulmonary tissue, and then into the lymphatic vessels and glands. In many trades the work must be done in a dusty atmosphere, and they are thus rendered detrimental to health.   Charcoal-burners, grinders, stone-cutters, file-cutters, weavers, spinners, tobacco-workers, sawyers, millers, bakers, and others suffer from various affections of the lungs, induced by the dust of their trades.   During a year's work a workman in a horse-hair mill inhales 15 grams of dust, in a saw-mill 27 grams, in a woolen mill 30 grams, in a grinding mill 37.5 grams, in an iron-foundry 42 grams, in a snuff-factory 108 grams, in a cement-factory 336 grams.

The ciliated epithelium is exceedingly sensitive to mechanical excitation.   The coördinated, continuous movement of the cilia on a larger surface does not depend wholly upon an external (mechanical) conduction of the stimulus, but also upon an internal conduction (as in the nervous system).

There is no doubt that with the inspired air the germs of infectious diseases are often taken into the respiratory organs, whence they gain entrance into the body.   Thus, the diphtheria-bacillus in the pharynx and the larynx, the glanders-bacillus in the nose, the germ of whooping-cough in the bronchi, the microbes of hay-fever and ozena in the nose, the influenza-bacillus in the air-

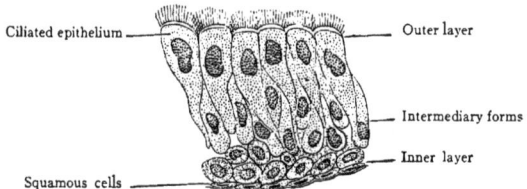

Ciliated epithelium — Outer layer

— Intermediary forms

— Inner layer

Squamous cells —

FIG. 91.—Stratified Ciliated Cylindrical Epithelium of the Larynx (Horse) (after Toldt).

passages, the pneumonia-bacillus in the air-vesicles.   The cause of tuberculosis, the bacillus tuberculosis, enters the air-filled pulmonary tissue with the dust of tuberculous sputa, and may spread from that focus through all of the tissues.   In a similar manner leprosy arises from the bacillus lepræ.   The cause of malaria, the plasmodium malariæ possessed of ameboid movement, reaches the blood partly through the respiratory organs, changes the hemoglobin within the red corpuscles into melanin, and causes their destruction.   In the same way the blood is invaded by the exciting agents of smallpox (micrococcus vaccinæ), the spirillum of relapsing fever, the still little known microbe of measles, and the as yet undiscovered germ of scarlet fever, etc.

Many disease-germs enter the mouth with the air, others with the food, and are swallowed, so that they undergo development in the intestinal tract.   This is true of cholera (comma-bacillus), dysentery, typhoid fever (bacillus typhosus), and amebic enteritis (amœba coli; the amœba coli mitis is less virulent, and the amœba intestina vulgaris is harmless).   In cattle, anthrax arises in the same way from bacterium anthracis.

RENEWAL OF THE AIR IN LIVING-ROOMS (VENTILATION). EXAMINATION OF THE AIR.

Fresh air is one of the most necessary conditions for salutary existence on the part both of the healthy and of the sick.   It may be assumed that a sufficient renewal of the air in living-rooms will be assured, if 800 cu. ft. of space be allowed for every inmate of a room, and about 1000 cu. ft. for every sick person.   The necessary space for the inmates of dwellings, schools, barracks, penal institutions, and hospital-wards should be measured accordingly, and the allotment of space to the individuals should be made only in this proportion.   However, this standard has been materially departed from in various countries.

In overcrowded spaces the amount of carbon dioxid at first increases.   The normal amount in the air (0.5 in 1000) has been found increased in comfortable living-rooms to from 0.54 to 0.7 in 1000; in badly ventilated sick-rooms to 2.4 in 1000; in overcrowded auditoriums to 3.2 in 1000; in pits to 4.9 in 1000; in school-rooms to 7.2 in 1000.   Although it is not the amount of carbon dioxid that makes the air of crowded spaces injurious, but rather the exhalations from the outer and inner surfaces of the body, which at the same time render the air offensive to the sense of smell, still the amount of carbon dioxid is an indication of the degree of vitiation of the atmosphere. To determine whether or not the ventilation is sufficient in spaces crowded with individuals, the carbon dioxid of the air should be estimated quantitatively at the time of occupation; hence, in school-rooms, if possible, shortly before the close of the school-session, or in sick-wards or dormitories (barracks) shortly before daybreak. As a good, comfortable room-atmosphere contains less than 0.7 of carbon dioxid in 1000, the ventilation of a space must be considered insufficient if more than 1.0 in 1000 is found.

The atmosphere contains only 0.0005 cubic meter of carbon dioxid in 1 cubic meter of air, and an adult produces hourly 0.0226 cubic meter of carbon dioxid. Therefore, it will be found on calculation that ventilation must supply 113 cubic meters (for a child 60 cubic meters) of fresh air hourly for each person if the carbon dioxid in the living-room is to be kept below 0.7 in 1000—0.7 : 1000 = (0.0226 + x × 0.0005): x; hence, x = 113. If the amount of carbon dioxid in the air of a room be allowed to reach 1.0 in 1000, then an hourly ventilation of 45 cubic meters is sufficient for an adult, and 24 cubic meters for a child.

The following method is employed to determine whether a living-room has sufficient ventilation. A large quantity of carbon dioxid is generated in the room, as much as 1 or 2 liters hourly for every cubic meter of space. The burning of stearin-candles may be employed as the source of carbon dioxid, each candle producing 12 liters of gas in one hour; a gas-burner supplies 100 liters an hour; an adult man produces 22.6 liters by respiration, and a school-child 12 liters hourly. If sufficient carbon dioxid has been produced at the end of an hour, the generator is removed, and the first estimation of carbon dioxid in the air is made, according to the method described later on. At the end of another hour, during which the windows and doors are kept closed, the second estimation of carbon dioxid is made. The amount of fresh air that has entered by ventilation during this hour is calculated by the following formula: $C = 2.3 \times m \times \log. \frac{p-a}{q-a}$, in which C represents the volume in cubic meters of fresh air that has entered by ventilation in one hour, m the volume of room-space in cubic meters, p the amount of carbon dioxid contained in 1 cubic meter of the air in the room at the first estimation, expressed in cubic meters, q the amount of carbon dioxid in each cubic meter, found at the second estimation and expressed in cubic meters, a the carbon dioxid in atmospheric air = 0.0005 cubic meter in 1 cubic meter of air.

Example: In a school-room, containing 40 children, the first estimation of carbon dioxid is made shortly before the close of school. If the result be 2 in 1000, it will indicate the presence of 0.002 carbon dioxid in 1 cubic meter of air. After the children have gone, the windows and doors are again closed, and the second analogous estimation is made at the end of an hour. If the result be 1 in 1000, there will be 0.001 carbon dioxid in 1 cubic meter of air. The size of the school-room is 600 cubic meters. The quantity of fresh air that has entered the space during the hour can be estimated according to the foregoing formula: $C = 2.3 \times$ $600 \times \log. \frac{0.002 - 0.0005}{0.001 - 0.0005} = 1380 \times \log. \frac{0.0015}{0.0005} = 1380 \times \log. 3 = 1380 \times$ $0.4771213 = 658.3$ cubic meters. Hence, 658.4 cubic meters of fresh air have entered the school-room by ventilation. As one child requires 60 cubic meters of fresh air hourly, the 40 pupils require 40 × 60 = 2400 cubic meters of fresh air in one hour; but, as a matter of fact, the ventilation of this space amounts to only 658.4 cubic meters; therefore, 1741.6 cubic meters are still wanting. Hence, either a better ventilation must be provided, or fewer children should be allowed to attend the school. A ventilation that amounts to more than three times the room-space hourly will be found to give rise to an unpleasant draft, and is, therefore, often directly harmful in winter. For the school-room in question containing 600 cubic meters of space, only 1800 cubic meters of ventilation hourly would be permissible; hence, there is only space in that room for at most 30 pupils (30 × 60 = 1800). As the space receives only 658 cubic meters of ventilation hourly, provision must be made by better ventilation for the addition of 1142 cubic meters more of fresh air; but without further ventilation place could be found in the school for only 11 children (658 ÷ 60).

In ordinary living-rooms, in which the necessary space (800 cu. ft.) is allowed for every inmate, the air is sufficiently renewed by the numerous pores possessed by the walls of the rooms, as well as by the going in and out, and further, in winter, by stoves (a well-heated stove providing a ventilation of from 40 to 90 cubic meters of air hourly). That this ventilation is sufficient is proved by the fact that the amount of carbon dioxid in the room remains constant. When there is a more considerable difference between the temperature in the room and that outside (as in winter), the ventilation is more than sufficient.

If, however, the cubic space allotted to each inmate is too small, as in over-crowded hospitals, narrow ship-quarters, etc., then the necessary change of air must be provided for by means of contrivances for artificial ventilation. The same must be done if noxious exhalations are given off by the sick. Above all, however, it is to be noted that the natural ventilation through the pores of walls is greatly limited if they be damp. At the same time, damp walls are prejudicial to health by reason of their greater conduction of heat, and also because the germs of infectious diseases can develop in them, as in moist ground generally.

Ventilation may be accomplished either by aspiration, the exchange of air being brought about by suction-power; or by pulsion, the fresh air being pumped into the room.

The carbon dioxid contained in the air of a living-room may be estimated as follows: A baryta-solution is prepared, containing 10 grams of crystallized barium hydrate and 0.5 gram of barium chlorid in 1 liter of water. A large, dry, accurately graduated, 6-liter flask is filled with air from the room to be in-vestigated, by blowing the air for some time down to the bottom of the flask by means of a bellows. Then, by means of a pipet 100 cu. cm. of the baryta-solution are allowed to run into the flask, naturally displacing 100 cu. cm. of the air. The flask is then closed with a rubber cap, and is allowed to stand for two hours, being shaken occasionally. In this way all the carbon dioxid is absorbed by the baryta-solution. Then, 25 cu. cm. of the clear, supernatant fluid are withdrawn into a medicine-bottle, and are titrated with a normal oxalic-acid solution from a graduated buret, until a drop of the mixture, when placed upon turmeric paper, does not form a brown stain; that is until the reaction is neutral. A few drops of a solution of 0.2 gram of rosolic acid in 100 cu. cm. of dilute alcohol may also be added to the baryta-solution in the medicine-bottle, producing a red coloration. When oxalic acid is added, the mixture is decolorized by the slightest excess of this acid. To prepare the normal oxalic-acid solution, 2.8636 grams of pure, crystallized, undecomposed oxalic acid, dried by having stood over concentrated sulphuric acid under a glass bell-jar for four hours, are dissolved in 1 liter of water; 1 cu. cm. of this solution is equivalent to 1 mgm. of carbon dioxid. The number of cubic centimeters of acid-solution added to the baryta-solution is noted. Now, 25 cu. cm. of the original baryta-solution, with which nothing has been done, are titrated in like manner with the normal acid-solution to the point of neutralization; here also the amount of the acid-solution added is noted. By subtraction the difference is found between the amounts of normal acid-solution added in both titrations. Each cubic centimeter of this difference is equivalent to 1 mgm. of carbon dioxid, and the resulting value must be multiplied by 4, in view of the fact that only 25 cu. cm. of the 100 cu. cm. of baryta-solution were titrated. The result gives the milligrams of carbon dioxid in six liters minus 100 cu. cm. of air.

The milligrams of carbon dioxid thus determined are converted into cubic centimeters by multiplying them by 0.508 (as 0.508 cu. cm. of carbon dioxid, at 0° C. and 760 mm. of barometric pressure, weighs 1 mgm.). The volume of the air is further reduced to 0° C. and 760 mm. of barometric pressure. This is done according to the formula $V_1 = \dfrac{V.B}{760.(1 + 0.003665.t)}$, in which V_1 represents the re-duced volume desired, V the volume of air taken in the flask for the experi-ment, B the barometer-reading taken at the time of the experiment, and t the temperature in the investigated room. By this reduction-procedure the results can be obtained in percentages for possible comparisons.

Example: Twenty-five cu. cm. of the baryta-solution are neutralized by means of 24.6 cu. cm. of the oxalic-acid solution; 25 cu. cm. of the baryta-solution after the absorption of carbon dioxid (taken from the experiment-flask) are neutralized by means of only 21.5 cu. cm. of the oxalic-acid solution. The difference between them, 24.6 — 21.5 = 3.1, represents 3.1 mgm. of carbon dioxid, which have been absorbed in the 25 cu. cm. of baryta-solution. Accordingly, there are contained in the 100 cu. cm. of baryta-solution employed 12.4 mgm. of carbon dioxid (4 × 3.1). If it be assumed that the large flask of air contains 4100 cu. cm., of which 100 cu. cm. have been displaced by an equal volume of baryta-solution that has been run in, so that there remains a volume of air equal to 4000 cu. cm.;

and if, at the time of the experiment, the temperature of the living-room was $20°$ C., and the barometer-reading 750 mm., then the reduced volume of air corresponding to the 4000 cu. cm. is $V_1 = \dfrac{4000 \times 750}{760 \times (1 + 0.003665 \times 20)} = 3678$ cu. cm., in which are contained 12.4 mgm. carbon dioxid. One mgm. of carbon dioxid, however, equals 0.508 cu. cm.; hence, there were in 3678 cu. cm. of air 6.299 cu. cm. of carbon dioxid (12.4 $\times$ 0.508). In 1000 cu. cm. air this amounts to 1.7 cu. cm. (according to the formula x : 1000 = 6.299 : 3678), or 1.7 of carbon dioxid in 1000.

NORMAL SECRETION OF MUCUS IN THE AIR-PASSAGES.
THE EXPECTORATION (SPUTUM).

The mucous membrane of the respiratory tract is covered by a thin layer of mucus. This mechanically hinders further formation of mucus by preventing the usual irritation of the air and dust. Additional mucus is secreted only in so far as is rendered necessary to replace that lost by evaporation. As a rule, increased circulation of blood in the tracheal mucous membrane is attended with increased secretion. Division of the nerves on one side (in the cat) gives rise to redness on the same side, with increased secretion.

On "catching cold" (for instance, as a result of covering the abdomen with ice) the mucous membrane first becomes completely pale, and then deep red, with marked increase in the secretion. Injection of sodium carbonate and ammonium chlorid restricts the secretion. The local application of alum, silver nitrate, or tannic acid dries the mucous membrane, so that the epithelium is cast off. Apomorphin, emetin, and pilocarpin actively stimulate the secretion; atropin and morphin limit it.

Even under normal conditions hawking and coughing will cause the expectoration of slimy, viscid material, which may be derived from the entire respiratory tract, and is always mixed with a little saliva. In the presence of catarrhal conditions or of more serious disease the expectoration becomes more profuse, and is often mixed with characteristic products. It contains:

1. *Epithelial cells*, especially squamous cells from the mouth and the throat (Fig. 92, 8), more rarely alveolar epithelium (2), still more rarely ciliated epithelium (7) from the larger air-passages. Not rarely changes are found in the epithelium as a result of maceration, including the cylindrical cells that have already lost their cilia (6) and contain swollen nuclei.

Alveolar epithelium (2), with a diameter from two to four times that of a leukocyte, is found especially in the morning-sputum, but only in that from persons over 30 years of age. In younger persons its presence indicates diseased conditions of the pulmonary parenchyma. Alveolar epithelium is found also in a state of fatty degeneration and filled with pigment-granules (3); also in the form of myelin-degenerated cells (4), that is, cells filled with clear refractive droplets of varying size, some being colorless, and some having absorbed pigment-granules (dust-particles). Also mucin in myelin-forms, that is, in the form of coagulated nerve-substance, is found constantly in the sputum (5). Mucus is stained yellow by safranin, while albumin is stained red.

2. *Leukocytes* (9) are present in large number in yellow sputum, and in smaller number in clear sputum. They are to be looked upon as white blood-corpuscles that have wandered from the blood-vessels. They also are often found in changed forms and in a state of dissolution; they may be shrivelled up, filled with fat-granules, or they may appear as conglomerations of granules; and, finally, isolated nuclei indicate the destruction of their cell-body.

Eosinophile cells are found in the sputum from cases of asthma, and also in the nasal secretion from cases of acute coryza and of nasal polyps. Leukocytes containing hemosiderin are found after capillary hemorrhages in the air-passages.

The fluid substance of the sputum contains much mucus, derived from the mucous glands and the goblet-cells, also some nuclein and lecithin, and the constituents of the saliva, according to the amount mixed with the sputum. Albumin is found in the sputum only in cases of inflammation of the air-passages; its amount increases with the degree of inflammation. Urea has been found in the sputum in cases of advanced nephritis.

Pathological.—In the presence of catarrhal conditions the sputum is usually at first glairy and slimy (sputa cruda); later, it becomes more consistent and yellow (sputa cocta).

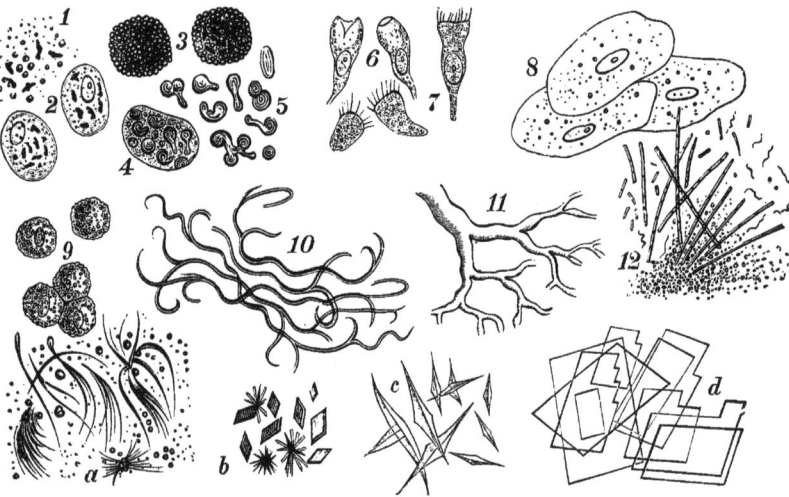

FIG. 92.—Objects Found in the Sputum: 1, detritus and dust-particles; 2, pigmented alveolar epithelium; 3, fatty degenerated and partially pigmented alveolar epithelium; 4, alveolar epithelium showing myelin-degeneration; 5, free myelin-forms; 6, 7, desquamated ciliated epithelium, partly changed and deprived of its cilia; 8, squamous epithelium from the mouth; 9, leukocytes; 10, elastic fibers; 11, fibrinous cast of a small bronchus; 12 leptothrix buccalis, together with cocci bacilli, and spirochetæ; a, fatty-acid crystals and free fatty granules; b, hematoidin; c, Charcot's crystals; d, cholesterin.

Under pathological conditions there may be found in the sputa:

(a) Erythrocytes, always from rupture of a blood-vessel.

(b) Elastic fibers (10) from destroyed pulmonary alveoli. Usually they occur in small bundles of delicate fibers, which at times suggest the rounded walls of the alveoli by their curved arrangement. Naturally, they always indicate destruction of pulmonary tissue.

(c) Much more rarely, in the presence of rapid and extensive disintegration of the lungs, there occur larger fragments of pulmonary débris, embracing several alveoli; likewise small pieces of fibro-cartilage or unstriated muscle-fibers from the small air-passages.

(d) Colorless coagula of fibrin (11) may be found, and are usually to be recognized as casts of the smaller or larger air-passages. They are formed in connection with inflammatory processes in the lungs or bronchi that are attended with a fibrinous exudation into the tubules. They are thus found frequently in cases of pneumonia in adults, in cases of bronchial croup, and also, rarely, in cases of severe influenza.

(c) Crystals of various kinds: Fatty-acid crystals (a), arranged in bundles of fine needles, usually lying in whitish, cheesy, fetid lumps of sputum. They indicate a more profound process of decomposition affecting the stagnating secretion and the underlying tissue. Crystals of leucin and tyrosin are rarely found as decomposition-products of the albuminates. Tyrosin is found more abundantly after rupture of an old abscess into the lungs. Colorless, octahedral or rhombic platelets with elongated points—Charcot's crystals (c)—have been found in the expectoration in cases of asthma, lying in and on peculiar, spirally wound plugs of exudate from the narrow air-passages; they have also been found in connection with other exudative affections of the bronchi. These structures, also called Curschmann's spirals, are produced when the respiratory air, in passing by, draws out parts of the secretion into threads, and rolls them spirally to and 'fro. Hematoidin-crystals (b), from old effusions of blood in the lungs, occur rarely; likewise cholesterin-crystals (d), arising from broken-up collections of pus.

(f) Fungi and other low organisms are found in the sputum, being taken in during inspiration. The threads of leptothrix buccalis (12) occur frequently, having been detached from deposits on the teeth. Mycelial threads and spores are found in the sputum in cases of thrush, which occurs frequently in the mouths of nursing infants as white, spreading deposits (oïdium albicans). Among the bacteria, the mucous-membrane streptococci (mostly diplococci) are constantly found, and frequently the micrococcus albus liquefaciens and harmless saprophytes; pyogenic cocci usually occur only in cases of pulmonary tuberculosis. In the presence of gangrene of the lungs monads and cercomonads have been found, in cases of pneumonia at times the bacillus pneumoniæ of Friedländer, in cases of influenza the influenza-bacillus of Pfeiffer and Canon, in cases of whooping-cough a minute diplococcus (according to Czaplewski and Hensel a non-motile bacillus), in cases of mumps a bacterium similar to the gonococcus, in cases of measles the bacillus causing that disease, in cases of pulmonary tuberculosis without exception the tubercle-bacillus. Rarely the sarcina is found; this is encountered more frequently in the stomach in the presence of gastric catarrh, and also in the urine.

With regard to its external appearance sputum may be described as mucous, muco-purulent, or purulent. When heated at 60° C. all sputa are reduced to a uniform degree of fluidity.

The sputum may have an abnormal coloration. Thus, it may be red from blood-pigment; if it remains long in the lungs, the blood-pigment may run through a whole scale of colors (as in external, visible blood-tumors), and it may thus give the sputa a dark-red, bluish-brown, brownish-yellow, deep-yellow, yellowish-green, or grass-green color. The sputum is sometimes yellow in cases of jaundice. Colored dust, if accidentally inspired, may also color the expectoration.

The odor of the sputa is usually stale, and more or less unpleasant. It becomes ill-smelling when it has remained for some time in pathological cavities in the lungs; it is stinking in the presence of gangrene of the lungs.

EFFECTS OF ATMOSPHERIC PRESSURE.

At the normal pressure of the atmosphere, with the barometer registering 760 mm. of mercury, a pressure is exerted on the entire surface of the body amounting to from 15,000 to 20,000 kilos, corresponding to the extent of surface—103 kilos to each square decimeter. This pressure acts on the body equally from all sides, and in those internal air-spaces as well which are in direct communication with the outer air either constantly—as the respiratory tract, the sinuses of the frontal, superior maxillary, and ethmoid bones—or only temporarily —as the digestive tract and the tympanic cavity. If an air-filled space, for example the tympanic cavity, be closed off from the outer air for some time, a rarefaction of the gases in the space occurs, as a result of the consumption of oxygen and its replacement by a smaller volume of carbon dioxid. As the fluids of the body (blood, lymph, secretions, parenchymatous juices) are practically incompressible, their volume may be regarded as unchanged by the prevailing pressure. These fluids, however, absorb gases from the atmosphere in accordance with the prevailing pressure—that is, the partial pressure of the several gases—and also with their temperature. The solid constituents of the body are composed of innumerable and minute elementary parts, such as cells and fibers, of which each presents only a microscopic extent of surface to the influence of the pressure. Hence, the prevailing atmospheric pressure for every cell can be estimated only at a few milligrams, a pressure under which even the most delicate histological structures develop with ease. As an example of the action of atmospheric pressure on larger masses, attention may be called to the fact that, as a result of the adhesion of the smooth, sticky, articular surfaces of the shoulder-joints and the hip-joints, the arm and the thigh are supported without the aid of muscular activity; so that, for example, the thigh is still held in the acetabulum after all of the soft parts around the neck of the femur, including the articular capsule, are divided.

An ordinary increase in barometric pressure has an influence on the respiratory activity in that it stimulates slightly the respiratory movements, while a fall in barometric-pressure has the opposite effect. The absolute amount of carbon dioxid remains the same; but in connection with the lessened frequency of respiration attending a low barometer, the percentage is naturally somewhat increased.

Marked diminution in the atmospheric pressure, such as occurs in ascending mountains or in balloon-voyages (the highest known ascension, without loss of consciousness having been made by Berson of Berlin, to a height of 9145 meters at a temperature of —47.7° C.), causes a series of characteristic phenomena: (1) As a result of great diminution in the pressure on surfaces in direct contact with the air, they undergo marked congestion. Hence, there occur redness and swelling of the skin and exposed mucous membranes, even to the extent of causing hemorrhages from the more delicate parts, as the nose, the lungs, the gums; turgidity of the cutaneous veins, profuse sweating, marked secretion from the mucous membranes. The arteries become more empty; at one-half the atmospheric pressure the blood-pressure in the radial artery begins to fall. (2) Other direct effects of diminished pressure are a feeling of weight in the legs, as the atmospheric pressure alone is not sufficient to keep the head of the femur in the acetabulum; bulging of the tympanic membrane by the air in the tympanic cavity, until the difference in tension is equalized through the Eustachian tube, and as a consequence pain in the ears and even impairment of hearing. (3) The diminution in the tension of oxygen in the surrounding air causes difficulty in breathing and oppression of the chest, as a result of which the respirations become more rapid (also the pulse), deeper, and irregular. At an elevation of from 3000 to 4000 meters the respiration and pulse are increased one-fourth; when the atmospheric pressure is reduced from one-third to one-half, the blood loses oxygen, and in consequence there is incomplete removal of the carbon dioxid from the blood and a considerable diminution in the oxidation-processes in the body. When the atmospheric pressure is one-half or less, the amount of carbon dioxid in the arterial blood is diminished, and the amount of nitrogen decreases in proportion to the diminution in atmospheric pressure. Rabbits kept under a pressure of from 300 to 400 mm. of mercury die on the third day, and present widespread fatty degeneration, especially of the heart.

In men and in animals, residence in high, mountainous regions appears to increase in the course of a few days the amount of hemoglobin in the blood and the number of red corpuscles. This effect should be favorable for the absorption of oxygen. A noteworthy phenomenon is the appearance of numerous microcytes in the first few days. Dyspnea from various causes also has a similar effect in man. (4) In consequence of the diminution in the density of the air, the latter is not able to produce loud tones in the larynx through the vibrations of the vocal bands; hence, the voice appears faint and altered. (5) In consequence of the determination of blood to the external parts in contact with the air, the internal parts become relatively poor in blood; hence result diminution in the secretion of urine, muscular weakness, digestive disturbances, dulness of the senses; fainting spells, all of which phenomena are intensified by the conditions mentioned in paragraph (3). According to the observations made by Whimper on himself during the ascent of the highest peak in the Andes, the body can, to a certain extent, accustom itself with respect to these latter phenomena. At an elevation of from 7000 to 8000 meters loss of consciousness occurs at times; the aëronauts Crocé-Spinelli and Sivel lost their lives at a height of 8600 meters, where the rarefied air contains only 72 per cent. of oxygen (the air-pressure being 241 mm. of mercury). In dogs a marked fall in the blood-pressure occurred first at 200 mm. of mercury, accompanied by a small, slow pulse.

The inhabitants of high, mountainous regions are sometimes attacked by an illness (mountain-sickness), which consists essentially of symptoms similar to those described, especially anemia of the internal organs, and which is accompanied by a diminution in the amount of hemoglobin in the blood. Alexander von Humboldt found remarkable roominess of the thorax in the inhabitants of the high Andes. This phenomenon has been attributed to a diminution in the carbon dioxid of the blood, which serves as a stimulant to the respiratory center. At an elevation of from 6000 to 8000 feet above the sea, water contains only about one-third the amount of air absorbed; therefore fish cannot longer live in it.

Animals can be subjected to a still greater rarefaction of the atmosphere under the receiver of an air-pump. Under such conditions birds die when the air-pressure is reduced to 120 mm. of mercury; mammals at 40 mm. of mercury. Frogs endure repeated evacuation, and as a result they become much distended by escaping gases and aqueous vapor; after the entrance of air, however, they collapse completely. Hoppe-Seyler ascribes the cause of death in warm-blooded animals to the development of gas in the blood, the bubbles obstructing the capillaries. Landois has often been able to confirm this phenomenon, and as far back as 1879 he suggested that the development of gas-bubbles in the parenchymatous juices, especially of the nervous system, might act injuriously through mechanical laceration of the tissues. Sudden reduction of a previously high air-pressure may act in a similar manner. The free gas that forms in the blood is almost pure nitrogen. The presence of air in the arteries of the spinal cord produces anemic paralysis, and later local destruction of the nerve-elements. Redi and Wepfer, in 1685, were the first to observe death from blowing air into the veins, as a result of mechanical obstruction to the circulation.

Local diminution of the air-pressure results in marked congestion and swelling of the tissues in the affected part; this is shown in the simplest manner by cupping. Under the name of the "cupping-boot" Junod described an apparatus for the rarefaction of air, made to include a whole extremity; this apparatus rendered possible a reduction to one-third in the air-pressure surrounding the leg. By this means from 2 to 3 kilos of blood may be aspirated into the leg, thus producing a temporary withdrawal of blood from other parts of the body, without causing a permanent loss of blood to the body. The energetic application is exceedingly painful, and the after-effects persist for 48 hours.

Marked increase of the atmospheric pressure is accompanied by phenomena that may for the most part be explained as the reverse of those described in the discussion of diminution of the air-pressure. They have been observed many times, partly in so-called pneumatic cabinets, in which, for therapeutic purposes, the pressure is gradually increased to one and one-fifth, two and two-fifths atmospheres and more; partly in closed reservoirs used in construction under water, and out of which the water is forced by pumping air in. Under such conditions men work at times even under a pressure of four and one-half atmospheres. The following phenomena are worthy of attention: (1) Pallor and dryness of the external surfaces, collapse of the cutaneous veins, reduction in perspiration and the secretions from mucous membranes, greater supply of blood to the abdominal organs. (2) Pressing inward of the tympanic membrane (until the Eustachian tube allows the compressed air in the tympanic cavity to escape, often with a noise); considerable pain in the ears and even impairment of hearing. (3) A feeling of lightness and freshness during respiration. The respirations become slower (from 2 to 4 in a minute), inspiration is made easier and shortened, expiration is lengthened, and the pause is distinct. The capacity of the lungs is increased, owing to freer movement of the diaphragm, in consequence of diminution in the gases contained in the intestine. G. v. Liebig has noted an increase in the absorption of oxygen; Panum found that with the same volumes of air interchanged, the excretion of carbon dioxid is increased; the venous blood appears to be reddened. (4) Difficulty in speaking, a nasal metallic tone to the voice, inability to whistle. (5) Increased secretion of urine; on account of the more rapid oxidation in the body, there is increased activity of metabolism, increase in muscular energy, increased appetite, subjective feeling of warmth. The pulse is slower, and the pulse-curve lower.

On account of the invigorating and stimulating effect of a sojourn in moderately compressed air, the employment of the latter has been practised for therapeutic purposes; and it has been found that repeated applications have produced favorable after-effects of considerable duration. Unduly rapid increase of pressure is to be avoided and likewise unduly rapid removal of the pressure.

Waldenburg and others have constructed apparatus in the form of a spirometer; either compressed air may be inspired from its bell-jar, or the bell-jar may be filled with rarefied air, into which the expirations are made. Both methods are used in suitable cases for therapeutic purposes.

Paul Bert has found at an excessively high, artificial atmospheric pressure, over 30 vol. per cent. oxygen in the arterial blood of animals .(investigated at 700 mm. of mercury). If the amount of oxygen reaches 35 vol. per cent., death occurs, accompanied by convulsions. At a somewhat lower point the bodily temperature falls, the oxidation-processes in the body are reduced, strange to say, and as a result of this the formation of carbon dioxid and urea is diminished. Greatly compressed oxygen also produces the effect of a relative deficiency of oxygen; animals die in it, exhibiting signs of suffocation with greatly reduced metabolic processes.

Frogs exhibit in compressed oxygen (up to 14 atmospheres) the same phenomena as they would in a vacuum or in pure nitrogen. There occurs paralysis of the central nervous system, at times preceded by convulsions. Then the heart stops beating (but not the lymph-hearts), and at the same time the motor nerves lose their irritability; finally, the direct muscular irritability disappears.

. Under exceedingly high pressure of oxygen (thirteen atmospheres) an excised frog's heart beats scarcely one-fourth the time that it remains active in the air. If the quiet heart be brought into the air, the pulsations may return. Under a pressure of 100 atmospheres the frog's muscles still contract normally, and only at 400 atmospheres do they become paralyzed.

Phosphorus ceases to be luminiferous under high pressure of oxygen, but not, however, the phosphorescent organisms,—for example, the lamprey,—or the phosphorescent bacteria, such as those of meat (micrococcus Pflügeri). Exceedingly high atmospheric pressure is injurious to plants also.

COMPARATIVE. HISTORICAL.

Mammals have lungs similar to those of man. Those of birds exhibit a spongy structure; they are fused with the inner surface of the chest-wall, and have, on their outer surface, openings that lead into large, thin-walled air-sacs, lying among the viscera. These air-sacs further communicate with the cavities in the bones, which contain air instead of marrow, in order to provide greater lightness (pneumaticity of the bones). There is no diaphragm. In reptiles the lungs are divided into larger and smaller divisions of vesicles; in snakes one lung atrophies, while the other becomes greatly drawn out and elongated, in accordance with the form of the body. Frogs pump air into their lungs by contraction of the pharyngeal sac, the nostrils being closed and the larynx opened. Turtles fill their lungs with air by a sucking movement. Amphibia (frog) possess two simple lungs, each of which in its structure to a certain extent represents an enormous infundibulum with its alveoli. When young (until their metamorphosis) they live as aquatic animals, and breathe by means of gills; the perennibranchiates (proteus) indeed, like the fishes, breathe in this manner throughout life. Among fishes the dipnoi, besides their gills, possess a swimming-bladder, abundantly supplied with afferent and efferent vessels, constituting an internal respiratory organ remotely comparable to the lungs. By the term "gills" is meant an organ for respiration in water, constructed in the form of numerous, vascular, plate-like diverticula. Among the fishes, the mud-fish (cobitis) exhibits an intestinal respiration, when there is lack of water and it buries itself in mud; in this process air is swallowed on the upper surface of the water, the oxygen is abstracted in the intestines, and carbon dioxid is discharged through the anus. Insects and centipedes respire through tracheas, which consist of numerous air-canals distributed throughout the body and communicating with the atmosphere on the outer surface of the body by means of openings (stigmata) that can be closed. As insects possess no true circulatory movement of the blood, the air conducted through tubes penetrates from all sides into the blood-filled body-cavities; while in the lung-breathing vertebrates the blood conducted through tubes is brought from the whole body to the respiratory organ. The stigmata on the outer surface of the body, constituting the entrances to the tracheas, are provided with peculiar contrivances for closing, and can be employed for the emission of sounds. Arachnids respire by means of tracheas and lung-like air-sacs (tracheal pouches); crabs, by means of gills. Mussels and cephalopods possess fully developed gills; snails have partly gills, partly lungs. Among the lower animals, gill-like formations are still found among the round worms and in the echinoderms; intestinal respiration occurs in the tunicates and many of the mites. Respiration by means of a water-vascular system, a system of canals through which water flows, is peculiar to the medusæ and the flat worms. The lowest animal forms—protozoa, sponges, polyps—do not possess a special respiratory organ; in them the surfaces in contact with water carry on the respiratory interchange of gases.

Historical.—Aristotle (384 B. C.) regarded the object of respiration to be the cooling of the body, in order to moderate the internal heat. He observed correctly that the warmest animals also respire most actively, but in the interpretation he reversed cause and effect; for the warm-blooded animals do not respire on account of their heat (for cooling purposes), but they are warm as a result of their more active respiration (combustion).

Galen (203–131 B. C.) already observed the purifying action of the respiratory organ, assuming that the "soot" was removed from the body with the expired air, together with the expired water. The most important experiments concerning the mechanics of respiration date from Galen. He maintained that the lungs passively follow the movements of the thorax, that the diaphragm is the most important respiratory muscle, that the external intercostals are inspiratory muscles, and the internal intercostals expiratory. He divided the intercostal nerves and muscles, and observed that loss of voice followed. After dividing the spinal cord at progressively higher levels, he found that successively higher thoracic muscles became paralyzed. Theophilus Philaretus taught that the circulation could be improved by loud crying, singing, or speaking. Oribasius (360 A. D.) observed that both lungs collapsed in the presence of double pneumothorax. Vesalius (1540) first described artificial respiration as a means of reanimating and stimulating the heart's action. Malpighi (1661) described the peculiar structures of the lungs. Lower (1669) saw the blood become bright red in the lungs. Borelli (died 1679) first explained most thoroughly the mechanism of the respiratory movements.

The chemical processes attending respiration were already suspected by Mayow (1679): "Ignis et vita iisdem particulis aëreis sustinetur." However, more accurate knowledge could be obtained only after the discovery of the several gases coming under observation. J. B. van Helmont (died 1644) discovered carbon dioxid, and found that the air was vitiated by respiration; but Black (1757) first discovered the excretion of carbon dioxid during respiration. In 1774 Pristley and Scheele discovered oxygen. Lavoisier, in 1775, found the nitrogen, and at the same time ascertained the composition of the atmosphere. The same investigator also represented the formation of carbon dioxid and water during respiration as being the result of combustion within the lungs. J. Ingenhousz (1779) discovered the respiration of plants—the absorption of carbon dioxid and the giving off of oxygen during that process. Senebier (1785) found that this exhaled oxygen arose from decomposition of the carbon dioxid. Vogel and others definitely proved the existence of carbon dioxid in venous blood. Hoffmann and others demonstrated the presence of oxygen in arterial blood. Lavoisier with Séguin, in 1789, made the first communication concerning the quantitative absorption of oxygen and excretion of carbon dioxid during respiration. More complete insight into the interchange of gases during respiration could be obtained only after Magnus extracted and analyzed the gases from arterial and venous blood.

PHYSIOLOGY OF DIGESTION.

THE MOUTH AND ITS GLANDS.

The mucous membrane of the mouth contains sebaceous glands at the red edge of the lips. It consists of fibrillar connective tissue intermixed with fine elastic fibers. Toward the free surface it forms papillæ, of which the largest (0.5 mm.) are found on the lips and the gums, including some with double points—twin papillæ. The smallest are on the palate and in the fold-like duplicatures of the mucosa. The submucous tissue, which passes directly over into the mucosa, is thickest and most dense where the mucous membrane is immovably attached to the periosteum of the maxilla and the palate, and also in the vicinity of glandular involutions; while it is most delicate over movable and folded parts. The surface is lined by stratified nucleated squamous epithelium (Fig. 92, 8), and it is, as a rule, strongest and consists of the largest number of layers in regions where the papillæ are longest. A diplosoma is found in the deeper cells of the surface of the tongue.

All of the glands of the mouth, including the salivary glands, are divided, with reference to their secretion, into three groups: (1) *albuminous* or *serous glands*, whose secretion contains albumin; (2) *mucous glands*, whose ropy secretion contains mucin, together with some albumin; (3) *mixed glands*, whose acini secrete partly albumin and partly mucin, as, for example, the submaxillary gland in man. For a description of their structure reference may be made to page 258.

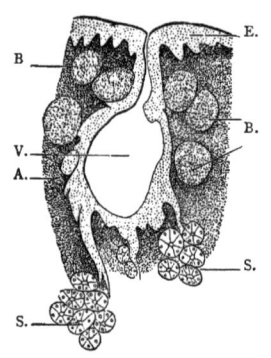

FIG. 93.—Section through Lymph-follicles of the Root of the Tongue (after Schenk); B, lymph-follicles; V, depression; A, adenoid connective tissue; S, mucous glands; E, epithelium.

Numerous mucous glands—termed buccal, palatine, lingual or molar muciparous glands, in accordance with the region in which they occur —are present in the tissue of the mucosa, their bodies appearing macroscopically as tiny white nodules. They represent the type of simple branched tubular glands. The contents of their secreting cells are partly mucus, which is expelled at the time of secretion. The excretory duct, formed of connective and elastic tissues, with a narrow outlet, is lined by a single layer of cylindrical epithelium. One duct often receives that of a neighboring gland. The labial glands are mixed glands.

The small glands of the tongue deserve special consideration. Two morphologically and physiologically distinct glands can be distinguished, namely (1) *mucous glands* (E. H. Weber's glands), situated especially near the root of the tongue; compound alveolar glands, with bright, transparent, secreting cells and mural nuclei, and a rather thick membrana propria; and (2) *serous glands* (von Ebner's glands), situated about the circumvallate papillæ (and the foliate papillæ in animals), and consisting of convoluted and branched tubules, characterized by small, narrow cells, filled with droplets of secretion, containing a centrosome and yielding an albuminous secretion. Halfway up between the cells the intercellular secretory ducts are found. (3) The Blandin-Nuhn glands, within the tip of the tongue, consist of glandular lobules secreting mucus and saliva, and are, therefore, mixed glands. Delicate varicose nerve-fibers pass up to the cells.

Of the *blood-vessels*, which are abundant, the larger lie in the submucosa, while the smaller penetrate into the papillæ, in which they form either capillary networks or simple loops.

Of the *lymph-vessels* the larger trunks, which form a coarse meshwork, lie in the submucosa, while the smaller, forming a finer network, pass through the mucous membrane itself. The cutaneous follicles or lymph-follicles constitute a part of the lymphatic apparatus. They form an almost coherent layer on the back of the tongue at its root. Several of these lymph-follicles always collect into a round mass, surrounded by connective tissue, and raising the mucous membrane somewhat. In the center of every such collection is a depression (Fig. 93) into the bottom of which mucous glands empty and fill the small crater with mucous secretion.

The *tonsils* exhibit on the whole the same formation,—crypt-like depressions, into the sinuses of which small mucous glands empty, and surrounded by masses of from 10 to 20 lymph-follicles. Layers of firm connective tissue form a sheath about the tonsils. The pharyngeal and tubal tonsils exhibit a similar structure.

Many medullated nerve-fibers, coming from the submucous tissue, ramify in the mucous membrane and terminate in part in separate papillæ in the form of Krause's end-bulbs, in larger number on the lips and the soft palate, in smaller number on the cheeks and the floor of the mouth. Probably the nerves also spread out in the form of fine terminal nodules between the epithelial cells, according to the Cohnheim-Langerhans mode of distribution. Functionally these are sensory nerves and nerves of touch.

THE SALIVARY GLANDS.

The salivary glands and also the pancreas are compound tubular glands. The excretory ducts, formed of connective and elastic tissues (Wharton's duct contains also unstriated muscle-fibers) are lined with

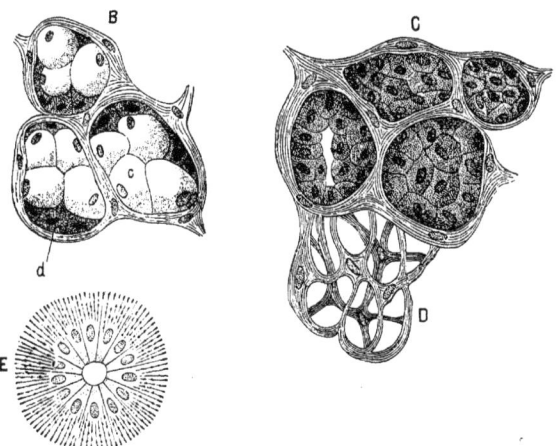

FIG. 94.—Histology of the Salivary Glands: B, alveoli of the rested submaxillary gland of the dog; c, the distended, glistening mucous cells; d, Gianuzzi's crescents; C, the alveoli after active secretion, showing the connective tissue of the alveoli isolated at D; E, section of a salivary tubule, lined with cylindrical epithelium.

cylindrical epithelium. Into the structureless membrane of the acinus that gives it its form, is incorporated a layer of star-shaped, anastomosing cells (Fig. 94, D). Next to the outer wall of the acinus lie fissure-like

17

lymph-cavities, and beyond these the blood-capillaries run in a net-like meshwork. The lymph-vessels leave the gland at the hilum.

The secreting cells are of varying structure, accordingly as the salivary gland secretes mucus (sublingual gland in man, submaxillary gland in the dog) or albumin (parotid gland in man), or is a mixed gland (submaxillary gland in human beings).

Two kinds of cellular elements are found in the acini of the submaxillary gland of the dog and the sublingual gland of human beings: (1) The so-called mucous cells (Fig. 94, B, c), which bound the secretory cavity. They possess a membrane and are filled with a flattened nucleus turned toward the acinus-wall. Centrosomes are difficult to recognize. The cell-body is abundantly impregnated with mucin, which gives it a bright, highly refractive appearance. On account of their mucous contents the cell-bodies hardly stain with carmine at all, while the nucleus takes up the stain. A process given off by the cell applies itself in a curved manner to the inner wall of the acinus. The true protoplasm of the cell is drawn out in a thread-like network from the nucleus through the mucin-mass. (2) The other variety of cellular elements form crescent-shaped complex bodies (Fig. 94, B, d)—Gianuzzi's crescents, Heidenhain's composite marginal cells—that lie in direct contact with the wall of the acinus. Each crescent consists of a number of small, closely packed, angular cells, with albuminous contents and nuclei and separated with difficulty. They are granular, darker, without mucous contents, easily impregnated by stains, and exhibit secreting spaces between the cells.

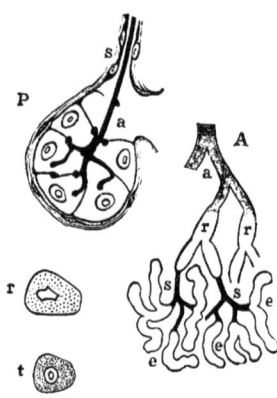

FIG. 95.—Diagrammatic Representation of a Salivary Gland: a, excretory duct; r, r, salivary tubules; s, intercalary portion; e, e, terminal portions. P, terminal portions of the parotid gland, with intercellular secretory ducts (stained black), passing over into the excretory duct (a) of the intercalary portion (s); r, parotid cell at rest; t, the same cell after secretion.

The parotid gland (Fig. 95), secreting albumin in man and in mammals, contains but one kind of secretory cells, namely, cubical cells, with a coarse-meshed protoplasm, staining little with pigments, without a membrane, with serrated, readily stained, centrally situated, highly refractive nuclei, without nucleoli, with secretory ducts between them. The smaller cells of the salivary tubules bear a diplosoma near their free surface. The salivary glands of animals that secrete saliva free from mucus present similar features.

By means of fine ducts, the so-called intercalary pieces, the terminal portions of the glands communicate with the thicker salivary tubules. The cells of these tubules, which, in their outer portions, appear fibrillated, and at times contain yellow granules (Fig. 94, E), bear a diplosoma near the surface. These salivary tubules empty into the excretory ducts. It is not improbable that these different portions of the gland also secrete different constituents of the saliva.

THE SECRETORY ACTIVITY OF THE SALIVARY GLANDS.

If the submaxillary gland of a dog is excited to active secretion by stimulation of its nerves, the mucous cells are after a while no longer seen, but in their stead only smaller protoplasmic cells, devoid of mucus, within the acini. The mucous cells have discharged their mucus into the secretion of the gland, while their shrunken, dark-granular protoplasmic cell-bodies remain (Fig. 94, C). These are capable, after a certain period of rest, of producing new mucus.

In regard to the crescents Stöhr believes that they are produced mechanically by inequality in the secretory phases in adjacent acinus-cells. The cells reduced in size after having discharged their mucus are pressed to the wall by other cells filled with mucus and therefore much swollen, and thus the flattened composite marginal cells are formed. Recently R. Krause and others, differing from this, state that the composite marginal cells secrete only serum and have no relation with the mucous cells.

In the parotid gland of the rabbit, after secretion induced by stimulation of the sympathetic nerve, the gland-cells assume a more shrunken appearance, and their contents become more granular and more readily stained. The nuclei appear rounder and exhibit a nucleolus (Fig. 95).

Ranvier observed in the secretion of the albuminous glands (submaxillary gland in the rat) that, after stimulation, many motile vacuoles were formed in the gland-cells. The water of the secretion is formed in the vacuoles, and in its excretion, carries with it the soluble ferment of the cells. A similar phenomenon occurs within mucous cells and also in goblet-cells. Morphological changes occur also in the cells of the salivary tubules.

THE NERVES OF THE SALIVARY GLANDS.

All the salivary glands derive their nerve-supply from two sources, namely from the sympathetic nerve and from a cranial nerve. The nerve-fibers, chiefly medullated, in part also non-medullated, pass in at the hilum and form a plexus rich in ganglion-cells between the lobules of the gland.

The sympathetic nerve sends branches (a) to the submaxillary and sublingual glands, derived from the plexus surrounding the external maxillary artery (Fig. 243); (b) filaments pass to the parotid gland from the sympathetic plexus, which, piercing the parotid, surrounds the external carotid artery.

Of the cranial nerves, (a) the submaxillary and sublingual glands are supplied by filaments from the chorda tympani branch of the facial nerve. (b) To the parotid gland fibers pass from the glosso-pharyngeal nerve in the dog, especially from its tympanic branch, which sends fibers through the tympanic plexus to the lesser superficial petrosal nerve. Together with the latter the former pass downward over the anterior surface of the petrous portion of the temporal bone, then through the sphenoidal fissure to the otic ganglion. With the latter they continue through communicating branches to the auriculo-temporal nerve (from the third division of the trigeminal nerve), which, covered by the parotid gland, on its way to the temple, sends the fibers to the gland.

The submaxillary ganglion, which gives off fibers to the submaxillary and sublingual glands, derives its roots from the tympanico-lingual plexus, as well as from the sympathetic plexus about the external maxillary artery.

With regard to the terminal distribution of the nerves to the salivary glands, two varieties are to be distinguished: (1) the vasomotor nerves, which give branches to the muscular walls of the blood-vessels, and (2) the true glandular nerves.

According to Arnstein the latter form a surrounding network outside of the gland-tubules. From this plexus fine filaments pierce the membrana propria and terminate on the surface of the secreting cells with a peculiar end-apparatus: namely, branched twigs possessing tiny bulbs or mulberry-shaped masses. The same condition exists in the sebaceous, sudoriferous, and mammary glands and in the pancreas.

THE INFLUENCE OF THE NERVOUS SYSTEM ON THE SECRETION OF SALIVA.

The Submaxillary Gland.—Stimulation of the facial nerve at its root causes profuse secretion of limpid saliva deficient in the specific constituents. At the same time the blood-vessels of the gland undergo dilatation. The capillaries, in the presence of increased blood-pressure in them, undergo such a degree of dilatation that the pulsating movement of the arteries is transmitted into the veins. More than four times as much blood flows back from the vein, which, besides, appears almost bright red in color and contains more than one-third as much oxygen as the venous blood of the unstimulated gland. In spite of the relatively large amount of oxygen in venous blood, the secreting gland consumes absolutely more oxygen than the inactive gland.

The facial nerve contains two sets of functionally different fibers: (1) true secretory nerves and (2) vasodilatator nerves. It is not permissible to regard the phenomenon of secretion as a simple result of increased circulatory activity.

Stimulation of the sympathetic nerve causes the scanty secretion of a viscid, gelatinous, ropy saliva, containing the specific constituents, particularly mucus and the salivary corpuscles, in abundance, and having a specific gravity of from 1007 to 1010. At the same time, with decrease in the blood-pressure, the blood-vessels of the gland undergo contraction, so that the small amount of blood escapes from the veins with a dark-blue color.

The sympathetic nerve likewise contains two sets of functionally different nerve-fibers, (1) true secretory fibers and (2) vasoconstrictor nerves. Continued stimulation of the chorda tympani and the sympathetic nerve alters the secretions, making them more nearly alike, and thus teaches that, essentially, the saliva produced by stimulation of the chorda tympani and that produced by stimulation of the sympathetic nerve differ not specifically, but only in degree. With increasing nerve-stimulation the secretion increases, and with it the amount of contained salts. The organic constituents depend, in addition to the intensity of the stimulation, upon the condition of the gland, whether at rest or exhausted. The constitution of the blood and the circulatory conditions in the gland likewise influence the composition of the saliva.

That the secretion of the glands cannot be considered as a simple filtration as the result of changes in blood-pressure, but that it occurs as an independent function in conjunction with changes in the blood-vessels, will appear from the following considerations:

1. The secretory activity of the gland, on stimulation of the nerves, continues for some time even after all blood-vessels have been ligated.

2. Atropin and daturin destroy the activity of the secreting fibers in the chorda tympani, but not that of the vasodilator fibers.

3. The pressure in the excretory ducts of the salivary glands, which can be measured by means of a manometer tied in the duct, may be almost twice as great as that in the arteries of the gland, having reached about 290 mm. of mercury in the excretory duct of the submaxillary gland. With increase in the pressure the amount of saliva diminishes, as does likewise the amount of work performed by the gland.

4. The salivary glands, in the same way as nerves and muscles, also fatigue, especially after injection of acids or alkalies into the excretory duct. This indicates that the secretory structure is independent of the circulation and under the influence of the nerves.

5. That in the secretion of saliva the cellular activity of the glands also is evident is shown by the researches of Zerner, who, after intravenous injection of indigo-carmine, found this substance within the mucous cells and the rod-cells.

It must, therefore, be inferred that the nerves exert a direct influence on the secreting cells of the glands, independent of any mediation on the part of the blood-vessels. As the direct anatomical connection between the nerve-fibers and the secreting cells appears proved, so, also, is the physiological connection to be accepted.

During the process of secretion the temperature of the submaxillary gland rises about $1.5°$ C. The gland, as well as the venous blood leaving it, is not rarely warmer than the arterial blood. Between the irritation of the nerve and the beginning of secretion, from 1.2 to 24 seconds elapse.

Paralytic Secretion of Saliva.—By this term is understood the persistent secretion of limpid saliva from the submaxillary gland, which sets in twenty-four hours after division of the cerebral nerves, whether the sympathetic nerve is also injured or is preserved. It increases for perhaps eight days; then, with degeneration of the gland, it decreases. The injection of small amounts of curare into the artery of the gland also produces the condition, which is prevented by apnea, while dyspnea favors it. After a unilateral lesion both glands are said to take part in the secretion. According to Langley, after division of the chorda tympani, its central end acquires increased irritability. This exerts a centripetal effect upon the salivary center on both sides. At the same time, soon after the division, a ganglionic local secreting center, situated in the gland of the same side, also is stimulated, so that, if all of the nerve-fibers passing to the gland are later separated, the salivary secretion from the gland still continues.

The Sublingual Gland.—Probably the conditions existing here entirely resemble those found in the submaxillary gland.

The Parotid Gland.—Stimulation of the sympathetic nerve alone does not cause the secretion of saliva in the parotid in the dog. This occurs only when the branch from the glosso-pharyngeal nerve to the parotid, which is accessible in the tympanic plexus within the tympanic cavity, is also stimulated at the same time. Then a viscid saliva, rich in organic elements, is poured out. Stimulation of the cerebral branch alone produces a clear, watery saliva, with few organic constituents, but containing salivary salts.

According to Langley, the sympathetic nerve also contains independent secretory fibers, which can be demonstrated only by stimulation soon after the termination of the irritation of the tympanic nerve. After destruction of the tympanic plexus, the parotid gland atrophies. Stimulation of the glosso-pharyngeal nerve in the rabbit causes secretion also in the glands of the tongue, with redness of the foliate papillæ.

In the intact body excitation of the nerves causing secretion of saliva occurs through reflex influences, a watery (cerebral) saliva being secreted under normal conditions. The nerve-fibers conveying the impulse centripetally are: (1) the gustatory nerves; (2) the sensory fibers of the trigeminal and glosso-pharyngeal nerves of the entire buccal cavity. These seem also to cause the secretion of saliva by mechanical irritation through the movements of mastication. Pflüger found that, on the side upon which mastication took place, one-third more saliva was secreted. Cl. Bernard observed the secretion to cease in horses while drinking. (3) The olfactory nerves, excited by certain exhalations. (4) The gastric branches of the pneumogastric nerve, especially in association with strangling movements. (5) Even the irritation of distant sensory nerves, such as those of the conjunctiva, by the application of irritating fluids in carnivora.

Further, stimulation of the central extremity of the divided sciatic nerve causes the secretion of saliva. In this category is probably to be included also the salivation sometimes observed in pregnant women. By irritation of distant sensory nerves both centers are excited reflexly; when nearby nerves are irritated, the center on the same side is especially excited.

The reflex center for the secretion of saliva is situated in the medulla oblongata, at the origin of the seventh and ninth cranial nerves. The center for the sympathetic fibers also is situated here. If the center is directly irritated mechanically, as by pricking, salivation occurs; suffocation has the same effect. This reflex may be inhibited by irritation of certain sensory nerves, as by drawing forward loops of intestines.

The reflex center is in direct communication with the cerebral hemispheres, as is evident from the fact that, with the thought of savory substances, especially during the state of hunger, watery salivation takes place. Irritation of the cerebral cortex, in the region of the cruciate sulcus (Fig. 258) also causes a flow of saliva in the dog. Also central disease in human beings may induce abnormalities in the secretion of saliva through their influence upon the intracranial center.

As long as all nerve-irritation is suppressed, no secretion of saliva takes place, as, for instance, during sleep. Secretion likewise ceases immediately after division of all of the glandular nerves.

Inflammations of the buccal cavity, neuralgia involving the nerves of the mouth, irruption of the teeth, ulcers of the mucous membrane, spongy conditions of the gums (as from the long-continued use of mercury) often induce active secretion of saliva (salivation, ptyalism), which rarely is unilateral.

The parotid gland in the sheep (ruminant) secretes continually. Division of all of the afferent nerves does not affect this secretion. Perhaps this gland contains a center through which secretion is excited.

Certain poisons also cause salivation by direct nerve-irritation, especially pilocarpin. Some, particularly atropin, paralyze the cerebral salivary nerves, and thus cause a cessation of secretion. Administration of muscarin under these conditions causes resumption of the secretion. Pilocarpin acts by irritation of the chorda tympani. Administration of atropin during the resulting salivation causes this to cease. Conversely, in the condition of abolished secretion of saliva following the administration of atropin, pilocarpin or physostigmin causes a resumption of the secretion. Curare acts as a sialogog by irritation of the center.

THE SALIVA FROM THE INDIVIDUAL GLANDS.

Method.—For obtaining the isolated saliva from the individual glands a thin metal tube is introduced into the excretory duct. If masticatory movements are then performed, or if a pungent substance be placed upon the tongue, the saliva will flow from the tube, drop by drop.

Parotid saliva is not ropy, dropping readily, of alkaline reaction, with a specific gravity of from 1003 to 1006. It contains 6.84 per cent. of total solids, of which 3.40 per cent. are inorganic. On standing it becomes cloudy and precipitates, together with some globulin, calcium carbonate, which is dissolved in fresh saliva as bicarbonate.

Through the precipitation of calcium, salivary calculi may be formed in the excretory ducts; dental calculi likewise may form, enclosing leptothrix-threads and bacteria.

Of the organic constituents of parotid saliva the most important is ptyalin; mucin is absent. Saliva contains, further, small amounts of a globulin-like body, alkali-albuminate and albumin, together with some urea, traces of volatile acid, and it appears never to be free from potassium or sodium sulphocyanid, which is wanting in some animals.

This substance is recognized, after acidulating the saliva slightly with hydrochloric acid, by adding a solution of ferric chlorid, when, with the formation of ferric sulphocyanid a dark red color results. Potassium sulphocyanid reduces hydriodic acid when added to saliva, with the development of a yellow color, and the formation of iodin, which can be recognized by the addition of starch. It is absent when the flow of bile into the intestine is prevented. It is formed through proteid metabolism, perhaps from the contained cyanogen. As potassium sulphocyanid is toxic for plants and microörganisms, it may be concluded that it acts, within certain limits, as a disinfectant for the buccal cavity.

The inorganic elements in the saliva are mainly potassium and sodium chlorids, with calcium bicarbonate, and calcium and sodium sulphates, phosphates, and chlorates.

The *submaxillary saliva* is alkaline, sometimes strongly alkaline. On standing for some time it precipitates fine crystals of calcium carbonate, together with an amorphous, albuminoid substance. It always contains mucin, and it is, therefore, as a rule somewhat ropy; also ptyalin—less than in the parotid secretion; and only 0.0037 per cent. potassium sulphocyanid.

In the submaxillary saliva of the dog there were found 1.755 of organic matter, of which 0.662 was mucin; from 2.604 to 3.662 of inorganic salts; and from 0.263 to 1.123 of soluble salts.

Pflüger investigated the gases of the submaxillary saliva and found, in 100 cu. cm. of saliva, 0.6 of oxygen, 64.7 of carbon dioxid, partly removable by exposure to a vacuum and in part capable of being expelled by phosphoric acid; and 0.8 of nitrogen; or of gases in 100 volumes 0.91 of oxygen, 97.88 of carbon dioxid, and 1.21 of nitrogen. Külz found in human parotid saliva as much as 1.46 volumes per cent. of oxygen and 3.2 of nitrogen, 4.7 of carbon dioxid removable by suction, and 62 of combined carbon dioxid.

The *sublingual saliva*, more viscous and more coherent than the submaxillary saliva, is strongly alkaline in reaction. It contains much mucin, numerous salivary corpuscles and some potassium sulphocyanid; but its composition has, on the whole, not been determined accurately.

THE MIXED SALIVA, THE SECRETION OF THE MOUTH.

The buccal fluid is a mixture of the secretions of the salivary glands and the small glands of the mouth.

Physical Properties.—It is an opalescent, tasteless and odorless, somewhat ropy fluid, with a specific gravity of from 1002 to 1006, and an alkaline reaction, due to alkaline phosphates.

Between midnight and morning the reaction may be faintly acid. The decomposition of epithelium, of salivary corpuscles or of remains of food by microbes may also cause the reaction to be acid temporarily, particularly after long fasting and after much talking. In the presence of digestive disturbances and of fever the reaction is not rarely acid, in consequence of stagnation and insufficient secretion; therefore, also, the mouth is dry.

The amount in twenty-four hours is between 200 and 1500 grams, according to Bidder and Schmidt between 1000 and 2000 grams. The total solids in the secretion amount to 5.8 per cent.

The solids are: 2.2 of epithelium and mucus, 1.4 of ptyalin and albumin, 2.2 of salts and 0.04 per cent. of potassium sulphocyanid in 1000. The ash contains especially potassium, phosphoric acid and chlorin.

Microscopical Constituents.—(a) The salivary corpuscles, from 8 to 11 μ in size, are nucleated, protoplasmic, spherical cells, without a limiting membrane. They exhibit as a vital phenomenon so-called molecular movements on the part of their numerous dark granules, which are embedded in the protoplasm, through whose internal, flowing movement they are set into a tremulous, dancing locomotion, which ceases with the death of the cells. Salivary corpuscles can be easily brought into view by slight pressure upon the excretory ducts beneath the tongue.

(b) Desquamated squamous epithelium is never absent, and is present in abundance in association with catarrh of the buccal cavity (Fig. 92, 8).

(c) Living organisms, which grow as saprophytes upon the buccal fluid and remains of food, at times in carious teeth, consist of the threads of the leptothrix buccalis (Fig. 92, 12), which turn blue, as a rule, on addition of iodin, and multiply with enormous rapidity. Leptothrix vegetations enter the dental tubules and cause caries of the teeth. The zooglea-form of the leptothrix appears as a cream-like, yellowish, smeary deposit on the teeth. Miller found in all healthy human beings, in addition to the ordinary leptothrix buccalis, another variety, the lepto-thrix buccalis maxima, also the iodococcus vaginatus, the bacillus buccalis maximus, the spirillum sputigenum and the spirochæta dentium. Further. pathogenic bacteria may be present, as, for instance, those of pneumonia, of diphtheria, etc.

Chemical Properties.—(a) Organic constituents: a globulin-like albu-minous substance, mucin, ptyalin; fats and urea are present only in traces; about 130 mg. of potassium or sodium sulphocyanid in twenty-four hours.

(b) Inorganic constituents: sodium chlorid, potassium chlorid, potas-sium sulphate, alkaline and earthy phosphates and ferric phosphate.

According to Schönbein, saliva contains traces of nitrous salts, which are recog-nizable from the yellow color produced by metadiamidobenzol in saliva diluted five times with water after addition of a few drops of dilute sulphuric acid; also traces of ammonia. Fresh saliva is said to contain hydrogen dioxid, which oxidizes the ammonia to nitrous acid; though when the reaction of the saliva is acid nitric acid is formed.

Abnormal Constituents of the Saliva.—In cases of diabetes lactic acid has been found as a result of decomposition of the sugar. It dissolves the calcium of the teeth and may thus give rise to caries, as in cases of diabetes. v. Frerichs found leucin, and an increased amount of urea and albumin were observed in cases of nephritis, and uric acid in cases of uremia. Of foreign substances that are admin-istered there appear in the saliva mercury, potassium, metallic and free iodin and bromin, the last displacing an equivalent amount of chlorin from the salivary chlorids. lead, morphin, lithium, and sodium chlorid.

Of the salivary glands in the new-born infant only the parotid contains ptyalin. In the submaxillary gland and in the pancreas the diastatic ferment appears to be formed not earlier than the end of the second month. Therefore the nourish-ment of infants with starches is not advisable. It is a remarkable fact that in new-born infants suffering from thrush (due to oïdium albicans) no ptyalin is demonstrable in the saliva. For the infant that takes milk alone, the diastatic action of the saliva is not indispensably necessary. Therefore, the mucous mem-brane of the mouth appears to be but slightly moistened during the first two months, though an abundance of saliva is secreted later . Also, the glands usually attain a considerable size only after the first half-year of life. The irruption of the first teeth causes the secretion of much saliva in consequence of the irritation of the buccal mucous membrane.

PHYSIOLOGICAL ACTIONS OF THE SALIVA.

The most important action of the saliva is amylolytic or diastatic, that is, the conversion of starch into sugar and dextrin. This is due to the *ptyalin*, an unformed, hydrolytic ferment or enzyme which, even when present in small amounts, causes the starch to take up water and become soluble, with absorption of heat, although the ferment itself undergoes no material change. Ptyalin is not present in the saliva of true carnivora.

According to Dubrunfaut, O'Sullivan, Musculus and v. Mering, maltose and dextrin, both soluble in water, are formed from starch (or glycogen) by the dias-tatic ferment of the saliva (and of the pancreas):

$$10(C_{12}H_{20}O_{10}) + 8(H_2O) = 8(C_{12}H_{22}O_{11}) + 2(C_{12}H_{20}O_{10})$$
$$\text{Starch} \quad + \quad \text{Water} \qquad = \qquad \text{Maltose} \quad + \quad \text{Dextrin.}$$

The exact course of events is as follows: At first with liquefaction of the starch-paste amylodextrin is formed. This does not reduce Fehling's solution; it is colored blue by iodin and is the principal constituent of the preparation formerly called soluble starch or amydulin. It is transformed into three molecules of erythrodextrin, which reduces Fehling's solution feebly, and is colored red by iodin. The erythrodextrin is transformed into three molecules of achroödextrin, which reduces Fehling's solution, but is not stained by iodin. From this iso-maltose and maltose are formed, the latter being formed from the former by the action of ptyalin. Isomaltose undergoes fermentation with greater difficulty than maltose. Finally all the starch is changed into maltose and dextrose.

When little ferment is present and the action is of short duration, the saliva or the pancreatic juice produces isomaltose principally; when much ferment is present and the action is of longer duration, the formation of maltose and of some dextrose is favored. The maltose subsequently may be changed in the intestine into dextrose, but the greater part is absorbed unchanged.

Kirchof, in 1811, showed that dextrose is formed from starch, by boiling with dilute sulphuric or hydrochloric acid.

Demonstration of Ptyalin.—This depends, as in the case of all hydrolytic fer-ments, upon the fact that a voluminous precipitate formed in the saliva carries the

FIG. 96.—Potato Starch.

ferment down with it mechanically, and from it the latter is then isolated by simple means. For this purpose the saliva is strongly acidulated with phosphoric acid, and lime-water is added until the reaction is rendered alkaline. As a result, a heavy precipitate of basic calcium phosphate forms, carrying the ptyalin down with it. This precipitate is collected upon a filter and the ptyalin is dissolved out with the aid of a little water. Alcohol precipitates the ptyalin in this watery extract as a white powder. By repeated solution in water, and subsequent precipitation with alcohol, the ptyalin is obtained in an absolutely pure state.

The cells of the glands first contain ptyalin in a preliminary stage, namely a ptyalinogenic substance, from which ptyalin is formed only during secretion. Ptyalin contains nitrogen, is free from ash, but yields no xanthoproteic reaction. It is precipitated from solution by neutral or basic lead acetate. It actively decomposes hydrogen dioxid.

v. Wittich taught that ptyalin could be extracted with glycerin containing water from the salivary glands of human beings or swine, cleansed, minced, placed in strong alcohol and then dried. After standing for several days, the glycerin is poured off and to it alcohol is added, precipitating the ptyalin. This is collected on a filter and then dissolved in water. In order to free it from any albumin that may be adherent to it, the aqueous solution is rapidly heated to 60° C., with the result that the albumin is precipitated, while the ptyalin remains unimpaired in solution in the filtrate.

The following details are worthy of consideration with respect to the action of the saliva in the process of saccharification:

(a) The process of saccharification is recognized: (1) from the disappearance of the starch. The addition of a little iodin to a thin solution of starch produces a blue color. If, now, saliva is added and the liquid is shaken, the blue color quickly disappears. (2) Directly by demonstration of the presence of sugar by appropriate tests.

(b) The process pursues a most favorable course at a temperature between 35° C. and 46° C.; it is slower in the cold; at 55° C. the action of the ferment becomes weaker, and at 75° C. it is destroyed. Ptyalin is distinguished from diastase, that is the diastatic ferment formed in germinating grain, by the fact that the latter exhibits its saccharifying action only at a temperature between 60° and 69° C. Ptyalin also breaks up salicin into saligenin and grape-sugar.

(c) The ptyalin, as a ferment, remains unchanged in the process of saccharification. Nevertheless, when once employed, it will not possess the same activity in a second experiment.

(d) The diastatic activity is greatest in the morning. It then declines, rising again toward noon and falling once more toward evening. It declines also after every ingestion of food.

(e) The action of the saliva is most intense when its reaction is feebly acid, though it takes place also when the reaction is alkaline or neutral. Ptyalin causes the production of sugar in the acid gastric juice of human beings only when the acidity is due to organic acids, such as lactic or butyric acid, but not when it is due to free hydrochloric acid. The production of dextrin occurs in either event. In the former case, therefore, saccharification may be continued in the stomach, although the ptyalin is destroyed by the hydrochloric acid or digested by the pepsin. The presence of peptone is said to be necessary for the production of sugar. The production of butyric and lactic acids in considerable amount may exert an inhibitory effect on the formation of sugar. Neutralization of these acids, however, permits the process to begin anew.

(f) The addition of sodium chlorid, ammonium chlorid, or sodium sulphate (in about 4 per cent. solution) increases the fermentative activity of ptyalin, as do also the acetates of quinin, strychnin, and morphin; further, curare and 0.625 per cent. sulphuric acid.

(g) Much alcohol and potassium hydroxid destroy the ptyalin; exposure to the air for a considerable time weakens it; sodium carbonate and magnesium sulphate delay its action; while salicylic acid inhibits saccharification, as does also much atropin.

(h) Ptyalin acts but feebly and gradually on unboiled starch only after the lapse of 2 or 3 hours; while it acts rapidly upon starch swollen by boiling (starch-paste).

(i) The different kinds of starch are transformed with varying rapidity in accordance with the quantity of cellulose contained in each: unboiled potato-starch (Fig. 96) in not less than 2 or 3 hours; unboiled corn-starch within 2 or 3 minutes; wheat-starch more quickly than rice-starch. When rubbed up into powder or boiled, all starches act in the same way.

(k) The mixture of saliva from all of the glands is more effective than that from any one gland alone; the mucus is inactive.

Ptyalin produces free hydrogen sulphid from radishes, onions, garlic, and the like, which contain sulphur. This fact explains the presence of the gas named in the intestines after the ingestion of the foregoing substances.

The saliva takes part in dissolving in the mouth articles of food soluble in water.

The saliva moistens articles of food ingested in a dry state, renders possible, by its viscosity, the formation of the bolus and facilitates deglutition through the slipperiness afforded by the mucus it contains. The mucus is later evacuated with the feces.

Recently the presence of peptone-producing ferments in the saliva has been discovered, but they are perhaps merely absorbed from the intestine and again excreted in the saliva (as occurs in the urine).

TESTS FOR SUGAR.

Trommer's test, like several others, depends upon the fact that sugar in hot alkaline solution acts as a reducing agent; here a metallic oxid is transformed into a suboxid. To one-half as much potassium-hydrate or sodium-hydrate solution, of a specific gravity of 1.25, is added the fluid to be tested. Then a weak solution of cupric sulphate is added drop by drop until the bluish precipitate that appears at first and consists of cupric oxid, is again dissolved by agitation. If sugar is present, the precipitate again forms a deep-blue solution after agitation. If heat is applied gradually almost up to the boiling-point a yellowish or reddish cloud is formed from above, which is finally precipitated as brownish-red cuprous oxid or as yellowish-red cupric oxid: $2CuO — O = Cu_2O$.

Cuprous oxyhydrate is dissolved also by other organic substances, though only certain sugars—maltose, grape-sugar, fruit-sugar and milk-sugar, but not cane-sugar—cause final reduction. Fluids previously turbid must be filtered and possibly treated with basic lead acetate. In the latter event the excess of lead is precipitated by sodium phosphate; then filtration is practised. When the amount of sugar is exceedingly small, concentration of the fluid over the water-bath may be necessary. If small amounts of sugar, less than 0.5 per cent., are present, together with ammonia, uric acid, and kreatinin, instead of a yellow precipitate, merely a yellow solution of cuprous oxid may result. The addition of an excess of cupric sulphate, which should always be avoided, causes confusion by the precipitation of black cupric oxid.

Böttger's test is made with an alkaline solution of bismuth oxid, best prepared according to Nylander as follows: Bismuth subnitrate 2 grams, sodio-potassium tartrate 4 grams, and sodium hydrate (8 per cent.) 100 grams. One cu. cm. of this mixture is added to 10 cu. cm. of the fluid to be tested. Upon boiling for several minutes the sugar present causes reduction to metallic bismuth, with the formation of a black precipitate.

Moore's and Heller's test: Sufficient sodium or potassium hydrate is added to the fluid to give it a strongly alkaline reaction. On boiling, a yellowish, brownish or brownish-black color results from the formation of humus-substances. If, after cooling, one drop of concentrated sulphuric acid is added, the odor of burnt sugar (caramel) and formic acid develops.

Mulder's and Neubauer's test: If a solution of indigo-carmin, made alkaline by sodium carbonate, is added to a fluid containing sugar until a pale-blue color is produced, and heat is applied, the color becomes successively green, purple, red and yellow. Agitated with atmospheric air the fluid again acquires the blue color.

Molisch's tests: To ½ cu. cm. of the fluid to be tested, 2 drops of a 17 per cent. alcoholic solution of *a*-naphthol or of a solution of thymol are added; with dilute solutions of sugar a small quantity of solid *a*-naphthol may be used instead of the solution. Then 1 or 2 cu. cm. of concentrated sulphuric acid are added and the fluid is rapidly shaken. In the presence of sugar the *a*-naphthol mixture becomes deep violet in color, the thymol-solution deep red. Subsequent dilution with water causes a precipitate of the same color, which is insoluble in concentrated hydrochloric acid. Albumin, casein and peptone also yield this reaction, but the precipitate appearing upon the addition of water is soluble in concentrated hydrochloric acid.

Phenylhydrazin test: To 7 cu. cm. of the fluid in a test-tube a small amount of phenylhydrazin chlorid (0.2) and also of sodium acetate (0.3) are added. Heat is applied until solution takes place, water being added if necessary. The tube is kept in boiling water for an hour. The contents are then poured into a conical glass, at the bottom of which characteristic yellow, microscopical tufts of fine, long needles of phenylglucosazone are found, which are almost insoluble in water; while maltose produces an analogous substance, phenylmaltosazone, which is soluble in hot water.

From all fluids to be tested for sugar, any albumin present should first be removed; from the urine by boiling, after slight acidulation with acetic acid; from the blood, by the method described on page 73; the alcohol is driven off by heat.

QUANTITATIVE ESTIMATION OF SUGAR.

By Fermentation. (An illustration of yeast is given in Fig. 140.) For this purpose the apparatus illustrated in Fig. 97 is employed. In the glass flask a is measured a quantity (as, for example, 20 cu. cm.) of fluid containing sugar, to which yeast is added. The flask b contains concentrated sulphuric acid. The entire apparatus is weighed immediately after being filled. At ordinary temperature (between 10° and 40° C.), most energetically at 25° C., the sugar breaks up into 2 molecules of alcohol and 2 molecules of carbon dioxid :

$$C_6H_{12}O_6 = 2(C_2H_6O) + 2(CO_2)$$
$$\text{Sugar} = 2 \text{ Alcohol} \div 2 \text{ Carbon dioxid.}$$

In addition some glycerin and succinic acid are formed. The carbon dioxid escapes through the flask b, and returns to the sulphuric acid any water that it may have taken with it. If the decomposition is concluded in the course of about two days, the apparatus is again weighed. From the loss in weight the amount of sugar that was contained in the 20 cu. cm. of fluid is estimated, in accordance with the fact that 100 parts by weight of sugar free from water are equal to 48.89 parts of carbon dioxid, or that 100 parts of carbon dioxid by weight correspond to 204.54 parts of sugar.

By Titration, with Fehling's alkaline cupric-oxid solution based on Trommer's test. The deep-blue titration-fluid, composed of cupric sulphate, potassium acetate, sodium hydrate and water, is so prepared that all of the cupric oxid in 10 cu. cm. of the solution will be reduced to yellowish-red cuprous oxid by just 0.05 gram of grape-sugar. For example, as in determining the amount of sugar in urine, 10 cu. cm. of Fehling's solution are placed in a porcelain dish, and gradually

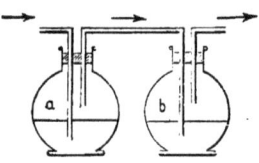

FIG. 97.—Apparatus for the Quantitative Estimation of Sugar.

diluted with 40 cu. cm. of water and heat applied almost up to the boiling-point. The urine, previously diluted to from 10 to 20 times its volume, is dropped from a burette into the hot titration-solution, and stirred until every trace of blue color has disappeared or until one drop of the fluid no longer produces a red color on blotting-paper saturated with acetic acid and potassium ferrocyanid. The amount of urine needed is now read from the scale of the burette, making allowance for the dilution, and it will then be known that the amount of urine used for reduction contained 0.05 gram of grape-sugar. From this the amount of sugar in the entire quantity of urine excreted can be readily estimated.

By Polarization.—Sugar possesses the peculiarity of turning the plane of polarized light to the right, just as albumin turns it to the left. Specific polarizing power is the term applied to the degree of rotation that 1 gram of the substance in question, dissolved in 1 cu. cm. of water, forming a layer 10 cm. thick, the length of the tube of the apparatus, effects with yellow light. For dextrose this is +56. As the rotatory power is directly proportional to the quantity of the substance dissolved in the fluid, the degree of deflection affords information as to the amount of the optically active substance contained in the fluid. In making the observation, the Soleil-Ventzke polarization-apparatus (Fig. 98) shows on its scale to the right directly the percentage of sugar; to the left, that of albumin.

The light derived from the lamp encounters a crystal of calcspar at a. Two Nicol's prisms are placed at v and s; that at v can be rotated about the visual axis, while the other is fixed. The Soleil double plate of quartz is attached at m; one-half of this deflects the plane of polarized light as far to the right as the other deflects it to the left. At c the field of vision is covered by a plate of levorotatory quartz. At b c is placed a compensator formed of two dextrorotatory prisms of quartz, which can be moved laterally by means of the screw g in such a way that the polarized light sent through the apparatus must pass through a thinner or thicker layer of the dextrorotatory quartz in accordance with the degree of rotation. With these dextrorotatory prisms in a certain position, the deflection of the levorotatory quartz at n is exactly neutralized. In this position the scale and vernier placed upon the compensator will be exactly at 0, and both halves of the double plate at *m* appear of the same color to the observer, who looks from *v* through the telescope introduced at *e*. By appropriate rotation of the Nicol's prism at *v*, a bright rose color is preferably selected. In this position the telescope must be so adjusted that the vertical dividing line of the double plate is plainly visible. Thus adjusted the instrument is ready for use. The tube, 10 cm. long, is filled with the fluid to be examined, which must be perfectly clear—should it contain albumin, this must be removed by boiling and filtering—and the tube is introduced into the apparatus between *m* and *n*. By rotating the Nicol's prism at *v*, the rose-red color is again brought into view. Then the com-

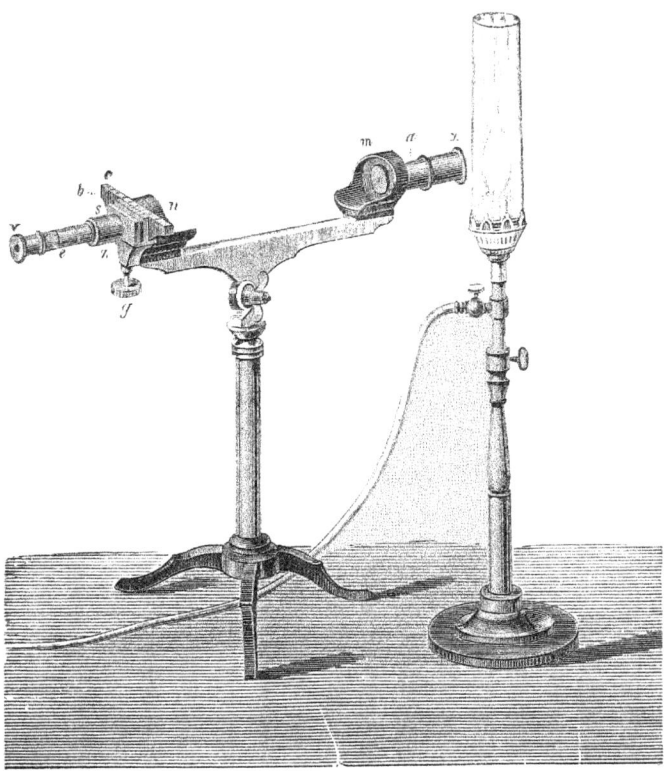

FIG. 98.—The Soleil-Ventzke Saccharimeter.

pensator at *g* is turned until both halves of the field of vision are exactly of the same color. When this has been done, the number of divisions the zero-mark of the vernier has been moved to the right—in the case of albumin to the left—can be read directly from the scale. The number of divisions read off shows directly the number of grams of the rotatory substance in 100 cu. cm. of the fluid. Turbidity that persists in spite of filtering often disappears after addition of a drop of acetic acid or a few drops of a solution of sodium carbonate or lime-water, with subsequent filtration. For a description of other apparatus employed for the same purpose—the polaristrobometer of Wild, the polarimeter of Zeiss, and the half-shadow apparatus of Laurent, Lippich, and others—the text-books on physics and chemistry should be consulted.

THE MECHANICS OF THE DIGESTIVE APPARATUS.

The mechanism of the digestive apparatus comprises:

1. The prehension of the food, the movements of mastication and of the tongue, insalivation, and the formation of the bolus.
2. The movements of deglutition.
3. The movements of the stomach and the small and large intestines.
4. The expulsion of fecal matter.

THE PREHENSION OF FOOD.

Liquid food is taken into the mouth (1) by suction. While the lips are applied hermetically about the utensil containing the fluid, the tongue, moving downward and at the same time flattened, often in conjunction with depression of the lower jaw, causes the fluid to enter the buccal cavity. Herz found that the negative pressure produced by the suction of infants equals from 3 to 10 mm. of mercury. (2) The liquid is sipped when it is brought directly in contact with the lips, and then is drawn by aspiration into the buccal cavity, together with air, with a characteristic sound. (3) Liquid can also gain entrance into the buccal cavity by being poured, the lower lip, as a rule, being applied to the containing vessel.

Among the solid articles of food, the smaller particles, supported by the lips, are picked up by the tongue; of the larger particles a piece is bitten off by the chisel-shaped incisor and sharp canine teeth, and then, for further comminution, it is brought between the rough surfaces of the bicuspid and molar teeth.

THE MOVEMENTS OF MASTICATION.

The articulation of the lower jaw is divided into two cavities, one above the other, by an interarticular cartilage, which also fulfils the duty of preventing mutual direct pressure of the articular surfaces during the energetic action of the muscles of mastication, in the act of chewing. The articular capsule, considerably strengthened ·by the external ligament particularly, is so capacious as to permit, in addition to elevation and depression of the lower jaw, also of displacement of the head of the inferior maxilla forward upon the articular tubercle, although the meniscus does not leave the head of the bone, which it covers like a cap.

The movements of mastication include: (a) Elevation of the jaw, which is effected by the united action of the temporal, masseter and internal pterygoid muscles. If the inferior maxillary bone had previously been greatly depressed, so that the condyles of the bone were moved forward upon the articular eminences, they now drop back into the articular cavity.

If, in raising the lower jaw, the bone is maintained in a particular position, the action of the muscle that would move the maxilla from this position ·is lost, as is shown by the following: (1) In elevating the lower jaw when it is pushed as far forward as possible, the action of the temporal muscles is lost, because these, in raising the jaw, draw it backward at the same time. (2) When the lower jaw is pushed as far backward as possible, the temporal muscles alone exert an elevating action, because the other muscles would tend also to draw it forward at the same time. (3) When the lower jaw is displaced laterally, the elevating action of the temporal muscles is lost.

(*b*) The downward movement of the lower jaw is partly due to its weight and partly to moderate contraction of the anterior bellies of the digastric and by the mylohyoid and geniohyoid muscles. These muscles act more powerfully when the mouth is opened widely and forcibly. The fixation of the hyoid bone necessary for this purpose is effected by the omohyoid and sternohyoid, as well as the combined action of the sternothyroid and thyrohyoid muscles.

As the articular heads of the bones move forward upon the articular tubercles when the inferior maxilla is greatly depressed, it has been assumed that, in this case, the external pterygoid muscles actively favor this displacement. When the mouth is opened to an especially marked degree, the head is bent backward, and, with the hyoid bone fixed, the posterior bellies of the digastric muscles, as well as the stylohyoid muscles, enter into action. Some animals possess upper jaws capable of movement upward and downward, as, for instance, parrots, crocodiles, snakes, and fish.

(*c*) Displacement of one or of both articular heads of the inferior maxillary bone forward or backward. (1) Projection forward of the lower jaw is caused by the action of the external pterygoid muscles. As under such circumstances the articular head of the bone slips upon the articular tubercle, and therefore also moves downward, the surfaces of the lateral teeth must separate from each other in this position. (2) Backward displacement is caused by the action of the internal pterygoid muscles. (3) The articular head on one side is drawn forward and then backward again by the external and internal pterygoid muscles of the same side, a transverse movement of the inferior maxilla taking place at the same time. The more the lower jaw is depressed, the more ineffective are these movements.

In the movements of mastication, with which both elevation and depression of the lower jaw, as well as with a transverse grinding movement are often combined, the food to be masticated is kept between the opposing surfaces of the teeth by the muscles of the lips (orbicularis oris) and the buccinators from without and by the action of the tongue from within. The sensibility of the masticatory muscles, together with the sensibility of the teeth and the mucous membrane of the mouth and lips, determines the amount of force to be expended by the muscles of the lower jaw for the purpose of mastication. By reason of simultaneous insalivation, the divided particles cohere, so that they can be readily formed into an oval bolus on the dorsum of the tongue.

The muscles of mastication, together with the mylohyoid and the anterior belly of the digastric, receive their motor nerves from the motor portion of the third division of the trigeminal nerve. The hypoglossal nerve innervates the geniohyoid, thyrohyoid, omohyoid, and sternohyoid muscles, as well as the sternothyroid. The buccinator, the posterior belly of the digastric, the stylohyoid and the muscles of the face that take part in opening and closing the mouth are supplied by the facial nerve. The common nervous center for the movements of mastication lies in the medulla oblongata.

When the mouth is shut, the permanent position of the jaws in contact with each other is due to atmospheric pressure, as the buccal cavity is made completely free of air, while the entrance of air is prevented anteriorly by the lips and posteriorly by the veil of the palate. The pressure of the atmospheric air corresponds to a column of mercury of from 2 to 4 mm. high.

STRUCTURE AND DEVELOPMENT OF THE TEETH.

The tooth is to be regarded as a modified papilla of the mucous membrane of the jaw, of exceptional size and peculiar structure. In its simplest form it appears as a horny tooth, as, for instance, in the lamprey and the duck-bill, in which the connective-tissue framework of the papilla is covered externally with layers of hard, horny epithelium, comparable with the formation of hair and of bristles. In the formation of human teeth a thick layer covering the papillary cone is transformed into the firm layer of calcified dentine. The epithelium of the papilla produces the enamel, while an accessory deposit takes place around the base of the cone in the form of a thin covering of bone (cement).

The *dentine*, ivory, or tooth-bone, which surrounds the cavity of the tooth (Fig. 99) and the root-canal, is firm, elastic and brittle. It appears, when subjected to special treatment, to be composed of fibrils, which unite to form lamellæ, and these in turn make up the dentine and are traversed perpendicularly by the dentinal tubules. These numerous, long, corkscrew-like, spiral dentinal tubules begin with free openings from $1.3\ \mu$ to $2.2\ \mu$ in diameter in the interior of the tooth, and traverse the dentine to its outermost layer. The tubules are bounded by an extremely resistent, thin cuticle-like layer, the dentinal sheath (Fig. 100), which is most unyielding to chemical agents. Within the cavities of the dentinal tubules and completely filling them lie soft fibers, the dentinal fibrils, which are to be considered as enormously elongated processes from the superficial pulp-cells, the odontoblasts.

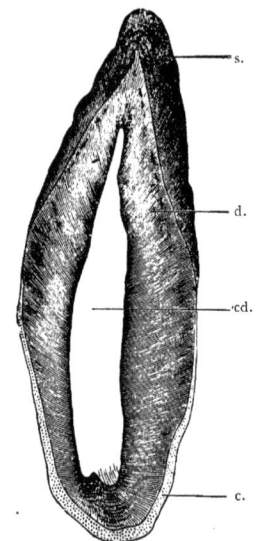

FIG. 99.—Longitudinal Section through an Incisor Tooth: s, enamel; d, dentine; cd, tooth-cavity; c, cement.

The dentinal tubules, and also their contents, the dentinal fibrils, anastomose throughout their entire course by means of processes. Most of them terminate near the enamel, or they penetrate by means of delicate processes into the cement substance between the enamel prisms. Only a few bend over, forming an arch and joining one another (Fig. 102, *A*, *c*), while others pass over into the interglobular spaces (Fig. 101). The latter are small, uncalcified areas of the ground-substance, or dilated tubules located in greater number particularly near the periphery of the dentine, and bound by spherical surfaces. With the naked eye peculiar lines can be seen in the dentine, particularly that of the elephant's tooth, running parallel with the contour of the tooth (Schreger's lines), which depend upon the fact that, at these points, all of the dentinal tubules pursue a similar course as respects their main curves. A special canal-system, rising from the root, lies between the dentine on one side and the enamel and cement on the other, and communicates with the other cavities of the tooth.

The *enamel* (vitreous substance), the hardest substance in the body, as hard as apatite or quartz, covers the free projecting crown of the tooth. It consists of perpendicular, hexagonal prisms (Fig. 102, *B*, *C*), arranged side by side like palisades, and united by cement-substance. These prisms are from $3\ \mu$ to $5\ \mu$ wide, varying in thickness throughout their course, at the same time arching in different directions, and they exhibit, after the action of acids, a coarse, transverse striation, which, however, is absent in entirely fresh prisms. As regards their nature, the enamel-prisms are elongated and calcified cylindrical epithelium of the dentinal papilla. Retzius described, in enamel, the presence of dark, brownish bands, running parallel to the outer border of the enamel, and due to the deposition of air in the enamel (Fig. 99). Fully formed enamel is in marked degree negatively doubly refracting and uniaxial, while developing enamel is positively doubly refractive.

The *cuticula*, the membranous capsule of the enamel, covers the free surface of the enamel as a structureless membrane, 1 μ or 2 μ thick, which, in the case of young teeth, exhibits an epithelium-like arrangement, and is derived from the outer epithelial layer of the enamel organ.

The *cement* (osseous substance) consists of a thin bony layer covering the root of the tooth, with a fibrillated ground-substance and provided with Sharpey's fibers (Fig. 103, *a*). Haversian canals and lamellæ are found only in the thick layers of cement at the apex of the root, the former at times leading into the tooth-cavity. Thin layers of cement may even be unprovided with bone-corpuscles.

Chemistry of the Solid Constituents of the Tooth.— The teeth consist of a framework of calcareous substance, infiltrated with calcium phosphate and carbonate, like the bones. (1) The *dentine* contains of organic matter 27.7, of calcium phosphate and carbonate 72.06, of magnesium phosphate 0.75, with traces of iron, fluorin and sulphuric acid, potassium, sodium, and carbon dioxid. (2) The *enamel* possesses as its organic basis a substance resembling the proteid of epithelial cells. It contains of inorganic matter—in addition to 3.60 of organic matter —calcium phosphate and carbonate 96.00, magnesium phosphate 1.05, with traces of calcium fluorid, an insoluble chlorin-combination, potassium, sodium, and carbon dioxid. (3) The composition of the *cement* is identical with that of true bone.

Fig. 100.—Transverse Section through Dentine. The light rings are the dentinal sheaths, the dark centers with the bright points are the dentinal fibrils lying in the dentinal tubules.

The Soft Parts of the Tooth.—The *tooth-pulp* in the adult tooth represents the remains of the dental papilla, about which the hardening layer of dentine has been deposited. It consists of connective tissue, at times not distinctly fibrous, and rich in capillaries, with connective-tissue cells and leukocytes. The most superficial layer of cells, which, not unlike epithelium, lie close together next to the dentine, is formed of unencapsulated odontoblasts,

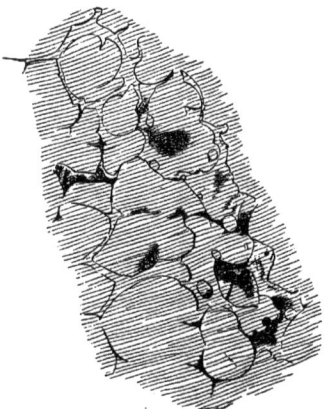

25 μ long by 2 μ wide, from which the production of the dentine proceeds. They send long processes into the dentinal tubules, while the nucleated cell-body, resting on the surface of the pulp, forms a connection with the pulp and with neighboring odontoblasts by means of other processes. Numerous medullated nerve-fibers, becoming non-medullated after repeated division, penetrate between the odontoblasts and end beneath the dentine in free extremities presenting nodular thickening in places. Other nerve-fibers lie partly in the dentinal tubules, in part in the substance of the dentine. Most of them appear to end free, in a brush-like radiation. A plexus of fine nerve-fibers lies beneath the enamel. The epidentinal canal-system is provided with a special nerve-apparatus, which in part penetrates into the enamel. The arteries of the tooth often lie in grooves in the nerve-branches. The capillaries even penetrate to the odontoblast-layer.

Fig. 101.—Interglobular Spaces in the Dentine.

The periosteum of the root of the tooth and, at the same time, of the alveolar cavity, is of a delicate structure, without elastic fibers, but rich in nerves and blood-vessels. The gums have no mucous glands and are characterized by their long, vascular papillæ.

The *development of the teeth* begins as early as the fortieth day (Röse). Throughout the entire length of the alveolar margin there is a projecting ridge, formed of a thick epithelial layer, the *dental ridge* (Fig. 104, *a*). From this epithe-

18

lial layer a furrow, also filled with epithelium, forms in the jaw, the *dental groove*, which thus runs beneath the base of the dental ridge. The dental groove grows deeper throughout its entire length, acquiring a form resembling the transverse sec. tion of elongated formative epithelial cells; this is the *enamel organ*. From the depth of the jaw there grows toward the enamel organ a conical papilla, formed of mucous tissue, the *dentinal papilla* (Fig. 104, *c*), in such a way that its apex sup.

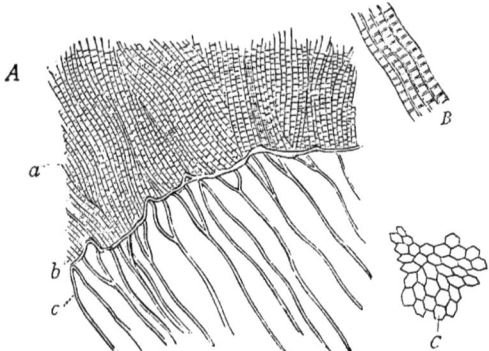

FIG. 102.—*A*, Section of a Tooth at the Junction (*b*) between Dentine and Enamel: *a*, enamel; *c*, dentinal tubules; *B* enamel prisms greatly magnified ; *C*, the same prisms in transverse section.

ports the enamel organ like a double cap. The connecting parts of the enamel organ, lying between the dentinal papillæ of the separate teeth, now disappear, through hyperplasia of the connective tissue, which next gradually surrounds the dentinal papilla and its enamel organ as the *dental sac* (Fig. 104, *d*, Fig. 105).

Of the epithelial cells of the enamel-organ those (Fig. 105, 3) that cover the head of the papilla as a continuous layer form cylindrical epithelium, which later,

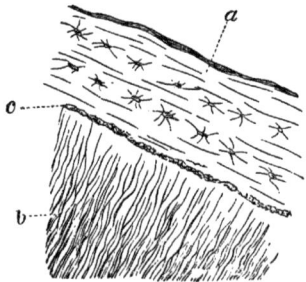

FIG. 103.—Transverse Section of the Root: *a*, cement, with bone-corpuscles; *b*, dentine with dentinal tubules; *c*, junction between the two.

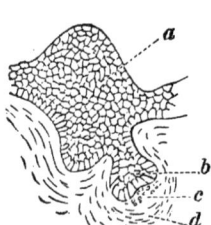

FIG. 104.—*a*, Dental ridge; *b*, enamel organ; *c*, site of the beginning dentinal papilla; *d*, first trace of the dental sac.

by calcification, hardens into the enamel prisms. The layer of cells of the double cap, however, which is turned upward toward the dental sac (Fig. 105, 1), flattens out, softens down and through a process of horny metamorphosis becomes the cuticula, while the epithelial cells lying between the two layers gradually under-go complete atrophy through a peculiar intermediary metamorphosis in which they resemble the star-shaped cells of mucous tissue (Fig. 105, 2). According to

v. Brunn the enamel extends along down the entire root of the tooth during the process of development, but is subsequently lost in this situation.

The *dentine* is formed on the uppermost surface of the protruding connective-tissue dentinal papilla, the odontoblasts (Fig. 105; Fig. 106, *k*) arranged here in a continuous layer becoming calcified, but in such a manner that uncalcified fibers, the dentinal fibrils, remain. "By means of the process of the pulp each odonto-blast is connected with the deeper lying, grad-ually growing cells of the young pulp, so that when an odontoblast is ossified down to the rudiment of its fibril another takes its place, without the continuity of the fibril being inter-rupted. Accordingly, each dentinal fibril with its anastomoses, must be considered as a rudi-ment of several communicating odontoblasts." In the hardening of dentine the same process occurs as in that of ossification by osteoblasts.

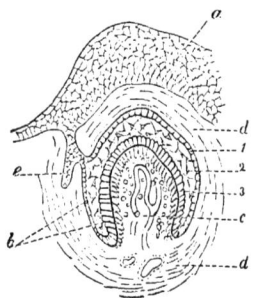

FIG. 105.—*a*, Dental ridge; *b*, enamel or-gan with (1) external epithelium, (2) middle reticular layer, and (3) en-amel cell layer; *c*, dentinal papilla, with blood-vessels and layer of elon-gated odontoblasts on the surface; *d*, dental sac; *e*, secondary enamel organ.

The *cement* is derived from the soft connec-tive tissue of the alveolar process by ossification. This connective tissue arises from the entire base of the dental sac.

The Shedding of the Teeth.—Even during the development of the milk-teeth, a special enamel organ (Fig. 105, *c*) for the permanent teeth forms by the side of that for the temporary teeth; but its growth is held in check until the time for the shedding of the teeth. The papilla of the permanent tooth is absent at the begin-ning. As the permanent tooth grows, its dental sac first breaks through the alveolar wall of the temporary tooth from below. The tissue of this dental sac, acting as an eroding granulation-tissue, causes absorption of the root of the temporary tooth and later also of its body, up to the crown, without its blood-vessels undergoing atrophy. The ameboid cells of the granulation-tissue engage in a process of undermining in the absorption of the temporary teeth by means of processes they send out, taking up, like phagocytes, calcareous fragments of the disintegrating tooth.

From the ninth month to the second year the twenty temporary teeth appear in the following order: lower internal incisors, upper internal incisors, upper external inci-sors, lower external incisors, first molar, canine and second molar teeth.

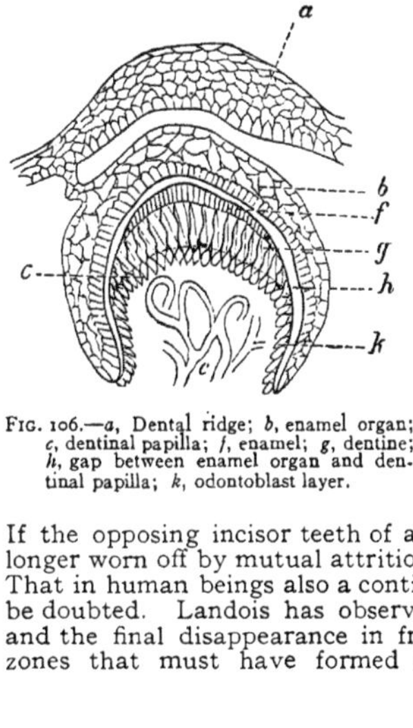

The shedding of the teeth begins in the seventh year, in the same order (the decidu-ous molars being replaced by the bicuspids). Then three new molars appear behind the bicuspid teeth, the most posterior at about the age of twenty years, therefore called "wisdom-teeth." They may, however, ap-pear as late as the eightieth year. Thus, the adult has thirty-two teeth.

According to Zuckerkandl, epithelial re-mains are found in the gums behind the last molar teeth, which must be regarded as the rudiment of a fourth undeveloped molar tooth. An analogous condition has been noted in animals.

The uninterrupted growth of the incisor teeth may be readily observed in rodents, replacing the free ends worn off by chewing.

FIG. 106.—*a*, Dental ridge; *b*, enamel organ; *c*, dentinal papilla; *f*, enamel; *g*, dentine; *h*, gap between enamel organ and den-tinal papilla; *k*, odontoblast layer.

If the opposing incisor teeth of a rodent are extracted, the remaining teeth, no longer worn off by mutual attrition, grow from the jaw in the form of a long arch. That in human beings also a continual replacement of the teeth must occur cannot be doubted. Landois has observed the advance toward the masticating surface and the final disappearance in from 8 to 9 years of rachitic, atrophic, circular zones that must have formed on the permanent teeth of a boy even before eruption. This proves the forward growth and the wear of the teeth at their free ends. Only when, in old age, the power of regeneration becomes diminished, do the teeth have worn-off surfaces. During the embryonal life of the baleen whale, dental sacs are noted in the jaws, which, however, undergo atrophy; in their place whalebone develops later. Tooth-bearing edentates, whose teeth are unprovided with enamel, nevertheless possess an enamel-organ, whose function it is, like a matrix, to insure for the developing tooth sufficient room for its formation. The edentulous armadillo possesses an embryonal dental ledge, which has also been found in birds and turtles as the last rudiment of a former dentition.

MOVEMENTS OF THE TONGUE.

The tongue keeps the food between the opposing surfaces of the teeth during mastication; it collects the finely divided particles of food, held together by the saliva and forms them into a bolus, and finally it transfers the bolus along its dorsal surface into the pharynx at the time of deglutition.

The course of the muscle-fibers in the tongue is three-fold: longitudinal, from the tip to the root of the tongue; transverse, originating mainly from the septum of the tongue stretched longitudinally; and vertical, traversing the thickness of the organ. The muscles of the tongue are in part confined to this organ alone; in part they pass to the tongue from other fixed points, namely, the hyoid bone, the lower jaw, the styloid process and the palate.

Microscopically the muscle-fibers are striated transversely. surrounded by delicate sarcolemma, and frequently divided like a fork at their extremities. The bundles interlace with one another to a considerable extent, and small deposits of fat are found in the spaces between them.

The movements of the tongue give rise in part to changes in form, in part to changes in position.

1. Shortening and widening, through the longitudinal lingual muscle, aided by the hyoglossus.

2. Elongation and narrowing, through the transverse lingual muscle.

3. Excavation of the dorsum of the tongue in the form of a longitudinal furrow, through contraction of the transverse lingual muscle, with simultaneous action of the median vertical fibers.

4. Arching the dorsum of the tongue: (a) transversely, through contraction of the lowermost transverse fibers; (b) longitudinally, through the action of the lowermost longitudinal muscles.

5. Protrusion of the tongue, through the genioglossus, aided somewhat by the geniohyoid, passing from the hyoid bone toward the chin; at the same time the tongue usually becomes elongated and narrowed.

6. Retraction of the tongue through the hyoglossus, chondroglossus and styloglossus; also as a rule with shortening and widening of the tongue.

7. Depression of the tongue upon the floor of the mouth is effected in the median line by the genioglossus; at the sides by the hyoglossus. By depression of the hyoid bone the floor of the mouth can be made even much deeper.

8. Elevation of the tongue to the palate: (a) at the tip, through the anterior portions of the upper longitudinal fibers; (b) in the center, through elevation of the entire hyoid bone by the mylohyoid muscle (trigeminal nerve); and (c) at the root, through the styloglossus and palatoglossus muscles, as well as indirectly by the stylohyoid muscle (facial nerve).

9. Lateral deflection of the protruded tongue is effected by the genioglossus (toward the opposite side); while similar deflection of the tongue, lying in the mouth, is effected by the styloglossus, hyoglossus, chondroglossus and palatoglossus muscles. Further lateral deflection of the tongue, so that the tip comes to lie behind the last bicuspid tooth, is effected through the combined action of the styloglossus and hyoglossus muscles on one side and the genioglossus on the other side.

The motor nerve of the tongue is the hypoglossal. In case of unilateral paralysis the tip of the tongue lying at rest in the mouth is directed toward the unaffected side, because the tone of the unparalyzed longitudinal fibers shortens the unaffected side to some extent. If, however, the tongue is protruded, the tip deviates toward the paralyzed side. This is dependent on the direction pursued by the genioglossus muscle from the middle line (internal mental spine) backward and outward, the direction of whose traction the tongue must naturally follow. The tongue in killed animals sometimes exhibits fibrillary twitchings for an entire day.

THE ACT OF SWALLOWING (DEGLUTITION).

The propulsion of the contents of the alimentary canal is effected by a motor process whereby the canal contracts upon the contained mass; and as this contraction progresses throughout the entire length of the tube, the contents are pushed on before it. This movement is called peristalsis.

The first and most complicated act of this movement is deglutition, in which the following stages can be distinguished:

1. The mouth is closed by the orbicularis oris muscle (facial nerve).

2. The jaws are pressed together by the muscles of mastication (trigeminal nerve); in this way the lower jaw becomes a fixed point, permitting the action of the muscles passing from the lower jaw to the hyoid bone.

3. The tip of the tongue, the back of the tongue, and the root of the tongue are successively pressed against the hard palate, and in this way the contents of the mouth (bolus or fluid) are forced toward the pharynx.

4. When the bolus has passed the anterior palatine arches, having been made slippery by the mucus of the tonsillar glands, its return to the mouth is prevented by the contraction of the palatoglossus muscles lying in the anterior palatine arches, which bring these arches firmly in contact with each other, like the scenes in a theater, and by the back of the tongue, which is elevated by the styloglossus muscle.

5. The bolus now lies behind the anterior palatine arches and the root of the tongue, within the pharynx and exposed to the successive action of the three constrictor muscles of the pharynx, which push it onward. The action of the superior constrictor muscle, which contracts first, is always combined with horizontal elevation, through the elevator of the veil of the palate (facial nerve), and tension of the soft palate, through the tensor of the veil of the palate (trigeminal nerve; otic ganglion). The superior constrictor, through the pterygopharyngeal muscle, presses the posterior and lateral pharyngeal wall firmly against the posterior edge of the veil of the palate horizontally elevated and made tense like a cushion (Passavant's cushion), while the edges of the posterior palatine arches are at the same time approximated through the palatopharyngeal muscles. In this way the nasopharyngeal cavity is closed, so that food is prevented from passing readily upward into the nasal cavity.

In persons with congenital or acquired defects of the soft palate, food can enter the nose during the act of deglutition.

The elevation of the veil of the palate can be readily demonstrated by introducing a fine probe through one nostril, along the floor of the nasal cavity, until its posterior extremity rests upon the veil of the palate. With every movement of deglutition the anterior extremity of the probe, projecting from the nostril, is depressed. A sensitive flame may also be employed, a T-shaped tube being introduced into one nostril, the other being closed. One arm of the tube is connected with a gas-pipe, the other with a burner. Every movement of deglutition is attended by movement of the flame.

6. Responding to the successive contractions of the superior, middle, and inferior constrictors of the pharynx and the esophageal muscles, the bolus is forced downward. During this time the entrance to the larynx must be kept closed, to prevent food from passing into the trachea.

7. According to Kronecker and Falk, semisolid foods and fluids in the mouth are forced through the pharynx and the esophagus by vigorous contraction of the muscles closing the mouth, particularly the mylohyoid muscles. If the act of swallowing is repeated several times in rapid succession, as in drinking, only the last is followed by movements of contraction in the pharynx and the esophagus, for every act of swallowing in the mouth exerts an inhibitory effect upon the lower portions of the esophagus, through stimulation of the glossopharyngeal nerve. That solid and semisolid articles of food are, however, pushed slowly through the esophagus, by peristalsis alone, has been demonstrated by the Röntgen rays on admixture of bismuth subnitrate with the bolus.

According to Meltzer and Kronecker, the duration of the act of deglutition in the mouth is 0.3 second. Then the constrictors of the pharynx contract; 0.9 second later the superior, 1.8 seconds later the middle, and 3 seconds later the inferior constrictor of the pharynx. The constriction of the cardiac orifice, after the food has passed into the stomach, is the final movement of the series.

On auscultation of the stomach two sounds are heard during deglutition: (1) the squirting sound, which is due to the fact that the material swallowed is forced into the stomach, and (2) the squeezing sound, due to peristalsis occurring at the end of deglutition forcing the contents of the esophagus through the cardia. The latter is a rale and, as such, is dependent on the presence of air in the mass swallowed.

Closure of the larynx is brought about as follows: (a) The lower jaw being fixed, the larynx is drawn upward and forward beneath the root of the tongue, which is arched over it. This is effected by a movement of the hyoid bone forward and upward through the action of the geniohyoid, the anterior belly of the digastric, and the mylohyoid muscles together with an approximation of the larynx to the hyoid bone, through the thyrohyoid muscle. (b) While the tongue, besides, is drawn somewhat backward by the styloglossus muscles, it presses the epiglottis over the entrance to the larynx, so that food can now slide over it. (c) The epiglottis, further, is pulled down over the entrance to the larynx by the action of the reflector epiglottidis and the aryepiglottic muscle.

Intentional injuries of the epiglottis in animals, or destruction of the epiglottis in human beings, cause choking readily from the entrance of liquids into the larynx, while solid foods can be swallowed with scarcely any trouble. In dogs, however, colored liquids pass directly from the back of the root of the tongue downward into the pharynx, without necessarily staining the upper surface of the epiglottis, hidden under the overhanging root of the tongue.

(d) Finally, closure of the glottis by the constrictors of the larynx prevents the entrance of swallowed substances into the larynx. This closure is brought about through reflex influences.

In order that the pharynx itself shall not be drawn down with the descending bolus it is drawn upward by the stylopharyngeal, salpingopharyngeal and basopharyngeal muscles during the activity of the pharyngeal constrictors.

Nervous Supply.—The nerves of the pharynx are comprised in the pharyngeal plexus, formed by branches from the pneumogastric, the glossopharyngeal and the sympathetic. The act of deglutition is voluntary only in so far as it takes place in the mouth. The passage of the bolus through the palatine arches on past the tonsils into the pharynx is involuntary, and entirely reflex in character. The pharynx takes up the movement only if its contents (food or saliva) mechanically excite reflex action. The sensory branches that excite this mechanical stimulation of the deglutition-reflex are, according to Schröder van der Kolk, the palatine branches of the trigeminal nerve, from the sphenopalatine ganglion, and the pharyngeal branches of the pneumogastric nerve. The center for the nerves in question for the striated muscles lies in the medulla oblongata. Deglutition is still possible in the state of unconsciousness, as well as after destruction of brain, cerebellum and pons. Irritation of the ninth cranial nerve prevents the deglutition-reflex.

Within the esophagus (Fig. 107), the stratified squamous epithelium of which is kept slippery by the mucus from small mucous glands opening at the edges of the folds of mucous membrane, the downward movement takes place also involuntarily through a coördinated muscular act, a peristaltic movement of the external (longitudinal) and the internal (circular) unstriated muscle-fibers.

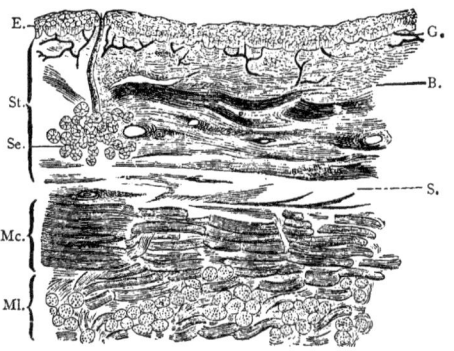

In the upper part of the esophagus, in which lie striated muscle-fibers, peristalsis is much more rapid than in the lower portion. The movements of the esophagus never originate spontaneously, but they always follow on a previous act of deglutition. Thus, if a bolus be introduced into the esophagus through an external wound, it remains where it was placed; it is carried downward only when movements of deglutition are initiated above. The peristalsis extends throughout the entire

FIG. 107.—Transverse Section through the Esophagus. E, epithelium; St, mucous membrane; Se, mucous gland; Mc, circular muscle-fibers; Ml, longitudinal muscle-fibers; G, capillaries; B, connective tissue; S, submucosa.

length of the esophagus, even if this be ligated or a portion has been excised. The peristalsis, likewise, continues downward in a dog, even after meat is withdrawn from the esophagus, though it has been halfway down.

Exceedingly large and exceedingly small masses of food are swallowed with greater effort than those of medium size. Dogs are able to swallow a bolus weighing 450 grams. Deglutition becomes difficult in consequence of great dilatation of the thorax, as in Müller's experiments; likewise in consequence of contraction of the thorax, as in Valsalva's investigations.

The motor nerve of the esophagus is the pneumogastric—after division of which on both sides food remains in the esophagus, particularly its lower part.

Goltz discovered the remarkable fact that the ganglionic plexuses situated in the esophagus and the stomach of the frog acquire greatly increased irritability when the brain and spinal cord or both pneumogastric nerves are destroyed. Esophagus and stomach contract vigorously like a string of pearls, even after slight irritation, while animals with an uninjured central nervous system swallow fluid introduced simply by peristalsis. It should be borne in mind that human beings with an enfeebled nervous system (hysteria) not rarely exhibit similar spasmodic contraction of the esophagus (globus hystericus). Schiff observed spasmodic contraction of the esophagus in dogs also after section of both pneumogastric nerves.

The heart-beats are accelerated with each act of swallowing, while the blood-pressure falls, the need of respiration diminishes and some movements, such as labor-pains and erection, are inhibited. All of these movements are brought about through reflex influences.

THE MOVEMENTS OF THE STOMACH.—VOMITING.

Three methods are employed for determining the position of the stomach: (a) the introduction of a rubber bougie through the esophagus, whose passage along the greater curvature of the stomach can be palpated; (b) electric transillumination of the stomach by means of a small round incandescent light attached to the extremity of a stomach-tube. The stomach is previously suitably dilated by the development of carbon dioxid from sodium bicarbonate administered; the interpretation requires great care; (c) the Röntgen rays have also been employed after filling the stomach with meat mixed with bismuth subnitrate, the latter being impervious to the x-rays.

For registering the gastric movements, a rubber bulb, introduced through an external gastric fistula in animals, and applied in various situations in the interior of the stomach, is employed. The bulb is connected with a writing-apparatus by means of a column of air. Einhorn has used the gastrograph in human beings. This consists of a metallic capsule attached to the extremity of a rubber tube, which is swallowed. With every movement of the stomach the metallic parts in the interior of the capsule are brought into contact, and thus employed to effect an electrical registration. A series of photographs taken with Röntgen rays also affords information as to the course of the movements and the evacuation of the gastric contents.

The anterior surface of the empty stomach lies in a frontal position, with a slight tendency to the right and upward, while the posterior surface accordingly occupies the opposite position. When the stomach is moderately distended, the anterior surface rises about the lesser curvature as an axis, so that it forms an angle of from 45° to 48° with the horizon. When the distention is more marked, the stomach comes progressively to occupy more nearly the horizontal position, so that its anterior surface gradually becomes the superior surface.

The muscular coat of the stomach consists of an external or longitudinal layer of fibers, a middle or circular layer, and an internal or oblique layer, one layer passing over into another in many places. At the pylorus the musculature forms a circular sphincter-muscle (sphincter of the pylorus), whose fibers continue into the pyloric valve. At the cardiac orifice also the muscle-fibers are grouped into a sphincter muscle.

The movements of the stomach are of two kinds: 1. The rotatory-rubbing movement, by means of which the walls of the stomach lying in immediate contact with the ingesta move to and fro with a slow displacing action. These movements succeed one another periodically, each cycle occupying several minutes.

These movements can be imitated by slowly rolling or molding a ball between the palms of the hands by means of rotatory movements of the hands in opposite directions. Indeed, hair swallowed by cattle and dogs is formed into a regular ball in the stomach. The object of this rotatory movement is thoroughly to moisten the surface of the stomach-contents with the secretion of the gastric glands, and at the same time to favor its escape by the pressure and the continuous passage of ingesta, as well as to detach the already loosened and softened superficial layers of the food. Further, the admixture of the ingesta with the gastric juice is effected in this way. This movement may be either diminished, in the presence of gastric disease, such as gastric ulcer, or increased, as when there is stenosis or dilatation.

2. The other kind of movement is a peristalsis of periodic recurrence, in conjunction with rhythmic opening and closure of the pylorus, as a result of which the partly dissolved gastric contents are little by little propelled into the duodenum, commencing after an interval of fifteen minutes and ending at about the fifth hour. Each wave lasts twenty seconds, with an interval of from fifteen to twenty seconds between waves. This peristalsis is most active from the pyloric antrum toward the pylorus. According to Rüdinger, the longitudinal fibers passing toward the pylorus, in contracting, especially when the pyloric antrum is full, cause dilatation of the pylorus.

Evacuation of the stomach occurs only when the intestine is free from contents. The following experiment will serve to determine when the ingesta enter the intestine. In the presence of an alkaline reaction in the intestine, salol is decomposed into carbolic acid and salicylic acid; the latter can be recognized in the urine from the violet color produced upon adding ferric chlorid. In healthy persons this reaction begins in from half an hour to an hour and disappears after twenty-four hours; while in the presence of motor insufficiency of the stomach it is delayed from three to twenty-four hours. Liquids are rapidly propelled from the stomach into the intestine.

The thick, muscular walls of the stomach in many grain-eating birds aid in triturating the ingesta. The energy of muscular action necessary for this purpose has often been measured by earlier investigators, who found that glass balls were broken, and lead pipes that could be flattened only by a pressure of 40 kilograms were compressed, in the stomach of the turkey. The masticating stomach of many insects also is capable of similar activity.

Mechanical stimulation causes contraction of the muscular layers directly affected; as does also application of potassium-salts, segmentary contraction of the circular muscles often taking place at the same time. Sodium-salts, on the contrary, usually cause semicircular contractions or contractions progressing toward the cardiac orifice. At the pyloric antrum the stimulations as a rule spread more rapidly. Electrical stimulation of the internal surface of the stomach causes no movement. The contraction induced by stimulation of the intestinal mucous membrane is always less than that due to stimulation of the external surface of the intestine. In human beings both endogastric and percutaneous electrical stimulation are without demonstrable effect on the evacuation and the secretion of the stomach.

Nervous Activity.—Openchowski and his pupils make the following statements with respect to the influence of the nerves upon the movements of the stomach: The cardia contains automatic ganglion-cells. which are connected with the pneumogastric and the sympathetic nerves. A center for the contraction of the cardiac orifice is situated in the posterior quadrigeminal bodies, whence the paths pass downward, mainly through the pneumogastric, and in lesser degree through the splanchnic nerves. The center for opening the cardia lies in the corpus striatum. and in connection therewith one in the cruciate sulcus of the central cortex, in the dog; the pneumogastric nerves constitute the conducting paths. Dilatation centers are situated also in the upper portion of the spinal cord, whence the path passes through the sympathetic nerve (aortic plexus, lesser splanchnic nerve). Reflex opening of the cardiac orifice can be induced by irritation of the sensory splanchnic nerves, and of the sciatic also.

The body of the stomach contains also automatic ganglia, connected with the pneumogastric and the sympathetic nerves. A center for contraction is situated in the corpora quadrigemina, whence paths pass through the pneumogastric nerves and the spinal cord, and from the latter into the sympathetic. The upper cord contains inhibitory centers; the paths pass through the sympathetic and the splanchnic nerves.

The pylorus contains automatic ganglia. It exhibits a certain, varying degree of tone during closure: the splanchnic nerve may more fully open the pylorus. while the pneumogastric tends to close it. The center for opening the cardiac orifice inhibits the movement of the pylorus; the path passes through the spinal cord and the splanchnic nerves. Inhibitory pyloric centers are situated in the corpora quadrigemina and the olivary bodies; the path passes through the spinal cord. The cortical center for opening the cardia causes simultaneous contraction of the pylorus; the path passes through the pneumogastric nerves. Centers for the contraction of the pylorus are situated in the corpora quadrigemina; the path passes through the pneumogastric nerves, a few fibers through the spinal cord and the sympathetic nerve.

Stimulation of the peritoneum and also of the skin causes reflex immobility of the pylorus and of the small intestine. Stimulation of the central stump of one pneumogastric, the other being intact. gives rise to immobility of the pylorus, contraction of the stomach and dilatation of the cardiac orifice. Elevation of the temperature to 25° C. causes movements in the excised empty stomach.

Vomiting takes place in consequence of contraction of the walls of the stomach, the pyloric sphincter being at the same time closed. It occurs most readily when the stomach is distended. Dogs usually distend the stomach greatly before vomiting, by swallowing air. There is no doubt that in infants vomiting is due principally to contractions of the walls of the stomach, though without the slightest spasmodic coöperation of abdominal pressure. When the act of vomiting is attended with straining, abdominal pressure comes energetically into play.

The contractions of the walls of the stomach that cause a general diminution in the size of the viscus can be recognized when the stomach is exposed. The pylorus contracts; then wave-like contractions appear from the pyloric extremity upward to the body of the stomach. The uppermost portion of the stomach, including the cardia, does not contract, but the cardiac orifice is opened by the contraction of the longitudinal muscle-fibers, which pass toward the esophageal opening, and therefore must act as dilators when the stomach is full.

The actual ejection of the contents of the stomach is immediately preceded by an eructation-like movement, dilating the intrathoracic portion of the esophagus. This takes place in such a manner that, with the glottis closed, violent, jerky inspiration suddenly occurs, causing the esophagus to be distended by gas rising from the stomach. At the same time the larynx and the hyoid bone are drawn forcibly forward by the combined action of the geniohyoid and sternohyoid, together with the sternothyroid and thyrohyoid muscles, with obliteration of the laryngeal angle. As a means of support the lower jaw is even moved horizontally forward; as a result air passes from the pharynx downward to the upper portion of the esophagus. At the same time the projection and the inclination of the head favor dilatation of the esophagus. If, now, sudden abdominal pressure is exerted, supported by the intrinsic movements of the stomach, the contents of the viscus will be ejected. If the vomiting be long continued, there may even be antiperistalsis of the duodenum, as a result of which bile enters the stomach and becomes admixed with the vomited matters.

Children, in whom the fundus of the stomach is not sacculated, vomit more readily than adults, in whom the fundus must contract forcibly.

The center for the act of vomiting is situated in the medulla oblongata. It is connected with the respiratory center, as experience teaches that attacks of nausea can be overcome by rapid, deep respiration. The act of vomiting can be inhibited likewise in animals by means of artificial respiration. On the other hand, the administration of emetics does not permit the development of apnea.

The act of vomiting may be excited by chemical or mechanical irritation of the centripetal nerves of the mucous membrane of the palate, the pharynx, the root of the tongue and the stomach; also, under certain conditions (pregnancy) by irritation of the uterus, of the intestines (peritonitis), and also of the genito-urinary apparatus; finally by direct stimulation of the vomiting center.

The act of vomiting excited by repulsive conceptions appears to result from the transmission of stimuli from the cerebrum through conducting fibers to the vomiting center. The act of vomiting is also common in connection with cerebral disease. Irritation of the central stump of the pneumogastric nerve is capable of inducing vomiting.

The ruminating process in ruminants resembles the act of vomiting. Also in human beings eructation of food resembling morbid rumination has been observed as the expression of a gastric neurosis. There exists under such circumstances relative insufficiency of the cardiac orifice of the stomach : with the glottis closed, the contents of the stomach on attenuation of the air in the thorax rise into the mouth. Forced expiratory pressure is capable of preventing this phenomenon.

Emetics act (1) directly upon the vomiting center (as, for instance, apomorphin). Central vomiting ceases after destruction of the corpora quadrigemina, or division of the anterior columns of the spinal cord or destruction of all the spinal sympathetic fibers that pass to the stomach. (2) Other emetics act upon the vomiting center through reflex influences from the stomach or the intestine (copper sulphate, tartar emetic). The irritation reaches the gastric musculature through the pneumogastric nerves. (3) Both of these modes of action may be combined. Emetics may also remove mucus from the respiratory organs. It would appear that emetics exert a favorable influence upon the respiratory movements, through irritation of the respiratory center, as, for instance, in small children.

THE MOVEMENTS OF THE INTESTINES.

For observing the peristaltic movements in animals, the abdominal cavity is opened under a 0.9 per cent. sodium-chlorid solution at blood-temperature in order to avoid the entrance of air ; or the observations may be made through the shaved and uninjured abdominal walls.

The small intestine exhibits peristaltic movements in a classical manner. The progressive constriction of the canal, which forces the contents before it, always passes from above downward. After death and on exposure of the coils of intestine to the air, peristalsis is often seen to develop in several parts of the intestine at the same time, and as a result the intestinal loops acquire the appearance of a mass of crawling worms. In addition to these movements, pendulum-like movements of the intestine also occur, by which the contents are moved some distance first in one direction and then in the other. The advance of new intestinal contents and the resulting increased distention of the tube due to solid contents or gas causes renewed movement.

The large intestine exhibits less active and less extensive movements. When the abdominal walls are thin, or in the sac of a hernia, peristalsis may be felt and even seen. Herbivora exhibit more active peristalsis than carnivora. Perhaps the transmission of peristalsis takes place directly through the musculature, as in the heart and the ureter. The ileo-cecal valve, as a rule, does not permit the usually more consistent contents of the large intestine to pass back into the small intestine. During sleep, at night, the movements of the stomach and the intestines cease.

If fluid material is gradually introduced into the rectum from a height of one meter of water-pressure through an intestinal tube, it may pass upward through the ileo-cecal valve into the small intestine, and, with great care, it may reach the stomach and esophagus, and even escape from the mouth and nose. In this way the entire intestinal tract in the living subject can be irrigated, and with curative results; as, for instance, in cases of cholera (1 or 2 per cent. solution of tannic acid in 7.5 per cent. solution of sodium chlorid). Eight or nine liters are sufficient to fill the entire alimentary canal.

A crystal of sodium chlorid applied externally to the intestine causes contraction at that point, with upward peristalsis, while potassium chlorid induces only local contraction. Particles saturated with sodium-chlorid solution and introduced into the rectum are carried upward, in part even to the stomach, through the mediation of nervous irritation, perhaps of the muscularis mucosæ.

Pathological.—If an inflammatory or catarrhal condition of the intestinal mucous membrane develops rapidly in consequence of an acute inflammatory irritation, contractions of the inflamed portion, at first marked, occur in the full intestine. When the affected portion has been emptied the movements are no longer more marked than normal. If further contents reach the inflamed portion, the peristaltic downward movement takes place more rapidly than normal and diarrhea results. At times a greatly contracted piece of the intestine is pushed into a neighboring portion (invagination, intussusception). Reduction in the bodily temperature is followed by a decrease in the peristalsis.

That antiperistalsis, that is a movement upward toward the stomach, occurs was formerly considered proved by the appearance of fecal vomiting in connection with intestinal obstruction due to stenosis in human beings with occlusion of the bowel. The investigations of Nothnagel, however, throw doubt upon this conclusion, as he failed to observe effective antiperistalsis after artificial occlusion of the bowel. The fecal odor of the vomited matter may also depend upon its prolonged sojourn in the duodenum, whence, as the well-known bilious vomiting shows, ingesta may be returned into the stomach.

THE EVACUATION OF FECES (DEFECATION).

The contents of the intestine remain in the small intestine about three hours, and for a further twelve hours in the large intestine, where they become inspissated, and in the lower portion formed into the fecal mass. Through the peristaltic movement, the feces are forced onward to a point somewhat above that portion of the rectum which is surrounded by both sphincter-muscles, of which the upper or internal is formed of unstriated and the external of striated muscle-fibers.

Immediately after the act of defecation the external sphincter (Fig. 108, S; Fig. 109) usually contracts, and remains contracted for some time. When the muscle relaxes, the elasticity of the parts surrounding the anal opening, particularly of both the sphincter-muscles, is sufficient to insure closure of the anus. In the interval of rest or until the pressure of the feces again occurs, there is no evidence of a permanent contraction (tonic innervation) of the anal sphincters. As long as the fecal matters lie above the rectum, they give rise to no conscious sensation.

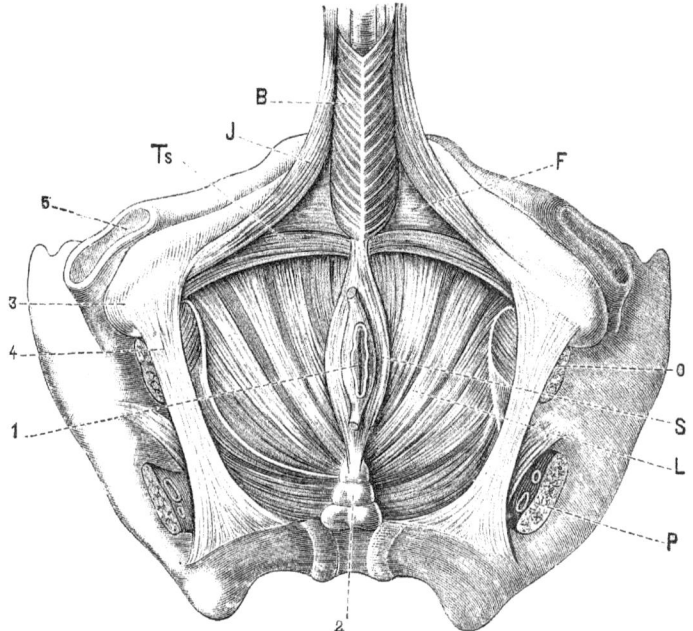

FIG. 108.—The Perineum and its Muscles: 1, anus; 2, coccyx; 3, ischial tuberosity; 4, tuberososacral ligament; 5, acetabulum; B, bulbocavernosus muscle; Ts, superficial transverse perineal muscle; F, fascia of the deep transverse perineal muscle; J, ischiocavernosus muscle; O, internal obturator muscle; S, external sphincter ani muscle; L, levator ani muscle; P, pyriformis muscle.

It is only their descent into the rectum that causes the feeling of tenesmus. At the same time the stimulation of the sensory nerves of the rectum causes reflex stimulation of the sphincters. The center for this reflex (Budge's anospinal center) is situated in the lumbar cord.

In animals, after division of the spinal cord above this center, the anal opening closes actively when touched; but soon after this reflex contraction the sphincters relax, and the anus may thus remain open for a time. This is due to the fact that the active voluntary contraction of the external sphincter-muscle, already mentioned, under the control of the will (cerebrum), which keeps the anus closed for some time after each evacuation of the bowel, is absent. In dogs, in which the posterior roots of the lower lumbar and the sacral nerves were divided, Landois observed that, while recovery was otherwise normal, the anus remained open. Not rarely a portion of the fecal mass protruded for a considerable time, as the sensibility in the rectum and anus was lost in such animals. Neither was reflex contraction of the sphincters possible, nor could voluntary closure of the anus, induced by the sense of feeling alone, take place, although this would other-wise have doubtless been possible.

An excitomotor as well as an inhibitomotor influence may be exerted upon the external anal sphincter, as upon any voluntary muscle, from the cerebrum. Nevertheless, closure can be maintained only for a certain time if the pressure from above is considerable. Finally ener-getic peristalsis overcomes even the strongest voluntary stimulation.

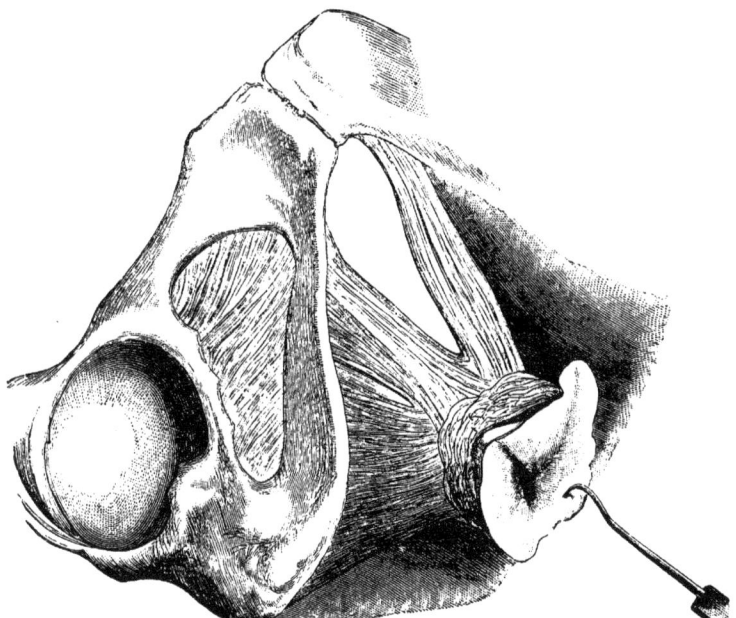

FIG. 109.—The Levator Ani and External Sphincter Ani Muscles.

The evacuation of feces, which takes place habitually in human be-ings at a definite interval, once or twice daily, rarely oftener, begins with active peristalsis in the large intestine which passes downward to the rectum. In order that the sphincter muscles may not be excited to reflex activity by the advancing column of feces, it appears that an inhibitory center for the sphincter-reflex, capable of voluntary stimula-tion, must become active. This is situated in the brain (Masius sup-poses in the optic thalamus), whence its fibers pass through the cere-bral peduncles to the anospinal center. During stimulation of this inhibitory apparatus, the column of feces passes through the anus without causing its reflex closure.

The active peristalsis necessary to cause defecation may be favored and to a certain extent excited, partly by pressure, partly by short voluntary movements of the external sphincter and the levator ani muscles, whereby the myenteric plexus of the lower portion of the large intestine is stimulated mechanically, with the result that active peristaltic movements of the large intestine are soon set up. The expulsion of feces is favored by active, voluntary abdominal pressure, principally with inspiratory depression of the diaphragm. The soft parts of the pelvic floor are forced downward conically with a strong effort at stool, whereby the anal mucous membrane, which coincidently becomes filled with venous blood, is at times everted. It is the function of the levator ani muscle (Figs. 108 and 109) voluntarily to elevate the soft parts forming the pelvic floor and thus, in elevating the anus, in a measure to slide it over the descending column of feces. At the same time it prevents relaxation of the soft parts of the pelvic floor, particularly the pelvic fascia. As the fibers of both levator ani muscles converge downward, and mix with those of the external anal sphincter, they coincidently aid the sphincter when energetic contraction takes place, as they bear approximately the same relation to the anus that the strings of a tobacco-pouch bear to its opening. When the desire for stool is marked the closure of the anus can be made more secure by pressure from without through forcible rotation of the thighs outward and the action of the gluteal muscles.

During the normal interval between evacuations of the bowel, the feces appear to descend only to the lower extremity of the sigmoid flexure. From this point to the anus the rectum normally is usually free from feces. The strong circular fibers of the muscularis, which Nélaton termed the third anal sphincter, appear, by their contraction, to arrest the further advance of the fecal matter.

NERVOUS INFLUENCES AFFECTING THE INTESTINAL MOVEMENTS.

The automatic center for the movements of the intestinal canal is the greatly developed myenteric plexus, which is embedded between the longitudinal and circular layers of the muscular coat. It is this that is responsible for the movements that continue for some time in an excised portion of intestine, just as they occur in the heart.

This plexus, constituted mainly of non-medullated nerves, distributes fibers that, after again forming a network, pass to the unstriated muscle-fibers. The cells of the plexus possess an axis-cylinder process and several protoplasmic processes. Nerve-fibers pass through the mass of ganglia, while others surround the ganglion-cells with their extremities. Special nerve-plexuses, containing ganglia, are found upon the blood-vessels and lymph-vessels of the intestinal wall.

When this center is free from all stimulation, the intestine remains in a state of rest, resembling the apnea that occurs with absence of stimulation of the medulla oblongata. This occurs during intra-uterine life, as it does also with respect to respiration, in consequence of the large amount of oxygen in the fetal blood. This condition may be termed intestinal rest—aperistalsis. It is observed also during sleep, perhaps in consequence of the greater amount of oxygen in the blood.

The circulation through the intestinal vessels of blood containing the usual amount of gases gives rise to the quiet peristaltic movement of the healthy individual—euperistalsis.

All stimuli transmitted to the myenteric plexus increase peristalsis, which finally may progress to violent movement, with rumbling in the intestines (borborygmus), and may even cause involuntary discharge of feces and spasmodic contraction of the intestinal musculature. This condition, which corresponds to dyspnea, may be designated dysperistalsis.

This condition may be caused (a) by interruption of the circulation in the intestines, it matters not whether anemia, as after compression of the aorta, or venous hyperemia is thereby induced. The exciting agent here is the deficiency of oxygen, or the excess of carbon dioxid. Even slighter circulatory disturbances in the intestinal blood-vessels, as, for instance, venous stasis in connection with abundant transfusion into the veins, whereby transitory overdistention of the venous system, and therefore stasis in the portal system occurs, give rise to increased peristalsis. This takes the form of noises and rumbling in the intestines, together with involuntary defecation, if, in consequence of transfusion of heterogeneous blood, stasis becomes marked, as a result of thrombosis of the intestinal blood-vessels. Landois explains in this way the irresistible inclination to stool and the increased peristalsis that attend certain forms of cardiac weakness of acute onset and sclerosis of the coronary arteries, in consequence of which the circulation in the intestines suddenly ceases. A similar state of affairs is observed even under normal conditions. Landois believed that the persistent pressure in constipated individuals induces the evacuation that eventually takes place, as much by exciting peristalsis through the venous stasis in the intestines as by mechanical pressure upon the intestinal canal. Also the increased peristalsis that constantly attends approaching death depends, undoubtedly, upon circulatory disturbances and thus upon an alteration in the amount of gases in the blood in the intestines. The same statement is applicable to the increased intestinal movement that attends certain emotional disturbances, as, for instance, fear. Here the stimulation of the brain passes through the medulla oblongata (containing the center for the vasomotor nerves) to the intestinal nerves and causes circulatory disturbances in the intestines (coincidently with pallor). Restoration of the normal circulatory condition restores the intestines to quiet peristalsis. Salvioli caused blood to flow artificially through excised pieces of intestine by means of cannulas introduced into the blood-vessels, and found that blood rich in oxygen caused intestinal rest, while interruption of the circulation caused contractions of the intestines. Bókai was able to overcome the dysperistalsis induced by the introduction of carbon dioxid into the intestines by introducing oxygen into the intestinal cavity.

(b) Direct irritation of the intestine causes movement not only of the part directly affected, but also of the neighboring part of the intestines, especially that lying toward the pylorus. The cumulative effect of stimuli is shown here; that is feeble stimuli, which are too weak to excite movement when applied but once, do so on persistent repetition, as exposure of the intestines to the air, in more marked degree in the presence of carbon dioxid and chlorin, the introduction of certain irritating substances into the intestine, marked distention of the intestinal canal, especially with coincident difficulty in or obstruction to defecation (which occurs frequently in human beings), or direct irritation of different kinds, also inflammatory processes involving the intestine either from within or from without. In this connection, the observation is of interest that induced currents applied to a hernial sac containing intestine excite active peristalsis in the hernia. Local irritation of a portion of the intestine with a tetanizing induced current causes a circular constriction, which advances especially toward the stomach when the current is of considerable strength. The shortening of the longitudinal fibers that are stimulated at the same time extends in both directions.

With increasing temperature intestinal rest first results—from irritation of the splanchnic nerves; when the temperature reaches 43° C. intestinal movement is resumed.

All persistent stimuli of moderate strength cause cessation of dysperistaltic intestinal movement from overstimulation. This condition may be designated intestinal exhaustion or intestinal paresis.

This state of rest of the intestine is thus widely different from that attending the condition of aperistalsis. Persistent stasis of blood in the intestinal vessels leads finally to intestinal exhaustion, as, for instance, when thrombosis occurs in the intestinal vessels after transfusion of blood from a different species. Distention of the vessels with indifferent fluids, after compression of the aorta had previously excited active peristalsis, likewise causes cessation of peristaltic movement. In the same category belongs also the condition of rest noted after the temperature of the intestine has been reduced to 19° C. Severe intestinal inflammation also has a similar effect. Under favorable conditions the intestine may recover from this stage of exhaustion after the irritation has ceased. This takes place, as a rule, through a transitional stage attended with active peristalsis. Thus the introduction of arterial blood into the vessels of the exhausted intestine causes at first active peristaltic movements, followed by normal peristalsis.

The continuous application of strong stimuli finally causes complete paralysis of the intestine in human beings as seen after inflammations, traumatisms, incarcerations, and the like. The intestine becomes greatly distended, as the paralyzed muscularis is no longer able to offer any resistance to the gases expanded by the heat (meteorism).

The Peripheral Intestinal Nerves.—Of the nerves passing to the intestine the pneumogastric nerve increases the movements of the small intestine and the upper portion of the large intestine, either by conveying the stimuli applied to it to the myenteric plexus, or by causing contractions of the stomach, which, in turn, as true mechanical impulses, excite the intestine to movement. The pneumogastric nerves also contain several inhibitomotor fibers.

The splanchnic nerve—the greater derived from the sixth to the ninth, and the lesser from the tenth and eleventh dorsal ganglia—is (1) the inhibitory nerve for the intestinal movements, but only so long as the blood in the capillaries has not become venous while the circulation in the intestine remains undisturbed. If the latter condition has arisen, irritation of the splanchnic causes increased peristalsis. If arterial blood be introduced, the inhibitory action is prolonged. Irritation of the origin of the splanchnic nerve in the dorsal cord also produces the inhibitory effect under analogous circumstances, even in the presence of irritation of the spinal cord as a result of strychnin-poisoning, with the occurrence of general tetanic convulsions. O. Nasse believes that it may be concluded from the experiments that, in addition to these readily exhausted inhibitory fibers, paralyzed by venosity of the blood, there are present (2) motor fibers that are excitable for a longer time, inasmuch as stimulation of the splanchnic nerve after death always causes peristalsis of the stomach and intestines, as does stimulation of the pneumogastric nerve. (3) The splanchnic nerve is also the vasomotor nerve of all of the arteries and veins of the intestines, including the portal vein, thus controlling the largest vascular area of the body. Stimulation of the splanchnic nerve causes contraction, its division dilatation, of all of the intestinal blood-vessels possessing muscle-fibers. In the latter event an enormous accumulation of blood takes place in the intestinal vessels, so that anemia of other parts of the body results, and in consequence even death may take place from anemia of the medulla oblongata. (4) The splanchnic nerve is, finally, the sensory nerve of the intestines, and, as such, it is extremely sensitive.

Almost all the cells of the solar plexus are included in the course of the fibers of the splanchnic nerve. Nicotin paralyzes these cells, while the peripheral fiber retains its irritability.

Stimulation of the nervi erigentes causes contraction of the longitudinal muscular fibers and relaxation of the circular fibers of the rectum; while irritation of the hypogastric nerves has the opposite effect according to Fellner.

Stimulation of the sigmoid gyrus on the cerebral cortex of the dog, as well as of parts lateral to and behind it, excites intestinal movements through the pneumogastric nerves, as does likewise stimulation of the optic thalamus. Inhibitory fibers pass from both of these situations through the spinal cord, from which they make their exit near the middle of the dorsal cord.

The drugs that affect the intestine are (1) those that diminish the irritability of the myenteric plexus, and thus decrease peristalsis, even to the point of intestinal rest, like belladonna; (2) those that stimulate the nerves inhibiting peristalsis, and paralyze in large doses, like opium or morphin. The drugs of these two classes cause constipation. Elevation of temperature (also during fever) diminishes intestinal peristalsis through irritation of the splanchnic nerve. (3) Other drugs stimulate the motor apparatus; such as nicotin, to the point of intestinal cramps, muscarin, caffein and some laxatives, which thus act as evacuants. The movement excited by muscarin can be neutralized by atropin. As, in consequence of the rapid movement of the intestinal contents, the contained fluid can be absorbed in but small measure, the frequent evacuations that follow are at the same time liquid. (4) Among purgatives, mention should be made of those that irritate the intestines directly, such as colocynth and croton-oil. It is supposed that agents of this kind cause a watery transudation from the blood-vessels into the intestine, just as croton-oil also causes vesicles on the external skin. (5) Certain laxative salts, sodium sulphate, magnesium sulphate and others, liquefy the intestinal contents by retaining for their solution in the intestine the water of the intestinal contents; if, therefore, they are injected into the blood-vessels of an animal, constipation may even result. (6) Calomel (mercurous chlorid) restricts the absorptive power of the walls of the intestine, and also putrefactive decomposition in the bowels. Therefore the fecal evacuations are thin, with little odor, and of a greenish color from admixture of unchanged biliverdin.

THE STRUCTURE OF THE GASTRIC MUCOUS MEMBRANE.

The surface of the mucous membrane of the stomach exhibits numerous small depressions, the gastric crypts (foveolæ gastricæ, Fig. 110), lined by a single layer of mucous goblet cells (Fig. 112, d). These cells are sharply delimited at the cardiac orifice from the stratified squamous epithelium of the esophagus; and at the pyloric extremity from the true cylindrical epithelium of the duodenum. The cells with almost homogeneous contents are provided with elliptical nuclei containing nucleoli. Between their narrowed, lower ends are scattered oblong or spindle-shaped, unencapsulated, nucleated elements, exhibiting mitosis, which are intended to replace desquamated cells. All cells are completely open upon their free surface, so that nothing prevents the escape of the mucus elaborated by mucous metamorposis from the cell-protoplasm. The simple tubular gastric glands, generally several in number, empty into the bottom of the gastric crypts. They occur in two different forms:

1. As true gastric glands, peptic glands of the fundus (Fig. 114), which number about five millions, the largest being present in the fundus. The structureless membrana propria of simple tubular form, has, on its internal surface, two different kinds of cells (Fig. 111, II, a), the chief cells (Fig. 111, II, a), the adelomorphous cells of Rollett; small, unencapsulated, nucleated, pale cells lying close together, lining the inner lumen of the glands, and (b) larger, mainly scattered, plainly projecting parietal cells (Fig. 111, II, h), the delomorphous cells of Rollett, ovoid or crescentic, without a membrane, darkly granular, readily stained with osmic acid and aniline-blue, containing, at times, several nuclei. They cause bulbous projections of the membrana propria. In human beings the parietal cells are thought to reach to the lumen of the spaces within the gland. They are even found scattered under the epithelium of the crypts and the surface of the mucous membrane, as well as in isolated pyloric glands. Between the chief cells secretory spaces are present, and likewise between neighboring parietal cells, while, at the same time, with the latter delicate branching and anastomosing passages in part lead from the excretory duct of the gland into the interior of the parietal cells and in part form a network surrounding them.

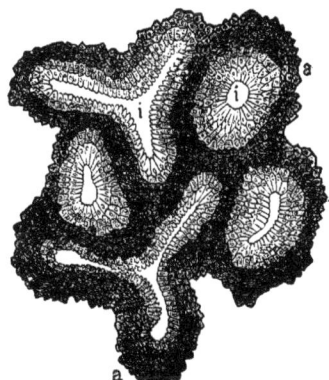

FIG. 110.—Sectional View of the Gastric Mucous Membrane, Showing the Crater-like Depressions of the Gastric Crypts: a, a, the most prominent projections of the mucous membrane (from a dog).

2. Only in the vicinity of the pylorus, where the mucous membrane has a

19

rather yellowish-white appearance, are the pyloric glands (Fig. 112, **A**) found, in general in smaller number. At their lower extremity their ducts are not rarely divided into two or more blind sacs. Their cellular contents consist, as a rule,

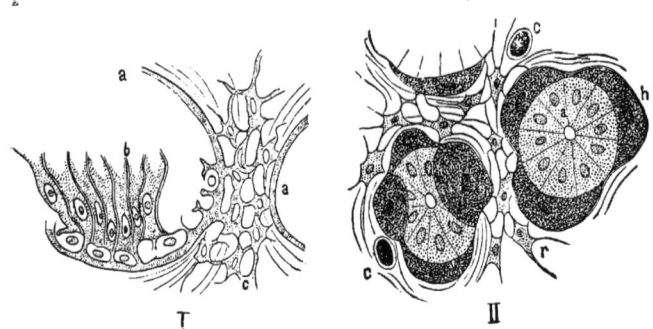

FIG. 111.—I, Transverse Section through the Duct of a Fundus-gland: a. membrana propria; b, goblet-cells; c, reticular connective tissue. II. Section through the Glands of the Fundus: a, chief cells; h, parietal cells; r, reticular tissue of the mucous membrane between the glandular tubules; c, divided blood-vessels.

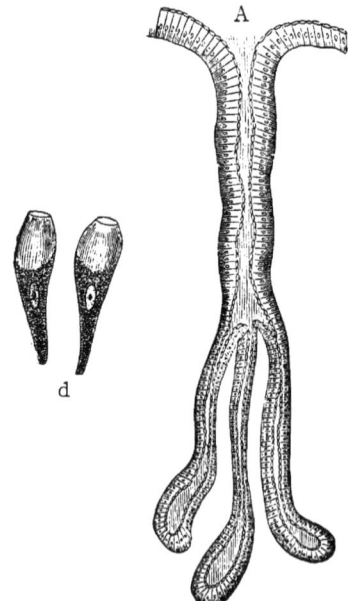

FIG. 112.—d, Isolated goblet-cells; A, pyloric gland of the stomach.

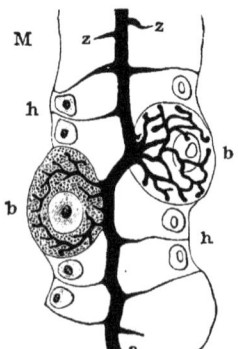

FIG. 113.—M, Portion of a gastric gland with chief cells (h h) and parietal cells (b b); the latter exhibit intracellular secretory canals. Between the chief cells intercellular secretory ducts (z z) penetrate for some distance; a, excretory duct of the gland.

of a variety of finely granular secreting cells, which most nearly resemble the chief cells of the lab-glands.

3. At the cardiac orifice, also, there lies a circular layer of tubules without parietal cells, which secrete diastatic ferment.

The scanty supporting structure of the gastric mucous membrane consists of reticular connective tissue with leukocytes, mixed with fibrillary connective tissue and elastic fibers. The mucous membrane possesses a special muscular layer, the muscularis mucosæ. This passes as a rather thick stratum under the base of the gland, often exhibiting an inner circular and an outer longitudinal layer. From

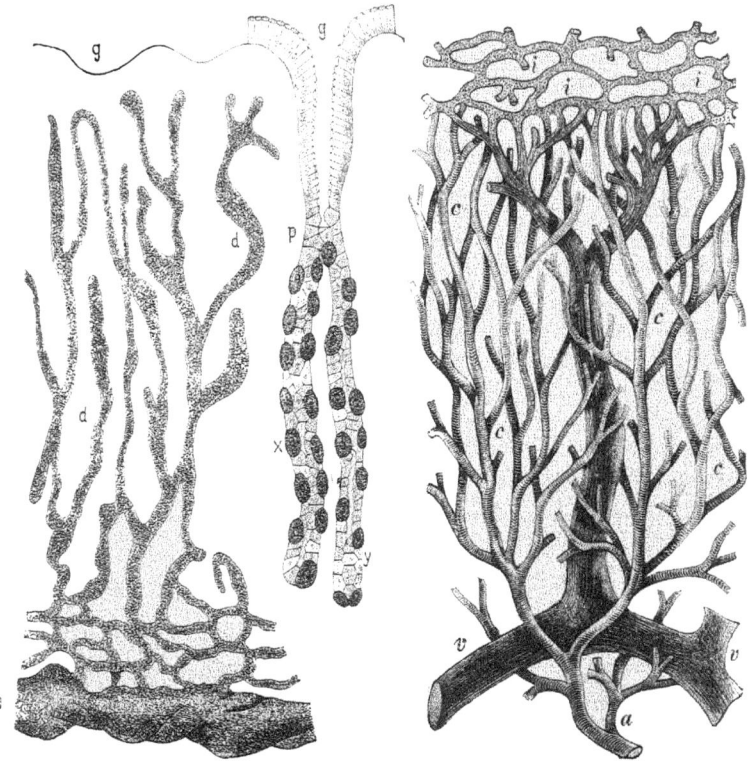

Fig. 114.—Vertical Section through the Gastric Mucous Membrane: g g, the crypts of the surface; p, the mouths of the peptic tubules (fundus glands) with parietal cells (x) and chief cells (y); a v c c, artery, vein and capillaries of the mucous membrane; i, capillary network for the passage of the mouth of the gland-duct; d d, the lymphatic vessels of the mucous membrane, passing over, at e, into a large trunk (semidiagrammatic representation).

this stratum a number of bundles of fibers pass upward between the glands and around them. They appear to be intended for active evacuation of the glandular tubules.

Numerous blood-vessels (Fig. 114) enter from the fibrillary connective tissue of the submucosa (a), spread out into a longitudinal capillary network (c c) between the glands, and reach the free surface, where they again form a fine meshwork (i i) just under the epithelium, and through the meshes of which the mouths of the ducts (g) make their appearance. Collecting at this point the veins unite in the submucosa to form trunks of considerable size (v).

The lymphatic vessels of the gastric mucous membrane begin rather close beneath the epithelium as bulbous or loop-like formations (d d), then pass perpendicularly to the submucosa, where they attain a considerable size (c) through the union of adjacent branches. The nerves are the same as those of the intestine. The submucosa consists of bundles of connective tissue with elastic fibers and embedded fat-cells.

THE GASTRIC JUICE.

The gastric juice is a fairly clear, colorless, levorotatory, readily filtered fluid, with a strongly acid reaction, an acid taste and a characteristic odor. From the presence of free hydrochloric acid, it counteracts putrefaction and, in part, fermentation. Its specific gravity, when the stomach is empty (fasting), ranges between 1004 and 1006.5; after the ingestion of food, from 1010 to 1020, and more than 1020 when the production of acid is diminished. Its amount was said by Beaumont, in 1843, from observations upon a human being with a gastric fistula, to be only 180 grams daily. According to Grünewald, in 1853, it was estimated in a similar case to be 26.4 per cent. of the body-weight in twenty-four hours. Finally it was placed by Bidder and Carl Schmidt, after comparative observations upon dogs, as 6½ kilograms in the day, corresponding to $\frac{1}{10}$ of the body-weight. The gastric juice contains:

1. **Pepsin**, the characteristic, nitrogenous, hydrolytic ferment or enzyme that dissolves proteids: from 0.41 to 1.17 per cent.

2. **Hydrochloric acid** occurs free in the gastric juice: from 0.2 to 0.3 per cent.

3. **Lactic acid** may also be found, either from fermentation of carbohydrates (fermentation lactic acid) or from being dissolved out of the meat of the food (sarcolactic acid).

Reactions.—Hydrochloric acid alone, and in the free state, is identified by Günzburg's reagent: To a few drops of filtered gastric juice an equal number of drops of a solution of 2 grams of phloroglucin and 1 gram of vanillin in 30 grams of alcohol are added, and the mixture is evaporated in a porcelain dish over the water-bath, with the development of a rose-red color. Resorcin, 2.5 grams, dissolved in 50 grams of dilute alcohol, with addition of 1.5 grams of cane-sugar, may be employed in a manner analogous to the foregoing reagent, likewise giving rise to a red color.

Reaction for Lactic Acid.—A freshly prepared blue mixture of 10 cu. cm. of a 4 per cent. solution of carbolic acid, with 20 cu. cm. of distilled water and one drop of ferric chlorid, is colored yellow by lactic acid. To 5 cu. cm. of the gastric juice to be tested 1 or 2 drops of hydrochloric acid are added, and the mixture is evaporated over a free flame to the thickness of sirup. The residue is extracted with a little ether, is then poured into a reagent glass containing 5 cu. cm. of water, one drop of a 5 per cent. solution of ferric chlorid is added, and the mixture is shaken. A greenish-yellow color appears even when 1 part of lactic acid in 1000 is present. The gastric contents, evaporated to the consistency of sirup, to expel the alcohol, are extracted by shaking with ether. The filtrate, on addition of an alcoholic solution of iodin and being heated, yields iodoform, in consequence of the formation of acetaldehyd from the lactic acid.

Hydrochloric acid and organic acids together yield the following reactions. To demonstrate the total free acids (those not combined with albumin), Congo-red is used, also in the form of reagent-paper. It indicates the presence of free hydrochloric acid or a considerable amount of free organic acids by becoming blue in color. The same information is afforded by dark-red benzopurpurin, which is changed to a violet color, and also by tropæolin OO. A little of a concentrated alcoholic solution of the latter, heated with 4 drops of gastric juice in a dish, yields a bluish-violet stain.

4. For a consideration of the milk-ferment, reference may be made to page 300.

5. The large amount of mucus adherent to the surface of the mucosa is a secretion of the mucous goblet-cells.

6. Inorganic matters are present in percentages for human beings (and for dogs, in parenthesis) as follows: Water, 994.40 (973.06); hydrochloric acid, 0.20 (2.84); calcium chlorid, 0.06 (0.96); sodium chlorid, 1.46 (2.82); potassium chlorid, 0.55 (1.09); ammonium chlorid (0.5); calcium, magnesium, and iron phosphates, 0.125 (2.7). Organic matters, principally pepsin, are present to the amount of 0.32 per cent. (1.71).

Of foreign substances, the following appear in the gastric juice after introduction into the body: potassium sulphocyanid, iron lactate, potassium ferrocyanid, sugar, etc. Ammonium carbonate is found in the presence of uremia.

THE SECRETION OF THE GASTRIC JUICE.

During the course of digestion characteristic changes take place in the chief cells, and in the parietal cells of the fundus glands and in the cells of the pyloric glands.

The chief cells contain granules that are consumed during the process of secretion. The granules contain the pepsin-forming substance, which is transformed into pepsin. The size of the chief cells diminishes also during secretion. At rest these cells take from the lymph, material for the production of the granules. The parietal cells, during the period of secretion, appear first to be swollen, then to become smaller. All of the cells, further, are darker, and the nucleus of the cells of the pyloric glands moves toward the center. The secretory ducts become more distended.

In some animals the chief cells, during secretion, bear a fringe of short, hair-like processes (Tornier's "brush-fringe"!), directed toward the lumen of the gland.

The pepsin is formed in the chief cells. If these are swollen, they produce much pepsin; if shrunken, they produce but little. The pyloric glands also secrete pepsin, though in much less amount. During the first stage of hunger the pepsin accumulates; while during the period of digestive activity it is eliminated, as it is also when hunger is protracted.

Klemensiewicz removed the pyloric portion of the stomach of a dog with two incisions; sutured the duodenum to the stomach, and allowed the pyloric portion, still in communication with its blood-vessels, to heal in the abdominal wound, after closure of its lower extremity by sutures. The secretion of this portion of the stomach was viscid and alkaline, containing 2 per cent. of solid matters, including pepsin.

The glands themselves contain no pepsin, but only a zymogen, namely, the pepsinogenic substance or propepsin, which occurs in the granules of the chief cells. The zymogen, of itself, exerts no influence upon proteids. If, however, it be treated with hydrochloric acid or sodium chlorid, it is transformed into pepsin. In addition to pepsin, the pepsinogenic substance may be extracted from the mucous membrane of the stomach by means of water free from acid. The milk-ferment also originates in the chief cells.

The hydrochloric acid is formed by the parietal cells. It is found on the free surface of the mucous membrane, as well as in the excretory ducts of the gastric glands. In the depth of the glandular tubules, however, the reaction is generally alkaline. The acid must, therefore, be advanced rapidly from the depth to the surface.

Sàrcolactic acid can be rapidly extracted as such from the chyme. For the production of lactic acid through fermentation in the stomach it is necessary that the carbohydrates have been present for a consider-. able time. This does not occur in the healthy individual, but in association with great diminution in the production of hydrochloric acid, stagnation of the ingesta in the stomach, and interference with gastric absorption, particularly in the presence of gastric carcinoma.

Lactic-acid bacteria are always present in the stomach, though they exhibit no activity in the presence of healthy gastric juice on account of the anti-fermentative influence of the hydrochloric acid. Lactic acid develops, however, only in the absence of free hydrochloric acid, which is particularly often the case in the presence of gastric carcinoma.

The hydrochloric acid first secreted at once combines in the stomach with the proteids to form acid albuminates. These do not yield the color-reactions of free acid. As the secretion progresses free hydrochloric acid makes its appearance. If the secretion of gastric juice be enfeebled it may, therefore, happen that the production of hydrochloric acid is not sufficient to permit of the appearance of free hydrochloric acid.

When the tests for hydrochloric acid in the stomach-contents are distinctly, even though feebly, positive, sufficient hydrochloric acid is present; an unusually strong reaction is indicative of abnormally increased production. If the reaction is wanting, a decinormal hydrochloric-acid solution is added to a measured amount of gastric contents until a distinct reaction is obtained by Günzburg's test. The amount of hydrochloric acid consumed is then proportional to the degree of the hydrochloric-acid insufficiency present.

In regard to the production of free acid, the following appears to be established. The parietal cells secrete hydrochloric acid from the chlorids that the mucous membrane takes up from the blood. Therefore, the production of hydrochloric acid ceases when the chlorids are withdrawn from the food, as well as in the state of hunger. The active agent in this connection has not been discovered; yet it is established that, if carbon dioxid acts continuously on the chlorids, nevertheless, hydrochloric acid is expelled by the much weaker carbon dioxid. Maly and others assume that the production of hydrochloric acid takes place within the parietal cells as follows:

$$2Na_2HPO_4 + 3CaCl_2 = Ca_3(PO_4)_2 + 4NaCl + 2HCl.$$

The bases set free by the separation of the hydrochloric acid are excreted in the urine, with the development of a slightly acid reaction.

When the stomach is empty the gastric juice contains some hydrochloric acid, but a more abundant secretion is, according to Pawlow, brought about in a most striking manner by the appetite, and also by the stimulation of the food under natural conditions, as well as by water, meat-extractives, and even by indigestible matters when introduced into the stomach. Under these circumstances the mucous membrane is reddened from increased activity of the circulation, so that the outflowing venous blood is lighter in color. The excitation of the secretion is a reflex process. The sensory nerves of the pharynx and the stomach excite, in a centripetal direction, the medulla oblongata, which contains the center for this reflex.   The centrifugal path to the mucous membrane traverses the pneumogastric nerves, after the division of which the reflex is abolished.   The mucous membrane subsequently furnishes a moderate amount of a feebly active, paralytic secretion. During sleep in the stage of digestion, the amount of acid increases.

Heidenhain found in experiments upon dogs—in which, in the same way as the pylorus, he isolated the fundus for the formation of a blind sac—that mechanical irritation induced only local secretion. If, however, absorption of digested substances took place at the point of irritation, the secretion spread out over a larger surface.

Small quantities of alcohol, introduced into the stomach, increase the secretion of the gastric juice, while large amounts abolish it and enfeeble the movements of the stomach. Fat inhibits the secretion of the gastric juice. Artificial digestion is somewhat disturbed by alcohol up to 2 per cent., and in greater degree by 10 per cent. alcohol; 20 per cent. alcohol retards, while still larger amounts abolish it. Beer and wine retard digestion, and undiluted they prevent artificial digestion. The administration of large amounts of sodium chlorid diminishes the secretion of hydrochloric acid, while the ingestion of much sugar only delays it. After two days of fasting the secretion of hydrochloric acid ceases (in the dog). Gastric ulcers cause reflex increase in the production of hydrochloric acid; jaundice, nervous gastric affections and anemias, a reflex diminution.

The gastric juice, which passes into the duodenum after digestion is completed, is neutralized by the alkalis of the intestinal and of the pancreatic juices.   The pepsin is absorbed as such, and can be found in small amounts in the urine and in the muscle-juice.

If the gastric juice is removed completely through a gastric fistula, the alkalies in the intestines become so abundant that the urine is rendered alkaline.

The acid gastric juice in the new-born is quite intensely active. It most readily digests casein, and next in order fibrin and other proteids. In consequence of excessive acidity of the gastric juice, large masses of casein, difficult of digestion, form in the stomach of infants, and are especially tough after the ingestion of cow's milk.

The following drugs are excreted by the gastric juice after introduction into the body-juices: Morphin, veratrin, caffein, quinin, antipyrin, chloroform, chloral hydrate, methyl-alcohol, ethyl-alcohol and acetone.

Comparative.—According to Klug, the parietal cells of grain-eating birds prepare also pepsin, in addition to hydrochloric acid. The gastric glands of the frog, which possess only parietal cells, likewise secrete pepsin; the pyloric glands of the dog, which contain only chief cells, nevertheless secrete acid. Accordingly both kinds of cells secrete hydrochloric acid.

METHODS OF OBTAINING THE GASTRIC JUICE.

THE PREPARATION OF ARTIFICIAL DIGESTIVE FLUIDS; DEMONSTRATION AND PROPERTIES OF PEPSIN.

To obtain the gastric juice Spallanzani had fasting dogs swallow bits of sponge enclosed in perforated leaden capsules, and withdrew them after they had become saturated with the gastric juice. In order to prevent admixture with the secretions of the mouth, the sponge is best introduced through an opening made in the esophagus ligated above.

Beaumont (1825–1833), an American physician, was the first to obtain gastric juice from a human being, in the case of the Canadian hunter, Alexis St. Martin, whose stomach had been opened by a bullet-wound, with the formation of a permanent gastric fistula. Various substances were introduced directly into the stomach through the opening, and examined from time to time as to their digestion.

Guided by this, Bassow, in 1842, was the first to establish an artificial gastric fistula in a dog. The wall of the stomach is opened below the xiphoid process, and the margins of the gastric opening are united by suture to the margins of the wound in the abdominal walls. A short tube with a terminal plate is placed in the fistula in such a manner that the plate lies in contact with the margin of the mucous membrane. The tube possesses a screw-thread, upon which an appropriate cannula can be so screwed that the terminal plate lies upon the abdominal wall outside of the margins of the wound. The parts are joined in the following manner ⊢ ⊣. As a rule the opening of the cannula is corked. If in such dogs the excretory ducts of the salivary glands are additionally ligated, unmixed gastric juice is secured.

According to C. A. Ewald and Leube, dilute gastric juice can be obtained from human beings by introducing water into the empty stomach through a tube that acts like a siphon, and withdrawing the fluid by siphonage after a short time.

An important advance was made by Eberle, in 1834, who taught that artificial gastric juice could be prepared by extracting pepsin from the gastric mucous membrane by means of dilute hydrochloric acid. Dilute hydrochloric acid serves for the extraction of the triturated gastric mucous membrane—0.088 per cent. for the digestion of fibrin, 0.16 per cent. for the digestion of coagulated albumin— being added anew, in quantities of a half liter every six or eight hours. The later extracts are even more active than the first. The fluid collected is filtered and in it are placed, at the temperature of the body, the substances to be digested. It is, however, necessary to add more hydrochloric acid from time to time. That degree of acidity affects digestion most favorably that most causes the proteids to swell. According to Klug, gastric juice containing 0.6 per cent. of hydrochloric acid and 0.1 per cent. of pepsin is most effective. Pepsin from dogs is especially active. Digestion pursues a favorable course between $37°$ and $40°$ C.; while it ceases in the cold, as well as at higher temperatures.

The hydrochloric acid employed may be replaced, to a certain extent, by other halogen-acids, whose activity is inversely proportional to their molecular weight; further by from six to ten times as much lactic acid; by nitric acid; in a much less effective manner, finally, by oxalic, sulphuric, phosphoric, acetic, formic, succinic, tartaric, and citric acids. In general, the acids with greater acidity act more powerfully, with the exception of sulphuric acid. The action of the different acids varies, however, accordingly as fibrin, casein, solid or liquid albumin is employed.

v. Wittich showed that pure pepsin can be extracted from the gastric mucous membrane by means of glycerin also. After cleaning the mucous membrane, it is left in alcohol for twenty-four hours, then dried, pulverized and sifted, and then extracted for a week in glycerin. On addition of alcohol to the filtered extract pepsin is precipitated, and this, dissolved in dilute hydrochloric acid, yields active gastric juice.

The preparation of *perfectly pure pepsin* has been effected by W. Kühne by exposing comminuted pigs' stomachs to autodigestion with dilute hydrochloric acid at the temperature of the body. The mass, which is for the most part liquefied, is saturated with ammonium sulphate, by which pepsin and albumoses still present are precipitated. The residue collected on the filter is again—and if necessary repeatedly—digested in the incubator, after addition of dilute hydrochloric acid. If, finally, all of the albumin has been converted into peptone, the pepsin alone is precipitated by repeated saturation with ammonium sulphate, and is collected on the filter. It is dissolved in water, its salts are removed by dialysis and it is finally precipitated in a pure state by alcohol. Brücke had previously prepared pure pepsin by causing a voluminous precipitate in the digestive mixture including the pepsin, and separating the latter. Pekelharing found that a strongly active artificial gastric juice, on dialysis with water, caused the separation of a precipitate of pepsin.

In all the processes of extraction, the yield of pepsin is greatest when the mucous membrane, protected from putrefaction, is exposed to the air for some time, as subsequently propepsin and pepsin are formed in the gland-cells.

Pure pepsin is a colloid substance. It does not yield the reactions of albumin to the following tests: It does not respond to the xanthoproteic test, is not precipitated by acetic acid and potassium ferrocyanid, by tannic acid, mercuric chlorid, argentic nitrate or iodin. In other respects it is to be included among the albuminoid substances. Pepsin, when heated to a temperature of from $55°$ to $60°$ C. or above, in acid solution, is rendered inactive.

THE PROCESS AND THE PRODUCTS OF GASTRIC DIGESTION.

The mixture of finely divided food and gastric juice is designated *chyme*. Upon this the gastric juice exerts its action.

ACTION UPON PROTEIDS.

The pepsin and the free hydrochloric acid are capable of transforming the proteids, at the temperature of the body, into a readily soluble modification that has been designated peptone. In this process the proteids are changed first into bodies possessing the character of syntonins, and in this condition the solid proteids are swollen. Syntonin is an acid-albuminate. By neutralization, with cautious addition of an alkali, the albumin is precipitated. Then, by combination with water and division into numerous small molecules, a product results, which is, to a certain extent, an intermediary body between albumin and peptone—the albumose of W. Kühne and Chittenden (propeptone of Schmidt-Mülheim). This is soluble in water, readily soluble in dilute acids, alkalies and salts. These solutions are not precipitated by boiling, but by acetic acid and potassium ferrocyanid, as well as by acetic acid and saturation with sodium chlorid or magnesium sulphate. Albumose is precipitated by nitric acid, but it is redissolved, with the production of an intense yellow color when heated, and it is again precipitated on cooling. Some albumoses possess diffusibility.

With the continued action of the gastric juice, the albumose passes over into soluble and readily diffusible peptone. The unchanged proteids behave toward the peptones as anhydrids with a large albumin-molecule. The production of peptone and its solution result, therefore, from decomposition with the taking up of water, brought about by the hydrolytic ferment, pepsin. This action takes place best at the temperature of the body.

According to W. Kühne, the proteid molecule contains two different substances, namely hemi-albumin and anti-albumin. By the action on these of hydrochloric acid syntonin is produced. This is next broken up into the two primary albumoses: protalbumose, soluble in water, and hetero-albumose, soluble in salt-solutions. Both are then transformed into deutero-albumoses, which, in contradistinction to the primary albumoses, are not precipitated in neutral solution by saturation with sodium chlorid. Deutero-albumose in contradistinction to protalbumose is not precipitated by copper sulphate. The deutero-albumoses are then decomposed into peptones: hemipeptone and antipeptone.

The pepsin enters into intimate relations with the proteid molecule. The greater the amount of pepsin present, the more rapidly, to a certain degree, does digestion take place. The pepsin itself undergoes almost no change, and if care is taken to keep the amount of hydrochloric acid always the same, it is able to digest new amounts of albumin (one part to about 500,000 parts). Nevertheless some pepsin is consumed in the process of digestion.

The proteids are introduced into the stomach either in a liquid or in a solid form. Of the liquid proteids only casein is at once coagulated in solid form and precipitated and then redissolved. The other liquid proteids remain liquid, are converted into the condition of syntonins, and then immediately into albumoses and finally into peptones, that is, actually digested.

Uncoagulated and coagulated proteids, globulins, fibrin, some forms of vitellin, chondrigen, collagen, and elastin, though with difficulty, are in the same way converted into albumoses and peptones; while neuro-keratin, keratin, and nuclein remain undigested.

During the digestion of albumin, absorption of heat takes place, demonstrable by the thermometer. Accordingly the temperature of the chyme in the stomach falls, in the course of two or three hours, from 0.2° to 0.6° C.

The coagulated proteids may be designated the anhydrids of the liquid proteids and the latter in turn the anhydrids of the peptones. Thus the peptones represent the highest possible stage of hydration of the proteid bodies.

Peptones may also be obtained from proteids with the aid of such agents as usually cause hydration, particularly by treatment with superheated steam vapor, through the action of strong acids, caustic alkalies, ferments of putrefaction and some other ferments, as well as by ozone.

The proteid anhydrids may be reconverted from this stage of hydration by the abstraction of water.

By heating with acetic-acid anhydrid at a temperature of 80° C. peptone is transformed into syntonin. Also by heating to a temperature of 170° C., through the action of the galvanic current in the presence of sodium chlorid, and through the action of alcohol together with salts, peptone is retransformed into albumin. Albumose was thus first seen to result from fibrin-peptone.

Properties of Peptones.—(1) They are readily and completely soluble in water. (2) They diffuse readily through membranes, more readily than propeptones. (3) They also filter much more readily than albumin through the pores of animal membranes. (4) From a mixture of pep-tone, propeptone, albumin and pepsin, first neutralized and then feebly acidulated with acetic acid, neutral ammonium sulphate added in excess precipitates everything except peptone. (5) Peptones are not precipi-tated by boiling, or by nitric acid, or acetic acid and potassium ferro-cyanid, or by acetic acid or by saturation with sodium chlorid. (6) They are precipitated by phosphotungstic, by phosphomolybdic acid, and by biliary acids; precipitated by tannic acid, they are redissolved in an excess. Other precipitating agents are mercuric chlorid and nitrate, mercuric iodid, potassium iodid. (7) They yield all of the color-reactions of albumin. (8) With sodium hydrate and copper sulphate in the cold, they give a purple-red color (biuret-reaction).

The biuret-reaction is yielded also by propeptone, as well as by a proteid body, the so-called *alkophyr*, formed coincidently in the process of artificial digestion and soluble in strong alcohol. Gelatin-peptone and gelatin are precipitable by tri-chloracetic acid, while albumin-peptone is redissolved in an excess of this acid. This is a useful means of differentiating these peptones. The peptones of the various proteid bodies are distinguished by the amount of sulphur they contain, with some of which this substance is but loosely combined, while with others it is firmly united. All have a disagreeable and bitter taste.

In order to demonstrate the rapidity with which fibrin is digested by the gastric juice, Grünhagen places in a funnel the fibrin that has been saturated with 0.2 per cent. hydrochloric acid, moistens it with digestive fluid and notes the rapidity with which the fibrin gradually melts away, drop by drop, and finally is entirely dissolved. Grützner stains the fibrin with carmine, saturates it with 0.1 per cent. hydrochloric acid, and places it in the digestive fluid. The more rapidly the latter becomes stained uniformly red, in consequence of digestion of the fibrin, the more energetic, naturally, is the digestive action.

Quantitative Estimation of the Activity of Pepsin.—Of a solution of egg-albumin (3 grams in 160 cu. cm. of 0.4 per cent. hydrochloric acid) two specimens of 10 cu. cm. are taken, 5 cu. cm. of gastric juice being added to the one and 5 cu. cm. of water to the other. The mixtures are poured into Esbach's tubes up to the mark U. Both tubes are then kept for one hour at a temperature of $37°$ C., after which Esbach's reagent is added up to the level of the mark R, and the amount of the precipitate in both tubes is noted after the lapse of twenty-four hours. Peptone is not precipitated. Chronic gastric catarrh and carcinoma yield low digestive values, while hypersecretion of the gastric juice may increase the digestive intensity.

Preparation of Pure Peptone.—The diluted digestive solution, freed from albuminates by boiling, and with an almost neutral reaction, is first saturated, while boiling, with ammonium sulphate, filtered when cool, again heated, after beginning to boil made strongly alkaline by adding ammonia and ammonium carbonate, again saturated in the heat with ammonium sulphate, filtered after cooling, again heated until the odor of ammonia has disappeared, again saturated with the salt, hot, and acidulated with acetic acid. The fluid, filtered in the cold, contains pure peptone.

The peptones are undoubtedly those modifications of proteids that are intended, after absorption from the digestive tract, and later through the blood, to be employed to replace the proteids consumed by the process of metabolism in the living organism.

If much albumin has already been digested by the gastric juice, the pepsin is precipitated and becomes inactive if some hydrochloric acid is not again added from time to time. Admixture with bile in the test-tube impairs the activity of pepsin; nevertheless the entrance of bile into the stomach causes no permanent derangement, as renewed amounts of pepsin are at once secreted by the gastric mucous membrane. The stomach digests less well food that has not been thoroughly masticated or properly insalivated. The presence of blood or of serum prevents the action of pepsin, as well as of trypsin and of the lab-ferment. Heated to a temperature of $65°$ C. the pepsin in the gastric juice becomes inactive, pure pepsin even at a temperature of $55°$ C. Concentrated acids, alum and tannic acid abolish the process of peptic digestion. Alkalinity of the gastric juice, as, for instance, from the presence of large amounts of saliva, also concentrated solution of alkaline salts, such as sodium chlorid, magnesium sulphate and sodium sulphate, have the same effect, as do also sulphurous and arsenous acids, and potassium iodid; while small amounts of sodium chlorid increase the secretion and favorably influence the action of the pepsin. The salts of the heavy metals, which form precipitates with pepsin, peptones and mucin, disturb gastric digestion. According to Langley and Eakins, alkalies rapidly destroy pepsin, and propepsin less rapidly. Acids (as lactic, acetic and hydrochloric) precipitate the gastric mucus and stimulate the secretion of pepsin, while the salts of the alkalies have exactly the opposite effect. Alcohol precipitates the pepsin, although this is redissolved on addition of water, so that digestion can then proceed again undisturbed. Agents that hinder thorough saturation of proteids, as, for instance, binding them tightly, or concentrated solutions of astringent salts, retard digestion.

The ingestion of half a liter of cool water does not disturb gastric digestion in the healthy individual, though it does when the function of the stomach is deranged, while the ingestion of a larger amount impairs the digestive activity of the stomach. The same effect is brought about by strong muscular action. In the horse moderate movement (trotting) assists the digestion of starches in the first hour. Warm compresses over the epigastric region favor gastric digestion.

According to Penzoldt, the digestibility of various proteid articles of food by the stomach is given in the following order. *Easily digestible:* boiled brain and thymus, pike, sea-fish, carp, oysters, chicken, boiled pigeon, raw scraped beef or veal, wheat-bread, cauliflower, soft-boiled egg (casein, alkali-albuminate). *Digestible with moderate ease:* boiled beef and veal, goose, pork, salt potatoes, rye-bread, rice, tapioca, asparagus, rape-cole, carrots, raw egg, purée of legumes. *Digestible with difficulty:* salmon, salt fish, highly salted caviare, string beans, hard-boiled egg. The digestibility of the different meats, from the more to the less readily digestible, is as follows: veal, lamb, mutton, pork, beef, rabbit, horse.

ACTION UPON OTHER FOODS.

Milk coagulates in the stomach, with the liberation of heat, as a result of precipitation of the casein, which encloses the fat globules. The free acid of the stomach is alone sufficient for precipitation, the alkali being withdrawn from the casein, which it holds in solution.

Hammarsten, in 1872, discovered a special rennet-ferment in the gastric juice, which coagulates the casein in either neutral or alkaline solutions. On this fact depends the preparation of cheese by means of calf's stomach rennet.

The rennet is formed in the chief cells of the gastric glands from a rennet-forming substance, by the action of an acid. The rennet-forming substance is present in the mucous membrane in much larger amount than rennet itself. One part of rennet-ferment is capable of precipitating 800,000 parts of casein. The addition of calcium chlorid hastens, while water retards, coagulation. An excess of alkali impairs the activity of rennet. The rennet-ferment is best assisted by hydrochloric acid, followed, in order, by lactic, acetic, sulphuric and phosphoric acids.

The casein, as well as the nucleo-albumin, is converted in the process of digestion, mainly into peptone rich in phosphorus; a residue poor in phosphorus, paranuclein, remaining as an insoluble product.

The rennet-ferment is destroyed by long-continued artificial digestion. To obtain rennet, Hammarsten agitates artificial gastric juice prepared from the calf's stomach, and after neutralization, with magnesium carbonate. The filtrate contains only rennet, which, after acidulation with acetic acid, is precipitated by the introduction of liquid stearic acid, to which it adheres. The acid is dissolved in ether, which can then be readily separated.

Finally, sugar of milk is converted in the gastric juice into lactic acid, by fermentative activity—lactic-acid ferment. Further, the milk-sugar in the stomach and intestines is, in part, transformed into grape-sugar.

Cane-sugar is gradually converted into grape-sugar, in which process, according to Uffelmann the gastric mucus, according to Leube the gastric acid, plays the most important part.

ACTION ON THE DIFFERENT TISSUES AND THEIR CONSTITUENT MATERIALS.

(1) The gelatin-yielding substance of the various supporting structures—connective tissue, fibrous cartilage and the matrix of bone as well as glutin itself, is peptonized and digested in the gastric juice. (2) The structureless membranes (membranæ propriæ) of the glands, sarcolemma, the nerve-sheath of Schwann, the capsule of the crystalline lens, the elastic layers of the cornea, the membranes of the fat-cells, are likewise digested, but scarcely the elastic, fenestrated membranes and fibers. (3) Striated muscular tissue forms after digestion of the sarcolemma and breaking up of the transversely striated contents into discs and fragments of fibrils, as well as unstriated muscular tissue, a true digested peptone, in consequence of hydration and the decomposition of the myosin. Remains of meat, however, always pass over into the intestine. (4) The proteid elements of the soft cellular structures of the glands, stratified epithelium, endothelium and lymphoid cells, are converted into peptone, while the nuclein of the nuclei cannot, apparently, be digested. (5) The horny portions of the epidermis, nails, hairs, as well as the chitin and the wax of lower animals, are indigestible. (6) The erythrocytes are digested, the hemoglobin decomposed into hematin and a globulin-like substance. The latter is peptonized; the former remains unchanged, and in part appears in the feces, and in part is absorbed and transformed into the coloring-matter of the bile. (7) The fibrin is easily digested into propeptone and fibrin-peptone by the taking up of water and the breaking up of the molecule. Mucin is digested in the stomach. (8) Of vegetable articles of food, vegetable fats are not changed by the gastric juice. The vegetable cells give up their protoplasmic contents for the production of peptone, while the cellulose of the cell-walls is undigestible in the stomach of human beings.

That the stomach is also capable of digesting parts of a living body is shown by the fact that the thigh of a living frog or the ear of a rabbit, introduced into a gastric fistula in a dog, will be partly digested. The edges of gastric ulcers and fistulæ in human beings are also eroded by the digestive activity of the gastric juice. The question was early asked, Why does the stomach-wall not digest itself? As, after death, the mucous membrane is, in fact, often rapidly softened by autodigestion (gastric softening), the opinion is justified that, so long as the circulation is maintained, the tissues are constantly protected against the action of the acid by the alkalinity of the blood. If the reaction of the gastric juice be alkaline, digestion cannot be inaugurated. Ligation of the blood-vessels of the stomach resulted, according to Pavy's investigations, in digestive softening of the gastric mucous membrane. In human beings morbid occlusion of the vessels causes, in an analogous manner, the development of gastric ulcers. Also the thick, firmly adherent layer of mucus may help to protect the uppermost layer of the mucous membrane against autodigestion. In general, however, the conditions, with respect to all peptonizing ferments, are such that fully living protoplasm, therefore also that of the epithelial cells of the stomach, possesses the property of being able to resist the action of enzymes, as it is capable of decomposing all, even the most complicated, molecules of inanimate substances. Amœbæ, bacteria, worms, larvæ and embryonal vegetable cells are not affected by artificial digestive juices, not even by trypsin.

After extirpation of the stomach, digestion is continued by the pancreas, the liver and the intestines. The stomach is a protective apparatus with respect to the intestine, as it removes various injurious influences, particularly of bacterial origin.

THE GASES OF THE STOMACH.

The stomach always contains gases, derived in part from air directly swallowed, as, for example, with the saliva, and in part from gases that pass backward from the duodenum.

If the larynx and the hyoid bone are suddenly drawn forcibly forward (as in vomiting), a considerable amount of air enters the space behind the larynx and when the latter returns to its position of rest, is carried down by the peristalsis of the esophagus. One can feel distinctly the downward passage of such a quantity of air. At times, even without any movement of deglutition, a number of small air-bubbles enter the stomach.

These masses of air constantly undergo change, owing to the absorption of oxygen into, and the elimination of carbon dioxid from, the blood. The rather abundant production of carbon dioxid in the stomach depends, however, on chemical processes resulting from the admixture of the pyloric secretion, containing sodium carbonate, with the secretion of the fundus, containing acid. According to Planer, the amount of oxygen is extremely small, while that of carbon dioxid is considerable.

A portion of the carbon dioxid in the saliva is set free by the acid of the gastric juice. The quantity of nitrogen is indifferent.

GASES OF THE STOMACH.   VOLUMES IN PER CENT.
(According to Planer.)

	HUMAN CADAVER AFTER VEGETABLE DIET.		DOG.	
	I.	II.	I. After a Meat diet.	II. After a Diet of Legumes.
CO_2	20.79	33.83	25.2	32.9
H	6.71	27.58	—	—
N	72.50	38.22	68.7	66.3
O	—	0.37	6.1	0.8

Abnormal development of gases, in cases of gastric catarrh, occurs only when the reaction of the gastric contents is neutral. Thus, in the presence of butyric-acid fermentation, hydrogen and carbon dioxid are produced, while acetic-acid and lactic-acid fermentation generate no gases. Marsh-gas (CH_4) also is found; though this can reach the stomach only from the intestine, as it can be produced only in the absence of oxygen. Traces of hydrogen sulphid generated by the bacterium coli commune are formed at times in connection with benign dilatation of the stomach and motor insufficiency. Yeasts and various bacteria are also found in the stomach.

STRUCTURE OF THE PANCREAS.

The pancreas is a compound tubular gland with terminal alveoli which constitute the chief portions of the gland. On the internal surface of the membrana propria, formed of fibrillar tissue, lie the somewhat cylindrical-conical secreting cells, which consist of two layers: (1) the smaller, parietal layer, which is transparent, lamellated, striated, and can be deeply stained by carmine, and (2) the internal layer (Bernard's granular layer), which is deeply granular, and stains but slightly. Between the two layers lies the nucleus. During the process of secretion a visible transformation takes place continually in the cell-substance; the granules in the granular layer undergo solution and form constituents of the secretion, while in the external layer the homogeneous substance is renewed, and is later again transformed into granular matter. This, in turn, again moves inward toward the lumen of the alveolus.

In detail there takes place in the first stage of digestion (from the sixth to the tenth hour) a consumption of the granular inner zone and a growth of the

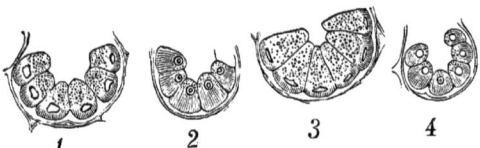

FIG. 115.—Changes in the Cells of the Pancreas in the Different Stages of Activity: 1, in the state of hunger; 2, in the first stage of digestion; 3, in the second stage; 4, with paralytic secretion.

striated outer zone (Fig. 115, 2). In the second stage (from the tenth to the twentieth hour) the inner zone of the swollen gland has increased greatly in size, while the outer zone is much diminished (Fig. 115, 3). In the state of hunger the latter again increases in size (Fig. 115, 1). In the pancreas, yielding a paralytic secretion, and reduced in size, the inner zone of shrunken cells is almost entirely lost.

In consequence of increased secretion, some of the secreting cells undergo a change, so that the acini form irregular collections containing many granules, and have lost all resemblance to glandular acini. Entire cells are also destroyed during the activity of the gland and new ones are again formed.

The finest excretory ducts of the acini begin as intercellular secretory spaces. With the alveolus there is connected an intercalary portion, constituted of flat cells, and which develops in the center of every acinus. Then a sort of salivary duct follows, without striated epithelium, as in the salivary glands. From the micro-center of the cells of the excretory-duct system a ciliated flagellum, the "outer thread," projects free into the lumen of the canal.

The pancreatic duct, which possesses an axial course and as a rule empties into the duodenum in common with the bile-duct, while a smaller branch of the duct makes its entrance at a special papilla at a higher level, consists of an inner, denser, and an outer, looser, wall of connective and elastic tissue, together with unstriated muscular fibers mainly pursuing a circular course, and lined internally by a single layer of cylindrical epithelium. Small mucous glands lie in the main duct and in its larger branches. Medullated and non-medullated nerves, which in their course are connected with ganglia, pass to the glandular acini; but their terminations are unknown. Blood-vessels surround the acini, in part of large size and in abundance, in part isolated. The fresh pancreas contains water, albuminates, ferments, fats and salts. The resting gland contains much leucin, isoleucin and tyrosin; further, butalanin, often xanthin and guanin; lactic acid, formic acid, fatty acids; most of these from autodecomposition.

THE PANCREATIC JUICE.

To obtain the pancreatic juice Regner de Graaf, in 1664, tied in the excretory duct of a dog a cannula provided with an empty bag at its extremity, in which the juice collected. Others passed the tube through the abdominal walls externally and thus made a transitory cannula-fistula, which closed in the course of a few days, with inflammatory expulsion of the extremity of the cannula that had been tied in place. In order to establish a permanent fistula, either a duodenal fistula is made, like a gastric fistula, through which the duct of Wirsung is catheterized by means of a thin tube; or the duct is opened in a dog and drawn toward the abdominal wound and an attempt is made to unite the wound in the duct with the abdominal wound so as to form a fistula. Heidenhain eliminates the portion of the duodenum in which the duct opens from the continuity of the intestine, incises it, and fixes it outside of the abdominal wound.

From such a permanent fistula an abundant, feebly active, watery secretion, rich in sodium carbonate, is collected. From a freshly made opening and before the onset of inflammatory processes, a scanty viscid fluid is obtained which exerts energetic and characteristic physiological actions.

Obviously, the scanty, viscid secretion is normal, while the watery, abundant secretion is abnormal and derived from the dilated blood-vessels, perhaps in consequence of paralysis of the vasomotor nerves, and as a result of increased transudation. The latter would thus in a certain sense be a paralytic secretion. The amount must vary greatly, accordingly as viscid or watery secretion is produced. During digestion a large dog secreted from 1 to 1.5 grams of viscid secretion; Bidder and Schmidt obtained from a permanent fistula from 35 to 37 grams of watery secretion in twenty-four hours, for each kilogram of weight.

While the resting, inactive gland is flabby, yellowish red in color, the secreting gland is turgescent and reddened from the dilatation of its blood-vessels.

Normal pancreatic juice is transparent, colorless and odorless, with a salty taste, and a strongly alkaline reaction from the presence of 0.4 per cent. sodium carbonate, and therefore effervescent from escape of carbon dioxid on addition of acid. It contains albumin and potassium albuminate (9.2 per cent.); like watery egg-albumin, it is viscid, flows with difficulty and coagulates at a temperature of 75° C. into a white mass. On standing in the cold a gelatinous coagulum of albumin separates, in which concentrated mineral acids, metallic salts, tannic acid, chlorin-water and bromin-water cause a precipitate; the precipitate produced by alcohol can be redissolved by water. The total solids in the pancreatic juice of human beings equal 13.6 per cent. Among the salts are sodium chlorid, 7.3; sodium bicarbonate, 0.4; sodium phosphate, 0.45; sodium sulphate, 1.1 in 1000, together with some lime and traces of magnesia, potassium sulphate and ferric oxid.

The more rapid and the more profuse the flow of the pancreatic juice, the more deficient is the secretion in organic constituents, the inorganic components remaining almost the same. Nevertheless, the total amount of solid constituents secreted is greater under such circumstances than when the secretion is scanty. The freshly discharged juice contains traces of leucin and soaps.

· In pancreatic juice that is no longer fresh, chlorin induces a red color, as does crude nitric acid in the putrefying juice, by the production of indol. Rarely the juice forms concretions in the pancreas, principally of calcium carbonate. In cases of diabetes dextrose has been found in the pancreatic juice; in cases of jaundice, urea.

THE DIGESTIVE ACTIVITY OF THE PANCREATIC JUICE.

The presence of four hydrolytic ferments, or enzymes (an amylolytic, a proteolytic, a lipolytic, and a milk-curdling ferment), makes the pancreatic juice a most important digestive fluid.

The *amylolytic activity* is due to the ferment amylopsin, which appears to be identical with the ptyalin of the saliva, though it acts more energetically, both upon raw and upon boiled starch and glycogen. At the temperature of the body almost immediately, but more slowly at a lower temperature, it converts the substances named into maltose, isomaltose and dextrin, as does the saliva. Even cellulose itself is said to be digested and gum to be transformed into sugar, but inulin remains unchanged.

The amylopsin is precipitated by alcohol and it remains dissolved in glycerin without material enfeeblement. All agencies that disturb the diastatic activity of the saliva also abolish that of the amylopsin, although admixture of acid gastric juice, as its hydrochloric acid is in combination, or of bile, is without injurious effect.

The ferment is isolated by the same method as that by which salivary ptyalin is obtained, but in this process the peptic ferment is at the same time precipitated with it.

In addition to this diastase, the pancreatic juice contains a second diastatic ferment, by which maltose and isomaltose are transformed into dextrose. Saliva contains hardly a trace, and blood-serum more of this ferment than of diastase.

The addition of bile, as well as of various neutral salts (in about 4 per cent. solution), increases the diastatic activity, and in the following order: potassium nitrate, sodium chlorid, ammonium chlorid, sodium nitrate, sodium sulphate, potassium chlorate. ammonium nitrate and ammonium sulphate.

The *proteolytic activity* is due to the ferment trypsin, which at the temperature of the body transforms the albuminates, in the presence of an alkaline medium, without previous swelling, first into albumoses (hemi-albumose and anti-albumose), also designated propeptones, and finally into true peptones, also designated tryptones. Previous swelling of the proteids by means of hydrochloric acid, as well as an acid reaction in general, have a tendency to prevent this transformation.

The albumoses of tryptic digestion have the character of the deutero-albumoses. Two kinds of peptones are formed, namely hemi-peptone, which later breaks up into the amido-acids, and antipeptone, which does not undergo further decomposition.

Trypsin peptonizes all proteids, casein, vitellin, elastin, mucin, and nuclein, while neurokeratin, keratin and amyloid remain insoluble. Glutin and the gelatin-yielding substance, swollen by acids are changed into gelatin-peptone, and the latter is not further changed. Oxyhemoglobin decomposes into albumin and hemochromogen. Pancreatic extract first affects milk-casein in such a manner that it is coagulated by heat, after which it is peptonized.   In other respects, trypsin has an action like that of pepsin upon tissues containing albumin.

Casein is almost wholly digested by trypsin. The tryptic ferment, which is also present in the pancreas of new-born infants, is carried down mechanically from the pancreatic juice diluted with water, by the production of a voluminous precipitate, with collodion. The precipitate is washed and dried, and then the collodion is dissolved out in a mixture of ether and alcohol. The residue is soluble in water, and represents the ferment. Kühne further separates with especial care the albumin still combined with the ferment in the aqueous extract of the gland, and thus secures the ferment in a purer form. It is soluble in water, insoluble in alcohol and in pure glycerin.

As trypsin is destroyed by hydrochloric acid, it is not advisable, as in the presence of weakened digestion, to administer trypsin by the mouth. In a dried state it can be heated to a temperature of 140° C. without injury; in a moist state, if pure, to 50° C.; and mixed with salts or with albumoses and peptones, to 60° C.

Method: For testing trypsin, gelatin is especially useful, being liquefied in a test-tube at the temperature of the body: 7 grams of gelatin boiled with 93 grams of an aqueous solution of thymol. For antiseptic purposes thymol should be added also, after filtration, to the fluid to be tested for the presence of the ferment.

Trypsin results through the taking up of oxygen within the pancreas, from a mother-substance, zymogen, which collects in the interior of the secreting cells in smallest amount between the sixth and the tenth hour, and in largest amount, on the other hand, sixteen hours after eating. It can be extracted from fresh glands by glycerin or by water. In aqueous solution this body yields the ferment. Within the excised pancreas the same result occurs on treatment with strong alcohol.

The addition of bile, sodium chlorid, sodium glycocholate and carbonate, as well as carbon dioxid, increases the activity of the ferment, while magnesium sulphate and sodium sulphate enfeeble its action.

With continued action of the trypsin upon the hemipeptone produced, this is converted in part into the amido-acids: leucin ($C_6H_{13}NO_2$), tyrosin ($C_9H_{11}NO_3$), aspartic or amidosuccinic acid ($C_4H_7NO_4$) in the digestion of fibrin and glutin, glutamic acid ($C_5H_9NO_4$), and butalanin or amidovalerianic acid ($C_5H_{11}NO_2$). Gelatin-peptone, according to Nencki, on further decomposition yields glycin and ammonia. The amido-acids produced may be partly absorbed as such and may be consumed in the circulation.

The following bases also occur: xanthin-bases, lysin, lysatinin, arginin, together with ammonia and a body that becomes reddened by chlorin-water or bromin-water.

If the action be continued still further, matters having a fecal odor result, and with especial rapidity when the reaction is alkaline, also indol (C_8H_7N), skatol (C_9H_9N), and phenol (C_6H_6O), volatile fatty acids with the development of hydrogen, carbon dioxid, hydrogen sulphid, marsh-gas and nitrogen. These products of decomposition, however, result wholly from putrefaction of the preparations. This can be prevented by the addition of salicylic acid or thymol, which destroys the putrefactive organisms that are always present.

Prolonged boiling of the albuminates with dilute sulphuric acid, like the action of trypsin, produces first peptone, then leucin and tyrosin, and glycin from gelatin. Hypoxanthin and xanthin result in this way on boiling fibrin, the former also from long-continued boiling of fibrin with water.

Leucin, tyrosin, glutamic and aspartic acids, together with xanthin-bodies,

20

result also in the germination of certain plants, by reason of which there is a resemblance between the transformation and the consumption of nutritive materials in seeds and the digestive effects of ferments.

The *lipolytic activity* depends on the presence of a ferment termed steapsin or pialyn, which exerts its action more especially on the neutral fats. This action is two-fold: (1) they are transformed into a fine, permanent emulsion, and (2), by taking up water, they undergo a cleavage into glycerin and fatty acids.

$$C_{57}H_{110}O_6 + 3H_2O = C_3H_8O_3 + 3(C_{18}H_{36}O_2)$$
Tristearin + Water = Glycerin + Stearic Acid.

The addition of bile increases this action in the rabbit very considerably. This cleavage action is due to a ferment, especially decomposed by acids, but which has not yet been isolated. Lecithin is split up by this ferment into glycerinphosphoric acid, neurin and fatty acids.

After decomposition is complete, the fatty acids are in part united with the alkalies of the pancreatic juice and the intestinal fluid to form fatty-acid alkalies, or soaps; and in part emulsified in the alkaline intestinal juice. Both the emulsion and the soap-solution are capable of being absorbed. After extirpation of the pancreas in the dog, the digestion and absorption of fats are correspondingly diminished.

If the fat to be emulsified contains free fatty acids, as is the case with all of the fats of the food, and if the fluid at the same time has an alkaline reaction, emulsification takes place with extraordinary rapidity. A drop of cod-liver oil, which likewise always contains some free acid, placed in a 0.3 per cent. soda-solution, is at once broken up into fine emulsion-granules. First a hard soapy membrane is formed on the surface of the oil-drop; this, however, is quickly dissolved and small drops are thereby torn away. The fresh surface becomes again covered with a layer of soap and the process is continually repeated. The soaps produced themselves in turn act as emulsifiers. If the amount of oleic acid contained in the oil and the concentration of the soda-solution are increased, so-called "myelinforms" are produced, that is, forms like those that appear when fresh nerve-fibers are teased in aqueous liquids. Animal fats furnish an emulsion more readily than vegetable fats, castor-oil not furnishing any at all.

The fatty acids also may undergo still further decomposition through the action of the fat-splitting ferment, with the production of carbon dioxid and hydrogen even, in the absence of microörganisms.

Danilewsky isolated the four pancreatic ferments in the following manner: If an acid infusion of dog's pancreas is super-saturated with magnesium oxid, the precipitate carries the fat-ferment down with it. Collodion added to the filtrate precipitates the trypsin; the precipitate is collected; and the collodion is dissolved out by a mixture of ether and alcohol. The diastatic ferment is contained in the filtrate from the collodion-precipitate.

For testing the digestive activity of the pancreas an extract of the swollen and reddened gland may be prepared after trituration with the aid of concentrated solution of sodium chlorid. Triturated pancreas, which has lain for a day, can also be extracted with glycerin or chloroform-water. Alcohol precipitates the ferments in these extraction-fluids. Kühne renders the minced pancreas free from water and fat by means of alcohol and ether, and pulverizes it. The powder, to which 10 parts of 0.1 per cent. salicylic acid solution at blood-heat are added, exhibits the activity of the ferments. An extract of the pancreas, prepared rapidly and at a high temperature with a 0.7 per cent. solution of sodium chlorid, contains almost alone the sugar-forming ferment, which is absent from the gland in the state of hunger. After long-continued maceration at a later period trypsin principally is obtained.

To demonstrate the effects of the pancreas Setschenow proceeds as follows: Minced calf's pancreas is infused with less than double its volume of water and is kept at a temperature of 38° C. for five hours. The decanted fluid is strained, shaken with ether, and alcohol is added until a precipitate forms. The latter is spread uniformly upon filter-paper by filtration, and the paper is dried at a temperature of 40° C. A strip of this paper about the length of a finger immersed in 3 or 4 cu. cm. of water yields a fluid capable of acting upon starches, albumin and fat.

The pancreas of new-born infants contains no diastatic ferment, but both peptic and fat-splitting ferments. Diseases of infants, diarrhea at times, appear to have a marked effect on the activity of the pancreas. Slight diastatic power is exhibited after the second month of life, complete activity only after the lapse of the first year.

The *milk-curdling activity* depends on the presence of a ferment, according to W. Kühne and W. Roberts, which can be extracted by means of a concentrated solution of sodium chlorid.

The pancreas also prepares a sugar-splitting ferment. If a solution of sugar is digested with an aqueous or glycerin extract of pancreas, the amount of sugar diminishes.

THE SECRETION OF THE PANCREATIC JUICE.

In the case of the pancreas, a resting stage, in which the gland is flabby and pale yellow, and a stage of secretory activity, in which the organ appears swollen and pale red, can be distinguished. The latter occurs only after the ingestion of food, and results probably in consequence of reflex excitation through the nerves of the alimentary canal, and apparently in consequence of the moistening of the intestinal mucous membrane with the acid gastric contents, for acids are the most powerful excitants of this secretion. W. Kühne and Lea found that all the lobules did not take part in the secretory activity at the same time. The pancreas in herbivora secretes continuously.

According to Bernstein and Heidenhain, the secretion begins to flow with the entrance of the food into the stomach, the quantity reaching its maximum in the second or third hour. After this the amount decreases between the fifth and the seventh hour; then, in consequence of the passage of all of the dissolved matters into the duodenum, it rises again between the ninth and the eleventh hour, and finally falls gradually between the seventeenth and the twenty-fourth hour, to the point of complete cessation.

During the act of secretion the blood-vessels behave like those of the salivary gland after stimulation of the facial nerve; they are dilated, the venous blood being bright red. It is, therefore, probable that a similar nervous mechanism is active here. In general, the activity of the gland is in large measure dependent upon an adequate blood-supply; anemic conditions impair the secretory processes. The secretion, in the rabbit, is under a secretory pressure of over 17 mm. of mercury.

The nerves are derived from the hepatic, splenic and superior mesenteric plexuses, to which the pneumogastric and splanchnic nerves send branches. The secretion of the gland is excited by stimulation of the medulla oblongata, of the splanchnic nerves (feebly), of the peripheral stump of the pneumogastric nerve, in consequence of which the amount of ferment in the juice is increased, as well as of the gland itself by means of induction-currents. Reflex increase in the secretion is brought about by stimulation of the central stump of the lingual nerve, at times also by that of the central stump of the pneumogastric nerve. The secretion is suppressed by atropin, by excitation through the act of vomiting, as well as by stimulation of the pneumogastric nerve or its central stump, as well as of other sensory nerves, as, for example, the crural and sciatic nerves. Destruction of the accessible nerves of the pancreas surrounding the blood-vessels renders the stimulation mentioned ineffective. On the other hand the secretion of a watery, paralytic, slightly active secretion becomes permanent; and the amount is then no longer modified by the ingestion of food.

Fat and water, further pilocarpin and physostigmin, excite pancreatic secretion. Solutions of neutral and alkaline salts of the alkaline metals exert an inhibitory action. Animals tolerate ligation of the pancreatic duct. It is a remarkable fact that the duct may regenerate spontaneously. This operation may, however, be followed by cyst-formation in the ducts and atrophy of the glandular structure. After total extirpation of the pancreas, the digestion of albumin, fat and starches is impaired. The severe diabetes that develops immediately after extirpation of the pancreas and which has been observed also in human beings after degeneration of the pancreas, is of obscure origin.

THE STRUCTURE OF THE LIVER.

The liver is included among the compound tubular glands. Its development shows that with its excretory ducts it evolves in the form of a reticulated tubular gland. The globular, polygonal hepatic acini (lobules, islands), flattened one against the other, from 1 to 2 mm. in diameter, are considered as the ultimate macroscopic units of the gland. They show the following histological peculiarities:

The liver cells (Fig. 116, II, a), 34 or 35 μ in diameter, are irregularly polyhedral, consisting of soft, friable protoplasm, filled with pigment-granules. They have no membrane, and contain one or more spherical nuclei, with nucleoli, and are so arranged that they radiate from the centre of the acinus in longer or shorter connected lines toward the surface of the lobule. Thus arranged they are in part surrounded by the more delicate bile-ducts (Fig. 116, I, x), in part separated one from the other in rows by the coarse network of blood-capillaries (d d). In the state of hunger the liver-cells are finely granular and deeply clouded (Fig. 117, 1). About thirteen hours after suitable nourishment the cells contain coarse, glistening flakes of glycogen (2). At the same time the protoplasm is condensed on the surface, whence a network extends toward the center of the cells, in which the nucleus is suspended. The liver-cells often contain fatty granules.

The Blood-vessels of the Lobule.—(a) Ramifications of the venous system. If the branches of the portal vein, well supplied with muscular fibers, and entering through the transverse fissure, be followed, small vessels will finally be found, after free dendritic branching, that, approaching from various directions, converge at the limits of the acini, and here enter into communication through capillary anastomoses, forming the interlobular veins (Fig. 116, V, i). From these veins capillary vessels (c c) pass from the entire periphery of the acinus toward its center. They are relatively large (from 10 to 14 μ in diameter) and form a longitudinal network in a radiating direction; and between them rows of connected hepatic cells, liver-cell columns (d), are always lodged. The capillaries are so arranged that they run along the edges of the rows of cells, and never between the surfaces of two adjacent rows. The radiating course of the capillaries necessarily brings it about that these vessels must unite at the center of the acinus to form the beginning of a larger vessel. This is the central or intralobular vein (V. c) which, in turn, piercing the lobule vertically, makes its exit at one point and, reaching the surface unites, as the sublobular vein (V. a), with similar vessels from neighboring acini, to form larger trunks that (100 μ in diameter) represent the roots of the hepatic veins. The trunks of this great system of venous radicles leave the gland at the blunt edge of the liver.

(b) *Ramifications of the Hepatic Artery.*—The branches of the hepatic artery, throughout their entire course, accompany the larger branches of the portal vein, to which, as well as to the adjacent larger bile-ducts, they supply nutrient capillaries. These branches enter into numerous anastomotic communications among themselves. The small capillaries pass mainly from the periphery of the acinus into the capillaries of the portal system (Fig. 116, i i). Those arterial capillaries, however, that lie in the thicker connective tissue upon the larger venous and biliary branches (r r) pass over chiefly into two venous trunks that, accompanying the corresponding arterial branches for some distance, empty into branches of the portal vein.

Individual arterial branches pass up to the surface of the liver, where they form a wide-meshed nutritive network, particularly under the peritoneal covering. The small venous radicles collecting from this point also reach the ramifications of the portal vein.

The Biliary Passages.—The finest biliary passages, bile-capillaries, originate from the center of the acinus, and likewise within its entire interior, as membraneless, regularly anastomosing straight ducts, 1 or 2 μ in diameter. They form a polygonal mesh about each liver-cell (Fig. 117, 3). The ducts almost always lie mid$_w$a$_y$ between the surfaces of two adjacent liver-cells (Fig. 116, II, a) as true intercellular passages or secretory spaces. When the cells fall apart in the process of maceration, they retain only semicircular depressions. The finest ducts of the bile-capillaries have been observed to penetrate the interior of the liver-cells and to communicate here with round, secretory vacuoles containing bile (Fig. 117, 3). As the blood-capillaries run along the edges of the rows of liver-cells, while the biliary ducts run along the surfaces of the cells, both systems of ducts are always at a definite distance from one another (Fig. 118).

In human beings individual bile-ducts sometimes run also along the edges of the cells, so that they must then act as intercellular ducts of 3 or 4 cells. This arrangement is said to predominate in the embryonal liver. In addition to injection, the capillaries can be made visible by staining by Golgi's method.

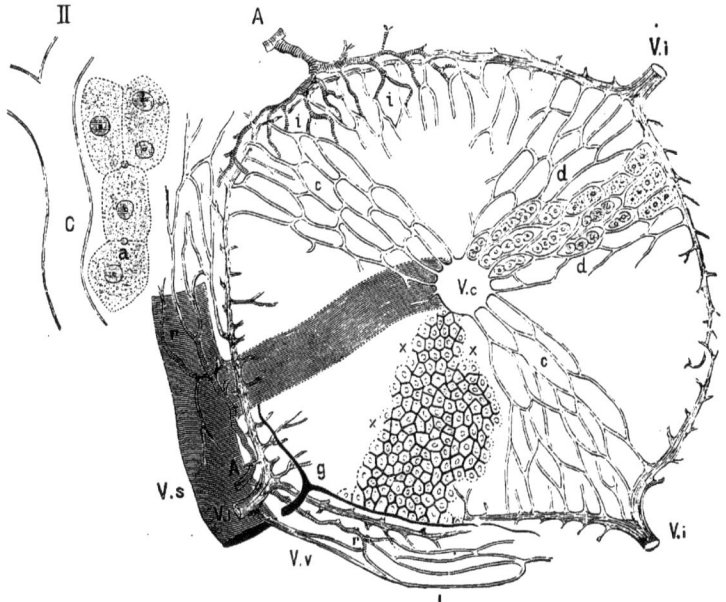

Fig. 116.—I. Diagrammatic Representation of an Hepatic Lobule: V. i., V. i, interlobular veins; V. c, central vein; c, capillary between the two; V. s, sublobular vein; V. v, vascular vein; A A, branches of the hepatic artery, approaching the capsule of Glisson and the larger blood-vessels at r r, and forming the vascular vein further on, entering the capillaries of the interlobular veins at i i; g, branches of the bile-duct, dividing at x x between the liver-cells; d d, situation of liver-cells in the capillary network. II. Isolated liver-cells, at c lying upon a capillary blood-vessel and forming a fine bile-duct at a.

Within the peripheral, cortical portion of the lobule the ducts, without walls, increase in size by anastomosis of neighboring ducts. They then leave the acinus, in order, from this point, uniting between the lobules (Fig. 116, g) with adjacent ducts, to form larger bile-ducts, with numerous anastomoses. These, in company with the branches of the hepatic artery and the portal vein, finally leave the transverse fissure of the liver as a collecting duct, the hepatic duct. The finer interlobular bile-ducts possess a structureless membrana propria with low epithelium. The larger (Fig. 119) exhibit a double membrane constituted of connective tissue and elastic fibers, the internal layer being especially supplied with blood-capillaries and bearing a single layer of cylindrical epithelium. Only in the largest branches, and in the gall-bladder, does this internal layer become an independent mucous membrane, with submucosa. Unstriped muscle-fibers are found in isolated bundles in the main ducts (longitudinal and circular especially in the lower portions of the bile-ducts), as well as in a delicate longitudinal and circular layer in the gall-bladder. The movements here are slowly rhythmic and peristaltic. The mucous membrane of the gall-bladder is provided with folds and comb-like depressions. The epithelium is a single layer of cylindrical epithelium with a basal membrane and intervening mucous goblet-cells. Small mucous glands are found in the mucous membrane of the large bile-ducts and of the gall-bladder.

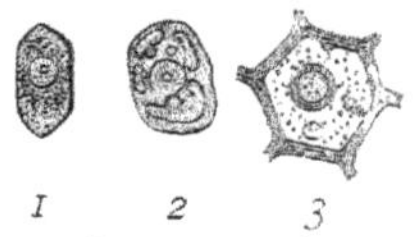

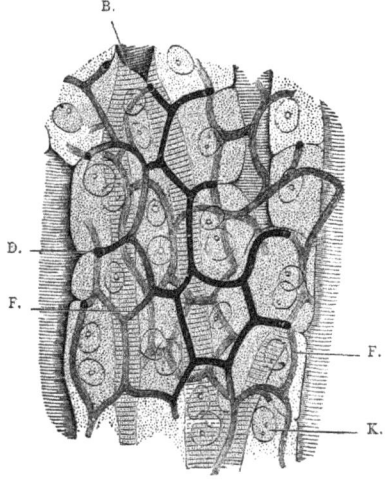

Fig. 117.—A, Liver-cell, in the state of hunger; 2, filled with masses of glycogen; 3, surrounded by bile-capillaries.

The *connective tissue* of the liver enters the portal fissure as a sheath (capsule of Glisson) for the vessels, and, mixed with elastic tissue, finally reaches the periphery of the acini, where in the pig, the camel and the polar bear it forms a clearly demonstrable capsule, but in human beings is inconspicuous. Delicate elements can, however, be followed even into the acinus, nucleated star-cells and a network of delicate reticular fibers, which effect the fixation of the elements.

The connective tissue of the acini not rarely undergoes considerable increase in drunkards, and its hyperplasia may even cause destruction of the contents of the acinus by pressure (cirrhosis of the liver). In this thickened, interacinous connective tissue newly formed bile-ducts have been found, and likewise in the cicatricial connective tissue of the "corset-liver."

The *lymph-vessels* begin as pericapillary ducts in the interior of the acinus. Further on they run within the walls of the hepatic veins and the branches of the portal vein; then they surround the venous branches. The larger vessels, formed from the union of the interlobular passages, leave the organ in part at the transverse fissure, in part with the hepatic veins, and in part at different points on the surface. At the blunt edge of the liver they form a close meshwork and pass through the triangular, hepato-renal and suspensory ligaments.

The *nerves* of the hepatic plexus, constituted in part

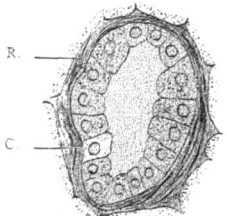

Fig. 118.—Blood-capillaries, Finest Biliary Ducts, and Liver-cells, in Their Mutual Relations in the Rabbit's Liver (after E. Hering): B, blood-vessel; D, finest biliary duct, in cross-section; F, finest biliary duct; K, nucleus of liver-cell.

Fig. 119.—Interlobular Bile-duct from the Human Liver (after Schenk): R, circular fibrous layer; C, cylindrical epithelium.

of Remak's fibers, in part of medullated fibers from the sympathetic and pneumogastric nerves, follow the ramifications of the hepatic artery. Ganglia are placed in their course in the interior of the organ. The nerves are in part vasomotor in nature. According to Pflüger, other nerve-fibers enter into direct connection with the liver-cells, as is the case in the salivary glands. The muscle-cells of the bile-ducts contain motor filaments.

The celiac plexus sends trophic and vasomotor nerves to the liver. Destruction of this plexus therefore causes degeneration of the liver-cells, and dilatation of the hepatic artery. The pneumogastric nerve supplies dilator-fibers to the vessels, and the greater splanchnic motor branches to the muscles of the bile-ducts.

CHEMICAL CONSTITUENTS OF THE LIVER-CELLS.

Proteids.—The fresh, soft liver-parenchyma has an alkaline reaction. After death, coagulation takes place, with cloudiness of the cell-contents; the tissue becomes friable and gradually acquires an acid reaction. This process is suggestive of rigor mortis, and is due to a myosin-like, post-mortem coagulating albuminous substance. The liver contains, further, a proteid body coagulable at 45° C., another coagulable at 70° C., and one slightly soluble in dilute acids and alkalies. The nuclei contain nuclein. The connective tissue yields gelatin.

Glycogen, $6C_6H_{10}O_5 + H_2O$, or animal starch, from 1.2 to 2.6 per cent., is a carbohydrate closely allied to inulin, soluble in water, and diffusible with difficulty, which surrounds the nuclei of the liver-cells in amorphous granules (Fig. 117, 2), though not always present and not always found in equal amounts in all parts of the liver. The glycogen in the liver represents the excess of carbohydrate material, which, after the ingestion of suitable foods, is temporarily stored like the starch in the plants. It is subsequently transformed into sugar and consumed by the tissues.

Qualitative Determination.—Glycogen is stained deeply red by iodin (best dissolved by means of potassium iodid in a concentrated solution of sodium chlorid), like inulin, even in microscopic sections hardened in alcohol. Organs containing glycogen, boiled with an excess of sodium sulphate, yield an opalescent filtrate. If the organs, as, for example, the liver, still contain diastatic ferment, the glycogen, after being kept warm for several hours, will be converted into sugar, and, as already stated, the resulting filtrate remains clear.

Quantitative Estimation.—According to Külz's modification of Brücke's method, the coarsely minced liver is thrown into boiling water immediately after death and boiled for half an hour. It is then crushed and potassium hydrate (4 grams to 100 grams of liver) is added. Evaporation over a water-bath to double the weight of the piece of liver employed is permitted to take place until in the course of three hours all is dissolved. After cooling, the mixture is neutralized with hydrochloric acid, and the albumin, together with the lime, is precipitated by means of hydrochloric acid, and potassio-mercuric iodid. Filtration is now practised, the precipitate being taken from the filter four times, mixed with a few drops of hydrochloric acid and potassio-mercuric iodid in water to the consistency of broth and filtered. All of the glycogen is now contained in the filtrate, to which, with stirring, double the volume of 96 per cent. alcohol is added. The glycogen deposited in the course of twelve hours is placed upon the filter, washed with 62 per cent. alcohol, then with absolute alcohol, with ether, again with absolute alcohol and dried at 110° C. Should the fluid remain cloudy after addition of hydrochloric acid and potassio-mercuric iodid, two parts of 98 per cent. alcohol are added and the filtered precipitate is dissolved in 2 per cent. potassium hydrate, then neutralized with hydrochloric acid and now all of the albumin can be precipitated by repeated addition of hydrochloric acid and potassio-mercuric iodid again.

According to Seegen, dextrin is present in the liver in addition to glycogen. Rabbit's liver contains about three times as much glycogen in winter as in summer.

The following are to be considered as the sources of glycogen in the liver: (1) The carbohydrates of the food, after they have been converted into dextrose in the alimentary canal; only the sugars fermentable by yeast form glycogen, and not those incapable of fermentation;

and (2) the proteids, including gelatin. If the proteids are a source of glycogen, it must result from a non-nitrogenous derivative of them.

Pflüger considers the formation of glycogen from albumin a synthetic process. The molecular group CH_2, found in albumin, as well as in the fatty acids, must be transformed by oxidation into CHOH. The cells taking part in the formative process may, however, also utilize this group CHOH wherever it is found already prepared, as in sugar or in glycerin.

Also fats (olive-oil), glycerin, taurin and glycin (the latter through decomposition into glycogen and urea), have been designated as the source of glycogen.

In rabbits, the production of glycogen is increased by the administration of asparagin, ammonium carbonate or urea. The excessive production of acid in cases of diabetes, demonstrated by Stadelmann, fixes the ammonia and thus materially diminishes the production of glycogen.

Ligation of the common bile-duct results in diminution of the glycogen in the liver. The liver after this operation appears to have lost the property of forming glycogen from suitable material brought to it. Also ligation of the hepatic artery renders the liver free from glycogen. After excluding the portal circulation the amount of sugar contained in the blood decreases. With reference to the occurrence of glycogen elsewhere reference may be made to p. 466.

If large amounts of starch, grape-sugar, cane-sugar, levulose and maltose are added to the proteids of the food, the amount of glycogen in the liver is greatly increased, while on a pure albuminous or fatty diet it is considerably decreased; the state of hunger may cause it to disappear entirely. Injection of grape-sugar or of glycerin into a mesenteric vein of a fasting rabbit causes the appearance of glycogen in a liver previously free from it.

The living liver-cell is capable of producing glycogen in considerable quantities only from the two kinds of sugar capable of direct fermentation, namely dextrose and levulose. The non-fermentable sugars are not converted into glycogen, and cane-sugar and maltose only in so far as they are transformed in the intestine into dextrose. As the infant consumes milk-sugar, it must form glycogen from albumin.

Forced muscular movement rapidly renders the liver of the dog free from glycogen. Reduction of temperature diminishes the amount of glycogen in the liver. The rigid liver after death contains dextrin and grape-sugar. Glycogen is also present in the liver for a considerable time after death, as well as in the muscles.

Under normal conditions, the glycogen in the liver is gradually transformed in small amounts into grape-sugar. The amount of sugar normally present in the blood is from 0.5 to 1 in 1000. The blood in the hepatic veins may contain somewhat more. Increased transformation into sugar occurs only in connection with marked circulatory disturbances in the liver, as a result of which the blood of the hepatic veins comes to contain a larger amount of sugar. The glycogen undergoes this transformation, likewise, soon after death, when the liver is always found to contain a larger amount of sugar and a smaller amount of glycogen.

The active ferment necessary for this process can be obtained from an extract of the liver-cells, by the method employed to obtain ptyalin. Nevertheless, it is said not to be formed in the liver-cells, but only reaches the liver to be quickly stored up, through the blood, within which the ferment is always formed with rapidity so soon as the movement of the blood undergoes marked disturbance. This transforming ferment develops also as a result of the solution of red blood-corpuscles; and as a constant slight destruction of red blood-corpuscles must surely be assumed to take place within the liver, a source is thus provided for the production of the ferment through the action of which small quantities of sugar are continually formed in the liver. As the liver is thus the seat for the production of sugar, extirpation of this organ or ligation of its vessels is followed by disappearance of the sugar contained in the blood.

The grape-sugar formed in the liver is destroyed in part in the blood-stream, on its way through the tissues, in part by a special ferment, which appears to be derived principally from the pancreas, and to be carried by the blood-corpuscles. A portion of the sugar in the blood is converted in the muscles into glycogen.

According to Külz and Vogel, the same process takes place in the liver in the formation of sugar from glycogen as results from the action of the saliva and the pancreatic juice, with the production likewise of maltose and isomaltose. According to E. Cavazzani, irritation of the celiac plexus causes the production of sugar in the liver, in connection with which the liver-cells undergo morphologic change.

Further, *fats* are observed in the liver-cells, in the form of granules, as well as free in the bile-ducts; occasionally when the diet is rich in fat (in greater amount in drunkards and tuberculous patients), olein, palmitin, stearin and volatile fatty acids are found. Further, sarcolactic acid, traces of cholesterin, jecorin, finally small amounts of urea (in increasing amount in the warm, "surviving" liver), uric acid; and leucin, tyrosin (guanin?), sarcin, xanthin, and cystin pathologically in conjunction with putrefactive disorders, may be present.

The liver-cells contain pigments, which are partly soluble in feebly alkaline water, partly in chloroform.

The pigment soluble in water, designated ferrin, varies from yellow to red in color and contains almost all of the iron of the liver. The latter can be demonstrated directly by means of potassium ferrocyanid or ammonium sulphid. The pigment soluble in chloroform, designated cholechrome, can be extracted from pulverized dried liver. It stands midway between bile-pigment and the lipochromes.

The *inorganic constituents* of the liver are potassium, sodium, calcium, magnesium and manganese. Iron in organic combination with albumin (in ferratin) is present in the liver to the amount of about 6 per cent. Abstraction of blood together with albumin-hunger causes its disappearance. It is utilized in the production of new blood. Chlorin, phosphoric, sulphuric, carbonic and silicic acids may also be present; and copper, zinc, lead, mercury and arsenic have been found deposited in the liver accidentally.

DIABETES MELLITUS.

The formation of large amounts of grape-sugar by the liver and their entrance into the blood and into the urine (glycosuria, diabetes mellitus) have been brought into relation with the normal conditions already mentioned. Extirpation of the liver in the frog or destruction of the liver-cells (fatty degeneration from phosphorous or arsenical poisoning) does not cause the appearance of this phenomenon. It occurs a few hours after injury to a particular spot (center for the vasomotor nerves of the liver) on the floor of the lower portion of the fourth ventricle (Cl. Bernard's sugar-puncture, piqûre); further, after division of the vasomotor paths in the spinal cord from above downward to the exit of the nerves for the liver, that is to the lumbar portion, in the frog to the fourth vertebra.

Division or paralysis of the vasomotor conducting paths from the center to the liver results in glycosuria. According to recent researches by François Franck and Hallion, the vasomotor nerves of the liver (for the hepatic artery and the portal vein) arise between the sixth dorsal and the second lumbar nerves and pass through the communicating branches into the splanchnic nerves. According to the opinions of earlier investigators, all of the paths, however, do not pass through the spinal cord alone. A number of vasomotor fibers for the liver leave the spinal cord at a higher level, and pass further on in the course of the sympathetic nerve to the liver. Thus, destruction of the uppermost, as well as of the lowest, cervical ganglion, and of the first dorsal ganglion, of the abdominal ganglia, often also of the splanchnic nerves, is followed by glycosuria. The paralyzed, dilated vessels render the liver exceedingly vascular, and the blood-stream in them is slowed. This disturbance of the circulation gives rise to the presence of a large amount of sugar in the liver, as the blood-ferment has time to effect transformation of the glycogen. Irritation of the sympathetic nerve at the last cervical and first dorsal ganglia causes contraction of the hepatic vessels at the periphery of the acini, with anemia. It is a remarkable fact that glycosuria when present can be removed by division of the splanchnic nerves. This is explained by the circumstance that the enormous hyperemia of the intestines occurring after this operation renders the liver anemic.

Also a number of poisons that paralyze the vasomotor nerves of the liver cause diabetes in the same manner, namely curare, when artificial respiration is not maintained: carbon monoxid, amyl nitrite, orthonitrophenyl-propionic acid and methyldelphinin: less constantly morphin, chloral hydrate and others. The toxic products of some of the infectious diseases also act in the same way at times. Blood-stasis of other sort in the liver also appears capable of causing glycosuria, as, for example, after mechanical stimulation of the liver. In this category probably belongs the glycosuria following the injection of dilute saline solutions into the blood, as a result of which the changes in the shape of the red corpuscles cause stasis. Also the fact that repeated venesection makes the blood richer in sugar may, perhaps, be explained by the slowing of the circulation.

Persistent irritation of peripheral nerves may also be active through a reflex influence upon the center for the vasomotor nerves of the liver. The appearance of sugar in the urine has sometimes been observed as a result of irritation of the central stump of the pneumogastric nerve, likewise after irritation of the central stump of the depressor nerve. Even division and central irritation of the sciatic nerve may cause the appearance of sugar in the urine: in this way is explained the occurrence of glycosuria in cases of sciatica and other nervous disorders.

According to Schiff, stagnation of the blood in various extensive portions of the body is said to increase the development of the ferment in the blood to such a degree that diabetes results. Of this character must be considered the glycosuria that occurs after compression of the aorta or the portal vein, although the pressure exerted under such circumstances may, perhaps, paralyze nerve-paths concerned. According to Eckhard, injury to the vermis of the cerebellum, in the rabbit, is said to bring about diabetes. In human beings, also, affections of the nervous structures mentioned may cause diabetes.

Various explanations have been assigned in elucidation of the ultimate cause of these symptoms:

(a) The glycogen of the liver may without interference be converted into sugar, as ferment may be conveyed to the liver-cells from the blood-mass, in consequence of its stagnation. Therefore the normally functionating vasomotor system of the liver, and especially its center, is, in a certain sense, to be designated an inhibitory system controlling the production of sugar.

(b) If it be assumed that, under normal conditions, a certain, even though small, amount of sugar flows continually from the liver into the blood, through the hepatic veins, diabetes might be explained as depending on the abolition of those metabolic processes (deranged combustion of sugar) that constantly remove this sugar from the blood under normal conditions.

The following experiments appear to confirm this latter view: Independently of one another, v. Mering and Minkowski, as well as de Dominicis, observed that dogs become diabetic after total removal of the pancreas. According to Minkowski, it is the function of the pancreas to consume the sugar of the blood. Lépine and Barral state that a ferment is produced in the pancreas that destroys the sugar in the blood; so that after extirpation of the pancreas, sugar must accordingly accumulate in the blood. The ferment is contained in abundance within the leukocytes in the portal vein; some is derived from the lymph, perhaps also from other abdominal glands. After extirpation of the pancreas, the blood contains little sugar-destroying ferment. Kolisch and von Stejskal found much jecorin.

Pflüger expresses himself as follows as to the development of diabetes mellitus: The sugar formed by the liver in excessive amount, in consequence of abnormalyl increased nervous excitation, stimulates the pancreas—for it is possible that this gland takes part in the synthetic production of fat from sugar—or the fat-forming organs to the production of an increased amount of fat, so that often fat-formation takes place at the beginning of the disease. As soon as the fat-producing organs, exhausted and paralyzed from over-activity, are no longer capable of disposing of the sugar wholly or in part (which may also be the result of excessive ingestion of sugar), this is excreted by the kidneys, because even the healthy body cannot assimilate the greater portion of the sugar as such, but only after it has been transformed into fats or into soaps. The living body strives to make good the resulting great loss in nutritive material by the assimilation of larger amounts of albumin and fat. Naturally a variety of diabetes is conceivable without hepatic disease as the result of paralysis of the pancreas, or of the fat-producing organs. Lépine's discovery of a glycolytic ferment yielded to the blood by the pancreas, which decomposes the sugar in the blood in some as yet unknown manner, and which is absent or diminished in cases of diabetes, would readily accord with the foregoing hypothesis.

In the presence of pancreatic diabetes, puncture of the floor of the fourth ventricle increases the excretion of sugar; likewise, remarkably, the addition of raw pancreas to the food.

(c) Phloridzin, a glucosid from the bark of the roots of cherry-trees and apple-trees, after ingestion causes the sugar normally present in the blood to pass rapidly over into the urine, so that the latter contains a larger and the former a smaller amount of sugar.

(d) According to Biedl, diabetes occurs after ligation of the thoracic duct in the dog.

The enormous need of food and drink, together with the signs of consumption of the bodily tissues, is characteristic of diabetic patients. Not rarely, in severe cases, collapse-like coma is observed, which has been designated also diabetic coma, and during the existence of which the breath often smells of acetone, which can also be demonstrated in the urine. Diabetic patients living on an exclusive meat-diet exhibit diacetic acid in the urine, in addition to acetone. Neither acetone nor its antecedent, diacetic acid (which can be recognized by the reddening of the urine when dilute ferric chlorid is added drop by drop), after the administration of which the urine contains much acetone, is, as direct feeding-experiments show, the cause of this coma; which is perhaps the result of excessive acid-production in the body, therefore an acid intoxication. To neutralize the acid, increased elimination of ammonia takes place from the body. The urinary tubules often exhibit signs of coagulation-necrosis, which can be recognized by a bright and swollen appearance of the necrotic cells of the tubules. v. Frerichs found, further, glycogenic degeneration in Henle's loops, in the liver, the heart, the leukocytes and the lungs. The urine of diabetic patients is discussed on p. 501.

THE CONSTITUENTS OF THE BILE.

The bile is a transparent fluid varying from yellowish brown to dark green in color, of a sweetish, bitter taste, feeble musk-like odor, and feebly acid or neutral reaction. The specific gravity of human bile from the gall-bladder is between 1026 and 1032, while that collected from a fistula varies from 1010 to 1011. The constituents of the bile are as follows:

Mucus, and in addition a considerable amount of mucoid nucleo-albumin, which together make the bile ropy, are products of the mucous glands and the goblet-cells of the mucous membrane of the bile-ducts. They are precipitated by alcohol, or dilute hydrochloric acid or dilute acetic acid. They cause rapid putrefaction of the bile.

The two biliary acids : glycocholic acid and taurocholic acid, the so-called conjugate acids, combined with sodium (and with potassium in traces) to form sodium glycocholate and taurocholate, have a bitter taste and are dextrorotatory. In human bile, as in that of cattle, glycocholic acid predominates; in carnivora, the sheep, the goat, tauro·cholic acid.

(*a*) *Glycocholic acid*, $C_{26}H_{43}NO_6$, is decomposed by boiling with potassium or barium hydrate or with dilute mineral acids, and by taking up water splits into—

$$\underset{\substack{\text{Glycin (glycocoll,} \\ \text{gelatin-sugar, amido-} \\ \text{acetic acid)}}}{C_2H_5NO_2} + \underset{\substack{\text{Cholalic or} \\ \text{cholic Acid}}}{C_{24}H_{40}O_5} = \underset{\text{Glycocholic Acid}}{C_{26}H_{43}NO_6} + \underset{\text{Water.}}{H_2O}$$

(*b*) *Taurocholic acid*, $C_{26}H_{45}NSO_7$, decomposes with similar treatment and addition of water into—

$$\underset{\substack{\text{Taurin (amido-ethyl-} \\ \text{sulphuric acid, pris-} \\ \text{matic crystals)}}}{C_2H_7NSO_3} + \underset{\text{Cholic Acid}}{C_{24}H_{40}O_5} = \underset{\text{Taurocholic Acid}}{C_{26}H_{45}NSO_7} + \underset{\text{Water.}}{H_2O}$$

Demonstration of the Biliary Acids.—The bile is evaporated to one-fourth its volume, triturated to a pasty mass with animal charcoal to remove the coloring-matter, and dried at 100° C. The black mass is extracted with absolute alcohol, which passes colorless through the filter. After a portion of the alcohol has been driven off by evaporation, the addition of an excess of ether causes at first a resinoid precipitate of salts of the biliary acids, which later pass over into a crystalline mass of brilliant needles (Platner's crystallized bile). The alkaline salts of the biliary acids obtained in this way are readily soluble in water or alcohol, but are insoluble in ether. From the solution of both salts neutral lead acetate precipitates a portion of the glycocholic acid as lead glycocholate. The latter is collected on a filter, dissolved in hot alcohol, and lead sulphid is precipitated by hydrogen sulphid. After removal of the precipitate, the addition of water causes separation of the isolated glycocholic acid. If, after precipitation of the lead glycocholate, basic lead acetate is added to the filtrate, a precipitate of lead taurocholate forms, uncontaminated, however, by lead glycocholate, from which the free acid is subsequently obtained by analogous treatment.

According to Schotten and others, human bile contains, in addition to cholic acid, still another acid, fellic acid ($C_{23}H_{38}O_4$); the bile of cattle contains cholic acid ($C_{24}H_{40}O_5$).

Of the products of decomposition of the biliary acids, glycin does not occur as such in the body, but only in the bile in combination with cholic acid, in the urine in combination with benzoic acid as hippuric acid, and finally in gelatin in complete combination.

Cholic acid is dextrorotatory, insoluble in water, soluble in alcohol; it is soluble with difficulty in ether, separating out in prisms. Its crystalline alkaline salts are readily soluble in water, like soap. With iodin, in direct light, it yields a yellow, in transmitted light a blue, crystalline combination. It occurs free only in the intestine.

Cholic acid is replaced in the bile of some animals by a related acid, as, for example, in the bile of swine, by hyocholic acid; in the bile of geese, chenocholic acid is present.

By boiling with concentrated hydrochloric acid or heating, dry, to 200° C., cholic acid is changed into an anhydrid dyslysin.

Dyslysin is only an artificial product and never occurs in the intestines. When fused with potassium hydrate, it is changed back to potassium cholate.

Pettenkofer's Test.—The biliary acids, the cholic acids and their anhydrids, when dissolved or broken up in water, and on addition of two-thirds concentrated sulphuric acid (drop by drop, without permitting the temperature of the fluid to rise above 70° C.), and a few drops of a 10 per cent. solution of cane-sugar, yield a purplish-red transparent color, which shows two absorption-bands in the spectrum, at E and F.

Before examining a solution for the presence of biliary acids, the albumin must always be first removed, as the latter yields a similar reaction, although the red solution here is characterized by only one absorption-band. If only small amounts of biliary acids are present, the fluid must first be concentrated by evaporation. Cholesterin, stearic and oleic acids, as well as phenol and pyrocatechin, exhibit a similar reaction. Pettenkofer's test, therefore, is absolutely reliable only when the salts of the biliary acids in alcoholic extract are precipitated and thus isolated. It depends on the production, from the reaction between sugar and sulphuric acid, of furfurol, which is stained red in the presence of the biliary acids. Instead of sugar a 0.1 per cent. aqueous solution of furfurol may be employed with advantage for this reaction.

The biliary acids are formed in the liver, as extirpation of this organ is not followed by their accumulation in the blood.

The manner in detail in which the production of the nitrogenous biliary acids takes place, is unknown, although they are supposed to result from albumin. A generous proteid diet increases the secretion of bile. Taurin contains the sulphur of the proteid; the biliary acids contain from 4 to 6 per cent. of sulphur. Probably the substance of the red blood-corpuscles broken up in the liver takes part in their production.

The Biliary Pigments.—Fresh human bile and that of some animals is yellowish brown in color, due to the *bilirubin* present which is combined with an alkali. Under the influence of oxygen, heat and light, bilirubin is transformed by oxidation into a green pigment, *biliverdin*. This predominates in the bile of herbivora and of cold-blooded animals, and likewise often in the state of hunger.

(*a*) *Bilirubin*, $C_{32}H_{36}N_4O_6$, from 0.15 to 0.25 per cent. in human bile, according to Städeler and Maly in combination with an alkali, crystallizes in transparent, sorrel, clinorhombic prisms. It is insoluble in water, but soluble in chloroform, by means of which it can be separated from biliverdin, which is insoluble in chloroform. It combines with alkalies as a monobasic acid and is thus soluble. It is identical with hematoidin.

It is most easily prepared from red gall-stones formed of bilirubin and lime, which are triturated, the lime being dissolved out by means of hydrochloric acid. On agitation with chloroform the bilirubin is taken up. The derivation of bilirubin from hemoglobin is not to be doubted, on account of its identity with hematoidin. Probably red blood-corpuscles are broken up in the liver, and their hemoglobin is converted into bilirubin.

In normal bile from a dog, a pigment is not rarely present having the spectral properties of methemoglobin, and which perhaps represents a body intermediate between the hemoglobin and the coloring-matter of the bile.

(*b*) *Biliverdin*, $C_{32}H_{36}N_4O_8$, is an oxidation-stage of bilirubin, from which it can be obtained by various oxidizing processes. It is readily soluble in alcohol, with great difficulty in ether, and not at all in chloroform. It is present in large amount in the placenta of the dog. It has not as yet been possible to reconvert it into bilirubin by means of reducing agents.

Gmelin's Test.—Bilirubin and biliverdin, which, in addition to the bile, are occasionally found also in other fluids, at times in the urine, are demonstrated by Gmelin's test. If to the fluid containing the substances named are added several cubic centimeters of nitric acid and one drop of nitrous acid, which are permitted to flow carefully from the edge down the sides of a conical glass, without agitation, a play of colors results as follows: green (biliverdin), blue, violet, red and yellow.

(*c*) If the addition of acid is stopped when the color becomes blue, thus preventing further oxidation, a stable transformation-product remains, namely *bilicyanin*. This has a blue color in acid solution, a violet color in alkaline solution, and it exhibits two ill-defined absorption-bands at D. Haycraft and Schofield were able to change this back by reduction with ammonium sulphid.

Fluids containing biliary pigment, if boiled for from three to five minutes with one-third formalin, acquire an emerald-green color, which is changed to amethyst violet on addition of hydrochloric acid.

(*d*) Small amounts of *bilifuscin* (bilirubin + water) have also been found in gall-stones and putrid bile.

(e) *Biliprasin* (bilirubin + water + oxygen) has also been found under like conditions.

(f) The yellow pigment finally obtained by the continued oxidizing effect of the nitric-acid mixture upon all of the biliary pigments is the *choletelin* of Maly, $C_{16}H_{18}N_2O_6$; it is amorphous, and soluble in water, alcohol, acids and alkalies.

(g) With addition of hydrogen and water in the intestine through the agency of bacteria bilirubin passes over into the *hydrobilirubin* of Maly, $C_{32}H_{40}N_4O_7$. The same result can be brought about artificially by treating an alkaline aqueous solution of bilirubin with actively reducing sodium-amalgam. Hydrobilirubin is but slightly soluble in water, more readily in salt-solutions or alkalies, alcohol, ether and chloroform, and it exhibits an absorption-band at F. This body, which, according to Hammarsten, occurs even in normal bile, is a constant pigment of the feces, from which, after acidulation with sulphuric acid, it can be extracted by absolute alcohol. Probably it is identical with the pigment of the urine, the urobilin of Jaffé. Hydrobilirubin is formed in the intestine from ingested bile, being in part absorbed and excreted from the portal circulation through the bile.

Hydrobilirubin to which a drop of sulphuric acid and some potassium nitrate are added again yields Gmelin's reaction. Fresh fecal matter, broken up in a porcelain dish in a concentrated solution of mercuric chlorid, yields a red color as the reaction of hydrobilirubin, while admixture of bilirubin causes a green color.

Cholesterin forms transparent rhomboid plates (Fig. 92, d), is insoluble in water, but soluble in hot alcohol, in ether or chloroform. In the bile it is kept in solution by the salts of the biliary acids. Cholesterin is not a secretory product of the liver, but a product of the disintegration of the epithelial cells of the biliary passages.

It is most easily obtained from the so-called white gall-stones, which not rarely consist principally of almost pure cholesterin, by boiling the triturated calculi with alcohol. The crystals that separate on evaporation of the alcohol become red in color from the edges on addition of sulphuric acid (five volumes to one volume of water), and blue, like cellulose, on addition of sulphuric acid and iodin. Dissolved in chloroform, one drop of concentrated sulphuric acid produces a deep-red color. Moistened with a deep wine-yellow, alcoholic solution of iodin, the crystals exhibit green, blue and red coloration after addition of sulphuric acid. Dissolved in glacial acetic acid, addition of sulphuric acid produces first a rose-red, then a blue color.

Other Organic Substances.—Lecithin, or its decomposition-products, neurin and glycerin-phosphoric acid; palmitin, stearin, olein, as well as their sodium-soaps; diastatic ferment; traces of urea, at times ethereal sulphuric acids; acetic and propionic acids and traces of myristinic acid in the bile of cattle.

Fat reaches the bile from the liver and, conversely, fat is in turn absorbed from the bile in the biliary passages (epithelial cells of the gall-bladder). Fresh unboiled bile decomposes hydrogen dioxid. Bacteria introduced into the blood-stream are in part eliminated by the bile.

The inorganic constituents of the bile (from 0.6 to 1 per cent.) include sodium chlorid, potassium chlorid, 0.2 per cent. soda, alkaline sodium phosphate, calcium and magnesium phosphate, and an abundance of iron. The last yields the usual reactions of iron even in fresh bile, so that iron must be present in the bile in one of its oxygen-combinations. Finally, some manganese and silica are present. Freshly secreted bile from the dog contains more than 50, from the rabbit 109, volumes per cent. of carbon dioxid, in part combined with alkalies, in part absorbed, the latter being almost completely absorbed within the bladder.

Analysis of Human Bile.—Water, from 82 to 90 per cent., salts of the biliary acids, from 6 to 11 per cent., fats and soaps, 2 per cent.; cholesterin, 0.4 per cent.; lecithin, 0.5 per cent.; mucin, from 1 to 3 per cent.; ash, 0.6 per cent. The amount of sulphur contained in dry bile from a dog is from 2.8 to 3.1 per cent.; the amount of nitrogen, from 7 to 10 per cent. The sulphur of the bile is not oxidized into sulphuric acid, but it appears in sulphur-containing compounds in the urine.

SECRETION OF BILE.

The secretion of bile is not a simple filtration of already prepared materials from the blood through the liver, but a chemical production, attended with oxidation, of the characteristic biliary matters in the liver-cells, which exhibit histological change during the process of digestion, and to which the blood of the gland only supplies the raw material. It takes place continuously, the bile being in part temporarily stored in the gall-bladder, and only discharged in considerable amount at the time of digestion. The higher temperature of the blood in the hepatic veins, as well as the large amount of carbon dioxid in the bile, indicates the occurrence of oxidation-processes in the liver. Even the water of the bile is not simply filtered out, since the pressure in the biliary passages may exceed that in the portal vein. It appears that the bile is derived from proteid only, and that the excretion of carbon dioxid in the act of respiration bears a certain relation to its production. In animals (birds) deprived of their livers the constituents of the bile are not produced.

After an albuminous diet the liver-cells undergo increase in size, and in still greater degree after administration of carbohydrates, in connection with which they contain glycogen; while after ingestion of fat they likewise become larger and contain fatty granules, principally at the periphery of the liver-lobules. Irritation of the celiac plexus causes reduction in the size of the cells, with deficiency in glycogen, and it appears to spur them on to secretion.

The experiments of Kallmeyer and Jul. Klein, performed under the direction of Alex. Schmidt, have yielded the interesting result that a paste of fresh, "surviving" liver-cells produces the glycin and the taurin of the biliary acids from a mixture of hemoglobin (or serum) and glycogen (or dextrose) and that addition of soda or 0.6 per cent. sodium-chlorid solution favors this production. In addition to this production, a body resembling urea is formed. It is now established that the source of the latter is to be referred to the liver.

Anthen, under Alex. Schmidt's direction, found that "surviving" liver-cells possess the ability to take up dissolved hemoglobin in their cell-bodies, and, in the presence of glycogen, to transform this into a pigment closely related to the biliary coloring-matter.

The Amount of Bile.—Copemann and Winston found the amount of bile to be from 700 to 800 cu. cm. in twenty-four hours, in a small woman with a biliary fistula, in whom the common bile-duct was completely closed, so that no bile could flow into the intestine; Mayo Robson found the amount to be 862 cu. cm. in a similar case; Paton found it to be as much as 680 grams, with 2.2 per cent. solid matter.

Older estimates are: 533 cu. cm. by v. Wittich; from 453 to 566 grams by Westphalen; 652 cu. cm. by Ranke, in 24 hours. Analogous estimates for animals are, to one kilogram of dog 32 grams (1.2 per cent. solid matter); to one kilogram of rabbit 137 grams (2.5 per cent. solid matter); to one kilogram of guinea-pig 176 grams (5.2 per cent. solids).

The *flow of bile* into the intestine exhibits two maxima during a digestive period, one from the second to the fifth, and the other from the thirteenth to the fifteenth hour after the meal. The cause resides in reflex stimulation of the hepatic vessels, which in consequence become greatly distended with blood.

The *influence of the food* is most striking. The most abundant secretion takes place after free ingestion of meat; on addition of fat or carbohydrates scarcely any more is formed. In a state of hunger the quantity is reduced from one-third to one-half, and even more with a pure fat-diet. The ingestion of water increases the amount, with simultaneous relative reduction in the solid constituents.

The influence of the circulation. The portal vein furnishes especially the material for the production of the bile, and in greater degree than the hepatic artery. The latter is at the same time the nutrient vessel of the tissues of the liver. This is shown by the following observations:

(a) Simultaneous ligation of the hepatic artery (diameter, $5\frac{1}{2}$ mm.) and of the portal vein (diameter, 16 mm.) abolishes the secretion of bile.

(b) If the hepatic artery is ligated, the portal vein alone maintains the secretion. According to Kottmeier, Betz, Cohnheim and Litten, ligation of the artery or of one of its branches is said, further, to result in necrosis of the parts supplied, and possibly of the entire liver, as the artery is the nutrient vessel of this organ. After ligation of the artery the production of urea diminishes greatly; while after ligation of the portal vein this is said to remain almost normal.

(c) If the branch of the portal vein for a lobule of the liver is ligated, only slight secretion takes place in this lobule through the agency of the artery.

Thus neither exclusive ligation of the hepatic artery nor exclusive gradual obliteration of the portal vein (rarely observed as a morbid condition) results in cessation of the secretion. Only diminution in the secretion takes place. The observation that the secretion ceases after sudden ligation of the portal vein (which, besides, is rapidly fatal) is to be explained by the fact that, in addition to the diminution in the secretion, the enormous blood-stasis in the abdominal viscera after this operation makes the liver intensely anemic and therefore unsuited for secretion.

(d) If the blood of the hepatic artery is introduced directly into the lumen of the opened portal vein, ligated peripherally, the secretion continues.

(e) The passage as rapidly as possible of large amounts of blood through the liver acts most favorably upon the secretion. In this connection the prevailing blood-pressure is not of primary importance, for after ligation of the inferior cava above the diaphragm, in consequence of which the highest degree of blood-pressure due to stasis develops, the secretion ceases. The transfusion of considerable quantities of blood always increases the production of bile, although excessive pressure in the portal vein, from the introduction of blood from the carotid artery of another animal restricts the production.

(f) Profuse loss of blood has a tendency to cause cessation of bile-production before the function of the muscular and nervous apparatus is abolished. A more abundant blood-supply to other organs, as, for example, to the muscles of the body engaged in hard labor, diminishes the secretion.

(g) *The influence of the nerves.* All procedures that cause contraction of the arteries of the abdomen, such as irritation of the valve of Vieussens, of the inferior cervical ganglion, the hepatic nerves the splanchnic nerve, the spinal cord, whether directly, as by strychnin, or reflexly, by irritation of the sensory nerves, diminish the secretion. All procedures that induce stagnation of blood in the hepatic vessels, such as division of the splanchnic nerves, diabetic puncture, division of the cervical cord, have a like effect. Paralysis (ligation) of the hepatic nerves is said at first to increase the secretion of bile, with reddening of the liver.

(h) With regard to the raw material brought to the liver by the blood-vessels for the production of bile, the difference in the composition of the blood in the hepatic veins and that in the portal vein is noteworthy. The blood in the hepatic veins contains somewhat more sugar, lecithin, cholesterin, and blood-corpuscles, but, on the contrary, it is deficient in albumin, fibrin, hemoglobin, fat, water and salts. The liver is capable of excreting unchanged in the bile biliary pigments circulating in the blood.

The production of bile is dependent preëminently upon the transformation of the red blood-corpuscles, as they furnish the material for the formation of some of the constituents.

All procedures, therefore, that induce increased destruction of red blood-corpuscles make the liver rich in hemoglobin and, as a result, cause increased production of bile, also pathologically, as, for example, in the presence of malaria and blood-degenerations.

Naturally, a normal condition of the liver-cells is necessary for normal secretion.

For observing the secretion of bile in animals, a biliary fistula is established, the fundus of the gall-bladder being opened somewhat to the right of the xiphoid process, and then being sutured into the abdominal wall, with the aid of a cannula kept constantly open. As a rule, all of the bile will then be discharged externally. If absolute certainty in the latter connection be desired, the common bile-duct should be ligated in two places and divided. Soon after the establishment of a fistula, the secretion of bile diminishes. This is dependent upon the removal of the bile from the body. Introduction of bile in the body from some other source again increases the secretion. Various investigators have been able to observe directly biliary fistulæ developed pathologically in human beings. In dogs regeneration of the divided bile-duct may take place.

EXCRETION OF BILE.

This takes place:

1. Through the constant advance of fresh amounts of bile from the seat of production toward the excretory ducts.

2. Through the periodic compression of the liver by the diaphragm from above, with each inspiration. In addition, each inspiration accelerates the blood-current in the hepatic veins; each respiratory increase in abdominal pressure hastens the blood-current in the portal vein.

Whether the diminution in the secretion of bile following bilateral division of the pneumogastric nerves is to be explained in this manner has been decided in the affirmative. Nevertheless it is to be borne in mind that the pneumogastric nerve sends branches to the hepatic plexus. Whether the excretion of bile is also decreased after paralysis of the phrenic nerves and relaxation of the abdominal pressure is undetermined.

3. By the peristaltic contraction, every fifteen or twenty seconds, of the unstriped muscle-fibers of the large biliary ducts and the gall-bladder, the secretion is forced onward. Stimulation of the region of the spinal cord, from which the motor nerves for these structures are derived (through the splanchnic nerves), for this reason induces acceleration of the discharge, which is later followed by retardation. Under normal circumstances this stimulation appears to be due to reflex action, excited by the entrance of the ingesta into the duodenum, in conjunction with stimulation of the movement of this portion of the intestine.

The movement of the biliary ducts can be in part excited, in part inhibited reflexly by stimulation of the central end of the pneumogastric or of the sciatic nerve. According to Oddi, the common bile-duct is provided with a sphincter at its duodenal orifice, which is affected by reflex influences: gastro-intestinal irritation is believed to cause spastic contraction, which would not be unimportant in the explanation of attacks of jaundice of nervous origin.

4. Direct stimulation of the liver or reflex stimulation of the spinal cord retards the excretion. On the other hand, extirpation of the hepatic plexus, as well as injury to the floor of the fourth ventricle, has no disturbing influence. The splanchnic nerve is the motor nerve of the bile-ducts and the gall-bladder. Stimulation of its central extremity causes relaxation of ducts and bladder, while stimulation of the central end of the pneumogastric nerve causes their contraction, together with relaxation of the sphincter of the duodenal orifice.

21

5. Stasis of bile occurs in the bile-ducts even from relatively slight resistance.

A manometer fastened in the gall-bladder of a guinea-pig balanced a column of water more than 200 mm. high. Up to this pressure, therefore, secretion took place. If this pressure were increased or maintained for an excessively long time, absorption of the water of the bile into the blood took place on the part of the liver, up to about four times the weight of the liver, as a result of which solution of red blood-corpuscles by the bile absorbed took place at the same time, with the passage of hemoglobin into the urine.

Various substances that enter the circulation readily pass over into the bile, particularly the metals, which are also deposited in the hepatic tissue. Further, potassium iodid, bromid, and ferrocyanid, potassium chlorate, arsenic, oil of turpentine, bile injected into the blood (also that from other animals), indigo-carmine and xanthophyllin pass over; less readily, cane-sugar and grape-sugar, sodium salicylate and carbolic acid. Sugar has been found in cases of diabetes, leucin and tyrosin in cases of typhoid fever, altered hemoglobin in the presence of blood-degeneration, lactic acid and albumin under other pathological conditions.

Some substances promote the secretion of bile, olive-oil most intensely; further, oil of turpentine, sodium salicylate, alkalies and laxatives, bile and salts of the biliary acids (particularly from other species of animals), which, after absorption, are again secreted by the liver. Pilocarpin and atropin diminish the secretion. The so-called lymphagogues induce marked secretion of bile in consequence of increased hepatic activity; the increase of lymph, on the part of the liver, is thought to depend upon the latter.

RESORPTION OF BILE.

Symptoms of Jaundice (Icterus; Cholemia).—If an obstruction occurs to the discharge of bile into the intestine,—as, for example, a plug of mucus or a gall-stone occluding the common bile-duct, or a tumor or pressure from without, rendering the duct impervious,—the biliary passages become distended, and, through their distention, cause enlargement of the liver. The pressure in the biliary passages is naturally increased under such conditions. As soon as this pressure has reached a certain point, in the dog up to 275 mm. of a column of the excreted bile—as must soon take place with the continued production of bile—resorption of the bile from the greatly distended bile-ducts of larger size into the lymph-vessels (not into the blood-vessels) of the liver occurs. In this way the biliary acids and the biliary coloring-matter enter the blood. Ligation of the thoracic duct therefore prevents the entrance of the substances into the blood. Also when the pressure within the portal vein is abnormally low, it is thought that bile can pass over into the blood without occlusion of the bile-ducts. This is said to be partly the case in the presence of icterus neonatorum, as blood no longer enters the portal vein from the umbilical vein after the umbilical cord has been tied; further, in the presence of the "hunger-icterus" observed during the state of hunger, as in the stage of inanition, the distribution of the portal vein is relatively empty, on account of deficient absorption from the intestine.

Cholemia may, however, result also from the excessive production of bile (hypercholia), which cannot be completely discharged into the intestine, and thus is resorbed. This takes place when erythrocytes, which furnish the material for the manufacture of the bile, are destroyed in excessive amount. From this material only the liver can elaborate bile. Under such circumstances a plug of inspissated secretion at times forms in the bile-ducts, as a result of which, in consequence of the stagnation of the bile, its resorption is in turn favored. The transfusion of heterogeneous blood acts in this way, in consequence of destruction of the red blood-corpuscles. Therefore icterus is a frequent symptom under such conditions. The author has encountered the same phenomenon after excessive transfusion of blood from the same species, the blood being in part likewise dissolved later. Such a solvent effect upon the erythrocytes is exerted also by the injection of some heterogeneous sera, of salts of the biliary acids, of water, of various acids, as, for example, phosphoric acid, and by the administration of large amounts of chloral, chloroform. and ether. Further, injections of hemoglobin in solution into the blood-stream or into the intestine, from which it is absorbed, have the same effect. (The subject is further considered on p. 341.)

If, as a result of compression of the placenta in the uterus, too much blood has been carried to the new-born infant, a portion of this excess of blood in the body may be dissolved during the first days of life, the hemoglobin being transformed into bilirubin, with symptoms of icterus. Under such circumstances also there is excessive destruction of erythrocytes, as, indeed, of all of the tissues, because in the new-born infant, with insufficient nourishment the metabolic processes must be more active for the maintenance of respiration, heat-production and digestive activity.

The jaundice that is exemplified by the foregoing symptoms is also designated hepatogenic, or resorption-icterus, because it is due to the absorption of bile already prepared in the liver.

Cholemia is accompanied by a series of characteristic symptoms:

1. Biliary coloring-matter and the biliary acids enter into the tissues of the body, giving rise to the most striking objective symptom (and therefore designated also jaundice). The external integument, particularly the sclera, acquires an exquisitely yellow color. In pregnant women the fetus also is discolored. Hematoidin-crystals have been found in the kidneys, the blood and the fatty tissue of icteric children. In exceptionally rare cases, as in the presence of hemiplegia, only one-half of the body has been found jaundiced.

2. The biliary acids and the biliary coloring-matters appear in the urine, though not in the saliva, the tears or in mucus. When the coloring-matter is present in large amount the urine acquires a deep yellowish-brown color, while its foam is intensely lemon-yellow. Immersed strips of paper or linen are stained the same color. Occasionally crystals of bilirubin are present.

3. The feces become clay-colored, because of the absence of hydrobilirubin derived from the bile-pigment; extremely hard, because the diluting bile does not reach the intestine; rich in fat, because the fats, particularly the more solid, are not sufficiently digested in the intestine in the absence of bile (so that even as much as 78 per cent. of the fat ingested passes out in the feces; principally fatty acids and soaps appear in the feces, and but little neutral fats); and highly offensive, because, under normal conditions, the bile poured out into the intestine inhibits putrid decomposition of the intestinal contents. The evacuation of the feces takes place sluggishly, partly on account of their hardness, partly because of the absence of bile, which excites peristaltic movements in the intestines.

4. The heart-beats are reduced to about 40 in the minute. This is due to the salts of the biliary acids, which at first stimulate the heart and then enfeeble it. Injection of the salts of the biliary acids into the heart causes, therefore, at first, transitory increase in the heart-beats, followed by slowing. The same result is brought about if these substances are injected directly into the blood, although under such circumstances the brief stage of stimulation is much less marked. Division of the pneumogastric nerve has no influence on this phenomenon. In addition to the action on the heart, there is marked dilatation of the smallest blood-vessels, slowing of the respiration and lowering of the temperature.

5. An influence on the nervous system, either through the salts of the biliary acids or through the cholesterin accumulated in the blood, perhaps also upon the muscles, is shown by the great general relaxation, fatigue, weakness and somnolence, finally deep coma; at times by insomnia, pruritus, even delirium and convulsions. In experiments on animals Löwit observed symptoms, after injections of bile, indicative of stimulation of the respiratory, cardio-inhibitory and vasomotor centers. Direct application of bile or its salts to the cerebrum causes convulsions.

6. Jaundice of marked degree is attended with yellow vision, in consequence of impregnation of the retina with yellow biliary coloring-matter.

7. The biliary acids in the blood dissolve the erythrocytes, and this leads to the further formation of bile. The dissolved hemoglobin is transformed into new bile-pigment, while the globulin-body of the disintegrated hemoglobin may form casts in the renal tubules, which later are washed into the urine. Should dissolution not take place, the erythrocytes become swollen and exhibit increased solubility.

After ligation of the bile-duct, the protoplasm of the liver-cells disappears, and according to some observers partial necrosis of the hepatic tissue occurs, with secondary reactive inflammation, connective-tissue hyperplasia, cell-multiplication of the epithelial cells of the biliary passages. The stagnating bile diminishes in amount and exhibits further an increase of mucus and cholesterin, but on the other hand a reduction in taurocholic acid (in the dog).

ACTION OF THE BILE.

The bile is a metabolic product largely destined for excretion, and participating in but small measure in the digestive process.

Bile plays an important part in the absorption of fat. It forms a fine emulsion of the neutral fats, in consequence of which the fatty granules, in addition to chemical division, are especially rendered capable of passing through the cylindrical epithelium of the small intestine. It does not effect the chemical decomposition of the neutral fats into glycerin and fatty acids, as does the pancreatic juice, but it is capable of dissolving the fatty acids through the salts of the biliary acids. .

The soaps present in the intestine are soluble in the bile and are capable in turn of greatly increasing the emulsifying power of the bile. The bile itself, however, is capable of converting the fatty acids directly into an acid solution that exerts an active emulsifying influence.

As the bile, like a soap solution, bears a certain relation to aqueous fluids as well as to fats, it may conduce to diffusion between the two, as the membrane can be moistened and can imbibe both fluids.

From the foregoing it follows that the bile is of great importance for the preparation and absorption of fats. This can also be demonstrated by experiments on animals, in which the bile is entirely conveyed externally through a fistula. Dogs thus treated absorb, at the most, 40 per cent. of the fat ingested, while normal dogs absorb 99 per cent. The chyle of such animals is, accordingly, deficient in fat, and is not white, but transparent. The feces, however, contain more fat and are greasy. The animals eat greedily; the tissues of the body show great deficiency of fat, even when the nutrition in general has not suffered much. In human beings suffering from derangement in the secretion of bile, a diet rich in fat is, for this reason, contraindicated.

Fresh bile contains some diastatic ferment, as starch and glycogen are converted into sugar.

This ferment is, however, absorbed from the walls of the alimentary canal and is then excreted as ptyalin by the bile, as by the urine also.

The bile acts as a stimulant to the intestinal musculature and thus contributes to absorption in general.

Perhaps through its biliary acids, acting as irritants, it causes the muscles of the intestinal villi to contract from time to time, in consequence of which these propel the contents of their lymph-spaces into the larger lymph-trunks, and thus are capable of absorbing renewed amounts.

Also the musculature of the intestinal wall itself appears to undergo excitation, probably through the agency of the myenteric plexus. In favor of this view is the fact that intestinal peristalsis is greatly impaired in animals with biliary fistulæ and in the presence of obstruction of the biliary passages, as well as the fact that the salts of the biliary acids, administered by the mouth, cause diarrhea and vomiting. As, however, intestinal contractions aid absorption, the bile is, in this connection also, active in taking up the dissolved food.

The presence of bile is necessary for the normal vital activity of the intestinal epithelium in the absorption of the fatty globules.

Through its excretion the bile supplies a sufficient amount of water for the feces. Animals with biliary fistulæ and human beings with obstructed biliary passages are markedly constipated. Besides, the slippery mucus of the bile facilitates the advance of the ingesta through the intestinal canal.

The bile diminishes putrefactive decomposition of the intestinal contents, especially with a fatty diet.

On the entrance of the strongly acid gastric contents into the duodenum, the glycocholic acid is precipitated by the acid of the stomach and carries the pepsin with it. Further, the albumin and the gelatin, still in solution, but not the peptones and propeptones, are precipitated by the taurocholic acid, salts of the biliary acids having already been decomposed by the acid of the stomach. If, however, the mixture is again rendered alkaline by the pancreatic and the intestinal juice and the alkali of the bases derived from the salts of the biliary acids, the pancreatic ferments enter energetically into action.

If bile enters the stomach, as, for instance, in the act of vomiting, the acid of the gastric juice combines with the bases of the salts of the biliary acids. There thus results principally sodium chlorid and free biliary acids. At the same time the acid reaction is diminished. The biliary acids are not effective as acids in gastric digestion, in place of the combined hydrochloric acid, the neutralization causing also precipitation of the pepsin and the mucin. As soon, however, as the wall of the stomach secretes additional acid, the pepsin is again dissolved. The bile entering the stomach has a disturbing effect on gastric digestion also, by causing contraction of the albuminates, as these can be peptonized only when swollen.

FINAL FATE OF THE BILE IN THE INTESTINAL CANAL.

Of the constituents of bile, some are evacuated with the feces, while others are again absorbed through the intestinal walls.

The mucin passes into the feces unchanged.

The biliary coloring-matters are mostly reduced in the large intestine and are partly evacuated with the feces as hydrobilirubin; a small portion of them is absorbed and finds its way into the urine as urobilin. The reduction may proceed beyond the formation of hydrobilirubin to that of a colorless material, which may, however, upon admission of oxygen, be again oxidized to hydrobilirubin.

Hydrobilirubin is absent from meconium, but bilirubin and biliverdin are present together with an unknown red oxidation-product derived from them. Therefore the process that takes place in the fetal intestine is not a reducing but an oxidizing one.

Cholesterin is in part evacuated with the feces; in part it is reduced to the form of hydrocholesterin (coprosterin), crystallizing in needles.

The biliary acids are, for the most part, again absorbed through the walls of the jejunum and the ileum, and are utilized anew in the production of bile. Tappeiner found them in the thoracic duct; small amounts find their way from the blood into the urine. Only a small portion of glycocholic acid appears unchanged in the feces. Taurocholic acid, in so far as it is not absorbed, is readily decomposed in the intestine by putrefactive processes into cholic acid and taurin. The former is found in the feces, the latter is not infrequently absent. Cholic acid is, however, in part resorbed and may again unite in the liver with glycin or taurin.

As putrefactive decomposition is absent from the fetal intestine, unchanged taurocholic acid is accordingly present in the meconium. Glycocholic acid, when administered, is found again in the bile from animals (dog) which normally excrete but little thereof.

The feces certainly contain merely traces of lecithin.

As, therefore, the largest part of the biliary acids is returned to the blood, it is clear that animals from which all the bile is lost through a biliary fistula, without their licking it up, lose considerably in weight. This is due partly to the impaired digestion of fat, in part to the direct loss of the biliary acids. If dogs are nevertheless to maintain the same weight, they must consume almost double their former nourishment. Under such conditions, carbohydrates are especially serviceable as a substitute for fat in the diet. If their digestive apparatus is in other respects intact, the animals may, by reason of their voracity, even gain in weight. Under such circumstances, however, it is the muscles almost alone and not the fat that is increased.

The fact that bile is secreted during fetal life, while none of the other digestive fluids are produced, indicates that the bile is in part a product of retrogressive tissue-metamorphosis, and is intended for the constant elimination of certain excrementitious matters.

The cholic acid, which is absorbed through the intestinal wall, is finally burned up in the body into carbon dioxid and water. The glycin gives rise to the production of urea, as well as hippuric acid, as, after the ingestion of that substance, the amount of urea is greatly increased. The fate of the taurin is not known. Considerable amounts administered to human beings by the stomach appear again in the urine principally as taurocarbamic acid, together with a small amount of unchanged taurin. When injected subcutaneously into a rabbit, it almost all appears in the urine.

THE INTESTINAL JUICE.

The human intestine is ten times as long as the length of the body from the vertex to the anus. In this it resembles that of fructivorous apes. It is relatively longer than that of omnivora. Its minimum length is 507 cm.; its maximum length, 1149 cm. Its capacity is relatively greatest in children, in whom also it is relatively longer. The intestine is somewhat longer in males than in females.

The *intestinal juice* is the digestive fluid secreted by the numerous glands of the intestinal mucous membrane. The largest amount is furnished by Lieberkühn's glands; the duodenum receives, besides, the scanty secretion of the compound alveolar grands of Brunner.

Brunner's glands, which occur singly in human beings, but in the sheep constitute a continuous layer in the duodenum, are present, in part, at the pylorus. Their cylindrical cells have a middle, darker zone; the flat nucleus lies near the base of the cell, with a diplosome nearer its free surface. During the state of hunger, the cells are turbid and small, and, like the pyloric glands of the stomach, they contain fatty granules, while during digestive activity they are large and clear. The glands contain nerve-filaments from Meissner's plexus in the mucous membrane.

The Secretion of Brunner's Glands.—The usually granular contents of the secretory cells consist, in addition to albuminous materials, of mucin and ferment-substances of unknown nature. It is not improbable that these glands are related to the pancreas, and perhaps are even to be regarded as detached portions of the pancreas. Their activity seems to favor this view. An aqueous extract (1) dissolves albumin slowly and feebly, at the temperature of the body. (2) It possesses diastatic activity. The secretion appears to have no effect on fats.

It should be especially emphasized that, as on account of the small size of the glands they must be viewed individually, with a magnifying glass, from the under surface of the intestinal mucous membrane, digestive experiments are exceedingly difficult.

Lieberkühn's crypts or glands are simple tubular glands, resembling the finger of a glove, that lie close to one another in the intestinal mucous membrane, and in greatest number in that of the large intestine (on account of the absence of villi). They possess a membrana propria, constituted of most delicate fibrils, and a single layer of cylindrical protoplasmic cells, between which goblet-cells also occur, in small number in the small, and in large number in the large intestine. The glands in the small intestine yield a watery secretion principally; those of the large intestine. from their numerous goblet-cells, ropy mucus. Both kinds of gland-cells multiply by mitosis, and the new products move from situations where active division is going on to places where the production is less active. The mucus in the goblet-cells encloses usually a single central body.

The secretion of Lieberkühn's glands is, from the duodenum downward, the chief source of the intestinal juice.

The intestinal juice is obtained, by Thiry's method. in the following manner: From a loop of the intestine of a dog. withdrawn from the abdomen, a piece of the length of a hand is so divided by two incisions that only the continuity of the intestinal canal is severed but not the mesentery. Then one end of this piece is ligated; the other, left open, is sutured in the abdominal wound, after the ends of the intestine, between which the piece has been removed, have been carefully united by suture. Vella permits both ends of this horseshoe-shaped portion of intestine to open on the abdominal wall. In this way, after the operation has been completed, the animal can continue to live with its but slightly abbreviated intestine. The intestinal fistula, with a free external opening, yields, however, an intestinal juice that is not contaminated by any other digestive secretion.

The intestinal juice derived from such a fistula flows spontaneously in very small amount; during digestion it is largely increased. Mechanical, chemical and electrical stimulation increase the secretion, especially of mucus, with reddening of the mucous membrane, so that 100 square centimeters yield from 13 to 18 grams of juice in an hour. The administration of pilocarpin also increases the secretion. The juice is light yellow in color, opalescent, watery, strongly alkaline,

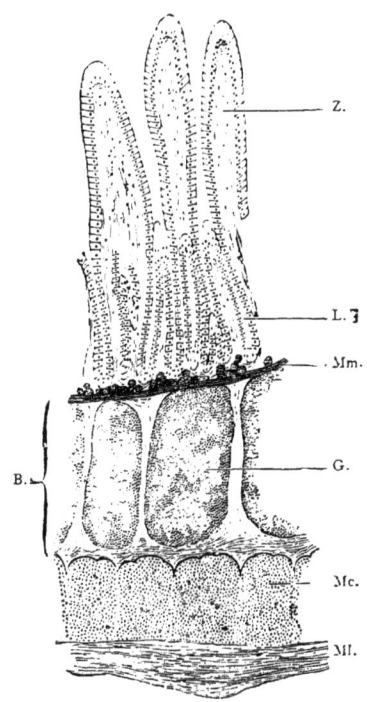

Fig. 120.—Longitudinal Section through the Small Intestine of a Dog: B, connective-tissue layer; Z, intestinal villi covered with cylindrical epithelium; L, Lieberkühn's glands; Mm, muscularis mucosæ; G, crowded lymph-follicles; Mc, circular muscular layer; Ml, longitudinal muscular layer.

effervescing on addition of acids, and has a specific gravity of 1010. It contains, in human beings, proteid (0.80 per cent.), ferments, mucin, especially in the large intestine (0.73 per cent.), and salts (0.88 per cent.), of which 0.34 per cent. is soda and 0.5 per cent. sodium chlorid.

The amount of intestinal juice secreted is least with the presence of dissolved grape-sugar in the intestines, greater with the presence of cane-sugar, and still greater with the presence of starch and peptone. It increases in the second hour of digestion.

Biedermann found the production of mucus in the goblet-cells of the intestine, in the frog, to take place in such a manner that droplets of mucus first appear in the cell-contents. These enlarge into vacuoles, which soon become confluent; then the mucus escapes from these and is discharged from the cell.

The digestive activity of the juice of the small intestine is still in many respects unexplained. The juice has been found most active in the dog, while it is more or less inactive in other animals.

It possesses less diastatic activity than the saliva and the pancreatic juice. It forms maltose, which rapidly passes over into dextrose. The glands of the large intestine are said to be wanting in this property. von Wittich has extracted the ferment by means of glycerin diluted with water.

The intestinal juice is capable of transforming maltose into grape-sugar. It, therefore, continues the diastatic action of the saliva and the pancreatic juice, which principally are active only up to the production of maltose.

According to Bourquelot, this action is due to intestinal bacteria, and not to the intestinal juice as such, nor to the saliva, the gastric juice or the invertin. The larger part of the maltose, however, seems to undergo absorption unchanged.

No action upon proteids is recognizable, or, at least, only traces. The peptonizing properties described are in part dependent upon putrefactive processes. According to earlier statements, fibrin is slowly peptonized by trypsin and pepsin; albumin, fresh casein, raw or cooked meat and vegetable albumin less readily. Gelatin is probably also brought into solution by a special ferment.

FIG. 121.—Transverse Section through Lieberkühn's Glands: H, cavity of the glandular tubule; D, glandular epithelium; B, connective tissue; G, blood-vessels.

The intestinal juice is capable of acting on fat, which it partially emulsifies in the presence of free acid. Whether the neutral fats are also decomposed, in small measure, has not as yet been determined with certainty.

The intestinal juice contains invertin, an unorganized ferment, which decomposes disaccharids (cane-sugar, milk-sugar and maltose) into monosaccharids (dextrose, levulose and galactose), with the taking up of water and the production of heat:

$$\underset{\text{Cane-sugar}}{C_{12}H_{22}O_{11}} \; \dotplus \; \underset{\text{Water}}{H_2O} \; = \; \underset{\text{Dextrose}}{C_6H_{12}O_6} \; + \; \underset{\text{Levulose.}}{C_6H_{12}O_6}$$

Milk (casein) is coagulated.

With regard to the ferments of the alimentary canal, Langley upholds the view that they undergo destruction: the diastatic ferment of the saliva is destroyed by the hydrochloric acid of the gastric juice; pepsin and the rennet-ferment succumb to the action of the alkaline salts of the pancreatic and intestinal juices and the trypsin; the diastatic and peptic ferments of the pancreas are rendered inert by the acid fermentation in the large intestine. Nevertheless some ferment is absorbed and passes over into the urine.

Of the *influence of the nerves* upon the secretion of the intestinal juice but little has been ascertained with certainty. Stimulation or division of the pneumogastric nerves is without apparent effect. On the other hand, destruction of the nerve-filaments passing to the intestinal loops and accompanying the blood-vessels is followed by distention of the intestinal canal with an abundance of watery fluid. This result is explained in part by paralysis of the vasomotor nerves of the intestinal tract. As the nerve-filaments for a limited portion of intestine, ligated in two places, can be completely separated, the watery intestinal contents will be found only in the corresponding loop of intestine. According to Hanau, the condition in this experiment of Moreau is one of paralytic secretion, which, with regard to time, pursues a typical course.

The following substances are after ingestion excreted by the intestinal mucous membrane of isolated fistulæ: iodin, bromin, lithium, metallic ferrocyanogen, salts of iron and others.

FERMENTATIVE PROCESSES IN THE INTESTINES DUE TO MICROBES; INTESTINAL GASES.

Wholly different from the peculiar digestive processes just described, which are brought about by definite unorganized ferments or enzymes, are those processes which are to be considered as fermentative or putrefactive decompositions. These are caused by microbes, the so-called excitants of fermentation or putrefaction, or organized ferments; and they may, therefore, take place outside of the body, in suitable media. Lower forms of organisms, which maintain fermentative processes in the intestinal tract, are often swallowed with food and drink, as well as with the buccal fluid. Upon the introduction of these the processes of decomposition begin, with simultaneous production of gas. On a pure milk-diet intestinal putrefaction is much less marked.

Fermentation, therefore, cannot occur in the intestine during fetal life. For this reason gases are always absent in the intestine of the new-born. The first bubbles of air reach the intestine through frothy saliva swallowed, even before food is taken. As, however, micro-organisms may enter the intestinal tract with the air swallowed, the development of gas by fermentation must soon follow. The development of the intestinal gases thus goes hand in hand with the fermentative processes. As, however, gases from the air swallowed are exchanged in the intestinal canal, the composition of the intestinal gases will be found to be dependent upon various factors.

Kolbe and Ruge collected intestinal gases from the human anus and found in 100 volumes:

Food.	CO_2.	H.	CH_4.	N.	H_2S.
Milk	16.8	43.3	0.9	38.3	Amount unknown.
Meat	12.4	2.1	27.5	57.8	
Peas	21.0	4.0	55.9	18.9	

Moreover, it should be noted: 1. That oxygen is rapidly absorbed by the walls of the canal from the air-bubbles swallowed with the food, so that, in the lower part of the large intestine, even traces of oxygen are absent. In exchange the blood-vessels of the intestinal wall give up into the intestine carbon dioxid, so that, therefore, a portion of the carbon dioxid in the intestines is derived from the blood by diffusion.

2. Hydrogen, carbon dioxid and ammonia, as well as marsh-gas, are also developed from the intestinal contents by fermentation, which may take place even in the small intestine.

Bacteria as Excitants of Fermentation. The organisms that especially cause fermentation, putrefaction and other forms of decomposition are bacteria (schizomycetes), namely, minute, unicellular structures, chiefly having the shape of a sphere (micrococcus), or a short rod (bacterium), or a long rod (bacillus), or a spiral thread (vibrio, spirillum, spirochæta). Their power of reproduction is beyond all conception. Through their vital phenomena they cause profound

chemical changes in the matters containing them.   As for their growth and
metabolism, they abstract certain substances from the nutritive fluid in which
they live, they decompose the chemicals contained therein.   In this process
some of them form certain substances that may subsequently act as ferments
upon matters in the nutritive fluid.

The microbes are destroyed by antiseptics, such as carbolic acid, salicylic
acid, etc., although the ferments are not destroyed.   Therefore, these substances
afford a means of distinguishing and separating the fermentative from the micro-
biotic decompositions.

The bacteria consist of a capsule and protoplasmic contents.   Some possess
flagella as organs of locomotion, which, perhaps, are possessed by all capable
of motion.   The organisms multiplying by division are sometimes collected
together in extensive colonies, united by a gelatinous mass, often visible to the
naked eye, and which are designated zoogleæ.   These appear in the form of
nodules, branches, patches, flakes, layers of mold, or ropy, creamy or greasy
deposits.   In the case of some micro-organisms, principally bacteria, multiplication takes place by spore-formation, especially if the nutritive fluid becomes
deficient in nutrient material.   The rods then grow into threads of considerable
size, which become jointed; and globular, strongly refractive granules, from 1 to 2 μ
in size, develop in the individual parts (Fig. 122, 8, 9).   In the case of some, as

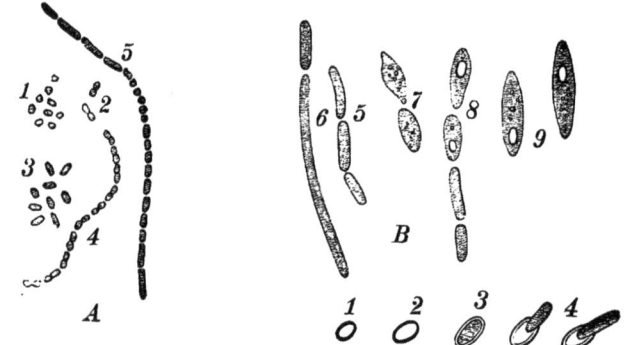

Fig. 122.—A, bacterium aceti, in the form of cocci (1), diplococci (2), short bacilli (3), and jointed threads (4, 5).
B, bacillus butyricus: 1, isolated spore; 2, 3, 4, germinating stage of the spore; 5, 6, short and long bacilli;
7, 8, 9, spore-formation in the bacteria.

the butyric-acid germ, the bacilli, before spore-formation, acquire the shape of
an enlarged spindle, within which the spores form.   After death of the mother-
cells, the spores become free, and from them, transplanted to a suitable soil,
the newly formed cells of the microbes again germinate.   The processes of spore-
formation (7, 8, 9) and of germination of the butyric-acid micro-organism (1, 2,
3, 4) are illustrated in Fig. 122, B.   The spores are extremely resistant, being
capable, in the dry state, of surviving for a long time, and some even withstanding
boiling.

Among bacteria, a distinction is made between those that exhibit their vital
activity in the presence of oxygen, aerobes, and others that thrive only when
oxygen is excluded, anaerobes.   In accordance with the products to which they
give rise by decomposition in their nutrient media, they may be divided into
those that induce decomposition in the form of fermentations (zymogenic schizomycetes), those that form pigments (chromogenic), those that generate bad
odors, as in the putrefactive processes (bromogenic), and, finally, those that,
developing in the living tissues of other organisms, cause morbid conditions,
even death itself (pathogenic).   Some also elaborate poisons (toxicogenic).   All
of these have been found in and upon the human body.

If it be borne in mind that a large number of bacteria are introduced into
the alimentary canal with foods and drinks, as well as, in part, also with the inspired air; that, further, the temperature of the intestine is especially favorable

to their development; and, finally, that sufficient material of the most varied kind, not entirely disposed of by the digestive processes, furnishes nutrient matter for the vegetation of the germs, it is not surprising that a rich formation of these organisms is found in the alimentary canal and that they cause numerous forms of decomposition in the intestinal contents. Knowledge of these processes is, at the present time, still highly deficient; and the formulæ proposed for the decompositions can, therefore, only approximately explain the processes. For this reason, the following statements can only be considered provisionally as aphorisms in the study of the mycotic intestinal decompositions.

Fermentation of Carbohydrates, which takes place principally in the small intestine. 1. Bacillus acidi lactici (bacterium lacticum), whose biscuit-shaped cells, from 1.5 to 3 μ in length, are arranged in groups or rows or are isolated, causes fermentative decomposition of sugar into inactive lactic acid:

$$C_6H_{12}O_6 \quad = \quad 2(C_3H_6O_3)$$
$$\text{1 Grape-sugar} \quad = \quad \text{2 Lactic acid.}$$

Milk-sugar ($C_{12}H_{22}O_{11}$) may be decomposed by the same bacterium, with the addition of water, first into two molecules of grape-sugar, $2(C_6H_{12}O_6)$, and this in turn into four molecules of lactic acid, $4(C_3H_6O_3)$.

This micro-organism, whose germs float in the air everywhere, causes the spontaneous souring and curdling of milk. It develops further in sour-crout, sour pickles, and the like. It induces fermentation of cane-sugar, mannite, inosite, and sorbite, as of the sugars mentioned. In addition to lactic acid, carbon dioxid also results. There are, besides, other lactic-acid-producing bacteria that are capable further of transforming starch into sugar. van de Velde obtained lactic, butyric and succinic acids as products of the fermentative activity of the bacillus subtilis (Fig. 123), and mannite as a reduction-product.

2. Bacillus butyricus, which is often stained blue by iodin in a starch-containing medium, transforms lactic acid into butyric acid, together with carbon dioxid and hydrogen.

$$2(C_3H_6O_3) \quad = \quad C_4H_8O_2 \quad + \quad 2CO_2 \quad + \quad 4H.$$
$$\text{2 Lactic Acid} \quad = \quad \text{1 Butyric Acid} \quad + \quad \text{2 Carbon Dioxid} \quad + \quad \text{4 Hydrogen.}$$

This bacterium (Fig. 122, B) is a true anaerobe, which vegetates only in the absence of oxygen. The lactic-acid bacillus, which actively consumes oxygen, is therefore its natural predecessor. Butyric-acid fermentation completes the transformation of many carbohydrates, chiefly starch, dextrin and inulin. It takes place constantly in the feces. There are a number of other bacteria with similar activity. The butyric-acid bacillus produces also dextrin from starch.

3. Certain micrococci are capable of developing alcohol as the chief product from sugar.

In the human small intestine there are present besides: bacterium Bischleri (short rods), which produces alcohol, inactive lactic acid and acetic acid from sugar; bacterium ilei (short rods), which transforms sugar into alcohol, succinic acid and some active paralactic acid, together with carbon dioxid and hydrogen; bacterium ovale ilei (almost spherical), which transforms sugar into alcohol, paralactic acid and traces of the fatty acids; bacillus gracilis ilei (delicate long rods), which has a similar action; bacterium lactis aerogenes, which transforms sugar into alcohol and succinic acid, together with lactic acid and some acetic acid.

The presence of yeast also may result in the production of alcohol in the intestine, in both instances likewise from milk-sugar, which at first passes over into dextrose. Only traces are found in the intestine.

4. Bacterium aceti (Fig. 122, A) is capable, outside of the body, of transforming alcohol into acetic acid.

$$C_2H_6O \quad + \quad O \quad = \quad C_2H_4O \quad + \quad H_2O$$
$$\text{Alcohol} \quad + \quad \text{Oxygen} \quad = \quad \text{Aldehyd} \quad + \quad \text{Water.}$$

Aldehyd is changed by oxidation into acetic acid ($C_2H_4O_2$). According to Nägeli, the same micro-organism is capable of producing small amounts of carbon dioxid and water. As acetic fermentation ceases at 35° C., it will not take place in the intestine, so that the acetic acid constantly met with in the feces must result from other fermentative processes. Thus, it is produced in considerable amount in herbivora as a product of the fermentation of cellulose; being, after absorption, burned up in the fluids of the body. Acetic acid is formed also as a result of the putrefaction of albuminates with exclusion of air.

5. Also partial solution of starch and of cellulose is caused by schizomycetes (bacillus butyricus, bacterium termo, vibrio rugula) in the intestines; for cellulose, mixed with cloacal discharge or the intestinal contents, is transformed into a sugar-like carbohydrate, which then breaks up into equal volumes of carbon dioxid and marsh-gas. The neurin produced by the pancreas also yields marsh-gas (CH_4), in addition to carbon dioxid.

The solution of the cellulose of the cell-walls then permits the action of the digestive juices upon the enclosed digestible portions of the

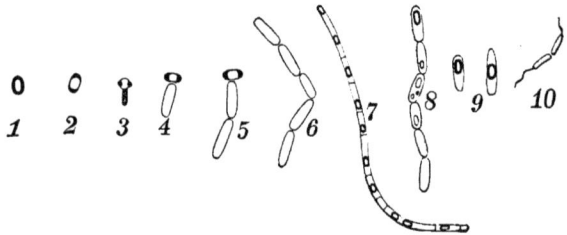

FIG. 123.—Hay-bacillus (Bacillus subtilis): 1, spore; 2, 3, 4, germination of the spore; 5, 6, short bacilli; 7, jointed filament with spore-formation in each cell; 8, short bacilli, in part with spore-formation; 9, spores in individual short bacilli; 10, bacteria with flagella.

vegetable food. In human beings the metabolism of cellulose is always slight, while in herbivora it is digested in considerable amounts.

6. Bacillus subtilis, cheese-spirilli and others are capable of transforming starch into sugar.

7. Micro-organisms (lactic-acid bacilli?) that produce invertin also occur in the intestinal canal. This substance can be obtained also from brewer's yeast by agitation with water and ether and subsequent filtration.

Fermentation of Fats. Putrefaction is capable, with the aid of as yet unknown micro-organisms, of decomposing neutral fats into glycerin and fatty acids, after taking up water. Glycerin is susceptible of varied fermentations with different microbes, as, for example, the bacillus Fitzianus. When the reaction is neutral, hydrogen and carbon dioxid are formed, together with succinic acid and a mixture of fatty acids.

Fitz observed alcohol, together with caproic, butyric and acetic acids, develop as a result of the action of the hay-bacillus (bacillus subtilis, Fig. 123), while in other cases butyl-alcohol principally resulted. van de Velde found butyric and lactic acids, together with traces of succinic acid, and also carbon dioxid, water and nitrogen.

The fatty acids yield, chiefly as calcium-soaps, material suitable for fermentation. Calcium formate, in fermentation with cloacal discharge, yields calcium carbonate, carbon dioxid and hydrogen; calcium acetate yields calcium carbonate, carbon dioxid and marsh-gas. Of the oxyacids, the fermentation of lactic, glyceric, malic, tartaric and citric acids is known.

According to Fitz, lactic acid, in combination with calcium, yields propionic acid, acetic acid, carbon dioxid and water. Valerianic acid in considerable amount is produced by other excitants of fermentation. Glyceric acid yields especially acetic acid, in addition to alcohol and succinic acid. Malic acid forms succinic acid and some acetic acid; as a result of other fermentative processes, propionic acid, and of still other fermentative processes, butyric acid, together with hydrogen; or it is decomposed into lactic acid and carbon dioxid. Tartaric acid breaks up into acetic acid, propionic acid, carbon dioxid and water; as a result of the action of other microbes, into butyric acid; and of that of still others, into acetic acid, together with some butyric and succinic acids and alcohol. Citric acid yields finally acetic, with some butyric and succinic acids.

Fermentations of Proteids. In the fermentation of the undigested proteids in the intestine and their derivatives, which takes place principally in the large intestine, micro-organisms likewise appear to take part. In the first place it should be emphasized that some schizomycetes are capable of producing peptonizing ferment, as, for example, the bacillus subtilis, bacillus liquefaciens ilei, the cheese-spirilli, the micro-organisms of pickled herring, etc., so that assistance to the peptic enzyme, even though slight, on the part of these microbes appears to be not wholly excluded.

It has been found that pancreatic digestion of albuminates does not proceed beyond the production of amido-acids: leucin, tyrosin and others. Putrefactive fermentation in the large intestine causes still further and more profound decompositions. Leucin ($C_6H_{13}NO_2$), by taking up two molecules of water, forms valerianic acid ($C_5H_{10}O_2$), ammonia, carbon dioxid and four molecules of hydrogen. Glycin behaves in a similar manner. Tyrosin ($C_9H_{11}NO_3$) breaks up into indol (C_8H_7N), which is constantly encountered in the intestine, together with carbon dioxid, water and hydrogen. If the admission of oxygen is possible, still other decompositions take place. These products of putrefaction are wanting in the intestine of the fetus and the new-born. In the putrefactive decompositions of proteids, as well as upon boiling them with alkalies, carbon dioxid and hydrogen sulphid develop, together with hydrogen and marsh-gas. Under such circumstances, gelatin yields, in addition to abundant leucin, much ammonia, carbon dioxid, acetic, butyric and valerianic acids and glycin. Mucin and nuclein undergo no decomposition. Artificial digestive experiments with the pancreas disclose an extraordinary tendency to putrefactive decomposition.

The body giving rise to the fecal odor, which likewise results from putrefaction, has not as yet been discovered. It is intimately related to indol and skatol, but these are odorless when prepared in the pure state.

Among the solid matters in the large intestine produced only by putrefaction, *indol* (C_8H_7N) is especially to be pointed out. This is a substance that results also from heating albuminates with alkalies, or in small amount by superheating them with water to 200° C. It is the forerunner of indican in the urine. If the products of the digestion of albuminates, the peptones, are rapidly absorbed in the intestine, only a small amount of indol is formed.   If, on the other hand, with a lesser degree of absorption, the putrefactive process can exert a profound effect chiefly upon the products of pancreatic digestion still present in large amount, considerable indol will be formed, and much indican subsequently appears in the urine.

Thus Jaffé found an abundance of indican in the urine in the presence of incarcerated hernia and obstruction of the bowel.   After transfusion with heterogeneous blood, in connection with which the walls of the intestine are often the seat of extravasation of blood and thrombosis, and paralytic conditions of the intestinal vessels and musculature itself are not rarely encountered, the author has often found the amount of indican contained in the urine to be large.

Test for indol: The fluid to be tested is acidulated with considerable hydrochloric acid and is well shaken after addition of a few drops of oleoresin of turpentine. If an intense red color results, the pigment is removed by agitation with ether. The pigment resulting from fibrin in the process of tryptic digestion, and becoming violet with bromin-water, can be isolated by agitation with chloroform.   In addition to the latter pigment, there is still a second pigment that passes over in the process of distillation, and can be extracted from the distillate by ether. Both appear to belong to the indigo-group.

A. v. Bayer was able to produce indigo-blue artificially from orthonitrophenol-propionic acid by boiling with dilute sodium hydrate and after addition of some grape-sugar.   From indigo-blue he obtained skatol, in addition to indol.   G. Hoppe-Seyler observed an abundance of indican in the urine after feeding rabbits upon sodium orthonitrophenol-propionate.

Further, some *phenol* (C_6H_6O) is formed in the intestine by the putrefactive process.   Baumann observed the same substance as a result of the putrefaction of fibrin with pancreas outside of the body, and Brieger found it constantly in the feces.   It appears to undergo an increase under conditions analogous to those attending an increase in the amount of indol, as an increase in the amount of indican in the urine is accompanied by an increase in the amount of phenyl-sulphuric acid.

Amidophenyl-propionic acid also can be obtained from putrefying meat and fibrin as a product of the decomposition of tyrosin.   Part of this is changed by putrefactive ferments into phenylpropionic acid (hydrocinnamic acid), which is completely oxidized in the organism to benzoic acid, and appears in the urine as hippuric acid.   In this way is explained the formation of hippuric acid when a pure proteid diet is taken.

Skatol (C_9H_9N, methylindol), a constant constituent of human feces, has been prepared artificially by Nencki and Secretan by protracted putrefaction of egg-albumin under water.   In this way results skatol-carbonic acid, which, when heated, readily decomposes into skatol and carbon dioxid.   Skatol also appears in the urine in combination with sulphuric acid.

Milk inhibits the decomposition of albumin and intestinal putrefaction through the presence of casein and thus also diminishes the amount of ethereal sulphates in the urine.

According to the brothers Salkowski, both skatol and indol result from a common substance preformed in albumin, which, when decomposed, at one time yields a larger amount of indol, and at another time a larger amount of skatol, accordingly as to whether the hypothetical indol-bacterium or the skatol-bacterium active under such conditions prevails in the development.

It is of great importance in the process of putrefactive fermentation whether this takes place with the exclusion of oxygen or not.   In the former case reduction occurs: oxy-acids are reduced to fatty acids, and there are developed, especially hydrogen, but also marsh-gas and hydrogen sulphid; the hydrogen, in turn, may cause further reduction.   If, however, oxygen is still present, the nascent hydrogen divides the molecule of ordinary free oxygen into two atoms of active oxygen; there forms, thus, on the one hand, water, and on the other hand, the second atom of oxygen brings about active oxidation.

The remarkable fact should yet be mentioned here that the putrefactive processes, after the development of phenol, indol, and skatol, and also of cresol, phenyl-propionic and phenylacetic acids, are again inhibited, and after a certain concentration in their production cease completely. Thus, the putrefactive process itself generates antiseptic substances even to the point of causing the death of the micro-organisms; for, as with highly organized beings, the excrementitious products of the bacteria themselves are poisons for them. It is, therefore, to be inferred that, in the intestinal canal also, the formation of the substances mentioned in turn inhibits the putrefactive decompositions to some extent. Ptomains are not formed normally in the intestines.

The *reaction* of the contents of the small intestine is alkaline, due principally to carbonates, and in less degree to phosphates. The contents are, however, rich in carbon dioxid, the presence of which causes, on one hand, the acid reaction of the indicators reacting to carbon dioxid, while, on the other hand, it ensures the maximum efficiency on the part of the ferments in the intestine. In the large intestine the reaction is generally acid, in consequence of the acid fermentation and decomposition of the ingesta and the feces.

PROCESSES IN THE LARGE INTESTINE. FORMATION OF THE FECES.

Within the large intestine the putrefactive and fermentative decompositions of the ingesta greatly exceed the fermentative or true digestive transformations, as only small amounts of the ferments of the intestinal juice are found in it. In addition, the absorptive activity of the walls of the large intestine is greater than the secretory activity, whence the consistency of the contents, which at the commencement of the large intestine are still semi-liquid, but become more consistent in the further course of the intestine. The absorption includes not only the water and the products of digestion in solution, but also, under certain circumstances, even unchanged fluid proteids. Also toxic substances are decidedly more readily absorbed here than from the stomach. The feces begin to be formed only in the lower portion of the large intestine. The cecum in some animals, as, for example, the rabbit, is of considerable size; fermentative decompositions appear to take place in it with great activity, with the development of an acid reaction. In human beings the cecum is principally an organ of absorption, as the abundance of lymphatic follicles indicates. From the lower portion of the small intestine and from the cecum onward, the ingesta acquire the fecal odor.

Observations on Thiry's intestinal fistulæ permit the conclusion that a considerable portion of the feces is derived from the secretion of the mucous membrane and from epithelial desquamation.

The *amount* of feces evacuated equals, on an average, 170 grams in twenty-four hours (from 60 to 250 grams), although, when large amounts of food, especially if difficult of digestion, are taken, even more than 500 grams may be discharged. After a diet of animal food the amounts of feces and of solid residue therein are less than after a vegetable diet. The consistent feces are broken up by the development of gas, and therefore float on water.

The *consistency* of the feces depends on the amount of water contained in them, which usually reaches 75 per cent. A pure meat-diet causes rather dry feces; food rich in sugar, rather watery feces; while the amount of fluid ingested is without influence. The more rapidly peristalsis takes place, however, the more watery are the feces, because there is not sufficient time for the absorption of fluid from the rapidly advancing ingesta.   Paralysis of the intestinal blood-vessels and lymph-vessels, after transection of the nerves, is likewise accompanied by liquefaction of the feces.

The *reaction* of the feces is often acid, particularly in consequence of lactic-acid fermentation of large amounts of carbohydrates ingested. Numerous other acids generated by fermentation are also present. If, however, considerable amounts of ammonia are produced in the lower portion of the intestine, a neutral and even an alkaline reaction may preponderate.   The secretion of large amounts of mucus in the intestine favors a neutral reaction.

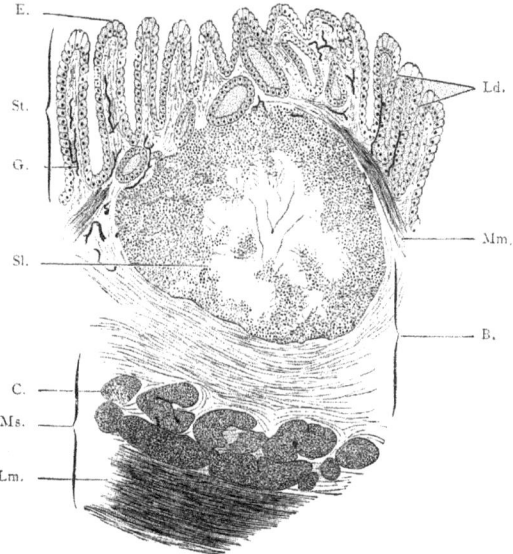

Fig. 124.—Longitudinal Section through the Large Intestine: E, epithelium; St, mucous membrane; G, blood-capillaries; Sl, solitary follicles; C, circular muscular layer; Ms, muscular layers; Lm, longitudinal muscular layer; Ld, Lieberkühn's glands; Mm, muscularis mucosæ; B, connective tissue.

The *odor* of the feces, which is more pronounced with a meat-diet than with a vegetable diet, is dependent upon the fecal-smelling products of putrefaction not yet prepared in an isolated state; further upon the volatile fatty acids, as well as upon traces of methylmercaptan.   The last-named substance can be prepared from proteid by means of fused potassium hydroxid, and it develops in traces on boiling varieties of cabbage, and it is also formed from hydrogen sulphid (as from eggs).

The *color* of the feces varies in accordance with the amount of altered biliary pigment present, hence shades vary from light yellow to dark brown.

In addition, the color of the food has considerable effect. Thus the presence of much blood in the food renders the feces almost brownish black, from hematin; green vegetables render them brownish green, from chlorophyll; bones, in dogs, render the feces white, from the calcium contained; bluish-red vegetable juices render them bluish black; iron-preparations stain them black in part, from the production of iron sulphid.

The feces contain (Fig. 125):

1. The secreted juice of the intestinal mucous membrane, together with desquamated and digested epithelial cells. After almost complete absorption of the digested food, the feces still contain from 8 to 9 per cent. of nitrogen, from 12 to 18 per cent. of ethereal extract and from 11 to 15 per cent. of ash. Certain articles of food stimulate these excretions more vigorously than others.

If a loop of the lower portion of the small intestine and the upper portion of the large intestine be excluded, as in a Thiry's fistula, and it be replaced in the

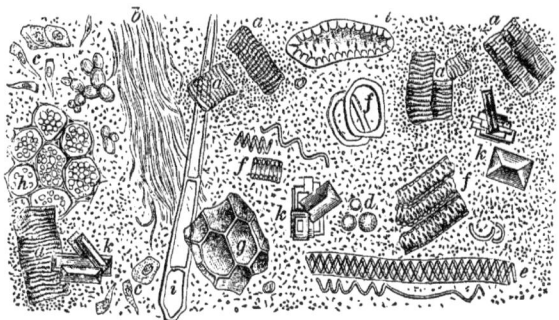

FIG. 125.—Feces: *a*, muscle fibers; *b*, tendon; *c*, epithelial cells; *d*, leukocytes; *e–i*, various forms of plant-cells, among which everywhere large numbers of bacteria (1) are scattered; between *h* and *b* are yeast-cells; *k*, ammoniomagnesium phosphate.

abdominal cavity after being closed by a circular suture, a mass of fecal character will be found in it. A loop of colon, thus excluded, will contain only a watery transudate, rich in salts.

2. The indigestible residue of the tissues of animal or vegetable food: hairs, horny tissue, elastic tissue; most forms of cellulose, wood-fibers, fruit-stones, spiral vessels from plant-cells, gum.

3. Fragments of otherwise readily digestible substances, particularly when they were ingested in excessive amount, or when not sufficiently comminuted by mastication; thus, the remains of meat (up to 1 per cent.), pieces of ham, shreds of tendon, bits of cartilage, flakes of fatty tissue, small pieces of hard albumin; further, plant-cells, starch in vegetable cells, firm-walled cells of ripe pulses, unground adhesive cells of grain, and the like. The presence of meat and starch is suggestive of an existing intestinal catarrh.

Of all articles of food certain remnants pass over into the feces: of wheat-bread, 3.7 per cent.; of rice, 4.1 per cent.; of meat, 4.7 per cent.; of potatoes, 9.4 per cent.; of cabbage, 14.9 per cent.; of rye-bread, 15. per cent.; of carrots, 20.7 per cent.

22

4. The metabolic products of the biliary coloring-matter, which are especially abundant in all diseases that cause increased destruction of erythrocytes, and which now no longer yield the Gmelin-Heintz reaction, as well as the altered biliary acids.   In diarrheal stools, as, for example, the green stools, the reaction, however, can often be readily demonstrated. It indicates accelerated peristalsis.   The meconium contains unaltered bilirubin, biliverdin, glycocholic and taurocholic acids.

5. Unaltered mucin and nuclein and, as a metabolic product of the latter, xanthin-bases; nuclein especially after a diet of bread; in addition, cylindrical epithelial cells from the alimentary tract in various stages of digestion; further, fat-globules at times.   Crystals of cholesterin and of coprosterin are rare.   The less intimately the mucus is admixed with the feces, the lower down in the intestine is its source.

6. After the ingestion of a large amount of milk, as well as after a diet of fat, crystalline needles of calcium-salts of the fatty acids, thus calcium-soaps, are found constantly in the feces, even in infants.   When courses of treatment with milk have been pursued undigested masses of casein and fat have besides been observed to be present.   Further, combinations of ammonia with the acids resulting from putrefaction already mentioned are among the substances constantly present in the feces. Larger masses of fat in the feces indicate accelerated peristalsis.

7. Among the inorganic residue, the readily soluble salts, which. therefore are readily diffused, are rare in the feces; thus sodium chlorid and other alkaline chlorids, the phosphoric as well as the sulphuric combinations.   On the other hand, the insoluble combinations, principally ammoniomagnesium phosphate, neutral calcium phosphate, yellow-colored calcium-salts, calcium carbonate and magnesium phosphate, constitute 70 per cent. of the ash.   The large amount of alkalies and earths contained in the feces is noteworthy, three-quarters of which are in combination with carbon dioxid and organic acids.   These are derived only in smallest part from the secretions of the intestinal mucous membrane.   By far the greatest part of the ash, however, is derived from the constituents of the food.   According to Rey, from 20 to 50 per cent. of solutions of calcium-salts, injected into the blood or subcutaneously, is excreted by the glands of the large intestine in the dog; 0.2 gram of iron is present daily.

In the presence of a fistula in the large intestine. Kobert and Koch observed, in the feces: sodium, calcium, magnesium, iron; phosphoric, sulphuric, hydrochloric acids; soaps, neutral fat, fatty acids, mucin, albumin, epithelium, traces of ethereal sulphates, together almost one gram daily.   At times the excretion of inorganic substances is so abundant as to form incrustations upon other fecal matter.   Under such circumstances either ammoniomagnesium phosphate is present alone, in large crystals, or magnesium phosphate is mixed with it.   Particularly the ingestion of rye-bran, in bread, which contains these substances in large amount, causes this result.   Charcot's crystals are found in the presence of entozoa.

8. Bacteria are present in abundance; yeasts are seldom absent.

For the identification of the individual bacteria, Escherich has developed pure cultures from the intestinal contents of infants, Bienstock from those of adults.   In the intestine of infants, fed upon mother's milk exclusively, the bacterium lactis aerogenes (Fig. 126, 2) produces, particularly in the upper portion, where milk-sugar is still unabsorbed, acetic acid, together with carbon dioxid, hydrogen and marsh-gas.   Lactates are transformed into butyrates.   The bacterium also produces acetic acid from starch.   A characteristic feature of the feces is the slender bacterium coli commune (Fig. 126, 1), provided with from one to three flagella, which forms lactic and formic acids, together with acetic acid, and at times exerts a pathogenic action.

In the feces of adults Bienstock found first of all two varieties of large bacilli (Fig. 126, 3, 4) resembling the bacillus subtilis in size and appearance, differing from the latter only in the form of its pure culture, by its manner of sporulation and by an absence of independent movement. These two bacilli are distinguishable macroscopically only by the form of this culture, which takes the shape either of a grape, or of a mesentery. Neither possesses any fermentative activity. A third, micrococcus-like, small, slowly multiplying bacillus (bacillus coprogenus parvus) was present in three-quarters of all of the stools. The fourth variety is the specific bacterium of proteid decomposition (bacillus putrificus coli), which is wanting in the feces of infants, and which with the production a fecal odor gives rise to the putrefactive products of proteids. Only this and no other causes these processes in the intestine; yet it does not decompose casein and alkali-

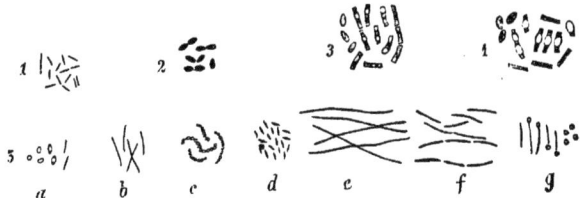

FIG. 126.—1, Bacterium coli commune; 2, Bacterium lactis aërogenes; 3, 4, the two large Bienstock bacilli with partial endogenous spore-formation; 5, the various stages of development of the bacillus of proteid putrefaction.

albuminate. The evolution of this bacterium is represented in Fig. 126, 5, a–g; of which the stages c and g are, however, wanting in the feces and are encountered only in artificial cultures.

If the feces are simply examined microscopically, without special precautions, the following are found as normal saprophytes: the bacterium coli commune, the staphylococcus aureus; frequently, also, varieties of proteus, at times with infective properties; in addition, other bacteria, whose entrance in part through the anus is possible: the bacillus butyricus, often staining blue with iodin, in feces rich in starch, and other small, spherical and rod-shaped schizomycetes, staining similarly. After the ingestion of uncooked food of various kinds, Lembke was able to verify the presence of as many as 73 different bacteria in the intestine.

In human beings, with accidentally acquired intestinal fistulæ or an artificial anus (intestinal fistula involving the colon), opportunity is afforded to study the changes in the intestinal contents with greater precision.

MORBID ALTERATIONS IN DIGESTIVE ACTIVITY.

The ingestion of food may be prevented by spasm of the muscles of mastication (usually as a symptom of general convulsions), by strictures of the esophagus, either from corrosive cicatrices (after the swallowing of caustic fluids) or from neoplasms, especially carcinoma. Inflammatory affections of any kind in the mouth and pharynx may also seriously interfere with the ingestion of food. Inability to swallow occurs as a symptom of disease of the medulla oblongata, in consequence of paralysis of the center for the motor nerves (facial, pneumogastric and hypoglossal) and of that for the sensory nerves through which pass reflex impulses (glossopharyngeal, pneumogastric and trigeminal). Irritation or abnormally heightened stimulation of this area may cause spasmodic swallowing and a feeling of constriction in the throat (globus hystericus).

The secretion of saliva is diminished in conjunction with inflammation of the salivary glands, occlusion of their ducts by concretions (salivary calculi), etc.; further, under the influence of atropin and daturin, in consequence of which the secretory (not the vasomotor) fibers of the chorda tympani appear to become paralyzed. Slight fever may increase the amount of saliva, though the amount of ferment may be lessened; fever of more marked degree diminishes both, while in the presence of high fever no saliva at all is secreted. The saliva secreted with lower grades of fever is cloudy, viscous and it usually becomes acid. With increase in fever the inertness of the diastatic action also increases. After the crisis the amount of saliva and the activity of the ferment become subnormal; likewise in the presence of diseases of the kidneys. After chronic illness of long standing the production of ferment frequently diminishes. The secretion of saliva is increased by morbid irritation of the nerves of the mouth, as from inflammations, ulcers, trigeminal neuralgia, so that enormous quantities may be poured out. Mercury and jaborandi-leaves cause salivation, the former with the simultaneous occurrence of a stomatitis that induces reflex secretion of saliva. Diseases of the stomach also may increase the secretion of saliva, in conjunction with paroxysms of nausea and retching. Viscid, ropy saliva, due to irritation of the sympathetic nerve, is secreted, together with some vascular disturbance, in consequence of active sexual excitement, but also as a result of certain psychical impressions. The reaction of the buccal secretion becomes acid in the presence of catarrhal conditions of the mouth and, further, as a result of the decomposition of accumulated epithelial cells in the mouth during the prevalence of fever, as well as in cases of diabetes mellitus, in consequence of acid fermentation of the sugar contained in the saliva. Diabetic patients therefore suffer frequently from carious teeth. The secretion of the mouth in infants also has a slightly acid reaction unless the greatest cleanliness is observed.

Disturbances in the activity of the gastric musculature may appear, as a paralytic phenomenon, with distention of the stomach, and a protracted sojourn of the ingesta. With more marked grades of the disorder decomposition and the production of gas take place. Diminution in muscular activity may give rise to dilatation of the entire stomach. Incompetency of the pylorus represents a special form of gastric paralysis. Derangement of innervation, central or peripheral in nature, may be the cause; further, actual paralysis of the pyloric sphincter or anesthesia of the mucous membrane of the pylorus, which exerts a reflex effect upon the sphincter muscle; finally, also, interference with the transmission of the reflex within the center. Abnormally increased activity of the gastric musculature will, as gastric diarrhea, hasten the ingesta into the intestine; often vomiting occurs. In nervous individuals so-called peristaltic unrest of the stomach is at times present, in conjunction with dyspeptic disorders. Spasm of the cardiac orifice or paresis of the inhibitory nerves of the cardia also occurs. Rarely, in the presence of stricture of the pylorus, true antiperistalsis of the stomach has been observed.

Gastric digestion is delayed by all severe physical and mental exertion and, if this be of more marked degree, digestion may even be inhibited. Also sudden emotional disturbance, as well as reflex influences from other organs (uterine dyspepsia), may have this effect. Probably these factors exert an influence upon the vasomotor nerves of the stomach. Impairment and abolition of the secretion of the gastric juice may, under certain conditions, be purely nervous in nature, as in cases of nervous dyspepsia and gastric neurasthenia. Complete absence of the gastric juice is found in connection with atrophy of the mucous membrane, principally in cases of pernicious anemia. Also excessive secretion of the gastric juice, continuous flow of the juice, and likewise excessive production of acid may depend upon derangement of nervous activity: nervous gastroxynsis, chiefly observed in women. Excessive production of hydrochloric acid occurs in association with round ulcer of the stomach.

Inflammatory or catarrhal affections of the stomach, as well as ulcers and neoplasms, disturb normal digestive activity, as does also the excessive ingestion of foods difficult of digestion, of sharp spices in considerabl‧ amount, or much alcohol. Grützner observed in a dog that the mucous membrane secreted continuously under the influence of a chronic gastric catarrh, but the gastric juice was deficient in pepsin, cloudy, viscous, less acid, even alkaline. The introduction of food did not modify the secretion; the stomach, therefore, never actually comes to rest. At the same time the chief cells of the gastric glands are turbid. Accordingly it would seem o‧ advantage for patients suffering from gastric catarrh to eat frequently, but only a little at a time, and in addition use a 0.4 per cent. hydrochloric-acid solution as a beverage. Small doses of sodium chlorid appear to aid gastric digestion.

In the presence of enfeebled digestion, the cause may be deficien‧ formation either of hydrochloric acid or of pepsin. Both substances may therefore be administered as remedial agents. In the presence of enfeebled gastric digestion and motor insufficiency decomposition of the contents of the stomach into lactic, butyric and acetic acids often takes place as a result of the action of lower organisms. Small doses of salicylic acid are advisable under such circumstances, together with some hydrochloric acid (notwithstanding possible heart-burn or acid eructation). The administration of pepsin probably is but rarely imperative, as this ferment is only seldom absent even from the diseased gastric mucous membrane. In the presence of marked dilatation and a protracted sojourn, the proteids in the stomach, notwithstanding the hydrochloric acid, undergo putrefaction, which, however, does not as a rule have an injurious effect. In cases of gastric catarrh and cholera, albumin has been observed to appear in the gastric juice.

Gastric Digestion in Patients with Fever and Anemia.—Beaumont, from observations made upon the man with the gastric fistula examined by him, found that only scanty secretion of gastric juice takes place in the presence of fever. The mucous membrane was deficient in secretion, red and irritable. Dogs, which Manassein had made febrile from septicemia or profoundly anemic by venesection, elaborated a fairly active gastric juice, characterized especially by a deficiency of hydrochloric acid. Hoppe-Seyler examined the gastric juice from a patient with typhoid fever—in which disease van de Velde found no free hydrochloric acid (for the parietal cells are destroyed under such conditions); as well as in cases of gastric carcinoma also, in which disease there is, as a rule, no excess of free hydrochloric acid—and found it absolutely inactive for artificial digestion, even after hydrochloric acid had been added. This investigator properly emphasizes the fact that the diminution in hydrochloric acid after such conditions favors the development of a neutral reaction of the gastric contents, by reason of which, on the one hand, digestion in the stomach can no longer take place; while, on the other hand, abnormal fermentative processes must take place, with the aid of developing micro-organisms and sarcinæ ventriculi (?). Uffelmann found that, in patients with fever, the secretion of a peptone-forming gastric juice ceases if the fever sets in violently, if a condition of great weakness develops, or if a high temperature persists for a long time. In any event, also the amount of gastric juice secreted is diminished. In the presence of fever the irritability of the mucous membrane is increased, so that vomiting is readily induced. Also the increased excitability of the vasomotor nerves of patients with fever is evidently detrimental to the secretion of active digestive juices. Gluzinski found an absence of hydrochloric acid in the acute febrile infectious diseases. Beaumont observed that fluids were rapidly absorbed from the stomach of a febrile patient, while, on the other hand, the absorption of peptones was diminished, on account of the frequently accompanying gastric catarrh and the disturbed activity of the muscularis mucosæ.

Many salts disturb gastric digestion, if added in considerable amount, particularly the sulphates. Of the alkaloids, morphin, strychnin, digitalin, narcotin and veratrin likewise have a disturbing influence. A small amount of quinin accelerates gastric digestion.

As the digestive activity of the stomach can be replaced by the pancreas, it is evident that dogs may continue to live without profound disturbance of nutrition after extirpation of the stomach. Langenbach observed a similar result in human beings after operation.

The secretion of bile undergoes a change in the presence of acute disease, as, for example, fever, in that it becomes scanty and at the same time more watery, and that it is poorer in its specific constituents. Should the liver undergo profound structural changes as a result of the morbid process, the secretion of bile may cease completely.

As a result of the decomposition of bile (acid fermentation?) gall-stones form within the gall-bladder or biliary passages. These calculi may be white or brown. The former consist almost entirely of laminated cholesterin-crystals. They are generally about 1 cm. in diameter, but they may be the size of a walnut or even larger. The brown gall-stones consist of bilirubin-lime, together with biliverdin, bilicyanin and choletelin, and also calcium carbonate and phosphate, often mixed with iron, manganese, copper and other precipitated heavy metals. All gall-stones, like urinary calculi, possess an organic supporting structure. Some are rather spherical, often studded with mulberry-shaped nodules. Those packed together in the gall-bladder become polished, from mutual attrition in consequence of the contraction of the walls of the gall-bladder. The white gall-stones often contain lime and biliary coloring-matter as a nucleus, together with a nitrogenous residue, probably derived from desquamated epithelium, mucus, salts of biliary acids and some fat. Gall-stones may cause obstruction of the bile-ducts and then give rise to symptoms of cholemia. Smaller stones, impacted in the ducts, may cause intense pain (biliary colic) and, by means of their sharp edges, they may even bring about fatal rupture of the ducts. The formation of biliary calculi is probably due ultimately to local stagnation and decomposition of bile in the gall-bladder, caused, for example, by tight lacing, in consequence of which kinking of the gall-bladder takes place. Cholemia and jaundice have already been discussed.

In the presence of high fever the *pancreatic secretion* appears to be diminished and its activity enfeebled. Cessation of secretion is attended with the appearance of fat in the form of globules and crystalline fatty acids in the feces. Degeneration of the pancreas may cause diabetes.

Among the *disturbances in the activity of the intestinal tract*, constipation (obstipation) is first to be considered. The causes of this condition may reside in: (1) Obstructions that occlude the normal passage. In this category belong constrictions of the intestinal canal, due to cicatricial strictures, as, for example, in the colon often after dysentery; neoplasms; further. axial torsion of a loop of intestine (volvulus), or invagination of one portion into another (intussusception), or into a hernial sac (hernia); also the pressure of tumors or exudates from without. Finally, congenital absence of the anus may constitute the cause. (2) Excessive dryness of the intestinal contents may cause obstipation. Under such circumstances the following factors may be operative: Excessive dryness of the food; further, diminution of the digestive juices, as, for example, of the bile in cases of icterus; or in consequence of great loss of fluid through other organs of the body, as after profuse perspiration or secretion of milk, or, finally, during fever. (3) Derangement of the activity of the muscles and of the motor nerve-apparatus of the intestine may induce constipation through insufficient peristalsis. This is caused especially by paralytic conditions, as in the presence of inflammation, degeneration, chronic catarrh and peritonitis. Spinal paralysis is generally attended with sluggish defecation; central affections often also. Whether the phenomena of mental impairment and hypochondriasis are the accompaniment or the sequel of constipation has not yet been demonstrated. Spasmodic contraction of certain portions of the intestine may give rise to transitory retention of the intestinal contents, with great pain (colic); as may also spasm of the anal sphincter, which may also take place reflexly, from irritation of the lower portion of the intestine. The feces are almost always hard and deficient in water, when constipation exists, because during their long sojourn in the intestines fluid is absorbed from them. In consequence, the fecal masses form large pieces (scybala) within the large intestine and these may, in turn, constitute a new obstacle to the onward movement (coprostasis). Diminution in the intestinal and gastric secretion occurs also as a sign of general nervous affections (hysteria, hypochondria, mental disorders), although increased secretion may also take place under such circumstances.

The agents that cause constipation are, in part, those that paralyze the motor apparatus temporarily, such as opium or morphin; and, in part, those that diminish the secretions of the intestinal mucous membrane, and exert a constringent effect upon the blood-vessels and the mucous membrane, such as tannic acid, alum, lime, lead acetate, argentic and bismuth nitrates.

Increase in the intestinal discharges is usually accompanied by a greater degree of fluidity of the feces (diarrhea). The causes are as follows:

1. Unduly rapid propulsion of the contents through the intestinal canal, particularly through the large intestine, so that absorption from this part cannot take place in a normal manner. The increased peristalsis is due to irritation of the motor-nerve apparatus of the intestine, and is principally reflex in character, Rapid passage of the ingesta through the intestinal canal results in the presence in the discharges of substances that could not be completely or at all digested in the short time afforded (lientery). This will also occur if portions of the intestine, situated high up, communicate with lower portions of the intestine, through abnormal openings.

2. The feces may be of the consistency of paste from the admixture of water, mucus and fat, in considerable amount; further, from the residue of fruits and vegetables. In rare cases in which the feces contain a good deal of mucus, so-called Charcot's crystals are present (Fig 92, c). In the presence of ulceration of the intestine, leukocytes (pus-cells) are found.

3. Diarrhea may develop in consequence of disturbances of the processes of diffusion through the intestinal walls. Affections of the epithelial cells should be mentioned in this connection: swelling in association with catarrhal or inflammatory conditions of the mucous membrane. As, further, in the process of absorption independent activity on the part of the cylindrical cells is to be taken into consideration, controlled, perhaps, by the nervous system, it is plain how sudden agitation, from fright, anxiety, etc., may cause diarrhea.

4. Diarrhea may be the result of increased secretion. In its simplest form this occurs through capillary transudation, when salts, as, for example, magnesium sulphate, introduced into the intestine, remove water from the blood by endosmosis. In this category belong the copious watery discharges that take place in consequence of alteration of the intestinal epithelium, as in cases of cholera, in which such excessive transudation takes place into the intestine that the blood becomes inspissated and may even stagnate in the veins. In addition, transudation into the bowel may take place in consequence of paralysis of the vasomotor nerves of the intestine. The diarrhea due to cold appears to belong in this group. Certain substances appear directly to irritate the secretory organs of the intestine or their nerves; among these are the drastic purgatives. Pilocarpin injected into the blood also induces marked secretion.

In the presence of febrile disorders, the secretion of the intestinal glands appears to undergo quantitative and qualitative changes, with simultaneous derangement in the activity of the intestinal musculature and the organs of absorption and increased irritability of the mucous membrane.

With respect to fermentations in the intestine, the fact should be emphasized that all, in excess, as, for example, the butyric or the acetic, give rise to pathological manifestation. With regard to the pathogenic schizomycetes acting from the intestinal canal (cholera, typhoid, dysentery, and others) reference may be made to p. 246 Flagellated trichomonads are exceedingly rare.

Finally, attention should be directed to the fact that, in consequence of abnormal decompositions in the intestinal canal, substances may be formed that exert a toxic effect upon the organism and thus give rise to auto-intoxications.

COMPARATIVE PHYSIOLOGY OF DIGESTION.

Among mammals, herbivora possess larger salivary glands than carnivora, while omnivora occupy an intermediate position. Whales have no salivary glands at all; the pinnipeds have a small parotid, the echidna none at all. The dog, like some carnivora, has an additional zygomatic gland situated in the orbit. In birds the salivary glands empty at the angle of the mouth; the parotid gland is wanting. Among snakes the parotid glands are in some species transformed into poison-glands; tortoises have sublingual glands; in addition, reptiles have labial glands at the margin of the lips. Amphibia and fish have only small, disseminated buccal glands. In insects the salivary glands are widely distributed, partly unicellular (as, for example, two glands in lice), partly compound; several pairs of them are usually present. In some the secretion contains formic acid, for which reason the stings of these animals cause burning and inflammation: in others the secretion is strongly alkaline, as that from the large salivary glands of the bed-bug. In bees and ants the lower salivary glands secrete a sort of cement-substance. The web-glands on the lower lip of caterpillars secreting the silky material, principally those of the silk-worm, should not be confounded with the salivary glands. Among vermes, leeches have unicellular salivary glands. In snails the salivary glands are also widely disseminated, and the saliva from dolium galea contains more than $3\frac{1}{2}$ per cent. sulphuric acid, which also is present in murex. cassis, and aplysia. Cephalopods have a double set of salivary glands. In the octopus the saliva digests fibrin, but not starch, and it is poisonous.

Crop-like formations are wanting in all mammals; the stomach appears to be single, as in human beings, or divided into halves, as in many rodents, into a cardiac portion and a pyloric portion.

The stomach of ruminants consists of four portions: the first and largest is the paunch (rumen), the next the honeycomb-bag (reticulum). In these two portions, principally in the paunch, the ingesta undergo maceration and fermentation. They are now returned to the mouth by the action of the voluntary muscular fibers passing to the stomach, again thoroughly masticated, and, by the closure of a special semicircular groove (esophageal groove), the bolus is carried into the third stomach, the manyplies (psalterium), which is absent in camels, and thence to the true, fourth stomach, the rennet-stomach (abomasum). In the two first stomachs starch and cellulose are digested, the sugar formed in part passing over into lactic acid. The third stomach performs chiefly mechanical work, while the fourth really digests albumin. In the small intestine proteids and carbohydrates are further digested.

The intestine is divided into the small and the large intestine. It is short in carnivora, and considerably longer in herbivora. The cecum, which in herbivora attains considerable size as the most important organ of digestion, and in some rodents is even multiple, represents in human beings an insignificant, typical remnant, and is wholly absent in carnivora. In birds the esophagus, especially in birds of prey and granivora, often possesses a diverticular appendix, the crop, for the maceration of the food. In the crop of pigeons there occurs, at the breeding-season, the secretion of crop-milk, the product of a special gland, which is also used as food for the young. The stomach consists of the proventriculus well supplied with glands, and the thick-walled muscle-stomach, which, with the aid of the inner horny plates, effects the crushing especially of grain. In the intestine, at the junction with the short large intestine, there is almost constantly present a pair of ceca shaped like a glove-finger. The intestinal mucous membrane exhibits principally longitudinal folds. The alimentary canal of fish is usually simple. The stomach frequently represents only a dilatation. Less commonly the pylorus possesses one, more frequently a large number of diverticular appendices, containing a large number of glands (appendices pyloricæ, as, for example, in the salmon). The mucous membrane of the usually short intestine exhibits longitudinal plication, as a rule, or the so-called spiral valve, as in the sturgeon, resulting from a spiral arrangement. The alimentary canal of fish, from the esophagus to the rectum, possesses peptonizing power. The short rectum is provided, in sharks and rays, with a diverticular appendage (bursa entiana).

In amphibia and reptiles the stomach is generally a simple dilatation. The intestine is longer in herbivora than in carnivora. Especially interesting in this connection is the fact that the vegetable-eating frog-larvæ acquire a much shorter intestine with the metamorphosis that makes them carnivorous, terrestrial animals. The intestinal mucous membrane of reptiles exhibits numerous plications. The liver is not wanting in any vertebrate, and is especially large in fish. The amphioxus has only a diverticulum indicative of the liver. The gall-bladder is wanting occasionally in all classes, in accord with which is the experimental observation that extirpation of the gall-bladder is unattended with appreciable influence on digestion and absorption. The pancreas is wanting only in some fish. One opening (in the amphioxus) or two openings (in the shark, the ray, the sturgeon, the eel and the salmon) lead from without freely into the abdominal cavity; the same conditions prevail also in crocodiles.

Among the molluscs, snails and cephalopods only have true organs of mastication. Some herbivorous land-snails have a movable, horny grinding plate situated in the upper pharyngeal wall. Horizontal maxillary plates, with hard edges working one upon the other, are present particularly in carnivorous snails with uncovered gills. A horny grinding plate, placed like a tongue, whose peculiar form serves for the systematic differentiation of various snails, is frequently present in others. Cephalopods possess a strong biting apparatus in the form of a large, horny pair of jaws, resembling a parrot's beak in shape. They also have a grinding plate upon a tongue-like prominence, studded with spines. The alimentary canal is divided into esophagus, stomach and intestine, at times provided with diverticula. In many mussels the rectum pierces the heart and the pericardium. In snails the anus is usually in the vicinity of the respiratory organs. The liver is, as a rule, large. The vineyard-snail has a cellulose-splitting ferment in the secretion of the liver. In the cephalopods the ink-bag opens into the rectum or near the anus.

Among vertebrates crustaceans have a masticating apparatus transformed from feet; in some, true masticating feet are still present; in parasitic crabs there are also sucking mouth-organs. Among arachnids the mites have sucking mouth-organs; in true spiders, there are, in addition to the sucking mouth-organs, horizontally acting clutching jaws, in part connected with poison-glands. Centipedes possess a strong pair of jaws, acting horizontally. Of insects, those provided with masticating mouth-organs possess, between the upper and lower lips, two pairs of jaws, acting horizontally against each other, of which the upper (mandibulæ) exceed the lower (maxillæ) in strength. In sucking-insects the four jaws are transformed into a long tube with a longitudinal slit (the stinging proboscis of the bed-bug), which lies in the semicircularly grooved lower lip as in a case. The proboscis of the butterfly consists of the greatly prolonged lower jaws, lying side by side, and capable of being rolled up, while the development of the upper jaws has been arrested. Bees have a sucking tongue, which lies in a groove formed in the lower jaws; in addition, the feeble upper jaws still persist as organs of mastication.

In crustaceans the esophagus is short; in some the stomach is a simple dilatation, in others it possesses diverticula, in which are situated the bile-producing glands. The fresh-water crab and its relatives possess a strong chitinized intima in the stomach, which is capable of acting as a masticating organ. This membrane is expelled when the skin is shed. Among arachnids, scorpions have a simple alimentary canal. True spiders possess a narrow esophagus and a circular stomach; in addition diverticula on all sides, at the base of which liver-tissue is present, and which may extend even down into the feet. In insects, in addition to the esophagus and the chyle-stomach, generally rich in glands, and at times serrated, there are present various portions, such as the crop in the cricket for instance, the sucking stomach in the butterfly, the masticating stomach in the beetle, in varying manner. The intestinal canal is usually shorter in carnivorous than in herbivorous insects. In the intestine of the flour-worm (tenebrio) ferments are present resembling those of the pancreatic juice. It is remarkable that, in the larval state, as, for example, of most bees, the tract is closed below the chyle-stomach. The rectum, with its auxiliary apparatus, exists by itself and empties, as an excretory duct, into the anus. Peculiar long, tubular excretory organs, the Malpighian vessels, several of which are present, open at the junction of the small and the large intestine.

Of the vermes, tape-worms, as well as the acanthocephala (echinorhynchus) among round worms, have no special digestive organ, but are nourished by endosmosis, through absorption on the part of the skin. The anus is wanting in trematodes (distomum), thread-worms, and almost all turbellaria. In the first, as well as in leeches (sanguisuga), the buccal orifice is surrounded by a sucking-cup, which, in leeches, possesses, in its depth, three dentated cutting organs. Some leeches, as well as the planaria, have a protrusile proboscis. The intestine of turbellaria, unprovided with an anus, is shaped simply like a glove-finger. It is variously branched in liver-flukes (distomum). In the annulate worms the intestine extends from the anterior to the posterior extremity of the body; both mouth and anus are present. Among them, the earth-worms possess a muscular pharynx, while leeches have a highly distensible stomach, provided with many lateral diverticula, which, when the animal has sucked itself full, can be incised through the skin of the back, so that the blood flows continuously from the wound, while the animal continues to take up blood through its sucking mouth (bdellotomy). All vermes are unprovided with a liver.

All echinoderms possess an intestinal canal. The mouth is often provided with a biting mechanism, which appears in sea-urchins in the form of five enamel-teeth connected with a movable, complicated maxillary apparatus (Aristotle's lantern). Many of the starfish are unprovided with an anus; a bile-like secretion is found in diverticula of their stomach. Salivary glands have been found in sea-urchins.

The aquatic celenterates possess no intestinal tract provided with independent walls. The abdominal cavity is the digestive cavity; mouth and anus are represented by the same central orifice, which often is surrounded by tentacles (medusæ, polyps). A system of canals, passing through the body (medusæ), and connected with the digestive cavity, conveys the nutritive fluid and, at the same time, the oxygen-containing water. It is, therefore, the water-vascular-system, and at the same time the nutritive, respiratory and excretory organ.

Among the protozoa, the gregarines are nourished by endosmosis through the skin. Infusoria possess mouth and anus, although their abdominal cavity is bounded only by the protoplasm of their body-substance. Rhizopods surround their food with their body-substance and excrete the indigestible material at another portion of the body. In sponges this process takes place from the interior of their numerous canals, which penetrate the colonies of their protoplasmic bodies.

Digestive Phenomena in Plants.—The observations upon the digestion of proteid on the part of a number of plants are highly remarkable. The sundew (drosera) possesses, upon the surface of its leaves, numerous tentacle-like processes, provided with glands. As soon as an insect lights upon the leaf, the former is suddenly seized by the tentacles. The glands discharge a juice of acid reaction and digest the animal with the exception of its insoluble chitinous remains. The juice contains a pepsin-like ferment and formic acid. The secretion, as well as, later, the absorption of the dissolved substances, takes place in conjunction with movement of the protoplasm of the leaf-cells. Venus' fly-trap (dionea) and butter-wort (pinguicula) exhibit similar processes, as well as the cavities of the transformed leaves of the nepenthe. Altogether, about 15 species of such carnivorous dichotyles are known.

The juice escaping from incisions in the green fruit of the papaw-tree (carica papaya) possesses peptonizing properties due to a ferment closely allied to trypsin. The milky sap from the fig-tree is likewise active, exerting a diastatic effect and also coagulating milk at 50° C. Albumin is dissolved also by some fungi (boletus, tuber), lichens (parmelia) and the sap of taraxacum, lactuca, agave and portulac. Artichokes, yellow or lady's bedstraw and other plants contain rennet-ferment. The sap of aloes and of sugar-cane, as well as dried figs, coagulates milk and has a peptonizing action; as does also ordinary flour-dough on admixture; further, the juice (containing peptone at the same time) from the seed of wheat, barley, poppy, beets and corn, after the addition of organic acids. Potatoes and rice have feeble, flour, grain and corn marked sugar-forming activity.

HISTORICAL.

Digestion in the Mouth.—The vessels of the teeth were known to the Hippocratic school. Aristotle ascribed an uninterrupted growth to the teeth. In addition, he directed attention to the fact that those animals that exhibit a development of horns and antlers, cloven-hoofed animals, possess an imperfect denture (absence of the upper incisor teeth). It is a remarkable fact that, in human beings with excessive formation of horny substance, in consequence of the presence of superfluous hair, imperfect development of the teeth (absence of the incisors) has also been observed. The muscles of mastication were recognized early. Vidius (died 1567) described the maxillary articulation, with the meniscus. The epiglottis, according to Hippocrates, prevents the entrance of food into the larynx. The ancients considered the saliva only a solvent and a means for moistening the food. In addition, in consequence of a knowledge of the saliva of rabid animals and the parotid secretion of venomous snakes, various poisonous properties were ascribed to the saliva, especially from fasting animals—a view that Pasteur again confirmed in part, referring the action to pathogenic bacteria in the secretions of the mouth. Aretæus (81 A. D.) emphasizes the muscular nature of the tongue. The salivary glands had been discovered in ancient times. Galen (131-203 A. D.) was familiar with Wharton's duct and Ætius (270 A. D.) with the submaxillary and sublingual glands. Regner de Graaf established salivary fistulæ in dogs in 1663, by tying tubes in Stenon's duct. Hapel de la Chenaye obtained in 1780 for examination large amounts of saliva from a salivary fistula established in a horse. Spallanzani in 1786 stated that insalivated articles of food are more readily digested than those moistened with water. Hamburger and Siebold investigated the reaction, consistency and specific gravity of the saliva and found mucus, proteid and salts present. Berzelius introduced the term ptyalin for the characteristic substance in the saliva, though Leuchs in 1831 first discovered its diastatic fermentative action.

Gastric Digestion.—The ancients compared digestion to cooking, through which solution is effected. Aristotle supposed that, through this "pepsis" chyle (ichor) first developed from the food, and then reached the heart. He also knew of the rennet-action of the stomach. According to Galen, only dissolved masses pass through the pylorus into the intestine. He described the movement of the stomach and the peristalsis of the intestines. Ælian recognized the four stomachs of ruminants and gave their names. Vidius (died 1567) observed the numerous small glandular openings in the gastric mucous membrane. van Helmont (died 1644) expressly mentions the acid of the stomach. He as well as Sylvius (died 1672) compared the action of the stomach with fermentation, in connection with which, according to Descartes (died 1650) and Willis (died 1675), the action of the acid predominates. Reaumur (1752) recognized that a juice was secreted by the stomach that effects solution and with which, together with Spallanzani (1777), he undertook digestive experiments outside of the stomach. Carminati (1785) then found that the stomach of carnivora, especially when engaged in digestion, secretes an actively acid juice. Prout discovered in 1824 the hydrochloric acid of the gastric juice and Sprott and Boyd in 1836 found the glands of the gastric mucous membrane, among which Wassmann and Bischoff distinguished the two different kinds. After Beaumont (1825-1833) had made his observations upon a man with a gastric fistula, Bassow (1842) and Blondlot (1843) established the first artificial gastric fistulæ in animals. Eberle subsequently (1834) prepared artificial gastric juice. Mialhe designated the albumin modified by digestion as albuminose; while Lehmann, who examined this more thoroughly, introduced the name of peptone. Schwann (1836) first prepared pepsin and defined its activity in combination with hydrochloric acid.

Pancreas, Bile, Intestinal Digestion.—The pancreas was known to the Hippocratic school. Moritz Hofmann demonstrated in 1641 its excretory duct in the turkey to Wirsung, who (1642) described it in human beings as his discovery. Regner de Graaf collected in 1663 pancreatic juice from fistulæ, and which Tiedemann and Gmelin found to be alkaline, while Leuret and Lassaigne found it to resemble saliva. Bouchardat and Sandras in 1845 discovered its diastatic, Eberle in 1834 its emulsifying, Purkinje and Pappenheim in 1836 its peptic, and Cl. Bernard in 1846 its fat-splitting properties, to the last of which Purkinje and Pappenheim had already directed attention.

Aristotle designates the bile as a useless excrementitious product. According to Erasistratus the bile is conveyed from the liver to the gall-bladder through most minute, invisible ducts. Aretæus attributed the cause of icterus to occlusion of the bile-ducts. Benedétti in 1493 described gall-stones. According to Jasolinus (1573) the gall-bladder is emptied by its own contraction. Sylvius de le Boë (1640) observed the hepatic lymph-vessels, Walæus (1641) the connective tissue of the so-called capsule of Glisson. Albr. v. Haller pointed out the utility of the bile in the digestion of fat. Henle, Purkinje and Dutrochet (1838) described the liver-cells. Heynsius discovered urea, Cl. Bernard (1853) sugar, in the liver, and with Hensen (1857), he found glycogen in the liver. Kiernan (1834) described the blood-vessels more thoroughly. Beale injected the lymph-vessels, Gerlach (1854) the finest biliary passages, Schwann (1844) established the first biliary fistula. Gmelin discovered cholesterin, taurin and the biliary acids. Demarcey pointed out the combination of the biliary acids with sodium (1838). Strecker found the sodium-combinations of both biliary acids and isolated them.

Corn. Celsus mentioned nutritive enemata (3–5 A.D.). Laguna (1533) and Rondelet (1554) knew of Bauhin's valve. Fallopia (1561) described the folds and villi of the intestinal mucous membrane, as well as the nerve-plexuses of the mesentery. J. Conrad Brunner (1687) discovered the duodenal glands that bear his name. Severinus (1645) knew of the agminated follicles (Peyer's patches, 1673) and Galeati (1731) knew of Leiberkühn's (1745) glands in the intestine.

PHYSIOLOGY OF ABSORPTION.

STRUCTURE OF THE ORGANS OF ABSORPTION.

The mucous membrane of the entire intestinal tract, so far as it is lined by a single layer of cylindrical epithelium, that is, from the cardiac orifice to the anus, is capable of absorption. The buccal cavity and the esophagus can take part in this process only to an exceedingly limited extent, on account of their thick, many-layered squamous epithelium. Nevertheless, poisoning, as, for example, with potassium cyanid, may take place by absorption from the mouth alone. The capillary blood-vessels, as well as the chyle-vessels, of the mucous membrane act as the absorbing channels of the intestinal tract. The former convey the materials absorbed almost wholly through the portal vein to the liver, while the latter, uniting in their further course with lymph-vessels, discharge the absorbed chyle or milky juice through the thoracic duct into the blood at the junction of the subclavian and internal jugular veins.

From the stomach are absorbed aqueous salt-solutions (within six minutes), sugar (namely, grape-sugar, milk-sugar, cane-sugar and maltose) in aqueous solution in moderate amount, in alcoholic solution in somewhat larger amount; dextrin and peptone, chiefly in concentrated solutions, in lesser amount; and poisons, especially when dissolved in alcohol. Klemperer and Scheurlen observed that, in the dog, neither fat nor the fatty acids were absorbed. The empty stomach absorbs more rapidly than that filled with food. Diseases of the stomach and fever cause delayed absorption.

In addition to absorption, an active secretion of water into the stomach, takes place, in general, in greater degree in proportion as the amount of absorbed substances is greater.

The small intestine constitutes the principal field of absorption, presenting, especially in its upper half, through its many folds of mucous membrane and through the innumerable cone-shaped villi projecting from them, an extraordinary expanse of surface for absorption. The villi are close together at their bases, so that the entire surface of the mucous membrane appears to be covered with them. In the spaces between their bases the numerous simple tubules of Lieberkühn's glands empty. Each villus is to be regarded as a projection of the entire mucous membrane, for it contains all of the elements comprised within it.

The cloak-like covering of the villi consists of a single layer of cylindrical epithelium with intervening isolated mucous goblet-cells. The surface of the cells directed toward the lumen of the intestine is polygonal (Fig. 127, D) and, viewed from the side (C), exhibits a broad seam-like outline, which was formerly considered the thickened wall of the cell-membrane and was designated by the term "lid-membrane." This seam exhibits a delicate longitudinal striation, which was interpreted in part as the expression of the constitution of the lid, of rods

348

arranged as a mosaic, in part as pore-canaliculi, intended for the passage of the finest fat-granules. As a matter of fact, however, this seam belongs only to the longitudinal surfaces of the epithelial cells and is comparable to the thickened edge of a cylindrical vessel, open above.

The protoplasmic cell-contents, which enclose a large elliptical nucleus with nucleolus in the lower portion of the cell, end approximately on a level with this edge, although at the same time, they contain, at the level of the thickness of this marginal seam, many pseudopod-like protoplasmic processes, which, standing side by side, and arranged in bundles, are surrounded by the edge of the marginal border. Thus, when viewed from the side, the lid-membrane appears striated, while, as a matter

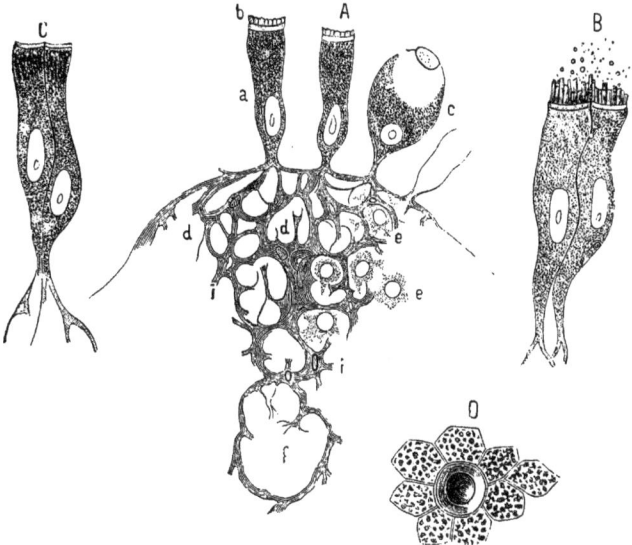

FIG. 127.—Structure of the Absorption-apparatus of a Villus: A, transverse section of a villus, in part; a, cylindrical epithelium, with thickened border (b); c, a goblet-cell; i, i, framework of the adenoid tissue of the villus; d, d, cavity within this, in which lie lymphoid cells (e, e); f, central lymph-space in transverse section. B, two cylindrical epithelial cells with extended pseudopod-like processes of the cell-protoplasm, participating in absorption of the fat-granules. C, cylindrical epithelium after absorption of the fat-granules has been completed. D, cylindrical epithelium of the villus, viewed from the surface, with a goblet-cell in the center.

of fact, neither the lid nor the mosaic plates or pores attributed to it exist. The cells are, therefore, open toward the intestinal surface. The protoplasmic processes, standing close together, and resembling the cilia of ciliated epithelium, are directed from the interior of the cell toward the periphery of the intestine. In their midst, near the free surface, lies a diplosoma.

These protoplasmic processes are rapidly extended from the cell-body beyond the edge of the cell-membrane, and in a manner comparable to the pseudopods of amœbæ, they seize the finely granular fat and draw it into the cell-body. Moistening with bile appears especially to promote their activity, as the movement is not observed in villi not moistened with bile.

In addition, the medulla oblongata, the spinal cord or the dorsal nerves must have been divided for about a day previously. This apparently depends upon the fact that, in the preparation of an uninjured animal (frog), the fresh division of nerves that becomes necessary acts as an irritant, as a result of which the cells settle down to rest, like irritated amœbæ or like the corneal cells after irritation of their nerves. This fact points to an influence of the nerves upon absorption.

When the epithelial cells are filled with fat-granules, the processes are withdrawn into the interior of the cell. The border then appears unstriated, and a transparent zone lies between it and the cell-protoplasm. The goblet-cells appear to be engaged principally in the secretion of mucus; although small fat-granules are also occasionally seen within them.

Pathological: In cases of cholera, as well as after poisoning with arsenic and muscarin, enormous desquamation of intestinal epithelium takes place.

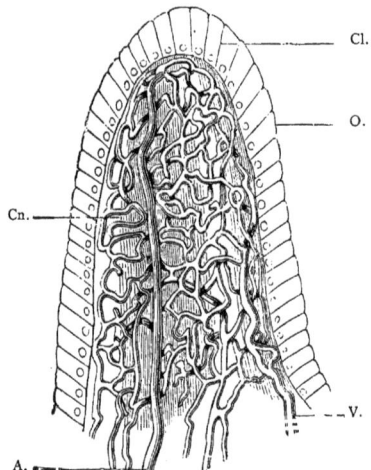

FIG. 128.—Blood-vessels of an Intestinal Villus: Cn, capillaries; A, artery; Cl, cylindrical epithelium; O, surface of the epithelium; V, vein.

According to the views of Eimer, Heidenhain, v. Thanhoffer and others, the constricted root-ends of the epithelial cells communicate with anastomosing connective-tissue corpuscles of the villous tissue. Into these the fat-granules are believed to migrate from the interior of the epithelial cells. The soft connective-tissue cells, finally, are thought to communicate with the central lymph-vessel; and in this manner a communication is established between the epithelium and the latter. Thus, the fat-granules would migrate through the body of the connective-tissue cells, as through lymph-canaliculi, to the central lymph-vessel. The author is able to agree with this conception with a modification, which approaches the views of His, Brücke and v. Basch. As a result of his investigations he believes that the epithelial cell narrows toward its lower extremity, like a funnel; the cell-membrane entering, in various directions, directly into communication with the supporting cells of the adenoid tissue of the villus, as well as with the subepithelial branching layer of the villus, which, accordingly, must be perforated in many places. The supporting cells of the villous tissue surround a spongy system of cavities within which lie protoplasmic, nucleated stroma-cells (Fig. 127, A) of varying appearance. The latter at times contain fat-granules in suspension. According to v. Davidoff, these cells are formed by constriction from the lower extremities of the epithelial cells, which, in time, develop a nucleus within themselves.

These cells, like ameboid cells without capsules, communicate with one another and with the protoplasm of the epithelial cells, and in them, through active movement of the protoplasm, wander the fat-granules, which the cells take up and again give up within the villus. Thus, the epithelial sheath, with the connective-tissue corpuscles of the villus, forms the supporting apparatus; the contents of the epithelial cells and the numerous stroma-cells are the active propellers of the fat-granules taken up. Through appropriate interstices in the tissues the cavities containing the stroma-cells communicate with the axial lymph-vessel, which is lined by endothelial cells. It is not improbable that leukocytes frequently migrate from the capillary blood-vessels of the villus into the tissue of the villus and, in part containing absorbed fat-granules, pass over into the central lymph-vessel. According to Schäfer, Zawarykin, Wiedersheim, Stöhr, Preusse, Heidenhain and others, the ameboid cells probably migrate from the parenchyma of the villi toward the epithelial layer and perhaps even between the epithelial cells, and return toward the axis of the villus, laden with the substances absorbed.

A small artery enters every villus and, lying excentrically, passes to the summit of the villus without division, to give off branches from this point. In human beings this division begins at the middle. The ramifications form a dense capillary network, which lies superficially in the parenchyma of the villus, almost directly beneath the epithelial layer, and from which, either at the apex of the villus or further downward, a vein, running backward, is constituted.

The villus is provided with unstriated muscular fibers, both deep-seated, their bundles accompanying the central lymph-vessel longitudinally, and also superficial, running rather transversely.

The connective tissue of the small intestine has two layers, a deeper, composed of thick, interwoven, mainly collagenous fibers (stratum fibrosum), and lying above this a reticular layer intermixed with elastic fibers (stratum granulosum), entering into the villi also.

Nerves enter the villi from Meissner's mucous-membrane plexus, are provided with small, granular ganglion-cells in their course, and end in part in the muscles of the villi and of the arteries, while in part they appear to communicate with the contractile protoplasm of the epithelial cells.

Nerve-filaments pass from Meissner's mucous-membrane plexus to the vessels of the submucosa. Meissner's plexus communicates, by numerous fibers, with a nerve-plexus that spreads throughout the entire thickness of the mucous membrane, extends into the villi and supplies the muscularis mucosæ, the vessels of the mucosa and Lieberkühn's glands.

The epithelial cells of the large intestine possess no seam-like marginal thickening.

The serous coat of the alimentary tract is provided with special lymph-vessels, at first distinct from the chyle-vessels.

ABSORPTION OF THE DIGESTED FOOD.

PHYSICAL FORCES: ENDOSMOSIS, DIFFUSION, FILTRATION.

Endosmosis and diffusion take place between two liquids that are capable of admixture, as, for example, hydrochloric acid and water, but never between two fluids that are opposed to admixture, as, for instance, oil and water. If two miscible dissimilar liquids are separated from each other by a membrane provided with physical pores, such as may be present even in apparently homogeneous membranes, an interchange of the constituent parts takes place through the pores of the membrane, until finally both fluids have the same composition. This process is designated endosmosis or diosmosis. The endosmotic passage of a substance through the membrane takes place if a solvent liquid having an attraction for the substance is present on the other side of the membrane.

If both miscible fluids are simply placed over one another in a vessel, without the intervention of a porous septum, an interchange of particles of the liquids also takes place, until the entire mass has undergone homogeneous admixture. This interchange is designated diffusion.

The rapidity of diffusion is influenced: 1. By the nature of the fluids. Acids pass over most rapidly, alkaline salts more slowly; liquid albumin, gelatin, gum, dextrin, and starch-solutions most slowly. The latter, in part, do not crystallize, and also in part do not represent true solutions, but only suspensions. 2. The more concentrated the solutions, the greater is the diffusion. 3. Heat promotes, cooling retards, diffusion. 4. If the solution of a body difficult of diffusion is mixed with a readily diffusible solution, the former diffuses with even greater difficulty. 5. Dilute solutions of various substances diffuse into one another without difficulty, while concentrated solutions mutually retard diffusion. 6. Double salts, of which one constituent diffuses more readily, and the other with greater difficulty, may even be separated chemically by diffusion.

In the endosmotic interchange of fluids, the passage of the fluid-particles takes place independently of the hydrostatic pressure. Fig. 129 is a simple illustration of endosmotic exchange. A glass cylinder is filled with distilled water (F). A flask (J) is kept immersed in the water to a suitable height, and closed by a membrane (m) replacing its broken bottom. From the neck of the flask, in which it is tightly corked, projects a glass tube (R). The flask is filled with concentrated salt-solution up to the level of the lower extremity of the tube. The flask is introduced into the glass cylinder to such a distance that both fluids stand at the same level (x). In a short while the fluid rises in the tube (R), because particles of water pass through the membrane into the concentrated salt-solution in the flask, and independently of the hydrostatic pressure. The fluid rises in the tube as high as the attraction of the water causes it to. The height of the fluid thus indicates the osmotic pressure.

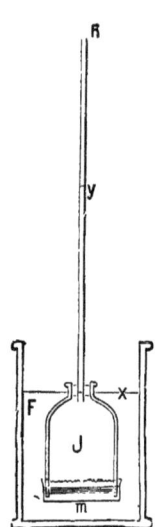

Conversely, also, particles of the concentrated salt-solution pass from the flask into the interior of the cylinder, mixing with the water (F). This interchange of current continues until an entirely uniform mixture is present in the flask and in the cylinder. Under these circumstances the level of the fluid will to the last always have risen higher in the tube (to y).

The circumstance that the level of the liquid within the tube can rise so high and be kept at such a height depends upon the fact that the pores of the membrane are too fine to permit of the action of hydrostatic pressure through them. Therefore endosmosis is defined as an interchange of particles of fluid independently of the hydrostatic pressure.

Reflection will show that if, in an endosmosis-experiment of similar kind, the water in the cylinder is renewed from time to time, the solution in the flask must become progressively more dilute, until, finally, the flask (J) and the cylinder (F) contain only pure water.

FIG. 129.—Apparatus for Diosmosis.

Endosmotic Equivalent.—It has been found that in endomosis-experiments, equal parts by weight of different fluids or soluble substances (which soon coalesce on the moist surface of the membrane within the flask to form concentrated solutions, as, for example, sodium chlorid) being present in the flask, a varying amount of distilled water passes through the membrane, so that, finally, if the water in the cylinder is constantly renewed, a variable amount of distilled water will be present in the flask. In other words, it has been found that a definite part by weight of a soluble substance in the flask has been exchanged by endosmosis for a definite part by weight of distilled water. The figure that indicates how many parts by weight of distilled water pass over in the endosmosis-flask for a definite part by weight of a soluble substance has been designated by Jolly as the endosmotic equivalent. For 1 gram of alcohol, 4.2 grams of water are exchanged; for 1 gram of sodium chlorid, 4.3 grams of water. The endosmotic equivalents for the following substances are:

Acid potassium sulphate.....	= 2.3	Magnesium sulphate.......	= 11.7
Sodium chlorid..............	= 4.3	Potassium sulphate........	= 12.0
Sugar	= 7.1	Sulphuric acid	= 0.39
Sodium sulphate	= 11.6	Potassium hydrate........	= 215.0

The amount of the substance passing through the membrane into the water of the cylinder within an equal time is proportional to the degree of concentration of the solution. If, therefore, the water within the cylinder is frequently renewed, the course of the endosmotic equalization is the more rapid. Further, the larger the pores of the membrane and the smaller the molecules of the substance in solution, the more quickly endosmosis takes place. It thus results that the rapidity with which endosmosis takes place varies for different substances. Thus the rapidity for sugar, sodium sulphate, sodium chlorid and urea is, as $1 : 1.1 : 5 : 9.5$.

The endosmotic equivalent for each substance, however, is not constant. It is influenced by: 1. The temperature, with increase in which, in general, the endosmotic equivalent increases. 2. C. Ludwig and Cloëtta have demonstrated that the endosmotic equivalent varies with the degree of concentration of the penetrating solutions; it is larger for dilute solutions of substances.

Should a solution of another substance be present in the cylinder instead of water, an endosmotic current takes place from both sides, until complete equalization is effected. In this process it is seen that these counter-currents of concentrated solutions have a disturbing influence on each other. If, however, two substances in solution are present in the flask at the same time, both diffuse toward the water, without interfering with each other. 3. The endosmotic equivalent varies with the employment of different membranes of different porosity. Sodium chlorid, which has an endosmotic equivalent of 4.3 when pig's bladder is used, possesses an equivalent of 6.4 when a cow's bladder is employed; 2.9 with a swimming-bladder, and 20.2 with a collodion membrane.

There are a number of fluids that, on account of the considerable size of their molecules, are capable of passing with difficulty, if at all, through the pores of a membrane impregnated with gelatinous substances, diffusible with difficulty. These consist in part of fluids that contain substances, not in true solution, but in a greatly diluted state of imbibition. Among such substances are the liquid albuminates, solutions of starch, dextrin, gum, mucus and gelatin. They are capable of gradually passing over into and mixing with other fluids by diffusion, in the absence of an intervening porous membrane-wall; they pass by endosmosis with difficulty, if at all, through the pores of membranes impregnated with gelatin. Nevertheless, the nature of the outside liquid must be taken into consideration; egg-albumin, it is true, passes through membranes into salt-solutions, but not into water; the transudate, under such conditions, becomes more concentrated. Graham has designated the substances in question colloids, because in considerable concentration they become gelatinous. They also possess the property of not crystallizing, as a rule, while crystalline substances, designated crystalloids, are exchanged by endosmosis. The endosmotic apparatus thus constitutes a mechanism for effecting a separation from mixtures of crystalloids and colloids, which by Graham is designated dialysis. If mineral salts are added to the colloid substances, their ability to pass through membranes is increased.

That endosmosis takes place within the alimentary canal, through its mucous membrane and the delicate membranes of the capillary blood-vessels and lymphatics, cannot be denied. On the one side of the membrane, within the tract, there are relatively concentrated aqueous solutions of salts, sugar, soaps, and peptones, all of which possess slight diosmotic power. On the inner side of the vessels is the colloid, albuminous solution of the blood and the lymph, practically incapable of osmosis, and deficient in the matters in solution within the alimentary canal, particularly in the state of hunger. The vital properties, however, probably in consequence of the motility of the protoplasmic structure within the membranes, also appear to exert some influence upon endosmosis. Thus, Reid observed that the exfoliated frog's skin is less permeable than living skin, and the latter, in turn, more so after irritation had been applied.

23

Filtration is the passage of fluid through the coarser intermolecular pores of a membrane dependent upon pressure. The higher the latter and the larger and more numerous the pores, the more rapidly will the filtrate pass through the pores of the membrane. Increase in temperature likewise accelerates filtration. Further, those fluids filter most readily that most rapidly soak into the membrane in question. Therefore, different fluids vary in the readiness with which they pass through different membranes. Further, the greater the concentration of the solutions, the more slowly, in general, is their passage. The filter has the property of retaining in part matters from the solutions passing through, either substances dissolved in the fluid (particularly colloid substances), or water (from dilute solutions of potassium nitrate). In the former case the filtrate is more dilute, in the latter more concentrated, than the fluid was before its passage through the filter. Other substances pass through without material change in concentration. Should the filtrate enter another fluid, the concentration of the transudate increases with the pressure under which filtration takes place. Some membranes exhibit a difference according as filtration takes place from their different surfaces; thus the membrana testacea of the egg permits of filtration only in the direction from without inward. The mucous membrane of the stomach and intestine also exhibits a difference in this respect.

It was formerly believed that filtration of substances in solution could take place from within the digestive canal into the vessels: 1. If the intestine contracted and thereby exerted pressure directly on the contents. This alone, however, could scarcely have any noteworthy influence, even in case the canal were contracted in two places and the intervening musculature, through contraction, compressed the fluid intestinal contents. 2. Filtration under negative pressure may be effected through the villi, which on contracting forcibly evacuate the contents of the blood-vessels and lymphatics in a centripetal direction. The latter particularly will remain empty, as the chyle in the fine lacteals is prevented from passing backward by numerous valves. When the villi are again relaxed, they will by suction be able to fill themselves with the fluids of the digestive tract capable of filtration. On the other hand, the fact must especially be emphasized that, according to Spee and Heidenhain, the muscles of the villus actively dilate the central lymph-vessels.

ABSORPTIVE ACTIVITY OF THE WALL OF THE ALIMENTARY CANAL.

The process of digestion prepares from the food in part true solutions, in part finely divided emulsions, whose small globules are surrounded by an albuminoid capsule.

Absorption of Solutions.—It cannot be denied that true solutions can pass over into the blood and the lymph of the intestinal canal by endosmosis, but some observations indicate that the cellular elements of the digestive tract also participate in the process of absorption through the functional activity of their protoplasm. It has not as yet been possible to refer the forces effective in this connection to simple physical or chemical processes. When Heidenhain introduced methylene-blue in solution into the intestine, he was convinced that the path of its absorption was in part through, in part between, the epithelial cells.

The *Inorganic Substances:* Water, and the dissolved salts necessary for nutrition, are generally easy of absorption, and in large measure by the blood-vessels. In the absorption of salt-solutions by endosmosis, water must naturally pass from the intestinal vessels into the intestine, while the salt-solutions enter the vessels. The amount of water, however, is but slight on account of the small endosmotic equivalent of the salts to be absorbed. Salts are absorbed in larger amount from concentrated than from dilute solutions. If, however, considerable amounts of salts with a high endosmotic equivalent are introduced into the intestine, as, for example, magnesium or sodium sulphate, these salts retain the water for their solution, and in addition more fluid escapes from the vessels of the intestinal wall, and diarrhea results. Conversely, it is evident that, on injecting these substances into the blood, a large amount of water passes from the intestine into the blood, so that constipation results, in consequence of the great dryness of the interior of the intestine. It should, however, especially be pointed out that the absorption of solutions of various salts, isotonic with one another, takes place differently. The epithelial cells of the intestine behave like the erythrocytes with respect to the permeability of the solutions. Water is absorbed from the stomach only in small amount.

The absorption of fluids takes place best at moderate pressure within the intestinal canal (from 80 to 140.cm. of water-pressure), in connection with which the surface of the mucous membrane is best smoothed out. A greater degree of pressure would compress the intestinal vessels and would accordingly allow absorption to diminish. During digestion, on account of the dilatation of the blood-vessels, absorption takes place rapidly. For this reason warm solutions also are more quickly absorbed from the stomach than cold, the latter causing contraction of the vessels.

The fact that a 0.5 per cent. sodium-chlorid solution is better absorbed than water, further a potassium-solution less well than sodium-solutions, and also the extensive absorption of dog's serum in the dog's intestine, are opposed to the view that only physical forces (endosmosis) are concerned in absorption.

Some other inorganic substances also, which are not, as such, constituents of the body, are absorbed by endosmosis: potassium iodid, potassium chlorate, potassium bromid; further, iron-salts, as well as dilute sulphuric acid, etc.

Carbohydrates in solution have their chief representatives in the different varieties of sugar—and principally in dextrose and maltose, which have relatively high endosmotic equivalents, as cane-sugar is generally transformed by a ferment into invert-sugar. Absorption appears to take place relatively slowly, as, at this time, only small amounts of grape-sugar are found in the intestinal vessels and in the portal vein. According to v. Mering, the sugar is absorbed from the intestine by the portal vein. Dextrin is also present in the blood of the portal vein, as boiling with dilute sulphuric acid increases the amount of sugar in this blood. The amount of sugar absorbed depends upon the concentration of its solution in the intestine. Therefore, the amount of sugar contained in the blood is increased after a diet rich in sugar, so that it may even pass over into the urine. To this end approximately a 0.6 per cent. solution of sugar in the blood is necessary. Also cane-sugar in small amount has been found in the blood. When a large amount of sugar-solution is present in the intestine, a portion also enters the lymph-vessels. In a girl with a fistula of the receptaculum chyli, not more than $\frac{1}{2}$ per cent. of the sugar introduced into the alimentary canal was found to be absorbed by the lacteals. The sugar is in part consumed in the blood and in metabolism, perhaps principally in the muscles.

Peptones have an endosmotic equivalent, more than four times smaller than that of dextrose. They can be rapidly absorbed, on account of their ease of diffusion and filtration. Absorption takes place through the blood-vessels, unless excessive amounts are present in the intestine, as after ligation of the thoracic duct, ingested proteids are as well absorbed as under normal conditions. Peptones have been recovered from the blood, with certainty, in small amounts only. It is, therefore, to be inferred that they are quickly retransformed into true proteids. The mucous membrane possesses the property of retransforming peptone into albumin. Heidenhain regards the epithelial cells of the villi as the seat of this transformation. Peptone gains entrance into the blood unchanged only in minimal amount and it disappears from this after its passage through the tissues.

If blood containing peptone is kept warm in the presence of a small piece of small intestine, while air is passed through the mixture, the peptone soon disappears from the blood.

The peptones undoubtedly represent the principal contingent of the albuminates destined for absorption. Of all the proteids they alone suffice to maintain the body equilibrium, as animals fed upon peptone only (in addition to the necessary fat or sugar) are able to maintain their nutrition. They can do the same when fed with propeptone.

According to Pfeiffer, the diffusion of the peptones is promoted by a 1 per cent. solution of sodium chlorid or sulphate. The absorption of grape-sugar and peptone in the stomach and intestine is increased by the addition of certain substances, as, for example, sodium chorid, pepper, alcohol or ethereal oils. In dogs a peptone-solution (5 cu. cm. of a 20 per cent. solution in 0.6 per cent. sodium-chlorid for an animal weighing 8 kilograms) introduced into the blood, causes death.

Unchanged Proteids.—In spite of their slight power of filtration and (on account of their great endosmotic equivalent) of diffusion, it has been demonstrated with certainty that unchanged proteids, such as liquid casein and the proteids of milk, meat-juice, dissolved myosin, alkali-albuminate, egg-albumin mixed with sodium chlorid, syntonin, gelatin, can be absorbed; their absorption takes place, in part, even from the mucous membrane of the large intestine. The amount of absorbed unaltered albumin is, however, smaller than that of the peptones.

Egg-albumin without sodium chlorid, serum-albumin, hemoglobin and fibrin are not absorbed. Many years ago the author made the observation in a young man that after the ingestion of the white of between 14 and 20 raw eggs, with sodium chlorid, albumin was excreted in the urine after from 4 to 10 hours. The amount of albumin thus excreted increased up to the third day, then becoming less and ceasing on the fifth day. The more albumin ingested, the earlier the albuminuria appeared and the longer it lasted. In this case the condition was evidently one in which considerable absorption of unchanged egg-albumin took place into the circulation. If egg-albumin be injected directly into the blood-stream of animals, it likewise passes, in part, into the urine.

The soluble *soaps* form only a part of the fats absorbed, the largest portion of the fat being taken up in the form of a finely granular emulsion. Absorbed soaps have, on the one hand, been found in the chyle; on the other hand, from the circumstance that the blood of the portal vein is richer in soaps at the time of absorption than during the state of hunger, it has been inferred that absorption of the soaps takes place, to some extent, through the intestinal capillaries. Nevertheless, only a small portion of the soaps enters the blood.

The experiments of Lenz, Bidder and Schmidt render it probable that the organism can take up only a limited amount of fat within a certain time, and this may, perhaps, bear a definite relation to the quantity of bile and pancreatic juice. Beyond that amount no more fat is absorbed. Thus, in cats, 0.6 gram of fat an hour was found to be the greatest amount absorbed for every kilogram of body weight. I. Munk and Rosenstein found the absorption of fat greatest from 5 to 8 hours after ingestion, and earlier or later accordingly as the fat was more or less readily liquefiable.

The greater part of the soaps in the intestine, transformed into neutral fat, passes over into the chyle. It seems as if the soaps are capable of uniting with glycerin in the parenchyma of the villus to form neutral fat. Perewoznikoff and Will found neutral fat after the injection of both of these ingredients into the intestinal canal, and also C. A. Ewald observed fat to form when he brought soap and glycerin in contact with the fresh, living intestinal mucous membrane. Blood and chyle contain no free fatty acids. In the blood the fat is subsequently decomposed in the presence of oxygen.

Of other organic matters in solution that are introduced into the intestinal tract, some are absorbed, as, for example, alcohol, and many others. Other bodies may be in part absorbed, in part fermented: tartaric acid, citric acid, malic acid, lactic acid, glycerin and inulin ; gum and vegetable mucin, which give rise to the formation of glycogen in the liver; and it is probable that unknown products of metabolism are also absorbed.

Of pigments, alizarin, alkanna and indigo-carmine are absorbed: others are in part absorbed, such as hematin; chlorophyll is not absorbed. Metallic salts appear, in part, to be held in solution by an excess of albuminates, and to be absorbed at the same time with these (iron sulphate has been found in the chyle), and, in part, to be conveyed to the liver through the blood of the portal vein. Numerous poisons undergo rapid absorption, prussic acid in the course of a few seconds; potassium cyanid has been found in the chyle.

Moreover, the purely physical conception of the absorption even of true solutions by endosmosis and filtration alone is not sufficient. Here, also, the protoplasm of the cells takes at least an active part, for only in this way is it possible to explain how even a slight derangement in the activity of these cells, as, for example, after cold or excitement, may be followed by sudden serious disturbances of absorption, even the escape of fluid into the intestine. Only in this way, also, can the fact be explained that the presence of different spices, in small amount, actively increases absorption in the stomach. If, further, absorption took place solely and alone by endosmosis, water would pass over into the intestine after the injection of alcohol; but this never occurs. Further, salt is absorbed in the intestine from a solution that has less osmotic energy than blood-plasma. Moreover, Brieger observed, after the injection of from 0.5 to 1 per cent. solutions of metallic salts into ligated loops of the intestine, that transudation of water into the bowel failed to take place; although this occurred when injections of 20 per cent. solutions were made.

Absorption of the Smallest Granules.—The largest amount of the neutral fats and at the same time also of the fatty acids is absorbed in the form of a milky emulsion prepared by the bile and by the pancreatic juice and composed of minute granules. The individual fat-granules appear to be surrounded by a delicate albuminous membrane, the haptogenic membrane, which is derived in part from the pancreatic juice. In the absorption of fat-emulsions, the villi of the small intestine participate primarily and in greatest degree; but the epithelial cells of the stomach also, as well as those of the large intestine, take part in this process. In the villi the fat-granules are seen: (1) Within the epithelial cells, the protoplasm of which is dotted with them. The nucleus remains free from them, yet it is so beset by the innumerable fat-granules as to escape observation. (2) Within the tissue of the villus itself, the granules traverse in large numbers the intercommunicating course of the spaces in the reticular tissue. Not rarely, when absorbed in smaller amount, the granules arrange themselves in connected reticular paths. At times they appear to be collected in undivided, band-like lines; at other times, the entire parenchyma of the villus is completely filled with innumerable granules. (3) At a later period the central lymph-vessel in the axis of the villus appears filled with fat-granules.

The amount of fat in the chyle varies in the dog, after generous feeding of fat, from 8 to 10 per cent. The fat disappears from the blood within thirty hours. If chyle, rich in fat, is mixed with blood (even if lake-colored), and is agitated with air, the amount of fat in the mixture diminishes as a result of the action of a lipolytic substance present in the blood, in consequence of which a body, insoluble in ether, is formed.

The fat-granules are taken up out of the blood by the various tissues, particularly by the liver, and in smallest measure by the muscles. The consumption of fat in the tissues begins with a division into glycerin and fatty acids, which is followed by the final combustion.

With regard to the forces that effect absorption of the fat-granules, it appeared conceivable from observations made by v. Wistinghausen that moistening of the porous membranes with bile is capable of facilitating the passage of fat-granules; but this does not adequately explain the abundant and rapid absorption. It appears most probable that the protoplasm of the epithelial cells of the alimentary tract seizes the fat-granules by an independent movement, and then actively draws them within itself. The protrusion of delicate protoplasmic filaments from the cell-body would take place in a manner similar to that in which the absorption and the inclusion of granular articles of food takes place in the lower organisms, the amœbæ. Absorption is possible on the part of the goblet-cells also, because the entrance to the cell remains open. The protoplasm of the epithelial cells communicates directly with the protoplasmic lymphoid cells present in large number within the reticulum of the villus. Thus, the granules may be conveyed to these cells and finally from them, through the stomata between the endothelial cells, into the central lymph-vessel of the villus.

The process of the absorption of granules—and perhaps the same is in part true of proteids—is thus established as a wholly active, vital one. This view receives adequate support from the investigations of Brücke and of v. Thanhoffer and others, as well as the observation of Grünhagen that the absorption of fat-granules in frogs takes place most rapidly at a temperature at which the motile phenomena of the protoplasm are most active. In fact, the conception of a simple physical filtration of the granules into the tissue of the villus is scarcely any longer permissible. This is to be concluded also from the fact that the number of fat-granules present in the chyle is independent of the amount of water present in it. If absorption took place essentially through filtration, the constancy of a direct relation between the amount of fat and the amount of water present would at least be highly probable. The fatty acids, in their passage through the intestinal wall, are retransformed with fixation of glycerin into neutral fats. They pass, in part, through the blood-vessels.

The intestine of distomum hepaticum may be considered as a truly classical object-lesson for a study of the cells of the intestine in their functional activity and of the manner in which they accomplish the absorption of solid substances by means of their pseudopod-like processes. Sommer has admirably depicted the conditions, and the author convinced himself of the accuracy of the representation by personal observation of the preparations. Metschnikoff noted similar conditions in celenterates, Du Plessis in turbellaria, Greenwood in earth-worms.

If carmine or India-ink is mixed with the food of rabbits, a deposition of either granular pigment takes place in Peyer's patches and in the lymph-cells.

Pathological.—In the presence of severe intestinal disease, injury to and alteration in the epithelial cells of the intestine appear to be caused by a poison elaborated in the bowel, as, for example, in cases of cholera and cholera infantum.

INFLUENCE OF THE NERVOUS SYSTEM.

Little is known with certainty concerning the influence of the nervous system upon the processes of absorption in the intestinal tract. After division of the mesenteric nerve-filaments, the intestinal contents become abundant and watery. This may be due, in part, to deficient absorption, as well as to an increased, paralytic secretion of the intestinal juice, although it is as yet impossible to determine with certainty to what extent transudation into the intestine on the part of the vessels participates in this process. After extirpation of the sympathetic ganglia of the abdomen, symptoms of paralysis of the intestine appear, with exhausting diarrhea, finally terminating fatally; acetone is also present in the urine. Of especial interest is the observation of v. Thanhoffer, who noted the protrusion of filaments from the protoplasm of the epithelial cells of the small intestine only when the medulla oblongata or the dorsal nerves had been divided some time previously.

NOURISHMENT BY MEANS OF "NUTRITIVE ENEMATA."

In those desperate cases in human beings in which administration of food by the mouth is impossible, e. g., in the presence of stenosis of the esophagus or of persistent vomiting, resort has been had to the procedure adopted by Corn. Celsus, namely, rectal alimentation. As the large intestine is capable of scarcely any digestive activity it is best to introduce fluid material capable of absorption, which is permitted to flow slowly, by its own weight, into the anus, preferably through a long tube provided with a funnel. The recipient must endeavor to retain the material for as long a time as possible. By means of slow and gradual injection, the fluid at times may even pass beyond the ileo-cecal valve. Particles of proteid substances, saturated with a solution of sodium chlorid, may even pass through the small intestine into the stomach, where they may be digested.

Nitrogenous substances are to be recommended for this purpose: eggs rubbed up into an emulsion with an aqueous solution of sodium chlorid, peptone or propeptone; less well, milk and egg-albumin with sodium chlorid. The commercial preparations of peptone are made by digestion with pepsin, by vegetable ferments or by superheated water, and they often contain much propeptone. An adult should receive daily 120 grams, a child 50 grams of meat-peptone; Leube advises from 50 to 80 grams dissolved in 250 cu. cm. of water. In addition, as a stimulant and as food-sparer, tea with wine may be given. Leube introduces into the rectum a pasty mixture consisting of 150 grams of meat with 50 grams of reddened pancreatic tissue and 100 grams of water, and it is believed that proteids are peptonized and absorbed here. In addition, as much as 50 grams of grape-sugar dissolved so as to make 300 cu. cm., or starch-paste and dilute lakecolored blood may be employed; also fat-emulsions (not more than 10 grams of fat daily); mixed with pancreatic paste, as much as 50 grams of fat can be given.

Whether thin soap-solutions are advisable, however, has not as yet been determined. This mode of administering nutriment by means of nutrient enemata, must, however, always remain imperfect; at best only one-quarter of the amount of proteids necessary for the maintenance of the metabolic equilibrium is absorbed.

SYSTEM OF LACTEAL AND LYMPHATIC VESSELS.

Within the tissues of the body, and even in those without special blood-vessels (cornea) or with but a poor supply, there is present a system of vessels conveying fluid, and within which the movement is only centripetal. These vessels begin within the parenchyma of the organs in widely different ways, and unite in their course to form delicate, then thicker tubes, which empty into two trunks of considerable size at the junction of the common jugular and subclavian veins: the thoracic duct on the left side, the lymphatic trunk on the right.

The importance of the lymph and of its movement in the various organs is apparent in different ways at different points. (1) In some tissues the lymphatics represent the nutrient channels through which the nutrient fluid given off by adjacent blood-vessels is distributed, as in the cornea particularly and often within the connective tissues. (2) In some tissues, as in the glands, for example the salivary glands and the testicles, the lymph-spaces constitute the chief reservoirs for fluid, from which, at the time of secretion, the cellular elements derive their necessary fluid. (3) In addition, the lymphatic vessels everywhere have the task of collecting the fluid with which the tissues are saturated and of conveying it back again to the blood. If the network of capillary blood-vessels be regarded, from this standpoint. as an irrigation-system, which supplies the tissues with nutrient fluid, the lymphatic system can be considered as a drainage-mechanism, which, in turn, conducts away the excess of the transuded fluids. Metabolic products from the tissues, the products of retrogressive metamorphosis, are added to this return-current. The lymph-channels are thus, at the same time, absorbent vessels: substances that would otherwise be carried to the parenchyma of the tissues are thus also absorbed by the lymphatic system.

A consideration of these circumstances shows that the system of the lymph-channels represents in reality an appendix to the blood-vascular system; therefore, further, the lymphatic system cannot be active at all if the circulation of blood is totally interrupted; it operates only as a part of the whole and with the whole.

If the lacteals are contrasted with the true lymph-vessels, this is done chiefly for anatomical reasons, because the important and considerable paths of the former, which are derived from the entire intestinal tract, have especially attracted the attention of investigators since antiquity and are to a certain extent an almost independent division of the lymphatic system, with conspicuous absorptive activity. In addition their contents, of white color from the generous admixture of fat-granules, as chyle or lacteal fluid, appeared at first sight to be essentially distinct from the clear and watery fluid of the true lymphatics. From the physiological standpoint, however, the lacteals cannot be given an independent position. They are, functionally and structurally, lymphatics, and their contents are true lymph, mixed with a large amount of absorbed materials.

ORIGIN OF THE LYMPH-CHANNELS. LYMPHATICS.

Development by Means of Secretory Spaces. Within the supporting substances (connective tissue, bone) numerous star-shaped or polymorphous spaces
are found that are connected with one another by means of delicate tubular
processes. This system of communicating spaces contains the cellular elements
of the tissues. The cells, however, by no means completely fill the spaces,
an interval often existing between the cell-body and the wall of the space,

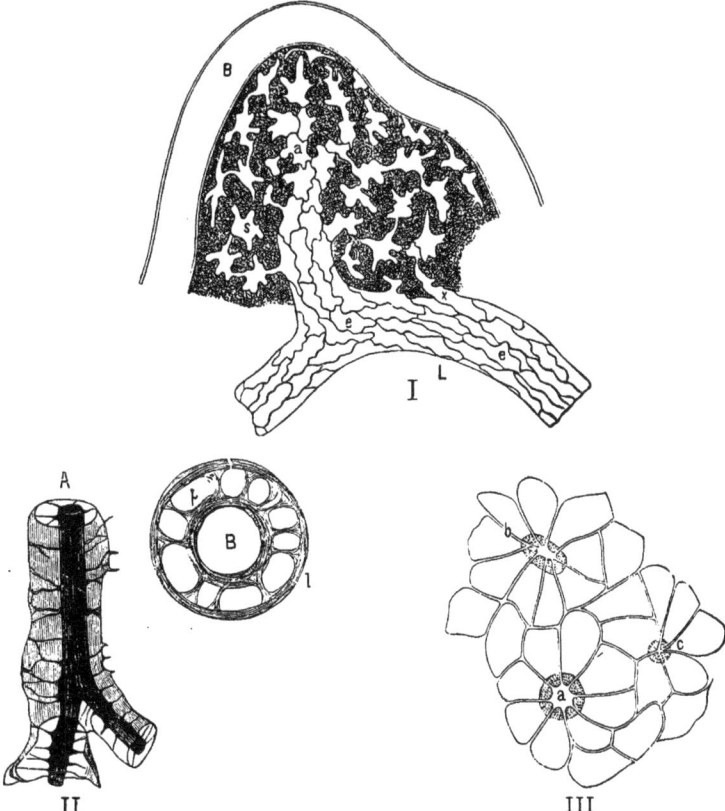

Fig. 130.—Origin of the Lymph-channels: I, from the central tendon of the rabbit (semi-diagrammatic); s,
secretory spaces, communicating with the lymphatic at x; a, commencement of the lymphatic from the
confluence of secretory spaces. II, perivascular lymphatics. III, lymph-stomata.

and varying in size, in accordance with the state of motility of the protoplasmic
cells. These spaces are the so-called secretory spaces, or secretory canals, and
they represent the commencement of the lymphatics. As adjacent spaces inter-
communicate, the propulsion of the lymph is provided for. The cells lying in
the secretory spaces are capable of ameboid movement. In part they remain
permanently in their spaces (fixed connective-tissue cells, bone-corpuscles); in
part they are capable of engaging in active migration through the secretory canal-

system (wandering-cells). At greater or lesser distances, these secretory clefts are connected with minute tubular lymphatics, which are designated lymph-capillaries (Fig. 130, I, L). Their commencement results from the more intimate approximation of secretory spaces (I, a).

The lymph-capillaries, generally exceeding the capillary blood-vessels in caliber, lie principally in the space midway between the arched loops of the blood-capillaries (B). They are composed of delicate nucleated endothelial cells (e), whose characteristic sinuous edges can be stained black by means of a solution of silver nitrate. Between the endothelial cells scattered spaces, stomata, are present. The endothelial cells constituting the wall are often united by bridges of protoplasm. According to Kolossow, the cells may recede from one another at their edges, and thus form spaces between them, while the connecting bands of protoplasm are capable, subsequently, of drawing them together again. Thus, the stomata would develop temporarily and again close.

It is to be inferred that the blood-vessel system communicates with the lymph-spaces; that the blood-plasma finds its way into the lymph-spaces from the thin-walled blood-capillaries through their stomata. In the lymph-spaces this fluid maintains the nutrition of the tissues, inasmuch as the necessary constituents are taken up independently by the tissues. The materials consumed are returned to the lymph-spaces and later reach the lymph-capillaries, which finally deliver them to the venous system.

To what extent the cellular elements within the lymph-spaces exert any action upon the discharge of blood-plasma and later upon its propulsion into the lymphatics can only be surmised. It can be conceived that, through contraction and diminution in size of their cell-bodies, as well as through partial change in position from the group of secretory spaces closer to the blood-vessel to that directed toward the lymph-capillary, they might exert suction upon the blood-plasma transuded. If the cells, themselves, then take up the transuded fluid, the conception is permissible, further, that by subsequent contraction they express this fluid in a certain direction, namely from secretory space to secretory space, toward the lymph-capillaries. In consequence of the independent migration of the cellular elements through the secretory spaces into the larger lymph-paths, small particles that may be contained in the secretory spaces (as, for example, pigment-granules that have been rubbed into the tissue of the irritated, horny skin in the process of tattooing, and also minute fat-granules, bacteria and the like), and which the lymph-cells are capable of taking up through ameboid movement, may be propelled onward.

After what has been said concerning the migration of leukocytes from the blood-stream through the stomata between the endothelial cells of the capillaries, or through the walls of smaller vessels, the migration of cellular elements from the blood-vessel system into the commencement of the lymph-channels may be regarded as a normal process. Granular pigments pass from the blood into the protoplasmic bodies of the cells in the lymph-spaces. Only when the granular substance is present in large amount is it distributed into the ramifications of the lymph-spaces as a granular injection.

The origin of the lacteals within the villi has been outlined in their description as organs of absorption.

Commencement of the Lymphatics in the Form of Perivascular Spaces.—In the tissue of bony substance, of the central nervous system and of the liver, the smallest blood-vessels are surrounded by wider lymph-vessels, so that the blood-vessels lie in the lymph-vessels like a finger in a glove. In the brain these lymph-vessels are in part constituted of delicate connective-tissue fibrils, which, partly traversing the lumen of the lymph-canal, are supported upon the surface of the blood-vessel. Fig. 130 II, B represents in transverse section a small blood-vessel (B), with a perivascular lymph-vessel, from the brain; p is the traversed lumen of the lymph-vessel. In addition to these so-called perivascular spaces of His, the cerebral vessels are provided also with lymph-spaces within the adventitia (Virchow-Robin spaces). In part these possess a well-developed endothelium. In their further course, where the vessels increase in caliber, the blood-vessel penetrates the wall of the lymph-vessel at one spot, and both continue separately side by side. Wherever the lymph-vessels serve as perivascular sheaths, the passage of blood-plasma and lymph-cells into the lymph-stream is greatly facilitated. It should be especially mentioned that, in tortoises, even the larger vessels are often covered by the lymph-vessels as a sheath. In Fig. 130, II, A, the bifurcation of the aorta, with the perivascular lymph-vessels, is shown according to Gegenbaur. The animals referred to exhibit macroscopically the same relations that warm-blooded animals present microscopically; and thus the illustration may serve also as the microscopical picture of small perivascular lymph-vessels in warm-blooded animals.

Commencement in the Form of Interstitial Spaces Within the Viscera.—In the testicles the lymphatics commence simply in the form of numerous spaces, which occur between the multifarious coils and convolutions of the seminiferous tubules. They will, therefore, here present the form of spaces bounded by the arched, cylindrical surfaces of the tubules. The limiting surfaces are, however, lined with endothelium. The lymphatics acquire independent tubular walls only beyond the parenchyma of the testicle. Similar conditions are found in the kidneys. In many other glands the glandular substance is likewise surrounded by lymph-spaces. Into these the blood-vessels first pour lymph, from which the secreting cells remove the material for the formation of the glandular secretion, as, for example, the salivary glands.

Commencement by Means of Free Stomata upon the Walls of the Larger Serous Cavities (Fig. 130, III). From the investigations of v. Recklinghausen, C. Ludwig, Dybkowsky, Schweigger-Seydel, Dogiel and others, it has been found that the old view of Mascagni, that the serous cavities communicate freely with the lymphatics, is correct. Upon examining serous membranes (most readily the peritoneal lining of the large lymph-cavity in the frog), best after moistening them with argentic nitrate, followed by exposure to the action of light, disseminated, relatively large, free openings of the stomata are found lying between the endothelial cells. Groups of the latter include a stoma among them. A portion of motile protoplasm lies in the cells surrounding the stoma, close to the edge of the opening. Upon the state of contraction of this protoplasm appears to depend the fact whether the stomata are widely open (a), half closed (b), or completely closed (c). These stomata are thus the beginnings of the lymph-capillaries. Fluids, introduced into the serous cavities, therefore readily reach the path of the lymphatics. The cavities of the peritoneum, the pleuræ, the pericardium, and the serous covering of the testicle, further the arachnoid space, the chambers of the eye, and the labyrinth of the ear have shown themselves to be true lymphatic cavities; the fluid in them is thus to be designated lymph. Fluids in the peritoneal cavity are absorbed, in part, also by the veins. The endothelial cells of the serous membranes are capable of movement and communicate with one another by means of connecting bridges of protoplasm. In the animal kingdom the free surfaces of the cells are frequently provided with cilia.

Even upon the free surface of a number of mucous membranes, it is stated, open pores have been observed as the commencement of the lymphatics: in the bronchi, in the nasal mucous membrane and in the larynx.

The larger lymphatics arising from the lymph-capillaries closely resemble veins of equal size in the structure of their walls. Especial stress is to be laid upon the presence of a large number of valves, which are placed so closely behind one another that the distended lymphatic is not unlike a string of pearls.

THE LYMPH-GLANDS.

The so-called lymph-glands are peculiar to the lymphatic apparatus. They are inappropriately designated glands, because they really represent only many-branched, lacunar, labyrinthine spaces, constituted of adenoid tissue, interposed in the course of the lymphatics. Simple and compound lymph-glands can be distinguished.

The **simple lymph-glands**, more correctly designated simple lymph-follicles or cutaneous follicles, are present either isolated (solitary follicle), as in the intestine, the bronchi, the spleen; or collected in masses (conglobate follicle), as in the tonsil, Peyer's patches, the follicles of the tongue. They are small, spherical vesicles, attaining approximately the size of a pin's head, and they consist throughout of delicate elements of the reticular connective tissue intermixed with elastic fibrils and arranged in a network (Fig. 131, C). In the meshes of this network, lymph and lymph-cells are present in abundance. Upon the surface the tissue becomes condensed into a somewhat more independent. conspicuous sheath. which, however, is variously traversed by small spongy spaces in the reticular tissue. Small lymphatics advance everywhere directly up to these lymph-follicles, often keeping considerable areas of their surface covered with a rich network. Frequently, also, the surface of the follicle is incorporated into the wall of the vessel, at times throughout a slight, at other times throughout a considerable, extent, so that the surface of the follicle is directly irrigated by the lymph of the vessel; and, if no direct canal-orifice of considerable size leads from the lumen of the lymphatic into the interior of the spherical follicle, a communication must, nevertheless. be assumed to exist between the small lymphatic and the lymph-follicle, and this is adequately provided by the innumerable spaces between the follicles. Thus, the lymph-follicle is a true lymphatic structure, whose fluid and lymph-cells can pass over into the stream of the adjacent lymphatics. The follicles are provided, upon their surfaces, with a network of blood-vessels, which also send numerous delicate ramifications and capillaries through the interior of the follicle (A), within which they are supported by the reticulum (B). It is to be inferred that leukocytes can pass from these capillaries into the follicle.

It should be mentioned as of special importance in connection with these follicles that, in the lymph-glands, the solitary as well as the

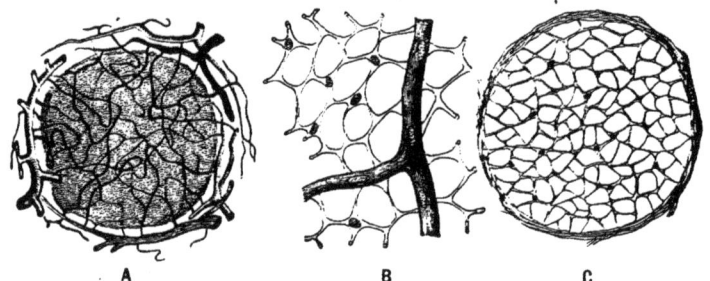

FIG. 131.—A, blood-vessels of the follicle; B, the reticulum and a branch of a blood-vessel; C, lymph-follicle with reticulum and sheath.

conglobate glands, an enormous migration of the leukocytes normally takes place uninterruptedly during life through the epithelium between the cells. The leukocytes insinuate themselves between the epithelial cells, but, by their enormous migration, as well as by the divisions that take place during this process, they impair the functions of the epithelium and may even destroy it. Thus, in a measure, physiological injuries result, which prepare the way for invading microörganisms. The cells that have thus migrated later undergo disintegration.

The compound lymph-glands (incorrectly designated lymph-glands) represent to a certain extent an aggregation of lymph-follicles of altered shape. Every lymph-gland is surrounded externally by a connective-tissue capsule traversed by numerous unstriated muscle-fibers, and from whose inner surface numerous septa and bands (Fig. 132, a a) penetrate into the interior of the body of the gland, and divide it into a large number of small compartments. The latter possess within the cortical substance of the gland a rather rounded shape (alveoli), in the medulla, a rather longitudinal sausage-shaped form (medullary spaces). All, however, are of the same significance and all are connected by communicating orifices. Thus, a rich network of cavities. connected in all directions, is formed within the lymph-gland by the septa. These spaces are traversed by the so-called follicular bands (f f). The latter represent to a certain extent the inner-most contents of the spaces, but in such a manner that they are smaller than the spaces and nowhere touch the walls of the cavities themselves. If the cavities of the gland be conceived as injected with a substance that at first has filled them all, but later, by contraction, is reduced to half its size, one will have an approxi-mate picture of the spatial relations of the follicular bands to the cavities of the gland. The follicular bands contain the blood-vessels (b) of the gland within them. About these there is deposited a rather thick cortex of reticular connective tissue, whose meshes (x) are extremely delicate and fine, whose spaces are rich in lymph-cells and whose surface (o o) is so constituted of the condensed reticulum-cells that a communication between the narrow meshes is still possible.

Between the surface of the follicular bands and the inner wall of all the cavities

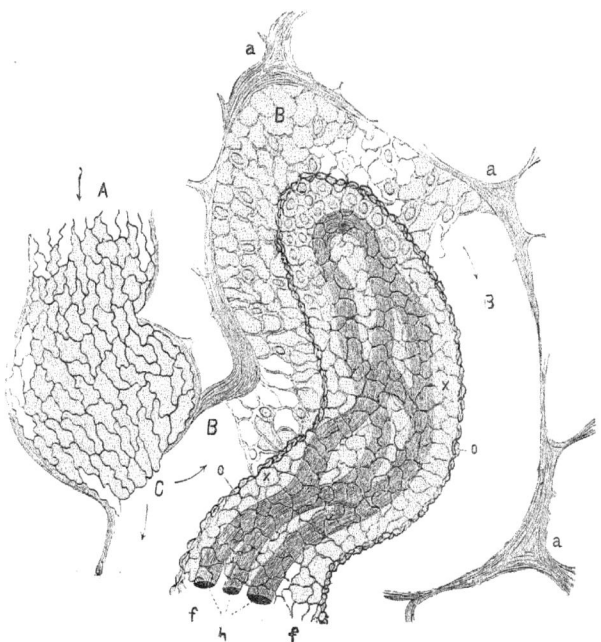

Fig. 132.—Part of a Lymph-gland: A, afferent vessel; B, B, lymph-path within the cavity of the gland; a, a, column and septa bounding the cavity of the gland; f, f, follicular band of the cavity; x, x, its reticulum; h, its blood-vessels; o, o, delicate reticular junction between the follicular band and the lymph-paths.

of the glands lie the paths of the lymphatics (B B). Perhaps they are lined by an endothelium; their lumina are traversed by a rather coarse reticulum.

The afferent-vessels (A), which spread out upon the surface of the gland, penetrate the external capsule and pass over into the lymph-paths of the glandular cavities (C). The efferent vessels, which exhibit large, almost cavernous anasto-moses and dilatations in the vicinity of the gland, arise at other parts of the gland directly from the lymph-paths. The latter, thus, to a certain degree represent a dense interlacing network of capillary vessels, lying within the glandular cavi-ties, arranged between the afferent and efferent vessels.

The movement of the lymph on its way through the many-branched and tortuous lymph-paths of the gland will be retarded and, on account of the resist-ance to the current that the cellular elements, arranged in the paths, must offer, will possess feeble propulsive power. The lymph-corpuscles, lying in the meshes of the reticulum, are carried onward by the lymph-stream, so that, after flowing through the glands, the lymph is richer in cells. The lymph-cells lying in the range of the follicular bands may again migrate through the narrow meshes of the reticulum (o) into the lymph-paths, to make good the loss. The formation of the lymph-cells in the follicular bands either takes place locally by division, or new cells migrate from the capillary blood-vessels into the follicular bands. Further on, the muscular activity of the capsule and of the trabeculæ should not be underestimated in the movement of the lymph through the glands. Such muscular contraction will express the gland like a sponge. The direction of the fluid thus discharged is governed by the presence of valves within the related lymphatics.

Of the chemical substances in the lymph-glands, in addition to those of the lymph, leucin and the xanthin-bodies are worthy of mention.

PROPERTIES OF THE CHYLE AND THE LYMPH.

Both chyle and lymph are colorless, albuminous, clear fluids, containing lymph-cells. The latter are in reality the same elements that enter the circulation with the lymph-stream, and within the former are designated white blood-corpuscles. The source of the lymph-cells is discussed on p. 370. As, in rare cases, isolated red blood-corpuscles also pass out through the walls of the vessels and into the commencement of the lymph-vessels again, the presence of erythrocytes in the lymph, rarely in the chyle, is readily explained. Red blood-corpuscles can also pass over from the veins into the central extremities of the large lymph-trunks when the pressure in the veins is high. Lymph and chyle contain also molecular granules, and fragments of disintegrated leukocytes; chyle contains, in addition, numerous fat-granules.

In the lymph a distinction is made between the lymph-plasma and the contained lymph-cells or leukocytes, whose chemical constituents are considered on p. 64. The lymph-plasma contains both of the fibrin-factors, derived from disintegrated lymph-cells. They cause coagulation of the lymph after withdrawal from the body, and in this process the soft, gelatinous, scanty lymph-clot, which forms but slowly, includes the still surviving lymph-cells within it. The fluid remaining, the lymph-serum, contains alkali-albuminates, serum-albumin and some diastatic ferment derived from the blood. Of the coagulable albuminates about 37 per cent. consist of paraglobulin.

The *chyle*, which is the sole fluid contained in the lymphatic vessels of the digestive tract (lacteals), can be obtained only in small amounts, before its admixture with the lymph, and it can, therefore, be examined only with great difficulty. A small number of lymph-cells are already present in the first beginnings of the lacteals in the villi; beyond the intestinal wall and, still more, after passing through the mesenteric glands, their number increases. On the other hand, the amount of the solid constituents of the chyle, which is increased after abundant good digestion, is decidedly diminished after the chyle has become mixed with lymph. After the ingestion of food rich in fat the chyle contains many fat-droplets (from 2 to 4 μ in diameter), which, however, decrease conspicuously in the further course of the current. The amount of fibrin-factors in the chyle increases with increase in the number of lymph-cells. In addition, chyle contains sugar (up to 2 per cent.), glycogen, peptone adherent to the leukocytes, diastatic ferment absorbed from the intestine, and lactates after ingestion of starches, traces of urea and leucin.

The chyle from the body of an executed person contained, together with 90.5 per cent. of water:

Solids .....9.5 { fibrin .... a trace / proteid........7.1 / fats..........0.9   Extractives.....1.0   Salts...........0.4

Carl Schmidt found the following inorganic constituents in 1000 parts of chyle from a horse:

Sodium chlorid......5.84	Sulphuric acid.....0.05	Magnesium phosphate 0.05
Sodium 1.17	Phosphoric acid....0.05	Iron ....a trace
Potassium 0.13	Calcium phosphate.0.20	

The *lymph*, at the beginnings of the lymphatics, is likewise deficient in cells, and clear and colorless. The fluid from the serous cavities and synovial fluid exhibit similar features. A variation in the lymph, in accordance with the tissues from which it is derived, can with certainty be assumed, although, up to the present time, this has not been established. After passing through the lymph-glands, the lymph becomes richer in cellular elements and, probably in consequence of this, also richer in solid constituents, particularly proteid and fat. In one cu. cm. of lymph from a dog, 8200 lymph-corpuscles were counted.

Hensen and Dähnhardt succeeded in collecting for examination pure lymph in considerable amount from a lymphatic fistula on the thigh of a human being. It had an alkaline reaction and a salty taste. The relative composition of pure lymph, cerebrospinal fluid and pericardial fluid is as follows:

	Pure Lymph. (Hensen and Dähnhardt.)	Cerebrospinal Fluid. (Hoppe-Seyler.)	Pericardial Fluid. (v. Gorup-Besanez.)
Water	98.63	98.74	95.51
Solids	1.37	1.25	4.48
Fibrin	0.11	—	0.08
Albumin	0.14	0.03—0.06	2.46
Alkali-albuminate	0.09	—	—
Extractives	—		1.26
Urea, leucin	1.05		
Salts	0.88		

Absorbed carbon dioxid, to 70 per cent. by volume, of which 50 per cent. could be obtained by extraction and 20 per cent. was obtained by addition of acid.

The cerebrospinal lymph contains a substance that reduces Fehling's solution, and that Nawratzki determined to be dextrose. This, however, disappears soon after death.

100 parts of lymph-ash contain:

Sodium chlorid....74.48	Calcium..........0.98	Sulphuric acid.......1.28
Sodium...........10.36	Magnesia..........0.27	Carbon dioxid........8.21
Potassium.........3.26	Phosphoric acid....1.09	Iron oxid...........0.06

Just as in the case of the blood, potassium and phosphoric acid, of the inorganic constituents, predominate in the cells; while in the lymph-serum, sodium preponderates, principally as sodium chlorid. Only in the cerebrospinal fluid are the potassium-combinations and the phosphates said to predominate. The amount of water in the lymph rises and falls in correspondence with that in the blood. Of gases, dog's lymph contains carbon dioxid in abundance (over 40 per cent. by volume, of which 17 per cent. can be pumped out and 23 per cent. can be removed by acids), traces of oxygen and 1.2 volumes per cent. of nitrogen.

QUANTITATIVE RELATIONS OF LYMPH AND CHYLE.

It is estimated that the total amount of lymph and chyle introduced into the circulation through the large lymph-trunks in twenty-four hours equals the total volume of the blood. Of this one-half will be contributed by the chyle, the other half by the lymph. The secretion of lymph in the tissues takes place without interruption. From a lymphatic fistula on a woman's thigh, about 6 kilograms of lymph were collected in twenty-four hours. In young horses, the amount of lymph collected from the large lymph-trunk of the neck in from one and one-half to two hours measured between 70 and more than 100 grams. The following influences affect the amount of chyle, as well as that of lymph.

The amount of chyle increases considerably during digestion, especially if the quantity of food taken has been large, so that the vessels of the mesentery and the intestine will at this time be constantly found filled with white chyle. In the state of hunger the vessels are collapsed and can be recognized with difficulty.

The amount of lymph increases especially with the activity of the organ from which it flows. Thus it was found that active and passive muscular movements increase the amount of lymph considerably, almost five-fold in the horse. Lesser obtained more than 300 cu. cm. of lymph in this manner from fasting dogs, in consequence of which, with inspissation of the blood, the animals became exhausted, to the point of death.

All agencies that increase the pressure to which the parenchymatous fluids of the tissues are subjected increase the amount of lymph secreted, and conversely. Of this the following observations are illustrative:

(a) An increase in blood-pressure, not alone in the entire blood-vessel system, but also in the vessels of the part in question, causes increase in the amount of lymph, and conversely.

(b) Ligation or compression of the efferent veins causes considerable increase in the amount of lymph given off by the parts in question, even more than double, because the escape of fluid is confined to the lymphatic vessels. The application of tight bands is also a cause for swelling of the parts to the peripheral aspect of the application, as copious effusion of lymph takes place into the tissues—hypostatic edema.

(c) An increased supply of arterial blood acts in a similar manner, but less powerfully. In this connection paralysis of the vasomotor or irritation of the vasodilator fibers may cause an increase in the amount of lymph by creating marked hyperemia. The process of dilatation favors the production of lymph in greater degree than permanent distention of the blood-vessels. Contraction of the arterial paths as a result of irritation of the vasomotor nerves or from other causes will naturally have the opposite result; but even after ligation of both carotids, the lymph-current in the large cervical trunk of the dog by no means ceases, as the head is still supplied with blood in small amount by the vertebral arteries.

If, after unilateral division of the sympathetic nerve, the blood-vessels of the ear are dilated, indigo-carmine, injected into the blood, enters earliest and in greater degree into the lymph of this ear; the latter also becomes decolorized earlier than the healthy ear. In this way the rare instances of unilateral or partial icterus are to be explained.

An increase in the total volume of blood as a result of injection of blood or serum into the veins causes increased formation of lymph, as, in consequence of the increased tension thus induced, blood-plasma passes over into the tissues in large amount. If water or a hypotonic salt-solution be infused, water passes out into the tissues.

After death and complete rest of the heart, the formation of lymph still goes on for some little time, although in slight degree. If fresh blood be then passed through the animal's body, still warm, increased lymph will in turn flow from the large lymph-trunks. It thus appears that the tissues are still capable of taking up plasma from the blood for the production of lymph for some time after cessation of the circulation. This fact may explain the circumstance that some tissues, as, for example, the connective tissue, appear to contain more fluid after death than during life, while, at the same time, the blood-vessels have after death given up much of the plasma from their interior.

Under the influence of curare an increase in the secretion of lymph takes place; the amount of the solid constituents of the lymph increasing. In the frog large amounts of lymph collect in the lymph-sacs, and this may be due in part to the fact that the lymph-hearts are paralyzed by curare. The production of lymph is increased also in the tissues of inflamed parts.

ORIGIN OF LYMPH.
SOURCE OF LYMPH-PLASMA.

The lymph-plasma is, in part, a filtrate from the blood-vessels, passing over into the tissues, in accordance with the prevailing blood-pressure. In this process, the salts (penetrating membranes most readily) pass through admixed in approximately the same proportions as the salts in the blood-plasma; the fibrin-factors, to about two-thirds; the albumin, about one-half. As in the case of filtration in general, the filtration of lymph also must increase with increased pressure.

C. Ludwig and Tomsa were able to demonstrate this by permitting blood-serum to pass through the blood-vessels of an excised testicle under varying pressure, with the result that the transuded fluid from the lymph-vessels was increased or diminished in amount. This artificial lymph exhibited a composition similar to that of natural lymph. The albumin contained in the lymph also increased with increasing pressure. In addition, the metabolic products of the tissues, concerning whose qualitative and quantitative conditions little is known, naturally undergo admixture with the lymph-plasma in the different tissues.

In part, however, the formation of lymph must be regarded as a secretory process of the cells of the blood-capillaries.

In favor of this view is the fact that materials injected into the blood (sugar, egg-albumin, peptone, urea and sodium chlorid) pass in concentrated form into the increased lymph ; further, that the blood is capable of maintaining the osmotic tension of its plasma. As a result of this secretory property on the part of the endothelium of the vessels, substances that would disturb the isotonia between the blood-corpuscles and the blood-plasma are quickly eliminated from the blood, including superfluous water. After the injection of peptone, the blood-pressure falls enormously, so that the passage into the lymph cannot be dependent upon this pressure. With increase in the lymph-current, the secretion of urine also is later increased. The lymph-paths may thus be considered as a reservoir that temporarily takes up out of the blood the substances to be eliminated, whence they are then gradually further consumed or excreted.

According to Heidenhain, there are materials that increase lymph-production, lymphagogues, which are in part effective by causing the passage of fluid from the blood into the lymphatic radicles. Among such agencies are injections into the blood of a decoction of leeches, crab-muscles, mussels, solution of nuclein, tuberculin, bacterial extracts, bile, physostigmin, pilocarpin and extract of helianthus. In part they increase the amount of lymph by causing the passage of water from the tissues into the lymph. In this category belong injections into the blood of sugar, urea and salts. Atropin diminishes lymph-production.

24

Muscular activity causes increased lymph-production, as well as a more rapid escape of the lymph. The tendons and fasciæ of the skeletal muscles, which possess numerous small stomata, absorb lymph from the muscular tissue. With alternate contraction and relaxation of these fibrous tissues, their lymph-ducts suck themselves full and propel the lymph onward. Even passive movements are effective in this direction. If solutions are injected beneath the fascia lata, they can be propelled onward by passive movements, contraction and relaxation, into the thoracic duct.

SOURCE OF THE LYMPH-CELLS.

A considerable portion of the lymph-cells are derived from the lymph-glands, out of which the lymph-stream washes them into the efferent vessel. Therefore it happens that the lymph-stream, after passing through the lymph-glands, is always found richer in lymph-cells. Within the lymph-glands there are large and small lymphocytes, the latter being the daughter-cells of the former, and arising by mitosis. In addition, new leukocytes are constantly migrating from the blood-capillaries of the follicular bands into the reticulum. The lymphatic follicles permit cellular elements to enter through the meshes of their limiting layer into the adjacent small lymph-vessels.

A second seat of lymph-cell production is found in the organs containing adenoid tissue as a basis, in the meshes of which lymph-cells are found in large number, such as the entire mucous membrane of the intestinal tract, the bone-marrow and the spleen. The cells reach the radicles of the lymph-vessels in these organs by ameboid movement.

Just as the lymph-cells reach the circulation through the large trunks and are there encountered as white blood-corpuscles, so, likewise, numerous leukocytes migrate in turn from the blood-capillaries into the lymph-vessels, especially in their small beginnings, and partly by active ameboid movement, partly by being forced by filtration-pressure exerted by the blood-column. In rare cases even a return movement of lymph-cells from the lymph-spaces into the blood-vessels has been observed.

Also particles of cinnabar or milk-globules introduced into the blood reach the lymph-vessels from the blood-capillaries in a short time; the nerves of the vessels having no influence in this condition. In case of venous stasis, in analogy with the processes attending diapedesis, this passage takes place more freely than when the circulation is unembarrassed. Inflammatory changes in the vessel-wall also favor the passage. The vessels of the portal system prove especially permeable.

New lymph-cells result also through multiplication by division of the lymph-corpuscles, and likewise of the so-called fixed connective-tissue cells, as has been demonstrated with certainty especially in the presence of inflammation of certain organs. If irritants which excite inflammation are applied to the excised cornea, kept in a moist chamber, a large increase in the wandering cells in the anastomosing lymph-passages of the cornea will be noted; and as, in the inflamed cornea, the corneal cells permit the recognition of a reproduction of their nuclei by division, the conclusion is probably justified that a division of the corneal corpuscles (fixed connective-tissue cells) is responsible for the increase in the wandering cells.

That a new-formation of leukocytes must take place by division, as well as by the setting free of divided connective-tissue cells, is shown by their often

enormous production in the presence of inflammations (pus-formation), particu.
larly in the case of extensive phlegmons and purulent effusions in the serous
cavities, when by reason of their enormous number, they cannot be regarded as
having resulted solely by migration from the circulation.

The destruction of the lymph-cells appears to take place in part at
the seats of origin of the vessels and in the vessels themselves. The
occurrence in the lymph of the fibrin-factors, which are derived from
disintegrated leukocytes, tends to support this view. Particularly in the
presence of severe inflammation, especially in connective tissue, the
new-formation of numerous lymph-cells appears to be attended with
their increased destruction. Therefore the lymph under such circumstances becomes especially rich in fibrin, and, naturally, also the blood,
through the lymph.

According to Hoyer, the lymph-glands are also filtering apparatus in which
degenerating leukocytes are intercepted and subjected to a destructive metamorphosis.

CIRCULATION OF CHYLE AND LYMPH.

The cause for the movement of the chyle and the lymph depends
ultimately on the difference in pressure prevailing between the lymphatic
radicles and the points at which the lymphatics empty into the venous
system.

In detail the following facts are noteworthy:

In the onward movement of the lymph, forces are primarily active
that are of influence at the points of origin of the lymphatics. These
forces must vary in accordance with the character of the points of origin.
(a) The lacteals receive the first motile impulse through the contraction
of the muscles of the villi. Inasmuch as these grow shorter and smaller,
they constrict the axial lymph-space, whose contents must move in a
centripetal direction. With the succeeding relaxation of the villus, the
numerous valves prevent the chyle from flowing backward. With constriction of the lumen of the intestine, through contraction of the intestinal muscles, the villi are forced more closely together longitudinally,
the evacuation of the central lymph-vessel being likewise favored. (b)
Within those lymph-vessels that originate as perivascular spaces, every
dilatation of the blood-vessels must cause a movement of the surrounding
lymph-stream in a centripetal direction. (c) Lymph enters the open
lymph-pores of the pleura with each inspiratory movement, which
excites suction upon the thoracic duct. A similar condition exists at the
orifices of the lymph-vessels on the abdominal aspect of the diaphragmatic peritoneum. The blood-vessels participate principally in absorption
from the abdominal cavity, the lymphatics relatively little. If serous
fluid or a solution of salt or sugar is introduced into the abdominal
or pericardial cavity, it will be absorbed, and, if isotonic with the blood-
plasma, without change; if it is not isotonic, it will first be made
isotonic by elimination from the blood. Accordingly, osmosis cannot
be alone the active agency in the process of absorption, as imbibition
contributes some influence. If the intra-abdominal pressure increases,
the blood-vessels absorb more freely, but with excessively high pressure
less freely, in consequence of compression of the abdominal veins. In
this manner is explained the clinical observation that, in the presence of
ascites, absorption is often promoted after the abdominal tension is

diminished through removal of a moderate amount of fluid. (*d*) In those vessels that arise by means of fine secretory canaliculi, the movement will depend directly on the tension of the parenchymatous fluids, and the latter, in turn, upon the tension in the blood-capillaries. Thus the blood-pressure will still be active as a force from behinde ven into the lymphatic radicles.

In the lymph-trunks themselves, the contractions of their muscular walls propel the current onward. Heller noted, in the lymphatics of the mesentery of the guinea-pig, that this movement was peristaltic in an upward direction. The large number of valves prevent a backward current. In addition, the contractions of the surrounding muscles, further, any pressure upon the vessels and the tissues as the seat of origin of the lymphatic radicles will force the current onward. If the escape of blood from the veins is rendered difficult, lymph is poured out more abundantly from the tissues in question.

The interposed lymph-glands offer considerable resistance to the current, as the lymph must flow through numerous spaces, traversed by fine meshes and partially filled with cells. Nevertheless the obstacles thus presented are in part compensated for by the numerous unstriated muscles that are present in the sheath and the trabeculæ of the glands. By means of these, compression of the glands (as of a sponge) can take place, the presence of the valves again determining the centripetal direction of the current. From this point of view electrical stimulation of swollen lymph-glands might be successful.

With the union of the vessels into a few of considerable size, and finally to form the main trunk, the sectional area of the current becomes diminished, and the velocity of the current correspondingly increased. Nevertheless, the velocity under such circumstances is always low, reaching only from 238 almost to 300 mm. in a minute in the main cervical lymph-trunk in the horse, a fact that is indicative of the exceedingly slow movement of the lymph in the small vessels. The lateral pressure in the same situation was from 10 to 20 mm.; in the dog only from 5 to 10 mm. of a dilute soda-solution, but in the thoracic duct of a horse it was 12 mm. of mercury.

The time required for the passage of the lymph through the walls of the capillaries of the abdomen or of the lower extremity, is about 2 minutes in the dog; for the propulsion of the lymph through the lymphatics of the lower extremity and of the trunk, 3.2 seconds.

The respiratory movements have an important influence upon the lymph-stream in the thoracic duct and the right lymphatic duct, as each inspiration conveys the current of lymph, together with venous blood, to the heart, and as a result the tension in the thoracic duct may even become negative.

Lymph-hearts.—The lymph-hearts containing valves found in some animals, particularly cold-blooded animals, are deserving of consideration. The frog possesses two axillary hearts (above the shoulder near the vertebral column) and two sacral hearts (above the anus near the apex of the sacrum). They beat, though not synchronously, about 60 times in the minute and contain about 10 cu. cm. of lymph. They contain striated muscle-fibers and are provided with special ganglia. The posterior hearts pump the lymph into the branches of the communicating iliac vein, the anterior into the subscapular vein.

Their pulsation depends in part on the spinal cord, for, as a rule rapid destruction of the cord causes cessation of the heart-beat, but pulsations are not rarely observed to continue after removal of the cord. A second normal source of excitation of the lymph-hearts is to be sought in Waldeyer's ganglia. Irritation of the skin, the intestine and the blood-heart gives rise to a reflex influence, partly acceleration, partly retardation of the beat, which does not affect the sacral heart if the coccygeal nerve, which connects the posterior lymph-heart with the spinal cord, is divided. Strychnin-convulsions accelerate the heat, as does also irritation of the spinal cord by heat, while it is diminished by cold. The heart that has ceased to beat in consequence of exposure or of the action of muscarin, but not resting in consequence of destruction of its nerves, can be excited to renewed pulsation by increased filling. Antiar paralyzes the lymph-hearts and the blood-heart; curare, the former only. In other amphibians, two lymph-hearts have been found; and one or two in the ostrich and the cassowary, in some web-footed birds, as well as in the chicken-embryo; in fish they have been found in the tail, as, for example, in the eel, where their pulsation visibly affects the adjacent veins.

The nervous system has a direct influence upon the movement of the lymph through innervation of the muscles of the lymphatics, the lymph-glands, and, when they exist, the lymph-hearts. In addition, there are still other special effects of the nerves upon the absorptive activity of the lymphatic radicles. Kühne noted, after irritation of the corneal nerves, that the corneal cells contracted within their secretory canaliculi. The following observation by Goltz is also interesting in this connection. When this investigator injected a dilute solution of sodium chlorid subcutaneously into the lymph-spaces, he saw that it was rapidly absorbed; it remained unabsorbed, however, after destruction of the central nervous system. Division of the nerves to the extremities also resulted, temporarily, in retarding the absorption.

If inflammation was excited in both posterior extremities of a dog, marked edema, together with acceleration of the lymph-stream, appeared in the one in which the sciatic nerve had been divided.

If the thigh of a frog is tightly constricted until the circulation ceases, the nerve being preserved, and the part is immersed in water, it becomes greatly swollen (the dead thigh does not swell); whence it follows that absorption takes place independently of the existence of the circulation. Division of the sciatic nerve or crushing of the spinal cord (though not mere transverse section or separation of the brain) abolishes absorption.

ABSORPTION OF PARENCHYMATOUS EFFUSIONS.

Fluids that transude into the tissue-spaces from the blood-vessels, or those that are injected into the parenchyma through a needle, undergo absorption. In this process the blood-vessels participate primarily, and the lymphatics also secondarily. Into the latter, there pass from the clefts and secretory spaces in the connective tissue, even small particles, as, for example, granules of cinnabar and India ink after tattooing of the skin, blood-corpuscles from hemorrhagic extravasations and fat-droplets from the marrow of fractured bones. If all the lymphatics of a part be ligated, absorption still takes place just as rapidly as before. Therefore, the absorbed fluid must have passed through the delicate membranes of the blood-vessels. The opposite observation, that no absorption of parenchymatous fluids takes place after ligation of all the blood-vessels, does not exclude a participation of the lymphatics in the process of absorption, because, after ligation of all of the blood-vessels, naturally all formation of lymph in the tissues, and consequently any lymph-current, must cease. The absorption of fluids introduced into the tissues artificially, particularly in the subcutaneous cellular tissue (parenchymatous and subcutaneous injection), generally takes place rapidly, as a rule more rapidly than after administration by the mouth. Therefore, subcutaneous injections of drugs in solution are much employed for therapeutic purposes. Naturally the substances to be injected should not have a destructive, corrosive or coagulating effect upon living tissues. In addition to the great rapidity of absorption, subcutaneous injection has the further advantage over the administration of a drug by the mouth that some agents that are ingested undergo decomposition in the stomach and intestine as a result of the digestive process, so that they cannot at all be absorbed unchanged.   Thus, particularly poisons that act through ferments, such as snake-venom, ptomains and putrid poisons, are destroyed by the stomach.   Emulsin also behaves in the same manner. If this ferment is introduced into the stomach while amygdalin is injected into a vein of the same animal, poisoning by hydrocyanic acid does not take place, because the emulsin is destroyed by the digestive process.   If, however, emulsin is injected into the blood and amygdalin into the stomach, hydrocyanic-acid poisoning takes place rapidly, because amygdalin is absorbed unchanged from the stomach.   Amygdalin is a glucosid ($C_{20}H_{27}NO_{11}$) that breaks up as a result of the fermentative activity of fresh emulsin with the taking up of water, $2(H_2O)$, into hydrocyanic acid (CHN), oil of bitter almonds (C_7H_6O) and sugar, $2(C_6H_{12}O_6)$. For observations on animals on the absorption of solutions from the parenchymatous structures, either poisons whose action gives rise to conspicuous tonic symptoms, or such harmless substances as are readily discoverable in the blood and subsequently especially in the urine are employed, as, for example, potassium ferrocyanid.

The author, in 1878, demonstrated that serum, injected into the subcutaneous tissue, is rapidly absorbed.   The serum, which must be obtained from an animal belonging to the same species or at least as indifferent as possible, undergoes decomposition in the circulation, so that the production of urea increases.   Infusions of serum may, therefore, be given for nutritive purposes. Febrile reaction is observed after such injections, as in the case of transfusion. Solutions of albumin and peptone, oil, butter, dextrose, levulose, galactose and maltose in solution have also been observed to undergo absorption.

LYMPH-STASIS AND SEROUS EFFUSIONS.

If obstruction to the efferent venous and lymphatic paths of an organ arises, stasis results, and later abundant effusion of lymph into the tissues.   This is most distinctly recognizable in the skin and the subcutaneous tissue, where the soft parts become swollen;   while, without redness and pain, tumefaction develops, with a doughy feeling, and pressure with the finger causes pitting.   These are the signs of lymph-stasis, which, if the fluid is especially rich in water, is designated by the term edema.

Also within the serous cavities, under like conditions, a similar collection of lymph takes place.   If numerous leukocytes migrate from the delicate bloodvessels into such serous cavities and undergo multiplication, the fluid, richer in cells, becomes more and more like pus.   The multiplication of these cells gives rise to the presence of a considerable amount of albumin, which may subsequently be increased by absorption of water from the effusion.   The latter will be made particularly easy when the pressure in the fluid exceeds that in the small bloodvessels.   These sero-purulent effusions not rarely undergo a change in constitution later on, for which no reason has been found.   The substances present are in part products of the decomposition of albumin, such as leucin and tyrosin, in part products of the retrogressive metamorphosis of nitrogenous substances, such as xanthin, kreatin, kreatinin (?), uric acid (?) and urea.   Further, endothelial cells from the serous cavities; sugar in pathological serous and chylous effusions and edematous fluid have been found; in the latter also diastatic ferment, often cholesterin; and in the fluid of serous hydrocele and echinococcuscysts, succinic acid.

Not only the pressure from without upon the lymphatics, but, in general, resistance of any kind that is present in the lymph-path may give rise to lymphstasis and serous effusions.   Thus, lymph-stasis results from occlusion of the lymphatics in consequence of inflammation and thrombosis (lymph-coagulation); further, as a result of impermeable, swollen, inflamed or degenerated lymph-glands. In these cases, however, the formation of new lymph-vessels is frequently observed, reëstablishing the former communication.   An effusion of lymph may also take place into the serous cavities of the abdomen or the thorax, from rupture of large lymph-paths, especially of the thoracic duct—chylous ascites or chylothorax.   Interference with or even removal of all those factors that have been found active in propelling the lymph onward will be capable of inducing lymph-stasis.

If stagnation of lymph can develop in this manner also on the part of the lymphatic apparatus, the appearance of considerable amounts of watery lymph, in the form of edema or anasarca, as well as of serous effusions, is often at the same time due to the fact that a copious transudate is furnished on the part of the blood-vessels. Obstruction in the distribution of the lymph-stream may then further increase such a collection of fluid. Particularly the vessels of the abdomen and, further, those that furnish a watery exudation also under normal circumstances appear, above all others, to be especially adapted to transudation. Such increase in transudation is favored by (1) any considerable degree of venous stasis. These hypostatic transudates are, as a rule, deficient in albumin and leukocytes, but, on the other hand, the richer in erythrocytes the greater the interference with the flow of venous blood. Ranvier induced hypostatic edema artificially in the lower extremity by ligation of the inferior vena cava and simultaneous division of the sciatic nerve. The paralytic dilatation of the vessels of the posterior extremity, induced by the latter, causes an increase in the amount of blood present and a rise in the blood-pressure, which, in turn, promotes edematous exudation. (2) Further, as yet unknown physical changes in the protoplasm of the endothelial cells of the blood-vessels and capillaries may render these capable of permitting the abnormal passage of albumin, hemoglobin and even blood-cells. This takes place when foreign matters are present in the blood in considerable amount, as, for example, hemoglobin in solution; further, when the blood is deficient in oxygen or albumin. Also after exposure to abnormal heat, a similar condition has been observed, and the tumefaction of the soft parts in the vicinity of inflamed tissues likewise appears to be due to an exudation of lymph through altered vessel-walls. Perhaps a nervous influence, which makes itself felt in a certain vascular area (by contraction or relaxation of the protoplasm of the blood-capil-laries ?), may even be capable of causing such a transitory change in the vessel-walls. Lymphatic transudates of this character are generally rich in cells and consequently also in albumin. (3) Further, the presence of a large amount of water in the blood will increase its capacity for transudation. Nevertheless, the fact should be considered in this connection that the large amount of water contained in the blood acts, in turn, by inducing changes in the protoplasm of the endothelium of the blood-vessels and capillaries, so that it is itself, when long continued, a factor that increases the permeability of the vessel-walls. Debilitated, poorly nourished, flabby individuals particularly exhibit watery lymphatic exudations from watery blood—cachectic edema.

There is no doubt that lymph-stasis (hydrops) may develop also under certain circumstances, and even through the action of microörganisms (bacterium lymphagogum), in consequence of the fact that irritation of the cells of the blood-capillaries (as by the products of metabolism of that organism) gives rise to increased exudation of fluid.

COMPARATIVE.

Extensive lymph-spaces, lined with endothelium, are present in the frog, beneath the entire external integument. In addition, a large lymphatic space, the cysterna lymphatic magna of Panizza, extends in front of the vertebral column, separated from the abdominal cavity by the peritoneum. Tailed amphibia, as well as many reptiles, have large lymph-spaces beneath the skin, occupying the entire length of the trunk in the lateral regions of the back. Further, all reptiles and the tailed amphibia possess, in the course of the aorta, large, longi-tudinal lymph-reservoirs. Tortoises likewise have an extensive lymphatic ap-paratus (Fig. 130, A, II). The bony fish have longitudinal lymph-trunks in the lateral regions of the back, from the tail to the anterior fins, and these are connected with dilated lymph-spaces at the root of the tail and the fins of the extremities. Within the interior of the body the extensive lymph-sinuses attain their greatest development in the region of the gullet. Many birds possess a sinus-like dilatation of a lymph-space in the region of the tail. In the carnivora the mesenteric lymph-glands are united to form a large, compact·mass, the so-called pancreas of Aselli. Naturally the lymph-spaces (provided with valves) always communicate with the venous system, and usually with the territory of the superior vena cava.

HISTORICAL.

Although the lymph-glands were known to the school of Hippocrates, espe-cially through their morbid enlargement, and although Herophilus and Erasistratus had observed the milk-white chyle-vessels in the mesentery, Aselli (1623) was the first to study the mesenteric chyle-vessels more thoroughly, together with their valves. Pecquet (1648), as a student, found the receptacle for the chyle, Rudbeck and then Thom. Bartholinus the clear, watery lymph-vessels (1650–1652). Eustachius (1562) was familiar with the thoracic duct, which Gassendus (1654) later claimed to have been the first to discover. Lister noticed that chyle was colored blue after the injection of indigo into the intestine (1671). Rudbeck (1652) observed the separation of fibrin in the lymph; Reuss and Emmert(1807) were the first to observe the lymph-corpuscles. The chemical examinations date from the first quarter of the nineteenth century, and were made by Lassaigne, Tiedemann, Gmelin and others, of whom the latter also recognized the fact that the white color was dependent upon the fat-granules.

PHYSIOLOGY OF ANIMAL HEAT.

SOURCES OF HEAT.

The heat of the body is a form of kinetic energy appearing without interruption and must be conceived as depending upon vibrations of the atoms of the body. In the last analysis every source of heat is contained in the mass of potential energy taken into the body as food, in combination with the oxygen obtained from the air in the act of respiration. The amount of heat liberated depends upon the amount of potential energy transformed.

The potential energy of nutrient matters may be appropriately designated as latent heat, inasmuch as it may be conceived that in their consumption in the body, which is essentially a process of combustion, kinetic energy is transformed only in the form of heat. As a matter of fact, mechanical energy and electricity are also developed from the potential energy supplied. However, in order to obtain a uniform measure for the forces transformed, it is advisable to express all potential energy in terms of heat-units. The calorimeter is an apparatus with the aid of which the amount of potential energy contained in food-stuffs can be converted experimentally into heat and the units of the latter can at the same time be measured.

Favre and Silbermann employed the so-called water-calorimeter (Fig. 133). A cylindrical box, the so-called combustion-chamber (K), serves for the reception of the substance to be burned. This box is suspended in a larger, cylindrical vessel (L), which is filled with water (w), so that the combustion-chamber is completely surrounded thereby. Three tubes enter into the upper portion of the chamber: one (O) is intended for the passage of air containing oxygen, which is necessary in the process of combustion. The second tube (a) in the middle of the cover is closed above with a thick glass plate, upon which is mounted at an angle a mirror (s), which permits the observer (B) to look into the interior of the chamber from a lateral point of view in the direction b b. in order to observe the process of combustion at c. The third tube (d) is employed only when it is desired to consume combustible gases in the chamber and through it these are then passed. Generally this tube is closed by a cock. A lead pipe (e e) also passes out of the upper portion of the chamber and in a convoluted arrangement traverses the volume of water, finally reaching the surface at g. Through this the gases of combustion escape, being cooled in the convoluted tube to the temperature of the water.

The cylindrical vessel containing the water is covered with a lid having openings for the four tubes that pass through it. The water-cylinder stands upon legs within a larger cylinder (M), which is filled with a poor conductor of heat. Finally this is placed in a still larger cylinder (N), which again contains water (W). This last layer of water is intended to prevent any heat from the exterior from raising the temperature of the water in the interior. A definite amount of the material to be examined is burned in the combustion-chamber. After combustion has been completed, during the progress of which the water in the interior is repeatedly stirred, the temperature of the water is determined by means of a delicate thermometer. If the amount of increase in temperature is noted, and if the amount of water in the inner cylinder is known, the number of heat-units furnished by the combustion of the measured amount of the substance under examination can be readily estimated.

Instead of the water-calorimeter the ice-calorimeter may be employed. In this instrument the inner container is surrounded with ice instead of with water. Around this in a second container is still more ice, which prevents any heat from without acting upon the ice in the interior. The body in the interior chamber gives off heat and causes a portion of the surrounding ice to melt, while the ice-water passes off below through a tube and is measured. In this connection it should be noted that 79 heat-units are required to melt 1 gram of ice into 1 gram of water at a temperature of 0° C. For animal experimentation the calorimeter has probably reached the highest grade of perfection at the hands of Rubner.

The air-calorimeter of d'Arsonval permits of measurement in human beings within a few minutes. A rigid cylinder of woolen material, within which a man may stand, is provided above with a chimney. If the man heats the air in the interior, this will escape through the chimney and set in motion a small windmill contained therein, whose revolutions can be counted. The amount of heat given off is proportional to the square of the velocity of the escaping current of air. A man in the nude state yielded 124, and in the dressed state 79 calories in an hour.

Just as in the calorimeter, though much more slowly, nutrient matters are consumed in the human body with a supply of oxygen, and as a consequence there takes place a transformation of potential into kinetic energy, which in a person at rest appears almost wholly as heat.

Favre and Silbermann, Frankland, Rechenberg, Stohmann, B. Danilewsky, Rubner and others have made calorimetric observations as to the amount of heat yielded by the combustion of many nutrient substances. One gram of water-free substance yields in heat-units as follows:

		CARBOHYDRATES.	
Proteids on the average,	5711	Galactose,	3722
Serum-albumin,	5918	Cane-sugar,	3955
Egg-albumin,	5735	Milk-sugar,	3952
Syntonin,	5908	Maltose,	3949
Hemoglobin,	5885	Glycogen,	4191
Milk-casein,	5858	Starch,	4183
Yolk of egg,	5841	Cellulose,	4185
Vitellin,	5745	Cow's milk,	5613
Meat, {	5663 / 5641	Woman's milk,	5786
Peptone,	5299	Rye-bread,	4471
Fibrin,	5637	Wheat-bread,	4351
Vegetable fibrin,	5942	Peas,	4889
Legumin,	5793	Buckwheat,	4288
Conglutin,	5479	Maize,	5188
		Alcohol,	6980
Animal fats on the average,	9500	Muscle-extractives,	4400
Butter,	9231	Liebig's meat-extract,	3216
Olive-oil, {	9467 / 9608	(Principally according to Stohmann).	
Rape-oil, {	9627 / 9759	Urea,	2537
Stearic acid,	2712	Glycin,	3128
Oleic acid,	2682	Leucin,	6533
Palmitic acid,	2398	Hippuric acid,	5678
Glycerin,	397	Kreatinin,	4275
Alcohol,	7100	Uric acid,	2741

As the proteids in the body are not transformed beyond urea the amount of heat resulting from the combustion of urea is to be deducted from that resulting from the combustion of the proteids. As one gram of proteids (average calories 5711) yields 0.3428 gram of urea, and 1 gram of urea yields 2537 calories, 870 calories are to be deducted.

Isodynamic food-stuffs, namely, those that yield the same amount of heat in the process of combustion, are as follows: 100 grams of animal proteid, after deduction of the heat resulting from the combustion of urea, equal 52 grams of fat, 114 grams of starch, 128 grams of dextrose. One hundred grams of fat are isodynamic with 243 grams of dry meat or 225 grams of dry syntonin, or with 256 grams of dextrose. According to Pflüger, 1 gram of nitrogen in meat equals 2.79 grams of fat; 1 gram of animal fat equals 0.364 gram of nitrogen in meat; 1 gram of starch equals 0.424 gram of fat or 0.154 gram of nitrogen in meat; 1 gram of grape-sugar equals 0.390 gram of fat or 0.42 gram of nitrogen in meat; 100 grams of vegetable albumin likewise equals 55 grams of fat or 121 grams of starch or 137 grams of dextrose.

Rubner estimates in human beings on a mixed diet the available heat-producing energy for 1 gram of proteid at approximately 4100 calories, for 1 gram of fat 9300 calories, for 1 gram of carbohydrate 4100 calories. For the dog Rubner determined that 1 gram of nitrogen in the excreta of the fasting animal had caused the production of 25,000 calories; further, that 1 gram of nitrogen in the excreta with a meat-diet had produced 26,000 calories; and 1 gram of carbon, formed from 1.3 grams of fat, had yielded 12,300 calories.

If it be known, therefore, how many parts by weight of the foregoing substances a human being takes up with his food during twenty-four hours, the calculation can be made as to how many heat-units he may generate therefrom through oxidation. In this connection the utilization of the nutrient materials must be taken into consideration, in accordance with which a certain, even though small, percentage of the food cannot be disposed of by the digestive and absorptive organs, and therefore is excreted unused.

Rubner found that, however abundant the administration of food, a larger amount of heat can be shown to be produced immediately on the first day of feeding, as compared with the

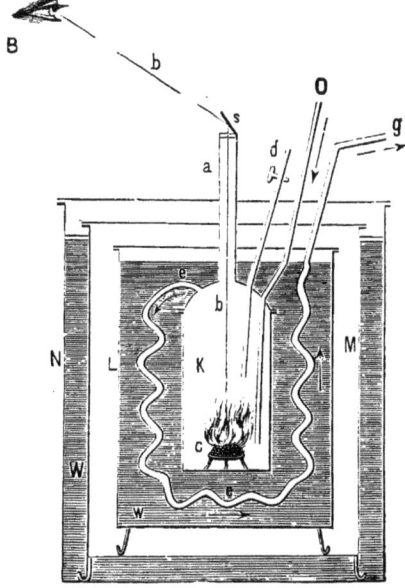

FIG. 133.—Water Calorimeter (after Favre and Silbermann).

preceding days of fasting. The bodily temperature under such circumstances remains unaltered. The greatest amount of heat is produced as a result of excessive administration of proteids, less from carbohydrates and least from fats.

In detail the *sources of heat* are as follows:

1. The *transformation of chemical combinations* of foods with high potential energy into those of lesser or completely exhausted potential energy. As the organic articles of food, exclusive of the inorganic accompaniments, consist of C, H, N and O, it is especially through (*a*) the combustion of C into CO_2 and of H into H_2O that heat is produced. In this connection it is to be noted that the combustion of 1 gram of C into CO_2 yields 8080 heat-units, while that of 1 gram of H into H_2O yields 34,460 heat-units, though the C and H in the molecules of the food-stuffs must not already be saturated with O. The amount of O necessary for this purpose is taken up in the act of respiration. There-

fore an approximate estimate may be made as to the quantity of heat produced by an organism from the amount of oxygen consumed in a unit of time. An equal consumption of O corresponds with an equal production of heat, whether it served for the oxidation of H or of C. As a matter of fact, a relation exists between heat-production in the animal body and the consumption of O, as between cause and effect. Thus, cold-blooded animals, which consume little O, have a low bodily temperature. Among warm-blooded animals 1 kilogram of living rabbit takes up 0.914 gram of O within an hour and by this means maintains its bodily temperature on the average at 38° C.; 1 kilogram of living hen, on the other hand, consumes 1.186 grams of O in an hour and maintains as a result an average temperature of 43.9 C. The amount of heat produced is equally large whether the combustion takes place slowly or rapidly. The activity of metabolism has, accordingly, an influence only upon the rapidity, but never upon the absolute amount, of heat-formation. Also, the combustion of inorganic substances in the body, such as that of sulphur into sulphuric acid, that of phosphorus into phosphoric acid, constitutes a source of heat, although it be but slight. According to Rubner this amounts to but 0.47 per cent. of the heat.

(b) In addition to the processes of combustion, however, all of those chemical processes in the human body, as a result of which the total amount of potential energy present is diminished, in consequence of greater saturation of affinities of the atoms previously present, are attended with the development of heat. Wherever the atoms combine with saturated affinities for greater stability in their ultimate position of rest, chemical potential energy is transformed into kinetic thermal energy, as, for instance, in the alcoholic fermentation of grape-sugar and other similar processes.

Heat is produced also in the following chemical process:

(α) The union of bases with acids. Here the character of the base determines the amount of heat formed, while the character of the acid is without any influence. Only when the acid, as, for instance, carbon dioxid, is not capable of neutralizing the alkaline reaction, is the production of heat smaller. Also, the formation of chlorin-combinations, as in the stomach, generates heat.

(β) The transformation of a neutral into a basic salt. In the blood the sulphuric and phosphoric acids resulting from the combustion of sulphur and phosphorus combine with the alkalies of the blood to form basic salts. The decomposition of the carbonates of the blood by lactic and phosphoric acids constitutes a double source of heat, namely through the formation of a new salt, as well as through the release of carbon dioxid, which is in part absorbed by the blood.

(γ) The combination of hemoglobin with oxygen. According to Berthelot the amount of heat produced in this way is equal to one-seventh of the total amount formed in the body.

In the chemical processes through which the body is provided with heat there not rarely occur heat-absorbing intermediate transformations of the bodies. At times, in order to bring about more complete saturation of the affinities, intermediary atom-groups in themselves firmly united must first be broken up. In this process thermal energy is consumed. Also in the breaking up of stable aggregate states in processes of retrogressive metamorphosis heat is bound up. All of these intermediary losses of heat, however, are extremely slight as compared with that due to the development of the end-products.

2. *Physical processes* may be mentioned as a second source of heat. (a) The transformation of the kinetic mechanical energy of the viscera furnishes heat, as the work done cannot be conveyed to the outside.

Thus, all of the kinetic energy of the heart is transformed into heat through the resistance opposed to the blood-stream. The same may be said of the kinetic energy of certain muscular viscera. Thus, the torsion of the costal cartilages and the friction of the current of air in the respiratory organs and of the contents of the digestive tract yield a certain amount of heat.

Small amounts of the mechanical energy of the heart are transmitted through the apex-beat and the superficial pulse to surrounding parts, but these are exceedingly small. Also, in the respiratory movement, in the expulsion of the respiratory gases, the expectorated and other matters, a small amount of energy is conveyed to the outside, which is not converted into heat. Joule has attempted to determine the amount of heat generated in consequence of the kinetic energy lost by a flowing fluid. According to him the amount of heat produced in this way as a result of the friction must stand in a relation to the product of the difference between the initial and the terminal pressure in the weight of the flowing fluid mass. If it be assumed that the daily work of the circulation equals more than 86,000 meter-kilograms, it will be seen that the resulting amount of heat in 23 hours will be about 204,000 calories, which is sufficient to raise the temperature of the body of a medium-sized person about 2° C.

(b) If the body through muscular activity does work transmitted to the outside, as, for instance, if an individual throws a heavy weight or ascends a tower, a portion of the kinetic energy is converted into heat through the friction of the muscles, the tendons, the articular surfaces, further through concussion and pressure of the ends of the bone upon one another.

(c) The electrical currents generated in muscles, nerves and glands, apart from the small amounts that pass outside of the body with suitable conduction, are most probably transformed into heat. Thermogenic chemical processes also generate electricity, which likewise is transformed into heat. This source of heat is, however, quite insignificant.

(d) As a further slight source of heat from physical causes there should yet be mentioned heat-production through absorption of carbon dioxid, through the condensation of water in its passage through membranes, and in the process of imbibition, the formation of stable aggregate states, as, for instance, of calcium in the bones. It is true, heat is again in part lost through the involution of solid parts at advanced age. After death, at times also under pathological conditions during life, coagulation of blood and the rigidity of muscles constitute in this manner a source of heat.

ANIMALS WITH CONSTANT AND WITH VARIABLE TEMPERATURE.

Instead of the older division of animals into cold-blooded and warm-blooded (mammals and birds), it is advisable to base their classification upon another characteristic, namely, the uniformity or the variability of the bodily temperature with respect to external influences. For the class of warm-blooded animals the name *homoiothermic animals* has been introduced by Bergmann, because they are capable of maintaining their bodily temperature with remarkable uniformity notwithstanding considerable variations in the surrounding temperature. He designated cold-blooded animals *poikilothermic animals* because their bodily temperature rises and falls within wide limits in accordance with the temperature of the surrounding medium. Accordingly, heat-production must be increased in homoiothermic animals if exposed for a long time in a cold atmosphere and diminished on exposure for a long time in a warm atmosphere.

An instance of this great constancy of the temperature in the human body was furnished by Fordyce, who died in 1792. After a man had been for ten minutes in a room filled with hot, dry air, the temperature of the interior of his closed hand, the cavity of his mouth beneath the tongue, as well as the urine, was raised only a few tenths of a degree. When Becquerel and Brechet were investigating by means of the thermo-electric needle the temperature in the middle of the biceps muscle in a man whose arm had been immersed for a whole hour in ice-water, they found the temperature of muscular tissue reduced only 0.2° C. The same muscle exhibited either no increase in temperature or a reduction of only 0.3° C. after the man had immersed the arm in water at a temperature of 42° C. for a quarter of an hour.

If marked alteration in temperature be brought about by powerful agents, namely, by vigorous abstraction of heat or by considerable addition of heat, great danger to the continuance of life results.

Poikilothermic animals react differently, the bodily temperature following in general the surrounding temperature, though with variations. On the basis of numerous observations Soetbeer therefore states that the poikilothermic vertebrates have no special temperature in the ordinary sense of the term, but their bodily temperature, like that of inanimate objects, is dependent upon that of the physical conditions of their surroundings.

The following may suffice as illustrations of the bodily temperature in the animal kingdom: Birds: sea-gull, 37.8° C.; swallow and titmouse, 44.03°; mammals: dolphin, 35.5°, mouse, 41.1°, echidna from 26.5° to 36°; arthropods: from 0.1° to 5.8° above the surrounding temperature; in bees aggregated in the hive from 30° to 32°, and in bees in swarms as high as 40°. The following animals raise their temperature above the surrounding temperature: cephalopods 0.57°, molluscs 0.46°, echinoderms 0.40°, medusæ 0.27°, polyps 0.21° C.

METHODS OF ESTIMATING THE TEMPERATURE: THERMOMETRY.

Thermometry.—By means of thermometric apparatus information is obtained as to the temperature of the body subjected to examination. For this purpose there are employed:

The *thermometer* (Galileo, 1603). Sanctorius was the first in 1626 to make thermometric measurements in human beings. It is advantageous to employ instruments graduated in 100 parts according to Celsius, each degree being subdivided into ten parts. The apparatus should be compared with a normal thermometer before being used. The column of mercury should be slender and the spindle neither too small nor too large, and preferably cylindrical in shape. A large bulb increases the sensitiveness and also the period of observation, because the large amount of mercury is influenced through and through by heat with greater difficulty. If the spindle be smaller the observation can be made more rapidly, but it is less trustworthy. The scale should be of porcelain.

All thermometers acquire an error after use for a considerable time, registering too high. Therefore, they should be compared from time to time with a normal instrument. At every observation the bulb should be completely surrounded and kept at rest for at least fifteen minutes and during the last five minutes no movement in the column of mercury should be noticeable. Minimal and particularly maximal thermometers, for the measurement of febrile temperature, are of the greatest convenience to the physician.

For delicate comparative measurements Walferdin's metastatic thermometer (Fig. 134) is especially useful. The tube is exceedingly narrow in proportion to the bulb. In order that on this account the instrument should not be drawn out to an extraordinary length, an arrangement is provided by which the necessary amount of mercury can be increased or diminished at will. So much mercury is taken that at the expected temperature the column reaches about to the middle of the tube. This end is attained by having at the upper extremity of the tube an expansion in which the superfluous mercury is received. If, for instance, a temperature is to be taken that is likely to be between 37° and 40° C., the bulb is first heated to somewhat above 40° C.; then it is cooled quickly and at the same time shaken, so that the column of mercury is broken below the upper expansion. Thus the play of the column is from about 40° downward. The tube is so fine that 1° C. comprises about 10 cm. in length, and $\frac{1}{100}$° C. is still 1 mm. long. A reading of even as little as $\frac{1}{1000}$° C. has been made possible. The scale is graduated arbitrarily. The value of the graduation must be determined by comparison with a normal thermometer, and also the temperature when the column of mercury reaches a certain level.

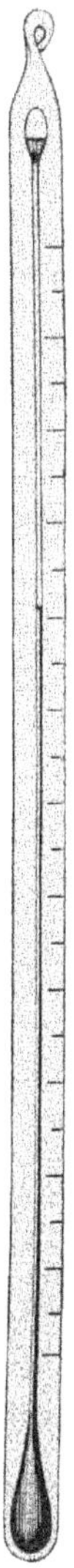

Kronecker and Mayer caused small maximal thermometers to be passed through the digestive canal or through vessels of considerable size. The small instruments are so-called outflow thermometers, whose mercury escapes through the short open tube, and in greater amount naturally when the temperature is highest. After removal, examination is made by comparison with a normal thermometer for the purpose of determining the temperature at which the mercury rises exactly to the free extremity of the tube.

The *thermo-electric apparatus* permits rapid and accurate measurement of the temperature (Fig. 135, I). The thermo-electro-galvanometer of Meissner and Meyerstein employed for this purpose contains a circular magnet (m) suspended from a silk thread (c) to which by means of a hook a small mirror (s) is attached. Near this magnet a fixed bar-magnet is fixed, with its poles similarly directed, and in such proximity that the free magnet is capable of turning to the north with the slightest degree of force. About the latter a thick copper wire (b b) is wound several times (in the diagrammatic representation but one turn is shown), and with the prolonged extremities of this two needle-like thermo-elements (a f, f a) made of different metals—German silver and iron—and soldered together, are connected. The free ends of these needles of similar name are, further, connected by means of a wire (b). Thus the two thermo-elements are incorporated into the closed circuit. At a distance of three meters from the mirror a horizontal scale (K K) is fixed, the numbers on which are reflected in the mirror. The scale itself is supported upon a telescope (F), which is directed toward the mirror. The observer (B), looking through the telescope, sees in the mirror the figures of the scale, which can be accurately adjusted. If the magnet swings out of the magnetic meridian, and with it the mirror, other figures on the scale appear to the observer in the mirror. If one of the thermo-elements is heated, an electric current results, which is directed in the warmer element from the German silver to the iron, and at the same time causes deflection of the movable magnet. If the observer conceive that he is swimming in the direction of the current within the conducting wire the north pole of the magnet is deflected to the left.

The tangent of the angle, through which the freely movable magnet is deflected from its position of rest in the magnetic meridian by means of a galvanic current passed before it, is equal to the relation of the galvanic energy G to the magnetic energy. Therefore, the tangent is as G is to D. In order, thus, to keep the tangent as large as possible, while G remains the same, the magnetic energy must be reduced as much as possible. If the magnetism of the swinging magnet be designated m and the magnetism of the earth T the magnetic energy D equals Tm. From this it appears that D can be diminished in two ways, namely (1) by reduction of the magnetic force of the swinging magnet, as may be done through the astatic pair of needles of the Nobili multiplicator, and (2) by lessening the magnetism of the earth by means of a fixed auxiliary magnet (Hauy bar) applied in the neighborhood of the swinging magnet with the same object.

Of importance for the rapid and accurate adjustment of the magnet is the employment of the so-called damping arrangement of Gauss, which is not indicated in the illustration. This consists of a thick,

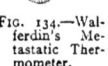

Fig. 134.—Walferdin's Metastatic Thermometer.

copper. hollow cylinder, upon which the wire of the coil is wound.   This mass of copper may be considered as a closed multiplicator of a single winding with a large cross-section.   The magnet set into oscillation induces in this closed copper mass a current whose intensity is greatest when the rapidity of oscillation of the magnet is greatest. and which takes the opposite direction as soon as the magnet is reversed.   In lesser degree the multiplicator itself as soon as it is closed operates in the same manner as a damper.   The currents thus induced cause a reduction in the oscillations of the magnet in such a way that the arc of move-

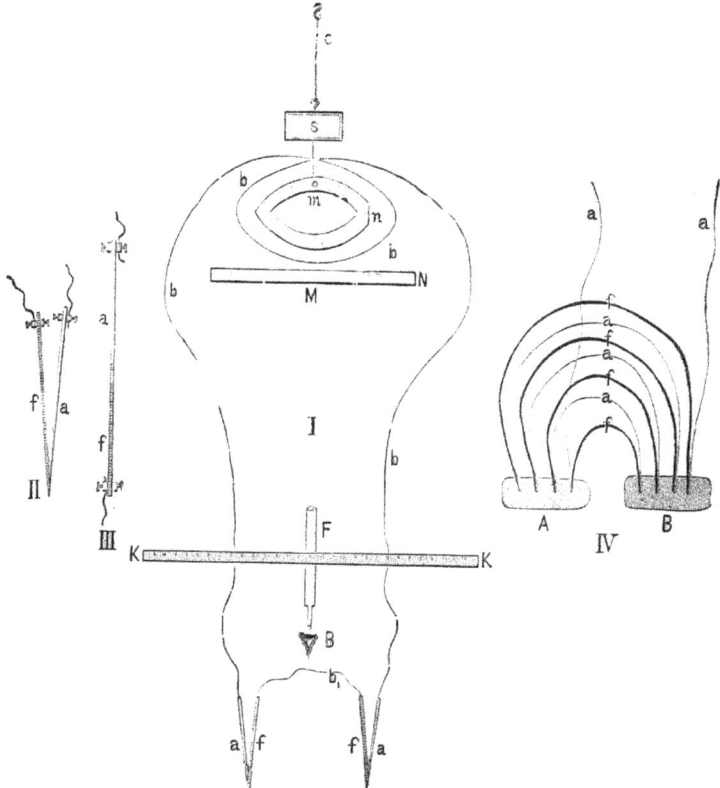

FIG. 135.—Diagrammatic Representation of Thermo-electric Apparatus for the Measurement of Temperature.

ment diminishes in rapid and almost geometric progression.   The induced, damping current is the stronger the less the resistance in the closed circuit, in the presence of the damper therefore the greater the transverse section of the copper ring.   By means of this damping arrangement the monotonous oscillation of the magnet to and fro is limited and the latter comes to rest rapidly and promptly after three or four small oscillations while the observation is sharp and made without loss of time.

So-called Dutrochet needles (II) are introduced as thermo-electric elements. These consist of German silver and iron and are soldered together longitudinally at their points. Becquerel needles (III) also may be employed. These are made of the same metals, which are soldered together in continuity. The needles must be well covered upon their surface with brown varnish in order that the currents resulting from the moistening of different metals with the parenchymatous fluids may not interfere with the thermo-currents obtained. Before the investigations are undertaken the extent of deflection on the scale to which a definite difference in temperature in the needles gives rise, thus about 1° C., must further be determined. In order to do this a sensitive thermometer is fastened by means of a loop to each of the thermo-needles, which are placed in separate oil-baths of a constant temperature, though differing by 1° C., as can be seen from the thermometers. If the circuit is now closed the deflection on the scale will naturally correspond to 1°. If, thus adjusted, the instrument exhibited a deflection of 150 mm., every displacement of the scale of 1 mm. would equal $\frac{1}{150}$ ° C. If this has been determined, either the two thermo-needles can be introduced into the different tissues or organs of animals at the same time, and in this way information be gained as to the prevailing differences in temperature in these portions of the body; or one of the thermo-needles is placed in a bath of constant temperature—approximately that of the body—in which at the same time there is a sensitive thermometer, while the other needle is introduced into the viscus to be examined. In this event the difference in temperature between the tissue and the constant source of heat is learned. For slight differences in temperature, such as usually exist in the tissues of the body, the thermo-electric energy is always proportionate to the difference in temperature between the two needle-elements.

It is obvious that instead of one pair of needles a multiplicity may be employed. By this means the delicacy of the apparatus naturally is materially increased. Thus, v. Helmholtz was able to increase the delicacy of the apparatus to the detection of differences of $\frac{1}{1000}$ ° C. by the employment of 16 antimony-bismuth elements. Schiffer constructed a thermopile of four pairs of needle-elements in a simple manner (Fig. 135, IV) by soldering together alternately wires of iron and German silver. It is intended that four such elements should be introduced into two substances (A and B) to be examined for differences in temperature.

Thermopalpation is the name given by Benczúr and Jónás to the following method of examination: If the finger be moved over an uncovered portion of the trunk it will be found that the skin is warmer over parts containing air, such as lungs and intestines, than over parts, normal or pathological, not containing air. The boundaries are said to agree with those determinable by percussion, but this has been disputed. Naturally this difference can be established also by thermometric examination.

TEMPERATURE-TOPOGRAPHY.

Although a powerful influence must be ascribed to the blood, on account of its constant movement, in the equalization of the temperature in the different parts of the body, nevertheless an exact equalization is never attained, but noteworthy differences exist in different parts of the body.

The *temperature of the skin* has been found to be as follows:

In the middle of the sole of the foot....32.26° C.
In the vicinity of the Achilles tendon...33.85° C.
In the middle of the anterior aspect of
   the leg.........................33.05° C.
In the middle of the calf..............33.85° C.
In the popliteal space.................35° C.
In the middle of the thigh............34.40° C.
In the inguinal fold...................35.80° C.
Over the apex-beat of the heart.......34.40° C.
On the face in a man..................31° C.
At the tip of the nose and on the lobule
   of the ear ....from 22° to 24° C.

J. Davy made these measurements immediately on arising without dressing, with the temperature of the room at 21°. Only the inferior surface of the bulb of a thermometer otherwise covered came in contact with the different portions of the skin.

In the closed axillary cavity, the temperature ranges, according to Wunderlich, from 36.49° to 37.25°; according to C. v. Liebermeister it is 36.89° C.

The skin overlying muscles is warmer than that covering bones and tendons. The cutaneous temperature is somewhat lower in the aged, while in children it ranges between 25° and 29° C.

The skin of the cranial vault has a higher temperature in the frontal and parietal regions than in the occipital region. Further, the left side is warmer than the right. The temperature of the skin is increased by dyspnea.

v. Liebermeister employs the following method in taking the temperature of free cutaneous surfaces: The bulb of the thermometer is heated to a point slightly above that of the temperature expected. Then the fall of the column of mercury is observed as the instrument is held in the air, and then at the apparently appropriate moment the bulb is applied to the surface of the skin. If the skin has the same temperature as the bulb, the mercury must remain stationary for a time. For the measurement of the cutaneous temperature, it is useful to employ a specially constructed thermometer with a flat vessel.

The temperature of the cavities of the body :

Cavity of the mouth beneath the tongue................37.19° C.
Rectum ....38.01° C.
Vagina ....38.03° C.
The temperature of the uterus is somewhat higher, while
     that of the cervical canal is somewhat lower.
Urine ....37.30° C.

The temperature of the stomach falls during the process of digestion. Cold rectal injections (11° C.) rapidly lower the temperature of the stomach 1° C.

The *temperature of the blood* is on the average 39° C. In the internal portions of the body venous blood is warmer than arterial blood, while the reverse condition prevails in the peripheral portions.

Blood of the right heart..............38.8° C. ⎫
Blood of the left heart...............38.6° C. ⎪
Blood of the aorta....................38.7° C. ⎬ Claude Bernard.
Blood of the hepatic veins............39.7° C. ⎭
Blood of the superior vena cava.......36.78° C. ⎫
Blood of the inferior vena cava.......38.11° C. ⎬ G. v. Liebig.
Blood of the crural vein..............37.20° C. ⎭

The lower temperature of the blood in the left heart is due to the fact that the blood is cooled in the lungs in the process of respiration. According to Heidenhain and Körner the temperature of the right heart is somewhat higher because it lies upon the warm liver, while the left heart is surrounded by the air-containing lung. This fact, observed by Malgaigne in 1832 and by Berger and G. v. Liebig, is disputed by others, who attribute the somewhat higher temperature of the left heart to the fact that more active processes of combustion take place in arterial blood and that heat is generated in the formation of oxyhemoglobin. In adjacent veins or in those of the same name the temperature of the blood is generally lower than in the corresponding arteries, on account of the greater amount of heat given off in the slower movement. Thus, the temperature of the blood of the jugular vein is from 0.5° to 2° lower than that of the carotid; that of the blood of the crural vein is from 0.75° to 1° lower than that of the crural artery. Superficial veins, particularly in the skin, give off much heat and therefore the contained blood has a lower temperature. The hepatic veins contain the warmest blood, 39.7° C., not alone on account of the glandular activity of the liver, but also on account of the extraordinarily protected situation of the organ.

The Temperature of the Tissues.—The temperature of the individual tissues is the higher: (1) the more they contribute to the production of heat through the transformation of potential energy, that is, the greater their metabolic activity; (2) the more blood they contain; and (3) the more protected their situation.

The muscles are the chief seat of heat-production, principally during contraction, but also during rest. The temperature of the blood in the aorta is from 0.1° to 0.6° lower than that of muscle at rest. In the second place, the glands generate heat, especially during activity, particularly the liver, the salivary glands, the glands of the stomach and the intestines.

Berger took the temperature of different tissues in the sheep and obtained the following results:

Subcutaneous connective tissue.........................37.35° C.
Brain ....40.25° C.
Liver ....41.25° C.
Lungs ....41.40° C.
Rectum ....40.67° C.
The right heart.......................................41.40° C.
The left heart..40.90° C.

In man, Becquerel and Brechet found the temperature of the subcutaneous connective tissue 2.1° C. lower than that of the adjacent muscles. The temperature of the cornea and of the aqueous humor depends in part upon the state of the iris. The smaller the pupil, the more heat must they receive from the vessels of the iris.

INFLUENCES AFFECTING THE TEMPERATURE OF INDIVIDUAL ORGANS.

The temperature of the individual organs is by no means constant, but there are numerous influences that at times cause it to rise and at other times cause it to fall.

1. The more heat a portion of the body generates independently within itself, the higher will be its temperature. As the production of heat depends upon the metabolic changes in the organs, it follows that with the activity of the latter the degree of heat-production must keep pace.

(a) The *glands* during secretion produce much heat, which they impart either to their secretion or to the outflowing venous blood.

C. Ludwig found the temperature of the escaping saliva on irritation of the tympanico-lingual nerve 1.5° C. higher than that of the blood passing through the glandular artery to the secreting organ. The temperature of the venous blood in the secreting kidney is higher than that of the arterial blood. The secreting liver in particular produces a large amount of heat. Claude Bernard studied the temperature of the blood in the portal vein and of the blood in the hepatic veins during hunger, at the beginning of digestion and at the height of digestion, and found

Temperature of portal vein......37.8° C. } After fasting for four days. Blood of
      "   hepatic veins....38.4° C. }   right heart during fasting 38.8° C.
Temperature of portal vein......39.9° C. } At the beginning of digestion.
      "   hepatic veins.....39.5° C. }
Temperature of portal vein.......39.7° C. } At the height of digestion. Blood of
      "   hepatic veins.....41.3° C. }   right heart during digestion 39.2° C.

In dogs feeding or chemical or mechanical irritation of the gastric mucous membrane, and even the holding of food before the animal, brought about elevation of temperature in the stomach and the intestines.

(b) The *muscles* produce heat in their contraction. J. Davy found the temperature of active muscle higher by 0.7°. Becquerel observed in 1835, by means of the thermo-galvanometer, an increase of 1 °C. in the temperature in the interior of a contracting muscle in man after five minutes.

Therefore the temperature in fast runners may rise above 40°. The increase in temperature following vigorous muscular activity disappears about one and a half hours after the commencement of rest. The lower temperature of paralyzed muscles is due only in part to the absence of muscular contractions.

(c) With reference to the influence of the *sensory nerves* upon the temperature it should in the first place be noted whether the circulation is increased or diminished as a result of their stimulation, whether respiration is slowed or accelerated, and whether the musculature of the body is relaxed, or is stimulated to activity through reflex influences. In the first place the temperature, in the interior of the body and the rectum, will be increased, and in the latter diminished. From this point of view the conflicting statements not rarely made can be reconciled.

(d) The bodily temperature rises also (about 0.3°) as the result of *mental activity*. The brain itself acquires a higher temperature in consequence of sensorial or sensory stimulation.

(e) The parenchymatous fluids, the serous fluids and the lymph generate but little heat within themselves by reason of the slight metabolic changes that take place in them, and accordingly their temperature is that of their environment. The epidermoidal and horny tissues produce no heat at all, and therefore maintain their temperature from the subjacent tissues.

2. The temperature of an organ depends upon the amount of blood it contains, as well as upon the time within which the volume of blood is renewed.

This is seen most distinctly in the differences in temperature between cold, pale skin, and warm, reddened skin.

When Becquerel and Brechet compressed the axillary artery in a man, the temperature in the interior of the biceps muscle of the arm fell several tenths of a degree. After ligation of the crural artery and vein in dogs Landois observed the temperature decline several degrees. Long-continued elevation of the extremities deprives them of blood and causes them to become colder.

Attention should be called at this point to a difference between the internal and external portions of the body, which is especially emphasized by v. Liebermeister. The external portions of the body give off more heat to the exterior than they generate within themselves. They will, therefore, be the cooler the more slowly the blood flows into them; and the warmer the more rapid the blood-current. Acceleration of the blood-current, therefore, will cause greater uniformity in temperature between the peripheral portions and the interior of the body, while retardation of the blood-current causes greater uniformity in temperature between the peripheral portions of the body and the surrounding medium. The internal portions of the body react in exactly the opposite manner. Here active production of heat takes place, while heat-dissipation occurs almost solely through the blood-current. The temperature in these parts must, therefore, fall when the blood-current is accelerated, and the reverse. From this it follows that the greater the difference in temperature between the periphery and the interior of the body, the less is the rapidity of the circulation.

3. If the situation of an organ causes it to lose much heat by conduction and radiation, or if other conditions bring about the same result, the temperature of the organ declines.

In the first place the skin is again to be mentioned in this connection, as it must exhibit a different temperature accordingly as it is exposed to colder or warmer surroundings, or is covered or not, or is dry or moistened by perspiration (in the evaporation of which heat is lost). The ingestion of considerable amounts of cold food and drink must cause the temperature of the stomach, and the inhalation of cold air must cause that of the respiratory tract down to the bronchial tree, to fall.

MEASUREMENT OF THE VOLUME OF HEAT: CALORIMETRY.

The calorimeter furnishes information as to the amount of heat that the body to be examined possesses or is capable of producing. The heat-unit or calory, that is the amount of kinetic energy that is capable of raising the temperature of one gram of water 1° C., is employed as the unit of measure.

Experiment has shown that equal amounts of different bodies require unequal amounts of heat in order to attain the same temperature. For instance 1 kilo of water requires nine times as much heat as 1 kilo of iron to attain the same temperature. Wherever, therefore, different materials with equal temperatures are found, each will be endowed with different amounts of heat. The same amount of heat imparted to different bodies will, thus, also produce different temperatures in them. On the other hand, bodies naturally of different tempera-ture may possess equal amounts of heat. The amount of heat that a definite amount (as, for instance, 1 gram) of a body requires in order to have its tempera-ture raised a definite amount (as, for instance, 1° C.), is designated the specific heat of that body. The specific heat of water, which possesses the greatest of all bodies, is placed at 1. Heat-capacity is the term applied to that property of bodies by means of which they are required to take up a varying amount of heat in order to maintain the same temperature.

Calorimetry is employed:

For the determination of the specific heat of the different organs of the body. But few observations in this connection have as yet been recorded.

The specific heat of a number of animal parts, as compared with that of water as 1, is as follows:

Blood from man, on the average. 1.02   (?)   Meat from man, on the average....0.741
(it is in proportion to the num-   Compact bone..................0.3
ber of erythrocytes)   Spongy bone. 0.71
Arterial blood, on the average. . 1.031 (?)   Fat 0.712
Venous blood, on the average . .0.892 (?)   Striated muscle................0.825
Cow's milk, on the average ....0.992   Defibrinated blood.............0.927

The specific heat of the human body as a whole is thus only ap-proximately that of an equivalent weight of water.

For the method of determining the specific heat of solid or liquid bodies works on physics should be consulted.

More important is the employment of calorimetry for the estima-tion of the amount of heat that either the entire body or an individual portion is capable of producing in a definite period of time.

Lavoisier and Laplace made the first calorimetric observations on animals in 1780, with the aid of the ice-calorimeter. A guinea-pig melted 13 ounces of ice in 10 hours. Crawford in 1779 and later Dulong and Despretz in 1822 employed for this purpose the water-calorimeter of Rumford—after which that of Favre and Silbermann (Fig. 133) is modeled. Small animals were placed in the interior chamber (K) made of thin copper and this was immersed in a large volume of water surrounded by a poor conductor of heat. The amount of the surrounding water and its initial temperature were known. From the elevation of temperature at the termination of the experiment, which lasted several hours, the number of calories furnished could be directly estimated. The air for breathing was supplied to the animal through a special tube from a gasometer. The expired gases were examined chemically for carbon dioxid.

According to Despretz, a small bitch generated 14,610 heat-units in an hour— 393,000 in twenty-four hours. The taking of the temperature of the animal before and after the experiment was carelessly omitted. Assuming equal metabolic activity, a human being about seven times heavier would, on the basis of this observation, produce in the neighborhood of 2,750,000 calories in twenty-four hours. Senator found that a dog weighing 6330 grams produced 15,370 calories, with a loss of 3.67 grams of carbon dioxid.

An adult man produces at rest in twenty-four hours 2,400,000 calories, therefore 100,000 in an hour. One kilogram of body-weight produces in twenty-four hours approximately 34,000 calories, therefore 1417 in an hour. These figures increase with increase in the total metabolism and also with functional activity.

The first calorimetric observations on man were made by Scharling in 1849. Leyden introduced the leg alone into the chamber of the calorimeter. This raised the temperature of 6600 grams of water 1° C. in an hour. If it be assumed that the total superficies of the body is about fifteen times as great as that of the leg the human body, assuming equal loss, would produce 2,376,000 calories in twenty-four hours.

HEAT-CONDUCTION OF ANIMAL TISSUES.

EXPANSIBILITY OF ANIMAL TISSUES BY HEAT.

The heat-conduction of animal tissues is principally of importance in relation to the arrangement of the external integument and the sub-cutaneous fatty tissue. The latter especially serves as a protecting shield in warm-blooded animals living in cold water (whale, walrus, seal) and through this abstraction of heat by means of conduction from the interior of the body is practically impossible. Few investigations have been made upon this question. Greiss in 1870 determined the con-ductivity of the following tissues by noting the distance from a central source of heat introduced into the tissues at which was melted a layer of wax. He studied the stomach of sheep, the bladder of oxen, the skin of cattle, calves' feet, the hoofs of oxen, the bones of oxen, the horns of buffaloes, the antlers of deer, ivory, mother of pearl and haliotis-shell (sea-snail). He found that fibrous tissues conduct better in the direction of their fibers than at right angles to their course. The figures formed by the melting wax upon tissues spread out over a wide area were therefore generally elliptical. Landois has made observations upon a number of human tissues by determining the melting-distance of a layer of paraffin from a thin test-tube filled con-stantly with boiling water and applied intimately to tissues in layers of equal thickness, and subsequently applied on the flat and supported by threads. Desiccation was avoided, and also the effect of radiant heat. Landois was able to confirm the fact of the better conduction in the direction of the fibers. Next to bone the best conductor was found to be blood-clot; then there followed successively spleen, liver, cartilage, ten-don, muscle, elastic tissue, nails and hair, anemic skin, gastric mucous membrane, washed fibrin-fibers. The great thermic conductivity of the blood as compared with the much lower conductivity of bloodless skin is of particular interest. In this way is explained the fact that but little heat is dissipated by anemic skin, while hyperemic skin conducts and gives off a much larger amount of heat.

Like all bodies the human body undergoes expansion at elevated temperatures. A man, weighing 60 kilos, will expand about 62 cu. cm. with an increase of his bodily temperature from 37° C. to 40° C. Of the different tissues, connective tissue (tendon) is expanded by heat, while elastic tissue and skin are contracted like rubber.

VARIATIONS IN THE MEAN BODILY TEMPERATURE.

General Climatic and Somatic Influences.—The bodily temperature remains on the whole constant within different climates. This is noteworthy if it be considered that a human being at the equator and in the polar regions is exposed to surrounding temperatures that differ from each other by more than 40° C. Further, it has been observed that when a person passes from a warm to a cold climate his temperature declines but little, but that when an individual passes from a cold to a hot region his temperature rises relatively in more considerable degree. In the temperate zone the bodily temperature in the cold winter-season is usually from 0.1° to 0.3° C. lower than on hot summer days. The elevation of a region above the level of the sea has no demonstrable influence upon the temperature. Race and sex cause no difference. Persons of vigorous constitution are believed to have a somewhat higher temperature in general than debilitated, flabby, anemic persons.

Influence of the General Metabolism.—As the production of heat is related to the breaking up of chemical combinations, from which, in addition to the formation of water, carbon dioxid and urea finally result as the most important excrementitious products, the amount of heat generated will keep pace with the total production of those bodies formed. The increased metabolic activity that sets in after a heavy meal causes an elevation of several degrees in temperature. As the general metabolism is naturally much less on days of fasting than on days on which a normal amount of food is taken, it is clear that in human beings the temperature will be found to be on the average 36.6° on fasting days and 37.17° C. on ordinary days.

Also Jürgensen found in human beings on the first day of inanition a reduction in the temperature, although on the second day a transitory elevation occurred. In experiments on fasting animals it was found that the temperature declined much at first, then for a considerable time remained pretty constant, and finally in the last days declined still further. Schmidt subjected a cat to starvation, and found that up to the fifteenth day the temperature was 38.6° C.; on the sixteenth day it was 38.3°, on the seventeenth, 37.64°, on the eighteenth, 35.8°, on the nineteenth, the day of death, 33° C. Chossat found the temperature of mammals and birds 16° C. lower on the day of death from starvation than under normal conditions.

Influence of Age.—The activity of the general metabolism must be in part responsible for the temperature of the body at different ages, but other influences of undetermined origin may also in part be contributory.

AGE.	MEAN TEMPERATURE AT ROOM-TEMPERATURE.	NORMAL LIMITS.	PLACE OF MEASUREMENT.
New-born	37.45° C.	37.35°–37.55° C.	Rectum.
5– 9 years	37.72° C.	37.87°–37.62° C.	Mouth and rectum.
15–20 "	37.37° C.	36.12°–38.10° C.	Axillary cavity.
21–25 "	37.22° C.		" "
26–30	36.91° C.		" "
31–40	37.10° C.	36.25°–37.5° C.	" "
41–50	36.87° C.		" "
51–60	36.83° C.		" "
–80	37.46° C.		Mouth.

According to Chelmonski the bodily temperature is somewhat lower in the old, and the evening temperature lower than the morning temperature.

The temperature in the new-born exhibits special peculiarities, such as would be readily explicable from the sudden change in the conditions of life. Immediately after birth the temperature of the child is on the average 0.3° higher than that of the vagina of the mother, namely 37.86° C. In the first hours after birth the temperature declines about 0.9° C., in conjunction with the reduction in gaseous interchange. After from nine to thirty-six hours it will again have risen to the average temperature of the infant, which is about 37.45° C. Certain irregular fluctuations occur during the first week of life. During sleep the temperature in infants declines between 0.34° and 0.56° C. Persistent crying may cause the temperature to rise several tenths of a degree. Less heat is produced in the aged on account of their lesser activity of metabolism, so that they suffer more readily from cold and therefore need warmer clothing.

Periodic variations in the course of the day are constant at all periods of life. In general, it may be stated that the temperature rises continuously by day, the maximum being reached between 5 and 8 p. m.; while it declines continuously by night, the minimum being reached between 2 and 6 a. m. The mean temperature of the body is found in the third hour after breakfast. The average level of all the temperatures observed in a person in the course of a day is designated the daily mean, which, according to Jäger, is 37.13° in the rectum. If the daily mean is above 37.8° it must be considered as febrile, and if below 37° as an evidence of collapse.

As the daily variations in temperature occur also during the state of hunger, although the elevations are somewhat less after the time for meals, the ingestion of food cannot alone be the cause of the variations, but these appear to reside essentially in the varying degree of muscular activity.

Hour	v. Bären-sprung	J. Davy	Hallmann	Gierse	Jürgensen		Jäger
a. m. 5					36.7	36.6	36.9
6	36.68				36.7	36.4	37.1
7		36.94*	36.63	36.98	36.7*	36.5*	37.5*
8	37.16*		36.80	37.08*	36.8	36.7	37.4
9		36.89			36.9	36.8	37.5
10	37.26		10½ = 37.36	37.23	37	37	37.5
11		36.89			37.2	37.2	37.3
m. 12	36.87				37.3*	37.3*	37.5*
p. m. 1	36.83			37.13	37.3	37.3	37.4
p. m. 2		37.05	37.21	37.50*	37.4	37.4	37.5
3	37.15*			37.43	37.4*	37.3*	37.5
4		37.17			37.4	37.3	37.5*
5	37.48	37.05*	5½ = 37.31	37.43	37.5	37.5	37.5
6		6½ = 36.83		37.29	37.5	37.6	
7	37.43	7½ = 36.50*	37.31*		37.5*	37.6*	37.3
8					37.4	37.7	37.1*
9	37.02*				37.4	37.5	36.9
10				37.29	37.3	37.4	36.8
11	36.85	36.72	36.70	36.81	37.2	37.1	36.8
m. 12					37.1	36.9	36.9
a. m. 1	36.85	36.44			37	39.9	36.9
2					36.9	36.7	36.8
3					36.8	36.7	36.7
4	36.31				36.7	36.7	36.7

* Indicates ingestion of food.

The excretion of carbon dioxid from hour to hour, also the daily variation in the pulse-frequency, almost coincides with the temperature. v. Bärensprung found that the mid-day maximum temperature somewhat preceded the maximum pulse.

If a person sleeps by day and performs all of his other daily duties by night, the typical course of the temperature-curve described may be inverted. The variations are, therefore, dependent upon the state of activity. With respect to the state of activity or of rest of the individual, the temperature of persons active during the day appears in general higher and during the night in general lower than in a person at rest.

The peripheral portions of the body also exhibit more or less regular variations in temperature. In the palm of the hand the course is somewhat as follows: After a relatively high temperature during the night a rapid fall sets in in the morning at six o'clock, which reaches its lowest between 9 and 10 o'clock. Then there follows a slow ascent, which reaches its maximum after the midday meal. Between 1 and 3 o'clock the temperature begins to decline, and the lowest level is reached in the course of two or three hours. Between 6 and 8 there is again a rise, and finally a decline toward morning. A rapid fall of the temperature at the periphery corresponds with a rise in the interior of the body.

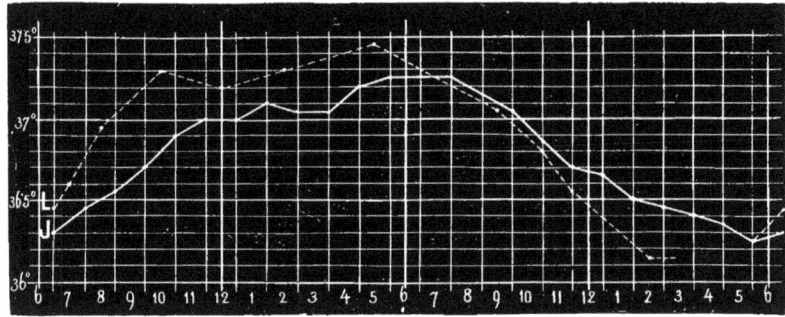

Fig. 136.—Variations in the Bodily Temperature during Health within Twenty-four Hours. L........according to v. Liebermeister. J———— according to Jürgensen.

Certain operations upon the body cause variations in temperature. After venesection the temperature at first falls. Then it rises several tenths of a degree with chilliness. In the first days it falls again to the previous level and even somewhat below this. Profuse acute hemorrhage causes a reduction in temperature of from 0.5° to 2° C., while long-continued, extensive hemorrhage may cause in dogs a reduction to as low as 31° and 29° C.

Here the reduction in oxidation-processes in the tissues the seat of lessened metabolic activity in consequence of the hemorrhage and the enfeebled circulation obviously constitute the cause of the reduction in temperature. Analogous conditions of diminished metabolism can be brought about if the peripheral extremity of the divided vagus is irritated for about an hour, so that the heart-beat becomes extremely slow, and in conjunction with it the entire circulation. Thus Landois was able to reduce the temperature in rabbits several degrees within a short time.

After every transfusion of any considerable amount of blood, beginning about half an hour after the operation, the temperature rises to a marked febrile attack, which will have subsided in the course of several hours. Direct transfusion from an artery to an adjacent vein in the same animal excites the same phenomenon.

Certain poisons, particularly chloroform, chloral and other anesthetics, as well as alcohol; further, digitalis, quinin and others, cause reduction in temperature. These substances appear in part to render the tissues less suited for the molecular decomposition necessary for the generation of heat. In the case of the anesthetics it is possibly a condition of the latter kind within the structure of the nerve that furnishes the cause. In part, however, they may also have an influence upon those processes that control the dissipation of heat from the body. Other poisons cause elevation of temperature from opposite causes.

Strychnin, nicotin, picrotoxin, veratrin, laudanin, cause elevation of the bodily temperature. The lowest temperature terminating in recovery observed was 24° (!) C. in the rectum of a profoundly intoxicated individual.

Reduction in temperature in connection with *disease* is due either to diminished heat-production (reduction in metabolic activity), or to increased heat-dissipation.

Marked reduction in temperature in individual instances (between 31° and 27.5° and down to 22° C. in the anus) has been observed particularly in cases of paralysis, in one of which Reinhard found a rectal temperature of as low as 22.5° C. four and one-half hours before death. The lowest temperature observed one day before death was 23° C. in the anus in a case of hemorrhage into the medulla oblongata. Also in cases of diabetes a reduction in temperature below 30° C. has been observed.

Elevation of temperature is exhibited as a rule in connection with fever, the highest temperature being observed by Wunderlich before death, 44.65° C.

REGULATION OF THE TEMPERATURE.

As human beings and other homoiothermic animals are capable of maintaining their bodily temperature at the same level under varying conditions, the body must possess special mechanisms by means of which the heat-economy is subjected to constant regulation. The latter can obviously make itself effective in two directions: either by control of the amount of molecular transformation through which potential energy is transformed into the kinetic energy of heat, or by influencing the dissipation of heat from the body in accordance with the production or the effects of external agencies.

Regulatory Mechanisms Governing Heat-production.

C. v. Liebermeister estimated the heat-production of a medium-sized person as 1800 calories per minute. It is in the highest degree probable that mechanisms are operative in the body upon whose stimulation the amount of heat-producing molecular transformation is dependent. It should especially be borne in mind that this stimulation is of reflex origin. Irritation from the peripheral extremities of the cutaneous nerves, through thermic excitation, or of the nerves of the intestines and of the digestive glands, through mechanical or chemical stimulation during the process of digestion or during inanition, may be transmitted to a heat-center, from which an influence is exerted through centrifugal fibers upon the reservoir for potential energy, for the purpose of stimulating either increased or diminished metabolism. Little is as yet known, however, concerning the nervous apparatus and channels necessary for the maintenance of this hypothesis. Nevertheless, numerous phenomena indicate that such a view is not unjustifiable.

Investigation has as yet furnished no adequate evidence as to the existence of a heat-center. Tschetschehin and Naunyn, as well as Ott and Wood recently, assume the existence in the brain (according to Ott in the anterior portion of the optic thalamus) of a center that is supposed to exert an inhibitory effect upon the combustion-processes in the body through fibers that descend through the pons, medulla and cord; and accordingly destruction of this center or its conducting paths would cause increased heat-production. Aronsohn and Sachs observed transitory rise of temperature, with increased metabolism, after deep puncture of the rabbit's brain several millimeters to one side of and behind the anterior fontanel. Injuries of the caudate nucleus, optic thalamus, corpus callosum, septum lucidum and trigone also cause similar phenomena. Confirmatory evidence is given by Richet, who attributes this elevation of temperature to increased heat-production. The animals eat more, yet lose flesh. Repeated cerebral puncture eventually induces marasmus, reduction in temperature, as low as 26°, and death. Centers with an opposite function, namely, stimulating heat-production, are said to be situated in the caudate nucleus, in the gray substance of the septum lucidum and in the gray matter in front of and below the caudate nucleus and in the tuber cinereum. After high division of the spinal cord heat-regulation, heat-production and heat-dissipation are disturbed.

The regulatory mechanisms governing heat-production can be recognized from the following phenomena:

1. As a result of the moderate, transitory influence of cold the bodily temperature rises, while as a result of the like influence of heat upon the external integument the temperature declines.

2. Heat-production is increased by reduction of the surrounding temperature, while the excretion of carbon dioxid and the consumption of oxygen are at the same time increased. Heat-production is diminished by increase of the surrounding temperature. The production of carbon dioxid takes place principally in the muscles, without contraction necessarily taking place at the same time.

Human beings generate at 0° C. about twice as much heat as at a surrounding temperature of 30° C. D. Finkler found in experiments on guinea-pigs that the production of heat is more than doubled in vigorous animals in consequence of a reduction in the surrounding temperature of about 24° C. Thus, during the winter the metabolism of the guinea-pig was increased about 23 per cent. as compared with the summer. It thus caused an alteration in heat-production in general that is entirely analogous to that resulting from lowering of surrounding temperature of shorter duration.

C. Ludwig and Sanders-Ezn have observed in rabbits, when the surrounding temperature was reduced from 38° C. to 6° or 7°, a rapid increase in the elimination of carbon dioxid. Conversely, this was diminished in animals as the surrounding temperature was raised from between 4° and 9° to from 35° to 37°. The thermic stimulation of the surrounding temperature must thus have had an effect also upon the combustion-processes. In accordance with this fact is the observation of Pflüger, who found increased consumption of oxygen and increased elimination of carbon dioxid in rabbits that had been immersed in cold water. When the influence of the cold was so profound that the bodily temperature fell as low as 30°, the gaseous interchange also diminished, and if the exposure continued until the temperature fell to 20° the interchange became only half of the normal. If mammals are placed in a warm bath whose temperature exceeds that of the body by 2° or 3° the elimination of carbon dioxid and the consumption of oxygen increase in consequence of a stimulation of the metabolic processes. The elimination of urea also increases from the same cause.

3. The application of cold to the external integument causes in part involuntary muscular movement (shivering), in part voluntary muscular movement. As a result of both heat is produced.

Cold thus stimulates muscular activity, which is attended with oxidation-processes. In human beings muscular activity induces, in addition to increased heat-production, also increased heat-dissipation. The latter, however, becomes less on conclusion of the activity than it had been before. After administration of curare, which paralyzes the voluntary muscles, this regulation of temperature falls to a minimum.

Strychnin increases heat-dissipation and heat-production, and the bodily temperature may be either increased or diminished in accordance with the preponderance of production (convulsions) or of dissipation. Cocain increases the bodily temperature, while the anesthetics have the reverse effect.

4. Change in the surrounding temperature has an influence upon the need for food. Ingestion of food increases the elimination of carbon dioxid, principally in consequence of increased activity on the part of the digestive glands. In winter, as well as in cold regions, the sense of hunger and the need for fats, whose combustion yields much heat, are increased.

Regulatory Mechanisms Governing Heat-dissipation.

The average dissipation of heat from the skin of a human being weighing 82 kilos is between 2,092,000 and 2,592,000 calories in twenty-four hours—therefore, between 1450 and 1798 calories in the minute.

1. Elevation of temperature causes dilatation of the cutaneous vessels. The skin becomes vividly reddened, soft, and full of fluid, so that it serves as a better conductor of heat and is swollen. The epithelium becomes moistened and sweat exudes from the surface. In this way provision is made for augmented heat dissipation, evaporation of the sweat playing an important part in the abstraction of heat.

The greater the increase in the moisture of the air, the less becomes the evaporation from the skin. Accordingly, heat-dissipation must be increased by conduction and radiation. The same amount of heat that is capable of transforming 1 gram of water at a temperature of 100° C. into steam is equal to that which will raise the temperature of 10 grams from 0° to 53.67° C. The sweat secreted is of the same temperature as the body; if it be completely converted into vapor it will require first sufficient heat to raise it to the boiling-point and then additionally the amount of heat that will convert it from this point into steam. For purposes of more precise determination there would be required a knowledge of the heat-capacity and of the boiling-point of the sweat.

The action of cold is to cause contraction of the cutaneous vessels. The skin becomes pale, less soft, deficient in fluid and collapsed. The epithelium becomes dry and permits the escape of no fluid for evaporation. In this way dissipation of heat through the skin is diminished. Through the contraction of the muscles of the skin and of the cutaneous vessels, with the displacement of well-conducting fluid and blood from the skin and the subcutaneous connective tissues, loss of heat from the periphery is diminished and heat-conduction transversely through the tissues is rendered difficult. The cooling of the body is lessened through the marked interference with the flow of blood through the skin, in the same way as is the case with a cooling apparatus made of convoluted tubing if the current passing through it is greatly lessened. If, however, the cutaneous vessels undergo dilatation, the temperature of the surface of the body rises, and the difference in temperature between it and the surrounding cooler medium is increased, and thus the loss of heat is augmented. Tomsa has shown that anatomically the arrangement of the fibrillation of the skin is such that every stretching of the fibers effected by the muscles of the skin gives rise to a reduction in the thickness of the skin, as a result of which an influence is exerted principally upon the readily displaceable blood present. When the author, in conjunction with Hauschild, ligated in dogs either the arteries alone or at the same time the axillary and crural arteries and veins, the carotids and the jugular veins, the temperature of the interior of the body rose several tenths of a degree within a short time. Chlorotic and anemic individuals, with pale, bloodless skin, at times exhibit, for the same reason of failing circulation through the skin, elevation of the bodily temperature.

By means of systematically employed stimuli, which, like cold baths and cold sponging, cause contraction of the muscles and vessels of the skin, the latter may be so invigorated and be maintained in such a state of irritability as to oppose vigorously loss of heat when the body or individual parts thereof are threatened with sudden abstraction of heat. Thus, cold spongings and baths constitute in a measure gymnastics for the muscles of the skin, which under the conditions indicated are capable of protecting the body against cold.

2. Elevation of temperature accelerates the heart-beat, while reduction of temperature diminishes the number of contractions of the heart. Through the action of the heart the blood that is relatively the warmest is pumped from the interior of the body to the surface of the skin; and in this way it may readily give off heat upon the extensive surface. The oftener the same amount of blood courses through the skin, the more will be the amount of heat given off, and the reverse. Therefore, the frequency of the heart-beat is in direct relation to the rapidity with which cooling takes place. Thus, the pulse has been observed to rise to more than 160 per minute in air of an excessively high temperature—above 100° C. This is true not alone of the range of normal conditions, but also of the pathological variations in temperature during the febrile state. C. v. Liebermeister records the following figures for the pulse and the temperature respectively in adults :

Pulse—in the minute: 78.6, 91.2, 99.8, 108.5, 110, 137.5.
Temperature in ° C.:   37°, 38°, 39°, 40°,   41°, 42°.

If the number of heart-beats is permanently diminished it might be anticipated that elevation of temperature would occur. When the author, in conjunction with Ammon, caused reduction for about one and one-half hours in the number of heart-beats by irritation of the peripheral extremity of the vagus in rabbits, the temperature in the rectum fell on the average from 39° to 34.5° C. The enfeebled circulation diminishes also metabolism and oxidation in the body; in fact, this diminution must therefore even over-compensate for the accumulation of heat resulting from the diminished circulation.

3. Elevation of temperature increases the number of respirations, so that a much larger amount of air passes through the lungs in a given time and in them is raised almost to the temperature of the body. In addition a certain amount of water undergoes evaporation in the expired air with every respiration, and in this way heat is taken up. Therefore, it is to be borne in mind that vigorous respiratory movement materially sustains the circulation, so that the respiration operates indirectly in the manner outlined in 2. Forced respiratory movement exerts a cooling effect even if. air heated to a temperature of 54° C. and saturated with watery vapor is inhaled.

4. Nature provides many animals with furs during the winter and with lighter covering in the summer. Many animals living in air and water of a low temperature are protected against excessive heat-dissipation by heavy layers of fat. In the same manner man provides for a more uniform dissipation of heat on the part of the skin by means of a difference in clothing for winter and summer. The attitude of the body

also is not without influence upon the temperature. Thus, a cowering position and drawing together of the head and the extremities help to retain heat, while spreading of the extremities, erection of the hair, ruffling of feathers, permit the escape of a greater amount of heat. Landois found that in rabbits suspended in air with their extremities spread out the rectal temperature declined from 39° C. to 37° C. in the course of three hours. Exposure in heated or cooled rooms, ingestion of hot or cold food and drink, hot or cold baths, exposure to a quiet atmosphere or to air in active motion (fanning) are measures employed by man for regulating the temperature at will.

In the cooling of the body from its surface, radiation, conduction (also through the air) and convection (as the layer of air in contact with the body is constantly being displaced by the heat) take part in addition to evaporation. The radiating power of the skin has been carefully studied by Eichhorst and Masje. It is increased after irritation and friction of the skin, after muscular effort, and in still greater degree—up to three or four times the initial amount—through the action of cold air or after a cold bath. After marked abstraction of heat radiation becomes small, while it is increased during the febrile process and after the employment of antipyretics. The amount of heat radiated by a naked man from each square centimeter of superficies is equal to 0.001 calory in the second. This would make for the entire body, weighing 82 kilos. approximately 1,782,000 calories in twenty-four hours. Stewart found the loss of heat through radiation for a clothed man. weighing 70 kilos, 700,000 calories; for a man, weighing 82 kilos, 820,000 calories. In a clothed person the radiation, according to Rubner, with a weight of 82 kilos and a superficies of 22,430 square centimeters. is 1,181,000 calories.

In estimating the influence of climate upon the regulation of heat of the body chief importance is to be attached to the rapidity of evaporation, which is proportional to the square root of the velocity of the wind.

CLOTHING.

The effect of the clothing is yet to be taken into consideration. A warm dress is an equivalent for food, for as the dress is intended to preserve the heat of the body generated by the latter from the combustion of food, it may be stated that the body has a direct income through the food. while by means of clothing it protects itself against unnecessary expenditure. The clothing thus at room-temperature saves 20 per cent. From this its importance in the heat-economy is obvious. Summer-clothing weighs from 3 to 4 kilos and winter-clothing from 6 to 7 kilos. The radiation of heat from the body through a full suit of clothing is only about one-third of that from the naked skin. At a low temperature this reduction in heat-radiation is greater than when the surrounding temperature is high.

With respect to the usefulness of clothing the following considerations are to be borne in mind: (1) Its *conductivity*. Those materials that are the poorest conductors of heat keep the body the warmest. The following is a list of conductors arranged successively from the poorest to the best: Hare-skin. down, beaver-skin, raw silk. taffeta. sheep's wool, cotton, flax. twisted silk. (2) The *radiating power*. Rough substances radiate heat more readily than smooth substances. The radiating power is, however. equal for different colors. (3) The relation to the *sun's rays*. Dark materials absorb more heat from the sun than light materials. (4) The degree in which materials are *hygroscopic* is of great importance, that is whether they are capable of taking up much moisture from the skin, and at the same time yield this up gradually by evaporation, or the reverse. Wool of the same weight takes up twice as much water as linen. but the latter permits its more rapid evaporation. Wool upon the skin, therefore, less readily permits accumulation of moisture and also the development of cold through rapid evaporation, and therefore affords protection against catching cold. (5) The degree of *permeability* for air—ventilation—is of importance with respect to clothing, but it bears no relation to heat-conduction. Thus the application of a coat of varnish to materials increases the heat-conduction, but destroys the ventilation. The permeability depends—apart from the thickness of the material—upon the specific gravity and the character of the thread. The following is a list of substances beginning with the more permeable and passing to the less permeable: Flannel, buckskin, linen, silk, leather, oil-cloth. (6) Clothing that is in direct contact with the skin naturally also takes up the excrementitious products of the skin, linen and cotton in greatest amount, and wool least of all transmitting the waste matters to the overlying clothing. The drawers take up the least material, the shirt twice as much, the socks eight times as much.

The temperature of the surface of the body beneath winter-clothing is on the average 18° C. and beneath summer-clothing 20° C. Heat is given off principally by conduction when clothing is worn. Clothing appears comfortable when the temperature of its surface is 5° or 6° C. higher than that of the air.

HEAT-BALANCE.

As the temperature of the body is capable of remaining constant within narrow limits, it is obvious that the amount of heat taken up must be equivalent to the amount of heat given off, that is exactly so much potential energy must within a given time be converted into heat as heat is given off from the body. Attempts have been made from different points of view to set up heat-balances which while partly at least without a reliable foundation, are nevertheless of great interest in the elucidation of the heat-economy of the animal organism. An adult produces on an average enough heat to raise the temperature of his body almost 1° C. in half an hour. If no heat were given off, the body would in a short time become enormously heated—in thirty-six hours to the boiling-point—providing the production of heat continued uninterruptedly.

HEAT-BALANCE ACCORDING TO H. v. HELMHOLTZ.

Hermann von Helmholtz was the first, in 1846, to determine numerically the amount of heat produced by man.

1. **Heat-income.**—(a) A healthy adult, weighing 82 kilos, expires in twenty-four hours 878.4 grams of carbon dioxid. The combustion of the carbon thereof into carbon dioxid generates.. 1,730,760 calories

(b) The man, however, takes up more oxygen than is present in the carbon dioxid given off. This excess is employed for purposes of oxidation, particularly for the formation of water through the combustion of hydrogen. In consequence of the excess of oxygen thus taken up 13,615 grams of hydrogen can be additionally burned up, yielding......... 318,600 "

2,049,360 calories

(c) About 25 per cent. of heat must be derived from other sources than combustion-processes, so that in round figures there will be..................................... 2,732,000 calories

This amount would in fact suffice to raise the temperature of a human body weighing from 88 to 90 kilos from an average temperature of 10°, 28° or 29°C., thus to 38° or 39° C., that is the normal temperature.

2. **Heat-expenditure.**—According to v. Helmholtz the following debits must be set against the heat-income:

(a) For the heating of food and drink, which have an average temperature of 12° C.... 70,157 calories = 2.6 per cent.

(b) For heating the inspired air = 16,400 grams, assuming the temperature of the air to be 20° C.. 70,032 " = 2.6 per cent.

[If the temperature of the air were 0° the number of calories would be 140,064 = 5.2 per cent.]

(c) 656 grams of water evaporated through the lungs................................... 397,536 = 14.7 per cent.

(d) The remainder through radiation and evaporation from the external integument ... from 77.5 per cent. to 80.1 percent.

ESTIMATION OF HEAT-INCOME ACCORDING TO FRANKLAND'S METHOD.

Frankland in 1866 burned food directly in the calorimeter (Fig. 133) and obtained the following results:

1 gram of proteids yielded ....4998 heat-units ⎫ These figures may be
   "   grape-sugar yielded 3277   "   ⎬ compared with ·Rub-
   "   beef-fat yielded 9069   "   ⎭ ner's results, p. 379.

The proteids are decomposed only to the stage of urea; therefore the heat yielded by the combustion of the latter is to be deducted from 4998, thus leaving 4263 heat-units for 1 gram of proteids.   If the number of grams of the individual foods taken by man has been determined by weight the number of heat-units taken up can be readily estimated.

When the amount of food is sufficient the production of heat, under otherwise like conditions, is always the same.   If the amount of food is insufficient, the amount of heat produced is but little diminished, as the body must then consume some of its own tissues.   This is naturally the case in the state of hunger especially. The character of the food, providing it is sufficient in other respects, is of subordinate importance.

VARIATIONS IN HEAT-PRODUCTION.

According to v. Helmholtz the average heat-production in a healthy adult, weighing 82 kilos, in twenty-four hours is 2,732,000 calories.

Influence of the Superficies of the Body.—Rubner found that heat-production is dependent not upon the weight of the body, but upon its size and the related superficies.   Small, and also young, animals have a relatively larger superficies than larger, and also older, animals. As, however, the dissipation of heat takes place principally from the external surface, accordingly greater heat-production will have to take place in animals with a greater superficies—heat-dissipating surface. Thus a relatively greater consumption of oxygen was accordingly observed in smaller animals.   Rubner's investigations have shown that for dogs of various sizes the heat-production for each square meter of body-surface uniformly equaled 1,143,000 calories.   If the body-weight was compared with the body-surface in different animals, he found that for every kilogram of weight there was in the rat 1650, in the rabbit 946, in man 287 square centimeters of surface.

According to J. Rosenthal the production of heat is to be estimated in the following manner: If n represents the amount of heat produced in an ·animal in one hour, g the body-weight, and A a factor that remains nearly constant in the same species and under like nutritive conditions (for the body of the child, 11.97; for that of the adult, 12.31; for that of the dog' 49; for that of the rabbit, 33), then $n = A\sqrt[3]{g^2}$.

Age and Sex.—In the earliest period of life, as well as in old age, the production of heat is less than at mature age.   It is likewise so in women as compared with men.

Daily Variation.—The production of heat exhibits a course similar to that of the bodily temperature at different hours of the day.

Bodily Functions.—During waking, with physical and mental exertion, as well as during digestion (on account of the greater glandular activity), the production of heat is greater than under the opposite conditions.

RELATION OF HEAT-PRODUCTION TO THE WORK ·PERFORMED BY THE BODY.

The potential energy supplied to the body can be transformed by the latter into heat and into kinetic energy.   In the resting body almost the entire amount of potential energy is transformed solely into heat, for the work of the muscles of the circulatory, digestive, and respiratory organs is transformed within the body into heat, and therefore is not work transmitted outward.   A man at work, however, in addition to the production of heat, transforms potential energy into work.   An equivalent measurement will serve for the comparison of both activities, namely, 1 heat-unit, that is, the energy that will raise the temperature of 1 gram of water 1° C., which equals 425.5 grammeters.

The following illustration will serve, first of all, to make clear the relation between heat-production and work.   If a small steam-engine, in which a given amount of coal is burned, is placed within the inner chamber of a capacious calorimeter, heat alone will be produced from the coal so long as the engine is not brought into working activity.   The water in the calorimeter will indicate exactly through the elevation of its temperature the number of heat-units furnished by the burning coal.   If this has been determined, the same amount of coal is burned in the steam-engine in a second experiment, but at the same time by means of a suitable device outside of the calorimeter work is performed by the engine, such as the raising of a weight.   This work must naturally be furnished by the potential energy of the fuel and be transformed.   If now the elevation of temperature at the end of the experiment is noted it will be found that a smaller number of heat-units have been transmitted to the water than in the first experiment, in which the engine was heated, but performed no work.   Comparative experiments of this kind have demonstrated beyond doubt that in the second experiment the useful working effect is almost proportional to the heat-deficit observed.

If the processes in the organism be compared with this illustration it will be seen that the resting human being generates between 2½ and 2¾ million calories from the potential energy contained in the ingested food, while the amount of work performed by a laborer is estimated at 300,000 kilogram-meters.   If the organism were exactly comparable with the engine, just so much less heat would have to be formed within the body as corresponds to the amount of work done.   As a matter of fact, the organism naturally can transform only a lesser amount of heat from the same measure of potential energy when work is performed.   One point, however, should be taken into consideration in which the laborer differs from the working engine.   The laborer consumes in the same time a far larger amount of potential energy than the resting individual.   A greater amount of combustion takes place in his body, and it therefore comes about that the loss through the increased combustion is not alone made good, but is even over-compensated.   The laborer is, by reason of his greater muscular activity, warmer than the resting individual.   The following may serve as an example of the relation indicated: Hirn in 1858 took up at rest in the calorimeter-chamber 30 grams of oxygen in an hour, and produced 155,000 calories.   When subsequently he undertook in the chamber work transmitted outward, to the amount of 27,450 kilogram-meters, he consumed 132 grams of oxygen and furnished only 251,000 calories.

In estimating the amount of work done only that transmitted outward as heat-equivalent is to be considered, as, for instance, the lifting of a load, the throwing of weights, the displacement of masses.   Also the lifting up of the body is to be included here.   In ordinary walking the overcoming of the resistance of the air and the activity of the muscles must be taken into consideration.   In descending from a height an increase in heat of the body is not to be looked for, for muscular activity is required to prevent the body from falling down and from collapsing, and to avoid a too precipitate descent.

26

The organism is superior to the engine in the fact that more work in proportion to heat is transformed from the same measure of potential energy. While the best gas-engine is capable of converting 10.82 per cent. of the potential energy of illuminating gas into work and the remainder into heat, the human being is capable of furnishing 35 per cent. of work—in making ascents and in doing work of other character only 25.4 per cent.—from the chemical transformation in its muscular tissue, Pflüger's experimental dog as much as 48.7 per cent., and an excised bit of frog's muscle even 50 per cent. Work alone, without simultaneous production of heat, can never be transformed from chemical potential energy in an inanimate or animate motor.

ACCOMMODATION TO VARIATIONS IN TEMPERATURE.

All bodies possessing great heat-conductivity, when brought in contact with the skin, appear much cooler or warmer respectively than poor conductors. The reason for this lies in the fact that they abstract more heat from the body or supply more heat to the body than the latter. Thus the water of a cold bath will always feel colder than the air at the same temperature, because it is a better conductor of heat. In the temperate zone, for example:

AIR	WATER
At 18° C. feels moderately warm,	Up to 18° C. appears cold,
From 23° to 28° C., hot,	From 18° to 29° C., cool.
Above 28° C., extremely hot.	From 34° to 35° C., indifferent.
	Above 35.5° C., warm,
	At 37.5° C. and above, hot.

So long as the temperature of the body is higher than that of the surrounding medium, the body gives off heat, and in greater amount and more rapidly the better the conductivity of the surrounding medium. As soon, however, as the surrounding temperature becomes higher than that of the body, the latter takes up heat and in greater amount and more rapidly as the medium is a better conductor. Therefore, hot water appears to be of a higher temperature than air at the same temperature.

A human being may remain for eight minutes in a bath at a temperature of 45.5° C., but not without risk to life. The hands tolerate immersion in water of a temperature of 50.5° C., but not of a temperature of 51.65° C. At a temperature of 60° C. intense pain is felt in the integument. On the other hand, a human being may tolerate air at a temperature of 127° C. for eight minutes. Girls have remained for as long as twenty minutes in air at a temperature of 132° C. Under these circumstances the bodily temperature rises but little, namely, to 38.7° or 38.9° C. This depends upon the fact that the air, acting as a poorer conductor of heat, does not convey so much heat to the body as does water. Further, and this is the most important fact, the body exposed to hot air is capable of losing heat at its surface through abundant sweating and evaporation, and to this end the increased evaporation of water due to the increased activity of the lungs contributes. The enormous acceleration of the heart-beat—up to above 160—causes constantly renewed volumes of blood to be sent to the skin through its greatly dilated blood-vessels, for the secretion of sweat and evaporation. In the degree in which these diminish the body becomes less capable of withstanding the surrounding heat, and thus is readily explained the fact that the human being is by far less able to withstand air rich in watery vapor than dry air at the same temperature, as heat must, under such circumstances, accumulate within the body. Thus in the Russian steam-bath at a temperature of from 53° to 60° C. the normal rectal temperature rises to between 40.7° and 41.6° C. A human being is just able to work in an atmosphere at a temperature of 31° C. almost completely saturated with watery vapor.

In water at the temperature of the body the normal bodily temperature rises 1° C. in one hour; about 2° C. in one and one-half hours . Gradual elevation of the temperature of the water from 38.6° to 40.2° C. caused an increase in the axillary temperature to 39° C. within fifteen minutes.

ACCUMULATION OF HEAT IN THE BODY.

As under normal conditions the constancy of the bodily temperature is the result of a constant relation between heat-production and heat-dissipation it is obvious that heat must be stored up in the body when heat-dissipation is lessened. The chief organ regulating heat-dissipation is the external integument. Contraction of the skin and its vessels diminishes heat-dissipation, while relaxation with dilatation of the vessels increases heat-dissipation. Accumulation of heat may, accordingly, be effected:

(a) By intense and extensive cutaneous irritation, through which a transitory influence is exerted, causing contraction of the skin and its vessels. (b) Also through other forms of restriction of loss of heat through the skin. (c) Through increased activity of the vasomotor center, as a result of which contraction of all vessels, and naturally also those of the external integument, is brought about. In this way the elevation of temperature following transfusion of blood from an animal of the same species is to be explained—direct transfusion of arterial blood from the crural artery into the adjacent vein in the same animal will suffice, as Landois was able to confirm by experiments on the carotid and the external jugular vein—as well as that following venesection after a preceding decline in temperature. In both events abnormal blood-distribution takes place. In the first the venous system is abnormally overloaded, in the second abnormally empty. For the restoration of the normal distribution vigorous activity on the part of the musculature of the vessels is necessary, excited through the vaso-motor center. The marked contraction of the cutaneous vessels hereby brought about exerts an inhibitory influence on heat-dissipation and heat-accumulation thus takes place. The elevation of temperature observed after sudden abstraction of water from the body must apparently be explained in the same way. The inspissated blood requires less vascular blood and the contracted vessels permit the escape of little heat into the skin. (d) If the circulation through the cutaneous vessels in considerable areas is retarded by mechanical means, as by occlusion of small vessels by viscous masses of stroma or coagula, which form after transfusion of blood from an animal of a different species, accumulation of heat takes place likewise in consequence of diminished dissipation. Perhaps a number of other pyrogenic agents act in the same manner. In dogs in which both carotids and both axillary and crural arteries were ligated at one time, with or without the related veins, the temperature was observed to rise almost 1° C. within two hours.

It is obvious that increased heat-production in the presence of normal heat-dissipation must give rise to accumulation of heat. In this category belongs the elevation of temperature following muscular and mental activity, and attending digestion. Finally, the elevation of temperature that appears several hours after a cold bath and is brought about by increased heat-production through reflex influences from the cooled skin is probably of the same character.

If the temperature of the body as a whole is raised about 6° C. death results, as in the case of heat-stroke or sunstroke. At this temperature molecular decomposition of the tissues appears to take place. With long-continued, though less marked, elevation, distinct fatty degeneration of many tissues occurs. If animals whose temperature is raised artificially to 42° or 44° C. are subsequently placed in a cooler atmosphere, the temperature at first becomes subnormal (36° C.) and it may remain so for days.

FEVER.

In many ways related to the accumulation of heat largely confined within the limits of physiological phenomena fever occurs as the most common pathological derangement in the bodily economy and to it some reference may be made. Fever consists essentially in increased metabolism, chiefly in the muscles, together with elevation of temperature. Under these circumstances a disturbance in the regulation of the heat-balance must naturally take place, for if provision be made that with the increased heat-production also increased heat-dissipation shall take place, there can then be no elevation of temperature, or accumulation of heat. According to v. Liebermeister heat-regulation is placed upon a higher temperature-level during the febrile process. As in the state of fever the body appears to be in large measure incapacitated for mechanical activity, the transformation of this larger amount of decomposing potential energy in the body almost wholly into heat, and the failure to utilize this for mechanical activity, must moreover be especially emphasized as characteristic. Malarial intermittent fever may be considered as the prototype of fever. It is attended with severe paroxysms of fever lasting several hours in alternation with wholly afebrile periods, so that its symptoms may be readily analyzed. Among the individual phenomena of fever there are encountered:

1. *Elevation of bodily temperature* (to 38° or 39° C. constitutes mild, and from 39° to 41° C. and above, severe fever). Not only the febrile patient with a burning, reddened skin (calor mordax), but also the shivering patient in a chill with an apparently cold skin may exhibit elevation of temperature. The reddened skin, however, is a good conductor, the pale skin a much poorer conductor of heat. Therefore, the former appears the warmer to the touch.

2. *Increased heat-production*, which had already been assumed by Lavoisier and Crawford, can be recognized indubitably by calorimetric measurement. This can be attributed only in smallest part to transformation of the increased circulatory activity into heat, but in largest part it is dependent upon heat generated in the processes of combustion.

3. *Increased metabolism*, to which the wasting character of fever is due. This was known to Hippocrates and Galen and was thus described by v. Bärensprung in 1852: "All so-called fever-symptoms indicate that during the febrile process tissue-consumption is abnormally increased. The increased metabolism is evidenced by augmented carbon-dioxid elimination (from 70 to 80 per cent.). In addition to carbon-dioxid elimination there is increased absorption of oxygen, at most 20 per cent. in a patient with acute fever, while the respiratory quotient remains unchanged. According to D. Finkler the production of carbon dioxid is susceptible of greater variation than the consumption of oxygen. The state of the nutrition is an index of the size of the respiratory quotient. The increase in gaseous interchange is not the result, but the cause of the increased bodily temperature. The former takes place also when the bodily temperature is reduced by

a cold bath. The elimination of urea is increased between one-third and two-thirds. In dogs suffering from septic fever Naunyn observed increased elimination of urea even before the temperature rose—prefebrile elevation. At times, however, the urea is in part retained during the febrile process and is eliminated in large amount after the termination of the febrile attack—epicritical elimination of urea. The uric acid also is increased. At the same time the urinary pigment derived from the hemoglobin may be increased twenty times and the elimination of calcium be increased seven times. The urinary water is diminished (in typhoid fever) and is excreted in greater amount during convalescence. The fact that the combustion-processes in the body of the febrile patient are exceptionally increased if he be placed in a warmer atmosphere appears especially noteworthy. During the febrile process there is also an increase in oxidation-processes under the influence of colder surroundings, but the increase of combustion in warm surroundings is much greater than in cold.

4. *Diminished Heat-dissipation.*—That in some cases febrile temperature may actually result from diminished heat-dissipation is shown, for instance, by the sudden attacks of fever that occur after catheterization or with the passage of a gall-stone through the bile-duct. These are brought about solely through reflex irritation of the vasomotor center, which greatly interferes with heat-dissipation in consequence of contraction of the cutaneous vessels. In other forms of fever in man diminished heat-dissipation is only in part a causative factor, as the following analysis will show:

(a) *The Stage of Chill, or the Cold Stage.*—Here the loss of heat through the pale, anemic skin, by conduction, radiation and evaporation of water, is diminished in greatest degree, but also heat-production is from one and one-half to two and one-half times greater. The rise of temperature in the febrile stage, which often is rapid and marked, alone establishes the fact that the diminished heat-dissipation is not the sole cause of the elevation of temperature. (b) In the *hot stage* loss of heat from the reddened, hyperemic skin is increased, but the increased heat-production still preponderates. v. Liebermeister estimates that a temperature-elevation of $1°$, $2°$, $3°$ or $4°$ C. corresponds with an increase in heat-production of 6 per cent., 12 per cent., 18 per cent., or 24 per cent., respectively. (c) In the *sweating-stage* heat-dissipation from the reddened, moist skin, and evaporation, are most pronounced, being more than two or three times the normal loss. Under these circumstances heat-production is either increased or normal or subnormal, so that under such conditions the temperature of the body likewise may become subnormal, down to 36° C. In case of fatal collapse the production has fallen to three-fourths or one-half of the normal, without simultaneous increase in heat-dissipation.

Plethysmographic examinations of the vessels of the arm in febrile patients have shown, in accordance with the temperature-variations during the febrile process, that the blood-vessels begin to contract before any elevation of temperature is evident. With the progress of the contraction the temperature then rises, and both reach their maximum at the same time. The decline in temperature is subsequently preceded by dilatation of the vessels, and with marked dilatation of the vessels the temperature again falls to the normal level.

5. *Deranged Heat-regulation.*—High surrounding temperature may increase that of the febrile patient more than that of a non-febrile individual. The reduction in heat-production that permits normal animals to maintain their normal temperature in warm surroundings is far less during fever.

Among the *accessory phenomena of fever* the following are especially noteworthy: Increase in the intensity and number of the heart-beats, and respiration—in adults to 40, in children to 60 in the minute. Both are compensatory phenomena of the elevated temperature. There are, further, diminished digestive activity and intestinal movement, derangement of cerebral activity, of the secretions, of muscular activity, interference with elimination, as for instance of water, or of administered potassium iodid, through the urine. Febrile pyrexia is by some considered as having a curative influence on the body, it being reasoned that the body is cleansed and purified by the heat of the fever. In the presence of high fever molecular degeneration of the tissues has often been found.

With respect to the blood-corpuscles during fever reference may be made to p. 50, 1, to the amount of carbon dioxid in the blood on p. 81, to the vascular tension on p. 142, to the saliva on p. 339, to the digestion on p. 341 D. The utilization of the food throughout the entire tract has not been found interfered with in marked degree.

In experiments on animals Krehl and Mathes found an increase of 10 per cent. in heat-production in conjunction with elevation of temperature, and diminution in heat-dissipation. At the height of the fever heat-production was likewise increased, while heat-dissipation was increased only when heat-production was considerable. With decline of temperature heat-production is generally diminished while heat-dissipation varies.

According to Filehne, Hildebrand, Richter, Stern, and others, antipyretics act by restoring heat-regulation to a lower level. Quinin reduces temperature by limiting heat-production. Toxic doses of metallic salts act similarly, diminished carbon-dioxid formation being at the same time demonstrable. According to others the influence of antipyretics is exerted principally upon the increase of heat-dissipation through the dilatation of the vessels, while heat-production is but little diminished—about 15 per cent.

The course of heat-production in infected cold-blooded animals follows that in febrile warm-blooded animals. It rises at the height of the disease and falls during collapse. Even in plants—injured bulbs—Pfeffer observed phenomena analogous to fever.

ARTIFICIAL ELEVATION OF THE BODILY TEMPERATURE.

Elevation of the bodily temperature, in addition to causing disturbances of the general condition, influences, first of all, consciousness, so that mental confusion, vertigo, insomnia and loss of consciousness occur. The functions of the medulla oblongata and the spinal cord are affected only later.

If mammals are kept constantly in air at a temperature of 40° C. escape of heat from the body ceases, and accordingly accumulation of the heat produced must take place. At first the bodily temperature declines somewhat for a short time, but later a distinct elevation sets in. Respiration and pulse are accelerated and the latter becomes weaker and irregular. Absorption of oxygen and elimination of carbon dioxid diminish in the course of from six to eight hours, and death takes place amid signs of great exhaustion, convulsions, salivation and loss of consciousness, even when the temperature of the body is not increased more than 4° or at most 6° C. Death is due not to the rigidity of the muscles, as the coagulation of their myosin does not take place in mammals at a temperature below 49° or 50° C., in birds at a temperature of 53° C., in frogs at a temperature of 40° C., but probably to a derangement of the heat-regulating functions of the nerves. If mammals are exposed suddenly to air of a high temperature, 100° C., death takes place amid similar phenomena, but much more rapidly—in fifteen or twenty minutes. The temperature of the body rises only 4° or 5° C. under such circumstances. Under like conditions a loss of 1 gram in body-weight is observed in rabbits within a minute. Birds tolerate the high temperature somewhat better, dying only after the temperature of their blood reaches 48° or 50° C. Man also is capable of surviving for a short time in air having a temperature between 100° and 132°, although the greatest danger to life sets in in the course of ten or fifteen minutes. At the same time the skin becomes burning red, copious sweating takes place, and the cutaneous veins are greatly distended and of a brighter red appearance. Pulse and respiration are greatly accelerated. Severe headache, vertigo, exhaustion and failure of sensory activity are indicative of great danger. At the same time the temperature taken in the rectum will have risen but 1° or 2° C. According to the observations of C. A. Koch, v. Voit and Simanowsky, artificial elevation of temperature in man and animals is not followed by increased proteid metabolism, whence it is to be concluded that the increased proteid metabolism attending the febrile process cannot be dependent upon the elevation of temperature, but must be brought about by the inadequate nutritive state of the tissues, or by bacterial poisons. Fever may also endanger life through elevation of the bodily temperature. If the temperature remains at 42.5° C. for a considerable time death is unavoidable. If the artificial elevation of temperature is not increased to the point of causing death, beginning in from thirty-six to forty-eight hours fatty infiltration and degeneration will take place in the liver, the heart, the kidneys and the muscles.

In cold-blooded animals the temperature can be raised from 6° to 10° C. within a short time, by exposure to hot water as well as to hot air. As the heart of the frog ceases to beat at a temperature of 40°, and as the muscles in the interior of the body begin to become rigid at the same temperature, the maximum temperature for the continuance of life is in this animal considerably lower. Actual death is preceded by a condition of apparent death, from which resuscitation is possible. Insects live in the desert at a temperature of 64° R. and arctisca and anguillulæ die in water at a temperature of 45°, while in a dry medium they can be heated to a temperature of 70°, and rotifers, after careful desiccation, can be heated to 125° C. Most juicy plants die after exposure for half an hour to air at a temperature of 52° C. or to water at a temperature of 46° C. Desiccated seeds (oats) may retain their germinating activity after exposure for a considerable time to air at a temperature of 120° C. Lowly organized plants, such as the algæ, can live in warm springs at temperatures up to 60° C. Some bacteria tolerate the boiling temperature.

EMPLOYMENT OF HEAT.

Brief and not intense heat applied to the surface of the body causes at first a transitory, slight reduction in the bodily temperature, partly because the production of heat is thereby diminished through reflex influences, partly because more heat is given off in consequence of dilatation of the cutaneous vessels and expansion of the skin. Baths at a temperature above that of the blood cause at once elevation of the bodily temperature. Following the bath a slight reduction in temperature takes place after a time. Apart from the changes in bodily temperature brought about by changes in circulation and in respiration, Oppenheimer estimates the elevation of temperature, t, brought about by a bath of 400 liters (kilos) at a temperature of 40 C. and of half an hour's duration (the time required to warm the body thoroughly), in a man weighing 75 kilos, with a bodily temperature of 37° C., assuming equal heat-capacity for the body and the water of the bath:

$$(400 + 75) t = 400.40 + 75.37; \text{ therefore } t = \frac{18,775}{475} = 39.5.$$

The temperature of the body, thus, rises from 37° to 39.5° C., an increase of 2.5° C., representing 187,500 heat-units.

General application of heat to the entire body is indicated when the bodily temperature has fallen extremely low, or when danger is threatened thereby, as in the algid stage of cholera, and in the case of a premature human fetus. General supply of heat is effected by means of warm baths, packs (beds), vapors, insolation, and copious hot drinks. Heat is applied locally by means of hot compresses, partial baths, placing of a part in hot earth or sand, introduction of a part into the body of a recently killed animal (animal bath), introduction of injured parts into receptacles containing heated air. After removal of the heating agent, the greater amount of heat-dissipation caused by the dilatation of the vessels is to be taken into consideration.

POST-MORTEM ELEVATION OF TEMPERATURE.

R. Heidenhain found as a constant phenomenon in dogs that were killed that a transitory elevation of temperature took place before the cooling of the cadaver set in, and this slightly exceeded the normal temperature of the body. Similar, and in part remarkable, elevations of temperature had been observed

previously in human bodies immediately after death, particularly when this resulted from violent muscular spasm.   Thus, for instance, Wunderlich found a
temperature of 45.375° C. in a body fifty-seven minutes after death from tetanus.
The causes of post-mortem elevation of temperature reside:

   1. In a transitory increase in heat-production after death, and especially
through the conversion of the viscid contents of the muscles (myosin) into the
solid form of coagulation (muscular rigidity).   The muscle in the process of becoming rigid produces heat.   All causes that excite rapid and intense muscular
rigidity—including transitory spasm—therefore favor post-mortem elevation of
temperature.   Rapid coagulation of the blood must also contribute to the production of heat.

   2. Further, a series of chemical processes take place in the interior of the
body soon after death that produce heat.   When Valentin placed dead rabbits
in a chamber at the temperature of the body, and in which loss of heat from
the body was impossible, the internal temperature of the body rose constantly.
The processes that thus give rise to the production of heat after death take place
more rapidly in the first hour than in the second.   The higher, further, the bodily
temperature at the moment of death, the more considerable will be the post-mortem
generation of heat.

   3. Diminished heat-dissipation after death is a third cause.   As the circulation
is abolished within a few minutes, but little heat is given off from the cutaneous
surface of the cadaver, because in order that rapid loss of heat should take place
constantly renewed filling of the cutaneous vessels with warm blood is necessary.

THE INFLUENCE OF COLD UPON THE BODY.

   Transitory slight cooling of the external integument causes either no change
in the bodily temperature or a slight elevation.   The latter is dependent upon the
fact that both through reflex influences a more rapid molecular transformation
for purposes of heat-production is stimulated, as well as through contraction
of the small cutaneous vessels and the skin itself heat-dissipation is diminished.
The long-continued action of more intense cold, however, causes reduction in
temperature, particularly through conduction, in spite of increased heat-production
at the same time.   Thus, after cold baths the temperature may be 34° or 32° C.,
and even as low as 30° C.   Cold baths at a temperature below 25° cause the
cutaneous temperature to fall as low as 19°.   Within the interior of the body
the temperature, after remaining stationary for a moment, declines in proportion
to the intensity of the cooling.   If the cooling be continued the body is placed
in the condition of that of a cold-blooded animal.

   As an after-effect of marked abstraction of heat, it is found that the bodily
temperature remains for some time lower than it had been before—primary after-
effect.   It was, for example, only 22° C. in the rectum at the end of an hour.
The designation secondary after-effect is applied to the elevation of temperature
that takes place after the primary after-effect has been neutralized.   This begins—
after cold baths—at the end of from five to eight hours, and reaches in the rectum
about 0.2° C.   In an analogous manner Hoppe-Seyler observed in the course of
a short time a decline in the bodily temperature after the action of heat upon
the body.

   Catching Cold.—If the body of a rabbit is suddenly cooled after exposure to
a surrounding temperature of 35° C. transitory diarrhea occurs at times, in addition
to shivering.   In the course of one or two days the temperature rises 1.5° C. and
albuminuria sets in.   Kidneys, liver, lungs, heart, nerve-sheaths, exhibit microscopic traces of interstitial inflammation; the dilated arteries, particularly in the
liver and the lungs, contain thrombi, and the adjacent veins migrated leukocytes.
In pregnant animals even the fetus exhibited the same conditions.   In explanation
of the phenomenon the question may be discussed whether increased destruction
of the cellular elements does not take place in the greatly cooled blood.

   Freezing.—As a result of the long-continued action of high degrees of cold
upon the skin, the musculature of the skin and its vessels contracts at first, in consequence of the stimulating influence of the cold, and pallor of the integument
develops.   If the action be maintained, paralysis of the walls of the vessels takes
place, and the skin becomes reddened, with dilatation of the vessels.   As the passage of fluids through the capillaries is seriously embarrassed in consequence of
the action of the cold, stagnation of the blood results.   This soon makes itself
manifest as livid discoloration, as the oxygen is almost consumed in the small

vessels in consequence of the slowness of the current. Thus the circulation at the periphery is slowed. If the intense effect of the freezing be continued the movement of blood at the periphery ceases entirely and principally in the thinnest parts, namely the ears, the nose, the toes and the fingers. The functions of the sensory nerves become impaired, and numbness and anesthesia develop. Later there may be even complete freezing throughout. If the peripheral parts become anemic, the internal organs naturally become hyperemic and the heart is distended with blood.

As the retardation of the circulation must naturally be transmitted from the surface of the body to the other circulatory areas, increased venosity of the blood develops in consequence of diminished circulation through the lungs, notwithstanding the larger amount of oxygen in the cold, dense air, and as a result the activity of the nerve-centers is affected. Great disinclination to movement, a distressing feeling of fatigue, a peculiar irresistible tendency to sleep, an inability to think logically, uncertainty in sensorial activity, and finally complete loss of consciousness are the symptoms of this condition. At a temperature of —0.56° C. the blood freezes, while the fluids of the superficial portions of the body become rigid somewhat earlier. The protoplasm, as, for instance, of the muscles, may be cooled on careful experimentation down to a temperature of 18° C. without becoming solid. In making attempts at resuscitation or at thawing, all bending or breaking movements of the rigid parts is to be avoided, in order that the crystals of ice do not perforate the tissues. Further, too rapid heating is to be avoided, as hereby sudden expansion of the tissues might be brought about, and give rise to their molecular destruction. Simple rubbing with snow in order if possible to set the blood gradually in motion from the parts that are not frozen toward those that are rigid, with gradual warming, will yield the best results. Often complete freezing is followed by partial death of the affected part.

ARTIFICIAL REDUCTION OF THE BODILY TEMPERATURE IN ANIMALS.

In consequence of reduction of the temperature of the body the activity of the most highly developed nerve-centers (cerebrum) is diminished first and only later that of the medulla oblongata. If the functions of the latter are beyond restoration death must result.

Artificial reduction of the temperature in warm-blooded animals by exposure to a cold atmosphere, or to cold-mixtures, is followed by a series of characteristic phenomena. If the temperature of the animals—rabbits—is lowered to 18° C.—rectal temperature—they are overcome by great prostration, without abolition, however, of voluntary and reflex movements, although these are lost at a temperature of 17° C. The pulse is reduced in frequency from 100 or 150 to 20 beats in a minute, and at the same time the blood-pressure falls to a few millimeters of mercury. Respirations are infrequent and superficial, and breathing, therefore, becomes inadequate (at 25° C., in rabbits). Asphyxia is no longer capable of exciting convulsions; the secretion of urine ceases; and the liver exhibits excessive hyperemia. In this condition the animal may remain for twelve hours; then, after the muscles and the nerves exhibit signs of paralysis, coagulation of the blood has taken place following the destruction of large numbers of blood-corpuscles, and the eye-ground has become pale, death takes place amid symptoms of paralysis of the heart, convulsions and asphyxia.

If left to itself, an animal whose temperature has been reduced to 18° C. is incapable of recovery when the surrounding temperature is the same. If, however, artificial respiration is practised the bodily temperature rises 10°. If in conjunction with the latter, heat is in addition supplied from without, the animals recover completely, even when they have been apparently dead for about forty minutes. Walther was in this way able to resuscitate full-grown animals whose temperature had been reduced to 9° C. and Howarth young animals even when the temperature had been reduced to 5° C. Mammals born blind and birds born without feathers, if left to themselves, suffer reduction in temperature more rapidly than others. Morphin, and in still greater degree alcohol, accelerate the reduction in the temperature of mammals, the gaseous interchange at the same time falling considerably, and for this reason drunken persons are more readily exposed to the danger of death from freezing.

Knoll lowered the temperature of rabbits by means of intravenous infusion of an ice-cold indifferent solution of sodium chlorid. He found reduction in pulse-frequency, prolonged systole, paralysis of the cardiac branches of the vagus, primary increase and secondary reduction in blood-pressure, accelerated, superficial breathing and later diminished frequency of breathing.

Cl. Bernard made the remarkable discovery that the muscles of animals whose temperature had been reduced maintain their irritability for a longer time, with respect to direct stimuli, as well as to stimulation through the nerve. He found the same condition when the animals were asphyxiated through deficiency of oxygen. Artificial cold-bloodedness, that is, a condition in which the temperature of warm-blooded animals is reduced, with preservation of the irritability of muscles and nerves, can be developed in warm-blooded animals also by division of the cervical cord while artificial respiration is maintained, and further by application of a cool solution of sodium chlorid to the peritoneum.

Hibernation, which is due essentially to the lowering of the temperature of the animals, exhibits a series of analogous phenomena. Valentin found that the marmot begins to be only half awake when the bodily temperature reaches 28° C.; at a temperature of 18° C. it is soporose; at 6° it exhibits shallow and at 16° C. deep sleep. At the same time the heart-beats fall to 8 or 10 in a minute, with reduction in the blood-pressure. The respirations and the movements of the bladder and the intestine cease entirely, and only the cardio-pneumatic movement maintains the slight diffusion of gases in the lungs. The temperature does not fall as low as 0°, but the animals awaken before the temperature has fallen to this level. At a temperature of 0° C. no further dissociation of the oxyhemoglobin would take place. Hibernating animals, indifferently whether in the waking or in the sleeping state, may, however, survive an artificial reduction of temperature down to —1° C. and recover spontaneously. Hibernating animals, therefore, submit to a greater reduction of temperature than other mammals. Under such circumstances, they yield up their heat rapidly and they are able to renew their heat with rapidity even spontaneously. Newborn mammals more closely resemble hibernating animals in this respect than do adult animals. The animals can be awakened from their winter's sleep by sensory stimulation and increasing temperature through the agency of the nerve-centers.

In cold-blooded animals exposed to great cold the temperature can be reduced almost to the freezing-point—tenches can be frozen into ice. In the state of cold their metabolism is greatly lowered and the animals are apparently dead, although they recover rapidly when exposed to warmer surroundings. Under favorable conditions animals frozen into a mass of ice may be resuscitated—the frog. If, for example, however, the fluids throughout the body have been frozen into ice, the animals will die, for the reason that with the formation of ice in the tissues, the gases are expelled in the form of bubbles and the salts separate in the form of crystals. The germs and ova of lower forms of animal life, as, for instance, the eggs of insects, survive long-continued, severe cold. A moderate degree of cold only retards their development. Snakes tolerate an external temperature of —25°, frogs a temperature of —28°, myriapods and infusoria a temperature of —50°, snails for days a temperature of —120°. Germs, grains of seed and spores of fungi exposed to a temperature of —200° are capable of germinating after being again warmed, and also the seeds of wheat, oats, peas, etc., exposed for four or five days to a temperature of —192° C.

The application of a coat of varnish to the skin gives rise to a series of conditions similar to those due to reduction in temperature. The varnished skin readily gives off heat outward through radiation, particularly as the blood-vessels of the skin are enormously dilated. Therefore the temperature of the animals falls greatly and some even die. If the reduction in temperature be prevented by applications of heat and of external coverings the animals survive. The blood of such animals as die contains no toxic substances and no retained excrementitious matters that might have caused death, for other animals injected with such blood remain healthy. In human beings varnishing of the skin appears to have no injurious effect.

EMPLOYMENT OF COLD.

Applications of cold to a large part of the surface of the body may be made from the following points of view:

(a) By means of cold baths or packs of considerable duration to remove large amounts of heat from the surface of the body when the temperature in the presence of fever has attained a dangerous elevation. This effect can be produced in a most lasting manner if the temperature of the bath is at first moderate and is gradually reduced, because the skin is rendered anemic and becomes contracted in consequence of low degrees of temperature, so that a marked obstacle to the dissipation of heat at once arises. Also, the gradually cooled bath is borne for a considerable time. The addition of stimulating substances, as, for instance, salt, which effects dilatation of the cutaneous vessels, favors heat-dissipation, chiefly because the salt-water acts as a better conductor of heat. The reduction in temperature is favored by simultaneous administration of alcohol internally. Also, evaporation of water from the skin, through spraying with aqueous vapor, is adapted for the reduction of the bodily temperature.

(b) Local external reduction of temperature, as by means of an ice-bag, serves in the first place to cause contraction of the vessels and of the tissues, as in case of inflammation, with simultaneous local abstraction of heat. Whether, under such circumstances, the heat-generating molecular disintegration of potential energy is retarded locally or not is as yet undetermined.

(c) Local abstraction of heat through the rapid evaporation of volatile substances, such as ether and carbon disulphid, causes anesthesia of sensory nerves. The introduction of media of low temperature into the interior of the body, such as the inhalation of cold air, the ingestion of cold drinks, cold injections into the intestine, the bladder or the genital tract, in part acts locally and in part may cause general abstraction of heat if the action be long continued and intense. In connection with the action of cold it should be borne in mind that the contraction of the vessels and the collapse of the tissues after cessation of the effect are usually followed by increased fulness and turgescence.

THE TEMPERATURE OF INFLAMED PARTS.

Heat is considered one of the fundamental phenomena of inflammation, in conjunction with redness, swelling and pain. Nevertheless, the apparent increase in the temperature of inflamed parts is by no means dependent upon increase in the temperature above that of the blood, a condition that has never been observed. In consequence of the dilatation of vessels, which causes redness, and the increased amount of blood flowing through the inflamed parts, as well as through tumefaction of the tissues with well-conducting fluid, the external portions of the body, such as the skin, are usually of a higher temperature than normal, and at the same time they more readily give off heat through conduction. Whether or not increased heat-production takes place in the inflammatory focus itself, perhaps in accordance with the character of the inflammatory process, in consequence of accelerated molecular disintegration, has not as yet been determined.

HISTORICAL. COMPARATIVE.

Hippocrates—born 460 B. C.—considered the indigenous heat as the cause of life. According to Aristotle the heart prepares heat within itself and distributes it to all parts of the body, together with the blood. This doctrine, which is presented in a similar manner also in the writings of Hippocrates and Galen, was for a long time the dominating one, and is found last in the writings of Cartesius and Bartholinus (1667, "Flammula cordis"). The iatromechanical school attributed the heat to the friction of the blood in the walls of the vessels. The iatrochemical school, on the other hand, looked for the source of heat in fermentative processes taking place in the blood through the entrance of absorbed articles of food. Lavoisier was the first, in 1777, to make the combustion of carbon in the lungs the source of heat. After the invention of the thermometer by Galileo, Sanctorius in 1626 made the first thermometric observations on the sick, while the first calorimetric observations were made by Lavoisier and Laplace in 1780. Comparative observations have already been recorded on p. 382, and also with respect to hibernation on p. 410.

PHYSIOLOGY OF METABOLISM.

SCOPE OF METABOLISM.

By metabolism is understood the phenomenon common to all living organisms, sharply differentiating the organized from the unorganized, and consisting in the power of incorporating into their own tissues the substances obtained from food (in animals by means of digestion) and of forming them into component parts of their own animate bodies. This division of metabolism is designated *assimilation*. Moreover, out of these assimilated substances, which constitute a reservoir of potential energy, the organism is, by means of transformation-processes, able to develop activities in the form of kinetic energy, which are manifested most strikingly among the higher animals as muscular work and heat. The resulting transformation of tissue-constituents, which terminates in the formation of excrementitious substances, is thus an indirect object in the study of metabolism.

Normal metabolism requires, accordingly, food-material suitable both qualitatively and quantitatively; a storing up within the body, in proportion to the consumption; a regulated chemical transformation of the tissues; and the preparation of the waste-products to be thrown off by the organs of excretion.

SYNOPSIS OF THE MOST IMPORTANT SUBSTANCES USED AS FOOD.

WATER. EXAMINATION OF DRINKING-WATER.

When it is considered that the body contains in all of its tissues about 58.5 per cent. of water, that water is constantly being thrown off with the urine and the feces, as well as by the skin and the lungs, and that in the processes of digestion and absorption most substances must be dissolved in water, and likewise that numerous waste-products, especially in the urine, must leave the body in aqueous solution, the importance of a constant supply and continual renewal of water will be at once obvious. Hoppe-Seyler epitomized admirably the importance of water to life in the following words: "All organisms live in water, and indeed in running water," a saying that deserves a place by the side of the old one of "Corpora non agunt nisi fluida."

Leaving out of consideration its presence as a constituent of fluid food, water is used as a drink in different forms: (1) As *rain-water* (in some countries, where it is collected in suitable reservoirs, cisterns, etc.), which most closely approximates distilled (chemically pure) water, although it, nevertheless, always contains small amounts of carbon dioxid, ammonia, nitrous and nitric acids. (2) As *well-water* or *spring-water*, which is ordinarily rich in mineral matter, it results from atmospheric precipitations, which filter through the layers of earth rich in carbon dioxid, and with the aid of the absorbed carbon dioxid it is capable of dissolving out the alkalies, the alkaline earths and metals. These substances enter into solution as bicarbonates, for instance calcium carbonate and ferric carbonate. The water is either drawn from the wells by mechanical appliances or it gushes from the surface of the earth in certain localities in the form of springs. (3) The *running water* of streams, rivers and brooks is generally much poorer in mineral matter than that of wells or springs.

Flowing on the surface spring-water soon gives off much of its carbon dioxid. As the solution of many minerals is possible only in the presence of carbon dioxid, insoluble precipitates of these substances must result. The water of wells and springs is poor in oxygen, and on the other hand rich in carbon dioxid. The latter gives it its refreshing and stimulating properties. For the same reason a generous vegetable life is possible about springs, while on the other hand the existence in spring-water and well-water of animal organisms requiring oxygen is extremely limited. Freely running water, however, absorbs oxygen from the air, while giving off carbon dioxid, and thus supplies the necessary conditions for existence to fishes and other aquatic animals. River-water contains about from $\frac{1}{30}$ to $\frac{2}{30}$ of its volume of absorbed gases, which may be driven off by boiling or freezing.

The water from wells and springs chiefly is used for drinking-purposes. River-water (with which, unfortunately, some large cities must yet content themselves) demands first a careful removal of the clay and other accidental impurities suspended in it. It may be cleared and purified by means of large filter-beds made of thick layers of sand mixed with charcoal. On a small scale the commercial charcoal-filter can be used with advantage to clarify the water, the charcoal being in addition disinfectant. In this connection it is a noteworthy fact that alum in a dilution of 0.0001 per cent. is able to clarify turbid water.

EXAMINATION OF DRINKING-WATER.

Drinking-water (even when viewed in thick layers) should be perfectly colorless and clear, also without odor, which is best perceived by heating to 50° C., with or without addition of sodium hydroxid. Moreover, it should not be too hard, that is, not unduly rich in salts of calcium and magnesium.

By the term *degree of hardness* is designated the content of compounds of calcium and magnesium in 100,000 parts of water. A water of 20 degrees of hardness contains, therefore, in 100,000 parts, 20 parts of calcium (calcium oxid) in combination with carbon dioxid, sulphuric and hydrochloric acids (the small amounts of magnesium need not be taken into consideration). A good drinking-water should not greatly exceed 20 degrees of hardness. To determine the degree of hardness a titrated soap-solution may be used. This is shaken with the water to be examined, and the later that foam appears the harder is the water. The degree of hardness exhibited by unheated water is designated its *total hardness;* that of heated water its *permanent hardness*. By means of boiling, the calcium carbonate principally is precipitated, as a result of the escape of the carbon dioxid. It is on this account that boiled water becomes softer.

Turbidity of the water following the addition of hydrochloric acid and barium-chlorid solution indicates the presence of *sulphuric acid*, usually in combination with calcium.

As chlorin (always in combination with a metal) appears only in small quantities in pure spring-water, and as its presence in large amounts (apart from saline springs, the vicinity of the sea, or factory-sewers) generally indicates a communication with water-closets or manure-heaps, the estimation of this is of especial interest. For purposes of demonstration, 20 cu. cm. of water are mixed with a few drops of nitric acid, and silver nitrate is added; a precipitate of silver chlorid results. For quantitative estimation by titration there are necessary a solution A of 17 grams of crystallized silver nitrate in 1 liter of water (1 cu. cm. of this solution precipitates 3.55 mgm. of chlorin as silver chlorid); and also a cold saturated solution B of neutral potassium chromate. In testing, 50 cu. cm. of the water to be examined are placed in a beaker, 2 or 3 drops of the solution B are added, and then from a buret solution A is permitted to flow drop by drop until the precipitate, at first white, remains red, even after stirring. If the number of cubic centimeters of A used be multiplied by 7.1, the amount of chlorin contained in 100,000 parts of water will be obtained. *Example:* if 50 cu. cm. required 2.9 cu. cm. of silver-solution, then 100,000 parts of water contain 2.9 × 7.1 = 20.59 parts of chlorin. In good drinking-water the chlorin should not exceed 15 mgm. in 1 liter.

If 50 cu. cm. of water are acidulated with a little hydrochloric acid, then ammonia added in excess, and to this a solution of ammonium oxalate, the white precipitate obtained is *calcium oxalate*. According as the resulting turbidity is only slightly cloudy or markedly milky it is known whether the water is *soft* (poor in calcium) or *hard* (rich in calcium). After this calcium-precipitate has settled, the clear fluid is poured off and mixed with a solution of sodium phosphate and a little ammonia; the crystalline precipitate that now forms indicates the presence of *magnesia*.

The feebler the reactions for sulphuric acid, chlorin, calcium and magnesium, the better is the water. Good drinking-water, should, moreover, contain only traces of nitrates, nitrites and ammonium-compounds, as their presence points to organic substances containing nitrogen in a state of decomposition.

Nitric acid is indicated when 100 cu. cm. of water are acidulated with two or three drops of concentrated sulphuric acid, some bits of zinc are added and then a solution of iodin, zinc and starch, and a blue tint appears. The following test is exceedingly sensitive: some fragments of brucin sulphate along with a drop of the water to be examined are to be placed in a watch-glass; then a few drops of concentrated sulphuric acid are added. A rose-red color appears. Diphenylamin sulphate mixed with a few drops of concentrated sulphuric acid yields in the presence of nitrates, even when in great dilution, a blue color. This test is, therefore, recommended for the demonstration of well-water in milk.

Demonstration of *nitrous acid:* To 100 cu. cm. of water a few drops of pure concentrated sulphuric acid and a solution of zinc iodid and starch are added: a blue color appears. In addition naphthionic acid and pure β-naphthol, thoroughly mixed in a mortar, are recommended as a reagent. To 10 cu. cm. of the fluid to be examined for nitrites two drops of a concentrated solution of hydrochloric acid and as much of the mixture mentioned as can be taken up on the point of a knife are added and the whole is thoroughly shaken. If ammonia is added in a layer on top of this mixture a red ring appears. This test has a sensitiveness of 1 : 100 millions.

Ammonium-compounds in considerable amount render the water suspicious. To 150 cu. cm. of water 0.5 cu. cm. of solution of sodium hydrate and 1.0 cu. cm. of solution of sodium carbonate are added and the precipitate is allowed to settle. Of the supernatant clear fluid a column 15 cm. high is introduced into a narrow graduated cylinder and mixed with Nessler's reagent (a solution of mercuric iodid and potassium iodid in an excess of potassium hydroxid): Traces of ammonia in water thus yield a color between yellow and red, large amounts a brown precipitate of mercuric-ammonium iodid.

The contamination of water by decomposed animal substances will be recognized by the amount of contained nitrogen. In most cases it is sufficient to determine the amount of nitric acid. For this two solutions are necessary: A, containing 1.871 gm. of potassium nitrate to a liter of water; 1 cu. cm. of this contains 1 gram of nitric acid; B, a dilute solution of indigo; 1 part of pulverized indigotin is slowly added with stirring to 6 parts of fuming sulphuric acid; the mixture is allowed to settle, the blue liquid is poured into 40 times its amount of distilled water and then filtered. Finally, the fluid is diluted with distilled water until it begins to be transparent in layers of from 12 to 15 mm. thick. To test the efficiency of B, 1 cu. cm. of A is mixed with 24 cu. cm. of water, some table-salt and 50 cu. cm. of concentrated sulphuric acid are added, and so much of B is now allowed to flow from a buret until a faint green tint appears. The number of cubic centimeters of B used corresponds to 1 mgm. of nitric acid. Twenty-five cubic centimeters of the water to be examined are mixed with 50 cu. cm. of concentrated sulphuric acid and titrated with B until the green color appears. This titration must, however, be repeated, and in the second observation the number of cubic centimeters of indigo-solution be permitted to flow in a stream; a somewhat larger amount of fluid may be required to produce the green color. The number of cubic centimeters of solution B thus used (in proportion to the previously ascertained strength) indicates the amount of nitric acid present in 25 cu. cm. of water. In well-water as much as 10 mgm. nitric acid is found in the liter.

Hydrogen sulphid is recognized, apart from its odor, through the brown color imparted to a piece of filter-paper that has been saturated in an alkaline solution of lead, and is held over the water boiling in a flask. If hydrogen sulphid is present in combination in the water, some sodium hydroxid and a dilute solution of sodium nitro-prussid are added; a reddish-violet color appearing

It is of the greatest significance with respect to the excellence of drinking-water that it should be free from *putrefying or decomposing organic matter.* The latter, in conjunction with the lower organisms always to be found in it, when ingested with drinking-water, expose the body to serious dangers, as a number of infectious diseases, especially cholera and typhoid fever, can be spread in this manner.   The latter is especially the case if the wells in use lie near water-closets and manure-heaps, so that the products of decomposition can filter through into the reservoir for the water.

Qualitative Demonstration.—(1) **A** fairly large amount of water is evaporated in a porcelain dish to dryness, and is then subjected to greater heat.   If large amounts of organic matter are present a discoloration between brown and black takes place.   If the matters contain nitrogen the odor of burning hair appears at the same time.   Good water, so treated, yields but a faint brown color.   Micro-. scopic examination may also be made to determine the presence of microörganisms in water.   About 1 cu. cm. of water is allowed to evaporate upon a slide having an up-turned edge and kept in a place free from dust, and the dry spot is examined. (2) A solution of gold-potassium chlorid added to water produces, after standing for a long time, a black muddy precipitate.   (3) A solution of potassium permanganate added to the water placed under cover is gradually decolorized, with the formation of a brown muddy deposit.   The precipitates from 2 and 3 are the more abundant the greater the amount of organic substances present in the drinking-water.

Quantitatively the amount of organic substances is determined, according to Kubel, as follows.   Two solutions are required: A, containing 0.63 gram of pure crystalline oxalic acid in 1 liter of distilled water; B, containing 0.33 gram of potassium permanganate in 1 liter of purest distilled water.   For the determination of the efficiency of the latter 100 cu. cm. of distilled water are placed in a wide-necked bottle of 300 cu. cm. capacity, together with 5 cu. cm. of dilute sulphuric acid (1 volume of acid to 3 volumes of water) and heated to boiling.   Into this from 3 to 4 cu. cm. of solution B are allowed to flow from a buret provided with a glass stop-cock.   The mixture is boiled for ten minutes, the heat is then removed and 10 cu. cm. of solution A are added.   Finally, the fluid, which has become colorless, is mixed with solution B until a faint red tint appears. The number of cubic centimeters used corresponds to 6.3 mgm. of oxalic acid, which are present in the 10 cu. cm. of solution A, and contains exactly 3.16 mgm. of potassium permanganate, or 0.8 mgm. of oxygen available for oxidation, which is necessary for the transformation of the 6.3 mgm. of oxalic acid into carbon dioxid.

In order to test a given water for the amount of organic matter present, 100 cu. cm. of the sample are placed in a flask of 300 cu. cm. capacity, 5 cu. cm. of dilute sulphuric acid are added and so much of solution B that the fluid becomes an intense red and remains so even when heated.   After five minutes' boiling, 10 cu. cm. of solution A are added.   The fluid, thus made colorless, is then titrated with solution B until a faint red tint appears.

For purposes of calculation as many cubic centimeters of solution B as are necessary for the oxidation of 10 cu. cm. of solution A are subtracted from the total number of cubic centimeters of solution B used in the experiment.   The difference in cubic centimeters is multiplied by 3.16 : x if the proportion of potassium permanganate, by 0.8 : x if the proportion of oxygen, necessary for the oxidation of the organic substances present in 100,000 parts of water is desired (x represents the number of cubic centimeters of solution B that corresponds to 10 cu. cm. of solution A).

Example.—Nine and nine-tenths cubic centimeters of solution B correspond to 10 cu. cm. of solution A.   After acidulation with sulphuric acid, 100 cu. cm. of the water under examination is mixed with 15 cu. cm. of solution B and boiled.   The red fluid is decolorized by the 10 cu. cm. of solution A.   To restore a faint red tint 4.4 cu. cm. of solution B must be added.   Estimation: 15 + 4.4 = 19.4; 19.4 — 9.9 = 9.5.   Therefore, for the oxidation of the organic substances in 100,000 parts of water (9.5 × 3.16) : 9.9 = 3.008 of potassium permanganate, or (6.5 × 0.8) : 9.9 = 0.77 part of oxygen are necessary.   Bad drinking-water, especially when it contains much organic matter, should never be used in its native state, but particularly not at a time when epidemics of typhoid fever, of cholera or of dysentery prevail or threaten.   It should be urgently advised that the water be thoroughly boiled previously, as by this means the germs of infection are destroyed.   The resulting insipid taste can be readily improved by means of effervescent powder, sugar or fruit-juice.

STRUCTURE AND SECRETORY ACTIVITY OF THE MAMMARY GLANDS.

About twenty milk-ducts open separately on the tip of the nipple, and just in advance of their mouth present an oval dilation, the lacteal sinus, generally expanded laterally. Each undergoes dendritic ramification and passes to a special lobe of the gland, which is bound together by loose interstitial connective tissue. Only at the time of lactation do all of the terminal branches of the milk-ducts lead to round glandular acini arranged in groups. Each vesicle has a membrana propria, upon which externally is a network of star-shaped connective-tissue cells, and internally a single layer of somewhat flattened, polyhedral and nucleated secretory cells. According to the degree of secretory activity of the acinus its lumen, at times narrow, at other times wide, is filled with a fluid in which float round, shining fat-granules (milk). Fibrillary connective tissue, principally arranged in a circular manner, and transversed externally by fine elastic fibers, forms the wall of the glandular ducts, which are lined by cylindrical epithelium. In the smallest of these a membrana propria can yet be recognized, which is continuous with that of the terminal vesicle.

During the first days following delivery (as well as before it), the breasts secrete little milk of considerable consistency and yellowish color (colostrum), in which large cells completely filled with fat-granules are present (colostrum-corpuscles). The latter appear also later on, when the discharge of the milk has for a time been discontinued. Sometimes a nucleus is recognizable in the cells,

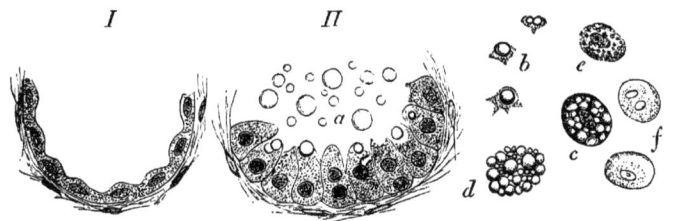

FIG. 137.—*I*, Acinus of the mammary gland, inactive; *II*, during the formation of milk: *a, b*, milk-globules; *c d e*, colostrum-corpuscles; *f*, pale cells (from the dog).

rarely ameboid movement (Fig. 137, *c, d, e*). The normal secretion of milk, appearing in the course of three or four days, is a productive activity of the gland-cells.

Heidenhain and Partsch found the secretory cells in the inactive gland (Fig. 137, *I*) to be flat, polyhedral and mononuclear; on the other hand, in the active gland often polynuclear, cylindrical, higher, and richer in albumin and granules (Fig. 137, *II*). The free edge, turned toward the cavity of the acinus, undergoes characteristic changes during secretion. There are formed in this part of the cells fat-granules, which are thrown off during secretion, together with the detached cell-margin. In part the nuclei also degenerate, their product likewise passing over into the milk (nuclein-content of milk). The same cells appear to be able to perform the secretory process repeatedly, undergoing regeneration during the period of rest. According to Bizzozero, Benda, Michaelis and Unger, the cells actively produce the fat-granules and throw them off.

Leukocytes containing fat-granules are, further, found in milk, representing colostrum-corpuscles, and isolated, pale cells (*f*). Some milk-globules still have shreds of cell-substance on their surface (*b*). With respect to the formation of the individual constituents of milk, H. Thierfelder, who exposed fresh mammary glands to digestive processes immediately after death, found that during the digestion of the gland at the temperature of the body a reducing substance, probably milk-sugar, was formed as a result of fermentative activity. The mother-substance (saccharogen) is soluble in water, but not in alcohol or ether; it is not destroyed by boiling and is not identical with glycogen. The ferment forming milk-sugar seems to be fixed in the glandular cell, as it does not pass over into the milk or into a watery extract of the gland. During the digestion

27

of the mammary gland at the temperature of the body casein also is formed as a result of a fermentative process, and probably from serum-albumin. This ferment is present in the milk.

The *mammillary areola and the nipple* are characterized by pigmentary deposits in the cells of the rete Malpighii (during pregnancy more abundant and of greater extent) and by large papillæ in the cutis, some of which contain tactile corpuscles. Numerous unstriated muscle-fibers in the deep layers of the corium and in the subcutaneous tissue (always free from fat) surround the milk-ducts of the nipples and in part also pass longitudinally to the tip of the nipple. The glands of Montgomery, about the size of a millet-seed, situated during lactation in the mammillary areola, are small, nodular, prominent, subcutaneous milk-glands, with a special duct of evacuation at the summit of each nodule.

Arteries enter the mammary gland from different directions. Their branches do not accompany the glandular ducts. Capillaries arranged in a network surround the glandular acini and anastomose by means of small arteries and veins with those of the neighboring vesicles. In the mammillary areola the veins are arranged in the form of rings (circles of Haller).

The *nerves* of the mammary gland arise from the supra-clavicular and the second, fourth and sixth intercostals. They pass in part to the skin of the gland and of the highly sensitive nipple, in part to the vessels, and in part to the unstriated muscle-fibers of the nipple and to the glandular vesicles themselves, in which their mode of termination is, however, as yet unknown. In connection with the investigations of the mammary glands great credit belongs to C. Langer.

Lymphatics are found close about the alveoli, often distended to their utmost, and from them material for the formation of milk appears to be derived.

Comparative.—From ten to twelve nipples are found in rodents, insectivora and carnivora; others among these animals have only four. Pachyderms and ruminants generally possess from two to four on the abdomen; the carnivorous whale has two at the side of the vulva. Apes, bats, herbivorous whales, elephants and sloths resemble man; the half-apes have from two to four nipples. The duck-bill (ornithorhynchus paradoxus) possesses tubes arranged in groups (similar to skin-glands), which open without nipples upon a hairless flat area of skin. The marsupials carry their undeveloped young in a muscular duplicature of the skin of the abdomen, in which the nipples are situated. In them and in the duck-bill there is a compressor muscle of the mammary gland, which promotes the evacuation of milk.

Development of the Breast.—The first indication of the breasts consists on each side in a transitory elevation passing downward on the lateral aspect of the thorax, and of which subsequently there remain only punctate and nodular formations, the precursors of the breasts. The further development of the latter begins in both cases as early as the third month; between the fourth and fifth a number of simple tube-like glandular ducts arranged radially are already present beneath the hairless, excavated mammillary areola. In the new-born the ducts already exhibit two or three branches, and they are provided with dilated extremities. In both sexes the ducts divide in a dendritic manner until the twelfth year, though without the development of actual acini. In girls who have reached puberty this branching proceeds rapidly and extensively, although also in them the gland, rich in connective tissue, exhibits the formation of acini only at the periphery, while only with the occurrence of pregnancy do characteristic acini develop also in the center of the body of the gland along with relaxation of the connective tissue.

In the climacteric period all of the acini and numerous small milk-ducts disappear. In the adult man the mammary gland usually resembles that of the new-born, having undergone involution after puberty. Supernumerary nipples on the breast are of interest as representing independent openings of individual milk-ducts. Supernumerary glands and nipples (*hypermastia* and *hyperthelia*) arranged in part irregularly and in part regularly in rows, like the dugs of the sow, point to their original multiple beginning and are worthy of note as points of resemblance among animals.

The situation of a breast in the axilla, on the back, on the acromion or on the tibia is a curiosity. A slight secretion from the breast of the new-born (witches' milk) is normal. On the other hand, suckling performed by a man is to be included among the greatest rarities. According to Aristotle buck-goats sometimes give milk (noted also by Schlossberger), as do also calves after their dugs have been frequently sucked, and goats that have not been covered, when their udders are irritated by nettles.

The *evacuation of milk* (from 500 to 1500 cu. cm. in twenty-four hours) is due not alone to the purely mechanical act of suction, but also to the functional activity of the mammary gland. The latter consists primarily in erection of the nipple, the smooth muscle-fibers of which exerting pressure on the sinuses of the ducts, so that the milk may spurt forth in a stream. Moreover, the glandular structure itself is reflexly stimulated to more active secretion through irritation of the sensory nerves of the nipple.

From the suddenly dilated glandular vessels a transudate pours abundantly into the gland, by which, admixed with the milk-corpuscles and transformed into milk, it is discharged. The amount of the secretion depends thus upon the degree of blood-pressure. Accordingly, not only the milk stored up in the breast is sucked out, but during the process of suction secretion is accelerated. "The breast is willing," as the nursing women say. Only in this manner can be explained the speedy arrest of the secretion of milk in connection with sudden emotional excitement, which (as anger, fright, etc.) acts, as is known from experience, on the vasomotor nerves. Laffont saw, following stimulation of the mammary nerve of a bitch, erection of the nipples, dilatation of the vessels and secretion of milk. After section of the external spermatic nerve in goats, Eckhard noted absence of erection of the dugs, although the formation of milk suffered no interruption, which appeared only after section of the nerves on both sides. Continued stimulation of sensory nerves diminishes the secretion. The rarely observed condition of galactorrhea is perhaps to be considered as a sort of paralytic secretion similar to the analogous secretion of saliva. Heidenhain and Partsch noted increased secretion in dogs when, following section of the glandular nerves, strychnin or curare was injected. Atropin decreases the amount of milk.

The slight *milk-fever* appearing with the commencement of the secretion of milk is probably due to increased stimulation of the vasomotors, whose activity must further be considered in relation with the transposition of the mass of blood out of the pelvic cavity after birth.

MILK AND MILK-PREPARATIONS.

Milk must be designated a complete food, in which all of the constituents are present in such proportion that the body can thrive upon it. According to Johannessen the proportions in human milk are as follows: Albumin 1, fat 2, sugar 4.2; in cow's milk: albumin 1, fat 1, sugar 1.43. Of the milk relatively more fat is absorbed in the intestine than albuminates. Human milk is utilized up to 91.6 per cent., cow's milk to 90 per cent.

Milk is opaque, bluish white, with a sweetish taste and a characteristic odor, probably due to peculiar odoriferous bodies in the cutaneous secretion of the gland. It has a specific gravity of from 1.030 to 1.032. On standing numerous butter-globules collect on the surface as cream, beneath which is a watery bluish layer. Human milk has always an alkaline reaction; cow's milk is at times alkaline, sometimes acid, and sometimes amphoteric. The milk of carnivora is always acid.

Milk consists of the fluid (*milk-plasma*, plasma lactis) and the morphological constituents suspended in it, among which the *milk-globules* predominate. If the milk be clotted, the *cheese-cake* (placenta lactis), which consists of coagulated casein, containing milk-globules, separates from the *whey* (serum lactis). The latter contains some dissolved albumin, milk-sugar and most of the salts.

Milk-globules or Butter-globules.—Microscopically, milk contains innumerable small globules (Fig. 137). Colostrum-corpuscles and epithelium from the milk-ducts are not common in mature milk. The milk-globules and the swollen casein cause, on account of the reflection of light, the white color and the opacity of milk. The milk-corpuscles consist of butter-fat and are apparently surrounded by a thin layer of casein (haptogenic membrane).

That the milk-corpuscles are actually surrounded by a capsule of casein has lately been definitely denied. Formerly, the following observations were offered in support of the presence of the capsule: if acetic acid, which dissolves the casein-capsule, be added to a microscopic preparation, the milk-globules run together in fat-droplets. Further, if cow's milk be shaken with potassium hydroxid, which destroys the casein-capsule, and then be mixed with ether, the milk becomes clear and transparent, as the ether dissolves all of the fat-granules. Previously to the treatment with potassium hydroxid or acetic acid, the ether is incapable of setting free the fat in cow's milk from its capsule; in the case of human milk the addition of ether and shaking alone suffice. Other investigators, however, deny the presence of the casein-capsule; according to them milk is a simple emulsion, permanently maintained as such by means of the colloid casein, which is simply swollen in milk-plasma.   The treatment of milk with potassium and ether renders the casein of the plasma unfitted to maintain the emulsion of the milk permanently.

The *fats of the milk-globules* (human) are the triglycerids of stearic, palmitic and oleic acids, in lesser amount of myristic, capric, caprylic, caproic and butyric acids.   In addition there are found traces of formic acid and cholesterin.

By means of long-continued beating of milk (churning), and even more readily of cream, the fat of the milk-globules (after rupture of the casein-capsule) is obtained as butter in coherent masses.   Butter is soluble in alcohol and ether, and is purified by melting at 60° C., or by washing with water at 40°.   On standing in the air it becomes rancid, from the glycerin of the neutral butter-fats being decomposed by the action of germs into acrolein and formic acid, which, with the volatile fatty acids, give the odor.

The *milk-fluid* (plasma lactis) is clear, somewhat opalescent and contains proteids, chief among which is *casein*, together with a small amount of *lactalbumin* and *lactoglobulin* and the opalescent *opalisin*, a little *nuclein*, phosphocarnic acid and a trace of *diastatic ferment* (in human milk).

Casein is retained in the filtration of milk by means of fresh animal membrane or by means of clay cylinders.   It can also be completely precipitated out of human milk by means of saturation with magnesium sulphate, from cow's milk by means of a little acetic acid.—*Quantitative Determination in Cow's Milk:* Twenty cubic centimeters of cow's milk are diluted with 60 cu. cm. of water, and 30 cu. cm. of a 1 in 1000 sulphuric-acid solution are added with stirring, precipitating the casein of cow's milk.   After five hours filtration is practised, the filter is washed with water, twice with alcohol and fifteen times with ether, and it is then dried and weighed.   The casein can be completely precipitated by addition of alum at a temperature of 37° C.   Magnesium sulphate then precipitates the globulin in the filtrate.   Globulin and albumin together are precipitated from the filtrate by means of tannic acid.

The albumin in milk coagulates on boiling; in addition the free surface becomes covered with a skin of insoluble casein.

The plasma contains, further, *milk-sugar*, a carbohydrate resembling dextrin, lactic acid?, lecithin (1⅔ times as much as in cow's milk), urea, traces of kreatin, kreatinin, xanthin-bodies (potassium sulphocyanid in cow's milk); sodium chlorid, potassium chlorid, alkaline phosphates, calcium and magnesium sulphates, alkaline carbonates and in addition traces of iron, metallic fluorids and silica, carbon dioxid, nitrogen, oxygen and ammonia.   Human milk contains numerous staphylococci.

The *curdling of milk* consists of a coagulation of the casein.   The latter is combined in the milk with calcium phosphate, and is therefore soluble.   Acids, which remove the calcium phosphate from the casein, cause coagulation of the casein; lactic acid acts most readily in this connection, then hydrochloric, nitric, sulphuric, acetic and phosphoric acids.   Acetic and tartaric acids, when added in excess, redissolve the precipitated casein.   Human milk is not curdled by all acids, but only by means of two or more drops of 0.1 per cent. hydrochloric or 2 per cent. acetic acid.   Heating above 130° C. coagulates the milk, acids being formed from milk-sugar as a result of the action of heat and the contained casein becoming more coagulable.   The spontaneous curdling of milk after standing for some time, especially in the heat, results from the formation of lactic acid by the bacillus acidi lactici, which transforms the neutral alkaline phosphate into acid phosphate, removes the calcium phosphate from the casein and thus precipitates it.   The sugar is transformed into lactic acid and carbon dioxid.   The bacillus furnishes the stimulation for this decomposition, while the casein of the milk is the actual fermenting body.

By means of the lab-ferment milk of alkaline reaction may be coagulated (sweet whey).   This ferment decomposes the casein in the precipitated cheese and the scanty but readily soluble whey-albumin.   The lab-coagulation is then quite different from the others.   It takes place only when calcium-salts are dissolved in the milk.   If these are precipitated by means of potassium oxalate, the lab-ferment no longer causes coagulation in the fluid; this latter, however, occurs again as soon as calcium chlorid is added.

Boiling (by destroying lower organisms), sodium bicarbonate ($\frac{1}{1000}$), ammonia, salicylic acid ($\frac{1}{1000}$), also glycerin and ethereal oil of mustard, prevent spontaneous coagulation.   Fresh milk renders tincture of guaiac blue; boiled milk does not. After standing for some time in the air milk gives off carbon dioxid and absorbs oxygen. At the same time, as a result of the activity of germs that develop rapidly in the milk(?), an increase of the fat, together with that of the alcoholic and ethereal extracts at the expense of the casein, is brought about. According to Schmidt-Mülheim some casein is transformed into peptone, but only in unboiled milk.

Milk-analysis.—Every 100 parts of milk contain:

	HUMAN.	COW.	GOAT.	ASS.
Water	88.3	88.0	86.25	89.01
Casein	0.9–1.2 ⎱ 1.1	3.0	2.53 ⎱	3.57
Albumin	0.3–0.5 ⎰	0.3	1.26 ⎰	
Butter	3.21	3.5	4.34	1.85
Milk-sugar	4.67	4.5	3.78 ⎱	5.05
Salts	0.2	0.7	0.65 ⎰	

Colostrum contains much serum-albumin and little casein, but, on the other hand, all other solid substances in larger amount, especially the butter.

Pflüger and Setschenow found in 100 volumes of milk the following substances by volume: carbon dioxid from 5.01 to 7.60; oxygen from 0.09 to 0.32; nitrogen from 0.70 to 1.41. The carbon dioxid can in part be displaced only by means of phosphoric acid.

Among the *salts*, those of potassium preponderate over those of sodium (as in the red blood-corpuscles and in meat); also a considerable amount of calcium phosphate is present, for the formation of the bones of the infant. Wildenstein found in 100 parts of ash from human milk sodium chlorid 10.73; potassium chlorid 26.33, potassium 21.44, calcium 18.78, magnesium 0.87, phosphoric acid 19, ferric phosphate 0.21, sulphuric acid 2.64, silica a trace. The amount of salts is influenced by that in the food.

Milk exhibits no difference in the amount of albumin before and after nursing. The amount of sugar, however, diminishes after nursing, while the fat increases considerably. With the progress of lactation, albumin appears most abundantly in the first six months, in lesser amount in the second six months, and after the first year it decreases still more. The amount of fat varies, but rather increases after the first year. The sugar exhibits a pretty uniform, inconsiderable increase. In primiparæ the amount of solid constituents (9.67 per cent.) is greater than in multiparæ (8.56 per cent.). Young mothers form more albumin and fat, older mothers more sugar. A starchy diet yields a fatter milk, while with a proteid and fatty diet the amount of albumin and of sugar increases. Camerer and Söldner found in milk a body resembling the basis of bone, together with hitherto unknown bodies consisting of carbohydrates combined with proteids.

If it be necessary to employ the *milk of animals*, it should be noted that cow's milk (best when containing much fat) must be diluted with water and mixed with milk-sugar. Heubner and Hofmann recommend for children from one to nine months old, as a general rule, only one mixture, consisting of 1 part of cow's milk and 1 part of a solution containing 69 grams of milk-sugar in 1 liter of water. Soxhlet recommends a mixture of 2 parts of cow's milk and 1 part of a 12.3 per cent. solution of milk-sugar. The casein of cow's milk varies qualitatively; further, it appears in larger flakes and is, therefore, more difficult of digestion than the small-flaked casein of human milk. The casein of human milk does not split off paranuclein in the process of digestion, as does that of cow's milk. Boiled cow's milk is somewhat more difficult of digestion than unboiled cow's milk, but, nevertheless, because sterilized, is to be preferred. The milk should be boiled for ten minutes, be cooled quickly to below 18° C. and be kept cool. In the case of children more than nine months old, the addition of water is progressively diminished. Cow's milk may also be diluted with advantage with beef-tea. For children that cannot take milk, v. Liebig has recommended especial soups prepared from cow's milk, water, wheat-flour, hop-flour and sodium bicarbonate. The starch is transformed, in the course of preparation, into sugar and dextrin. The more rapid the growth exhibited by mammals the richer is their milk in albumin.

Milk-tests.—The amount of *cream* is measured by permitting the milk to stand for twenty-four hours in a cool place in a high glass cylinder graduated into 100 parts. The cream that collects on the surface should amount to from 10 to 14 volumes per cent. The *specific gravity* of whole cow's milk is between 1029 and 1034, of skimmed milk between 1032 and 1040. It is determined by means of the areometer at a temperature of 15° C. Every degree Celsius more or less makes a difference of –0.1° or + 0.2° on the areometer. Should only an approximative estimation be desired, the amount of sugar both in the whey as well as in the whole milk diluted with water can be titrated directly by means of Fehling's solution (1 cu. cm. of which corresponds to 0.0067 gram of milk-sugar); or the amount in the whey may be determined by means of the polarization-apparatus. If the estimation is to be made with exactness, the proteid must be removed from the whey; and in addition the fat-globules dissolved out of the whole milk, and the fat extracted. The amount of *water*, as compared with the amount of milk-corpuscles (fat)—the latter should not be less than 3 per cent. in whole milk, and 1½ per cent. in half-skimmed milk—is determined by means of the lactoscope (the diaphanometer of Donné, modified by Vogel and Hoppe-Seyler). This consists of a glass vessel 1 cm. in diameter with plane parallel walls. A measured quantity of milk is introduced into the vessel and water added from a graduate until the flame of a lighted candle placed about a meter behind the apparatus can be distinctly seen outlined (in a dark room) with the eye placed directly in front of the apparatus. In such an experiment from 70 to 85 cu. cm. of water are needed for each cubic centimeter of good cow's milk. Feser's *galactoscope* also is serviceable in the examination of milk, even in the hands of the laity.

The following substances pass into the milk: fat taken with the food, numerous odorous substances (anise, vermouth, garlic, etc.); chloral hydrate, opium, indigo, salicylic acid, iodin, iron, zinc, mercury, lead, bismuth and antimony. In cases of osteomalacia the amount of calcium in the milk is increased. Potassium iodid diminishes the secretion of milk.

Abnormal admixtures include hemoglobin, biliary pigments, mucin, blood-corpuscles, pus, fibrinous clots, tubercle-bacilli and other bacilli. Numerous germs develop in evacuated milk, of which the bacillus cyanogenus, which occurs rarely, gives the milk a sky-blue color. It is the milk-serum that is blue, not the germ. There are also schizomycetes that produce bluish-black and green colors. Red and yellow milk are also observed as a result of similar action by other chromogenic schizomycetes. Red milk is due to the notorious micrococcus prodigiosus, which is itself colorless, and also to the bacterium erythrogenes; yellow milk, to the bacillus synxanthus. Some of the pigments formed seem to be related to the aniline dyes and others to those belonging to the phenol-group. As the possibility of the entrance also of other pathogenic germs cannot be excluded, the milk should be sterilized by boiling.

The rennet-like activity of bacteria is widespread, so that they coagulate and peptonize casein and finally cause further decompositions. Thus the butyric-acid bacilli first cause the coagulation of the casein, which they then peptonize and later decompose, with the development of ammonia. Milk becomes viscous from the action of the bacillus lactis viscosus, perhaps in other ways, just as beer and wine may become "long."

Preparations of Milk.—1. *Condensed Milk.* Eighty grams of cane-sugar are added to each liter of milk, the mixture evaporated to one-fifth its volume and then sealed in tin cans while hot. For the use of nursing infants a teaspoonful is dissolved in a pint of cold water and then boiled.

2. As a food replacing albumin A. Salkowski recommends the following preparation of casein: Casein 20 parts, sodium phosphate 2 parts, water 200 parts, or the soluble ammonia-compound prepared by conducting ammonia over casein (*eucasin*); Röhmann advises acid casein-calcium 3 grams, milk-sugar 4.5 grams, di-sodium phosphate 0.375 gram, monopotassium phosphate 0.153 gram, calcium chlorid 0.04 gram, potassium chlorid 0.3 gram, magnesium acetate 0.01 gram, water 100 grams.

3. *Koumiss and Kefyr.* The Kirghese are accustomed to produce alcoholic fermentation in mare's milk, and the Caucasian mountaineers do the same with cow's milk. As a result of the addition of sour milk, which contains the bacterium lacticum and the bacillus caucasicus, the unfermentable milk-sugar is transformed into fermentable glucose, and by the action of yeast, which is present in an addition of completed koumiss, alcoholic fermentation of the glucose takes place, the mixture being vigorously stirred. Koumiss contains from two to three per cent. of alcohol. The casein, which is precipitated at first and later is partly dissolved again, is transformed into acid-albumin and peptone. The kefyr-fungus (dispora caucasica) also gives rise to a similar preparation, in part containing peptones. In addition to the kefyr-fungus, there is also found the bacterium lacticum and a schizomycete that peptonizes casein, as well as a streptococcus that forms lactic acid and another organism that ferments milk-sugar. Koumiss and kefyr are also prepared at some health-resorts.

4. *Cheese* is prepared by coagulating skim-milk (poor cheese), or whole milk

(fat cheese), by means of rennet, permitting the whey to run off and well salting the coagulated mass. After some time the cheese ripens, the casein, probably with the formation of sodium albuminate, becoming soluble in water once more. In some kinds of cheese liquefaction occurs, with the formation of peptone and diastatic ferment as a result of the action of the cheese-spirillum (spirillum tyrogenum). On further decomposition, leucin and tyrosin appear. The amount of fat in the cheese increases, while that of casein diminishes. Later on, the fats decompose; the volatile fatty acids yield the characteristic odor. The formation of peptone, leucin, and tyrosin and the splitting up of the fats suggest the digestive processes. Cheese contains also saprophytic microbes.

EGGS.

Eggs also must be looked upon as a complete food, as the organism of the young bird is capable of developing from them. The yolk contains, as the characteristic albuminous body, vitellin; also an albuminate in the capsules of the yellow yolk-globules, the egg-casein, which is precipitated by means of a one per cent. solution of sodium chlorid, on combining with oxygen; nuclein from the white yolk; fats in the yellow yolk (palmitin and olein); cholesterin, much lecithin, and, as a product of decomposition, glycerophosphoric acid, grape-sugar; pigments (lutein), including one containing iron and related to hemoglobin; finally salts, qualitatively as in the blood, quantitatively as in the blood-corpuscles; gases.

The white of the egg contains egg-albumin as the principal constituent, in addition to some globulin, mucin-matter and albumose, also small amounts of palmitin and olein, partly saponified by sodium; grape-sugar; extractives; finally salts that resemble qualitatively those of the blood and quantitatively those of the serum. There are, besides, traces of fluorin. On a diet of eggs and also on a diet of roast meat, relatively more of the nitrogenous constituents are absorbed than of the fats.

MEAT AND MEAT-PREPARATIONS.

In the form in which it is consumed, meat contains, in addition to the proper muscular tissue, an admixture in greater or lesser amount of the elements of fatty, connective and elastic tissues. Beef freed from fat and dried contains, according to Argutinsky, carbon 49.6, nitrogen 15.3, hydrogen 6.9, ash 5.2, oxygen and sulphur 23 per cent. According to H. Schulz the amount of sulphur present is 1.1 per cent. of the dried muscle. The chemistry of muscle is fully discussed on p. 548. The proteids of muscle are contained in the contractile substance and in part in the saturating fluid. The fats are derived for the greater part from the inter-fibrillary fat-cells, the lecithin and cholesterin chiefly from the muscle-nerves. The gelatin-yielding substance is supplied by the connective-tissue fibers of the perimysium, the perineurium, the vessel-walls and the tendinous parts. The red coloring-matter, which is present in varying amount even in the muscles of the same animal (red muscles and white muscles), is hemoglobin. In addition, some muscles, for example the heart, contain the related substance, myohematin. Elastin is present in the sarcolemma, the neurilemma and the elastic fibers of the perimysium and the vessel-walls. Keratin, in small amount, is derived from the endothelium of the vessels. The following are to be considered as end-products of the retrogressive metamorphosis of the muscle-substance proper, in which also they occur in greatest amount: kreatin

25 per cent., kreatinin, sarcin, xanthin (especially in fasting pigeons), carnin (oxidizable into xanthin, present in meat-extract), uric acid (urea 0.01 per cent.). There are present further inosite (abundant in the muscles of alcoholics), inosinic acid (inconstant), phospho-carnic acid, resembling nuclein and decomposing into carnic acid ($C_{10}N_3O_5H_{15}$), and a carbohydrate, and also some non-coagulable albumin, taurin (especially in cold-blooded animals), some grape-sugar, glycogen (0.43 per cent.) abundantly in fetal muscles. Further, meat contains lactic acids together with volatile fatty acids. Among the salts the potassium-compounds with phosphoric acid predominate; magnesium phosphate predominates over calcium phosphate.

According to Schlossberger and v. Bibra 100 parts of meat contain the following:

	Ox.	Calf.	Deer.	Pig.	Man.	Chicken.	Carp.	Frog.
Water	77.50	78.20	74.63	78.30	74.45	77.30	79.78	80.43
Solids.	22.50	21.80	25.37	21.70	25.55	22.70	20.22	19.57
Soluble albumin, Coloring-matter.	2.20	2.60	1.94	2.40	1.93	3.0	2.35	1.86
Glutin.	1.30	1.60	0.50	0.80	2.07	1.20	1.98	2.48
Alcoholic extract	1.50	1.40	4.75	1.70	3.71	1.40	3.47	3.46
Fats.	..	..	1.30	..	2.30	..	1.11	0.10
Insoluble albumin, vessels, etc.	17.50	16.20	16.81	16.81	15.54	16.50	11.31	11.67

100 parts of meat-ash contain the following:

.	Horse.	Ox.	Calf.	Pig.
Potassium carbonate	39.40	35.94	34.40	37.79
Sodium.	4.86	..	2.35	4.02
Magnesium.................	3.88	3.31	1.45	4.81
Calcium.	1.80	1.73	1.99	7.54
Potassium.	..	5.56	..	..
Sodium.	1.47	4.86	10.59	0.40
Chlorin.				0.62
Iron oxid.	1.00	0.98	0.27	0.35
Phosphoric acid............	46.74	34.36	48.13	44.47
Sulphuric acid.	0.30	3.37	..	..
Silicic acid.	..	2.07	0.81	..
Carbon dioxid.	..	8.02	. ..	..
Ammonia.	..	0.15	..	..

Potassium and sodium may partially replace each other. The flesh of the pike contains almost twenty times as much calcium as does beef.

The amount of fat in meat is exceedingly variable in accordance with the state of nutrition of the animal. In 100 parts of human flesh, after the visible fat has been cut away, it is from 7 to 15; in beef from 11 to 12; in veal 10.4; in mutton 3.90; in wild goose 8.80; in chicken 2.50.

The amount of extractives is most abundant in the flesh of those animals that exhibit vigorous muscular activity, therefore especially in game. After severe muscular exertion the extractives increase, and at the same time sarco-lactic acid is formed, the meat as a result becoming tender and more pleasant to the taste. Among the extractives there are some that have a stimulating influence on the nervous system, such as kreatin, kreatinin, etc., and some that give to the flesh the pleasant characteristic taste (osmazome). The latter is due in part also to the various fats in the meat, and at times it appears distinctly only on preparation. In 100 parts of meat the amount of extractives is in man and in pigeons 3, in deer and in ducks 4, in swallows 7.

Preparation of Meat.—As a general rule the flesh of younger animals is more tender and more easily digested than that of older animals, because the sarcolemma, the connective tissue and the elastic constituents are less tough. Further, after being allowed to hang for a while, the flesh is still more tender, because the inosite is converted into sarcolactic acid, the glycogen into sugar, and the latter into lactic acid, so that the constituents of the meat undergo a sort of maceration. Meat is further always more easily attacked by the digestive juices when in a finely divided state than when in large pieces; and, finally, it should be noted that adequately boiled, steamed, broiled or roasted meat is easily digestible, although not so rapidly as raw meat. In the preparation the heat should not be too intense or too long continued, because in such an event the muscle-fibers become hard and much shrunken. On the other hand, those pieces of meat that are heated to about 60° or 70° C., such as the pieces from the middle of a large roast that yet have a rosy but not bloody appearance, are most easily digested, as this temperature is quite sufficient with the aid of the acid in the meat to transform the connective tissue into gelatin.

Thus, the meat falls apart and the separate fibers are readily isolated in the stomach. To obtain a piece of good, readily digestible meat, a large cubical block is taken and its surface is suddenly exposed to strong heat by frying in fat or immersion in boiling water. In this manner a firmly coagulated layer of albumin forms on the surface, which no longer allows the juices of the meat to escape from the interior. The reddish, juicy parts from the interior of a piece of meat thus prepared are the most nutritious and the most readily digested, while the hard and much shrunken outer crust resists the digestive juices for some time.

Meat-broth is most suitably prepared by permitting thoroughly chopped meat to stand for some hours in cold water and then boiling. v. Liebig found that out of 100 parts of chopped beef thus treated, but 6 parts pass over into the cold water. Of these, 2.95 parts are again precipitated as coagulated albumin and for the greater part skimmed off and thrown away, so that only 3.05 parts remain dissolved. Of 100 parts of chicken 8 parts are extracted, of which 4.7 are coagulated and 3.3 parts dissolved in the broth. By protracted boiling a part of the coagulated albumin may again pass into solution. The dissolved substances are: 1. Inorganic salts of the meat, of which 82.27 per cent. pass over into the broth. The boiled-out meat retains principally the earthy phosphates. 2. Kreatin, kreatinin, lactates and inosinates, which give to the meat-broth its stimulating and nerve-strengthening power, as well as a small amount of extractives of pleasant taste and some glycogen. 3. Gelatin, which is obtained in considerable amount from the flesh of younger animals.

In accordance with the facts and figures presented, meat-broth is thus to be considered really as merely a highly valuable stimulant, acting as a restorative to the muscles, but not as a food in the ordinary sense of the word, for the constituents of the meat-extract and the kreatin leave the body in the urine in a practically unchanged condition. From larger pieces of meat cooked in the broth even fewer constituents are obtained. Such cooked-out meat accordingly, provided it is not much shrunken by excessive boiling and therefore rendered difficult of digestion, possesses a high nutritive value, which is usually underestimated by the laity. On the other hand, the preparation of meat-broth at home is a real luxury, its so-called strength in the sense of the laity being a pure illusion.

J. v. Liebig's Meat-extract is a fat-free meat-broth containing, however, some gelatin and glycogen, and also about 30 per cent. of albumoses and peptone. It is prepared from finely chopped beef or mutton, in parts of South America and Australia where beef is plentiful, and is evaporated in wide dishes on a water-bath to the consistency of an extract. By solution in water a cheap meat-broth can thus be readily obtained: 1 kilogram of beef yields 31 grams. By boiling the solution with bones (gelatin), some beef-fat, pot-herbs and addition of salt a beverage completely replacing fresh broth is obtained. The so-called "bouillon-tablets" on sale consist almost entirely of desiccated gelatin, which is obtained to the extent of about 28 per cent. from bones boiled in Papin's pots under high pressure. These alone, when dissolved in hot water, naturally cannot replace meat-broth; they can, however, be advantageously employed in conjunction with v. Liebig's meat-extract. In boiling, meat loses weight, principally through loss of water, as follows: beef 15 per cent., mutton 10 per cent., chicken 13.5 per cent. In roasting the same kinds of meat the loss is respectively 19 per cent., 24 per cent., 24 per cent.

J. v. Liebig's "Infusum carnis frigide paratum" is prepared by mixing finely chopped meat with 1 : 1000 hydrochloric acid (3 cu. cm. of fuming hydrochloric acid to 1000 cu. cm. of water), stirring frequently and expressing after some hours. The almost tasteless fluid, which, in addition to the constituents of the broth, is also rich in albumin, is often useful in cases with enfeeble digestion. Albumin is, however, precipitated by the addition of sodium chlorid or by boiling. Leube and J. Rosenthal reduced such a mixture of hydrochloric acid and meat to a gelatinous spongy condition (containing but little peptone) by heating under high pressure in hermetically sealed vessels. The meat-solution thus obtained is employed advantageously in cases of weak stomach.

Of other methods of preservation there are yet to be mentioned: the canning in its own juices of meat boiled at a temperature of 100°; the drying of fat-free meat cut into long narrow strips (the pemmican of the Indians); dried, powdered, salted beef (carne pura). C. v. Voit discovered that the nutritive value of meat is not impaired to any great degree in the process of pickling. He found in pickled meat, in addition to increase in sodium chlorid, a loss of water of 10.4 per cent., of organic matters 2.1 per cent., of albumin 1.1 per cent., of extractives 13.5 per cent., of phosphoric acid 8.5 per cent. The practice of smoking is based upon the antiseptic action of the smoke.

Poor quality and decomposition of meat may result from the development of the alkaloids of putrefaction (ptomains), as well as from the action of bacteria. Such a condition should always cause rejection of the meat. Although it is often enough consumed without bad result, as the popularity of the "haut goût" or "gamey taste" demonstrates. At least, the meat, before being eaten, should always be thoroughly boiled. The decomposition of sausages and similarly prepared meat at times results in the development of a peculiar and even fatal poison, the sausage-poison. Occasionally the decomposition of meat, particularly also of fish, gives rise to a peculiar actively phosphorescent light, due to the development of lower organisms. The use of such meat. however, does not seem to be directly injurious. A knowledge of the occurrence of the trichina spiralis in pork is highly important; also of bladder-worms varying in size from a pea to a bean in pork and beef, the development of which into tapeworms is discussed under Reproduction. The cysticercus of bothriocephalus latus is found in the pike.

VEGETABLE FOODS.

The nitrogenous constituents of plants are less readily absorbed than those of animal foods. Nevertheless the former may completely replace the animal proteids, provided they contain an equal amount of nitrogen. Carbohydrates, starch and sugar are quite completely absorbed and even some cellulose is digested. The greater the amount of fat in vegetable food the less the carbohydrates are digested and absorbed.

Among the vegetable articles of food the *cereals* occupy the first place. They contain proteids, starch, and salts, together with water to about 14 per cent. The nitrogenous gluten is most abundant beneath the capsule, so that the use of finely ground bran in coarse bread seems plausible for good digestive organs; although the varieties of bread that contain a considerable amount of bran are digested with appreciably greater difficulty, as the cellulose-membrane of the gluten-layer is scarcely dissolved in the process of digestion. Rye-bread is assimilated with greater difficulty than wheat-bread. For commercial bread a mixture of both kinds of flour is advisable. Quantitative composition:

100 parts of dry flour contain:		100 parts of cereal ash contain:		
	Albuminates.	Starch.	Red Wheat.		White Wheat.
Wheat	16.52%	56.25%	27.87	*Potassium carbonate*	33.84
Rye	11.92	60.91	15.75	Sodium	....
Barley	17.70	38.31	1.93	Calcium	3.09
Corn	13.65	77.74	9.60	Magnesium	13.54
Rice	7.40	86.21	1.36	Iron oxid	0.31
Buckwheat	6.8–10.5	65.05	49.36	*Phosphoric acid*	49.21
			0.15	Silica	....

It is remarkable that in white wheat sodium is wanting and is replaced by other alkalies. Rye contains more cellulose and dextrin than wheat, but less sugar. Rye-bread is, as a rule, less porous. Barley and oats are much used as gruel, and in the North also mixed in bread.

Oats contain a crystalline globulin (avenalin), and a proteid soluble in alcohol and another soluble in alkalies. By admixture with water or neutral salts three other proteids are obtained as products of transformation. Rye and wheat yield one globulin (edestin); one albumin (leukosin); gliadin, forming gluten, and soluble in dilute alcohol; and glutenin (absent from rye), soluble in dilute acids and alkalies. Barley contains leukosin, edestin, hordein, corresponding to gliadin, and also other proteids.

In the preparation of bread, flour is kneaded, together with water, to form dough, in which the gluten acts as a cementing substance, and to which salt and also yeast (saccharomyces cerevisiæ) are added. Under the influence of heat the albuminates of the flour begin to undergo decomposition and the ferments act upon the swollen starch, which is partially transformed into sugar. The sugar undergoes further decomposition into carbon dioxid and alcohol, of which the first, forming bubbles in the stiff dough, causes this to become spongy. Certain bacteria also coöperate with the yeast to the same end. By the baking at 200° C. the alcohol is driven off and the dough is done. Much readily soluble dextrin is formed in the crust. In the preparation of sour bread, old sour dough, in which the sugar has partially undergone lactic-acid fermentation, is added instead of yeast, and as a result, in addition to the alcoholic fermentation, the lactic-acid fermentation of the grape-sugar in the dough is also initiated. As in the transformation of starch into sugar, and the latter into carbon dioxid and alcohol (which eventually escape), material is directly lost; ammonium carbonate, which escapes during the process of baking with the expansion of the dough, is added. This loss amounts to about one per cent.; with an average daily consumption of bread for each individual of 256

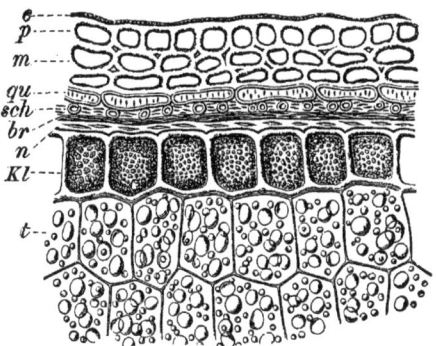

Fig. 138.—Section through a Grain of Wheat: p, epidermis, with c, cuticula, m middle layer, qu transverse, sch tubular cells, br and n seed-membrane, Kl gluten, t starch.

grams, the daily loss for 1,000,000 persons should equal 2500 kilograms of bread, or sufficient for 10,000 persons. J. v. Liebig proposes the use of sodium bicarbonate and hydrochloric acid for the same purpose; then the dough will not have to be salted, because of the formation of sodium chlorid. Horsford's baking-powder (calcium phosphate and sodium bicarbonate) is also used. It permits the escape from the dough of the expanding carbon dioxid, the phosphoric acid of which is also useful to the body.

The *legumes* contain much albumin. Beans contain two globulins readily soluble in salt-solutions—the phaseolin of Ritthausen and phaselin. Peas and vetches yield in considerable amount a globulin, designated legumin by Braconnot, which is soluble in a solution of sodium chlorid, and also three other proteids. Legumes contain also starch, lecithin, and cholesterin, together with from 9 to 19 per cent. of water. Peas contain 28.02 per cent. of albumin and 38.81 of starch; beans 28.54 and 37.50; lentils 29.31 and 40 respectively. The last are richer in cellulose. Because of the deficiency in gluten no dough can be made from them, and therefore no bread. As a food for the mass of the people, these plants deserve the greatest consideration, because of the large amount of albumin they contain, although they may be a source of intestinal discomfort in consequence of the development of gas, as well as of the presence of indigestible cellulose. Leguminous flour, when mixed in different proportions with the flour of cereals (for instance in the form of Hartenstein's leguminose), can be used with advantage in the feeding of children and debilitated persons.

Maize contains three globulins, several albumins, and a proteid—zeïn—soluble in alcohol.

Potatoes contain from 70 to 81 per cent. of water. In the juicy cellular tissue, which yields an acid reaction when fresh, from the presence of phosphoric, malic and hydrochloric acids, there is present from 16 to 23 per cent. of starch, 2.5 per cent. of dissolved proteids, consisting of one globulin (tuberin), soluble in potato-juice, and some albumin, together with a trace of asparagin. The cell-capsules become swollen when boiled, and are changed by dilute acids into sugar and gum. The "eyes" contain the poisonous substance solanin. One hundred parts of potato-ash contain: potassium carbonate 46.96, sodium chlorid 2.41, potassium chlorid 8.11, magnesia 13.58, calcium 3.35, phosphoric acid 11.91, sulphuric acid, derived from burned albuminates, 6.50, silica 7.17.

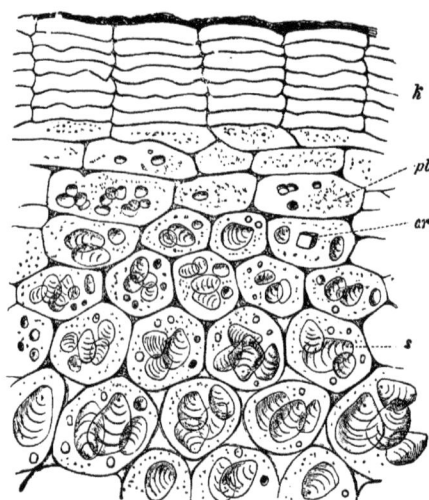

FIG. 130.—Section through Potato: *k*, cork; *pl*, plasma-containing cells, with small starch-granules; *cr*, protein crystalloid; *s*, starch.

Fruits contain as the principal food-constituents sugar and salts. Their characteristic taste is due to the organic acids. The gelatinizing substance of fruit-jellies is the soluble so-called *pectin* ($C_{32}H_{48}O_{32}$), which can also be obtained artificially by cooking from the pectose of unripe fruit, which is soluble with difficulty, and from carrots.

Green vegetables are especially rich in salts that resemble the salts in the blood. For instance, unseasoned lettuce contains 23 per cent. of salts, spinach much iron. Of less importance in them are starch, dextrin, sugars and small amounts of albumin.

CONDIMENTS.

COFFEE, TEA, CHOCOLATE, ALCOHOLIC DRINKS AND SPICES.

Since the time of v. Bibra the term *condiment* has been applied to such articles of food as are used less because of their direct nutritive properties, than because of their agreeable action and stimulation, in part upon the organs of taste and in part also upon the nervous system. Coffee, tea and chocolate are prepared as infusions or decoctions of the familiar vegetables. They contain respectively an active constituent, caffein or thein ($C_8H_{10}N_4O_2 + H_2O$—trimethylxanthin), or the closely related theobromin ($C_7H_8N_4O_2$—dimethylxanthin), which are classified among the alkaloids, or vegetable bases. These have recently been prepared artificially from xanthin.

These alkaloids and similar bodies in many other plants are present in the plants preformed. Their behavior is similar to that of ammonia. They have an alkaline reaction and with acids form crystalline, well-defined salts. All of these vegetable bases affect the nervous system, some feebly, as the preceding, or more actively stimulating, as quinin; others have a more powerful stimulating effect, to the point of paralysis, including active poisons, such as morphin, atropin, strychnin, curarin, nicotin, etc.

The alkaloids of coffee, tea and chocolate confer upon the infusions of these substances generally used as popular beverages the pleasantly stimulating effect upon the nervous system, refreshing the mind, animating movement and stimulating to increased activity. In this respect they resemble the stimulating extractives of beef-broth. Coffee contains about ⅛ per cent. of caffein, which is partially first set free in the process of roasting. Tea contains 6 per cent. of thein, green tea also 1 per cent. of ethereal oil, black tea ½ per cent.; green tea 18 per cent., black tea 15 per cent. of tannic acid. Green tea yields on the whole about 46 per cent., black tea scarcely 30 per cent., of extract.

In addition the inorganic substances in these beverages are to be considered. Tea contains 3.03 per cent. of salts, including considerable amounts of soluble compound of iron and manganese, which are important in the formation of hemogloblin, and also sodium-salts. In coffee, which yields 3.41 per cent. of ash, potassium preponderates. In all three beverages, however, the remaining inorganic substances that are found in the blood are present in suitable proportions. Cocoa is only inadequately utilized as a nutritive agent: of 50 grams only 5 grams of albumin, 16 grams of fat and 6 grams of starch.

Alcoholic beverages owe their activity primarily to the alcohol they contain. With regard to the latter the following is to be noted: 1. Alcohol is decomposed in the body, principally into carbon dioxid and water. In this respect it does not differ essentially from other articles of food, and it is thus to be regarded as a source of heat. As alcohol readily undergoes this combustion in the body, its use can, therefore, diminish to a certain degree the consumption of the constituents of the body itself. It has, however, been shown that with a mixed diet alcohol is not capable of protecting the albumin from decomposition, but solely the fat. With a mixed diet alcohol is not able to replace any of the carbohydrate of the food. Only from 1 to 2.5 per cent. of the alcohol passes over into the urine, from 5 to 6 per cent. into the breath. The odor of the breath is due, in addition, to other volatile substances in the alcoholic beverage, such as fusel-oil and others. Traces pass into the cutaneous secretions. 2. In small amounts alcohol has a stimulating effect, in large doses, through overstimulation, a paralyzant effect upon the nervous system. By means of this stimulating effect it is therefore capable of spurring the body temporarily to greater functional activity for achievement, at the

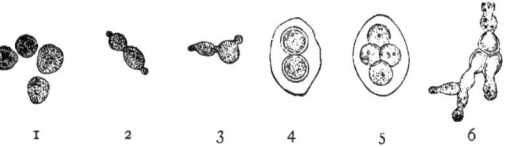

FIG. 140.—1, Isolated yeast-cells; 2, 3, formation of buds; 4, 5, endogenous cell-formation: 6, germination and bud-formation.

expense, it is true, of a subsequent depression. 3. When taken in small suitable doses before or after meals, it aids the digestion, while larger doses interfere with digestion. 4. It diminishes the sensation of hunger. 5. It induces more active respiratory movements, and stimulates the heart and the vascular system, and thus accelerates the circulation of bright-red blood, so that muscles and nerves become more capable of action. It also causes a subjective sensation of heat. In a larger dose, however, it paralyzes the vessels by overstimulation and they become dilated, for example, in the external integument. As a result heat is radiated in greater degree through the skin than it is generated in the body, and therefore the bodily temperature is lowered. Large doses also diminish the activity of the heart by the excitation of smaller, weaker and more rapid beats. In elevated regions the power of alcohol is greatly enfeebled, as, on account of the low atmospheric pressure, the alcohol is rapidly given off from the blood.

From the foregoing remarks it is clear that alcohol, when taken in small amounts, can be of incalculable benefit in conditions of temporary privation and want of food, in conjunction with which resistance to fatigue and an extraordinary amount of work are yet required. In a similar manner it is capable of protecting the tissues of the sick from too rapid consumption, with the exception of the albumin. When taken habitually, however, and especially in large amounts, it causes derangement of the nervous system by overstimulation, and undermines the forces of mind and body, partly in consequence of its poisonous properties, chiefly due to its volatile constituents (fusel-oil), affecting the nervous system permanently, partly through its direct action in causing injurious catarrhal and inflammatory conditions in the digestive organs, and partly finally through interference with and impairment of the normal metabolism.

Alcoholic beverages are prepared by fermentation of the sugar obtained from various carbohydrates, especially starch. Alcoholic fermentation is a result of the vital activity of a low order of fungus, namely the yeast-fungus—the saccharomyces cerevisiæ (in the fermentation of beer) and the saccharomyces ellipsoideus (in the fermentation of wine), the fungus removing directly from the saccaharine mixture the substances necessary for its existence, namely carbohydrates, albuminates and salts, chiefly calcium phosphate, potassium phosphate and magnesium sulphate and causing their decomposition into alcohol and carbon dioxid, together with some glycerin (from 3.2 to 3.6 per cent.) and succinic acid (from 0.6 to 0.7 per cent.). The yeast-liquor alone, in the absence of yeast-cells, also causes fermentation, through the presence of a ferment known as zymase, which acts like a chemical agent. Yeast belongs to the budding fungi, which multiply both by budding and by sporulation (ascospores). It is added directly to the fluids to be fermented, or its spores, which constantly float about in the air, fall into the uncovered mixture. Perfect exclusion of yeast-cells, or their destruction, as by boiling the syrup in sealed vessels, therefore, prevents the occurrence of fermentation. Alcoholic fermentation is, therefore, a result of the vital activity of a lower form of organism.

Wine contains on an average from 89 to 90 per cent. of water, from 7 to 8 per cent. of alcohol, together with the ethyl-alcohol, propyl-alcohol and butyl-alcohol. The color of red wine is derived from the skins during fermentation. If the skins be previously removed red grapes will yield whitish wine. The fine taste (flower, bouquet) develops during the storing of the wine. Enanthic ether is said to give rise to the characteristic odor of wine. The value of wine depends upon the as yet unknown stimulating, volatile substances that confer upon each wine its special character. Of great importance are, further, the salts, which in their composition resemble those of the blood.

Beer contains from 75 to 95 per cent. of water, alcohol from 2 to 5 per cent. (porter and ale as much as 8 per cent.), carbon dioxid from 0.1 to 0.8 per cent., sugar from 2 to 8 per cent., gum, dextrin from 2 to 10 per cent., cholin, the constituents of hops, some residue of protein-substances (gluten), fat, lactic acid, ammonia-compounds, the salts of barley and of hops.

In the ash the enormous amount of phosphoric acid and potassium carbonate, so important in the formation of blood, is noteworthy; one hundred parts of ash contain of potassium carbonate 40.8, phosphorus 20, magnesium phosphate 20, calcium phosphate 2.6, silica 16.6 per cent. The favorable influence of beer on the formation of blood, muscles and other tissues is due to the abundance of phosphoric acid and potassium carbonate. The obesity of the beer-drinker depends chiefly on the fat-sparing action of the alcohol. The potassium carbonate present in beer has a fatiguing effect after heavy drinking.

Spices are not consumed because of their nutritive value, but in part on account of their taste, in part because of their stimulating effect, through which they arouse the digestive organs to increased activity. In a certain sense sodium chlorid must also be regarded as a spice, being withheld at present apparently from only a few savage tribes. A similar fact was recorded by Homer. Also certain as yet unknown substances that have a marked effect on the organs of taste, and that develop only in the course of preparation of some dishes, as in the crust of a roast, and in the crust of pastry, may be included among spices.

PHENOMENA AND LAWS OF METABOLISM.

METABOLIC EQUILIBRIUM.

By *metabolic equilibrium* is understood that normal condition in which precisely the same amount of material for the maintenance and growth of the organism is taken up and assimilated from the digested nourishment as is removed from the body through the excretory organs in the form of waste-materials or end-products of retrogressive tissue-metamorphosis. The income must always balance the expenditure. During the period of growth of the body a certain excess of formative activity corresponding to the increase in size of the body must predominate. Thus, growing portions of the body exhibit from 2.5 to 6.3

times as active a metabolism as parts of the body already formed. On the other hand, in the years of senile debility a certain excess of bodily expenditure is to be considered as a normal phenomenon.

Method.—The normal metabolic equilibrium in the organism may be recognized: 1. By determining chemically that the sum of all the egesta, given off by the body within a certain period of investigation, corresponds to the sum of the ingesta furnished by the food. In this connection the amount of carbon, nitrogen, hydrogen, oxygen, and salts, together with the water of the food and the oxygen of the inspired air, must be equal to the carbon, nitrogen, hydrogen, oxygen, the salts and the water in the excretions (urine, feces, expired air, evaporated water) of the organism. 2. The physiological equilibrium of metabolism may further be recognized in a purely empirical way from the fact that with a suitably selected diet, the body performing its ordinary functions is able to maintain its normal weight. Thus, this simple procedure of weighing makes it possible for the physician to inform himself quickly and with certainty concerning the metabolism of his patient or convalescent. The tedious method of elementary metabolic analysis was first successfully undertaken particularly by the Munich investigators, v. Bischoff, v. Voit, v. Pettenkofer and others. It was soon apparent that of all the elements the greatest importance was to be assigned to the passage through the body of carbon and nitrogen.

The total amount of carbon taken with the food (which is ascertained by elementary analysis of a sample of each article of diet, or is calculated from reliable analyses of the articles of food) must, in complete metabolic equilibrium, correspond to the carbon in the carbon dioxid contained in the expired air (90 per cent.) from the lungs and the skin. To this there should also be added the relatively small amount of carbon in the organic excrementitious matters of the urine and the feces (10 per cent.), which is to be determined by elementary analysis. As all organic food and all the constituents of the body contain carbon, an increased loss of carbon (as compared with the income) indicates that organic matter in excess is being decomposed in the body; on the other hand, diminished excretion of carbon necessarily indicates an addition to the substance of the body. For the exact determination of the carbon dioxid in the expired air the Munich investigators employed v. Pettenkofer's respiratory apparatus.

With regard to the nitrogen, which should be determined in the ingesta as well as the excreta by the method of Kjeldahl, it was found that almost all the nitrogen of the ingested food is excreted again in the urine within 24 hours, principally in the form of urea. The remaining nitrogenous urinary constituents (uric acid, kreatinin, etc.) furnish only about 2 per cent. of the nitrogenous elimination. In addition, the nitrogen-content of the feces is to be taken into account (from 4 to 5 per cent. in dogs). A small amount of nitrogen also escapes from the organism in the expired air; also a portion with the desquamated epidermal structures (about 50 mg. of hair and nails daily) and in the sweat.

The opinion that practically all the nitrogen ingested with the food is excreted in the urine and the feces was established for carnivora by v. Voit and Gruber; for ruminants by Henneberg, Stohmann and Grouven, and for man by Ranke. Contrary to this view a number of the older as well as more recent investigators have made the assertion that the total amount of nitrogen cannot be recovered from the excretions mentioned, but that an appreciable nitrogen-deficit exists.

According to Leo about 0.55 per cent. of the albumin decomposed in the body yields its nitrogen (which may be assumed to amount to 15 per cent.) in the gaseous state. In making exact analyses of metabolism this gaseous excretion of nitrogen must naturally be taken into account.

In the food nitrogen occurs almost exclusively as a constituent of albuminous substances. In the excretions it indicates decomposition of the albuminous constituents of the body. As proteids contain on the average 16 per cent. of nitrogen the amount of albumin corresponding to the amount of nitrogen excreted is determined by multiplying the latter figure by 6.25. Nitrogenous equilibrium only indicates that the albuminous substances in the body are unchanged. If nitrogen is retained gain in weight takes place, principally in the form of muscle; if there is an excess of nitrogenous excretion, consumption of the albuminous constituents of the body must ensue.

The relative amount of nitrogen and carbon in albumin may be expressed as 1 : 3.3. Of the amount of carbon decomposed in the process of metabolism there are 3.3 parts for every part of nitrogen in the proteids subjected to the process. The excess is to be attributed to the decomposition of non-nitrogenous substances (fats or carbohydrates).

It· is believed that the greater portion of the 'proteids are decomposed in the tissues into carbamic acid, which is then transformed in large amounts, in the liver, into urea.

The excretion of nitrogen after the taking of food is, in animals, not uniform from hour to hour, but it increases rapidly at once, reaches its maximum after 5 or 6 hours and then gradually declines. The excretion of sulphur and phosphorus pursues a similar course, though the maximum excretion, on a meat diet, occurs as early as the fourth hour. On addition of fat to a meat-diet the excretion of nitrogen and of sulphur is more evenly distributed throughout the hours of the day. In human beings Rosemann found during the day an increase between 9 and 11 a. m., as a result of breakfast and the stimulation of all of the functions in the morning; a second increase between 3 and 4 p. m., as a result of dinner; a third, smaller one between 7 and 9 p. m., following supper; and a final increase between 9 and 11 p. m. The excretion diminishes during the night.

The nitrogenous constituents of the body become poorer in carbon as a result of the processes of metabolism, but richer in nitrogen and oxygen; for there are, in the albumins, 4 atoms of carbon for each atom of nitrogen, in gelatin $3\frac{1}{2}$ atoms of carbon, in glycocoll 2 of carbon, in kreatin $1\frac{1}{2}$ of carbon, in uric acid $1\frac{1}{4}$ of carbon, in allantoin 1 of carbon, in urea, finally, only $\frac{1}{2}$ atom of carbon.

The oxygen furnished by respiration is either determined directly from the reduction in its amount in the air supplied to the animal, or it is calculated from other data. This inspired oxygen, as well as the oxygen of the food, makes its appearance principally in the form of carbon dioxid and water; a small amount leaves the body in the excrementitious products. The amount of oxygen absorbed in respiration is the measure of the entire process of combustion in the body, by which carbon is oxidized into carbon dioxid and hydrogen into water. The respiratory quotient indicates the amount of inspired oxygen that is required alone for the combustion of the carbon. If the volume of carbon dioxid produced by the combustion of pure carbon is exactly the same as the volume of oxygen consumed for this purpose, the respiratory quotient is 1. This is the case in the decomposition of carbohydrates. As hydrogen and oxygen are present in these compounds in the proportion necessary to form water by combustion, practically all the oxygen is utilized for the oxidation of the carbon of the carbohydrates. For albumin the respiratory quotient is 0.8, for, on a purely albuminous diet, only 800 cu. cm. of carbon dioxid are excreted for every 1000 cu. cm. of oxygen. For fats the respiratory quotient is 0.7, for, on a diet of fat, only 700 cu. cm. of carbon dioxid are excreted for every 1000 cu. cm. of oxygen consumed. Thus for fats and albumin the respiratory quotient is smaller than for carbohydrates, the volume of carbon dioxid excreted is less than that of oxygen inspired, because in the combustion of albumin or fat a part of the oxygen taken up must be employed for the oxidation of hydrogen into water.

In case a larger volume of carbon dioxid is excreted than the amount of oxygen absorbed, the respiratory quotient rises above 1. This happens if, in addition to the inspired oxygen, a portion of the oxygen from the constituents of the food is converted into carbon dioxid in the process of combustion in the body. This is the case, for example, when nutritive materials rich in oxygen (for example, carbohydrates) are transformed in the body into those poor in oxygen (for example, fats).

The respiratory quotient may also, under certain circumstances, become even smaller than it is after an exclusive fat-diet, if, for instance, a portion of the inspired oxygen is employed in the body for the formation and deposition in the tissues of compounds rich in oxygen (for example, in the formation of glycogen).

The respiratory quotient may, however, exhibit certain periodic variations independently of the character of the diet. As the oxygen taken up is not always used in the formation of carbon dioxid immediately upon its entrance into the body, but as certain intermediate predecessors of carbon dioxid, rich in oxygen, may accumulate in the body and be excreted only later completely oxidized into carbon dioxid, it may happen that during one part of a period of dieting more oxygen may be taken up than is given off as carbon dioxid.

Hydrogen leaves the body principally oxidized into water, but it may also leave the body combined with organic excreta. The water is given off with the urine, the feces, through the lungs and by evaporation from the skin. As hydrogen is oxidized into water the amount of water given off is naturally greater than that taken up. The salts are excreted in various ways, the most soluble of them passing out with the urine, a few, particularly potassium-salts, and those that are soluble with difficulty, with the feces, and some, like common salt, in part also with the sweat. The salts contained in the ingesta and excreta are estimated by weight after incineration.

If the sulphur and the phosphorus are to be estimated separately the amount of each in the ingested food is oxidized by combustion, by addition of sodium hydroxid and potassium nitrate into sulphuric and phosphoric acids respectively. The same method is followed for their estimation in the feces, as well as for sulphur in the epidermal structures. The sulphuric and phosphoric acids so obtained, as well also as the sulphur and the phosphorus excreted in the urine in an already oxidized form, are estimated according to the method described on p. 491. The sulphur is derived principally from the albuminous food; about half of it is excreted with the urine as sulphuric acid, half with the feces (as taurin) and through the epidermal structures.

For every body, there is, according to its weight and activity, a minimum and a maximum limit of metabolic balance. If less food is supplied than is necessary for the first, loss of body-weight results. If, on the other hand, an excess of food is supplied, any amount exceeding the maximum limit will be passed unabsorbed as superfluous ballast with the feces, provided it cannot be utilized for increase of flesh. The more the body gains in weight on a generous diet, the higher the minimum limit rises. With marked increase of flesh, therefore, the necessary supply of food must be relatively much greater than in the case of thin persons, in order to cause a like increase in the tissues of the body. With continued increase in flesh there finally results a condition in which the digestive organs are able to prepare only sufficient material for the maintenance, but not for increase, of weight.

QUALITY AND QUANTITY OF NOURISHMENT FOR A HEALTHY ADULT.

The question as to the substances that are needed by man for his satisfactory nourishment, and the amount required, has naturally been answered in a purely empirical way by observation of the manner of nourishment of healthy individuals at different ages and with varying degrees of activity. As, for example, the infant flourishes and grows on a diet of milk, milk must undoubtedly include in its composition nutrient matter qualitatively and quantitatively appropriate.

In accordance with his entire organization man belongs to the omnivora, that is, to the class of beings that is adapted to a mixed diet. He possesses the canine tooth of the carnivora, but his intestine is shorter than that of the herbivora.

For his continued existence man requires the following four principal nutritive substances, none of which can be spared from the diet for any length of time:

1. *Water;* for an adult from 2700 to 2800 grams daily in food and drink.

Withdrawal of water increases the disintegration principally of the albuminous tissues. If thirsting cats are kept for a long time in hot air, a concentration of the blood becomes manifest, which, through chemical injury, leads to a fatal central narcosis by poisoning of the vital centers.

28

2. *Inorganic matters* as integral parts of all of the tissues, without which their structure could not be formed. These substances are present in sufficient amount in all the usual articles of diet, so that it is unnecessary to supply them separately (as the nutrition of animals shows). Increase in the supply of salt causes increase in the consumption of water, and this in turn causes increase of nitrogenous decomposition in the body. Withdrawal of certain necessary salts causes disturbances in the nutrition of the tissues containing them. Thus, the use of food free from calcium is followed by disturbance in normal bone-formation; withholding common salt causes albuminuria. The body absorbs the iron required for the formation of blood in part in the form of complex organic compounds from the vegetable and animal kingdoms, but in part also in an inorganic form; phosphorus principally from proteids containing phosphorus. The alkaline salts derived from vegetable food serve to neutralize the sulphuric acid formed by oxidation of the sulphur of the proteids. Food that has been artificially deprived of all salts causes rapid death in animals by acid intoxication.

Only as a matter of necessity does man occasionally resort to the use of considerable amounts of inorganic matter in order to obtain the organic nutrient material mixed with it, as A. v. Humboldt relates of the inhabitants along the shores of the Orinoco and the Meta, and who, in times of scarcity, when the catch of fish is low, are compelled to eat a certain kind of rich clay, containing an abundance of infusoria.

3. At least *one* animal or vegetable *proteid*. The albuminates are utilized to replace the consumed nitrogenous tissues, particularly the muscles. In addition, they are used as sources of energy and heat. The latter function of albuminous food can be fulfilled by non-nitrogenous food; the first, however, can not. The albuminates contain from 15 to 18 per cent. of nitrogen.

Different tissues of the body contain proteids in the following proportions: Blood 20.56 per cent., muscles 19.9 per cent., liver 11.74 per cent., brain 8.63 per cent., blood-plasma 9.0 per cent., milk 3.8 per cent., lymph 2.4 per cent. According to Pflüger and Bohland a growing youth weighing 62 kilos decomposes 89.9 grams of albumin daily.

It is a remarkable fact that asparagin—a nitrogenous amido-body, which is formed in sprouting plants from albumin and under certain circumstances may be again transformed into this in the plant—combined with gelatin is capable of replacing the albumin of the food. Asparagin alone is capable (only in herbivora) of diminishing the decomposition of albumin. Salts of ammonia, glycocoll, sarcosin and benzamid increase the destruction of proteid.

4. At least one form of *fat* or digestible *carbohydrate*. These serve principally for the replacement of the decomposed fat and non-nitrogenous constituents of the body. On account of the large amount of carbon they contain, they are, through their oxidation, the chief sources of heat-production. Fats and carbohydrates can replace each other in the diet in reciprocal amounts corresponding to the quantity of heat that they are able to produce by their combustion in the body. In the same way, the portion of the albumin of the food that does not serve for the restoration of the tissues can be replaced by equivalent amounts of fat or carbohydrates. In this connection 100 parts of fat are equivalent to 256 of grape-sugar, 243 of milk-sugar, 234 of cane-sugar, 221 of dry starch. In general the same amounts correspond to 227 parts of carbohydrate, as well as to 227 of albumin.

The compound sugars in the organism must first be decomposed into monosaccharids before they are oxidized, just as they are decomposed into monosaccharids in the process of fermentation. If the organism is unable to split a compound sugar into its components, it cannot oxidize the sugar. This splitting, for example, of cane-sugar and milk-sugar, is carried out in the intestine. If these substances are introduced subcutaneously they are not split up, and therefore are not made use of; that is, they are excreted again in the urine.

It is a remarkable fact that butter can be injected subcutaneously in considerable amounts and, thus introduced, is utilized either for combustion or for the deposition of fat.

With regard to the relative proportions in which these various nutritive materials should be combined, experience has taught that for man, a diet in which the nitrogenous and non-nitrogenous elements are mixed in the proportion of one nitrogenous to $3\frac{1}{2}$, or at the most $4\frac{1}{2}$, parts of the non-nitrogenous elements, must be considered as the most advantageous. If the customary articles of diet be considered according to this standard, it can easily be seen to what degree they conform with the requirement mentioned, and, furthermore, that a suitable diet may often be formed by a mixture of several of them.

The following table shows the proportion of nitrogenous and non-nitrogenous matters in various articles of food:

	Nitrog-enous.		Non-nitrog-enous.			Nitrog-enous.		Non-nitrog-enous.
1. Veal,	10	to	1	10. **Human milk,**	10	to	37
2. Hare,	10	"	2	11. **Wheat-flour,**	10	"	46
3. Beef,	10	"	17	12. Oat-meal,	10	"	50
4. Lentils,	10	"	21	13. Rye-flour,	10		57
5. Beans,	10	"	22	14. Barley-flour,	10	"	57
6. Peas,	10	"	23	15. White potatoes,	10	"	86
7. Mutton (fattened),	10	"	27	16. Blue potatoes,	10	"	115
8. Pork,	10	"	30	17. Rice,	10		123
9. **Cow's milk,**	10	"	30	18. Buckwheat-flour,	10	"	130

This survey shows that in addition to human milk, wheat-flour lies within the normal limits with regard to its proportional composition. On the other hand the articles of diet from 1 to 9 require an addition of non-nitrogenous, those from 12 to 18 of nitrogenous, substances in order to maintain the proportions from 10 : 35 to 10 : 45. A man who attempted to live on meat alone would, therefore, be just as irrational as one who took potatoes alone as food. Experience long ago impressed upon the mind of the people the fact that milk and eggs will indeed support life, but that a meal of meat requires potatoes or bread; a dish of beans a portion of bacon.

It should also be especially mentioned that the proportions of the diet vary in accordance with climate and season. As with a considerable degree of cold the organism must produce more heat, the inhabitants of higher latitudes take relatively more non-nitrogenous food (fat and sugar or starches), which, on account of its richness in carbon, is especially suited for the generation of heat in the body.

The graphic representation of the composition of the most important articles of food in Fig. 141 (after A. Fick) is especially clear.

If it be borne in mind that the nitrogenous bodies in the food must be in the proportion of $1 : 3\frac{1}{2}$ to $4\frac{1}{2}$ of the non-nitrogenous, a glance will show at once what articles of diet are suited for food without addition, as well as which of them may be suitably combined to supplement one another.

The *absolute amount of food* that an adult needs during twenty-four hours is influenced by various factors. As food represents the reservoir

of chemical potential energy from which the body generates on the one
hand heat and on the other kinetic energy, the absolute amount of food

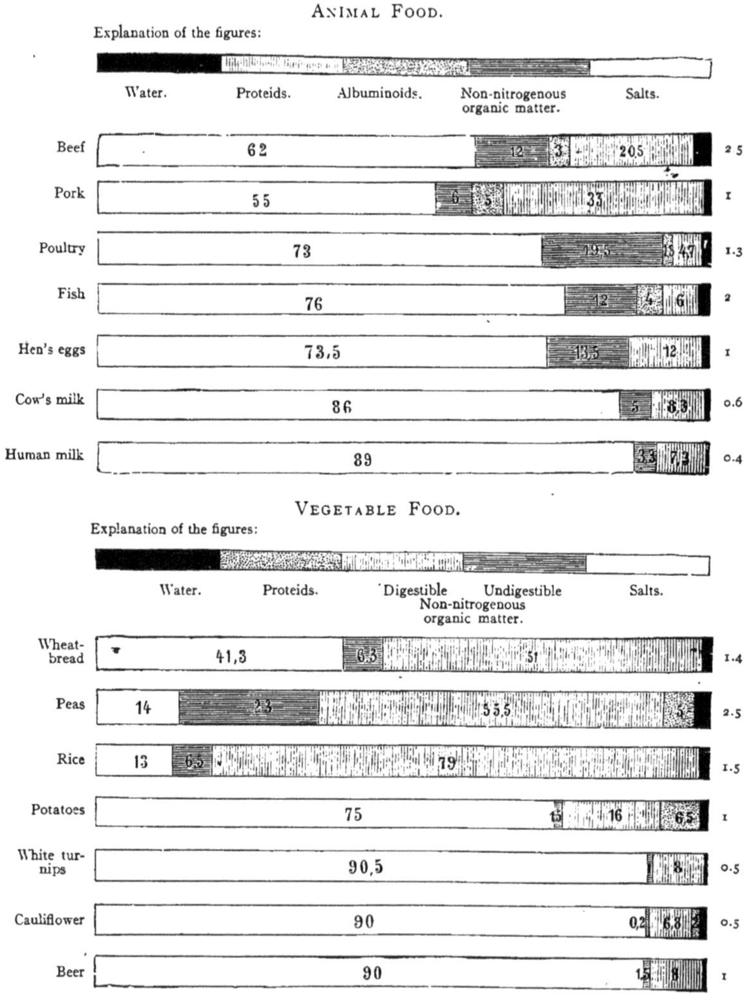

ANIMAL FOOD.

Explanation of the figures:

| | Water. | Proteids. | Albuminoids. | Non-nitrogenous organic matter. | Salts. |

Beef	62 ... 12 3 20.5	2.5
Pork	5 5 ... 6 5 33	1
Poultry	73 ... 19.5 4.7	1.3
Fish	76 ... 12 4 6	2
Hen's eggs	73,5 ... 13.5 12	1
Cow's milk	86 ... 5 8.3	0.6
Human milk	89 ... 3.3 7.3	0.4

VEGETABLE FOOD.

Explanation of the figures:

| | Water. | Proteids. | Digestible / Undigestible Non-nitrogenous organic matter. | Salts. |

Wheat-bread	41,3 ... 6.5 51	1.4
Peas	14 ... 23 53,5	2.5
Rice	13 ... 6.5 79	1.5
Potatoes	75 ... 1.5 16 6.5	1
White tur-nips	90,5 ... 8	0.5
Cauliflower	90 ... 0.2 6.8	0.5
Beer	90 ... 1.5 8	1

FIG. 141.

must be increased when the loss of heat from the body (winter) or its
muscular activity (work) increases.   On the average a man requires 130
grams of proteids, 84 grams of fat, and 404 grams of carbohydrates.

The following figures are average values derived from many individual observations:

An adult requires in 24 hours:

Amount of Food in Grams.	Resting (Playfair).	Moderate Work (Moleschott).	Hard Work (Playfair).	(v. Pettenkofer and v. Voit.)
Proteids,	70.87	130	155.92	137
Fats,	28.35	84	70.87	117
Carbohydrates (sugar, starch, etc.),	310.20	404	567.50	352

In an analogous example taken from C. v. Vierordt the elementary matters in the food will be estimated and at the same time the amounts ingested be compared with those excreted.

An adult with moderate activity consumes:

	C	H	N	O
120 grams of albumin, containing,	64.18	8.60	18.88	28.34
90 grams of fats, "	70.20	10.26	—	9.54
330 grams of starch, "	146.82	20.33	—	162.85
	281.20	39.19	18.88	200.73

In addition: 744.11 grams of oxygen from the air by respiration.
2818 grams of water.
32 grams of inorganic compounds (salts).

The whole amounts to about 3.2 kg. or about $\frac{1}{20}$ of the body-weight. Over 6 per cent. of the water, about 6 per cent. of the fat, about 1 per cent. of the albumin and about 0.4 per cent. of the salts in the body are thus daily replaced.

An adult with moderate activity excretes:

	Water	C	H	N	O
With respiration,	330	248.8	..	?	651.15
By transpiration,	660	2.6	..	..	7.2
In the urine,	1700	9.8	3.3	15.8	11.1
In the feces,	128	20.0	3.0	3.0	12.0
	2818	281.2	6.3	18.8	681.45

In addition 296 grams of water—not included in the 2818 grams of water ingested—are formed in the body by oxidation of the hydrogen of the food. These 296 grams of water contain 34.89 grams of hydrogen and 263.31 grams of oxygen. Further, 26 grams of salts are passed with the urine and 2 grams with the feces.

An adult at rest consumes during twenty-four hours 96.5 grams of proteid—equivalent to 1.46 grams for each kilogram; at hard work, 107.6 grams—equivalent to 1.6 grams for each kilogram. Three or four times as much fat as albumin is transformed daily.

Investigations, principally by the Munich school, have determined the following *minimum* figures for the diet at various ages:

Age.	Nitrogenous.	Fat.	Carbohydrate.
For a child up to 1½ years,	20–36 gms.	30–45 gms.	60–90 gms.
For a child from 6 to 15 years,	70–80 "	27–50 "	250–400 "
For a man, with moderate activity,	118 "	56 "	500 "
For a woman, with moderate activity,	92 "	44 "	400 "
For an old man,	100 "	68 "	350 "
For an old woman,	80 "	50 "	260 "

It is frequently asserted that, in case of necessity, a considerably smaller amount of proteid (55 gms. for a man) would suffice, providing that the amount of food were sufficient to supply the requisite number of calories for the body, that is, 45,000 calories for each kilogram of body-weight. The diet of the Japanese contains, for example, a much smaller amount of nitrogen than that of the European. Numerous experiments have demonstrated, however, that an adult weighing 70 kilograms can be sufficiently nourished only temporarily, and not for any length of time, on less than 80 grams of proteid.

The minimum amount of proteid requisite for preserving the nutrition must be so large that the nitrogen it contains will be equal to the nitrogen excreted by the individual in question in a fasting condition.

Small animals consume for each unit of body-weight decidedly more than large ones. This depends not so much on the fact, as was formerly believed, that the metabolism is more active in small animals, as on the fact that small animals, in proportion to their body-weight, possess a larger body-surface, and are, therefore, more exposed relatively to external influences, and especially to the cooling effect of the surrounding air. If the amount of the substances decomposed is compared, not to the body-weight but to the body-surface, for example to one square meter, almost the same values will be obtained for small is for large animals of the same species. On the other hand, the values for animals of different species are different.

The absolute amount of food that an adult requires in twenty-four hours is most conveniently expressed in the form of units of energy that it is capable of supplying, that is, in calories. An adult, with a moderate amount of fat, requires daily, for each kilogram of body-weight:

At complete rest .... from 32,000 to 38,000 calories.
At light work................... "   35,000 " 45,000   "
At hard work................... "   50,000 " 70,000   "

Therefore, a man weighing 70 kilograms at light work would require in the neighborhood of 70 × 40,000, or 2,800,000 calories. Any diet containing 2,800,000 calories is sufficient; but the diet must always contain proteid—and in no event less than 80 grams daily. As 1 gram of proteid yields 4100 calories, 1 gram of carbohydrate 4100 calories and 1 gram of fat 9300 calories, the following dietetic combinations may be considered as sufficient:

Grams.		Calories.	Grams.		Calories.
80 of proteid.........	=	328,000	80 of proteid.........	=	328,000
300 of carbohydrate	=	1,230,000	200 of carbohydrate	=	820,000
113 of fat	=	1,237,000	177 of fat	=	1,646,000
		2,795,000			2,794,000
			80 of proteid.........	=	328,000
80 of proteid.........	=	328,000	400 of carbohydrate	=	1,640,000
265 of fat	=	2,465,000	89 of fat	=	828,000
		2,793,000			2,796,000
			For a short time also:		
100 of proteid.........	=	410,000	60 of proteid.........	=	246,000
280 of carbohydrate	=	1,148,000	320 of carbohydrate	=	1,212,000
133 of fat	=	1,237,000	133 of fat	=	1,237,000
		2,795,000			2,795,000

In most of the ordinary articles of food nitrogenous and non-nitrogenous substances occur together, but, as the statements on page 435 show, in widely different proportions. Man requires a diet in which the proportion between nitrogenous and non-nitrogenous substances is between 1 : 3½ and 1 : 4½. If a person takes food in which this proportion does not hold, he must consume an excessive amount of it, in order to obtain a sufficient quantity of that substance in which the article of diet is relatively deficient. This, it is clear, must necessarily cause waste of the preponderating substance. Moleschott has, in this connection, grouped the principal articles of diet together. In order to obtain the necessary 130 grams of proteid a laborer must consume the following amounts of various foods:

Cheese........388 gms.   Beef........ 614 gms.   Rice........ 2562 gms.
Lentils........491 "   Eggs........ 968 "   Rye-bread... 2875 "
Peas..........582 "   Wheat-bread..1444 "   Potatoes ....10000 "

It is quite evident that, in using the last-named substances, the laborer must consume a useless excess of non-nitrogenous food. In order to obtain from his food the necessary 448 grams of carbohydrate (or the equivalent amount of fat) required for his subsistence, such a laborer would have to eat:

Rice..........572 gms.   Peas..........819 gms.   Cheese ....2011 gms.
Wheat-bread...625 "   Eggs..........902 "   Potatoes .....2039 "
Lentils........806 "   Rye-bread.....930 "   Meat ....2261 "

Thus, particularly with the exclusive use of cheese or meat, the laborer would be compelled to consume enormous quantities, which would be equivalent to a waste of nitrogenous material.

Finally, attention should be drawn to the fact that not all of the food is digested or absorbed in the digestive tract, but that there is always a certain residue that is unutilized and is voided with the feces. Calculated as dry substance this amounts in percentages: in rice to 4.1, in white bread to 4.5, in meat to 5.2, in eggs to 5.2, in milk to 9, in potatoes to 9.4, in peas to 11.8, in beans to 18.3, in black bread to 15. It is more advantageous to administer the amount of food required daily in several portions than to give it at infrequent intervals or all at once; the distribution of the food over several meals diminishes proteid decomposition.

For the herbivora a diet suffices containing one part of nitrogenous to eight or nine parts of non-nitrogenous material.

METABOLISM IN THE STATE OF STARVATION.

If a warm-blooded animal is deprived of all food, it must, naturally, decompose and utilize the energy stored in its own tissues in order to generate its bodily heat and to perform any mechanical labor demanded of it. Its body-weight, accordingly, steadily decreases till death from starvation occurs, the tissues and organs meanwhile becoming richer in water.

Method.—For an exact investigation of the state of inanition (1) the starving man or animal is weighed daily. (2) All of the carbon and nitrogen in the expired air, the urine and the feces is estimated daily. The nitrogen found can be derived only from the consumed proteids of the body, especially the muscles, and from the same source also a varying amount of carbon, in accordance with the composition of the muscles. The amount of carbon remaining after subtracting this amount is to be attributed to the decomposition of the non-nitrogenous tissues of the body, principally the fat. After the amount of muscle and fat broken down has been thus computed, the subtraction of this amount from the total loss of body-weight will yield the amount of water lost.

The following example, which deals with a cat starved to death by Bidder and Schmidt, shows the various excretions on the successive days of starvation:

Day.	Body-Weight.	Amount of Water Taken.	Amount of Urine.	Urea.	Inorganic Constituents of the Urine.	Dry Feces.	Expired Carbon.	Water in Urine and Feces.
1	2464		98	7.9	1.3	1.2	13.9	91.4
2	2297	11.5	54	5.3	0.8	1.2	12.9	50.5
3	2210		45	4.2	0.7	1.1	13	42.9
4	2172	68.2	45	3.8	0.7	1.1	12.3	43
5	2129		55	4.7	0.7	1.7	11.9	54.1
6	2024		44	4.3	0.6	0.6	11.6	41.1
7	1946		40	3.8	0.5	0.7	11	37.5
8	1873		42	3.9	0.6	1.1	10.6	40
9	1782	15.2	42	4	0.5	1.7	10.6	41.4
10	1717		35	3.3	0.4	1.3	10.5	34
11	1695	4	32	2.9	0.5	1.1	10.2	30.9
12	1634	22.5	30	2.7	0.4	1.1	10.3	29.6
13	1570	7.1	40	3.4	0.5	0.4	10.1	36.6
14	1518	3	41	3.4	0.5	0.3	9.7	38
15	1434		41	2.9	0.4	0.3	9.4	38.4
16	1389		48	3	0.4	0.2	8.8	45.5
17	1335		28	1.6	0.2	0.3	7.8	26.6
18† ...	1267		13	0.7	0.1	0.3	6.1	12.9
	−1197	131.5	775	65.9	9.8	15.8	190.8	734.4

The cat before death had lost 1197 grams in weight. This loss may be distributed, according to what has been said, as follows: Proteid 204.43 grams, or 17.01 per cent.; fat 132.75 grams, or 11.05 per cent.; water 863.82 grams, or 71.91 per cent. of the total loss of weight.

With regard to the general phenomena of inanition it is worthy of remark that strong, well-nourished dogs die of starvation only after four weeks, while man succumbs in twenty-one or twenty-two days. Six persons suffering from melancholia, who had taken water, lived for forty-one days, however. In recent years voluntary exhibitions of starvation have become the fashion. The most striking of these was given by the Italian painter, Merlatti, who, it is alleged, withstood starvation for a period of fifty days with the use of water only. Succi, according to unexceptionable testimony, fasted for thirty days.

Under such conditions the regulation of temperature, the circulation, the respiration, the muscular and the nervous activity were found to be within limits of normal variation; the secretions necessary for digestion were, on the other hand, almost abolished.

Small mammals and birds succumb within nine days, but frogs only after nine months. Full-grown, vigorous mammals, on the contrary, have lost as much as $\frac{1}{10}$ of their weight (from $\frac{1}{4}$ to $\frac{1}{2}$) before death. In man the decrease in weight is relatively greatest during the first few days. Young individuals die much earlier than adults. To outward appearance the emaciation is striking. The mouth is dry, the walls of the alimentary tract become remarkably thin, the digestive juices are no longer secreted, the action of the heart is enfeebled, the pulse, smaller and of lower tension, is less frequent, the respirations are increased in frequency and more superficial, the urine is highly acid on account of increase in sulphuric and phosphoric acids, and its chlorin-compounds soon disappear almost entirely. The blood is poorer in water, the plasma in albumin; the gall-bladder is greatly distended, a fact that points to uninterrupted destruction of red blood-cells in the liver. The liver is small and extremely dark. The muscles tire readily. Finally, great weakness of the wasted and friable muscles develops and death follows amid signs of the greatest prostration and coma.

The conditions of metabolism are apparent from the foregoing table, according to which the decrease in the excretion of urea is much greater than that of carbon dioxid. From this it may be concluded that a correspondingly greater breaking-down of fats than of proteids takes place. According to the calculations a tolerably constant amount of fat is broken down daily, while the proteids undergo much slighter destruction with the progress of the days of fasting. Drinking of water hastens the destruction of the proteids.

In the case of the fasting virtuoso, Cetti, Zuntz and Lehmann found that the consumption of oxygen and the production of carbon dioxid, as calculated for the unit of body-weight, rapidly reach minimal values, below which they did not fall with continued starvation. On the average, the consumption of oxygen from the third to the sixth day of fasting amounted to 4.65 cu. cm. for each kilogram of body-weight and for each minute. Absolutely, as regards the individual, the respiratory interchange decreased slowly, but this decrease failed to keep pace with the decrease in the weight of the body. At the beginning of starvation the amount of carbon dioxid diminished more than the consumption of oxygen. The respiratory quotient was 0.67. The urea from the first to the tenth day of starvation decreased from 29 to 20 grams.

In the case of another faster, Succi, Luciani found that a nitrogenous excretion of 16.23 grams had decreased on the first day of fasting to 13.8 grams, on the seventeenth day to 7.8 grams, on the twenty-second to 4.75 grams, on the twenty-eighth day to 5.6 grams. Also Johannson, Landgren, Sondén and Tigerstedt found that metabolic activity at first declined quickly and to a large degree, later slowly and slightly.

A consideration of the relative loss of weight of the various organs is also of great interest, as shown by comparison with a similar animal killed without preliminary starvation. It should be stated, however, in this connection, that many organs lose weight proportionately, for example, the bones (and as a result phosphoric acid, calcium, and magnesium increase in the urine), while other parts exhibit a disproportionately marked decomposition, for example the fat. The latter are broken down with especial rapidity and from them other organs are in part nourished during starvation. Finally, certain organs, like the heart and the nerves, suffer slight loss, as they are able to maintain themselves on the decomposition-products of other tissues. In the breaking down of the tissues the nuclei also suffer and certain glands undergo fatty degeneration.

A starved male cat lost, according to v. Voit:

		Percentage of the Amount Originally Present.	Percentage of the Total Loss of the Body.
1.	Fat	97	26.2
2.	Spleen	66.7	0.6
3.	Liver	53.7	4.8
4.	Testicles	40.0	0.1
5.	Muscles	30.5	42.2
6.	Blood	27.0	3.7
7.	Kidneys	25.9	0.6
8.	Skin	20.6	8.8
9.	Intestines	18.0	2.0
10.	Lungs	17.7	0.3
11.	Pancreas	17.0	0.1
12.	Bones	13.9	5.4
13.	Central nervous system	3.2	0.1
14.	Heart	2.6	0.02
15.	Remaining portions of the body together	36.8	5.0

The average resistance of the hemoglobin is increased by inanition.

Allusion should be made also to an important difference between animals that have been liberally fed with meat or fat before the beginning of the period of inanition, and those that have been kept on a barely sufficient diet. Liberally fed animals suffer much greater loss in weight in the early days of starvation than in the later. Furthermore, fat individuals exhibit from the first a greater decomposition of fat in proportion to proteids than thinner individuals. Animals liberally fed with proteid continue to decompose much albumin in the early days of fasting. Animals fed with little proteid, on changing to a liberal albuminous diet, likewise continue to decompose only a limited quantity of albumin in the first few days.

METABOLISM WITH AN EXCLUSIVE DIET OF MEAT, ALBUMIN OR GELATIN.

According to Pflüger the higher animal (as has been demonstrated experimentally for the dog) can be nourished and maintained almost exclusively on *proteids*, without impairment of its functional activity. Proteids are, therefore, to be designated as foods of the first order, as fundamental foods. Pflüger applies the term *nutritive requirement* to the smallest amount of lean meat that is capable of maintaining the metabolic equilibrium, without fat or carbohydrate of the body being utilized for decomposition. The amount of this nutritive requirement is determined by the weight of the flesh of the animal and increases with this on addition of flesh to the body. The decomposition of proteids increases also with the supply of proteid if the latter exceeds the nutritive requirement. Under such circumstances, however, a certain portion of the excess of proteid is conserved and deposited as flesh. The nutritive requirement of the dog is for 1 gram of animal nitrogen 0.0636 gram of nutrient nitrogen; or 1 kilogram of nitrogenous animal tissue requires 2.099 grams of nitrogen in the food.

Human beings provided exclusively with meat free from fat are, however, not able to maintain the metabolic equilibrium. Compelled to adhere to such a diet permanently they would certainly succumb. The reason for this is obvious. In beef the proportion of nitrogenous to non-nitrogenous elementary nutritive constituents is as 1 to 1.7. The healthy person gives off daily in the carbon dioxid of the respiration, in the feces and in the urine, about 280 grams of carbon. If he desired to obtain these 280 grams of carbon from the carbon of an exclusively meat-diet, he would be compelled to digest and assimilate more than 2 kilos of pure meat in twenty-four hours. His organs, however, would by no means suffice to accomplish this permanently. The man would under such conditions soon be compelled to consume less meat. This result would require the decomposition of the constituents of his own body, first of all the fat and then also the proteids.

Also in the following manner it can be clearly shown that a human being is unable to maintain himself on a meat-diet exclusively. A man weighing 70 kilograms and doing a moderate amount of work requires 40,000 calories daily for each kilo of body-weight, therefore a total of 2,800,000 calories. One thousand grams of lean beef yield 95,000 calories. Such a person would, therefore, be compelled to consume about 3 kilograms of beef, or 2,850,000 calories, daily, and this, naturally, is impossible.

The carnivora (dog), whose digestive organs are especially adapted to the digestion of meat, by reason of the short intestine and actively solvent influence of the digestive fluids upon proteids, cannot be maintained permanently on chemically pure albumin, although this is possible with the leanest meat, which, however, always contains not less than 0.59 per cent. of fat. Under such circumstances the animal consumes large amounts of meat, and as a result the elimination of urea is increased correspondingly. If it eat still larger amounts, it may even put on flesh, and then, in accordance with the maintenance of the newly deposited flesh, it requires naturally a constantly increasing amount of meat.

The herbivora are under no circumstances capable of subsisting upon a meat-diet exclusively, as their digestive apparatus, which is adapted for vegetable food, would by no means suffice for the disposal of the necessary amounts of meat.

Of *gelatin* it has been shown that it may replace the proteids in the food, in so far as these serve as sources of energy and heat, but not if bodily tissue is to be replaced. Under such circumstances two parts of gelatin take the place of one part of proteid. The carnivora, which can maintain their metabolic equilibrium with large amounts of meat, are capable of doing this with less meat and a corresponding addition of gelatin. According to Munk the dog is capable for a few days of replacing $\frac{3}{6}$ of its proteid requirement by gelatin. A diet of gelatin exclusively is, however, inadequate. In addition the animals soon lose their appetite for such food.

In consequence of its solubility the addition of gelatin (calf's-foot jelly) to the food of convalescents has been recommended. The absorbed products of the digestion of gelatin are conveyed to the connective tissues, which constitute a repository for it. After a long-continued diet of chondrin, together with meat, glucose has been found in the urine.

AN EXCLUSIVE DIET OF FATS OR CARBOHYDRATES.

If *fat* alone is supplied, the body is unable to maintain itself. In consequence of the deficiency of nitrogen, the animal must necessarily perish. The symptoms occurring with this form of diet are as follows: The animal in question secretes less urea than in a state of hunger. Therefore, the consumption of fat must restrict that of the flesh of the animal itself. This is due to the fact that the fat, being a readily combustible substance, is more readily oxidized in the body than the less readily combustible nitrogenous albuminates. If the amount of fat taken is exceedingly large, not all of the carbon of the fat can be recovered in the excreta, or as carbon dioxid in the expired air. Accordingly the body must accumulate fat, while naturally it destroys proteids in corresponding amount. The animal thus becomes fatter and at the same time poorer in flesh.

The result of administration of *carbohydrates* alone, which must first be converted into sugar by the digestive processes, exhibits marked similarity to that obtained with a pure fat-diet. It should, however, be noted that the sugar in the body more readily undergoes destruction than the fat, and, further, that with reference to the nutritive value, 256 parts of glucose are the equivalent of 100 parts of fat. Accordingly, a carbohydrate-diet restricts the decomposition of proteids even more readily than a pure fat-diet.

Just as it is necessary outside of the body for the fermentation of disaccharids and polysaccharids that these be first decomposed into monosaccharids, so also the combustion of sugar in the body can occur only on condition that a transformation into monosaccharids has previously taken place.

LAWS GOVERNING METABOLISM ON A MIXED DIET OF MEAT AND FAT OR CARBOHYDRATES.

If a dog in a state of metabolic equilibrium be given an amount of fat and starch exceeding its requirements, elimination through metabolism is not increased, but the excess of these non-nitrogenous foods administered is deposited in the body of the animal as fat.

If a dog fed with the leanest possible meat, and in a state of metabolic equilibrium, be given an additional amount of meat exceeding its requirements, the elimination through metabolism increases almost proportionately to the additional amount administered beyond the requirements. Only a small portion of the addition is conserved and increases the body-weight as a deposition of flesh. This augmentation of metabolism not only causes an increase in the nitrogenous excretion in general proportional to the supply of proteid, but also the carbon contained in the supply of proteid is again excreted, for of the proteid fed no portion is deposited in the body as fat or carbohydrate. From both of these statements it follows that neither fat nor carbohydrate is capable of increasing metabolism beyond the requirements, although proteid is.

The seat of active proteid metabolism after a diet rich in proteids is, according to Pflüger, not in the increased flow of fluid, but within the proteid-containing cells which have undergone a marked alteration (saturation) as a result of the entrance of the proteid into them. This view is confirmed by the experiments of Schöndorff, who found that if the blood of a fasting animal be forced through the tissues of a generously nourished animal the urea in the blood of the latter increases, while, on the contrary, if the blood of a well-nourished animal be forced through the tissues of a fasting animal the urea in the blood of the latter diminishes. As, on providing an adequate amount of proteid, muscular activity takes place only at the expense of proteid, and as in the decomposition of this proteid neither fat nor carbohydrate results, fat or carbohydrate cannot be the source of muscular activity (Pflüger). Other investigators are of the opinion, however, that with adequate nitrogenous nourishment energy as well as heat can be generated from fat and carbohydrate.

Nourishment with Carbohydrates and Meat.—The organism is capable of generating fat from carbohydrates. A deposition of fat in the body thus brought about takes place only if in addition to the proteid of the meat a nutritive excess of carbohydrates is present. Such an excess of starch may be present even when the supply of starch itself is small, while the excess may even be wanting when the supply of starch is large. The result depends upon the character of the food that is supplied in addition to the starch. The larger the amount of proteid, in addition to the starch, contained in the food, the more readily is an excess of starch to be attained without the necessity of supplying too much starch. If this condition of such an excess is not fulfilled, fat does not result even with generous administration of carbohydrates. The newly formed fat possesses the same potential energy as the nutritive excess resulting from the carbohydrates administered.

Deposition of fat in the body does not take place, however large the excess of proteid food, if carbohydrates or fat be not supplied at the same time. On feeding with meat and starch, or in general with a mixed diet, the amount of newly formed fat depends in no wise upon the amount of proteid decomposed, but only upon the amount of nutritive excess due to carbohydrates. Deposition of fat from carbohydrates takes place even when no proteid at all is supplied and the metabolism, therefore, must be maintained in part at the expense of a portion of the body-proteid.

While for the maintenance of the metabolic equilibrium on a pure meat-diet an enormous consumption (from $\frac{1}{15}$ to $\frac{1}{10}$ of the body-weight in the dog) is required, a third of the amount of meat suffices with an adequate addition of fat or carbohydrate. For 100 parts of fat, added to the meat, 245 parts of dry meat or 227 of syntonin can be conserved. If carbohydrates are selected instead of additional fat, 100 parts of fat correspond to from 230 to 250 parts of carbohydrate. It should, however, be borne in mind that, at least for a short time, the carbohydrates are superior to fat as a proteid-sparer, as the fat is less completely utilized in the process of metabolism than the carbohydrates.

It appears that, instead of fat, a corresponding amount of fatty acids has the same effect in the process of metabolism.

Glycerin is not capable of lessening the destruction of bodily proteid, although recently I. Munk has stated that moderate amounts of glycerin introduced into the circulation are consumed in the body and through their oxidation protect a portion of the bodily fat against oxidation. According to Lebedeff, v. Voit and Arnschink, glycerin, however, diminishes the decomposition of bodily fat and is therefore a food-material.

ORIGIN OF THE FAT IN THE BODY.

A portion of the bodily fat is derived directly from the food, being simply deposited in the tissues after absorption. In favor of this view is the observation that with a scanty proteid diet a generous addition of meat causes the deposition of large amounts of fat in the body. The administration of fatty acids alone may also contribute to the formation of fat, inasmuch as glycerin, formed by the body, must combine with them in the process of metabolism.

As a result of fattening experiments with different warm-blooded animals (pig, goose, dog), in which, in addition to a large excess of starch, only a small amount of fat and proteid is supplied, the conclusion has been reached that a direct transformation of the absorbed carbohydrates, rich in oxygen, into fatty tissue, poor in oxygen, takes place. Pflüger found that the sugar-molecule of the food, given in excess of the requirements for the development of fat in the animal, is in part oxidized and in part reduced, so that, on the one hand, carbon dioxid, and, on the other hand, the group of atoms concerned in the formation of fat, result, inasmuch as the molecular groups CH OH are reduced to CH_3. The carbon dioxid that is exhaled when fat-formation takes place in consequence of the administration of starch is thus derived from two sources, namely, in part from the process of decomposition described, and in part from the total combustion of starch. The excessive elimination of carbon dioxid in this process of fat-formation in consequence of an excessive starchy diet must naturally cause an increase in the respiratory quotient, even above 1.2.

If the carbohydrates be considered as decomposing into fat, carbon dioxid and water, 100 grams of starch or 111.1 grams of sugar will yield at most 41.1 grams of fat, 47.5 grams of carbon dioxid and 11.4 grams of water. Also the circumstance that bees utilize the sugar of honey in the formation of wax is in favor of the production of fat from carbohydrates. According to Pasteur and E. Voigt, glycerin can be formed from carbohydrates.

Does fat result from proteid metabolism? v. Pettenkofer and v. Voit reached the conclusion, as a result of their experiments, that fat can be formed in the animal body from proteids. They fed a dog with large amounts of meat, and although all of the nitrogen thereof was excreted in the urine and the feces, a portion of the carbon of the meat could not be recovered from the excreta. They concluded, therefore, that this carbon had been transformed into fat for accumulation in the body.

This statement is contradicted by Pflüger on the basis of his own investigations, which lead him to the conclusion that the doctrine of the development of fat from proteids in the bodies of animals is entirely groundless. If it were assumed that fat could be formed from proteid, such formation is not possible through simple decomposition of the proteid molecule, but rather it would be necessary for decomposition first to take place and then synthesis of the decomposed parts.

Earlier investigators, who accepted the formation of fat from proteids, believed that the proteids administered broke up into a non-nitrogenous and a nitrogenous atom-complex. The former, in case it did not leave the body completely decomposed into carbon dioxid and water when a rich proteid diet was taken, was believed to furnish the material for the formation of fat, while the latter was supposed to leave the body oxidized principally into urea.

The following experiments support the view that fat can develop from proteid furnished as food: (1) Ssubotin and Kemmerich fed nursing bitches with meat almost free from fat, and found that the greater the amount of meat eaten, the greater was the amount of milk produced and thus also of fat. In these experiments, however, the possibility is not excluded that the bitches utilized the fat of their own bodies in the preparation of the milk. (2) Radziejewski gave a lean dog meat almost free from fat and in addition pure rape-oil, one of whose constituents, erucic acid, does not occur normally in the animal body. When the animal, after a period of feeding of considerable length, had accumulated fat, chemical examination demonstrated that the tissues contained, in addition to erucin, also fat which otherwise is normally present in the dog. In an analogous manner Lebedeff found in a dog after feeding with lean meat and linseed-oil considerable amounts of linoleic acid, together with normal dog's fat. In both experiments, however, the normal dog's fat could have been derived from the fat of the meat fed. (3) The fat found within organs in a state of pathological fatty degeneration had previously often been considered as derived from the proteid protoplasm of the tissues. Even though it be admitted, says Pflüger, that the fat of the degenerated organs has developed within them, and has not gained entrance from without, it would still first be necessary to believe that the cells everywhere contain carbohydrates or their derivatives, which it is known with certainty can be transformed into fat by synthetic processes. Also the fatty degeneration produced in the animal body by phosphorus-poisoning affords no support for the view that fat is developed from proteid, for although a small amount of fat is found in the body after such poisoning, its development from proteid has not yet been demonstrated. In the case of fatty degeneration, there is primarily an injury of proteid bodies and in place of these fat from other sources appears in the cells in a certain measure as a reparative procedure. (4) Nägeli showed that lower forms of fungi, like other plants, are able to form proteid, fat and carbohydrates synthetically from various matters, in part exceedingly simple. Thus, for example, fungi generate fat synthetically in ripening cheese probably from the products of decomposed proteid. In the decomposition of entire cadavers and their transformation into a mass consisting almost wholly of palmitic and stearic acids (adipocere) in the presence of fungi, it cannot be concluded that a simple transformation of albumin into these fats takes place.

DEPOSITION OF FAT AND FLESH IN THE BODY (HYPERNUTRITION).

CORPULENCE AND THE MEANS FOR ITS CORRECTION.

Hypernutrition results if more food is supplied than the body is capable of decomposing and again eliminating. The digestive apparatus (collectively and in common activity) is probably capable of digesting twice as much as the requirements demand. The absorbed excess of food that is not decomposed is accumulated and forms the superfluous tissue. Higher animals are capable, although not in the strict sense, of surviving on an almost exclusively proteid diet. Pflüger was able to keep a dog engaged in hard work alive for an indefinite time on a diet exclusively of meat and almost free from fat. All of the vital phenomena, therefore, can be carried on by means of proteid alone. Albumin may, accordingly, wholly replace fat in the process of metabolism. The smallest amount of lean meat that thus maintains the metabolic equilibrium is designated by Pflüger as the *nutritive requirement*. The supply of fat or carbohydrates exclusively is never capable of maintaining life, as the animal under such circumstances is compelled to consume its own flesh. Therefore, a certain indispensable amount of proteid must absolutely be present in every diet.

If an amount of proteid be added to the food that is sufficient in itself to fulfil the requirement and if any desired amount of fat is added, almost all of the proteid will be decomposed and almost all of the fat will be deposited as such. The conditions are much the same if carbohydrate is supplied instead of fat, except that in this case the carbohydrate is transformed in the body into fat and is deposited as such. The greater the amount of non-nitrogenous food that is supplied in addition to the nutritive requirement of proteid the more favorable are the conditions for fattening, because all of the non-nitrogenous matters are transformed into bodily fat.

If proteid is not supplied in sufficient amount the deficiency may be made good by fat or carbohydrate, and in such proportion that two-thirds of the nutritive requirement may be supplied by non-nitrogenous matters. Under such circumstances the latter replace the deficiency of proteid in accordance with the amount of their potential energy as indicated by the number of calories yielded in their combustion. From these facts it follows that the greater or smaller amount of albumin supplied with such food is decomposed almost wholly in the process of metabolism, indifferently whether much or little fat or carbohydrate is supplied at the same time. In direct contrast to the proteid, the amount of fat or carbohydrate that is consumed in the process of metabolism is in nowise dependent upon the amount thereof contained in the food. Generally, the amount of carbohydrate or fat that undergoes decomposition is the smaller the larger the amount of proteid supplied. The nutritive requirement is satisfied first and foremost by proteid, but if the amount of proteid supplied is not sufficient, the fats and the carbohydrates are also utilized in so far as the requirements demand. In order to comprehend the laws of fattening by means of proteid and starch, it should be borne in mind that for the satisfaction of the nutritive requirement, in addition to almost the entire amount of proteid supplied, so much carbohydrate is decomposed as will wholly suffice for the nutritive requirement. The amount of carbohydrate left over is deposited as fat. In accordance with the foregoing statements, on supplying equal amounts of carbohydrate a proportionately larger amount will be conserved the larger the amount of proteid furnished.

The amount of nutritive requirement, that is, the smallest amount of fat-free meat that alone establishes metabolic equilibrium, is governed by the flesh-weight of the animal and increases in direct proportion to this. A fat animal has, therefore, apparently a smaller nutritive requirement only because the total amount of fat, acting as a similar amount of dead matter, consumes nothing.

The decomposition in the process of metabolism of the proteid taken with the food increases with the supply, even when this far exceeds the requirement, but a portion of the excess is always conserved. In this manner there is a gradual deposition of flesh in the body.

As the amount of proteid supplied with the food has practically no influence upon the deposition of fat in the body, and the carbohydrates are generally not so useful as proteid, fat will be produced most advantageously with the smallest amount of proteid possible, but with the largest possible amount of starch in the food. If an animal on a mixed diet in a moderate state of fattening be given a further supply of proteid, this will at once satisfy a portion of the nutritive requirement, which theretofore had been satisfied by non-nitrogenous matters. These therefore can be dispensed with and are deposited as fat.

With a diet of meat exclusively deposition of flesh is possible only when the proteid of the food exceeds the requirement. The largest portion of the excess of proteid is decomposed and some is deposited. With increase in the weight of flesh, the consumption of proteid soon increases, and, accordingly, the amount of excess diminishes. It is, therefore, one of the properties of proteid food that it tends speedily to neutralize the conditions necessary for the deposition of flesh if these are present.

With a mixed diet deposition of flesh can be attained only if the supply of proteid exceeds the amount indispensable. Under such circumstances only from 7 to 9 per cent. on the average, at most 16 per cent., of the proteid supplied, is conserved by the non-nitrogenous articles of food. The deposition of flesh is then the greater the larger the amount of proteid contained in the food. Of the proteid consumed the body can deposit only one part of proteid, while nine parts are decomposed. In addition, for two parts of decomposing proteid one part of fat is formed from the carbohydrate supplied in excess.

Excessive deposition in the body of man, *corpulence*, is to be considered an abnormal manifestation of metabolism, which to the subject may be a source not alone of inconvenience, but also of disorders or even of serious danger. With reference to the causes of obesity, a certain degree of congenital predisposition (in from 33 to 56 per cent. of the cases) cannot be denied, inasmuch as members of certain families increase more readily in weight (as is likewise true of certain breeds of animals), while others, even when supplied with an abundance of food, which may reach enormous amounts, remain thin. The principal cause, however, is an habitually excessive supply of food beyond the normal metabolic average, although almost every corpulent person will with complacent self-deception maintain that he really eats remarkably little.

The mistake should be avoided of considering the corpulent individual as always excessively fat. The process of overfeeding results at first in the deposition both of fat and of flesh. On continuance of the process the development of muscular tissue diminishes, because in consequence of his clumsiness and helplessness the corpulent individual is rendered inactive. As a result, the nutrition of the muscular structures is secondarily impaired. Some active corpulent individuals, however, retain their large deposition of flesh throughout life. If, however, those factors become especially operative later on that favor the production of fat, corpulence may be transformed into obesity exclusively, as, naturally, is frequently the case.

The following influences favor the development of corpulence: (1) An excessive diet of proteid, with a corresponding addition of fat or carbohydrate. The proteid of the food serves for the deposition of albuminates in the body, while the fat is produced by the ingestion of fat and carbohydrates. (2) Diminished consumption of nitrogen in the body, in consequence of (a) lessened muscular activity (little movement, much sleep). (b) Enfeeblement of the sexual functions, as shown by the fattening of castrated animals, as well as the circumstance that women readily become corpulent after cessation of menstruation, probably in consequence principally of withdrawal of the stimulating influence of vascular activity. (c) Diminished mental activity (obesity of idiocy), phlegmatic temperament, probably for the foregoing reason. Conversely, vigorous mental activity, an excitable temperament, anxiety and grief counteract the fattening process. (d) The corpulent individual need consume relatively less material for the generation of heat in his body, partly because his compact body, in consequence of the greater concentration of mass, gives off less heat from the external integument than a delicate slender body, and partly because of the thick layer of fat as a poor conductor of heat prevents direct loss of heat by conduction. As a result of the relatively lessened production of heat in the body thus required, there may be an increased deposition of tissue. (e) A reduction in the number of red blood-corpuscles, which stimulate oxidation-processes in the body, is generally followed by an increase in the amount of fat. Corpulent persons are, therefore, not rarely fat because they are anemic. Women with a reduced number of red blood-corpuscles are generally fatter than men. (f) The use of alcohol favors the conservation of fat in the body, because, on account of the readiness with which it is oxidized, it protects the fat in the body from combustion (the obesity of drunkards).

In addition to the great inconvenience due to the weight of the body, corpulence, and particularly obesity, is attended with certain disadvantages and dangers. Among these are dyspnea, readiness of fatigue, the development of intertrigo in the folds of the skin and of so-called fat-hernia, and finally the danger of fatty degeneration, of cardiac paralysis and of apoplexy.

For the *correction of obesity* the following measures should be adopted: (1) Uniform reduction of all of the articles of food to the proportions of the normal diet. The obese patient should weigh himself and his daily amount of food from week to week. So long as he observes no reduction in body-weight, the amount of food (in spite of the appetite) should be gradually and uniformly reduced. This course should be pursued slowly, without unduly sudden limitation. Almost all good resolutions fail in the face of the excellent appetite. A moderate reduction of the fat and the carbohydrates in the normal diet would at the same time result in consumption of the fat of the body itself. Such individuals as are still capable of muscular activity may be permitted 156 grams of proteid, 43 grams of fat, 114 grams of carbohydrates. Those in whom hypostasis, hydremia, and respiratory difficulty have developed may be permitted 170 grams of proteid, 125 grams of fat and 170 grams of carbohydrates.   It is, however, not advisable to restrict a corpulent person excessively as to fats and carbohydrates alone, as is customary in the so-called cure of Banting.   Such a violent modification of the normal diet is often attended with profound derangement of the entire metabolism.   Many persons have suffered greatly in health as a result of this procedure.   Every long-continued limitation of diet in one direction is deleterious and will accordingly result in emaciation, but not without danger, for it has a disturbing influence upon the entire metabolism and thus in a given sense is pathological.   (2) It is advisable during the principal meals to avoid as much as possible the use of fluids of all kind (until about three-quarters of an hour later), because by this means the absorption and the digestive activity in the intestine are rendered less effective. (3) Muscular activity should be increased by vigorous work, and also mental activity should be encouraged.   (4) Heat-dissipation should be favored by cold baths of long duration, followed by vigorous friction of the skin to the point of bright redness.   At the same time the clothing should be light.   The patient should sleep in a cool room and for not too long a time.   In this manner the increased ingestion of tea and coffee also is useful, actively stimulating the cutaneous circulation and thereby the dissipation of heat.   (5) Mild laxatives, such as acid fruits, cider, alkaline carbonates (Marienbad, Carlsbad, Vichy, Neuenahr, Ems, etc.), have a favorable influence in the correction of obesity by increasing the evacuations from the intestines and diminishing absorption.   (6) If, in the presence of marked deposition of fat, there is already danger of enfeeblement of the action of the heart an attempt should be made, with caution, by means of increased muscular activity (mountain-climbing and the like), to stimulate the heart and to strengthen its musculature.   By this means the circulation is improved and metabolism becomes more active, so that recovery may even yet be brought about with the aid of a sensible diet.

Entirely different from the process of fattening, which consists in the deposition of large droplets of fat in the fat-cells of the panniculus and about the viscera, as well as in the bone-marrow (but never in the subcutaneous connective tissue of the eyelids, the penis, the ears, the red margin of the lips, the nose), is the condition of *fatty atrophy* or *fatty degeneration*, which occurs in the form of fatty granules in the albuminous tissues, for example, in muscle-fibers (of the heart), glandular cells (liver, kidneys), cartilage-cells, lymph-corpuscles and pus-corpuscles, as well as in divided nerves.   If this process increases in the tissues to such a degree that the albumin is as a result made to disappear without being again restored, the fatty atrophy or degeneration is marked.   It is observed after severe fevers, marked (artificial) heating of the tissues, diminished absorption of oxygen into the body (as has been observed especially after phosphorus-poisoning), also in drunkards, after certain forms of intoxication (arsenic), and in connection with disorders of circulation and innervation.   Finally, some organs exhibit fatty degeneration in connection with special diseases.   In rare cases in the new-born the entire body may rapidly undergo fatty atrophy.

THE METABOLISM OF THE TISSUES.

All tissues require for their normal existence and for their functional activity the process of metabolism.   The chief medium for this is the blood-current, which, acting as the principal traffic-carrier in the metabolic process, conveys the material for the restoration of the tissues and removes the products of their vital activity.   Those tissues that, like the cornea and cartilage, possess no vessels in their structure, must receive the nutritive plasmatic fluid from the adjacent capillaries through their cellular elements, which thus act as channels for the conveyance of the fluid.   Therefore, interference with the normal circulation in the tissues, as, for example, through constriction or calcification of the walls of the vessels and the like, is attended with derangement of nutrition; complete occlusion, as, for example, by thrombosis, total compression, or artificially by ligature of all the afferent vessels, is followed by certain destruction of the tissues, which soon appears in the form of gangrene (necrosis).

Atrophy resulting from reduction in the normal supply of blood gradually disappears in the further course of time.

In accordance with what has been stated a double current can be recognized in the fluids of the tissues, the afferent current, which brings the materials for the restoration of the tissues, and the efferent current, which removes the effete products of metabolic activity. The former will convey the albuminates, fats, carbohydrates, as well as the salts in solution, as they are taken up by the organs of absorption, for the formation of the tissues. It is clear that obstruction of any sort in the arterial system of the tissue in question will diminish this supply. The metabolism is as a result restricted, in consequence of deficient formative activity.

This current can be recognized from the circumstance that after injection of a relatively indifferent, readily demonstrable substance, for instance potassium ferrocyanid, into the blood, that substance will be found in the blood within the tissues, whither it has been conveyed with the afferent current.

The efferent current removes the products of metabolism, particularly urea, carbon dioxid, water and salts, in order to convey these with the utmost rapidity to the excretory organs.

This current can be recognized from the circumstance that if a soluble substance be introduced into the tissues themselves, as with a syringe for subcutaneous injection, for example potassium ferrocyanid, this will be found in the urine in the course of a few (from two to five) minutes.

If the efferent current from the tissues is so strong and so large that the excretory organs are unable to eliminate the waste matters from it, these may again wander through the tissues. Such a condition is observed after subcutaneous injection of considerable doses of poisonous substances, which often enter the blood in such large amount that, before they can be eliminated, they are conveyed to other tissues, for example the nervous system, upon which they exert their effects before any considerable degree of elimination has taken place. If large amounts of foreign substances are injected they may even be temporarily deposited partly in other tissues, particularly in the liver and the bone-marrow. As the afferent current traverses two canal-systems, the veins and the lymphatics, it is clear that obstruction of these paths will disturb the metabolism as a result of interference with the normal removal of effete matters. On tight constriction of a peripheral portion of the body, in consequence of which veins and lymphatics are compressed, stagnation of the current takes place to so marked a degree that even swelling of the tissues may result.

In the propagation of the currents in the tissues the activity of the muscles is of great importance, inasmuch as not only do they favor the movement of the fluid in the vessels by pressure within the yielding tissues, but also where they are attached to the periosteum, the perichondrium and the joints they cause changes in the form of the interstices and thereby influence the movement of the fluid within the latter by alternate contraction and relaxation.

H. Nasse found the specific gravity of the blood in the jugular vein 0.225 in a thousand higher than that of the blood in the carotid artery, and containing 0.9 part more by weight in 1000 of solids. One thousand cu. cm. of blood yield in circulating through the head more than 5 cu. cm. of transudate to the tissues.

The activity of metabolism in the tissues and at the same time the intensity in the varying currents depends upon diverse factors:

1. Upon the *activity of the tissues themselves*. The increased activity of an organ can be recognized from the larger amount of blood contained in it and the increased activity of the circulation, which in turn are the media for the metabolism. If an organ is subjected to complete inactivity, for example a paralyzed muscle or the peripheral extremity of a divided nerve, the amount of blood and its interchange soon diminish. The organism sends its fluids only to active tissues. The affected part becomes pale and flaccid and finally undergoes fatty degeneration. For some organs increased metabolism in association with their activity has been demonstrated, for example the muscles. Langley and Sewall have been able to observe microscopically the metabolism in thin lobules of glands during life. The cells both of the serous and of the mucous and peptic glands become filled in the state of rest with coarse granules, dark in transmitted and white in reflected light, which are consumed during the period of activity. During sleep, in which most of the organs are at rest, metabolism is restricted. It is likewise diminished by darkness, while it is stimulated by light (obviously through nervous influences). The variations in total metabolism will be reflected in the elimination of carbon dioxid and urea, which in conformity with the activity of the organism yields a curve that is fairly parallel with that for the daily variations in respiration, pulse and temperature.

2. Also the *state of the blood* has a distinct influence upon the currents in the tissues on which the metabolism depends. A highly concentrated blood deficient in water (such as is observed after profuse sweating, copious diarrhea, for example in cases of cholera) renders the tissues dry; while, conversely, the taking up into the blood of large amounts of water renders the tissue more succulent, even to the point of dropsy. The presence of a considerable amount of sodium chlorid in the blood and a reduction in the amount of oxygen in the red blood-corpuscles, the latter in association with muscular exertion causing dyspnea, give rise to increased disintegration of albuminates and thus to increased production of urea. Therefore, exposure to rarefied air causes increased elimination of urea. Certain abnormal changes in the blood are noteworthy: Thus, carbon-monoxid blood is not capable of abstracting oxygen from the air and conveying carbon dioxid from the tissues. The presence of hydrocyanic acid in the blood immediately interrupts the chemical oxidation-processes carried on through the blood; the tissues no longer remove oxygen from the bright-red blood overladen with oxygen, and there thus results asphyxia from interference with the internal respiration. Fermentative processes also are interfered with in the same way by hydrocyanic acid. A reduction in the total volume of blood causes, on the one hand, the passage of a larger amount of water from the tissues into the vessels, while, on the other hand, it retards the absorption of substances from the tissues (for example, poisons or pathological exudates) or from the surface of the intestine. If the substances derived from the tissues are rapidly eliminated from the blood, or transformed therein, subsequent absorption takes place the more rapidly.

3. The *blood-pressure* has an influence upon the fluid-current, inasmuch as marked increase of pressure renders the tissues richer in fluid, but the blood itself more concentrated (up to from 3 to 5 in 1000). That pressure upon the afferent vessels causes the escape of blood-plasma through the walls of the capillaries can be demonstrated on a surface of corium denuded of its epidermis, as, for example, in a blister. Reduction of the blood-pressure will have the opposite effect. After administration of phosphorus, copper, ether, chloroform, and chloral, the oxidation-activity in the animal body is diminished.

4. *Elevation of the temperature of the tissues* (for several hours during the day) does not cause increased destruction of proteid and fat. This subject is discussed also on pp. 404, 406 and 409.

5. An *influence of the nervous system* upon the tissue-metabolism has also been observed. Doubtless this influence is a double one. In the first place it may be exerted indirectly through the intermediation of the vessels, the vascular nerves causing contraction or dilatation of the vessels, and thus increasing or diminishing the amount of blood passing through the vessels. In this connection attention should be called especially to pathological conditions, abnormal stimulation or paralysis of the vascular nerves or their centers. Independently of the vessels, however, certain special nerves that have been designated trophic control the metabolism in the tissues. Atrophy caused by nerve-paralysis increases the longer it persists. Examples of metabolism in tissues excited directly through the nerves are the secretion of saliva on nerve-irritation after exclusion of the circulation

and metabolism on contraction of bloodless muscles. Increased respiration and apnea are not followed by increased oxidation.

REGENERATION.

The power of regenerating parts that have been lost varies widely in different organs and tissues. It is much more marked in the lower animals than in warm-blooded animals. Division of the fresh-water polyp (hydra) is followed by the development of two new individuals. An entire being may even develop from every excised portion of the trunk of the body; only exceedingly small pieces give rise to incomplete reproduction. No animal regenerates portions of the arm. Also the planaria exhibit similar powers of regeneration. From every portion of the umbrella of certain medusæ (thaumantiades), if it contain only a portion of the margin, a new medusa may develop. From the surface of a piece of the trunk of a turbellaria directed downward a pedal extremity develops, from the upper surface a cephalic extremity, and if attached horizontally heads develop at both extremities. Artificial division is possible also in rhizopods and infusoria. Divided infusoria regenerate only if the divided portion contains a part of the nucleus. Transversely divided earthworms (lumbriculus variegatus) regenerate entirely to whole individuals. The decapitated head has been observed to regenerate five times. In starworms the excised snout, together with the pharyngeal ring of the central nervous system, regenerates. Spiders and crabs regenerate feelers, legs and claws; snails, parts of the head, including feelers and eyes, providing the central nervous system is uninjured. Some fish are capable of replacing repeatedly destroyed fins, principally the caudal fin. Salamanders and lizards exhibit regeneration of the entirely lost tail, with bones, muscles and even the posterior extremity of the spinal cord. In young frogs amputated legs regenerate, but only when the bones also are divided, and not after exarticulation. In tritons the lower jaw regenerates. In order, however, that this regeneration shall take place a stump at least must be left. Total extirpation of the parts mentioned destroys the power of regeneration.

Loeb designates as *heteromorphosis* the phenomenon that occasionally after injuries supernumerary parts appear that otherwise do not belong in such situations. Thus, for example, in young lizards lateral notching of the tail may cause the growth of a second tail from the wound; likewise supernumerary extremities develop in tailed amphibia after amputation. Planaria injured on the head exhibit the growth of a second head.

In amphibia and reptiles the regeneration of organs and tissues follows, on the whole, the type of embryonal development, and the histological processes in the growing caudal extremity and in regenerating portions of the body of earthworms take place in the same manner. In amphibia and reptiles only tissue of the same kind develops from injured tissue. The spinal cord regenerates from the epithelial cells of the central canal. In the process of tissue-formation the leukocytes assume only the function of nutrition and conveyance of material. It is a remarkable fact that tadpoles develop after destruction of the brain and the medulla and functional exclusion of the spinal cord.

The power of regeneration is much more restricted in warm-blooded animals and in man. In these also it is confined principally to early life. True regeneration is exhibited by—

1. The *blood;* first the plasma, then the white and finally also the red blood-corpuscles.

2. The *epidermal structures* and the *epithelium* of the mucous membrane regenerate by cell-division in the deepest layers after previous nuclear division. After direct loss they regenerate so long as the normal matrix upon which they grow and the deepest layer of cell-protoplasm capable of development is not also destroyed. In the latter event, regeneration ceases and restoration must take place from the margins of the deficiency. In the process of regeneration, therefore, growth takes place always either from the deep layers, or, after their destruction, from the margins. There develop under such circumstances protoplasmic wandering cells that in part become detached and help to close the deficiency, and in part the deepest layer of cells develops into large, multinucleated protoplasmic cells, which multiply by division into polygonal flat nucleated cells.

The *nail* grows from the posterior fold forward, on the fingers in the course of from four to five months, on the great toe in about twelve months (and more slowly in extremities with fractured bones). Its matrix extends as far as the lunula, and its total or partial destruction causes corresponding loss of the nail. The *eyebrows* are changed in from one hundred to one hundred and fifty days, the remaining hairs more slowly. Destruction of the papilla in a hair-follicle prevents regeneration. Cutting accelerates the growth of the hair, although cut hair does not grow longer than uncut hair. After attaining a certain length the hair falls out. The hair never grows at its free extremity. The *epithelial cells* of the mucous membranes and the glands appear to be subjected to a regular cycle in their utilization and in the regeneration of new cells. In the mammary gland and likewise in the sebaceous glands partial desquamation of secretory cells, and also their regeneration, are evident. The regeneration of spermatozoa takes place through spermatoblasts. In *catarrhal conditions* increased desquamation and regeneration of epithelial cells take place upon the mucous membranes, together with the appearance of indifferent cell-forms (leukocytes) in large number. The *crystalline lens*, which represents an invaginated epidermal sac that has become independent, regenerates like epithelial structures. Its matrix is the anterior wall of the capsule, with the single layer of cells present in this situation. If the lens is removed, but with preservation of these cells, regeneration takes place, the cellular elements becoming elongated into lenticular fibers and filling the entire cavity of the empty capsule. The removal of large amounts of water from the body may cause turbidity of the lens.

3. The *blood-vessels* exhibit extensive regeneration, which takes place in the same way as the formation of the vessels, but this has already been discussed. There always develop at first capillaries, about which, later on, the characteristic tissue-elements are deposited from without in places that subsequently are to become arteries or veins. In case of injury or permanent occlusion of a vessel, at least the portion to the next collateral vessel is always wholly obliterated, derivatives of the endothelial cells, connective-tissue corpuscles from the vessel-wall and wandering cells being transformed into the spindle-cells of the obliterating cicatrix. On the blood-vessels of young and adult animals blind and solid processes are present as an evidence of constant obliteration and regeneration of the vessels. The *lymphatics* behave in the same way as the blood-vessels. After removal of lymphatic glands regeneration may take place, especially when stasis of lymph is present.

4. The *contractile substance of the muscular fibers* may undergo regeneration if destroyed by injury or degenerative processes. The contractile, transversely striated contents of the sarcolemma undergo granular or fibrillar degeneration, or break up into discs or plates, the latter being observed in connection with waxy degeneration of the abdominal muscles in cases of typhoid fever. At the same time nuclei in large number appear within the sarcolemma, as well as in this itself, and the previously contractile contents are converted into cell-protoplasm. In the course of a few days mitotic cell-division is observed. The protoplasm exhibits at first fine fibrillary longitudinal striation. From this fibrous tissue of myogenous origin, transversely striated, nucleated fibers may be developed in the course of months. In case of considerable loss of muscular tissue or gaping wounds a fibrous cicatrix forms. In fibers injured through subcutaneous wounds Neumann observed, after from five to seven days, a budlike prolongation of the divided extremities, at first without transverse striation, which, however, appeared later. Unstriated muscle-fiber may regenerate after injury. The nuclei of the injured fibers divide by karyokinesis and about each newly formed nucleus a new muscle-fiber develops in consequence of the differentiation of the surrounding protoplasm. The fibers divide in the middle of their length.

5. Immediate reunion of a divided *nerve* never takes place with immediate restoration of function. If a portion of a nerve-trunk be excised, the peripheral extremity of the nerve degenerates first, the medullary sheath and the axis-cylinder being transformed into cells. The deficiency is soon filled with juicy connective tissue. The process pursued later in the regeneration of nerve-fibers is fully considered on p. 636. It is an especially noteworthy fact that in the peripheral nerves a constant loss by fatty degeneration, associated with consecutive regeneration of fibers, takes place. Regeneration of peripheral ganglion-cells does not occur. On the other hand, v. Voit observed in a decerebrated pigeon, after the lapse of five months, a regenerated nerve-mass in the skull, consisting of medullated fibers and central ganglia. Also, Vitzou has reported the regeneration of destroyed cerebral ganglion-cells after the appearance of karyokinesis in the adjacent cells. Eichhorst and Naunyn found in young dogs in which the spinal cord was divided between the thoracic and the lumbar portion that an anatomical

and functional regeneration takes place, so that voluntary movements again occur. Vaulair observed in frogs and Masius in dogs first motility, then sensibility, return. Regeneration of the spinal ganglia did not take place. According to Stroebe a formation of fibers takes place in a small, limited area at the site of injury to the spinal cord of the rabbit, but not complete regeneration of the actual spinal tissue.

6. In some *glands* the regeneration of their cells during normal activity is exceedingly active, for example the sebaceous glands, the mucous follicles of the stomach, the glands of Lieberkühn, the uterine glands, the mammary glands during pregnancy; in others regeneration is less active. The removal of considerable portions of various glands is not followed, as a rule, by regeneration, while after injury of glands regeneration of the affected parts does not take place if suppuration occurs. Regeneration of the biliary passages, the bile-duct, and of the pancreatic duct, is remarkable. After injury of the liver Tizzoni and Colluci, as well as Griffini, observed the regeneration of liver-cells and biliary passages even beyond the normal limits of the liver. Pisenti reports similar observations upon the kidney.

After injury to the liver Podwisotzky observed the deficiency disappear completely through partial multiplication of the liver-cells and partial hyperplasia of the epithelial cells of the biliary passages, which are likewise transformed into true liver-tissue (resembling the embryonal development of the liver). Ponfick extirpated even three-quarters of the liver, and regeneration set in within a few days after the operation and was complete in the course of a few weeks.

According to Philippeaux and Griffini regeneration may take place after partial removal of the *spleen*, according to Laudenbach, in the dog, even after almost complete removal. After mechanical injury to the secretory cells of certain glands (liver, kidney, salivary, mammary, Meibomian) hyperplasia and division of adjacent cells take place for the purpose of regeneration. The nipple of which half has been extirpated undergoes regeneration.

7. Of the connective tissues, *cartilage*, providing its perichondrium remains intact, appears to regenerate by division of the cartilage-cells, although, probably, loss of tissue is most frequently replaced by connective tissue.

8. After incised wounds of *tendons* reunion takes place through the agency of the tendon-cells themselves. These multiply considerably by utilizing the matrix for the formation of cells and by mitotic division of the latter. If the extremities of the divided tendon are widely separated, granulation-tissue forms for the development of a cicatrix, as a result of marked reaction in the surrounding connective-tissue tendon-sheath.

9. The regeneration of *bone* is remarkable. If the articular extremity, together with the adjacent portion of the bone, be resected, it may be regenerated, although an appreciable shortening results. Pieces of bone that have been broken or sawed off reunite if replaced; likewise teeth that have been removed and replaced in the alveolus. An isolated piece of periosteum, even if transplanted to another bone, gives rise to a piece of bone of corresponding size. Defects in bone are readily filled by bony tissue if the periosteum be preserved. For this reason the surgeon in resecting diseased bones carefully preserves the periosteum, in the hope that the bone will regenerate from it. The medulla of bone may also regenerate. The internal medullary membrane is capable, if transplanted, of producing bony tissue in small amount from the osteoblasts present.

If a bone, for example a long bone, has been fractured, a circular thickened deposit, at first of rather gelatinous, vascular and cellular, later of firmer cartilaginous, character, forms from the periosteum upon the external surface at the site of fracture—the *external callus*. A similar process takes place at the same time within the medullary cavity, which is thereby diminished in size—*internal callus*. These formations are due to cell-multiplication, in part from the periosteum, in part from the medulla and the bone-tissue itself. The callus generally resembles tissue, and is often cartilaginous.

In the external and internal callus calcification of the cartilage later takes place, as well as the deposition of osseous lamellæ, which, acting as rings, fix the fractured extremities. Later (up to the fortieth day) a thin layer of the same material forms between the fractured extremities, and this subsequently undergoes ossification—*intermediate callus*. With the final solidification of the latter, the bony matter of the external and internal callus gradually disappears. Externally, the swelling disappears, internally the medullary canal becomes again of uniform size and the intermediate callus eventually acquires the same architecture as the adjacent portions. Bone-fractures toward which the course of the nutritive vessels of the bone is directed are said to heal relatively more readily and more rapidly.

With reference to the growth and the metabolism of bones a number of interesting observations may be recorded: (1) Exceedingly small amounts of phosphorus or arsenous acid, added to the food, cause marked thickening of the bones. This appears to be due to the fact that the portions of bone that undergo absorption in the process of normal growth—for example, the walls of the medullary cavity—are not absorbed, but persist, while new growth continues to take place. Small doses of phosphorus are employed for the correction of rachitic softening of bone. In cases of osteomalacia Neumann found an increased elimination of phosphoric acid with the urine. (2) Complete exclusion of lime from the food does not impair the growth of the bones, but makes them thinner, all parts, even the organic matrix of the bone, undergoing uniform atrophy. (3) The ingestion of madder (rubia tinctorum) makes the bones red, the pigment being deposited in the osseous tissue together with the calcium-salts. In birds the egg-shell likewise is stained. (4) Long-continued administration of lactic acid has a solvent influence upon the osseous tissue. The ashy constituents of the bones are diminished. The changes in the bones in youth induced by the withdrawal of calcium-salts are increased by administration of lactic acid. The bones resemble rachitic bones. Osteomalacia in women can be relieved by castration. (5) Artificial hypostatic hyperemia is capable of increasing the growth of bone. The normal growth of bone is considered in connection with its embryological development.

At all portions of the body where considerable amounts of tissue have been lost, with secondary inflammation, such defects heal by the formation of a cicatrix of the structure of connective tissue that fills the defect.

After injury to permanent connective tissue there occurs in the course of three hours an abundant multiplication of the nuclei, which are derived from the matrix, followed by the formation of cells (awakened slumbering cells), while the fixed connective-tissue corpuscles undergo increase in size. After the formation of the previously slumbering cells from elastic and gelatinous fibers has continued for one or two days, mitotic division is observed particularly early in the cells of the adjacent capillaries, then also in the tissue-cells themselves. This often persists for more than eight days. The spindle-cells form blood-vessels, which bridge over the wound-defect, and soon also bundles of fibers, that is, a young cicatrix. The larger the number of cells that become fibers, the firmer becomes the cicatrix; the vessels atrophy and the old cicatrix is firm and deficient in vessels.

The formative process described occurs in all situations where lost tissue is replaced by connective tissue. On the free surface of the body the newly formed vascular tissue not rarely grows (from wounds and ulcers) above the adjacent level—proud flesh. This soon returns, however, to the normal level (after the application of astringents to the vessels), becoming pale, and, finally, after a protecting layer of epidermal cells has developed upon the free surface, forms the *cicatrix*.

If the continuity of a tissue has been severed by a wound, as, for example, an incision, the divided surfaces may, after careful apposition, unite directly, without inflammation—union by primary intention. The surfaces are at first held together by blood-plasma, and later on direct union of the parts takes place. Divided blood-vessels, however, never reunite to form a blood-channel. The cut surfaces of nerves often unite directly, but direct physiological restoration does not take place. Wherever direct union does not take place, cicatricial connective tissue forms in the sequence of inflammation and suppuration—union by secondary intention.

TRANSPLANTATION AND ADHESION.

Parts of the body, such as the nose, the ears, and even the fingers, if severed by means of a sharp and clean-cutting surface, may unite, even after the lapse of hours, an evidence that the life of severed tissues may persist for a time. As a matter of fact, some tissues detached from the body may continue to live for a considerable time, for example leukocytes for three weeks, ciliated epithelium for eighteen days.

The transplantation of flaps of skin is often practised by surgeons to effect closure of existing defects. The flap of skin intended for transplantation, and detached from the subjacent tissues, is permitted to remain for a time attached by means of a pedicle in its original position, and its margins are united accurately by suture to the freshened margins of the deficiency, the pedicle being divided only after the approximated margins have united firmly.   In this way, a new cutaneous covering for the nose can be formed from the skin of the back from another person, or from the skin of the patient's own arm, or from the skin of the forehead.   It is possible also to transplant even large, entirely detached flaps of skin, without a pedicle, even after they have been preserved for fifty hours in 0.6 per cent. sodium-chlorid solution at room-temperature.

To form a cutaneous covering for large granulating (previously carefully cleansed) ulcerous surfaces Reverdin and Thiersch apply under pressure numerous rapidly detached bits of cutis the size of beans upon the granulations, or after removal of the latter upon the freshened wound-surface, where they become adherent.   From the margins of these fragments newly formed layers of epidermis extend over the entire surface of the ulcer.   Enderlin was able to employ successfully such fragments after preservation for four days moistened with physiological salt-solution.   The excised spur of the cock can be made to grow upon the comb.   Bert transplanted the denuded tails and feet of rats beneath the skin of the back of other rats.   The transplanted parts became adherent and formed vascular communications with adjacent tissues, and even their bony parts increased in size.   Parts excised as long as three days previously exhibited similar phenomena.   Detached portions of periosteum transplanted to other situations likewise heal in place and even develop bone.   Extracted teeth may be replaced and even in a second person.   v. Hippel transplanted successfully a piece of a rabbit's cornea 4 mm. square in a defect in a human eye, the clear membrane of Descemet being preserved as a foundation, but the transplanted structure subsequently became turbid.   Also blood and lymph can be transfused.

All of the transplantations mentioned succeed almost solely between individuals of the same species.   Most tissues, however, are not susceptible of transplantation, for example muscles, nerves, glands and organs of special sense.   In the lower animals, even entire parts can be transplanted; for example two pieces of different earthworms may unite, and also of hydra.

The union of two higher animals (rats and others) was successfully effected first by Bert in 1862, who divided the skin of the trunk and united the margins of the wounds in the respective animals by suture.   Union had taken place in the course of five days.   When atropin was administered to one of the animals the pupils of both dilated.   Post-mortem injection demonstrated the existence of anastomoses between the vessels of both.   That such union may take place also in man is shown by the experiments related on p. 454.   The procedure might be of therapeutic significance, as the possibility does not appear excluded that the union of the skin, for example along the extensor aspect of the two forearms, might result in an influence of the one individual upon the other, whether to the end of conveying nutritive juices, or for the removal of certain substances from the body of the one (as for example in case of insufficiency on the part of certain excretory organs), or for the transmission of antitoxins and the like.

INCREASE IN SIZE AND IN WEIGHT IN THE PROCESS OF GROWTH.

In the first period after birth the *length of the body*, which on the average is $\frac{1}{3.5}$ of that of an adult, exhibits the most rapid increase; in the first year about 20 cm., in the second 10 cm. more, in the third about 7 cm.; from the fifth to the sixteenth year the annual increase (about $5\frac{1}{2}$ cm.) is pretty much the same. From the twentieth year on, only slight growth takes place.   From the fiftieth year on, the size of the body diminishes, principally in consequence of attenuation of the intervertebral discs.   The reduction may reach 6 or 7 cm. up to the eightieth year.

The *weight of the body* (about $\frac{1}{20}$ of that of the adult) diminishes constantly in the first five days or week after birth in consequence of evacuation of meconium and of the small amount of food taken at first, together with increased functional activity (generation of heat, respiration, digestive activity), as a result of which the metabolic products are considerably augmented.   Not before the tenth day does the weight of the child again equal that of the newborn.   Later on, the increase in weight exceeds that of the increase in length of the body during corresponding periods.   In the first year the weight is trebled.   In man, the maximum is reached at about the fortieth year.   At about the sixtieth year reduction in weight sets in, in consequence of the retrogressive nutritive processes of age, and this may reach about 6 kilos up to the eightieth year. The detailed figures are given in the following table:

Age.	Length (cm.) Male.	Female.	Weight (kilos) Male.	Female.	Age.	Length (cm.) Male.	Female.	Weight (kilos) Male.	Female.
0	49.6	48.3	3.20	2.91	15	155.9	147.5	46.41	41.30
1	69.6	69.0	10.00	9.30	16	161.0	150.0	53.39	44.44
2	79.6	78.0	12.00	11.40	17	167.0	154.4	57.40	49.08
3	86.0	85.0	13.21	12.45	18	170.0	156.2	61.26	53.10
4	93.2	91.0	15.07	14.18	19	170.6	—	63.32	—
5	99.0	97.0	16.70	15.50	20	171.1	157.0	65.00	54.46
6	104.6	103.2	18.04	16.74	25	172.2	157.7	68.29	55.08
7	111.2	109.6	20.16	18.45	30	172.2	157.9	68.90	55.14
8	117.0	113.9	22.26	19.82	40	171.3	156.5	68.81	56.65
9	122.7	120.0	24.09	22.44	50	167.4	153.6	67.45	58.45
10	128.2	124.8	26.12	24.24	60	163.9	151.6	65.50	56.73
11	132.7	127.5	27.85	26.25	70	162.3	151.4	63.03	53.72
12	135.9	132.7	31.00	30.54	80	161.3	150.6	61.22	51.52
13	140.3	138.6	35.32	34.65	90	—	—	57.83	49.34
14	148.7	144.7	48.50	38.10					

In the first three days the newborn child loses from 170 to 222 grams in weight. Nourished with mother's milk, the child doubles its weight in the first five months and trebles it in the first year. The weight of a five-year-old child is double that of a child one year old, and that of a twelve-year-old child double that of a child five years old. Between the twelfth and the fifteenth year, the weight and the size of girls are greater than those of boys, on account of the earlier advent of puberty in girls. Growth is most rapid in the last months of fetal life; then from between the sixth and the ninth year to between the thirteenth and the sixteenth year. At about the thirtieth year the length of the body is complete, while the weight is not.

Normally developed individuals weigh as many kilos as their length measures in centimeters after subtraction of the first meter. As compared with the growth of the entire body, the individual parts exhibit wide variations. The brain grows least, namely, only to the third year, and from this time on scarcely at all. Also the liver and the intestines grow little, while the heart, the spleen and the kidneys grow only in slightly lesser measure than the entire body. Fat, and particularly muscles, grow more than the entire body.

SUMMARY OF THE CHEMICAL CONSTITUENTS OF THE ORGANISM.

INORGANIC CONSTITUENTS.

Water constitutes 58.5 per cent. of the entire body and is present in the different tissues in widely varying amounts. The tissues of the kidneys contain the largest amount of water, namely 82.7 per cent., while the bones contain 22 per cent., the teeth 10 per cent. and the enamel at least 0.2 per cent. Schönbein found some hydrogen dioxid in the urine.

Gases: Oxygen, ozone, hydrogen, nitrogen, carbon dioxid, methane, ammonia, hydrogen sulphid.

Salts: Sodium chlorid, potassium chlorid, calcium chlorid, ammonium chlorid, calcium fluorid, sodium carbonate, sodium bicarbonate, calcium carbonate, sodium phosphate, alkaline disodium phosphate, acid monosodium phosphate, neutral potassium phosphate, acid potassium phosphate, tribasic calcium phosphate, acid calcium phosphate, magnesium phosphate, neutral sodium sulphate, potassium sulphate, calcium sulphate.

Free acids: Hydrochloric acid (and sulphuric acid in the saliva of some snails, for example dolium galea).

Silicon (as silicic acid), manganese, iron in the blood (and combined with a proteid as ferratin, which aids in blood-formation), iodin (in the thyroiodin of the thyroid gland, diminished in the presence of goiter, increased after administration of iodid), copper (?).

On the whole, a man weighing 70 kilograms consists of thirteen elementary substances, namely, 44 kilograms of oxygen, 7 kilograms of hydrogen, 1.72 kilograms of nitrogen, 0.8 kilogram of chlorin, 0.1 kilogram of fluorin, 22 kilograms of carbon, 800 grams of phosphorus, 100 grams of sulphur, 1750 grams of calcium, 80 grams of potassium, 70 grams of sodium, 50 grams of magnesium, 45 grams of iron.

ORGANIC CONSTITUENTS.

THE PROTEID BODIES OR PROTEIN-SUBSTANCES.

THE TRUE ALBUMINOUS BODIES.

The albuminous or proteid bodies, consisting of C, H, N, O and S, are the fundamental and principal constituents of the animal body, to which they are supplied through vegetable food. They are present in almost all animal and vegetable fluids and tissues, partly in liquid form, partly in more consistent, semi-solid form as constituents of the tissues. Their chemical constitution is unknown; their percentage-composition is described on p. 26. The nitrogen is combined in them in two different ways, in part loosely, in which form it can be separated on treatment with dilute hot potassium hydroxid, with the formation of ammonia; and in part firmly. According to Pflüger a portion of the nitrogen of the living proteid portions of the body is combined in the form of cyanogen. Also the sulphur in the proteid molecule is combined in part firmly, in part loosely. The loosely combined sulphur can be split off by hot potassium hydroxid as potassium sulphid. With lead acetate it forms lead sulphid. The firmly combined sulphur can be prepared only after destruction of the albumin. In serum-albumin the proportion of the loosely to the firmly combined sulphur is as 3 to 2.

The proteid molecule is exceedingly large, and is probably complex. A small portion of it belongs to the group of aromatic substances (which appear especially in connection with putrefaction); the larger portion of the molecule to the series of fatty bodies (in the oxidation of proteids, fatty acids especially develop). Also carbohydrates may appear as decomposition-products, not being entirely wanting in any form of albumin studied by Krukenberg. The decompositions in the process of digestion that are of physiological interest are discussed on p. 304, those occurring in the putrefactive processes on p. 333.

The proteids form a large group of related substances, which perhaps represent only modifications of the same body. If it be borne in mind that the infant prepares from the casein of milk the majority of all the proteids of its own body this last view will be clear. The proteids are generally soluble in water or dilute salt-solutions, but with the exception of the peptones, are incapable of diffusing through membranes on account of the large size of their molecule. They are insoluble in alcohol or ether. They are in general not crystallizable, so that they can be prepared in a pure state only with difficulty. They rotate the plane of polarized light to the left and in the flame they yield the odor of burned horn. They are transformed into a solid modification, that is coagulated, by heat and the long-continued action of alcohol, and are then insoluble in neutral solvents. Coagulated albumin is soluble only (1) in dilute alkalies, alkali-albuminate resulting, having lost a portion of nitrogen and sulphur; (2) in dilute mineral or strong organic acids, acid-albumin (syntonin) developing; and (3) by the process of digestion, albumoses and peptones being formed. By neutralization of alkali-albuminate and acid-albuminate, these substances are rendered insoluble. As a result of long-continued boiling with dilute mineral acids or alkalies, as well as of the action of steam under high tension, the proteids take up water and break up into amido-acids, with the formation of ammonia and hydrogen sulphid; on boiling with alkalies, splitting off also carbon dioxid, oxalic acid and acetic acid.

Color-reactions: (1) Coagulated and heated with nitric acid proteids are stained yellow—xanthoproteic acid. Supersaturation with ammonia makes the color orange. (2) If heated above 60° with Millon's reagent (mercuric nitrate with nitrous acid) a red color results. (3) Boiled with potassium hydroxid, then cooled and copper sulphate added, proteids become deep violet-blue. (4) Concentrated hydrochloric acid (pure) dissolves them on boiling and produces a violet color. (5) Solid proteids are made blue by sulphuric acid containing molybdic acid. (6) The solution of thoroughly desiccated albumin in glacial acetic acid is made violet by concentrated sulphuric acid and exhibits the absorption-band of hydrobilirubin. (7) Iodin may be employed as a microscopic reagent, staining proteids brownish-yellow; also sulphuric acid and cane-sugar, which stain them purple-violet.

Precipitation: (1) By boiling. (2) By strong alcohol. (3) By "salting." Most proteids are precipitated by the addition of neutral salts to their solutions to the point of complete saturation, especially if the reaction be acid. If the addition of salt be made gradually, some of the albumin can thus be separated in crystalline form. (4) Nitric acid precipitates albumin, as does also metaphosphoric acid. (5) Further precipitants are the salts of the heavy metals (iron chlorid, lead acetate, copper sulphate, platinum chlorid, mercuric chlorid in solution with hydrochloric acid). (6) Precipitation is caused by acetic acid and potassium ferrocyanid, also by tannic acid, picric acid or trichloracetic acid. (7) Mercuric-iodid, potassium-iodid on addition of hydrochloric acid, phosphotungstic and phosphomolybdic acids also precipitate albumin.

Animal proteids.

Albuminous bodies can be divided into several characteristic groups: The first group comprises albuminous substances in the strict sense, designated *genuine albuminous substances* or *proteins*, which are soluble in water or in dilute saline solutions and are levorotatory. This **first group** comprises the *albumins* and the *globulins.*

The **albumins** are soluble in water and precipitable by complete saturation with ammonium sulphate, but not by means of sodium chlorid or magnesium sulphate.

Serum-albumin has been prepared in crystalline form by Gürber. By diffusion almost all of its salts, and thereby its coagulability by heat, can be removed. It is precipitated by strong alcohol. It is readily soluble in concentrated hydrochloric acid, acid-albumin, which is soluble in water, being precipitated on addition of water.

Egg-albumin, $C_{80}H_{122}N_{30}SO_{24} + H_2O$, has been prepared in crystalline form by Hofmeister. It occurs in the white of birds' eggs and exhibits a specific rotation of polarized light of $-37.8°$. After injection into the veins or beneath the skin, or even after introduction into the intestine in large amount, it appears partly unchanged in the urine. It is precipitated by agitation with ether. Its composition is $C_{53.28}H_{7.26}N_{15}S_{1.09}$.

Lactalbumin.

Muscle-albumins, that is, the proteid bodies in the aqueous extract of muscle.

The **globulins** are insoluble in water, the majority soluble in dilute saltsolutions. They contain less sulphur and yield a more marked xanthoproteic reaction than the albumins. In solution they are coagulated by a temperature of 75° C. and they are precipitated by abundant addition of water. Dilute acids convert them into acid-albumins. They are precipitated by saturation of the solution with magnesium sulphate and also by semisaturation with ammonium sulphate, by very dilute acids, as well as by carbon dioxid. The globulins include:

Serum-globulin, the presence of which in the urine is described on p. 496.

Fibrinogen, from which fibrin results. The substances from which this is produced are described on p. 69. Stroma-fibrin is considered on p. 72.

Myosinogen.

Vitellin, which occurs in the yolk of birds' eggs and likewise in the crystalline lens, perhaps also in the chyle and in the amniotic fluid, is not precipitable by saturation of a neutral salt-solution with sodium chlorid. Crystalline vitellins occur as yolk-plates in the eggs of fish, frogs, tortoises. In the eggs of birds and in tissues the vitellins are amorphous.

Alkali-albuminates.—Potassium and sodium, also calcium hydroxid and barium hydroxid, form combinations with proteids, and the more rapidly the more concentrated the alkaline solution and the higher the temperature. These combinations are designated alkali-albuminates. They exhibit especially marked circumpolarization, are not coagulated on boiling and are precipitated from solutions by acids, which combine with the alkali. If, for example, egg-albumin be mixed with a solution of potassium hydroxid, potassium albuminate is formed as a gradually developing jelly, which is soluble in boiled water.

Acid-albuminates.—If proteids are dissolved in strong acids, for example hydrochloric acid, they acquire the properties of so-called acid-albumin, which exhibits great similarity to alkali-albuminate (also the specific rotation). This body is insoluble in water and neutral salt-solutions, readily soluble in dilute hydrochloric acid. They are thrown out of solution by the addition of much salt (sodium chlorid or sodium sulphate). Also neutralization by alkali causes precipitation, though boiling does not. On cooling, the boiled (concentrated) fluid becomes gelatinous and again fluid when heated. The syntonin from muscle is an acid-albuminate. It is converted into myosin by milk of lime and ammonium chlorid. The **second** group comprises the complex albuminous bodies. These are proteins combined with bodies of complex composition and they are also designated proteids. They are precipitated by alcohol, which coagulates them after long-continued action. Heat does not cause coagulation. They are generally precipitated from their solutions by slight acidulation. They are readily soluble in dilute alkalies. The second group comprises:

Chromoproteids, that is combinations of protein with pigment. These include:

Hemoglobin, whose combinations and derivatives are described on pp. 55–63.

Glycoproteids, that is combinations of protein with carbohydrates. These include:

Mucin, probably present in various slightly different modifications. It is richer in oxygen, but poorer in nitrogen and carbon, than albumin, free from phosphorus, and contains up to 1.79 per cent. of sulphur and up to 13.5 per cent. of nitrogen. It is liquefied in water into a ropy mucous mass, but it is insoluble in water. On addition of alkali it is converted into a neutral ropy solution. It serves as a protecting substance against the entrance of injurious agents. It is precipitated by a small amount of acetic acid and is redissolved by a larger amount of the same acid. It is precipitated also by alcohol, the resulting precipitate being soluble in water. Acetic acid and potassium ferrocyanid cause no precipitation, although nitric acid and other mineral acids do. Mucin yields all the color-reactions of the albuminous bodies. It is present in saliva, bile, the mucous glands, the secretions from mucous membranes, in "mucous" tissue and in the tendons. In addition it is occasionally found pathologically in cysts (in the lower animals, especially in snails and in the skin of holothurians). On boiling with water or on standing in alcohol it is transformed into coagulated albumin. Alkalies and lime-water transform it into alkali-albuminate, acids into acid-albuminate. On decomposition it yields leucin and 7 per cent. of tyrosin. The mucins yield like glucosids. At high temperatures they break up under the influence of dilute mineral acids into a proteid and a carbohydrate, namely, animal gum.

Peptone and **propeptone** are discussed on p. 298; their demonstration in the urine on p. 496. Peptone is found also in dry lupins, in oats, etc., and less in germinating seed. There may yet be mentioned *proteic acid,* precipitated from the meat-juice of animals (fish) by Limpricht with the aid of acids; and finally *amyloid,* encountered partly in the form of laminated granules on the brain and in the prostate gland, partly (pathologically) as a glistening infiltration of the liver, spleen, kidneys, coats of the vessels, and recognizable from the blue discoloration on addition of iodin and sulphuric acid (like cellulose), and the red discoloration on adding iodin. It can with difficulty be converted into albuminate by alkalies and acids.

APPENDIX : VEGETABLE PROTEIDS.

Plants contain, although in distinctly smaller amount than animals, proteids of various kinds. These occur either in liquid (swollen) form, particularly in the juices of living plants, or in solid form. They resemble the animal albuminates in composition and reaction. There are distinguished:

I. The **vegetable albumins.**

II. The **vegetable globulins.** Of the globulins forming crystals or spheroids that were formerly grouped together under the names *conglutin* and *vitellin,* together with *legumin,* the following may be mentioned: *Edestin* in grain, *amandin* in almonds, *corylin* in nuts, *excelsin* in the Para nut, *avenalin* in oats, *conglutin* in lupins. The globulins include as a decomposition-product *glutin,* an important constituent of wheat, whose glutinous property makes it possible to convert a mixture of flour and water into a coherent dough. Gluten can be obtained from wheat-flour, which may contain as much as 17 per cent., by washing the dough repeatedly with water. Thus prepared, it is viscid, gray, insoluble in water and alcohol, soluble in dilute acids (for example 1 in 1000 parts of hydrochloric acid) and in alkalies. Gluten results from a myosin-like globulin-substance, which is transformed by a ferment in the presence of water into gluten.

III. The **nucleins,** which comprise a special group of readily decomposed complex proteids, containing phosphoric acid in firm combination. They form the chromatin-substance of the cell-nucleus (whence the name), as well as the tingible constituents of the cell-body, and accordingly they are widely distributed in the animal and vegetable kingdoms. The nucleins have a strongly acid character. They are divided into the following two groups:

1. *Paranucleins*, which consist of albumin plus phosphoric acid. If more albumin is added to paranuclein, nucleoalbumin is formed. Casein is such a body, in which, besides, calcium, is present for the neutralization of the acid. It occurs in solution in the milk of all mammals, from which it can be precipitated by addition of acid or of rennet, but not by heat. In the process of gastric digestion nuclein is gradually separated from casein. On boiling casein with hydrochloric acid and stannous chlorid *lysatin*, $C_6H_{13}N_3O_2$, results, which yields urea when boiled with baryta-water.

2. The *true nucleins*, which consist of albumin plus nucleinic acid. Nucleinic acids are decomposed by hydration into phosphoric acid and xanthin-bases (nuclein-bases). The latter include xanthin, guanin, adenin, hypoxanthin, cytosin. The true nucleins may combine with more albumin and yield *nucleoproteids*. A carbohydrate is derived from nucleinic acid, namely *pentose*.

The nucleins are insoluble in water or dilute acids, readily soluble in dilute alkalies, with which they unite by reason of their acid character to form neutral combinations. They swell in solution of sodium chlorid, and yield all the color-reactions of albumin. In alkaline solution they are readily decomposed into proteids and nucleinic acids (or phosphoric acid). The nucleins resist the solvent action of the gastric juice, which is capable of dissolving and digesting only the proteids of the nucleoalbumins and nucleoproteids. Upon the latter property depends the possibility of isolating the nucleins. The nucleinic acids occur also uncombined with albumin in certain cellular structures of the animal kingdom (salmon-spawn). Nuclein-bases have been found free in animal and vegetable tissues.

The yolk of the egg contains a nuclein-like body containing iron that is utilized in the formation of blood from the yolk (hematogen), and that also aids in hemogenesis on a diet of eggs. From a body, phosphosarcic acid, closely related to the paranculeins, can be prepared a ferruginous body, carniferrin, which contains iron in similar firm combination as in hematogen.

Nucleohiston, a combination of nuclein and histon, which can be prepared from the erythrocytes of the goose, is readily decomposed into nuclein and histon. The latter prevents coagulation of the blood.

Nucleoalbumin is prepared by Halliburton in the following manner: Kidneys are rubbed up with powdered sodium chlorid and some water. The expressed extract is poured into distilled water, in which the remains of tissue and the globulins fall to the bottom, while the mucoid nucleoalbumin floats on the surface. This is collected and washed repeatedly with distilled water.

Injected into the veins nucleoalbumin causes coagulation. According to Pekelharing, the zymogen of the fibrin-ferment is a nucleoalbumin. *Histon*, a base consisting of protamin and albumose, is present in the nuclei of the erythrocytes of birds and in leukocytes, thymus, spleen, testicles, in combination with nuclein. It is coagulable by ammonia, not by boiling, and can be extracted by means of dilute acids. *Reticulin*, the ground-substance of reticular connective tissue, is a related body. It contains phosphorus and sulphur, is indigestible and insoluble, and on heating with alkalies splits off the phosphorus-containing group, and is then soluble with difficulty. With hydrochloric acid it splits off amidovalerianic acid (but no tyrosin). *Plastin* is similar to nuclein and occurs in the nuclei and in the protoplasm of spermatozoa. It is formed in the process of peptic digestion, and is insoluble in sodium carbonate as well as in hydrochloric acid 4 to 3 of water.

THE ALBUMINOID BODIES.

These resemble the true albuminous bodies with reference to their composition and source. They are uncrystallizable; some of them are free of sulphur; while most cannot be prepared in an ash-free state. Their reactions and decomposition-products resemble those of the albuminous bodies. Some of them yield, in addition to much leucin and tyrosin, also glycin and alanin (amidopropionic acid), although in physiological, chemical and physical respects they exhibit considerable differences from albuminous bodies. They occur in the tissues both as organized constituents as well as in liquid form. Whether they are formed by oxidation from the albuminous bodies or by synthesis is not known. They are in part indigestible, in part digestible, although the products of their digestion can replace the decomposed albumin in the body not at all or but incompletely. They are contained principally in the connecting and protecting structures of the body. They can enter into combination with acids or alkalies.

1. *Keratin* is present in all horny and epidermal structures. It is soluble only in boiling caustic alkalies, while it swells in cold alkalies and in concentrated acetic acid. It contains from 2 to 5 per cent. of sulphur, a large part of which can be split off by alkalies. It is indigestible; decomposed by hydrolysis it yields 10 per cent. of leucin and 3.6 per cent. of tyrosin. *Neurokeratin* is described on p. 627.

2. *Fibroin* is soluble in strong alkalies and mineral acids, as well as in cupric-ammonium sulphate. Boiled with sulphuric acid it yields 5 per cent. of tyrosin, leucin and glycin. It is the principal ingredient of the web of insects and spiders. By long boiling silk-gelatin (*sericin*) is obtained from silk. This body is richer in oxygen and water than fibroin. Treated with sulphuric acid it yields, in addition to leucin and tyrosin, also *serin*, a crystalline amidoacid.

3. *Spongin*, a body resembling fibroin, and derived from sponges, yields leucin and glycin as decomposition-products.

4. *Elastin*, the ground-substance of all elastic tissue-elements, is soluble only when boiled in concentrated potassium hydroxid. It yields from 36 to 45 per cent. of leucin, together with one-half per cent. of tyrosin. It yields the reactions of albumin and its decomposition-products. It contains sulphur only in loose combination. It is peptonized by trypsin, but not by the gastric juice.

5. *Glutin* or *bone-gelatin* can be prepared from all connective or gelatin-yielding substances (which contain *collagen*) in the form of gelatin by boiling with water. This gelatin on cooling forms a jelly. Collagen is soluble by boiling with acids or alkalies. Glutin is strongly levorotatory. It is transformed by long boiling and digestion into a peptone-like state, in which it does not become gelatinous. A glutin-like body is present in leukemic blood and in splenic juice. Glycin, leucin and tyrosin, also *serin*, a crystalline amidoacid. result on hydrolytic decomposition. Glutin contains 0.7 per cent. of sulphur.

6. *Chondrin* or *cartilage-gelatin* is obtained by boiling hyaline cartilage. It becomes gelatinous in the cold. It is precipitated by acetic acid and by small amounts of mineral acids. It is dissolved in an excess of the latter as well as by neutral salts.

The true characteristic substance of hyaline and elastic cartilage is a mono-basic acid, namely, *chondroitin* ($C_{18}H_{27}NO_{14}$), which as an ethereal sulphate, namely, as chondroitin-sulphuric acid, is contained in cartilage. This acid is present in cartilage only in exceedingly loose combination with albuminous or gelatinous substances. Alkalies separate the albuminous bodies from the chondroitin-sulphuric acid by forming alkaline salts with the latter. The *chondrin* (of the earlier writers) is a gelatinizing solution consisting of a mixture of ordinary gelatin and the last-mentioned chondroitin-sulphates of the alkalies. It can, therefore, be prepared artificially from gelatin and potassium or sodium chondroitin-sulphate. True hyaline cartilage is, therefore, distinguished from (gelatin-yielding) osseous cartilage by the circumstance that the ground-substance of the former contains chondroitin-sulphates.

On decomposition of chondroitin (as well as of chitin) *glycosamin* ($C_6H_{11}O_5NH_2$) is formed, the latter on treatment with nitrous acid being transformed into glucose —an example of the manner in which non-nitrogenous carbohydrates may be derived from nitrogenous albuminous bodies.

7. The *hydrolytic ferments*, also designated *enzymes* (in order to distinguish them from the organized ferments, for example yeast and bacteria). The characteristic of all organized ferments is that they are active only in the presence of water and in such a manner that they cause a decomposition of the body upon which they act as a result of which the latter takes up water. All of the ferments likewise decompose hydrogen dioxid into water and oxygen. Their activity is greatest at a temperature between 30° and 35° C. They are destroyed by boiling. In the dry state they may tolerate exposure to a temperature of 100° C. without attenuation. The addition in considerable amount of antiseptics that destroy lower organisms does not check their activity. During periods of protracted inactivity their solutions undergo destruction in greater or lesser degree. The following hydrolytic ferments are distinguished:

(*a*) *Sugar-forming ferments* in the saliva, the pancreatic juice, the intestinal juice, the bile, the blood, the lymph, the chyle, the liver, the urine, the milk. and *invertin* in the intestinal juice.

Almost all dead tissues, organic fluids, and even albuminous bodies, may exert a feeble diastatic action.   Diastatic ferment is found also in grain and leguminous fruits, in hay and other vegetable foods.

(b) *Proteolytic ferments:* In the gastric juice (pepsin), the muscles, also in germinated seeds, for example vetches, malted barley, and in the myxomycetes; in the pancreatic juice (trypsin), the intestinal juice, the urine.   Pepsin and trypsin diffuse through membranes like peptone.

(c) *Fat-splitting ferments:* in the pancreatic juice.

(d) *Milk-coagulating ferments:* in the stomach, the pancreatic juice, the urine.

NITROGENOUS GLUCOSIDS.

The following nitrogenous glucosids, which on hydrolytic treatment take up water and are decomposed into sugar and other atom-groups, may be considered here:

Cerebrin, $C_{57}H_{110}N_2O_{25}$.

Protagon in the medullary substance of nerves ($C_{66 \cdot 30}N_{2 \cdot 39}H_{10 \cdot 68}P_{1 \cdot 068}$ per cent.)

Chitin, $2(C_{15}H_{26}N_2O_{10})$, a nitrogenous glucosid or amin of a carbohydrate in the cutaneous covering of all arthropods, also in the intestine and the trachea of these animals; soluble in concentrated hydrochloric or nitric acid.   The *hyalin* of the bladder-worms is closely allied to chitin.   Among the glucosids of the vegetable kingdom are also solanin, amygdalin and salicin.

NITROGENOUS PIGMENTS.

These are of unknown constitution and occur only in animals.   In all probability they are all derivatives of hemoglobin.   They are: (1) *Hematin* and *hematoidin.*   (2) The *biliary pigments.*   (3) The *urinary pigments.*   (4) *Melanin* or the black pigment contained partly in epithelial cells (choroid, iris, deep epidermal cells in colored races), partly in connective-tissue corpuscles (lamina fusca of the choroid), in hairs and in pathological neoplasms.   Schmiedeberg produced melanin by boiling albumin for a long time with concentrated mineral waters.   The melanin prepared from a melanosarcoma had the following composition: $C_{68}H_{64}N_{10}SO_{26} + H_2O$.

ORGANIC NON-NITROGENOUS ACIDS.

The fatty acids, constructed according to the formula $C_nH_{2n-1}O(OH)$, are present in the body in part free, in part combined.   In the free state the volatile fatty acids are found in decomposing cutaneous secretions (sweat), also in the large intestine.   In combination, acetic acid and caproic acid will appear as amido-combinations in glycin (amido-acetic acid) and leucin (amido-caproic acid). Particularly, however, the fatty acids are combined with glycerin to form neutral fats, from which, in the process of pancreatic digestion, the fatty acids are again decomposed.

The acids of the acrylic-acid series, constructed according to the formula $C_nH_{2n-3}(HO)$, yield the animal organism but one acid, namely oleic acid.   This, also, forms with glycerin the neutral fat, olein.   It will be advisable at this point to discuss the neutral fats, in the formation of which both the fatty acids and oleic acid are utilized.

THE FATS.

The fats occur abundantly in the animal body, but probably also in all plants, in the latter particularly in the seeds (nuts, almond, cocoanut, poppy), less commonly in the pericarp (olive), or in the root.   They are obtained by expression, by melting or by extraction with ether or boiling alcohol.   They contain a smaller amount of oxygen than the carbohydrates.   On paper they produce characteristic fat-spots; agitated with colloidal substances they yield an emulsion.   If neutral fats are superheated with water or are heated with certain ferments or are permitted to undergo decomposition, they take up water and break up into glycerin and free fatty acids, of which the latter, if volatile, diffuse a rancid odor.   Treated with caustic alkalies they likewise take up water and are decomposed into glycerin and fatty acids.   The fatty acids form salt-like combinations (soaps) with the alkali, while the glycerin is set free.   The soap-solutions in turn dissolve fats. Glycerin, a triatomic alcohol, $C_3H_5(OH)_3$, combines (1) with the following mono-basic fatty acids:

1. Formic acid, CH_2O_2,	12. Laurostearic acid, $C_{12}H_{24}O_6$,
2. Acetic acid, $C_2H_4O_2$,	13. Tridecylic acids, $C_{13}H_{26}O_2$,
3. Propionic acid, $C_3H_6O_2$,	14. Myristic acids, $C_{14}H_{28}O_2$,
4. Butyric acids, $C_4H_8O_2$,	15. Pentadecylic acids, $C_{15}H_{30}O_2$,
5. Valerianic acid, $C_5H_{10}O_2$,	16. Palmitic acids, $C_{16}H_{32}O_2$,
6. Caproic acids, $C_6H_{12}O_2$,	17. Margaric acids, $C_{17}H_{34}O_2$,
7. Enanthylic acids, $C_7H_{14}O_2$,	18. Stearic acids, $C_{18}H_{36}O_2$,
8. Caprylic acids, $C_8H_{16}O_2$,	19. Arachinic acid, $C_{20}H_{40}O_2$,
9. Pelargonic acid, $C_9H_{18}O_2$,	20. Hyenic acid, $C_{25}H_{50}O_2$,
10. Capric acid, $C_{10}H_{20}O_2$,	21. Cerotic acid, $C_{27}H_{34}O_2$,
11. Undecylic acids, $C_{11}H_{22}O_2$,	22. Melissic acid, $C_{30}H_{60}O_2$, etc.

The acids form an homologous series according to the formula $C_nH_{2n-1}O(OH)$. With each additional CH_2 the boiling-point is raised $19°$. The acids containing a larger amount of carbon are consistent and do not volatilize; those containing a lesser amount of carbon (to 10 inclusive) are oleaginous and volatile, with a pungent acid taste and a rancid odor. The earlier may be produced from the later in the series by oxidation, CH_2 disappearing, with the formation of CO_2 and H_2O: for example, butyric acid results from propionic acid. Human and animal fat contain 16 and 18, in smaller amount and inconstantly 14, 12, 6, 8, 10, 4. Some are contained in the sweat and in the milk. Many develop from albumin and gelatin in the process of putrefaction. The majority, with the exception of those from 19 to 22, are present in the contents of the large intestine.

2. In addition, glycerin combines with the monobasic oleic acids, which likewise form a series and stand in an intimate relation to the fatty acids. Their general formula is $C_nH_{2n-3}O(OH)$; they all thus possess 2H less than the corresponding members of the fatty-acid series. By suitable procedures the corresponding fatty acids can be obtained from the oleic acids, and conversely oleic acids develop from the corresponding fatty acids. *Oleic acid* (elaic acid), $C_{18}H_{34}O_2$, is the only member found in the organism; combined with glycerin it yields fluid olein. The fat in the new-born contains more glycerids of palmitic and stearic acids than that of the adult, which contains more glycerids of oleic acid. In addition, oleic acid occurs in combination with alkalies (in soaps), and, like a number of fatty acids, in the lecithins. Lecithin is considered as a glycerophosphate of neurin, in which two atoms of H in the radicle of glycero-phosphoric acid are replaced by two atoms of stearic, palmitic, or oleic acid. If barium hydrate is added to lecithins, insoluble barium stearate or oleate or palmitate + oleate is produced, together with neurin in solution and barium glycero-phosphate. There appear to be different lecithins, of which those combined with the stearic-acid and that with the palmitic-acid + oleic-acid radicle are the most frequent. Lecithin is present in the blood-corpuscles, in larger amount in the semen, in the yolk of birds, in the nervous tissue, in traces in all animal cells. Neurin also is a constant constituent of bacteria and of the seeds of vetch and peas.

The neutral fats, the glycerids of the fatty acids and of oleic acid, are triple ethers of the triatomic alcohol, glycerin. Fat in the ordinary sense of the word consists of palmitin (with a melting-point of $62°$), stearin ($71.5°$), olein ($0°$). Related to the neutral fats is glycero-phosphoric acid, an acid glycerin-ether, resulting from the combination of glycerin with phosphoric acid, with the giving off of 1 molecule of water—$C_3H_9PO_6$; it is a product of the decomposition of lecithin. Spermaceti (cetaceum), obtained from the cranial cavity of certain whales, contains principally palmitic-acid cetyl-ether.

3. The *glycolic acids*, acids of the lactic-acid series, are constructed according to the formula $C_nH_{n-2}O(OH)_2$. They result from the fatty acids by oxidation, if 1 atom of H in the fatty acids is replaced by OH (hydroxyl). Conversely also fatty acids can be obtained from the glycolic acids. Those fatty acids that (from propionic acid downward) contain more than 2 atoms of C may form various isomeric glycolic acids, in accordance with the C-atom in which the other hydroxl-group enters. There occur in the body

(a) *Carbonic acid*, hydrooxyformic acid, $CO(OH)_2$, in this form, however, forming only salts. Free carbonic acid is the anhydrid, namely CO_2.

(b) *Glycolic acid*, oxyacetic acid, $C_2H_3O(OH)_2$, does not occur in the body in the free state. A combination of this, glycin—glycocol, amido-acetic acid, gelatin-sugar —occurs as a conjugate acid, namely as glycocholic acid in the bile and as hippuric acid in the urine. Glycin exists in gelatin in complex combination.

(c) *Lactic acid*, oxypropionic acid, $C_3H_4O(OH)_2$, is contained in the body in two isomeric forms: (1) Ethylidene lactic acid, which occurs in two modifications, namely as dextrorotatory sarcolactic acid, paralactic acid, a metabolic product of muscle and also in the thymus and thyroid glands; it develops also from the action of bacteria on grape-sugar, and as ordinary optically inactive or fermentation-lactic-acid, which is present in the gastric juice, in sour milk, sour-crout, sour pickles, and can also be obtained from sugar by fermentation. (2) The isomer ethylene lactic acid is likewise present in muscle in small amounts.

(d) *Leucic acid*, oxycaproic acid, $C_6H_{12}O_3$, does not occur independently, but only as a derivative, namely as leucin, amidocaproic acid, as a metabolic product in certain tissues, as well as a product of pancreatic digestion. By treatment with nitrous acid leucic acid can be produced from leucin, and glycolic acid from glycin.

4. *Acids of the oxalic-acid or succinic-acid series*, with the formula $C_nH_{2n-4}O_2(OH)_2$, dibasic acids, which develop from fatty acids and glycolic acids as completed oxidation-products by taking up oxygen and giving off water. Their development from bodies rich in carbon, particularly fats, carbohydrates and proteids, is, therefore, noteworthy.

(a) *Oxalic acid*, $C_2O_2(OH)_2$, results by oxidation from glycol, glycin, cellulose, sugar, starch, glycerin and many vegetable acids, and occurs normally in the urine in combination with calcium.

(b) *Succinic acid*, $C_4H_4O_2(OH)_2$, has been found by some in small amounts in dead animal tissues and fluids, urine, echinococcus-fluid, hydrocephalus-fluid, hydrocele-fluid. It is present in large amount in the urine of the dog after a diet of fat and meat, in the urine of the rabbit when fed with carrots. It is generated by micro-organisms and is wanting in fresh, living tissues. It develops in small amounts in the process of alcoholic fermentation.

5. The *cholalic acids* are present in bile and in the intestine.

6. *Aromatic acids*, containing the benzol-nucleus: benzoic acid (phenylformic acid) occurs in the urine in conjunction with glycin as hippuric acid.

THE ALCOHOLS.

Alcohols are bodies that develop from carbohydrates by the substitution of hydroxyl (HO) for one or more atoms of hydrogen. They can also be viewed as water, $^H_H\}O$, in which half of the hydrogen is replaced by a CH-combination. Thus, for instance, C_2H_6, ethyl hydrid, is transformed into $^{C_2H_5}_H\}O$, ethyl-alcohol.

(a) *Cholesterin*, $_{C_{27}H}^{H_{45}}\}O$, is a levorotatory alcohol that occurs in blood, yolk, brain and bile, and, besides, quite generally in vegetable cells. It is present also in tissues of man and animals containing keratin. Liebreich considers cholesterin as a necrobiotic fat. By oxidation cholesteric acid $(C_8H_{10}O_5)$ is developed from cholesterin, appearing also as an oxidation-product of cholic acid.

(b) *Glycerin*, $C_3H_5O\{^{OH}_{OH}_{OH}$, is considered as a triatomic alcohol. It occurs in combination with fatty and oleic acids in neutral fats. It is formed in the process of pancreatic digestion by decomposition of the neutral fats. It is developed in small amount as a result of the fermentation of fats in the intestine, as well as in the process of alcoholic fermentation.

(c) *Phenol* (phenylic acid, carbolic acid, oxybenzol).

(d) *Pyrocatechin* (dioxybenzol).

(e) The *sugars* may be considered advantageously in connection with the alcohols, as they behave like polyatomic alcohols. Their exact constitution is as yet unknown. The sugars form, together with a series of closely related bodies, the large group of *carbohydrates*, which will be considered collectively. Although many of these do not occur in the animal body, their consideration is justified by the fact that they occur largely as constituents of vegetable food.

THE CARBOHYDRATES.

These bodies occur in the animal and vegetable kingdoms and have received their designation because they contain in their molecules, in addition to at least six atoms of C, the atoms of H and O always in the proportions present in water, namely, H_2O. All are solid, chemically indifferent, without odor. They either have a sweet taste (sugars) or may at least be readily transformed into sugar by the action of dilute acids. They deflect the ray of polarized light either to the right or to the left. Heated dry they give off the odor of caramel. They stain red with thymol and sulphuric acid. According to their constitution, they may be considered as fatty bodies, as hexatomic alcohol in which 2 atoms of H are wanting. In small amounts the carbohydrates are constituents of almost all animal tissues. In the presence of special nutritive disturbances decomposition of complex organic constituents of the viscera appears to take place. Nitrogenous products that are readily decomposed into urea are split off from the albuminates, and in addition to these the non-nitrogenous portion appears as carbohydrate. The formation of carbohydrates (sugar) from fats occurs in the germination of seeds containing oil, with the taking up of oxygen.

The carbohydrates can be divided into the following groups:

I. The **monosaccharids** or **glucoses,** which contain only one molecule of simple sugar: $C_6H_{12}O_6$. 1. *Grape-sugar* (glucose, dextrose, lump-sugar, starch-sugar, liver-sugar or urinary sugar) has been produced synthetically by E. Fischer. It occurs in the animal body in small amounts in the blood, chyle, muscle, liver, urine; in large amounts in the urine in cases of diabetes mellitus. It is formed in the process of digestion by diastatic ferments from other carbohydrates. In the vegetable kingdom it is distributed in the sweet juices of many fruits and flowers, and thence into honey. It is formed from cane-sugar, maltose, dextrin, glycogen and starch on boiling with dilute acids. It crystallizes in cauliflower-like, warty masses, with one molecule of water of crystallization, combines with bases, salts, acids and alcohols, but is readily decomposed by bases. It exerts a reducing action on many metallic oxids. Phenylglucosazone melts at a temperature of 204° C. As a result of its oxidation there develop first the monobasic glyconic acid and then the monobasic glucic acid. A fresh solution has a rotatory power of $+106°$, which falls to $+56°$. On fermentation with yeast it is decomposed into alcohol and carbon dioxid; by putrefactive bacteria it is broken up into two molecules of lactic acid. The qualitative and quantitative estimation of grape-sugar is discussed on pp. 267, 268, and 501. In alcoholic solution it enters into almost insoluble combinations with calcium, barium or potassium; also with sodium chlorid it crystallizes into a combination.

2. *Galactose* is formed by hydrolytic decomposition of milk-sugar (lactose), but also by hydrolysis of gum and mucoid substances; and as a decomposition-product of the glucosid, cerebrin. It forms needles and plates, soluble in water, is dextrorotatory $+88.08°$ and fermentable and yields the reactions of dextrose. Its phenylosazone melts at 193° C. When oxidized it forms galactonic acid and later mucic acid.

3. *Levulose* (levorotatory fruit-sugar, invert-sugar or mucous sugar) is formed from inulin by the action of acids together with levulin, which is the analogue of dextrin. It occurs in the acid juices of some fruits and in honey as a colorless syrup, crystallizable with difficulty, fermentable with greater difficulty, insoluble in cold alcohol, with a rotatory power of $-106°$. It reduces like dextrose and has the same osazone. It is formed normally in the intestine and is rarely found abnormally in the urine.

II. **Disaccharids** or **saccharoses** contain two molecules of simple sugar, and having the formula $C_{12}H_{22}O_{11}$, are the anhydrids of the members of the first division.

1. Milk-sugar (lactose = dextrose + galactose) occurs only in milk, crystallizes in crusts, with 1 molecule of water, from whey evaporated to a syrupy consistence, is dextrorotatory $+52.5°$, soluble with greater difficulty than grape-sugar in water and particularly in alcohol. On boiling with dilute mineral acids it is decomposed into galactose and dextrose; it can be transformed directly into lactic acid only by fermentation, and the resulting galactose is, however, susceptible of alcoholic fermentation with yeast (preparation of koumiss). Its quantitative estimation is discussed on p. 422. It occurs rarely in the urine. Lactosazone melts at 200° C.

2. *Maltose* (malt-sugar), $C_{12}H_{22}O_{12} + H_2O$, contains one molecule of water less than two molecules of grape-sugar ($C_{12}H_{24}O_{12}$) and results in the diastatic decomposition of starch. It is soluble in alcohol, but is precipitated from an alcoholic solution by ether in the form of needles (dextrose is not). It is dextrorotatory 138.4° and crystallizable, and 100 parts reduce equally with 66 of dextrose. Dextrose reduces cupric acetate, while maltose does not. Maltosazone melts at 208° C. Isomaltose is not susceptible of fermentation.

3. *Saccharose* (cane-sugar or beet-sugar = dextrose + fruit-sugar) is present in cane-sugar and a number of plants, but it does not reduce copper. It is soluble in alcohol with difficulty, is dextrorotatory and not fermentable. In the intestine, and also when boiled with dilute acids, it is transformed into a mixture of glucose

30

and levulose.   Oxidized with nitric acid it is decomposed into glucic acid and oxalic acid.

III. **Polysaccharids** or **amyloses,** result, as anhydrids of members of the first division, from the union of many molecules of the monosaccharids. Many of them do not undergo fermentation. They yield colloid solutions, do not diffuse and do not crystallize.

1. *Glycogen,* with a rotatory power of $+211°$, devoid of reducing action, occurs in the liver, the muscles, many embryonal tissues, the fetal membranes, the rudimentary embryo of the chick and in part in normal and pathological epithelium. It is present in small amounts in many organs: testicle, lung, skin, and in pus and inflammatory foci. It has been found in considerable amount in the body of the diabetic, in the brain, the pancreas, and cartilage. It occurs also in oysters and other molluscs, but it may be present in the cells of all of the tissues and of all classes of animals. Errera found glycogen in yeast.

2. *Dextrin* is dextrorotatory $+138°$, forms a viscid solution with water, from which it is precipitable by alcohol and acetic acid, and is discolored feebly red by iodin. It results from scorched starch (and is therefore abundant in bread-crusts) through the action of dilute acids, and in the body through the action of ferments. It is formed from cellulose by treatment with dilute sulphuric acid. It occurs also in beer. In the vegetable kingdom it is present in most vegetable juices.

3. *Starch* is present in the mealy portion of many vegetables, partly consisting of organized granules in layers forming within the vegetable cells, with a generally excentric nucleus, and partly, though less commonly, occurring unformed in vegetables. The diameter of the starch-granule varies considerably in different plants. It is, for instance, from 0.14 to 0.18 mm. in the potato and only 0.004 mm. in the seed of the red beet. In water at a temperature of $72°$ C. it swells as a paste. It is stained blue by iodin only in the cold. The starch-granules contain further always more or less cellulose, as well as a body stained red by iodin (erythrogranulose). The transformation of starch and glycogen takes place through the action of the saliva, the pancreatic and the intestinal juice; both are transformed into dextrose by dilute sulphuric acid.

4. *Gum,* $C_{10}H_{20}O_{10}$, occurs in man in organs containing mucus, such as the salivary glands, the mucoid tissues, the lungs, and in bile, occasionally in albuminous fluids, and in small amounts in the urine. It is susceptible of fermentation and is decomposed by boiling with dilute acids into a body reducing copper oxid. In the vegetable kingdom gum is found in the juices particularly of acacias and mimosas, partly soluble in water (arabin), partly swelling up like mucus (bassorin). It is precipitated by alcohol. Wood-gum (pentosan, $C_5H_3O_4$) occurs abundantly in fibrous vegetable matters consumed by herbivora as food. On heating with dilute acids there result by hydration pentoses, in the same way as dextrose is formed from starch. There are two pentosans: xylan, which yields xylose, and araban, from which arabinose results. Pentosan results from the oxidation of cellulose and starch, 1 atom of C being transformed into CO_2.

5. *Inulin,* a crystalline powder found in the root of chicory and dandelion, and in the bulbs of dahlia variabilis, is not stained blue by iodin.

6. *Lichenin,* lichen-starch, the intercellular substance of lichens, especially of Iceland moss (cetraria Icelandica), and of algæ. It can be transformed into glucose by dilute sulphuric acid.

7. *Cellulose,* $C_6H_{10}O_5$, the cellular tissue of all vegetables, is found also in the integument of tunicates, the exoskeleton of arthropods and the skin of snakes. It is soluble only in ammoniated cupric oxid and is colored blue by sulphuric acid and iodin. On boiling with dilute sulphuric acid dextrin and glucose are formed. It is transformed (cotton) by concentrated nitric acid mixed with sulphuric acid into nitro-cellulose (gun-cotton, $C_6H_7(NO_2)_3O_5$,), which, dissolved in a mixture of ether and alcohol, forms collodion. *Tunicin,* a body similar to cellulose, occurs in the integument of tunicates (molluscs). Cellulose is dissolved in the intestine of herbivora with the aid of bacteria.

For the sake of completeness, *inosite,* $C_6H_{12}O_6$, hexahydrohexaoxybenzol, muscle-sugar, phaseomannite, bean-sugar, may be discussed at this point. This is not a true sugar, but it has a sweet taste. It occurs in the muscles, the lungs, the liver, the spleen, the kidneys, the brain of the ox, and the kidneys of man; pathologically in the urine and in echinococcus-fluid. It is widely distributed in the vegetable kingdom, especially in beans (Leguminosæ) and in grape-juice. It is optically inactive, generally crystallizes like cauliflower, with two molecules of water, in long monoclinic crystals, insoluble in alcohol or ether; it does not respond to Trommer's test and is susceptible only of sarcolactic-acid fermentation. Evaporated to dryness with nitric acid, then with ammonia and calcium chlorid, it leaves a rosy-red stain.

AMMONIA-DERIVATIVES AND THEIR COMBINATIONS.

The ammonia-derivatives are products of proteids, decomposition-products of the tissue-metamorphosis of proteids.

1. **Amins,** that is compound ammonias, which may be derived from ammonia (NH_3) or from ammonium hydroxid (H_4N, OH) by replacing one or all of the atoms of H by carbohydrate-groups (alcohol-radicles). The amins derived from a single molecule of ammonia are designated monamins. Among these are *methylamin*, $\begin{matrix} H \\ H \\ CH_3 \end{matrix} \Big\} N$, and *trymethalamin*, $\begin{matrix} CH_3 \\ CH_3 \\ CH_3 \end{matrix} \Big\} N$, known only as decomposition-products of cholin (neurin) and of kreatin. Neurin occurs in lecithin in complex combination. The lecithins are described on p. 463 and the diamins are discussed on p. 305.

2. **Amids,** that is derivatives of acids in which NH_2 is substituted for the hydroxyl (HO) of the acids. *Urea*, $CO(NH_2)_2$, the diamid of CO_2 is the principal end-product of the tissue-metamorphosis of the nitrogenous constituents of the body. Carbon dioxid containing water is $CO(OH)_2$, in which both OH-atoms are replaced by NH_2, thus $CO(NH_2)_2$.

3. **Amido-acids,** that is nitrogenous combinations exhibiting partly the character of an acid, and partly that of a feeble base, in which H-atoms of the acid-radicle are replaced by NH_2 or substituted ammonia-groups.

(a) *Glycin* (amido-acetic acid, glycocol, gelatin-sugar) results on boiling gelatin with dilute sulphuric acid. It is present in the cornea, which contains, besides, chondrin. It has a sweet taste (gelatin-sugar), behaves like a feeble acid, but unites also as an amin-base with acids. It occurs as glycin + benzoic acid = hippuric-acid in the urine (it has also been prepared artificially), and as glycin + cholic acid = glycocholic acid in the bile. (b) *Leucin* (amidocaproic acid) has been found pathologically in pus and in the atheromatous matter of sebaceous cysts, generally in combination with tyrosin. (c) *Serin* (amidolactic acid) is obtained from silk-gelatin. (d) *Blood-alinin* (amidovalerianic acid). (e) *Aspartic acid* (amidosuccinic acid). (f) *Glutamic acid* (amidopyrotartaric acid) is obtained in the decomposition of albuminates. Aspartic acid can be obtained from asparagin by boiling with acids and the splitting off of ammonia. Asparagin is formed largely in the vegetable kingdom from albumin and has been prepared artificially, while in the animal body it is transformed into urea and uric acid. (g) *Cystin* (amidolactic acid), in which O is replaced by S, is strongly levorotatory. (h) *Taurin* (amido-ethyl-sulphuric acid) occurs, besides in a number of glands, particularly in combination with cholic acid as taurocholic acid in bile. It has also been prepared artificially. (i) *Tyrosin* (paraoxyphenylamidopropionic acid, prepared synthetically) occurs together with leucin in the presence of pancreatic digestion. It may occur pathologically in the urine. It is abundant in dahlia-bulbs.

There are related to the amido-acids further: (a) *Kreatin*, methylguanidin-acetic acid, $CH_9N_3O_2$, which is present in muscle, brain, blood, and urine, and has been prepared artificially. Boiled with baryta-water it takes up water and is decomposed into urea; (b) *Sarcosin* ($C_3H_7NO_2$, methylamido-acetic acid). When boiled with water or heated with strong acids in the presence of decomposing substances kreatin is transformed into *kreatinin* with the loss of water; kreatinin occurs in the urine. This strong base can be retransformed into kreatin by the action of alkalies.

4. **Ammonia-derivatives in Part of Unknown Constitution.**—*Uric acid. Allantoin* results from the oxidation of uric acid by means of potassium permanganate. It has been found, together with guanin and sarcin, also in the buds of Platanaceæ. *Cyanuric acid* has been found in dog's urine. *Inosinic acid* is present in muscle. *Guanin* ($C_5H_5N_5O$), together with adenin, xanthin and hypoxanthin, 'a decomposition-product of nuclein, occurs in traces in normal blood, in larger amount in leukemic blood, and in considerable amount in embryonal muscle, as well as in the liver, the spleen and the pancreas. It occurs, pathologically, in rapidly growing neoplasms rich in nuclei, and in the muscles of swine suffering from guanin-gout. It is transformed by nitrous acid into xanthin, by oxidation into urea and when fed to animals it increases the elimination of urea. It occurs, further, in guano, in the excrement of spiders, in the skin of amphibia and reptiles, in the silver gloss of some fish (for example the herring). *Hypoxanthin* or *sarcin*

in association with xanthin occurs in many organs and in urine. Kossel succeeded in preparing hypoxanthin from nuclein by prolonged boiling. It can be obtained from fibrin by putrefaction and the action of gastric and pancreatic juice. Xanthin can be produced from hypoxanthin by oxidation and is convertible into caffein. Xanthin and guanin have been produced synthetically by Gautier. *Paraxanthin* and *heteroxanthin* occur in the urine; *carnin*, which resembles them, in meat. An intermediate stage between nuclein and hypoxanthin is represented by the adenin ($C_5H_5N_5$) of Kossel, which has been found in the spleen, the pancreas, the thymus, the seminal fluid, and also in tea and in yeast. It appears to occur as an amorphous powder, or in six-sided columns disintegrating in the air, and as a decomposition-product of nuclein in all animal and vegetable cell-tissues.

AROMATIC BODIES.

1. *Monatomic phenols:* (a) The *phenol* (hydroxl or benzol) in the intestine; phenylsulphuric acid in the urine. (b) *Kresol* in the form of orthokresol, metakresol and parakresol combines with sulphuric acid in the urine. (2) *Diatomic phenols:* (a) *Pyrocatechin* combined with sulphuric acid in the urine. (3) *Aromatic oxyacids:* (a) *Hydroparacumaric acid;* (b) *Paroxyphenylacetic acid* in the urine. (4) *Substituted carbohydrates:* (a) *Indol* (also prepared artificially) and (b) *skatol* both occur in the intestine and combined with sulphuric acid in the urine. Stoehr has prepared skatol artificially by distillation of strychnin with calcium.

HISTORICAL.

According to Aristotle the body requires food for three purposes, namely for growth, for the generation of heat and to compensate for loss from the body. The generation of heat was thought to take place in the heart by a process of concoction, the heat being carried with the blood to all parts of the body, while the act of respiration was considered as a means for dissipating the excess of heat generated in the process of combustion. In a somewhat modified form also Galen held this view. According to him the metabolism is comparable to the conception of a lamp, the blood representing, to a certain degree, the oil, the heart the wick, and, finally, the lungs the draft. According to the view of the iatrochemical school, metabolism takes place in the body in the form of fermentative processes in which the substances introduced are decomposed in conjunction with the bodily juices. There thus result refined, useful juices, and fermentative waste products intended for excretion. Since the middle of the seventeenth century knowledge of the metabolic processes has progressed hand in hand with the development of chemistry. A. v. Haller ascribed the heat to chemical processes. He believed that the nourishment must make good to the body the constant loss of excrementitious matter. Anabolism takes place through a lymphatic fluid, which is poured out for the reconstruction of the used-up animal fibers between the latter. Mayow believed in 1679 that metabolism was essentially a process of combustion, the blood becoming bright red in the lungs. After the discovery of oxygen Lavoisier formulated the theory of combustion in the lungs, in which he believed carbon dioxid and water were formed. He compared the relatively slow course of physiological combustion with the heating of dung that takes place at a lower temperature. Mitscherlich compared the metabolic processes in the living body directly with putrefactive phenomena. Magendie first pointed out the difference between nitrogenous and non-nitrogenous foods and showed that the latter alone are not capable of sustaining life. Also gelatin alone is insufficient for this purpose. His results were less precise with respect to the nutritive value of albuminates, which he nevertheless gave foremost rank, and among which he was willing to recognize only meat as adequate as nutritive material.

The greatest advance in the principles of nutrition is due to J. Liebig, who laid the foundation of the present knowledge of metabolism. According to his view food-stuffs subserve two purposes, namely as plastic, for the growth of the body, and, as respiratory, for the generation of heat. The former includes especially the albuminates, the latter especially the non-nitrogenous carbohydrates and fats.

Among recent investigators the Munich school deserves especial mention as advancing knowledge: v. Bischoff, v. Pettenkofer, v. Voit and others. Most recently Pflüger has made important contributions.

THE SECRETION OF URINE.

STRUCTURE OF THE KIDNEY.

The kidneys are compound tubular glands (Fig. 142).

All of the *urinary tubules* arise within the cortex of the kidney from Bowman's capsules, which are globular in shape, measure from 200 to 300 μ in diameter, are constituted of endothelioid cells (k), and whose inner surface is lined with a single layer of epithelial cells (Fig. 142, II). In the interior of the capsule lies the convolution of vessels known as the glomerulus or Malpighian body. Each capsule passes by means of a narrowed portion into the convoluted urinary tubule, which has a diameter of 45 μ (I, x). This possesses a membrana propria constituted of extremely fine fibers and passes through the cortical structure in a devious course. It is lined by characteristic epithelium, the cells containing a turbid protoplasm that swells readily and is occasionally filled with fat-globules. That portion of the protoplasm which is directed toward the relatively narrow lumen of the tubule contains a distinct globular nucleus, while the portion adjacent to the membrana propria, and different also chemically, presents a fibrillated appearance, as if constituted of rods. Where the rods are in direct contact with the membrane they diverge like the bristles of a brush pressed against a surface. The free extremities of the rods of adjacent cells touch one another, so that the attached surface of the cells acquires an irregular radiating appearance. When secretion takes place the free surface has a brush-like margin.

Landauer describes the cells of the convoluted tubules and of the wider portion of Henle's loop as provided with lateral folds (and not with the rod-like structure), by means of which adjacent cells are brought in direct contact with one another.

At the junction of the medullary and the cortical tissue the convoluted tubule becomes suddenly constricted and passes over as Henle's loop in the form of an elongated arch into the medullary structure (t, t). A distinction is made in the loop between the small descending limb, with a relatively large lumen (14 μ) and clear, flat epithelium arranged in alternating order and bulged out at the middle by its nucleus (IV, S), and the wider, ascending limb. The transition from the one to the other takes place in man, as a rule, in the lowermost portion of the descending limb. The ascending limb becomes dilated to a diameter of from 20 μ to 26 μ, its lumen is relatively large and its epithelium is essentially the same as that of the convoluted tubules, except that the rods are shorter. Where the ascending limb penetrates into the cortical structure the canal at first becomes smaller again. It then passes into the intercalated portion (n, n), which has a diameter of 40 μ and in structure most nearly resembles the convoluted tubules, than which, however, it is shorter, though lined with similar cells. After a second constriction the intercalated portions pass over into the collecting tubules (o), which within the medullary rays projecting into the cortex have a diameter of about 45 μ. In their further course downward in the papilla adjacent collecting tubules unite and form tubes having a diameter of from 200 to 300 μ, the papillary ducts or excretory tubes (O), of which from 24 to 80 open at the apex of each of the 12 or 15 papillæ (foramina papillaria or cribrum benedictum).

In the lowermost and widest portion the membrana propria of the duct is surrounded and fortified by a layer of delicate connective-tissue fibers. The cells are large cylindrical epithelia, with well-defined, spherical nuclei (VI) and diplosomata. Further upward the constricted portion of the collecting tubules is lined by low, cylindrical, rather cubical cells, with large nuclei (V) supported upon a structureless membrana propria. Within the cortical structure the cells assume an inclined position, so that they overlie one another like the shingles on a roof. In the cells of all of the urinary tubules, excepting those of the collecting tubules, a ciliated process projects from the centrosoma into the lumen of the tubule. The same peculiarity is present in the epithelium of the seminal vesicles.

The Blood-vessels of the Kidney.—The renal artery with its branches reaches the junction of the medullary and the cortical structure after repeated division. From this point the interlobular arteries (a) arise at equal distances apart and traverse the cortex vertically. Throughout their entire course they give off laterally the afferent vessels (i), each of which enters into a capsule formed by the urinary tubule at a point exactly opposite to that from which the tubule

itself passes off.   By breaking up into numerous capillary loops the vascular tuft
or glomerulus is formed within the interior of the capsule.   The glomerulus is
provided toward the wall of the capsule with a covering of flat, nucleated cells

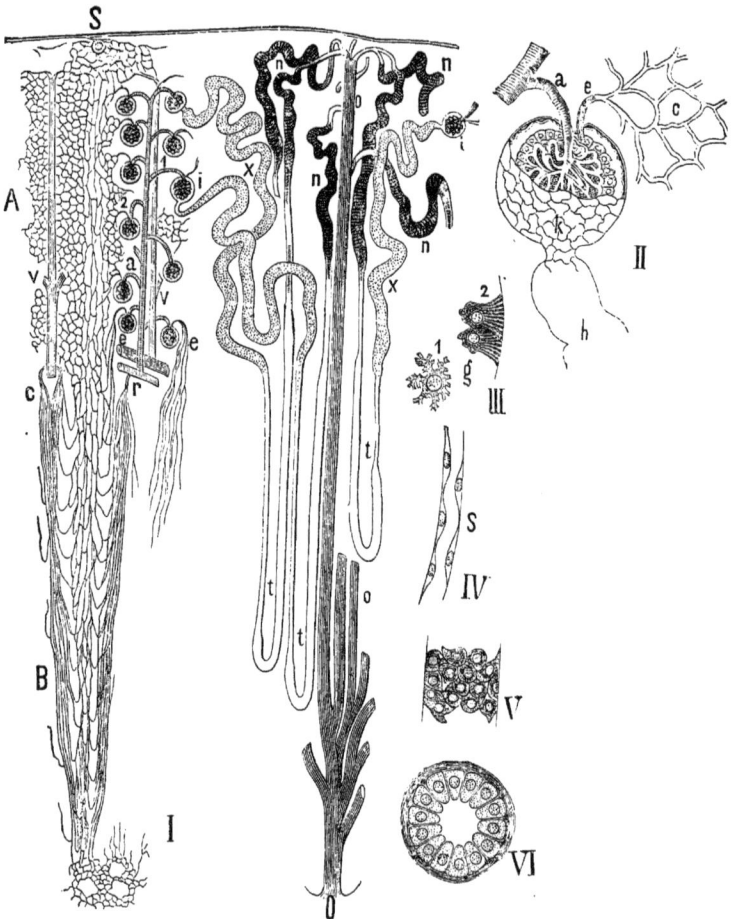

FIG. 142.—Structure of the Kidneys: I, the vessels and urinary tubules in semi-schematic arrangement; A, cortical
capillaries; B, medullary capillaries; a, interlobular artery; I, afferent vessel; 2, efferent vessel; r e, straight
arterioles; c, straight venules; v v, interlobular veins; S, origin of a stellate vein; i i, capsule enclosing a
glomerulus; x x, convoluted tubule; t t. Henle's loops; n n, intercalated portion; o, collecting tubules;
O, excretory duct; II, capsule and glomerulus: a, afferent vessel; e. efferent vessel; c, capillary network of
the cortex; k, endothelioid structure of the capsule; h, origin of the convoluted tubule; III, rod-cells from
the convoluted tubules; 2, viewed from the side (g, internal area containing the nuclei); I, viewed from the
surface; IV, cellular lining of Henle's loop; V cells in collecting tubule; VI, section of the excretory duct.

(Fig. 142, II), which are present also between the capillary loops of the tuft.
From the loops there passes, from the center of the tuft, the efferent vessel (2),
which is always of smaller size and makes its exit from the capsule close by

the side of the afferent vessel and in structure and further course resembles a small artery. Throughout the entire cortex all of the efferent vessels enter into the formation of a fine capillary network (A and II, c), which surrounds the urinary tubules. Within the range of the medullary rays of the cortex the fibers of the network, in accordance with the straight course of the urinary tubules, are arranged rather longitudinally, while in the remainder of the cortex their arrangement is polygonal. From this capillary network of the cortex venous radicles arise to form the interlobular veins (v). These originate just beneath the fibrous capsule of the kidney from the union of the radicles of the smallest venules arranged in a stellate manner (stellulæ Verheynii or stellate veins) and then pass each in the company of an interlobular artery to the junction of the medulla and the cortex.

The *vessels of the medullary structure* arise from the straight arterioles. These either begin at the junction of the cortical and the medullary structure of the kidney as individual, direct branches (r) of the interlobular arteries, still provided with muscular fibers, or they are formed from those efferent vessels (e) that lie adjacent to the medullary structure of the kidney. The latter are said to be unprovided with muscular fibers. Finally it is stated that a number of these vessels are formed from the union of the capillaries of the medullary rays. All of the straight arterioles, accompanying the straight urinary tubules, pass into elongated brush-like capillary bundles, which surround the urinary tubules. From these capillaries there collect throughout the entire extent of the medulla loops curving upward and downward, representing the beginning of the veins. The latter pass back toward the junction between the medullary and the cortical structure and gradually constitute the straight venules (c), which empty into the lower portion of the interlobular veins. On the papillæ the capillaries of the medulla communicate with vascular branches in garland-like arrangement surrounding the papillary ducts.

The *vessels of the fibrous capsule of the kidney* are derived in part from penetrating branches arising from the extremity of the interlobular arteries and in part from branches of the suprarenal, phrenic and lumbar arteries, between which anastomoses take place. The capillary network is a simple mesh-arrangement. The venous radicles pass over in part into the stellate veins and in part into veins of the same name as the arteries. A number of venous radicles also pass out of the cortex. The communication between the distribution of the renal artery and other arteries in the capsule explains the fact that after ligation of the renal artery within the kidney the blood-stream may enter from the capsule. Arterial blood also is sent to the kidney and this may even give rise to a slight secretion.

Lymphatics are present within the fibrous capsule as a wide-meshed network and beneath the capsule in the form of spaces of considerable size. In the parenchyma of the kidney itself the lymph is said to circulate between the urinary tubules and the blood-vessels, in tissue-spaces without walls which are found in larger number around the convoluted tubules than around the straight tubules. The spaces reach to the surface of the kidney and are distributed extensively beneath the capsule. Marked distention of the lymph-spaces compresses the urinary tubules and the vessels. Large lymphatics, provided with valves, are visible at the hilus of the kidney, while others pass through the fibrous capsule, both communicating with the lymph-spaces of the capsule of the kidney.

Of the *nerves*, branches provided with ganglia accompany the afferent vessels. Non-medullated fibers penetrate to the surface of the capsule and between the urinary tubules. It is established physiologically that motor fibers are present for the unstriated muscular fibers, also vasomotor fibers and sensory branches in the capsule and the pelvis of the kidney. The existence of vasodilator and secretory fibers is also probable.

The *connective tissue* of the kidney forms in the papillæ fibrillated, concentric layers about the excretory ducts (VI). Further upward star-shaped cells of reticular tissue appear in addition and these are found alone in the cortex. The outer layers of the fibrous capsule of the kidney are formed of dense bundles of fibrils, while the inner layers are looser and send processes into the cortical layer. The fatty capsule of the kidney is connected with the organ itself, in part through vessels and in part through bands of connective tissue.

Unstriated muscular fibers are contained in the kidney in three forms: (1) A sphincter-like layer surrounding each papilla; (2) a wide-meshed network on the surface of the kidney; (3) fibers that arise from the depth of the pelvis of the kidney and pass through the pyramids with the blood-vessels

H. Kostjurin found at the junction of the cortical and the medullary structure, in the dog, a layer of muscle-fibers from which bundles pass in each direction.

THE URINE.*

THE PHYSICAL CHARACTERS OF THE URINE.

The *amount* of urine in men is between 1000 and 1500 cu. cm. in twenty-four hours; in women between 900 and 1200. There is a minimum between 2 and 4 a. m., a maximum in the morning and a second maximum between 2 and 4 p. m.

The amount of urine is diminished by profuse perspiration, diarrhea, thirst, food deficient in nitrogen, reduction in the general blood-pressure, after profuse hemorrhage, as a result of the action of certain poisons, such as atropin and morphin, and in the presence of certain diseases of the structure of the kidney. The minimum that may still be considered normal is between 400 and 500 cu. cm.

The amount is increased by increase in the blood-pressure in general, or in the distribution of the renal artery alone, by copious drinking, contraction of the cutaneous vessels from the action of cold, the elimination of soluble diuretic substances, such as urea, salts, and sugar through the urine, a diet rich in nitrogen, as well as by certain medicaments, such as digitalis, juniper, squill, alcohol, etc. Carbonated beverages increase the urine in the succeeding hour.

The direct influence of the nervous system upon the amount of urine is also familiar. In this category belongs the polyuria suddenly developed after nervous perturbation, as, for instance, in hysterical persons, following epileptic attacks, and also after pleasurable excitement, and finally the remarkable increase in urinary secretion after injury of the floor of the fourth ventricle of the brain. Nocturnal polyuria occurs in persons suffering from disease of the heart and the kidneys, in cachectic states and in the presence of arterio-sclerosis. Neurasthenic anuria of neurotic origin lasting from twelve to fifty-six hours is extremely rare. The urine can be measured in graduated cylinders or flasks.

The *specific gravity* of the urine varies between 1015 and 1025. The minimum is observed after abundant ingestion of water, 1002; the maximum after profuse sweating and marked thirst, 1040. In the new-born the specific gravity falls considerably in the first few days, in conformity with the progressive increase in the amount of nourishment taken. The adult discharges per diem on the average 1 gram of solids through the urine for every kilogram of body-weight.

The determination of the specific gravity is made, with the urine at a temperature of 16° C., by means of the urinometer (Fig. 144). If but a small amount of urine is obtainable and it does not sufficiently fill the urinometer-cylinder the urine is diluted with twice or thrice its volume of distilled water, and then the last two figures on the urinometer are multiplied by two or three respectively. By means of the formula of Trapp or Haeser the amount of solids contained in 1000 parts of urine can be estimated approximately from the specific gravity. Of the number indicating the specific gravity, as, for instance, 1018, the last two figures are taken, in this instance therefore 18, and multiplied by 2.33. The estimation of the total solids can be made in a more trustworthy manner by evaporating about 15 cu. cm. of urine in a weighed porcelain-dish over the water-bath and subsequent complete drying in the air-bath at a temperature of 100° C. and cooling over concentrated sulphuric acid. In this way some urea is decomposed into carbon dioxid and escaping ammonia, in consequence of which the result is somewhat too low.

The specific gravity depends naturally upon the amount of water in the urine. The urine of the morning (urina noctis) is most concentrated, that is, heaviest, because water is absorbed from the bladder after the urine has been

---

* The illustrations are taken in part from Ultzmann and Hoffmann's Atlas of Urinary Sediments.

present for a considerable time during sleep and the urine thus becomes inspissated. The most dilute urine is encountered after copious drinking (urina potus). Hunger and laxatives diminish, while physical exertion increases, the amount of solids in the urine. Among pathological conditions, highly concentrated and copious urine, up to 10,000 cu. cm., is observed in cases of diabetes mellitus (p. 313), when the specific gravity may be from 1030 to 1060. Concentrated, scanty urine is encountered in the presence of fever. Simple, for instance, nervous, polyuria is characterized by extremely dilute and copious urine, and the specific gravity may be as low as 1001.

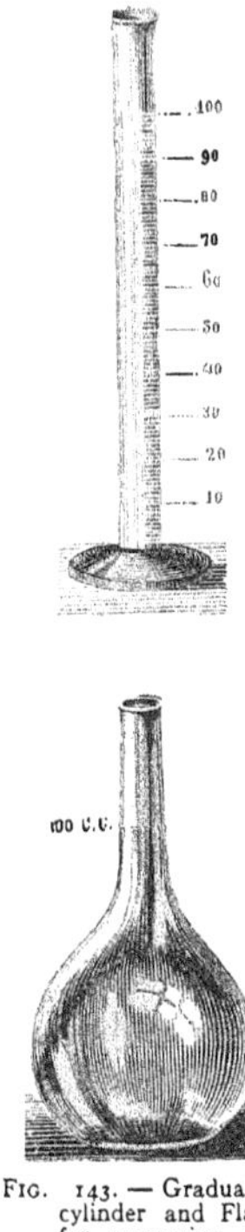

The *color* of the urine exhibits various gradations principally in accordance with the amount of water contained. Highly diluted urine is likely to be pale yellow in color. Urine of watery clearness has even been observed in association with sudden polyuria—as, for instance, the spastic urine of the hysterical. Concentrated urine, particularly after a generous meal, varies from dark yellow to brownish red in color. Urine of similar tint in association with fever is commonly designated high-colored.

Fetal urine, as well as that passed immediately after birth, is as clear as water. Admixture of blood gives rise, in accordance with the degree of disintegration of hemoglobin, to a color varying from red to deep brownish-red; biliary pigment to a deep yellowish-brown color, with an intense yellowish foam; senna, taken by the mouth, causes the urine to have a deep-red color, rhubarb a brownish-yellow color, carbolic acid a black color. Urine in a state of ammoniacal decomposition may present a dirty-blue appearance from the formation of indigo. For uniform estimation of the color of the urine a urinary color-scale has been devised empirically.

The urine, especially if in a state of ammoniacal decomposition, exhibits fluorescence, which disappears on addition of acid, and reappears on addition of alkali. Normal urine precipitates in the course of a few hours a cloud or nubecula of vesical mucus that settles slowly. The froth of normal urine is white and it disappears rather quickly, though persisting for a longer time when albumin is present. Not rarely the urine contains a number of epithelial cells.

Normal urine flows in a limpid stream like water.

The presence of considerable amounts of sugar, albumin or mucus diminishes its fluidity. So-called chylous urine from patients in the tropics may even present a white, gelatinous appearance.

The *taste* of urine is saline and bitter, the *smell* characteristically aromatic, approximating that of beef-broth, particularly after the ingestion of roast meat.

FIG. 143. — Graduated cylinder and Flask for measuring the Amount of Urine.

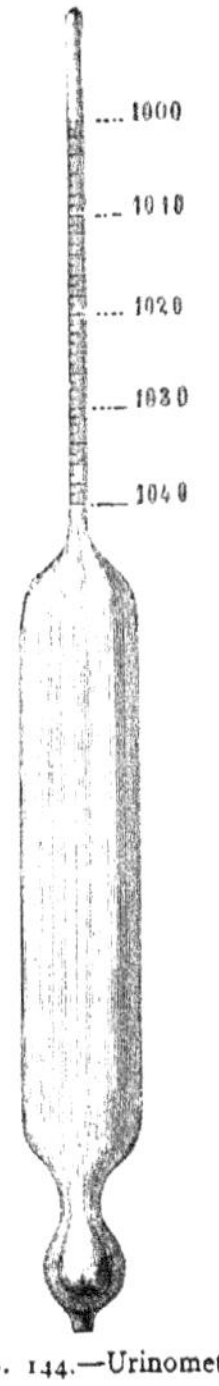

FIG. 144.—Urinometer.

Urine in a state of ammoniacal fermentation exhibits the odor of ammonia. Of substances taken by the mouth, turpentine gives rise to the odor of violets, copaiba and cubebs to an aromatic odor, and asparagus to a disgusting odor due to methylmercaptan. Valerian, garlic and castor yield up some of their odorous constituents to the urine.

The *reaction* of normal urine is acid from the presence of acid salts, especially acid monosodium phosphate (PO_4H_2Na). The latter results from alkaline disodium phosphate (PO_4HNa_2), uric acid, hippuric acid, sulphuric acid and carbon dioxid each taking up one atom of sodium, so that the phosphoric acid must be displaced to form the acid salt. After a meat-diet acid potassium phosphate especially causes the acid reaction. That the urine contains no free acid is shown by the fact that no precipitate takes place on addition of sodium hyposulphite.

Night-urine exhibits the highest, morning-urine the lowest degree of acidity. Sometimes the reaction of the morning-urine is alkaline.

The acid reaction becomes increased after ingestion of acids, such as hydrochloric acid and phosphoric acid; as well as of ammonium-salts, which are transformed in the body into nitric acid; after active muscular exercise; after a milk-diet; and pathologically in the presence of hyperacidity of the gastric juice. The absolute elimination of acid is increased by marked diuresis, while the relative elimination is diminished.

The acidity of the urine is lessened and its reaction may even be rendered alkaline: (1) By the ingestion of caustic alkalies, alkaline carbonates, or alkaline salts of the vegetable acids—the last being oxidized in the body into alkaline carbonates. (2) By the presence of calcium or magnesium carbonate. (3) By admixture of blood or pus of alkaline reaction. (4) By drainage of the acid gastric juice outside the body through a fistula; further, in from one to three hours after digestion, in consequence of the formation of acid in the stomach. (5) By the absorption of alkaline transudates, such as serum or blood. (6) In consequence of profuse secretion of sweat and hot baths. If the surface of the body is kept at a temperature of 31° C. and 30 per cent. of relative humidity, alkaline urine will be excreted in the morning-hours, on account of the fixed alkaline carbonates, while the evening-urine exhibits a strongly acid reaction. (7) The urine has rarely been observed to be alkaline in anemic persons, from deficiency of phosphoric and sulphuric acids.

The reaction is tested by means of strips of violet litmus-paper, which become red when dipped in acid urine and blue in alkaline urine. In order to determine the degree of acidity of the urine it is necessary to learn the amount of sodium hydroxid required to render exactly neutral the reaction of 100 cu. cm. of urine. For this purpose a solution of sodium hydroxid is employed of which each cubic centimeter contains 0.0031 gram of sodium; 1 cu. cm. of this solution neutralizes exactly 0.0063 gram of oxalic acid. From a graduated buret (Fig. 145) the sodium-solution is permitted to escape drop by drop into a beaker containing 100 cu. cm. of urine, with constant stirring, until violet litmus-paper no longer becomes either red or blue. The amount of sodium-solution in cubic centimeters is read from the scale of the buret, and as each cubic centimeter corresponds to 0.0063 gram of oxalic acid, the amount of oxalic acid that is the equivalent of the acid in the 100 cu. cm. of urine can be readily estimated. The degree of acidity of the urine is therefore expressed in terms of the equivalent amount of oxalic acid that is fully neutralized by the same amount of sodium hydrate.

The urine of carnivora varies in color from pale to golden yellow. It has a high specific gravity and exhibits a strongly acid reaction. The urine of herbivora has an alkaline reaction and therefore exhibits precipitates of earthy carbonates (so that it effervesces on addition of acid) and of earthy basic phosphates. In the state of hunger it acquires the character of the urine of carnivora, as under these conditions the animal in a certain measure lives upon its own tissues.

THE ORGANIC CONSTITUENTS OF THE URINE.

UREA: $CO(NH_2)_2$.

Urea, the diamid of CO_2 or carbamid must be considered as the principal end-product of the oxidation of the nitrogen-containing constituents of the body. It has the following extremely simple composition: 1 atom of carbon dioxid + 2 atoms of ammonia — 1 atom of water. It crystallizes in silky-glistening, four-sided prisms, with oblique ends, belonging to the rhombic system (**Fig.** 146, 1, 2), without water of crystallization; when, rapidly crystallized it forms delicate, white needles. It has no influence upon litmus, is odorless, and of a feeble bitter, cooling taste like that of potassium nitrate. It is readily soluble in water and in alcohol, but almost insoluble in ether. It is isomeric with ammonium cyanate, from which it develops on evaporation through atomic displacement. Numerous other modes of artificial preparation are known.

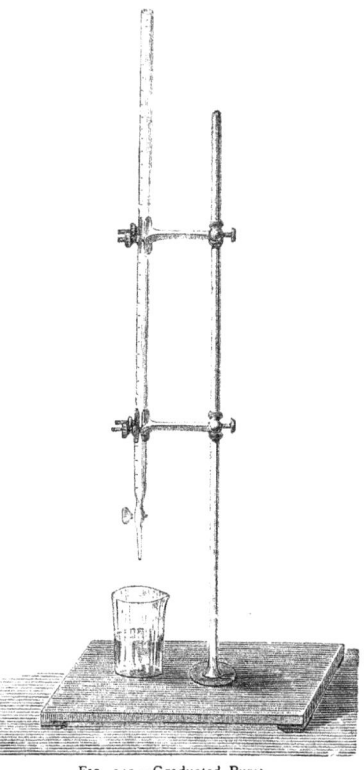

Fig. 145.—Graduated Buret.

Heated to a temperature above 120° it is decomposed, with the development of vapors of ammonia, and leaving a vitreous mass of biuret and hydrocyanic acid. In the process of ammoniacal putrefaction and as a result of treatment with strong mineral acids, of boiling with alkaline hydrates and of heating with water at a temperature of 240° C., it takes up two atoms of water and yields ammonium carbonate: $CO(NH_2)_2 + 2H_2O = CO(ONH_4)_2$. Brought in contact with nitrous acid it is decomposed into water, carbon dioxid and nitrogen. The last two forms of decomposition have been employed for the quantitative estimation of urea.

The amount of urea in normal urine is between 2.5 and 3.2 per cent. Adults excrete daily about from 30 to 40 grams; women less; children relatively more. In accordance with the more active metabolism in the latter, the amount of urea furnished by the weight-unit of the child's body, as compared with that of the adult, is as 1.7 to 1. If the body is in a condition of metabolic equilibrium almost as much nitrogen is eliminated in the form of urea as is introduced into the body with the food.

The amount of urea increases with the amount of proteids in the food, as well as with the degree of disintegration of the nitrogen-containing tissues in the body. As the latter is increased by withholding oxygen and by hemorrhage, these also cause an increase in the amount of urea. The administration of large amounts of water—by more thorough washing out of the tissues—and also of salts, frequent micturition and exposure to compressed air likewise increase the amount of urea. In diabetics who partake of large amounts of food, the amount of urea occasionally exceeds 100 grams daily, while in the state of hunger it falls to 5.6 grams. In the state of inanition a maximum of elimination has been observed toward noon, and a minimum toward morning.

Daily variations in the amount of urea pursue a course parallel with the amount of urine. Three or four hours after digestion begins the formation of urea reaches its maximum, subsequently falling again and reaching its minimum during the night. The excretion of urea, and in the same proportion that of the total nitrogen, with the urine is materially augmented in consequence of increased muscular activity. This excretion

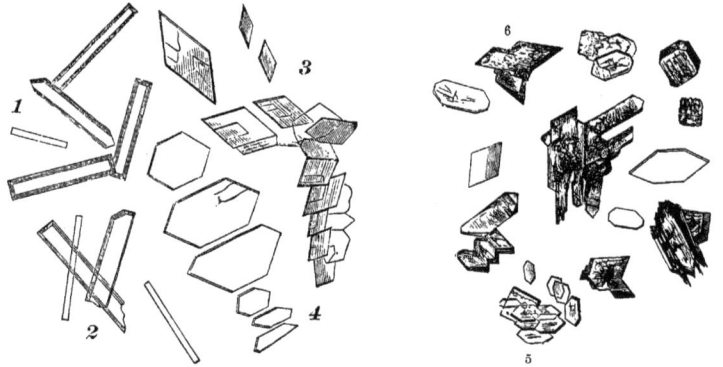

FIG. 146.—1, 2, Prisms of pure urea; 3, rhombic plates; 4, hexagonal tablets; 5, 6, irregular scales and plates of urea nitrate.

is less on the first working day, as observed in dogs, than on the second and third, but it is still increased on the two resting days succeeding the work.

Pathological.—In the presence of acute febrile inflammatory processes and of fever in general, the excretion of urea increases to the height of the morbid process, in association with which it again declines. After the cessation of the process the excretion is often subnormal. At times the formation of urea may be increased in association with high fever, but the excretion may be checked and retention of urea takes place. In the further course of the disorder the excretion may be greatly increased. In chronic diseases the amount of urea varies with the state of the nutrition, the metabolism of the patient and in accordance with the height of the accompanying fever. Degenerative disorders of the liver, as, for instance, from phosphorus-poisoning, may be attended with diminished excretion of urea and increased excretion of ammonia.

Substances that increase the proteid disintegration in the body, as, for instance, arsenic, antimony-combinations, and small amounts of phosphorus, increase the formation of urea; while those that conserve proteids, as, for instance, quinin, diminish the production. Increased formation of bile in the liver gives rise at the same time to increased formation of urea.

Urea represents the end-product of the metabolism of proteids. Next in order there stand, as lower stages of oxidation, uric acid, guanin, xanthin, hypoxanthin, alloxan and allantoin. Uric acid administered as urates appears in the urine as urea, being transformed by the liver, with increase in the secretion of bile. Muscle-extractives have the same effect, and in general increased formation of bile is attended with augmented formation of urea. After administration of leucin, glycin, aspartic acid or of ammonium-salts an increase in the excretion of urea takes place.

The liver is the principal, but not the sole seat of the formation of urea. The correctness of the supposition of Schmiedeberg that the urea is derived from ammonium carbonate through loss of water was demonstrated by v. Schröder, who found urea in large amount in blood to which ammonium carbonate had been added, and made to pass through a recently removed liver. It is, therefore, to be concluded that ammonium-combinations derived from nitrogen-containing tissues as metabolic products pass over into the circulation, through which they are conveyed to the liver for the formation of urea. The organism is capable of converting considerable amounts of ammonia, as, for instance, in the form of lactate or acetate, into urea. The liver forms urea also from the ammonia in the blood of the portal vein. In the metabolism especially of proteids there is formed in many organs by oxidation carbamic acid, CO_2NH_3, which likewise is transformed principally in the liver into urea, and also the amido-acids. If acids are taken into the body before the ammonium-combinations are transformed into urea, there result ammonium-salts, with a corresponding reduction in the amount of urea in the urine. Under pathological conditions the urea-forming activity of the liver may be diminished.

After extirpation of the liver, the urine no longer contains urea, and likewise after exclusion of the hepatic circulation, but on the other hand large amounts of ammonium-salts.

Eck, in the dog, diverted the blood of the portal vein directly into the inferior vena cava, by establishing an artificial communication between the two vessels, and then ligated the portal vein close to the liver. The dogs were seized with severe nervous attacks and convulsions. As, according to v. Schröder, ammonium-salts are transformed in the liver into urea, this transformation is thus almost wholly prevented, and the substances named now exert a toxic effect upon the nervous system.

By injection of a 6.2 per cent. solution of sulphuric acid into the bile duct, in the dog, all of the liver-cells became necrotic, and the animal died in one or two days with signs of prostration, mental derangement, loss of sensibility, central narcosis and finally convulsions. From this it has been concluded that the liver serves the purpose of converting a toxic metabolic product, carbamic acid, into an innocuous one, urea.

In birds the liver thus produces the largest amount of uric acid from the ammonium supplied. As birds readily tolerate ablation of the liver, Minkowski observed after this operation reduction in the amount of uric acid and increase in the amount of ammonium-salts in the urine.

Urea is present in the following parts of the body: Blood (1 : 10,000); lymph, chyle (2 : 1,000); liver, lymphatic glands, spleen, lungs, brain, eye, saliva, amniotic fluid; by Schöndorff it was found in the muscles and the erythrocytes and in almost all of the organs of the dog; besides, pathologically, in the sweat, as, for instance, in cases of cholera, as well as in the vomitus and in dropsical fluids of uremic patients.

The preparation of urea can be accomplished directly from dogs' urine, after generous feeding with meat, the fluid being evaporated to a syrupy consistency, extracted with alcohol, the filtered extract again evaporated, the crystals thus separated freed of the adherent extractives by means of alcohol and then dissolved in absolute alcohol. Filtration is practised again and evaporation is permitted to take place until crystallization occurs. A given volume of human urine is evaporated to one-sixth of its original volume, is reduced to a temperature of 0° and an excess of strong, pure nitric acid is added. Urea nitrate contaminated with coloring-matter is precipitated. The precipitate is filtered, expressed, dissolved in a little boiling-water, mixed with animal charcoal for the removal of the coloring-matter, and filtered hot. On cooling, decolorized crystals of urea nitrate separate from the filtrate. These are again dissolved in hot water, and barium carbonate is added so long as effervescence takes place. Barium nitrate and free urea are thus formed. Evaporation to dryness is now practised, followed by exhaustion with absolute alcohol, filtration and evaporation, after which the urea separates in crystals.

Combinations of Urea.—Urea is capable of entering into combination with acids, as nitric, oxalic or phosphoric, or with bases, or with salts, as sodium chlorid, mercuric nitrate. The most important combinations are:

1. *Urea nitrate:* $CH_4N_2O.NO_3H$, whose mode of preparation from the urine has just been described. The preparation of urea nitrate is employed with advantage for the microscopic demonstration of urea. If there are but a few drops of watery fluid in which the presence of urea is suspected—and this must be so prepared that the urea present is in concentrated watery solution—one drop of this fluid is placed upon a glass slide, a thread is laid through the middle of the drop and over both is placed a cover-slip. From the extremity of the thread a drop of concentrated nitric acid is permitted to find its way beneath the cover-slip. The characteristic crystals appear upon either side of the thread (Fig. 146, 3, 4, 5, 6). Urea nitrate is readily soluble in water, soluble with difficulty in water acidulated with nitric acid. Less commonly, when crystallization takes place slowly, it yields six-sided prisms.

2. *Mercuric-nitrate urea* is obtained in the form of a white, cheesy precipitate, when mercuric nitrate is introduced into a solution of urea. If, on the development of the precipitate, the nitric acid set free is neutralized by sodium carbonate, all of the urea eventually combines with the mercuric salt. When this point has been reached, all excess of mercuric nitrate gives rise, on addition of sodium carbonate, to the production of sodium nitrate and yellow basic mercuric carbonate. The titration-method of J. v. Liebig for urea is based upon this reaction.

QUALITATIVE AND QUANTITATIVE ESTIMATION OF UREA.

The qualitative estimation of urea aims (1) at the preparation of this substance directly as such. If its presence be suspected in an albuminous fluid mixed with blood or pus, the following course is pursued: Three or four times its volume of alcohol are added to the fluid, and filtration is practised after the lapse of several hours. The filtrate is evaporated over the water-bath, and the residue is dissolved in a few drops of water. (2) This aqueous solution is employed for the microchemic preparation of urea nitrate, which has important diagnostic significance. (3) By means of a solution of sodium hypobromite, the urea in the fluid submitted to examination is decomposed into carbon dioxid, water, and nitrogen. The nitrogen rises in the mixture in the form of small bubbles. The Knop-Hübner method of quantitative estimation is based upon this reaction. (4) A crystal of urea is cautiously fused in a dry test-tube, and yields an odor of ammonia. On cooling, it is dissolved in a small amount of water, and sodium hydrate, together with one drop of dilute copper sulphate, is added, with the development of a red-color—biuret-reaction.

Quantitative estimation of urea in the urine, according to the method of Mörner and Sjöqvist:

To 2.5 cu. cm. of urine are added 2.5 cu. cm. of baryta-mixture (1 volume of a cold saturated solution of barium hydrate and 2 volumes of cold saturated barium nitrate) and 7.5 cu. cm. of ether-alcohol (the alcohol must be 70 per cent.). The mixture is preserved for a day sealed. It is now filtered, and the filtrate, which contains of the nitrogenous substances only the urea, is evaporated at a temperature of 55° C. after the addition of 0.5 gram of magnesium oxid. After the addition of 10 cu. cm. of sulphuric acid it is further evaporated upon a boiling water-bath, until no further reduction in volume takes place. Then it is placed in a Kjeldahl boiling-flask, and the examination is continued according to the Kjeldahl method.

The method of Kjeldahl is employed for the **estimation of the total amount of nitrogen in the urine.** It is based upon the fact that all of the nitrogen is transformed into ammonia, and this is estimated quantitatively. Five cu. cm. of urine of moderate concentration are measured by means of a pipet and introduced into a flask having a capacity of about 200 cu. cm., with 20 cu. cm. of pure English sulphuric acid (to one liter of which 200 grams of phosphoric anhydrid are added), and one drop of metallic mercury; and this is boiled over the sand-bath until the fluid, which at first was dark, is entirely decolorized. On cooling, the fluid is rinsed with about 200 cu. cm. of water into a flask having a capacity of half a liter, and 100 cu. cm. of sodium hydrate (of a sp. gr. of 1.34), a few cu. cm. of an aqueous solution of potassium sulphid, and some powdered zinc are added. The flask is then quickly closed with a stopper and the ammonia set free is distilled into a receiver containing 50 cu. cm. of one-tenth normal sulphuric acid. The extremity of the tube from which the ammonia escapes must be immersed in the normal sulphuric acid. In order to determine whether all of the ammonia is present in the receiver, the stopper of the receiver is cautiously removed, a strip of litmus-paper is placed by means of a pair of forceps in front of the tube conveying the ammonia, and note is made whether the escaping distillate causes the strip to turn blue. The amount of sulphuric acid in the receiver not saturated by ammonia is determined by titration with one-tenth normal sodium hydrate.

According to Pflüger and Bohland, the amount of nitrogen in the urine can be estimated approximately by the following simple method: To 10 cu. cm. of urine, Liebig's urea-titrating solution is added from a buret, and the mixture is tested upon a dark glass plate with sodium bicarbonate drop by drop, as in the estimation of urea. If the stirred stain remains yellow, the number of cubic centimeters of titration-fluid employed is multiplied by 0.04 and in this way the percentage of nitrogen present is obtained. The total amount of nitrogen in the urine is to the nitrogen in the urea approximately as 5 to 4.

URIC ACID—$C_5H_4N_4O_3$.

Next to urea, the greatest amount of nitrogen is eliminated as uric acid, namely, 0.5 gram in 24 hours (in the state of hunger, 0.24 gram; after a generous meat-diet, 2.11 grams). The amount of uric acid is to that of urea on the average as 1 to 46, though with many variations. In the mammalian body the uric acid is formed from the nuclein of the disintegrating leukocytes. With increase in the latter, there is increase in the amount of

Fig. 147.—Graduated Pipet.

uric acid formed. Ingestion of nuclein—as, for instance, after the eating of thymus gland—increases the number of leukocytes in the blood and the excretion of uric acid. Xanthin-bases (guanin, xanthin, hypoxanthin) occur in the intestines as products of the digestion of nucleins. If they be increased in amount, an increase in the amount of uric acid results.

In birds, reptiles and insects, uric acid is the principal nitrogenous excrementitious product; while it appears in but small amount in the urine of herbivora.

The products of the decomposition of leukocytes present in surviving splenic pulp (nuclein) yield, when treated with fresh blood at the temperature of the body, an abundance of uric acid, together with xanthin and hypoxanthin. Also the nuclein of the nuclei of many other tissues has also shown itself to be a source of uric acid. In addition to uric acid, xanthin-bodies are formed in the same way. When animals are fed with nucleinic acid and hypoxanthin, the elimination of uric acid is increased.

Uric acid fed to mammals passes into the urine in part further oxidized into urea, together with an increase in the amount of oxalic acid. In hens there is increased elimination of uric acid after the administration of leucin, glycin, aspartic acid, hypoxanthin, or ammonium carbonate. The urea administered to hens is, however, eliminated chiefly reduced to uric acid.

Uric acid is dibasic, tasteless, odorless, and colorless, soluble with
great difficulty in water (in 15,000 parts of warm, or 18,000 parts of cold
water, though in 2,000 parts of a 2 per cent. solution of urea), insoluble
in alcohol or ether. It crystallizes in various forms (Fig. 148), the
basic type of which is the rhombic plate (1). Enlargement of the opposed larger angles causes the formation of the whetstone-shape frequently observed (2). If the longer sides of the latter are flattened,
six-sided plates result. Large, golden-yellow crystalline rosets (6, 8)
often separate spontaneously from diabetic urine. Precipitated by
addition of hydrochloric acid (25 cu. cm.) to urine (one liter) or of
acetic acid, the crystals usually assume the form of a barrel (9) or a
bundle of spears that are tinged brownish violet by adherent urea.

Uric acid is readily soluble in alkaline carbonates, borates, phosphates, lactates,
and acetates. Removing a portion of the base from these salts, there result,
on the one hand, acid urates; and on the other hand, acid salts from the neutral

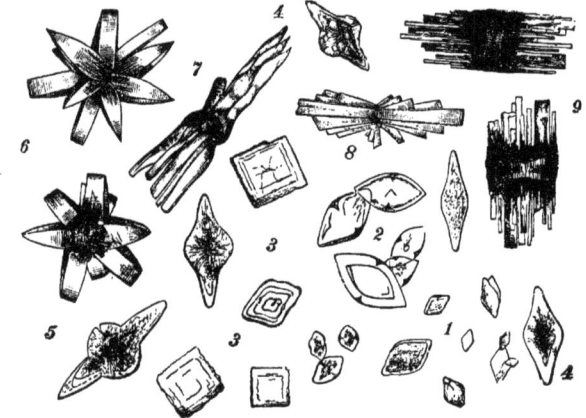

FIG. 148.—Different Forms of Uric Acid: 1, rhombic plates; 2, whetstone-shape; 3, quadratic shape; 4, 5, elongated forms with two pointed extremities; 6, 8, arrangement of several crystals in the form of a roset; 7,
crystals drawn out into the shape of a lance; 9, so-called barrel-shape obtained from human urine by means
of hydrochloric acid, in part darkly discolored.

salts. Among alkalies, lithium (citrate) is especially noteworthy as a solvent of
uric acid.

According to v. Noorden and Strauss, a favorable composition of the urine
will be obtained if calcium carbonate or calcium-salts of the vegetable acids (from
2 to 10 grams) are administered. Phosphoric acid leaves the body with the calcium through the intestines. In consequence, the monosodium phosphate in the
urine is diminished, as it gives up the phosphoric acid and thus disodium phosphate results. The latter, however, is capable of dissolving uric acid, inasmuch
as sodium urate and monosodium phosphate are formed. Uric acid is soluble in
concentrated sulphuric acid, from which it is reprecipitated by water. Plumbic
oxid converts it into urea, allantoin, oxalic acid and carbon dioxid; ozone produces the same substances, together with alloxan. Reduced by hydrogen in a
nascent state, xanthin and sarcin are produced. Horbaczewski has prepared
uric acid synthetically by fusing one part of glycin and seven parts of urea.

In the urine, the uric acid is dissolved principally in the form of acid
sodium and potassium urate. These salts are present also in urinary

sediments, urinary sand, and urinary calculi. Ammonium urate is contained in lateritious sediment in but small amount, being formed in large amount only as a result of ammoniacal decomposition of the urine (Fig. 154). Free uric acid occurs in normal urine only in the smallest amount. It is, however, not rarely precipitated subsequently on standing (Fig. 153), and it is present, further, also in urinary sand and calculi. Deficiency of neutral phosphates in the urine favors the formation of uric-acid sediment.

The urine of the new-born is rich in uric acid (uric-acid infarct of the kidneys). The uric acid, together with its salts, is increased by marked muscular activity attended with perspiration, also in the presence of catarrhal and rheumatic fevers and such as are attended with derangement of respiratory activity; further, in cases of leukemia with increased leukocyte-destruction and splenic tumor, granular liver; and, finally, quite generally in connection with gastric and intestinal catarrh following excessive indulgence in alcohol, after generous ingestion of cheese and salt fish or salt meat, after administration of glycerin, and a diet containing nuclein. Hypoxanthin fed to birds is eliminated in part transformed into uric acid.

The amount of uric acid is diminished after generous ingestion of fresh fruits (strawberries, cherries, grapes) or of quinic acid or alkaline salts of the vegetable acids contained in them; further, after hot baths; also after ingestion of proteids in large amount and after the administration of caffein, potassium iodid, sodium chlorid, sodium carbonate, lithium carbonate, sodium sulphate, inhalations of oxygen, gentle muscular exercise, though not after copious ingestion of water. In cases of gout in which uric acid is deposited in the gouty nodules, its elimination is slight. In the presence of chronic splenic tumor, anemia, and chlorosis, it is diminished, particularly if no respiratory disorder is at the same time present; and likewise in cases of epilepsy in advance of an attack.

The Urates.—With various bases uric acid forms principally acid urates, which are soluble with difficulty in cold water and readily in hot water. Neutral urates are transformed by carbon dioxid into acid salts. Hydrochloric and acetic acids dissolve the combinations and the uric acid separates in the form of crystals.

Acid sodium urate, sodium biurate, has a neutral reaction, and appears as a uratic sediment (lateritious sediment) generally of a brick-red color from uroerythrin (according to Hoppe-Seyler from urobilin)—less commonly it is between light gray and whitish in color—in the presence of catarrhal digestive disorders and of rheumatic and febrile affections. Microscopically it appears as amorphous granules (Fig. 153, b). The sediment is dissolved by heating the urine. Not rarely the sediment contains also the potassium-salt, which is entirely similar

Acid ammonium urate is soluble with difficulty in water, is always present, as a sediment, in ammoniacal urine, appears in reflected light in the form of yellowish spheres of thorn-apple or morning-star shape—in transmitted light of a darker color—and is frequently accompanied by triple phosphates (Fig. 154, a).

Acid sodium urate and acid ammonium urate are recognized by the separation of free uric-acid crystals in microscopic preparations, after addition of a drop of hydrochloric acid.

Acid calcium urate, occasionally present in urinary calculi, is a white amorphous powder soluble with difficulty in water. Fused upon a platinum plate, it leaves a residue of calcium carbonate. Rarely magnesium urate occurs in urinary calculi.

31

QUALITATIVE AND QUANTITATIVE ESTIMATION OF URIC ACID.

Qualitative Estimation.—1. The *microscopic demonstration* of uric acid and the urates is based upon the characteristics that have been described. Uric acid is precipitated from urine by addition of acetic or hydrochloric acid.

2. The *murexid test.* Uric acid or urates are heated in a shallow dish with nitric acid at a low temperature. Decomposition takes place, with the development of a yellow color. Nitrogen and carbon dioxid escape, while urea and alloxan ($C_4H_2N_2O_4$) remain behind. Evaporation is now cautiously carried further, and the resulting yellowish-red stain is permitted to cool. The addition of a drop of dilute ammonia produces a purple-red color (murexid = ammonium purpurate: alloxantinamid). This red color becomes blue on further addition of potassium hydrate. If, at the outset, potassium or sodium hydrate is added to the stain, instead of ammonia, a violet color results.

3. If upon a strip of filter-paper saturated with a solution of silver nitrate is dropped uric acid or urate dissolved in an alkaline carbonate, a black stain at once appears through reduction of the silver.

Quantitative Estimation.—1. The *method of Hopkins,* by means of which the uric acid is precipitated as ammonium urate. If 100 cu. cm. of urine are thoroughly saturated with ammonium chlorid (about 30 grams are necessary), all of the uric acid is precipitated as ammonium urate, particularly if some ammonia is added besides. After the lapse of two hours the precipitate is collected upon a filter, where it is washed several times with a saturated solution of ammonium chlorid. The precipitate is now rinsed from the filter with boiling water, and exposed to the action of hydrochloric acid with heat. The uric acid that separates is collected upon a dry filter, and is again dried and weighed.

2. The *method of Salkowski,* modified by E. Ludwig, is based upon the precipitation of the uric acid by silver nitrate and its subsequent separation and weighing. The following solutions are necessary: *A.* Twenty-six grams of silver nitrate dissolved in water, and admixed with ammonia, until complete solution takes place; then addition of water to make 1 liter. *B.* Magnesia-mixture: 100 grams of crystallized magnesium chlorid dissolved in water; ammonia added until a strong odor is developed; then ammonium chlorid to the solution; and, finally, addition of water to make 1 liter. *C.* Ten grams of pure sodium hydrate dissolved in 1 liter of water. One-half of this is completely saturated with hydrogen sulphid, and then both halves are mixed.

Mode of Procedure.—One hundred cubic centimeters of filtered non-albuminous urine (if necessary freed of albumin) are placed in a beaker. In another glass, 10 cu. cm. of the solution *A* are mixed with 10 cu. cm. of the solution *B*, and ammonia is added, if necessary also ammonium chlorid to the point of complete saturation. This solution is poured with stirring into the urine, and the mixture is permitted to stand for one hour. The precipitate is then collected upon a filter, is washed with water containing ammonia, and is brought, by means of a pipette and a glass rod, without injury to the filter, back again into the beaker. Now 10 cu. cm. of the solution *C*, diluted with 10 cu. cm. of water, are heated to the boiling-point, and this solution is passed through the used filter into the beaker which contains the silver-precipitate; the filter is then washed with hot water, and the beaker is heated for some time over the water-bath with stirring. On cooling, the solution is filtered into a dish; the filter is washed with hot water; the filtrate is acidulated with hydrochloric acid; and the product is evaporated to about 15 cu. cm., when 15 drops of hydrochloric acid are added, and the solution is permitted to stand for twenty-four hours. The uric acid separated out is collected upon a previously weighed filter, washed with water, alcohol, ether, and hydrogen sulphid; dried at a temperature of 100° and weighed. For every 10 cu. cm. of the watery filtrate, 0.00048 gram of uric acid are to be added.

KREATININ, XANTHIN-BASES, OXALURIC, OXALIC, AND HIPPURIC ACIDS.

Kreatinin ($C_4H_9N_3O_2$) is derived in part from the kreatin present in the muscles by loss of water, and in part from the meat in the food. Its amount daily is from 0.6 to 1.3 grams.

The amount of kreatinin is diminished in cases of progressive muscular atrophy, of tetanus, and of marantic, anemic, or paralytic conditions of the musculature.

It is increased particularly by greatly augmented muscular activity, after the ingestion of food rich in nitrogen. It is wanting in the urine of infants.

Kreatinin yields an alkaline reaction, is readily soluble in water and in hot alcohol, and it forms colorless, oblique rhombic columns. It combines with acids, but also with salts. Kreatinin-zinc chlorid is prepared for the detection of kreatinin.

Demonstration.—A few drops of a slightly brown, watery solution of sodium nitroprussid and then dilute sodium hydrate added to 5 cu. cm. of urine cause a Burgundy-red color that soon disappears. Addition of acetic acid changes the color to yellow. Acetone yields a similar reaction, though in the case of this substance the red color becomes still darker, almost purple, after addition of acetic acid. Acetone can first be driven off from the urine by boiling, and then the reaction of kreatinin is certain.

Xanthin-bases : Alloxuric Bases.—Under the names xanthin-bases or nuclein-bases, or, alloxuric bases, are comprised a group of bodies, including xanthin, hypoxanthin, adenin, guanin, carnin, which are related genetically to uric acid, and, together with it, are also designated alloxuric bodies. The mother-substance of all alloxuric bodies, including uric acid, is purin ($C_5N_4H_4$), from which are derived: hypoxanthin, as oxypurin; xanthin, as dioxypurin; uric acid, as trioxypurin; adenin, as 6-aminopurin; guanin, as 2-amino-6-oxypurin. By the entrance of one methyl-group into the xanthin-molecule, there result the isomers, 1-methylxanthin, 3-methylxanthin, 7-methylxanthin (heteroxanthin). If two methyl-groups enter, there are formed theobromin, paraxanthin, and theophyllin. If 3 methyl-groups enter, caffein is formed.

Salomon and Krüger found in the urine hypoxanthin, xanthin, adenin, heteroxanthin, paraxanthin, 1-methylxanthin, 7-methylguanin; and of the foregoing, respectively, in 10,000 liters of urine, 8.5 grams, 10.1 grams, 3.5 grams, 22.3 grams, 15.3 grams, 31.3 grams, 3.4 grams.

Alloxuric bases are prepared from the urine as combinations with silver or copper, and these are decomposed by hydrogen sulphid. The crude bases, treated with dilute hydrochloric acid, exhibit varying solubility. The vegetable alkaloids of coffee, tea, and cocoa are the antecedents of heteroxanthin. Paraxanthin is derived from caffein. Studies of the alloxuric bodies, therefore, are of value only after protracted abstinence from the beverages named.

Xanthin, $C_5H_4N_4O_2$, is present in small amounts only; according to E. Salkowski, it may, however, under some circumstances, be as much as one-eighth of the weight of uric acid. It is an amorphous, yellowish-white powder, quite readily soluble in boiling water. It is said to be present in the urine in somewhat greater amount after courses of treatment with sulphur, in leukemic patients and in conjunction with nephritis in children. Rarely it forms urinary calculi. It represents an intermediate link between sarcin and uric acid. Guanin and hypoxanthin can be converted into xanthin. In contact with water and ferments, xanthin is transformed into uric acid. Evaporated with nitric acid, it leaves a yellow stain that becomes yellowish red with potassium and violet red when further heated.

Hypoxanthin, sarcin, $C_5H_4N_4O$, can be prepared in the form of needles or exfoliating scales from meat, milk, bone-marrow, liver, blood from the cadaver. It is present in normal urine in smaller amount. Hypoxanthin exhibits great resemblance to xanthin, into which it can be transformed by oxidation. Hydrogen in the nascent state conversely reduces uric acid to xanthin and hypoxanthin. Evaporated with nitric acid, it yields a light-yellow stain, which becomes more intense on addition of sodium hydrate, but not reddish yellow. It is more readily soluble in water than xanthin, and a means of differentiating the two is thus afforded. Guanin is wholly insoluble in water.

Paraxanthin has proved toxic in moderate amount to dogs. Rachford found it in the urine in considerable amount in cases of severe migraine with convulsive conditions.

Oxaluric acid, $C_3H_4N_2O_4$, occurs in the urine in but small amount as an ammonium-salt, is but slightly soluble in water and appears as a loose white powder. Ammonium oxalurate can be prepared from uric acid. Perhaps there is a physiological connection between uric acid and oxaluric acid.

Oxalic acid, $C_2H_2O_4$, occurs, though not constantly, as calcium oxalate, to an amount varying from 10 to 25 mg. daily. It is recognizable from its envelop-shaped clear octahedra (Fig. 153, *d*), which are insoluble in acetic acid; biscuit-shaped or hour-glass shaped crystals (Fig. 159, *c*) are less common. The genetic relation between oxalic acid and uric acid appears demonstrated by the fact that dogs after being fed with uric acid excrete much calcium oxalate. It should, how-ever, be pointed out that the oxalic acid may also result as an oxidation-product from derivatives of the fatty-acid series. Oxalic acid is formed from oxaluric acid by the taking up of water, together with the appearance of urea.

Oxalic acid is wanting on a pure milk-diet. Almost all vegetable articles of food contain it. The ingestion of substances that contain a considerable amount of calcium oxalate, such as sorrel and tea, increases the excretion. Citric acid, treated with ozone, yields carbon dioxid and oxalic acid. The presence of calcium oxalate after the use of lemons is thus explained. Increased elimination of oxalic acid in the urine is designated oxaluria, and is considered in part a sign of retarded metabolism, as, for instance, from deficiency of oxygen, in the dog; and in part as dependent upon hyperacid-ity of the gastric juice. It may become dangerous in conse-quence of the formation of cal-culi. In conjunction with oxa-luria, the uric acid has often been found increased. The amount of oxalic acid is in-creased in the urine of jaun-diced persons. According to Neubauer, dissolved calcium oxalate, held in solution by acid sodium phosphate, also occurs in the urine. The elim-ination of this substance takes place in crystalline form the more completely, the more nearly the urine approaches a neutral reaction.

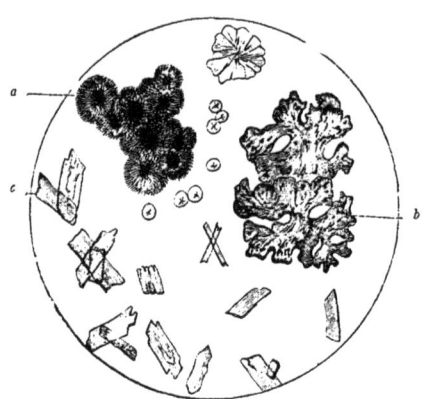

Hippuric acid, $C_9H_9NO_3$, benzoylamidoacetic acid, oc-curs in the urine of herbivora, and as the principal represen-tative of the nitrogenous prod-ucts of metabolism; and in human urine only in small amount—from 0.3 to 3.8 grams in a day. It is an odorless, monobasic acid, with a bitter taste, crystallizing in color-less four-sided prisms; and it is readily soluble in alcohol. but only in 600 parts of water. It is a conjugate acid and results in the body from benzoic acid (or from the cuticular substance of plants, which is closely related to it), or from oil of bitter almonds, cinnamic acid, quinic acid (in hay), which are readily transformed by reduction (quinic acid) or by oxidation (cinnamic acid) into benzoic acid, with which glycin combines with the giving up of water.

$$C_7H_6O_2 \quad + \quad C_2H_5NO_2 \quad = \quad C_9H_9NO_3 \quad + \quad H_2O$$
$$\text{Benzoic Acid} \qquad \text{Glycin} \qquad \text{Hippuric Acid} \qquad \text{Water.}$$

The formation of hippuric acid is, accordingly, dependent principally upon the food. It is, therefore, wanting in the urine of nursing calves, as well as after the ingestion of such vegetables as possess no cuticula; as, for instance, earthy bulbs and peeled vegetables. Similar syntheses of glycin occur in the organism also after ingestion of many other substances, as, for instance, after administration of substituted benzoic acids or of aromatic acids. As the albuminates also are capable of yielding benzoic acid and oil of bitter almonds through oxidizing agents, the hippuric acid may be formed in the body from disintegrating albuminates. This explains the fact that it is found also in the urine of fasting persons.

In the dog, the conjugation of hippuric acid takes place in the kidneys; in frogs, also outside these organs. Kühne and Hallwachs refer the formation to the liver, Jaarsfeld and Stokvis to the kidney, the liver and the intestines. The observation of Salomon that hippuric acid was present in the blood and in the liver of nephrectomized rabbits after injection of benzoic acid into the blood indicates that the formation does not take place exclusively in the kidneys. Fur-ther, the hippuric acid formed in human beings may, under pathological condi-tions, particularly when the reaction is alkaline and albuminuria is present. be again decomposed in the urine, in consequence of a fermentative process. Whether the hippuric acid formed is already decomposed in the blood and the tissues of man is doubtful. In the kidneys of swine and of the dog, fermentative decomposition of hippuric acid takes place.

After ingestion of pears, prunes, cranberries, unpeeled apples, the amount of hippuric acid increases greatly. It is increased also in the presence of jaundice, diseases of the liver, and diabetes. If it be contained in the urine in large amounts, it appears in the sediment, from which it can be isolated by boiling with alcohol. Boiled in strong acids or alkalies, or in combination with putrid substances or the micrococcus ureæ, it is decomposed again into benzoic acid and glycin, with the taking up of water.

The urine of the dog contains, in addition to uric acid, *kynuric* acid, $C_{10}H_{11}N_3O_6 + H_2O$, and *uroprotic* acid, $C_{66}H_{116}N_{20}SO_{34} + H_2O$.

Allantoin, $C_4H_6N_4O_3$, a constituent of the amniotic fluid of the cow, in lesser degree of that of human beings, is normally present in the urine in traces, especially after the eating of meat; in larger amount in the first week of life and in pregnant women, as well as after administration of thymus gland and pancreas. The amount increases after the ingestion of considerable amounts of tannic acid; in the dog, from the oxidation of uric acid fed.

Allantoin forms glistening, prismatic crystals. It crystallizes in transparent prisms from the urine of nursing calves on evaporation to a sirupy consistence, and standing at rest for a day. It is decomposed by ferments into urea, ammonium oxalate and carbonate, and another substance whose identity has not yet been established. It is readily soluble in water, with difficulty in alcohol, and not at all in ether. For its preparation, the urine is precipitated by means of basic lead acetate, the lead being removed from the filtrate

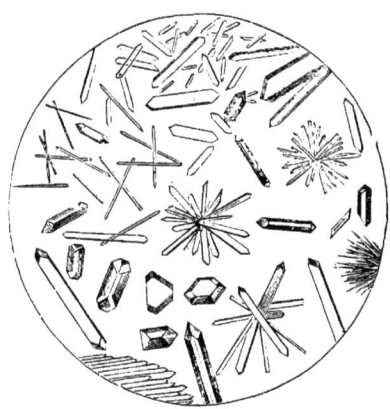

FIG. 150.—Hippuric Acid.

by means of hydrogen sulphid. The fluid is then evaporated to a sirupy consistence, and the crystals separate in the course of days. These are washed with water and recrystallized out of hot water.

Oxyproteic acid is an oxidation-product of albumin containing nitrogen and sulphur. It can be prepared as a baryta-combination, is soluble in water but not in alcohol, and is precipitable by mercuric nitrate and sulphate. On a mixed diet, it constitutes from 2 to 3 per cent. of the total nitrogen, thus somewhat more than the uric acid. It is greatly increased in cases of phosphorus-poisoning, and perhaps also in conjunction with other forms of proteid decomposition.

COLORING-MATTERS OF THE URINE.

Urobilin is present in considerable amount in highly colored febrile urine, often also in normal urine, particularly after the ingestion of readily digestible food and after the termination of gastric digestion; in small amount in the state of hunger and during the process of gastric digestion. It is a derivative of hematin, or of the biliary coloring-matter resulting therefrom. It closely resembles the hydrobilirubin of Maly, from which, however, it differs in the greater amount of nitrogen it contains. It gives the urine a red or reddish-yellow color, which becomes yellow after admixture with ammonia.

Urobilin can be extracted from some urine by agitation after addition of an equal volume of ether or chloroform. The urobilin passes over into the latter, and if this be permitted to evaporate, it remains as a residue.   It is soluble in ammonia-water or in dilute soda-solution.   If urobilin be dissolved in dilute sodium hydrate and a small amount of calomel be added, the yellow solution becomes rose-red (urorosein).   If a chloroform-extract be prepared by agitation of urine containing urobilin, and if iodin be added, and be combined by agitation with dilute potassium-solution, the solution acquires a color varying from yellow to brownish yellow, with a beautiful fluorescence in green.   This reaction can also be applied directly to any urine containing urobilin.   At times the urobilin, on standing, undergoes a modification, and then the usual reactions fail.

If sodium or potassium carbonate be added to the urine, the characteristic absorption-band at F approaches b and becomes much darker and more sharply defined.   According to Hoppe-Seyler and Saillet, urobilin develops in the urine only after evacuation by the taking up of oxygen on the part of another body forming urobilin (Jaffe's chromogen).   If acetic ether be added to recently discharged urine acidulated with acetic acid and agitation be practised, the chromogen passes over into the ether.   If the acetic ether be agitated in sunlight with water, urobilin is formed; and this can be again shaken out by means of chloroform.   If zinc chlorid be added to urine containing urobilin and rendered alkaline by addition of ammonia, the urine exhibits marked fluorescence, with a distinct green luster, particularly in reflected rays of sunlight.   The isolated urobilin is fluorescent also without addition of zinc chlorid.   Phosphotungstic acid precipitates all urobilin as a rose-colored deposit, which is soluble in water, and, after addition of hydro-

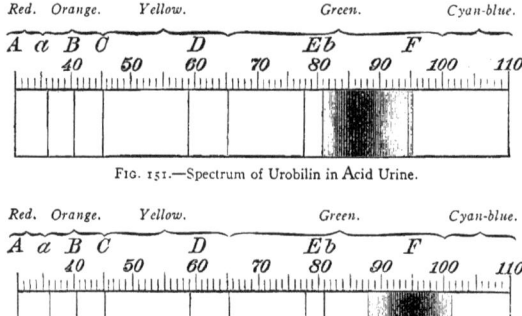

FIG. 151.—Spectrum of Urobilin in Acid Urine.

FIG. 152.—Spectrum of Urobilin in Alkaline Urine.

chloric acid, also in chloroform.   By the employment of reducing agents (sodium-amalgam) a colorless reduction-product is formed from urobilin; but this, on standing in the air, is retransformed into urobilin, with the taking up of oxygen. The colorless body is identical with the chromogen that Jaffe found in urine. In many cases of jaundice, in which, at times, Gmelin's test for biliary pigment fails to develop, urobilin is present, particularly when incomplete biliary stasis exists.   This urobilin-icterus occurs especially after the absorption of considerable extravasations of blood.   According to Cazeneuve, the urobilin is increased in all diseases that are attended with increased destruction of red blood-corpuscles.

Urochrome is considered by Thudichum as the peculiar yellow coloring-matter of the urine.   It can be isolated in yellow crusts that are soluble in water, as well as in dilute acids and alkalies.   The watery solution oxidizes in the air, with the development of a red color through the formation of uroerythrin.   Treated with acids, further decomposition-products appear; among them, uromelanin. The uroerythrin often gives the urates a red color, the urochrome a yellow color. The latter, however, is by many not considered a well-characterized chemical body.   Human urine saturated with ammonium sulphate yields, on agitation with 90 per cent. phenol, all of its coloring-matter to the latter.   If this solution of phenol be mixed with ether and water, the water is stained yellow (urochrome), the phenol-ether mixture red (urobilin and hematoporphyrin).

In the presence of melanotic neoplasms, black urine has from time to time been observed, due to *melanin* or a pigment containing iron.

A brown pigment containing iron is carried down by the uric acid precipitated on addition of hydrochloric acid. By repeated addition of sodium urate to urine and precipitation of the uric acid by hydrochloric acid, this pigment can be obtained in considerable amount.

SUBSTANCES FORMING INDIGO, PHENOL, KRESOL, PYROCATECHIN, AND SKATOL. OTHER SUBSTANCES.

Indican, or the indigo-forming substance, is derived from indol, C_8H_7N, the mother-substance of indigo, which is formed in the intestine as a result of the pancreatic digestion of proteids, and as a putrefactive product. The indol, conjugated with the sulphuric-acid residue, SO_3H, and combined with potassium, represents the indican, or indigogen of the urine ($C_8H_6NSO_4K$, potassium indoxylsulphate). It forms white glistening tables and plates, readily soluble in water, slightly in alcohol. By oxidation it forms indigo-blue:

$$2 \text{ indican} + O_2 = C_{16}H_{10}N_2O_2 \text{ (indigo-blue)} + 2\,HKSO_4 \text{ (acid potassium sulphate)}.$$

Jaffe found between 4.5 and 19.5 mg. of indigo in 1500 cu. cm. of normal human urine. Indigo is more abundant in the urine of inhabitants of the tropics, less abundant on a milk-diet, and it is wanting in the new-born. The urine of horses contains 23 times as much as human urine. Subcutaneous injections of indol increase the amount of indican in the urine. E. Ludwig obtained indican by heating hematin or bilirubin with potassium hydrate and powdered tin. It has been found also in the sweat.

Demonstration.—One-half of a test-tubeful of urine is mixed with an equal amount of hydrochloric acid, and 2 drops of a freshly prepared solution of chlorinated lime are added. The mixture becomes at first clear, then grayish blue. Now a few drops of chloroform are added, and the mixture is persistently agitated, the pigment being dissolved by the chloroform. If the mixture is permitted to stand, the blue chloroform-layer is deposited at the bottom. *For quantitative estimation*, the indican is transformed into indigo and further into sulphoindigotic acid, and this is titrated with a solution of potassium permanganate. Certain bacteria may produce indigo-blue in the evacuated urine, but also in the urinary passages; therefore, a lustrous bluish-red coating of microscopic rhombic crystals of indigo-blue upon the surface of putrid urine, or a precipitate thereof, is occasionally observed (Heller's uroglaucin).

Pathological.—Indican is increased in the urine when the formation of indol is increased in the intestines in consequence of active putrefactive fermentation; as, for instance, in cases of typhoid fever, lead-colic, trichinosis, gastro-intestinal catarrh, hemorrhage from the stomach or bowel, diseases of the small intestine, cholera nostras, carcinoma of the liver and the stomach, strangulated hernia, peritonitis. As indican is developed as a result of the decomposition of pus, an increased amount in the urine may indicate the presence of suppuration, when the intestinal conditions are normal.

Urine boiled with hydrochloric acid yields to the ethereal extract, together with indigo-blue, a garnet-red pigment, crystallizing in rhombic plates, namely *indigo-red*, urorubin, urorosein, which is developed by oxidation from an unknown chromogen. Its amount depends upon the same conditions as does that of indican. The urine thus extracted yields a brownish-black pigment to amylic alcohol, namely *uromelanin*. All urinary pigments that are produced through the activity of acids are contaminated by dark-colored, nitrogenous humin-substances, which are formed, in part, from the carbohydrates of the urine.

Reaction for Indigo-red.—One-quarter of a test-tubeful of urine is boiled continuously, with addition of nitric acid, drop by drop, until a red color is produced; it is then cooled and rendered alkaline with ammonia. If now it be gently agitated with 2 cu. cm. of ether, indigo-red dissolved in the ether passes over. The red reaction takes place in the presence of insufficiency of the intestine and its glands, in conjunction with severe diarrhea and most profound nutritive disorders.

Phenol, C_6H_6O, carbolic acid, occurs, according to Baumann, likewise united with sulphuric acid, as phenol-sulphuric acid, $C_6H_5OSO_3H$, which is found in the urine in combination with potassium. It is present in large amount in horses' urine.

Phenol results from the decomposition of albuminates through pancreatic digestion, and especially through putrefactive processes. The mother-substance is tyrosin. The formation of phenol-sulphuric acid is, therefore, entirely analogous to the formation of indican. Phenol, as well as kresol, is increased in the urine in the presence of infectious and suppurative diseases, as well as of diabetes, If phenol is employed internally or externally, the amount of phenol-sulphuric acid in the urine increases greatly. Therefore, sulphuric acid must unite with it. For this reason, alkaline sulphate is decomposed in the body, so that it may be entirely wanting in the urine. Living muscular structure or liver, digested for seven hours in a current of air with blood, with addition of phenol and sodium sulphate, forms phenol-sulphuric acid. Likewise, under these circumstances, pyrocatechin forms ether-sulphuric acid.

The dark discoloration of the urine often observed in human beings after the internal or external employment of phenol depends upon oxidation of the phenol into hydroquinone (orthodioxybenzol, $C_6H_6O_2$), which appears in the urine in large part as ether-sulphuric acid.

Parakresol (hydroxyltoluol, C_7H_8O) is present in larger amount than phenol, together with the isomers, orthokresol and metakresol, the latter in traces; also these, combined with sulphuric acid as kresol-sulphuric acids.

For the demonstration of phenol, and also of kresol, 150 cu. cm. of urine are distilled with dilute sulphuric acid. The distillate yields, with bromin-water, a precipitate of tribrom-phenol, which soon crystallizes; as well as a red color with Millon's reagent. The hydroxylbenzols—pyrocatechin and hydroquinone—are given off after protracted heating of urine to which hydrochloric acid is added. Resorcin, which is isomeric with hydroquinone, leaves the body in the urine as ether-sulphuric acid when ingested. Toluol and naphthalin react in a similar manner. When benzol is administered, it is first oxidized into phenol.

Pyrocatechin, $C_6H_6O_2$, metadihydroxylbenzol, is formed, together with hydroquinone, from phenol; and it is likewise isomeric with hydroquinone. In an analogous manner to indol and phenol, it is united with sulphuric acid. Infinitesimal amounts occur normally. It has been observed in larger amount by Ebstein and Müller in the urine of a dyspeptic child. It can be recognized by the dark discoloration of the urine that results from putrefaction.

Possibly pyrocatechin is formed in the body from decomposed carbohydrates, from which Hoppe-Seyler observed it to develop by heating with water under high pressure, as well as by treatment with alkalies.

Skatol, which appears in crystalline form in the presence of intestinal putrefaction, likewise appears in the urine as an ethereal sulphate. Brieger found potassium skatoxyl-sulphate after feeding dogs with skatol.

Demonstration.—The skatol-combination can be recognized by addition of dilute nitric acid, in consequence of which a violet color results; or of fuming nitric acid, in consequence of which red flakes are precipitated. Its amount varies with the same causes as does that of indican.

Also **hydroparacumaric acid** and **paraoxyphenylacetic acid,** which belong to the aromatic oxyacids, are encountered in the urine in increased amount, together with a large amount of indican, in the presence of urticaria, acne, and senile pruritus, as signs of increased intestinal putrefaction. The first is a putrefactive product of meat, while the second has been obtained by E. and G. Salkowski from putrid albumin.

Demonstration.—If the urine, to which a mineral acid has been added, is agitated with ether, the latter then evaporated, and the residue dissolved in water, it will yield a red color with Millon's reagent. This is the reaction of the aromatic oxyacids.

Baumann has named the following list of substances that result from tyrosin by decomposition and oxidation, of which most members develop both as a result of the putrefaction of proteids in the intestine, and pass thence into the urine.

Tyrosin, $C_9H_{11}NO_3 + H_2 = C_9H_{10}O_3$ (hydroparacumaric acid) $+ NH_3$.
$C_9H_{10}O_3 = C_8H_{10}O$ (parethylphenol, not yet demonstrated) $+ CO$.
$C_8H_{10}O + O_3 = C_8H_8O_3$ (paraoxyphenylacetic acid) $+ H_2O$.
$C_8H_8O_3 = C_7H_8O$ (parakresol) $+ CO_2$.
$C_7H_8O + O_3 = C_7H_6O_3$ (paraoxybenzoic acid, not yet demonstrated) $+ H_2O$.
$C_7H_6O_3 = C_6H_6O$ (phenol) $+ CO_2$.

Potassium sulphocyanate or **sodium sulphocyanate** is present in the urine in the proportion of from 0.02 to 0.08 gram to the liter, in larger amount in the urine of smokers. It is derived from the saliva and can be recognized by the ferric-chlorid test after acidulation with hydrochloric acid.

Succinic acid, $C_4H_6P_4$, occurs particularly after the ingestion of meat and fat, and in infinitesimal amounts after the taking of vegetable food. It occurs in considerable amount as a product of the decomposition of asparagin, after the eating of asparagus. Also, as a product of alcoholic fermentation, it finds its way into the urine through ingestion of spirit; or, administered internally, it passes undecomposed into the urine.

Lactic acid, $C_3H_6O_3$, is a constant constituent of the urine. Fermentation lactic acid has been found principally in cases of diabetes, sarcolactic acid in cases of phosphorus-poisoning and of trichinosis.

Traces of volatile fatty acids are inconstant. They occur particularly in connection with destructive diseases of the liver.

Ferments.—Diastatic, peptic, and rennet-like ferments have been found by Grützner principally in urine of high specific gravity. Fat-splitting ferment is not present normally. Trypsin is much attenuated.

Traces of **grape-sugar** occur up to between 0.01 and 0.05 per cent. After the ingestion of milk-sugar, cane-sugar, or grape-sugar (50 grams and more), these varieties of sugar appear unchanged in the urine in small amounts. Baisch found some isomaltose.

Reducing substances (yielding Trommer's reaction) are always present in the urine. Normal human urine effects reduction almost like a 0.3 or 0.4 per cent. solution of grape-sugar, in larger measure in the presence of fever. Almost five-sixths of these substances are probably combinations of glycuronic acid, while one-sixth consists of uric acid and kreatinin. There is present a dextrin-like carbohydrate and one soluble in alcohol, as well as some *animal gum.* Bechamp's nephrozymose consists principally of gum. This substance is prepared by precipitating the urine with thrice its amount of 90 per cent. alcohol. It is not a simple body, and it transforms starch into sugar at a temperature of between 60° and 70° C.

Acetone, C_3H_6O, appears after an exclusive diet of meat and fat, according to v. Noorden only on digestion of the flesh of the body. As soon as carbohydrates are taken, it is no longer observed. Also the digestion of the muscle and fat of the body occasions its appearance. Vicarelli found it in pregnant women with dead fetuses.

Demonstration.—One-half liter of urine is acidulated with hydrochloric acid and is distilled. On addition of tincture of iodin and ammonia, iodoform appears in the distillate as a cloudiness and is recognizable by its peculiar odor.

Optically inactive urine that becomes discolored brown or black on exposure to the air after addition of alkali, with the taking up of oxygen and a powerful reducing activity, contains *alkapton*, homogentisic acid, which occurs but rarely, and is produced from tyrosin (by the action of microörganisms?) within the body and then passes over into the urine.

THE INORGANIC CONSTITUENTS OF THE URINE.

The inorganic constituents of the urine either are taken into the body as such with the food and pass unchanged into the urine, or they are formed independently, inasmuch as the sulphur and the phosphorus of the food are consumed and unite with bases to form salts. From 9 to 25 grams of *salts* are eliminated daily.

During sleep, the chlorin, potassium, and sodium in the urine are greatly reduced, sulphuric acid and the solid constituents of the urine generally are somewhat reduced, while the acidity is considerably increased.

Sodium chlorid, table-salt, 12 grams (from 10 to 13 grams) daily, is increased after meals as the result of movement, of the copious drinking of water, of increase in the amount of urine generally, of generous administration of sodium chlorid, but also of potassium-salts. It is diminished principally under the reverse conditions.

Under abnormal conditions the elimination of sodium chlorid is diminished in large measure in association with pneumonia and other affections attended with inflammatory exudation; further, in conjunction with most febrile disorders, except malaria, as well as with persistent diarrhea and sweating; constantly also when the urine contains albumin and when dropsy is present. Destruction of red blood-corpuscles increases the chlorids in the urine, while, on the contrary, the amount of chlorin in the urine (as well as in the gastric juice) is diminished in the presence of anemia, although the blood contains more chlorin than normal.

Qualitative Estimation.—Urine is acidulated in a test-tube with nitric acid, and a solution of silver nitrate is added, with the result that a white, cheesy deposit of silver chlorid takes place. The albumin must first be removed from albuminous urine by boiling. On microscopic examination, attention should be given in the evaporated preparation to the terrace-like arrangement of the cubes of sodium chlorid; and, at the same time, also to the rhombic prisms of sodium-chlorid urea.

Quantitative Estimation, according to the method of Habel and Fernholz: 15 cu. cm. of the mixture of urine and baryta are acidulated, after neutralization, with 10 drops of dilute nitric acid having a specific gravity of 1119, and a solution of silver nitrate, of which 1 cu. cm. fixes 10 mg. of sodium chlorid or 6.065 of chlorin, is added so long as a precipitate of silver chlorid is observed to take place. Then a small portion is filtered into a test-tube, and a test is made to determine whether turbidity results from addition of one or two drops of the silver-solution. If this be marked, the whole amount is poured back into the beaker, 0.1 cu. cm. of the silver-solution is added, and the test is repeated until the turbidity produced by two drops of the silver-solution is no longer particularly distinct. Now an equal amount is filtered into a second test-tube, and two drops of a one per cent. solution of sodium chlorid are added. If the turbidity is equally marked with that produced by two drops of the silver-solution, the correct point has been reached. Next, exactly so many cubic centimeters of the silver-solution are added to a new specimen acidulated with 10 drops of the nitric acid; and the intensity of the turbidity in the filtrate induced by two drops of silver-solution is compared with that induced by two drops of the one per cent. solution of sodium chlorid. If the turbidity caused by the sodium chlorid is the greater, 0.05 cu. cm. less of the silver-solution is added, and the turbidity of the filtrates is compared. Then so much more or less of the silver-solution is added as represents the difference between the two points last found; and this is continued until an equal amount of silver nitrate and sodium chlorid produce equal turbidity in the filtrate.

Titration of the chlorids according to the Freund-Töpfer modification of Mohr's method: Ten cubic centimeters of urine are diluted to 25 cu. cm., and 2.5 cu. cm. of a mixture of 3 parts of acetic acid, 10 parts of sodium acetate, and 100 parts of water are added. Next, a few drops of a 10 per cent. solution of potassium.bichromate are added, and titration is practised with the silver-solution (14.63 grams in 500 cu. cm. of water), until the well-stirred yellow fluid retains a reddish tint. Every cubic centimeter of silver-solution used corresponds to 10 mg. of sodium chlorid, or 0.00607 gram of chlorin.

Phosphoric acid, about 2 grams daily, occurs in the form of mono-potassium and monosodium phosphate, and acid potassium and magnesium phosphate. It is present in larger amount after the ingestion of animal than of vegetable food. Its amount increases from the midday meal till evening, and it then declines in the night until the next morning. It is increased by muscular activity. It is derived in largest part from the alkaline and earthy phosphates of the food; and it is in part a metabolic product of lecithin and nuclein.

In the presence of fever, the increased elimination of potassium phosphate is indicative of consumption of blood and muscle. When abnormal destruction of blood takes place suddenly in the body, the phosphoric acid, together with the urea, is greatly increased. In the state of hunger, the phosphoric acid is derived principally from the breaking down of the bones, which contain thirty times as much as the muscles. Also in the presence of cerebral meningitis, softening of the bones, diabetes and oxaluria, the elimination of phosphorus is said to be increased; likewise, after administration · of lactic acid, morphin, chloral, or chloroform. It is diminished during pregnancy, on account of the formation of bone in the fetus. It is diminished also in consequence of the ingestion of ether and alcohol, and likewise of inflammation of the kidney.

Qualitative Estimation.—Potassium hydrate is added to urine in a test-tube, and heat is applied. The earthy phosphates are thus precipitated in a cloud, while the alkaline phosphates remain in solution. For qualitative estimation, there are necessary a titrated solution of uranic acetate, of which 1 cu. cm. unites with exactly 0.005 gram of phosphoric acid. To 50 cu. cm. of urine are added 5 cu. cm. of a solution of sodium acetate containing 100 grams of the latter salt and 100 cu. cm. of strong acetic acid diluted to 1 liter with water; and the mixture is heated. The titrating solution is permitted to flow with stirring so long as precipitation is apparent. As soon as free uranium oxid is present in the fluid, one drop of the mixture, to which a solution of potassium ferrocyanid is added upon a porcelain plate, yields a brownish-red reaction of uranium ferrocyanid.

In addition to phosphoric acid, phosphorus occurs in the urine in an incompletely oxidized form, namely as glycerin-phosphoric acid, about 0.05 gram daily, in larger amount in the presence of nervous diseases and after chloroform-narcosis.

Sulphuric acid is united in part with alkaline metals, in part with indol, phenol, skatol, and pyrocatechin, in the form of aromatic ethereal sulphates, both in the proportion of 1 to 0.1045 on the average. All factors that favor the formation of indol, phenol, skatol, or pyrocatechin increase the conjugate ethereal sulphates. The total amount of sulphuric acid eliminated is from 2.5 to 3.5 grams daily. It is increased after the ingestion of sulphur. The sulphuric acid is derived principally from the decomposition of albuminates. It is increased by muscular activity; and, therefore, its amount is always parallel to that of the urea eliminated. The amount of alkaline sulphates administered with the food is, as a rule, exceedingly small.

Increased excretion of sulphuric acid in febrile urine indicates increased tissue-metabolism in the body. In the presence of inflammation of the kidney a diminution has been observed, in cases of eczema a marked increase in the amount of sulphuric acid in the urine. In rabbits, but not in carnivora and human beings, administration of taurin, which contains sulphur, causes the presence of an increased amount of sulphuric acid in the urine. According to Zülzer, the relative amount of sulphuric acid in the urine is small when the secretion of bile in the intestine is large.

The **qualitative demonstration** is made by addition of barium chlorid to the urine, which yields a fine, white, insoluble precipitate of barium sulphate.

For **quantitative estimation,** 50 cu. cm. of urine are strongly acidulated with acetic acid and an equal volume of water and barium chlorid is added. After three-quarters of an hour's warming upon the water-bath, the precipitate will have been deposited. This is collected upon an ash-free filter, first washed out with water, then with warm dilute hydrochloric acid, and finally again with water. The barium sulphate thus purified is fused and weighed. It contains all of the sulphuric acid united with salts. The filtrate and the wash-water contain besides the conjugate sulphates. The combined fluid is mixed with one-eighth of its volume of concentrated hydrochloric acid and heated for a considerable time. Barium sulphate and a resinous mass separate out. The fluid is filtered, and the resinous mass is dissolved and washed from the filter with hot alcohol, and finally, again washed with hot water, then dried and fused. One part of barium sulphate corresponds to 0.3433 sulphuric acid.

In addition to sulphuric acid, sulphur (one-fifth) occurs in the urine in an incompletely oxidized form (potassium sulphocyanate, cystin, and a sulphurous substance derived from the bile). Sulphurous acid, constant in carnivora, occurs in normal human urine only when hydrogen sulphid is formed in the intestine in considerable amount. Hydrogen sulphid, which is less commonly observed, is abnormal. It is recognizable by the black discoloration of paper moistened with lead acetate and ammonia when held over the urine. It results principally through fermentation by bacteria (bacterium coli), and is rarely absorbed from the intestine or from pathological putrid foci.

Small amounts of silicic acid and nitric acid, derived from drinking-water, the latter, however, in part produced in the body itself, are present. In the fermentation of urine, the nitrates are reduced to nitrites. After the administration of salts of the vegetable acids, car-bonates appear in the urine, which then effervesces upon addition of an acid.

Sodium in the urine is principally united with chlorin, and in lesser degree with phosphoric acid and uric acid. Potassium, equaling about one-third of the sodium, is combined principally with chlorin. During fever, more potassium is excreted than sodium, while the reverse occurs during convalescence. Calcium and magnesium are present in normal acid urine dissolved as chlorids or acid phosphates. If the urine be-comes neutral, neutral calcium phosphate and magnesium phosphate are precipitated. The latter has been found also in alkaline urine in association with disorders of the stomach, in the form of large, trans-parent, four-sided prisms. If the urine becomes alkaline, calcium car-bonate (Fig. 172, a), or amorphous tribasic calcium phosphate is pre-cipitated, the magnesium, however, in the form of ammonio-magnesium phosphate (triple phosphate). The calcium is derived from food, and its amount varies in accordance with the digestive and absorptive capability of the digestive tract. In the presence of pulmonary tuber-culosis and diabetes, the excretion of calcium is increased.

Free ammonia, from 0.06 to 0.88 gram in the day, occurs also in quite fresh urine, and in larger amount with animal than with vegetable food. After administration of mineral acids, the excretion of com-bined ammonia likewise increases. The appearance of an increased amount of ammonia indicates a predominance of acids in the body and a deficiency of alkalies. *Demonstration:* A strip of red litmus-paper held over a mixture of urine and milk of lime in a covered glass becomes blue. The alkaline combinations of the organic acids diminish the excretion of ammonia. Inorganic ammonia-combinations are trans-formed into organic combinations, perhaps into an ammonium albu-minate.

Iron is never wanting in the urine, from 2.5 to 10 mg. being excreted daily. Further, some hydrogen dioxid is present, and is recognized by discoloration of a solution of indigo on addition of an iron sulphate.

One liter of urine contains 24.4 cu. cm. of gases; 100 volumes of urinary gases obtained by exhaustion contain 65.40 volumes of carbon dioxid, 2.74 volumes of oxygen, and 31.86 volumes of nitrogen. After vigorous muscular activity the amount of carbon dioxid may be doubled. The act of digestion also causes an increase, while drinking in large amount causes a reduction.

SPONTANEOUS ALTERATIONS IN THE URINE ON STANDING; ACID AND AMMONIACAL URINARY FERMENTATION.

When kept in a cool place, normal urine often exhibits a formation of newly developed acid—acid urinary fermentation. This results in consequence of the development of peculiar fermentative microörganisms, both budding-fungi, as well as fission-fungi, and is accompanied by excretion of uric acid (Fig. 153, c), acid sodium urate (b), and calcium oxalate (d). The nature of the fermentative process is not as yet entirely known. According to Brücke, lactic acid is formed from the sugar of the urine; according to Scherer, the germs decompose the vesical mucus and some urinary pigment into lactic and acetic acids. According to Röhmann, who observed acid fermentation develop only exceptionally, this is due to acids that result from the decomposition of sugar and of alcohol accidentally present. Also the occurrence of butyric and formic

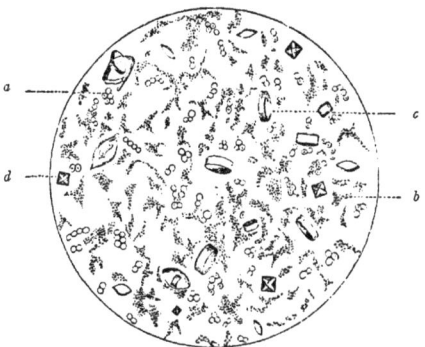

FIG. 153.—Sediment due to Acid Urinary Fermentation: a, fermentative budding-fungi; b, amorphous acid sodium urate: c, uric acid; d, calcium oxalate.

acids as products of the decomposition of other constituents of the urine has been observed. These newly formed acids expel the uric acid from the simple sodium urate, so that free uric acid and neutral sodium biurate (acid sodium urate) must be formed. With the commencement of acid fermentation, the urine appears to absorb oxygen. Even while the urine has an acid reaction, it becomes turbid and exhibits the presence of nitrous acid, whose source is as yet undetermined. The presence of nitrites is disclosed by the development of an intensely yellow color on addition of potassium ferrocyanid and acetic acid. According to v. Voit and Hoffmann, phosphoric acid is detached from acid sodium phosphate, with the formation of the basic salt, and partly displaces uric acid from sodium urate and partly causes its transformation into biurate.

On standing for some time, and more readily when exposed to heat, the urine eventually undergoes ammoniacal fermentation (Fig. 154), the urea being decomposed, with addition of water, into carbon dioxid and ammonia, as a result of the development of the micrococcus and the bacterium ureæ (Fig. 155), at times arranged like a string of pearls. The ammonia is recognizable both from its odor and from the vapor that forms when a rod moistened with hydrochloric acid is held over the urine.

The capability of decomposing urea is possessed besides by various other bacteria, including the staphylococci and the pulmonary sarcinæ, whose germs float everywhere in the air. These organisms produce a soluble ferment that decomposes urea. Miquel describes ten microörganisms that decompose urea and uric acid.

In consequence of the presence of the ammonia formed in the urine, the latter becomes turbid because substances are precipitated that are no longer capable of being held in solution, namely amorphous tribasic calcium phosphate; acid ammonium urate (Fig. 154, a) in the form of thorn-apple or morning-star spherules; and, finally, the large, clear, coffin-lid shaped crystals of ammonio-magnesium phosphate (b). There are formed, also, volatile fatty acids, principally acetic acid (from the carbohydrates of the urine). In the presence of catarrhal and inflammatory conditions of the bladder, the fermentative process may take place within this viscus. Under such circumstances, however, leukocytes (pus-corpuscles, Fig. 160), and desquamated epithelial cells are admixed in considerable amount. When pus is present in large amount, the urine becomes albuminous. Rarely, free gases form in the bladder (pneumaturia), as, for

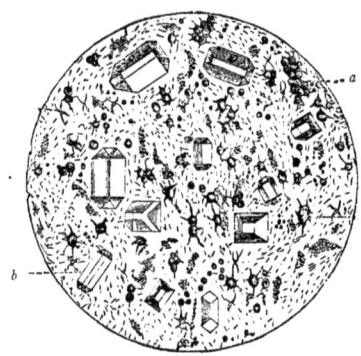

FIG. 154.—Sediment due to Ammoniacal Urinary Fermentation: a, acid ammonium urate; b, ammonio-magnesium phosphate.

FIG. 155.—Micrococcus ureæ.

instance, in consequence of the entrance of the bacterium lactis ærogenes (Fig. 126, 2). A bacillus generating hydrogen sulphid (bacterium coli commune) and one generating methylmercaptan, have also been found.

ALBUMIN IN THE URINE: PROTEINURIA, ALBUMINURIA.

For the physician, albumin is a most important abnormal constituent of the urine.

(1) **Serum-albumin** (whose properties are described on pp. 73 and 458) may appear in the urine in the absence of anatomical alteration in the structure of the kidney, and the condition has been designated by Leube as *physiological albuminuria*. Albumin has often been found normally in the urine in minute traces, particularly in consequence of the presence of a considerable amount of albumin in the blood-plasma (as, for instance, when the secretion of milk is suppressed) and after a meal containing an excess of proteid. Albumin is common, also, in the urine of the fetus and the new-born. (2) When the *pressure in the distribution of the renal vessels is increased* (as, for instance, after a cold bath or after excessive

drinking of fluids), either for a considerable time or as a transitory phenomenon, particularly in association with hypostatic hyperemia attending diseases of the heart, emphysema, chronic pleural effusions, infiltrations of the lungs; and after compression of the chest that causes stasis in the pulmonary circulation, and finally extends into the renal veins. (3) After *division or paralysis of the vasomotor nerves of the kidney*, in consequence of which intense hyperemia of the kidney is brought about. In this category belongs the albuminuria following severe and protracted painful affections of the abdominal viscera, as, for instance, strangulated hernia, in consequence of which reflex paralysis of the nerves of the renal vessels is induced. After severe *muscular exertion*, as in marches, parturition, or convulsive seizures, in cases of epilepsy, eclampsia, the convulsions attending suffocation and strychnin-poisoning. The albuminuria observed in conjunction with concussion of the brain, apoplexy, and spinal paralysis, severe emotional disturbances, excessive mental activity, and morphinism, is possibly attributable to a disorder of the vasomotor centers. (4) *Inability on the part of the epithelial cells to restrain the albumin* may cause albuminuria; and, as it appears, in consequence of defective nutrition and functional debility of the secretory elements. In this category belongs the albuminuria attending ischemia, and that following hemorrhage and attending anemic conditions, scorbutus, icterus, diabetes, and the death-agony. (5) In association with many *acute febrile diseases*, especially the acute exanthemata (as, for instance, scarlet fever); further, typhoid fever, pneumonia, and pyemia. It is probable that under such circumstances the secretory apparatus of the kidney has undergone changes (cloudy swelling of the epithelial cells of the urinary tubules, inflammation of the glomeruli) that render these incapable of preventing the escape of the albumin. (6) *Degeneration of the kidneys*, such as contraction of the kidneys, amyloid degeneration, further inflammatory processes in their various stages, are generally attended with albuminuria. Semmola has shown that the albuminuria attending nephritis is not rarely dependent rather upon the state of the blood than upon the disease of the kidneys. He believed that the renal lesion occurs in general as a secondary phenomenon, while the albuminuria is primary, except the form that is a result of the inflammation of the kidney itself. (7) Finally, *inflammatory and suppurative processes in the urinary passages*, from the pelvis of the kidney to the extremity of the urethra, may cause albuminuria. Under such circumstances, however, leukocytes are always found in the urine; not rarely, also, erythrocytes or the products of their solution, and fibrin-coagula. Certain substances that give rise to irritation and inflammation of the urinary apparatus should finally be mentioned, such as cantharides and carbolic acid. (8) The appearance of albumin in the urine after *sodium chlorid has been entirely eliminated from the food* is noteworthy. The albumin disappears when the salt is resumed.

Demonstration of Albumin in the Urine.—(a) After strong acidulation with acetic acid, a few drops of a concentrated solution of *potassium ferrocyanid* causes a precipitate.

(b) Urine to which is added one-third its volume of pure *nitric acid* exhibits a precipitate. A resulting turbidity may be due, apart from albumin, to the precipitation of urates. Slight heat, however, causes solution of the latter, while albumin remains turbid.

(c) Urine to which a few drops of acetic acid are added and which is then mixed with an equal volume of *concentrated sodium sulphate* and boiled yields a precipitate.

(d) The urine is acidulated with a few drops of concentrated acetic acid, and filtration is practised for the removal of mucin. The urine is then cautiously overlaid, drop by drop, in a test-tube held obliquely, by the following mixture: Mercuric chlorid, 8; tartaric acid, 4; glycerin, 20; water, 200. Turbidity results at the line of contact. Albumose is disclosed by the same reaction, but it is redissolved by heat. Jolles recommends the following mixture: Ten parts of mercuric chlorid, 20 parts of succinic acid, 10 parts of sodium chlorid, and 500 parts of water. Five cubic centimeters of filtered urine are acidulated with 1 cu. cm. of 30 per cent. acetic acid and 4 cu. cm. of the reagent described are added.

(e) A few drops of 30 per cent. sulphosalicylic acid are added to filtered urine. This reaction discloses also the presence of albumoses, but the precipitate due to the latter is cleared up on heating.

Boiling, by driving off the carbon dioxid, may cause a precipitate of earthy phosphates in alkaline urine, and this may simulate albumin. If, however, a small amount of acetic acid be added, the phosphates are redissolved, while albumin would be coagulated. Only small amounts of clear urine should be employed in making the tests for albumin. Turbid urine, therefore, should first be filtered.

The **quantitative estimation of albumin** is made as follows: 100 cu. cm. of urine, if necessary after addition of a small amount of acetic acid, are heated in a dish to the boiling-point, with the result that the albumin is precipitated as a flocculent deposit. The precipitate is collected upon a weighed, ash-free filter, dried at 110°, and it is washed repeatedly with hot water, then with alcohol, and is thoroughly dried in the air-bath at 110°. The dried filter is now weighed, and the weight of the filter is deducted. Finally, the filter with the albumin is reduced to ash in a weighed platinum crucible, and the weight of the ash is subtracted.

For the estimation by the *polarization-apparatus*, reference may be made to p. 268.

By means of Esbach's *albuminimeter*. A glass cylinder is filled with urine to the mark U, and with the albumin-precipitating reagent (20 parts of citric acid, 10 parts of picric acid, 970 parts of water) to the mark R, and is then closed with a stopper and agitated. After the lapse of twenty-four hours (at room-temperature) the coagulated albumin will have settled to the bottom. The divisions of the scale on the glass indicate the number of grams of albumin in 1000 grams of urine. The urine must have an acid reaction, be fresh, and its specific gravity should not be too high. The presence of an excessive amount of albumin also may therefore require dilution of the urine with from 2 to 4 times as much water. The amount of albumin obtained is then naturally to be multiplied by 2 or 4.

Globulin has been found almost exclusively in albuminous urine; and, indeed, in the majority of cases. To demonstrate its presence, 50 cu. cm. of albuminous urine are rendered feebly alkaline with potassium hydrate, and powdered magnesium sulphate is added to an amount approximating somewhat more than 24.11 per cent. If exposed to a warm temperature, all of the globulin is precipitated in the course of twenty-four hours, and it can be filtered out, dried, and weighed. With this the total amount of albumin should be compared. The presence of globulin is of unfavorable prognostic significance. Its amount is diminished by favorable circulatory conditions in the kidney.

Propeptone (Albumose).—Peptone does not occur in the urine. What has previously been described as such is propeptone. The latter occurs sometimes in acid albuminous urine: rarely, also, in urine free from albumin. Maixner found it constantly in connection with all suppurative disorders, empyema, peritonitis, pneumonia, meningitis, ulcerative affections of the digestive tract, etc.—*pyogenic propeptonuria.* Albumose is always present also in pus, and propeptonuria is a sign of the destruction of pus-corpuscles. It occurs further in connection with increased retrogressive or destructive processes in tissues rich in albumin; as, for instance,

FIG. 156.—Esbach's Albuminimeter.

in the presence of carcinoma and of fever. In the same category probably belongs also its constant occurrence in the puerperium; often, also, during pregnancy, when the fetus has died and is undergoing putrefaction—*puerperal propeptonuria.* Propeptone is found, also, when the urine contains semen.

Demonstration.—Ten cu. cm. of urine are heated with 8 grams of ammonium sulphate until the latter is dissolved. Then the hot fluid is centrifuged for a minute. The fluid is decanted, the residue rubbed up with 97 per cent. alcohol for the removal of the urobilin, then dissolved in a small amount of water, and boiled and filtered. The filtrate is subjected to the biuret-test. When the urine contains hematoporphyrin, it is advisable to precipitate this first with barium chlorid.

Egg-albumin appears after generous ingestion of fluid egg-albumin, as well as after injection into the tissues or into the blood-stream.

Mucus is present in association with catarrhal conditions of the urinary organs, particularly of the bladder. Microscopically, the presence of numerous leukocytes is noteworthy. As these contain albumin the intensity of the reaction for albumin will vary with their abundance. The characteristic reagent for mucus, however, is acetic acid, which produces a flocculent sediment also in clear filtered urine. Mucin, however, is not precipitated by boiling. The mucoid substance, *nucleo-albumin*, which is precipitated by an excess of acetic acid in dilute urine, occurs as a sign of renal irritation.

In the presence of disorders of the bladder, there rarely occurs in the urine an admixture of a peculiar ropy, gum-like substance, consisting of transformed mucus, which is thought to be the product of an anaërobic bacterium gliscrogenum. *Nucleoalbumin* also has been found, derived partly from the bladder, partly from the urinary tubules of the medullary structure; it is precipitable by acetic acid. Kolisch and Burián found *histon* in a case of leukemia, and Jolles *nucleohiston.* According to Mörner, the urine contains substances that precipitate albumin, such as chondroitin-sulphuric acid, nucleinic acid, rarely taurocholic acid, in larger amount in association with jaundice. If acetic acid be added to normal urine, these substances are eventually precipitated out.

BLOOD AND HEMOGLOBIN IN THE URINE: HEMATURIA, HEMOGLOBINURIA.

In case of hematuria the blood may be derived from any portion of the urinary apparatus. (1) In case of *hemorrhage from the kidney*, the blood is generally admixed with the urine in small amount and is well distributed. The erythrocytes under such circumstances often exhibit peculiar alterations in shape, and processes of division, which may be brought about by the action of the urea, and which have been attributed by Friedreich to independent ameboid movement (Fig. 159). The *blood-cylinders* present in the sediment are pathognostic of renal hemorrhage, that is, elongated microscopic coagula of blood, which must be considered as actual casts of the collecting tubules of the kidneys, and which are washed thence into the urine (Fig. 166). (2) In case of *hemorrhage from the ureters*, long, worm-like strings of coagulated blood are occasionally observed in the urine as casts

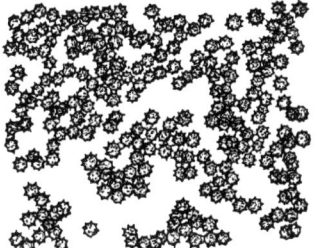 

FIG. 157.—Thorn-apple shaped Blood-corpuscles in the Urine.

FIG. 158.—Peculiar Changes in the Shape of the Red Blood-corpuscles in Case of Renal Hematuria (after Friedreich).

of the ureter. (3) Relatively the largest coagula of blood occur in cases of *hemorrhage from the bladder*. (4) Blood is present in the urine as an admixture at every *menstrual period*.

Urine containing blood should always be examined microscopically for blood-corpuscles. In addition, attention should be given to fibrin-coagula. In acid urine, erythrocytes can be recognized for as long as two or three days; though never arranged in rouleaux. If the hemorrhage has been considerable, the corpuscles are generally normal in shape. If, however, the urine is concentrated, they appear mulberry or thorn-apple shaped (Fig. 157).

The blood-corpuscles always settle gradually to the bottom in urine at rest. If the blood is slowly admixed with the urine and in small amount from ruptured capillaries, the erythrocytes appear of variable size, some not larger than between one-eighth and one-half of the normal (Fig. 159). At the same time, their pigment has become brownish yellow in color (methemoglobin). If, in a case of hemorrhage of this kind, there exists catarrhal inflammation of the bladder, numerous leukocytes, at times adherent to one another (Fig. 160), which,

32

in fresh preparations, often exhibit distinct ameboid movement, are found among the erythrocytes, which often are greatly shrunken. If the urine, as is usual, is of alkaline reaction, crystals of ammonio-magnesium phosphate will be present (Fig. 160). If the erythrocytes have already become pale, they are not rarely rendered more distinct by addition of a wine-yellow solution of iodin and potassium iodid.

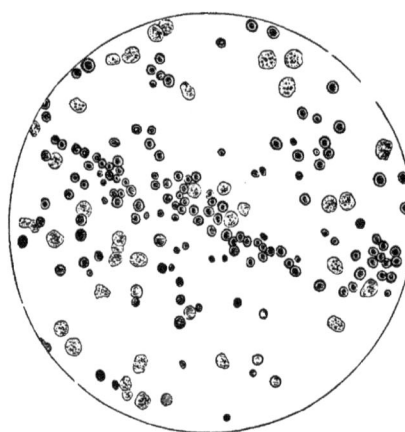

FIG. 159.—Red and White Blood-corpuscles of Varying Size.

Hemoglobinuria, that is, the elimination of hemoglobin through the urine, is entirely distinct from true hematuria. It occurs only when a considerable amount of hemoglobin has already been set free in the vessels from dissolved red blood-corpuscles (hemocytolysis). This is observed in its purest form after transfusion of blood from an animal of a different species, and also from lambs' blood in human beings. The foreign blood-corpuscles are dissolved in the blood-stream of the recipient and the hemoglobin appears in the urine. In addition, microscopic casts of the urinary tubules of coagulated globulin-like substance, stained yellow by hemoglobin, are present. Hemoglobin has been found in the urine, also, after extensive burns; after decomposition of blood in the body in cases of pyemia, scorbutus, purpura, severe typhoid fever; after the ingestion of unboiled toad-stools, and of lupins by sheep; after inhalation of hydrogen arsenid; after the entrance of azobenzol, naphthol, pyrogallic acid, toluylendiamin, potassium chlorate, chloral, phosphorus, or carbolic acid, into the circulation, as these bodies dissolve the erythrocytes; and, finally, periodically in attacks (in the horse also) of as yet unexplained nature, in which the condition appears to depend upon undue solubility of the erythrocytes, particularly from the action of external cold upon the skin.

Demonstration of Blood in the Urine.—1. The *color* of urine containing blood has been observed to be of all shades, from light red to dark brownish-black, in accordance with the amount of blood present. Often the urine is turbid.

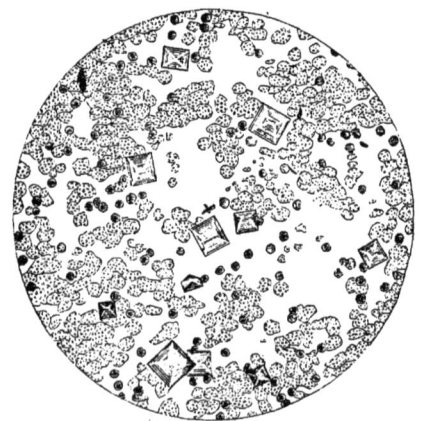

FIG. 160.—Greatly Shrunken Red Blood-corpuscles in the Urine from a Case of Catarrh of the Bladder, in the midst of numerous Leukocytes and small Crystals of Triple Phosphates.

2. Urine containing blood or hemoglobin must always exhibit the reactions of *albumin*.

3. *Heller's blood-test:* To urine in a test-tube, one-third potassium hydrate is added and moderate heat is applied. The earthy phosphates are precipitated,

and carry down with them hemochromogen, so that garnet-red flakes are deposited. When the urine contains but a small amount of blood, these flakes appear red in reflected light and greenish in transmitted light, the distinction being clear when as little as one part of hemoglobin is present in a thousand. If the earthy phosphates are already precipitated in alkaline urine, deposition is effected artificially by addition of a few drops of magnesium sulphate and ammonium chlorid, and the same change in color is apparent.

4. From the earthy phosphates thus obtained, containing hemoglobin, and collected upon a filter, *hemin-crystals* can be prepared. For this purpose the same procedure may be followed as is described on p. 62.

5. The reaction may be tested, also, with tincture of guaiac and oil of turpentine; the blood acting as a *carrier of ozone*. The urine should not lose the property of developing a blue color as a result of previous heating.

6. Urine containing blood when examined with a spectroscope exhibits characteristic appearances. The arrangement of the apparatus is shown in Fig. 161. The urine is placed in the chamber D (hematinometer) 1 cm. thick, with parallel glass walls. Through this pass the rays of light from a lamp, E, while

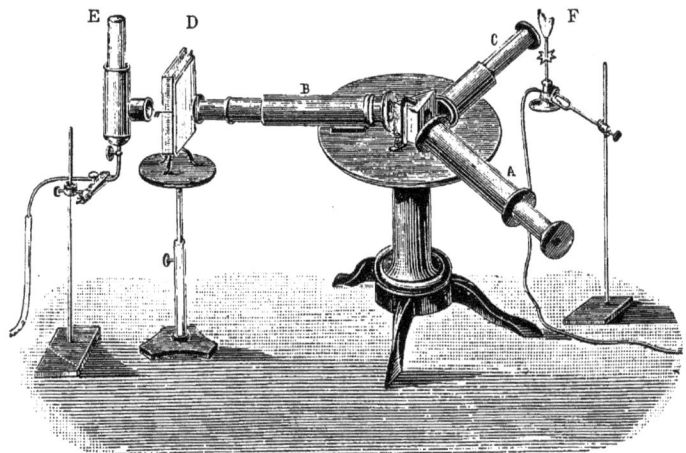

Fig. 161.—Spectroscope for Examination of the Urine as to the Presence of Hemoglobin.

another, F, illuminates a scale, and the observer makes his observation through the telescope, **A**. The examination yields the following results:

(*a*) Fresh urine containing blood exhibits the spectrum of *oxyhemoglobin* (Fig. 15). Under some circumstances, it is necessary, in this connection, to dilute the urine with distilled water and to secure perfect clearness by filtration. To confirm the observation, the oxyhemoglobin may be exposed to the action of reducing substances, which produce reduced hemoglobin.

(*b*) If concentrated urine containing blood is permitted to stand for a somewhat longer time, especially at the temperature of the blood, it acquires a deep, dark-brown color, like coffee-grounds, in the presence of an acid reaction. The hemoglobin is thus converted into *methemoglobin*. Methemoglobin in solution is, in contradistinction from oxyhemoglobin, precipitable by lead acetate. The acid solution of methemoglobin in urine thus resulting exhibits in the spectroscope a close resemblance to hematin in acid solution (Fig. 15). If the urine is now rendered alkaline, the absorption-bands of methemoglobin in alkaline solution appear. The spectra of oxyhemoglobin and methemoglobin are also found combined in the urine. When treated with reducing substances methemoglobin is transformed into hemoglobin. Later on, also hematin is present in acid solution in the urine. If such urine is treated with reducing substances, alkaline hematin appears.

Traces of *hematoporphyrin* are constant in the urine, but in considerable amount this substance is, however, rare (in cases of lead-poisoning, intestinal hemorrhage, administration of sulfonal).

Demonstration.—To 500 cu. cm. of urine are added 100 cu. cm. of a ten per cent. sodium-hydrate solution. The precipitate is washed upon a filter, dissolved in hydrochloric-acid alcohol, and exhibits spectroscopically acid hematoporphyrin.

(*c*) If urine containing blood is coagulated by boiling and the brownish-black coagulum is washed out and dried, and then extracted at gentle heat with alcohol containing sulphuric acid, a brown fluid is obtained, which, if sufficiently concentrated, proves on spectroscopic examination to be hematin in acid solution (Fig. 15, 5).

BILIARY CONSTITUENTS IN THE URINE: CHOLURIA.

The physiological factors that are of importance in connection with the presence of biliary matters in the urine have been in part already discussed (p. 319). If bilirubin is formed from hemorrhagic extravasations through the activity of the connective-tissue cells, bile-pigment may pass over into the urine, while the tissues acquire a yellow color. Cases presenting this peculiarity have been designated instances of *hematogenous* or *anhepatogenous icterus.*

The biliary coloring-matters are demonstrated by the Gmelin-Heintz test (p 317.); the appearance of the green color-ring of biliverdin can be considered as characteristic. The method has received several modifications. (1) If a considerable amount of icteric urine is passed through filter-paper, one drop of nitric acid with nitrous acid yields the color-rings upon the inner surface of the yellow-colored, and, if necessary, warmed, filter. (2) If 50 cu. cm. of icteric urine, acidulated with acetic acid, be agitated with 10 cu. cm. of chloroform, bilirubin passes over into the latter. If bromin-water be added, beautiful color-rings appear. If to the chloroform-extract oil of turpentine containing ozone be added, together with a little dilute potassium hydrate, a green color due to biliverdin appears in the watery solution. (3) Tincture of iodin diluted ten times with alcohol and overlaid on the urine gives rise to a grass-green ring. (4) According to Jolles, the following procedure yields the most distinct results: To 50 cu. cm. of urine are added 5 cu. cm. of a ten per cent. solution of barium chlorid and 5 cu. cm. of chloroform, agitation being practised for four minutes in a vessel closed with a glass stopper. After the lapse of ten minutes chloroform and precipitate are pipetted into a dish, placed over the water-bath at a temperature of 80° until evaporation takes place. The mixture is then permitted to cool. Now one or two drops of concentrated nitric acid are permitted to flow upon the precipitate at several places, and the color-rings appear. (5) The urine is rendered alkaline with soda, and calcium chlorid is added drop by drop until the fluid overlying the precipitate appears normal. The precipitate is filtered off and washed, over it is poured alcohol, and it is dissolved by means of hydrochloric acid. If the solution be boiled, a color varying between green and blue develops. When cooled, it yields a play of colors from blue to violet to red with nitric acid.

In the presence of protracted high fever, the urine at times contains only *biliprasin.* If it contains only *choletelin*, the urine, to which hydrochloric acid has been added, is examined with the spectroscope, and a pale absorption-band will be found between b and F.

Hematoidin-crystals (Fig. 92, b) are present in the urine when erythrocytes are destroyed in the blood-stream in large number. After these had been found first by v. Recklinghausen and Landois, after transfusion of heterogeneous blood, they were observed in conjunction with various infectious diseases that exercise a destructive effect upon the erythrocytes; in cases of scarlet fever; in lesser degree in cases of typhoid fever; and Landois with Strübing observed them in the urine in association with attacks of periodic hemoglobinuria. Landois refers the biliary acids often observed by him in the urine after solution of the erythrocytes to the hemoglobin of the destroyed corpuscles. If old collections of blood rupture into the urinary passages, as in cases of pyonephrosis or in conjunction with the perforation of necrotic areas, the appearance of the crystals is comparable to that in the sputa in analogous cases. In cases of hypostatic icterus, bilirubin, which is identical, was found in crystalline form.

The biliary acids, which Dragendorff demonstrated to the extent of 0.8 gram in 100 liters of normal urine, appear in larger amount in connection with resorption-icterus, although even under such circumstances never in considerable amount.

Landois observed them, also, in association with the passage of biliary matters in consequence of marked destruction of erythrocytes. Their properties and reaction have already been described (p. 315), a solution of cane-sugar 0.5 gram to one liter of water being employed for the latter. Urine of low specific gravity should be concentrated upon the water-bath. To insure absolute certainty, a portion of urine is evaporated over the water-bath almost to dryness, and the residue extracted with alcohol. The alcoholic extract is again carefully evaporated in a porcelain dish, and the residue dissolved in a few drops of water and subjected to Pettenkofer's test. If the test is applied directly to the urine, one must previously have convinced himself that the urine is free from albumin, as this substance yields a similar reaction. In such an event the albumin should be removed by boiling and filtration. If filter-paper is dipped into urine to which cane-sugar has been added, and the paper is dried and brought in contact with sulphuric acid, a violet-red color results, which is particularly pretty in transmitted light.

SUGAR IN THE URINE: GLYCOSURIA.

Normal urine contains traces of dextrose. Small amounts of sugar are present after ingestion of sugar in large amounts (alimentary glycosuria), and also in the presence of fever, after the drinking of beer supplemented by alcohol, occasionally in the exceedingly obese, in neurasthenics, in association with cerebral disease, and in advanced age. Glycosuria occurs also as a result of intestinal activity in ill-nourished individuals; and, artificially, after ligation of the mesenteric arteries. Dextrosuria of considerable degree is a sign of diabetes mellitus. In this connection, the large amount of urine, up to 10,000 cu. cm., as well as the high specific gravity, from 1030 to 1040, are striking. The diabetic patient excretes a relatively larger amount of water through the kidneys; and, on the other hand, a relatively smaller amount through the skin (and the lungs?) than a healthy person. Also the elimination of the water ingested takes place later and more uniformly than in health. The urine is pale yellow in color, although the amount of coloring-matter is, in the aggregate, by no means diminished; and the nitrogenous matters are increased. A diet of carbohydrates generally increases the excretion of sugar; while a proteid diet may reduce it. Uric acid and calcium oxalate are often found increased at the commencement of the disease. On standing for a considerable time yeast-cells constantly develop in the urine.

For quantitative estimation the tests for sugar already described (p. 268) are appropriate, although the urine must be free from albumin or be rendered so. The following tests are most to be recommended:

(a) The fermentation-test is the most reliable. A test-tube inverted over mercury is filled with the saccharine urine and a piece of yeast, living and free from sugar, as large as a pea, and also one drop of tartaric acid, are added, and the mixture is kept in a warm place. Carbon dioxid collects at the bottom of the inverted tube, and disappears after the introduction of potassium hydrate.

(b) A 2.5 per cent. solution of copper sulphate and a solution containing 10 parts of sodiopotassic tartrate in 100 parts of a 4 per cent. solution of sodium hydrate are employed. Five cubic centimeters of urine are boiled in a test-tube, and from 1 to 3 cu. cm. of the copper-solution and 2.5 cu. cm. of the tartaric-acid solution in a second test-tube. The boiling of both fluids is interrupted simultaneously, and after the lapse of from 20 to 25 seconds, the contents of the one tube are poured without agitation into the other; reduction then takes place spontaneously.

(c) Böttger's test with Nylander's modification (p. 267).

(d) In the application of the phenylhydrazin-test, 5 cu. cm. of urine are diluted with 5 cu. cm. of water, and 0.5 of phenylhydrazin hydrochlorate and 1 gram of sodium acetate are added. The mixture is boiled for two minutes over the water-bath, is permitted to cool slowly and to stand for four hours in the cold. Combinations of glycuronic acid form similar, though plumper, crystals, more like thorn-apples.

(e) In applying Molisch's test, a-naphthol dissolved in chloroform, instead of in alcohol, is employed. The test discloses the presence of all of the carbohydrates in the urine, under normal circumstances 0.96 per cent. altogether, of which 0.1 is grape-sugar. Urine containing sugar should be diluted 100 times.

(f) If to 10 cu. cm. of diabetic urine in a test-tube 0.5 mg. of powdered gentian-violet are added, the urine is colored, while normal urine is not.

Quantitative estimation is made by fermentation or by the titration-method. The estimation by circumpolarization is, according to Worm-Müller, almost valueless for the estimation of the amount of sugar in diabetic urine, as the urine often contains in part as yet unknown optically active substances. If, however, it be desired to employ this method, the urine must be previously agitated with commercial animal charcoal and filtered, in consequence of which it becomes colorless. Small amounts of *glycogen* derived from urinary tubules that have undergone glycogenic degeneration have been found by Leube in diabetic urine.

After ingestion, the sugars that are most readily decomposed pass with greatest difficulty, while those that are not at all decomposable pass most readily, into the urine. If, therefore, considerable amounts of dextrose are administered, a portion thereof passes into the urine; and a larger amount in cases of diabetes than in health. Ingested levulose does not increase the amount of sugar in the urine of a diabetic patient. The use of starch in considerable amounts never gives rise to the presence of sugar in the urine in health, although it increases the amount of sugar in cases of diabetes. The ingestion of *cane-sugar* or of *milk-sugar* in considerable amount causes the passage of small amounts of each into the urine during health. The diabetic, under such circumstances, excretes an increased amount of dextrose. According to Külz, the cane-sugar ingested by a diabetic patient is decomposed into grape-sugar and fruit-sugar; the latter is consumed in the body, the former in part excreted. The same takes place with milk-sugar.

Levulose is rarely present in the urine, constituting *levulosuria*.

In severe cases of diabetes mellitus, Külz found *levorotatory β-oxybutyric acid,* the next higher analogue of lactic acid, in the urine, from the oxidation of which diacetic acid is produced. The latter, in its turn, is readily decomposed into carbon dioxid and acetone. β-oxybutyric acid is never wanting when diabetic coma is present. *Acetone* is present in the urine of diabetics often in considerable amount, principally in association with progressive loss of strength, and often even in spite of administration of carbohydrates. From oxybutyric acid there results, by dehydration, a-cro-

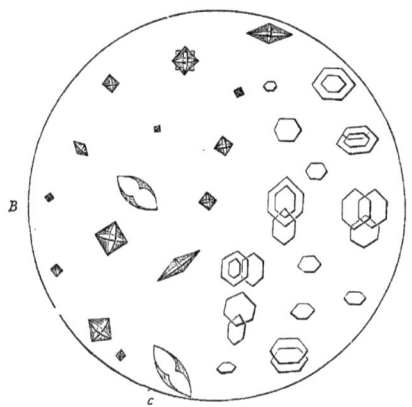

FIG. 162.—*A*, crystals of cystin: *B*, of calcium oxalate; *c*, hour-glass shaped crystals of calcium oxalate.

tonic acid, which Stadelmann found in diabetic urine. As albuminuria results from administration of acetone, the complication of albuminuria with diabetes is clear.

Milk-sugar—lactosuria—is present in the urine of puerperal women, together with glucose and isomaltose, chiefly in connection with milk-stasis. The condition is thus due to absorption from the breasts. Milk-sugar likewise appears in the urine of infants with derangement of digestion.

Pentose has, on several occasions, been observed in the urine: pentosuria. This substance contains 5 atoms of carbon, is not susceptible of fermentation, and is capable of causing reduction. It may possibly be due to disease of the pancreas. Phloroglucin and hydrochloric acid yield a red color. Pentose is present in coffee, in many wines, and in varieties of milk and sugar. Ingested pentoses—arabinose, xylose—pass over into the urine.

Reichart has called attention to the simultaneous appearance of *dextrin* in urine containing sugar. *Inosite* has been found both in cases of diabetes and in cases of polyuria and albuminuria. Traces of it are contained even in normal urine. Occasionally, "sugar-puncture" in animals is followed by the appearance of inosite instead of dextrose in the urine. For the detection of inosite, the dextrose is removed by fermentation; and albumin by boiling after addition of a few drops of acetic acid and sodium sulphate. Of the filtrate, a few cubic centimeters are evaporated in a porcelain dish down to a few drops; then 2 drops of a solution of mercuric nitrate (titration-solution according to J. v. Liebig) are added. A yellow precipitate takes place. If this is spread out and further carefully heated to a point beyond desiccation, a dark-red color appears, which on cooling gradually disappears.

The sugar may, in rare cases, also give rise to *pneumaturia*, fermentation by microbes causing the development of carbon dioxid.

CYSTIN.

Cystin, $C_6H_{12}N_2S_2O_4$, is a levorotatory body that occurs normally in [traces in the urine and but rarely in considerable amount. It appears in the form of colorless, six-sided plates (Fig. 162, A), in children also forming concretions. Cystin is insoluble in water, alcohol, and ether; readily soluble in ammonia, after the evaporation of which it crystallizes out. According to Baumann and Preusse, there exist intermediary products of metabolism that contain the material necessary for the formation of cystin. When the metabolism is normal, these, however, undergo further change; and their sulphur appears in the urine oxidized as sulphuric acid. In rare cases this oxidation fails to take place; and then the sulphur appears in the urine as cystin. In cases of phosphorus-poisoning the cystin is increased.

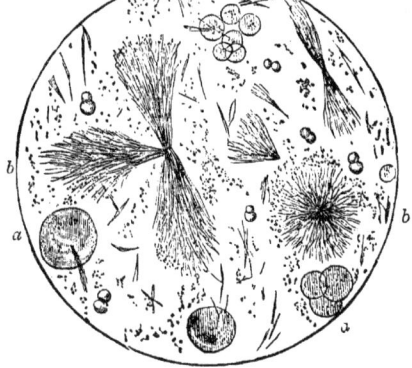

FIG. 163.—*a a*, Leucin-spheres; *b b*, tyrosin-sheaves; *c*, double spheres of ammonium urate.

LEUCIN, $C_6H_{13}NO_2$, AND TYROSIN, $C_9H_{11}NO_3$.

Both of these bodies, whose development has been referred to in the consideration of pancreatic digestion, are present in traces in normal urine. They occur together in somewhat larger amount in association with derangements in the function of the liver-cells (acute yellow atrophy of the liver, phosphorus-poisoning). As the elimination of urea is generally diminished at the same time, it may be concluded that the liver is the seat of the formation of urea.

Leucin, which separates either spontaneously in the precipitate or only after evaporation of an alcoholic extract of the inspissated urine, appears in the form of yellowish-brown spheres (Fig. 163, *a a*), occasionally with concentric radiation or provided with fine points at the periphery. When heated dry leucin sublimes without fusing.

Tyrosin forms silky, colorless sheaves of needles (Fig. 163, *b b*). If a solution of tyrosin be boiled with Millon's reagent, there results at first a pretty red color, and shortly afterward a deep brownish-red precipitate. If tyrosin is gently heated with a few drops of concentrated sulphuric acid, it is dissolved with the development of a transitory deep-red color. If it now be diluted with water, and barium carbonate be added to the point of neutralization, the mixture boiled and filtered, and dilute iron chlorid added to the filtrate, a violet color appears. Dissolved in hot water, addition of quinone produces a red color.

SEDIMENTS IN THE URINE.

Both in normal, as well as in pathological urine, precipitates may form at the bottom of the vessel; and these are designated *sediments*. They are either *organized* or *unorganized*.

ORGANIZED SEDIMENTS.

(A) *Sediment of blood:* derived from erythrocytes and leukocytes (Figs. 157, 158, 159, 160), occasionally also shreds of fibrin (Figs. 6, 7).

(B) *Pus-corpuscles,* in greater or lesser amount in association with catarrhal or inflammatory processes in the urinary passages, entirely resemble the leukocytes (Figs. 6, 7). Marked, persistent admixture of pus is indicative of profound parenchymatous suppuration; numerous mononucleated leukocytes, of disease of the kidneys. **Demonstration.**—If the supernatant fluid be poured off and a bit of potassium hydrate be dissolved in the sediment, the pus is converted into a vitreous, ropy mass, later becoming more consistent (alkali-albuminate). Mucus treated in this manner is dissolved into a thin fluid admixed with flakes.

(C) *Epithelial cells* of varied shape and not always distinguishable as to the source whence they are derived. They are more abundant in the presence of catarrhal conditions in the parts in question. In the urine of women, pavement epithelial cells from the vagina are also present. The spermatozoids likewise are included among epithelial structures.

(D) *Lower forms of organisms.* The freshly collected urine from healthy

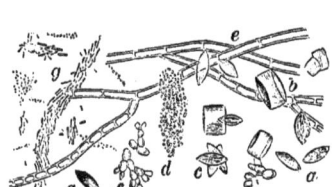

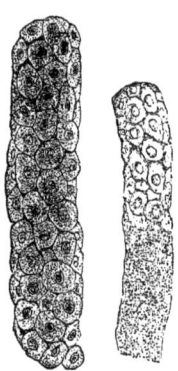

FIG. 164.—*e,* Molds; *f,* budding-fungi (yeast); *d g,* bacteria (micrococci and bacilli); *a b c,* uric acid (after v. Jaksch).

FIG. 165.—Epithelial Tube-casts.

persons always contains many microörganisms, which, however, have probably been washed away from the urethral mucous membrane. They are principally large or small diplococci. In cases of gonorrhea, gonococci thus gain entrance into the urine. Lower forms of organisms may also appear in the urinary passages, as, for instance, in the bladder, when their germs have been introduced by means of unclean catheters. The following varieties may be distinguished:

1. *Schizomycetes* (fission-fungi). In pathological cases bacteria may gain entrance into the urinary tubules and the urine from the blood. Bacterial cultures injected artificially into the vessels are in part eliminated through the kidneys. In urine undergoing alkaline fermentation, both micrococci and rod-shaped bacteria or bacilli appear (Fig. 164). The sarcinæ are further included among schizomycetes.

2. *Saccharomycetes* (fermentative germs): (a) The germ of acid fermentation of urine (saccharomyces urinæ): small vesicular cells, arranged partly in groups, partly in rows (Figs. 153, a; Fig. 164, f). (b) Yeast (saccharomyces fermentum, Fig. 140) is present in diabetic urine.

3. *Phycomycetes* (molds) appear in putrid urine as mold-formations (Fig. 164, e). They are without significance.

(E) Of great significance in the diagnosis of certain diseases of the kidney is the occurrence of so-called *urinary cylinders,* that is, casts of the urinary tubules. If these structures are relatively thick and rather straight, they are probably derived from the collecting tubules of the kidney; while if they are thinner and tortuous, their source is suspected to be the convoluted tubules.

Various kinds of tube-casts can be distinguished: 1. *Epithelial casts* (Fig. 165), which consist of coherent and desquamated cells of the urinary tubules. They indicate that no profound change has as yet taken place within the kidney, but that, as in catarrhal inflammatory states of mucous membranes, the epithelial cells are in process of desquamation. 2. *Hyaline tube-casts* (Fig. 171) are completely homogeneous and transparent. They are most readily demonstrated by addition of a solution of iodin to the preparation. They are generally long and narrow; occasionally, they present fine disseminated points, or fat-granules (finely granular tube-casts, Fig. 169). They appear not to be derived from a transudation from the blood, but as a result of the secretory activity of the epithelial

FIG. 166.—Blood-casts.

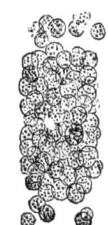

FIG. 167.—Casts of Leukocytes (after v. Jaksch).

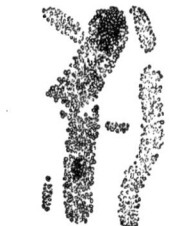

FIG. 168.—Acid Sodium Urate in. the Form of Tube-casts.

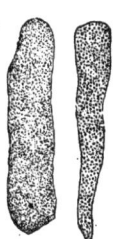

FIG. 169.—Finely Granular Tube-casts.

FIG. 170.—Coarsely Granular Tube-casts (after v. Jaksch)

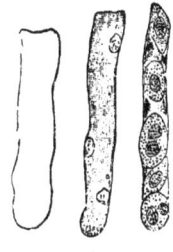

FIG. 171.—*a*, Hyaline tube-cast; *b*, hyaline tube-cast with leukocytes; *c*, hyaline tube-cast with renal epithelium (after v. Jaksch).

cells of the urinary tubules. 3. *Darkly granular tube-casts* (Fig. 170), brownish yellow, opaque, and consisting wholly of a granular mass, are usually somewhat wider than hyaline tube-casts. Marked variations of the latter occur. Not rarely, they exhibit fattily degenerated or atrophic epithelial cells of the urinary tubules. 4. *Amyloid tube-casts* occur in cases of amyloid degeneration of the kidneys. They have a waxy luster, are completely homogeneous (Fig. 171, *a*) and yield, with sulphuric acid and solution of iodin, the blue color of amyloid reaction. 5. *Blood-casts*, consisting entirely of coagulated blood, with distinct blood-corpuscles, occur in association with capillary hemorrhage into the tissue of the kidney (Fig. 166). These are allied to the casts found in connection with hemoglobinuria; as, for instance, after transfusion of heterogeneous blood. They consist of hemoglobin or of its globulin tinged with hematin. The tube-casts stained yellow that have been observed in conjunction with icterus probably also result from the albumin of dissolved blood-corpuscles. Urine containing tube-casts is always albuminous.

Tube-casts of leukocytes are observed in connection with suppurative processes in the urinary tubules (Fig. 167). *Urates* arranged in the shape of tube-casts are without significance (Fig. 168); as well as *cylindroids*, formed of mucus, with which short strands of mucus arising in the ureter, the bladder, the prostate, the uterus, and the vagina, may be confounded.

UNORGANIZED SEDIMENTS.

The unorganized sediments, in part crystalline, in part amorphous, have already received consideration in the discussion of the individual constituents of the urine.

SCHEMATIC RÉSUMÉ FOR THE RECOGNITION OF ALL OF THE SEDIMENTS IN THE URINE.

I. In acid urine there may be found—

1. An *amorphous crumbling sediment*,

(*a*) Which is soluble in the heat and is again precipitated in the cold, and which, on addition of a drop of acetic acid to the microscopic preparation, forms crystals of uric acid, which often has a reddish color (brick-dust powder).

This sediment consists of *sodium* or *potassium biurate* (Fig. 153).

(*b*) The sediment is not dissolved by heat, but on addition of acetic acid, and without effervescence. This is probably *tribasic calcium phosphate.*

(*c*) Highly refracting granules, occurring occasionally and soluble in ether.

FIG. 172.—*a*, Finely granular calcium carbonate; *b* and *c*, crystalline neutral calcium phosphate.

are *fat-globules*. Fat occurs in the urine particularly in conjunction with the presence of a round-worm (filaria sanguinis hominis) in the blood (only in foreigners or travelers); further, occasionally together with sugar in the urine, in tuberculous patients; in cases of phosphorus-poisoning, of yellow fever, of pyemia; after protracted suppuration; and, finally, after injections of fat or milk into the circulation. Fatty degeneration in some portion of the urinary apparatus, admixture of pus from old abscesses, and severe injuries to bones, should further be taken into consideration. In this connection, attention should be given also to cholesterin and lecithin. Rarely, the amount of fat in the urine may be so marked as to give rise to a creamy appearance—chyluria.

2. A sediment consisting of crystals:

(*a*) *Uric acid* (Fig. 148 and Fig. 153—whetstone-shaped crystals).

(*b*) *Calcium oxalate* (Fig. 153, Fig. 162, *B*)—envelop-shaped crystals, insoluble on addition of acetic acid.

(*c*) *Cystin*—extremely rare (Fig. 162, *A*).

(*d*) *Leucin* and *tyrosin*—of great rarity (Fig. 163).

II. In alkaline urine there may be present:

1. The sediment is wholly *amorphous* and *crumbling;* it consists of *tribasic calcium phosphate.* It is soluble on addition of acids without effervescence.

2. The sediment is *crystalline*, or, at least, of *characteristic form.*

(a) *Ammonio-magnesium phosphate* (Figs. 173, 160, 154): Large coffin-lid crystals, immediately soluble on addition of acids.

(b) Small globules, yellowish in reflected light, dark in transmitted light, often provided with points; thorn-apple or morning-star shaped, together with amorphous granules (Figs. 154 and 175). These consist of *acid ammonium urate.*

(c) *Calcium carbonate:* Small whitish globules, biscuit-shaped or arranged side by side in irregular masses. together with amorphous granules. Efferves-

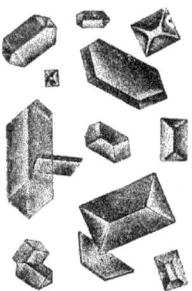

FIG. 173.—Ammonio-magnesium Phosphate.

FIG. 174.—Imperfectly Developed Crystals of Ammonio-magnesium Phosphate.

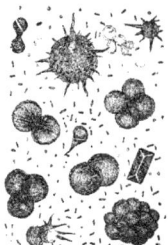

FIG. 175.—Acid Ammonium Urate (after v. Jaksch).

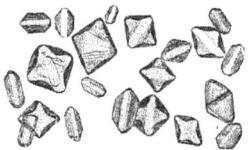

FIG. 176.—Basic Magnesium Phosphate.

cence a es place on addition of acids, also in the microscopic preparation (Fig. 172, a) k

(d) *Leucin* and *tyrosin* are extremely rare (Fig. 163). Crystals of *neutral calcium phosphate* (Fig. 172, c), with their spear-shaped extremities in contact, are also rare, as well as plates of *basic magnesium phosphate* (Fig. 176).

Organic sediments may occur both in acid, as well as in alkaline, urine. Among them, *pus-corpuscles* are present especially in alkaline urine, and the *lower forms of vegetable organisms* likewise predominate under such circumstances.

URINARY CONCRETIONS.

Urinary concretions vary in size from that of a grain of sand or a pebble to that of a fist. They are encountered in the bladder, also in the pelvis of the kidney, in the ureters, and in the prostatic sinus. All urinary concretions contain a framework of organic structure uniting the particles of the formation into a coherent mass. They are divided, according to Ultzmann, as follows:

1. Concretions whose nucleus consists of the sediment formed in acid urine—primary calculus-formation. All of these arise primarily in the kidney and pass thence into the bladder, where they undergo enlargement in accordance with the development of the crystals in the urine.

2. Calculi that have for a nucleus either the sediments found in alkaline urine

or a foreign body—secondary calculus-formation. These develop in the bladder itself.

Primary calculus-formation takes place from free uric acid in the form of sheaves as a nucleus (Fig. 148, 7), and surrounded by layers of calcium oxalate. *Secondary calculus-formation* takes place in neutral urine from calcium carbonate and crystalline calcium phosphate, in alkaline urine from acid ammonium urate, ammonio-magnesium phosphate, and amorphous calcium phosphate.

Chemical examination next determines whether or not the particles of the concretion are combustible upon a platinum plate.

I. *Combustible concretions* can consist only of organic matter.

(*a*) If the murexid-test yields a positive reaction, the concretion contains *uric acid*. Uric-acid calculi are common, often of considerable size, smooth, rather hard, and in color from yellow to reddish brown.

(*b*) If another specimen on boiling with potassium hydrate yields an odor of ammonia, and if moist turmeric-paper held in the vapor becomes brown, or a glass rod moistened with hydrochloric acid and held over the vapor yields fumes of ammonium chlorid, the concretion contains *ammonium urate*. If this test yields a negative result, the concretion contains pure uric acid. Calculi of ammonium urate are rare, generally small, of earthy consistence, and in color between clay-yellow and whitish.

(*c*) Should the *xanthin-reaction* be positive, this substance is present, though it is rare. In one instance, *indigo* has been found in a calculus.

(*d*) If *cystin-crystals* (Fig. 162, *A*) are detected after solution in ammonia and evaporation of the latter, the presence of this rare substance is demonstrated.

(*e*) Concretions composed of *blood-coagula* or *fibrinous flakes*, without any crystallization whatever, are rare. If burned, they yield an odor of singed hair. They are insoluble in water, alcohol, and ether. They are soluble in potassium hydrate, out of which they are reprecipitable by acids.

(*f*) *Urostealith* is the name that has been given to the substance composing rarely found concretions which in the fresh state are soft and elastic, resembling India rubber. On drying, they become brittle and hard, and in color between brown and black. Warmth causes them to become softer again, and they melt when heated. Solution takes place in ether, the residue of the evaporated ethereal solution becoming violet in color on further heating. Urostealith is dissolved by heated potassium-hydrate solution, with saponification. Concretions containing *fat* or *cholesterin* are rare.

II. If concretions are only *in part combustible*, with a residue, they contain organic and inorganic matters.

(*a*) A portion of the calculus is reduced to powder, and this is boiled with water and filtered hot. Urates that may be present undergo solution. In order to determine whether the uric acid is combined with sodium, potassium, calcium, or magnesium, the filtrate is evaporated and fused. The ash is examined spectroscopically (flame-spectra), and by this means *sodium* and *potassium* are recognized. *Magnesium* urate and *calcium* urate are transformed by fusing into carbonates. In order to separate the two, the ash is dissolved in dilute hydrochloric acid, and filtration is practised. The filtrate is neutralized with ammonia; then again dissolved with a few drops of acetic acid. Addition of ammonium oxalate precipitates *calcium oxalate*. Filtration is now practised, and to the filtrate are added sodium phosphate and ammonia. By this means the magnesia is separated as *ammonio-magnesium phosphate*.

(*b*) *Calcium oxalate* occurs principally in children, either as small, smooth, pale hempseed-calculi, or in dark, nodular, hard mulberry-calculi. It is not affected by acetic acid, is soluble in mineral acids, without effervescence; and is reprecipitated by ammonia. When fused upon a platinum plate, the specimen becomes black; it is then burned white to calcium carbonate, which undergoes effervescence upon addition of acid.

(*c*) *Calcium carbonate* occurs generally in whitish-gray, earthy, chalk-like, rather rare calculi that usually are multiple. It is soluble in hydrochloric acid with effervescence. When fused, it becomes at first black, from admixture of mucus; but soon afterward white.

(*d*) *Ammonio-magnesium phosphate* and *basic calcium phosphate* are usually united in soft, white, chalky stones, which at times attain quite considerable size. Such calculi imply a long sojourn in ammoniacal urine. The first substance yields an odor of ammonia when heated, and more distinctly when heated with potassium hydrate. It is soluble in acetic acid without effervescence, and is precipitated in crystalline form from this solution on addition of ammonia. When fused, the specimen melts to a white, porcelain-like mass. *Basic calcium phosphate* does not effervesce with acids. The solution in hydrochloric acid is precipitated by ammonia. The solution in acetic acid yields calcium oxalate on addition of ammonium oxalate. In order to isolate calcium and magnesium from such stones, the process described in paragraph (*a*) should be followed.

(*e*) *Neutral calcium phosphate* is rarely found in calculi, but, on the other hand, not rarely in urinary sand. Such concretions resemble the earthy phosphates in physical and chemical properties, except that they contain no magnesia.

THE PHYSIOLOGICAL PROCESS OF URINARY SECRETION.

The two older and most important theories of secretion will be mentioned: (1) Bowman held that the glomeruli secrete only water, and that the epithelial cells of the urinary tubules through their glandular activity furnish the specific urinary elements, which the onflowing urinary water washes out of the cells. (2) C. Ludwig assumed that a dilute urine is secreted in the capsules. Passing through the urinary tubules, this, by endosmosis, returns water to the blood, which is more deficient therein, and to the lymph of the kidney, and thus becomes reduced to normal consistence.

The secretion of the urine in the kidneys depends, however, not alone upon physically definable influences, but it must rather, in accordance with a series of acquired facts, be assumed that in addition the vital activity of special secretory cells plays a prominent rôle. The physical or chemical forces obviously underlying the latter have not as yet been determined. The secretion includes (1) the urinary water, and (2) the urinary elements dissolved therein. Both together constitute the totality of the secretion. The amount of urinary water secreted in the glomeruli determines principally the amount of urine, while the amount of substances dissolved in the urinary water determines the concentration of the urine.

The amount of urinary water, which is secreted principally in the capsules, depends, in the first place, upon the blood-pressure in the distribution of the renal artery; and, accordingly, is governed by the laws of filtration. The amount of urinary water furnished is, however, not dependent upon the hydrostatic pressure alone, but the functional activity of the cells lining the glomerulus is also of influence. In addition to the water, a certain amount of the salts occurring in the urine is secreted in the glomerulus; albumin, however, is retained. In consideration of the functional activity of the cells, the amount of urinary water must depend also in part upon the rapidity with which new blood conveying the material for secretion passes to the glomeruli; and, in part, upon the amount of urinary elements and water contained in the blood.

The independent activity of the secretory cells is present only when their vitality is intact. It is paralyzed in consequence of transitory occlusion of the renal artery. For this reason, the kidney no longer secretes under such circumstances, even when the circulation is restored after removal of the compression. The observation that the urine is not rarely found to have a higher temperature than the arterial blood is also indicative of this activity.

The dependence of the secretion upon the blood-pressure will be made clear by the following observations:

1. Increase of the total contents of the vessels, in consequence of which the tension in the vascular system must increase, increases the

amount of filtered urinary water.   Injections of water directly into the
vessels, or the ingestion of considerable quantities of fluid, operates in
this direction.   If the increase in blood-pressure exceeds a certain
level, albumin may even pass into the urine.   Conversely, loss of water
in consequence of profuse sweating or diarrhea, or copious venesection,
as well as prolonged thirst, will cause diminution in the amount of
urinary secretion.   The circumstance that the blood-pressure does not
rise constantly after free drinking is evidence of the functional activity
of the cells of the glomeruli, as is also the fact that the amount of urine is
not increased after large transfusions.

2. Diminution in the vascular capacity will operate in a similar
manner: contraction of the cutaneous vessels under the influence of
cold, stimulation of the vasomotor center or of considerable areas of
the vasomotor nerves, ligation or compression of arteries of large size,
envelopment of the extremities in tight bandages.   Naturally the opposite conditions will be followed by a reduction in the amount of urine:
the influence of heat upon the skin to the point of redness and dilatation of the vessels, enfeeblement of the stimulation of the vasomotor
center, or paralysis of considerable areas of the vasomotor nerves.

3. Increased cardiac activity, in consequence of which the tension
and the rapidity of the current in the arterial distribution are increased,
augment the amount of urine.   Conversely, enfeeblement of the heart's
action (paresis of the motor nerves of the heart, disease of the heart-
muscle, valvular lesions) diminishes the amount of urine.   Artificial
irritation of the vagi, in consequence of which, with slowing of the heart-
beats, the average blood-pressure fell in animals from 130 to 100 mm.
of mercury, with slowing of the pulse, was followed by a reduction in
the amount of urine to about one-fifth,   At 40 mm. of aortic pressure
the secretion of urine ceases.

4. The amount of urine secreted rises or falls with increasing or
diminishing fulness of the renal artery.   Even moderate compression
of the artery in animals is followed by distinct reduction.

Pathological.—In the presence of fever, there is diminished fulness of the
renal vessels, with consecutive reduction in the amount of urine.   The observation
is of especial significance for the pathogenesis of certain diseases of the kidney
that ligature of the renal artery, even if continued for only two hours, causes
necrosis of the epithelium of the urinary tubules.   In case of arterial anemia of
longer duration, necrosis of the entire renal structure takes place.   Ribbert found
the epithelial cells of the convoluted tubules greatly altered after compression of
the renal artery for some time.

Most diuretic medicaments act in one or another of the directions
indicated.   In case of increased diuresis, the lumen of the urinary
tubules is increased.

The pressure within each afferent vessel must be relatively large, because (1)
the duplicate capillary arrangement in the kidney offers considerable resistance,
and because (2) the efferent vessel has a much narrower lumen than the afferent
vessel.   In accordance with these facts, an excretion from the blood into the
capsules of the urinary tubules will take place from the capillary loops of the
glomerulus in consequence of the filtration-pressure.   Dilatation of the afferent
vessels, as, for instance, from the action of the nerves upon the unstriated muscular
fibers, will increase the filtration-pressure; while constriction will diminish the
secretion.   If the reduction in the pressure has become so considerable that the
blood-current in the renal vein is distinctly slowed, the secretion of urine begins
to diminish.   It is a remarkable fact that occlusion of the renal veins completely
suppresses the secretion.   C. Ludwig has concluded from this that the secretion

of fluid accordingly can not take place from the true renal capillaries, because the blood-pressure in these must be increased by occlusion of the veins, and this would cause increased filtration. On the other hand, the observation mentioned would indicate that the secretion takes place from the capillaries of the glomerulus. The venous stasis in the efferent vessel distends this vessel, which arises in the center of the convolution, to such a degree that the capillary loops are pushed together against the wall of the capsule and compressed, so that no filtration can take place from them. Whether some fluid is given off through the urinary tubules, especially the convoluted tubules, is as yet undecided.

The amount of urine and the amount of contained urea are diminished by venous stasis in the kidneys. The amount of sodium chlorid remains constant, while that of albumin in pathological urine increases.

As the blood-pressure in the renal artery equals between 120 and 140 mm. of mercury, and the urine in the ureter is propelled under exceedingly slight pressure, so that it is no longer capable of escaping against a counter-pressure of from 10 to 40 mm.—provided by a manometer introduced into the ureter divided transversely—it must be clear that the blood-pressure is also capable, as a *vis a tergo*, of forcing the stream of urine through the ureter.

The degree of concentration of the urine depends upon the amount of the constituents in solution passing out of the blood into the urinary water. The cells of the convoluted urinary tubules appear to take up these substances from the blood by means of an independent activity. The urinary water passing through the urinary tubules from the glomerulus, and containing only readily diffusible salts, later takes up these substances out of the cells of the convoluted tubules by a process of extraction. The independent activity of the cells is indicated by the following facts:

1. Sulphindigotate of sodium (indigocarmin), which, when injected into the blood, passes into the urine, can be recognized in the interior of the cells of the urinary tubules, but not in the capsules. Further on, this substance is visible in the lumen of the urinary tubules, whither it is washed by the current of urinary water from the glomerulus. If, in such an experiment, the cortical layer containing the capsules has been removed two days previously by cauterization or with the knife, the blue pigment will have remained in the convoluted tubules. It will not have advanced onward, as the current of water from the destroyed glomeruli is wanting. This observation thus indicates that the glomeruli furnish principally the urinary water, and the convoluted tubules the specific urinary elements. Heidenhain and Sauer observed also urates (injected into the blood) secreted by the convoluted tubules. Nussbaum has also demonstrated that urea is not secreted by the capsules, but by the urinary tubules. Möbius found the same with respect to the biliary pigment, Glaevecke with respect to the iron salts of the vegetable acids when injected subcutaneously, and Landois first described the same condition with respect to hemoglobin. After infusion of milk into the vessels, Landois encountered numerous fat-globules within the cells of the urinary tubules.

It appears that the capsules may also take part in the process only after abundant secretion. After infusion of large amounts of sodium sulphindigotate and after the observation has been continued for some time, the epithelium of the Malpighian capsules also exhibits the blue discoloration. Likewise in the presence of albuminuria, the abnormal elimination of albumin takes place first in the urinary tubules and later in the capsules. Also hemoglobin occurs in part in the capsules. Egg-albumin is believed by Nussbaum to be excreted through the capsules.

Disse studied the alterations in the secretory cells during their activity. With the commencement of this activity the cells become larger, and bright areas of the protoplasm, infiltrated with secretion, appear as halos about the nucleus. The discharge of the secretion into the lumen of the tubules takes place through filtration. The brush-border indicates only the empty cell; it disappears while the cell is being filled with secretion.

Henle, H. Meckel, Leydig, and Bial observed in snails constituents of the urine (guanin) within the cells of the kidney.

2. Also when, either after ligation of the ureter or in consequence of marked reduction in blood-pressure in the renal artery (after division of the cervical cord or venesection), urinary water is no longer secreted, the substances named are, nevertheless, after introduction into the blood, seen to pass over into the urinary tubules. Injection of urea likewise again stimulates the secretion. This indicates that the secretory activity takes place independently of the filtration-pressure.

The independent vital activity of the glandular cells of the urinary tubules not explainable by physical processes makes it impossible to consider the glandular tubules as a simple apparatus resembling physical membranes. This is shown also by the following experiment: Abeles permitted the circulation of arterial blood to continue artificially through fresh, living, extirpated kidneys. Pale urinous fluid escaped from the ureter drop by drop. If urea or sugar were added to the circulating blood, the vessels became dilated and the secretion contained the admixed substances in greater concentration. Thus, also the surviving kidney excretes, in concentrated form, substances that circulate in the blood in a dilute state. The same observation was made by I. Munk in analogous experiments with sodium chlorid, potassium nitrate, caffein, grape-sugar, and glycerin, with an increase in the total amount of the secretion. Addition of caffein or theobromin to the circulating blood induces an increase in the secretion, stimulates, thus, the secretory cells themselves to increased activity. The assumption of vital activity alone explains, also, why the serum-albumin of the blood does not pass into the urine, although egg-albumin or dissolved hemoglobin, introduced into the blood, does so rapidly.

Among the salts that occur in the total blood, also in the blood-corpuscles, naturally only those in solution can pass over into the urine. Those that are united to proteids or in the cellular elements cannot pass over, or at least only after decomposition. This fact explains the difference between the salts of the total blood and those of the urine. The urine can, likewise, take up only the gases absorbed into the blood; and not those in chemical combination.

Should stagnation of the secretion take place in the ureter, as after ligation, and, later, in the urinary tubules, a return of the secretion into the tissue of the kidney and, later, into the blood will be observed. The kidney becomes edematous, in consequence of distention of its lymph-spaces. The secretion is altered, inasmuch as, first, water is reabsorbed into the blood; then the sodium chlorid in secretion is diminished, likewise sulphuric acid and phosphoric acid, and, finally, also the urea. Kreatinin will still be present in considerable amount. A true secretion of urine, later on, no longer takes place.

The circumstance, further, is noteworthy that the two kidneys never secrete symmetrically. The condition is one of alternation in activity and hyperemia. The one kidney secretes a fluid containing a larger amount of water, and, at the same time, more sodium chlorid and urea. It may even be more acid. v. Wittich observed that the excretion of uric acid in the kidneys of birds does not take place uniformly in all of the urinary tubules, but only in varying areas. The extirpation of one kidney or its morbid destruction in human beings does not diminish the secretion. There occurs increased activity, with enlargement of the remaining organ; and this is due to the increased functional demands upon the secretory cells of this kidney.

THE PREPARATION OF THE URINE.

The question has often been raised, whether the urine is really secreted through the kidney, or whether the urinary constituents are not in part prepared by the kidney itself. The following experiments will shed light upon this subject:

1. The blood already contains one part of urea in from 3000 to 5000 parts; but the blood in the renal vein contains less urea than that of the artery. This fact indicates that urea is excreted from the blood.

2. After extirpation of the kidney—nephrectomy—or ligation of its vessels, urea accumulates in the blood and progressively with the lapse of time to between $\frac{1}{100}$ and $\frac{1}{300}$. At the same time, fluids containing urea and ammonia are vomited and discharged with the stools. Animals die after such profound operations, moreover, within from one to three days.

3. If the ureters are ligated, the actual secretion of the kidneys soon ceases. After this, the accumulation of urea in the blood likewise increases, and, indeed, as it appears, not in greater amount than after nephrectomy. Nevertheless, it is possible that the kidney, in its metabolic activity, does, like other portions of the body, prepare some urea in its tissues.

4. The blood of birds contains uric acid even under normal conditions. Ligation of the ureters, as well as of the renal vessels, or gradual destruction of the secreting renal epithelium by means of subcutaneous injections of neutral potassium chromate, is followed in birds by a deposition of uric acid in the joints and tissues; so that the serous membranes particularly acquire a whitish incrustation therefrom. The brain remains free. Also acid combinations of uric acid with ammonia, sodium, and magnesium are thus deposited. Extirpation of the kidneys in serpents gives rise to the same phenomena in lesser degree.

From these experiments it may be concluded that the urea, and with it probably most of the organic constituents of the urine, are excreted principally through the kidneys, but are not prepared in them. The seat for the formation of all of these substances is probably to be referred to the tissues. The urea is formed from decomposed proteid, and principally in the liver. As a result of experiments with birds and serpents, v. Schröder and Colasanti come to the conclusion that the formation of uric acid cannot be assumed to take place exclusively in any definite organ. Urobilin is formed from hemoglobin.

Little is known concerning the physiological-chemical processes in the kidneys themselves. Hippuric acid is formed in part in the kidney, for the blood of herbivora contains no trace thereof; but the synthesis of this substance in rabbits takes place also in other tissues. If blood to which sodium benzoate and glycin have been added is passed through the vessels of a fresh kidney, hippuric acid is formed. If, further, phenol and pyrocatechin are digested with fresh renal tissue, the corresponding sulphuric-acid combinations that occur in the urine are formed. The latter, it is true, are formed also by similar digestion with hepatic and pancreatic tissue and with muscle. From these observations it may be concluded that in the body the substances in question are prepared within the kidneys and the organs named.

The kidneys are extremely rich in water, and they yield an alkaline reaction. In addition to serum-albumin, globulin, nucleo-albumin, albumin soluble in sodium carbonate, a gelatin-yielding substance, fat in the epithelial cells (principally after

33

the ingestion of milk and meat), the elastic sarcolemma-like substance of the membrana propria of the urinary tubules, and the tissue-elements of the vessels and their unstriated muscles, the kidneys contain leucin, xanthin, hypoxanthin, kreatin, taurin, inosite, cystin (the last is present in no other tissue); and of these, the majority pass into the urine either not at all or only in small amount. The presence of these substances indicates, probably, active metabolism in the kidneys; and this is suggested also by the great vessels of the kidney. During the secretion of the kidneys, the blood of the renal vein is said to become bright red, and to be deprived of its fibrin. If alkaline blood-serum be filtered through a layer of nucleo-albumin or lecithin-albumin, an acid filtrate passes through. Liebermann explains in a similar manner the development of acid urine on passing blood-plasma through the renal epithelium containing lecithin-albumin.

THE PASSAGE OF VARIOUS SUBSTANCES INTO THE URINE.

The following substances pass unchanged into the urine: Alkaline sulphates, borates, silicates, nitrates, carbonates; alkaline chlorids, bromids, and iodids; potassium sulphocyanate, potassium ferrocyanid; salts of the biliary acids; urea, kreatinin; cumaric, oxalic, camphoric, pyrogallic, sebacylic acids; further, many alkaloids, as, for instance, morphin, strychnin, curarin, quinin, caffein; among the pigments, sodium sulphindigotate, carmine, gamboge, madder, logwood, the coloring-matter of huckleberries, mulberries, cherries, rhubarb; further, santonin; and, finally, the salts of gold, silver, mercury, arsenic, bismuth, antimony, iron (but no lead), which, however, pass in largest amount into the bile and into the feces.

Inorganic acids appear in human beings and carnivora as neutral ammonium-salts; in herbivora, as neutral alkaline salts.

Certain substances that generally undergo decomposition, even when they gain entrance into the blood in small amounts, pass in part into the urine when they accumulate in the blood in considerable amount, because they are not completely decomposed, such as sugar, hemoglobin, egg-albumin, alkaline salts of the vegetable acids, alcohol, chloroform.

Many substances appear in the urine as oxidation-products: moderate amounts of alkaline salts of the vegetable acids as alkaline carbonates; uric acid in part as allantoin; sodium sulphite acid and hyposulphite in part as sodium sulphate; potassium sulphid as potassium sulphate. Many oxids appear as sub-oxids, benzol as phenol.

Those bodies, such as glycerin and the resins, that are completely consumed, exhibit no special derivatives in the urine.

Some substances undergo synthesis with metabolic products, and appear in the urine as conjugated combinations. In this category belongs the development of hippuric acid by conjugation, the formation of the conjugate sulphates, as well as the formation of urea by synthesis from carbamic acid and ammonia. After administration of camphor, or of chloral and butyl-chloral, a conjugated combination with glycuronic acid, an acid closely allied to sugar, appears in the urine. Taurin and sarcosin undergo conjugation with sulphamic or carbamic acid. Phenyl bromid, when administered, enters into conjugation with mercapturic acid, a body allied to cystin.

Tannic acid, $C_{14}H_{10}O_9$, takes up water, and is thus decomposed by hydrolysis into two molecules of gallic acid — $2C_7H_6O_5$.

Potassium iodate and bromate are reduced to potassium iodid and bromid; malic acid, $C_4H_6O_5$, in part to succinic acid, $C_4H_6O_4$; indigo-blue, $C_{16}H_{10}N_2O_2$, takes up hydrogen to form indigo-white, $C_{16}H_{12}N_2O_2$.

Finally, many substances do not pass into the urine at all, such as serum-albumin, oils, insoluble metallic salts, and metals.

INFLUENCE OF THE NERVES UPON THE SECRETION OF THE KIDNEYS.

As yet, only the influence of the vasomotor nerves upon the filtration of the urine from the renal vessels is known, and these nerves appear to be derived from both halves of the spinal cord for each kidney. In general, it should be borne in mind that dilatation of the branches of the renal arteries, particularly of the afferent vessels, must increase the pressure in the glomeruli, and therefore the amount of filtered fluid increases. The greater the measure in which the dilatation of the vessels is confined to the distribution of the renal artery alone, the greater will be the amount of urine. The lower dorsal nerves, in the dog principally the twelfth and thirteenth, contain the largest number of the vasomotor fibers for the kidney.

Division of the renal plexus is, as a rule, followed by increase in the amount of urine. Occasionally, in consequence of the increased pressure, albumin is observed to pass into the Malpighian capsules; and with rupture of the vessels of the glomeruli even blood may appear in the urine. The center for these renal vasomotor fibers is situated on the floor of the fourth ventricle, in front of the origin of the vagus. Injury, as by puncture, in this situation is, therefore, followed by increase in the amount of urine (diabetes insipidus), occasionally with the simultaneous appearance of albumin and blood. Naturally, any injury of the active nerve-path from the center to the kidney has a similar effect. The center for the vasomotor nerves of the liver is situated close to this center, and injury of the former gives rise to the production of sugar in the liver. Eckhard observed hydruria develop after irritation of the vermiform process of the cerebellum lying upon the medulla. A similar result is brought about in human beings also as a result of irritation in this situation by tumors, inflammatory processes, and the like.

If, in addition to the distribution of the renal artery, an adjacent extensive vascular area be paralyzed simultaneously, the blood-pressure in the distribution of the renal artery will be less high; as, at the same time, much blood finds its way into the other paralyzed area. Under such conditions, therefore, either a slight or only a transitory polyuria will be observed. In this way, there results a moderate increase in the amount of urine for a few hours after division of the splanchnic nerve, which contains the vasomotor fibers for the kidney. These leave the spinal cord in part through the first dorsal nerve, and pass into the sympathetic nerve. The splanchnic contains, at the same time, also the fibers for the extensive distribution of the intestinal vessels. Irritation of this nerve is, naturally, attended with the opposite effect.

If, with paralysis of the renal nerves, the overwhelming majority of all of the vasomotors of the body are at once paralyzed, the pressure throughout the entire arterial distribution falls in accordance with the extensive dilatation of all of these vascular paths. In consequence, the secretion of urine diminishes, even to the point of complete cessation. This last effect is seen after division of the cervical cord down to the seventh cervical vertebra. This fact explains the observation that the polyuria that occurs after injury to the floor of the fourth ventricle disappears as soon as the spinal cord down to the twelfth dorsal nerve is divided.

The presence of a large amount of urea in the blood causes contraction of the vessels of the body, but dilatation of the renal vessels.

Contraction of the vessels, and, therefore, at the same time of the volume of the kidney, are caused by asphyxia and strychnin-poisoning; also irritation of sensory nerves has a similar reflex effect. Extirpation of the nerves of the kidney has the opposite effect. During fever, the vessels of the kidney are contracted, probably in consequence of irritation of the center by the abnormally heated blood.

Repeated inhalation of carbon monoxid is said occasionally to be attended with polyuria, perhaps in consequence of paralysis of the center for the vasomotor nerves of the kidneys.

According to Cl. Bernard, irritation of the vagus at the cardia causes increased secretion of urine, with reddening of the blood in the renal veins. Possibly this nerve contains vasodilator fibers that behave similarly to the corresponding fibers in the facial nerve for the salivary glands. The vagus innervates the intrinsic unstriated musculature of the kidney.

According to Arthaud and Butte and others, irritation of a peripheral extremity of the vagus, conversely, diminishes the secretion of urine and the circulation in both kidneys. Atropin renders the experiment impossible. The vagus thus appears to be the vasomotor nerve of the kidney. According to Boeri, it possesses trophic functions, as albuminuria occurs after division of the vagus. Irritation of the cervical sympathetic likewise diminishes the secretion. This irritation appears to be reflex, being transmitted through the spinal cord to the splanchnic nerve.

UREMIA; AMMONIEMIA; URIC-ACID DYSCRASIA.

After extirpation of the kidneys, nephrectomy, or ligation of the ureters, which renders further secretion of urine impossible; further, also, in human beings, as a result of extreme urinary stasis, as well as in consequence of morbid alterations in the kidneys (inflammation, fatty degeneration, and desquamation of the epithelial cells of the urinary tubules, cicatricial contraction of the kidney, amyloid degeneration), there develop a series of characteristic phenomena that resemble an intoxication, and, if of marked degree, cause death, with degeneration of the ganglia in the cerebral cortex and the spinal cord. This condition is designated *uremic intoxication* or *uremia*. Among the phenomena, the following are conspicuous: Mental prostration, somnolence, even loss of consciousness to the point of deep coma, and, in addition, from time to time, the occurrence of twitching or even widespread, severe convulsions. Occasionally, there are delirium and general excitement. At the same time, the occurrence of the so-called Cheyne-Stokes respiratory phenomenon is often observed. Occasionally, transitory, invariably bilateral, blindness occurs, from toxic paralysis of the psycho-optic center. There may, however, take place, quite independently, hemorrhagic extravasations into the retina, causing, rarely permanent, blindness—apoplectic retinitis. Also impairment of hearing is observed. Vomiting and diarrhea are common. Ammonium carbonate, formed in the digestive tract from urea, as well as kreatin, causes uremic diarrhea. Also the breath and the emanation from the skin may exhale the odor of ammonia. The alkalinity of the blood and the amount of oxygen in the blood are diminished. The retention of substances that are normally excreted by the urine must be considered as the cause of these symptoms, although it has not, as yet, been possible to designate with certainty the substances that must be considered as the agents upon which the toxic phenomena depend.

Suspicion was first directed to urea. v. Voit observed that even healthy dogs exhibited uremic manifestations when they partook of urea for a considerable time with their food if, at the same time, the use of water, which would have carried off the urea rapidly through the kidneys, was prevented. Further, Meissner found that death amid uremic manifestations could be hastened in nephrectomized animals, if urea was at the same time injected into the blood. An injection of moderate amounts of urea into the blood of entirely healthy animals was not, it is true, followed by uremic symptoms, although one or two grams caused a comatose state in rabbits. Dogs died after subcutaneous injection of urea to an amount equaling one per cent. of the bodily weight. Hippuric acid is said to have an entirely similar effect in frogs. Although urea, when introduced into the blood in large amounts, causes death with convulsions, this condition should not be confounded with uremic attacks of intermittent occurrence.

As injection of ammonium carbonate causes symptoms similar to those of uremia, v. Frerichs and Stannius believed that the decomposition of urea in the blood causes the intoxication—ammoniemia. However, after nephrectomy or ligation of the ureters, even on simultaneous injection of urea into the blood, careful chemical investigation fails to disclose the presence of ammonia in the blood. Therefore, spontaneous formation of ammonia in the blood cannot be the cause of the uremic symptoms.

As in birds and reptiles, which eliminate principally uric acid, ligation of the ureters likewise induces a comatose state, it was necessary to think of other substances as possibly causing the toxic symptoms. Meissner observed prostration and twitchings develop in dogs after injection of kreatinin. Cl. Bernard, Traube, Ranke, Astaschewsky, Feltz and Ritter, and others attribute the phenomenon to an accumulation of the neutral potassium-salts; Schottin and Oppler suggest the accumulation of normal or abnormally decomposed extractives, Thudichum that of the oxidation-stages of the urinary pigment. Possibly many substances and their decomposition-products act in conjunction. R. Fleischer found a reduction in the elimination of sulphuric and phosphoric acids in advance of the uremic attack in man.

On placing various substances occurring in the urine—kreatinin, kreatin, acid potassium phosphate, uratic sediment from human urine—directly upon the surface of the cerebrum, Landois observed the development of all signs of uremia. There occurred, particularly, fully developed convulsive seizures, with intervals of rest, in dogs, with subsequent coma. Also, many other secondary phenomena of uremic eclampsia could be thus induced. Urea is inactive in this direction, ammonium carbonate, leucin, sodium carbonate, sodium chlorid, potassium chlorid, feebly active.

After long-continued excessive ingestion of food, together with the use of spirit, and slight activity, there occurs, principally in conjunction with respiratory disorders, derangement of metabolism, and not rarely a marked accumulation of uric acid in the blood. The latter is deposited in the joints and their ligaments and cartilages, principally of the foot and the hand, and gives rise to inflammatory and painful attacks—gouty nodules, uric arthritis. Rarely, the kidneys, the heart, and the liver are involved. In the vicinity of the foci, the tissues undergo necrosis. Food containing nuclein is to be avoided; also meat-broths, meat-extract, sodium chlorid; while cheese, peptone, legumins, and aleuronat are to be commended. As to the amins, piperazin, lysidin, lycetol, urotropin, the investigations are not as yet concluded. As uric acid is more readily soluble in solutions of urea, the administration of this substance has been advised. Uric acid introduced into the blood or into the lymphatic system causes changes in the renal epithelium, in the form of uric-acid spheroliths between and within the cells of the convoluted tubules. Administration of adenin, while it does not increase the excretion of uric acid, favors its deposition in the kidney amid inflammatory symptoms. In birds, long-continued administration of oxalates, sugar, acetone, phenol, gives rise to deposition of urates in the urinary tubules, as well as in the serous and the synovial membranes, and these have disappeared after administration of piperazin.

Human urine, when injected beneath the skin or into the veins of animals, has a toxic and even fatal effect, particularly in the case of infectious diseases, diseases of the liver, carcinoma, exophthalmic goiter, and, in accordance herewith, after extirpation of the thyroid gland. The toxic properties are due to organic (toxins) and inorganic constituents, principally potassium-salts. Pregnant animals are especially susceptible to this poison.

STRUCTURE AND FUNCTIONS OF THE URETERS.

The pelvis of the kidney and the ureter possess a mucous membrane constituted of delicate connective-tissue fibers with many embedded cells, upon which a laminated transitional epithelium is situated. The deepest layer of the latter is provided with small, round, soft cells. Then follows a layer of more nearly vertical, club-shaped and bulbous cells, whose attenuated extremities ramify between the cells of the deepest layer; the free surface is covered by cubical cells, which finally are surmounted by a homogeneous cuticular border. Beneath the epithelium there is a layer of adenoid tissue, containing disseminated lymph-follicles. In the pelvis of the kidney, the mucous membrane contains isolated small grape-like mucous glands, which are present also in the ureter. The muscular coat consists of an internal somewhat thicker longitudinal layer and an external circular layer, to which, in its lower third, a number of disseminated bundles of longitudinal fibers are added. All of these layers are rather freely interwoven with connective tissue. The external connective-tissue sheath forms a sort of adventitia, in which the larger vessels and the nerves, together with the ganglia, are situated. The layers of the ureter may be followed upward to the pelvis of the kidney and to the calices. They finally line the pelvis itself only with mucous membrane, passing over upon the base of the pyramids, while the muscle-fibers cease at the foot of the pyramids, where they form a sort of sphincter about the pyramids by means of circular bundles. The blood-vessels supply the various layers and form a capillary network beneath the epithelium. The relatively scanty medullated nerves, in the vicinity of which ganglia are found, in part supply the muscles as motor fibers, while in part they penetrate toward the epithelium. These are reflex and sensory, as indicated by the severe pain attending impaction of calculi. The ureter penetrates the thickness of the bladder-wall, passing obliquely through it for a considerable distance. The internal opening is a slit in the mucous membrane directed obliquely inward and downward, and provided with a sharp, valve-like process (Fig. 177).

The propulsion of the urine through the ureter takes place (1) in consequence of the fact that the urine constantly secreted in the kidney under considerable pressure forces onward the urine in the ureter, which is under lower pressure. (2) In the erect posture, the urine flows by gravity down the ureter. (3) The muscles of the ureter through their peristaltic movement propel the urine into the bladder. This movement occurs only as a reflex phenomenon in response to the entrance of the urine, a few drops every three-quarters of a minute, or in consequence of direct irritation. It always passes downward with a velocity of from 20 to 30 mm. in

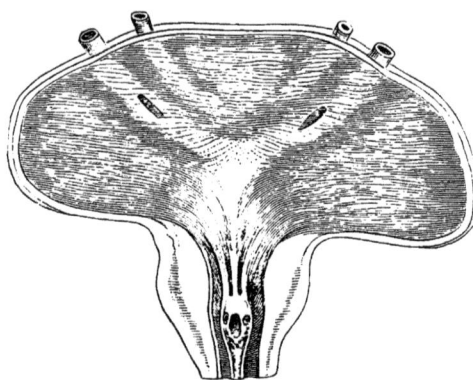

FIG. 177.—Lower Portion of the Male Bladder, with the Commencement of the Ureter, Opened through a Median Incision in the Anterior Wall, and spread out (after Henle). The clear lines of the trigone, the slit-like openings of the ureters, the ureters divided above and the seminal vesicles can be recognized. On the colliculus seminalis there appear in the middle the large opening of the prostatic sinus, and on either side the small circular orifice of the ejaculatory duct, and below both the numerous punctate openings of the excretory ducts of the prostate gland.

a second. The greater the distention of the ureter by the urine, the more rapidly does this peristaltic movement take place. Asphyxia, venous hyperemia, and irritation of the splanchnic increase the number of contractions; while rapid ligation of the renal vessels, as well as ligation of the ureter, diminishes them.

In case of local irritation, the contraction takes place in both directions. As Engelmann observed these movements also in excised portions of ureter in which neither nerve-fibers nor ganglia were visible, he believes that the movements are due to direct muscular conduction in the unstriated muscles, just as takes place in the heart.

The stagnation of urine toward the kidney is prevented (1) by the fact that the secretion collecting in the pelvis of the kidney and in the calices under high pressure presses upon the pyramids from all sides, so that the urine cannot pass back into the urinary tubules closed by pressure. (2) If when the urine has accumulated in the ureter in considerable amount, as from occlusion by concretions, the musculature engages in increased activity for the propulsion of the urine, the portion of the muscular fibers surrounding the pyramids so compresses the urinary tubules that the urine cannot pass back into the excretory ducts of the tubules. The return of urine from the bladder into the ureter is rendered difficult in part by the fact that with marked stretching of the bladder-wall the ureter, in so far as it is contained therein, is likewise compressed; and in part by the fact that the stretching of the mucous membrane of the bladder firmly approximates the margins of the slit-like openings of the ureters (Fig. 177).

In case of retention of urine in the bladder, a return of urine into the ureters may, it is true, take place.

STRUCTURE OF THE URINARY BLADDER AND THE URETHRA.

The mucous membrane of the bladder is not unlike that of the ureter. The laminated epithelium exhibits flatter cells in the upper layer. When the bladder is distended, the epithelial cells become stretched and thinner. The unstriated muscular fibers are arranged in bundles that form an external longitudinal layer and an internal circular layer. In addition, fibers pass in various directions and cross one another, forming a wide-meshed trabecular network. Between the muscular coat and the mucous membrane there is a layer of delicate, fibrillar, cellular connective tissue, with an intermixture of elastic fibers. An excessively minute dissection of the individual layers and bands of the musculature of the bladder has given rise to erroneous physiological interpretations. In this category belongs the establishment of a special detrusor urinæ muscle, which is said to consist of fibers pursuing a vertical direction from the vertex to the fundus, principally upon the anterior and posterior surfaces. The conception of a special internal sphincter vesicæ is likewise unjustified as constituted of a circular layer of unstriped muscles, from 6 to 12 mm. thick, surrounding the commencement of the urethra, and in its form helping to give rise to the funnel-shape of the outlet of the bladder. This layer, also designated annulus urethralis vesicæ, is no sphincter at all. In the trigone of Lieutaud there are, at times, between the orifices of the ureters, numerous muscular bundles, attached in part to the circular, in part to the longitudinal fibers of the wall of the bladder. Waldeyer believes, particularly of the trigone, that it facilitates the distention of the bladder, favors its complete evacuation and aids its closure.

From the physiological standpoint, it should be borne in mind that the entire musculature of the bladder represents a continuous hollow muscle whose sole function it is, in contracting, to diminish the cavity of the bladder from all directions and to expel its contents.

The vessels of the bladder resemble those of the ureter in their distribution. The nerve-fibers are provided with ganglia, as is the case generally at all parts of the urinary passages outside the kidney. These are situated in part in the mucosa, in part in the muscularis, and they communicate with one another by means of filaments. In the mucous membrane and its epithelium, the nerves terminate in end-bulbs. In accordance with their functions, the nerves are motor, sensory, reflex, and vascular.

In women, the urethra serves only as the excretory duct of the urinary bladder. The mucous membrane, formed of a large amount of fibrillary connective and elastic tissue and supplied with papillæ, is lined by laminated pavement epithelium. In addition, a number of Littré's mucous glands are embedded in it. Next to the mucous membrane is a layer of longitudinal unstriated muscular fibers, and next to the latter a layer of circular fibers. These layers contain an abundance of connective-tissue and elastic fibers, and, besides, extensive venous plexuses, suggestive in their structure of cavernous spaces.

The true sphincter muscle of the bladder is a striated muscle, which undergoes contraction and relaxation under the influence of the will, and consists in part of transverse, completely circular fibers, which extend downward to the middle of the urethra and lie next to the unstriated circular fibers, and in part of longitudinal fibers, which pass upward to the base of the bladder only on the posterior wall of the urethra, and downward between the circular fibers. Additional circular fibers are situated below the middle of the urethra, and only in isolated distribution on its anterior surface.

In the male urethra, the epithelium of the prostatic portion still resembles that of the bladder, in the membranous portion it becomes laminated, and in the cavernous portion a simple cylindrical epithelium. The mucous membrane beneath the laminated epithelium, provided with papillæ, contains, principally in the posterior portion, the mucus-secreting glands of Littré. Unstriated muscle-fibers are present in the prostatic portion as a longitudinal layer, especially at the colliculus seminalis; while the membranous portion contains principally circular fibers, with intervening longitudinal fibers. The cavernous portion contains posteriorly delicate circular fibers, anteriorly only isolated insignificant oblique and longitudinal fibers.

With respect to the mechanism for closure of the male urethra, it should be pointed out that the so-called internal sphincter vesicæ of the anatomists, which consists of unstriated muscular fibers, and, as an integral portion of the musculature of the bladder surrounds the commencement of the urethra down to within the prostatic portion of the urethra, above the colliculus seminalis, is not a sphincter muscle at all. The true striated sphincter of the urethra, or external sphincter of the bladder, is situated below the former. It is a completely circular muscle, surrounding the urethra, just above the entrance of the urethra into the urogenital septum, at the apex of the prostate gland, where its fibers anastomose with those of the subjacent deep transverse peroneal muscle.

This sphincter muscle includes, also, longitudinal fibers, which pass downward from the bladder along the upper border of the prostate. Isolated transverse bundles are derived anteriorly from the surface of the neck of the bladder. The sphincter muscle includes, besides, certain transverse fibers that lie within the prostate even opposite the apex of the colliculus seminalis, passing like a thick transverse column in advance of the commencement of the urethra into the structure of the prostate—prostatic muscle.

In the male urethra, the blood-vessels form a rich capillary network beneath the epithelium, in the midst of which a wide-meshed lymphatic vascular network is situated.

COLLECTION AND RETENTION OF THE URINE IN THE BLADDER. EVACUATION OF THE URINE.

After the evacuation of the bladder, urine reaccumulates, with gradual distention of the viscus. As long as the amount of urine is but moderate, the elasticity of the elastic fibers surrounding the urethra and of the sphincter muscle of the urethra—in men, in addition, that of the prostate—suffices perfectly to retain the urine in the bladder. This is indicated by the fact that in the cadaver the urine does not escape from the bladder. The movements for the evacuation of the bladder, as well as for the retention of the urine in the bladder, exhibit, in many respects, an agreement with the motor mechanism at the rectum. In the first place, it should be pointed out that the walls of the bladder are capable of independent contraction. Whether these are due to the ganglion-cells in the bladder that are found in the course of the nerves has not been demonstrated. It is rather more likely that the musculature of the bladder is capable of rhythmic movement without nervous aid.

The urinary bladder, especially when considerably distended, exhibits the occurrence of intermittent slight contractions, which can be compared with the peristaltic movements of the intestines. Even the excised frog's bladder, and even portions thereof without ganglia, exhibit similar rhythmic contractions, which can be increased by heat. After division of all of the nerves of the bladder, bleeding with asphyxia is still followed by contractions as a result of direct stimulation of the muscles of the bladder. The contractions occur, further, more actively in the presence of derangement of the circulation in the bladder, or of venosity of the blood, in the same way as the movements of the intestine are brought about in marked degree by like influences. In this category belongs the evacuation of the urine when the action of the heart ceases in cases of sudden asphyxia or protracted suppression of respiration. As emotional disturbances also influence the contraction of the walls of the bladder, the evacuation of the urine in connection with sudden fear can be explained in this manner. In the state of apnea, as well as in apneic intervals after persistent deep respiratory movements, the independent contractions of the bladder cease.

In order to comprehend the mechanism of the retention of the urine in the bladder, as well as of its evacuation, a description is necessary of the following nervous apparatus which participates in these processes:

1. The sensory nerves of the walls of the bladder are derived from the first, second, third, and fourth posterior sacral roots. A number of sensory fibers pass into the spinal cord through the intermediation of the hypogastric plexus. The sensory nerves pass upward in the cord to the cerebral cortex.

2. The center for reflex stimulation of the unstriated musculature of the wall of the bladder—vesicospinal center—is situated in the neighborhood of the fourth lumbar vertebra, in the dog.

3. The motor tracts pass from this center to the unstriated musculature of the wall of the bladder through the nerves between the second lumbar—by way of communicating branches of the sympathetic—and the fourth sacral by way of the nervi erigentes. Irritation of the sensory nerves of the wall of the bladder causes reflex contraction of the bladder-wall.

In addition to the sensory nerves of the bladder, the reflex described may be excited also by irritation of other sensory nerves; thus, active tickling, or warming of the region of the knee during sleep at times causes evacuation of urine, likewise the hearing of splashing and whistling sounds. In animals, stimulation of certain sensory nerves likewise causes contractions of the bladder.

Omitting consideration of the sphincter muscle of the urethra, the sensation of a distended bladder will become apparent as soon as the bladder is moderately distended. Then the mechanical irritation of the sensory nerves of the bladder in the mucous membrane excites in the vesicospinal center the reflex through the motor nerves of the unstriated musculature of the bladder, and in consequence of this the walls of the bladder undergo contraction. This constitutes the process as it takes place, for instance, normally always in infants, who do not as yet have control of the urethral sphincter. Also voluntary evacuation of the bladder, whatever the degree of distention, is always effected only through excitation of the reflex described. The will is incapable of influencing directly the unstriated musculature of the bladder; and this is emphasized particularly by the author, in opposition to the statements of many other observers. To induce reflex stimulation of this movement of the bladder, principally in the presence of considerable degrees of distention, the direction of the attention to the sensations in the urinary apparatus alone suffices. When the distention of the bladder is only moderate or slight, the sensory, excito-reflex nerves of the bladder must first be stimulated, and either through irritation of the sensory nerves -by voluntary contraction of the striated muscles of the urethra and the floor of the pelvis, or of the nerves of the bladder as a result of abdominal pressure.

As electric stimulation from the cerebral peduncle downward through the motor paths of the spinal cord to the motor nerves of the unstriated musculature of the bladder causes contraction of the bladder, many investigators have concluded that the will is capable of exciting spontaneous contractions of the bladder directly in this way. The author considers this view as incorrect. In his opinion, voluntary evacuation of urine is always induced by reflex influences, in the excitation of which the will participates only in a secondary manner. With the vesical center situated in the spinal cord still other nervous apparatus coöperates. As painful irritation of sensory nerves in different parts of the body also is capable of causing reflex contraction of the bladder—the involuntary discharge of urine that occurs frequently in children.suffering from disorders of dentition may be of this character; as, further, as has already been pointed out, sensory nerves situated at a higher level, even cerebral nerves, are capable of exciting the vesical reflex, it must be concluded that the vesical center extends for a considerable distance upward, perhaps to the anterior portion of the optic thalamus, and that from these higher levels descend motor paths that are susceptible of— possibly reflex—stimulation in the spinal cord. Irritation of the medulla from the cerebral peduncle downward causes contraction of the walls of the bladder.

With respect to the mechanism for the retention of the urine in the bladder through the sphincter muscle of the urethra, consideration should be given to the following facts:

4. The motor nerves for the striated sphincter muscle are contained in the pudendal nerve, derived from the anterior roots of the third and fourth sacral nerves. Irritation causes contraction of the muscle; paralysis, inability to close the urethra, with the result that dribbling or incontinence of urine takes place. The nerves may be both stimulated—voluntary interruption of the stream of urine—and inhibited through the action of the will.

5. The sensory nerves of the urethra pass into the spinal cord through the posterior roots of the third, fourth, and fifth sacral nerves. These stimulate, on the one hand, the reflex for the urethral sphincter, so that as soon as urine escapes from the bladder into the commencement of the urethra the sphincter muscle contracts; as, for instance, in adults, during sleep, when the bladder becomes distended. On the other hand, they transmit sensory impressions from the urethra, particularly also when urine forces its way into the canal.

6. The center for the urethral-sphincter reflex—urethrospinal center—is situated, in the dog, at the level of the fifth, and, in the rabbit, at the level of the seventh lumbar vertebra.

7. From the cerebral cortex the voluntary motor paths course downward through the spinal cord to the sphincter muscle of the urethra, within the pyramidal tracts.

8. The inhibitory paths for this muscle likewise pass from the brain through the spinal cord, and through them the muscle may voluntarily

be relaxed into inactivity. It has not yet been possible to stimulate this center experimentally.

With respect to the mutual relations between the activity of the musculature of the bladder—expulsion of urine—and of the sphincter of the urethra—retention of urine—the action of the sphincter muscle preponderates, as a rule, when the distention of the bladder is not excessive. In other words, as soon as urine is forced into the urethra by contraction of the musculature of the bladder, reflex closure of the urethra takes place. The action of the sphincter muscle, however, predominates only to a certain degree; and neither the reflex nor the voluntary contraction of the sphincter is capable of resisting strong pressure by the urine. In the act of micturition, as it takes place when the bladder is moderately distended, the sphincter of the urethra must always be voluntarily inhibited in its contraction during the contraction of the walls of the bladder.

The foregoing description of the innervational conditions of the bladder is based upon the published experiments of Budge, all of which were performed in collaboration with Landois. Division of the sacral nerves, in the dog, causes degeneration of the nerves of the bladder and of the rectum, but not of the internal genitalia—some fibers of the urethral and vulvar nerves undergo degeneration. Bilateral division renders micturition and defecation impossible, while unilateral division renders these difficult. In addition, there is complete anesthesia at the anus, of the vagina, and on the posterior aspect of the thighs, together with weakness at the ankle-joint.

Normally, the bladder is completely evacuated. The residual urine that collects abnormally in greater or lesser amount is a source of danger, on account of the tendency to decomposition. The urine undergoes alterations during its sojourn in the bladder. According to Kaupp, retention is attended with an increase in the amount of sodium chlorid, and a diminution in the amount of urea and of water. The reduction in the latter is much more marked in conjunction with simultaneous sweating. The question whether the mucous membrane of the bladder absorbs soluble matters has been answered in the affirmative by Cl. Bernard, for the dog. Under such circumstances, water is again excreted into the bladder. Maas and Pinner noted absorption also on the part of the urethral mucous membrane, Lewin and Goldschmidt also on the part of the ureter, and the pelvis of the kidney, as well as the prostatic vesicle (strychnin).

As the ureters empty rather toward the base of the bladder, the urine most recently secreted is always the lowermost. Under varying conditions of secretion the urine may therefore (in a resting posture) form layers in the bladder, so that when evacuated the different layers may be clearly distinguishable. In quiet dorsal decubitus, the pressure in the bladder is from 13 to 15 cu. cm. of a column of water. The pressure is naturally increased by increase of the intra-abdominal pressure, especially in consequence of coughing and expulsive efforts. The erect posture has a similar effect, in consequence of the pressure of the viscera from above. In the evacuation of the urine, the amount expelled is at first small; this increases later in the same interval of time, and toward the end of the act it again diminishes. In men, the last portions are expelled from the urethra through voluntary contraction of the bulbo-cavernous muscle. Adult dogs constantly accelerate the stream of urine rhythmically through the action of this muscle.

MORBID DERANGEMENT OF URINARY RETENTION AND OF MICTURITION.

Derangement in the mechanism of retention and evacuation of urine may be referred by the physician to its cause from a consideration of the physiological conditions described. *Retention of urine—ischuria*—results (1) from occlusion of the urethra by foreign bodies, concretions, strictures, prostatic enlargement; (2) from paralysis or exhaustion of the musculature of the bladder, the latter also following parturition in consequence of the pressure of the child's parts against the bladder; (3) primarily, after division of the spinal cord. Under such circumstances, retention of urine takes place (*a*) because the division of the spinal cord gives rise to increased reflex activity on the part of the urethral sphincter, and (*b*) because inhibition of this reflex cannot take place.   If, with increasing distention of the walls of the bladder, the urethral orifice is finally dilated mechanically, dribbling of urine takes place.   Nevertheless, the urine escapes only drop by drop, as it overcomes the maximum tension at which the urethra still closes. Therefore, the bladder becomes more and more distended, as the tone of the continuously stretched walls lessens progressively, and the bladder may be distended to an enormous size.   In consequence of the entrance of bacteria into the bladder, ammoniacal decomposition of the long-retained urine may readily take place; and, as a result, catarrhal and inflammatory conditions of the bladder may be excited.   (4) From interference with the voluntary control of the inhibition of the reflex of the urethral sphincter, as well as from increased reflex excitability of the urethral center.

Incontinence of urine—stillicidium urinæ—occurs as a result (1) of paralysis of the urethral sphincter; (2) of anesthesia of the urethra, in consequence of which the reflex of the sphincter must be lost; (3) incontinence of urine is, sec-ondarily, always a result of division of the spinal cord or of abnormal degeneration. *Strangury* is observed as an excessive reflex of the walls of the bladder and the sphincter muscle, in consequence of irritation of the bladder and the urethra, as observed in association with inflammation, irritation, and neuralgia.   So-called *nocturnal enuresis*, nocturnal involuntary discharge of urine, may be a result of increased reflex activity of the walls of the bladder, or of enfeeblement of the reflex of the sphincter muscle.   Nothing of a definite nature is known as to the influence of deranged action of the will, principally in connection with unilateral injury, apoplexy, and the like.   In patients suffering from disease of the spinal cord, there is impairment of the sensation of a distended bladder, as well as of the contractile power of the walls of the bladder.   In neurasthenic patients, the latter is diminished, while the sensation of distention is increased.   In patients with prostatic disease, there is, at first, likewise increased sensitivity with a dis-tended bladder.

COMPARATIVE.   HISTORICAL.

In vertebrates, with exception of the bony fishes, there is often a union of the urinary and the generative organs.   The primitive kidney (Wolffian body), which serves during the first period of embryonic life as an excretory organ, assumes this function throughout life in fish and amphibia.   The myxenoids (cyclostomata) possess the simplest kidneys: On either side there is a long ureter, upon which are situated capsules with short pedicles containing glomeruli, and arranged in rows. Both ureters empty into the genital pore.   In the remaining fishes, the kidneys often extend longitudinally, lying as more compact masses on either side of the vertebral column.   The two ureters unite to form the urethra, which always opens behind the anus, either united with the genital orifice or behind this.   In the sturgeon and the shark the anus and the urethral orifice together form a cloaca.   Bladder-like formations, which, however, do not resemble the urinary bladder of mammalia morphologically, occur in fish, either at each ureter (ray, shark) or at the junction of the two.

In amphibia, the efferent vessels of the testicles unite with the urinary tubules. The testicular-renal duct unites, in the frog, with that of the other side; and both, united, open into the cloaca, while the capacious urinary bladder opens through the anterior wall of the cloaca.

From the reptiles upward, the kidney in all vertebrates is no longer the per-sisting Wolffian body, but a newly formed organ.   In reptiles, it is generally flattened longitudinally.   The ureters open separately into the cloaca.   Saurians and tortoises possess a bladder opening into the anterior wall of the cloaca.   In birds, the ureters remain separate and open into the urogenital sinus emptying into the cloaca internally to the excretory ducts of the generative glands.   The bladder is constantly wanting.   In mammalia, the kidneys often consist of many small lobules, reniculi, as, for instance, in the seal, the dolphin, the ox.

Among invertebrate animals, molluscs possess excretory organs in the form of canals provided with an external opening and an internal opening, communi-cating with the cavity of the body, and occasionally functionating also as oviducts. In mussels, this canal is expanded into a spongy organ (organ of Bojanus), situated at the base of the gills, often possessing a central cavity of considerable size, and provided with ciliated secretory cells. The internal (ciliated) excretory duct opens into the pericardial cavity; the outer, occasionally united with the sexual orifices, opens upon the external surface of the body. In the analogous, generally unpaired, often contractile organ of snails, guanin has been demonstrated. The organ is capable, in a remarkable manner, not alone of excreting water from the blood, but also of conveying water into the blood. Cephalopods possess sacculated excretory organs, provided with glands and opening into the mantel-cavity lying on the vascular trunks of the gills.

Insects, spiders, and centipedes have so-called Malpighian vessels, partly as uric-acid forming excretory organs; partly, also, as biliary organs. These vessels are long tubes that open into the commencement of the large intestine. In crabs, the blind tubes of the digestive tract probably have similar functions. In cestodes, the excretory organs are longitudinal tubes; in tape-worms two that extend throughout the entire chain, in the teniæ anastomosing at the junction of the segments by means of a large communication. In trematodes (distomum) the branching organ opens at the posterior extremity of the body. Also in most round-worms the excretory organ is formed of tubes, which, united, open at a pore in the abdominal line. Earth-worms possess, almost in all segments of the body in pairs, the so-called ·nephridia-canals, that is, tubes, often much convoluted, that commence in the abdominal cavity with an inner, ciliated orifice, and communicate upon the ventral aspect of the body with the external surface. In the sea-urchin, the star-fish, and the medusæ, the water-vascular system is, at the same time, the excretory organ. Also in sponges, the canals passing through the body and conveying water may be considered as such.

Historical.—According to Aristotle the urine is derived from the blood passing through the kidneys, and then flows through the ureters into the bladder; the venous blood of the kidneys does not undergo coagulation. He pointed out the relatively large size of the human bladder. Berengar (1521) observed, on injecting water into the renal vessels, that fluid escaped from the papillæ. Massa (1552) discovered lymphatic vessels in the kidneys. Eustachius (died 1580) ligated the ureters and subsequently found the bladder empty. Cusanus (1450) studied the color and the specific gravity of the urine. Rousset (1581) pointed out the muscular nature of the walls of the bladder, in which Sanctorius (1631) was unable to recognize any special sphincter muscle; while Vesling (1641) had already described the trigone of Lieutaud (1753). The first more important chemical investigations were made by van Helmont in 1644. He demonstrated the solid constituents of the urine, and found among them sodium chlorid. He noted the higher specific gravity of febrile urine, and explained the development of urinary calculi from the solid constituents of the urine. With respect to the discovery of individual urinary constituents, it may be noted that Scheele, in 1776, discovered uric acid; Bergmann calcium phosphate; Brand and Kunckel phosphorus; Rouelle, in 1773, urea, which was named by Fourcroy and Vauquelin in 1799; Berzelius lactic acid; Seguin albumin in pathological urine; J. v. Liebig hippuric acid; Heintz and v. Pettenkofer kreatin and kreatinin; Wollaston, in 1810, cystin; Marcet, in 1817, xanthin; Lindbergson magnesium carbonate. The more recent histological, physiological, and chemical investigations are discussed in the text.

FUNCTIONS OF THE EXTERNAL INTEGUMENT.

STRUCTURE OF THE SKIN.

The external integument, from 2.3 to 2.7 mm. thick, with a specific gravity of 1057, is constituted of the *cutis vera*, corium, cutis, and the overlying *epidermis*.

The corium (Fig. 178, I, C) forms upon the entire surface numerous papillæ, from 0.1 to 0.5 mm. high, of which the largest are encountered upon the palmar aspect of the hand and the plantar aspect of the foot, as well as upon the nipple and the glans penis. The majority of the papillæ contain loops of capillary blood-vessels (g), and in circumscribed areas of the skin also so-called tactile corpuscles (Fig. 179, a). The papillæ are arranged upon the skin in groups in the small areas bounded by the delicate furrows in the skin that are still macroscopically visible. On the palmar aspect of the hand and the plantar aspect of the foot they follow the characteristic cutaneous lines. The horny skin consists of a dense, uniformly woven network of elastic fibers, more delicate in the papillæ, and coarser in the deeper layers, with which fibrillary connective tissue, with connective-tissue corpuscles and lymphoid cells, are intermixed. In the deepest layers, the connective tissue predominates, and, by the interlacing of its bundles, forms longitudinal-rhombic reticular spaces (a a), generally filled with fatty tissue, whose longitudinal expansion corresponds with that of the greatest degree of

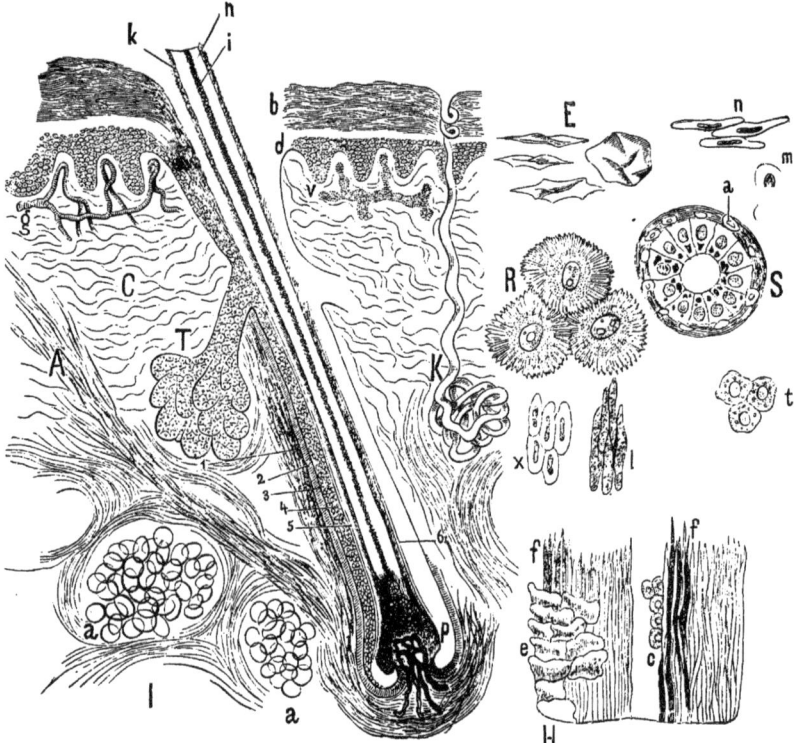

FIG. 178.—Histology of the Skin and the Epidermoidal Structures: I, transverse section through the skin, with hair and sebaceous glands (T), corium and epidermis are shown in reduced size; 1, external; 2, internal fibrous layer of the hair-follicle; 3, cuticula of the hair-follicle; 4, external root-sheath; 5, Henle's layer of the inner root-sheath; 6, Huxley's layer of the inner root-sheath; p, hair-root attached to the vascular hair-papilla; A, arrector pili muscle; C, corium; a, subcutaneous fatty tissue; b, horny layer; d, Malpighian mucous layer of the epidermis; g, vessels of the cutaneous papillæ; v, lymphatics of the cutaneous papillæ; h, horny substance; i, medullary canal; k, epidermis of the hair; K, sudoriferous gland; E, epidermal scales from the horny layer, viewed partly from the side, partly from the surface; R, prickle-cells from the Malpighian layer; n, superficial, deep nail-cells; H, hair, more highly magnified; e, epidermis; c, medullary canal with medullary cells; f f, fiber cells of the hair-substance; x, cells of Huxley's layer; ı, cells of Henle's layer; S, transverse section through a sudoriferous gland of the axillary cavity; a, adjacent unstriated muscular fibers; t, cells of a sebaceous gland, in part with fatty contents.

tension of the skin at the part of the body in question. Beneath the corium lies the subcutaneous connective tissue, which, however, is without fat-cells in some places. At certain points, firm fibrous bands of connective tissue unite the skin to the underlying fascia, ligaments, or bones (tenacula cutis). In other situations, principally over projecting bony parts, there are subcutaneous mucous bursæ filled with a synovial-like fluid, their interior partly lined by endothelium.

Unstriated muscle-fibers are present in the uppermost layers of the corium, principally on the extensor aspects; further, particularly on the nipple, the mammillary areola, the prepuce, the perineum, and in especial abundance in the tunica dartos of the scrotum.

The arteries of the skin in the palm of the hand and the sole of the foot, which must sustain the greatest amount of pressure, possess the thickest walls for the propulsion of the blood-stream. In silver-workers, the elastic fibers of the skin of the hands are discolored black in places from the deposition of reduced silver, and the same condition exists in cases of medicamentous argyria.

The **epidermis** is a layer of pavement epithelium, from 0.08 to 0.12 mm. thick, united by cement-substance. The deepest layer, the mucous layer (d), rete Malpighii, consists of several layers of protoplasmic nucleated prickle-cells (R), without membrane, pigmented in the colored races, as well as on the scrotum and at the anus, and of which the deepest are rather cylindrical and vertical. Among these cells scattered lymphatic wandering cells are encountered, which convey important constructive and nutrient material to the epithelial cells. On high magnification the cells are found to be provided with a fibrillar structure. The interstices between the prickles serve as lymph-paths. The more superficial layers (b), stratum corneum, consist of flat, horny, non-nucleated, epidermic scales (E) that swell up in sodium hydrate. The division between these two layers is constituted by a layer especially distinct when the epidermis is thick—of bright transitional forms of cells—stratum lucidum (between b and d). The uppermost layers of the epidermis are being continually desquamated, while new layers of cells resulting from division of the rete cells are constantly brought up from the depth. In this process, the cells that are elevated acquire the microscopic and chemical character of the horny layer, inasmuch as the nucleus undergoes atrophy.

Wherever pigment is present in the epidermis itself and likewise in the epidermoidal structures, it is conveyed, in many situations, from the underlying connective tissue by the stellate wandering cells. In this way is explained the fact that pieces of epidermis transplanted from a white person to a negro soon become dark. In certain other situations, however—as, for instance, on the mammilla—

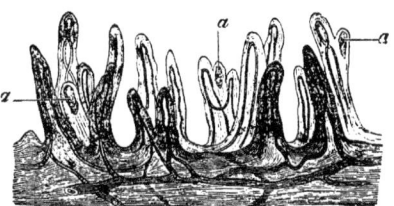

FIG. 179.—Cutaneous Papillæ Deprived of their Epidermis and the Vessels Injected: q a a, tactile papillæ, each containing a Meissner corpuscle.

it can be shown that the pigment is formed in the deep epidermal cells themselves. Finally, the pigment in connective-tissue cells is said to be derived in part from that formed in the epidermal cells.

In the layer of the epidermis in which the process of cornification takes place, therefore, from the upper layers of prickle-cells down to the actual cornified epidermis, the cells contain two varieties of granules—the albuminoid, intracellular, hyaline granules, and the fat-like, extracellular granules of eleidin, which are exhibited in an analogous manner by all horny structures at the boundary of the process of cornification. The granules of eleidin can be stained with henna, the hyaline granules with hematoxylin. Both structures are said to be allied to chitin.

Between the prickle-cells of the epidermis, and between the laminated epithelial cells of the mucous membrane, Herxheimer observed peculiar, spiral, solid fibers, which appeared to consist of fibrin-like masses. The elastic fibers of the horny skin undergo hyaline swelling and scaly or granular disintegration as a phenomenon of age.

THE NAILS AND THE HAIR.

The **nails** consist of numerous layers of firmly united cornified prickly epidermal cells, which can be isolated by caustic alkalies, and at the same time undergo swelling and display a nucleus (Fig. 178, n, m). The entire inferior surface of the nail rests upon the nail-bed. The posterior and the lateral borders are situated in a deep groove, the nail-fold (Fig. 180, e). The corium beneath the nail is provided throughout the entire extent of the nail-bed with longitudinal rows or bands of papillæ (Fig. 180, d). Immediately above these, as upon the skin in other situations, is the laminated, prickle-cell layer of the Malpighian mucous network (Fig. 180, c). Over this the nail is spread, thus representing the horny layer of the nail-bed (Fig. 180, a). The posterior nail-fold and the semilunar, brighter portion of the nail, the lunula, constitute the root of the nail. With the exception of a small surrounding area, they form at the same time the matrix, from which the growth of the nail takes place. The whitish crescent, present also on isolated nails, is due to the lessened translucence of this posterior portion of the nail, and this is a result of the special thickness and the uniform distribution of the cells of the mucous layer in this situation.

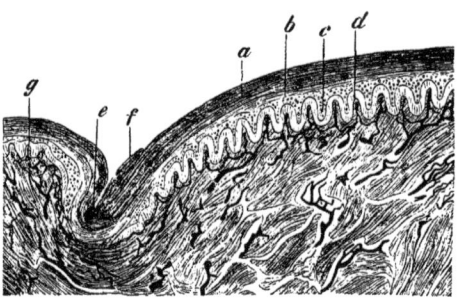

FIG. 180,—Transverse Section of One-half of a Nail, through the True Nail-bed (after Biesiadecki): a, nail-substance; b, subjacent loose horny layer; c, mucous layer; d, nail-ridge divided transversely; e, nail-fold without papillæ; f, the horny layer of the nail-fold, which has pushed itself over the nail; g, papillæ of the skin of the dorsum of the finger.

Growth and Development.—According to Unna, working under Waldeyer, the matrix of the nail is formed only by the floor and not also by the roof of the fold up to the anterior border of the lunula. The nail grows continuously from behind forward, and it is formed in layers by separation of the matrix. These layers are parallel with the surface of the matrix, though not with that of the nail. They pass obliquely from above and behind, downward and forward, through the thickness of the nail-structure. The nail is of uniform thickness from the anterior border of the lunula to the free margin. It, therefore, no longer grows in thickness in this area, except by the deposition of new cornified layers of cells from the mucous layer on the under surface of the nail. In the course of a year, the fingers yield about 2 grams, the hands and feet, 3.43 grams of nail-substance—in the summer relatively more than in the winter.

In the *development* of the nail, Unna observed the following stages: (1) Between the second and the eighth month of fetal life, the situation of the nail is occupied by a partial increase of the cornification of the epidermis on the dorsal aspect of the terminal phalanx—the eponychium. As the remains of this, there persists throughout the whole of life the normally formed, epidermal, horny layer that separates the subsequently developed, definitive nail from the roof of the fold. (2) The definitive nail develops in the fourth month beneath the eponychium. The base of the nail is situated, at first, at the extremity of the terminal phalanx, and subsequently moves further toward the dorsum. In the seventh month, the actual thin nail, itself still covered with eponychium, covers the entire extent of the nail-bed, and in the eighth month it penetrates the fold wholly.

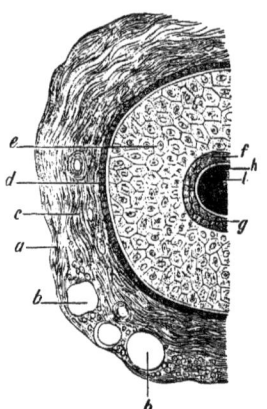

FIG. 181.—Transverse Section of a Hair below the Neck of the Hair-follicle: a, external sheath of the hair-follicle, with (b) blood-vessels in transverse section; c, internal sheath of the hair-follicle; d, vitreous layer of a hair-follicle; e, external, g, internal root-sheath; f, external layer (Henle's sheath); g, inner layer of the latter (Huxley's sheath); h, cuticula; l, hair.

(3) When, subsequently, the eponychium is exfoliated, the nail is disclosed. After birth, the papillæ develop upon the nail-bed, and, at the same time, the matrix extends to the most posterior portion of the fold.

The Hair.—With the exception of the palm of the hand, the sole of the foot, the dorsal aspect of the third phalanges of the fingers and toes, the external surface of the eyelids, the glans penis, the inner surface of the prepuce, a portion of the labia, and the lips, the skin of the entire body is covered with in part large and in part small hairs (lanugo). The hair is embedded by means of its root in a depression in the skin—hair-follicle (Fig. 178, I)—which passes obliquely through the thickness of the skin, at times down into the subcutaneous connective tissue. In the hair-follicle the following parts are distinguished: (1) The external fibrous layer (Fig. 178, 1, and Fig. 181, a), constituted of nucleated connective-tissue bundles pursuing principally a longitudinal course, and in which the vessels and nerves are distributed. (2) The inner fibrous layer (Fig. 178, 2, and Fig. 181, c), which contains connective-tissue fibers pursuing especially a transverse course. Toward the orifice of the hair-follicles, this layer passes over into the portion of the cutis vera forming the papillæ. At the bottom of the hair-follicle there is formed from the inner fibrous sheath the bulbous, vascular hair-papilla—comparable to a papilla of the cutis—the matrix of the hair, from which the growth of the hair takes place. (3) The innermost layer of the hair-follicle proper forms, besides, a vitreous layer (Fig. 178, 3, and Fig. 181, d). It terminates at the neck of the hair-papilla; above, its prolongation passes to the junction between the cutis vera and the epidermis. In addition to these layers, the hair-follicle has an epithelial lining, which must be looked upon as related to the epidermis. Thus, the external root-sheath (Fig. 178, 4, and Fig. 181, e), consisting of several layers of soft cells of fibrillar appearance, separated by spaces, and lying in contact with the vitreous layer, appears as a direct continuation of the Malpighian mucous layer, and its outermost layer exhibits cells stretched laterally. At the bottom of the hair-follicle it becomes narrower, and on fully developed hairs it is delimited from the root of the hair itself. The horny layer of the epidermis, passing down into the hair-follicle to the orifice of the sebaceous glands, retains the properties that it possesses upon the external skin. Below the orifice, however, its continuation forms the so-called internal root-sheath. This consists (1) of the outer single layer (Fig. 178, 5, and Fig. 181, f) of longitudinal, flat, homogeneous, nucleated cells (Fig. 178, magnified at 1)—Henle's layer—lying next to the outer root-sheath. Internal to this, there lies (2) the layer of Huxley (Fig. 178, 6, and Fig. 181, g), constituted of nucleated, rather longitudinal, polygonal cells (Fig. 178, x); and, finally (3) the cuticula of the inner root-sheath, a layer formed of cells in a manner analogous to the superficial covering of the hair, separates the inner root-sheath from the hair itself. Toward the hair-bulb this triple layer becomes ill defined, its cells mingling with those of the hair-bulb, without distinct limitation. All hair-bulbs are provided with nerve-cells and nerve-fibers, the latter having a bifid termination.

The *arrector pili muscle* (Fig. 178, A) is a flat, expanded layer of unstriped muscle-fibers passing from the outer fibrous layer of the bottom of the hair-follicle to the upper layer of the true skin, and always subtending the obtuse angle formed by the obliquely directed hair-follicle with the surface of the skin. Therefore, its contraction must cause the hair to become erect (goose-flesh). As a sebaceous follicle is usually present in the angle mentioned, the contraction may, by pressure, cause evacuation of the secretion of the gland. In addition, the muscle exerts a compressing effect upon the vessels of the papillary body. Goose-flesh never occurs upon the ear, the hand, or the foot. Occasionally it is only unilateral or confined to circumscribed areas. The pilomotor nerves are described on p. 719.

The arrectores pilorum receive their nerves (pilomotor nerves) from branches that pass from the spinal cord and thence into the sympathetic. The irritation of certain ganglia of the sympathetic has caused erection of the hair in definite circumscribed areas of the skin in the ape. The muscles are stimulated by reflex influences, which either extend over the entire body or remain strictly unilateral or local.

The *hair*, which remains firmly attached to the surface of the hair-papilla by means of its swollen, lowermost portion, the head of the hair, consists of three parts: (1) The medullary substance (Fig. 178, I, i), which is wanting in the lanugo and in the hair of early childhood, consists of a central row of cells, from two to eight in number, lying side by side (H, c). (2) Surrounding this is the thicker cortical layer (h), which consists of long, rigid, cornified hair-fiber cells (H, f, f), containing the pigment-granules of the hair. Nevertheless, the hair-fibers at times possess, besides, a diffuse tint. These fibers consist of minute longitudinal horn-fibrils, and exhibit a longitudinal nucleus when boiled with

34

caustic alkalies. (3) Upon the surface of the hair is the cuticula (k), consisting of laminated and non-nucleated scales arranged like the shingles on a roof (H, e).

The *graying of the hair* in late life is dependent upon a deficiency in pigment-formation in the cortical structure. The silvery luster of white hair is further increased by the development of numerous air-bubbles, in large number in the medulla, but also in small number in the cortex, which reflect the light. Occasionally pigment develops in the growing hair, at times not, so that, accordingly, it appears discolored in places and not so in others. Sudden graying of the hair, of which well-authenticated records exist, and which has also been observed upon one side of the body, was found by Landois in the case of a man who during an attack of delirium tremens was harassed by frightful hallucinations and became gray during a single night, to be dependent upon the presence of many air-bubbles throughout the entire medulla of the blond hair, and in smaller numbers, also, in the cortical structure, while the pigment was preserved. These air-bubbles imparted an exquisite gray luster to the hair. In rare cases, intermittent graying of the hair of the scalp has been observed; so that the hair exhibited alternately light and dark curls at intervals of about 1 mm. In such a case, Landois found the bright areas to be due to an abundant development of small air-bubbles in the medullary canal and the surrounding cortical area, while the pigment was well preserved.

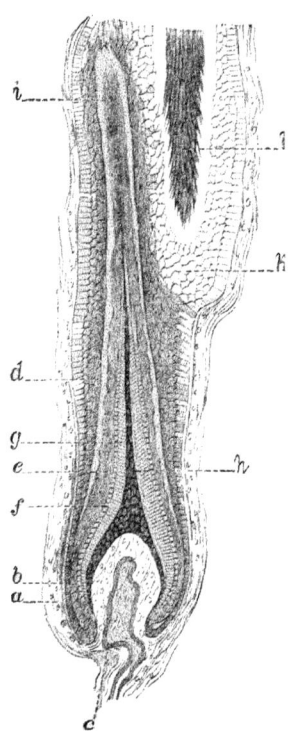

As to the *development of the hair*, Kölliker has discovered that, first, about the twelfth or thirteenth week, depressions like the finger of a glove take place from the epidermis into the corium. They are bounded externally by a vitreous membrane, and internally are occupied by soft homogeneous cells of the Malpighian mucous network. As these depressions subsequently enlarge downward and acquire a flask-like shape, the cells, arranged axially, acquire a rather longitudinal form and constitute a conical body, rising from the bottom of the recess. On this body there can be recognized an inner, darker portion, the primitive hair; and a thin, light, overlying cover, the inner root-sheath. The outermost cells, in contact with the wall of the fold, become the external root-sheath. Even before this, the papilla grows from below toward the hair-root; while, at the same time, the fibrous layers of the hair-follicle develop externally. Later on, the apex of the hair grows toward the horny layer of the epidermis. Here the apex penetrates the inner root-sheath, which is reflected upon the constantly growing hair like a sleeve. In the nineteenth week the hairs appear upon the forehead and the brow; between the twenty-third and the twenty-fifth week the lanugo-hair appears free, having a characteristic direction or grain on all parts of the body,

FIG. 182.—Longitudinal Section through a Hair-follicle, with the Hair in Process of Change (after y. Ebner): *a*, external and middle hair-follicle sheaths; *b*, vitreous layer; *c*, hair-papillæ with vascular loop; *d*, external, *e*, internal root-sheath (differentiated into Henle's and Huxley's layer); *f*, cuticula of the inner root-sheath; *g*, cuticula of the hair; *h*, young, non-medullated hair; *i*, conical tip of the new hair; *l*, hair polyp of the exfoliated hair with, *k*, the remains of the exfoliated external root-sheath.

just as is the case in animals. According to Kölliker, children are born only with lanugo-hair.

Of the *physical properties of the hair*, its great elasticity (tension, 0.33 of its length), marked cohesion (traction of from 1½ to 3 ounces), great resistance to putrefaction, as well as its high hygroscopic power, should be pointed out. The last property is possessed, also, by the epidermal cells, as indicated by the pains of clavi and cicatrices in damp weather.

The *growth of the hair* takes place by the constant formation by cellular division of new cells, at first soft, upon the surface of the papilla, which represents the matrix of the hair. These cells are situated upon the lower surface of the hair-bulb, acquire the shape characteristic of the different portions of the hair to which they become attached, and eventually undergo cornification. Thus, every newly formed layer raises the hair to a higher level out of the follicle. Human beings, between the eighteenth and twenty-sixth year, produce daily 0.20 gram of hair-tissue—corresponding to a loss of nitrogen represented by 0.0615 gram of urea—and even more in summer; and when frequently cut, according to Beneke, 14.6 grams of hair-tissue from the scalp annually. Iodin or bromin, ingested into the body, passes into the tissue of the hair.

As to *changes in the hair*, the statements made are by no means unanimous. According to one view, after the hair has attained its typical length, the formative process upon the surface of the hair-papilla is uninterrupted. The hair-bulb rises from the papilla, becomes cornified, remains generally free from pigment, and is finally raised more and more from the surface of the papilla, while its bulbous lower extremity becomes fibrillated like a broom (Fig. 182). The lower portion of the hair-follicle, thus made empty, diminishes in size; and upon the old papilla a new hair is formed through resumption of the formative processes, while the old soon becomes detached and falls out. In opposition to this view, Steinlein, Stieda, and others, contend that the papilla of the old hair is destroyed, while a new one forms in the hair-follicle, from whose surface the formation of the new hair takes place. Finally, Kölliker and Waldeyer believe both that new hair forms upon the old papilla and that its formation may take place upon a new papilla. The statement that hairs may be newly formed in adults, as in the fetus, is denied by v. Ebner.

THE GLANDS OF THE SKIN.

The **sebaceous glands** (Fig. 178, I, T) are simple acinous glands that in the case of large hairs empty laterally by from one to three openings into the hair-follicle, while in the case of small hairs the follicle projects free through the excretory duct of the gland (Fig. 183). The glands upon the labia minora, the glans penis, the prepuce (Tyson's glands), and those upon the red surface of the lips bear no relation to hair-follicles. The largest are present upon the nose and the labia; they are entirely wanting upon the palm of the hand and the sole of the foot. The glands contain polyhedral or circularly flat, nucleated, secretory cells (Fig. 178, t), through whose proliferation several layers of epithelium result, the elements of which undergo fatty degeneration as they advance toward the lumen of the gland, where they are broken up into fatty detritus. The membrane that gives form to the gland-vesicle is a structureless vitreous skin.

The **sudoriferous glands** (Fig. 178, I, K), also designated sweat-glands, each consist of a long, intestine-like, diverticular tube, whose extremity is rolled into a convoluted mass in the subcutaneous connective tissue; while the somewhat smaller excretory extremity passes through the corium and the epidermis in a spiral manner—in the illustration it is shown in abbreviated form. The cells of the sweat-glands are more compact, and are provided with intercellular and intracellular secretory passages and a rod-shaped central body. The glands are numerous and large on the palm of the hand, the plantar surface of the foot, in the axilla, the groin, the forehead, and about the nipple; scanty on the dorsum of the trunk; and are wanting on the glans penis, the prepuce, and the margin of the lips. Modifications are seen in the glands about the anus, the wax-glands of the ears (ceruminous glands), and the glands of Moll at the margin of the lids (which empty into the hair-follicles of the eyelashes).

The glandular tube is lined within the convolution, in the smaller part of the tube by a single layer of nucleated pavement-epithelium, and in the larger part by cylindrical epithelial cells (Fig. 178, S) without membrane, and in part containing fatty granules. The membrana propria is structureless and surrounded by delicate connective-tissue fibrils. Unstriated muscular fibers pass in a longitudinal direction on the larger glands (Fig. 178, S, a). The excretory duct (sweat-canal) contains no muscular fibers and is lined by a laminated epithelium of flat cells, whose surface possesses a thick, cuticular border. Within the epidermis, the canal pursues an intercellular course, without an independent membrane, between the epidermal cells. A network of capillaries surrounds the convolution. Before the vessels become capillary, the arteries form an intricate network surrounding the convolution. This bears a remarkable re·emblance to the network forming the glomerulus in the Malpighian capsule of the kidney. Finally, a plexus of nerves passes to the glands.

The total number of sudoriferous glands may be about two and one-half millions, representing a secretory superficies of approximately 1080 square meters. With respect to their function, it should be borne in mind that they secrete sweat. Nevertheless, an oily fat is admixed with their secretion, possibly from special cells. and this may predominate in the secretion in animals, as in the hoof-glands of the frog of a horse's foot, the glands on the sole of the dog's foot, and those of birds' feet. Meissner attributes only a secretion of fat to the convoluted glands, and Unna also believes that the sweat is produced from the intercellular spaces of the prickle-cells, which communicate with the penetrating sweat-ducts.

Tubular and reticular lymphatics without valves (Fig. 178, I, v) are present in the cutis, in part with blind terminations in the papillæ. Neumann observed them arranged in the form of a network about the hair-follicles and their glands. A coarser network of larger lymph-trunks is found in the subcutaneous tissue.

The blood-vessels appear principally in two layers; namely, in a superficial layer, from which the loops for the cutaneous papillæ arise; and a deep subcutaneous layer. Both vascular areas anastomose by means of processes. In addition, the glands of the skin are surrounded by a network of vessels.

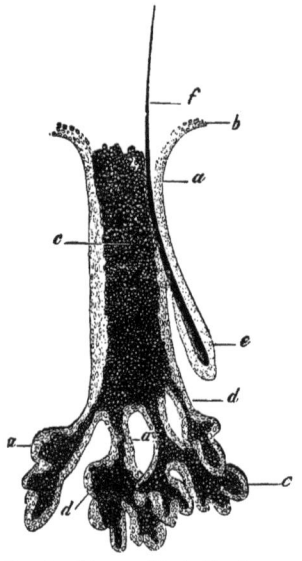

Fig. 183.—Sebaceous Gland with a Lanugo-hair: *a*, glandular epithelium; *b*, rete Malpighii. continued into the glandular epithelium; *c*, fat-containing cells and free fat as glandular contents; *d*, acini; *e*, root sheath with the hair.

THE SKIN AS AN EXTERNAL COVERING.

It is the function of the subcutaneous fatty tissue to fill the depressions between the different parts of the body, as well as to round off projecting portions, so that the rounded fulness of the body-form, agreeable to the eye, results. The fatty tissue acting as a soft cushion, affords protection from excessive pressure, as on the sole of the foot, in the palm of the hand, on the buttocks; and it encloses various more important parts that may be readily injured, as, for instance, the vessels and nerves in the axilla, the inguinal fold, and the popliteal space. As a poor conductor of heat, the subcutaneous fat shields the body against excessive loss of heat; the cutis vera and the epidermis exert a similar influence The firm, elastic, readily movable cutis is capable of affording protection against external mechanical injuries, and in this it is aided by the epidermis, whose dry, impervious, horny tissue, without nerves and vessels, is especially adapted to afford protection against poisons in solution; and is capable of offering considerable resistance even to thermic and chemical influences. A thin layer of sebum protects the free surface of the epidermis from maceration by fluids and from the destructive action of the air. The epidermal layer is, further, important in the fluid-economy of the body. It exerts pressure upon the cutaneous capillaries, and thus affords protection against excessive loss of fluid from the vessels of the skin. Portions of skin deprived of epidermis, therefore, appear reddened, and exude droplets of moisture. Large weeping areas of skin are capable of impairing considerably the nutritive state of the body through loss of albumin. The epidermis and the epidermoidal structures are, further, when dry, poor conductors of electricity. The passage of a strong current diminishes this resistance to one-thirtieth, in consequence of cataphoric infiltration. Finally, it may be stated that the presence of the uninjured epidermis protects adjacent parts against adhesion.

As the epidermis is but slightly extensible, it is drawn tensely over the folds and papillæ of the corium, which are obliterated on stretching the skin. Even the papillæ disappear in this way, if the tension is considerable.

The hairs serve in various situations as tactile organs—eye-lashes, lanugo-hair of the face; and upon the head, as a poor conductor of heat, they regulate the taking up and the giving off of heat and afford protection against direct radiation from the sun.

CUTANEOUS RESPIRATION. CUTANEOUS SECRETION.

SEBUM. SWEAT. PIGMENT-FORMATION.

The secretory activity of the external integument, whose extent exceeds more than one and a half square meters, comprises (1) the respiratory excretion; (2) the secretion of the cutaneous fat; and (3) the secretion of sweat.

Cutaneous respiration has already been discussed (p. 241).

Suppression of the activity of the skin by varnishing is followed, in warm-blooded animals, at first by no reduction in the total gaseous interchange. Probably increased respiratory activity on the part of the lungs compensates for the loss of the respiratory activity of the skin. In certain mammals, especially in rabbits, death results from varnishing of the skin, probably in consequence of excessive loss of heat. Strong animals die later than weak; horses only in the course of several days, with trembling and emaciation. The greater the area of skin that is not varnished, the later does death take place. Rabbits die after one-eighth of the surface of their body has been varnished; and after total covering of the skin the temperature at once declines, to as low as 19°. Pulse and respiration generally become less frequent; but with circumscribed varnishing, increased respiratory frequency and increased excretion of urea have been observed. Swine, dogs, and horses are said to exhibit only transitory depression of temperature and languor after one-half of the surface of the body has been varnished, though life is preserved. Varnishing of the skin is not injurious to human beings.

The **sebum of the skin.** The fat secreted by the sebaceous glands is fluid when discharged, but stagnating within the excretory duct of the gland it is transformed into a white, tallowy mass, which, principally on the alæ of the nose, can be expressed in sausage-shaped comedones. Its function is to keep the epidermis and the hair pliable and to protect the skin against excessive desiccation. Microscopically, the secretion contains innumerable fat-globules, a few gland-cells filled with fat and rendered visible on addition of sodium hydrate, and in almost all human beings microscopic mite-like animals—demodex folliculorum.

Chemical examination demonstrates the presence principally of fats, particularly olein (fluid) and palmitin (solid), together with fatty soaps, and some cholesterin; in addition, a small amount of albumin and unknown extractives. Among the inorganic constituents, the insoluble earthy phosphates preponderate; while the alkaline chlorids and phosphates are subordinate. There is some doubt as to the occurrence of sodium and ammonium phosphate and of ammonium chlorid.

The *vernix caseosa*, which covers the skin of the new-born, is a greasy mixture of cutaneous fat and macerated epidermis. It contains 35 per cent. of water and 14 per cent. of ethereal extracts, together with traces of albumin, chlorin, calcium, magnesium, and phosphoric acid. Examination for fats disclosed the presence of cholesterin, isocholesterin, oleic and palmitic acids (salts of fatty acids), together with glycerin. The *preputial smegma* (52.8 per cent. fat) is a similar product, in which an ammonium-soap occurs. *Ear-wax* is a mixture of the secretion of the ceruminous glands, which resemble the sudoriferous glands, and of the glands of the hair-follicles of the auditory canal. It contains, in addition to the constituents of the cutaneous fat, a brown pigment, soluble in alcohol and fat; a bitter yellow extractive; albumin; lecithin; cholesterin; potassium-soaps; and a special fat. The *secretion of the Meibomian glands* is cutaneous fat. The production of the fatty coating necessary for the oiling of the epidermis takes place, together with the formation of keratin, in part within the epidermis itself. The presence of cholesterin-fats in this situation has also been demonstrated in the layer of beginning cornification.

The **sweat** is secreted by the convoluted glands, the nuclei of the secretory cells acquiring a more nearly circular outline, and the cells, in the horse, becoming granular. So long as the secretion is confined within narrow limits, the water secreted, together with the volatile constituents, evaporates at once from the surface of the skin. As soon, however, as the secretion increases or evaporation is inhibited, the sweat appears in pearly drops at the orifices of the sweat-glands. The former has been designated *insensible*, the latter *sensible perspiration*.

The insensible perspiration varies widely. Generally, the right side of the body perspires more freely than the left. The palm of the hand sweats in greatest measure. Then, in order, follow the sole of the foot, the cheek, the breast, the thigh, and the forearm. Sweating increases slowly from the morning onward, in greater degree in the afternoon, and declines after the evening meal; then, increasing, it reaches its maximum before midnight. The presence of a large amount of moisture in the surrounding air diminishes the perspiration, as do also copious sweating previously and increased diuresis. Children have a relatively greater insensible perspiration. Ingestion of water increases, and withholding of water diminishes, the sweat; alcohol also diminishes it. The smallest measure of dissipation of watery vapor takes place at 15° C., while both above and below this temperature-level the dissipation increases. The ordinary temperature beneath the clothing is about 32° C. At this temperature the insensible perspiration equals 1500 grams of water. When the temperature of the surrounding atmosphere is 33° C. and above, sweating begins. Generous nutrition, warm clothing, and work cause greater excretion of water.

Pathological.—The insensible perspiration is increased in the presence of diseases of the skin, principally the acute erythemata. It is diminished in cases of scarlet fever, especially in association with uremia.

Sweat can be collected in largest amount from human beings by exposure in the steam-bath at a high temperature in a metallic tub in which the subject lies and into which the secretion of the skin flows. In this way Favre collected 2560 grams of sweat in one and a half hours. It is convenient, also, to obtain thus the partial secretion of sweat from the arm, which is placed in a glass cylinder hermetically sealed by rubber bandages about the arm.

In animals, sweating takes place in the horse, less in cattle, on the palm and the sole of the foot of the ape, the cat, the hedgehog. Swine sweat (?) on the snout, cattle about the mouth (?), while goats, rabbits, rats, mice, and dogs do not sweat at all.

Microscopically, sweat contains epidermal scales and fatty granules from the glands of the skin accidentally present. The sweat is colorless and slightly turbid, with a specific gravity of 1005. It has a salty taste and a characteristic odor in different portions of the body, due to volatile fatty acids.

The moist epidermis, including the hair and the nails, has an acid reaction, while the cutis has an alkaline reaction. The sweat secreted during rest has an acid reaction, while if the secretion is increased, the acidity diminishes and the reaction may even become alkaline. The sweat is composed of a glandular secretion having an alkaline reaction and an acid epidermal secretion. The reaction will vary in accordance with the preponderance of the one or the other of these constituents.

The constituents of the sweat:—Water, together with volatile substances, and it increases after copious drinking, 991 parts in 1000. E. Harnack found the solids on the average 8.5 in the thousand, including organic matters, 2 in the thousand, and inorganic matters, 6.5 in the thousand. Among the organic substances there should be mentioned some neutral fats, palmitin, stearin, found also in the sweat of the palm of the hand, which contains no sebaceous glands; in addition, cholesterin, volatile fatty acids, principally formic acid, together with acetic, butyric, proprionic, caproic, and capric acids, probably varying qualitatively and quantitatively in different portions of the body. They are present in largest amount in the acid sweat first secreted. Further, there are traces of sulphocyanid-combinations, of albumin (resembling casein), considerable urea, more than 0.1 per cent., and also ammonium-salts as decomposition-products of the latter in the air. Also sulphuric acid, in conjugation with skatol and phenol, and oxyacids were found by Kast in the sweat, uric acid by Tichborne. In the uremic state—anuria attending cholera—urea is even found upon the skin in crystalline form.

Marked increase in the secretion of sweat in healthy persons and in uremic patients diminishes the amount of urea in the urine. The reddish-yellow pigment that alcohol extracts from the residue of sweat and that is colored green by oxalic acid, is of unknown composition.

Among the inorganic substances, those that are readily soluble preponderate over those that are soluble with difficulty. There have been found sodium chlorid, 0.2; potassium chlorid, 0.02; sulphates, 0.01 in 1000, together with traces of earthy phosphates and sodium phosphate. Of gases, the sweat contains carbon dioxid absorbed together with some nitrogen.

Of ingested substances, the following appear again in the sweat: readily, benzoic acid, according to H. Meissner, also hippuric acid; cinnamic, tartaric, succinic acids; with greater difficulty, quinin, potassium iodid, mercuric chlorid, arsenous and arsenical acids, potassium and sodium arsenate. After the ingestion of iron arsenite, iron is found in the urine and arsenous acid in the sweat. Mercuric iodid is found transformed into chlorid, the iodin passing over into the saliva. When ingested, sweat has toxic effects.

Pigment-formation takes place in the form of a granular deposition, principally in the deeper, and less in the upper layers of the Malpighian network. It thus occurs particularly in the anal fold, on the scrotum, and the nipple; as well as universally in the colored races. The horny layer of the epidermis contains a diffuse yellowish-white pigment, which becomes darker in old age. This pigment-formation is supposed, like the process of cornification, to depend upon a chemical process, in consequence of which reduction takes place. This process is increased by light. In addition, the prickle-layer contains granular pigment. The dark discoloration of the epidermis can be removed and the process of cornification can be prevented by means of free oxygen.

Among pathological pigment-formations is that which occurs in liver-spots, freckles, and in conjunction with Addison's disease.

INFLUENCES AFFECTING THE SECRETION OF SWEAT.

The secretion of the skin, which on the average equals about $\frac{1}{84}$ of the weight of the body, or twice the elimination through the lungs, may be increased or diminished as a result of various influences. The tendency to sweating varies greatly in different individuals. Among the influences affecting the secretion of sweat the following are known: 1. Elevation of the surrounding temperature causes marked redness of the skin and profuse secretion of sweat. Cold and a temperature of the skin above 50° C. suppress the secretion. 2. The presence of an increased amount of water in the blood, principally after the ingestion of warm fluid in large amount, increases the sweat. 3. Marked activity on the part of the heart and the vessels, in consequence of which the blood-pressure in the capillaries of the skin is increased, has a similar effect. In this category belongs, also, the increased sweating in consequence of violent muscular activity. Under such circumstances the excretion of nitrogen through the sweat is increased. 4. Certain agents—hydrotics—increase sweating, such as pilocarpin, physostigma, strychnin, picrotoxin, muscarin, nicotin, camphor, and ammonium-combinations. Others, such as atropin and morphin in large doses, diminish the sweat. 5. The antagonism that exists between the secretion of sweat and the secretion of urine and the intestinal discharges, probably in consequence principally of mechanical influences, is especially noteworthy in so far as abundant micturition, as, for instance, in cases of diabetes, and thin stools are associated with dryness of the skin.

If the amount of sweat is increased, the proportion of salts, urea and albumin present increases; while the remaining organic substances diminish. The more saturated the air with watery vapor, the more readily does the secretion appear in drops upon the surface; while in dry air in active motion the secretion appears as fluid later in consequence of the rapid evaporation.

NERVOUS CONTROL AFFECTING THE SECRETION OF SWEAT.

As in the secretion of saliva, vascular nerves are principally active in the secretion of sweat, in addition to the true secretory nerves; and most frequently the vasodilators, as indicated by the sweating when the skin is reddened. The observation of sweating when the skin is pale (the sweating of fear and of death) shows, however, that also in the presence of vasoconstriction, the sweat-fibers may at the same time be active.

Under certain conditions an increase in the amount of blood present appears alone to be sufficient for the occurrence of sweating. In favor of this view is the observation of Dupuy, who noted unilateral sweating of the neck in a horse after division of the cervical sympathetic; and in opposition to this view is the statement of Nitzelnadel, who observed diminution of sweating in human beings after percutaneous galvanization of the cervical sympathetic.

Independently of the circulation, sweat-nerves of independent activity control the secretion from the surface of the body. Irritation of the appropriate nerve-trunk still causes transitory secretion of sweat even if the extremity has been previously amputated; and therefore the circulation no longer exists. In addition, the secretion of sweat may take place under higher pressure than that of the blood. In the healthy body, profuse secretion of sweat, it is true, appears usually to be associated with vascular dilatation, like the secretion of saliva after irritation of the facial nerve. Indeed, the sudoriferous and the vascular nerves appear to pursue almost identical paths.

For the hind extremity, in the cat, these fibers are contained in the sciatic nerve. Luchsinger was able to excite constantly renewed secretion of sweat for half an hour by irritation of the peripheral stump, if the paw was constantly kept dry. This nervous activity is destroyed by atropin. If a young cat, whose sciatic nerve on one side has been divided, is placed in a room filled with hot air, the three intact members soon sweat, but not that with the divided nerve, not even when excessive hyperemia of the member is induced by ligation of the veins. The sweat-fibers pass centripetally from the sciatic nerve, in the abdominal sympathetic, in order to reach the upper lumbar and lower dorsal cord (twelfth dorsal and first, second, and third lumbar roots in the cat), through the communicating branches of the sympathetic and through the anterior roots. The center for the secretion of sweat in the hind extremities is situated in the ganglia of the anterior horns in the lower dorsal and upper lumbar portions of the spinal cord. According to Langley, non-medullated sweat-fibers pass in the cat to the nerves from the eleventh dorsal to the fifth lumbar, and are derived from the sixth and seventh lumbar and the first and second sacral ganglia of the sympathetic. The origin and course of the vasomotors are, on the whole, the same.

This spinal center may be irritated directly (1) through marked venosity of the blood; therefore through dyspneic stimulation. In this category belongs probably also the sweat of the death-agony. (2) Through overheated blood (45° C.) passing through the center. (3) By certain poisons (see p. 536). Reflex stimulation of this center is effected, though with varying result, through irritation of the crural or peroneal nerve of the same side, as well as of the sciatic nerve of the opposite side.

For the fore-paws, in the cat, the sweat-fibers pass in the ulnar and median nerves. These pass from the dorsal roots between the fourth and the tenth to the dorsal division of the sympathetic, and then pass downward through the stellate ganglion, and thence into the nerves of the anterior limb.

An analogous center for the anterior extremities is situated in the lower half of the cervical cord. Irritation of the central stump of the brachial plexus causes reflex sweating of the paw of the opposite side. Under such circumstances, the hind paws also sweat at the same time.

Pathological.—Degeneration of the motor ganglia of the anterior horns of the spinal cord induces loss of the secretion of sweat, together with paralysis of the striated muscles of the trunk. Perspiration is increased in enfeebled as well as in edematous extremities. Nephritic patients exhibit great variations in the amount of water given off by the skin. Dieffenbach observed that sweating reappeared in transplanted bits of skin only after the return of sensitivity.

The sweat-fibers for the head (man, horse; snout in swine) are derived from the upper dorsal sympathetic, pass through the stellate ganglion and ascend in the cervical sympathetic. The observation is probably appropriate here that in human beings percutaneous galvanization of the cervical sympathetic causes sweating on the same side of the face and the arm, as well as the pathological observation that in association with unilateral sweating of the head, neck, and upper extremity, the corresponding pupil is dilated and the skin is pale. In the cephalic portion of the sympathetic the sweat-fibers enter the branches of the trigeminus, and this fact explains the circumstance that irritation of the infra-orbital nerve excites the secretion of sweat. Some fibers, however, arise directly from the trigeminal roots and the facial nerve.

Undoubtedly, the cerebrum must also exert a direct influence either upon the vasomotor nerves or upon the sweat-fibers, as is shown by the sweating that attends emotional disturbances, fright, etc.

An observation of Adamkiewicz and Senator tends to support this view. They noted that in a human being with an abscess in the motor area of the cerebral cortex for the arm, convulsions and sweating occurred in this member.

According to Adamkiewicz, all of the four paws of the cat sweat on irritation of the medulla oblongata, in which the dominating center for the secretion of sweat appears to be situated, even three-quarters of an hour after death.

Nerve-fibers that pass to the unstriated muscular fibers of the sudoriferous glands, and are wanting in the smaller glands, must have an influence upon the discharge of the secretion.

Pilocarpin and other diaphoretics, when injected subcutaneously, even after division of the nerves, cause sweating first at the site of injection. Atropin, in the same way, causes first local suppression of sweat-secretion. If the sweat-nerves are divided, in the cat, the irritability of the fibers (sciatic) to electrical stimulation is lost in the course of four days. In cats operated on, delayed sweating after injection of pilocarpin occurs during the course of three days, and this may be prolonged, after the lapse of six days, even to a delay of ten minutes. At a later period, the sweating may remain entirely in abeyance. The familiar phenomenon of the dry skin of paralyzed members is in accord with this observation.

If in man, a motor nerve, such as the tibial, median, or facial, be irritated, sweat appears in the distribution of the active musculature and in the corresponding distribution on the unirritated half of the body; and both when the circulation is free, as well as when it is arrested. On sensory and thermic irritation of the skin, there likewise occurs reflex sweating always upon both sides, independently of the circulation. The seat of the sweating is independent of the site of cutaneous irritation. In the case of the author himself, cold sweat appeared immediately upon the forehead as soon as the mucous membrane of the mouth was irritated by strong vinegar.

PHYSIOLOGICAL CARE OF THE SKIN.

PATHOLOGICAL ABNORMALITIES IN THE SECRETION OF SWEAT AND SEBUM.

In order to maintain the normal secretion of the skin, the care of this organ by means of frequent ablution and baths, soap being used to remove the fatty accumulation upon the skin, is of the greatest significance, as in this way the pores are kept open. By friction of the epidermis, baths aid metabolism, by an action upon the cutaneous vessels influence the circulation and the heat-economy of the body, and have a stimulating effect upon the nervous system. The establishment of public bath-houses must be considered among the most beneficent measures for the preservation of the public health.

Diminution in the secretion of sweat, anidrosis, occurs in cases of diabetes and carcinomatous cachexia; further, together with other nutritive disorders of the skin, in connection with certain nervous diseases, as, for instance, paralytic dementia. It has been observed in circumscribed areas of the skin as one of the phenomena of certain trophoneuroses; as, for instance, unilateral atrophy of the face, and in paralyzed parts. In some of these cases there may be paralysis of the nerves in question, or of their spinal centers.

Increase in the secretion of sweat, hyperidrosis, occurs in part in readily excitable persons, in consequence of irritation of the nerves in question. In this category belongs the sweating that attends debilitated states, and that occurs also in hysterical persons, principally upon the head and the hands; and the so-called epileptoid sweats that occur paroxysmally. Unilateral sweating, principally of the head, long known to earlier physicians, is especially noteworthy. This has been observed in conjunction with other nervous disorders, in part among the symptoms of irritation of the cervical sympathetic—dilatation of the pupils, exophthalmos. Landois has, however, observed unilateral sweating without other evidence of sympathetic disorder, probably as a manifestation of irritation of the true sweat-fibers.

Qualitative alterations in the secretion of sweat, paridrosis. In this category belong the rare cases of *blood-sweating, hematidrosis*, which may also be unilateral, and in which, at times, the bloody discharge from the pores of the skin appears to take place vicariously for absent menstruation. More commonly, however, the condition has been one of the symptoms of a profound nervous disorder, especially convulsive seizures. Blood-corpuscles, rarely blood-crystals, have been found in the escaping drops of red sweat. Yellow fever also is, at times, attended with bloody sweats. Biliary pigment has been found in the sweat of jaundiced persons; a bluish-black discoloration, further a blue color from indigo, from pyocyanin (the rare blue pigment of pus), produced by the bacillus pyocyaneus, or from ferric phosphate, are among the rarest exceptions. Such colored sweating is designated *chromidrosis*.

Between the epidermal scales and upon the hair there live numerous micrc-organisms, which, however, must be designated as innocuous: Two varieties of saccharomyces; the leptothrix epidermidis and various bacteria on surfaces the seat of intertrigo, namely, five varieties of micrococci; and between the toes, the bacterium graveolens and bacillus saprogenes, which generate the odor of the sweat of the foot. Yellow, blue, and red sweat are likewise caused by bacteria, the last by the micrococcus hæmatodes. Red sweat and black sweat may be caused also by a variety of torula. Within the lesions of acne and in comedones, there vegetates a thick bacillus, which Hodara considers as the cause for the formation of the pustule.

Grape-sugar has been found in the sweat in cases of diabetes; rarely uric acid, in individuals with calculi; cystin, in cases of cystinuria. In the fetid sweat of the feet, leucin, tyrosin, valerianic acid, and ammonia are present. This condition can be corrected only by the most systematic and scrupulous cleanliness. To the foot-baths, anti-fermentative and bactericidal substances should be added, such as salicylic acid or potassium permanganate. Odorous secretion of sweat is designated as *osmidrosis*, fetid sweating as *bromidrosis*. In the sweating stage of intermittent fever, considerable calcium butyrate has been found; in cases of puerperal fever, lactic acid. The viscid sweat of acute articular rheumatism is said to contain a greater amount of albumin, as does also the sweat attending enforced diaphoresis.

With respect to **abnormalities in the secretion of the cutaneous sebum,** there should be mentioned the pathological increase in secretion—*seborrhea*—which occurs either locally or disseminated over the entire skin. In cases of premature baldness, there is increased production of sebum on the scalp. Diminished secretion of sebum—*asteatosis of the skin*—causes the skin to become brittle and rough, in part locally and in part extensively. Often, as upon the bald head in the aged, the sebaceous glands undergo atrophy. If the excretory ducts of the sebaceous glands become obstructed, the sebum accumulates, in greater or lesser amount. Not rarely, the excretory ducts are occluded by particles of dirt, granules of ultramarine derived from washing blue, and vegetable fibers from the clothing. By pressure, the fatty, worm-like comedo is discharged.

ABSORPTION THROUGH THE SKIN. GALVANIC CONDUCTIVITY.

After prolonged exposure to water, the epidermis becomes moist and swollen. On the other hand, the skin is incapable of absorbing substances, either salts or vegetable poisons, from watery solutions, such as baths. This inability is due to the fat normally present in the epidermis and the pores of the skin. If, therefore, substances dissolved in such fluids as dissolve and extract the cutaneous sebum, as alcohol, ether, and particularly chloroform, are applied to the skin, they may be absorbed in small amounts—in larger measure in rabbits. Volatile substances, such as carbolic acid, that exert a corrosive effect upon the epidermis, are capable of absorption through the injured areas. Absorption does not take place from ointments applied simply to the skin. In the case of persistent vigorous inunction, there occurs, at times, a forcible introduction into the pores of the skin, not rarely in association with mechanical lesions in the continuity of the layers of epidermis.

Under such circumstances, absorption, as of potassium iodid, may take place from ointments.   Thus, v. Voit observed globules of mercury between the layers of epidermis and even in the corium of an executed individual, to whom, while still warm, he had given vigorous inunctions.

In courses of treatment with inunctions of mercurial ointment, globules of mercury penetrate, on rubbing, also into the hair-follicles and excretory ducts of the glands, where under the influence of the glandular secretion they may be transformed into a combination susceptible of absorption.   In addition, mercury, in the form of vapor, reaches the respiratory mucous membrane, where, likewise, it is transformed into an absorbable combination.   The inflamed skin, especially, however, when covered with fissured or injured epidermis, absorbs rapidly, like a wound-surface.   As all substances that irritate the skin sever the continuity of the latter when the effect is long continued, it can readily be understood that they are eventually absorbed from the wounded areas.

As the skin, under normal conditions, absorbs oxygen from the atmosphere, it may also absorb gases, such as hydrocyanic acid, hydrogen sulphid, carbon monoxid, carbon dioxid, vapors of ether and chloroform.   From a bath that contains absorbed hydrogen sulphid, this gas is absorbed; conversely, carbon dioxid is given off to the bath-water.

In frogs, active absorption of watery solutions takes place through the skin, the epidermal cells undergoing enlargement and exhibiting motor phenomena. These phenomena may also be induced artificially by electric stimulation.   The frog also absorbs much water through the skin even when the circulation is eliminated and the central nervous system is destroyed; though more, however, when the circulation is maintained.   The skin of the frog exhibits, in the process of absorption, a vital cellular activity, in consequence of which penetration takes place from without inward.

The transfer of watery solutions through the skin by means of the constant galvanic current, cataphoric action, is a matter of especial interest.   Both electrodes are impregnated with a watery solution of the substance in question, and the direction of the current is altered from time to time.   Thus, H. Munk was able to introduce through the skin of rabbits within several minutes strychnin, from the effects of which they died.   In man, the introduction of quinin and potassium iodid into the body was thus effected, these substances being subsequently demonstrated in the urine.   In the introduction, the compound bodies are (always?) decomposed by the current; thus, for instance, the positive pole of the current introduces the calcium of calcium chlorid; the negative, only chlorin.

COMPARATIVE.   HISTORICAL.

In all vertebrates, the skin consists of corium and epidermis.   In reptiles, the cornification of the epidermis occurs in large plates (scales of the snake, shell of the tortoise).   Among mammals, the armadillo exhibits a similar formation.   In addition to hair and nails, there occur in animals, as epidermoidal structures, prickles, bristles, feathers, claws, hoofs, horns (the antlers of the deer are bony formations arising from the frontal bone), spurs (cock), the horny covering of the beak of turtles and of birds, and the horn of the rhinoceros.   The scales of fish, on the other hand, consist of ossified portions of skin.   Some fish possess considerable portions of bone upon the skin.

The skin is provided with a large variety of glands.   In the amphibia, they secrete either mucus alone or poisonous substances.   Serpents and tortoises possess no cutaneous glands at all.   In lizards, the thigh-glands extend from the anus to the popliteal spaces.   In crocodiles the glands open beneath the margin of the cutaneous osseous scales.   Birds have no cutaneous glands.   The coccygeal gland, situated above the coccygeal vertebra, furnishes a secretion for lubricating the feathers.

The civet-glands at the anus of the civet-cat, the preputial glands on the musk-bag of the musk-deer, the inguinal glands of the hare, the pedal glands of ruminants are peculiarly developed sebaceous glands.   The strongly odorous castoreum is the secretion of the prepuce in both sexes of the beaver.

In molluscs, the skin, consisting of epidermis and corium, is intimately united with the underlying muscles to form the musculo-cutaneous tube of the body. Cephalopods have, in their skin, the so-called chromatophores; that is, round cells filled with granular pigment, at the periphery of which muscular fibers are attached in a radiate manner, so that their contraction must increase the colored surface. Through the play of these muscles there result the color-variations observed in cuttle-fish. Chromatophores are present, also, in other classes of animals, such as amphibia (frog) and fish (pike). In these animals, they appear as connective-tissue cells, within which pigment-granules either collect toward the center or swarm toward the periphery, while the processes of the cells themselves do not change their place. Every cell is provided with numerous nerve-endings, which surround the pigment-mass in the form of garlands, with free terminal radiations. Special glands furnish the material for the formation of the scales of the snails. In all invertebrates, the development of the scales takes place from a portion of the surface of the body of the animal that has been designated the mantle.

In articulates, the entire surface of the body is covered by a more or less solid shield, which is to be considered as a cuticular structure consisting of chitin, which is separated from the underlying matrix. It extends for some distance into the digestive tube and the trachea. In the formation of the skin it is thrown off and replaces itself anew from the matrix. This shield, which affords protection to the body, serves, at the same time, for the attachment of the muscles. It thus becomes a passive motor organ comparable to the skeleton of the vertebrates.

The echinoderms exhibit deposits of lime in their skin, in consequence of which they often acquire a cutaneous skeleton. The deposits of lime are either united to form large immovable plates, as in the scale of the sea-urchin, or united together in segments, as in the arms of the star-fish. In holothurians alone, the significance of calcification with respect to the cutaneous skeleton is of subordinate importance. In them, only isolated plates of lime have remained in various forms. In worms, the skin forms with the underlying muscles the musculo-cutaneous tube. The epidermis is, in some, provided with cilia; in others (tape-worms) it is traversed by pores; while in still others it is without any appendage. The hooklets on the head of teniæ, the rod-shaped motor bristles on the body of earth-worms, are cuticular formations. Cutaneous glands are present in the more highly developed worms, such as the leech.

The integument of cœlentrates (zoöphytes) is characterized by the forerunners of disseminated nettle-cells; that is, cells provided with whip-like processes, which contain a corrosive fluid and serve as organs of capture. Cilia are present in many; in some a tubular, external, chitin-like skeleton is formed. The integument of sponges is suggestive of that of zoöphytes. Infusoria possess numerous cilia, which in part are even subject to voluntary stimulation. Rhizopods are wholly unprovided with a true skin. Nevertheless under these circumstances the formation of silicious (radiolaria) or calcareous structures (monothalamia and polythalamia) is noteworthy.

Historical: Hippocrates (born 460 B. C.) and Theophrastus (born 371 B. C.) distinguished perspiration from sweat. According to the latter the secretion of sweat stands in a certain antagonistic relation to the secretion of urine and to the amount of water in the feces. Individuals suffering from fright were believed to sweat more freely from the feet. Father Augustinus stated that he knew an individual who was able to sweat voluntarily. According to Cassius Felix (97 A. D,) the skin absorbs water in the bath. He made investigations into the evaporation from the skin. Sanctorius (1614) measured more accurately the insensible perspiration and the loss of weight on the part of a fasting individual. The hair-follicle and the root of the hair are mentioned in the Talmud. Alberti (1581) recognized the hair-bulb. Donatus (1588) made the first report of sudden graying of the hair. Riolan (1626) discovered the cutaneous pigment of the negro in the epidermis. De Heyde (1684) and Leeuwenhoeck described the ciliated movement on the beard of mussels (1694).

PHYSIOLOGY OF THE MOTOR APPARATUS.

STRUCTURE AND ARRANGEMENT OF THE MUSCLES.

The striated (voluntary) muscles are covered on their outer surface by a connective-tissue sheath, the *external perimysium*. From this sheath septa extend into the interior of the muscle, the *internal perimysium*, carrying vessels and nerves, and dividing the muscle into bundles of fibers, which are sometimes finer (eye muscles) and sometimes coarser (gluteals). Each compartment thus formed contains a number of muscle-fibers lying close together.

Each muscle-fiber is surrounded by a rich meshwork of blood-capillaries, with neighboring lymphatics; it also has a nerve-fiber leading to it. These structures are held on the surface of the muscle-fiber by means of an extremely delicate connective tissue with a scarcely recognizable fibrillar structure, representing to a certain extent a perimysium for each separate fiber.

The individual *muscle-fibers* or *primitive muscular bundles* may be isolated by means of a 35 per cent. solution of potassium hydroxid, or of nitric acid containing an excess of potassium chlorate. They are from 10 to 100 μ in diameter, and are of limited length, in man from 5.3 to 9.8 cm. Within short muscles (the stapedius among others and the small muscles of the frog) the fibers, therefore, traverse the entire length of the muscle; within longer muscles, however, each fiber tapers to a point and is attached obliquely by cement-substance to the succeeding, similarly pointed fiber. Each muscular spindle is completely enclosed in a structureless, transparent sheath, the *sarcolemma* (Fig. 184, 1, S).

The muscle-fiber exhibits at intervals of from 2 to 2.8 μ a *transverse striation* due to alternate light and dark layers (1, Q). As a result of the action of hydrochloric acid (1 : 1000) or of the gastric juice, or after freezing, the fiber not rarely undergoes a solution of continuity in the region of the light bands, so that it breaks up into plates or *discs* (5) resembling an overthrown pile of coins, the discs always corresponding to the dark parts of the fiber. In addition to the transverse striation, a *longitudinal striation* may be observed in the fiber. This is due to the fact that the muscle-fiber is made up of numerous, fine, contractile threads (from 1 to 1.7 μ in diameter), the *primitive fibrils* (Fig. 184, 1, F), lying side by side. Each separate fibril is striated transversely, and all are bound together by a small amount of a fluid, finely granular, cement-substance (Rollet's *sarcoplasm*), in such manner that the transverse striations of all fibrils are situated at the same level. The sarcoplasm embeds all of the fibrils uniformly, and occurs also in a thin layer between the sarcolemma and the muscle-substance; it contains minute *interstitial granules* (fat and lecithin). The fibrils are prismatically flattened against one another; hence, a cross-section of a fresh frozen muscle exhibits a design consisting of polygonal figures—*Cohnheim's fields* (2).

The study of an isolated fibril under high magnification shows it to be a columnar structure, made up of numerous parts superposed in layers. These sections, which may be termed *muscular elements*, exhibit individually a complicated structure. Each muscular element (4) is a prismatic body, from 2 to 2.8 μ in height, with plane terminal surfaces. The entire middle layer is occupied by the darker and more highly refractive, true contractile substance, the *transverse disc* (Bowman's sarcous elements, Kühne's muscle-prisms). This is doubly refractive (anisotropic) and contains a bright layer, the *median disc* (4 c), which can be recognized as a bright line bisecting the dark field. On the upper and lower surfaces of the darker, contractile substance is a layer of light, singly refractive (isotropic) substance (4 d). Where this lighter disc comes in contact with that of the adjacent element, a dividing band can be recognized, the *terminal* or *intermediate disc* (4 a), which appears as a dark line.

542

In the muscles of arthropods there lies within the isotropic layer, at a short distance from the terminal disc, still another narrow layer of doubly refractive substance, the *accessory disc*, which contains chromatin. Every muscle-fiber is closed off toward its extremity by a layer of singly refractive substance. When the tube of the microscope is lowered, the doubly refractive discs appear dark, the singly refractive, light; when the tube is raised, the conditions are reversed.

The fibrils are readily obtained singly from the muscles of insects; in mammalian muscles they may be isolated after the action of dilute alcohol or Müller's fluid, especially at the torn ends of the fibers (Fig. 184, 3).

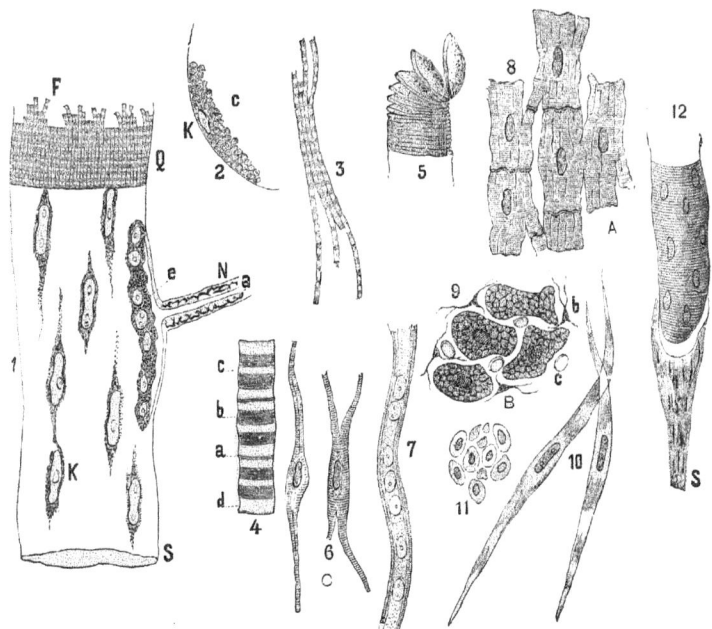

FIG. 184.—Histology of Muscular Tissue: 1, Diagrammatic representation of the parts of a striated muscle-fiber: S, sarcolemma; O, transverse striation; F, fibrils, further on giving rise to longitudinal striation; K, nuclei of the muscle-fiber; N, motor nerve leading to the fiber, with the axis-cylinder a, which passes over into the motor end-plate, seen in profile, the latter lying upon a nucleated, protoplasmic layer e. 2, Part of a cross-section of a striated muscle-fiber with Cohnheim's fields c; K, a muscle-nucleus in contact with the sarcolemma. 3, Isolated fibrils from a striated muscle-fiber. 4, Part of a fibril from an insect's muscle, highly magnified: a, Krause-Amici line limiting the muscular elements; b, the dark, doubly refractive substance; c, Hensen's line; d, the singly refractive substance. 5, Striated muscle-fiber breaking up into discs. 6, Striated muscle-fiber from the heart of the frog. 7, Structure of a striated muscle-fiber from a three-months' human embryo. 8, Reticulated muscle-fibers of the heart. 9, Cross-section of the heart-muscle: c, capillaries; b, connective-tissue corpuscles. 10, Unstriated muscle-fibers. 11, Unstriated muscle-fibers in cross-section. 12, Striated muscle-fibers with the related tendon S (detached).

In all fibers there are encountered several nuclei, from 9 to 13 μ long and from 3 to 4 μ wide, directed longitudinally, which become more evident on addition of dilute acetic acid. They are surrounded by a thin layer of protoplasm or sarcoplasm (1 and 2 K), and are designated *muscle-corpuscles*. Each nucleus contains one or two nucleoli; the protoplasm sends to adjacent corpuscles distinct, delicate processes, at times containing refractive granules, so that a continuous network of cells is formed beneath the sarcolemma.

Histogenetically the muscle-corpuscles are the remains of cells, from the body of which the muscle-fibers were formed (7); the striated substance is the differentiated parietal or intracellular substance that has separated from the corpuscles.

The latter probably represent the natural source of nutrition for the muscle-fibers. In amphibia, birds, fish, and insects, the muscle-corpuscles are situated in the axis of the fiber between the fibrils.

The protoplasm of the muscle-corpuscles is further connected with the protoplasm that throughout the entire muscle-fiber forms longitudinally and transversely a network of fibers on the fibrils—the *sarcoplasm.* The transverse fibers follow the course of Hensen's and the Krause-Amici lines; the longitudinal fibers pass in the interstices between Cohnheim's fields. The amount of sarcoplasm is greater in the lower than in the higher animals.

The *relation of the muscle-fibers to the tendons* varies. According to Toldt, the delicate connective-tissue elements that surround the individual muscle-fibers pass over the extremity of the latter directly into the elements of the tendon. Apart from this it may happen that the extremities of the muscle-fibers are attached by a special cement-substance to the plane surface or in shallow depressions at the extremity of the independently formed tendon (Fig. 184, 12 S). In arthropods there is doubtless also a direct transition of the sarcolemma into the substance of the tendon.

The tendons consist of parallel bundles of fibrillar connective tissue, containing connective-tissue corpuscles and elastic fibers. They are covered by a loose sheath of connective tissue, the *peritendineum,* which contains the blood-vessels, the lymphatics, and the nerves. The tendons pass through sheaths, whose slippery synovial fluid favors the gliding movement. In some situations the extremities of the muscle-fibers are attached directly to the fixed point; in other situations, such as the face, they terminate among the tissue-elements of the skin or the fasciæ.

Motor Nerves.—The trunk of the nerve, as a rule, enters the muscle at its geometrical center; hence, the point of entrance in long muscles with parallel fibers or in spindle-shaped muscles is situated near the middle. If the muscle with parallel fibers is more

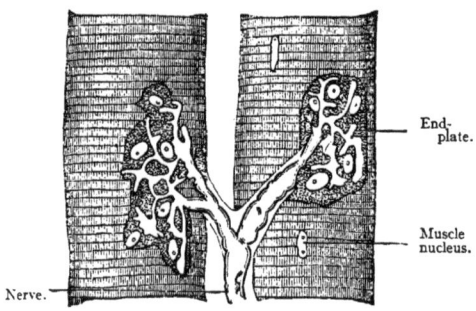

End-plate.

Muscle nucleus.

Nerve.

FIG. 185.—Muscle-fibers with Nerve-ending, from the lizard (after W. Kühne).

than 2 or 3 cm. wide, several branches enter side by side at its middle. In triangular muscles the point of entrance of the nerve is displaced toward the tendinous point of convergence of the muscle-fibers, the amount of displacement varying with the degree of convergence and the consequent thickness of the pointed extremity of the muscle. In general, the entrance of the nerve-trunk into a muscle may be suspected to take place at that point where the least displacement of muscular tissue occurs during the contraction of the muscle.

The motor nerve destined for a certain muscle does not originally contain as many fibers as there are muscle-fibers; in the eye-muscles of man about seven muscle-fibers correspond to three nerve-fibers in the trunk; in other muscles, in the dog, there is one nerve-fiber to from forty to eighty-three muscle-fibers. Hence, it is necessary, in the course of their ramification in the muscle, for the separate nerve-fibers to subdivide dichotomously.

In warm-blooded animals each *muscle-spindle* has only one point of innervation, while the longer spindles of cold-blooded animals have several. The medullated fiber enters the muscle-fiber and forms at its point of entrance a nodular prominence, the *nerve end-bulb* (Fig. 184, 1, e). In this transition the neurilemma fuses directly with the sarcolemma, the medullary substance disappears, and the axis-cylinder becomes transformed into a flattened ramification, the *nerve end-plate,* or *nerve-arborization,* which rests upon a finely granular accumulation of sarcoplasm (Fig. 185), in which nuclei are present. According to Kühne the connection of the nerve with the muscle-fiber is established only through the coalescence of the end-plate with the substratum of sarcoplasm, and through the connection of the latter with the interstitial rows of nuclei in the sarcoplasm, or with the protoplasm surrounding the muscle-nuclei, and further with the interstitial cement-substance of the fibrils.

In the arthropods (crawfish) each muscle-fiber possesses two nerve-terminations, arising from separate axis-cylinders. There are no end-plates at the point of entrance; but the nerve-fibrils are distributed for a great distance between the muscle-fibrils, and the nerve-endings appear to extend as far as the terminal or intermediate discs. Some investigators have assumed the existence of a similar arrangement in higher animals also.

The muscle is supplied also with **sensory fibers,** which subserve the muscular sense. Their existence is demonstrated physiologically by the fact that stimulation of a muscle will cause reflex variations in the blood-pressure and dilatation of the pupil, also an increase in the respiratory movements and muscular reflexes; and, further, by the fact that inflamed muscles are painful.

At first slender and medullated, these fibers finally become non-medullated. It appears that they are distributed on the outer surface of the sarcolemma, as they wind around the muscle-fibers after undergoing dendritic ramifications. According to others, the sensory nerves, after branching dichotomously, terminate only in the overlying connective tissue or in the aponeuroses, either abruptly or by means of a small swelling. Bremer designates their termination as unbelliferous; while according to Landauer they pass longitudinally along the muscle-fibers, in the frog, in the form of filaments provided with oval, nuclear formations. According to still other observers, they terminate as muscular buds or muscular spindles, or as free endings. In the horse the sterno-maxillary muscle receives sensory branches from a separate nerve.

Red and Pale Muscles.—In some fish, such as the sturgeon; birds, such as the turkey; and mammals, such as rabbits, two kinds of striated muscles can be distinguished, namely *red* or dark, for example the soleus and semitendinosus of the rabbit, and *pale* or light, for example the crural of the rabbit. The fibers of pale muscle are usually wider, and poorer in protoplasm than those of red muscle; their transverse striation is closer, and their longitudinal striation less prominent; their fibrils are placed at regular intervals; their muscle-nuclei, lying directly in contact with the sarcolemma, are less numerous than those of the fibers of red

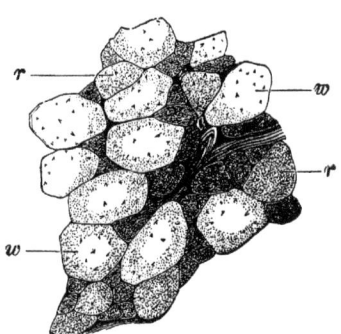

FIG. 186.—Cross-section through the Gastrocnemius Muscle of the Frog (after Grützner): *w*, the pale muscle-fibers; *r*, the red fibers.

muscle, within which they are situated between the fibrils; and they contain less glycogen, myosin, and water.

Julius Arnold found in man white fibers extensively distributed among the red. Even in the same muscle, in fact in almost every muscle, in the frog and in mammals, red and white fibers are intermingled (Fig. 186). Nevertheless it should be pointed out that their color is not always clearly differentiated. The physiological differences are considered on p. 563.

The heart of the frog, as well as that of invertebrates, contains transition-forms between striated and unstriated muscle-fibers (Fig. 184, 6). The spindle-shaped, mononuclear cells have the form of unstriated fibers, but the transverse striation of voluntary muscles.

Development.—Each striated muscle-fiber develops from a mononuclear, mesodermal cell without a wall, which becomes elongated in the form of a spindle. As it progressively increases in length, the nuclei multiply by mitosis. At a more advanced stage the peripheral or parietal substance of this structure is transformed into the fibrillar, striated mass of the fiber (Fig. 184, 7), while the nuclei with a scanty covering of protoplasm (muscle-corpuscles) form a continuous line in the axis of the fiber, where they remain permanently in some animals. In man the nuclei advance later toward the surface of the fiber, where a structureless cuticle, the sarcolemma, separates. The muscle-corpuscles serve in a certain sense as trophic centers for the striated parietal substance; they may bring about degenera-

35

tion, or restitution of the latter may arise from them. The muscles of the young have fewer fibers than those of the adult, although the former are on the whole smaller. In growing muscles, both in the new-born and in later life, the number of fibers is increased by the separation from a fiber of a band of sarco-plasm, together with a continuous row of muscle-corpuscles. This, as a "myo-blast," develops into a new fiber according to the embryonal type. The new fiber also receives its nerve-fiber, which develops from the nuclei of the sheath of Schwann. The individual fibers increase in thickness by an increase in the number of fibrils. In the growth of the muscle as a result of continuous in-creased exertion, the individual fibers become thicker, but not more numerous; the sarcoplasm is increased, while the fibrils and the nuclei are not changed.

Degeneration.—An active degeneration of fibers prob-ably takes place in the muscles, in accordance with their active metabolism. As an introduction to this process (rigid ?) muscular substance accumulates on the fibers in the form of nodules or rings; at such situations the fiber disintegrates into fragments, termed sarcolytes, which un-dergo absorption.

Comparative.—In addition to the parts of vertebrates analogous to human muscular tissues, striated muscle-fibers are found also in the iris and the choroid of birds.

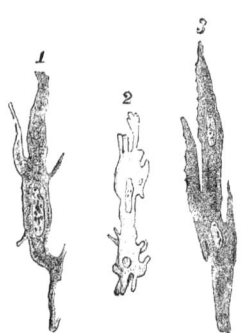

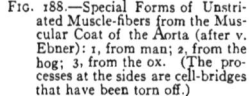

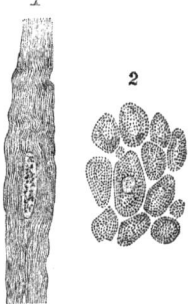

FIG. 187.—Unstriated Mus-cle-fibers, isolated by means of Diluted Alco-hol: 1, from the intes-tine; 2, from the radial artery of man.

FIG. 188.—Special Forms of Unstri-ated Muscle-fibers from the Mus-cular Coat of the Aorta (after v. Ebner): 1, from man; 2, from the hog; 3, from the ox. (The pro-cesses at the sides are cell-bridges that have been torn off.)

FIG. 189.—Muscle-cells from the Frog's Stomach with Distinct Fi-brils (after Engelmann): 1, por-tion of a fiber treated with am-monium bichromate; 2, cross-section of cells that have been treated with 8 per cent. sodium-chlorid solution.

Arthropods have only striated muscle-fibers, while molluscs, worms, and echino-derms have chiefly unstriated fibers. The latter possess also special, energetically contracting fibers with double oblique striation, formed of crossed, oblique lines. In cephalopods the muscle-fibers exhibit spiral lines at the periphery. Among vertebrates fish have the thickest muscle-fibers, then follow with decreasing width, toads, lizards, mammals, and birds.

The unstriated, involuntary muscle-fibers, or contractile fiber-cells, can be isolated by means of a 35 per cent. solution of potassium hydroxid. They are unencapsulated, unicellular, spindle-shaped, flattened fibers, in some places pre-senting an appearance of longitudinal fibrillar striation, from .45 to 230 μ long, and from 4 to 10 μ wide. They are occasionally forked at one or both extremities, and contain at the middle a solid, rod-shaped nucleus, which may be sharply brought out by the addition of dilute acetic acid. The nucleus is surrounded by a small amount of protoplasm, and encloses one or two bright nucleoli within a rich network (Fig. 184, 10 and 11). The fibers either lie singly, or are joined together in continuous layers or reticular columns, being arranged longitudinally with their tapering extremities in contact with one another. The fibers are not held together by a viscid interstitial cement-substance, as was formerly supposed, but they are universally connected by so-called cell-bridges. The conduction of stimuli through unstriated muscles is at the same time thus explained.

Where fibrils are visible in the fiber-cell (Fig. 189), they lie embedded in a rather homogeneous, granular substance, the sarcoplasm. According to Engelmann, the disintegration of the substance of unstriated muscle into the separate spindle-shaped elements is a postmortem change in the tissue. The transverse, thickened areas occasionally observed are not due to transverse striation, but to partial contraction or fold-formation (Fig. 184, 10). Unstriated muscle-fibers also have tendinous insertions at times. The blood-capillaries pass in longitudinal meshes between the fibers, as do also the numerous lymph-capillaries that surround the cells.

The **motor nerves,** according to J. Arnold, form a plexus of medullated and non-medullated fibers, partially supplied with ganglion-cells, and situated in the connective-tissue of the envelop surrounding the unstriated muscle-fibers—the *ground plexus.* From this arises a second non-medullated plexus, with nuclei at the nodal points—the *intermediate plexus.* This is situated either immediately upon the musculature or in the connective tissue between the individual bundles. The delicate fibrils (from 0.2 to 0.3 μ) given off by this plexus unite to form still an-

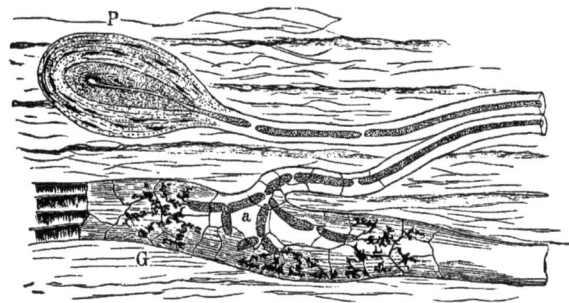

FIG. 190.—Sensory Nerve in a Tendon. One fiber terminates in a Pacinian corpuscle (P), the other in a tendon-spindle of Golgi (G).

other network, the *intermuscular plexus,* and pass to each fiber, running along its border and terminating in a pear-shaped thickening. According to Frankenhäuser the fibrils terminate in the nucleolus; according to Lustig, in the vicinity of the nucleus; according to J. Arnold they traverse both fiber and nucleus and re-enter the plexus. P. Schultz describes also sensory nerves, connected with ganglion cells and provided with terminal nodules.

In tendons the sensory nerves, after subdividing repeatedly, become non-medullated fibers (Fig. 190, a), which at the junction of muscle and tendon twine around or spread out over the bundles. This situation is covered with endothelium. The non-medullated fibers terminate finally in a tuft of delicate ramifications, designated Golgi's *tendon-spindle.* Terminations in the form of Pacinian corpuscles (P) or end-bulbs are also found in the tendons.

PHYSICAL AND CHEMICAL PROPERTIES OF MUSCULAR TISSUE.

The **consistency** of muscular tissue is similar to that of living protoplasm; it is semi-solid, that is, not fluid to such a degree as to be diffluent, nor is so solid that confluence of separated parts would not be possible. The consistency, therefore, may be compared to that of a jelly at the moment of liquefaction.

The view expressed is supported by the following facts: (1) The analogy between the function of the muscle-substance and that of the contractile protoplasm of cells, the latter surely possessing this semi-solid property, as must be inferred from the movement of the protoplasm. (2) The observation of the course of the contractile wave-movement through the length of the muscle-fiber. In the same category belongs the wave-like movement first observed by W. Kühne, when a strong, constant current is passed through the muscle. The phenomenon depends upon the occurrence of slow contraction-waves within the fibers in the direction of the galvanic current, which are increased by heat, and disappear when the muscle is tightly stretched, or when its extremities are forcibly pushed together. (3) Under the microscope, the progression of a parasitic round-worm (Myoryctes Weismanni) has been observed to take place by means of the serpentine movements through the contractile substance, the separated, semi-solid masses becoming again confluent behind it.

Refraction of Light.—The contractile substance refracts the light doubly (anisotropic), while the ground-substance is singly refractive (isotropic). The contractile substance behaves like a doubly refractive, positively uniaxial body, whose optical axis corresponds with the longitudinal axis of the fiber. Under the polarization-microscope, with the Nicol's prisms crossed and the fiber so placed that the longitudinal axis intersects the vibration-planes of the Nicol's prisms at an angle of 45°, the doubly refractive substance can be recognized by its appearing bright in a dark field of vision, while in a colored field (purple-red from the interposition of a mica plate) it appears of another color (blue, yellowish-red, to yellow). Although the doubly refractive contractile substance undergoes change in form during contraction, its double refraction nevertheless persists unaltered. Catherine Schipiloff, A. Danilewsky, and O. Nasse believe that the contractile, anisotropic mass consists of myosin. According to the observations of Engelmann all contractile elements possess the property of double refraction, and the direction of shortening always corresponds with that of the optical axis. With respect to the actual cause of the anisotropy, the comprehensive investigations of v. Ebner have demonstrated that as a result of the processes of growth in the tissue, tensions are produced (for example, the tension-phenomena of bodies subject to imbibition) that give rise to double refraction.

During sustained contraction in degenerating muscle-fibers the refractive index of the muscle-substance is increased as a result of loss of water from the tissue and the consequent increased concentration of the dissolved parts of the muscle.

The **chemical composition** of muscle undergoes rapid and profound changes after death. As, however, the muscles of the frog, when thawed after freezing, again become capable of contracting, they are, therefore, not altered chemically by the freezing. W. Kühne cooled to 10° C. frogs' muscles rendered bloodless by means of a 1 per cent. sodium-chlorid solution, triturated them in an ice-cold mortar, and expressed the juice (which thaws at —3°) through linen. The fluid thus expressed is filtered in the cold and appears as a slightly opalescent juice of a neutral or generally alkaline reaction and light yellowish tint, and designated *muscle-plasma*. In common with blood-plasma it coagulates spontaneously. The muscle-plasma becomes at first uniformly gelatinous. Later, turbid, opaque, doubly refractive flakes and threads undergo contraction in the jelly, and like the fibrin of the contracting blood-clot express a juice, *muscle-serum*, which has an acid reaction. Cold prevents the coagulation of muscle-plasma; above 0° it takes place but slowly, then more rapidly with increasing temperature, finally with great rapidity at 40° C. for the muscles of cold-blooded animals, or at 55° C. for those of warm-blooded animals. The addition of water or of a little acid to the muscle-plasma causes immediate coagulation. This coagulated proteid, the most abundant in the muscles, is derived from the doubly refractive substance, and is designated *myosin*. Its chemical formula is $C_{108}H_{172}N_{30}SO_{33}$.

Myosin forms from 3 to 11 per cent. of moist muscular tissue. It can be extracted from muscle-juice by means of a 5 to 10 per cent. solution of ammonium chlorid. Myosin belongs to the globulins; Halliburton has prepared it also from the muscles of warm-blooded animals. It is precipitated from its solutions by saturation with sodium chlorid or magnesium sulphate. When dissolved in a 10 per cent. solution of sodium chlorid, it is coagulated by heat. It is dissolved by 2 per cent. hydrochloric acid, with the formation of acid-albumin (syntonin), and by alkalies or alkaline carbonates, with the formation of alkalialbuminate. Like fibrin, myosin actively decomposes hydrogen dioxid. A. Danilewsky has succeeded in reconverting syntonin in part into myosin. Myosin is not present in unstriated muscles.

Muscle-serum contains further small amounts of *myoalbumin* (C_{114}-$H_{174}N_{30}SO_{30}$), which is coagulable at 73° C., but is not precipitated by saturation of the serum with magnesium sulphate; also *myoglobulin*, which is precipitated by this last procedure, and is coagulable at 63° C.; and a little *nucleoalbumin*.

Halliburton distinguishes the following proteids in muscle: (1) *Paramyosinogen*, or musculin, a globulin-like body, forming 20 per cent. of the total proteids, and coagulating at 47° C. (2) *Myosinogen*, forming 77 per cent. of the total proteids, coagulating at 55°. Both of these bodies are coagulable spontaneously, forming myosin. (3) According to v. Fürth myosinogen gives rise to myogenfibrin, which is soluble, is coagulable at 35°, and, like paramyosinogen, is readily transformed into a fibrin-like modification that is dissolved with difficulty. Certain salts or organic substances (caffein, veratrin) accelerate this process, while it is inhibited by blood-serum, and also by egg-albumin. (4) *Myoalbumin*, which is similar to serum-albumin. The coloring-matter of muscle (*myohematin*) appears to be different from hemoglobin. The absorption-bands are situated somewhat nearer to the red end of the spectrum. According to Levy, myohematin is identical with hemochromogen. There is an oxidized and a reduced myohematin (by ammonium sulphid). The muscle-nuclei yield some nuclein. The sarcolemma contains a substance resembling keratin. Several ferments are present in traces: pepsin, diastatic, lactic-acid (?), glycolytic, and coagulating (fibrin-)ferments. Proteic acid is a proteid substance in the flesh of fish.

The other chemical constituents of muscle have already been mentioned in the consideration of meat (p. 423). It will suffice to add a little more here. (1) In addition to volatile fatty acids (formic, acetic, and butyric acids), two isomeric lactic acids are found in muscle having an acid reaction: (*a*) Ethylidene-lactic acid in the modification of dextrorotatory paralactic or sarcolactic acid. (*b*) Ethylenelactic acid in small amount, which Maly also observed develop as an occasional fermentation-product of carbohydrates (glycogen, etc.). The formation of lactic acid during the rigidity of death is discussed on p. 552. Acid potassium phosphate also contributes to the acid reaction. (2) Glycogen is found to the amount of 1 per cent. after an abundant meat-diet, and of 0.5 per cent. during fasting. During digestion it is stored up in the muscles, as well as in the liver, but it disappears in the state of hunger. It is formed in the muscles themselves, probably from albuminates. (3) Dextrose, 0.02 per cent. (4) Of gases, there are present carbon dioxid (from 15 to 18 vol. per cent., partly absorbed, partly in chemical combination, the latter probably being formed as a result of decomposition), some absorbed nitrogen; but no oxygen, although muscle continually absorbs oxygen from the blood. The muscles contain a substance that yields carbon dioxid on decomposition; exercise consumes this substance, so that muscles that are greatly fatigued are capable of generating less carbon dioxid.

METABOLISM IN MUSCLE. THE SOURCE OF MUSCULAR ENERGY.

The **resting muscle** continuously abstracts oxygen from, and returns carbon dioxid to, the capillary blood passing through it. Nevertheless, the muscle excretes less carbon dioxid than corresponds to the amount of oxygen it absorbs. Excised muscles deprived of blood exhibit an analogous but diminished interchange of gases. Further, as such muscles retain their irritability longer in oxygen or in air than in indifferent gases free from oxygen, it is to be assumed that this gaseous interchange is a vital phenomenon connected with normal metabolism, and to which the functional activity of the muscle is due.

The excised, resting, surviving muscle gives off carbon dioxid, which in part has been present in the muscle preformed, and in part is subsequently generated by processes of decomposition that accompany the development of rigidity. A small part of this carbon dioxid arises only when oxygen is supplied. Bacterial putrefaction of the muscles causes marked excretion of carbon dioxid.

In **active muscle** the blood-vessels are always dilated, and the amount of blood passing through them is increased three or four times, a circumstance that obviously indicates increased metabolic activity. Accordingly, active is distinguished from passive muscle by a series of chemical changes:

1. The contents of living passive muscle have an alkaline, or, more correctly, a neutral reaction, changing red litmus to blue, but acid to turmeric paper. The reaction becomes acid in active muscle (not of the unstriated variety), and, indeed, the degree of acidity increases, to a certain limit, in proportion to the amount of work performed. The acidity is due to phosphoric acid resulting from the decomposition of lecithin and nuclein.

The earlier view, that the acidity is due to the development of lactic acid produced from glycogen, has not been substantiated. Pflüger and Warren, and also Astaschewsky and Heffter, even found the quantity of lactic acid in active muscles diminished, as compared with passive muscles. Other investigators, however, still adhere to the theory of lactic-acid formation, especially if there is a deficiency of oxygen during the work.

2. The active muscle excretes considerably more carbon dioxid than the resting muscle: (a) Active muscular exertion in man or animals increases considerably the excretion of carbon dioxid from the body. (b) Venous blood flowing from the tetanized muscles of an extremity contains an increased quantity of carbon dioxid; and, indeed, under these conditions more carbon dioxid is excreted than corresponds to the amount of oxygen simultaneously absorbed. (c) Also, excised, contracted muscles excrete an increased amount of carbon dioxid.

3. Active muscle consumes a greater amount of oxygen: (a) During work the entire body takes up much more oxygen, even four or five times as much. (b) Venous blood flowing from the active muscles of an extremity contains a diminished amount of oxygen. Nevertheless, the increase in the consumption of oxygen by an active muscle is not so great as the increase in the excretion of carbon dioxid. The increase in the interchange of gases continues in the period of rest immediately following the activity.

The consumption of oxygen can also be demonstrated volumetrically in excised muscles deprived of blood. It is true, oxygen is not absolutely necessary for muscular activity of short duration, as the excised muscle is capable of contracting for some time in a vacuum or in a gaseous mixture free from oxygen, and no free oxygen can be obtained from its tissue. The muscle must, therefore, contain a supply of oxygen in chemical combination, which is consumed during activity. Frogs' muscles abstract the oxygen from easily reducible substances; thus, they may decolorize a solution of indigo. Muscles that have rested act less energetically than those that have been active.

4. An active muscle contains less extractives soluble in water, but, on the other hand, more of those soluble in alcohol. It also contains less of the substances that form carbon dioxid, less fatty acids, kreatin, kreatinin, and sarcophosphoric acid.

5. During contraction the amount of water in muscular tissue is increased, while that in the blood is correspondingly diminished. The solid matters of the blood are increased, while those of the lymph (albumin) are diminished.

6. The question as to the extent to which the proteids of muscular substance generate the kinetic energy of muscular activity, by the transformation of their chemical potential energy, has been answered by Pflüger with the statement that albumin, if given in sufficiently large amounts, may be the exclusive source of muscular force.

This albumin represents a special variety, and is thought to be formed synthetically from ordinary living albumin by the absorption of alcohol-radicals, which may be withdrawn either from another proteid, or from fat and sugar if there is a deficiency of proteids. The living albumin is transformed into a readily decomposable, living proteid, which contains a greater amount of carbon, and represents the immediate source of muscular energy.

If a lean dog, fed only with lean meat and in a state of metabolic equilibrium during muscular rest, is subjected to a period of several days' work, it must receive a definite excess of lean meat in order to maintain its bodily weight. During the period of activity, the animal, therefore, decomposes more proteid, in accordance with the extent of the activity, and the metabolic equilibrium is thus maintained. Undoubtedly the work performed is accomplished at the expense of an increased consumption of proteids. If the dog does not receive an increased quantity of proteids on beginning to work, it loses in bodily weight.

Even though sufficient quantities of fat and carbohydrates, in addition to the proteid, be administered to the active dog, there will still be an increased consumption of proteids during work.

As on administration of a sufficient amount of proteid the muscular work is performed with the aid of this alone, and as in the decomposition of this proteid neither fat nor carbohydrate results, the fat and carbohydrate cannot be the true source of muscular force (Pflüger).

The carbon dioxid resulting from the decomposition of proteid leaves the body quickly through the pulmonary respiration; while the nitrogenous products of decomposition are excreted slowly, even for as long as two days after the completion of the work.

One and the same readily decomposable proteid is thus oxidized slowly and continuously in the muscular tissue, with the generation of heat, while under the influence of innervation it is consumed rapidly and in larger amount, and is then the source not only of heat, but also of kinetic energy.

Pflüger estimated that in his experimental dog one gram of nitrogen in the proteid, decomposed within the body, produced 7456 kilogrammeters of work. Of the total supply of energy contained in the proteid (measured by means of the calorimeter in calories), the dog converted 48.7 per cent. into kinetic energy, the remainder being transformed into heat. This 48.7 per cent. represents the *mechanical equivalent* of the proteid.

At an earlier period Fick and Wislicenus, as well as v. Voit and v. Pettenkofer, had reached the conclusion as a result of their experiments that the daily excretion of nitrogen is not increased to any considerable extent by forced work, whereas the consumption of oxygen and the excretion of carbon dioxid are increased, provided that the body has at its disposal sufficient material containing carbon, such as glycogen and fat, in its tissues or in the food. Hence, the proteid cannot be the source of muscular energy. Increased elimination of nitrogen takes place only when the activity gives rise to dyspnea, for deficiency in oxygen causes decomposition of albuminates.

Also the increased excretion of sulphuric acid resulting from work is indicative of a more active decomposition of albuminates.   The excretion of sulphur is increased by muscular exertion, and indeed the non-oxidized sulphur is at first excreted more rapidly than the oxidized.   The excretion of phosphoric acid also is increased.

7. In the muscles of animals the amount of glycogen (0.43 per cent.) has been observed to diminish as a result of activity, and even to disappear completely in consequence of strychnin-convulsions.   The same observation has been made with respect to the glycogen of the liver. Luchsinger maintains that muscles can still contract when completely free from glycogen; so that the latter cannot be the source of muscular energy.   Also, the sugar of the blood undergoes a decrease in the muscles as a result of activity.

There is a difference of opinion as to whether the muscle-glycogen is carried by the circulation from the liver into the muscles, or whether it is produced in the muscular tissue itself as the result of an as yet unknown decomposition of the albuminates.   Külz observed an increase in the amount of glycogen in the muscles of frogs that had been deprived of their livers after subcutaneous injections of sugar.   Likewise, the muscles retained their glycogen for a much longer time than the liver during the state of hunger.   These facts indicate the formation of glycogen in the muscular substance itself.   In any event, the normal circulation is a requisite for the production of glycogen in muscle, for this diminishes after ligature of all of the vessels.   Surviving muscle converts glycogen into sugar.

Some investigators, however, assume also that not only proteid but, in part, also fat and carbohydrate may be the source of muscular energy in the body.

MUSCULAR RIGIDITY (CADAVERIC RIGIDITY, RIGOR MORTIS).

Excised muscles, striated as well as unstriated, and also the muscles of the intact body some time after death, pass into a state of rigidity, described more fully later on, that is designated *muscular rigor*.   If the muscles of the dead body become involved, the entire cadaver becomes completely stiff (cadaveric rigidity).   The cause of this phenomenon resides in a spontaneous coagulation of the myosin within the muscle-fibers, with the development of a small amount of acid.   During this process of coagulation, heat is liberated owing to the transition of the fluid myosin into the solid condition, and, also, owing to the thickening of the tissue that takes place at the same time.

Myosin, dissolved in a 5 per cent. solution of magnesium sulphate diluted with water, separates after a time in the form of solid flakes, with the development of an acid reaction.   Warming hastens this process.

The rigid muscle exhibits the following properties: It is shortened, thickened, and somewhat denser; stiff, firm and solid; turbid and opaque, in consequence of the coagulation of the myosin; incompletely elastic, less extensible, and less readily torn.   It is completely unresponsive to stimuli, and its electrical potential has disappeared.   The amount of glycogen present is diminished.   Striated muscle has an acid reaction, on account of increased formation of the two varieties of lactic acid (unstriated has not), and it develops free carbon dioxid.   If incisions be made into rigid muscles, a fluid exudes spontaneously, the muscle-serum.

The view was formerly held that during rigidity, partial or complete transformation of the glycogen occurred, first into sugar and then into lactic acid. This view, however, has been contested by Böhm, who asserted that during digestion a transitory accumulation of large amounts of glycogen takes place in the muscles, as in the liver; so that approximately as much can be found in the former as in the latter. Rigidity causes no diminution of glycogen, provided putrefaction is prevented; hence, the lactic acid of rigid muscles cannot arise from glycogen, but probably from decomposition of albuminates.' Heffter maintains that lactic acid is not formed at all during postmortem rigidity.

The amount of acid does not vary, whether the rigidity develops slowly or rapidly. With the onset of acidification, the rigidity becomes more marked, on account of the coagulation of the alkali-albumin in the muscle. The less carbon dioxid there is generated by the rigid muscle the more it had already given off previously during activity.

Fibrin-ferment is present in muscle in a state of cadaveric rigidity. It is in general a product of protoplasm, and is never wanting where the latter is present. There is thus an analogy between coagulation of blood and muscular rigidity.

Two stages of rigidity are to be distinguished: In the first stage the muscle is already somewhat stiff, but still excitable; the myosin in this stage acquires a gelatinous consistency. Restitution is still possible from this stage. In the second stage the rigidity is fully developed in all of the characteristics mentioned.

Rigidity appears in man in from ten minutes to seven hours; the duration is likewise variable, from one to six days. After its disappearance, the muscles again become soft, owing to the onset of further decomposition and an alkaline reaction; the rigidity yields. The onset of rigidity is always preceded by a disappearance of nervous activity. Therefore, the muscles of the head and the neck are first affected, and then the others in a descending order. Likewise those muscles that usually degenerate earliest are the first to become rigid; for example, in the frog the flexors before the extensors. Rigidity disappears earliest also in those muscles that first became rigid. Great muscular activity before death, for example during the convulsions of tetanus, cholera, strychnin-poisoning, or opium-poisoning, causes rapid and intense rigidity. Therefore, the heart becomes strongly rigid and with relative rapidity. White muscles become rigid later than red muscles. Wild animals, hunted to death, may become rigid in a few minutes. Usually the rigidity lasts the longer the later it sets in. Rigidity never occurs in the fetus before the seventh month. Frogs' muscles cooled to $0°$ C. become rigid only after from four to seven days.

Stenson's Experiment.—The influence of the amount of blood in the muscles upon the onset of rigidity is especially worthy of notice. Ligation of the muscular arteries in warm-blooded animals causes first increased irritability of the muscular tissue, lasting a few minutes, then rapid diminution in the irritability, followed by the onset of both stages of rigor in succession. If the arteries of the muscles were ligated, Stannius observed that the irritability of the motor nerves disappeared in the course of an hour, that of the muscular tissue itself in from four to five hours; then rigidity sets in.

Pathological.—Thrombotic occlusion of the muscular vessels will also cause rigidity. Excessively tight bandaging may give rise to true rigidity in man by cutting off the circulation. The muscles become paralyzed and stiff, and later break up into flakes, and the contents of the fibers are subsequently absorbed. The circulatory disturbances, arising in muscles under the influence of cold, also cause paralyses that are often designated rheumatic. Also in cases of trichinosis the affected muscle-fibers are the seat of rigidity, and the stiffness in the muscles is thus explained. The contractures occurring in cases of cholera should probably be included in the class of muscular contractions resulting from circulatory disturbances, the inspissated blood giving rise to stagnation; as should also certain contractions occurring in the presence of atheroma and in the agonal period. The sensory nerves in completely anemic extremities retain their irritability for from five to ten hours.

If the circulation be restored in the first stage of rigidity, the muscle soon recovers. If, however, the second stage has set in, restitution is impossible. In cold-blooded animals rigidity does not set in for several days after ligature of the vessels. Brown-Séquard, by the injection of fresh blood containing oxygen, succeeded in restoring softness and irritability to a human cadaver in the first stage of rigidity even four hours after death. Heubel obtained the same result with the frog's heart as long as fourteen and one-half hours after death. On passing blood containing oxygen through excised muscles, C. Ludwig and Al. Schmidt found that the onset of rigidity was retarded for a long time; this did not occur, however, with blood deprived of oxygen. After considerable loss of blood, rigidity sets in relatively early. If an artificial circulation be kept up in the dead muscles of a frog by means of feebly alkaline fluids, rigidity does not occur.

Previous section or paralysis of the motor nerves results in delayed onset of rigidity in the relaxed muscles. The reason is found in the greater abundance of blood in these muscles, in consequence of associated paralysis of the vasomotors, the alkaline blood remaining in the muscles even after death, while the arteries in other parts of the body become empty. This view is supported by the fact that rigidity appears much later in fish whose medulla oblongata is suddenly destroyed than in those that die slowly. According to Ewald and Willgerodt the labyrinths of the ear, as organs controlling tone, likewise have an influence on the course of rigidity.

Freezing and thawing cause rigidity to set in more rapidly, and it is favored likewise by mechanical injury.

Continuous passive movements may retard the onset of rigidity, but on their cessation their rigidity sets in all the more rapidly. Rigidity that has already developed may be overcome by forced movements, but it may set in again.

Rigidity may be induced artificially:

1. By heat (heat-rigor), which causes coagulation of the myosin in cold-blooded animals at 40°, in mammals at from 45° to 47° C., and in birds at about 53° C. Under such circumstances there is marked excretion of carbon dioxid, but less after previous tetanization. Protoplasm, for example of the amœba, is similarly subject to heat-rigor.

The degree of heat required to bring about rigidity is the higher the longer the muscles have been excised. If the muscles of a frog in a state of cadaveric rigidity be heated, the remaining proteids undergo coagulation successively, and the muscle becomes still more rigid as a result of these coagulative processes.

2. Saturation with water induces water-rigor, with the development of an acid reaction, in consequence of the coagulation of the globulin-substances, the excretion of carbon dioxid not being increased.

If the thigh of a frog be ligated, and the muscles, deprived of their skin, be immersed in warm water, they will become rigid. On loosening the ligature a slight degree of rigidity may disappear through restoration of the circulation. On the other hand, a more marked degree of rigidity can be removed only by placing the leg in a 10 per cent. solution of sodium chlorid, which will dissolve the myosin-coagulum.

3. Acids, even weak acids such as carbon dioxid, induce rapid acid-rigor. This is probably different from normal rigidity, as the muscle does not develop free carbon dioxid. Injection of from 0.1 to 0.2 per cent. solutions of lactic or hydrochloric acid into the vessels of frogs' muscles causes immediate rigidity, which can be overcome by 0.5 per cent. acid, and also by a neutralizing solution of sodium bicarbonate, or 13 per cent. solution of ammonium chlorid. The acids enter into combination with the myosin.

4. Among poisons and other substances, the following promote rigidity: Caffein, quinin, digitalin, veratrin, hydrocyanic acid, also oils of mustard, fennel, and anise, and, when placed in direct contact with the muscles, potassium sulphocyanid, ammonia, metallic salts, alcohol, ether, chloroform. Chloroform, acetic acid, and heat induce rigidity with shortening; ammonia, on the other hand, rigidity without shortening.

The position of the entire body during rigidity is usually that which it occupied at death. The position of the limbs corresponds to the resultant of the various degrees of muscle-tension. If the limbs occupied another position before death, they are frequently seen to move during the onset of rigidity. The arms and fingers especially are readily flexed. If the rigidity develops with especial firmness and rapidity in certain groups of muscles, an unusual position may be assumed, for example the fencing attitude of cholera-cadavers. If the rigidity occurs rapidly, the body at times remains in the same position that it occupied at the moment of death, for example on the battle-field. Under such circumstances, however, the contracted muscle never passes immediately into a condition of rigidity, a period of relaxation intervening, even though short.

Muscles scalded by immersion in boiling water do not become rigid; neither do they become acid, nor evolve free carbon dioxid. Muscles coagulated by concentrated alcohol or by immersion in concentrated solutions of sodium chlorid, potassium nitrate, sodium and magnesium sulphate, do not yield an acid reaction.

Attention has repeatedly been directed to the analogies between muscle in active contraction and in the state of rigidity. The form of the contracted and of the rigid muscle is shortened and thickened; both are denser, of changed elasticity, and evolve heat; the contents of the contracted as of the rigid muscle are negative electrically as compared with resting or non-rigid contents; both evolve free carbon dioxid and the remaining acid from the same source. A contraction may, therefore, be regarded as a temporary rigidity, disappearing physiologically, just as earlier investigators, and recently Bernstein, designated rigidity as being, to a certain extent, the final vital act of the muscles.

A muscle in process of becoming rigid will lift a weight, like a living, contracting muscle. The height to which the weight is lifted by a rigid muscle is greater in the case of small weights and less for heavy ones than if the living muscle be stimulated to a maximum degree. If a muscle, in which heat-rigor has been induced, be at first prevented from contracting, and if later (for example after ten minutes) it be set free, its elastic energy will cause it to contract, and it must lose heat at the same time.

The disappearance of cadaveric rigidity takes place at first as a result of increased formation of acid in the muscle, by which the myosin is redissolved. Subsequently, with the development of microörganisms putrefaction sets in, with the associated evolution of ammonia, hydrogen sulphid, nitrogen, and carbon dioxid.

The loss of irritability in the muscles that precedes the onset of rigidity occurs in the following order in man (beheaded criminal): Left ventricle, stomach, intestine (fifty-five minutes), urinary bladder; right ventricle (sixty minutes); iris (one hundred and five minutes); muscles of the face and the tongue (one hundred and eighty minutes); the extensors of the extremities about one hour before the flexors; the muscles of the trunk (from five to six hours). The esophagus remains irritable for a long time.

IRRITABILITY, STIMULATION, AND DEATH OF THE MUSCLE.

By the *irritability* of a muscle is understood its ability to contract in response to stimuli applied directly to it (not to its nerves). *Stimulation* is the state of functional activity in which a muscle is placed by stimuli. At the moment of activity the stimulation causes the chemical potential energy of the muscle to be converted into work and heat; stimuli thus act as liberating forces. The mean temperature of the body is most favorable for the manifestation of irritability. Each muscle appears to possess a special degree of irritability peculiar to itself, as do likewise the nerves.

So long as the current of blood in the muscle is uninterrupted, stimulation first causes an increase in its functional activity, partly because the circulation becomes more active in association with dilatation of the vessels; later, however, the functional activity diminishes.

This diminution in functional activity is a sign of fatigue. If the same stimulation be continued, the muscular activity will exhibit a periodic variation, in such manner that after a series of weaker contractions stronger ones will again set in, followed in turn by weaker, and so on. This phenomenon depends upon periodically recurring improvement in the nutrition of the muscle, as a result of analogous variations in its circulation.

In excised muscles also, especially if the large nerve-trunks have already undergone degeneration, the irritability is at first somewhat increased after each stimulation, so that with a uniform series of stimuli the contractions at first exhibit an increase in extent. Thus, it may happen that, while the first weak stimulus is still ineffectual, the second will give rise to a contraction. The unstriated muscles exhibit, under certain conditions, automatic and rhythmic movements without the intervention of nerves.

Frogs' muscles that have been cooled, or those in which desiccation has begun, exhibit an excessively increased irritability, especially to mechanical stimuli. This fact may explain the remarkable muscular movements that often take place in cholera-cadavers. Cooled muscles from the frog or the tortoise may preserve their irritability for as long as ten days, but the muscles of warm-blooded animals often degenerate in from one and one-half to two and one-half hours. The irritability of the heart-muscle is considered on p. 118. Curarized, isolated frogs' muscles exhibit the least amplitude of contraction at 0°, the greatest at 30°; if heated beyond the latter temperature, the contraction gradually diminishes, until the point is reached where rigor sets in. The duration of contraction and the latent period are also shortest at 30°.

Since the time of Alb. v. Haller (1743) it has been thought necessary to attribute to muscle a peculiar irritability (even without the intermediation of the motor nerve). In more recent times attempts have been made to adduce further support in favor of this specific *muscular irritability:* (1) There are chemical irritants that induce no movement when applied to the motor nerves, but cause contraction when applied directly to the muscle; for example ammonia, lime-water, carbolic acid. (2) The extremities of the sartorius muscle of the frog, in which no nerve-endings can be demonstrated by means of the microscope, nevertheless react to direct stimulation by contractions. (3) Curare paralyzes the motor nerves, while the muscle itself remains irritable. The action of cold, or the arrest of the circulation in the muscle of an animal, will likewise abolish the irritability of the nerve, but not of the muscle at the same time. In general, the directly stimulated muscle will still contract for some time after its motor nerve has degenerated. (4) After section of the nerves, the muscles still remain irritable, even though the nerves have undergone total fatty degeneration. (5) At times electrical stimuli act only upon the nerves, and not upon the muscles themselves.

In lower animals (hydra, medusa) unicellular structures, neuro-muscular cells, have been found in which nervous and muscular tissue are represented in one and the same cellular structure.

With regard to the stimuli that act upon the muscles, the following are to be noted:

1. The **normal** stimulus under ordinary circumstances acts upon the muscle by way of its nerve, as in voluntary movement, the automatic motor impulse, reflex excitation. Its nature is unknown. The irritation of a muscle through the intermediation of its nerve is designated indirect stimulation. Pseudomotor effects are considered on p. 559.

2. **Chemical Stimuli.**—All chemical agents that alter the chemical constitution of muscular tissue with sufficient rapidity act as muscle-stimuli. According to Kühne, the mineral acids (0.1 per cent hydrochloric acid), acetic and oxalic acids, the salts of iron, zinc, copper, silver, and lead, bile, all act as stimuli to the muscle in dilute solution, and only on the nerves in much stronger solutions. Lactic acid and glycerin, when concentrated, excite only the nerve (?); when dilute, only the muscle. The neutral alkaline salts act equally on muscle and nerve. Alcohol and ether both act feebly. Water, especially if injected into the muscular vessels, causes fibrillary contractions. Solutions of sodium chlorid, from 0.6 to 0.9 per cent., or normal solutions of other sodium-salts, act indifferently toward the muscular substance, even after the latter is exposed to their influence for days; this is especially true after the addition of a trace of calcium chlorid or calcium phosphate. A 6 per cent. solution of sodium chlorid causes the sartorius, when deprived of its nerve, to contract much more strongly than when its nerve is preserved, and especially in its active, thick fibers. Acids, potassium-salts, and meat-extract diminish the irritability of the muscle, while other muscle-stimuli, such as alcohol, sodium-salts, some metallic salts, in small doses increase the irritability. Gases and vapors also have a stimulating influence on the muscles, either exciting simple contractions or immediately causing contracture. Protracted exposure to the gases causes rigidity. Only the vapor of carbon disulphid has an irritating effect on the nerves, while most vapors (for example, of hydrochloric acid) destroy without causing excitation.

In comparative observations on the influence of chemically related substances, only chemically equal quantities, for example normal solutions, should be employed. Thus, among the halogens, sodium iodid, with its high molecular weight, has the strongest effect; while sodium chlorid, with its low molecular weight, has the feeblest effect. The combinations of the metals act in like manner; also the salts of the alkaline earths. Those with the highest molecular weight cause the greatest excitation and the least injury. The following substances cause injury in the order of their sequence, arranged from those with stronger to those with weaker effects: ammonia, potassium, sodium, hydrochloric acid, nitric acid, sulphuric acid, phosphoric acid (in accordance with their avidity); the fatty acids with larger molecules as compared with those with smaller; the higher alcohols as compared with the lower.

In making experiments upon the chemical irritation of muscles, it is inadvisable to immerse the transverse section of the muscle in the solution. The substance in solution should rather be applied to a limited area on the uninjured surface of the muscle. The stimulation will then be manifested in a few seconds by contraction or fibrillary motion of the superficial muscular layers.

If the sartorius of a curarized frog be immersed in a solution of 5 grams of sodium chlorid, 2 grams of alkaline sodium phosphate, and 0.5 gram of sodium carbonate in 1 liter of water at 10° C., the muscle will be thrown into rhythmic contractions, which may persist even for days. These contractions suggest, to a certain degree, the rhythmic action of the heart (Biedermann).

The following act as chemical irritants upon unstriated muscles: ergot, aloes, colocynth, the alkalies; atropin and nicotin paralyze the nervous elements in such muscles, as does also ether; chloroform also destroys the muscle-fibers themselves. Carbon dioxid in small amounts acts as an irritant to the nerves, in larger amounts as a paralyzant, and in still larger amounts it irritates and finally paralyzes the muscle-fibers themselves.

3. Thermal Stimuli.—If a frog's muscle be rapidly heated, a gradually increasing contraction begins at about 28° C., becomes more pronounced at 30° C., and attains its maximum at 45° C.; following this, further heating rapidly leads to heat-rigor. Local cooling of the muscle increases its irritability for all kinds of stimuli. Frog's muscle cooled to 0° is exceedingly responsive to mechanical irritation, and it may be stimulated by degrees of cold below 0°, until freezing takes place. Heat has a relaxing effect on unstriated muscle (frog), while cold has a moderately stimulating effect. Variations in temperature, however, also affect the nerves of these muscles, each fluctuation causing reflex contraction, which does not occur if the nerves are paralyzed.

Cl. Bernard made the remarkable observation that the muscles of artificially cooled animals retain their irritability for many hours after death. Heat causes rapid disappearance, with temporary increase of the irritability.

4. Mechanical stimuli of all kinds cause a contraction at each separate, sudden blow; and tetanus if repeated. Strong, local stimuli induce an elevated contraction of considerable duration at the point of application. Moderate stretching of a muscle increases its irritability. Mechanical stimulation of a muscle poisoned with veratrin causes a heaving movement of its fibers, which may persist for as long as one minute.

5. Electrical stimuli are discussed in conjunction with nerve-stimuli (p. 631).

Curare, the arrow-poison of the South American Indians, is the dried juice of the root of Strychnos Crevauxi. When introduced into the blood or injected subcutaneously, it first causes paralysis of the intramuscular termination of the motor nerves, the muscles themselves retaining their irritability, while the sensory nerves and those of the central organs and the viscera (heart, intestine, and vessels) remain for a time unaffected. In warm-blooded animals the paralysis of the respiratory muscles naturally causes early asphyxia, which is unattended with convulsions. Frogs, whose skin is their most important respiratory organ, on receiving a suitable dose, may recover completely, after remaining motionless for days, during which the poison is eliminated through the urine. Larger doses paralyze also the cardiac inhibitory and vasomotor nerves. In electrical fish paralysis of the nerve transmitting the electrical shock occurs. In frogs the lymph-hearts also are paralyzed. If the doses that are fatal when administered subcutaneously be given by mouth, poisoning does not result, because the poison is eliminated by the kidneys at the same rate that it is absorbed by the gastric mucous membrane. For the same reason the flesh of an animal killed by a poisoned arrow is harmless. If, however, the ureters be ligated, the poison accumulates in the blood, and intoxication results. Large doses, however, will kill uninjured animals also by way of the intestinal tract.

Atropin appears to be a specific poison for unstriated muscle-fibers, although different muscles are variously affected by it.

The irritability of the muscles after lesions of the nerves deserves especial attention. After three or four days the irritability of the paralyzed muscle is diminished for direct or indirect (nerve) stimuli. There then follows a stage in which a constant current has an abnormally excessive effect, while induced currents are almost or completely without effect; irritability to direct, mechanical stimuli is also increased. This increased irritability is observed at about the seventh week. It then diminishes gradually, until it completely disappears at about the sixth or seventh month. Beginning with the second week, the muscle begins to undergo progressive fatty degeneration, to the point of complete atrophy. In experiments on animals Schmulewitsch found, immediately after section of the sciatic nerve, that irritability was increased in the muscles innervated by it.

After death the muscles degenerate (excised muscles more rapidly), and the earlier if they have been exhausted and exposed to stimuli of considerable intensity. Thick muscles survive longer (in their interior) than thin muscles. It would appear that there is a definite stage of early or late death for each individual muscle; for example, the extensors in man degenerate earlier than the flexors.

The muscles of the frog degenerate in twenty-four hours at summer temperature, in the course of two or three days at moderate temperature, and only after about twelve days at 0°. The muscles of warm-blooded animals degenerate on an average in the course of from one-sixth to twelve hours. The degeneration of the heart is considered on p. 113.

CHANGE OF SHAPE IN ACTIVE MUSCLE.

Macroscopic Phenomena.—1. Active muscle becomes shorter and at the same time increases in thickness.

The degree of shortening, which in exceedingly irritable frogs may amount to as much as from 65 to 85 per cent. (on an average 72 per cent.) of the entire length of the muscle, depends upon various factors: (*a*) To a certain degree an increase in the strength of the stimulus gives rise to a greater amount of shortening. (*b*) With increasing exhaustion after continuous, vigorous activity, the same strength of stimulus causes less shortening. (*c*) Elevation of temperature up to 30° C. causes stronger contractions in the frog's muscles. If the temperature be further elevated the degree of shortening is again diminished.

2. The contracting muscle is somewhat diminished in volume. Consequently, the specific gravity of contracting muscle is somewhat increased, the ratio to that of non-contracting muscle (in the marmot) being as 1062 : 1061. The diminution in volume amounts to only $\frac{1}{1310}$.

Method.—Swammerdam placed a frog's muscle in a glass tube containing air and drawn out into a thin tubule containing a small drop of fluid. The nerve was conducted to the exterior through a small lateral opening. Mechanical stimulation of the exposed nerve caused contraction of the muscle and descent of the small drop. In an analogous manner Ermann placed irritable fragments of an eel in a similar tube, filled with an indifferent fluid. The fluid rises to a certain level in a thin tubule communicating with the glass container. When the musculature of the eel was made to contract, the fluid sank. Landois demonstrated the diminution in volume of contracted muscle by means of the manometric flame. The cylindrical glass vessel containing the muscle receives two electrodes passing through its walls in an air-tight manner. It communicates at one point with a gas-supply pipe, and at another point it gives off a thin tubule, at the extremity of which a small flame is ignited at low gas-pressure. The muscular contraction following each electrical stimulus causes a reduction in the size of the flame. If a pulsating heart, naturally containing no air, be placed in the gas-chamber, each pulsation will be attended with a reduction in the size of the flame.

3. Under normal conditions, all stimuli applied to the muscle, as well as to the motor nerve, will cause contraction in all of its fibers. The muscle thus conducts to all of its fibers the impulses communicated to it. Deviations from this rule are observed, however, in two directions: (a) When the muscle is greatly exhausted, or when it is about

to degenerate, a violent mechanical, and also a chemical or electrical stimulus, applied to a circumscribed portion of the muscle will cause contraction in this portion alone; so that an elevated thickening of the fibers (Schiff's *idiomuscular contraction*) is observed at this point. The same phenomenon may be induced in the muscles of a healthy person, and especially in weakened and poorly nourished individuals, if the fibers be struck with a blunt edge at right angles to their course. (b) Under certain conditions, as yet not fully known, the muscles will be seen to exhibit so-called *fibrillary contractions*, that is the various bundles of fibers in the muscle are from time to time traversed by short contractions. Such a condition is observed in the tongue-muscles of the dog after section of the hypoglossal nerve, and in the face-muscles after section of the facial nerve.

According to Bleuler and Lehmann, section of the hypoglossal nerve in the rabbit is followed in the course of from sixty to eighty hours by fibrillar contractions that persist for months, even when stimulation of the healed nerve above the point of union again excites movements in the corresponding half of the tongue. Stimulation of the lingual nerve increases the fibrillar contractions. This nerve contains vasodilator fibers from the chorda tympani. Schiff believes that the cause of the contractions resides in the increased blood-supply to the tongue. Sigm. Mayer also observed contractions in the facial muscles in rabbits, after restoration of the circulation in the carotids and subclavians, previously compressed. Section of the motor nerves in the face does not abolish the phenomenon, while repeated compression of the arteries does so. The cause of the contractions resides, accordingly, in the musculature itself. This motor phenomenon is designated *pseudomotor*. It may be compared to the paralytic secretion of the salivary glands. Similar phenomena have been observed also in man under pathological conditions, but at times fibrillar contractions may be observed even in the absence of other evidence of pathological disturbances.

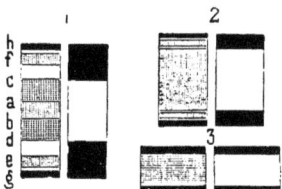

Fig. 191.—The Microscopic Phenomena of Muscular Contraction in the Individual Elements of the Fibrils (after Engelmann).

Microscopic Phenomena.—1. The separate fibrils of the muscle exhibit the same phenomena as does the entire muscle, in that they become shorter and thicker. 2. The observation of the individual muscle-elements is attended with especial difficulties. In the first place, it is certain that during contraction they become collectively shorter and thicker, so that the transverse striations appear to be pushed more closely together. 3. Opinions are not fully in accord as to the behavior of the constituent parts of each muscle-element during contraction.

Fig. 191, 1 represents, according to Engelmann, on the left a muscular element at rest; from c to d extends the doubly refractive, contractile substance, in the middle of which the median disc a b is situated; h and g are the terminal discs. In addition, there is in each singly refractive light layer an accessory disc, f and e, which is doubly refractive in but slight degree, and occurs only in the muscles of insects. Fig. 1 shows on the right the same element in polarized light, the middle portion of the element, so far as the actual contractile substance extends, appearing light on account of the double refraction; while the remainder of the muscle-element appears black on account of the single refraction. Fig. 191, 2 represents the transition-stage, and 3, the actual contractile stage of the muscle-element, both on the left as viewed in ordinary light, and on the right in polarized light.

According to Engelmann, during contraction (3) the singly refractive layer becomes on the whole more highly refractive, and the doubly re-

fractive layer less so.   As a result, the fiber may with a certain degree
of shortening (2) appear homogeneous and only faintly striated when
observed in ordinary light, the *homogeneous* or *transitional stage* (Merkel's stage of dissolution).   If the shortening be more pronounced (3),
distinct dark striæ again appear, corresponding to the singly refractive
discs.   At every stage of shortening, including, therefore, the transition-
stage, the singly and doubly refractive layers may be demonstrated, by
means of the polarizing apparatus, as sharply defined, regularly alternating layers (in 1, 2, 3, to the right).   They do not exchange places in the
muscle-compartment during contraction.   The height of both layers is
diminished during contraction, that of the singly refractive much more
rapidly than that of the doubly refractive layer.   The total volume of
each element is not appreciably changed during contraction.   Therefore, the doubly refractive layers increase in volume at the expense of the
singly refractive layers.   Hence it follows that during contraction fluid
passes from the singly into the doubly refractive layer; the former
shrinks, the latter swells.

Method.—The phenomena described can be best observed by instantaneously
coagulating the living muscle-fibrils of insects in the various stages of rest or
contraction by suddenly applying alcohol or dilute perosmic acid to the muscles,
and thus fixing the different stages.   The movement itself may be followed under
the microscope, either by stimulating the thin, outspread muscle electrically, or,
still better, by observing the independent muscular contractions in the transparent parts of an insect, for example in the head of the gnat's larva.

A thin, extended muscle, for example the sartorius of the frog, yields a double
spectrum (like a Nobert's glass screen), if light be allowed to pass through a
narrow slit, held closely in front of the fibers and at right angles to them.   If the
muscle be made to contract, for example by mechanical stimulation, the spectrum
broadens, an evidence that the intervals between the transverse striæ become
smaller.   At the same time the transparency of the muscle becomes greater than
during rest.

THE TIME-RELATIONS OF MUSCULAR CONTRACTION.
MYOGRAPHY. SIMPLE CONTRACTION. TETANUS.
ISOTONY. ISOMETRY.

Isotonic muscular activity is the term applied to the contraction in
which the tension of the muscle remains the same, while the fibers become shorter.

Method.—The time-relations of the contraction in the isotonic muscular act
may be shown by v. Helmholtz's myograph (Fig. 192).   The suspended muscle
(M), fastened at its upper extremity (K), is attached by its lower extremity to
a lever constructed like a balance, which can be weighted by means of the
weights (W) as desired, and is raised by the shortening of the muscle.   From the
free extremity of the arm of the lever is suspended by means of a hinge-joint
a style (F), which records the movement of the lower extremity of the muscle
on the smoked surface of a cylinder made by means of clockwork to rotate at a
uniform speed in front of the style.   In this way the contracting muscle itself
records its contraction-curve, in which the abscissas represent the units of time
calculated from the known rapidity of rotation of the cylinder, and the ordinates
represent the degree of shortening at any particular moment.

Fick improved the myograph materially by making the writing lever exceedingly light, and applying the weight close to the rotation-axis of the balance.
In this way the swinging movement accompanying the muscular contraction is
reduced to a minimum, as is also the change in tension brought about by such
movements.

The surface intended for the reception of the myogram must be moved rapidly,
as the process of movement takes place rapidly.   Therefore, either a plate fastened

to the rod of a pendulum (Fick's pendulum-myograph), or a surface set in motion by gravity (Jendrássik's gravity-myograph) or by means of a spring (Du Bois-Reymond) or a rotating convex surface (Rosenthal's rotating myograph), may be employed. Under the myogram a time-curve is traced by means of a vibrating tuning-fork. The apparatus is, in addition, provided with an arrangement for indicating in the tracing the moment of stimulation.

The curve may be traced advantageously on the vibrating plate of a tuning-fork (Fig. 194, I). The time-units are thus registered in all parts of the curve, each complete vibration being equal to 0.01613 second. The moment of stimulation coincides with the beginning of the vibration of the fork, which is at first moved to one side for a time, without vibrating. This is accomplished by releasing a clamp, which at the same time opens a galvanic circuit, and sends an induction (opening) shock of the secondary coil through the muscle. The tuning-fork can also be set in vibration by a blow on one of its prongs. If under such circumstances the nerve is so placed upon the fork as to be struck by the blow, the latter acts at the same time as a mechanical nerve-stimulus.

The balance, together with the imposed weights, is jerked upward at the commencement of the contraction. As a result the curve is distorted, because the muscle is no longer weighted after the moment of occurrence of the jerk. For this reason the muscle has been made to draw up an elastic spring. In this way, however, a stronger pull must be made on the muscle as the spring is raised higher and higher. To avoid this Grützner constructed a spring that exerts a steadily diminishing tension on the apparatus as the muscle progressively contracts.

If it be desired to record only the extent (height) of the contraction, the tracing is made on a stationary surface, which is displaced slightly after each movement (Pflüger's myograph).

Muscular contractions may also be recorded in the case of man. It is best to transfer the increase in thickness attending contraction either to a lever or to a drum covered with rubber, for example that of Brondgeest's pansphygmograph (p. 101).

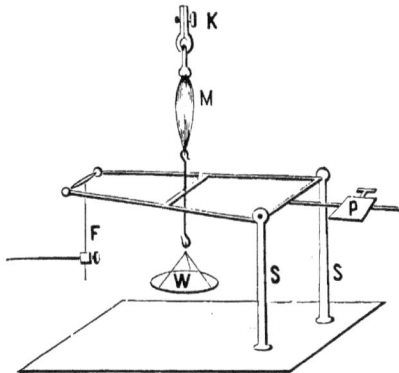

Fig. 192.—Diagrammatic Representation of v. Helmholtz's Myograph: M, the muscle, fastened at K; F, the writing-style, suspended from the arm of the lever that is to be raised; P, a counter-weight for maintaining equilibrium; W, scale-pan for weighting the muscle as desired; S S, posts supporting the balancing lever.

If a single stimulus of momentary duration be applied to a freely movable muscle, the latter executes a *simple contraction*, that is it shortens rapidly and also returns quickly to the relaxed condition. Under such circumstances the internal tension of the muscle remains the same during the course of the entire contraction, and for this reason the resulting curve is designated an *isotonic myogram*.

The following details can be noted in an *isotonic contraction-curve* described by a muscle that has to lift only the light writing lever, and is not overweighted by any other attached weights: 1. The *stage of latent stimulation* (Fig. 193, a b), which arises from the fact that the contraction of the muscle does not begin at the moment of stimulation, but always somewhat later. If the momentary stimulus, for example an induction-shock, be applied directly to the entire muscle, the latent period is about 0.01 second.

In man the stage of latent stimulation varies from 0.004 to 0.01 second.

36

If provision is made in the experiment for the muscle to contract immediately, so that no time is lost between the act of the relaxed muscle becoming tense and the commencement of the contraction, the latent stage may fall below 0.004 second. For the excised frog's muscle, Bernstein and Engelmann found the shortest period to be 0.0048 second. If the animal's muscle remains attached to the body, protected as well as possible from external injuries and supplied with circulating blood, then the latent stimulation may be shortened to 0.0033 second, and even to 0.0025 second.

Influences Affecting the Duration of the Latent Period.—The latent period is diminished by increase in the strength of the stimulus and by heat, and increased by fatigue, cooling, and increase in the weight. The latent period of an opening contraction is also longer (even 0.04 second) than that of a closing contraction. Before the muscle contracts as a whole, individual muscle-elements within it must already have undergone contraction. It is, therefore, assumed that the latent period of the individual muscle-elements is shorter than that of the entire muscle. The latent period is shorter after direct muscle-stimulation than after indirect stimulation through the nerve, as the transference of the stimulus through the motor end-organ requires some time. The transmission of the nerve-stimulus is considered on p. 667.

2. From the beginning of the contraction to the height of the shortening (b d), the muscle contracts at first somewhat slowly, then more rapidly, and finally toward the end of the shortening more slowly again; so that the ascending limb of the curve has the form of an $\int$. This is termed the *stage of increasing energy;* it lasts about 0.03 or 0.04 second.

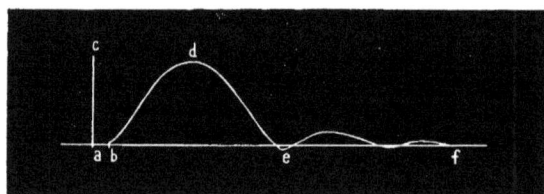

Fig. 193.—Myogram of an Isotonic Contraction.

Its duration is the shorter the smaller the contraction (weaker stimulus), the smaller the weight to be raised, and the less fatigued the muscle.

3. After the height of contraction has been reached, the muscle again becomes extended, at first slowly, then more rapidly, and finally more slowly again; so that the descending limb has the form of an inverted $\int$. This is the *stage of diminishing energy* (d e); it is usually of shorter duration than that of increasing energy.

4. After the descending limb of the curve has been recorded, there follow several after-vibrations (from e to f), due to the elasticity of the muscle, and disappearing gradually. These constitute the *stage of elastic after-vibrations*. The latter are, however, regarded as factitious, and due to the after-vibrations of Helmholtz's apparatus.

If the stimulus is applied to the motor nerve instead of the muscle, the contraction is the greater and lasts the longer the nearer to the spinal cord the nerve is stimulated.

It has, until now, been assumed that the muscle is weighted only with the light writing lever that it has to raise in recording the curve. If, however, it be after-loaded, that is if additional weights be hung on the lever that, supported during rest, must be lifted during contraction, then the commencement of the contraction is delayed as the after-loading is increased. This is due to the fact that the muscle, from the moment of stimulation on, must first accumulate so much contractile force as is necessary to raise the weight. The greater the weight the longer is the period of time that must elapse before the act of lifting begins. Finally, a degree of after-loading is reached at which it is no longer possible to raise the weight. This indicates the limit to which the lever-force may operate.

If a muscle, during contraction, be subjected to a temporary increase in tension, it will be found that a short, quick, and considerable increase in tension immediately diminishes the contraction; while a more prolonged and slow increase somewhat later increases the contraction.

The temperature of the muscle also has some influence. The duration of the contractile force diminishes with increasing temperature, increasing with increase in weighting. The rapidity with which the contractile force develops increases with increasing temperature, diminishing with increased weighting. The height to which an unweighted muscle may lift a weight increases with its temperature. A frog's muscle, supplied with circulating blood, exhibits the greatest contraction in response to stimuli at o° C. As the temperature rises, the extent of contraction diminishes progressively.

If the muscle becomes fatigued as a result of repeated stimulation, the stage of latent stimulation becomes longer and the curve remains lower, because the contraction of the muscle is less; while the abscissa becomes longer, because the muscle contracts more slowly (Fig. 194, I). Cooling of a muscle has like effects. Also the muscles of the new-born behave in a similar manner. The contraction-curve has a flat apex, and is considerably prolonged, especially in the descending limb.

If the nerve of the muscle is stimulated by the closing or opening of a constant current, the muscular contraction corresponds exactly to that already described. If, however, the current is applied directly to the muscle itself, and is closed and opened, a certain degree of persistent contraction, though often but slight, takes place during the period of closure, so that the curve assumes the form shown in Fig. 194, IV, in which the current was closed at S and opened at O.

According to Cash and Kronecker, the individual muscles have a special form of contraction-curve. Thus, the omohyoid of the tortoise contracts more rapidly than the pectoral. The flexors of the frog contract more quickly than the extensors. The muscles of tortoises, the adductors of mussels, the muscles of the bat, and the heart contract slowly. The muscles of flying insects contract with great rapidity, those of the fly 350 times, and of the bee 400 times in a second. There are, however, slowly contracting muscles among beetles also, for example in the water-beetle, hydrophilus.

White muscle-fibers are more irritable, have a shorter latent period, and are more readily fatigued than red fibers; their contraction-period is shorter. They are therefore more active, and the contraction-wave is propagated more rapidly in them. They also produce more acid and heat during their activity. The red fibers execute protracted, continuous movements; hence, moderate, physiological tetanus. They intermediate the adjustment of the muscular force to the resistance to be overcome. Red fibers, or those rich in protoplasm, are further present, especially in the continuously active muscles—respiratory, masticatory, ocular, and cardiac. The white fibers execute the rapid, single movements. Muscles that contain principally white fibers have a greater lifting capacity and a more marked absolute power in the single contraction, but they are inferior to the red muscles in tetanic contraction. The contraction-curves of a mixed muscle containing white and red fibers may exhibit two elevations in the ascending limb, the first being due to the contraction of the active white fibers, and the second to that of the more sluggish red fibers. These are observed especially after the action of veratrin on the muscle-substance. The nerves supplying the white and red muscles also exhibit differences in their irritability.

Action of Poisons.—Small doses of curare, as well as quinin and cocain, increase the size of the contractions induced by stimulation of the nerve; larger doses reduce the size to the point of complete paralysis. Suitable, small doses of veratrin likewise increase the size of the contractions, while the stage of relaxation is conspicuously lengthened. Acids accelerate the relaxation. Veratrin, antiarin, and digitalin in large doses induce such changes in the muscle-substance that the contractions become greatly prolonged and similar to a continuous, tetanic contraction. In muscles poisoned with veratrin or strychnin, the latent stage of contraction is at first shortened, but later lengthened. The gastrocnemius of a frog will contract more rapidly if supplied with circulating blood containing sodium bicarbonate. Kunkel believes that the essential factor in the action of the muscle-poisons consists in their control of the imbibition of water by the muscle-substance. As the muscular contraction depends on imbibition, the form of contraction of the poisoned muscle will be influenced by the state of imbibition produced in it by the poison.

The contraction-curves of unstriated muscles are similar to those of striated muscles, but the contraction takes place, after a latent period of as much as several seconds, visibly later and more slowly.

The contraction in a preparation of a frog's stomach lasts 600 times as long as that of a striated muscle, and the latent stage amounts to 1.5 seconds. The curve ascends more steeply than it descends, and its apex is flattened. Warming increases the height of the curve, and shortens the latent period and the duration of contraction; above 39° C., however, the conditions are reversed.

A muscle contracted as a result of stimulation returns to its original length only if a sufficient extending force is applied to it, as by weights suspended from it. Otherwise it will remain somewhat shortened for a considerable time, the resulting condition being designated *contracture* or *contraction-remainder*. This is especially well marked in muscles that have been previously subjected to strong, direct stimulation, or are greatly fatigued, or more strongly acid, or approaching

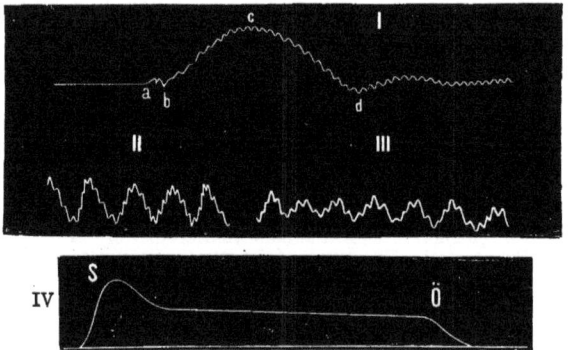

Fig. 194.—I, Contraction of a fatigued calf-muscle from the frog, recorded on a vibrating plate attached to a tuning-fork. Each dentation represents 0.01613 second; a b, latent irritation; b c, stage of increasing energy; c d, stage of diminishing energy. II, The most rapid writing movement of the right hand, recorded on the vibrating plate of a tuning-fork. III, The most rapid tetanic tremor-movement of the right forearm, recorded on the same plate. IV, Myographic curve on closing and opening a current applied to the muscle itself (after Wundt).

a condition of rigor, or have been obtained from animals poisoned with veratrin. The phenomenon of contracture is also observed in man.

In man, single twitching movements of the muscles may be executed with great rapidity. The determination of the time-relations of such movements may be made most simple by recording the movement in question upon the vibrating plate of the tuning-fork. Fig. 194, II, represents the most rapid movement that Landois could execute voluntarily with the right hand in writing the letters n n in succession. Each ascending and descending part of the movement comprises 3.5 vibrations (1 = 0.01613 second) = 0.0564 second. In III the right arm was made to vibrate laterally to and fro on the tuning-fork plate in tetanic tremor; here the to-and-fro movement comprised from 2 to 2.5 vibrations—from 0.0323 to 0.0403 second.

v. Kries found that a simple muscular contraction excited by an induction-shock lasts longer than a single, momentary, voluntary movement. The direct registration of the muscular thickening during a single voluntary contraction shows that the contraction within the muscle lasts longer than the movement developed in the passive motor organ itself. This shorter duration of the resulting movement, which at first appears paradoxical, is due to the fact that, shortly after the primary voluntary muscular contraction, a contraction of antagonists takes place, and as a result a part of the intended movement is cut off. Even with the most rapid voluntary movements in man, v. Kries found that about four impulses in the muscle were effective, so that they really represented short tetanic contractions.

Pathological.—In the presence of secondary degeneration of the spinal cord following apoplexy, of atrophic muscles associated with ankylosed extremities, of muscular atrophy, of progressive ataxia, and of paralysis agitans of long standing, the latent period is increased. On the other hand, it is diminished in the presence of the contractures attending senile chorea and spastic tabes. The entire curve appears to be lengthened in cases of icterus and diabetes. In cases of cerebral hemiplegia in the stage of contracture the muscular contraction resembles the veratrin-curve, as it does likewise in cases of spastic spinal paralysis and amyotrophic lateral sclerosis. In cases of pseudohypertrophy of the muscles, the ascending limb is short and the descending limb greatly lengthened. In the presence of muscular atrophy following cerebral hemiplegia and tabes, the height of the curve is reduced, ascent and descent take place gradually, and the contraction of the atrophic muscle resembles that of a fatigued muscle. In cases of chorea the curve is short. The reaction of degeneration is described on p. 669. According to Goldscheider contraction takes place sluggishly also in conjunction with affections of the nerves, without any change in the irritability of the muscles themselves. In rare cases the observation has been made in man that spontaneous motor stimuli give rise to prolonged muscular contractions, followed by aftercontractions (Thomsen's disease). The muscle-fibers of such patients are broad, the nuclei increased in number, and the fibrils hypertrophied; it has been suggested that the white fibers are wanting. Fr. Schultze and others have observed a peculiar muscular undulation.

If two momentary shocks be applied successively to the muscle in such a way that each would alone have induced a maximal contraction, that is the greatest possible contraction, the effect will vary in accordance with the time that elapses between the two shocks. If the second shock be applied after the muscle has already become relaxed from the contraction of the first stimulus, then a second maximal contraction simply results. If, however, the muscle is still in a phase of contraction or relaxation from the influence of the first stimulus, the second shock gives rise to a new maximal contraction from the phase of contraction existing at that time. If, finally, the second shock follows so quickly upon the first that both occur during the period of latent stimulation, only one maximal contraction results.

If both stimuli are only of moderate strength, not sufficient to induce maximal contraction, a summation of the effects of both takes place. At whatever stage of contraction the muscle may be as a result of the first stimulus (Fig. 195, I, b), the second shock will have an effect (b c) as if the phase of contraction brought about by the first shock were the natural passive form of the muscle. Thus, under favorable conditions, the contraction may be even twice as large as that induced by the first stimulus alone. The most favorable condition is the application of the second stimulus $\frac{1}{20}$ second after the first. The effects of both are also produced if the second shock is applied within the period of latent stimulation.

The second contraction of a summated contraction reaches its height in a shorter period of time than the first contraction alone would have done. The time for b c (Fig. 195, I) is, thus, shorter than that for a b.

If a series of shocks be applied to the muscle in rather rapid succession, the muscle will have no time to relax in the intervals. It, therefore, in accordance with the rapidity with which the stimuli follow one another, remains in a state of continuous, shock-like, tremulous contraction that is designated *tetanus*. The condition of tetanus, or rigid spasm, is, thus, not a state of continuous, uniform contraction, but a discontinuous form of movement, resulting from accumulated contractions. If the stimuli succeed one another with only moderate rapidity, the separate shocks may still be recorded in the curve (*II*). If, however, the stimuli are applied in more rapid succession, the curve has an uninterrupted appearance (*III*). As a single contraction takes place more slowly during fatigue, it is obvious that a fatigued muscle will be more readily thrown into tetanus by a smaller number of single stimuli than a fresh muscle.

All movements of considerable duration excited in the human body are thus to be regarded as tetanic, for they are constituted of a series of single contractions in rapid succession. Accordingly, every movement, however steady, will on close observation be found to exhibit intermittent vibration, which reaches its climax in tremor and becomes so conspicuous in cases of paralysis agitans.

The number of single impulses sent to the muscles of the body in the execution of voluntary movements varies considerably—when the contractions are slow from 8 to 14 in a second, when the contractions are rapid from 18 to 20, the average being 12.5 in a second. Fig. 196, I, represents a myogram of the left flexor brevis pollicis and the abductor pollicis during a continuous contraction of moderate intensity, recorded on the vibrating plate of a tuning-fork. The wave-like elevations indicate the separate impulses, each dentation being equal to 0.01613 second. II represents a similarly recorded curve made by the extensor digiti tertii.

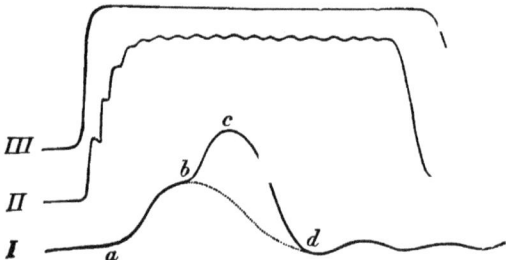

Fig. 195.—*I*, Two successive submaximal contractions. *II*, A series of contractions induced by 12 induction-shocks in a second. *III* Marked tetanus induced by rapid shocks.

By the summation of single stimuli, the muscle voluntarily excited slowly to contraction is gradually brought to the desired degree of shortening. It is customary to effect an exact adjustment of the extent of movement by the development of resistances through antagonistic muscles, as observations on lean, muscular persons show.

The tetanic contraction that occurs under normal conditions in the intact body has also been shown to be composed of single, successive contractions, as secondary tetanus may result from it; the latter may be induced also from a muscle in a state of strychnin-tetanus.

If a muscle be connected with a telephone whose wires are attached to two pins, one of which is inserted into the tendon and the other into the tissue of the muscle, a sound will be heard when the muscle is thrown into tetanus, indicating that periodic motor processes, that is, successive contractions, are taking place in the muscle. The sound is most distinct when the tetanizing Neef's hammer vibrates about fifty times a second.

The rapidity with which the successive stimuli must follow one another in order to induce tetanic contraction varies for the different muscles of the body, as well as for those of different animals.

In the case of the muscles of the frog 15 successive shocks in a second are required on an average to induce tetanus (in the hyoglossus muscle only 10, in the gastrocnemius 27 shocks). If the shocks are feeble, more than 20 in a second are required. The muscles of the tortoise are thrown into a state of tetanus by only 2 or 3 shocks in a second; the red muscles of the rabbit by 10, the white muscles by more than 30, human muscles by from 8 to 12, the sluggish abductor minimi digiti of man by 6 shocks in a second. The muscles of birds are not thrown into a state of tetanus even by 70 shocks, and the muscles of insects not even by from 350 to 400 in a second. In the muscles of the crab's claw, rhythmic contractions or rhythmically interrupted tetanus (in the astacus and hydrophilus) are observed as a result of tetanic stimulation.

O. Soltmann found that the white muscles of new-born rabbits are tetanized by 16 shocks in a second, and that the tetanus thus induced resembles that of fatigued adult muscles. This fact explains the readiness with which tetanus occurs in the new-born.

Curarized muscles are at times thrown into a state of tetanic contraction by a momentary stimulus.

The extent of shortening in a muscle in a state of tetanic contraction is, within certain limits, dependent upon the strength of the individual stimuli, and also upon their frequency. The steepness of the tetanus-curve increases with increase in the strength of the stimuli rather than with increase in the frequency of the individual stimuli. With feeble stimuli the muscle exhibits greater continuity in its contraction; intensification of the stimuli then causes a greater discontinuity in the curve (tendency to clonic spasm); and if the intensity of the stimuli be still further increased the curve becomes again more nearly continuous. The contracture that may remain after tetanus is the more marked the stronger and longer the stimulation and the weaker the muscle. The height of the contraction and that of tetanus are the same for an unweighted muscle. Only in the case of the weighted muscle is the height of the single contraction less

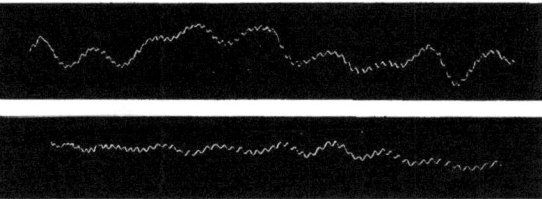

FIG. 196.—I, Fluctuations during a continuous contraction of the flexor brevis pollicis and the abductor pollicis. II, of the extensor digiti tertii.

than that of the tetanic contraction. At times a stimulus applied immediately after tetanus has a greater effect than one applied before tetanus.

The tetanized muscle cannot maintain the same degree of contraction indefinitely if the succession of shocks remains the same. On the contrary, it will lengthen somewhat as fatigue sets in, at first rapidly, but later more slowly. If the tetanizing stimulus is withdrawn, the muscle does not immediately regain its natural length, but a certain contraction-remainder persists for some time, especially after long-continued induction-shocks.

Muscle may also enter into a state of permanent contraction, which has not been definitely determined to be due to fusion of single contractions; for example the transient contraction induced by certain chemical agents (such as ammonia and others), the elevations attending idiomuscular contraction, and that induced by the passage of a constant current.

If rapid, weak induction-shocks (more than 224 and 360, even as many as 5000 in a second for frogs' muscles) be applied to the muscle or its motor nerve, the tetanus may cease after the initial contraction. This occurs with the least frequency of stimulation when the nerve is cooled; the higher the temperature of the nerve the greater the frequency of stimulation that may still be effective in inducing a long-continued tetanus.

This initial contraction is a short tetanus; increase in the strength of the current renders the tetanus continuous. On the other hand, Kronecker and Stirling, however, observed tetanus occur with more than 24,000 shocks in a second. According to these investigators, the upper limit of frequency for the muscle that will still cause tetanus appears to lie near the limit at which fluctuations in the current can no longer be appreciated, even with other rheoscopes.

Isometric Muscular Activity.—While the experiments discussed in the foregoing are concerned with the determination of the changes in the length of a muscle on stimulation and the movement of a weight supported by it, Fick has investigated the changes that take place in the tension of a muscle under the influence of stimuli, when its length is kept constant. Fick designates this process the *isometric muscular act.*

The following apparatus will serve to demonstrate the isometric muscular act (Fig. 197): The angular frame R is provided at its base with a long writing-lever S (abbreviated in the illustration), which is movable at the hinge-joint p. The muscle M, suspended from above, is connected with the writing-lever near its point of attachment. A strong spiral spring F is connected with the other arm of the writing-lever, and during the activity of the muscle, permits only the slightest degree of shortening to take place. This, however, is sufficiently magnified by the great length of the lever. A momentary electric stimulus is applied to the muscle by means of two elec-trodes (r r), and the writing-lever records the isometric curve.

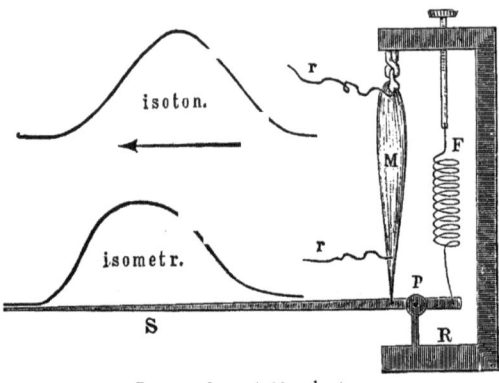

FIG. 197.—Isometric Muscular Act.

The isometric contraction-curve is, on the whole, similar to the isotonic curve, as a comparison of the curves in Fig. 197 will show. Nevertheless, the following differ-ences exist: (1) The contracting muscle attains its maximum tension in the isome-tric muscular act more rapidly than it attains its maximum shortening in the iso-tonic act. (2) The contracting muscle in the isometric act maintains the degree of highest tension somewhat longer, while in the isotonic act it recedes more rapidly from the highest degree of shortening.

In the isometric muscular act in man, voluntary excitation gives rise to a higher degree of tension than electric stimulation. In the frog the tension of the muscle in a state of tetanus is about twice as great as it is in a maximal contraction; in human muscle it may be even ten times as great. During extension of the tetanized muscle, as during its contraction, equal degrees of tension correspond to smaller lengths.

In the case of unstriated muscles, the entire curve is much shorter in the isometric act than during the isotonic act, and its form is almost symmetrical.

RAPIDITY OF PROPAGATION OF MUSCULAR CONTRACTION.

If a muscle of considerable length is stimulated at one extremity a contraction occurs at that point, and rapidly traverses in a wave-like manner the entire length of the muscle to its other extremity. The excitation is therefore communicated to each successive part of the muscle by virtue of a special conductive capacity on the part of the muscle to enter into a state of contraction. In the frog the wave of contraction has an average velocity of from 3 or 4 to 6 meters in a second, in the rabbit of from 4 to 5 meters, in the lobster of only 1 meter, in unstriated muscles and in the heart of only from 10 to 15 millimeters. These figures, however, apply only to excised muscles, for in the striated muscles of living human beings the rapidity of propagation is much greater, namely from 10 to 13 meters.

Method.—For the demonstration of this motor phenomenon, Aeby placed a writing-lever transversely across the origin of a muscle of considerable length and another across its insertion. Both were raised by the thickening resulting from the contraction of the respective parts of the muscle, and the movements were recorded one above the other on the drum of a kymograph. If one extremity of the muscle is now stimulated, the contraction-wave that rapidly traverses the muscle lifts first the proximal and then the distal lever. As the velocity with which the drum revolves is known, the rapidity with which the contraction-wave is propagated through the portion of muscle under examination can readily be calculated from the distance between the elevations of the two levers.

The time corresponding to the length of the abscissa of the curve inscribed by each recording lever is equal to the duration of the contraction in that part of the muscle; according to Bernstein this is from 0.053 to 0.098 second. By multiplying this value by the ascertained rapidity of propagation of the contraction-wave through the muscle, the wavelength of the contraction-wave is obtained; this equals from 206 to 380 millimeters.

The rapidity and the height of the contraction-wave are diminished by cold, fatigue, gradual degeneration, and some poisons. On the other hand, the strength of the stimulus and the extent to which the muscle may be weighted have no influence on the rapidity of the wave. In excised muscle the wave diminishes in size during its course through the muscle, but not in the muscles of a living human being or animal. The contraction-wave never passes from one fiber to an adjacent fiber; neither does it leap over an interposed tendon or a transverse tendinous septum.

If a muscle of considerable length be stimulated locally at its middle, a contraction-wave is propagated from the point of stimulation toward both extremities, and in other respects possesses the properties previously described. If two or more points in the muscle are stimulated at the same time, the wave-movement sets out from each, and one may even pass over another.

If a stimulus be applied to the motor nerve of a muscle, it will be conducted to each muscle-fiber, whose contraction-wave must develop at the motor end-plate and be propagated thence in both directions along the fiber, which is only from 5 to 9 centimeters in length. In accordance with the obvious inequality in the length of the motor fibers from the nerve-trunk to the end-plates, the contraction will not commence at absolutely the same moment in all of the muscle-fibers, as the conduction through the motor nerves likewise occupies a certain amount of time. The difference, however, is so small that the muscle, stimulated through its nerve, appears to contract simultaneously as a whole.

An absolutely simultaneous, momentary contraction of all of the fibers of a muscle can occur only if all are stimulated at the same time.

MUSCULAR WORK.

The muscles are most perfect machines not only because they utilize most completely the substances consumed in their activity, but because they are distinguished from all machines of human construction by the fact that, as a result of repeated exercise, they become stronger, better developed, and capable of increased activity.

According to the usual method of estimation, the amount of work performed by a muscle is equal to the product of the weight lifted (**P**) and the height to which it is lifted (s); hence, $A = s\,P$. From this it follows, first, that if the muscle is not at all weighted, therefore, if P equals o, then A must equal o; that is, no work is performed if there is no weighting. Further, if the muscle is burdened with an excessively heavy weight so that it is no longer able to contract, therefore, s equals o, then, likewise, no work is performed. Between these two extremes the active muscle is able to execute work.

Strictly speaking, the contracting muscle lifts, in addition to the suspended weight P, half of its own weight p, which should be added to P as $\frac{1}{2}$ p; hence, $A = (P + \frac{1}{2}\,p)\,S$.

With the strongest possible stimulation, or maximal stimulus, that is, a strength of stimulus that causes the maximum degree of contraction in the unweighted muscle, the work performed increases progressively with each contraction as the weight increases to a certain maximum. If, as the weight is increased, the muscle can raise it to a gradually diminishing height, the amount of work diminishes progressively; and, finally, as already noted, it becomes o, when no elevation is effected.

The following table will illustrate the work performed by a frog's muscle, according to Edward Weber:

Weight Lifted, in Grams.	Height in Milli- meters.	Amount of Work Performed in Gram-millimeters.
5	27.6	138
15	25.1	376
25	11.45	286
30	6.3	189

If the weight be increased at any given moment during the contraction of the muscle, more work can be performed, but only if the stimulus applied does not fall below a certain minimum. The duration of the contraction is longer.

If a muscle has contracted as much as possible for the purpose of lifting a heavy weight, it can be made to perform still more work by gradually diminishing the weight. It contracts still further and performs additional new work by raising the diminished weight.

If the amount of work performed by the muscle be diminished by raising the weight before the contraction to a part of the height to which it would have been lifted by the muscle stimulated to the maximum, then the muscle will raise the weight to a still higher level.

The investigations concerning muscular work yield the following results:

1. The muscle is capable of lifting a greater weight the larger its transverse section, that is the more fibers it contains arranged side by side.

2. The muscle is capable of lifting a weight the higher the longer it is, that is the more muscle-fibers it contains arranged in succession.

3. The muscle is capable of lifting the greatest weight at the commencement of contraction; as the contraction progresses, it is capable of lifting only a progressively smaller weight, and near the maximum contraction only a relatively light weight.

4. By the term absolute muscular energy is meant, according to Ed. Weber, the weight that the muscle stimulated to the maximum is no

longer capable of raising while in its natural passive form, without being stretched by the weight at the moment of stimulation.

In order to obtain a standard for the comparison of the absolute muscular energy in different muscles and also in different animals, an estimation is made of the absolute muscular force for one square centimeter of cross-section. The mean cross-section of a muscle is determined by dividing its volume by its length. The volume is equal to the absolute weight of the muscle in question divided by the specific gravity of muscle-substance (1058). Thus, the absolute muscular energy for one square centimeter of a frog's muscle is from 2.8 to 3 kilos; for one square centimeter of human muscle from 7 to 8 or even from 9 to 10 kilos. Analogous figures for crustaceans are from 1.8 to 3.2; for beetles from 3.4 to 6.9; for mussels from 4.5 to 12.4 kilos. The transverse section of the muscles tested in man is estimated from cadavers having the same constitution and muscular development as the person under observation.

In conformity with proposition 3 it is evident that a muscle during contraction will develop the greater absolute muscular energy the more it is extended before contraction.

5. If a muscle in a state of tetanic contraction maintains a weight in an elevated position, it performs no work during the time, but only in the act of elevation. Nevertheless, the muscle in the state of tetanus requires continued stimuli, and it exhibits metabolic changes and fatigue. The transformation of its potential energy is applied to the generation of heat.

When the maximal stimulus is applied, a muscle is not capable of lifting as heavy a weight at one contraction as when tetanic stimulation is applied. During tetanic stimulation, moreover, the muscle develops the greater energy (even as much as twice the ordinary) the more frequent the stimulation, as has been observed with increasing frequency up to 100 stimuli in a second.

If only moderate stimuli that do not excite the maximal contraction are applied to the muscle two possibilities present themselves. If the feeble stimulus remains constant, while the weight changes, the amount of work performed follows the same law that is operative during maximal stimulation. If the weight remains the same, while the strength of the stimulus varies, then, according to Fick, the height to which the weight is raised varies in direct proportion to the strength of the stimulus.

The stimulus that sets a muscle into activity must, naturally, attain a certain strength before it becomes effective—*liminal intensity of the stimulus.* This is independent of the weight attached to the muscle. With a minimal stimulus a small weight is raised to a higher level than a large one; but as the stimulus is increased, the contractions increase in greater proportion with a heavy weight.

A contracting muscle is capable of performing considerably more work if the weight to be lifted is attached to an inert mass that acts like a fly-wheel, or if the weight is swung to a considerable height. Starke was able almost to quadruple the work corresponding to a maximal contraction by a proper selection of materials for this purpose. Also the production of heat is increased under such conditions, although in much less degree, and it is much more quickly diminished on fatigue.

If the resistance applied to prevent the movement of a limb whose muscles, strained to the utmost degree, be suddenly removed, the limb will, with the greatest energy and rapidity, assume the position brought about by the muscles. Such springing movements are observed especially in grasshoppers, leaping beetles, and cheese-mites.

Under special conditions a muscle may perform considerable work through its increase in thickness.

In the intact body the vessels of a muscle dilate during muscular contraction, so that the amount of blood circulating thröugh it is increased. Evidently the vasodilator nerve-fibers contained in the same nerve-trunks as the motor nerves are stimulated at the same time as the latter.

In estimating the absolute muscular energy of single muscles or groups of muscles in man, close attention should always be paid to the physical relations, for example leverage, effects, direction of the traction, degree of shortening, and the like.

The absolute energy of certain groups of muscles may practically be measured readily by means of the dynamometer. This is constructed in part on the principle of the spring-scales, upon which the pressure or pull of the muscles in question is allowed to act. Quetelet has determined statistically the strength of certain groups of muscles. The pressure of both hands in man equals .70 kilos. The pull amounts to double this weight. The strength of the hands of a woman is about one-third less. Further, a man can carry more than twice his own weight, a woman only half her weight; boys are able to carry about one-third more than girls.

In estimating the work done by man, not only the amount of work he is able to perform in any one moment should be taken into consideration, but also the number of times in succession he can perform this work. In accordance with practical experience, the mean value of the daily work performed by a man during eight hours' activity has been estimated at from 6.3 to 10 (at most from 10.5 to 11) kilogrammeters in a second, hence a daily usefulness of 288,000 (in round numbers 300,000) kilogrammeters. The work performed by a horse in a second is assumed to be 75 kilogrammeters—*horse-power* or *dynamic horse*.

This average performance of work may, it is true, be temporarily increased, but the organism then requires a prolonged rest after the work is done, so as not to suffer in health as a result of the overexertion. The amount of work performed in walking and in bicycling is discussed on p. 595.

Some substances, when introduced into the body, impair and eventually abolish muscular activity; for example mercury, digitalin, helleborin, potassium-salts, apomorphin, and others. Others have been shown to increase the functional activity of muscular tissue; for example caffein, theobromin, veratrin, muscarin in small doses, glycogen, kreatin, and hypoxanthin; extract of meat likewise causes rapid recovery of the muscles after fatigue.

Unstriated muscles are capable of performing a great amount of work, for example the uterus during labor, the craw of granivorous animals. The longitudinal musculature of the earth-worm is capable of raising more than 15 kilos, the frog's intestine of overcoming the pressure of a column of water of 1½ meters.

THE ELASTICITY OF PASSIVE AND ACTIVE MUSCLE. MYOTONOMETRY.

Preliminary Physical Considerations.—Every elastic body has its *natural form*, that is the outer form that it possesses when no external force (traction or pressure) operates upon it. Thus, the passive muscle also possesses a natural form, when no traction or pressure is exerted upon it. If traction in a longitudinal direction be made on a muscle its connected parts must be somewhat separated from one another, and the natural form will be stretched under the influence of the elastic energy. If the extending force be removed, the elastic body will return to its natural form. A body is said to be *completely elastic* if it returns entirely to its natural form after the tension ceases. By *amount of elasticity* (modulus) is meant the weight, expressed in kilograms, by which an elastic body having a cross-section of one square millimeter would be stretched the equivalent of its own length, provided it did not previously rupture, as, naturally, often it does. For passive muscle this equals 0.2734, for bone 2294, for tendon 1.6693, for nerves 1.0905, for the coats of the arteries 0.0726. The amount of elasticity of passive muscle is, thus, small, as the latter yields readily to tractile force; hence its elasticity is not great. The *coefficient of elasticity* is that fraction of the length of an elastic body to which it is stretched by the unit of weight applied to it. This is large for muscle at rest. When the traction reaches a certain degree the elastic body finally ruptures. The *carrying capacity* of muscular tissue, just short of the point of rupture, varies for youth, middle and advanced age approximately in the proportions 7 : 3 : 2.

In the case of unorganized elastic bodies the amount of extension is always directly proportional to the extending weight. In that of organized bodies, therefore also of muscle, this is not the case, however, as with continued increase in weight in equal amount they are extended less and less in the further course of observation than at first. At the same time, they may for days or even weeks gradually undergo a still further increase in length after the primary extension, corresponding to the suspended weight, has been attained, if the same weight be continued. This is designated elastic after-effect.

The *elasticity of passive muscle* is small but complete, and is comparable to that of India rubber. The muscle can be greatly elongated by means of small weights. With the uniform addition of further units of weight, uniform extension, however, no longer takes place, but a slighter increase in length corresponds with equal increments of weight the greater the load. This phenomenon may also be expressed as follows: the amount of elasticity of passive muscle increases with its increased extension.

Method.—For the purpose of studying elasticity, the muscle is suspended free from a support provided with a scale, and the lower extremity is loaded successively with different weights placed in a small attached weighing-pan. The resulting elongation is measured in each instance. To construct a *curve of elongation*, the units of weight added successively are taken as abscissas, and the lengths corresponding to each load as ordinates. The following is an example from the hyoglossus of the frog:

Weight in Grams.	Muscle-length in Millimeters.	Elongation in Millimeters.	Elongation in Percentages.
0.3	24.9	—	—
1.3	30.0	5.1	20
2.3	32.3	2.3	7
3.3	33.4	1.1	3
4.3	34.2	0.8	2
5.3	34.6	0.4	1

The curve of elongation is not a straight line, as in the case of unorganized bodies, but it resembles a hyperbola in form. The stretched muscle has a somewhat diminished volume, as have the contracted and the rigid muscle.

Muscles permitted to retain their connections in the living animal with their vessels and nerves are more extensible than excised muscles. Perfectly fresh muscles elongate at first proportionately to the weight if the increase in the latter be uniform and be kept within narrow limits, therefore like unorganized bodies. If the weights be heavy the observations are not made without consideration of the elastic after-effect.

Dead, and especially rigid, muscle possesses a greater elasticity than fresh, living muscle; that is a greater weight is required to stretch the former than is needed to stretch the latter to the same length. On the other hand, the elasticity of dead muscle is less complete; that is, after being stretched, it regains its natural form only within narrow limits.

In contradistinction from the elastic after-extension of muscle when weighted, after the tension has become constant, Blix recognizes an *after-contraction* of muscle, which comes into play after removal of the weight. Further, he distinguishes an *after-relaxation* in muscle that has been stretched, its tension increasing with the increase in length, but diminishing when the length has become constant; and, finally, an *after-tension* in a previously stretched muscle, whose length is diminished, the previously low tension again increasing, when the length has become constant.

In the intact body the muscles are already stretched to a slight extent, as indicated by the moderate retraction that usually takes place when the muscular insertion is detached. This slight degree of extension is of importance with the occurrence of contraction; as otherwise the muscle would first have to contract, without immediately entering into activity, before it could exert traction upon the bones. The elasticity of the muscles becomes evident on the contractions of its antagonists. The position of a passive limb depends upon the resultant of the elastic traction of the various muscle-groups.

The *elasticity of active muscle* is diminished as compared with that of passive muscle; that is it is lengthened by the same weight to a greater extent than is resting muscle. For this reason active muscle is softer, as can be demonstrated in an excised, contracted muscle. The apparently increased hardness of tense, contracted muscles is due only to their tension. When an active muscle becomes fatigued, its elasticity is still further diminished. During the latent period, in which the development of electrical phenomena and of heat points to metabolic activity in the muscle, no change in elasticity has as yet been demonstrated.

Method.—Ed. Weber made observations in the following manner. The hyoglossus muscle of the frog was suspended vertically, and its length was measured in the passive state. The muscle was then tetanized by induction-shocks and again measured. Progressively increasing weights were then attached to it, in succession, and the amount of stretching of the passive and then the length

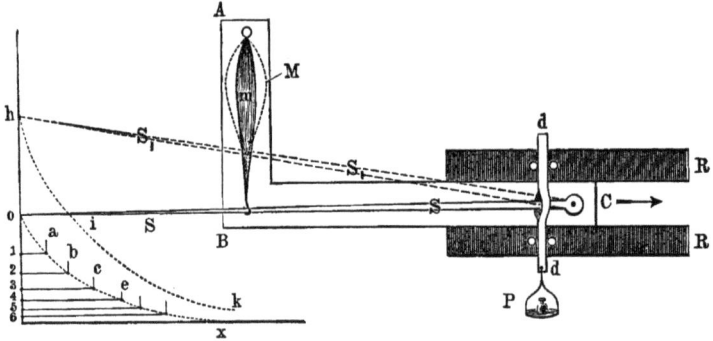

Fig. 198.—Blix's Elasticity Recorder.

of the tetanized muscle (supporting the same weight) ascertained. The extent to which the active, weighted muscle contracted from the passive, weighted condition is termed the *height of the lift.* This becomes steadily less as the weight increases, until finally the heavily weighted, tetanized muscle can no longer contract; that is the height of the lift is zero. Indeed, if the weight be exceedingly heavy it may happen that the muscle, when stimulated, not only can contract no further, but it may even elongate. According to Wundt, however, the elasticity of the muscle does not change under such conditions. In these observations the length of the active, weighted muscle is equal to the length of the equally weighted, passive muscle, minus the height of the lift.

Tracings of the length-curves recorded by passive or contracting muscle stretched by weights can be conveniently made by means of the apparatus of Blix, as shown in Fig. 198. The rectangular piece (A B C) is movable horizontally between two strips (R R). To the vertical portion of the former is attached the freely suspended muscle (m), which is connected with the writing-lever (S S), the latter being attached to the horizontal portion near C by means of a hinge-joint. The writing-lever is provided with a small movable rod (d d), from which a weight is suspended. When the rectangular piece (A B C) is moved in the direction of the arrow, the weighted rod (d d) more closely approaches the muscle, which thus becomes constantly more and more heavily weighted.

With the muscle at rest (m) the curve o a b c e is first recorded by means of the displacement described. Then a similar curve is recorded while the muscle is tetanized (M) by electrical stimulation; and the curve h i k is thus traced. With the aid of the apparatus both the extension-curve with increasing weight and the contraction-curve with diminishing weight can be recorded. Both curves are necessarily analogous, except that their form is reversed.

The elasticity of muscle may also be measured by its rate of oscillation when twisted about its longitudinal axis. Kaiser found that the elasticity of active muscle depends upon its length at the time. It is least when the muscle has the same length in the active as in the passive state. If shortening occurs in a muscle stretched by a weight, its elasticity is diminished, and this reaches its minimum when the muscle becomes of the same length as the passive, unweighted muscle. If the active muscle contracts still further, its elasticity increases.

Under the influence of certain *poisons* the elasticity of the muscles is altered as a result of changes in the condition of the contractile substance. Potassium causes shortening of the muscle, with simultaneous increase in its elasticity. Digitalin causes elongation of the muscle, together with increased elasticity. Physostigmin also increases the elasticity, while veratrin diminishes it and interferes with its completeness. Tannic acid renders the muscles less extensible, but more elastic.

Ligation of the vessels causes first a diminution, and later an increase in the elasticity. *Separation of the nerves* from the muscle results in a diminution of the elasticity. The influence of *temperature* on the extensibility is as follows: As the temperature increases—from 0° to 30°—the muscle elongates, as its extensibility increases. The increase in length is proportional to the load. At 34° contraction occurs as a result of the thermal stimulation; above 47° the muscle-proteid coagulates.

Unstriated muscles possess an exceedingly small amount of elasticity; at the same time the elastic after-effect lasts much longer, and immediately follows the primary stretching. Fibrous connective tissue possesses the greatest elasticity, elastic tissue less, and unstriated muscular tissue the least. The elasticity of a complex organ, made up of these tissues, depends, accordingly, upon the relative abundance of these elements.

As a result of his experiments Edward Weber has reached the following conclusions as to the nature of the contractile energy of muscle. He assumes the existence of two states in muscular tissue—the passive and the active. Each of these is characterized by a special natural form. The passive muscle possesses the longer, thinner form; the active muscle the shorter, thicker form. Both the active and the passive muscle tend to maintain their respective form. If, now, the passive muscle be thrown into activity, the passive form suddenly changes into the active form, by virtue of its elastic energy. It is this latter that is capable of performing the work of the muscle. Schwann has already alluded to the similarity between the energy of an active muscle and that of a long, elastic, tense spiral spring. Both are able to lift the greatest weight only from the form in which they are most stretched. The greater the shortening they have already undergone, the smaller is the weight that they are further able to raise.

Observations on elasticity can also be made on the muscles of living human beings. Under such circumstances, however, not alone the simple physical law of elongation is to be taken into consideration, for the elongation at the same time causes in the muscle changes in its irritability and in the blood-supply, as well as direct or reflex stimuli, all of which must necessarily modify its extensibility. If the extremity of the foot in man be raised vertically by means of a cord passing over a pulley and having weights attached to it, the muscles of the calf will be stretched. Mosso and Benedicenti found that, as the weight increased, the muscles became longer at the same or at an increasing rate, if the weight were continuous and increasing. If, however, the muscle is completely released, before the new, heavier weight is applied, then the length of the stretched muscle diminishes as the weight is increased. Further, the curve of elongation exhibits individual differences; it exhibits fluctuations in association with the respiratory curves; it may exhibit after-extensions and after-contractions; it changes with frequent repetition, with heat and cold. Strong, sudden stretching, and previous voluntary contraction and fatigue likewise have an effect. Investigations of this sort are designated *myotonometry*.

HEAT-PRODUCTION IN ACTIVE MUSCLE.

Method.—The increased temperature of a muscle during contraction may be determined either by means of delicate thermometers introduced between the muscles, or thermo-electrically. The passive muscle on the opposite side of the body, or the blood within a vein, will serve for purposes of comparison. As the resistance to conduction in metals (platinum wire, lead strips) is increased by heat, the observation may also be made in this way.

After Bunzen, in 1805, had observed the development of heat during muscular activity, v. Helmholtz demonstrated in 1848 that also excised frogs' muscles, tetanized for two or three minutes, exhibit a rise in temperature of from 0.14° to 0.18° C. R. Heidenhain even succeeded by thermo-electrical means in demonstrating an increase in temperature of·from 0.001° to 0.005° C. for each individual contraction. A similar condition exists in the beating heart, whose temperature rises with each systole. The production of heat in the muscle exhibits a latent stage, which is, however, of shorter duration than the latent period of movement.

A contraction of a frog's muscle, weighing one gram, will produce an amount of heat equal to about three microcalories, which will raise the temperature of three milligrams of water from 0° to 1° C.

The following facts have been ascertained concerning heat-production:

1. *It bears a relation to the amount of work performed.* (a) If the muscle during contraction carries a weight that during rest extends it again, it performs no work that is communicated externally. All of the transformed, chemical, potential energy is, therefore, converted into heat during this movement. Under these conditions the generation of heat corresponds with the activity; that is it increases at first with the weight and the height of the lift to the maximum point, and then, as the weight is further increased, the generation of heat diminishes. The heat-maximum, however, is attained with a smaller weight than the maximum of work.

(b) If the muscle at the height of its contraction is relieved of its weight, then it will have performed some active work communicated externally. Under such circumstances the amount of heat generated is less than in the previous case; and, indeed, the amount of work performed and the lesser amount of heat evolved, are the same in accordance with the law of the conservation of energy.

(c) If the same amount of work is performed in the one case by many small contractions, and in the other by fewer but larger contractions, the amount of heat generated is greater in the latter instance. This fact indicates that large contractions are attended with a relatively greater metabolism than smaller ones, and experience is in accordance with it. Thus, the ascent of a tower by steps with a high tread causes much more fatigue (that is requires more metabolism) than ascent by .steps with a low tread.

(d) If a weighted muscle executes single contractions in succession, by means of which it performs work, the amount of heat thus generated is greater than if it carries the weight constantly in tetanic contraction. The transition of the muscle into the shortened form thus develops a greater amount of heat than the maintenance of that form. Also summated contractions are, accordingly, attended with the generation of a smaller amount of heat than corresponds to that developed by two single successive contractions.

As the stimulus becomes stronger, heat-production increases, in the case of isometric contractions proportionately to the degree of tension; that of isotonic contractions at first more rapidly than the height of the lift, but with strong stimuli proportionately to the latter. Even if the height of the lift, the strength of the stimulus, and the tension of the contracting muscle remain the same during successive contractions, the muscle nevertheless generates more heat during the first than during the following contractions. The amount of heat generated also depends upon the character of the stimulus employed; thus, a muscle tetanized by slow shocks generates more heat than one contracted by rapid shocks.

2. *The development of heat depends upon the tension of the muscle;* it increases with increase in tension. If the muscle be prevented from contracting by fixation of its extremities, the maximum of heat-production takes place during stimulation, and the more quickly the more rapidly the stimuli succeed one another. Such a condition arises during tetanus, in which the violently contracted muscles mutually oppose each other. Therefore, a marked development of heat has been observed in connection with this disease. Dogs thrown into a state of continuous tetanus by electrical stimulation or by the induction of spasm die in consequence of elevation of their bodily temperature to a fatal height ($44°$ or $45°$ C.). This large production of heat is attended with a considerable degree of acidity and the formation of alcoholic extractives in the muscular tissue.

In the case of isometric tetanus the metabolism and heat-production increase more rapidly than the tension as the stimulus becomes stronger. The continuous maintenance of tension in the muscle on the one hand, as well as the contraction of the muscle with a small amount of work without considerable tension, nevertheless requires only relatively little metabolism for the generation of heat, as compared with the work, which is essentially proportional to the consumption of combustible material. If the stimulated muscle be so fixed that it cannot contract, and if it then by releasing its lower extremity be permitted to contract and lift a weight, an additional amount of chemical potential energy will be transformed for the performance of this latter task.

3. Heat-production diminishes as fatigue increases, and it again increases during recovery. The muscle becomes fatigued earlier in its production of heat than in its performance of work.

4. In a muscle normally supplied with circulating blood the production of heat, and also the mechanical performance of work, takes place much more energetically than in a muscle whose circulation is interrupted. Recovery following fatigue also takes place under such conditions more rapidly and completely.

The total amount of work and heat in a muscle must always be equivalent to the transformation of a corresponding amount of chemical potential energy. Of this the portion that is transformed into work will be the larger the greater the force that is opposed as a result of the contraction of the muscle. In the latter event this equals about one-fourth of the transformed potential energy. If the resistance be less, the work performed is a smaller fraction of the transformed potential energy.

At a high temperature, therefore probably in the febrile state, muscle exhibits greater metabolism for the generation of increased amounts of heat, without increase in the work performed.

In man, the production of heat in muscles made to contract by electrical stimulation can be appreciated through the skin. It was observed by Landois also when voluntary movements were executed. Venous blood flowing from a contracting muscle acquires a higher temperature than the arterial blood—by as much as $0.6°$ C.—as a result of energetic action.

37

The statement made by some that a rise in temperature, amounting to about $\frac{1}{30}°$ C., occurs also in a nerve in action is denied by others; but an increase in temperature does occur in a nerve in process of degeneration.

5. As the muscle is an elastic body, thermal phenomena will occur in it as a result of purely physical influences, as in inanimate, elastic bodies, such as India rubber. Thus, heat is set free on stretching living or dead muscle; and, conversely, the temperature of the muscle falls on elastic shortening.

THE MUSCLE-MURMUR.

If a contracted muscle be at the same time maintained in a state of tension by the application of resistance to it, a sound or murmur will be audible, arising from intermittent variations in tension within the muscle.

Method.—For purposes of observation, auscultation is practised either by means of the ear applied directly, or with the aid of a stethoscope, over a tetani-cally contracted muscle in another person. Some individuals are able to appre-ciate the murmurs of their own muscles of mastication on closing the external auditory canals, and pressing the jaws forcibly together.

If one external auditory canal be closed, and into the other there be inserted a small rod from the end of which is suspended a tetanized, weighted frog's muscle, the sound of this isolated muscle can be readily heard.

If the contracting muscle is attached to an elastic spring, whose rate of vibra-tion can be varied, and if the rate of vibration is determined that must be im-parted to the spring in order that it shall be energetically set into vibration by the sounding muscle, the rate of vibration of the muscle-sound can be readily deter-mined for each case after a few trials. A writing-style may be fastened to the tip of the vibrating spring, and record the vibrations upon a smoked surface.

A muscle, thrown into contraction by the will, vibrates from 19.5 to 20 times a second. The deep tone corresponding to such a small number of vibra-tions is, however, not audible, but the first overtone, corresponding to twice this number, is heard. The sound has the same rate of vibration when the muscle is contracted in animals, by stimulation of the spinal cord, and also when the motor nerve of a muscle is irritated by chemical means. If, however, the motor center in the cerebral cortex be stimulated, the contracting muscle will generate a tone that is the higher the stronger the stimulus.

If a tetanizing induced current be applied to the muscle (also in man), the rate of vibration of the muscular sound corresponds exactly with the rate of vibration of the spring-hammer in the induction-apparatus. The sound can, therefore, be raised or lowered by changing the tension of the spring.

Lovén found that the muscle-sound is relatively the strongest when the weakest current is employed that will induce tetanus. The sound will then have the vibration-rate of the next lower octave. As the current is increased in strength, the muscle-sound disappears, and with a strong current it reappears with the same rate of vibration as that of the interrupter of the induction-apparatus.

If the induction-shocks are sent through the nerve, the sound is not so loud, but otherwise it is of the same vibratory duration. By means of rapid induction-shocks sounds have been produced up to 704 and 1000 vibrations in a second.

The first sound of the heart is in part a muscle-sound.

Landois, in 1873, first reported the observation that the rumbling murmurs emitted by many fish (Cottus, sea-scorpion) are due to the loud sounds generated by the spasmodically contracted muscles of the shoulder-girdle, and still fur-ther intensified by the resonance of their large oropharyngeal cavity sur-rounded by a firm bony framework. He found at that time that even a single induction-shock that excited the muscles was able to generate the muscle-sound. Herroun, Yeo, and MacWilliam also noted a like condition in the contracting mus-cles of man. It must, accordingly, be considered as doubtful whether the muscle-sound can be regarded as evidence that tetanus is made up of a series of fluctua-tions in the density of the muscle. According to Bernstein, the sound heard during contraction occurs in the latent period. Hence, the cause of the muscle-sound is not to be sought in a displacement of the mass of the muscle.

which is stationary during the latent stage, but in molecular processes that are responsible also for the process of negative variation in the current.

FATIGUE OF MUSCLE.

The term *fatigue* is applied to that condition of diminished functional capacity in which the muscle is placed as a result of prolonged activity. This condition is recognized during life by a peculiar sensory perception localized in the muscles. In the intact body the fatigued muscle is capable of *recovery*, as is also the excised muscle to a slight degree. A muscle is more readily fatigued than its motor nerve.

The *cause* of fatigue is the accumulation in the muscular tissue of the products of metabolism, *fatigue-bodies*, that are formed as a result of muscular activity. Among these products are: phosphoric acid, free or combined in acid salts; acid potassium phosphate; glycerin-phosphoric acid(?); and carbon dioxid. The accuracy of the foregoing explanation is indicated by the fact that the fatigued muscle becomes again more capable of activity if the substances named are washed away by the passage of a normal solution (0.6 per cent.) of sodium chlorid or of a weak solution of sodium carbonate through the blood-vessels of the muscle. The consumption of oxygen on the part of the active muscle also promotes fatigue; for the passage of arterial (but not venous) blood through the vessels removes the fatigue by replacing substances that have been consumed by the muscle. Conversely, a muscle that is capable of activity may be rapidly fatigued by the injection of dilute phosphoric acid, acid potassium phosphate, or dissolved meat-extract into its vessels. An animal may be fatigued also by the transfusion of blood from a completely fatigued animal. A muscle fatigued by work absorbs less oxygen while in this condition, and it also generates only a small additional amount of acid and of carbon dioxid. The activity that gave rise to fatigue has thus induced considerable metabolic activity in the muscle.

The fatigued muscle requires a stronger stimulus than the fresh muscle in order to perform the same amount of work, that is, to lift a weight the same distance. The fatigued muscle is no longer able to raise heavy weights; its absolute muscular energy is therefore diminished. If the muscle is loaded with the same weight throughout the experiment, and if the stimulus is a maximal one (strong induced opening shock), then, from one contraction to the other, the height of the lift steadily diminishes by a constant fraction of the shortening. The fatigue-tracing is, thus, a straight line. The more rapidly the contractions follow one another, the more marked is this diminution in the height of the lift, and conversely. The excised muscle becomes fatigued to the point of exhaustion after a certain number of contractions. Under such circumstances it is a matter of indifference whether the stimuli follow one another in rapid or in slow succession. Analogous conditions are also observed in connection with submaximal stimuli.

The fatigued muscle requires, further, a longer period of time for its contraction, which, therefore, takes place more sluggishly. Finally, the period of latent stimulation is also lengthened in a state of fatigue. The fatigued muscle is said to be more extensible.

If the muscle is loaded with a weight so heavy that it cannot be lifted at all when contraction takes place, the muscle, nevertheless, becomes fatigued, and, indeed, in a still higher degree than if it were able to lift the weight. The metabolism and the formation of acid are, thus, greater in a stimulated muscle maintained in an extended position than in one that contracts when stimulated. If a muscle loaded with no weight is made to contract by stimulation, it becomes fatigued but ·gradually. If the muscle is weighted only during the contraction, but not during the extension, it tires more slowly than if it were continuously weighted; as it does also if it is required to lift its weight only in the course of its contraction, instead of raising it at once at the beginning of the contraction. The suspension of weights from a muscle that is continually at rest does not cause fatigue.

If the arteries of a warm-blooded animal are ligated, complete fatigue will result after from 120 to 240 contractions, in from two to four minutes, on irritation of the nerve. Direct irritation of the muscle, however, will still be able to excite an additional series of contractions. The fatigue-tracings in both cases are straight lines.

If the circulation in the muscles of a warm-blooded animal be uninterrupted, the contractions first increase in height, and then diminish, to pursue a straight line on stimulation of the nerve. Accordingly, it is found in persons that have used their muscles to the point of fatigue that the muscles and their nerves respond more actively to galvanic and faradic stimulation in the beginning, but to a steadily diminishing degree in the further course of the work.

Novi has demonstrated with greater detail the course of the contraction to the point of fatigue. According to him, the isolated muscle stimulated to the point of fatigue exhibits several phases in its action. The first phase exhibits a period in which the contractions occur rapidly and increase in size—an indication that the repetition of the stimulus causes an increase in the irritability of the muscle. In the second phase, of longer duration, the rapidity of the contractions is maintained, but their height diminishes—a sign that the irritability of the muscle is now decreasing. The third phase, again shorter, embraces contractions of slower course, the height remaining unchanged. In a fourth phase the contractions become still slower, but again increase in height. Finally the fifth phase exhibits uniform diminution in the height of the contractions and increase in their duration, until exhaustion occurs. Only this last phase corresponds to Kronecker's law.

According to v. Kries a fatigued muscle tetanized in maximum degree behaves like a fresh muscle tetanized in submaximum degree. Both exhibit an incomplete transition from the passive to the active state.

Recovery from the condition of fatigue may be brought about by the passage of a constant galvanic current through the entire length of the muscle, likewise by the injection of fresh arterial blood into its vessels, or of small doses of veratrin.

Relatively small amounts of sugar (30 grams) increase the muscular energy. Cocoa, coffee, tea, and other substances exert a stimulating influence on muscular activity.

Among the poisons, curare and the putrefaction-toxins (ptomains) cause the fatigue-curve to pursue an irregular course.

A. Mosso and Maggiora made observations on living persons, by having a weight lifted by the flexors of the middle finger, with the arm in a fixed position. Mosso found that the muscle tires sooner when stimulated directly than when excited indirectly through its nerve. Only for medium weights is the fatigue-tracing a straight line; for smaller weights it is S-shaped, and for larger ones hyperbolic. If a tetanizing, electrical stimulus be continued until the muscular power is exhausted, there will still be left in the muscle a residue of energy that can be utilized by the will; and, conversely, a muscle finally exhausted by voluntary contractions can still perform some work when impelled by an electrical stimulus. If both forms of excitation be employed in immediate succession, they will exhaust the muscle completely. Mental exertion diminishes the muscular energy in a marked degree, as do likewise hunger and high temperature, especially in conjunction with marked humidity and diminution of atmospheric pressure; also local artificial elevation or diminution of the muscle-temperature. The strongest muscular contraction induced by the will cannot be further increased by strong electrical stimulation of the motor nerve.   On the other hand, if the motor nerve be stimulated so that a less powerful contraction results, the will is unable to strengthen this contraction.   The work performed by a muscle already fatigued is much more exhausting than a greater amount of work performed when it has been rested. Anemia gives rise to symptoms similar to those of fatigue, up to the point of inability to contract; while an abundant supply of blood rapidly refreshes the muscle.   Fatigue of the legs, as after marching, hastens fatigue in the arms.   Long-continued watching and fasting favor fatigue.   Massage exerts a favorable influence on fatigued muscles.

If a muscle be completely exhausted by voluntary movement, and if, nevertheless, the will be allowed to act as if to excite a contraction, the muscle will actually begin to contract again after a period of rest, until it becomes again exhausted, and so on.   Mosso and Brandis assume that involvement of the central nervous system, including the psychic centers, is, in part, to be taken into account in connection with fatigue in man.   If a sensory stimulus be applied at the commencement of a voluntary contraction, the movement will be intensified and accelerated.

Pathological.—In rare cases a morbid increase in the liability to muscular fatigue (myasthenia) has been observed without muscular atrophy or sensory or reflex disturbances.

MECHANISM OF THE BONES AND THEIR ATTACHMENTS.

The *bones* exhibit in their spongy structure an internal architecture, consisting of lamellæ arranged for pressure and traction exactly in accordance with those lines that would be constructed by graphic statics in the representation of the forces in weighted beams of the same form.   This architecture is, therefore, so completely adapted to the function of bone that it combines the greatest capability as a supporting apparatus with the least expenditure of material.

The *joints* are covered with a layer of cartilage, which moderates, by means of its elasticity, the concussions communicated to the bones.   The surface of the articular cartilage is smooth, and thus permits the articular extremities to move freely upon each other.   At the outer boundary of the cartilage arises the capsule of the joint, which encloses the cartilaginous extremities like a sac.   The inner surface of the capsule is lined by synovial membrane, which secretes the viscid, slippery synovial fluid, and this facilitates considerably the free movement of the surfaces.   The outer surface of the capsule of the joint is covered with numerous fibrous bands, which act partly as fortifying and partly as restraining ligaments.   The bony processes also are included among the restraining contrivances, for example the coronoid process of the ulna, which permits the forearm to be flexed only to an acute angle; also the olecranon, which prevents hyperextension at the elbow-joint.   The continuous apposition of the articular surfaces is made possible (1) by the adhesion of the smooth cartilaginous surfaces, covered with synovial fluid and sliding on each other; (2) by the external capsular ligament; and (3) by the elastic tension and the contraction of the muscles.

The articular cavities must be regarded as cleft spaces, bounded by free connective-tissue surfaces, and unprovided with endothelium.   The articular cartilage and also the adjacent connective tissue are bare.   The intima of the synovial membrane does not consist of endothelium, but of protoplasmic cells provided with processes, together with a fibrous interstitial substance.   It is almost everywhere separated from the articular cavity by a thin layer of fibrillar tissue.

The *synovial membrane* is composed of delicate bundles of connective tissue intermixed with elastic fibers; it is provided on its inner surface in part with folds containing fatty tissue and in part with villi containing blood-vessels.   The internal articular ligaments or cartilages are not lined by synovial membrane. The points of attachment of the synovial membrane to the bones are termed insertion-zones.

The colorless, stringy, synovial fluid has an alkaline reaction and the composition of transudates.   In addition, it contains a substance resembling mucin, as well as albumin and traces of globulin, lecithin, cholesterin, fat, soaps, lutein, and also salts.   Excessive movement diminishes its amount and increases its density and also the amount of mucin, but diminishes the amount of salts.

With regard to the mode of movement, joints may be divided into the following classes:

1. *Joints with a Rotatory Movement about One Axis.*—(a) The *hinge-joint* or *ginglymus.* The one articular surface represents a section of a cylinder or cone, about one axis of which the other surface, with a corresponding concavity, moves on flexion or extension at the joint. Examples: the joints of the fingers and the toes. Strong lateral supporting ligaments are always present, to prevent lateral flexion of the joint.

The *screw-hinge joint* is a modification of the hinge-joint. The humero-ulnar articulation belongs to this class. Strictly speaking, simple flexion and extension do not take place at the elbow-joint; but the ulna is rotated on the trochlea of the humerus like a nut on a bolt; on the right humerus the screw is wound to the right, and on the left humerus to the left. The ankle-joint also belongs to this class; the nut is the articular surface of the tibia; the right joint resembles a left-handed screw, the left joint the reverse.

(b) The *pivot-joint* (rotatio), with a cylindrical articular surface; for example, the articulation between the atlas and the odontoid process of the axis, which represents the axis of rotation. The axis of rotation of the articulation at the elbow-joint for pronation and supination extends from the middle of the cotyloid cavity on the head of the radius to the styloid process of the ulna. Accessory joints for this pivot-joint are, above, the articulation between the articular circumference of the head of the radius and the lesser sigmoid cavity of the ulna; and, below, the articulation between the head of the ulna and the sigmoid cavity of the radius.

2. *Joints with a Rotatory Movement about Two Axes.*—(a) The joints exhibit in the two axes, which intersect at right angles, a curvature that is different in degree, but the same in direction: for example, the atlanto-occipital articulation, or the wrist-joint, in which both flexion and extension, as well as lateral inclination, are possible. (b) The joints have a surface of curvature that pursues a different direction in the two axes, which intersect at right angles. To this class belongs the saddle-joint, the surface of which is concave in the direction of the one axis, and convex in that of the other; for example, the articulation between the trapezium and the metacarpal bone of the thumb. The principal movements here are (1) flexion and extension, and (2) abduction and adduction. Further, movement is possible to a limited degree in all other directions, and a cone-shaped space can be circumscribed by the thumb. In this manner the saddle-joint resembles a limited arthrodial joint.

3. *Joints with a Movement on a Spiral Articular Surface (Spiral Joints).*—To this class belongs above all the knee-joint. The condyles of the femur, curved from before backward, exhibit, on sagittal section of their articular surface, a spiral the center of which lies toward the posterior portion of the condyle, and whose radius vector increases from behind downward and forward. The joint permits, first of all, extension and flexion. The strong lateral ligaments on both sides arise from the condyles of the femur, at a point corresponding to the center of the spiral, and are inserted on the head of the fibula and the internal condyle of the tibia respectively. When the knee-joint is strongly flexed, the lateral ligaments are relaxed; they become tense as extension increases, and in complete extension they form tense bands, which ensure lateral fixation of the knee-joint. In accordance with the spiral form of the articular surface, flexion and extension do not occur about one axis, but the axis constantly shifts with the points of contact; the axis traverses a path that likewise is spiral. The greatest flexion and extension cover about 145°. The anterior crucial ligament is made more tense during extension, and acts as a check-ligament for excessive extension; the posterior crucial ligament is made more tense during flexion, and is a check-ligament for excessive flexion. The movements of extension and flexion at the knee are, however, rendered more complex by the screw-like movement of the joint, with the result that the leg deviates outward during extreme extension. Accordingly, the thigh must be rotated outward during flexion, if the leg is fixed. Pronation and supination further are observed in the knee-joint, amounting to 41° in extreme flexion, but being entirely absent in extreme extension. They are due to rotation of the external condyle of the tibia about the internal condyle. In all positions of flexion the crucial ligaments exhibit a fairly uniform degree of tension, as a result of which the articular extremities are held in apposition. It is owing to their arrangement that with increase in the tension of the anterior ligament during extension the condyles of the femur must roll more on the anterior portion of the articular surface of the tibia; while with increase in the tension of the posterior ligament during flexion they must roll more on the posterior

portion. Braune and Fischer found in the course of their investigations that flexion at the knee-joint is attended with rotation of the tibia. The transition from a position of extension to one of flexion of 20° is attended with an internal rotation of 6°. From this point further flexion is attended with an external rotation, which amounts to 6° at a flexion of 90°.

4. *Joints with Rotation about One Fixed Point.*—These are the freely movable ball-and-socket joints (arthrodia). Movement is possible about innumerable axes, all of which intersect at the point of rotation. The one articular surface has an approximately spherical shape, while the other has that of a hollow sphere. The shoulder-joint and the hip-joint are types of this articulation. Instead of the numerous axes, about which movement is possible, three may be substituted, intersecting at right angles in space. Therefore, these joints have also been designated tri-axial. The movements possible are: (1) pendulum-like movements in any desired plane; (2) rotation about the longitudinal axis of the extremity; (3) movements circumscribing the surface of a cone, the apex of which corresponds to the center of rotation of the joint, and whose surface is circumscribed by the extremity itself.

Limited arthrodial joints are ball-and-socket joints with a more limited range of movement, and in which, moreover, rotation about the longitudinal axis is wanting; for example, the metacarpo-phalangeal joints.

5. *Rigid joints (amphiarthrosis)* are characterized by the fact that movement is possible in all directions, but is limited in extent, owing to short and unyielding external articular ligaments. The articular surfaces are usually of the same size, and are almost flat. Examples are afforded by the articulations of the carpal and tarsal bones with one another.

With regard to the mechanical origin of the articular forms of two bones movable upon each other, it is to be noted that the articular extremity to which the muscles are inserted near the joint becomes the acetabulum; while that extremity to which the muscles are inserted at a greater distance becomes the head.

Symphyses, synchondroses, and *syndesmoses* represent the junction of bones without the formation of an articular cavity. They are movable in all directions, but only to an extremely limited extent. Physiologically, they are thus closely related to the amphiarthroses.

Sutures unite bones without permitting any yielding. The physiological significance of sutures resides in the fact that the bones may grow at their margins, so that distention of the cavity enclosed by the bones is possible.

ARRANGEMENT AND FUNCTION OF THE MUSCLES IN THE BODY.

The muscles form 45 per cent. of the total mass of the body. The musculature on the right side of the body is heavier than that on the left. If the muscles are considered with regard to their function from the mechanical standpoint, the following categories may be distinguished:

A. Muscles without Definite Origin and Insertion.

1. The *hollow muscles*, enclosing spherical, oval, or irregular cavities, such as the urinary bladder, the seminal vesicle, the gall-bladder, the uterus, the heart; or forming the walls of more or less cylindrical canals, such as the intestinal tract, the muscular ducts of glands, the ureters, the oviducts, the vasa deferentia, the blood-vessels, and the lymphatics. Under such circumstances the muscle-fibers frequently are arranged in several layers, for example in longitudinal, circular, and oblique directions. During activity all of the layers contract to effect diminution in the size of the enclosed cavity. It is inadmissible to ascribe different individual mechanical effects to the various layers, for example to maintain that the circular fibers of the intestine narrow the tube, while the longitudinal fibers dilate it. Both sets of fibers rather act together in diminishing the enclosed cavity, namely by narrowing and shortening it. If, however, the wall of a hollow organ is pushed or folded inward either

by pressure from without or by partial contraction of a number of circular fibers, muscle-fibers that pass through the valley of the excavation to the surrounding borders may obliterate the depression by partial contraction, thus partially dilating the enclosed cavity, and converting the concave aspect of the depression into a smaller, plane one. The various layers are innervated from the same motor source, a fact that likewise supports the view of their homologous action.

2. The *sphincters* encircle an opening or a short canal, which is either narrowed or firmly closed by their action; for example the iris, orbicularis palpebrarum, orbicularis oris, sphincter pylori, sphincter ani, sphincter vulvæ, sphincter urethræ.

B. Muscles with Definite Origin and Insertion.

1. The origin is completely fixed when the muscle is in action. The course of the muscle-fibers to their insertion is such that during contraction the insertion approaches the origin in a straight line; for example the attollens, attrahens, and retrahens auriculæ, and the rhomboids. In the case of some of these muscles, the insertion is lost in soft structures, which then follow the line of traction; for example the azygos uvulæ, the elevator of the soft palate, most of the facial muscles arising from the bones and inserting into the skin, the styloglossus, stylopharyngeus, and others.

2. *Both Origin and Insertion are Movable.*—Under such circumstances the movements of both points are inversely as the resistances that have to be overcome by the movement. In this connection it should be borne in mind that these resistances can often be voluntarily increased either at the origin or at the insertion. Thus, for example, the sterno-cleido-mastoid may act either as a depressor of the head, or, if the head be fixed, as an elevator of the chest; the pectoralis minor may act either as an adductor and depressor of the shoulder or, if the shoulder be fixed, as an elevator of the third, fourth, and fifth ribs.

3. Some muscles with a fixed origin undergo a deviation from the straight line in the further course of their fibers or their tendons. This may be merely a slight curving, as in the occipito-frontal or the elevator of the upper eyelid; or it may be an angular deflection of the tendon around a firm prominence, so that the muscular traction is made in an entirely different direction, namely as if the muscle acted from this process directly on its insertion. Examples of the latter are the superior oblique muscle of the eyeball, the tensor tympani, tensor veli palatini, obturator internus.

4. Many muscles of the extremities act upon the long bones as upon levers: (a) The muscle may act upon a lever with a single arm, the insertion of the muscle and the weight being situated upon the same side of the point of support, or fulcrum, for example the biceps, the deltoid. The point of application of the force, under such circumstances, is often situated close to the fulcrum. By this means, the rapidity of the movement during contraction of the muscle is greatly increased at the extremity of the arm of the lever; for example, in throwing, the hand may move at a rate exceeding 22 meters a second; but force is lost. This arrangement, however, has the advantage that with lesser contraction of the muscle its force is diminished less than it would be if the contraction were more marked. (b) The muscles may act upon the bones as upon levers with two arms, the point of application of the force (muscular insertion) being situated upon the other side of the fulcrum than the

point of application of the weight; for example the triceps, the muscles of the calf. In both instances the muscular force necessary to overcome a given resistance is calculated according to the laws of the lever. Equilibrium will be established when the static factors—that is the product of the force in its vertical distance from the fulcrum—are equal; or when the force and the weight are inversely proportional to their vertical distances from the fulcrum.

In determining the amount of muscular force and the weight, especial attention should be given to the direction in which these act on the arms of the lever. Thus, it often happens that the direction that was perpendicular to the arm of the lever in a certain position may act obliquely upon it during movement. The static factor of a force or weight acting obliquely on the arm of the lever is obtained by multiplying the force by the perpendicular dropped from the fulcrum upon the line of direction in which the force is acting.

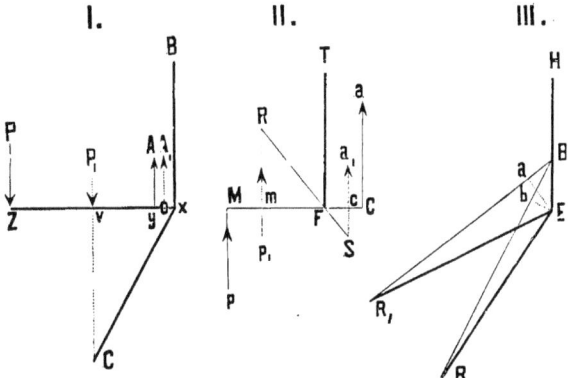

FIG. 199.—Diagrammatic Representation of the Action of Muscles on the Bones.

In Fig. 199, I, B x represents the humerus, and x Z the radius; A y the direction of traction of the biceps. If the biceps acted only in the rectangular position, as in holding horizontally a weight (P) attached to the forearm or the hand, then the force exerted by the biceps (A) could be determined by the formula A . y x = P . x Z; whence A = (P . x Z) : y x. It is evident that in the depressed position of the radius x C, the conditions are different; then the force of the biceps A_1 = (P_1 . v x) : o x.

In Fig. 199, II, T F represents the tibia, F the ankle-joint, M C the foot in the horizontal position. The force (a) of the calf-muscles necessary to neutralize a force p directed from below against the anterior extremity of the foot would be: a = (p . M F) : F C. If the position of the foot is changed to the direction R S, then the force of the calf-muscles a_t = (p_1 . m F) : F c.

From the foregoing the amount of force with which muscles that, like the coraco-brachialis, are stretched over the angle of a hinge-joint, act on the arms of their levers is also evident. Here also the static factor is equal to the force multiplied by the perpendicular dropped from the fulcrum upon the line of direction of the force.

In Fig. 199, III, H E represents the humerus, E the elbow-joint, E R the radius, B R the coraco-brachialis muscle. The factor in this position is A . b E.

If the radius is raised to the position E R_1, the factor is A . a E.   It should, how-ever, be noted here also that B R_1 < B R; hence the absolute muscular energy must be less in the more flexed position, as every muscle is able to lift less weight with increasing contraction.   What is thus lost in energy is made up in elongation of the arm of the lever.

5. Some muscles have a *double motor effect*, which they usually exe-cute combined; for example, the biceps muscle is a flexor and a supi-nator of the forearm.   If one of these movements is prevented by other muscles, the muscle does not participate in the execution of the other movement.

Examples.—If the forearm be strongly pronated and then flexed, the biceps does not participate; or if the elbow be tensely extended, supination is effected by the supinator brevis alone, not by the biceps.   The muscles of mastication furnish another example.   The masseter raises the lower jaw and at the same time pulls it forward.   If the depressed jaw, however, be kept drawn strongly backward, the masseter does not participate in the succeeding elevation of the jaw.   The tem-poral muscle raises the jaw, and at the same time draws it backward.   If the de-pressed jaw be raised when drawn forcibly forward, the temporal does not par-ticipate in its elevation.   The muscles of this group execute this partial movement only on the strongest exertion, or when the position of the bones is specially in-fluenced by other mechanical factors.   The flexors of the leg also exhibit interesting, analogous relations.

A muscle connected with one joint as a rule causes in a neighboring joint a movement opposite to that to which it gives rise in the joint over which it passes. For example, the brachialis anticus causes, in addition to flexion at the elbow-joint, also backward extension at the shoulder-joint.

6. *Diarticular or polyarticular muscles* are those that pass over two or more joints in their course from origin to insertion.   In these muscles the tendons may deviate from a straight line in certain positions, for example the extensors and flexors of the fingers and toes in flexion of the latter; or the direction remains constantly straight, for example the gastroc-nemius.   The muscles of this group exhibit also the following interest-ing conditions: (*a*) The phenomenon of so-called *active insufficiency*. If the origin and insertion of a muscle are too closely approximated as a result of certain positions of the joints over which it passes, it may happen that the muscle is compelled to contract to such a degree before its action becomes effective that further active contraction is not possible beyond the point at which its effect may first become manifest.   For example, when the knee is flexed at an acute angle, the gastrocnemius is no longer able to accomplish plantar flexion of the foot; the traction on the Achil-les tendon is made by the soleus alone.   (*b*) *Passive insufficiency* is exhibited by the polyarticular muscles under the following conditions: In certain positions of the joints a muscle may already be so stretched and made tense as from this position to limit certain movements of other muscles like a rigid restraining band.   For example, the gastroc-nemius is too short to permit complete dorsal flexion of the foot when the knee is extended.   The long flexors of the leg arising from the tuber-osity of the ischium are too short to permit complete extension at the knee-joint when the hip-joint is flexed at an acute angle.   The extensor-tendons of the fingers are too short to permit complete flexion of the joints of the fingers when the wrist-joint is completely flexed.

In the dependent upper extremity movement of the forearm at the elbow-joint is attended with a change in the position of the upper arm. The long head of the biceps tends to rotate the upper arm backward with the elbow-joint in a position between extension and flexion at a right angle; with the elbow-joint in a position of greater flexion, however, the rotation is forward.

A diarticular muscle when sufficiently contracted will move the bone situated between the two joints in the same manner as that on which it is inserted. This associated movement impairs the strength of the principal movement; and, conversely, the latter is strongest when the former is inhibited. The muscles that effect this inhibition have been designated by H. E. Hering *pseudo-antagonists*. They take part involuntarily in every movement, in order to limit the associated movement.

7. *Synergists* is the designation applied to those muscles that, collectively, serve to exercise a certain kind of movement; for example the flexors of the leg, the calf-muscles, and others. Also the abdominal muscles, including the diaphragm, in contracting to diminish the size of the abdominal cavity, as in the act of straining; also the inspiratory and the expiratory muscles may be regarded as synergists. The different heads of a muscle, or the two bellies of a digastric muscle, may also be considered from this point of view.

Antagonists, on the other hand, is the designation applied to those muscles that in contracting have an effect opposite to that of other muscles. Thus, flexors and extensors, pronators and supinators, adductors and abductors, elevators and depressors, sphincters and dilators, inspirators and expirators, are antagonists.

When it is desired to develop the action of a muscle in its full force, it is customary to place it involuntarily first in a state of greatest possible extension, as it is from this condition that the muscle is really capable of developing the greatest amount of force. Conversely, in the execution of delicate movements, requiring the smallest possible amount of force, a position is chosen in which the muscle in question is already contracted to a considerable extent.

All of the fascias of the body are attached to muscles, and are made tense by corresponding movements of the latter (tensors of fasciæ).

GYMNASTIC EXERCISES AND THERAPEUTIC GYMNASTICS. PATHOLOGICAL VARIATIONS IN THE MOTOR FUNCTIONS.

Gymnastic exercises are of great importance in the development of muscular function and of strength, and they should be practised by both sexes from early youth. The systematic activity increases the size of the muscles, and renders them capable of doing more work; in addition, the bodily fat is consumed in greater degree. With the increase in the size of the muscles the amount of blood is increased, and at the same time the bones, tendons, and ligaments are rendered more resistent. As the circulation is greatly increased in active muscle, exercise causes a general improvement in the circulation and in cardiac activity. As a result a favorable influence is exerted on the movement of the fluids of the body in persons especially of sedentary habits, who suffer from stagnation of blood in the abdominal organs (hemorrhoids, etc.). As, further, active muscle consumes a good deal of oxygen and generates much carbon dioxid, respiration is thus actively stimulated by gymnastic exercises. The general increase of metabolism gives rise to the feeling of well-being and of vigor, limits abnormal irritability and the tendency to fatigue. The whole body becomes more solid, firmer, and of heavier specific gravity.

Swedish therapeutic gymnastics are employed to strengthen systematically the muscles in persons suffering from weakness of certain muscles or muscle-groups, and in consequence not infrequently exhibiting deformities in the position of the skeleton. The movements of these muscles are practised especially, being opposed by suitable resistance, which should be overcome by the subject, or be opposed by him without overcoming them.

Kneading, pressing, and stroking the muscles (massage) also promote the circulation of blood through them. These procedures may, therefore, be applied with advantage to muscles that are so enfeebled by disease that independent, systematic training by exercises or gymnastics can no longer be successfully pursued.

Derangement of normal movements may occur in the apparatus concerned in passive movements, namely the bones, joints, ligaments, and aponeuroses, or in apparatus concerned in active movements, namely the muscles with their tendons and motor nerves.

Fractures, caries, and necrosis, and also inflammatory processes, which render movements of the bones painful, impair such movements or even render them wholly impossible. A similar result is caused by dislocations or inflammations of the joints, relaxation of the articular ligaments, or firm adhesions between the articular surfaces (ankylosis) or between the ligaments and soft parts surrounding the joint. Deviations from the normal function may further be caused by abnormal curvatures of the bones, enlargements (hyperostosis), or outgrowths (exostosis). Among the abnormal positions of the skeletal parts that occur frequently are to be included curvature of the spinal column laterally (scoliosis), backward (kyphosis), or forward (lordosis). These also give rise to disturbances of the respiratory movements. In the lower extremities, which have to bear the weight of the body, genu valgum (knock-knee) develops, especially in flabby, tall, young persons engaged in trades requiring much standing. The opposite curvature of the legs, genu varum (bowlegs), is usually the result of rachitic disease. Flat-foot (pes valgus) is due to depression of the arch of the foot, which then no longer rests upon its three normal points of support. This condition is often due to the same causes as genu valgum. The ligaments of the small tarsal joints are stretched, and the longitudinal axis of the foot is usually directed outward in marked degree. The inner border of the foot is brought closer to the ground. Pains in the foot and the malleoli render walking and standing difficult. Club-foot (pes varus) is the condition in which the inner border of the foot is raised, and the point of the foot is turned upward and inward; it is caused by a fetal arrest of development. All children are born with a slight degree of this position. Pointed toe (pes equinus) is the condition in which the point of the foot touches the ground; pes calcaneus, that in which the heel touches the ground. Both are usually dependent upon contracture of the muscles causing these positions, or upon paralysis of their antagonists.

Persistent absence of earthy salts from the food results in a deficiency of these in the skeleton; the bones become thin, transparent, and even flexible. Rickets in children and the identical lameness in young domestic animals are caused by the fact that the calcium-salts of the food cannot be absorbed, on account of persistent disturbances of digestion. Analogous disturbances of the motor functions develop if the fully developed bones subsequently lose their calcium-salts to the extent of one-third or one-half (halisteresis), and thus become brittle and soft—osteomalacia. A certain minor degree of fragility of the bones and halisteresis occurs in old age.

With regard to pathological alterations in the muscles, it should first be pointed out that the normal nutrition of muscular tissue can only take place if a sufficient supply of sodium chlorid and of potassium-salts is provided in the food, as these are integral constituents of muscular tissue. Otherwise, the muscles atrophy, and their reconstruction is prevented. Under such conditions, further, the central nervous system and the digestive apparatus also suffer, and the animals perish. The extent to which the muscles suffer in a state of inanition is described on page 440. Muscles and bones that for any reason are thrown out of function also undergo atrophy. In the atrophic muscles associated with ankylosis there is often found an enormous proliferation of the muscle-corpuscles, occurring as an "atrophic proliferation" at the expense of the contractile substance. A certain degree of muscular atrophy takes place normally in old age.

The great reduction (from 1000 to 350 grams) in the muscular structure of the uterus after parturition is especially noteworthy. This is due in part to the diminished vascularization of the organ. In cases of lead-poisoning the extensors and interossei especially undergo atrophy. Atrophy and degeneration of the muscles give rise to secondary shortening and thinning of the bones to which they are attached.

Section and paralysis of the motor nerves cause inactivity and finally degeneration of the muscles. Inflammation, softening, and sclerosis of the ganglion-cells in the anterior horns or in the motor nuclei of the medulla oblongata also give rise to atrophy of the muscles connected with them. Spinal paralysis and acute bulbar palsy (paralysis of the medulla oblongata) thus have an acute onset, while progressive muscular atrophy and progressive bulbar paralysis pursue a chronic course. Under these conditions the muscles and their nerves become thin and wasted. The muscles exhibit many nuclei, their contractile substance is partly in a state of fatty degeneration, and later disappears altogether. The intramuscular connective tissue is increased, often also the interstitial fat. According to Charcot, the central nerve-cells are also the trophic centers for the nerves arising from them and for the related muscles. According to Friedreich, however, progressive muscular atrophy is a primary disease of the muscles, a primary interstitial myositis resulting in atrophy and degeneration, the central nervous system becoming involved in the degenerative processes only secondarily; just as after amputation of an extremity corresponding parts of the spinal cord degenerate secondarily.

Finally, mention should be made of pseudo-hypertrophy or lipomatous muscular atrophy, in which the muscle-fibers are completely atrophied, in association with an abundant development of fat between the fibers, without, however, degeneration of the nerves or the spinal cord. The muscular substance may also undergo amyloid degeneration, the amyloid substance penetrating and infiltrating the tissue. At times atrophic muscles exhibit a deep brownish-red color, probably due to alteration of the muscle-pigment. Muscles constantly compelled to perform a large amount of work, such as the heart-muscle, the bladder, the intestine, undergo hypertrophy. If the mechanism of the skeleton becomes altered, for example as a result of rigidity of a number of joints, the muscles adapt themselves more or less completely to the altered mechanical conditions by changes in their growth, expenditure of energy, and manner of movement.

SPECIAL MOVEMENTS.

STANDING.

Standing is the vertical position of equilibrium of the body, secured by muscular action, in which the line of gravitation—that is a perpendicular dropped from the center of gravity of the body—strikes the ground within the supporting area of the soles of both feet. Of the various positions, that of "standing erect" will be analyzed here. In this position, muscular activity is exercised in two directions: (1) to fix the articulated body into an inflexible column (to "stiffen"); and (2) in case of a variation of the equilibrium to neutralize the disturbance by suitable muscular contractions.

The following muscular activities are observed in standing:

1. Fixation of the head on the vertebral column. The occiput may move in various directions on the atlas, whose two concave articular surfaces converge anteriorly. The act of nodding is the most readily performed. As the center of gravity of the head lies in front of the supporting points on the atlas, relaxation of the muscles, as in sleep or in death, causes the chin to fall upon the chest. The strong muscles of the neck, which pull from the spinal column upon the occiput, fix the head on the vertebral column.

In addition to the nodding movement directly forward, a similar movement is also possible obliquely forward and to the side. Rotation of the head in the articulations of the atlas is possible only to an inappreciable extent around the sagittal axis, likewise around the vertical axis, the latter occurring only when the head is flexed. No special muscular activity is necessary to prevent these movements in standing. When the head is rotated to the side, the contralateral vertebral artery is compressed in the vertebral sulcus, while that on the same side is enabled to carry more blood.

The chief rotatory movement of the head occurs about the vertical axis of

the odontoid process of the axis. The articular surfaces on the pedicles of the
first and second vertebræ are convex toward each other in the middle, becoming
somewhat lower anteriorly and posteriorly. The head is, therefore, highest in
the erect position; if it is rotated on the odontoid process, it undergoes a slight
spiral movement downward. In this way distortion of the medulla is avoided
when the head is strongly rotated. In standing, no muscular action is required
to fix these vertebræ, as rotation cannot occur when the muscles of the neck and
the flexors and extensors of the head are at rest.

2. The vertebral column requires fixation by muscles in those sections where its mobility is the greatest; these are the cervical and lumbar
regions. Here fixation is secured by the numerous and strong muscles of
the cervical vertebræ, especially those of the neck, and the lumbar muscles, especially the strong origins of the extensor dorsi communis, supported by the quadratus lumborum.

The least movable vertebræ are those from the third to the sixth dorsal; the
sacrum is completely immovable. For a definite length of the column the mobility depends upon the following factors: (a) The number and the thickness of
the elastic intervertebral discs. These are most numerous in the cervical region,
and are thickest in the lumbar region and relatively also in the lower cervical
region. They permit movement in every direction. The intervertebral discs
together form one-fourth the entire length of the spinal column. They are compressed somewhat by the weight of the body; hence, the body is longest in the
morning and after recumbency of some duration. The smaller circumference of
the bodies of the cervical vertebræ must be more favorable for their movement
on the discs than is the greater size of the lower vertebræ. (b) The position of
the processes also materially influences the mobility. The greatly depressed
spines of the dorsal vertebræ prevent hyperextension. The articular processes
of the cervical vertebræ are so situated that their surfaces are directed obliquely
from before and above backward and downward. By this means relatively free
movement is rendered possible in rotation, lateral inclination, and flexion. In
the dorsal region the articular surfaces of the superior articular processes are
directed vertically and directly forward, while those of the inferior articular
processes are directed directly backward; in the lumbar region the corresponding
position is almost vertical and sagittal. In the act of bending backward as far
as possible, the most movable points of the spinal column are the lower cervical
vertebræ, from the eleventh dorsal to the second lumbar vertebra, and the two
lower lumbar vertebræ.

3. The center of gravity of the part of the body thus stiffened (the
head and the trunk with the arms) is situated on the anterior border of
the inferior surface of the eleventh dorsal vertebra. The perpendicular
line dropped from the center of gravity passes behind a line joining both
hip-joints. Hence, the trunk would fall backward at the hip-joints; but
this is prevented by the ilio-femoral ligament, 14 mm. thick, stretched
between the anterior inferior spine and the anterior intertrochanteric
line, and also by the anterior tense layer of the fascia lata. As ligaments
alone are never able to withstand continuous traction, they are materially supported by the ilio-psoas muscle, which is inserted on the lesser
trochanter, and also in part by the rectus femoris, whose origin extends
upward over the acetabulum to the anterior inferior spine. A lateral
movement of the hip-joint, in which one thigh would be abducted and
the other adducted, is prevented especially by the large mass of the
gluteal muscles, which fix the thigh on the pelvis posteriorly and laterally. When the thigh is extended, the ilio-femoral ligament also is
able to prevent adduction, aided by the tense fascia lata.

4. The part of the body that has thus far been made rigid, including
the head and the trunk, with the arms and the thighs, and whose center of
gravity is situated somewhat lower and only to such a slight degree further

forward that the line of gravity passes through the line connecting the posterior borders of the knee-joints, must now be fixed at the knee-joints. Falling backward is prevented by the strength of the quadriceps femoris, supported by the tension of the fascia lata. Indirectly, the ilio-femoral ligament is believed also to aid in preventing falling backward, because in this act the thigh must be rotated outward, and this is prevented by the tension of the ligament named in the upright position. Lateral flexion at the knee-joint is impossible on account of the arrangement of the hinge-joint, strengthened by the strong lateral ligaments of the knee. Rotation at the knee-joint is impossible in the extended position.

5. The center of gravity of the entire body is situated 4.5 cm. in a vertical line below the promontory of the sacrum. A perpendicular dropped from this point strikes the ground a little in front of the line connecting both ankle-joints. The body would, therefore, fall forward at the latter joints. This is prevented by the muscles of the calf, aided by the muscles of the deep layer, namely the tibialis posticus, the flexors of the toes, and the peroneus longus and brevis.

The following additional factors have also been considered worthy of mention: (a) As the longitudinal axes of the feet form an angle of 50° at the heels, falling forward can take place only if the feet have taken a position more nearly parallel to their longitudinal axes. (b) Falling forward is opposed also by the form of the articular surfaces of the foot, as under such circumstances the anterior, broader part of the astragalus would have to be pressed between the two condyles. This last factor is actually of little importance, as falling forward does not require such a marked change of position as would be necessary to bring this mechanism into play.

6: The tarsal and metatarsal bones, united by tense ligaments, form the arch of the foot. This touches the ground at three points, the tuberosity of the os calcis, the head of the first metatarsal, and the head of the fifth metatarsal bone. Between the last two points, however, the heads of the other metatarsal bones also form points of support. The weight of the body falls upon the highest point of the arch, the head of the astragalus. The arch of the foot is maintained only by ligaments. The toes are able materially to aid in balancing the body by means of their muscle-play. Standing erect causes more fatigue than walking.

Braune and Fischer have recently distinguished the following varieties of station, for which, in contradistinction from the foregoing older exposition, a different form of muscular activity is required. (1) The "normal position" is characterized by the fact that the line of gravity passes downward through the lines connecting the central points of both hip-joints, knee-joints, and ankle-joints, and passes upward through the centers of gravity of the trunk and the head. Accordingly, the body need only be stiffened; no muscular activity at all is required to prevent falling forward or backward. (2) In the "comfortable position" the line of gravity strikes the ground in front of the line connecting the centers of both ankle-joints at a point corresponding approximately to the anterior border of the ankle-joint. Hence, muscular action is necessary to prevent falling forward at the ankle-joints. (3) In the "military position" the line of gravity falls in front of the knee-joints and ankle-joints, striking the ground at a point corresponding approximately to the middle of the sole. Hence, falling forward must be prevented at both joints, and this induces great fatigue on account of the considerable and continuous muscular exertion.

The position of the center of gravity in the living person is determined as follows: The body is placed on a narrow board the length of the body. A balancing edge is placed beneath the board, and first the upper and lower halves are balanced, then the right and left halves. Finally, the body is balanced when standing upright on a small board. The center of gravity is situated at the intersection of the three planes, passing in each instance at right angles to and along the balancing edge. The center of gravity of individual parts of the body may be determined in a similar manner on sections of a frozen cadaver.

Pathological.—The security of firm station is recognized from the swaying of the body, which may be easily registered with the aid of a small rod placed vertically on the top of the head, the swaying being recorded by means of a pen or a brush on a surface stretched horizontally above the head. Disturbances of sensation, such as occurs in tabes and the like, cause marked swaying; as do also muscular weakness, tremor, fatigue, coldness of the feet, the action of anesthetics on the soles of the feet.

SITTING.

Sitting is the position of equilibrium in which the body is supported on the tuberosities of the ischia, on which a to-and-fro rocking movement can take place, as upon the rockers of a rocking-horse. The head and the trunk together are made rigid so as to form an immovable column, as in standing. The essential purpose of sitting is to place the lower extremities out of service from time to time, in order that their muscles may recover from fatigue. The following varieties of the sitting posture have been distinguished: 1. The *forward sitting posture*, in which the line of gravity passes in front of the tuberosities. In this position the body is supported either against a firm object, for example by means of the arms on a table, or on the upper surface of the thigh, which is either held horizontally or is flexed to an acute angle at the hip by a support placed under the feet. 2. The *backward sitting posture* is characterized by the passage of the line of gravity behind the tuberosities. Falling backward is prevented under such circumstances by the back of a chair (if the latter extends upward as far as the head the neck-muscles also may undergo relaxation during rest), or by the counter-weight of the legs, kept extended by muscular action. In the latter event the sacrum may serve as a further point of support, while the trunk is fixed on the thigh by the ilio-psoas and the rectus femoris, and the leg is kept extended by the extensor quadriceps. Usually the center of gravity is so situated that the heels form additional points of support. This latter sitting posture is naturally not adapted for resting the muscles of the lower extremities. 3. In the *median sitting posture* (sitting erect) the line of gravity passes between the tuberosities. The muscles of the lower extremities are relaxed; the rigid trunk requires only slight muscular action to balance it, falling backward being prevented by the ilio-psoas and the rectus femoris, and falling forward by the lumbar portion of the strong dorsal muscles. Usually, the balancing of the head is sufficient to maintain equilibrium.

WALKING, RUNNING, JUMPING.

By walking is understood horizontal progression effected with the least possible muscular exertion by alternate activity of the two legs.

Method.—The brothers William and Edward Weber, in 1836, analyzed the various positions of the body during the movements of walking, running, and jumping, and recorded these positions in continuous series, which thus represent a true picture of all the successive phases of locomotion. Marey, in 1872, determined the time-relations attending change of position by connecting the motor organs in man and animals with apparatus that registered by means of air-transference. He also further developed Weber's original idea, and has recorded the various phases of movement in walking, running, and jumping, and in moving animals by means of complete series of instantaneous photographs taken by a camera working on the principle of the revolver. The duration of exposure in each instantaneous photograph equals $\frac{1}{1000\pi}$ of a second. When placed in a stroboscope, these series reproduce the natural movements; and by projection with the aid of a kinematograph they may also be shown as "moving pictures." Figs. 201, 202, and 203 represent such series of instantaneous photographs obtained in the manner described. Braune and O. Fischer, between 1895 and 1899, introduced a new method of recording the motor process in walking by means of bilateral chronophotographic exposures on an extensive coördinating system.

In the act of walking the legs are alternatively active. While one, the "supporting" or "active" leg, carries the body, the other, the "hanging," "swinging," or "passive" leg, is inactive. Thus, each leg in regular alternation goes through an active and a passive phase. The motion of walking may be divided into the following acts:

First Act (Fig. 200, 2).—The active leg is vertical, slightly flexed at the knee, and supports alone the center of gravity of the body. The passive leg is fully extended, and touches the ground only with the tip of the great toe (z). This position of the legs corresponds to a right-angle triangle, in which the active leg and the ground form the two sides (catheti), and the passive leg the hypothenuse.

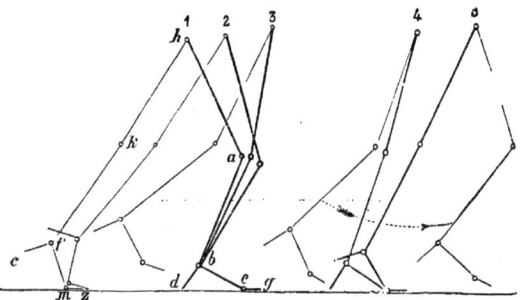

FIG. 200.—Phases of the Movement of Walking. The thick lines represent the active, the thin lines the passive leg: *h*, hip-joint; *k a*, knee-joint; *f b*, ankle-joint; *c d*, heel; *m e*, ball of the metatarso-phalangeal joint; *z g*, tip of the great toe.

Second Act.—To advance the trunk, the active leg tilts from its vertical position (cathetus) into an oblique position (3) inclined forward (hypothenuse). In order that the trunk may remain at the same height, it is necessary for the active leg to be lengthened. This is accomplished first by complete extension of the knee (3, 4, 5), then by elevation of the heel from the ground (4, 5), so that the foot rests on the ball formed by the heads of the metatarsal bone, and finally by elevation of the foot on the joint of the great toe (2, thin line). As both sections of the foot are successively raised from the ground, like the links of a measuring chain that is lifted from the ground ("unwound"), the elevation of the foot from the ground has also been termed "unwinding" of the foot. During the extension and forward inclination of the active leg the tips of the toes of the passive leg have been compelled to leave the ground (3).

While this leg now becomes slightly flexed at the knee for the purpose of shortening, it executes at the same time a "pendulum-like" movement (4, 5), by means of which its foot is moved just as far in front of the active foot as it was previously behind the latter.

When it attains this position, the foot is placed flat upon the ground
38

(1, 2, thick line).   The center of gravity is transferred to this, the hence-forth active leg, which at the same time assumes a vertical position, somewhat flexed at the knee.   The first act is now begun again.

In walking, the trunk also exhibits some characteristic secondary movements: (1) It inclines each time toward the active leg, as a result of traction of the glu-teal muscles and the tensor vaginæ femoris, with the object of transferring the center of gravity.   In heavy, short persons with broad pelves this produces the "waddling" gait.   (2) In order to overcome the resistance of the air, especially

FIG. 201.—Slow Walking, Photographed in Instantaneous Pictures (after Marey).   Only the side direct toward the observer is represented.   From the vertical position of the right active leg (*I*) the entire phase of the movement of this leg follows in six pictures (from *I* to *VI*); after *VI* the vertical position is again reached. The Arabic numerals denote the simultaneous corresponding positions of the left leg, thus 1 = *I*, 2 = *II*, etc., so that, for example, during position *IV* of the right leg the left leg at the same time has the position *I*.

FIG. 202.—Instantaneous Photographs of a Runner (after Marey).   Ten pictures in a second; the base line repre-sents the distance traversed in meters.

in rapid walking, the trunk is balanced at a forward inclination.   (3) During the "pendulum"-motion the trunk rotates slightly about the head of the active femur.   This rotation is compensated, especially in rapid walking, by the arm on the same side as the oscillating leg swinging in the opposite direction; while that on the other side at the same time swings in the same direction as the oscil-lating leg.

O. Fischer has accurately determined the movement of the center of gravity of the body.   The external forces to be considered are the weight, the resistance of the ground, the friction on the latter, and the resistance of the air.

The time-relations of walking are influenced by the following conditions: (1) The duration of the step.   As the rapidity of the pendulum-motion depends upon the length of the leg, it is evident that each individual, in accordance with the length of his leg, has a certain natural time of oscillation, which especially influences his accustomed rate of walking. In addition, however, the duration of the step depends upon the length of time during which both feet touch the ground simultaneously. Naturally, this can be increased voluntarily. With a "rapid pace" the period of time is zero; that is, at the same moment that the active leg is placed on the ground the passive leg is raised. (2) The length (or stretch) of the step, which amounts to six or seven decimeters on the average, must be the greater, the more the length of the hypothenuse of the passive leg exceeds the cathetus of the active leg. Hence, in the longest steps the active leg is markedly shortened by flexion at the knee, so that the trunk is carried at a lower level. Similarly, long legs are especially able to make greater steps.

According to Marey, Carlet, and H. Vierordt the pendulum-movement of the passive leg cannot be regarded as a true pendulum-oscillation, because it possesses a more nearly uniform rapidity, owing to muscular action. During the pendulum-movement of the whole limb, the leg oscillates independently at the knee-joint, as is especially evident in women. According to Ed. and Wm. Weber the head of the femur of the passive leg is held in the acetabulum chiefly by air-pressure, so that no muscular activity is necessary to carry the whole extremity. If all the muscles and the joint-capsule be divided, the head still remains attached to the acetabulum. By pulling on the thigh the borders of the cartilaginous rim of the acetabulum are closely applied in a valve-like manner to the margin

FIG. 203.—Instantaneous Photographs of a High Jump (after Marey). The pictures partly overlap as soon as the velocity of the forward movement diminishes on the descent after the jump. In the upper, left-hand corner is a dial, the white radius of which has moved forward one division in one-twelfth of a second. The base line represents the distance traversed, in meters.

of the cartilage on the head of the femur. According to the statements of the brothers Weber, the thigh is released from the acetabulum as soon as air is allowed to penetrate the articular cavity by perforating the bottom of the socket.

The brothers Weber showed that in walking on level ground an appreciable amount of mechanical work is performed, as the weight of the body must be lifted several centimeters with every step. Marey and Demery estimated that the work performed by a person weighing 64 kilos, when walking slowly, is equal to six kilogrammeters in a second; when running rapidly, it amounts to 56 kilogrammeters. The performance consists in raising the whole body and extremities, in imparting rapidity of motion to them, and in maintaining the center of gravity. According to Rziha the work performed in each second in walking slowly is 3.5 kilogrammeters, in walking at a medium gait 5.46, in walking rapidly 7.87, in a short run 21.87, in a brisk run 42.87, and in a fast run 87.50 kilogrammeters.

A bicycle-rider going at the rate of two meters in a second, performs 1.12 kilogrammeters, at a four-meter pace 4.51 kilogrammeters, at a five-meter pace 7.05, and at a six-meter pace 10.15 kilogrammeters. The normal capability of a bicycle-rider is three and one-half minutes for each kilometer, or a rate of 4.73 meters a second, with a daily capability of from 90 to 100 kilometers. The normal capability of a workman is in this connection assumed by comparison to be 6.3 kilogrammeters a second. A bicycle-rider, going at an average rate, traverses the same distance in half the time and with half the expenditure of energy that a pedestrian requires. With the same metabolic consumption of muscular tissue, the exertion and the degree of fatigue are greater in walking than in cycling. In long-continued cycling, likewise in long marches, there is an increase in the consumption of energy for the successive units of distance covered; at a moderate pace this increase amounts to about 20 per cent.

The pressure on the ground in walking is distributed in the following manner: The supporting leg always presses more firmly on the ground than the other; the longer the step the stronger the pressure. The heel attains the maximum pressure more rapidly than the point of the foot.

The length of the step varies not inconsiderably even when a voluntary attempt is made to have the steps of equal length; as do also the degree of spreading of the legs and the duration of the various phases of walking.

Running (Fig. 202) differs from rapid walking in the fact that a moment exists in which both legs are off the ground, so that the body hovers in the air. The active leg, in being forcibly extended from a more flexed position, must each time give the body the necessary impetus.

In *jumping* (Fig. 203) the body is suddenly raised by the most rapid and powerful contraction possible of the muscles in the lower extremities, care being taken at the same time to maintain the equilibrium by appropriate muscular action.

Pathological.—Variations in the walking movements depend primarily upon diseases of the bones, joints, ligaments, muscles. and tendons. Then the motor nerves must be taken into consideration, irritation and paralysis of which give rise to disturbances of the normal movements. The extent to which the sensory nerves and the reflex apparatus in the spinal cord influence the gait is pointed out on pages 716 and 728.

H. Vierordt has applied the graphic method to the analysis of pathological varieties of gait. Among these are, for example, the spastic, the oscillating or zig-zag gait, the gait of tabes and that of paralysis agitans. Abasia and astasia are the terms applied by Blocq in 1888 to the inability to walk and stand, arising from cerebral affections (hysteria, hypochondria, violent emotions, imperative conceptions, vertigo), while all other movements, even those of the legs, can be executed with full force and coördination.

COMPARATIVE STUDY OF MOTION.

The *absolute muscular energy* in animals is not, generally speaking, appreciably different from that of man. The greater exhibitions of force encountered in the animal kingdom arise from the thickness and number of the muscles, as well as from differences in the arrangement of their leverage or in the means for the transference of force. Thus, for example, insects are capable of exerting a great amount of force; some of them being able to drag 67 times their own weight, while a horse can scarcely drag its own weight. While further, for example, a man, by pressure on a dynamometer with one hand, overcomes a weight equal to 0.70 time his own body-weight, a dog by lifting his lower jaw can overcome a weight 8.3 times that of his body; a crab by closing its claw overcomes 28.5 times its weight; a mussel in closing its shell, 382 times its body-weight.

Standing is made easier in quadrupeds by reason of the much greater supporting surface; the springing animals assume, besides, more of a sitting position, and often use the tail as an additional support (kangaroo, squirrel), Birds possess a mechanical arrangement by means of which, in perching, their toes are flexed; in this way they are able to retain their grasp on twigs when asleep. In the stork and the crane, prolonged standing on one leg is made easy by the fact that no muscular action is required to render the leg rigid; fixation is secured by a process of the tibia fitting into a depression on the articular surface of the femur.

In *walking*, a gait can be distinguished in quadrupeds; the four feet are moved at different times, and always diagonally one after the other; for example, in the horse, right fore, left hind; left fore, right hind. In trotting there is an acceleration of this gait, so that the legs are moved together diagonally at two different times, while the body is at the same time raised higher. In the interval between both hoof-beats the body is in the air half the time in ordinary trotting, longer in an extended trot.

The *gallop:* When a (right) galloping horse moves horizontally through the air, the left hind foot comes down first. Shortly afterward the left fore foot and the right hind foot come down simultaneously; the right fore foot has not yet reached the ground, and is directed far forward. Up to this point the body has maintained its horizontal position. When, however, a few moments later, the left hind foot leaves the ground, it is at a higher level than the fore foot; at the same time, the right fore foot is also brought down and placed far forward; the right hind leg and the left fore leg are in extreme extension. At the next moment these limbs also leave the ground, and the hind foot acquires such an ascendency over the fore foot that it comes to be situated much higher than the latter. The body, therefore, is thrown forward and downward until the right fore leg, which alone still touches the ground, contracts actively, and pushes the body forcibly from the ground. When this has occurred, the horse again soars in air with the body directed horizontally. In galloping the longitudinal axis of the horse's body is placed obliquely to the direction of the movement, forming an acute angle. In an *extended gallop* (carrière), which is really a continuous jumping motion, the right hind leg and the left fore leg, for example, do not reach the ground simultaneously, the former striking first. The rapidity of this movement in the horse is 8$2\frac{1}{2}$ feet a second. Most beasts of prey, hares, etc., employ only the carrière for rapid movements.

The *amble* is a modification of the gait that is peculiar to many animals, for example the camel, the giraffe, the elephant. It occurs also in dogs and in horses, but it is not a favorite gait with the latter. It consists in advancing both feet on the same side simultaneously or almost so.

Marey fastened compressible ampullæ under the hoofs of the horse, connecting them with registering apparatus; and thus accurately recorded the time-relations of the various gaits. Muybridge, in 1872, was the first to obtain series of instantaneous photographs of running horses, which Schmidt-Mülheim placed together in the stroboscope.

In snakes the progression of the body is secured by elevation and depression of the ribs in a manner resembling rowing.

Swimming is an acquired art on the part of man. The specific gravity of the whole body is, on an average, somewhat higher than that of river-water, though somewhat lower than that of sea-water. In the quiet dorsal decubitus, with only the mouth and the nose above the surface of the water, sinking can be prevented by slight downward pressing movements of the hands; sometimes no movement at all may be necessary. In this position progression may be accomplished by simple extension and adduction of the legs. The movement may be accelerated by oar-like strokes with the arms. Swimming on the abdomen is more difficult, because the head, being held above water, increases the specific weight of the body. The body is advanced and held above water by movements divided into the following three phases: First phase, horizontal rowing movement of the extended arms from before backward to the horizontal position (forward movement); second phase, downward pressure of the arms toward the depth, with subsequent adduction of the elbows to the body (elevation of the body), together with a drawing up of the extended legs; third phase, forward thrust of the arms, in contact with each other, and at the same time extension and adduction of the legs obliquely backward and toward the depth, as a result of which both elevation of the body and forward progression are effected. Unduly rapid movements are exhausting and defeat their own purpose. Special attention should be paid to suitable respiratory movements.

Many land mammals, whose bodies are specifically lighter than water, move through it with a walking motion, especially of the hind legs; at the same time the feet, being directed downward, assure the normal position of the body, as they are specifically the heaviest parts of the body. Those mammals that live much in the water, as well as reptiles and amphibia, possess webbed feet and a propelling tail partly resembling that of fish. Whales resemble fish in the external appearance of their bodies.

Fish primarily make use of their tail as a motor organ, which is moved by the powerful lateral muscles. Usually the caudal fin is bent in two opposite directions above and below; in slight movements it is bent only in one direction. By sudden extension of the tail, the fish exerts a pressure against the water, and thrusts itself forward. Many fish, such as the salmon, can thus hurl themselves up out of the water. The dorsal and anal fins maintain the vertical position. The pectoral and abdominal fins, corresponding to the extremities, effect the smaller movements, especially upward and downward; during sleep the abdominal fins are spread out. Most fish possess a swimming-bladder. This is wanting, however, in many cartilaginei (cyclostomi), or is rudimentary, as in the shark. It either opens into the alimentary tract through the air-passage, or the latter is only a temporary structure that is later obliterated. The swimming-bladder is, in part, to be regarded as a respiratory organ with afferent and efferent vessels, while in part it serves for hydrostatic purposes. In the dipnoi the bladder is transformed into a lung. The body of swimming birds has a much lighter specific gravity than has water, while their feathers are lubricated by the coccygeal glands. They propel themselves forward with their webbed feet.

Flying, in mammals, is confined to the bat and its allied species. The bones of the upper extremities, including the phalanges, are greatly lengthened. Between the latter, as well as the hind limbs (except the feet), is stretched a thin membrane, which also partially includes the tail. The flying movement of this membrane is effected by the powerful pectoral muscles, which arise in part from a ridge-like elevation of the sternum and the strong clavicles. The so-called flying lemurs, squirrels, and opossums have merely a duplication of the skin, stretched laterally between the larger bones of the extremities, and serving as a parachute in jumping.

Man is unable to imitate flying movements successfully, for even though he were able to construct artificial wings, he would still lack the strength of the pectoral muscles that is necessary to effect elevation of the body.

In birds the body specifically is exceedingly light. Large air-sacs extend from the lungs into the thoracic and abdominal cavities; even the bones are connected with the lungs by special canals, so that all the spaces in the bones of the cranium, spinal column, bill, and extremities are filled with air instead of marrow. The upper extremities, transformed into wings, are supported by the powerful coracoid bone and the clavicles (furcula), the latter being fused in the middle. The wings are operated by the powerful pectoral muscles, which arise from the large crest of the sternum.

In flying upward the wings are half closed, and are moved with the anterior border directed obliquely forward and upward. The plane of the wings, without offering resistance to the air, follows in the same direction as the edge of the wings. Then the latter are spread out in a large arc downward and backward, with their surfaces pressed downward. While the under surfaces of the wings press against the air from above and forward, downward and backward, the bird moves forward and upward. Birds can rise only against the wind, partly because the wind striking horizontally against their backs would press them down, and partly because it would disarrange their feathers. By means of a revolving photographic camera, arranged in an apparatus resembling a musket, Marey obtained complete series of pictures of flying birds at which he directed the apparatus.

Among invertebrates, all insects possess six legs. In addition some of them (butterflies, bees) have two pairs of wings on the second and third thoracic segments. In beetles and earwigs the first pair is merely a covering; in the strepsiptera it is entirely rudimentary. Conversely, in the flies the second pair of wings is reduced to small swinging bulbs. Lice, fleas, and bedbugs have no wings at all. All spiders have eight legs, the moths having six in their youth. In the centipedes the first three body-rings carry each one pair of legs, while all the rest have either one or two pairs. The crustaceans also possess numerous feet, as a rule, some of them undergoing peculiar transformations, for example in the river-crawfish into mandibles, claws, ambulatory feet, abdominal swimming feet and fin-foot. In the arthropods all of the muscles are inserted on the inner surface of the chitinous covering. The muscles themselves are highly developed and capable of a great amount of energy and rapidity of movement.

Molluscs lack internal supporting organs, while external ones (shells) of simpler construction are present. The muscles, which are partly striated, form a musculocutaneous tube about the body that causes the changes in the form of the body. In mussels the strong single or double sphincter-muscle of the shells is noteworthy. In the pecten (scallops) this muscle effects a springing movement in the water by rapidly bringing the shells together. The molluscs provided with shells possess strong retractors.

In the worms likewise the integument forms with the muscles a musculo-cutaneous tube. The unstriated muscle-fibers pass either longitudinally only (round-worms), or longitudinally and transversely (scratching worms), or finally longitudinally, transversely and vertically through the body (flat-worms). Some worms possess muscular suckers, and others one or two pairs of motile stump-like feet. In round-worms the epidermal cells, and in some bristle-worms the intestinal epithelium. pass directly over into muscle-cells, both together being called "epithelio-muscular cells." In the echinoderms also the muscles are united with the integument; in the holothurians there is an external, continuous layer of circular fibers, beneath which is a longitudinal musculature, arranged in five separate bands.

In the star-fish and the hair-stars special muscles move the limbs of the radiating parts of the body. The sea-urchin, surrounded by a firm lime-capsule, has special muscles that move its spines, and by means of which it is capable of locomotion. The ambulacral feet also aid in locomotion.

In the celenterates the muscle-fibers are transformed sections of epithelial cells. Hence, there are present "epithelio-muscular cells," which are striated in the medusa, and unstriated in the anemone and hydroid polyp. The free epithelial part may be provided with cilia. In the medusa these elements lie partly on the umbrella and partly on the tentacles. Among the polyps, the actinia have a strong muscular base, and, in addition, longitudinal and circular fibers on the body and on the tentacles. In some polyps muscles also accompany the gastro-vascular apparatus.

Among the protozoa, striated muscle-fibers have been found in some infusoria, for example in the pedicle of the vorticella; while, in addition, the movements are executed by the movable protoplasm of the body, or by voluntarily motile cilia.

VOICE AND SPEECH.

SCOPE OF THE VOICE. PRELIMINARY PHYSICAL CONSIDERATIONS CONCERNING THE PRODUCTION OF SOUND IN REED-APPARATUS.

The current of expired air, and under certain circumstances also that of inspired air, can be employed to throw the tense true vocal bands of the larynx into regular vibration, as a result of which a sound is produced. This is termed the human voice.

The true vocal bands of the larynx are elastic, "membranous reeds." By "reeds" are meant elastic plates that almost completely fill the space (frame) in which they are spread out, leaving, however, a small space for their movement. If air be blown against the reeds from a tube below them (air-tube), they will yield at the moment that the tension of the air overcomes the elastic tension of the reeds. In this way a considerable quantity of air suddenly escapes, its tension rapidly diminishes, and the reeds return to their former position, to repeat again the movement described. From the foregoing it results that—

1. During the vibration of the reeds, alternate condensation and rarefaction of the air must take place. It is chiefly this that (as in the siren) produces the sound, not so much the reeds themselves.

2. The "air-tube," which conducts the air to the membranous reeds, consists in the human voice-apparatus of the lower section of the larynx, the trachea, and, below, the entire bronchial tree. The bellows is the thorax, diminished in size during expiration by muscles.

3. The air-passage above the reeds is called a "reinforcing tube," and consists of the upper section of the larynx, the pharynx, and also the oral and nasal cavities, which are arranged in two stories one above the other, and can be closed alternately.

The pitch of the tone depends upon the following factors:

(a) The length of the elastic plates. The pitch is inversely proportional to the length of the elastic plates; that is the fewer the units of length that enter into the elastic plates the more numerous will be the units of time (vibrations) entering into the tone produced. For this reason the pitch of the shorter vocal bands in children and in women is higher than that in adults and in men.

(b) The pitch of the tone is, further, directly proportional to the square root of the elasticity of the elastic plates. In the case of membranous reeds, and also in that of silk, it is directly proportional to the square root of the extending weight, which in the larynx corresponds to the force of the tensor muscles.

(c) In the case of membranous reeds a more powerful blast not only strengthens the tone by increasing the amplitude of vibration, but it also raises the pitch of the tone, because the greater amplitude of vibration increases the mean tension of the elastic membrane.

Among physical influences the following further are to be noted:

(d) The reinforcing tube, which is exceedingly variable in form, also resounds when the larynx is intonated; its primary tone is mingled with the sound of the elastic reeds, and, thus, it is able to reinforce certain overtones of the latter. This subject will be discussed in greater detail in the section on voice-formation. The individual characteristics of the voice depend essentially upon the form of the reinforcing tube. In reed-instruments the pitch of the tones can undoubtedly be influenced by varying lengths of the reinforcing tube; but this is not taken into consideration in the case of the larynx.

(e) During intonation of the reeds the strongest resonance takes place in the air-tube, as the latter contains compressed air. This causes the vocal resonance that is heard when the ear is applied to the chest-wall. Strong intonation may even cause an accompanying vibration of the thoracic wall. In weak individuals, and in cases of falsetto voice, the vocal resonance is exceedingly slight.

(f) Narrowing or widening of the glottis has no effect on the pitch of the tone; but with the glottis wide open, disproportionately more air must pass through it, thus materially increasing the work of the thorax.

ARRANGEMENT OF THE LARYNX.

Cartilages and Ligaments of the Larynx.—The fundamental framework of the larynx is formed by the cricoid cartilage, which is shaped like a seal-ring. The inferior cornu of the thyroid cartilage articulates with the cricoid in its postero-lateral region. This joint allows the plate of the thyroid cartilage to tilt forward, the inclination occurring as a rotatory movement about a horizontal axis connecting the two joints, the upper border of the cartilage moving forward and downward. The joints also permit a slight shifting of the thyroid cartilage on the cricoid upward and downward, forward and backward. The triangular, pyramidal arytenoid cartilages articulate on the upper border of the plate of the cricoid cartilage to one side of the median line, forming approximately a saddle-shaped joint with oval articular surfaces. The latter permit a double movement on the part of the arytenoids, namely rotation on their base about their vertical, somewhat oblique, longitudinal axis, by which the vocal process directed forward is rotated outward and upward, and the muscular process directed outward and overlapping the border of the cricoid cartilage posteriorly is rotated inward and downward, or conversely. In addition, the arytenoid cartilages may be displaced somewhat inward or outward on their bases.

The true vocal bands, or vocal ligaments, are composed principally of elastic fibers. They arise close together from about the middle of the internal angle of the thyroid cartilage, and are inserted on the vocal processes of the arytenoid cartilages directed forward. The "ventricles of Morgagni" allow free play for the vibrations of the bands, and separate them from the upper "false" bands, or ventricular ligaments, which are covered by a fold of mucous membrane. The latter take no part in phonation. Numerous mucous glands of the mucous membrane keep the vocal bands moist.

In accordance with the functions of the laryngeal cartilages in connection with the voice-apparatus, C. Ludwig has called the cricoid the "foundation-cartilage," the thyroid the "tension-cartilage," and the arytenoids the "position-cartilages."

Owing to the oblique downward inclination of their under surfaces the vocal bands readily come together when the glottis is narrowed during inspiration (for example in sobbing); and if the glottis is already closed, inspiration makes this closure still firmer. The false vocal bands exhibit the opposite relation, for when in mutual contact they are readily separated during inspiration; while during expiration they readily close, owing to the inflation of the ventricles of Morgagni.

Action of the Laryngeal Muscles.—*Dilatation of the glottis* is effected by the posterior crico-arytenoid muscles. In drawing the muscular processes of the arytenoid cartilages backward, downward, and toward the median line (Fig. 208), these muscles cause the corresponding vocal processes (*I*, *I*) to separate and move upward (*II*, *II*). A large isosceles triangle is thus formed between the vocal bands, and another between the inner borders of the arytenoid cartilages, having their bases in contact, so that the aperture assumes a rhomboidal form.

Pathological.—Paralysis of these muscles may cause intense inspiratory dyspnea, on account of the failure of the glottis to dilate. The voice remains unchanged. In a freshly excised larynx the dilators first lose their excitability.

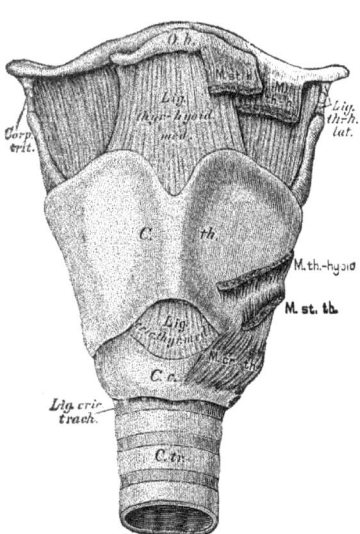 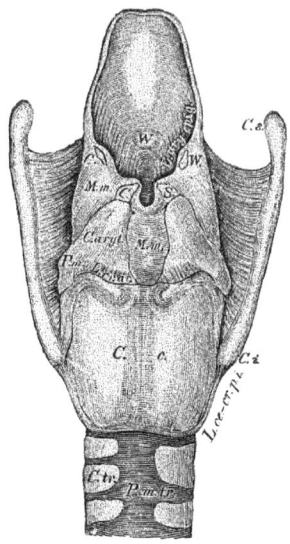

FIG. 204.—Anterior View of the Larynx, with its Ligaments and Muscular Insertions: *O. h*, hyoid bone; *C. th.*, thyroid cartilage; *Corp. trit.*, corpus triticeum; *C. c.*, cricoid cartilage; *C. tr.*, tracheal cartilages; *Lig. thyr.-hyoid. med.*, median thyrohyoid ligament; *Lig. th.-h. lat.*, lateral thyro-hyoid ligament; *Lig. cric.-thyr. med.*, median crico-thyroid ligament; *Lig. cric. trach.*, crico-tracheal ligament; *M. st.-h.*, sterno-hyoid muscle; *M. th.-hyoid*, thyro-hyoid muscle; *M. st.-th.*, sterno-thyroid muscle; *M. cr.-th.*, crico-thyroid muscle.

FIG. 205.—Posterior View of the Larynx, after Removal of the Muscles: *E*, epiglottis with the cushion (*W*); *L. ar.-ep.*, ary-epiglottic ligament; *M. m.*, mucous membrane; *C. W.*, cartilage of Wrisberg; *C. S.*, cartilages of Santorini; *C. aryt.*, arytenoid cartilages; *C. c.*, cricoid cartilage; *P. m.*, muscular process of the arytenoid cartilage; *L. cr. ar.*, crico-arytenoid ligament; *C. s*, superior cornu, *C. i.*, inferior cornu of the thyroid cartilage; *L. ce.-cr. p. i.*, postero-inferior kerato-cricoid ligament; *C. tr.*, tracheal cartilages; *P. m. tr.*, membranous portion of the trachea.

Also in the presence of organic disease in the distribution of the recurrent nerve, the branch to the posterior crico-arytenoid muscle is the first to be paralyzed. Likewise, in cooling the exposed recurrent nerve, this branch is always the first to fail in its function.

The *constrictor of the entrance to the larynx* is the transverse arytenoid muscle, which connects the two outer borders of the arytenoid cartilages by transverse fibers throughout their length (Fig. 209). On the posterior surface of this muscle are situated the crossed bundles of the oblique arytenoid muscles (Fig. 206), which have a similar action.

Pathological.—Paralysis of these muscles renders the voice feeble and hoarse, as much air escapes between the arytenoid cartilages during phonation.

The *intimate approximation of the vocal bands* is effected by bringing the vocal processes of the arytenoid cartilages close together.   To this end the latter must be rotated inward and downward by a forward and upward movement on the part of the muscular processes affected through the vocal or internal thyro-arytenoid muscles.   These muscles, which are applied to the elastic borders of the vocal bands, and in fact are embedded in their substance and whose fibers extend to the outer borders of the arytenoid cartilages, rotate the latter so that their vocal

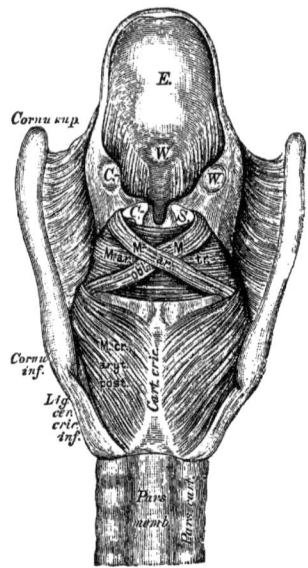   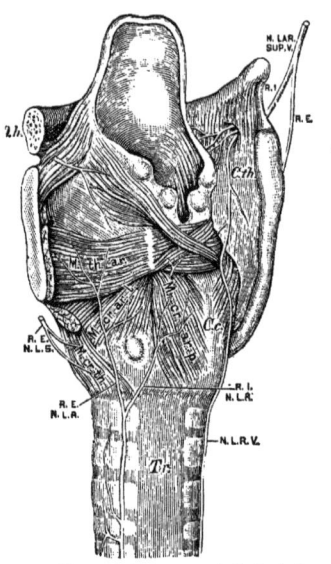

FIG. 206.—Posterior View of the Larynx, with the Muscles: *E,* epiglottis with the cushion (*W*); *C.-W.,* cartilages of Wrisberg; *C.-S.,* cartilages of Santorini; *Cart. cric.,* cricoid cartilage; *Cornu sup.,* superior cornu, *Cornu inf.,* inferior cornu of the thyroid cartilage; *M. ar. tr.,* transverse arytenoid muscle; *Mm. ar. obl.,* oblique arytenoid muscles; *M. cr. aryt. post.,* posterior crico-arytenoid muscle; *Pars cart.,* cartilaginous portion of the trachea; *Pars memb.,* membranous portion of the trachea.

FIG. 207.—Nerves of the Larynx: *O. h.,* hyoid bone; *C. th.,* thyroid cartilage; *C. c.,* cricoid cartilage; *Tr.,* trachea; *M. th.-ar.,* thyro-arytenoid muscle; *M. cr. ar. p.* posterior crico-arytenoid muscle; *M. cr. ar. l.,* lateral crico-arytenoid muscle; *M. cr. th.,* crico-thyroid muscle; *N. LAR. SUP. V.,* superior laryngeal branch of the vagus; *R. I.,* internal branch; *R. E.,* external branch; *N. L. R. V.,* recurrent laryngeal branch of the vagus; *R. I. N. L. R.,* its internal branch; *R. E. N. L. R.,* its external branch.

processes must move inward.   The glottis between the vocal bands is thus narrowed to a slit, while a broad, triangular opening remains between the bases of the arytenoid cartilages (Fig. 210).

The lateral crico-arytenoid muscle is inserted into the anterior border of the articular surface of the arytenoid cartilage; hence, it can only draw the cartilage forward.   Some investigators, however, believe that it also can effect a rotation of the arytenoid cartilage similar to that of the vocal or internal thyro-arytenoid muscle, with the difference that the vocal process are not brought so close together. .

Pathological.—Paralysis of the muscles effecting approximation of the vocal bands results in loss of voice.

The *tension of the vocal bands is effected* by the action of muscles in separating their two points of attachment from each other. To this end the thyroid cartilage is drawn forward and downward chiefly by the crico-thyroid muscles, the angle of this cartilage being at the same time somewhat enlarged. One can readily convince himself of this movement by feeling his own larynx during the emission of high tones. The same muscles also approximate the anterior arch of the cricoid cartilage to the inferior border of the thyroid cartilage; and as a result the posterior plate of the cricoid cartilage undergoes a backward inclination. At the same time the posterior crico-arytenoid muscles must draw both arytenoid cartilages somewhat backward, and hold them in that position. The tense vocal bands become longer and narrower.

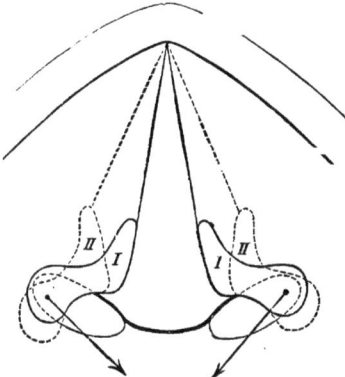
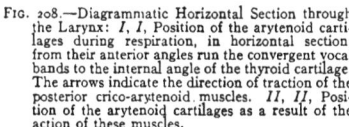
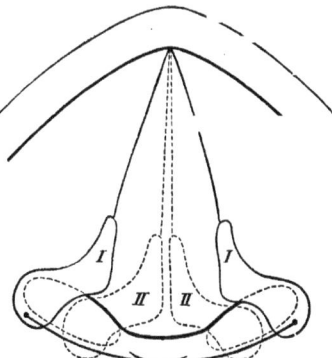

FIG. 208.—Diagrammatic Horizontal Section through the Larynx: *I, I,* Position of the arytenoid cartilages during respiration, in horizontal section; from their anterior angles run the convergent vocal bands to the internal angle of the thyroid cartilage. The arrows indicate the direction of traction of the posterior crico-arytenoid muscles. *II, II,* Position of the arytenoid cartilages as a result of the action of these muscles.

FIG. 209.—Diagrammatic Horizontal Section through the Larynx, to Illustrate the Action of the Arytenoid Muscle: *I, I,* Position of the arytenoid cartilages during quiet respiration. The arrows indicate the direction of traction of the muscle. *II, II,* Positions of the arytenoid cartilages produced by the action of this muscle.

The tension of the vocal bands is aided by the genio-hyoid and hyo-thyroid muscles, which together draw the hyoid bone, and thus indirectly the thyroid cartilage, upward and forward in the direction of the chin.

According to Harless, Schech, Kiesselbach, Hooper, and others, the crico-thyroid muscle effects elevation of the arch of the cricoid cartilage toward the thyroid cartilage. In this way the plate of the cricoid cartilage is directed backward and downward, thus causing increased tension of the vocal bands.

Pathological.—Paralysis of the crico-thyroid muscles renders the voice harsh and deeper, on account of insufficient tension of the vocal bands.

The tension thus induced is of itself by no means sufficient for phonation, for on the one hand the triangular aperture of the glottis between the arytenoid cartilages that would result from the isolated action of the internal thyro-arytenoid muscles must be closed. This is brought about by the transverse and oblique posterior arytenoid muscles. Then the vocal bands themselves, which, with the action of the crico-thyroid and posterior crico-arytenoid muscles, retain their concave border, so that the glottis between them appears as a space having the form of a myrtle leaf, must be fully stretched, so that the glottis assumes the shape of a linear slit (Fig. 214). This compensation likewise is brought about by the internal thyro-arytenoid muscle. It is this muscle, moreover, that effects those delicate gradations of tension in the vocal band itself that are necessary for the production of tones of slightly different pitch. It is especially adapted for this purpose, as it comes close to the edge of the vocal band and is firmly inserted into the elastic tissue of the latter. The contracting muscle in addition gives to the vibrating vocal band the resistance necessary for its vibrations. As some of the fibers of the vocal muscle terminate in the elastic tissue of the vocal band itself, they may impart increased tension to individual segments of the vocal band, as a result of which modifications in tone-formation are possible. It must, therefore, be assumed that the coarser variations in tension are caused by separation of the thyroid cartilage from the arytenoid cartilages, while the finer gradations of tension are induced by the vocal muscle. The usefulness of the elastic tissue in the vocal bands does not consist so much in its extensibility, as in its property of shortening without forming folds or creases.

Pathological. — When these muscles are paralyzed the voice can be produced only by powerful blasts, as much air escapes through the glottis. At the same time the tones are deep and impure. Unilateral paralysis results in flapping of the corresponding vocal band.

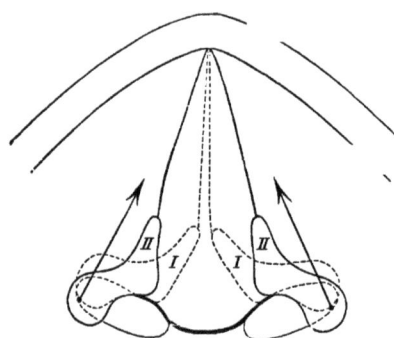

FIG. 210.—Diagrammatic Horizontal Section through the Larynx, to Illustrate the Action of the Internal Thyro-arytenoid Muscles in Narrowing the Glottis: *II, II,* Position of the arytenoid cartilages during quiet respiration. The arrows indicate the direction of traction of the muscles. *I, I,* Position of the arytenoid cartilages brought about by action of these muscles.

Relaxation of the vocal bands occurs spontaneously when the stretching forces cease to act, the thyroid cartilage drawn forward and the arytenoid cartilages fixed posteriorly returning to the position of rest in consequence of the elasticity that is peculiar to their arrangement. Relaxation of the vocal bands may result also from the action of the thyro-arytenoid and lateral crico-arytenoid muscles.

From the foregoing it follows that tension of the vocal bands and narrowing of the glottis are necessary for phonation.

The epiglottis, which becomes more erect with high tones and falls with low ones, has an influence on the timbre (clear or muffled) of the voice, but has no effect on the pitch.

The mucous membrane of the larynx, as well as the submucosa, is rich in delicate, elastic networks of fibers. The submucosa is loose and yielding in the region of the entrance to the larynx and the ventricles of Morgagni, a fact that explains the enormous swelling that often occurs in connection with so-called edema of the glottis. A clear, even, limiting layer lies beneath the epithelium. The epithelium is stratified, cylindrical, and ciliated, interspersed with goblet-cells, except on the true vocal bands and the upper surface of the epiglottis, where a stratified, squamous epithelium covers the mucous membrane; which in this situa-

tion bears papillæ. Racemose mucous glands are present in groups on the cartilages of Wrisberg, the cushion of the epiglottis, and in the ventricles of Morgagni;
and are scattered in the other situations, especially on the posterior wall of
the larynx. The blood-vessels form a dense, capillary network under the limiting
layer of the mucous membrane; beneath this are two more layers of vascular networks. The lymphatics form a superficial, narrower network beneath the blood-
capillaries, and a deeper, coarser network. The medullated nerves, which have

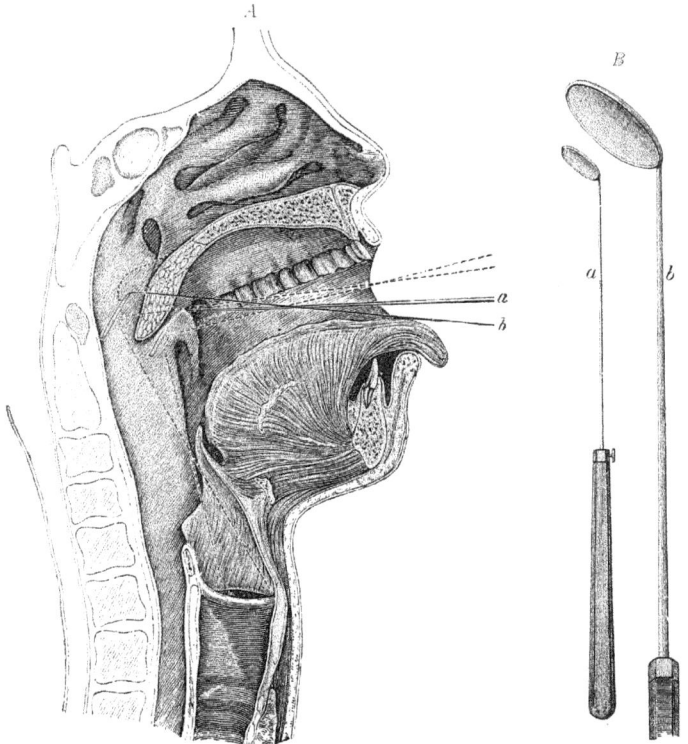

Fig. 211.—*A*, Vertical section through the head and neck as far as the first dorsal vertebra: *a* shows the position
of the laryngoscope in order to see the posterior part of the glottis, the arytenoid cartilages, the upper surface
of the posterior laryngeal wall, etc.; *b* shows the position of the laryngoscope in order to obtain a view of the
anterior angle of the glottis. *B*, Large (*b*) and small (*a*) laryngeal mirrors.

ganglia on their branches, are numerous in the mucous membrane; their terminations are unknown. The cartilage is hyaline in the thyroid, the cricoid, and
almost in the entire arytenoid cartilage, with a tendency to ossification. Fibro-
cartilage is found toward the apex and the vocal process of the arytenoid cartilage,
and also in all the remaining laryngeal cartilages.

The larynx grows until about the sixth year, then rests, but rapidly increases
in size again at puberty.

EXAMINATION OF THE LARYNX.
LARYNGOSCOPY. EXAMINATION OF THE EXCISED LARYNX.

After Bozzini, in 1807, had given the first impulse toward illuminating and examining the internal cavities of the body by means of the mirror, and Babington, in 1829, had viewed the glottis in this way, the singing-teacher, Manuel Garcia, in 1854, made investigations, by means of the laryngoscopic mirror, on himself and other singers, concerning the movements of the vocal bands during respiration and phonation. Türck and Czermak rendered the greatest service in the applica-

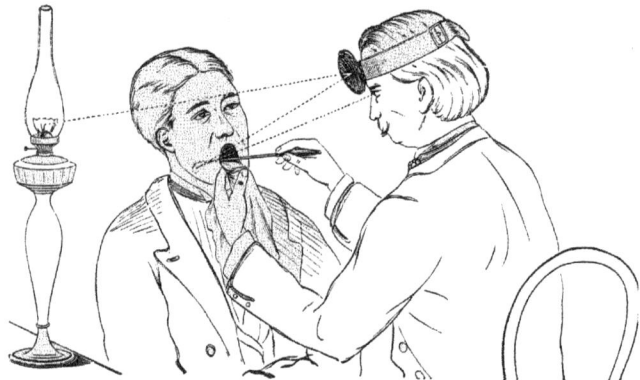

FIG. 212.—Method of Making a Laryngoscopic Examination.

tion of the laryngoscope to medical purposes, the latter being the first to use artificial light for illumination. Rhinoscopy was first attempted by Baumès in 1838, and was systematically developed by Czermak.

The laryngoscope consists of a small mirror, attached to a handle at an angle (Fig. 211, B), the instrument being introduced with the mouth wide open and the tongue drawn out (Fig. 211, A). The position of the mirror must be changed in accordance with the region to be reflected; and it may at times even be necessary to elevate the soft palate by means of the mirror (b). The mirror receives the picture of the larynx in the direction of the dotted line, and reflects it at the same angle through the oral cavity to the eye of the observer, which has taken its position in the line of the reflected rays. The illumination of the larynx is accomplished by collecting either sunlight or light from an artificial source in a concave mirror, and permitting the concentrated bundle of rays to fall on the laryngoscopic mirror held in the throat. The latter reflects the light against the larynx, which is thus illuminated. The observer looks in the same direction as the rays of light, either under the edge of the illuminating mirror, or through a central perforation in the latter.

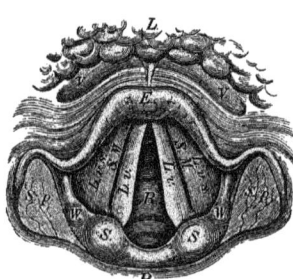

FIG. 213.—The Laryngoscopic Image During Respiration.

The laryngoscope received an important improvement at the hands of Oertel, who showed how the movements of the vocal bands could be followed directly· with the eye by means of rapidly intermittent illumination through the disc of a stroboscope (laryngo-stroboscope). By replacing the eye by a photographic camera, Ssimanowsky was able to photograph the movements of the vocal bands in an artificial larynx.

v. Ziemssen showed that long, thin electrodes could be introduced as far as the larynx under the guidance of the laryngoscope, and that the vocal bands could be stimulated to activity by irritation of the muscles. Rossbach succeeded in stimulating the muscles and nerves of the larynx externally through the skin.

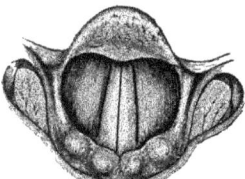
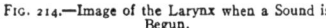

FIG. 214.—Image of the Larynx when a Sound is Begun.

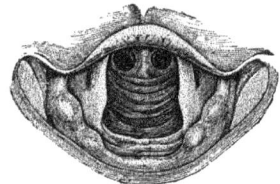

FIG. 215.—View of the Trachea as far as the Bifurcation.

In this way physiological information may be gained, or therapeutic applications may be made to the parts.

Autolaryngoscopy was first employed by Garcia, and then by Czermak especially for the study of the movements of the larynx. If one introduce an illuminated laryngoscopic mirror into his own throat, while placing the mouth opposite a plane mirror, he may easily see the picture of his own larynx reflected in the latter.

The *laryngoscopic picture* (Fig. 213) exhibits the following details: *L*, the root of the tongue, from the middle of which the glosso-epiglottic ligament passes downward; on each side of the latter are the so-called valleculæ (*V V*). The epiglottis (*E*) appears as an arch, shaped like the upper lip; beneath it in quiet respiration is seen the lancet-shaped chink of the glottis (*R*), and on either side the bright, yellowish vocal ligament (*L. v.*). This vocal band is from 6 to 8 mm. long in children, from 10 to 15 mm. long in women when relaxed, and from 15 to 20 mm. when tense. In men it measures from 15 to 20 mm. and from 20 to 25 mm. respectively. The whole chink of the glottis is 23 mm.

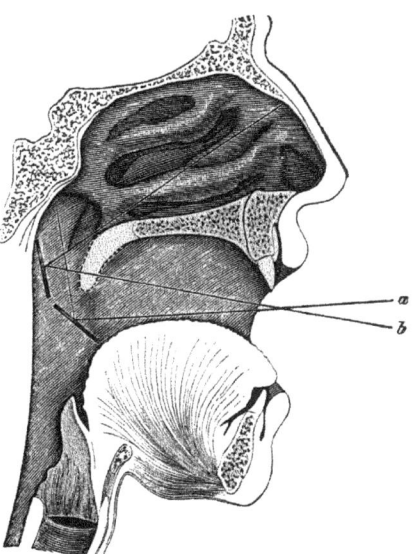

FIG. 216.—Position of the Laryngeal Mirror in the Practice of Rhinoscopy.

long in men and 17 mm. in women; when the vocal bands are tense 27.5 and 20 mm. respectively.

The width of the vocal bands varies from 2 to 5 millimeters. External to the vocal band is the entrance (rima vestibuli) to the sinus of Morgagni (*S. M.*), represented by a dark band. Still further outward, and on a higher plane, may be seen the fold of mucous membrane (plica

ventricularis) covering the false vocal band or the ventricular ligament
(*L. v. s.*).   On the lower, lip-shaped border of the entrance to the larynx
may be distinguished the posterior lower notch of the ostium pharyn-
geum laryngis (above *P.*); and on either side of this the apices of the cartilages of Santorini (*S. S.*) are visible, resting on the apices of the arytenoid cartilages; immediately behind is the adjacent pharyngeal wall (*P.*).
In the ary-epiglottic ligaments (*IV. II'.*) are the cuneiform cartilages of
Wrisberg, and finally, external to these, may be recognized the depressions of the sinus piriformes (*S. p.*).

Special attention should be given to the condition of the glottis and
the vocal bands during respiration and phonation.   During quiet respiration the chink of the glottis (Fig. 213) appears as a lancet-shaped slit,
which is wider during life than in the cadaver.   If deep respirations are
taken, the chink widens considerably (**Fig. 215**), and if the mirror is
favorably placed, it may be possible to see the rings of the trachea, and
even the bifurcation.   When the voice is produced, the glottis closes
each time to a narrow slit (Fig. 214).

Appendix.—Rhinoscopy.—The nasal cavity has important relations to speech
and to respiration.   By the introduction of a mirror bent at an angle, with the
reflecting surface directed upward, it is pos-
sible gradually to survey a field such as is
reproduced in Fig. 217.

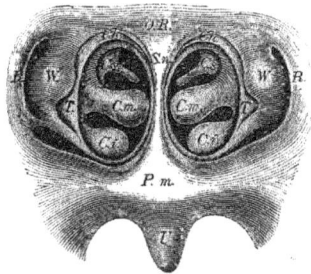

In the middle appears the nasal septum
(*S. n.*), on either side the longitudinally oval
choanæ (*Ch.*), and further below the soft
palate (*P. m.*) with the pendant uvula (*U.*).
On the borders of the choanal openings may
be recognized the posterior portions of the
inferior (*C. i.*), middle (*C. m.*) and superior
(*C. s.*) turbinated bones, with the corresponding nasal meatus beneath each one.
Least distinct are the upper turbinated bone
and the lower meatus.   At the uppermost
part a strip of the roof of the pharynx
(*O. R.*) may yet be seen, with the more or
less developed pharyngeal tonsil.   This
latter structure is composed of lymphatic

FIG. 217.—The Rhinoscopic Image.   This illustration is more or less diagrammatic, in so far as
a repeated shifting of the mirror is necessary
in order to obtain the entire image as is given
here.

glandular tissue, and extends in an arch-
like manner over the roof of the pharynx
between the openings of the two Eustachian
tubes (*T. T.*).   External to the mouth of
the Eustachian tube on each side is the so-called tubal eminence (*II'.*), and still
more external the fossa of Rosenmüller (*R.*).

For the study of the larynx experimentation on the excised larynx is further
of importance, as carried out by Ferrein in 1741 and especially by Johannes
Müller in 1839.   The latter conducted the air into an excised human larynx
through a tracheal tube the air-tension of which was measured by a communicating
mercurial manometer.   The bases of the arytenoid cartilages were held in a fixed
position against each other by means of a suture; while a cord passing over a
pulley and carrying weights drew the thyroid cartilage forward.   By increasing
the tension the tones could be raised about 2½ octaves.   When the tension remained the same, stronger blasts of air raised the tone to the fifth.   The tone
was not lowered by placing tubes over the larynx to increase its length, but these
measures modified the timbre and increased the resonance of the note.

Landois employed the fresh, living, excised larynx from the dog or the sheep;
the muscles being stimulated by various pairs of electrodes, while a bellows supplied
the air through a tracheal tube.   In this way the most reliable information concerning the action of the various muscles can be obtained.

The Röntgen rays have recently been applied with success to the study of
the position of the laryngeal cartilages and the hyoid bone, and also of the soft
palate.

CONDITIONS INFLUENCING THE SOUNDS OF THE VOCAL APPARATUS.

The *pitch of the voice-tone* depends upon the following factors:

1. The *tension of the vocal bands;* hence upon the degree of contraction of the crico-thyroid and posterior crico-arytenoid muscles, with the assistance of the vocal or internal thyro-arytenoid muscles.

2. The *length of the vocal bands.* In this connection it should be noted: (*a*) That children and women, with shorter vocal bands, produce higher tones. The voices of women are about one octave higher than those of men. (*b*) If the arytenoid cartilages are pressed tightly together by the action of the transverse and oblique posterior arytenoid muscles, so that only the vocal bands themselves can vibrate, while the intercartilaginous parts between the vocal processes cannot, then the tone will be higher. To produce deeper tones, the vocal bands, and also the margins of the arytenoid cartilages, must vibrate. At the same time the space above the exit of the larynx enlarges, so that the throat becomes more prominent. (*c*) Each individual has a certain medium pitch of voice, which corresponds to the least possible muscular tension within the larynx.

3. The *strength of the blast.* That the strength of the blast is able to raise the pitch of the tone in the human larynx is shown by the fact that the highest tones can be emitted only in a loud voice. With medium tones the air-tension in the trachea amounts to 160 mm., with high tones to 200 mm., with exceedingly high notes to 945 mm., in whispering only to 30 mm. of water measured through a tracheal fistula. In changing the intensity of a tone from loud to soft, or conversely, while maintaining the same note, the muscular action must undergo a change in force. When the note is loud the force diminishes, while it increases as the tone becomes soft. J. Müller called this process the "compensation of energy in the larynx."

The following accessory phenomena have been observed in the production of high notes, but no certain interpretation of them has been given: (*a*) As the pitch of the note increases, the larynx becomes elevated, partly because the elevating muscles of the larynx are brought into activity, and partly because the intra-tracheal air-pressure lengthens the trachea to such an extent that the larynx is raised up. The uvula also is raised higher and higher. (*b*) The upper vocal bands approach each other more and more, without touching or participating in the vibrations. (*c*) The epiglottis inclines more and more backward over the glottis. In explanation of *c* and *b* it is supposed that, in the production of high tones, all of those muscles are active that aid in shortening the vibrating section of the rim of the glottis and in constricting its opening. In this act the edge of the (external) thyro-arytenoid muscle displaces the upper vocal band inward; while the epiglottis is drawn downward by those fibers that pass upward toward it laterally—the thyro-ary-epiglottic muscle.

4. So-called *registers* can be distinguished in the voice. There is generally the *chest-register*, the thorax vibrating (pectoral fremitus), and the voice appearing to come from the depths of the chest; and also the *head-register*, the voice apparently coming from the throat. The latter, with its soft timbre and lack of resonance in the air-tube, is designated also a *falsetto voice* or *shriek*. Oertel observed under such circumstances that the vocal bands vibrated so as to form nodal lines across their width; at times only one nodal line is formed, so that the free border of the vocal band and the basal border vibrate, and are separated from each other by a nodal line parallel to the edge of the vocal

39

band. In high falsetto notes, as many as three such nodal lines may arise in succession. The formation of the nodal lines must be occasioned by a partial contraction of the fibers of the internal thyro-arytenoid muscle. At the same time the vocal bands must be stretched into the thinnest possible plates by the combined action of the crico-thyroid, posterior arytenoid, thyro-hyoid, and genio-hyoid muscles. The glottis is elliptical in form, while with the chest-voice it is bounded by the straight lines of the vocal bands. In the latter case more air passes out of the larynx.

Oertel found, moreover, that with the falsetto voice the epiglottis assumes a vertical position. The apices of the arytenoid cartilages are inclined somewhat backward; the entire larynx appears longer in its sagittal diameter and narrower in its transverse diameter; the ary-epiglottic folds are stretched tensely, with sharp edges; the entrance to the ventricles of Morgagni is constricted. The vocal bands are longer than in the production of the same tone with the chest-voice; further, they are narrower, and the vocal processes are in contact with each other. The rotation of the arytenoid cartilages necessary for this is brought about solely by the lateral crico-arytenoid muscle, while the thyro-arytenoid is to be regarded only as an accessory, aiding muscle. Elevation of the pitch with the falsetto voice is effected exclusively by increasing the tension of the vocal bands. In addition to the characteristic modification in the vibration of the vocal bands already described, still another series of partly transverse and partly longitudinal partial vibrations are superposed upon the former. In the case of the chest-voice a narrower edge of the vocal band vibrates than in that of the falsetto voice; in the production of the latter there is a feeling of less muscular exertion in the larynx. The uvula is raised horizontally. In the so-called chest-register the entire width of the vocal band vibrates, in the middle register only the inner narrower border. In the chest-voice the overtones in the note are richest and strongest, while in the falsetto voice they are less numerous and feebler.

Pathological.—By means of Oertel's laryngo-stroboscope important information can be obtained concerning variations in the vibrations of the vocal bands, such as unequal amplitude of vibration in the two vocal bands (laryngeal catarrh), with or without alternating vibrations; the formation of vibration-nodes in one band; the absence of vibrations in one or both bands.

In order that the voice may be produced, the following processes are necessary: (1) The required amount of air is accumulated in the thorax; (2) the larynx and its parts are fixed in the appropriate position; (3) then follows the "onset" of the voice, either the closed glottis being forced open by means of an expiratory effort, or some air being permitted to pass almost noiselessly through the glottis, and the vocal bands being then thrown into vibration as the blast of air is gradually increased.

RANGE OF THE VOICE.

The range of the human voice for the chest-register is shown in the accompanying diagram:

The figures indicate the number of vibrations in a second for the corresponding tone. It will be readily seen that the notes from c' to f' are common to all voices; nevertheless, each has a different timbre. The lowest note, which exceptionally is sung by bass singers, is the contra-F with only 42 vibrations; the highest note of the soprano voice is a''' with 1708 vibrations.

Hensen devised an especially ingenious method for determining with exactness the pitch of a sung note. The note is emitted against a König's capsule, with a gas-flame. Opposite this is a tuning-fork, vibrating horizontally, and provided at the extremity of one prong with a mirror, in which the flame is reflected. If the pitch of the voice is the same as that of the fork, the flame appears in the mirror as a single jet; at the octave two jets appear, at the twelfth three jets, and at the double octave four.

Each individual has his characteristic voice-timbre, which depends upon the configuration of all of the cavities belonging to the vocal organ. The so-called palatal tones arise from the approximation of the soft palate to the posterior pharyngeal wall. In the production of nasal tones the air in the nasal cavities vibrates more forcibly, as access to these cavities must be freer.

SPEECH. THE VOWELS.

The motor processes concerned in speech are carried out in the reinforcing tube—the pharyngeal, oral, and nasal cavities; they are directed toward the production of tones and noises. If the latter alone are developed, the voice-apparatus being passive, "whispering" results (vox clandestina); if, however, the vocal bands vibrate at the same time, "audible speech" results. Whispering itself may be made quite loud, but to bring about this result a strong blast is required; hence it is fatiguing. It can be practised during both inspiration and expiration, in contradistinction to audible speech, which sounds transient and indistinct when produced during inspiration. Whispering results from the sound that is generated, when the glottis is moderately narrowed, by the passage of the air over the blunt margins of the vocal bands. In the production of audible speech the vocal processes are so placed that the sharp margins of the vocal bands are directed toward the air-current, and are thrown into vibration by it.

The soft palate always participates in the production of speech. It is raised with every word, Passavant's transverse ridge being at the same time formed in the pharynx. The palate is raised highest during the utterance of u (oo) and i (ee); less high with o and e (a), and least high with a (ah). During the enunciation of m and n the palate is stationary; with the explosives it is about as high as with n; and it is lower with the fricatives. With l, s, and especially with the guttural r, it is thrown into a tremulous movement.

Speech is made up of vowels and consonants.

Vowels.—(Analysis and artificial formation are considered on page 905.)

In whispering, a vowel is the sound produced, during either expiration or inspiration, by the inflated characteristically shaped oral cavity, the sound having not only a definite pitch, but also a characteristic timbre. The characteristically shaped oral cavity may be designated the "vowel-cavity."

The *pitch of the vowels* may be determined musically, either by paying close attention to one's own whispered vowels, or in the case of another by blowing by means of a suitable air-tube from the opening of the mouth into its cavity placed in the position peculiar to the vowel in question. It is a remarkable fact that the fundamental tone of the "vowel-cavity" is almost constant for various ages and sexes. The differences in the internal capacity of the mouth can be compensated for by varying the size of the opening of the mouth. The pitch of the vowel-cavity may also be estimated by holding a series of vibrating tuning-forks of varying pitch in front of the mouth. When the one is found that corresponds with the fundamental tone of the vowel-cavity, the note of the tuning-fork will be strengthened considerably by resonance from the oral cavity. Finally, a membrane having the same rate of vibration as the vowel-tone may be held in front of the mouth, and the vibrations of the vowel-tone may be transferred to the membrane, the vibrations of the latter being recorded on smoked paper, as in the "phonautograph" of Donders.

König found the fundamental tones of the vowel-cavities to be as follows: U (oo) = b, O = b', A (ah) = b'', E (a) = b''', I (ee) = b''''.

If the vowels be whispered in this series, their pitch will at once be heard to increase. Otherwise, these fundamental tones of the mouth in the vowel-positions may vary within certain limits; hence it is better to speak of the region of a characteristic tone-position. This may be best demonstrated by placing the mouth in the characteristic position and percussing the cheeks. The sound of the vowel will then be heard, and its pitch will vary within certain limits in accordance with the position of the mouth.

In pronouncing *A* (ah), the mouth has the shape of a funnel dilating anteriorly (**Fig.** 218, **A**). The tongue lies on the floor of the mouth; the lips are wide open. The soft palate is raised moderately; being successively more elevated with O, E (a), U (oo), I (ee). The hyoid bone is in a position of rest when **A** (ah) is being uttered; but the larynx is somewhat raised, being higher than with U (oo), but lower than with I (ee).

When a transition is made from A (ah) to I (ee), the larynx and the hyoid bone retain their relative positions, but both ascend. During the transition from A (ah) to U (oo), the larynx sinks to the lowest possible level. At the same time the hyoid bone moves slightly forward. In pronouncing A (ah), the space between the larynx, the posterior pharyngeal wall, the soft palate and the root of the tongue is only moderately dilated; it becomes larger with E (a) and especially with I (ee), and is smallest with U (oo).

In sounding *U* (oo), the shape of the mouth is that of a capacious flask with a short, narrow neck. The entire reinforcing tube is under such conditions longest. Accordingly, the lips are protruded as far as possible, are corrugated, and are brought together so as to form a small opening. The larynx is at its lowest level. The root of the tongue is approximated to the posterior palatine arch.

In sounding *O*, the mouth, as with U (oo), resembles a wide-bellied flask with a short neck. The latter, however, is shorter and at the same time more widely open, the lips approaching more closely to the teeth. The larynx is somewhat higher than with U (oo). The entire reinforcing tube is, therefore, shorter than with U (oo).

In sounding *I* (ee), the mouth has the shape of a flask with a small belly in the posterior part, and a long, narrow neck. The belly has the fundamental tone f, while the neck has that of d''''. The reinforcing tube is shortest with I (ee), as the larynx is raised as far as possible, and the mouth is bounded anteriorly by the teeth, the lips being retracted. The oral canal between the hard palate and the back of the tongue is greatly constricted to a median, narrow channel. Hence, the air can pass through only with a clear, whistling sound, and even the vertex of the skull may be set into perceptible vibration; if the ears are stopped up, a shrill sound may be audible in them. It is impossible to pronounce I (ee) when the larynx is depressed and also when the lips are protruded, as for U(oo).

In pronouncing *E* (a), which stands next to I (ee), the cavity has likewise the form of a flask with a small belly (fundamental tone f') and a long, narrow neck (fundamental tone b'''). The neck, however, is wider, so that it does not give rise to a whistling sound. The larynx is somewhat lower for E (a) than for I (ee), but higher than for A (ah).

Fundamentally, Brücke is right in assuming that there are only three fundamental vowels, namely *I* (ee), *A* (ah), *U* (oo), between which the others, as well as the so-called diphthongs, are interpolated. The hieroglyphic, Indian, old Hebraic, and Gothic writings contain only these three vowels.

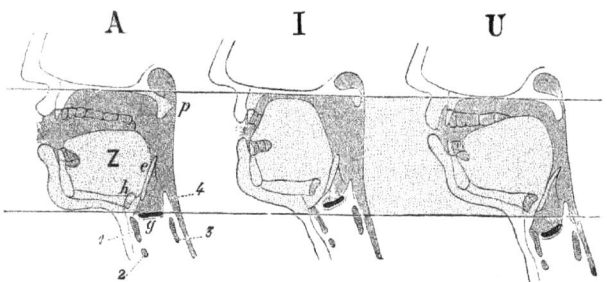

Fig. 218.—Sagittal Section through the Human Larynx in the Vowel-positions A (ah), I (ee) and U (oo): Z, tongue; *p*, soft palate; *e*, epiglottis; *g*, glottis; *h*, hyoid bone; 1, thyroid cartilage; 2, 3, cricoid cartilage; 4, arytenoid cartilage.

Diphthongs occur during the utterance of a sound, by passing from the position of one vowel into that of another. Distinct diphthongs are sounded only on passing from a vowel with a wider oral opening to one with a narrower opening; if the reverse occurs, the vowels appear separated to the ears.

Landois was especially successful in producing the vowels artificially. In the two halves of a head sawn through in a sagittal plane he arranged all of the parts in the positions that they would have to assume in enunciating a certain vowel, and then the entire space from the trachea to the lips was filled with paraffin. Both halves were then welded together. A paraffin cast was thus obtained of the vowel-cavity. The cast was covered with plaster-of-Paris, and then the paraffin was removed by melting. In this way a plaster reproduction of the vowel-cavity was obtained. A vocal apparatus was then introduced into the trachea from below. This apparatus was made of a thin, ivory reed, set in a wide frame, and having its pitch accurately adjusted to the fundamental tone of the plaster cavity. All the vowels, even I (ee), were thus produced with surprising success.

In addition to the pitch the characteristic timbre, or tone-color, of the vowel is worthy of notice. In this connection the mouth, characteristically shaped for the utterance of a vowel, may be compared to a musical instrument, which not only gives forth its sound in a certain pitch, but also allows it to ring with a characteristic timbre.

Thus, the vowel-sound U (oo), when whispered, has, besides its fundamental tone b, a soft, whistling timbre; I (ee), with its fundamental tone b''''', a hissing, whistling timbre; A, with its tone b'', an open, blowing timbre. This timbre depends upon the number and the pitch of the overtones peculiar to the vowel-sound, which are considered in the section on analysis of the vowels (auditory apparatus, p. 905).

The timbre of the vowels may, further, be modified in a special manner when they are uttered with a "nasal" twang, as is prevalent in the French language. The nasal timbre is produced when the soft palate does not close off the nasal cavity, as happens during utterance of the pure vowels, so that the air in the nasal cavity is set into sympathetic vibration. When a vowel is spoken with a nasal timbre, the air thus escapes through both the mouth and the nose; when the vowel is spoken purely, the air escapes only through the mouth. Hence, in the former case, a light held in front of the nostrils will flicker, or a cold glass or metal will be moistened; but not in the latter case.

In closure of the nasal cavity the soft palate is raised, in smallest measure when A (ah) is pronounced; then follow O, E (a), U (oo), I (ee). High and loud tones require more marked elevation, during which the velum presents a notch in the situation of the elevators of the palate. The opening of the Eustachian tube is constricted, but never entirely closed, by the elevation of the palate.

In uttering the pure, non-nasal vowels, the nasal cavity is so firmly closed off from the mouth that it can be sprung open only by an artificial increase in the pressure within the nose of from 30 to 100 mm. of mercury, with the development of a gurgling rhonchal sound.

The nasal twang occurs as a result of resonance in the naso-pharyngeal cavity; at the same time a portion of the cavity of the mouth is excluded by elevation of the dorsum of the tongue and depression of the palate.

Especially the vowels a (ah), ä (æ), ö (œ), o, e (a), are employed with a nasal accent. The nasal i (ee), however, does not appear to occur in any language. At all events it is hard to form, because in sounding it the oral canal is so narrow that if the nasal cavity be open at the same time the air will escape almost completely through the latter, while the small amount passing through the mouth is hardly sufficient to produce a sound.

In pronouncing the vowels it should also be observed whether they are uttered through a previously closed glottis, as is the case in German with all vowels placed at the beginning of words. Thus, the glottis is at first closed, but it is sprung open simultaneously with the intonation at the moment of commencing the word. Pronunciation of vowels in this manner was termed by the Greeks spiritus lenis. If, however, the vowel is pronounced after a preliminary breath has passed through the open glottis, and the sound of the vowel follows immediately, then the aspirate vowel results, corresponding to the spiritus asper of the Greeks.

If the vowels are pronounced audibly, therefore with a simultaneous sound of the voice, the fundamental tone of the vowel-cavity, with its constant, absolute pitch, strengthens in a characteristic manner the corresponding partial tone present in the sound of the voice. Accordingly, the vowels are intonated most purely from a musical point of view when the pitch of the tone is so adjusted as to contain overtones that correspond harmonically with the fundamental tone of the vowel-cavity when blown upon.

THE CONSONANTS.

The consonants are noises that are generated at certain parts of the reinforcing tube. They are classified as follows: I. According to their acoustic properties into (1) sounding or liquid consonants, that is those that are audible even without vowels (*m*, *n*, *l*, *r*, *s*); (2) mutes, including all the rest, which cannot be distinctly heard without the simultaneous pronunciation of a vowel. II. According to the mechanism of their formation, as well as the parts of the speech-apparatus by which they are produced.

1. *Mutes*, *stops*, *checks*, or *explosives*, the air being forced through an existing closure, with the production of more or less noise; or, conversely, the current of air may be suddenly interrupted, while at the same time the nasal cavity is closed off by elevation of the soft palate.

2. *Fricatives* or *spirants* or *sibilants*, the canal being constricted at one point, so that the air is forced through with a hissing noise, while the nasal cavity is closed off.

L and similar consonants are closely related to the fricatives, differing, however, from these in that the narrow passage through which the air is forced is not situated in the middle line, but to either side of the closed middle. The nasal cavity is closed off.

3. *Vibratives*, which result when air is forced through a narrow part of the canal, so that the margins of the constriction are thrown into vibration. The nasal cavity is closed off.

4. *Resonants*, also designated *nasals* or *semi-vowels*. The nasal cavity is entirely open, but the mouth is tightly closed anteriorly at one point. In accordance with the position of this closure of the mouth, the air in a larger or smaller portion of the oral cavity may be set into sympathetic vibration.

In addition to these possible forms of origin of the sounds the points at which they may be produced must be taken into consideration. These points may be designated *articulation-positions*. They are: A, between the lips; B, between the tongue and the hard palate; C, between the tongue and the soft palate; D, between both true vocal bands.

(A) *Consonants of the First Articulation-position.*

1. *Explosive Labials.*—*b* (bay): the voice is sounded before the soft explosion occurs; *p* (pay): the voice is sounded only after the much stronger explosion has taken place.

2. *Fricative Labials.*—*f*: between the upper incisor teeth and the lower lip (labio-dental); it is absent from all true slavic words; *v* (fow): between the two lips (labial); *w* (vay): results when the mouth is adjusted as for *f* (labial, as well as labio-dental), but instead of merely blowing air out, the voice is also sounded. There are really two different forms of *w* (vay), namely, one corresponding to the labial *f*, as in Würde (pronounced veerde), and the labio-dental, as in Quelle (pronounced kwelle).

3. *Vibrative Labials.*—The "burring" sound of drivers, not employed in civilized languages.

4. *Resonant Labial.*—*m* is formed when the voice is sounded, and the air in the oral and nasal cavities is thrown into resonance.

(B) *Consonants of the Second Articulation-position.*

Method.—In order to determine the extent to which the tongue and the palate are in contact in the formation of consonants in the second and third articulation positions, the tongue is sprinkled with a powdered dye. while the mouth is held wide open. When the consonant is formed, the palate receives a colored impression at those points where contact has taken place. Also in the case of the consonants, with the exception of m, n, ng, the soft palate is elevated.

1. The *explosives*, which are produced between the tongue and the hard roof of the oral cavity, are the hard *T*-sounds (also *dt* and *tt*), when enunciated sharply and without the voice; the soft *D*-sounds when uttered softly with simultaneous intonation of the voice. Variously designated and uttered modifications of these consonants occur in various languages, accordingly as the tip or the back of the tongue, on the one hand, and the teeth or the alveolar process or the hard palate, on the other hand, are employed in their formation.

2. The *fricatives* embrace the consonants allied to *S*, including the sharp s (also written *ss* and *sz*), which is produced without the sound of the voice, and the soft s, which can be produced only with intonation of the voice. Modifications occur also here, in accordance with the regions between which the aspirate consonant is formed. Thus, to the sharp aspirates belong also the sharp *Sch* and the hard English *Th*; to the soft aspirates the soft French *J* and the soft English *Th*. The sound *L* likewise belongs to this class, occurring in manifold modifications in various tongues, for example the soft *L* of the French. The sound *L* may also be uttered softly with the voice, or sharply without it.

3. The *vibratives* of the second articulation-position, or the lingual *R*-sounds, are usually enunciated with the sound of the voice, although they may also be formed without it.

4. The *resonants* are the *N*-sounds, which likewise may occur in various modifications.

(C) *Consonants of the Third Articulation-position.*

1. The *explosives* are the *K*-sounds, if hard and without the sound of the voice; or the *G*-sounds (gay), if the voice is also given. There are various modifications of both; for example, the explosive position of *G* (gay) and *K* preceding e (a) and i (ee) is situated farther forward on the palate than that of *G* (gay) and *K* before a (ah), o, u (oo).

2. The *aspirates* of these positions are the *Ch*-sounds, hard and without the voice; and *J* (y), if soft and without the voice. Following a (ah), o, u (oo), these consonants are formed farther back on the palate than those that follow e (a) and i (ee).

3. The *vibrative* is the palatal *R*, which results from vibration of the uvula.

4. The *resonant* is the palatal *N*. After e (a) and i (ee) the closure is displaced further forward, after a (ah), o, u (oo), further back. The nasal N of the French is, however, not a consonant at all, but only the nasal timbre of the vowel that results from the patulousness of the nasal cavity.

According to Saenger the participation of the nasal cavities in the production of *m*, *n*, and *ng* consists chiefly in affording a passage for the air expired during phonation.

(b) *Consonants of the Fourth Articulation-position.*

Logically, the glottis itself may further be considered as a fourth articulation-position.

1 An *explosive consonant* is not produced by forcing open the glottis, if a vowel has been loudly intonated from a previously closed glottis. If this occurs during whispering, a feeble, short sound may undoubtedly be heard, arising from the sudden opening of the glottis. As already noted, the Greeks applied the term spiritus lenis to the utterance of vowels from a previously closed glottis.

2. The *aspirates of the glottis* are represented by the *H*-sound (hah), which is produced with a moderately wide glottis. The Arabic Hha is emitted with especial sharpness from a still narrower glottis.

3. A *glottis-vibrative* occurs in the so-called laryngeal *R* of lower Saxony, and in the Arabic Ain. It can be produced by pronouncing a vowel with the deepest possible voice. This is followed by a distinct, shock-like, resounding vibration of the vocal bands, which represents the laryngeal *R*. The sound is represented especially in the low German dialect of Hither Pomerania, for example in Côarl (Carl), Wŭort (Wort).

4. A laryngeal resonant cannot be produced.

The *combination* of different consonants is accomplished by the rapid, successive execution of the movements necessary for each one. *Compound consonants* are those that are formed by adjusting the parts of the mouth for two different consonants at the same time, so that a mixed sound is formed from the simultaneous production of both sounds. Examples: *Sch, tsch, tz, ts, Ps* (Φ), Ks (X, Ξ).

PATHOLOGICAL VARIATION IN VOICE AND SPEECH.

Paralysis of the motor nerves of the larynx (from the vagus) as a result of wounds or of pressure by tumors results in loss of voice or aphonia. In the presence of aneurysm of the arch of the aorta the left recurrent laryngeal nerve is often paralyzed in consequence of being greatly stretched. Rheumatism, over-exertion. and hysteria may cause transitory paralysis of the laryngeal nerves.

Serous effusions into the laryngeal muscles in consequence of inflammatory processes will also cause paralysis of these muscles and thus aphonia. If the tensors chiefly are paralyzed, monotonia of the voice develops. The disturbances of respiration attending paralysis of the larynx are worthy of special notice. There may be no disturbance so long as the respiration is quiet, but as soon as the respiration becomes more active, a high degree of dyspnea, such as Landois has observed also in dogs, may set in, owing to the inability to dilate the glottis.

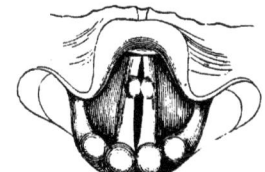

FIG. 219.—Tumors of the Vocal Cords, Causing Diphthongia.

If only one vocal band is paralyzed, the voice becomes impure and falsetto-like. The diminished vibration on the paralyzed side of the larynx may even be felt externally, but it may be still better recognized by means of the sensitive flame. It has been observed that unequal tension, from unequal innervation of the tensor-muscles, may give rise to alternating vibrations of the two bands, with opposite phases of movement. At times the vocal bands are paralyzed only to such an extent as not to move during phonation, but only on forced respiration and on coughing (phonetic paralysis). Mogiphonia, or premature fatigue of the voice, is the name given by Fränkel to a condition of paralysis of the laryngeal musculature that consists in failure of certain coördinated movements that have been acquired by practice. This corresponds to the paralytic form of writer's cramp.

Incomplete, unilateral paralysis of the recurrent laryngeal nerve results at times in double tone (diphthongia) of the voice, on account of the unequal tension of the two vocal bands. According to Türck and Schnitzler, diphthongia may develop also when the vocal bands are in contact at one point in their course (perhaps by reason of deposits or tumors), so that the glottis is divided into two sections, each of which produces the sound of the voice in a different pitch. If the glottis is suddenly closed by muscular spasm while the voice is being sounded, the rare condition of spastic aphonia results. In tabetic patients ataxic phenomena have been observed occasionally in the laryngeal musculature. Hoarseness is caused by accumulation of mucus on the vocal bands, or by roughness, swelling, or relaxation of the bands. If the bands suddenly come in contact while closely approximated during speaking, the voice "breaks," on account of the formation of nodal points.

Diseases of the pharynx, nasopharynx, and uvula may cause reflex nervous disturbances of the voice.

The reëstablishment of audible voice and speech has been observed even after total extirpation of the larynx, the individual breathing through a tracheal tube and no air escaping through the cavity of the mouth. The subject under such circumstances fills with air the cavity left by the removal of the larynx, and forces the air through a narrowed space above into the cavity of the mouth, thus producing a monotonous sound of stenosis that is remarkably like the voice.

Paralysis of the soft palate, as well as perforation or congenital fissure, gives a nasal timbre to all vowels; the former also causes difficulty in the normal formation of consonants of the third articulation-position. The resonants are well marked, while the explosives are enfeebled, on account of the escape of air through the nose.

Paralysis of the tongue causes difficulty in the pronunciation of I (ee); E (a), and A (ah) also are less easily pronounced. In addition, the formation of consonants of the second and third articulation-positions is disturbed. However, persons having even considerable defects of the tongue have reacquired intelligible speech. Aphthongia is that condition in which each attempt to speak results in spasmodic movements of the tongue.

In the presence of paralysis of the lips (facial nerve) the extent to which the consonants of the first articulation-position can be pronounced should be observed. Hare-lip also must be taken into consideration in this connection. In case of nasal obstruction speech assumes the so-called "obstructed oral tone." The formation of the resonants in the normal way is prevented. After extirpation of the larynx, an artificial larynx has been inserted between the trachea and the mouth, consisting of a metallic reed in a tube. All disturbances in the formation of consonants may be designated as "stammering" (dysarthria litteralis). Speech-derangements of cerebral origin are considered on p. 796.

COMPARATIVE.   HISTORICAL.

Speech may be included among the "movements of expression." Psychic excitement arouses in man characteristic movements, in which special groups of muscles constantly participate; for example laughing, crying, facial expression, and gestures in fear, anger, shame, discouragement, ambition, disgust, abhorrence, desire, joy, merriment, etc. Such movements constitute a medium by means of which related beings are enabled to communicate their inner thoughts to one another. In their origin these movements of expression are reflex motor phenomena; but when reproduced for purposes of explanation, they are voluntary imitations of these reflexes. In addition to emotional movements, impressions on the sense-organs also call forth characteristic reflexes, which are converted into movements of expression; for example, stroking or painful stimulation of the skin, movements following the perception of pleasant or unpleasant odors, likewise the influence of sound, also of light (bright or dark, and of colors), and the perception of objects of all kinds.

    In their simplest form movements of expression are manifested in sign-language. In a narrower sense speech may be designated as "sound-pantomime," the accompanying motor phenomena often taking the form of facial expression and gesture. Thus, articulate sound is caused, in the first place, by characteristic reflex motor phenomena in the speech-forming organs.

A second means of expression lies in the imitation of sound-phenomena by the organ of speech (onomatopoësis); for example the hissing of flowing water, the roaring of the storm, the rolling of thunder, ringing, howling, whistling, etc. If a further attempt is made to transform impressions depending on excitation of other senses into somewhat corresponding sound-perceptions, the term indirect onomatopoësis may be employed; for example the attempt to represent a sudden stab or a blinding flash of lightning by a short, shrill, whistling sound (Heise's principle of sound-metaphor).

Therefore, the primitive speech of man may have been a series of reflex sound-pantomimes and onomatopoetic imitations.

Moreover, expression in language is naturally related to the process of apperception. No idea can be expressed in language or gesture unless it be first apperceived, that is raised from the mass of ideas that fill the conscious mind to the psychic view-point.

Many different sounds occur in the various languages. Some tongues, for example that of the Hurons, have no labials: on some South Sea Islands no laryngeal consonants are spoken; f is wanting in Sanskrit, Finnish, etc.; the short e (a), o, and the soft sibilants in Sanskrit; d in Chinese and Mexican; s in many Polynesian tongues, r in Chinese; etc.

Movements of expression occur also among animals, especially the higher ones. The vocal organ of mammals is essentially like that of man. In some apes (orang-outang, mandril, pavian, macacus, mycetes) large sacs, which can be inflated with air, and which open between the larynx and the hyoid bone, serve as special resonance-organs. The whale has no voice.

Birds possess two larynxes, of which the lower is situated at the bifurcation of the trachea and is capable of producing the voice. Two folds of mucous membrane (in singing birds three) project one into each bronchus; they are rendered tense and are approximated by from one to five or six pairs of muscles, and they serve for the production of the voice.

Among reptiles, the tortoise can produce only a snorting noise, because it possesses no vocal bands; in the emys this may be increased to a peculiar whistling. The blind snakes are wholly voiceless; the chameleon and the lizard exhibit feeble voice-formation; the alligator and the crocodile are able to emit a roar, but in the adults of some species of crocodile the voice is lost, owing to changes in the larynx. Snakes lack special apparatus for voice-formation; in the act of forcing the air from their capacious lung through the entrance of the larynx, they produce a hissing sound, which at times may be surprisingly loud and harsh (puffing adder, hooded snake).

Among the amphibia, frogs possess a larynx with vocal bands and muscles. By blowing gently through this without muscular action, they produce deep, intermittent sounds; by blowing more forcibly and contracting the constrictors of the larynx, they produce a clear, continuous sound. The males of Rana esculenta possess at the angle of the mouth on each side a sounding-bag, which can be inflated and which intensifies the sound. In the tree-toad these sacs are united in the middle line to form a single laryngeal sac. Among toads, the sounds produced are usually weaker, and of these the bell-like note of the bombinator is worthy of notice; true toads emit feeble tones. The vocal organ of the Surinam toad (pipa) is peculiar: two cartilaginous rods project into the lumen of a large larynx; these are set into vibration by the current of air, and sound like vibrating rods or the branches of a tuning-fork. The salamander occasionally emits a sound resembling "uik." Among fish, utterance of sound occurs, as a result of friction of the upper and lower pharyngeal bones against each other, or of vibration of fins induced by muscular action, or of the escape of gas from the swimming-bladder, mouth, or anus. Finally, muscular sounds may be observed in fish.

Among invertebrates, insects are able to produce sounds partly by forcing the expired air through their stigmata, which are provided with reeds supplied with muscles (for example bees, many diptera, etc.). In addition, the wings often generate sounds by the rapid movement of their muscles (as in flies, beetles, bees). The death-head (Sphinx atropos) produces sound by forcing air from its sucking stomach. In others sounds are generated by rubbing the legs on the wing-cases (acridium), or the wing-cases on each other (gryllus, locust), or the breast (cerambyx), the leg (geotrupes), further the abdomen (necrophorus) on the margin of the wings, or the lower wing on the wing-case (pelobius). In the cicadas drum-membranes, pulled upon by muscles, are caused to vibrate. In some spiders (theridium) friction-sounds are produced between the cephalothorax and the abdomen, in some crabs (palinurus) also by the claws. In certain snails (helix) the escape of air produces a kind of voice.   Finally, some mussels (pecten) are able to produce sounds by beating their shells on each other.   In the animal world the utterance of sounds serves principally as a decoy.

Historical.—The Hippocratic school was aware of the fact that division of the trachea abolishes the voice.   Aristotle made numerous contributions regarding the voice and the venting of air in animals.   The true insight into the cause of voice-formation was, however, hidden from him, as well as from Galen.   The latter compared the vocal bands with the reeds of a shepherd's pipe.   Loss of voice in conditions of extreme weakness, especially after hemorrhage, was known to the ancients.   Galen observed loss of voice after establishment of double pneumothorax, further after section of the intercostal muscles or their nerves, also after destruction of the lower portion of the spinal cord, even when the diaphragm still performed its functions.   He gave the laryngeal cartilages the names that they still bear, recognized some of the laryngeal muscles, and asserted that the voice sounds only when the glottis becomes narrowed.   Dodart (1700) first attributed the development of the voice to the vibration of the vocal bands as a result of the air passing through the glottis; as the tension of the bands becomes greater the pitch of the voice increases.   The Paris professor Ferrein first declared correctly in 1741 that the width of the glottis is without influence on the pitch of the voice; he was the first to produce sounds in the excised larynx by blowing air through it.

The study of phonetics was practised already by the ancient inhabitants of India, less by the Greeks, but later by the Arabians.   Pietro Ponce (died 1584) was the first to give instruction in speech to deaf-mutes.   Later, Bacon (1638) studied the configuration of the mouth in the utterance of the various sounds; Johann Wallis (1653) did the same, partly for the instruction of deaf-mutes, and likewise Conrad Ammann (1692).   Kratzenstein (1781) first produced artificial vowels by fastening variously shaped resonators to a freely vibrating reed-apparatus.   Wolfgang v. Kempelen, of Vienna (1769–1791), constructed the first talking machine.   The voice-apparatus consisted of an ivory reed moved by means of a bellows and striking upon leather.   On the whole, the consonants were well produced; the aspirates were produced by whistling and hissing resonating tubes, the explosives by valve-like arrangements, R by a small rod dancing on the ivory reed, etc.   The vowels were produced by a megaphone, the cavity of which was altered by hand; A (ah), O, U (oo) were readily produced, E (a) with more difficulty, I (ee) incompletely.   Air was forced through the entire apparatus by means of a bellows, while the machine was "played upon" by the right hand raising valves and the left hand changing the megaphone; Wolfgang v. Kempelen stated correctly that tension of the vocal bands and narrowing of the glottis take place together; he is to be credited with many other accurate observations concerning the formation of articulate sounds.   F. H. du Bois-Reymond gave, in 1812, a natural system of consonants.   Robert Willis (1828) found that an elastic, vibrating spring yields the vowels in the series U (oo), O, A (ah), E (a), I (ee), in accordance with the pitch of its tone; also that the vowel-like sounds can be produced in the same order by lengthening or shortening an artificial resonating tube attached to a voice-apparatus.

GENERAL PHYSIOLOGY OF THE NERVOUS SYSTEM AND ELECTROPHYSIOLOGY.

GENERAL CONCEPTION OF THE NERVOUS SYSTEM.

STRUCTURE AND ARRANGEMENT OF THE ELEMENTS OF THE NERVOUS SYSTEM.

With relation to the general comprehension of the structure and function of the nerve-elements, two opposed views are held at the present time. According to the one the *neuron*, that is a ganglion-cell with all of its processes, is to be considered as the independent physiological unit of nervous tissue. The various neurons are not in immediate and direct connection with one another. The axis-cylinders of all nerve-fibers arise from ganglion-cells, and not from a network of fibers. All nerve-fibers terminate finally by means of terminal arborescences or telodendrites. It is only through these terminal filaments that the nerve-elements are connected by contact, the minute radicles being applied to one another. The nerve-cells and the nerve-fibers have each a distinct physiological importance, the cells acting as the physiological centers (for automatic or reflex movement, for sensation, perception, for trophic and secretory functions), the fibers, which always originate from the nerve-cells as processes, representing a conducting apparatus.

The more recent view rejects the neuron as the physiological unit and considers the fibrillary substance or the neuropile as the medium of nervous activity. The fibrillary substance is present in the great mass of gray matter, which represents a fine lacework or network of nerve-fibrils. It can be seen further in the nerve-cells and in the fibers passing off from them. The higher the plane of development in the animal scale the less numerous are the nerve-cells in proportion to the fibrillary structure, the ganglion-cells serving only as nutritional centers for the metabolism of the nerve-tissue. As Bethe has shown that reflex activity persists in crabs even after the ganglion-cells have been extirpated, conduction must obviously take place in the mass of fibers exclusively. The neuron thus loses its significance in the physiological sense and also from the histological standpoint.

The **nerve-fibers** are of several varieties:

1. The simplest form of nerve-fibers are the *primitive fibrils* or *axis-fibrils* (Fig. 220, 1), distinguishable only with high powers of the microscope. They occur as delicate fibers, presenting at varying intervals slightly varicose or spindle-shaped thickenings (postmortem change), and they can be recognized by the brown color that develops after the application of gold chlorid. They appear in part in the vicinity of the terminal distribution of the nerves, resulting from the fibrillation of the axis-cylinders, as, for example, in the layer of fibers of the optic nerve in the retina, in the terminal distribution of the olfactory fibers, in the net-like connection at the terminal distribution in unstriped muscle, and in part in the gray substance of the brain and the spinal cord as the most delicate processes of divided dendrites.

2. *Naked axis-cylinders* (Fig. 220, 2) represent bundles of primitive fibrils, which are characterized by most delicate longitudinal striation, separated by a number of fine granules. They are encountered in most exquisite form as neurites of central ganglion-cells (Fig. 220, I, z).

3. *Axis-cylinders surrounded by neurilemma* or the sheath of Schwann, from 3.8 to 6.8 μ in thickness, and designated *non-medullated* or *gray* nerve-fibers. The sheath of these fibers is a delicate, elastic cylinder composed of flattened cells

and covered here and there with oval nuclei. Dilute acids clear the fibers, without swelling. They are stained brownish red by gold chlorid. They occur in large numbers in the sympathetic nerves. In embryonal life all nerves, as well as the nerves of many invertebrates, are of this variety. In some situations several axis-cylinders are present within one sheath. These are designated Remak's fibers. They occur chiefly in the sympathetic system and in the olfactory nerves.

4. *Axis-cylinders* or nerve-fibrils surrounded only by a *medullary sheath* are present in the white and gray substance of the central nervous system, also in the optic and auditory nerves. After death they exhibit a tendency to undergo varicose and nodular thickening in certain areas in consequence of the coagulation of the medullary substance; hence they are also designated *varicose fibers*. Osmic acid acts upon these fibers only imperfectly; otherwise the medulla exhibits the same characteristics as the fibers of the following category.

5. The *medullated fibers*, with a sheath of Schwann (Fig. 220, 5, 6), occurring principally in the cerebrospinal nerves, but also in small number in the sympathetic nerves, exhibit the most complicated structure. They vary in width from 1 to 22.6 μ. The essentially nervous element of these fibers is the axis-cylinder (Fig. 220, 6, a), which occupies from one-fourth to one-sixth of the width and is surrounded by nerve-marrow like the wick of a candle. Generally it is somewhat flattened, and at times it is somewhat eccentric (Fig. 220, 7). Otherwise the axis-cylinder is composed of fibrils. Its consistence during life is that of semiliquid protoplasm or even more fluid. According to Kupffer a fluid (neuroplasm) is present between the fibrils.

Chloroform and collodion render the axis-cylinder visible. It is most readily isolated by means of nitric acid with an excess of potassium chlorid. On treatment with silver nitrate, Frommann noted the appearance in places of striation transverse to the axis-cylinder (Fig. 220, 8), the significance of which could not be determined.

The axis-cylinder is surrounded by the medullary sheath, which in the fresh state is homogeneous and strongly refracting. It is at the same time of fluid consistence, so that it exudes in globular drops (x) from the cut surfaces of the fibers. After death, however, or under the influence of heterogeneous fluids, the medullary substance at first retracts somewhat from the sheath, and as a result the fiber exhibits a double contour. Then the substance by a process of emulsification breaks up into larger and smaller globules, which tend to lie close together. Peculiar broken-up masses are thus formed in the nerve-fiber, giving it its characteristic appearance (Fig. 220, 6).

The medullary sheath is strongly refracting and positively uniaxially doubly refracting. The optically active body is lecithin. The substance of the medullary sheath is especially rich in cerebrin and lecithin, which swell up in warm water and assume similar forms, which have been well named myelin forms. Ether, chloroform, and benzine increase the transparency of the fibers (with disappearance of the double refraction) by solution of the fat-like constituents. Osmic acid stains them black.

Immediately surrounding the medullary sheath is the sheath of Schwann or *neurilemma* (Fig. 220, 6, c), a delicate, structureless membrane, resembling the sarcolemma. It contains disseminated oblong readily stained nuclei. On addition of acetic acid, and in chromic-acid preparations, this sheath appears in part isolated.

The sheath of Schwann exhibits, in the case of thick fibers at intervals of considerable length, in that of thin fibers at somewhat shorter intervals, the *nodes of Ranvier* (Fig. 220, 6, t t, and Fig. 221, *f s*). These are annular constrictions, about which the myelin is wanting. Between each two constrictions is a nucleus, so that such a segment of the fiber is equivalent to a cell from which it may be considered to have originated. At the annular constrictions the nutritive plasma probably enters the fiber for the axis-cylinder, as stains are able to penetrate at this point (8); probably also the products of metabolism are removed by the same channels. Apparently two segments of the sheath of Schwann are united by cement-substance at the annular constriction.

The axis-cylinder exhibits at the situation of the annular constriction regular preëxisting interruptions, as can best be demonstrated by treatment with silver nitrate. And although the discoverer Engelmann does not believe that in the living fiber a dividing layer of microscopic thickness is interposed in the annular constriction between each two adjacent segments of axis-cylinder, yet such an observation obviously gives material support to the view that the nerve-fiber is a chain of individual cells.

In the spinal nerves those fibers are the thickest that are the longest to their

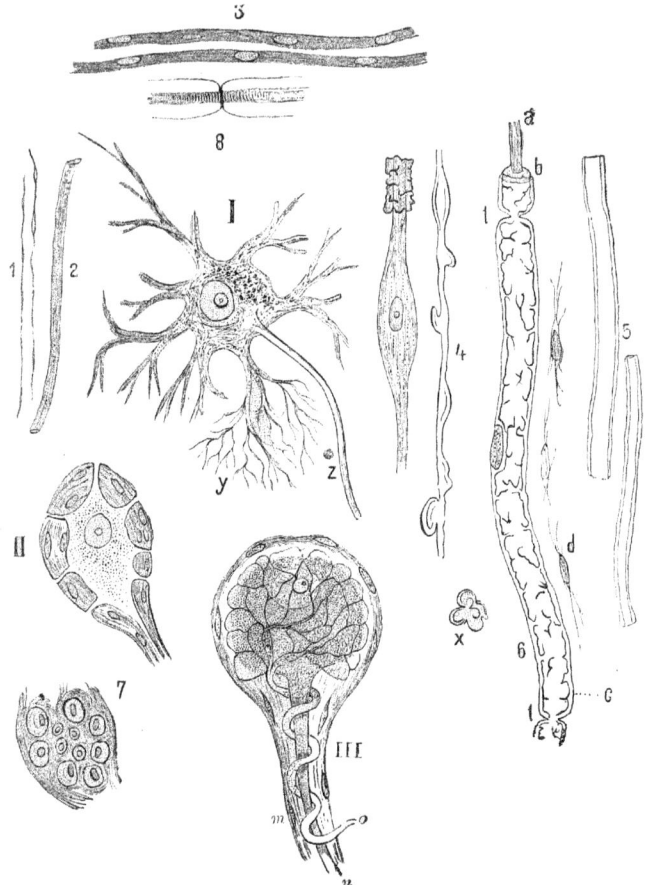

FIG. 220.—1, Primitive fibrils; 2, axis-cylinder; 3, fibers of Remak; 4, medullated varicose fibers; 5, 6, medullated fibers, with sheath of Schwann; c, neurilemma; t t, the annular constrictions of Ranvier; b, the medullary substance; d, cells of the endoneurium; a, axis-cylinder; x, medullary drop or myelin-globule; 7, transverse section of a nerve with distinct axis-cylinders, medullary sheaths and endoneurium; 8, nerve-fiber treated with silver nitrate; the axis-cylinder striated transversely (after Frommann). I, Multipolar ganglion-cell of the spinal cord; z, neurite; y, dendrites; to the right a bipolar ganglion-cell. II, Peripheral sympathetic ganglion-cell with connective-tissue capsule. III, Ganglion-cell with surrounding fibers; m, capsule; n, cellulifugal, o, cellulipetal fiber.

end-organ; and those ganglion-cells are the largest that give off the longest nerves.

According to Ewald and W. Kühne both the axis-cylinder and the medullary sheath are further surrounded by an exceedingly delicate horny sheath consisting of neurokeratin. Both are connected through the substance of the myelin by means of transverse or oblique fibers that divide the myelin between two annular constrictions into a number of successive segments. In this way are formed the oblique *clefts* or *indentations* of Schmidt, Lantermann, and Kuhnt in the myelin, as shown in Fig. 221.

According to Leydig and Joseph the axis-cylinders contain a delicate reticular framework, in the midst of which the fibrils of the axis-cylinder are embedded.

The nerve-fibers are undivided in their course in the nerve-trunk. As they approach their terminal distribution, they divide usually into two, less commonly into several, fibers that undergo no further change.

In animals the nerve-sheaths are sometimes still more complex. Thus, in the electrical nerve of the electrical catfish the sheath of Schwann of the individual nerve-fiber is constituted of such a large number of layers that the fiber may attain the thickness of a knitting needle. In the invertebrates the nerve-fibers as

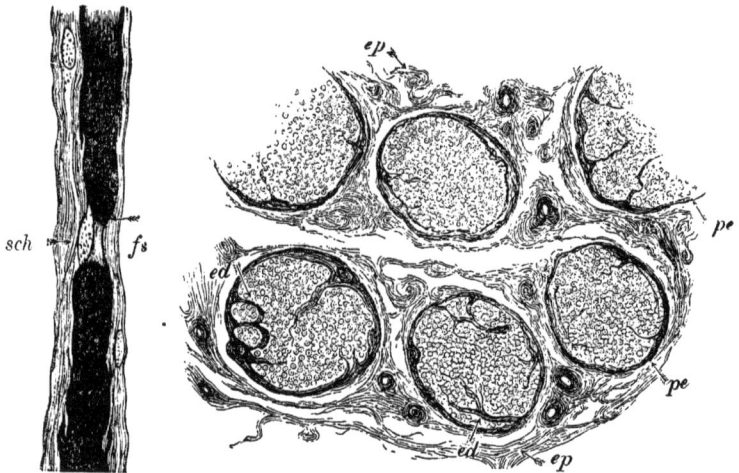

FIG. 221.—Medullated Nerve-fiber Stained Black by Osmic Acid: *fs*, annular constriction of Ranvier; *sch*, sheath of Schwann (after Eichhorst).

FIG. 222.—Transverse Section through a Portion of the Median Nerve: *ep*, epineurium; *pe*, perineurium; *ed*, endoneurium (after Eichhorst).

a rule have no medullary sheath. Retzius found them present, however, in the shrimp.

The connective tissue mixed with elastic fibers that surrounds a nerve-trunk is designated *epineurium* (Fig. 222, *ep*). The individual nerve-fibers are united in the nerve-trunk into bundles and each of the latter is surrounded by the *perineurium* (*pe*), from which the *endoneurium* (*ed*) passes inward between the nerve-fibers. These sheaths are at the same time provided with concentric layers constituted of smooth connective-tissue cells that can be stained with silver nitrate. Even the individual nerve-fiber, which results from the breaking up of a bundle of fibers, is further surrounded by a sheath composed of flat connective-tissue cells in mutual contact (Henle's sheath). The longitudinal network of blood-vessels of the nerve-fibers is supported by the connective tissue of the latter. The lymphatics form spaces, which reach to the individual fibers.

In their *development* the nerve-fibers first consist of fibrils that become surrounded by connective-tissue and finally by medullary sheaths. At birth the cerebral motor nerves and the auditory nerve alone are medullated. The longitudinal growth of the fibers takes place through elongation of the individual interannular segments and at the same time by the formation of new segments.

The ganglia have been considered partly as cells, partly as more complex structures. There are to be distinguished:

1. *Central ganglia* (Fig. 220, 1), occurring partly as large cells (up to 150 μ in diameter, visible to the naked eye, in the anterior horns of the spinal cord), partly as small cells (from 4 to 9 μ in diameter, deficient in protoplasm, in the posterior horns, in many parts of the cerebrum and cerebellum and in the retina), spherical, ovoid, or pear-shaped, with numerous processes that often give to the cells a star-shaped appearance. The brothers Landois found the ganglia of young insects much smaller than those of adult insects. A similar statement is made also by Schwalbe with respect to these cells and their nuclei. The cell-body is without a capsule, of soft consistence and exhibiting a reticular structure, or a finely fibrillar structure which extends into the processes. Between the fibrillæ is everywhere distributed yellow or brown finely granular pigment either heaped up at some particular part of the cell or disseminated throughout the entire cell. The relatively large nucleus is clear, granular or reticular, and in early life without a membrane. It contains little or no chromatin-substance. The nucleus contains a nucleolus, which in the fresh state is angular and motile, and after death is spherical, highly refractile, staining feebly, and often in turn contains a smaller granule.

On treatment with precipitating and coloring materials (alcoholic methylene-blue) the cell-substance can be demonstrated to contain chromophilic granular masses, so-called Nissl bodies, which are discernible with greater difficulty in the fresh and living state, and which appear with varying degrees of distinctness in different nerve-cells. In the motor ganglion-cells these bodies are arranged in concentric parallel layers. Electrical irritation induces contraction of the cell, causes it to appear darker and brings about a closer approximation of the Nissl bodies. Section of the motor fiber emanating from the ganglion-cell leads to granular degeneration and diminution in the number of the bodies. If the nerve-fiber undergoes regeneration the cell again acquires its normal appearance. According to Lugaro the cells of the spinal ganglia become altered and finally degenerate after division of their peripheral fibers, the nucleus becomes smaller, while the Nissl bodies increase in number and size and become grouped about the nucleus. Various poisons (strychnin, alcohol, tetanus, also uremia, autointoxications, inanition, anemia, prolonged high fever, heat-stroke, etc.) cause various changes in the ganglion-cells, mainly affecting the Nissl bodies. The cells of the cerebral cortex are affected in a peculiarly specific manner by each poison.

Of the processes given off by the ganglion-cells there are a large number that break up soon after their origin, like an intricately branching root; into numerous, delicate fibers presenting a varicose appearance, and designated *dendrites* or protoplasmic or secondary processes (Fig. 220, I, y). These dendrites conduct cellulipetally and form an intricate terminal network. The dendrites of adjacent cells do not anastomose with one another, but lie in close relation, entering merely into contact. Neither do the fibers of the dendrites give rise to nerve-fibers passing in a peripheral direction. The group of terminal filaments of a dendrite is designated a terminal network or a *telodendron*. In addition to the dendrites, the ganglion-cell gives off a process of considerable length, arising by a conical base, then pursuing a uniformly simple course, and conducting in a cellulifugal direction. This process is designated an *axone* or a *neurite* (an axis-cylinder or principal process) (I, z). The neurite is often continued peripherally into a nerve-fiber. Within the central nervous system it gives off here and there delicate branches that are designated *collaterals*. These break up into a fine terminal network, the radicles of which penetrate between the elements of the central organs. The neurites from the ganglion-cells of the cerebral and cerebellar cortex divide after a short course in a complex manner. Also these fine divisions come in contact with other nerve-elements in the central organ. Thus nerve-cells are connected only by the contact of their delicate processes. Moreover, a nerve-fiber never passes directly over into the histological elements of a nerve end-organ, the conducting fiber being only in telodendritic contact with that organ. If certain nerve-tracts are especially used and exercised, the altered function of the ganglion-cells in question may perhaps be explained by a further penetration of the dendrites into additional areas of the interstitial tissue, into which they had hitherto not penetrated. Demoor and Heger observed changes in the dendrites and neurites of the brain after irritation, cocain-anesthesia and the application of cold.

40

2. *Ganglia with connective-tissue capsules* (sheaths of Schwann) (II) occur (about 50μ in diameter) in the peripheral nerve-nodes. The soft cell-body, possessing two or more cell-processes, is surrounded by a firm capsule of cells closely applied to one another. The cell-body of the spinal ganglion cells is traversed by fine fibers; the capsule is later on connected with that of the nerve-fiber.

3. *Bipolar ganglia* are best seen in fish, for example in the spinal ganglia of the ray and the shark, as well as in the Gasserian ganglion of the pike. They appear as nucleated spindle-shaped swellings of the axis-cylinder (on the right, next to I). The nerve-marrow is often absent where the ganglion is interposed in the course of the fiber. Occasionally, however, the marrow, and always the sheath of Schwann, is continued over the ganglion.

4. *Ganglia surrounded with fibers* occur in the sympathetic system of the frog. From the pear-shaped cell (III, n) a process that remains non-medullated extends in one direction, and perhaps further on divides into two branches. In addition, on the surface of the cell, a second nerve-fiber is connected with an extremely delicate network of fine fibers. The second nerve-fiber winds around the first in a spiral manner and then proceeds in another direction (o) as a medullated fiber. Both cell and process are enclosed in a nucleated capsule (m). The straight fiber has been thought to conduct in a cellulifugal direction, the spiral fiber in a cellulipetal direction. It is possible, however, that the spiral fiber is derived from another ganglion-cell.

The cells in the peripheral ganglia are deserving of special consideration. They may be divided into two kinds: (1) the cells of the sensory ganglia, including the spinal ganglia, and on the cerebral nerves the Gasserian ganglion, the petrosum glosso-pharyngei, the jugular ganglion, and the nodose plexus of the vagus, the auditory ganglion, the geniculate ganglion. From the pear-shaped cells extend short prolongations that become gradually thinner and divide into two processes diverging in the shape of the letter T and becoming medullated nerve-fibers. Only the ganglia of the auditory nerve have cells with bipolar processes. The process from the peripheral sensory area to the ganglion is the *cellulipetal dendrite*. The neurite of the cell passes, however, in a cellulifugal direction into the central nervous system (in the case of the spinal ganglia it passes through the posterior roots into the spinal cord) and breaks up in a dendritic manner.

Every cell of the spinal ganglia has a connective-tissue capsule lined by a single layer of epithelium. The cell-body exhibits a granular chromophilic deposition and always pigment, while the ground-substance of the protoplasm has a reticular structure, and the nucleus is without chromatin.

2. The second group of peripheral ganglia contains sympathetic ganglion-cells. It includes in the course of the cerebral nerves the sphenopalatine, the otic, submaxillary, and ciliary ganglia, and all ganglia in the distribution of the sympathetic system. The cells of the sympathetic system have numerous dendrites and a single neurite. The latter becomes a non-medullated nerve-fiber with neurilemma, and it leaves the ganglion to surround the nerve-cells of other ganglia with a network or to terminate on a blood-vessel, while the dendrites ramify within the ganglion. The nerve-fibers passing from the central nervous system to the sympathetic ganglia surround the cells with a delicate network.

Numerous blood-capillaries surround the individual ganglion-cells, which also are provided with large lymph-spaces.

CHEMISTRY OF NERVOUS TISSUE.

MECHANICAL PROPERTIES OF NERVES.

Proteids.—Of the solid constituents of the gray matter about one-half are proteids; of the white matter one-third. There are present two phosphorus-free globulins and a nucleoalbumin (almost absent from the white matter).

One of the globulins is precipitable by a little neutral salt, and coagulates at 47° C. It is present also in leukocytes, muscles, and the liver. The other globulin is precipitated only on saturation with magnesium sulphate and coagulates at 70°. It is present also in liver-cells. The nucleoalbumin, containing 0.5 phosphorus, coagulates between 55° and 60° and is precipitated from a watery extract of brain-material by acetic acid.

Neurokeratin, a phosphorus-free substance rich in sulphur and closely related to keratin, occurs in the horny sheaths of the nerve-fibers, remaining after artificial digestion of the gray nervous substance by trypsin; treatment of the resulting product with potassium hydroxid yields pure neurokeratin. The material of the sheath of Schwann closely resembles elastin, although it is more readily soluble in alkalies. The connective tissue of the nerves yields gelatin.

Fats and fat-like substances soluble in ether occur principally in the white matter as follows:

(*a*) Liebreich's *protagon*, which resembles cerebrin, is readily decomposed, contains nitrogen and phosphorus, doubtfully sulphur, is the principal constituent of the brain-mass, but is wanting in the ganglion-cells, as well as in their decomposition-products.

It can be extracted from the white central nerve-mass by treatment with 85 per cent. alcohol at 45° C. It is readily soluble in ether, glacial acetic acid, and benzol, scarcely soluble in alcohol, and crystallizes in plates. It swells in water and becomes opalescent. When heated to 50°, the glucosid-like, phosphorus-free body, cerebrin, is separated. Boiled with baryta it yields the decomposition-products of lecithin. Protagon was considered by Diakonow and Hoppe-Seyler as a mixture of lecithin and cerebrin.

(*b*) *Cerebrin* occurs as a decomposition-product of protagon.

It is a white powder, consisting of spherical, transparent, smooth granules containing nitrogen, but free from phosphorus, soluble in hot alcohol, chloroform, and benzol, insoluble in ether or water. Boiled with dilute sulphuric acid it is decomposed into galactose and a fat. Farkus has separated from cerebrin homocerebrin (kerasin), a homologous readily soluble body, crystallizing in needles, and encephalin, which swells up in hot water like starch and contains an additional molecule of water.

(*c*) *Lecithin* is chemically combined in protagon. In addition there are present decomposition-products of lecithin, such as glycerophosphoric acid, oleophosphoric acid.

Lecithin is an ether-like combination of neurin in which the latter takes the place of the alcohol. It is of waxy consistence and it swells in water in myelin-forms; it is soluble in alcohol or ether. Neurin, $C_5H_{15}NO_2$, is a strongly alkaline, colorless fluid, forming crystalline salts with acids. It can be produced synthetically from glycol and trymethylamin; it is soluble in water and in alcohol. Neurin results by reduction from cholin; muscarin results by oxidation from cholin. Cholin is non-toxic, while neurin and muscarin are toxic. Cephalin closely resembles lecithin; it is precipitable from an ethereal solution by alcohol and is stained black by osmium.

(*d*) *Cholesterin* occurs partly free and partly in combination in the ganglia and in larger quantities in the white matter.

Whether *neutral fat* or *fatty acids* occur has not been positively determined.

3. The following products of retrogressive tissue-metamorphosis can be *extracted by water:* xanthin, guanin, and hypoxanthin, (?) adenin, kreatin, urea (in larger amount in case of retention of urine), (?) uric acid, jecorin, neuridin, a diamin occurring in connection with putrefaction. Further, W. Müller has found formic and acetic acids, taurin, much inosite, and in cattle leucin; v. Bibra, fermentation lactic acid, and Jaffe, a starch-like substance in human brains.

Nervous tissue in a state of rest has a neutral or slightly alkaline reaction, while when active and also when dead the reaction is acid.

The cerebral cortex in the fresh state yields an alkaline reaction, which is quickly changed to an acid reaction after death (?by fermentation lactic acid). The reaction of the nerve-fibers during life is variable. After the ingestion of methylene-blue Ehrlich found in living animals that the substance of the axis-cylinders stains blue, especially in those nerves that yield an alkaline reaction (cerebral cortex, cardiac nerves, the sensory and motor fibers of the unstriped muscles, gustatory and olfactory nerve-endings), while the endings of the voluntary motor nerves, which he considers have an acid reaction, remain unstained. According to Flesch the ganglion-cells exhibit differences in their reactions to stains, in accordance with their functions. The irritated nerve develops carbon dioxid.

As dead nerves exhibit increased consistence, it is probable that a condition of rigidity develops in them after death comparable to muscular rigidity and attended with the development of free acid. Fresh brain rapidly scalded at 100° C. remains alkaline (as do the muscles).

The gray matter contains more water (from 83 to 84 per cent.) than the white (from 68 to 70 per cent.). The dried material contains albumin (in the gray matter 30.89 per cent., without nuclein), partly soluble, partly insoluble (in the white matter 19.33 per cent., without nuclein and neurokeratin); lecithin 17.2 per cent. (9.9 per cent.); cholesterin and fats 18.7 per cent. (51.9 per cent.); cerebrin 0.5 per cent. (9.5 per cent.); extracts insoluble in ether 6.7 per cent. (3.3 per cent.); salts 1.5 per cent. (0.6 per cent.); the gray matter contains more phosphoric acid; neurokeratin (0.3 per cent. in moist peripheral nerves, 2.9 per cent. in moist white brain-matter). Breed found in 100 parts of ash: potassium 32, sodium 11, magnesium 2, calcium 0.7, sodium chlorid 5, iron phosphate 1.2, combined phosphoric acid 39, sulphuric acid 0.1, silicic acid 0.4.

Among the mechanical properties of nerve-fibers the absence of elastic tension in the various positions of the body is noteworthy. This is recognized from the fact that divided nerves do not retract and that the nerve breaks up on its surface into delicate macroscopically visible transverse folds (Fontana's transverse striation).

The marked cohesive resistance of the nerves to traction is responsible for the fact that when a part of the body is forcibly torn from the trunk, as in machinery accidents, the nerve-trunks often resist. The nerve, however, readily breaks up into its individual fibers.

If a constant electrical current be passed through an excised (or even dead) nerve there is a motor reaction of the contents of the fibers to the anode, of the sheath to the kathode.

METABOLISM IN NERVES.

Little is at present known concerning metabolism in nerves. The occurrence of various extractives has been determined, and these must be considered as decomposition-products. On the other hand, it has not yet been possible to demonstrate with certainty an interchange of oxygen and carbon dioxid. That, however, anabolism from the blood must take place in the nervous tissue is indicated by the fact that the irritability of the nerves diminishes after compression of the blood-vessels, and returns on restoration of the circulation. Thus, compression of the abdominal aorta is followed by paralysis of motion and sensation in the lower half of the body; occlusion of the cerebral vessels gives rise to almost immediate abolition of the functions of the cerebrum. Under such circumstances, the poverty of the nerve-trunks in blood-vessels is striking. As, however, the central nervous organs, especially the brain, undoubtedly receive a larger blood-supply the opinion seems justified that they are the seat of more active metabolism than the simple conducting apparatus. The ganglia form much lymph.

Hodge, Vas, Nissl, Mann, Beek, F. Pick, and others have studied the changes in the ganglion-cells that take place as a result of activity and exhaustion. It has been found that during rest the chromatic substance accumulates in the cells, while it is consumed during activity. Actively functionating cells are enlarged, as are also their nuclei and nucleoli. Exhaustion is indicated by contraction of the nucleus, probably also of the cell, and by the formation of a diffusely staining substance in the nucleus. The clearing up in the vicinity of the nucleus is due to disappearance of the chromatic substance. According to Levi numerous granules appear in the chromatic substance in rabbits as a result of activity of the spinal ganglia. These granules were wanting in the state of rest. Pick found the Nissl bodies in the spinal cord reduced in size. Demoor maintains that the active cells of the visual area stain less intensely, are diminished in size, and have an irregular nucleus. Poisoning with morphin gives rise to a granular condition of the protoplasmic processes of the cortical cells. The cells of the motor area of the cortex are shrunken after long-continued irritation, and their processes are granular.

IRRITABILITY OF NERVES. STIMULI.

Nerves possess the property of being thrown into a condition of irritability by stimuli, and they are, therefore, spoken of as irritable. Stimuli may be effective if applied at any point in the course of a nerve. An entirely uninjured and normally nourished nerve possesses the same degree of irritability at all points in its course. In new-born animals and in human beings up to the sixth week the nerves (and muscles) react less readily to electrical stimuli; the resulting contractions are slower and more extensive. The cause appears to reside in the incomplete development and evolution of these tissues. All stimuli, if powerful and long continued, soon cause paralysis by over-irritation of the nerve at the site of application. The nerve, therefore, loses its irritability at this point. Further on, toward the periphery, however, its irritability is still retained.

Mechanical stimuli affect the nerve when they induce a change in the form of the nerve-particles with a.certain degree of rapidity; for example, a blow, pressure, crushing, traction, puncture, section, concussion, sudden release of tension. In the case of sensory nerves pain occurs as a result ("falling asleep" of the extremities; pain on striking the ulnar nerve in its groove at the elbow); in the case of motor nerves, muscular contraction. If the mechanical injury to the fibers has resulted in interference with the continuity of their conducting elements (the axis-cylinders), the conductivity of the nerve ceases. If the molecular arrangement of the nerve-particles is permanently disturbed (for example by concussion), the irritability of the nerve is lost.

A light blow on the musculo-spiral nerve in the arm, on the brachial plexus in the supraclavicular fossa, causes in normal individuals contraction in the muscles supplied. The mechanical irritability of nerves may be abnormally increased under pathological conditions.

Tigerstedt discovered that the minimal value of the mechanical stimulation (induced by the falling of a weight upon the isolated nerve) is 900 milligram-millimeters, the maximal value from 7000 to 8000. More powerful stimulation causes exhaustion, but this does not extend beyond the irritated area. The mechanically irritated nerve does not acquire an acid reaction. A lesser degree of pressure or tension increases the irritability, which again diminishes after a short time. The work done by the irritated muscle as a result of this irritation was as much as 100 times greater than the kinetic energy of the mechanical nerve-irritation.

If a mechanical influence acts gradually the nerve may lose its conductivity or its irritability without any manifestation of irritation in the process. This class of phenomena includes the paralysis in the distribution of the brachial plexus as a result of long-continued pressure by a crutch and the paralysis of the recurrent laryngeal nerve by aneurysms.

As a result of pressure upon nerves by gradually increasing the weight applied there was observed at first an increase and then a diminution in the irritability.

In mixed nerves the reflex-conducting power is abolished earlier than the motor power.

Nerve-stretching is a mechanical procedure that has been employed for therapeutic purposes. If the exposed nerve is stretched, the tension acts as an irritant when it reaches a certain degree. After slight stretching the reflex irritability is at first increased; stronger stretching causes for a time diminution of irritability, as well as of reflex activity, and even temporary paralysis. The most extreme degree of stretching finally gives rise to permanent paralysis. It appears that the centripetal fibers (sciatic nerve) lose their conductivity earlier than the centrifugal fibers. In the process of stretching mechanical changes are induced in the nerve-tubes or in the end-organs that bring about alteration in irritability. The effect of the stretching may be propagated also to the central nervous system. Paralysis following forced stretching may undergo a marked degree of recovery. If, therefore, a nerve is in a state of excessive irritability, for example in a case of neuralgia, if this be due to inflammatory fixation or constriction of a nerve in its course, nerve-stretching may be useful partly by diminishing the irritability of the nerve, partly by breaking up the inflammatory adhesions. Nerve-stretching may be useful also in cases in which irritation of a centripetal nerve gives rise to reflex or epileptic convulsions by diminishing the peripheral irritability (in addition to the action described). In the case also of diseases of the spinal cord that have not yet advanced to a state of gross degeneration nerve-stretching is not to be neglected as a therapeutic agent.

For physiological purposes R. Heidenhain's tetanomotor is employed to induce mechanical nerve-stimulation. This consists of a vibrating ivory hammer attached to an extension of the Neef's hammer of the induction-apparatus, which by a rapid succession of blows upon the underlying nerve develops a condition of tetanus lasting up to two minutes.

Naturally, other mechanical stimuli of a similar nature will yield analogous results, such as contact with a vibrating tuning-fork, or with a sounding string, stroking with a bow-like apparatus, rhythmic stretching of the nerve (longitudinal traction).

Thermal Stimuli.—If a frog's nerve be heated to 45° C. its irritability at first increases and then declines. The higher the temperature the greater is the increase in irritability, but the shorter is its duration. Heated for a short time to 50° C. the irritability and the conductivity of the nerve are abolished; but on cooling, the frog's nerve is capable of recovering its irritability. Heat increased above 65° C. destroys the irritability, without preceding contraction, with degeneration of the myelin. A gradually frozen nerve retains its irritability when thawed. The cooled nerve retains its irritability for a long time. In motor nerves the irritability is increased, but the contractions are slighter and more prolonged and the conduction in the nerve continues for a longer time. Sudden cooling of nerves by a temperature of —5° C. and below excites contraction, as does also sudden warming by a temperature of from 40° to 45° C. and above. At still higher temperatures persistent tetanus occurs instead of the contraction. All such irritating variations in temperature if continued rapidly destroy the nerve, and probably give rise to chemical and mechanical alterations in the nerve-substance.

Of the nerves of mammals only the centripetal fibers and the dilators of the blood-vessels exhibit the effects of irritation by temperatures between 45° and 50° C. The remainder exhibit merely a change in irritability. Cooling to +5° C. diminishes the irritability of all of the nerve-fibers. Cooling of the ulnar nerve by immersion of the elbow-joint in cold water causes pain and contraction in the peripheral distribution of the nerve, such as is brought about by prolonged pressure. Local cooling of nerves increases the irritability to the constant current lasting for a considerable time (at least for $\frac{2}{100}$ second), and to mechanical and some chemical stimuli. A marked lowering of the temperature locally may abolish the conductivity of a nerve at the point of application. Local heating of the nerve to 35° C. increases its irritability to the faradic current, as well as to constant currents of shorter duration (of less than $\frac{1}{100}$ second).

According to Howell the extremes at which irritability is still present in motor nerves are 4° (cat), in the inhibitory nerve of the heart below 15°, in vasomotor nerves between 2° and 51°, in the sweat-nerves between 3° and 45°, in the respiratory fibers of the vagus 7°, and in the pressor fibers of the sciatic in rabbits about 2° C.

Chemical stimuli (chemical muscle-stimuli are discussed on p. 556) give rise to irritation in nerves when they cause alteration in the constitution of the latter with a certain degree of rapidity. As a result of the action of most of these irritants the irritability of the nerves is at first increased; then follows diminution to the point of abolition. Chemical irritants have, as a rule, less effect on sensory nerve-fibers than on motor fibers; so that chemical and thermal stimuli have to a certain degree opposite effects upon motor and sensory nerves. According to Grützner the failure of chemical stimuli to exert any effect on sensory nerves observed in most cases may, however, be due for the most part to want of simultaneousness of irritation of the individual fibers; and this view is supported by the circumstance that substances having a rapid and powerful action are capable under certain conditions of stimulating also centripetal fibers. Potassium and its salts exert a stronger action upon the sensory nerves (causing pain) than sodium; ammonium causes the most intense irritation. The painful effect of acids is proportionate to their degree of acidity. Of the monatomic alcohols the higher have a more intense action than the lower. Among stimuli of the motor nerves are: (a) rapid dehydration either by dry air (surrounding the nerve with filter-paper or suspending it over sulphuric acid) or by dehydrating fluids, such as concentrated solutions of neutral alkaline salts (sodium chlorid is said to stimulate only the motor nerves in mammals; sugar, urea, also concentrated glycerin and solutions of some metallic salts). Subsequent addition of water at times causes the contractions and spasm to disappear and the nerve may remain irritable. The dehydration at first increases the irritability, but later it is diminished. Imbibition of water decreases the irritability of the nerves. (b) Free alkalies, the mineral acids (not phosphoric acid), many organic acids (acetic, oxalic, tartaric, lactic), most salts of the heavy metals. While acids generally have irritant effects only in strong concentration, caustic alkalies have such effects in solutions down to 0.8 per cent. or even as low as 0.1 per cent. Neutral potassium-salts in concentrated form cause rapid destruction, but are less powerfully irritant than sodium-combinations. Neutral potassium-salts, when used in dilute solution, at first increase the irritability of the nerves and then diminish it, as can be demonstrated especially by stimulation with the opening shock of an induced current. (c) The anesthetics (ether, chloroform) and carbon dioxid in small amounts increase the irritability of isolated nerves; in larger quantities they diminish it. The haloid salts, especially the bromids, likewise diminish the irritability. Of the alkaloids and narcotics some (opium, cocain, curarin, chloral hydrate) diminish the irritability in part, while others (morphin, strychnin, muscarin, atropin) are indifferent in action. Other substances, such as dilute alcohol, bile, salts of the biliary acids, and sugar, generally excite contractions at first, after which the nerve rapidly dies. Ammonia, lime-water, solutions of some metallic salts, carbon disulphid, and the ethereal oils destroy the nerve without stimulating it (thus without inducing contractions in the frog-preparation). Carbolic acid (which excites convulsions on direct application to the spinal cord) has a similar action. These substances have a directly irritating effect upon the muscle.

Tannic acid has no irritating effect either upon the nerves or upon the muscles. In general, the irritating substances must be applied to the nerves in more concentrated solution than to the muscles in order that contractions may result.

Of chemically allied substances those act most intensely upon motor nerves that have a higher molecular weight; for example sodium iodid acts more intensely than sodium chlorid.

The Physiological Stimulus.—The nature of the physiological nerve-stimulus in the normal body is not known. It passes either in a centrifugal direction, from the central nervous system, as motor, secretory, or inhibitory impulses, or in a centripetal direction, from the specific terminal expansion of the organs of special sense and of the sensory nerves. The last-named class of stimuli are conveyed to the central nervous organs, where they are perceived as sensations, or they give rise by transference within the center to effects transmitted in a centrifugal direction and known as reflex processes. The individual physiological motor stimulus occupies a longer time than the transitory irritation of an induced current. It is not a uniform process, causing different effects in accordance with the varying intensity and the more or less frequent repetition, but it is rather a process exhibiting marked variability in the time of its occurrence and attaining a duration as great as one-eighth of a second.

Homologous and heterologous irritants and the law of specific energy are considered on p. 813.

Electrical Stimuli.—The electrical current exerts its strongest irritant effects upon a nerve at the time of its entrance into the nerve, and at the time of its disappearance. In like manner any rapid increase or decrease of the current passing through a nerve has a strong irritant effect. If, on the other hand, the current be allowed to pass gradually into the nerve-trunk, or to disappear gradually, or if the current passing through the nerve be gradually increased or diminished, the visible signs of nerve-irritation are much less marked. In general, the stimulation is most pronounced the more rapid the current-variation within the nerve, that is, the more suddenly the strength of the current passing through the nerve is increased or diminished.

This law applies, however, only to the rapidly moving muscles and their nerves (frog, warm-blooded animals). For muscles that move slowly (toad), the unstriated muscles and those of some invertebrates, slowly acting and slowly increasing electrical stimuli are the most effective.

If linear variation in the current (v. Fleischl's orthorheonome, v. Kries' spring rheonome) be employed as the stimulus, the intensity of the current must be the greater in order to obtain the same irritant effect the more slowly such linear variation takes place.

The electrical current must have a definite strength before it becomes active (threshold-value). With uniform increase in the strength of the current the size of the muscular contractions first increases rapidly and then more slowly. An electrical current must continue at least for 0.0015 second in order to stimulate the nerve; a current of shorter duration will have no effect. If the duration of the current be somewhat longer the stimulation on opening is wanting. The duration of closure of a constant current that continues for a time just short enough to be inactive need be prolonged only from 1.3 times to twice as long in order to attain the most complete effects.

The electrical current, further, is most effective when it is passed through the long axis of a nerve. It is ineffective when applied at right angles to the axis of the nerve. The muscle also is less responsive to electrical currents that pass transversely through its fibers than to such as pass longitudinally. The greater, further, the length of nerve through which the current passes the smaller need the electrical stimulus be.

When the constant current is applied to a motor nerve it exerts its most pronounced stimulating effect on closing and on opening. The stimulation, however, does not completely cease during the period the current is closed, for if the current be of moderate strength the muscle supplied by it remains in a condition of tetanus—galvanotonus or closing tetanus. The analogous reaction of the muscle on direct application of the constant current is considered on p. 563. When strong currents are employed this tetanus, it is true, subsides, but it does so because as a result of the action of the current in the nerve through diminution of its irritability resistances are developed that prevent the stimulation from reaching the muscle. According to Hermann, a descending current excites this tetanus more readily if the current is passed through the nerve at some distance from the muscle; while an ascending current excites the tetanus more readily when applied in the neighborhood of the muscle. The constant current manifests its stimulating influence upon sensory nerves in most marked degree at the moment of closing and opening. During the period of closure feeble stimulation is perceptible, but strong currents may under such circumstances give rise to unbearable sensations. Closing, opening, and the passage of the current stimulate all centripetal fibers, and also the vasodilators of the skin. The constant current has no effect upon vasoconstrictor and secretory fibers.

The following observation of Wedenskij is noteworthy: If the sciatic nerve of a frog be irritated by means of strong and frequent currents the tetanus induced soon disappears on account of exhaustion of the portion of nerve concerned, but begins again if the stimulus is either weakened or made less frequent. If the muscle is relaxed after the strong irritation, it contracts again if stimulated directly by a current of moderate intensity after the nerve-stimulation has been removed. The fact appears noteworthy that if two tetanizing stimuli are applied to the nerve of a frog-preparation, the two stimulations may counteract each other. If, for example, sodium chlorid be applied to the portion of nerve near the leg until the muscles are thrown into a tetanic contraction, this will cease if at the same time a portion of the nerve higher up is irritated. Both stimulating effects may, however, act together.

The phenomenon of deficiency is discussed on p. 664.

If the individual short current-shock occurs in rapid succession the related muscle is thrown into a state of tetanus.

The motor nerve possesses a greater specific irritability to electrical stimuli than the muscle-substance. This can be recognized from the circumstance that contractions take place on feebler stimulation if the nerve rather than the curarized muscle is stimulated.

Mention should yet be made of the remarkable fact that on irritation of a motor nerve the stimulating effect (contraction) is under certain circumstances the greater the nearer the point of stimulation to the central nervous system. According to v. Fleischl, however, the nerves are equally responsive to chemical stimuli at all points in their course. For electrical stimuli they are more responsive in their proximal portions only when the stimulating current is of the descending type. The reverse is said to be the case when the current is of the ascending type. Rutherford and Hallstén found that the reflex contractions induced by the irritation of a sensory nerve are the greater the more proximally the irritation is applied.

Nerve-fibers of like function in the same trunk do not always have the same degree of irritability. Thus, for example, feeble stimulation of the sciatic nerve of the frog causes contractions only of the flexors, while stronger stimulation is required to induce contraction also of the extensors. The effects of stimulation with long or with short intervals are analogous. According to Ritter, the nerves for the flexors also degenerate first. In a similar manner feeble stimulation of the hypoglossal nerve causes retraction of the tongue, while strong stimulation causes protrusion. In the facial nerve the fibers for the eyelids are more irritable than those for the mouth or the ear.

Also on direct irritation of the muscles (of curarized animals) the flexors contract in response to a weaker current than the extensors, but at the same time the former are more readily exhausted. Poisons generally injure the flexors earlier than the extensors. Treatment of a frog-preparation with ether causes flexion to occur on strong stimulation of the sciatic nerve. Increase of the stimulation, however, finally causes extension. Likewise, strong stimulation of the recurrent, laryngeal nerve during deep ether-narcosis causes dilatation, during slight intoxication constriction, of the glottis. Dilatation of the glottis has been induced by feeble stimulation. The adductor muscle of the crab's claw relaxes on feeble stimulation, while it undergoes contraction on strong stimulation.

Stimuli may be applied also by means of a single electrode of the induction-apparatus—*unipolar induced effect*. The cause resides in a movement of the electrical fluid to and from the free extremities of the open induced circuit at the moment of induction.

Upon *muscles* the action of electrical irritants is entirely similar to that upon nerves. The following, however, is noteworthy: currents of short duration have no effect upon muscles whose nerves are paralyzed by curare, as well as upon muscles enfeebled by intense exhaustion, degeneration, or pathological paralytic conditions.

A remarkable reaction designated *galvanotropism* is exhibited by entire animals when a current is passed through them. Feeble currents cause the animal to be thrown into a position with its long axis corresponding to the direction of the current, while strong currents have the opposite effect.

DIMINISHED IRRITABILITY; DEATH OF THE NERVE.

NERVE-DEGENERATION AND NERVE-REGENERATION.

The persistence of normal irritability in a nerve within the intact body depends first upon normal nutritive processes and the blood-supply of the nerve. In this relation it should be especially mentioned that insufficient nutrition is generally followed at first by an increase in the irritability. Only after advanced disturbance does the irritability diminish.

The physician should constantly bear in mind that whenever he encounters evidences of increased irritability of nerves under the influence of defective, or disturbed nutrition, as may be manifested in various ways, such as general nervousness and irritable weakness, etc., the condition is the beginning stage of a diminution of nerve-energy. Under such circumstances the nutrition should be improved by restorative remedies. Only the ignorant, misled by the signs of increased irritability of the nervous system, would employ depressing measures. In case of total obstruction of the blood-supply to the nerve-trunk, its irritability may persist for from five to ten hours.

If the terminal nerve-apparatus is exposed to a temporary disturbance of its

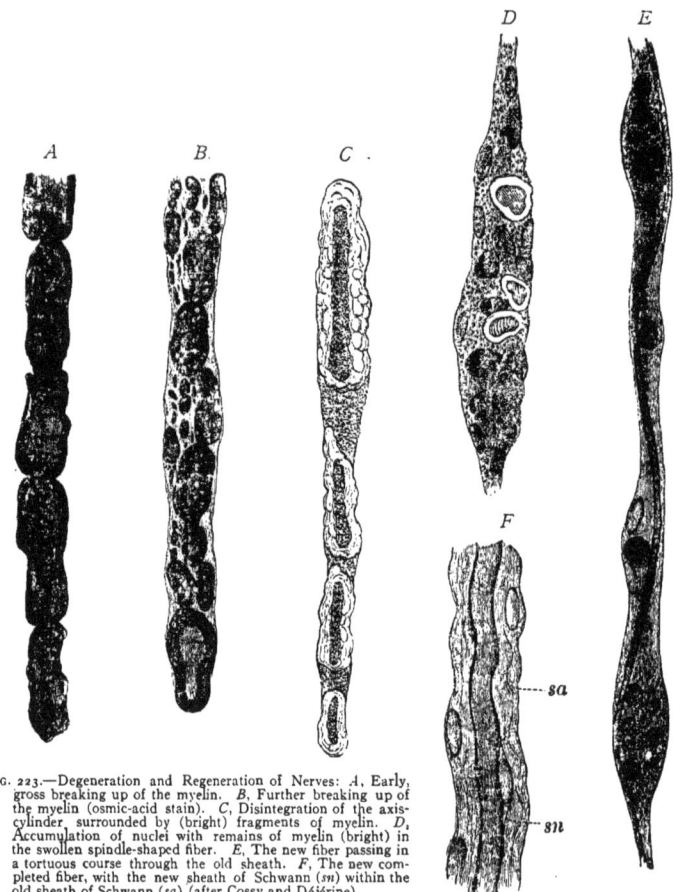

FIG. 223.—Degeneration and Regeneration of Nerves: *A*, Early, gross breaking up of the myelin. *B*, Further breaking up of the myelin (osmic-acid stain). *C*, Disintegration of the axis-cylinder, surrounded by (bright) fragments of myelin. *D*, Accumulation of nuclei with remains of myelin (bright) in the swollen spindle-shaped fiber. *E*, The new fiber passing in a tortuous course through the old sheath. *F*, The new completed fiber, with the new sheath of Schwann (*sn*) within the old sheath of Schwann (*sa*) (after Cossy and Déjérine).

normal nutrition, it responds to the restoration of normal nutritive processes by the development of a more or less intense irritative process. The effective disturbance of nutrition need exist a shorter time the more sensitive the nervous end-apparatus in question is to the nutritive disturbance, such as cutting off of the arterial blood-supply or interference with respiration.

Long-continued excessive irritation of a nerve without suitable intervals of rest for purposes of recuperation soon causes fatigue of the nerve and later on diminution of irritability through exhaustion.   Neverthe-less the nerve exhibits extraordinary resistance to various stimuli.   It may not be exhausted even after irritation continued for hours.

The enfeeblement and finally the cessation of muscular contraction after long-continued stimulation of the motor nerve connected with the muscle are due to exhaustion of the muscle and not of the nerve.   If while a nerve is being stimulated the muscle is prevented from contracting, by rendering the nerve incapable of conducting at a point distal to the site of stimulation (by anelectrotonus or curare), it will be found that even after twelve hours of constant irritation of the nerve, the muscle can again be made to contract if this obstruction (blocking of the nerve) be removed.   Also the observation that the negative ·variation in the nerve-current in an irritated nerve continues for a long time is interpreted in the same way.

The recovery of nerves takes place at first slowly, then somewhat more quickly, and finally again more slowly.   Should recovery not take place in the first half-hour in the frog, after long, intense irritation, the nerve does not recover at all.

Long-continued inactivity diminishes the irritability ·to the point of complete abolition.

The characteristic example of this is furnished by the degeneration of nerves after amputation of an extremity.   Not only the sensory nerves to the cutaneous area, etc., removed, but also the motor nerves to the muscles removed, undergo atrophy, and also their continuations in the spinal cord exhibit atrophic changes. The degeneration of the optic nerve after extirpation of the eye and of the auditory nerve after that of the internal ear is considered on pp. 679 and 699.

Excised nerves preserve for a time, as does muscle, their func-tional activity.   At first the end-apparatus of the nerve degenerates; then, in the case of motor nerves, the muscle; and finally the nerve itself.

Nerve-fibers are capable of maintaining their normal nutrition only when they are in uninterrupted connection with their trophic center, which controls the nutritive processes.   If, however, the nerve within the otherwise normal body is separated from its nutritional center, as by section or crushing, it loses its irritability in a short time and the per-ipheral end undergoes fatty degeneration, which begins in warm-blooded animals in the course of from four to six days, in cold-blooded animals after a longer interval.

The irritability of the nerve under these conditions—the so-called reaction of degeneration—is discussed on p. 672.   The degeneration after section of the roots of the spinal nerves is described on p. 716.

In the otherwise intact body both extremities at the point of division undergo *traumatic degeneration* in from one to two days in frogs, as a result of which the white substance of Schwann and the axis-cylinder can no longer be distinctly differentiated.   This degeneration extends, however, only to the next node of Ranvier.   Later, so-called *fatty degeneration* takes place simultaneously in the entire peripheral portion.

Fatty degeneration of nerves begins by a breaking up of the myelin (Fig. 223, *A*), which later becomes transformed into drop-like masses (*B*).   The axis-cylinder also swells up and disintegrates (seventh day) (*C*).   The nuclei in the sheath of Schwann become swollen and proliferate by mitosis (up to the tenth day) (*D*).   According to Ranvier, it is this nuclear proliferation and that of the protoplasm or neuroplasm lining the sheath of Schwann that first cause disinte-gration of the myelin and the axis-cylinder and that subsequently increase to such a degree as to convert the entire peripheral portion of the nerve into a connective-tissue strand, the fragments formed at the same time undergoing absorption. In the motor end-plates degeneration likewise takes place, at first in the non-medullated fibers, then in the terminal filaments and lastly in the nerve-trunk.

If regeneration takes place, the extremities of the divided nerve must have united, and for this purpose in the human being nerve-suture has been employed.

In the middle of the fourth week small, bright bands, developed from the proliferated protoplasm, appear within the sheath of Schwann, and these penetrate between the nuclei and the remains of the myelin (E). They are the new axis-cylinders, which thus develop in an endogenous manner within the old sheath of Schwann. Soon they become thicker, and receive myelin, with Lantermann's clefts, Ranvier's nodes and sheaths of Schwann (from the 2d to the 3d month) (F). According to Ziegler the new axis-cylinder, which develops independently, becomes only later connected with the central stump. The formation of the myelin takes place continuously. A portion of the nuclei disappear; the outer, with their protoplasmic portion, give rise to the sheaths of Schwann. Exactly the same process takes place in nerves ligated in continuity. It is a remarkable fact that several new fibers may develop within an old fiber. From the central extremity of the divided nerve-fiber the axis-cylinder after the fourteenth day grows toward that of the newly formed fiber and unites with it. The central extremity of a divided motor nerve may unite with the peripheral extremity of another nerve and still functionate. Langley united the central extremity of the vagus with the peripheral extremity of the sympathetic, and found after union took place that the vagus had acquired control of all structures supplied by the cervical sympathetic. According to Gessler restoration of the end-plate occurs first in the process of regeneration. In the case of non-medullated fibers the contents only and not the sheaths degenerate from the third day on. After two days perceptible regeneration begins.

The regeneration of nerves is under the influence of the nerve-centers acting as nutritive centers. If nerves be completely and permanently separated from these centers regeneration will not take place.

In the regeneration of mixed nerves sensation returns first, then voluntary muscular movement, and finally movement on irritation of the motor branches.

As the fatty degeneration involves the peripheral extremity of the nerve the observation of this process in a divided nerve affords a means of determining the central origin of fibers in a complex arrangement of nerves. The division of motor nerves results also in fatty degeneration of the related muscles in case restitution does not take place.

After division of the axis-cylinder the nerve-cell from which it originates undergoes alteration, the Nissl bodies swelling and disintegrating, though later being restored. In the process of degeneration the cell increases in volume and the nucleus assumes a peripheral position. This period covers from one and three-fourths to twenty days. Restitution occupies about ninety-two days. These changes are indications not of paralysis of the cells, but only of a certain impairment of their function. If the degeneration is permanent the cell disintegrates.

Under the influence of various procedures, such as crushing of the nerve-fiber, the remarkable observation has been made that voluntary impulses or irritating influences originating above the site of compression are conducted through the nerve to the muscle and give rise to contraction, while the irritability to stimuli below the site of compression is greatly diminished. Nevertheless Erb did not observe this difference with respect to mechanical stimulation. In an analogous way it will be found that the nerves of animals poisoned with carbon dioxid, alcohol, cocain, curare, or coniin, occasionally also the nerves in paralyzed parts of the body in man, are no longer responsive to local stimuli, although they still conduct impulses from the central areas. The injured segment of nerve thus loses its irritability earlier than its conductivity. The analogous phenomenon in muscle-fibers is discussed on p. 114.

After the administration of certain poisons to living animals, especially veratrin, the irritability of the nerves is at first increased, then diminished to the point of complete abolition, as indicated by the extent of the contractions in the muscle supplied by the affected motor nerve. In the case of other poisons the abolition of irritability takes place rapidly, as, for example, that induced by curare. Coniin, cynoglossum, methylstrychnin iodid, and ethyl-strychnin iodid.

If a frog-preparation consisting of nerve and muscle be placed in a poisonous solution results are occasionally manifested which are different from those produced if the poison is administered internally to the living animal. Atropin, for example, gives rise to a reduction in the irritability of the preparation, without preceding increase. Alcohol, ether, and chloroform, first increase and then diminish the irritability.

If the nerve is separated mechanically from its connection with its centers, as by section, or if the center has undergone degeneration, the nerve is first thrown into a state of increased irritability beginning in its central extremity and extending toward the periphery; then the irritability diminishes to the point of complete abolition. This process takes place more rapidly within the portions of the nerve nearer the center than in the more distal portions. This phenomenon is known as the Ritter-Valli law.

The rapidity of conduction of stimuli in nerves is increased in the stage of increased irritability and diminished in that of lowered irritability. In the latter stage the current, on electrical stimulation, must be continued for a longer time in order to be effective, therefore the rapidly successive shocks of the induced current are generally ineffective. Also the law of muscular contraction is modified in the various stages of the alteration in irritability during the process of degeneration.

Finally, attention should be called to the fact that some nerves possess a greater irritability at certain points, and that they retain this slightly longer at such points.

Thus, for example, the sciatic nerve of the frog in its upper third is more irritable to various stimuli, in both its sensory and its motor fibers, than at a more distal portion. Such inequalities in the irritability are due alone to accidental injuries inflicted in the course of preparation. A branch is given off from the upper third of the sciatic. According to Beck, in an uninjured nerve the more central portions require stronger stimuli than the more peripheral to induce the first minimal effect.

After section or crushing of a nerve all of the electrical currents employed for the stimulation of the nerve that pass in the nerve from the site of the lesion exert a much more active influence than those in an opposite direction. The cause for this resides in the fact that the current developing in the nerve after the injury is added to the electrical stimulating current. Also in an uninjured nerve, for example the sciatic of the frog, there are points at the central or peripheral termination of the nerve, or where large branches are given off, that react in a manner similar to the sites of lesion previously mentioned.

The dead nerve has lost its irritability completely. Death advances in accordance with the Ritter-Valli law from the central organs of the nervous system gradually to the peripheral paths. An acid reaction, which is present in dead muscles, can be demonstrated, though not con-· stantly, in dead nerves.

The functions of the brain cease immediately after the onset of death, as indicated by loss of consciousness and cessation of perceptive activity; so that reports of brain-activity after decapitation are to be relegated to the realms of fable. The vital functions of the spinal cord, however, persist for a somewhat longer time, particularly those of the white substance. Death occurs next in the large nerve-trunks; then in the nerves for the extensors and in those for the flexors (in three or four hours). The sympathetic fibers retain their irritability longest (up to ten hours in the intestine). The irritability of the nerves of frogs can be preserved for several days in the dead body if kept in the cold.

The irritability of the peripheral stumps of divided nerves is lost in pigeons and rodents in from two to three days; in from eight to ten days; in other warm-blooded animals in four days. In mixed nerves death of the fibers takes place at different intervals; for example. in the vagus, of the inhibitory fibers first, later of the accelerator fibers for the heart. The stumps of the cerebral nerves retain their irritability for a longer time than do those of the spinal nerves.

ELECTRO-PHYSIOLOGY.

The discussion of electrical phenomena will be preceded by a concise summary of the necessary preliminary physical considerations, without which a comprehension of the subject is impossible. This presentation will be made in a connected manner, the apparatus and methods devised for electro-physiological and electro-therapeutic purposes being described in their proper place. The student should familiarize himself thoroughly with this preliminary knowledge of physics.

PRELIMINARY PHYSICAL CONSIDERATIONS. THE GALVANIC CURRENT.

ELECTROMOTORS. CONDUCTION-RESISTANCE. OHM'S LAW. CONDUCTION THROUGH ANIMAL TISSUES. THE RHEOCORD.

If two of the bodies named later on are brought into direct contact with each other, positive electricity will be appreciable in the one and negative electricity in the other. The cause of this phenomenon is the electromotive force, which causes positive electricity to pass into the one body and negative electricity into the other. In accordance with the relations of the bodies to be discussed later on these are divided into *conductors* and *non-conductors*, and the conductors again into those of the first and those of the second class.

Conductors of the first class, chiefly the metals, can be arranged in such a series (tension-series) that, on contact of the first mentioned with one of the succeeding members of the series, the first body becomes electrically negative and the last positive. This tension-series is: Manganese, carbon, platinum, gold, silver, copper, iron, tin, lead, zinc. The intensity of the electrical excitation resulting from the contact of two of these bodies is the greater the farther the bodies are separated from each other in this tension-series. The contact of the bodies may take place indifferently at one or at several points. If several of the bodies in the tension-series are placed one upon the other, the electrical tension thus induced is the same as if the two terminal bodies alone were brought in contact, with omission of the intervening bodies.

If, on the other hand, conductors of the first and the second class are brought in contact, the result is different. Zinc in contact with water is strongly negative, the fluid positive. Zinc in contact with diluted acids is likewise negative, while other metals, such as copper and platinum, are less actively negative or even positive.

Experience has shown that those metals in contact with a fluid become in strongest degree electrically negative that are chemically most strongly acted upon by the fluid. Every combination. however, exhibits a constant difference in tension or potential. The density of the amounts of electricity set free from both bodies is dependent upon the size of the surfaces in contact. The fluids, for example the solutions of acids, alkalies or salts, are designated *exciters of electr.city of the second class*. They form no definite tension-series among themselves. Immersed in most of the fluids named, the metals nearer the positive side of the tension-series, particularly zinc, prove in greatest degree electrically negative, while those situated nearer the negative side are so in less degree.

If two different substances of the first class be immersed in a fluid, without coming in direct contact, for example zinc and copper, free negative electricity will appear at the projecting extremity of the (positive) zinc, and free positive electricity at the projecting extremity of the (negative) copper. Such a combination of two electromotors of the first class with an electromotor of the second class is designated a *galvanic circuit*. As long as the two metals remain separated in the fluid, the circuit is said to be *open;* as soon, however, as the projecting extremities are connected, for example by a wire arc, the circuit is *closed* and a *galvanic current results*. Both forms of electricity flow mutually into and neutralize each other, although in accordance with the degree in which the tensions neutralize each other new electricity is constantly generated in the circuit.

The galvanic current encounters resistances in its course that are designated *conduction-resistance* (W). This is (1) directly proportional to the length (l) of the circuit; (2) inversely proportional to the transverse section of the circuit (q), the length being the same; and (3) dependent upon the molecular peculiarities of the materials (*specific conduction-resistances*). Therefore, the conduction resistance $W = (s, l) \div q$.

The conduction-resistance increases in the case of metals and diminishes in the case of fluids with increase in temperature.

The *strength of the galvanic current* (S), or the quantity of electricity passing through the closed circuit, is thus proportional to the electromotive force (E), or the electrical tension, but inversely proportional to the total conduction-resistance (L). Therefore, $S = E \div L$ (Ohm's law).

The total conduction-resistance in the closed circuit, however, is made up (1) of the resistance in the closing arc, external resistance, and (2) of the resistance within the battery itself, internal resistance. The specific conduction-resistance of the different substances is thus variable. In the case of metals it is relatively slight, in that of fluids, however, quite marked. The specific conduction-resistance or rather the specific conductivity is at present generally indicated with reference to mercury as the unit. Accordingly, the conductivity of copper is 55, that of iron from 6 to 10, that of German silver from 3 to 6. In the case of fluids the resistance is exceedingly slight—for concentrated salt-solution 0.00002, for concentrated solution of copper sulphate 0.000004.

Conduction in Animal Tissues.—In animal tissues the conduction-resistance is exceedingly great, generally some millions of times as much as in metals. A constant current passing from the skin through the animal body encounters progressively diminishing resistance, on account of the galvanic conductivity of the water in the epidermis and the increased fulness of the vessels in consequence of the cutaneous irritation. Nevertheless, different portions of the surface of the body react differently, the least resistance being offered by the palms of the hands and the soles of the feet. The seat of the resistance is the epidermis, after removal of which, as by a cantharidal blister, the resistance is greatly reduced. The resistance is diminished, however, by increased superficies of the electrodes, and by increased moisture, heat, and intensity of their saturation. It is greatest in the extremities, least in the face. Dead tissue is usually a poorer conductor than living tissue. If the current is passed transversely through a muscle, it encounters nine times as much resistance as if it were passed longitudinally through the fibers of the muscle. In the longitudinal direction the resistance of the muscle is two and a half million times greater than that of mercury. Tetanus and cadaveric rigidity diminish the resistance in the muscles. If the conduction-resistance be tested with alternating currents much lower figures are obtained than if the constant current be employed, because the occurrence of polarization, especially internal polarization, can largely be omitted from consideration.

The resistance of the body varies between 260 and 1,250,000 ohms. It is high in cases of hysteria and melancholia, and low in cases of tetanus and exophthalmic goiter. Alt and Schmidt make the following statements as to the degree of conduction-resistance in various tissues: Nerve 0.17, muscle 1, blood 1, skin 1.25, brain 1.57, tendon 3.25, fat 3.92, muscle-sheath 4.41, bone 14.1.

From Ohm's law two laws of great importance in electro-physiology may be deduced: I. If there is great resistance in the circuit in the arc of closure, as is the case when a nerve or a muscle is intercalated in a closing arc, the strength of the current may be increased only by increasing the number of electromotive elements. II. If the conduction-resistance in the arc of closure, in comparison with that in the battery, is exceedingly small, an increase in the strength of the current cannot be brought about by increasing the number of elements, but only by an increase in the surface of the plates in the element.

It is important to differentiate exactly the terms *electromotive force* and *current-strength*. The electrical current may be compared with a current of water. The cause of the current in the water is the hydraulic pressure, that of the electrical current the electromotive force. The current-strength is the amount of water, or the amount of electricity, that passes in one second through the transverse section of the conductor. The pump that drives the water to the top of a high vessel, and thus generates the hydraulic pressure, corresponds to the electrical element. In the current of water a mass is set in motion, in the electrical current a force.

Since 1881 the electrical values, especially current-strength, electromotive force and resistance. have been indicated with reference to units that have a simple relation to the so-called absolute units. Those units are designated absolute that refer to the unit of length (cm.), the unit of time (sec.) and unit of weight (gr.). The *unit of resistance* is the *ohm* (= 10^9 absolute units). It is equal to the resistance of a column of mercury at a temperature of $0°$ C., having a transverse section of 1 square meter and a length of 1.026 meters. The ohm is, therefore, only a little larger than the earlier unit of Siemens (1 m. long and 1 square meter in transverse section).

The *unit of electromotive force* is the *volt* (= 10^8 absolute units). A Daniell's cell has an electromotive force of 1.1 volts.

The *unit of current-strength* is the *ampère*, that is the current that generates an electromotor force of 1 volt in a circuit having a resistance of 1 ohm (therefore 0.1 absolute unit). An ampère generates 0.174 cu. cm. of exploding gas in one second at a temperature of $0°$ and an atmospheric pressure of 760 mm.

An electrical current in passing through a wire generates heat, the amount of which is proportional to the product of the resistance multiplied by the square of the current-strength. or, according to the law of Ohm, of the current-strength multiplied by the electromotive force. The product of this from volt and ampère is equal to 10^7 work-units and is designated a *watt*.

According to the technical designation, 1 watt equals 0.00136 horse-power (1 horse-power equals 75 kilogrammeters).

As, in absolute measure. the mechanical heat-equivalent of the gram-calory equals 42,000,000 work-units, $\frac{42}{175}$ or 0.24 gram-calory is generated in a circuit with an electromotive force of 1 volt and a current-strength of 1 ampère in a second.

The *density of the current* must further be especially distinguished from the current-strength. As the same amount of electricity must always pass through any given transverse section of the circuit, the electricity must obviously be denser at the constricted portions and less dense at the wider portions if the transverse section varies in size. If S indicate the current-strength and q the transverse section of the part in question, the density (d) at this point will be $d = S \div q$.

If the arc of closure of the galvanic circuit be divided at the one pole into two or several circuits, which are reunited at the other pole, the total of the current-strengths is equal to the strength of the undivided current. If, further, the different circuits vary with respect to length, transverse section, and material, the current-strengths passing through the wires are inversely proportional to the conduction-resistances.

According to this principle, that of the derived circuit, the *rheocord* of du Bois-Reymond is constructed. With the aid of this instrument it is possible to pass from a galvanic current a derived current of any determined strength for the stimulation of a nerve or a muscle.

From each of the poles (Fig. 224. a b) of a galvanic battery are given off two wires, of which the one pair (a c and b d) pass to the nerve of the frog-preparation (F). The intercalated segment of nerve (c d) offers a high degree of resistance to this branch of the current (a c d b). The second branch of the current conducted from a and b (a A, b B) passes through a thick brass plate (A B), which is made up of seven pieces lying side by side (1–7) and united through the brass plugs (from S_1 to S_3) placed in the intervals, except between 1 and 2, so as to form an uninterrupted circuit. It will at once be clear that by means of this arrangement, as depicted in Fig. 224, only a minimal current passes through the segment of nerve (c d), which offers great resistance, while by far the greater portion of the galvanic current passes through the well-conducting brass plate (A – B). If increased resistance be introduced into this latter circuit. the branch current a c d b must naturally be increased correspondingly. These resistances can be interposed by means of the portions of fine wire indicated by the letters I a, I b, I c, II, V, X. If it be supposed that all of the brass plugs (from S₁ to S₅) are withdrawn, the branch current entering at A must pass through the entire system of fine wire. In this way a high degree of resistance is interposed and the branch current in the nerve must be increased correspondingly. If but one plug is withdrawn, the current passes only through the respective length of wire. The resistances offered by the various lengths of wire (from I a to X) are so related that I a, I b, and I c each represents a unit of conduction-resistance, II twice as much, V five times as much, and X ten times as much resistance. The distance I a may finally be lessened by the bridge (L), which can be moved upward, the scale (x y) indicating the length of the resistance-distance. It will be readily perceived that in accordance with the manner of applying the plugs and the bridge, the apparatus permits of a varied gradation in the branch current to be sent through the nerve. If the bridge L is pushed up close to 1, 2, the current passes directly from A to B, and not through the length of thin wire I a.

Other forms of apparatus intended for introduction into the closing arc of a circuit, in order to increase the conduction-resistance at will, are designated *rheostats*.

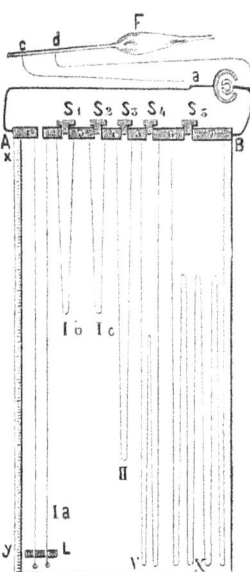

FIG. 224.—Diagrammatic Representation of the Rheocord of du Bois-Reymond.

THE ACTION OF THE GALVANIC CURRENT UPON THE MAGNETIC NEEDLE. THE MULTIPLICATOR.

If a galvanic current be passed (for example through a wire) parallel to a magnetic needle, the latter will be deflected from its position pointing to the north. If it be conceived that one is swimming in the positive current, the head in front and the abdominal surface directed toward the needle, the north pole of the magnetic needle will always be deflected toward the left (Ampère's rule). The deflecting force exerted by the galvanic current upon the needle always operates at right angles to the so-called *electromagnetic plane*, that is the plane passing through the north pole of the needle and two points in the conducting wire (passing in a straight direction parallel to the needle). If, for example, the conducting wire passes just above and parallel with the magnetic needle, whose plane of oscillation is formed by the horizontal surface, the electromagnetic plane will be vertical to the horizontal plane, and it will pass through the north pole of the needle and the conducting wire. The strength of the galvanic current that causes the deflection of the magnetic needle is proportional to the sine of the angle between the electromagnetic plane and the plane of oscillation of the needle.

This deflecting power of the galvanic current can be increased if the conducting wire is passed, instead of once, several times in the same direction in front of the magnetic needle. An apparatus constructed according to this principle is designated a *multiplicator*. In this the conducting wire passes in numerous turns at right angles to the horizontal plane around the magnetic needle suspended in the middle and swinging in the horizontal plane. The larger the number of turns the greater will be the angle of deflection of the needle, although not exactly in direct proportion, as the individual turns are at varying distances and occupy different positions with reference to the needle. The multiplicator is thus an apparatus by means of which a feeble current can readily be detected.

Experience has taught further that if the feeble galvanic current to be examined encounters great resistance in the closed circuit, such as in animal tissues through which the current is passed, then many turns of a thin wire are to be made about the needle. If, however, the conduction-resistance in the circuit is but slight, as is the case, for example, in the application of the thermoelectric apparatus, only a few turns of a thick conducting wire are made about the magnetic needle.

In order to render the multiplicator more sensitive in another manner the magnetic power of direction of the needle, by means of which it tends to turn toward the north, can be enfeebled. The extent to which this has been attained in the thermoelectrogalvanometer for the examination of feeble currents has been described and illustrated in connection with the study of feeble thermic currents (p. 333). It should be especially mentioned at this point that for the demonstration of electrical currents in animal tissues a coil consisting of a large number of turns of thin wire is to be attached to that instrument.

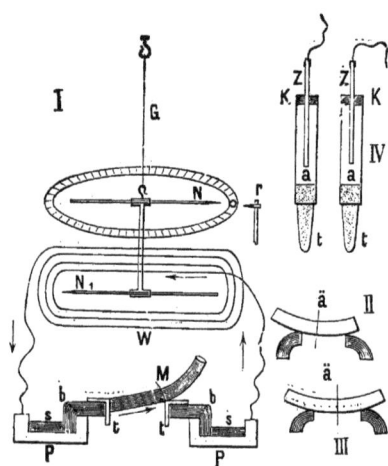

FIG. 225.—I, Diagrammatic representation of the multiplicator adjusted for the investigation of a muscle-current; N N₁, a static pair of needles suspended from a silk thread G; P P, the conducting vessels with the muscle M; II and III, other adjustments of the muscle; IV, non-polarizable electrodes.

In the multiplicator of Schweigger, employed for physiological purposes, the tendency of the needle to point toward the north has been materially enfeebled by the employment of the astatic pair of needles, as suggested by Nobili. Two identical magnetic needles are attached parallel one above the other by means of a fixed middle piece of horn, but in such a manner that the north poles point in opposite directions. As it is impossible to impart to each needle a magnetic strength of absolutely equal degree, one of the needles will thus be always somewhat stronger than the other. This difference in strength should, however, not be so great that the stronger needle is directed toward the north, but it should be sufficient only to cause the freely suspended pair of needles to assume a certain angle to the magnetic meridian, to which position it always returns after having been deflected therefrom, with the execution of a number of progressively diminishing oscillations. The angle assumed by the astatic pair of needles to the magnetic meridian is designated the *free deflection*. The greater the degree of astasia attained, the more nearly will the angle formed by the direction of the free deflection with the magnetic meridian approximate a right angle. The greater the degree of astasia, the fewer will be the number of oscillations made by the pair of needles in a given time, when they attempt, after deflection, to resume their original position. The duration of each of these periodic oscillations will then be quite long.

The multiplicator is so constructed that the direction of the needles is the same as that of the coils of wire. The upper needle oscillates above a scale graduated in degrees on which the extent of the deflection of the needle can be read. Even the purest copper wire in the coil always contains a certain admixture of iron, which exerts an attraction upon the magnetic needle. Therefore, a small fixed magnetic rod, designated a correcting rod or compensatory magnet (r), is attached to the multiplicator. This is directed toward the one pole of the upper needle and diminishes the strength of the astatic needles to such a degree that the attracting force in the coils of wire (in consequence of the iron present) is rendered ineffective with respect to the force of the earth's magnetism.

ELECTROLYSIS. TRANSITION-RESISTANCE. GALVANIC POLARIZATION. CONSTANT BATTERIES AND UNPOLARIZABLE ELECTRODES. INTERNAL POLARIZATION OF MOIST CONDUCTORS. CATAPHORIC ACTION OF THE GALVANIC CURRENT. SECONDARY RESISTANCE.

Every galvanic current that is passed through a fluid conductor causes decomposition of the fluid (*electrolysis*). The products of decomposition, designated *ions*, are deposited at the poles immersed in the fluid, the *electrodes* (of which the positive is designated the *anode* and the negative the *kathode*), *anions* collecting at the anode and *kations* at the kathode. If the products of decomposition are deposited upon the electrodes, they may mechanically, through their adhesion, either increase or diminish the difficulty of conduction through the electric fluid. This is designated *transition-resistance*. If by this means the conduction-resistance already present in the battery is increased, the transitional resistance is designated *positive*, while if it diminishes the conduction-resistance in the battery, it is designated *negative transition-resistance*.

The ions that collect at the electrodes may, however, modify the strength of the current also by the development between the anions and the kations (as between two different bodies connected by a conducting fluid) of a new galvanic current. This phenomenon is designated *galvanic polarization*. Thus, for example, water is decomposed by immersed platinum electrodes in such a manner that the negative oxygen collects at the positive pole, and the positive hydrogen at the negative pole. The polarization-current thus generated usually has a direction opposite to that of the original current, and, accordingly, is designated *negative polarization*. In rare cases, however, the polarization-current has the same direction as that induced by the decomposition and then the phenomenon is known as *positive polarization*.

Naturally, in the process of electrolysis both factors may be operative, namely transition-resistance, as well as polarization.

Polarization, when present, may be so slight as not to be recognizable with the naked eye, but it may then be demonstrated in the following manner: After the lapse of a short time the primary source of the current, for example the element with which the electrodes were connected, is excluded, and the extremities of the electrodes projecting out of the fluid are placed in communication with a multiplicator, which at once indicates even slight polarization by deflection of the needle.

The ions set free in the process of electrolysis cause, at times, at the moment of their development, further secondary decomposition. If, for example, platinum electrodes are immersed in sodium-chlorid solution, chlorin accumulates at the anode, and sodium at the kathode. The chlorin, however, immediately exerts a decomposing influence upon the water, the oxygen of which it takes up for oxidation, while the hydrogen is deposited secondarily at the kathode.

The degree of polarization increases (although in slighter measure) with the current-strength, while it diminishes almost proportionately with elevation of temperature. The endeavor to overcome the polarization, which, as can be seen, would soon modify the strength of the galvanic current present, has led to the invention of two important devices; namely, *constant galvanic batteries* and the so-called *unpolarizable electrodes*.

The *constant batteries* yield a constant current, that is a current of the same intensity, because the ions generated upon the electrodes are removed at the moment of their development, so that they are thus unable to give rise to a polarization-current. For this purpose the two bodies used for the tension-series are each immersed in a separate fluid, separated by a porous septum (porcelain cylinder). In the zinc-platinum cell of Grove the zinc is immersed in dilute sulphuric acid, the platinum in nitric acid. The oxygen deposited in the process of electrolysis at the positive zinc forms zinc oxid, which is at once dissolved in the dilute sulphuric acid. The hydrogen attracted to the platinum unites at once with the nitric acid to form water, the acid giving off oxygen and being converted into nitrous acid. The zinc-carbon cell of Bunsen acts in the same way, the negative carbon being immersed in nitric acid, the positive zinc in dilute sulphuric acid. In the cell of Daniell the positive zinc is immersed in dilute sulphuric acid and the negative copper in a concentrated solution of copper sulphate. The zinc undergoes the same change as in the Grove cell. The negative copper, however, attracts hydrogen, but the latter at once in the nascent state reduces the copper from its combination to metallic copper, which accumulates on the copper plate as a bright deposit. The electromotor force of a Daniell cell varies, in accordance with the degree of amalgamation of the zinc and the concentration of the fluid, between 0.909 and 1.35 volts, the internal resistance being 2.8 ohms.

If the electrodes of a constant element be conveyed to a moist animal tissue, for example nerve or muscle, electrolysis and. as a result, polarization must, naturally, at once take place. In order to avoid this, *unpolarizable electrodes* have been constructed (Fig. 225, IV). As a result of the studies of Regnauld, Matteucci, and du Bois-Reymond, it has been determined that such electrodes can be constructed if the conducting wire coming from each element be first connected with an amalgamated plate of zinc (z. z), the latter being secured (k, k) in a tube filled with a solution of zinc sulphate (a a), whose lower extremity is closed by means of an inverted cone of clay (t, t) moistened with 0.6 solution of sodium chlorid. If these clay points are applied to the tissues, no polarization takes place or at most only a very slight amount.

Exactly the same device is employed for examining the currents in muscles and nerves (Fig. 225, I). As these tissues when in direct connection with metals generate currents, a similar non-polarizable device is employed, but under such circumstances it has a somewhat different form. It consists of cups of zinc (P, P) filled with concentrated acid-free zinc-sulphate solution (s. s). In each cup is immersed a pad of blotting paper (b, b), which is saturated with the zinc-solution. Finally, this is covered with a thin layer of plastic clay (t, t) moistened with 0.6 per cent. sodium-chlorid solution, which protects the tissues from the direct caustic effects of the dissolved zinc salt.

Nerve-fibers and muscle-fibers, as well as moist vegetable tissues, fibrin, and similar bodies, which have a porous structure filled with fluid, likewise exhibit the phenomena of polarization on the application of currents of considerable strength. and this has been designated *internal polarization of moist conductors.* It is believed that the better-conducting solid particles in the interior of these bodies exert an electrolytic effect upon the particles of fluid in contact with them, as do metallic electrodes in contact with fluid. The ions resulting from the disintegration of the particles of the internal fluid would then give rise to the internal polarization in consequence of the tension existing between them. The conduction-resistance of muscle and nerve depends. according to Hermann, in part upon polarization. He considers the marked polarization of animal tissues (only comparable with that of the metals) as a specific vital property of protoplasm.

If the two electrodes of the cell are introduced into the divisions of a fluid separated into two halves by a porous partition, it will be observed that particles of fluid are conveyed in the direction of the galvanic current, from the positive to the negative pole, so that after the lapse of some time the amount of fluid in one half of the vessel has diminished, while that in the other half has increased. This phenomenon of direct transference has been designated the *cataphoric effect.* Upon it depends the galvanic transference of soluble substances through the external integument. Upon this depends, apparently, also the phenomenon of so-called *secondary external resistance.* If the copper electrodes of a strong constant cell are each introduced into a vessel filled with copper-sulphate solution. from which projects a pad saturated with this fluid, and if further over this pad is placed a bit of muscle, cartilage, vegetable tissue or a prismatic strip of coagulated albumin, it will be seen that after closure of the circuit the current undergoes considerable enfeeblement. If the current be now reversed, its strength is at first increased, but later it declines from the maximum. Thus, a constant alternating reversal of the current gives rise to similar alternation in the variation of the current. If a prismatic bit of albumin has been used in the experiment, it will be observed that simultaneously with the enfeeblement of the current the albumin has become deficient in water and presents a shrunken appearance in the vicinity of the positive pole, while, conversely, the albumin applied to the negative pole (probably through cataphoric action) is swollen and contains more water. If the direction of the current be altered the same phenomena is observed. but at the opposite poles. The contraction and loss of water in the albumin at the positive pole described must be the cause of the resistance in the circuit that explains the enfeeblement of the galvanic current. This phenomenon is designated that of *secondary external resistance.*

INDUCTION. THE EXTRA CURRENT. MAGNETIZATION OF IRON BY THE GALVANIC CURRENT. VOLTAIC INDUCTION. UNIPOLAR INDUCTION-EFFECTS. MAGNETO-INDUCTION.

If a galvanic element be closed by means of a short curved wire a feeble spark will be observed at the moment when the circuit is again opened. If, however, the closure is effected by means of a long wire wound into a coil a strong spark is observed on opening the circuit. If two handles are attached to the closing wire and held in the hands so that the current (through interruption of the wire-conduction between the two handles) at the moment of opening is conducted only by the body, a severe shock is felt at the moment of opening. This phenomenon is due to a current induced in the long, coiled spiral, which Faraday designated the *extra current*. The cause for its development is as follows: If the circuit is closed through the spiral wire, the galvanic current passing through the latter induces an electrical current in the adjacent turns of the same spiral. This induction-current is, at the moment of closure in the spiral, opposite in direction to the galvanic current in the circuit. Therefore, its effect is limited and it likewise causes no shock. At the moment of opening, this induction-current has, however, the same direction as the current in the circuit and, therefore, its effect is intensified.

Electrical apparatus, which, therefore, is so constructed that the irritation to which it gives rise results from interruption of the circuit in a spiral conductor is designated *extra-current apparatus*.

If a soft-iron rod be introduced into the cavity of a coiled wire spiral, it becomes magnetic so long as an electrical (galvanic) current passes through the spiral. If one extremity of the iron rod is turned toward the observer, and the other in the opposite direction, and if further the positive current passes through the spiral in the direction of the hands of a clock, the extremity of the rod turned toward the observer is the negative pole of the magnet. The strength of a magnet thus produced depends upon the strength of the galvanic current, the number of spiral turns and the thickness of the iron rod. As soon as the current is opened the magnetism in the iron bar disappears.

If a spiral roll be made of a long insulated wire, which may be designated the *secondary spiral;* if, further, a similar wire spiral designated the *primary spiral* be placed in the vicinity of the first, and the ends of the primary spiral are connected with the poles of a galvanic element, an electrical current is generated in the secondary spiral when the primary current is closed, or when opened after having been closed. A current, likewise, appears in the secondary spiral if this is brought closer to or removed further from a closed primary spiral (through which a current is constantly passing). The current appearing in the secondary spiral is designated the *induced* or *faradic current*. The process of this induction has been designated *voltaic induction* or *electrodynamic distribution*. The current developed in the secondary spiral on closure of the primary current or on approximation of the two coils to each other passes in the direction opposite to that of the primary current. On the other hand the current induced on opening the primary current or on separation of the two spirals from each other has the same direction as the primary current. While the primary current is closed or when the distance between the two spirals remains unchanged no current is demonstrable in the secondary spiral.

The currents developed in the secondary spiral on opening and closing the circuit differ from each other in the following particulars. Although the amount of electricity neutralized on opening and closing the current is the same, so that the same effect from both can be demonstrated by means of electrolysis, as also by means of a galvanometer, the electricity at once attains its maximum intensity and continues for a short time with the opening current, while the electricity increases but gradually, does not reach an equally high maximum and flows for a much longer time with the closing current. The reason for this important difference is as follows: With the closure of the primary circuit, there develops in the primary spiral the extra current, which passes in a direction opposite to that of the primary current. It, therefore, offers resistance to the more rapid development of the primary current to its full strength. The current induced in the secondary spiral therefore develops slowly. As, however, on opening the primary spiral the extra current in the latter passes in the same direction as the primary current, the disturbing influence mentioned disappears. The more rapid and profound action of the opening current is of great significance with relation to the physiological employment of induction-currents.

It may naturally be desirable under some circumstances to remove this inequality in the closing and opening shocks. This end can be attained by greatly weakening the extra current. This is accomplished simply by giving the primary spiral only a few turns. v. Helmholtz has attained the same object by introducing a secondary circuit in the primary circuit. By this means the current never disappears entirely in the primary spiral, but it is alternately weakened and strengthened by the alternate closing and opening of this secondary circuit of much less resistance.

If a current is made to appear or disappear in the primary coil with great rapidity, the induction-current develops in the secondary spiral not alone when the free extremities of the spiral wire, which may be connected with some part of an animal, are closed, but also when only one extremity of the wire is made to divert the current by the contact. There occur, therefore, on contact with only one extremity of the secondary spiral contractions in the frog-preparation that are designated *unipolar induced contractions*. Generally they appear only on opening the primary circuit. The occurrence of these contractions is favored by connecting the other extremity of the spiral in diverting contact with the earth, and also if the frog-preparation is not completely insulated.

Brief consideration may now be given to so-called *magneto-induction*. According to Ampère one may conceive of a magnetic bar as surrounded permauently by electrical currents in such a manner that if the south pole be directed toward the observer the currents pass around each transverse section of the bar like the hands of a clock. On this assumption it will be readily understood that a magnet will develop a current in a wire coil near by as soon as the two are approximated, and also if a piece of soft iron is suddenly rendered magnetic or suddenly loses its magnetism. The direction of the currents thus induced in the coil is the same as that of those induced on voltaic induction: that is the development of magnetism or the approximation of a coil of wire to a magnet gives rise to an induced current in a direction opposite to that of the current assumed to be present in the magnet; conversely, the disappearance of the magnetism or the separation of the coil from the magnet gives rise to a current in the same direction.

Approximation and separation of a magnet and a coiled wire may be effected in rapid succession if a magnetic bar that is fastened at one extremity is permitted to vibrate freely in the vicinity of the coil. The pitch of the note of such a rod will then naturally indicate the rapidity of the movement and thereby at the same time the number of induced shocks—Grossmann's acoustic current-shocks and the resulting acoustic tetanus in the frog-preparation.

DU BOIS-REYMOND'S SLIDING INDUCTION-APPARATUS. PIXII-SAXTON'S MAGNETO-INDUCTION MACHINE.

The sliding apparatus is an improved modification of the magneto-electromotor of Neef for physiological purposes. The apparatus is readily comprehensible from the accompanying sketch (Fig. 226). A wire passes from one pole (a) of the galvanic battery (D) to the metallic column (S), from the upper extremity of which an easily vibrating metallic spring (F) projects in a horizontal direction and is provided at its free extremity with a rectangular strip of iron (e). An adjustable screw (b) is approximated to the middle of the spring from above so that contact between the two takes place. From the screw (b) passes an insulated copper wire (c) to a hollow spiral (x x), within which are placed a number of rods of soft iron (i i) insulated by a coating of varnish. From the spiral the wire (d) passes on to a horseshoe of soft iron (H), which it surrounds in spiral turns, passing finally from this back again (at f) to the battery (g).

While the current is closed in this manner, it must effect the following results: It renders the horseshoe (H) magnetic and it in consequence at once attracts the movable strip of iron (e, Neef's hammer). By this means the contact of the spring (F) with the screw (b) is broken. The current is thus interrupted, the horseshoe accordingly loses its magnetism, and it releases e, which is drawn upward by the spring, so that contact takes place again at b. This new contact causes renewed magnetization of H and attraction and release are thus repeated in rapid succession, in consequence of which the primary current between F and b is alternately opened and closed with equal frequency.

A coil (K K), designated the secondary spiral, hollow within, and consisting of numerous turns of thin insulated wire, passes in the same direction as the spiral

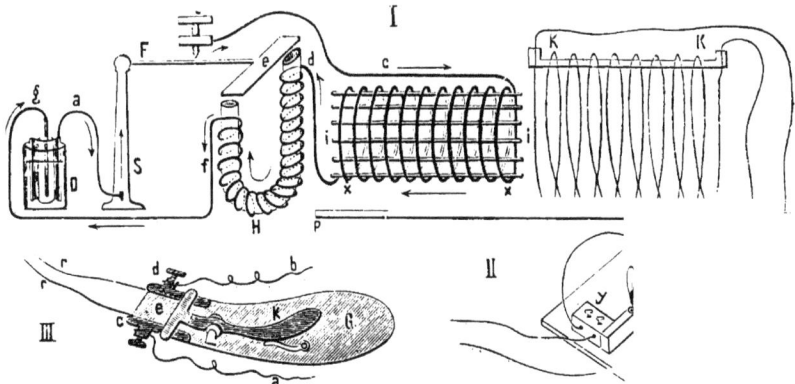

(x x) of the primary current. This is mounted upon a long board or slide (p p), provided with a scale upon which it may be moved over the primary spiral, which it then receives into its concavity (the induced current being then strongest), or it may be removed any desired distance from the primary spiral (the current then being feeblest). The degree of separation of the coils is thus an index of the strength of the stimulus. The measurement of the current-strength may naturally be made more accurately by means of graduated instruments. According to the laws of voltaic induction there develops in the secondary spiral (K K) on closing the primary current an induced current opposite in direction to that of the primary current, and on closing the primary current an induced current in the same direction. Moreover, according to the laws of magneto-induction the magnetization of the iron bar (i i) within the primary spiral (x x) through closure of the primary current causes the development of a current in the secondary coil (K K) in the opposite direction, and the demagnetization of the bar, by opening the primary circuit, an induced current in the same direction. These facts explain the more powerful effect of induced opening currents, as compared to closing currents. The removal of the inequality in the two currents has been discussed on p. 646.

The magneto-induction (or rotation) apparatus (Fig. 227), devised by Pixii, and improved by Saxton, and provided by Stöhrer with a commutator, consists of a powerful horseshoe steel magnet, opposite to whose two poles (N and S) is placed a horseshoe of soft iron (H), which can be rotated about a horizontal axis (a b).

FIG. 227.—Magneto-induction Apparatus with Stöhrer's Commutator.

The extremities of the horseshoe are surmounted by wooden spools (c d), around which an isolated wire is wrapped in numerous spirals. If the horseshoe is in a position of rest, as indicated in the figure, it is exposed to the influence of the large steel magnet, and it becomes magnetized itself. It turns to the poles of the steel magnet the opposite poles s and n. In the wire of the two wooden spools c and d an electrical current is developed whenever the horseshoe loses its magnetism or again acquires it. If half a rotation of the axis a b is made, so that the spool c is apposed to the poles, the magnetism in the horseshoe naturally changes its polarity, as the poles of the steel magnet N and S must always be in relation to the opposite poles of the horseshoe. This alternation in the poles of the horseshoe can naturally be brought about only when the original magnetism present disappears and the new magnetism of opposite polarity develops. The disappearance of the magnetism in the horseshoe and the development of the opposite kind gives rise to currents in the spiral in the same direction. With the second half-rotation the poles are restored to their original position. There must, therefore, be induced in the spiral a current of opposite direction from that of the current resulting with the first half-rotation. Each complete rotation of the horseshoe thus gives rise to two currents passing through the spiral in opposite directions, so that the conducting wires o and p are alternately positive and negative.

Stöhrer has by the application of his commutator succeeded in causing the two currents mentioned to pass in the same direction. For this purpose two metallic collars (m and n) well insulated from each other are placed upon the axis (a b) one over the other. Each collar is provided at both its upper and its lower extremity with a hollow metallic half-ring: thus, the collar n with the half-rings 3 and 4, and the collar m with the half-rings 1 and 2. The half-rings are arranged alternately in pairs. Of the two polar wires of the spiral one (o) is connected with the inner collar (m) and the other (p) with the outer collar (n). The divided metallic plates Y and Z are prolongations of the poles and act as conductors to the electrodes. It can be readily seen that in this position p passes to 3 of the outer collar and thence to Z. After a half turn, however, o is connected by 2 of the inner collar with Z. An analogous change in position takes place at Y. If, now, as has already been pointed out, o and p change their polarity with each half-turn, so that after every half-rotation first o and then p becomes positive, by means of the commutator Z remains constantly connected with the positive and, accordingly, Y constantly with the negative pole. The half-rings 1 and 4, as well as 3 and 2, project somewhat beyond each other at their extremities. By this means it results that, in a certain position, o and p are closed for a short time above and below by Z and Y. At this moment no current passes through the electrodes. The apparatus is most efficient and it is also available for electrolytic purposes.

The *key* (Fig. 226, II) is an adjunct to this apparatus. It consists of a device by means of which the current is made to pass through a wide metallic bridge (y, r, z) until it is sent through the parts to be stimulated. The latter takes place at the moment when the connecting metallic plate (r) is introduced between the two blocks y and z. The key-electrode (III) can be employed in the same manner for physiological purposes. This conveys the current to the tissues as soon as the spring connecting plate (e) is raised by pressure upon k. This instrument can be controlled with a single hand: a b are the polar wires, r r the insulated electrodes connected with the parts to be stimulated, and G the handle of the instrument.

ELECTRICAL CURRENTS IN RESTING MUSCLE AND NERVE. CUTANEOUS CURRENTS. GLANDULAR CURRENTS.

Method.—To test the law governing the muscular current there is required a muscle made up of parallel fibers and of simple structure, thus representing a prism or a cylinder (Fig. 228, *I* and *II*). The sartorius muscle of the frog may subserve this purpose. In such a muscle a distinction is made between its surface or the natural longitudinal section, its tendinous extremities or the natural transverse sections, and, if the latter are divided at right angles to the longitudinal axis, the artificial transverse sections; finally the designation equator (a b—m n) is applied to an imaginary line that exactly bisects the length of the muscle-fibers.

As the currents present are exceedingly feeble, a multiplicator (Fig. 225, I) is required for their demonstration or a tangent mirror-galvanometer, for example the electrogalvanometer (p. 384), with a damped periodic magnet. If the wires of the multiplicator were placed in direct communication with the moist animal tissue, they would give rise to a current by reason of their inequality, and, besides, polarization would develop on the surface of the wires on the passage of a current. Therefore unpolarizable electrodes, upon which the tissues may rest (Fig. 225, I, P, P), are always used in conjunction with the conducting wires.

The capillary electrometer of Lippmann (Fig. 229) has been advantageously employed for the demonstration of muscular currents. In this a thin column of mercury in a capillary tube lying in contact with a conducting fluid (dilute sulphuric acid) is displaced by the galvanic current, the constant of capillarity of the mercury undergoing alteration in consequence of the polarization at the surface of contact. The displacement, which the observer (*B*) recognizes with the microscope (*M*), takes place in the direction of the positive current. The image of the capillary tube can be projected upon a screen and the oscillations

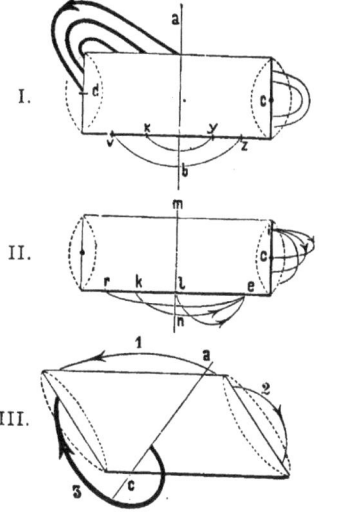

FIG. 228.

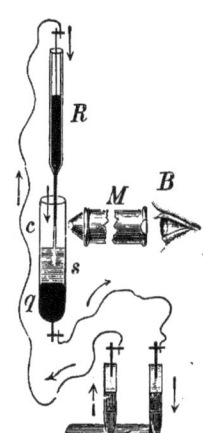

FIG. 229.—Diagrammatic Representation of the Capillary Electrometer.

of the mercury may be photographed. In Fig. 229, representing such an apparatus diagrammatically, *R* is a glass tube drawn out below to capillary fineness, and filled from above with mercury and from *c* downward with dilute sulphuric acid. The capillary tube extends downward into a wide glass tube, which has a platinum wire fused into it below and is filled with mercury (*q*) and dilute sulphuric acid (*s*). The conducting wires are connected with unpolarizable electrodes, which are applied to the transverse section and the surface of a muscle. On closing the current the column of mercury is displaced downward from *c* in the direction of the arrow. The electromotive force can be measured with the aid of the capillary electrometer from the extent of the displacement of mercury. On the other hand, when the electrical processes take place rapidly the movement of mercury cannot follow rapidly enough on account of the resistance.

The strength of the currents in animal organs is best measured by permitting another current of graduated and known strength to pass through the electrometer circuit in an opposite direction, so that the tissue-current present is reduced to zero—compensatory method of Poggendorf.

1. Perfectly fresh, uninjured muscles exhibit no current at all, nor do wholly dead muscles.

2. Strong electrical currents are observed if, as in Fig. 225, I M, the transverse section of the muscle is connected with one unpolarizable electrode, while the surface, or longitudinal section, is connected with the other. The direction of the current in the connecting wire is from the positive, longitudinal section to the negative, transverse section, therefore in the muscle itself from the transverse to the longitudinal section (Fig. 225, I, and Fig. 228, I). This current is the stronger the more one electrode is approximated to the equator and the other to the center of the transverse section. The strength diminishes the more the electrode applied to the surface approaches the extremity and the more the electrode applied to the transverse section approaches the margin of the section. The demonstration of the strong current may even be made on a single, isolated muscle-fiber. Unstriated muscles also exhibit similar currents between transverse section and surface.

3. Feeble electrical currents are obtained: (a) If the electrodes are applied at two points on the surface unequally distant from the equator. The current then passes from the positive point nearer the equator to the farther removed negative point, in the muscle naturally in the reversed direction (Fig. 228, II, k e and l e). (b) Equally feeble currents develop on applying the electrodes to points on the transverse section unequally distant from the center, the current passing from the point nearer the margin of the section to that nearer the center of the section, in the muscle itself in the opposite direction (Fig. 228, II, i c).

4. If the application be made to two points on the surface equidistant from the equator (I, x, y; v, z; II, r, e) or to two equidistant from the center of the transverse section (I, c) no current appears.

5. If the transverse sections of a muscle are made obliquely (III), so that the form of the section is rhombic, the conditions present will be the same as those described in paragraph 3. A point close to the obtuse angle of the transverse section or of the surface is positive with relation to one equally near the acute angle. The equator passes obliquely (a, c). These divergent currents are designated inclination-currents and their course is indicated by the lines 1, 2, and 3, III.

The electromotive force of a strong muscle-current, in the frog, is equal to from 0.035 to 0.075 of a Daniell cell, and in the case of the strongest inclination-currents even up as much as 0.1. The muscles and nerves of a curarized animal exhibit at first stronger currents. Exhaustion of the muscle diminishes the strength of the current, and it disappears entirely on the death of the muscle. Elevation of the temperature of a muscle increases the current, but a temperature above 40° C. again enfeebles it. Reduction of the temperature lessens the electromotive force. A current that has become feebler in the course of a short time can be made stronger by application of the electrodes to a new transverse section. Heated living muscular tissue and nerve-tissue are positive to cooler tissues of the same kind.

6. The resting nerve exhibits with reference to the conditions described in paragraphs 1, 2, and 3 effects analogous to those of the muscle.

The electromotive force of the strong nerve-currents, conducted from transverse section and surface, equals 0.02 of a Daniell cell. Heating the nerve to between 15° and 25° C. increases the strength of the nerve-current, while higher temperatures enfeeble it. In the development of a strong nerve-current the negativity of the transverse section rapidly diminishes with the death of the nerve.

This takes place only up to the next annular constriction, and after it has been completed, the nerve under such conditions is devoid of current. A new transverse section permits again of the development of a strong nerve-current.

7. If the electrodes are applied to the two transverse sections of an excised nerve or to two points on the surface equidistant from the equator, a feeble current appears and passes in a direction opposite to that caused by the physiological activity of the nerve-fiber (axial current); therefore in the case of centrifugal nerves in a centripetal direction and in that of centripetal nerves in a centrifugal direction. Perhaps this current depends upon differences in the time of death at the two extremities of the nerve.

The electromotive force of such a current increases with the length of the nerve-segment and with the size of the transverse section. It is enfeebled by exhaustion (by tetanization), especially in the case of motor nerves and less in that of centripetal nerves.

The muscle-current can be demonstrated also without the aid of a multiplicator:
1. By a sensitive frog-preparation, designated the *physiological rheoscope*. A moist conductor is applied to the transverse section and the surface of the gastrocnemius muscle from a frog. When the sciatic nerve of a frog-preparation connected with the leg is stretched over the muscle contraction takes place at once; likewise when the nerve is again removed. If at the lower extremity of the frog-preparation a transverse section is made through the gastrocnemius muscle, and the sciatic nerve (whose distribution in the muscle is connected with the surface of all of the fibers) is placed on this transverse section, the leg twitches as the muscular current, from the surface to the transverse section, enters the nerve. This observation was familiar to Galvani as *contraction without metals.*
2. An isolated muscle can be stimulated directly and made to contract by means of its own muscle-current. If unpolarizable electrodes are applied to the transverse section and the surface of a curarized frog-muscle and the circuit is closed by mercury the muscle contracts. In an analogous manner the nerve also may be stimulated by its own nerve-current. If the lower extremity of a muscle with a transverse section be immersed in a 0.6 per cent. solution of sodium chlorid, which itself is entirely indifferent, a secondary circuit is established through this fluid between the transverse section and the adjacent surface of the muscle. In consequence, the muscle contracts. Other indifferent conductors used to complete the circuit have a similar effect.
3. If the muscle-current be passed through potassium-iodid paste it causes by electrolysis a separation of iodin at the positive pole, as a result of which the starch-paste becomes blue.

The total current in the body should be the resultant of the electrical currents of the individual muscles and nerves, and, in the frog deprived of skin, it passes from the extremity of the limbs to the trunk and in the trunk from the anus to the head. This is the "corrente propria della rana" of Leopoldo Nobili, or the "frog-current." In mammals the corresponding current passes in an opposite direction.

If muscles or nerves have lost their irritability in the state of narcosis induced by ether or chloroform, the muscle-current may persist and even be increased. After death, the currents disappear earlier than the irritability. They persist longer in the muscle than in the nerve, in which they are abolished earlier in the more central portions. Also a motor nerve wholly paralyzed by curare still exhibits the current (spark), as did also a nerve in process of degeneration that had lost its irritability entirely for two weeks. In the divided nerves of a living animal the current-strength is at first increased after one or two days, while later it diminishes. Muscles that have become rigid at times exhibit currents in opposite directions in consequence of inequalities developed in the process of decomposition. The nerve-current is reversed by boiling water or by desiccation.

Of other tissues that exhibit electrical currents there may be mentioned the skin (frog), whose surface is positive, while its inner aspect is negative. The mucous membrane of the digestive tract exhibits the same relation, as does also the cornea, as well as the aglandular skin of fish and snails. Currents have been observed also in glands, principally in the unicellular and multicellular mucous glands of lower vertebrates (frog, eel).

CURRENTS OF STIMULATED MUSCLES AND NERVES AND OF SECRETORY ORGANS.

1. If a muscle that exhibits a strong electrical current is thrown into tetanic contraction, best by means of tetanization of its nerve, its current is enfeebled, at times even to the point of complete return of the magnetic needle to zero. This phenomenon is the *negative variation.* It is directly proportional to the primary deflection of the magnetic needle and to the energy of the contraction of the muscle.

After the tetanus the muscle-current is feebler than before. If the muscle be placed upon the electrodes in such a manner that the current is a feeble one, a diminution of this feeble current appears during tetanus in an analogous manner. In the ineffective arrangement the contraction of the muscle has no influence upon the magnetic needle. If the contraction of the muscle is prevented by making it tense, a somewhat slighter negative variation is observed. Therefore, it is also smaller in the isometric than in the isotonic act. If a contracted muscle be stretched, the negative variation present decreases. If, however, a resting muscle be stretched the resting current is diminished.

2. Excised frog-muscles thrown into a state of tetanus through their nerves exhibit electromotive force, *action-current.* A descending current is present, for example, in the tetanized gastrocnemius of the frog and a similar current in the entire hind leg. In wholly intact muscles of man, however, thrown into tetanic contraction through their nerves, such a current is wanting. Also wholly intact frog's muscles directly thrown into tetanus exhibit no current.

3. If a muscle is momentarily irritated directly at one extremity, so that the contraction-wave rapidly traverses the entire length of the muscle-fibers, every part of the muscle is successively negative electrically shortly before it undergoes contraction. A wave of negativity thus precedes the wave of contraction. The former therefore occurs during the period of latent irritation. Waves of negativity and contraction have the same velocity of about 3 meters in a second. The negativity, which at first increases and then diminishes, continues at each point for only about 0.003 second.

4. Also a single contraction indicates the development of an electrical current in the muscle. The pulsating frog's heart serves as an appropriate illustration, the observation being made with the aid of the electrogalvanometer. Each pulsation causes a deflection of the needle of the instrument, and this takes place earlier than the contraction of the heart-muscle itself. The electrical process in the muscle causing the negative variation precedes in general the contraction, and occurs, therefore, in the stage of latency. In the contraction of the wholly intact gastrocnemius muscle of the frog stimulated through its nerve there is at first a descending and then an ascending current.

Careful investigations of the electrical processes in the pulsating heart have shown that, with the cardiac pulsation, first the base and then the apex of the ventricle becomes negative. A brief latent period occurs in advance. If the heart-muscle is placed in a condition of relaxation by irritation of the vagus, a positive variation is naturally observed in the muscle-current. On the other hand, irritation of the accelerator nerve in the condition of arrest due to muscarin, even when the pulsation of the heart is not stimulated anew, gives rise to negative variation.

Also in man an electrocardiogram can be obtained if the two hands are connected with the capillary electrometer. The right arm exhibits the electrical tension of the base of the heart, the left arm that of the apex.   In correspondence with the different individual phases of a cycle of the heart the electrocardiogram is complicated.   Five variations appear in the course of the contraction: The first, third, and fifth indicate negativity of the base of the heart, the second and fourth negativity of the apex.   Also the two muscles of the iris exhibit negative variation in their contraction.   Naturally, the descending contraction-wave in the esophagus observed in the act of swallowing is attended with corresponding electrical phenomena.

The electrical processes in muscle on simple contraction are exhibited also by the frog-preparation.   If a segment of the nerve of such a preparation be placed upon a muscle the frog-preparation twitches whenever the muscle is made to contract.   If the nerve of a frog-preparation is placed upon a pulsating mammalian heart, a contraction takes place in the leg with every pulsation.   Thus, after division of the phrenic nerve, particularly on the left side, the diaphragm contracts synchronously with the heart-beat.   This contraction is designated the *secondary contraction*.   The moving muscle thus stimulates another muscle applied to it.   This phenomenon occurs readily if the muscles employed are in a state of beginning desiccation.

A muscle in a state of tetanic contraction from the action of an induced current causes *secondary tetanus* in a frog-preparation in contact with the muscle. The latter is considered an evidence that in the process of negative variation in the muscle many current-variations occurring in rapid succession must have been present, as only rapid variations of this kind have a tetanizing effect upon the nerve, and not long-continued current-variations.

Also when the muscle is in a state of tetanic contraction (toad) as a result of voluntary innervation, or of chemical irritation, or of strychnin-poisoning, secondary tetanus generally does not take place in an applied frog-preparation, although Lovén has observed secondary strychnin-tetanus, comprised of from 6 to 9 contractions in a second.   Lippmann's sensitive capillary electrometer (Fig. 229) also shows that both strychnin-spasm, as well as the voluntary contraction, are discontinuous processes.   The slight activity of chemical irritants is explained by the fact that they do not cause the muscle-fibers to enter promptly into a state of uniform contraction.   In case of voluntary tetanus and strychnin-tetanus the electrical process takes place perhaps with insufficient variation in the current. For this reason also muscles in a state of tetanic contraction in the normal body do not stimulate adjacent nerves or muscles.

Biedermann made the remarkable observation that striated muscle under the influence of ether-vapor is thrown into a state in which on irritation it exhibits no appreciable alteration of form or movement, while, on the other hand, changes demonstrable with the aid of the galvanometer appear at the point of irritation in the same degree as prior to the action of the ether, as an expression of the irritation, but in consequence of the abolished power of conduction they are capable of manifesting themselves only locally.

5. If a nerve resting with its transverse section and surface upon the electrodes be irritated electrically, chemically, or mechanically (or also, if this be possible, reflexly), its current diminishes. This negative variation, which may be propagated in both directions in the nerve, is made up of periodic interruptions of the original current occurring in rapid succession, as in the contracted muscle. Hering was able by this means, as in the muscle, to induce secondary contraction or secondary tetanus. The extent of the negative variation is dependent upon the extent of the primary deflection, upon the degree of irritability of the nerve and upon the strength of the irritant applied. The negative variation is demonstrable both on tetanizing with individual waves of irritation. The negative variation has not yet been observed in wholly intact nerves.

Hering found that the negative variation of the nerve-current induced by electrical tetanization is followed in general by a positive variation.   It increases to a certain degree with the duration of the irritation, with the strength of the stimulating currents with commencing desiccation of the nerves, and if the point on the longitudinal section to which the electrode is applied recedes from the transverse section. The negative variation after chemical or mechanical irritation is observed especially in winter-frogs exposed to cold, also on application of peripheral pressure-irritation to the skin, as well as in the fresh electrical nerve of the torpedo. Non-medullated nerves exhibit negative variation, just as they exhibit in general all electrophysiological phenomena in the same manner as medullated nerves.

The action of certain substances upon the isolated nerve of the frog gives rise to changes in the appearance of the electrical phenomena. If the isolated nerve is exposed to carbon dioxid, the negative variation remains in abeyance for a few minutes, after which it occurs with increased intensity. Chloroform and ether increase the electrical activity at first, later having an inhibiting effect. Potassium bromid has an inhibiting effect. The influence of electrotonus upon negative variation is discussed on p. 662.

The galvanic relation of the still irritable spinal cord is in general the same as that of the nerves. If longitudinal and transverse currents are established in the upper portion of the medulla oblongata, spontaneous intermittent variations, perhaps due to the intermittent stimulation of the centers situated in this locality, and particularly of the respiratory center, are observed. Similar variations occur also reflexly in response to individual electrical shocks applied to the sciatic nerve, while strong irritation by means of sodium chlorid or induced currents inhibits them. Also the surface of the cerebrum exhibits the development of currents, if the centers situated within it, for example the psychosensorial, are irritated by stimulation through the organs of special sense.

6. The same phenomenon that is exhibited by the muscle, as described in paragraph 3, is exhibited also by the nerve. The wave of negativity can be best followed if its rapidity of propagation is diminished by the action of severe cold. In its progress the wave of negativity does not decrease in extent, while it does so in the excised muscle.

The process of negative variation is propagated through the nerve-fiber with measurable rapidity, which is greatest at a temperature between 15° and 25° C. This is the same as that of the propagation of the stimulus itself, and in the normal average is from 27 to 28 meters a second. This rapidity exhibits the same variations as the rapidity of propagation of the nerve-stimulus. The duration of an individual variation, of many of which the process of negative variation is constituted, is only from 0.0005 to 0.0008 second. The length of the waves in the nerve is estimated at 18 mm.

J. Bernstein has by means of the *differential rheotome* found in the following manner the time required by the negative current-variation in the nerve to propagate itself from the point of irritation throughout the course of the nerve. A long nerve (Fig. 230. Nn) is so arranged that at one of its extremities (N) transverse section and surface are connected with the electrodes of a galvanometer (G). At the other extremity (n) are the electrodes of an induction-coil ($\mathcal{F}$). A disc (B), rapidly rotated about its vertical axis (A) by means of a pulley (S), is provided at one point of its periphery with a device (C) by means of which the current of the primary circuit (E) is rapidly closed and again opened with each revolution. In this way a stimulating closing and opening induction-shock is applied to the extremity of the nerve with each revolution of the disc. At the diametrically opposite side ($r\ r$) of the periphery of the disc is an arrangement (c) by means of which the galvanometer-circuit is closed and opened with each revolution. There thus take place at the same moment the stimulation and the closing of the galvanometer-circuit. When the disc is rapidly rotated, the galvanometer indicates the presence of a strong nerve-current, the magnetic needle being deflected to the point y. At the moment when stimulation takes place, the negative variation has not yet advanced to the other extremity of the nerve. If, however, the device that closes the galvanometer-circuit is so displaced at the periphery of the disc (for example to o) that the galvanometer-circuit is closed somewhat later than the nerve is stimulated, the current appears to be enfeebled by the negative variation, the needle being deflected only to x. If the rapidity with which the disc revolves is known it will be readily found that the time corresponding to the displacement of the closing must be equal to the rapidity with which the stimulus causing the negative variation is propagated from the one extremity of the nerve (n) to the other (N).

The negative current-variation in the nerve is wanting in degenerated nerves as soon as its irritability is abolished.

If light is permitted to fall upon a freshly extirpated eye the current from the positive cornea to the negative transverse section of the optic nerve exhibits at first an increase.

Yellow light has the most marked effect, while other colors have less marked effect. The inner surface of the resting retina is positive with relation to the posterior surface. On illumination of the inner surface a double variation occurs, namely, after a brief latent period, a negative preceded by a positive. On disappearance of the light a simple positive variation occurs. Retinas with the visual red bleached by light exhibit smaller variations. According to Beauregard and Dupuy the auditory nerve also exhibits similar manifestations of negative variation. One electrode is applied to the transverse section of the nerve, the other to the tympanic membrane, and a loud sound serves as the irritant.

Irritation of the secretory nerves of membranes containing glands gives rise to changes in the resting currents, with the formation of secretion. This secretory current in the skin of the frog and of warm-blooded animals has the same direction as the resting current. In the frog it is sometimes preceded by a current in the

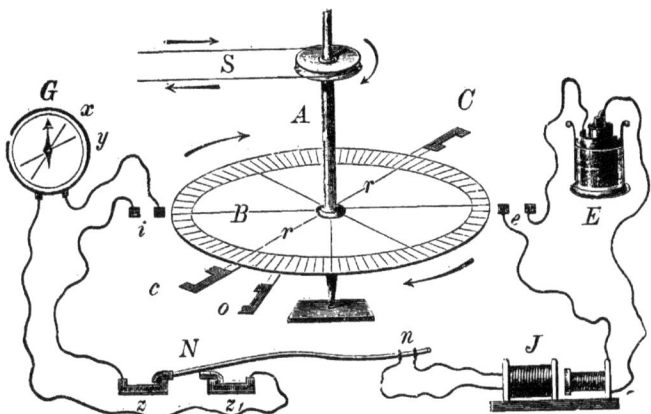

FIG. 230.—Diagrammatic Representation of Bernstein's Differential Rheotome.

opposite direction. Also, denuded portions of skin in cats exhibit analogous phenomena.

If in the cat a current is passed uniformly from the skin of both hind legs, and if one sciatic nerve is now irritated, a penetrating secretory current is set up, with secretion of sweat. If, in an analogous manner, the electrodes are applied uniformly to two points on the skin of the extremities in man, and the muscles of one extremity are contracted, a penetrating current is likewise set up. Tarchanoff observed in the skin of man feeble currents after irritation as by cold, tickling, and pain, and after other nervous stimuli, such as mental exertion and bright light. Destruction of the gland abolishes both the secretion and the secretory current, as does also atropin. Portions of the skin covered by hair, but without sweat-glands, have no secretory current. The current of the gastric mucous membrane during rest, which, as a rule, is penetrating, exhibits on irritation of the vagus, which exerts an influence upon the secretion in rabbits, a negative variation preceded by a slight positive variation. In the dog the external surface of the salivary glands is negative as related to the hilus. In case of abundant watery secretion, as from irritation of the chorda tympani, the surface exhibits a first phase of negative potential with respect to the hilus, which is at times followed by a second phase of feebler difference of potential in the opposite direction. In the presence of abundant watery secretion, the first phase preponderates, while when the secretion is less abundant and more viscid the second phase preponderates.

CURRENTS IN NERVES AND IN MUSCLES IN THE ELECTROTONIC STATE.

If a nerve be connected with non-polarizable electrodes in such a manner that its transverse section is applied to the one and its surface to the other (Fig. 231, I), the multiplicator will indicate the presence of a strong nerve-current. If a constant electrical current, designated the polarizing current, be now passed through the length of the extremity of the nerve projecting beyond the electrode, in a direction which coincides with that of the current in the nerve, the magnetic needle exhibits a still more marked deflection, as a sign of increase in the nerve-current—positive phase of electrotonus. This is directly proportional to the length of nerve traversed and the strength of the galvanic current, and inversely to the distance between the portion traversed and the portions of the nerve applied to the pads.

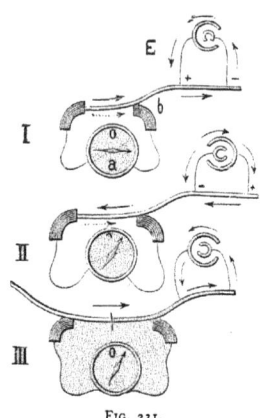

FIG. 231.

If, with the nerve in the same position, the constant electrical current is passed in a direction opposite to that of the nerve-current (II), there is a diminution in the electromotive force of the latter—negative phase of electrotonus.

If the electrodes are applied to two points on the surface of the nerve almost equidistant from the equator (III), the galvanometer at first exhibits no deflection with this ineffective arrangement. If, now, a constant current be passed through the free, projecting extremity of the nerve, the magnetic needle exhibits electromotive activity in the same direction as the constant current.

The foregoing experiments demonstrate that a nerve traversed by a constant electrical current undergoes, not alone within the directly traversed portion, but also beyond this, an alteration in its electromotive activity that is designated *electrotonus.* This alteration is attended with a change in the irritability of the nerve-segment in question.

The electrotonic current is strongest near the electrodes. It may be 25 times stronger than the resting nerve-current. Its strength increases with the strength of the constant, polarizing current, likewise with the length of the segment traversed. It is larger upon the side of the anode than upon that of the kathode. It appears with the closing of the constant current, while it reaches its maximum earlier at the kathode. It gradually increases at the anode and decreases at the kathode. On tetanization it undergoes negative variation like the resting nerve-current, while the polarizing current appears to be stronger. On the other hand, no noteworthy electrotonic increase in current between the electrodes can be observed beyond the polarizing current itself. Cold has a marked inhibiting influence upon the production of the electrotonic current.

The phenomena described occur only so long as the nerve is irritable. Ligation of the extremity of the nerve projecting beyond the galvanometer-circuit abolishes the phenomena in the segment so shut off. The galvanic electrotonic alterations in the extrapolar segments described—and due to a peculiar diffusion by physical means of the polarizing current—are wanting in the case of non-medullated nerves, which on the other hand exhibit physiological electrotonus. By treating medullated nerves with ether the physiological electrotonus may be abolished, while the physical phenomena referred to persist.

The negative variation appears more rapidly than the electrotonic increase in current, so that the former will have disappeared before the electrotonic increase in current is observed; for the rapidity of the electrotonic alterations in current is less than the propagation-velocity of the impulse in the nerve, namely only from 8 to 10 meters a second.

Upon the electrotonic process depends the *secondary contraction from the nerve.* If the sciatic nerve of a frog-preparation be applied to a divided nerve and then a constant current is sent through the free extremity of the latter (nonelectrical nerve-stimuli are ineffective), contraction takes place in the frog-preparation. This occurs because the electrotonizing current in the excised nerve irritates the adjacent nerve. On rapidly closing and opening the current secondary tetanus results. The same conditions are observed in connection with the *paradoxical contraction.* If the current is directed to one of the two branches into which the severed sciatic nerve of the frog divides, the muscles supplied by both nerves contract.

If the constant current is opened, *after-currents* appear, which according to du Bois-Reymond are due to internal polarization. In living nerve, muscle, and electrical organ this internal polarization-current is always positive, that is it has the same direction as the primary current, when a strong primary current of short duration is employed. If the primary current be of greater duration, negative polarization eventually results. Between the two there is a stage in which the preparation exhibits no polarization at all. Positive polarization appears particularly strong in the nerve when the primary current has the same direction as the course of the impulse in the nerve, in the muscle when the primary current passes from the point of entrance of the nerve to the extremity of the muscle. An analogous condition is observed in the electrical organ.

The muscle likewise exhibits the electrotonizing effect of the constant polarizing current. A constant current in the same direction intensifies the muscle-current, while a current in the opposite direction enfeebles the muscle-current. The effect is, however, relatively feeble.

THEORIES OF CURRENTS IN MUSCLES AND NERVES.

In explanation of the currents in muscles and nerves du Bois-Reymond proposed the so-called molecular theory. According to this, nerve-fibers and muscle-fibers contain minute molecules, of electromotive activity, arranged successively in series, and surrounded by a conducting indifferent fluid. The molecules are in a peripolar electrical state, namely, provided with a positive equatorial zone, directed toward the surface, and two negative polar surfaces, facing the transverse section. Each newly prepared transverse section exposes new negative surfaces, and each artificial longitudinal section new positive areas.

This arrangement explains the strong currents, for if the positive circuit be connected by means of a closing arc with the negative transverse section, a current must pass through this from the surface to the transverse section. On the other hand, the theory does not explain the feeble currents. To comprehend these it must be assumed that the electromotive activity of the molecules is enfeebled with varying rapidity on the one hand at unequal distances from the equator, on the other hand at unequal distances from the center of the transverse section. Then naturally differences in electric potential will develop between the molecules of greater activity and those that are already enfeebled. The muscles, however, show that their natural transverse section, that is the extremity of the tendon, does not become negative electrically, like an artificial section, but positive in greater or lesser degree. In explanation of this anomalous phenomenon du Bois-Reymond believes that a layer of electropositive muscular substance is still present at the extremity of the tendon. To facilitate comprehension he considers the peripolar elements of the muscle as consisting each of two bipolar elements, a layer of the half-element being so applied to the extremity of the tendon that its positive side is directed toward the free surface of the tendon. This layer he designates the *parelectronomic layer.* It is never entirely wanting. The better it is developed the greater is the absence of current on conduction from the surface and the tendon. If parelectronomy be well developed, the extremity of the

42

tendon may even become positive with respect to the surface. The parelectronomic layer is destroyed by cauterization.

The negative variation in current is explained by assuming that during the activity of muscle and nerve the electromotive force of all of the molecules is diminished. On partial contraction of the muscle the contracted portion assumes rather the character of an indifferent conductor, which is in simple conducting connection with the negative zones of the resting contents of the muscular fibers. The electrotonic currents beyond the poles, particularly in the nerve-fibers, require a special explanation, while the electrotonic state of the muscles extends principally to the intrapolar portion. In explanation of the electrotonic currents, it is assumed that the bipolar molecules have the property of rotation. The polarizing current, however, exerts a directive influence upon the molecules so that these turn their negative surface toward the anode and their positive surface toward the kathode. In consequence the molecules of the intrapolar segment are arranged like the voltaic pile. In the portions of the nerve lying beyond the pole the molecules are the less accurately arranged the farther removed they are. Therefore, the deflections of the needle become correspondingly feebler in the extra-polar portions.

The differential theory proposed by Hermann, which has recently been developed by Hering, explains the phenomena in a satisfactory manner. Any protoplasmic structure, such as muscle, nerve, or cell, develops no current that can be conducted outward so long as its metabolism, that is the internal chemical processes, remains the same in all parts. Every disturbance of this equilibrium in one part of the protoplasmic structure causes the development of currents that can be conducted away. Therefore, (1) the protoplasm at the point where death occurs, whether from injury of any kind or from degeneration, becomes electrically negative with respect to living and irritable protoplasm. (2) The protoplasm is negative at points that are irritated with respect to those that remain in an unirritated resting state. (3) The protoplasm becomes electrically positive in warmed situations, negative in cooled situations. In addition it may be stated (4) that protoplasm is strongly polarizable on its surface (nerve, muscle). The constant of polarity is diminished by irritation (and death).

In detail the following statements may yet be made in this connection. It has been shown that resting, uninjured and absolutely fresh muscles are entirely without current, as also are wholly intact nerves. The heart likewise is free from current, and also the muscles of fish still covered by skin. As the skin of the frog possesses currents of its own, it is possible, with special precautions, after destruction of the cutaneous currents through cauterants, to demonstrate also here the freedom from current on the part of the frog's muscles. Furthermore, L. Hermann found that the muscle-current always develops only after the lapse of a certain, though short, time after making a transverse section.

All injuries of muscles and nerves give rise at the site of injury (the demarcation-surface) to negative, dying tissue with relation to the positive, intact tissue. In this way is to be explained the negativity of the transverse section with relation to the surface. The current thus developed is designated by Hermann the *demarcation-current.* If potassium-salts or muscle-juice be applied to certain parts of a muscle these become electrically negative. If these substances are again removed the negativity of these parts disappears.

It appears to be a phenomenon peculiar to all living protoplasmic substances that after injury at one point this becomes negative on dying, while the remaining intact portion is electrically positive. Thus, all transverse sections of living vegetable structures are negative with relation to their surface. The same condition is observed in animal structures, for example glands and bones. The electrical organ of fish is discussed on p. 675.

Engelmann has made a remarkable observation. He found that the heart and the unstriated muscle-fibers again lose the negativity of their transverse section if the divided muscle-cells have died completely as far as the adjacent cement-substance of the neighboring cells; while nerves lose their negativity when the divided segments, each corresponding to a single cell, have died to the nearest annular constriction of Ranvier. Under such circumstances, all of these organs are entirely without current, for the entirely dead substance reacts essentially as an indifferent moist conductor. Also muscles divided subcutaneously likewise no longer exhibit negative cut surfaces after union of the wound-surfaces.

· Notwithstanding all of the foregoing observations the preëxistence of currents in resting living tissues cannot be assumed.

An alteration of the chemical processes in a part of the protoplasmic structure may, according to Hering, have such an effect that the portion that is decomposed (dissimilated) is negative with respect to the unaltered portion, but also that the portion that is replaced (assimilated) is positive to the remaining portion. Hering thus distinguishes negative and positive alterations, which must give rise to corresponding currents.

According to Grünhagen and others the electrotonic currents are due to internal polarization in the nerve-fibers between the conducting tissue of the nerve and that of the sheath. Matteucci had already found that if a wire be covered with a moist sheath and the latter be connected with the electrodes of a constant circuit, currents due to polarization appear, resembling the electrotonic currents in nerves. If either the wire or the moist covering be interrupted at a given point, the polarization-currents do not pass beyond the point of discontinuity. The polarization developed at the surface of the wire causes, through its transitional resistance, the conducted current to pass far beyond the electrodes.

Muscles and nerves consist in a similar manner of fibers surrounded by indifferent conductors. As soon as a constant current is closed on the surface, internal polarization develops between the two, and this gives rise to the electrotonic diffusion of the current. The polarization disappears on opening the current. It can be recognized from the fact that in the living nerve the galvanic resistance transversely through the fibers is five times as great and in muscles seven times as great as through their length. Recently also Boruttau states that all electrical phenomena of the nerve can be explained if it be considered in its capacity as a conductor.

With reference to the currents developed during the activity of the muscles, the currents of action, Bernstein established the doctrine that when a single wave of irritation (contraction) passes longitudinally through muscle-fibers that are connected at two points with the galvanometer, that point below which the wave passes is negative with reference to the other. Occasionally local points of contraction are present in muscle-preparations in certain situations and these are negative with relation to other resting points of the same muscle. In order to explain the currents that appear in connection with tetanus of frog's muscles it must be assumed that the extremity of the fibers takes part in lesser degree in the process causing the negativity than the middle of the fibers. This is, however, the case only in exhausted muscles or in those in process of dying.

As will be pointed out on p. 663 the contraction occurring on direct application of a constant current to the muscle takes place on closure of the current at the kathode, on opening the current at the anode. It will, thus, be clear that with the closing contraction the muscle exhibits negativity at the kathode, but with the opening contraction at the anode. These facts according to Hering and Biedermann explain the after-currents considered on p. 657.

If a muscle is made to contract by stimulation of its nerve, the wave of excitation passes from the point of entrance of the nerve in both directions and it is likewise negative to the resting muscle. In accordance with the situation of the entrance of the nerve into the muscle the ascending or the descending wave of excitation will reach the extremity (origin or attachment) of the muscle earlier. If, therefore, such a muscle be introduced by its upper and lower extremities into the circuit of the galvanometer, that extremity will at first be negative that is nearest the point of entrance of the nerve, for example in the gastrocnemius the upper, and later the lower. There thus appear in rapid succession first a descending, then an ascending current in the galvanometer-circuit, in the muscle naturally in the reverse order.

The same conditions are observed also in the forearm-muscles of man. If these are thrown into contraction from the nerve, the point of entrance of the nerve, 10 cm. below the elbow, is first negative, then the muscle-extremities if the wave of contraction has reached these points with a velocity of from 10 to 13 meters in one second. In this experiment the brachial plexus is stimulated in the axillary cavity. The conduction in the forearm (in the upper portion and above the wrist-joint) is established by surrounding the skin with strips of material saturated with zinc sulphate. The strips themselves come in contact with the paper pads of the non-polarizable electrodes.

If a wholly intact muscle free from current is made to contract entirely, no current is set up either with the individual contraction or in the state of tetanus, because at the same moment the entire muscular structure passes into a state of irritation and into a firmer condition. With respect to the nerves also it has in like manner been determined that everywhere dying and active contents are negative to resting normal contents.

Attention may yet be called to the following facts. If water passes through a capillary space an electrical movement takes place in the same direction. So also the advance of water in the capillary interstices of inanimate structures (pores of a porcelain plate) is attended with an electrical movement in the same direction as the stream of water. Exactly the same .conditions prevail in the movement of water that causes the swelling of a body. Landois has pointed out that imbibition and swelling take place at the demarcation-surface of an injured muscle or nerve; further that swelling occurs also at the contracted portion of a muscle in consequence of the absorption of fluid, and that in the process of secretion movement of fluid takes place from the blood into the glandular cells and from these to the excretory ducts. Mention should finally be made of the fact, as H. Munk found, that, at the moment of closing the current at the anode and beyond, loss of water and increase in resistance take place in the nerve; and at other situations and beyond the kathode the reverse. The total resistance of the distance traversed diminishes at first, then increases with accelerated rapidity. On opening the current neutralization of this difference rapidly takes place. In plants electrical phenomena are observed both on passive bending of portions of plants, as of the leaves or the stems, and also in active movements associated with bending of parts of the plants, for example in the movements of the mimosa, the dionea, and others. These electromotor effects are also to be explained in all probability by the movement of water in the parts of the plant, which must take place in their interior on movement. The tip of the root of germinating plants is negative with respect to the seed-covering, the cotyledons positive with relation to all other portions of the seedling. In the incubated bird's egg the embryo is positive, the yolk negative.

ALTERED IRRITABILITY OF NERVE AND MUSCLE IN ELECTRO-TONUS.

If a living nerve is traversed throughout a definite length by a constant electrical (polarizing) current it passes into the condition of *altered irritability* that is designated the electrotonic state or simply *electrotonus*. The condition of altered irritability extends not alone over the traversed (intrapolar) distance, but is communicated to the entire nerve.

Pflüger has discovered the following law of electrotonus. At the positive pole or anode (Fig. 232, A) the irritability is diminished and anelectrotonus prevails. At the negative pole or kathode (K) it is increased, and the increase in irritability prevailing here is designated katelectrotonus. These alterations in irritability are most pronounced near the poles.

In the intrapolar segment there must naturally be a point where anelectrotonus and katelectrotonus coincide and where therefore the irritability is unaltered. This point is designated the *indifferent point.* It is situated in the case of feeble currents near the anode (i), in that of strong currents, however, near the kathode (i_{11}); therefore in the first instance almost the entire intrapolar segment is more irritable and in the latter less irritable. Exceedingly strong currents greatly diminish the conducting power at the anode and they may even render the nerve wholly incapable of conducting.

Also at the kathode, but only after the current has been for some time flowing through the nerve, the irritability is diminished and the nerve becomes incapable of conducting.

Beyond the electrodes (extrapolar) the area of altered irritability is the more extensive the stronger the current. Further, the extent of extrapolar anelectrotonus is greater with the feeblest currents than that of extrapolar katelectrotonus. In the case of strong currents, this relation is reversed.

Fig. 232 exhibits diagrammatically the relations of irritability of a nerve (*N n*) that is traversed by a constant current in the direction of the arrow. The curves are so constructed that the degrees of increased irritability in the vicinity of the kathode (*K*) are represented as elevations above the line representing the nerve, and those of lowered irritability at the anode (*A*) as depressions. The curve m o i$_{11}$ p r represents the irritability of the nerve with strong currents, the curve e f i$_1$ h k that with currents of moderate strength and finally a b i c d that with feeble currents.

The electrotonic effects increase with the length of nerve traversed. The alteration of irritability in katelectrotonus appears at the moment of closure of the circuit. Anelectrotonus develops and extends slowly. Electrotonus is diminished by cold, and by heat up to 40°. At 30° katelectrotonus is increased and anelectrotonus diminished. Also with induced currents the anode diminishes the irritability.

If the polarizing current is opened there is at first a reversal of the conditions of irritability. There then follows a transition to the normal

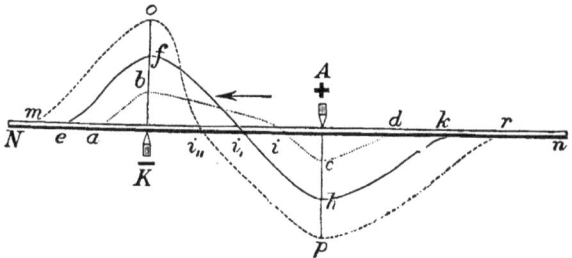

FIG. 232.—Diagrammatic Representation of the Electrotonic Relations of Irritability.

state of irritability of the resting nerve. At the initial moment of closure Wundt observed that the irritability of the entire nerve was augmented.

Testing Electrotonus in Motor Nerves.—In order to demonstrate the laws of electrotonus in motor nerves the frog nerve-muscle preparation (Fig. 233), consisting of the leg and the sciatic nerve, is employed. By means of unpolarizable electrodes (Fig. 225, IV) the current of a constant circuit is conveyed to the nerve throughout a limited distance. An irritant, such as an electrical shock, or chemical irritation by the application of sodium chlorid, or mechanical irritation, is now applied to the nerve at either the anode or the kathode, and note is made whether the contractions following upon the irritation vary in size when the polarizing circuit is opened or when it is closed. The contractions themselves may be recorded from the gastrocnemius muscle with the aid of the myograph. The following examples may be considered: (*a*) *Descending extrapolar anelectrotonus*, that is with a descending current the irritability at the anode within the extrapolar section is to be tested. If in such a case (**A**) the irritant, sodium chlorid, which is applied at R while the circuit is still open, gives rise to moderately large contractions in the leg, these become feebler, or are abolished, as soon as the constant current is passed through the nerve. After opening the current, the contractions appear again in their original strength. (*b*) *Descending extrapolar katelectrotonus* (**A**). The irritating salt is placed at R$_1$. The contractions induced increase immediately on closure of the polarizing circuit. On opening the circuit the contractions resume their previous activity. (*c*) *Ascending extrapolar anelectrotonus* (B). The salt is applied at r$_1$. The contractions of moderate intensity present before closure of the circuit become feebler after closure. (*d*) *Ascending extrapolar katelectrotonus* (B). The salt is placed at r. In this case a distinction must be made in accordance with the strength of the polarizing current: (1) If the current is extremely weak, and it can be appropriately regulated with the aid of the rheocord (Fig. 224), increase of the contractions is observed after closure of the polarizing circuit. (2) If, however, the current is stronger, the contractions become smaller, or they are even

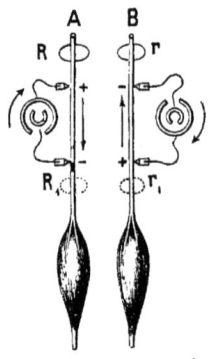

wholly abolished. The reason for this latter apparently abnormal relation lies in the fact that under the influence of stronger currents the conducting power at the anode is diminished or even abolished. Although in this case the salt acts upon an irritable segment of nerve the effect does not appear in the muscle, as the conduction of the stimulus to the latter is prevented.

The laws of electrotonus can be demonstrated also in an entirely isolated nerve. One extremity of the nerve is applied to the electrodes of a galvanometer for the production of a strong current. The polarizing circuit is applied to the nerve at some distance. If, now, the nerve with the circuit closed is irritated in the anelectrotonic segment, as, for example, by induction-shocks, the negative current-variation is feebler than if the polarizing circuit were open. Conversely, the variation is stronger if the irritation be applied to the katelectrotonic segment. Also the extrapolar currents appearing in electrotonus exhibit the negative variation when the nerve is irritated. Katelectrotonus is

FIG. 233.—Testing the Irritability in Electrotonus.

intensified by the action locally of elevation of temperature, of acids and by tetanization, while anelectrotonus is diminished by the same influences.

The *law of electrotonus has been demonstrated also in living human beings.* If it be desired to test electrotonus in living human beings the conditions of the distribution of the current in the part of the body are especially to be considered. If, for example, the two electrodes are placed in the course of the ulnar nerve (Fig. 234), it will be seen that the currents appearing in the nerve at the anode (+ *a a*) must diminish the irritability, although above and below the anode (at *c c*) the positive current emerges in part from the nerve and naturally causes katelectrotonus in these situations. In an analogous manner increased irritability

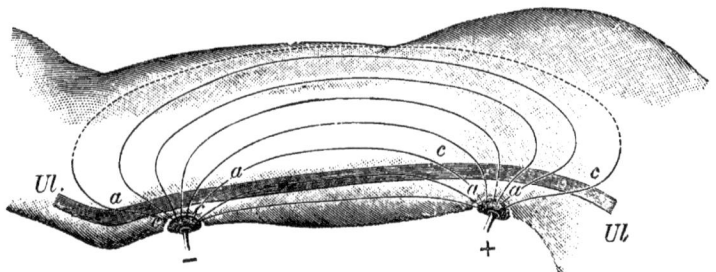

FIG. 234.—Diagrammatic Representation of the Distribution of the Electric Current in the Arm on Galvanization of the Ulnar Nerve.

prevails immediately at the point of application of the kathode (at — *c c*), but in the portions of the nerve above and below, where (at *a a*) the positive current (from +) enters the nerve-path, the irritability is diminished—anelectrotonus. If, thus, it be desired to apply irritation in the vicinity of an electrode, the application would not be made to a portion of the nerve whose irritability is influenced by that electrode. In order, therefore, to apply the irritation directly to the situation occupied by the electrode, it is necessary at the same time to apply the irritation through the electrode itself, for example mechanically, or by electrical irritation, passing the irritating current simultaneously through the path of the polarizing current.

Testing Electrotonus in Inhibitory Nerves.—In order to ascertain the action of the cardioinhibitory vagus fibers in electrotonus Landois proceeded as follows: If dyspnea be excited in rabbits, the number of heart-beats diminishes because the dyspneic state of the blood irritates the cardioinhibitory center in the medulla oblongata. If, under such conditions, a constant descending current applied to the vagus is closed, the nerve of the opposite side having been previously divided, the number of pulse-beats again increases—descending extrapolar anelectrotonus. If, on the other hand, the current is sent through the nerve in an ascending direction, the number of heart-beats diminishes still further with feeble currents, while with strong currents the number increases—ascending extrapolar katelectrotonus. From the foregoing it appears that the action of the inhibitory nerves in electrotonus is exactly the opposite of that of the motor nerves.

Testing Electrotonus in Sensory Nerves.—In a decapitated frog the sciatic nerve on one side is dissected free and isolated. If the nerve is irritated at one point with sodium chlorid reflex contractions take place in the other leg through the intact spinal cord. These disappear as soon as a constant current is closed on the nerve in such a manner that the salt is situated in the anelectrotonic segment.

Testing Electrotonus in Nerves of Special Sense.—Katelectrotonus at the central extremity increases the irritability in all nerves of special sense, in greatest degree in the eye for the shortest light-waves, on the tongue for acid taste. Anelectrotonus at the central extremity diminishes the electrical irritability, in the eye in least degree for the longest waves; at the tip of the tongue there develops a salty taste, on the posterior portion a bitter taste. At the moment of closure or of opening there occur alone in the eye and the ear so-called flashes. These result, however, only when muscular contractions take place at the same time. They are, therefore, caused in the eye solely by sudden movement of the eyeball, and in the ear by that of the muscles of the auditory ossicles, which are suddenly contracted strongly.

In the muscle the intrapolar segment is in a state of altered irritability during electrotonus. Also the delay in conduction extends only to this area.

To the question as to the real nature of the galvanic effects Loeb replies that probably all electrical effects upon living tissues are only indirect, that those effects that are designated electrical are in reality only the chemical and molecular effects of the ions or the combinations formed by them.

THE DEVELOPMENT AND THE DISAPPEARANCE OF ELECTROTONUS.

THE LAW OF CONTRACTION. THE LAW OF POLAR STIMULATION.

Both at the moment of development and at that of disappearance of electrotonus, therefore on closing and on opening the circuit, the nerve undergoes irritation. 1. On closing the circuit, this stimulation occurs only at the kathode, at the moment when katelectrotonus develops. 2. On opening the current, the stimulation takes place only at the anode, at the moment when anelectrotonus disappears. 3. Of these two stimuli that attending the development of katelectrotonus is stronger than that caused by the disappearance of anelectrotonus.

That the stimulation on opening the current occurs at the anode was demonstrated by Pflüger in the following manner with the aid of Ritter's opening-tetanus. The latter consists in the development of tetanus of some duration after the opening when a strong constant current is passed through a nerve-segment of considerable length. If the current is a descending one, this tetanus ceases immediately on division of the intrapolar nerve-segment, an evidence that the (tetanic) irritation emanates from the (now separated) anode. If the current is an ascending one the same operation fails to cause disappearance of the tetanus. Pflüger and v. Bezold found further evidence in favor of the view that the closing contraction is due to irritation at the kathode and the opening contraction from the anode in the fact that they observed with the descending current the closing contraction take place earlier after the moment of closure and the opening contraction later after the moment of opening in the muscle; and conversely, with the ascending current the closing contraction later, the opening contraction earlier. The difference in time observed corresponds to the time required for the propagation of the stimulus through the intrapolar segment. If a large portion of the intrapolar segment of the nerve of a frog-preparation be made inirritable by application of ammonia, only the electrode directed toward the muscle will have a stimulating influence, therefore with a descending current closure and with an ascending current opening.

The law of stimulation is applicable to all kinds of nerves.

The Law of Contraction.—The contractions occurring on closing and opening the circuit exhibit differences in accordance with the *direction* and the *strength* of the current.

Exceedingly feeble currents cause, in accordance with the third of the foregoing propositions, and whether descending or ascending, only closing contraction. The disappearance of anelectrotonus is such a feeble stimulus that the nerve does not react.

Currents of moderate strength, whether ascending or descending, cause contraction both on closing and on opening.

Exceedingly strong descending currents cause contraction only on closing. Contraction on opening is wanting because in the state of electrotonus with exceedingly strong currents almost the entire intrapolar segment has become incapable of conduction. Ascending currents cause contraction only on opening for the same reason. The muscle remains in contraction (closing tetanus) during the period of closure with currents of a definite strength.

Polar effects may be observed also in connection with rapid variations brought about by the induced current. These cause irritation to a certain degree only at the kathode in their development. At the anode the irritation is feebler on opening and the irritability of the nerve is diminished in this situation. From this point of view the phenomena of so-called *hypermaximal contractions* and *deficiency* are to be explained. If the nerve of a frog-preparation is stimulated by a descending current, which is gradually increased, the contractions at first increase in extent with increase in the intensity of the irritation. On further increase, however, the extent of the contractions no longer increases. If, now, the increase be continued still further the height of contraction again increases—*hypermaximal contraction.* It is only with this last intensity of action that the anodal opening stimulation manifests itself, and this is added to the kathodal closing stimulation, which at first is alone effective. If in an analogous manner the irritation be applied by means of an ascending current, the contractions will be observed to increase at first with increase in the strength of the current; then with further increase the contractions become smaller and they may for a time be entirely wanting— *deficiency.* This omission is explained by the action of anelectrotonus in rendering conduction difficult. If the increase be made still greater, the contractions appear again and they become still greater—*hypermaximal contractions.* This last phenomenon is explicable by the effects of anodal opening stimulation.

The nerve in process of dying, with change in irritability according to the law of Ritter-Valli, also exhibits a modified contraction-law. In the stage of increased irritability, feeble currents in both directions cause only closing contraction. In the succeeding stage of commencing diminution in irritability, feeble currents in both directions give rise to contraction on closing and on opening. Finally, in the stage of greatly diminished irritability the descending current causes contraction only on closing: the ascending, contraction only on opening.

As the various stages of irritability advance through the nerves centrifugally, the different stages may often be observed simultaneously in different segments of the nerve. According to Valentin, A. Fick, Cl. Bernard, Schiff. and others, the living, wholly intact nerve exhibits only closing contractions with the current in either direction, but also opening contractions with currents of considerable strength.

Eckhard observed in living rabbits, with currents of moderate strength, passing through the hypoglossal nerve, twitching of one-half of the tongue (instead of contraction) on opening the circuit of an ascending current and a similar manifestation on closing the circuit of a descending current.

Pflüger has represented the contraction-law diagrammatically. According to him the molecules of the resting nerve are in a state of a certain moderate degree of mobility. In katelectrotonus the mobility of the molecules is increased, while in anelectrotonus it is diminished. Accordingly, stimulation is produced if the nerve-molecules pass from the state of moderate mobility into that of free mobility, or if they pass from a state of difficult mobility into one of moderate mobility (of rest).

Analogous phenomena, such as are yielded by the contraction-law for the motor nerves, can also be established for the *inhibitory nerves.* Moleschott, v. Bezold, and Donders have examined the cardiac branches of the vagus in this connection. The results correspond entirely with those obtained with motor nerves, except naturally that inhibition of the heart's action occurs in this instance, instead of the contraction that takes place on stimulation of a motor nerve.

The *sensory nerves* likewise react in a similar manner, although it must be borne in mind that the reacting organ in this instance is situated at the central extremity of the nerve-tract, while in the case of the motor nerve it is situated at the peripheral extremity, in the muscle. Pflüger studied the influence of closure and opening on sensory nerves by observing the resulting reflex contraction. Feeble currents caused contraction only on closure; currents of moderate strength contraction on both closure and opening; strong descending currents contraction only on opening, and ascending currents contraction only on closing. Applied to the skin of man feeble currents give rise to sensation only on closing with the current passing in either direction; while strong descending currents give rise to sensation only on opening, and strong ascending currents only on closure. During the closure of the circuit there is a prickling, burning sensation, which increases with the strength of the current. The phenomena (sensations of light and of sound) observed in the nerves of special sense, are analogous to the foregoing.

In the *muscles* the contraction-law is tested by keeping one extremity stretched, so that it cannot shorten, and closing and opening the circuit in this situation. The movable extremity then exhibits the same law of contraction as if the motor nerve were stimulated. On closing the contraction begins at the kathode, on opening at the anode.

E. Hering and Biedermann demonstrated more thoroughly in this connection that contractions on closing and opening are purely polar effects. They found that when a feeble current is passed through the muscle, the first result that appears is a small contraction confined to the kathodal half of the muscle. Increase in the strength of the current causes greater contraction, which extends to the anode, but is feebler at this point than at the kathode. At the same time the muscle remains in a state of permanent contraction during the period of closure. On opening, the contraction takes place from the situation of the anode. Also after opening the muscle may remain for some time in a state of contraction, which ceases on closing the current passing in the same direction. The law of polar effects manifests itself also in the unstriated muscle of the excised uterus and intestine kept warm; also in the isolated ventricle of the frog, as well as in the musculocutaneous tube of worms and holothurians.

In some animals apparent deviations from the foregoing law of Pflüger occur, but these are only apparent. They are caused by the actions of ions induced in part by the internal and in part by the external polarization. In the protoplasmic current in chars Hörmann observed with each stimulation—attended with a wave of negativity—sudden arrest of the movement analogous to a muscular contraction. In this instance there is thus a law of arrest instead of a law of contraction.

Destruction of the extremity of a muscle by various procedures gives rise to diminution of irritability in the neighborhood of the portion destroyed. Therefore, the polar effects in such a situation are but feeble. Also moistening such a point with meat-infusion, potassium hydroxid, or alcohol diminishes the polar effects locally, while sodium-salts and veratrin increase them.

Under certain circumstances not only permanent irritation, but also contraction, may appear at both extremities of a muscle on passing a current longitudinally through it, for example after destruction of one of its extremities, or in case of peripheral muscular paralysis in man, during closure as well as after opening of a galvanic current.

The persistent moderate shortening of the muscle—continued closing contraction (Fig. 194, IV)—at times observed during the period of closure of the circuit, is due to the abnormal persistence of the kathodal closing excitation (with strong stimuli in dying muscles, or in the muscles of cooled winter-frogs). Also opening at times gives rise to a similar contraction originating at the anode. Treatment of the muscle with 2 per cent. sodium-chlorid solution containing sodium carbonate increases the permanent contraction considerably, and it appears occasionally as rhythmic shortening.

If the entire muscle is introduced into the circuit the closing contraction predominates when the current passes in either direction. During the period of closure a permanent contraction is most marked with the ascending current.

It is a remarkable fact that the constant current has an effect upon a muscle in the state of permanent contraction entirely opposite to that upon a relaxed muscle. If a constant current be passed, by means of unpolarizable electrodes, longitudinally through a muscle in a state of permanent contraction, for example as a result of poisoning with veratrin, or through the contracted ventricle, relaxation begins on closure at the anode and extends thence. On opening the current in the permanently contracted muscle the relaxation takes place from the kathode.

In correspondence with these remarkable phenomena, the currents in the muscular substance appear in accordance with the law that every contracted portion is negative with relation to every resting portion of a muscle. Perhaps the experiments of Pawlow throw light upon these observations. This observer found that the sphincters of mussels contain nerve-fibers, irritation of which causes the production in the muscle of a state of relaxation.

If a nerve or muscle has been traversed for a considerable time by a constant current, permanent tetanus often appears after the opening— so-called Ritter's opening tetanus. This is abolished by closing a current passing in the original direction, while the closing of a current in the opposite direction increases it—Volta's alternative. The persistent passage of the current increases the irritability for the opening of a current in the same direction and for the closing of a current in the opposite direction; while, conversely, it diminishes the irritability for the closing of a current in the same direction and the opening of a current in the opposite direction.

According to Grützner, Tigerstedt, and others, the cause for the opening contraction resides in part in the development of polarizing after-currents. The irritating effect of the kathode is dependent upon the escape of water at this point. Engelmann and Grünhagen explained the opening and closing tetanus in a different manner, namely as due to latent stimulation of the prepared nerve, as a result of drying and fluctuations in temperature, the stimuli being in themselves too feeble to cause tetanus, but becoming effective when increased irritability of the nerve is set up in the vicinity of the kathode after closing, in that of the anode after opening.

Biedermann showed that under certain circumstances two opening contractions could be observed in succession in the frog-nerve-preparation, of which the second or later corresponds to Ritter's tetanus. The first of these contractions is caused by the disappearance of anelectrotonus in the sense of Pflüger. The second is to be explained like Ritter's opening tetanus in the sense of Engelmann and Grünhagen.

Pathological.—The observation is rarely made in morbid conditions of the nervous system (hysteria) that after interruption of an electrical current through a nerve, tetanic contractions persist, and this has been appropriately designated the *neurotonic reaction.*

Simultaneous Action of the Constant Current and the Inherent Current.—The Action of Two Currents.—In the frog-preparation arranged for testing the contraction-law, a demarcation-current naturally occurs in the nerve. If a feeble, artificial, stimulating current be applied to such a nerve interference-phenomena may occur between these two currents. The closing of an exceedingly feeble constant current causes a contraction that is really no closing contraction, but is due to opening (conduction) of a branch of the demarcation-current. Conversely, the opening of an exceedingly feeble constant current may cause a contraction that is due really to closing of the nerve-current branch previously diverted in consequence of secondary closure (through the electrodes).

If a motor nerve is acted on simultaneously by two induced currents the following two results are possible. One induced current may be so feeble that the nerve is not irritated by it to the point of contraction, while the other induces only a feeble contraction. In this event the inframinimal current plays the part of a feeble constant current, and the size of the contraction depends only upon whether the effective irritating current is applied near the anode or the kathode of the inframinimal current. If, however, two stimulating currents of unequal strength, separated by a considerable distance from each other, in order to exclude electrotonic effects, are applied to a nerve, and each of them alone is effective, the same result occurs as if the stronger stimulus alone were applied. The feebler wave of stimulus is lost entirely in the stronger.

If in man a nerve is compressed and the affected portion of the body is rendered anemic by compression of its arteries, the opening contractions soon predominate greatly and kathodal opening contraction in greater degree than anodal opening contraction—*compression-reaction* of R. Geigl.

RAPIDITY OF CONDUCTION OF THE STIMULUS IN NERVES.

If a motor nerve is stimulated at its central extremity the impulse is propagated like a wave-movement through the course of the nerve to the muscle with great rapidity, which in the case of the sciatic nerve of the frog is equal to 27.25 meters in a second, for the motor nerves, and in that of man, 33.9 meters.

The rapidity of conduction is apparently less in the visceral nerves; for example, 8.2 meters in the pharyngeal fibers of the vagus. Fredericq and van de Velde found the rate in the motor nerves of the lobster to be 6 meters.

The propagation-rapidity of the wave of excitation is susceptible to various influences. It is retarded by cold and also by considerable heat applied to the nerve, by curare, and by anelectrotonus, while it is increased by katelectrotonus in the freely exposed nerve. It varies with the length of the conducting portion of nerve, but it increases with the strength of the stimulus, though not at first. The power of conduction is diminished in the anelectrotonic portion.

Method.—v. Helmholtz determined the velocity of propagation of the impulse for the motor nerves of the frog according to the method of Pouillet, which is based on the fact that the needle of the galvanometer is deflected by a constant current of short duration. The degree of deflection is proportional to the duration and the strength of the current, which is known in this instance. The method itself is so applied that the time-measuring current is closed at the moment that the nerve is irritated, and it is again opened when the muscle contracts. If, now, the nerve is irritated first at the central extremity and then close to its entrance into the muscle, the time between the beginning of the stimulation and the contraction will in the latter event be shorter, and therefore the deflection of the galvanometer will be less, than in the first case, as the stimulus must pass through the entire nerve to the muscle. The difference between the two periods of time is the propagation-time for the stimulus in the portion of nerve examined.

Fig. 235 is a diagrammatic representation of the arrangement of the experiment. The galvanometer G is introduced into the (still open) circuit a, b, c, d, e, f, g, h, yielding the time-measuring current. Closure is effected by depressing the lever S, the platinum plate c of the pivoted arrangement W being tilted down by d. At once, with the occurrence of closure the magnetic needle is deflected and the extent of the deflection is noted. At the same moment that the current between c and d is closed the primary circuit of the induction-apparatus, k, i, p, O, m, l, is opened by raising the extremity of the pivoted arrangement at i. By this means an opening current is induced in the induction-spiral R, which stimulates the nerve of the suspended frog's leg at n. The impulse is propagated through the nerve to the muscle (M); the latter contracts as soon as the impulse reaches it, and, by raising the lever H, which can be rotated about x, opens the time-measuring current by means of the double contact e and f. At the moment of opening. the further deflection of the magnetic needle ceases. The contact at f consists of a mercurial dome drawn out to a thread. If, after the contraction of the muscle, the lever H falls so that the point e rests upon the underlying fixed plate y, the contact at f nevertheless remains open, and, therefore, also the galvanometer-circuit. Another method is described on p. 654.

In man v. Helmholtz and Baxt determined the propagation-velocity of the impulse in the median nerve by having the musculature of the thenar eminence record its contraction by means of a lever upon a rapidly rotating cylinder. The irritation of the nerve was practised on one occasion in the axillary cavity and on the second at the wrist-joint. The contraction-curves, naturally, exhibited differences as to the moment of beginning. The difference in the time-value for these two gives the time for the conduction in the intervening nerve-segment. In the experiment the entire arm is enclosed in a plaster bandage in order to secure rest of the muscles of the arm.

According to Bernstein, the time necessary for the stimulus that passes from the motor nerve to the muscle to excite the motor nerve-endings is on the average 0.0032 second in the frog and 0.0015 second in warm-blooded animals.

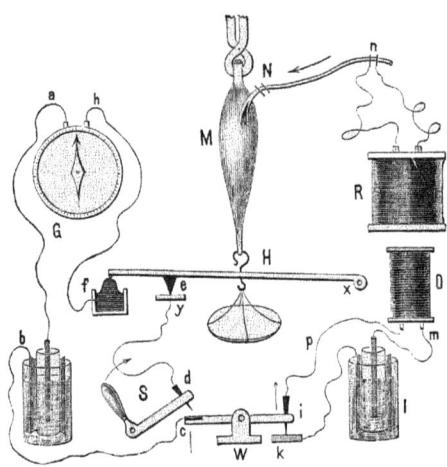

In the sensory nerves of man the impulse is probably propagated with the same rapidity as in the motor nerves. The figures obtained vary, it is true, between the wide limits of 94 and 30 meters in a second.

FIG. 235.—v. Helmholtz's Method for Determining the Propagation-velocity of the Nerve-stimulus.

Method of Examination.—In the person under observation two points at as widely unequal distances from the brain as possible are irritated for a moment in succession, for example the lobule of the ear and the great toe, as by an opening induced current. The moment of irritation is noted, for example by beginning the vibrations of a tuning-fork plate, the removal of the clamp from the tuning-fork at the same time opening the primary current. The person examined should in each instance indicate by an appropriate sign upon the registering surface the time when irritation is perceived. The reaction-time to be taken into consideration in this connection is discussed on p. 777.

A needle-prick of the skin causes at first the sensation of pricking. There then follows an interval free from sensation and finally again a pricking sensation, apparently originating from within.

Pathological.—In the presence of disease of the spinal cord the remarkable observation has occasionally been made of a striking retardation of conduction in the sensory nerves of the skin. The sensation itself may, under such circumstances, be unaltered. Occasionally only the conduction of painful sensations was found to be retarded, so that a painful impression upon the skin was first perceived only as a tactile sensation and then as pain, or conversely. If the interval of time between these two impressions is considerable, there may be well-defined double sensation.

In the sphere of the motor nerves retarded conduction for voluntary movement is observed, for example in cases of senile paralysis agitans. The observation has been made, further, in rare cases that with otherwise well-developed musculature, voluntary movements were executed much more slowly, as the interval between the impulse of the will and the contraction was prolonged, and, besides, the muscles occupied a longer time in contracting, a sort of tonic contraction thus resulting. In nerves exhibiting degenerative reactions the retardation of conduction in electrotonus was marked. In tabetic patients reflex movements have also been observed to be retarded, in the case of irritation by heat more than in that by cold.

DOUBLE CONDUCTION IN NERVES.

The property of living nerve by means of which it transmits an impulse throughout its course is designated *conductivity*. All procedures that injure the nerve in its continuity, such as division, ligation, crushing, chemical destruction, or that destroy its irritability at any point, such as absolute deficiency of blood, certain poisons, for example curare for the motor nerves, also marked anelectrotonus, abolish the conductivity. The conduction takes place only through fibers directly in communication, and it can never be transferred to an adjacent fiber—*law of isolated conduction*. By *double conduction* is understood the ability of the nerve to transmit in both directions a stimulus applied in its course. Naturally, as a result of the anatomical conditions in the intact body the motor nerves are capable of conducting only in a centrifugal direction and the sensory nerves only in a centripetal direction, but under suitable conditions it can be shown that each nerve is capable, in the same way as an inanimate conductor, of conducting in both directions.

The evidence brought forward in support of the presence of double conduction is as follows: If a nerve be irritated, alterations in its electrical properties appear both in an upward and in a downward direction. If the posterior free extremity of the electrical centrifugal nerve-fibers of the malapterurus be irritated, the branches arising at a higher level are also set in irritation, so that the entire electrical organ is discharged. If the lower third of the sartorius muscle in the frog be divided longitudinally and one division be irritated mechanically, the irritation passes in the nerve-fibers thus separated at first upward to the point of division and thence centrifugally in the unirritated muscular extremity whose individual fibers now contract. The gracilis muscle is divided by a tendinous line into two halves. The nerves to both are derived from a bifurcation of the individual fibers in the nerve-trunk. Every irritation of the nerve for one portion of the muscle causes contraction in both halves of the muscle. All of the earlier evidence in support of double conduction of nerves derived from experiments on division and reunion will not bear rigid scrutiny.

The intercentral association-fibers in the cerebrum must normally be assumed to conduct in both directions.

EMPLOYMENT OF ELECTRICITY FOR THERAPEUTIC PURPOSES.

DEGENERATIVE REACTIONS OF MUSCLE AND NERVE.

Electricity is much employed in medicine for therapeutic purposes. It may be used in the form of the rapidly interrupted current of the induction-apparatus (faradic current), or of the magneto-electrical machine, or of the extra-current apparatus, or in the form of the constant current. The employment of electricity is based upon its physical and physiological properties.

In cases of *paralysis* the faradic current is applied by means of suitable wet electrodes covered with sponge either to the muscle itself or to the point of entrance of the motor nerve (Figs. 236, 237, 238, 239). The motor points on the face are shown in Fig. 245, those of the neck in Fig. 244. In employing the faradic current the object is, by means of artificially induced movements, to protect the paralyzed muscle from secondary degeneration, which it would undergo as a result of long-continued inactivity. If in addition to its motor nerves also the trophic nerves of the paralyzed muscle are inactive, even long-continued faradization will have no noteworthy effect, as the muscle will nevertheless undergo atrophy. The application of the induced current may, however, have a good effect upon the paralyzed muscle by increasing the amount of blood sent to it and

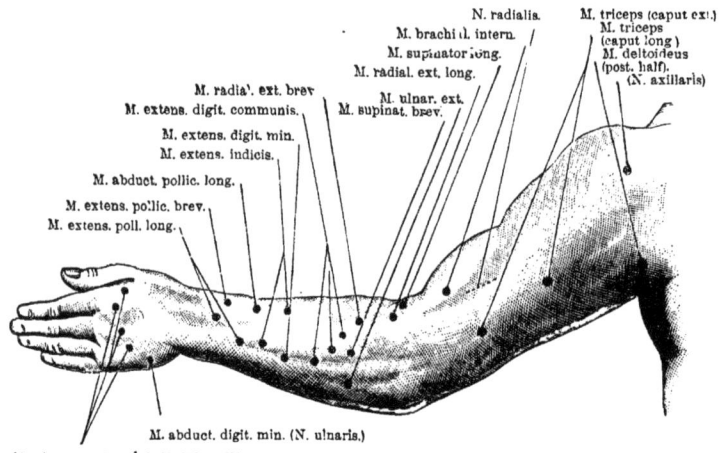

FIG. 236.—Motor Points of the Radial Nerve and of the Muscles supplied by it. Dorsal aspect of the upper extremity (after Eichhorst).

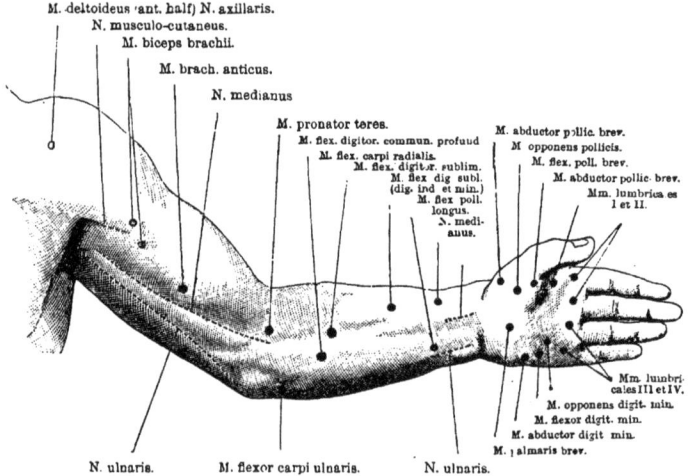

FIG. 237.—Motor Points of the Median and Ulnar Nerves, as well as of the Muscles supplied by them. Palmar aspect of the upper extremity (after Eichhorst).

reflexly influencing its metabolism. Feeble induction-currents are further capable of reviving the irritability of enfeebled nerves.

The constant current deserves consideration in cases of paralysis not only as a stimulus for exciting contraction, on closing, opening, reversing, increasing, and diminishing the current, but rather through its so-called polar effects. On closing the circuit the nerve is thrown into irritation at the kathode, and on opening the circuit at the anode. Then, during the period of closure of the circuit through the nerve the irritability is increased at the kathode and by this means a remedial influence may be exerted upon the nerve. In man, however, the special conditions described on p. 662 should be kept in mind in the employment of percutaneous galvanization. In the vicinity of the anode there is also increased irritability. This is observed chiefly on repeated reversal of the current, but also after closing and opening, or even on the uniform passage of the current. If the increase in irritability obtained by means of the current be tested, it will be found that as a result of the application of the current the irritability for the closing of a

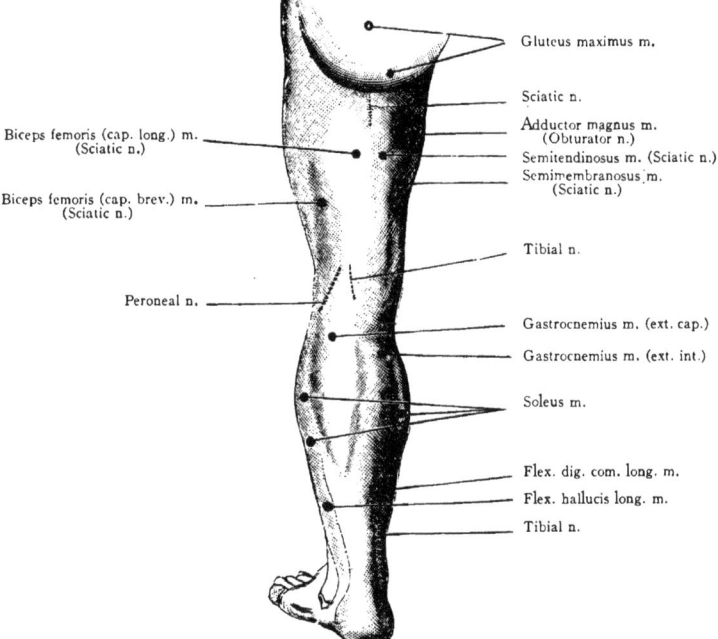

Gluteus maximus m.

Sciatic n.

Adductor magnus m. (Obturator n.)

Biceps femoris (cap. long.) m. (Sciatic n.)

Semitendinosus m. (Sciatic n.)

Semimembranosus m. (Sciatic n.)

Biceps femoris (cap. brev.) m. (Sciatic n.)

Tibial n.

Peroneal n.

Gastrocnemius m. (ext. cap.)

Gastrocnemius m. (ext. int.)

Soleus m.

Flex. dig. com. long. m.

Flex. hallucis long. m.

Tibial n.

FIG. 238.—Motor Points of the Sciatic Nerve and its Branches, the Peroneal and Tibial Nerves (after Eichhorst).

current in the opposite direction and for the opening of a current in the same direction is increased.

Furthermore, in the employment of the constant current its restorative effect should be taken into account, chiefly the ascending current, as R. Heidenhain has found that exhausted and enfeebled muscles can be refreshed by the passage of a constant current. Finally, the constant current must be conceded a therapeutic influence by reason of its catalytic or cataphoric effects, in consequence of which it exerts a solvent, decomposing, or dispersing action upon possible accumulated products of inflammation or stagnation in nerve or muscle. The current may, besides, exert a direct or reflex stimulating influence upon the nerves of the blood-vessels and lymphatics.

If the cause of the paralysis reside in the muscle itself, it is customary to apply the induced current by means of sponge-electrodes directly to the muscle. In case of primary lesions of the motor nerves the electrodes are applied to the latter. The currents used under such circumstances must be only of moderate strength. Strong tetanic contractions are to be avoided as injurious and likewise unduly prolonged action.

The galvanic current may likewise be applied either to the muscle alone or to the motor nerve or even to its center, or to both nerve and muscle at the same time. As a rule, under such circumstances, the kathode should be applied to the point whose irritability is lowered, as under its influence the irritability is increased. The anode is placed at some indifferent point, for example upon the sternum. Stroking along the nerve with the kathode, as well as variation in the

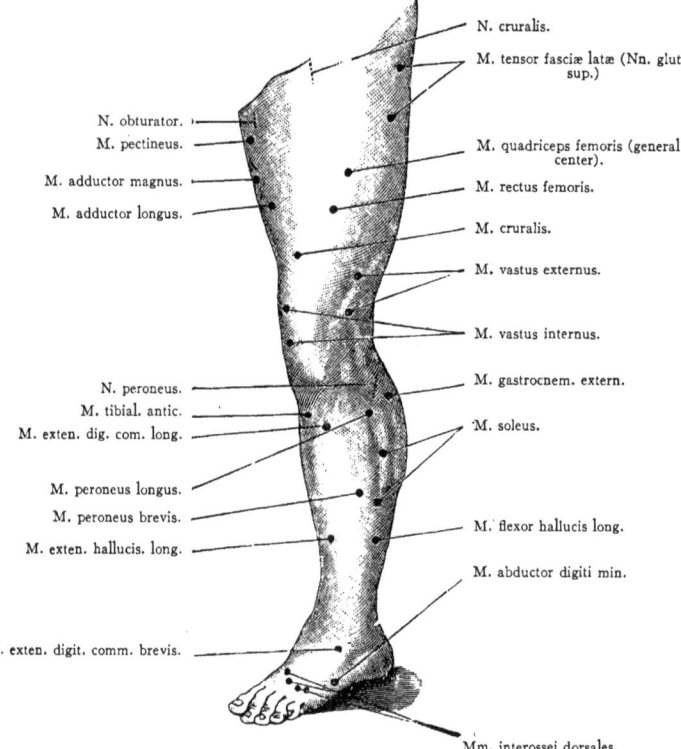

N. cruralis.

M. tensor fasciæ latæ (Nn. glut. sup.)

N. obturator.

M. pectineus.

M. adductor magnus.

M. adductor longus.

M. quadriceps femoris (general center).

M. rectus femoris.

M. cruralis.

M. vastus externus.

M. vastus internus.

N. peroneus.

M. tibial. antic.

M. exten. dig. com. long.

M. gastrocnem. extern.

M. soleus.

M. peroneus longus.

M. peroneus brevis.

M. exten. hallucis. long.

M. flexor hallucis long.

M. abductor digiti min.

M. exten. digit. comm. brevis.

Mm. interossei dorsales.

FIG. 239.—Motor Points of the Peroneal and Tibial Nerves on the Anterior Aspect of the Leg and Thigh. Peroneal nerve on the left, tibial nerve on the right (after Eichhorst).

strength of the current, is believed to increase the favorable effect. When the seat of the lesion is in the central organs, galvanization may be applied along the vertebral column, or to the vertebral column and the course of the nerve at the same time, or to the head (with care), or when possible at the suspected seat of the disease, for example the speech-center or the central convolutions. Care should be taken to avoid currents of undue strength and applications of prolonged duration.

The varied reaction of the paralyzed nerves and muscles to the induced constant current is especially noteworthy. This relation has been well designated the *reaction of degeneration*. In the first place, the physiological fact should be noted that the muscles supplied by dying nerves and also the muscles of a curarized animal respond less readily to a rapidly interrupted faradic current than fresh non-curarized muscles. According to Neumann, it is the longer duration of the constant current, as contrasted with the momentary closing and opening of the induced current, that permits the possibility of contraction. If the constant current be interrupted with the same rapidity as the faradic, it also will be ineffective. On the other hand, the induced current can be made effective if it be permitted to continue in action for a longer time. This can be accomplished in the sliding apparatus by keeping the primary circuit closed, and raising and depressing the induction-coil upon the slide. By this means slowly increasing and diminishing induced currents are generated that act energetically in causing contraction of curarized muscles. Therefore, in the stimulation of muscle and nerve, not alone the strength, but also the duration, of the current must be taken into consideration, just as the deflection of the magnetic needle is dependent upon both factors. According to E. Remak, however, the muscle reacts, nevertheless, to individual induced shocks, when the reaction of degeneration is present, and with a contraction of slower evolution, but this reaction is no longer demonstrable after the lapse of a short time. The muscle is, therefore, not inirritable to the faradic current, but only exhausted with extreme readiness. It must constantly recuperate after its exhaustion.

The *typical reaction of degeneration* is characterized essentially by the following points. For the muscle there is, on direct irritation, diminution to the point of abolition of faradic irritability, with increase of galvanic irritability (from the third to the fifty-eighth day), the latter diminishing, although with considerable variations, from the seventy-second to the eightieth day. There is also a preponderance of anodal closing contraction as compared with kathodal closing contraction. The contraction in the affected muscle takes place slowly; it is pseritaltic and limited locally, in contradistinction to the lightning-like contraction of normal muscles. In the stage of diminished galvanic irritability the latent period is prolonged fourfold, the duration of contraction twofold. For the nerve there is diminution to the point of abolition of faradic and galvanic irritability. If the reaction of the nerve is normal, while the muscle on direct stimulation with the constant current exhibits the reaction of degeneration, the condition is described as *partial reaction of degeneration*, which is constant in cases of progressive muscular atrophy. In the presence of deranged sensibility in cases of tabes, the sensory nerves have been observed to react in a manner analogous to the motor nerves in connection with the reaction of degeneration.

In rare cases the contraction of the muscle from the nerve on application of the induced current exhibits also a sluggish vermicular course—*faradic reaction of degeneration*. In the case of paralyzed and degenerated muscles the motor points may be found displaced further toward the periphery. The lessened thickness of the muscle and the resulting increase in density of the current may be the cause for this.

Nerve-degeneration and nerve-regeneration are considered on p. 663.

In the different forms of *spasm*, contracture or electrical spasm, the constant current especially has been found useful. Under such circumstances pathologically increased irritability of the nerves or muscles is diminished through the effects of anelectrotonus. Therefore, the anode should be applied to the nerve or the muscle, or in case of reflex spasm upon those points that are discovered to be the actual source of the pathological irritation. Uniform, feeble currents are especially effective under such circumstances. Also the relaxing (inhibiting polar) action is to be considered. The constant current may, however, exert a beneficial influence also through its catalytic action, by means of which it removes irritants at the seat of disease. Finally, it has often been observed, since the time of Remak, that with the application of the constant current, voluntary control of the affected motor apparatus is increased. In cases of spasm of central origin the constant current may be applied even to the central organ.

Faradization may be employed in cases of spasm to strengthen possibly enfeebled antagonists. Under such circumstances faradized muscles in a state of contracture are said to acquire increased extensibility, as the muscle is more extensible in a state of active physiological contraction.

In the treatment of *cutaneous anesthesia*, stimulation should be applied first to the skin itself, the induced current being often applied by means of wire-brush electrodes. In the employment of the constant current the kathode should be applied to the insensitive area. It is even possible with strong currents to cause vesication of the skin. When the lesion has possibly a central situation only the

43

constant current should be employed. The question should be raised as to the extent to which the suppression of sensation could be aided by the establishment of katelectrotonus in the central focus.

In cases of *hyperesthesia* and *neuralgia*, faradic currents are applied with the object of obtunding to a certain extent irritated areas of skin by hyperirritation by means of active applications. For this purpose strong currents passed through a wire brush cause a sort of flagellation, and the brush on long-continued application may act as an electrical moxa. In addition to this local effect, feeble currents excite, reflexly, acceleration of the circulation, with increased action of the heart and contraction of the vessels, while strong currents have the opposite effect. Both may, under certain circumstances, be of therapeutic value.

The employment of the constant current in cases of neuralgia is intended in the first place to induce diminution in the irritability of the morbidly irritated portion of the nerve by causing anelectrotonus. In accordance with the character of the case, the anode may be applied to the nerve-trunk or even to the center, and the kathode to an indifferent portion of the body. The catalytic and cataphoric effects should be taken into consideration, as through them, especially in cases of recent rheumatic neuralgia, irritating inflammatory products may be dissolved and dispersed. Descending currents kept closed permanently in the course of the nerve are especially recommended, and often prove surprisingly effective, especially in recent cases. Finally, the constant current, acting as a cutaneous irritant, may, like the faradic current, exert a reflex influence upon the activity of the heart and the vessels.

To determine definitely whether the irritability from the nerve or the muscle is normal, it is necessary to have an absolute current-meter, preferably Edelmann's unit-galvanometer, with an electrode having a section of 3 sq. cm.—unit-electrode. On application of this, the normal irritability in the same individual exhibits a galvanic variation of 2.3 milliampères. The differences in irritability between different healthy persons in the same nerve are smaller (1.2 m. a.) than between the different nerves of the same individual (2.3 m. a.). Kathodal closing contraction usually occurs earlier than anodal opening contraction. Stronger currents are required in the new-born to cause contraction on irritation of nerves and muscles than in adults.

v. Ziemssen and Edelmann have established an accurate dosage for the induced current in the treatment of diseases of the nervous system.

Recently sparks from the electrical machine or charges from the same source have been employed successfully by Charcot and Ballet in the treatment of anesthesia, facial paralysis, paralysis agitans. According to the former, isolated contraction of muscles can be induced in cases of spinal paralysis by the spark, even if they no longer react to the faradic current.

Mention should finally be made here of the fact that electricity is also employed for the production of thermic effects in various forms of the cautery—Mitteldorpf's galvanocautery.

The electrolytic properties of the electrical current have been employed for the purpose of causing coagulation in aneurysms or varices (arterial and venous tumors filled with blood)—galvanopuncture.

Under the influence of currents of high tension and extraordinary frequency, d'Arsonval observed increased respiratory activity, increased elimination of urine, increased combustion in the body, an influence upon the vascular nerves and the skin, and also effects upon the protoplasm of the cells, attenuation of toxins, and immunization, through which perhaps an enlarged view is opened into the treatment especially of disorders of metabolism.

ELECTRICAL CHARGING OF THE ENTIRE BODY AND OF INDIVIDUAL PORTIONS.

The elder Saussure investigated by means of the electroscope the charge in many persons placed upon an insulated stool. He attributed the irregular phenomena observed by him to the electricity generated through the friction of the clothing upon the skin. Later Gardini and others contended that the presence of a positive charge in the body is normal, while Sjösten and others held that the charge is negative. It is, however, probable that all of these charges, as well as those observed by Meissner, are purely due to friction-phenomena, modification in the effects of the distribution of the air, and to the contact of heterogeneous conductors.

Strong charges, to the point of causing a spark, have frequently been described. The earliest statement that Landois could find is made by Cardanus (1553), who makes mention of the appearance of sparks from the hair of the scalp. According to Hosford (1837) a nervous woman of Oxford exhibited sparks more than 4 cm. long at the fingers while standing on insulated carpet. Sparks, on combing the hair or on stroking cats, horses, etc., are often observed when the air is dry.

Of the various constituents of the body, recently voided urine has been found to be electrically negative; likewise the freshly drawn threads of spiders' webs; while the blood has been found to be positive. Also feathers and hairs become charged with electricity if rubbed.

COMPARATIVE. HISTORICAL.

Among the most interesting phenomena in the domain of animal electricity are exhibited by the electrical fish, of which about fifty varieties are known. The electrical eel, gymnotus electricus, is found in the fresh waters of the Orinoco district, and attains a length of 2.5 meters. The electrical rays include torpedo marmorata, from 30 to 70 cm. long; torpedo ocellata; nascinæ, found in the Mediterranean Sea; and a number of related species. The electrical catfish, malapterurus electricus, is found in the Nile. Finally there is mormyrus, or Nile-pike. By means of a special electrical organ these animals are able, in part voluntarily (eel, catfish), in part on reflex stimulation (ray), to give severe electrical shocks. The electrical organ consists of variously formed compartments bounded by connective tissue and filled with a mucoid, gelatinous substance designated torpedomucin by Weyl, to one surface of which the nerves pass and form a plexus. The latter gives rise finally to a cellular plate representing the terminations of the telodendrites and designated the electrical plate. By stimulation of the afferent electrical nerves the shock-like discharge of the organ takes place.

In the gymnoti the organ, which is comparable to a Voltaic pile arranged longitudinally in a series of rows, extends on each side of the vertebral column downward to the tail beneath the skin and receives from the anterior aspect several branches from the intercostal nerves. In addition to the larger organ, there is situated above the anal fins on each side a smaller one. The plates in this situation are vertical, and the direction of the electrical current is, in the fish, an ascending one, and in the conducting arc of closure, therefore, in the surrounding water, a descending one.

In the electrical catfish the organ, which surrounds the body of the fish like a mantle, is similarly situated, and contains a single nerve-fiber whose axis-cylinder arises in the vicinity of the medulla oblongata from a huge giant-cell, and is constituted of dendritic processes. The plates in this animal also are vertical and receive the nerves from the posterior aspect. The direction of the current when the shock is given is descending in the fish.

In the ray the organ is situated just beneath the skin to one side of the head, extending to the thoracic fins. It receives several nerves, which arise from the special portion of the brain, the electrical lobe, situated between the quadrigeminate bodies and the medulla oblongata. The plates, which do not increase in number with the growth of the animal, occupy a horizontal position. The nerve-filaments pass from these plates from the ventral aspect. The current passes in the fish from the ventral to the dorsal aspect. Torpedo occidentalis, of the eastern coast of America, may attain a length of 1.5 meters and is capable of throwing down a robust man by its discharge.

It is believed that the electrical organs are modified muscles, in which histologically the nerve-endings are highly developed, while the contractile substance has disappeared, and in whose physiological activity the chemical potential is transformed into electricity. In favor of this view is the circumstance that in the process of development the organs are preformed in a manner analogous to the muscles; further, that the organs in the resting state are neutral and in the active or degenerated state are acid in reaction; finally, that they contain an albuminous substance related to myosin, and that both after death exhibit signs of rigidity. Stimulated organs, as well as muscles, exhibit an increase in phosphoric acid, resulting from decomposition of lecithin or nuclein. Both further become exhausted, and moreover in both a period of latent irritation, lasting 0.016 second, follows actual irritation of the nerve, while a shock of the organ, which thus resembles the current in an active muscle, lasts 0.07 second. About 25 such shocks together constitute a discharge, which lasts about 0.23 second.

The discharge is thus, like tetanus, a discontinuous process. However, isolated individual discharges also take place, in the torpedo 0.006 second in duration, which thus would correspond to single muscular contractions. Veratrin causes marked discharges, comparable to the veratrin muscular tracings (p. 263). Mechanical, thermal, chemical, and tetanic-electrical stimuli give rise to discharge-shocks. During the occurrence of the electrical shock in the fish, a number of currents pass also through the muscles of the animal. In the ray the muscles are thrown into contraction, while in the eel and the catfish they remain at rest. An electrical ray may give fifty shocks in a minute; it then becomes fatigued and must recuperate; it is capable also of discharging the organ but partially. The activity of the organ is enfeebled by cooling and increased by heating it to about 22°. The organ is thrown into a state of tetanus by strychnin and it is paralyzed by curare. Irritation of the electrical lobe of the ray causes discharge; cold retards the discharge. Division of the electrical nerve paralyzes the organ. The electrical fish are themselves only slightly sensitive to strong faradic currents passed into the water surrounding them.

The substance of the electrical organ is simply refracting; excised portions exhibit a resting current that has the same direction as the shock and is increased by heat. Tetanus of the organ enfeebles the current. Mormyrus, raja, and gymnarchus are among the "feebly electrical fish," whose discharge is incomparably feebler, but which possess an organ, formerly improperly designated "pseudo-electrical," analogous in construction to that of the "strongly electrical fish" previously mentioned.

Historical.—The ancients were familiar with the shocks of the electrical fish of the Mediterranean Sea. Richer (1672) made the first reports upon the electrical eel. Walsh (1772) investigated experimentally the discharge and the power of the rays to give shocks. J. Davy was able to magnetize bits of steel by means of the shocks, to deflect the magnetic needle, and to induce electrolysis. In addition to the investigators named, Becquerel, Brechet, and Matteucci studied the direction of the discharging current, from which the last named and Linari obtained from 8 to 10 sparks. Al. v. Humboldt described the mode of life and the action of the gymnoti ("trembladores") of South America, which are able to throw down even horses by their shock.

Hausen (1743) and de Sauvages (1744) assumed the active force in the nerves to be electricity. The actual investigations into animal electricity begin—after Caldini (1756) had first observed that the muscles of the frog move on applying an electrical current—with Luigi Galvani (1789-92), who observed contractions in the frog's thigh as a result of the return stroke on discharge of the electrical machine and likewise when the muscle was placed in contact with two different metals. He believed that the nerves and the muscles possess the power of generating electricity independently. Alessandro Volta, on the other hand, attributed the contraction in the second experiment to an electrical current, whose source is situated outside of the frog-preparation at the point of contact of the heterogeneous metals. The contraction without metals of Galvani and Aldini (1794) appeared at first to contradict this view. Then, the latter showed that the animal parts themselves must contain sources of electricity. Pfaff (1793) was the first to observe the influence of the direction of the current upon the contraction of the frog's leg stimulated from the nerve. Bunzen prepared an effective pile from muscles of the frog. The subject entered upon a new phase as a result of the discovery of the galvanometer and of the classical methods introduced by du Bois-Reymond in 1843.

PHYSIOLOGY OF THE PERIPHERAL NERVES.

CLASSIFICATION OF NERVE-FIBERS ACCORDING TO FUNCTION.

As the nerve-fibers when stimulated possess the property of conducting impulses in both directions, their physiological activity is essentially dependent upon their relation to their peripheral end-organ and to their central connection. In this way the individual nerve is distributed to a definite area, within which, under normal conditions, its function is exercised in the uninjured body. This activity of the individual nerve, due to its anatomical arrangement and connections, is designated its *specific energy*.

I. CENTRIFUGAL NERVES.

(a) **Motor.**—The center consists of central or peripheral ganglia; the end-organ is a muscle.

  1. Motor fibers of transversely striated muscles.
  2. The motor nerves of the heart.
  3. The motor nerves of unstriated muscle-fibers, for example of the intestine. The peculiarities of the movement induced by these nerves has been discussed in the section on the Physiology of the Movement of the Digestive Apparatus (pp. 280 and 547). The vasomotor nerves are deserving of especial consideration in this group.

(b) **Secretory.**—The center is a central or peripheral ganglion, the end-organ the glandular cell.

Examples are furnished by the salivary secretion, the secretion of sweat, etc.

(c) **Trophic.**—The as yet unknown end-organ is situated in the tissues themselves, whose normal metabolism, growth, and uninterrupted intact existence they control.

In some tissues a direct connection with nerves is known to exist that is capable of influencing their nutritive processes. Anatomically or physiologically, the connection of the nerves with corneal cells, with the pigment-cells of the frog's skin, the connective-tissue corpuscles of the serous coat of the frog's stomach, with the cells that surround the stomata of the lymph-spaces, is known.

The statements which are to follow with reference to the trophic functions of certain nerves should be consulted, particularly the influence of the trigeminus upon the eye, upon the mucous membrane of the mouth and the nose, upon the face; of the vagus upon the lungs; of the motor nerves upon the muscles; of the nerve-centers upon the conservation of nerve-fibers, and of certain central organs upon individual viscera.

Furthermore, a description will be given here of the influence of the division of nerves upon the growth of bone. H. Nasse found that the bones after such an operation exhibited a diminution in the absolute amount of all their individual constituents, but, on the other hand, an increase of fat. After division of the spermatic nerve, degeneration of the testicle has been observed; after destruction of the secretory nerve, degeneration of the submaxillary gland; after division of the related nerve, interference with the nutrition of the cock's comb; after division of the second cervical nerve (in cats and rabbits), loss of hair from the ear; changes in the skin of the frog after injury of the spinal ganglia; after division of the cervical sympathetic (which is attended with hyperemia of the corresponding half

677

of the head), enlargement of the ear and increased rapidity in the growth of the hair were observed, together with hypertrophy of the muscular coat of the veins, of the cartilage, and of the horny skin, with atrophy of the epidermis; further, diminution in the size of the cerebral hemisphere of the corresponding side, perhaps in consequence of the pressure exerted by the dilated vessels. Lewaschew observed hypertrophy of the leg and foot in the sequence of long-maintained chemical irritation of the sciatic nerve in dogs, and also the development of aneurysmal dilatation of the vessels.

In man, irritation or paralysis of the nerves or degeneration of the gray matter of the spinal cord is not rarely attended with alterations in the pigment of the skin, and of the nails and hair and in their growth, as well as cutaneous eruptions, for example herpes zoster after inflammation of the spinal ganglia or nerves, and a tendency to bed-sores; further, rare affections and degenerations of the joints (in cases of tabes). Local diseases of the brain have been observed to be attended with unilateral derangement in the growth of the hair and the nails.

(d) **Inhibitory nerves**, which suppress or diminish a movement or secretion already present.

Examples are found in the vagus as the inhibitory nerve of the movement of the heart, the splanchnic as that of the movements of the intestine, the vaso-dilators as inhibitory nerves of the unstriated muscle of the vessels.

II. CENTRIPETAL NERVES.

(a) **Sensory nerves**, which convey sensory impressions to the central organ by means of special end-apparatus.

(b) **Nerves of special sense.**

(c) **Reflex or excito-motor nerves**, which, when stimulated at the periphery, conduct the irritation to the center, within which this excitation is transmitted to the centrifugal fibers (I, a, b, c, d), so that the activity of the latter is manifested as reflex movement, reflex secretion or reflex inhibition.

III. INTERCENTRAL NERVES.

These connect ganglionic centers one with another for the communication of the excitation among them, for example in the coördinated movements, for instance of the eyes and of widespread reflexes.

THE CEREBRAL NERVES.

All cranial motor nerves arise from their cerebral nuclei of origin as neurites of ganglion-cells in the same way as the fibers of the anterior roots of the spinal cord arise from the ganglia of the anterior horns. The sensory cerebral nerves have their actual origin in the bipolar cells of the peripheral ganglia of the sensory nerves. Into each of these cel's a cellulipetal dendrite enters from the region endowed with sensation, while a cellulifugal neurite passes from the cell to the brain, where it comes into contact with the terminal ramifications of the sensory nucleus of origin.

I. OLFACTORY TRACT AND BULB.

The strand-like triangular-prismatic olfactory tract, situated upon the inferior surface of the frontal lobe, becomes enlarged on the cribriform plate of the ethmoid bone to form the olfactory bulb, which is the analogue of a special portion of the brain that exists in different vertebrates with a well-marked power of smell. From the bulb there pass through the cribriform plate between 15 and 20 olfactory filaments, which continue first between the mucous membrane and the periosteum and into the mucous membrane itself only in the lower third of the olfactory region. The structure of the bulb, as well as the relations of the olfactory nerves, are discussed on p. 914.

The origin of the olfactory fibers may be traced as follows: (1) To the fornicate gyrus, median root (Fig. 262). (2) The lateral root passes through the anterior perforated plate to the internal capsule (sensory path of the cerebrum), and further through the uncinate gyrus (sensory cortical center), where its fibers enter into contact with the ganglion-cells by means of telodendrites. Possibly the fibers of origin decussate within the cerebrum. (3) Fibers may be traced also in the head of the caudate nucleus and thence into the anterior commissure, in which there is a communication between the two olfactory bulbs.

The olfactory nerve is the nerve of smell, the physiological excitation of which takes place only through volatile odorous substances. Congenital deficiency or division of both nerves destroys the sense of smell.

Pathological.—The designation *hyperosmia* is applied to a condition in which the acuity of the sense of smell is abnormally exaggerated, for example in hysterical individuals. Purely subjective impressions of the sense of smell, olfactory hallucinations, for example in the insane, probably depend upon abnormal excitation of the cortical center. In some persons the ingestion of antifebrin, which is odorless and tasteless, excites a subjective sense of smell even when the existence of marked coryza renders the nose incapable of smelling. *Hyposmia* and *anosmia*, diminution and abolition of the sense of smell, occur as the result of catarrhal conditions of adjacent cavities, through the action of injurious gases or fluids, as one of the phenomena of general intoxication or disease, and in consequence of absence of the pigment in the olfactory region. Strychnin increases and morphin occasionally diminishes the sense of smell.

II. OPTIC NERVE AND TRACT.

The *optic tract* arises from the anterior quadrigeminal body, from the lateral geniculate body, from the pulvinar and the zonal stratum of the optic thalamus (Fig. 242) and from the tuber cinereum. By means of a broad bundle of fibers (visual fibers of Gratiolet), which pass directly outward from the posterior horn, the origin in the parts named is connected with the cortical psycho-visual center in the occipital lobe of the same side. Fibers pass from the cerebellum through the crura.

The two tracts unite to form the *chiasm*, from which on each side arises the *optic nerve*, the fibers of which are medullated but without neurilemma.

In the chiasm a semidecussation of the fibers takes place as a rule (Fig. 240), so that the left tract sends fibers into the left half of each retina and the right tract fibers into the right half.

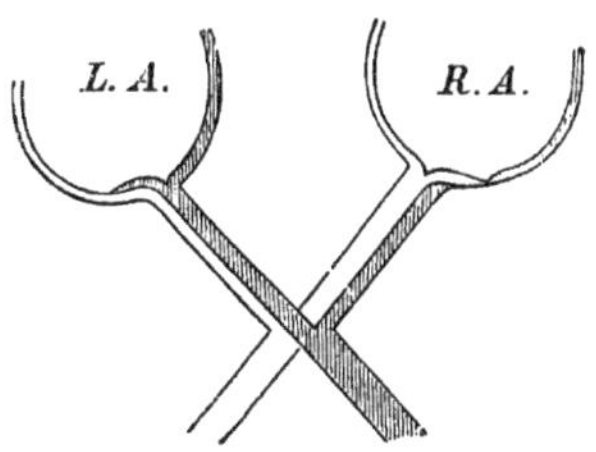

FIG. 240.—Diagrammatic Representation of the Semidecussation of the Optic Nerves.

It can thus be understood that in man destruction of one tract causes so-called homonymous hemiopia, that is blindness of the two corresponding retinal halves in the sense described. As the left half of each retina receives impressions from the right half of the visual field, and conversely, all fibers intended for seeing objects in the right half of the visual field are situated in the left tract, and vice versa. The same effect is produced by destruction of its origins, as by destruction of the optic tract, according to Bechterew also by that of the external geniculate body and the anterior brachium alone. Sagittal division of the chiasm has caused in man in one case blindness of the nasal half of each retina. In exceedingly rare cases the decussation has been wholly wanting in man.

Among animals partial decussation occurs in the following : Ape, cat, dog; complete decussation in the rabbit, mouse, guinea-pig, pigeon, owl. In the bony fish both optic nerves cross separately; in the cyclostomata there is no decussation at all. Two commissures lying upon the optic chiasm, that of Meynert and that of Gudden, have really nothing whatever to do with the optic nerve.

After extirpation of an eye in man there degenerate centripetally the fibers that enter the optic nerve from the organ, therefore in man half of the fibers in each tract. The degeneration extends to the origins in the quadrigeminal

bodies, the geniculate bodies and the pulvinar, but not into the conducting path
to the psycho-visual center. The secondary degenerations following destruction
of the cortical visual center are discussed on p. 787.

The optic nerve is the nerve of vision the physiological stimulation
of which occurs only through conveyance of the vibrations of the
luminiferous ether to the rods and cones of the retina. Every other
form of irritation of the nerve, either in its course or at its center,
causes a sensation of light. Division or degeneration of the nerve gives
rise to blindness. Irritation of the optic nerve causes also reflex contraction of the pupils through the oculomotor nerve, and marked irritation, also closure of the lids and flow of tears.

As the optic nerve has separate connections both with the psycho-visual center
and with the pupil-contracting center it will be readily understood that under
pathological conditions, on the one hand, blindness with preservation of the iris-
reaction, and, on the other hand, loss of the movement of the iris, with preservation
of vision, have been observed.

Gudden, in 1882, found two different kinds of fibers in the optic nerve, namely
fine or visual fibers, whose center is situated in the quadrigeminate body, and
coarse or pupillary fibers, whose origin can be traced to the external geniculate
body. Destruction of the visual fibers causes blindness, that of the pupillary fibers
gives rise to marked dilatation of the pupils.

Pathological.—Irritation in the range of the entire nervous apparatus may
cause excessive sensitiveness of the visual organs (*optic hyperesthesia*), and also
visual sensations of varied kind (*photopsia, chromopsia*), which, in case the irritation extends to the psycho-visual center, may even become actual *visual hallucinations*. Material alterations and inflammatory processes in the nervous apparatus are often followed by nervous impairment of vision (*amblyopia*) or even by
blindness (*amaurosis*). Nevertheless, both conditions may occur as the signs of
disorder in other organs, so-called sympathetic symptoms, being often probably
due to alterations in the circulation of the blood through irritation of the vasomotor nerves, and readily undergoing retrogression. Remarkable intermittent
forms of amaurosis are the *day-blindness* (*hemeralopia*, for example in connection
with diseases of the liver) and the *night-blindness* (*nyctalopia*). Disorders of the
cortical visual center are considered on p. 787.

III. OCULOMOTOR NERVE.

The fibers of the oculomotor nerve arise as neurites of the ganglion cells of
the oculomotor nucleus situated in the gray matter beneath the aqueduct of
Sylvius. Several groups of cells can be distinguished in this nucleus: (1) The *lateral chief nucleus*, consisting principally of large ganglion cells and passing below
the aqueduct of Sylvius on each side close to the middle line. (2) Between the two
lateral nuclei lies the single, smaller, large-celled *central nucleus*, and (3) in front
of this on each side, a smaller, small-celled nucleus. The fibers from the posterior
portions of the lateral and central nuclei decussate. In apes the nerves for the
external ocular muscles arise from the chief nucleus of the same and the opposite
side, those for the internal muscles from the accessory nuclei.

From the angular gyrus of the cerebral cortex, the psychomotor center for
the voluntary movements of the eyes, and probably also from the visual sphere
(for the involuntary adjustment of the eyes for direct vision), fibers that undergo
partial decussation in the raphé of the tegmentum pass to the oculomotor nucleus,
with whose cells they come in contact by means of terminal branches. Not far
from the pons the nerve appears in the midst of the inner bundle of fibers of the
peduncle as a median and a posterior lateral group of fibers.

The oculomotor nerve contains: 1. The voluntary motor fibers for all
of the external ocular muscles, with the exception of the external rectus
and the superior oblique, and for the elevator of the upper lid. The
coördinated movement of both eye-balls is, however, independent of the
will. 2. The fibers for the sphincter muscle of the pupil that are active
through reflex stimulation from the retina. 3. The fibers for the muscle

of accommodation. The fibers mentioned under 2 and 3 are given off from the branch for the inferior oblique muscle as the short root of the ciliary ganglion (Fig. 343, 3) and pass from the latter through the short ciliary nerves into the bulb. v. Trauwetter, Adamük, Hensen, and Völckers observed on irritation of the nerve that the eye underwent change as in near vision, and the pupil diminished in size. Details as to the origin of the individual portions of the nerve are given on p. 833

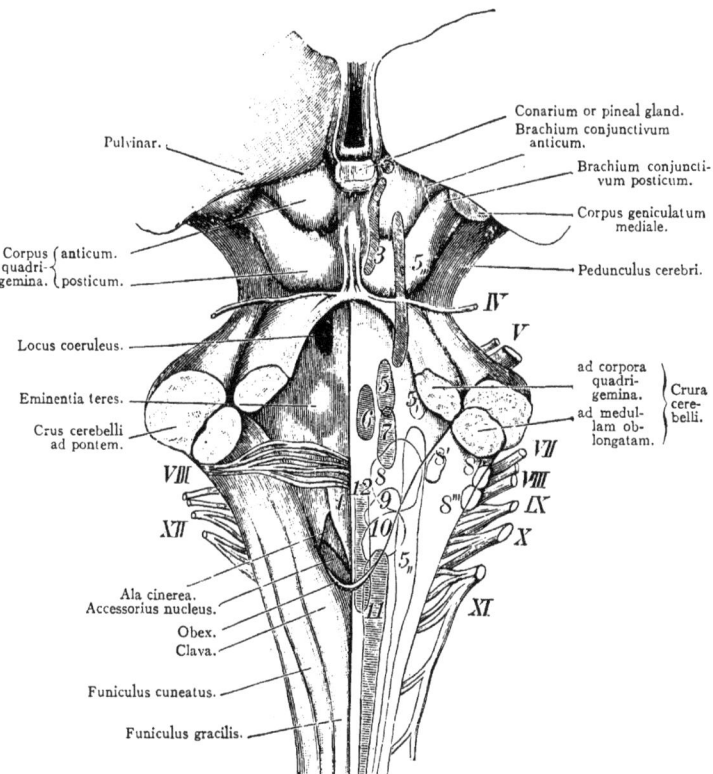

FIG. 241.—Medulla Oblongata and Quadrigeminate Bodies, Magnified: The figures from *IV* to *XII* indicate the superficial origin of the cerebral nerves; the figures from 3 to 12 indicate the position of their nuclei of origin; *t*, funiculus teres.

The *center* for reflex stimulation of the fibers of the pupillary sphincter by light is situated in the quadrigeminate bodies near the aqueduct of Sylvius. A detailed description is given on p. 842. The contraction of the pupil that occurs in conjunction with the act of accommodation is to be looked upon as an associated movement.

In man, the nerve anastomoses at the cavernous sinus with the first branch of the trigeminus, in this way receiving muscle-sense fibers; further with the sympathetic through the carotid plexus and indirectly through the abducens, in this way receiving vasomotor fibers. The rare cases in which fibers for the sphincter have been found in the abducens or even in the trigeminus must be considered as examples.of variations in the course of the pupillary fibers.

The intraocular fibers of the oculomotor nerve are paralyzed by atropin and stimulated by physostigmin (or the sympathetic is paralyzed, or both).

Contraction of the pupils on irritation of the nerve can be best demonstrated in the severed and opened head of a bird. Asphyxia, sudden cerebral anemia (from ligature of the carotid arteries or beheading), and likewise sudden venous stasis, cause dilatation of the pupils, as in death, through paralysis of the oculomotor nerve.

Pathological.—Complete *paralysis* of the oculomotor nerve gives rise: (1) To drooping of the upper lid (*paralytic ptosis*). (2) To immobility of the eyeball. (3) To rotation of the eye outward and downward (*strabismus*), and as a result to diplopia. (4) To slight protrusion of the bulb, because the superior oblique, which draws the eye forward, is unopposed by the action of its antagonists, the three paralyzed rectus muscles, which draw the eye backward. In animals, which have a retractor muscle of the bulb, this symptom is more conspicuous. (5) To moderate dilatation of the pupil (*paralytic mydriasis*). (6) To inability on the part of the pupil to contract upon stimulation by light. (7) To loss of the power of accommodation of the eye for near vision. The paralysis naturally may be confined to individual portions or be incomplete. Destruction of the posterior portion of the oculomotor nucleus causes only paralysis of the external ocular muscles (external ophthalmoplegia).

Irritation of the branch for the elevator of the lid causes spastic lagophthalmos in man; of the other muscular branches, corresponding spastic strabismus. These latter irritations may be induced also reflexly, as, for example, during dentition and in association with the diarrheas of childhood. Clonic contractions manifest themselves bilaterally as involuntary oscillation of the eyes (*nystagmus*) in consequence of irritation of the quadrigeminate bodies. Tonic spasm of the sphincter of the pupil is designated *spastic myosis*, clonic spasm *hippus*. Spasm of accommodation is also observed, and in conjunction with it not rarely macropia in consequence of imperfect estimation of distance.

IV. TROCHLEAR NERVE.

The trochlear nerve arises by means of neurites from the ganglion-cells of the trochlear nucleus, which is situated immediately behind the lateral chief nucleus of the oculomotor nerve, and really forms a continuation of the anterior horn (constituted of two sections joined together), below the gray matter surrounding the aqueduct of Sylvius. It then passes to the lower border of the posterior quadrigeminate body, and further on into the superior medullary velum, and decussates with the root of the opposite side in the velum and then appears free (Fig. 241). Like the third and sixth cerebral nerves it is probably connected by fibers with the cortical motor center for the ocular muscles.

The trochlear is the voluntary motor nerve of the superior oblique muscle. Its coördinated innervation, however, is involuntary.

Its connections with the carotid plexus of the sympathetic and the first branch of the trigeminus have the same significance as the analogous connections of the oculomotor nerve.

Pathological.—*Paralysis* of the trochlear nerve causes only slight loss of the mobility of the eyeball outward and downward, with the development of slight rotation inward and upward and diplopia. The images are placed obliquely one above the other, approach each other when the head is turned toward the unaffected side, and are separated when the head is turned toward the affected side. The patient at first inclines the head forward, but subsequently rotates it about the vertical axis toward the unaffected side. When the head is rotated, the healthy eye retaining the primary position, the eye makes a similar movement. *Spasm* of the trochlear nerve causes rotation of the eye outward and downward.

V. TRIGEMINAL NERVE.

The trigeminal nerve (Fig. 242) arises, like a spinal nerve, by two roots (Fig. 241). The smaller, anterior, motor root originates as a bundle of neurites from the motor nucleus of the trigeminus (nucleus masticatorius and locus cœruleus, Fig. 241, 5), rich in large cells, on the floor of the fourth ventricle close to the middle line, some of the fibers coming from the opposite side. From the cortical motor center of the cerebrum fibers from the opposite side pass through the cerebral peduncle to this nucleus. In addition, the descending root furnishes motor fibers.

This root (5,,,) extends from the anterior quadrigeminate body laterally along the aqueduct of Sylvius downward to the point of exit of the nerve. The large sensory, posterior root receives fibers (1) from the gray matter of the sensory nucleus of the trigeminus (5₁), situated to the side of the motor nucleus, and the analogue of the posterior horn. (2) From the gray matter of the posterior horn of the spinal cord down to the second cervical vertebra. These fibers give off collaterals to the origins particularly of the hypoglossal and facial nerves, participate in the reticular formation and enter the white posterior column and then, as the spinal or ascending root, the sensory branches of the trigeminus. (3) From the cerebellum (undecussated) fibers passing through the crus were described by Meynert.

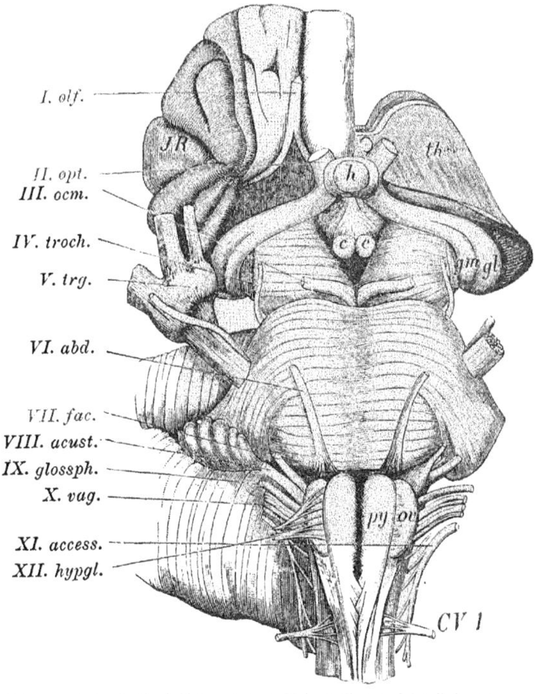

The origins of the sensory root are connected with the motor nuclei of all of the nerves arising in the medulla oblongata, with the exception of the

FIG. 242.—The Cerebral Nerves, *I* to *XII* (according to Schwalbe): *JR.* island of Reil; *h*, hypophysis; *th*, optic thalamus; *c, c*, corpora albicantia; *gm, gl*, mesial and lateral geniculate bodies; *py*, pyramid; *ov*, olivary body; *CVI*, first cervical nerve.

abducens. This fact explains the various reflex effects.

The thick trunk appears laterally between the fibers of the pons; then its posterior root forms the Gasserian ganglion (Figs. 242 and 243) at the apex of the petrous bone. In the ganglion the bipolar ganglion-cells are actually the seats of origin of the sensory root. Filaments of the sympathetic from the cavernous plexus pass to the Gasserian ganglion. Then the nerve divides into its three large branches.

The first, or ophthalmic, division (Fig. 243, d) receives sympathetic (vasomotor) fibers from the cavernous plexus, then passes through the sphenoidal fissure into the orbit. Its branches are:

1. The small *recurrent nerve*, which gives off sensory branches to the tentorium cerebelli. To it are added fibers from the carotid plexus of the sympathetic as vasomotors for the dura mater.

2. The *lacrimal nerve* gives off (*a*) sensory branches to the conjunctiva, the upper lid, the adjacent skin of the temple (Fig. 243, a); (*b*) true secretory fibers to the lacrimal gland. Accordingly, irritation of the nerve excites the secretion of tears, while division causes paralytic flow of tears. The secretion can be excited reflexly by the irritation of strong light and by irritation of the first and second branches of the trigeminus, and even of all of the sensory cerebral nerves and some of the nerves of the trunk. The reflex center for the secretion of tears is situated in the optic thalamus.

3. The *frontal nerve* (f) gives off, through its supratrochlear branch, sensory fibers to the upper lid, the brow, the glabella, and fibers reflexly stimulating the secretion of tears; and, through its supraorbital branch, analogous fibers to the upper lid, and the skin of the forehead and of the adjacent temple to the vertex.

4. The *nasociliary* nerve (n c), through its infratrochlear branch, supplies fibers analogous to those just mentioned to the conjunctiva, the lacrimal caruncle and sac, the upper lid, the brow, the root of the nose. Its ethmoid branch supplies the tip and the alæ of the nose externally and internally with sensory fibers and also the anterior portion of the septum and the lower turbinates with tactile fibers (which also in part excite reflexly the flow of tears) and perhaps also with vasomotor fibers, which may possibly arise through anastomosis with the sympathetic. From the naso-ciliary branch arise also the long roots (Fig. 243) of the ciliary ganglion (c) and the first, second, and third long ciliary nerves.

The ciliary ganglion (Fig. 243, c), which really belongs rather to the third than to the fifth nerve, has three roots: (*a*) the short root, from the oculomotor (3), (*b*) the long root (1) from the nasociliary, and (*c*) the sympathetic root (s) from the carotid plexus, occasionally united with b. From the ganglion there arise between six and ten short ciliary nerves (t), which, together with the long ciliary nerves, penetrate the sclera near the entrance of the optic nerve and pass forward between this and the choroid. They contain:

1. The *motor fibers* for the sphincter muscle of the pupil and the tensor of the choroid from the oculomotor root.

The oculomotor root is connected in the ciliary ganglion by terminal ramifications with dendrites of the ganglion-cells (not the sympathetic and those of the trigeminus). After division of the oculomotor nerve, degeneration of its fibers takes place only into the ciliary ganglion, but not further in a peripheral direction. Small doses of nicotin paralyze the oculomotor nerve from its origin to the ciliary ganglion and this portion rapidly loses its function after death, while the ciliary nerves that cause contraction of the pupil retain their irritability for a considerable time.

2. *Sensory fibers* for the cornea, which are distributed by means of most delicate filaments throughout the epithelium; and for the bulbar conjunctiva, which penetrate the sclera. These stimulate also reflexly the flow of tears (lacrimal nerve) and closure of the eyelids (facial nerve). The iris, which is the seat of pain in the presence of inflammatory processes and as a result of operation; the choroid, which is the seat of painful tension on contraction of the tensor of the choroid; and the sclera also receive sensory fibers.

3. *Vasomotor nerves* for the vessels of the iris, the choroid, and the retina. These, however, are derived in part only from the sympathetic

root and the connection of the sympathetic with the first division. The iris and the retina probably receive the larger number of vasomotor fibers from the trigeminus itself, and a small number from the sympathetic. According to Klein and Svetlin, the retinal vessels are not at all influenced through the sympathetic.

4. *Motor fibers* for the dilator muscle of the pupil, which are derived in largest part from the sympathetic, particularly the sympathetic root of the ganglion, and from the anastomosis of the sympathetic with the trigeminus. The first division itself, however, also contains pupil-dilating fibers, passing directly from the medulla oblongata into the first branch.

After division of the trigeminus the pupil in the rabbit and the frog, therefore, contracts after a brief antecedent period of dilatation; and after destruction of the superior cervical ganglion of the sympathetic the power of the pupil to dilate is not wholly abolished. The contraction that disappears in the course of half an hour in the rabbit can be looked upon as caused by reflex stimulation of the oculomotor fibers of the sphincter, in consequence of the painful irritation attending division of the trigeminus.

Whether dilator-branches pass in man through the sympathetic root of the ciliary ganglion and further on through the ciliary nerves has not been demonstrated with certainty. In the dog and in the cat at least, these fibers do not pass through the ciliary ganglion, but directly along the optic nerve to the eye, all passing through the Gasserian ganglion, the first division and finally through the long ciliary nerves. The center for the motor fibers of the dilator of the pupil is described on p. 749.

The phenomena brought about by *irritation or paralysis of the cervical sympathetic* or its path upward to the eye may be discussed now. Irritation causes, in addition to dilatation of the pupil, also an effect upon the unstriated muscle in the orbit and the eyelids. The orbital membrane, which separates the orbit from the temporal fossa in animals, contains numerous unstriated muscle-fibers (orbital muscle). The corresponding membrane of the spheno-maxillary fissure in man is also provided with a muscular layer 1 mm. thick, generally passing in a longitudinal direction through the fissure. Further, both eyelids contain unstriated muscle-fibers, which cause narrowing of the palpebral fissure. In the upper lid they continue as a prolongation of the elevator of the upper lid, in the lower they lie just beneath the conjunctiva. Also the capsule of Tenon contains unstriated muscle-fibers. All of these muscles are innervated by the sympathetic (the orbital muscle in part from the spheno-palatine ganglion), as is also the retractor of the third eyelid at the inner canthus of the eye in some animals. Irritation of the sympathetic therefore causes dilatation of the pupil, enlargement of the palpebral fissure, and protrusion of the eyeball. This irritation may also be induced reflexly by intense stimulation of sensory nerves. Also active stimulation of the nerves of the sexual organs gives rise to the signs mentioned in the eye in moderate degree as an accompanying manifestation. Perhaps the dilatation of the pupils in small children in connection with the presence of intestinal irritation from worms also belongs in this category. Also irritation of the spinal cord (sympathetic origin) in cases of tetanus causes dilatation of the pupils. Division of the sympathetic causes narrowing of the palpebral fissure and permits retraction of the eyeball (and projection of the relaxed third eyelid in animals). The division causes in dogs internal strabismus because the external rectus muscle receives in part motor fibers from the sympathetic. The origin of these fibers from the cilio-spinal region is described on p. 734.

5. It is as yet undetermined whether *trophic fibers* also arise from the trigeminus through the ciliary nerves. If the trigeminus be divided in the cranial cavity, there result in the course of from six to eight days inflammation, necrosis of the cornea, and finally destruction of the eyeball.

The results described take place, however, only when the nerve is divided in the Gasserian ganglion or peripherally (but not centrally) from it. The cause of the nutritive disturbances is dependent upon the ganglion-cells. In an estimation of the views as to the trophic fibers the following points must be taken into consideration: (1) Division of the trigeminus renders the entire eye anesthetic. The animal, therefore, is not conscious of direct injury and makes no effort to escape it. Also, adherent dust and mucus are no longer removed reflexly by closure of the eyelids. In general, in consequence of absence of the reflex, the palpebral fissure is wider and the eye is thus exposed to many injurious influences and to desiccation. Reflex secretion of tears also is wanting. When Snellen attached in front of the eye the sensitive auricle of the rabbit, through whose sensibility it avoided injury, the inflammation of the eye occurred much later. On placing a perfectly secure protecting capsule in front of the eye the inflammation was entirely prevented. The same result was observed when Gudden sutured the freshened margins of the lids in rabbits and permitted them to grow together. The cornea can be kept intact also by scrupulous cleanliness. There can, therefore, be no doubt that the loss of the sensibility of the eye favors the occurrence of inflammation. Efforts were made, further, to discover the trophic nerves and to divide them separately. As Meissner, Büttner, and Schiff observed the eye to become the seat of inflammation after dividing only the trophic (innermost) fibers of the trigeminus, the eye retaining its sensibility, the existence of trophic fibers might be considered as demonstrated; but Cohnheim and Senftleben contradict these facts. Conversely, the sensibility of the eye may be lost in consequence of partial injury of the nerve, and the eyeball does not become inflamed. Ranvier, who denied the existence of trophic nerves, incised the cornea in a circular manner through its superficial layers, at the same time dividing the nerves, all of which are present in this situation. There resulted anesthesia, but never keratitis. Further, in men and animals in which inability to close the eyes exists, redness with flow of tears or slight desiccation and cloudiness of the surface of the eyeball (xerosis) set in, but never such a destructive inflammation as that described.

(2) The following factors, to which hitherto little reference has been made, should further be taken into consideration: Division of the trigeminus paralyzes the vasomotors in the interior of the eyeball, and in consequence derangements in the circulation of the blood must take place. According to Jessner and Grünhagen the trigeminus also transmits vasodilator fibers to the eye, irritation of which causes increased flow of blood to the eye, with consecutive elimination of fibrin-factors and increase in the amount of albumin in the aqueous humor.

(3) After division of the nerve the tension of the eyeball is diminished. Conversely, irritation is followed by considerable increase in the intraocular pressure. The reduction in intraocular pressure must, naturally, alter the normal relation between the fulness of the blood-vessels and the lymphatic channels and the movement of the fluids within them, upon which the normal nutrition in large measure depends.

(4) W. Kühne observed movement of the corneal corpuscles on irritation of the corneal nerves. It does not appear impossible that the movement of these corpuscles has an influence upon the normal movement of the fluid in the canalicular system of the cornea. If, however, it were dependent upon the nervous system destruction of the latter would be followed also by nutritive disturbances. As a matter of fact, Gaule observed after division of the nerve that the corneal corpuscles were partly shrunken, partly enlarged, and that the epithelium of the cornea was partly necrotic, partly in a condition of proliferation.

Pathological.—In man, inflammation of the conjunctiva, ulceration and perforation of the cornea and finally panophthalmitis, which is designated *neuroparalytic ophthalmia*, have been observed after trigeminal anesthesia and less commonly in association with profound irritative states of the fifth nerve. Samuel was able to induce the same effects in animals by electrical stimulation of the Gasserian ganglion.

The affections of the eye due to disorders of the vasomotor nerves are entirely different than those described, as they never give rise to degenerative processes, as does division of the trigeminus. In this category belongs *intermittent ophthalmia*, a condition of unilateral, intermittent marked injection of the ocular vessels, due to malarial influences, associated with flow of tears, photophobia, often also with iritis and suppuration in the chambers of the eye, which was first considered by Eulenburg and Landois as a vasoneurotic disorder of the ocular vessels. Pathological observations, as well as experiments on animals, have demonstrated that an intimate physiological connection exists between the vascular distribution in the two eyes, so that affections in the vascular distribution of one eye readily excite analogous affections in the other eye. This fact explains why inflammatory processes chiefly in the interior of one eyeball give rise to so-called *sympathetic ophthalmia* in the other eyeball. Thus, irritants affecting the ciliary nerves or the fifth nerve upon one side cause at the same time dilatation of the vessels in the other eye, together with its sequelæ. The pathological excessive tension of the eye, with its sequelæ (*simple glaucoma*), is worthy of mention and has been attributed by Donders to irritation of the trigeminus. Unilateral flow of tears has been observed repeatedly in conjunction with irritative states of the first division of the trigeminus and unilateral suppression of tears, but rarely in association with paralytic states.

The second, or superior maxillary, division (Fig. 243, e) gives off:

1. The slender *recurrent nerve*, a sensory branch to the dura mater, which in the distribution of the middle meningeal artery accompanies the vasomotor nerves of this vessel derived from the superior cervical ganglion of the sympathetic. Irritation of this nerve causes also reflex closure of the lids in the frog.

2. The *subcutaneus malæ*, or orbital nerve (c), supplies with its two branches, the temporal and the orbital, the external canthus of the eye and the adjacent cutaneous area of the temple and the cheek, with sensory fibers. Some of the filaments of the nerve are said to be true secretory nerves for the tears.

3. The posterior and median superior *alveolar nerves*, and with them the anterior from the infraorbital nerve, give off sensory fibers to the teeth of the upper jaw, the gums, the periosteum and the maxillary antrum. The vasomotor nerves for all of these parts are supplied by the superior cervical ganglion of the sympathetic.

4. The *infraorbital nerve* (R), which, after its exit from the infraorbital foramen, distributes sensory fibers to the lower eyelid, the bridge and alæ of the nose and the upper lip to the angle of the mouth. The accompanying arteries receive vasomotor fibers from the superior cervical ganglion of the sympathetic. The sweat-fibers in swine are described on p. 537.

The **sphenopalatine, or basal ganglion** (n), is connected with the second branch of the trigeminus. It contains cells arranged like those of the sympathetic ganglia. To it pass, first, with one or several filaments, short sensory root-fibers from the second branch itself, which are designated *sphenopalatine nerves*. Motor fibers pass from behind into the ganglion through the greater superficial petrosal nerve from the facial (j) and finally gray vasomotor fibers from the sympathetic plexus of the carotid (greater deep petrosal nerve). The motor and vasomotor fibers form the Vidian nerve, which passes through the Vidian canal to the ganglion. The ganglion gives off the following fibers:

1. The *sensory fibers* (N) supply the roof, the lateral walls, and the septum of the cavity of the nose (posterior superior nasal nerves). The nasopalatine nerve passes through the incisor canal with its terminal filaments to the hard palate behind the incisor teeth. The sensory posterior inferior nasal nerves for the lower and middle turbinates and the two lower nasal passages are derived from the anterior palatine nerve of the ganglion, which descends in the pterygopalatine canal. Finally, the sensory branches for the hard (p), and the soft palate (p₁) and the tonsil are derived from the descending posterior palatine nerve. Irritation of any of the sensory fibers of the nose causes reflex sneezing, which is

always preceded by a sense of tickling in the nose.   The same result may be brought about, in addition to direct irritation, also by dilatation of the vessels of the nose.   The latter occurs readily from the action of cold upon the external integument.   The vascular dilatation is later on associated with increased secretion from the nasal mucous membrane. Irritation of the nasal nerve excites also flow of tears reflexly.   Irritation of the nasal branches causes finally also expiratory cessation of the respiratory movements.   2. The *vasodilators* of the nose pass with the sensory fibers of the ganglion; they are derived principally from the sympathetic root.   3. The *motor branches* descend through the posterior palatine nerve in the pterygopalatine canal and give off motor fibers (h) to the elevator of the veil of the palate and the azygos uvulæ. The muscle-sense fibers are supplied by the trigeminus.   Spasmodic conditions in these muscles are said to cause paroxysmally crackling sounds in the ear.   4. Filaments representing *gustatory* fibers pass also from the intermediate portion of the facial nerve to the gums.   5. The *vasomotors* of this entire area are derived from the sympathetic root, therefore from the cervical sympathetic.   6. The trigeminus-root furnishes the *secretory* nerves for the mucous glands of the nasal mucous membrane.   Irritation excites secretion, while resection of the trigeminus diminishes secretion and causes at the same time atrophic degeneration of the mucous membrane.   Accordingly trophic functions for the mucosa have also been attributed to the trigeminus.

Feeble electric irritation of the exposed ganglion causes abundant secretion of mucus and elevation of temperature in the nose, together with dilatation of the vessels.   After division of the trigeminus, redness of the nasal mucous membrane on the same side occurs.   This is probably due to the fact that penetrating dust or secreted nasal mucus is not removed from the nose through reflex influences, but remains and causes irritation and inflammation.

The third, or inferior maxillary, division (g) unites all of the motor filaments of the fifth nerve, with a number that are sensory, into a plexus, from which are given off:

1. The *recurrent nerve*, which arises from the sensory root and enters the skull through the spinous foramen and further on with the recurrent nerve of the second division supplies the dura with sensory filaments. From it pass also filaments through the petrososquamous fissure to the mucous membrane of the mastoid cells.

2. *Motor branches* for the muscles of mastication; the masseteric nerve, the two deep temporal nerves, the external and internal pterygoid nerves.   The muscle-sense fibers are probably derived from the sensory fibers.

3. The *buccinator* is a sensory nerve for the mucous membrane of the cheek and the angle of the mouth as far as the lips.

According to Jolyet and Laffont it contains besides, probably in the last instance derived from the sympathetic, vasomotors for the mucous membrane of the cheek and the lower lip, and their mucous glands.

As after division of the trigeminus this region of the mucous membrane becomes ulcerated, it has been thought that the buccinator contains also trophic fibers.   Rollet, however, called attention to the fact that section of the third division causes paralysis of the muscles of mastication on the same side, and as a result the teeth do not come in vertical apposition, but are pushed against the cheek.   In addition, in consequence of the anesthesia in the mouth, remnants of food, often insufficiently comminuted, remain in contact with the cheeks, and irritate the mucous membrane both mechanically and, as a result of decomposition, also chemically. Subsequently, by reason of the abnormal attrition of the teeth, ulcers form also on the healthy side. The assumption of trophic fibers is, therefore, not justified.

4. The *lingual nerve* (k) receives at an acute angle the chorda tympani (i i), a branch of the facial nerve, after its exit from the tympanic cavity. The lingual contains no motor fibers; it is the sensory and tactile nerve of the tongue, the anterior palatine arch, the tonsil, and the floor of the mouth. Irritation of this nerve, as well as of all of the remaining sensory fibers of the cavity of the mouth, excites reflex secretion of saliva. In addition, the lingual is the gustatory nerve for the tip and margins of the tongue (which are not supplied by the glossopharyngeal nerve), for after division of the lingual nerve in man, Busch, Inzani, Lussana, and others observed abolition of tactile sensation upon the entire half of the tongue and of the sense of taste upon the anterior portion of the tongue. These fibers, however, are, as a rule, derived from the chorda tympana, as has been pointed out in the description of the facial nerve.

According to Schiff, the lingual nerve itself contains gustatory fibers, and this view is supported by cases of Erb, Senator, Ziehl, Schreier, and others. These are probably exceptional cases. A. Schmidt believes that the gustatory fibers reach the brain through the trunk of the fifth nerve in the following manner (Fig. 243): chorda, facial trunk, connection with the lesser superficial petrosal nerve (B), otic ganglion, third division, trunk of the fifth nerve. In the interior of the tongue the lingual filaments are supplied with small ganglia. The lingual appears to receive vasodilators for the tongue and the gums from the chorda. After division of the trigeminus animals often bite the tongue, whose position and movement in the mouth they are unable to feel, and in consequence injuries and inflammations often result.

5. The *inferior alveolar nerve* is the tactile nerve of the tongue and the gums; the vasomotors pass through the superior cervical ganglion. Before entering the alveolar canal, it gives off the mylohyoid nerve, which supplies the motor fibers for the mylohyoid muscle and the anterior belly of the digastric, and likewise filaments for the triangularis menti and the platysma; muscle-sense fibers also probably are contained in these filaments. The mental nerve, which makes its exit from the mental foramen, is only the tactile branch for the chin, the lower lip, and the skin at the margin of the jaw.

6. The *auriculotemporal nerve* (A) sends sensory fibers to the anterior wall of the external auditory canal, the tympanic membrane, the anterior portion of the ear, the adjacent temporal region, and to the inferior maxillary articulation.

In Fig. 244 the area of distribution of the trigeminal branches to the head, as well as that of the cervical nerves, is indicated, and from this the nerves involved can be determined in the presence of morbid affections (neuralgia, anesthesia) involving the parts mentioned.

The otic ganglion is situated beneath the oval foramen upon the inner aspect of the third division. There enter into it as roots: 1. Motor filaments from the third division itself. 2. Vasomotor fibers from the plexus of the middle meningeal artery (therefore passing through the superior cervical ganglion of the sympathetic). 3. From the tympanic branch of the glossopharyngeal nerve filaments pass to the tympanic plexus (Fig. 243, λ), thence through the petrosal canal in the lesser superficial petrosal nerve into the cranial cavity, then through the sphenoidal fissure into the otic ganglion (m). Through the chorda tympani the facial nerve is in constant connection with the ganglion, just below which it passes (Fig. 243, m, i).

44

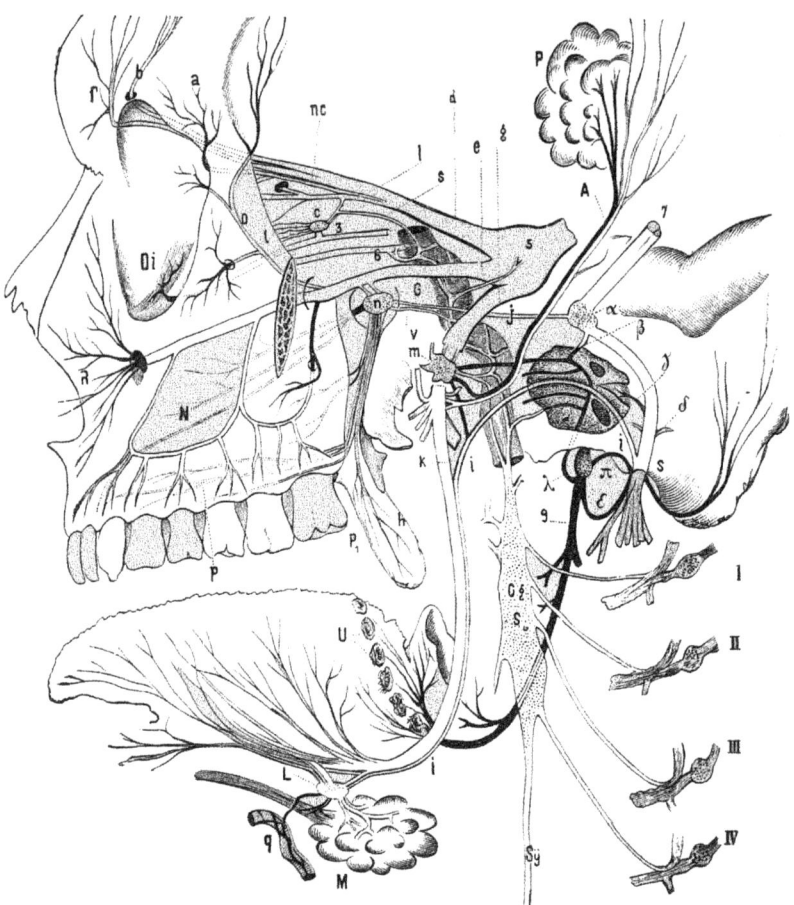

Fig. 243.—Semidiagrammatic Representation of the Ocular Nerves, the Connections of the Trigeminus and Its
Ganglia and Those of the Facial and Glossopharyngeal Nerves: 3, branch to the inferior oblique muscle
(Oi) from the oculomotor nerve, with the thick, short root to the ciliary ganglion (c); t, ciliary nerves; l, long
root to the ganglion from the nasociliary nerve (nc); s, sympathetic root from the plexus of the sympathetic
(Sy) surrounding the internal carotid artery (G); d, first division of the trigeminus (s), with the nasociliary
nerve (nc) and the terminal branches of the lacrimal (a), supraorbital (b) and frontal (f); e, second division
of the trigeminus; R, infraorbital nerve; n, sphenopalatine ganglion, with the roots (j) from the facial, and
(v) from the sympathetic; N, the nasal branches p p, the palatal branches of the ganglia; g, third division
of the trigeminus, k lingual nerve; i i, chorda tympani; m, otic ganglion wth the rnoots from the tympanic
plexus, the carotid plexus, and from the third division, and with its branches to the auriculotemporal (A)
and the chorda (i i); L, submaxillary ganglion, with the roots from the tympanicolingual and the sympathetic
plexus of the external maxillary artery (q). 7, Facial nerve—j, its greater superficial petrosal nerve; α, geniculate ganglion; β, branch to the tympanic plexus; γ, stapedius branch; δ, anastomoses with the auricular
branch of the vagus. s, Stylomastoid foramen. 9, Glossopharyngeal nerve—λ, its tympanic branch; π and
ε, connections with the facial; U, termination of the gustatory fibers of the ninth nerve in the circumvallate
papillæ. Sy, Sympathetic, with (Cg S) the superior cervical ganglion. I, II, III, IV, the four upper cervical
nerves. P, Parotid gland; M, submaxillary gland.

The otic ganglion gives off (as a continuation of 1): 1. Motor branches for the tensor tympani muscle and the tensor of the veil of the palate (with which muscle-sense fibers probably are also admixed). 2. One or several connecting branches of the ganglion to the auriculotemporal nerve are probably conveyed through the root-fibers (2 and 3) from the sympathetic and the glossopharyngeal nerve, which the nerve in question (Fig. 243, A) gives off to the parotid gland (P) in its passage through it. These branches control the salivary secretion of the parotid, as has been pointed out on p. 259.

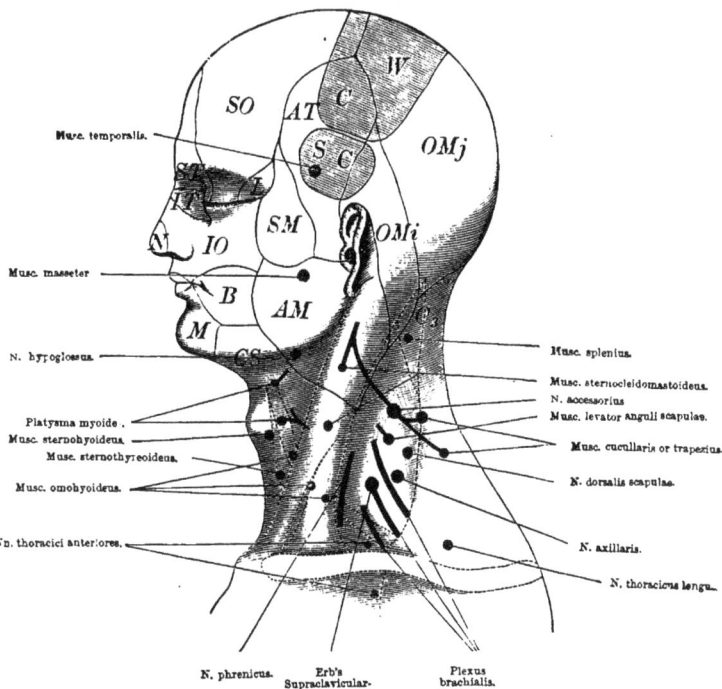

FIG. 244.—Distribution of the Sensory Nerves of the Head, together with the Situation of the Motor Points on the neck.
SO, Distribution of the supraorbital nerve; ST, supratrochlear nerve; IT, infratrochlear nerve; L, lacrimal nerve; N, ethmoid nerve; IO, infraorbital nerve; B, buccinator nerve; SM, subcutaneous malar nerve; AT, auriculotemporal nerve; AM, great auricular nerve; OMj, greater occipital nerve; OMi, lesser occipital nerve; C₃, third cervical nerve; CS, cutaneous branches of the cervical nerves; CW, situation of the central convolutions of the cerebral hemisphere; SC, situation of the speech-center (third frontal convolution).

Division of the trigeminus causes inflammatory changes in the mucous membrane of the tympanum in all possible degrees (in the rabbit). Lesions of the sympathetic or the glossopharyngeal are ineffective.

The **submaxillary or lingual** ganglion (Fig. 243, L) lies upon the convex arch of the united tympanicolingual nerve and the excretory duct of the submaxillary gland (M), and receives as root-fibers: 1. Branches of the chorda tympani (i i). These are related to the salivary secretion

of the submaxillary and sublingual glands, inasmuch as they contain secretory nerves, yielding a limpid saliva, and vasodilators. In addition,
they give branches to the unstriated muscular fibers of Wharton's duct.
Not all of the fibers of the chorda, however, pass to the gland; some are
distributed to the tongue. 2. The sympathetic root of the ganglion
arises from the plexus of the submental branch of the external maxillary
artery (q), thus from the carotid plexus of the sympathetic. It passes
to the glands and is the secretory nerve yielding concentrated saliva.
It, further, transmits the vasoconstrictors to the vessels of the glands.
3. Sensory root-fibers derived from the lingual nerve in part send filaments to the glands and their excretory ducts, and in part, again entering
the tympanicolingual from the ganglion, pass peripherally to the tongue.

Pathological.—*Spasm* of the muscles of mastication occurs as a pathological
manifestation in the distribution of the third division. It is, as a rule, bilateral,
and it may be either clonic (chattering of the teeth) or tonic (trismus). The
spasm is generally one of the manifestations of widespread convulsions; rarely it
is isolated as a symptom of cerebral focal disease of the medulla oblongata, the pons,
or the cortex in the situation of the motor center of the trigeminus. The spasm
may, naturally, be also reflex in origin, principally in consequence of irritation
of sensory nerves of the head.

Degeneration of the motor nucleus or affections of the root in the skull cause
paralysis of the muscles of mastication, which is rarely bilateral. Paralysis of the
tensor tympani muscle is said to have caused impairment of hearing or roaring
in the ears. In this connection, as well as with respect to paralysis of the tensor
of the veil of the palate, further observations are desirable.

With reference to all of the branches of the trigeminus, mention must be
made first of *neuralgia*, which is attended paroxysmally with intense radiating
pain in the peripheral distribution of the nerve. Generally unilateral, the disorder usually involves only a few branches or even fibers. Points of radiation
for the pain are often constituted by the bony canals from which the branches
make their exit. Rarely the ear, the dura mater, and the tongue are involved.
Occasionally twitching occurs in the corresponding group of facial muscles in
conjunction with the attack, either being excited reflexly or developing directly
as a result of peripheral irritation of the facial fibers connected with the terminal
fibers of the trigeminus. The reflex contractions if marked may extend even to
the muscles of the arms and the trunk.

Marked *redness* of the affected area occurs as an accompanying manifestation
of pain in the face, and in some cases increased or diminished secretion from the
conjunctiva and the nasal and buccal mucous membrane. The condition is certainly a reflex phenomenon of sympathetic origin. The *derangement in cerebral
activity* often observed in consequence of the altered distribution of blood is probably due to reflex vasomotor stimulation. C. Ludwig and Dittmar found that
irritation of sensory nerves causes contraction of the arterial blood-stream and
increased pressure in the cerebral vessels. Thus, melancholia and hypochondriasis
are often found in marked degree. Landois had cognizance of a case in which
during the severe attacks, involving the third division, marked visual hallucinations appeared. Disorders of the fifth nerve are capable, in general, of causing
varied reflex affections.

The *trophic disorders* that occur in association with affections of the trigeminus
are of great interest. Among these are brittleness and splitting of the hairs,
graying and falling out of the hair, circumscribed inflammation of the skin and
vesicular eruptions on the face (zoster) and also the cornea (neuralgic herpes of
the cornea).

Finally, there should be mentioned *progressive facial atrophy*, which is almost
always unilateral, but may also be bilateral. It is probably due to derangement
in the trophic activity (of the descending root?) of the trigeminus, although the
vasomotor activity of the sympathetic may also be invoked reflexly. Landois
found on sphygmographic examination of the famous case of Romberg (in a man
named Schwahn) that the pulse-tracing from the carotid on the atrophic side was
distinctly smaller than that from the vessel on the unaffected side. Intracranial
division of the sensory root of the fifth nerve causes in dogs a similar trophic disturbance. The antithesis of this obscure disorder, which is dependent upon the

trophic relations of the nerves to the tissues, is the rare condition of *unilateral hypertrophy of the face*, which bears some resemblance to the analogous manifestations of so-called *partial giant-growth* (acromegaly).

Reference should be made here also to the extremely remarkable observation of Urbantschitsch, who found that irritation of branches of the trigeminus, and principally those that pass to the ear, causes increase in the light-sense of the individual in question. Blowing on the cheek or the nasal mucous membrane, electrical irritation, the snuffing of tobacco, the smelling of strong odors may temporarily increase the sensibility to light. Also the gustatory and the olfactory sense, as well as the sensibility of certain cutaneous areas, may be thus increased reflexly through slight irritation of the trigeminus. In the presence of severe affections of the ear, in consequence of which fibers of the trigeminus may be seriously involved, the sense-functions mentioned may be impaired. Local improvement in the aural disorder is then attended with an increase in the activity of the special senses mentioned.

After extirpation of the Gasserian ganglion, together with its roots, in man the entire distribution of the trigeminus has been found completely and irremediably anesthetic. All parts remained intact so far as their trophic state was concerned, but the anesthetic eye was less resistant to influences exciting inflammation, and Keen and Laguaite observed keratitis after extirpation of the ganglion, and Scheier ulceration of the cornea and the mucous membrane of the mouth and the nose after injury to the nerve. The secretion of tears was in some instances diminished, in others abolished. The skin of the cheek and of the eyebrow exhibited slight trophic change. Immediately after the operation the skin exhibited signs of abnormal distribution of the blood and later a sense of heat on the forehead and in the eye. The sense of taste was impaired in the distribution of the lingual nerve and likewise that of smell in the corresponding nasal cavity. The muscles of mastication are paralyzed, and the delicacy of movement in the muscles of the face was impaired in consequence of absence of the muscle-sense. In the course of time the anesthetic area becomes smaller, as branches from adjacent nerves grow into it.

VI. ABDUCENS NERVE.

The abducens nerve arises from the abducens nucleus with neurites from large cells that correspond to those of the anterior horns of the spinal cord. The nucleus lies below the eminentia teres on the floor of the fourth ventricle (Fig. 241) and below the knee-shaped flexure of the facial nerve. Probably some oculomotor fibers arise from the abducens nucleus, and from the left those fibers of the right oculomotor that rotate the right eye inward (this explains the synergistic action of the two eyes in lateral movement). Physiologically, connecting fibers should pass between the nucleus of origin and the contralateral cortical center in the cerebrum for the ocular movements. The nerve makes its appearance at the posterior margin of the pons (Fig. 242).

The abducens is the voluntary nerve of the external rectus muscle, although in the coördinated movements of the eye it is stimulated involuntarily.

Branches of considerable size pass from the sympathetic in the cavernous sinus to the abducens nerve (Fig. 243, 6); smaller branches from the trigeminus, whose significance is the same as that of analogous branches of the trochlear and oculomotor.

Pathological.—Complete *paralysis* causes internal strabismus, and as a result diplopia. In dogs, division of the cervical sympathetic causes slight rotation of the eyeball inward. This must be due to the fact that the abducens receives motor muscle-nerves from the cervical sympathetic. Spasm of the abducens causes external strabismus.

With reference to *strabismus*, it should be mentioned that it may be caused, in addition to irritation or paralysis of the nerves, also by primary muscular affections, such as congenital shortness, contractures, injuries. Finally, strabismus occurs in connection with cloudiness of the transparent media of the eye, the individual involuntarily rotating the eye so that the visual rays, so far as possible, pass through the portions of the media that are still clear. Lesions of the retina at the yellow spot give rise to similar results. The affected eye, further, may be rotated involuntarily, in order that it may not interfere with the vision of the unaffected eye, the patient thus unconsciously placing himself in the position of a person with but one eye.

VII. FACIAL NERVE.

The facial nerve arises with centrifugal fibers from the ganglion-cells of two nuclei on the floor of the fourth ventricle. The anterior, smaller nucleus, which is situated immediately behind the most posterior cells of the oculomotor nucleus, is the origin for the muscular nerves of the eyelids and the structures about the orbit. The posterior, larger nucleus is situated in the most ventral portion of the tegmentum to the inner side of the ascending root of the fifth nerve. The fibers that arise here surround the nucleus of the abducens and contain the nerve for the muscles of the mouth and of the remainder of the face. In their passage from the facial center in the cerebral cortex to the nuclei the fibers for the mouth decussate, while the fibers for the eyes in part do not decussate.

The nerve makes its appearance at the posterior margin of the pons to the inner side of the auditory nerve. Between the two arises the thin intermediate portion of Wrisberg, which sends most of its fibers to the facial nerve and the remainder to the auditory. The filaments of origin of the intermediate portion arise from the glossopharyngeal nucleus. The gustatory and the tactile fibers possessed by the chorda tympani appear to enter the facial through these filaments. These fibers have ganglion-cells in the geniculate ganglion. The intermediate portion would thus be a separate division of the gustatory nerve, which unites with the facial and passes through the tongue with the chorda. With the auditory nerve the facial first enters the internal auditory canal and at the bottom of this, separated from the former, it enters the facial or Fallopian canal. At first it has a transverse direction as far as the hiatus of this canal. It then turns at a right angle at the geniculate ganglion (Fig. $_{243}$, a), containing ganglion-cells, passing over the tympanic cavity, to descend into the bone on the posterior aspect of this cavity. Finally, it makes its exit at the stylomastoid foramen, penetrates the parotid gland, and divides into its terminal branches, to be distributed in a fan-shaped manner (pes anserinus major).

The branches of the facial nerve (Fig. 243) are:

1. The motor *greater superficial petrosal* nerve (j). It passes from the geniculate ganglion through the hiatus out of the facial canal into the cranial cavity, then downward upon the anterior surface of the petrous bone, next through the sphenoidal fissure to the inferior surface of the base of the skull, and finally through the Vidian canal to the spheno-palatine ganglion. It is also possible that the nerve transmits sensory fibers to the facial from the second division of the trigeminus.

2. From the geniculate to the otic ganglion *connecting fibers* (β) whose course and function are described on p. 689.

3. The *motor branch* to the stapedius muscle (γ).

4. The *chorda tympani* nerve (i i) arises before the exit of the facial from the stylomastoid foramen (s), passes through the tympanic cavity, above the tendon of the tensor tympani, between the handle of the malleus and the long process of the incus, then through the petrotympanic fissure externally to the base of the skull and downward at an acute angle into the lingual nerve. In advance of this union an exchange of fibers takes place between the chorda and the otic ganglion (m). Both these, as well as the connection of the chorda with the lingual nerve, may transmit sensory fibers to the chorda and later on to the facial nerve. The chorda contains sensory fibers, for irritation of the nerve, which is possible in man when the tympanic membrane is destroyed, causes a sticking and prickling sensation in the anterior lateral portion and at the tip of the tongue. After division of the chorda O. Wolf found sensibility for tactile and thermic irritations abolished in man in the same distribution, and also gustatory sensation.

The chorda contains secretory fibers for the sublingual and sub-maxillary glands and vasodilators for these and the tongue. From the observations of numerous investigators it has, further, been estab-lished that the chorda tympani contains also gustatory fibers for the margin and the tip of the tongue, which, further on, it gives off to the tongue in the course of the lingual. Urbantschitsch observed a man whose chorda was exposed and in whom irritation of the nerve in the tympanic cavity caused gustatory sensations, together with sensory im-pressions.

It must, therefore, be accepted as established that the gustatory fibers of the chorda originate in the glossopharyngeal nerve. They may enter the chorda: 1. Through the intermediary portion of Wrisberg, and this view has recently received general acceptance. 2. A further pos-sible means of communication is afforded beyond the stylomastoid fora-men, namely, through the communicating branch with the glossopharyn-geal nerve (Fig. 243, e), which passes from the nerve last mentioned into that branch of the facial that at the same time contains the motor fibers for the stylohyoid muscle and the posterior belly of the digastric (Henle's styloid nerve). This nerve gives off also, perhaps, muscle-sense fibers for the stylohyoid muscle and the posterior belly of the digastric. In addi-tion, it is assumed that by means of this anastomosis motor fibers from the facial are brought to the glossopharyngeal. A third point of union be-tween the ninth and seventh nerves is situated in the tympanic cavity: the tympanic branch of the glossopharyngeal (λ) that enters the tympanic cavity is connected in the tympanic plexus with the lesser superficial petrosal nerve ($\jmath$), which is derived from the geniculate ganglion of the facial. The lesser superficial petrosal nerve may thus transmit gustatory fibers to the geniculate ganglion of the facial. It may, however, convey the gustatory fibers first to the otic ganglion, which is constantly con-nected with the chorda tympani. Finally, a fourth connection has been described as taking place through a filament (π) from the petrous gan-glion of the ninth nerve directly to the facial trunk in the Fallopian canal.

The chorda contains vasodilators for the anterior two-thirds of the tongue.

Mention should be made here of the remarkable fact that from 1 to 3 weeks after division of the hypoglossal nerve, irritation of the chorda causes movements in the paralyzed tongue. These movements are feeble and tardy, in comparison with those resulting from hypoglossal stimulation. The phenomenon is explained on p. 559. It depends essen-tially upon an increased supply of blood, in conjunction with an aug-mented secretion of lymph, as a result of which the corresponding half of the tongue becomes edematous. Heidenhain designates this action *pseudomotor*.

With respect to Heidenhain's interpretation it should be recalled that mus-cular contraction depends on swelling through the taking up of fluid. The pseudo-motor contraction has a latent stage ten times as long as that of hypoglossal irritation. A single moderate induction-shock is ineffective, as is also chemical irritation; nevertheless reflex stimulation may occur through various sensory nerves. Nicotin first stimulates, then paralyzes, movement excited through the chorda. The chorda transmits motor impulses even for a short time after sup-pression of the circulation. The pseudomotor contraction gives rise to no muscular sound.

5. Even before the chorda is given off the trunk of the facial enters

into direct relations with the auricular branch of the vagus (δ), which crosses its path in the mastoid canal and from which it may receive sensory fibers.

6. After making its exit from the canal the facial nerve gives off only motor branches to the stylohyoid muscle and the posterior belly of the digastric, to the occipital muscle, as well as to all of the muscles of the external ear and of the face, to the buccinator and to the platysma. It contains also sweat-fibers for the face.

Although the facial nerve, in most of its branches on the face, is under the control of the will, most persons are unable to move voluntarily the muscles of the nose and the auricle. Landois was able to contract the transverse and oblique muscles of the auricle, a rumbling sound being at the same time audible in

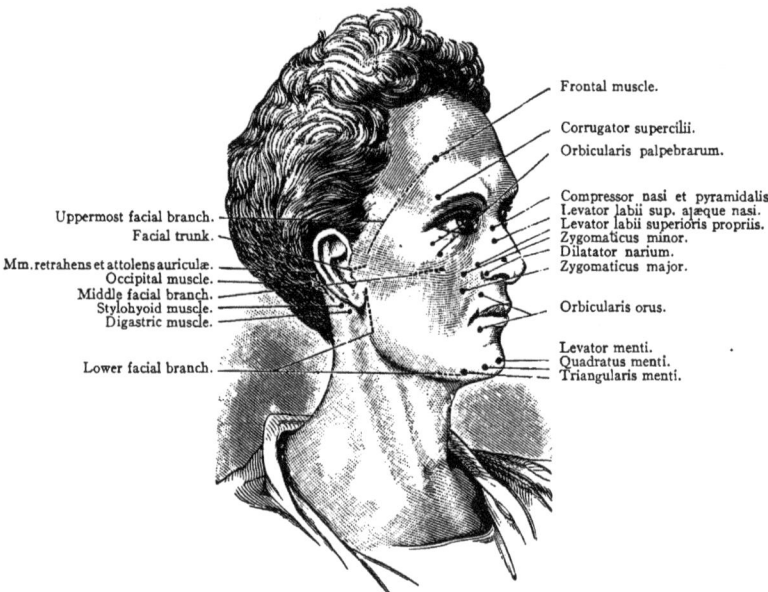

Frontal muscle.

Corrugator supercilii.
Orbicularis palpebrarum.

Compressor nasi et pyramidalis.
Levator labii sup. alæque nasi.
Levator labii superioris propriis.
Zygomaticus minor.
Dilatator narium.
Zygomaticus major.

Orbicularis oris.

Levator menti.
Quadratus menti.
Triangularis menti.

Uppermost facial branch.
Facial trunk.

Mm. retrahens et attolens auriculæ.
Occipital muscle.
Middle facial branch.
Stylohyoid muscle.
Digastric muscle.

Lower facial branch.

FIG. 245.—Motor Points of the Facial Nerve and of the Muscles Supplied by It (after Eichhorst).

the corresponding ear from the flexion of the cartilage of the external ear. He was able also to contract one-half of the orbicularis oris of the lower lip. According to Mendel the fibers of the facial for the orbicularis take their origin from the posterior extremity of the oculomotor nucleus.

On the face the facial branches unite regularly with those of the trigeminus. In this way the latter furnish also muscle-sense fibers to the muscles. The peripheral anastomoses of the sensory branches of the auricular nerve of the vagus and the great auricular nerve have the same significance for the muscles of the ear, as well, finally, as the anastomoses of the sensory filaments from the third cervical nerve for the facial fibers of the platysma. Division of the facial at the stylomastoid foramen is painful, but still more painful is that of the peripheral facial branches, as will be obvious from what has been stated.

The foregoing illustration shows accurately the course of the trunk of the facial nerve and its superior, middle and inferior branches on the face, as well as the points where the individual motor fibers pass into their muscles. By the application of one electrode at these points, the other being applied to any indifferent part of the body, the individual muscles can be made to contract electrically. The electrodes are applied in the same way in employing electricity for therapeutic purposes.

Pathological.—In connection with *paralysis* of the facial nerve it is above all important to determine whether the seat of the affection is a peripheral one, in the neighborhood of the stylomastoid foramen, or in the course of the long Fallopian canal, or, finally, central (cerebral). A careful analysis of the symptoms will lead to a conclusion in this respect. A frequent cause for paralysis at the stylomastoid foramen is designated rheumatic and probably depends upon exudation paralyzing the nerve by compression (perhaps at the situation of the lymph-space discovered by Rüdinger at the inner side of the Fallopian canal between the periosteum and the nerve, an evagination of the arachnoid sac). Other causes are inflammation of the parotid, direct traumatism, pressure of the obstetric forceps in the new-born. In the course of the canal fractures of the petrous bone, effusions of blood into the canal, syphilitic deposits, caries of the petrous bone, principally in connection with inflammation of the middle ear, are to be mentioned as causes of the paralysis. Among intracranial causes there should finally be mentioned affections of the cerebral membranes and the base of the skull in the vicinity of the nerve, disease of the facial nucleus, and finally of the cortical center for the nerve and the connections between this and the nucleus.

The symptoms of *unilateral facial palsy* are as follows: 1. Paralysis of the muscles of the face: the forehead is smooth, free from furrows; the palpebral fissure is open (paralytic lagophthalmos), with the external canthus at a lower level. The anterior surface of the eye readily becomes dry, and the cornea appears dull, chiefly because the distribution of tears is interfered with by absence of winking, and, in consequence of the dryness, slight inflammatory irritation may result (xerotic keratitis). According to some observers the facial nerve is believed to be the secretory nerve for the lacrimal gland (so that the secretion of tears is interfered with when the nerve is paralyzed) and the vasomotor nerve for the conjunctiva. Its course is believed to be as follows: facial, greater superficial petrosal nerve, sphenopalatine ganglion, second division of the trigeminus, orbital nerve. In order to protect the eye from exposure to light, the patient generally rotates the globe upward and outward beneath the upper eyelid, and relaxes the elevator of the upper eyelid, so that the lid droops somewhat. The nose cannot be moved, and the nasolabial fold is obliterated. In consequence, the sense of smell may be impaired, because the nasal orifice can no longer be dilated. The derangement of smell, however, is due principally to the defective distribution of tears (in consequence of paralysis of winking and of the muscle of Horner), which leaves the corresponding side of the nasal cavity dryer than normal. Horses, which in breathing visibly dilate the nostrils, are said either to die after bilateral division of the facial nerve from interference with respiration or at least to suffer from marked respiratory difficulty. The entire face is drawn toward the unaffected side, so that the nose, the mouth, and the chin generally occupy an oblique position. In consequence of paralysis of the stylohyoid muscle and the posterior belly of the digastric, the base of the tongue on the paralyzed side may occupy a lower level, and on forced movement of the base of the tongue this organ may undergo a deviation toward the unaffected side. Paralysis of the buccinator interferes with the normal formation of the bolus of food, which collects in the concavity of the relaxed cheeks from which the patient must eventually remove it with the fingers. Saliva and fluid readily escape from the angle of the mouth. In strong expiration the cheek is distended like a sail. Speech may be interfered with in consequence of difficulty in forming the labial consonants (particularly when the paralysis is bilateral) and also the vowels o ü ö. Speech becomes nasal in the presence of (bilateral) paralysis of the branches to the muscles of the palate. Whistling, suckling, blowing, expectoration are interfered with. *Bilateral paralysis* causes many of these symptoms in exaggerated degree. Others, such as the oblique position of the face, naturally are wanting. The face is completely relaxed, without any play of expression, and the patients cry and laugh "as behind a mask." 2. In the presence of *paralysis of the palate*, the uvula is deflected toward the unaffected side, and the paralyzed half of the palate is depressed and relaxed and cannot be elevated (greater superficial petrosal nerve). It has not as yet been determined whether and to what extent it affects the movements of swallowing and the formation of consonants. 3. *Impairment of the sense of taste* (either absence upon the anterior two-thirds of the tongue or delay and alteration in the sensation) results in accordance with what has been stated concerning the chorda tympani. 4. *Diminution in the secretion of saliva* upon the paralyzed side was first described by Arnold, although it will have to be determined to what extent any impairment of taste that may be present at the same time may give rise to interference with the reflex secretion of saliva, or whether, possibly, increased evaporation of saliva from the separated lips and the angle of the mouth may result in greater dryness of the affected side of the mouth. 5. Since the time of Roux increased attention has been called to *acuity of hearing* (oxyakoia or hyperacusis of Willis). The paralysis of the stapedius muscle causes oscillation of the stapes in the oval window, so that impulses from the tympanic membrane are strongly transmitted to this bone, which in turn gives rise to marked oscillations in the labyrinthine fluid. Less commonly, in consequence of paralysis of the stapedius muscle, it is observed that deeper notes are heard at a greater distance than upon the unaffected side. 6. As the facial nerve in man appears to contain sweat-fibers, it is clear that with the occurrence of atrophy of this nerve *loss of sweating* in the face must result. 7. Derangement of sensibility naturally cannot occur in connection with pure central affections of the facial nerve. As, however, numerous sensory fibers enter the peripheral portion of the nerve, peripheral paralysis will be attended with a certain limited impairment of sensibility (principally affecting the muscle-sense) in the face.

Division of the facial nerve in young animals causes atrophy of the related muscles. Therefore, the bones of the face are retarded in their growth. They remain smaller and the bones of the opposite side finally extend beyond the middle line toward the affected side. Also the salivary glands remain smaller.

Irritation of the facial nerve gives rise to *circumscribed* or *diffuse, direct* or *reflex, tonic* or *clonic* spasm. The diffuse form of spasm is designated mimetic facial spasm. Among the forms of circumscribed spasm *tonic spasm of* the eyelids, blepharospasm, is the most frequent, being caused by stimulation of the sensory nerve of the eye, principally in connection with scrofulous inflammation of the eye or in consequence of excessive irritability of the retina (photophobia). Less commonly the irritation is transmitted from a remote point, for example in one case in consequence of inflammatory irritation of the anterior palatine arch. The center for reflex stimulation is the facial nucleus. The clonic form of spasm, *abnormal winking* (spasmus nictitans), is generally of reflex origin through irritation of the eyes, the dental nerves or even remotely situated nerves. In marked cases the disorder is bilateral, and the spasm may extend to the muscles of the neck, the trunk and the upper extremities. Twitching of the muscles of the lips is caused in part by emotional influences, in part by reflex influences. *Fibrillary twitching* appears also in the sequence of paralysis of the facial nerve as a degenerative phenomenon. In dogs Schiff observed for years fascicular twitching in the paralyzed facial area, which in contradistinction to fibrillary twitchings, could be excited reflexly, and to which Schiff attributes the oblique position of the face in man. Intracranial irritation of the most varied form, affecting the cortical center or the nucleus of the nerve, may likewise cause spasm. Finally, facial spasm may occur as part of general convulsions, such as attend epilepsy, eclampsia, chorea, hysteria, and tetanus. Aretæus (81 A. D.) made the interesting observation that the muscles of the auricle take part in the convulsions of tetanus. With respect to the influence of irritation of the facial nerve upon the sense of taste information must be derived from future careful investigations. Rarely, spasmodic elevation of the palate and increased salivation have been described in connection with irritation of the facial nerve. Moos observed profuse secretion of saliva on irritation of the chorda in consequence of an operation in the tympanic cavity. Aristotle had already observed transitory impairment of hearing during the act of yawning and this has been attributed by Landois to spasm of the stapedius. This is the antithesis of the hyperacusis of Willis. In conjunction with this there occurs a feeble droning sound, due to the vibrations of the labyrinth induced by the contraction of the muscle named. Gottstein observed in one case this stapedius droning to occur paroxysmally in addition to blepharospasm.

VIII. AUDITORY NERVE.

Two roots serve for the origin of the auditory nerve, an anterior median root with coarse fibers, and a posterior lateral root with fine fibers. The vestibular nerve arises from the former, the cochlear nerve from the latter. The two are entirely distinct in the sheep and the horse. Each vestibular and cochlear nerve arises from a peripheral ganglion (the vestibular ganglion in the internal auditory canal and the spiral ganglion in the cochlea), constituted like the spinal ganglia, and at the same time the trophic center for the fibers. Into each ganglion-cell there enters a cellulipetal dendrite passing from the sensory epithelium in the labyrinth, while on the other hand each cell sends to the medulla oblongata a cellulipetal neurite to the nuclei of origin of the auditory nerve, with whose cells it comes in contact by means of terminal filaments and collaterals. The vestibular nerve is essentially connected with gray matter that is in relation with the cerebellum and probably subserves the purpose of maintaining the equilibrium. From the origin of the cochlear fibers the main portion passes on the opposite side to the posterior quadrigeminate body and the internal geniculate body and further (particularly through the lower fillet, the upper olive and the trapezoid body) to the temporal lobe of the cerebrum, in which the psycho-auditory cortical center is situated. After extirpation of the temporal lobe its fibers through the corona radiata atrophy into the internal capsule, as well as fibers in the posterior quadrigeminate body and the internal geniculate body. The striæ acusticæ represent a central path for the lateral auditory root. They form a secondary projection-system of the auditory nerve, decussating somewhat like a chiasm. The nuclei of origin of both auditory nerves are connected in the brain by commissural fibers. In the internal auditory canal root-fibers pass from the intermediate portion into the auditory nerve.

The auditory nerve has a double function: in the first place it is the nerve of hearing. Every irritation at its origin, in its course or in its terminal distribution causes auditory impressions; every injury, in accordance with its intensity, impairment of hearing to the point of deafness; also destruction of the labyrinths, the end-organs of the auditory nerves, causes complete deafness.

As animals after removal of both cochleæ still react to coarse sounds, the ampullæ must serve for the perception of the sounds, and the cochlea for the appreciation of the remaining auditory qualities. After extirpation of the labyrinth the auditory nerve undergoes atrophy in an upward direction.

Entirely distinct from the auditory is the function of the nerve that is localized exclusively in the semicircular canals, namely that govering the necessary movements for the maintenance of the bodily equilibrium, through stimulation of the peripheral distribution in the ampullæ.

Of especial importance is the behavior of the auditory nerve in response to the galvanic current. If an electrode is placed in a healthy person upon the tragus on each side, it will be found that upon the anodal side with closure of the current silence occurs, on opening the current a sense of sound, while the opposite takes place upon the kathodal side (Brenner's normal formula). If one electrode is placed on the tragus and the other is held in the hand, the same result is observed, except that the sound upon the unarmed ear is much feebler. The sound agrees exactly with the resonance fundamental tone of the sound-conducting apparatus of the ear itself.

The appearance of this sound is to be explained in the following manner: In the middle ear there exists a permanent blood-murmur, to which the system of cavities of the middle ear resonates with its fundamental tone. In consequence of habituation this tone is, as a rule, not noted, but it appears at once if the auditory nerve is placed in a condition of increased irritation, namely (in the sense of electrotonus) on kathodal closure and anodal opening.

According to Gradenigo, Pollak, and Gärtner, the auditory nerve in healthy persons does not react at all to currents of moderate strength. Only in the presence of hyperemic and irritative states of the auditory apparatus does a reaction take place, and then in both ears even when only one side is affected. The reaction-formula resulting under such circumstances conforms entirely with Pflüger's law, namely kathodal closure causes ringing in the ears and anodal opening deeper roaring. While the current is closed, a permanent reaction exists even when the strength of the current is slight. Even in the presence of complete deafness this typical reaction may persist.

Pathological.—*Increased irritability* of the auditory nerve at any point in its course, its centers, or its terminal distribution causes nervous acuity of hearing (*hyperacusis*), which is generally a symptom of widespread increase in nervous irritability, for example in hysterical persons. If present in particularly marked degree it may give rise to a distinctly painful sensitiveness, which may be designated *acoustic hyperalgia*. *Irritation* of the area named causes auditory perceptions, among which nervous roaring or ringing in the ears (*tinnitus*) is due to the fact that either the vascular noises in the ear are abnormally loud or the auditory nerve is hyperesthetic. In this way is explained the tinnitus following large doses of quinin or salicylates in consequence of vasomotor influences upon the labyrinthine vessels, which may increase to the degree of causing rupture of a vessel. Frequently, in the presence of roaring in the ears, the reaction is increased upon applying the galvanic current. Less commonly there is a so-called *paradoxical reaction*, that is, upon applying the galvanic current to one ear there appears, in addition to the reaction in this ear, the opposite in the ear through which the current is not passed. This phenomenon can be explained in the sense of transference. In other cases of lesions of the auditory nerve noises rather than musical notes can be excited by the current. In addition, various deviations from the formula of Brenner have been observed, and even complete reversal of this formula. Excitation, particularly of the cortical center of the auditory nerve, especially in the insane, may cause *auditory hallucinations*. If the irritability of the auditory nerve is diminished or even destroyed, nervous impairment of hearing (*hypacusis*) and nervous deafness (*anacusis*) develop. Often disease of one ear is attended by a compensatory relation to the other.

The Semicircular Canals of the Labyrinth.—After division of the canals, especially if bilateral, marked disorders of equilibrium appear. The oscillating movement of the head in the direction of the plane of the injured canal is characteristic. If the horizontal canal is divided the head (of the pigeon) is rotated alternately to the right and the left. The rotation appears chiefly when the animal attempts to make movements; during rest, the movements cease. The phenomenon may persist for months. Injury to the posterior vertical canal causes marked upward and downward nodding movements, in connection with which the animal occasionally falls forward or backward. Injury, finally, of the upper vertical canal causes likewise oscillatory vertical movements of the head, frequently with falling forward. Destruction of all of the canals is frequently followed by various oscillatory movements of the head that often render standing impossible. Breuer observed on mechanical, thermic and electrical irritation of the canals analogous rotation of the head. On applying salt-solution with a brush to the exposed canals Landois likewise observed the oscillatory movements described, which occasionally disappeared after having persisted for some time. The instillation of a 25 per cent. solution of chloral into a rabbit's ear will in the course of fifteen minutes have an effect similar to that due to destruction of the canals. Division of the auditory nerves in the skull has the same effect.

Goltz considers the canals as the sensory mechanism for maintaining the equilibrium of the head. Mach considered them as a mechanism for appreciating the movements of the head. According to Goltz, the endolymph, with every position of the head, exercises a maximum degree of pressure upon a given portion of the semicircular canals, and in this way stimulates the nerve-terminations in the ampullæ in varying degree. According to Breuer there occur in the semicircular canals on rotation of the head currents in the endolymph that stand in fixed relations to the direction and extent of the movement of the head, and which, therefore, if perceived constitute a delicate means for estimating the movement of the head. The nervous end-organs of the ampullæ are adapted to execute this perception. If, therefore, the semicircular canals act as a mechanism, in a measure as a static sense-organ, for the sense of equilibrium, the appreciation of the position or of the movements of the head, their destruction or irritation will modify these perceptions and thus give rise to abnormal oscillations of the head. Breuer, as a result of his experiments, reaches the conclusion that the labyrinth is intended as a means of orientation in space, and, particularly, that the semicircular canals bring rotatory and angular movements to perception, while the nerve-terminations in the saccule, with the otoliths, do the same for the position of the head with relation to the vertical and the existence of straight translational movements. Vertigo (with nystagmus) cannot be induced in deaf-mutes and animals whose labyrinths have been destroyed, nor in a labyrinthine invertebrates, and young tadpoles, which are as yet unprovided with semicircular canals.

The feeling of vertigo, of deception as to spatial relations of the surroundings, and at the same time of oscillation of the body, occurs particularly in connection with acquired alterations in the normal movements of the eyes, whether these consist either in involuntary lateral movements of the eyeballs (nystagmus), or in paralysis of these movements. On active or passive movement of the head or of the body, synchronous movements of the eyeballs take place normally, and these are definite for each position of the body. The general characteristic of these bilateral ocular movements, which may be designated as compensatory, consists in the fact that through them both eyes in the various changes in the position of the head and of the body tend to retain their primary position of rest. Division of the aqueduct of Sylvius at the level of the anterior quadrigeminate bodies, the cerebral portion on the floor of the fourth ventricle, the auditory nuclei, both auditory nerves, as well as destruction of the membranous labyrinth on each side, cause loss of these movements. Irritation of the same parts, conversely, causes bilateral associated ocular movements. It thus appears that compensatory ocular movements are under normal conditions excited reflexly from the membranous labyrinth. Both labyrinths are connected with both eyes by means of reflex nerve-paths, nerve-fibers passing to each eye from both labyrinths. These pass through the auditory nerve to the center (which extends from the interbrain to the commencement of the spinal cord) and from the center centrifugal fibers pass to the ocular muscles. Destruction of the semicircular canals thus causes change in the normal compensatory ocular movements and in this way gives rise to vertigo.

Chloroform and other poisons exhaust the compensatory ocular movements. Nicotin and others, as well as asphyxia, suppress them through an action upon the center. Cyon found that irritation of the horizontal semicircular canal causes horizontal nystagmus, irritation of the posterior canal vertical nystagmus and irritation of the anterior canal oblique nystagmus. Irritation of one auditory nerve causes rotatory nystagmus and axial rotation of the animal toward the irritated side.

The thought suggests itself that the disorders of equilibrium, attacks of vertigo and the feeling of apparent movement of external objects that are observed on the passage of a galvanic current through the head between the ears or between the two mastoid processes are due to influences acting on the semicircular canals of the labyrinth. Under such circumstances also oscillation of the eyes takes place, as well as a movement of the head on closure of the circuit at the anode.

Pathological.—The attacks of vertigo of sudden onset occurring in the course of disorders of the labyrinth and of so-called Menière's disease, the latter not rarely being attended with roaring in the ears, vomiting, staggering gait and marked impairment of hearing, must be referred to an affection of the ampullar nerves or their central organs or of the semicircular canals. The labyrinthine nerves may be affected also reflexly, or in the form of a pure neurosis. Even irritative phenomena upon one side may cause vertigo. Forcible injections into the ears of rabbits also cause attacks of vertigo, with nystagmus and rotation of the head toward the side treated. Also in workmen exposed to greatly increased atmospheric pressure, analogous phenomena appear. In the presence of deficiencies in the tympanic membrane in man Lucae observed, on application of the air-douche to the auditory canal, rotation of the eyes and vertigo. Inflammation of the middle ear in man may likewise cause nystagmus with vertigo. In this way is explained the vertigo observed in connection with spasm of the tensor tympani, as a result of which excessive pressure is exerted upon the labyrinth. Urbantschitsch found that even certain tones are capable of causing in persons occupying the vertical position disturbance of equilibrium and apparent movement. Also transitory derangement of the circulation in the nuclei for the nerves to the ocular muscles is, according to Mendel, often a cause of vertigo. Strabismus, paralysis of ocular muscles, pupillary changes are rare as reflex phenomena from the ear. It is a remarkable fact that occasionally a tendency to attacks of vertigo occurs in association with chronic disease of the stomach

FIG. 246.

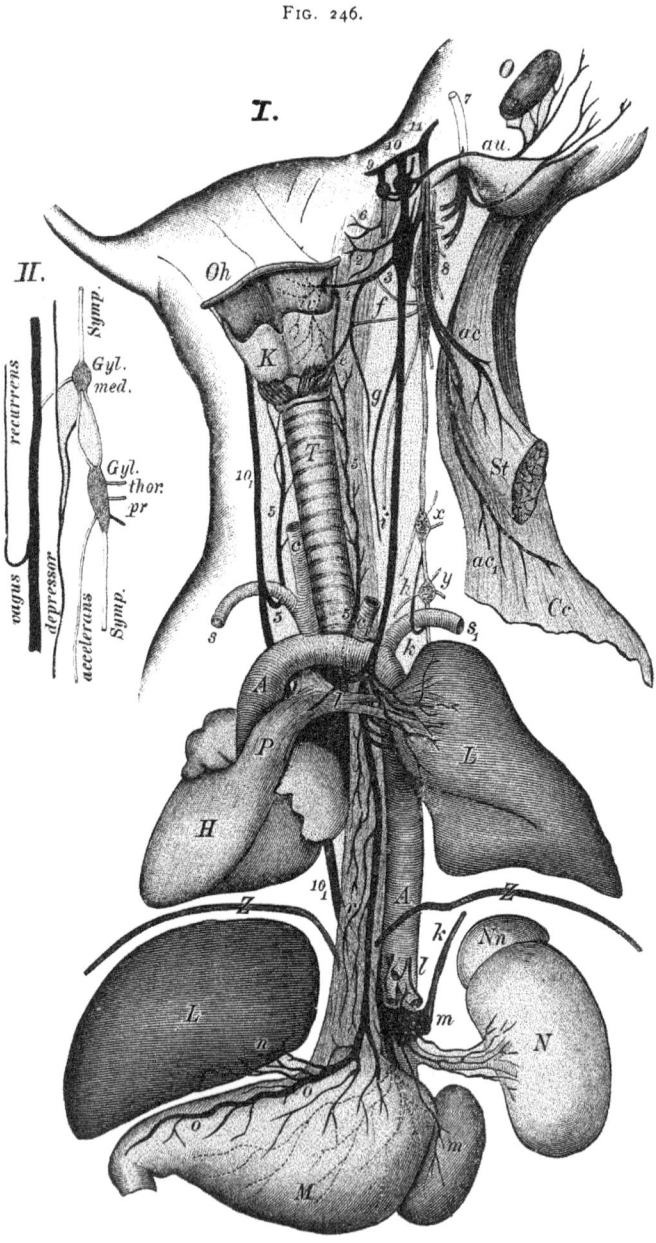

(Trousseau's gastric vertigo). This condition may result from irritation of the vasomotor nerves of the labyrinth secondary to that of the gastric nerves, producing an influence upon the pressure-relations of the endolymph. Intestinal vertigo, laryngeal vertigo and urethral vertigo have been described as occurring in an analogous manner.

IX. GLOSSOPHARYNGEAL NERVE.

The glossopharyngeal nerve arises from three nuclei:

1. The sensory, gustatory nucleus, constituted of small cells, is situated near the ala cinerea to the side of the hypoglossal nucleus just beneath the floor of the fourth ventricle. The fibers related to this nucleus arise actually from peripheral ganglion-cells (ganglionic plexus of the lingual branch). From these peripheral ganglion-cells the cellulifugal neurites enter into contact in the gustatory nucleus; the cellulipetal dendrites are derived from the neighborhood of the sense-cells of the tongue. 2. The motor nucleus is constituted of larger cells and is more deeply situated. It passes without sharp limitation into the motor nucleus of the vagus and sends as neurites motor fibers to the ninth and also to the tenth nerve. 3. The sensory, descending root is situated at the side of the solitary bundle, and with it also filaments from the vagus are associated. The cells of the jugular and petrosal ganglia serve for its origin; from them neurites pass into the medullary nucleus, while the dendrites are derived from the mucous membrane of the pharynx. The most anterior portion of the sensory nucleus of origin is considered as the root of the portio intermedia of Wrisberg.

The filaments unite to form two nerves, which subsequently coalesce, and leave the medulla in front of the vagus (Fig. 241). Close to the point of exit it forms the jugular ganglion, then in the petrosal fossa the petrous ganglion (Fig. 246). In the jugular ganglion the nerve anastomoses with the trigeminus, the facial (Fig. 243, ϵ and π), the vagus (Fig. 246) and the carotid plexus. From this ganglion there ascends vertically the tympanic nerve (Fig. 243, λ) into the tympanic cavity to unite with the tympanic plexus. This branch gives sensory branches also to the tympanic cavity and the Eustachian tube. Further, through the lesser superficial petrosal nerve it transmits fibers for the salivary secretion of the parotid gland (in the dog).

Functionally, the glossopharyngeal is: 1. The gustatory nerve for the posterior third of the tongue, the lateral portion of the soft palate and the glossopalatine arch.

The gustatory activity of the anterior two-thirds of the tongue has been discussed in connection with the consideration of the lingual nerve and of the chorda tympani. The lingual branches are provided with ganglia, principally at the plexuslike points of division and at the base of the vallate papillæ. The terminal branches can be traced to the circumvallate papillæ (Fig. 243, U), whose taste-buds they surround as telodendrites.

2. The glossopharyngeal is the motor nerve for the stylopharyngeus muscle. Nevertheless, the motor fibers of origin later pass also through the pharyngeal branches of the vagus.

FIG. 246, p. 702.

I. Diagrammatic Representation of the Distribution of the Vagus and Accessory Nerves: 10, exit of the left trunk of the vagus from the cranial cavity. (10₁, right vagus.) 9, Glossopharyngeal nerve. 7, Facial nerve. 1, Deep posterior auricular branch of the facial. 2, Pharyngeal branch of the vagus. 6, Pharyngeal branch of the glossopharyngeal 3, Superior laryngeal nerve, with its anastomoses (f) with the sympathetic and its division (4) into the internal branch (v) and the external branch (e). 5, Inferior or recurrent laryngeal. au, Auricular branch of the vagus. Cardiac nerves: g, cardiac branches from the trunk of the vagus and from the superior laryngeal; i, h, the three cardiac branches from the superior (8), middle (x), and inferior (y) cervical ganglia of the sympathetic. k, Ansa Vieussenii. l, Cardiac branch from the recurrent nerve L, Lung with the anterior and posterior pulmonary plexuses. r, Esophageal plexus. o, o, Gastric branches of the left vagus, together with the hepatic branches (n). m, Celiac plexus. h, The splanchnic nerve. 11, Accessory nerve of Willis, which sends its inner branch into the gangliform plexus of the vagus; its outer branch supplies with fibers (ac) the sternocleidomastoid muscle (St) and (ac_1) the trapezius (Cc). O, External auditory canal. Oh, Hyoid bone. K, Thyroid cartilage. T, Trachea. H, Heart. P, Pulmonary artery. A, A, Aorta. c, Right carotid. c_1, Left carotid. S, Right subclavian. s, Left subclavian. Z, Z, Diaphragm. N, Kidney. Nn, Adrenal body. M, Stomach. m, Spleen. Ll, Lung and liver. (The viscera are reduced in size.)

II. Diagrammatic Representation of the Course of the Depressor Nerve (Its origin from the Vagus is situated at a higher level), as well as of the Accelerator Branch of the Sympathetic Nerve (of the cat).

3. The glossopharyngeal is the sensory nerve for the posterior third of the tongue, the anterior aspect of the epiglottis, the tonsils, the anterior palatine arches, the soft palate and a portion of the pharynx. These nerves exert an inhibitory influence upon the act of deglutition and that of respiration. They cause, as do likewise the gustatory fibers, reflex secretion of saliva.

4. The salivary fibers are described on p. 259.

5. A branch accompanies the lingual artery. This is vasodilator for the posterior third of the tongue.

Definite pathological observations in man referable to pure and isolated affections of the ninth nerve are wanting.

X. VAGUS NERVE.

The origin of the vagus in connection with that of the ninth and eleventh nerves consists of: 1. A sensory nucleus, constituted of small cells, situated to the dorsal aspect of the hypoglossal nucleus (Fig. 241). 2. Other fibers of origin arise from a solitary bundle of longitudinal fibers (Lenhossék's bundle, W. Krause's respiratory bundle) situated on the outer side of the nucleus and extending downward into the cervical enlargement of the spinal cord. 3. Finally, a motor nucleus (nucleus ambiguus), situated further inward and a continuation of the anterior horn of the spinal cord, gives off fibers from either side.

The vagus leaves the medulla oblongata behind the ninth nerve (Fig. 242) by means of from 10 to 15 filaments between the pyramidal and lateral columns and forms at the jugular foramen the jugular ganglion, which, together with the gangliform plexus, behaves like a spinal ganglion with reference to the fibers of origin. Its branches contain fibers of varied function.

The sensory *meningeal branch* (from the jugular ganglion), in association with vasomotor fibers from the sympathetic, follows the posterior branch of the middle meningeal artery, and also sends branches to the occipital and transverse sinuses.

In cases of marked cerebral congestion and inflammation of the dura mater irritation of this branch may cause vomiting.

The *auricular branch* (Fig. 246, *au*), from the jugular ganglion, receives a communication from the petrous ganglion of the ninth nerve; then, passing through the mastoid canal, it crosses the path of the facial (7), which it is supposed to supply with sensory fibers. In its further course, it gives sensory fibers to the posterior portion of the auditory canal and the adjacent portion of the auricle. A branch passes with the posterior auricular nerve of the facial, to which it gives muscle-sense fibers for the muscles.

Irritation of this branch, as by inflammation or from the presence of foreign bodies in the external auditory canal, may cause vomiting. Irritation in the depth of the external auditory canal in the area of innervation of the auricular branch also excites reflex cough, less commonly symptoms of cardiac inhibition. Finally irritation of the auricular nerve causes reflex contraction of the vessels of the ear.

The *anastomotic branches* of the vagus are as follows: 1. A branch that connects the petrous ganglion of the ninth directly with the jugular ganglion of the tenth nerve. Its function is unknown. 2. Just above the gangliform plexus of the vagus, the entire inner half of the accessory nerve enters the trunk of the vagus. This transmits to the latter motor fibers for the larynx (through the recurrent branch of the vagus), for the pharynx (?) and the cervical portion of the esophagus and the stomach (?), as well as the cardiac inhibitory fibers. 3. In the gangliform plexus, nbers of unknown function from the hypoglossal, the superior cervical ganglion of the sympathetic and the cervical plexus unite with the vagus.

According to Grossmann the fibers for the cricothyroid arise in rabbits from the glossopharyngeal, as do also the Hering-Breuer pulmonary fibers. According to Grabower, the fibers for the muscles of the larynx come from the vagus itself. According to Kreidl, the fibers for the esophagus are situated in the glossopharyngeal in the rabbit, but enter the trunk of the vagus.

To the *pharyngeal plexus* the vagus (2) sends from the upper portion of the gangliform plexus one or two branches that at the level of the middle constrictor of the pharynx, together with the pharyngeal branches of the ninth nerve and the superior cervical ganglion of the sympathetic, form the *pharyngeal plexus* at the side of the ascending pharyngeal artery. The posterior portion of the trunk of the vagus itself supplies through this plexus the three constrictors of the pharynx, as well as the palatoglossus and palatopharyngeus muscles (according to observations on the ape) with motor fibers. Filaments from the middle of the anterior accessory root innervate the elevator of the veil of the palate. Sensory fibers from the vagus to the pharyngeal plexus supply the pharynx from a point below the level of the veil of the palate downward. These fibers stimulate reflexly the constrictors of the pharynx in the act of deglutition. In case of considerable abnormal irritation they are also capable of inducing vomiting. The sympathetic fibers of the pharyngeal plexus give vasomotor fibers to the vessels of the pharynx. The pharyngeal branches of the ninth nerve are described on p. 703.

The vagus sends two branches to the *larynx:* (a) The *superior laryngeal nerve* (3), which after receiving a vasomotor filament from the superior cervical ganglion of the sympathetic divides into an external and an internal branch. (1) The *external branch* receives from the same source vasomotor fibers (which later on accompany the superior thyroid artery) and it supplies the cricothyroid muscle with motor fibers (which in the ape are derived from the posterior fibers of the trunk of the vagus) and the inferior lateral portion of the laryngeal mucous membrane with sensory fibers. (2) The *internal branch* gives off only sensory fibers: to the glottoepiglottic fold and the adjacent lateral portion of the root of the tongue, to the aryepiglottic fold and to the entire interior of the larynx (in so far as it is not supplied by the external branch). Irritation of these sensory branches causes reflex cough, although irritation of the vocal bands does not, but only that in the vicinity of the respiratory glottis. The same effect is brought about through the sensory branches of the vagus to the trachea, particularly at the point of bifurcation, also through those of the bronchial mucous membrane, as well as those of the pulmonary tissue and of the pleura when altered by disease (inflammation). The cough-center is supposed to be situated on either side of the raphé in the neighborhood of the ala cinerea. Severe attacks of coughing may be attended with vomiting in consequence of irritation of the pharynx or as an associated movement. Hédon found in the superior laryngeal nerve vasodilator and secretory fibers for the mucous membrane of the larynx and Kokin in both laryngeal nerves secretory fibers for the mucous glands of the larynx and the trachea.

45

It is a noteworthy fact that in some persons coughing can be induced by irritation of even remotely situated sensory nerves, for example of the external auditory canal (auricular branch of the vagus), the nasal mucous membrane (trigeminal cough of Schadewald), the liver, the spleen, the stomach and intestine, the uterus, the mammary glands, the ovaries, and even some portions of the skin. Whether under such circumstances the perhaps abnormally irritable cough-center is directly stimulated centripetally through the irritated nerve, or whether in consequence of the nerve-irritation the vascularization and the secretion of the respiratory organs are first affected, and in turn cause the cough-reflex, must be submitted to future investigation, although to the writer the latter appeared the more probable.

The cough induced through irritation of the trachea and the bronchi (dog, cat) occurs immediately and persists as long as the irritation continues. On irritation of the larynx there occurs first inhibition of breathing, with accompanying movements of deglutition, cough occurring only on cessation of the irritation.

The superior laryngeal contains further centripetal fibers, irritation of which causes arrest of breathing, with closure of the glottis; and also fibers that excite movements of swallowing; and, finally, centripetal fibers, irritation of which stimulates the vasomotor center to increased activity, therefore designated *pressor fibers*.

(*b*) The *inferior laryngeal or recurrent nerve* passes on the left around the arch of the aorta, on the right around the subclavian artery, and, ascending in the interval between the trachea and the esophagus, gives off motor fibers to these structures and the inferior constrictor of the pharynx, and then passes to the larynx, to whose muscles it distributes motor fibers (with the exception of the cricothyroid muscle). In apes these fibers are derived from the most posterior fibers of the internal branch of the accessory nerve. The muscles of the epiglottis (aryepiglottic and thyroepiglottic) are innervated at times by the superior and at other times by the inferior laryngeal nerve. Irritation of the latter nerve also exerts an inhibitory effect upon the respiratory center.

The fibers of the nerves that subserve the respiratory functions pass isolated from those that control the phonetic activity of the muscles, from the origin to the muscle. From the superior laryngeal nerve an anastomotic branch passes to the inferior laryngeal (the so-called anastomosis of Galen), and it gives off sensory branches to the upper half of the trachea, to the larynx, perhaps also to the esophagus, and the muscle-sense fibers (?) for the laryngeal muscles supplied by the recurrent nerve.

Exner describes a middle laryngeal nerve, derived from the pharyngeal nerve of the vagus and its anastomoses in the pharyngeal plexus, which takes part in the innervation of the cricothyroid muscle (present only in rabbits) and the anterior and inferior portions of the laryngeal mucous membrane. According to Onodi fibers from the inferior cervical and the superior thoracic ganglion of the sympathetic take part in the innervation of the laryngeal muscles. On the other hand, the accessory is not believed to participate in this.

Irritation of the superior laryngeal nerves is painful and causes movement of the cricothyroid muscles, as well as, reflexly, of the remaining laryngeal muscles. Division of these nerves is said to cause slight slowing of the respiration in consequence of the paralysis of the cricothyroids. At the same time the voice in the dog becomes deeper and rough in consequence of deficient tension of the vocal bands. Further, the larynx is anesthetic, so that fluid from the mouth and particles of food (without causing reflex closure of the larynx or coughing) gain entrance into the trachea and the lungs, in consequence of which so-called deglutition-pneumonia results, with a fatal termination.

Irritation of the recurrent nerves causes spasm of the glottis. Division paralyzes the laryngeal muscles supplied by these nerves and the voice becomes toneless and rough (in the pig, in man, the dog, the cat, while rabbits retain their clear, shrill voice). The glottis is small. With each inspiration the vocal bands approach each other considerably in their anterior portions. In expiration they are blown apart. Therefore, inspiration (particularly in young individuals having a narrow respiratory glottis) is labored and noisy, while expiration takes place with perfect ease. In the course of a few days the animal (carnivora) becomes quieter, breathes less laboriously and the passive, flabby movement of the vocal bands disappears. If, however, in the further course of events, even after a considerable time, the animal is actively stimulated, there occurs in the presence of the marked need for air an attack of extreme dyspnea, which subsides only when the animal (dog) gradually becomes quieter. In consequence of the paralysis of the larynx, foreign bodies may gain entrance into the trachea, particularly as the paralysis of the uppermost portion of the esophagus renders swallowing difficult. In this way bronchopneumonia may develop.

The *depressor nerve*, which in rabbits arises from the trunk of the superior laryngeal and occasionally, with a second root from the trunk of the vagus itself, passes with the sympathetic downward in the neck, descends into the stellate ganglion and enters thence into the cardiac plexus. It is a centripetal nerve, irritation of which, and also of its central stump, diminishes the energy of the vasomotor center, so that the blood-pressure falls. At the same time this irritation is conveyed to the cardiac inhibitory center, so that the number of pulsations of the heart diminishes.

The depressor nerve is present also in the cat (Fig. 246, *II*), the hedgehog, the rat. and the mouse. In the horse and in man fibers analogous to the depressor nerve pass back again into the trunk of the vagus. Also in the rabbit fibers having a depressing effect may pass in the trunk of the vagus itself. The depressor fibers of the rabbit enter the oblongata through the upper root-filaments of the vagus. The inhibitory reflex for the heart is effective only upon the same side.

The branches of the vagus for the *cardiac plexus* (g, l), as well as the latter itself, have already been described. They contain the inhibitory fibers for the movement of the heart (they are derived from the most anterior root-filaments of the inner branch of the accessory nerve), also sensory fibers for the heart (in the frog and in part in mammals). Finally, the heart receives also through the vagus a portion of its accelerator fibers; feeble irritation of the vagus causes at times acceleration of the heart-beat. In cases of poisoning with atropin and nicotin, which paralyzes the inhibitory fibers, irritation of the vagus causes acceleration of the heart-beat. The following experiment tends to support the existence of vasomotor fibers in the cardiac branches: Persistent irritation of the peripheral stump of the vagus causes extravasation of blood into the endocardium (long-continued poisoning with digitalin or strychnin has a similar effect), in consequence of spasmodic contraction of the endocardial vessels, with secondary paralytic relaxation and rupture.

The *pulmonary branches* of the vagus are grouped together in the anterior and posterior pu monary plexuses. The former supplies sensory and motor branches to the trachea and passes then on the anterior surface of the bronchial ramifications into the lungs (L). The posterior plexus, formed of from three to five large branches derived from the trunk of the vagus at the side of the bifurcation, anastomoses with branches from the inferior cervical ganglion of the sympathetic and with fibers of the cardiac plexus, and, after fibers from each side have interchanged by decussation, passes with the branches of the bronchial tree into the lungs. The pulmonary branches are supplied with ganglion-cells, as are also the larynx, the trachea, and the bronchi. From the pulmonary plexus filaments pass to the pericardium and the superior vena cava.

The function of the pulmonary branches of the vagus is a varied one: (1) They supply the motor branches for the unstriated muscles of the entire bronchial tree. (2) They supply the sensory fibers (exciting cough) to the entire bronchial tree and the lungs. (3) They supply, in smaller part, vasomotor nerves to the pulmonary vessels, although these are in largest part, if not wholly, derived from the anastomosis with the sympathetic (in animals from the superior thoracic ganglion). (4) They contain, in the ape situated in the posterior portion of the trunk of the vagus itself, centripetal fibers passing from the parenchyma of the lungs to the medulla oblongata, irritation of which stimulates the respiratory center. Division of both vagi is, accordingly, followed by marked reduction in the number of respirations, which at the same time are deepened, so that the animals for a time exchange the normal volume of air containing normal amounts of oxygen and carbon dioxid. Irritation of the central stumps of the vagi causes acceleration of respiration. This labored and embarrassed breathing is explained as due to elimination of these reflex-stimulating fibers, which maintain the normal easy play of reflex breathing. After division of the nerves, the stimulation of the respiratory movements must take place in the medulla oblongata itself. (5) They contain centripetal fibers, irritation of which has a depressant effect upon the vasomotor center, as shown by fall of the blood-pressure on forced expiratory pressure. (6) Also fibers, irritation of which has an inhibitory influence upon the cardiac inhibitory fibers of the vagus, thus accelerating the pulse. Simultaneous irritation of the last two sets of fibers mentioned is capable of altering the rhythm of the pulse.

Carbon dioxid, as well as the vapors of ammonia and chloroform, introduced into the air-passages, cause (from the mucous membrane of the large bronchi) inspiration, while, acting upon the entrance to the respiratory tract situated above the trachea, they cause reflex expiration.

Pneumonia after section of both vagi has attracted the interest of investigators since the time of Valsalva (died 1723), Morgagni (1740) and Legallois (1812). In explanation of this condition the following facts are to be taken into consideration: (a) In the first place, section of both vagi is attended by loss of the motility of the larynx, as well as of the sensibility of the larynx (if the section has been made above the origin of the superior laryngeal nerve), the trachea, the bronchi, and the lungs. Therefore, the larynx fails to close during the act of swallowing, and reflex closure of the larynx when foreign bodies threaten to enter (fluids in the mouth, particles of food, irritating gases) does not take place, and reflex cough for the expulsion of substances that have entered is suppressed. Thus, foreign bodies enter the lung without hindrance, and all the more readily as the associated paralysis of the esophagus permits the food to remain lodged in this tube for a time, and thus readily enter the larynx. That herein resides an essential exciting factor for the inflammatory process Traube was able to show by demonstrating that the inflammation could be prevented by permitting the animals to breathe through a tracheal cannula introduced through an external wound in the neck. If, however, only the motor filaments of the recurrent nerves were divided, and the esophagus was ligated, so that foreign bodies necessarily entered the air-passages, so-called *foreign-body pneumonia* resulted in an analogous manner, with a fatal termination. (b) A second factor resides in the fact that in consequence of the extensive and labored snoring and noisy breathing, the lungs must become hyperemic, as during the protracted and marked dilatation of the chest the pressure of the air in the lungs must be abnormally low. As a result, serous transudates (pulmonary edema) result, or even extravasation of blood and dilatation of the pulmonary vesicles at the margins of the lungs. Through this influence the entrance of foreign bodies, particularly of fluid, into the glottis is facilitated. A tracheal cannula introduced from without will likewise prevent the inflammation under these circumstances. (c) Perhaps partial paralysis of the pulmonary vasomotors takes some part in the inflammation, as the hyperemia thus induced affords an inviting field for the complication. (d) Finally, it is still to be determined whether trophic fibers in the vagus subserve the normal preservation of the lung-tissue. According to Michaelson the pneumonia developing immediately after section of the vagus is seated principally in the middle and lower lobes, while the catarrhal inflammation that develops more slowly after section of the recurrent nerves is situated usually in the upper lobes. Rabbits die amid symptoms of pneumonia as a rule within twenty-four hours; when the precautions mentioned are taken, in the course of several days. Dogs may survive for a considerable time. It is doubtful whether the paralysis of the intra-abdominal fibers of the vagus favors the occurrence of death. In rabbits, extirpation of the ninth, tenth, and twelfth nerves on one side causes death from pneumonia. After section of both vagi in rabbits acute fatty degeneration of the heart and abnormal friability of the smaller coronary vessels develop in consequence of loss of the trophic functions of the vagus. In birds the lungs remain free from inflammation after section of both vagi, because the upper portion of the larynx retains its faculty of reflex closure. Nevertheless, death results in a week from inanition in consequence of paralysis of the crop, in which the food undergoes putrefaction. At the same time the heart is in a state of fatty degeneration, as are also the liver, the stomach, and the muscles. According to Wassilieff the heart exhibits parenchymatous swelling and slight waxy degeneration. In ruminants considerable tympanitic distention of the stomach results, because eructation is impossible. Frogs, which with each inspiration open the glottis, which is closed during rest, die of asphyxia after section of the trunks of the vagi. Section of the pulmonary branches is unattended with injurious effect. If the vagi are divided below the origin of the recurrent nerves the lungs remain healthy in the dog, although disorders of secretion and of the movements of the stomach set in. As a result putrefactive decomposition occurs in the stomach, in consequence of which death takes place.

The *esophageal plexus* (r) is formed by branches above from the inferior laryngeal, then from the pulmonary plexus, and below from the trunk of the vagus itself (being derived from the posterior root-fibers of the trunk). The plexus endows the esophagus with motility and with indistinct sensibility (also that of muscular contraction) only in its upper portion and it supplies it with reflex fibers.

The *gastric plexus* (o o) consists of the anterior (left) extremity of the vagus, which also sends fibers to the esophagus and passes along the lesser curvature and in part sends fibers through the transverse fissure to the liver. The posterior (right) vagus, after giving off a few fibers to the esophagus, takes part in the formation of the gastric plexus, which receives sympathetic fibers at the pylorus. The vagi supply the stomach with motor fibers, derived from its root (not from the accessory nerve), and also inhibitory or relaxing fibers for the cardia. Further, the vagus supplies the secretory fibers for the gastric mucous membrane. These contain vasomotor fibers, for division of the trunks of the vagi causes hyperemia of the mucous membrane of the stomach. The gastric fibers, however, receive the centripetal filaments, through which the secretion of saliva is stimulated. Whether also vomiting can be excited through them is still doubtful.

After section of both vagi below the diaphragm death results, at the latest after an interval of three months, preceded by emaciation, inflammatory altera-tions in the mucous membrane of the stomach and perivascular hyperplasia in the liver and in the kidneys.

About two-thirds of the right vagus, however, passes at the stomach into the celiac plexus (m) and thence, accompanying the arteries, to the liver, the spleen, the pancreas, the small intestines, the kidneys, and the adrenal glands. The influence of the vagus upon the movements of the intestine has been discussed in the consideration of the intestinal nerves. According to some observers, irritation of the vagus causes movements in the small as well as in the large intestine. Irritation of the peripheral stump of the vagus causes in the spleen contraction of the unstriped muscular fibers in the capsule and in the trabeculæ (in the dog and the rabbit). With respect to the kidneys, irritation of the vagus at the cardia causes increased secretion of urine, with dilatation of the renal vessels and redness of the blood in the renal veins. In dogs and rabbits a number of vasomotor fibers are said to be supplied to the abdominal viscera by the vagus, while the overwhelming majority are derived from the splanchnic.

The trunk and the branches of the vagus contain also fibers (in part already mentioned), irritation of which acts in a centripetal direction upon certain nervous stuctures:

(a) The *vasomotor center* is affected through (n) pressor fibers (especially in the two laryngeal nerves), irritation of which causes reflex contraction of the arteries and thus increase in blood-pressure; (β) depressor fibers (in the depressor nerve or in the vagus itself), which exert a contrary effect. This subject is discussed in connection with the vasomotor center.

(b) The *respiratory center* is affected through (n) accelerator fibers (pulmonary branches), irritation of which accelerates the respiration; and (β) inhibitory fibers (in the two laryngeal branches), irritation of which inhibits respiration. This subject is discussed in connection with the respiratory center.

(c) The *cardio-inhibitory system* is influenced through fibers in the trunk of the vagus, irritation of which acts on the center in a centripetal direction and places the heart in a condition of diastolic rest. Irritation of the central stump of the vagus, therefore, causes arrest of the heart. In conformity with these facts is the observation of Mayer and Pribram that sudden dilatation of the stomach causes slowing and even arrest of the heart, the arteries of the medulla oblongata contracting at the same time with increase in blood-pressure.

(d) The *vomiting center* is excited by irritation of the central stump of the vagus and of a number of centripetal fibers of the vagus.

(e) The *secretion of the pancreas* is influenced by irritation of the central stump of the vagus, the secretion being arrested in this way; therefore, probably through the intermediation of certain pancreatic nerves.

(f) According to Claude Bernard the pulmonary branches contain fibers, irritation of which causes reflex increase in the *formation of sugar in the liver*, perhaps through the intermediation of the hepatic branches of the vagus; for after section of both vagi the formation of glycogen in the liver ceases. Conversely, irritation of the peripheral stump of the vagus causes an increase in the formation of sugar in the liver.

The various branches and paths of the vagus possess an unequal degree of irritability. If irritation, at first feeble, be applied in a centrifugal direction, the muscles of the larynx move first, then the heart-beat is slowed. If the central stump is stimulated the excito-respiratory fibers become exhausted, even on feeble irritation, and only later the inhibito-respiratory fibers. According to Steiner, the various fibers are so arranged in the vagus of the rabbit that the centripetal are situated in the outer and the centrifugal in the inner half of the cervical trunk.

Pathological.—Irritation or paralysis in the distribution of the vagus will present a varying clinical picture accordingly as the lesion involves the entire trunk or only individual branches, and accordingly, also, as the affection is unilateral or bilateral. *Paralysis of the pharynx and the esophagus*, which is generally of central or at least intracranial origin, renders difficult or abolishes movements of deglutition, so that stagnation in the esophagus, entrance of foreign bodies into the larynx, dyspnea, and the passage of food into the nares are observed. In drinking, a rumbling murmur is at times audible in the relaxed tube (deglutitio sonora). When the paralysis is incomplete, the act of swallowing is only delayed and rendered difficult, and large masses of food are most readily swallowed. *Increased contraction*, even spasmodic constriction, is observed in association with the symptoms of general nervous irritability.

Spasm of the muscles of the larynx causes especially spasmodic closure of the glottis, so-called spasm of the glottis. The latter is peculiar to childhood, and occurs paroxysmally with dyspnea, tight, whistling inspiration, with which twitching of the muscles of the eyes, the jaw, the fingers, the toes, etc., may be associated.

The condition is probably one of reflex spasm that can be excited from the sensory nerves of various areas (such as the teeth, the intestine, the skin) in the medulla oblongata. Spasm of the dilators of the glottis and of other laryngeal muscles also occurs.

Irritation of the sensory nerves of the larynx, as is well known, causes cough. If the irritation is intense, for example in cases of whooping-cough, the fibers in the laryngeal nerves, having an inhibitory influence upon the respiratory center, may also be irritated; there occurs diminution in the respiratory frequency and finally arrest of respiration with relaxation of the diaphragm; in the presence of the most intense irritation, spasmodic arrest in expiration occurs, with closure of the glottis, even for as long as fifteen seconds. The condition is an inhibitory neurosis of the respiratory apparatus. Paralysis of the laryngeal nerves causing alterations in the voice has already been considered. Paralysis of both recurrent nerves, due, for example, to stretching in consequence of dilatation of the aorta and the innominate artery, is attended with great waste of air as a result of the fruitless efforts at phonation; expectoration is rendered difficult, and forcible cough impossible. In addition severe attacks of dyspnea may occur on exertion, entirely like those that can be induced experimentally in animals. The increased irritability of hysterical persons is associated with hyperesthesia and anesthesia of the larynx, the upper air-passages, aphonia, a tendency to vomiting, a slow and irregular heart-beat as signs of a neurosis of the vagus. Attacks of extreme dyspnea lasting for from one-quarter of an hour to several hours have been referred to irritation of the pulmonary plexus, which is supposed to cause *spasm of the bronchial muscles*, bronchial asthma. Physical examination of the lungs discloses in addition to rhonchi, no indication as to the cause of the severe attack. If the condition is really one of spasm, this is probably in most instances of reflex origin, the centripetal nerves of the respiratory passages, or of the skin (cold), or of the genitalia (sexual asthma) being involved. Landois, however, was of the opinion that in many cases considered as instances of asthma, the condition is one of transitory paresis of the pulmonary nerves, exerting a stimulating effect upon the respiratory center. The attack would, then, be a reproduction of the labored breathing following section of both vagi. Whether the acute pulmonary emphysema constantly observed in connection with this disease is due to irritation or to paralysis of the muscular fibers in the lungs is as yet a matter of doubt. According to Biermer this is due to slight obstructions to expiration in the small bronchi that are more readily overcome in inspiration than in expiration. Such obstructions comprise catarrhal swelling of the mucous membrane, accumulation of mucus or of blood, or spasm of the bronchi.

Irritation in the distribution of the cardiac branches of the vagus may, by direct excitation, cause attacks of diminished or even temporarily suspended contraction of the heart, together with a feeling of extreme prostration and of abolition of the functions of life, occasionally also with pain in the precordium. Such attacks may likewise be induced reflexly through irritation of the abdominal viscera, in conformity with the percussion-experiment of Goltz. Landois first analyzed these symptoms in 1865 on the lines of a physiological experiment and designated them pneumogastric or reflex angina pectoris. Extirpation of the larynx is occasionally followed by circulatory disturbances that may eventually prove fatal. These are due to a persistent state of irritation of the laryngeal nerves, eventually with extension to the vagus itself. Hennoch and Silbermann observed slowing of the heart in children presenting irritative phenomena referable to the stomach, and Landois, intermission of the heart-beat even in adults. Through the same reflex action a derangement in the respiratory functions of the vagus that Hennoch has designated *dyspeptic asthma* may be brought about. A similar condition may be brought about reflexly also through other sensory nerves (uterine asthma). Rarely, intermittent paralysis of the cardiac branches of the vagus is attended with marked acceleration of the heart-beat, up to between 160 and 240 beats, the rhythm and the strength at times exhibiting great irregularity, and dyspnea in part setting in at the same time. Under such circumstances a careful analysis is necessary in each case in order to determine to what extent irritation of the heart-muscle, the heart-centers or the accelerator cardiac fibers are concerned. Little of a trustworthy nature is known with regard to abnormal affections of the intra-abdominal fibers of the vagus. If the trunks of the vagi or their centers are paralyzed, the most conspicuous symptom is labored, deep, slow breathing, exactly as occurs after section of both vagi.

XI. ACCESSORY NERVE OF WILLIS.

The single elongated nucleus of origin (Fig. 241) comprises the dorsolateral group of cells of the anterior horn of the cervical cord, which begins below at the level of the seventh cervical nerve and extends upward without interruption in the medulla oblongata to the upper extremity of the pyramidal decussation. The nucleus of origin approaches at its highest point the hypoglossal nucleus, then is situated above the first cervical nerve in the middle of the anterior horn, next passes laterally, and between the second and fourth nerves is situated at the lateral margin of the anterior horn. Still further downward, to below the sixth cervical nerve, it is situated at the base of the lateral horn. All fibers arise from the ganglia as neurites. From the cortical center on the opposite side there must pass to the nucleus fibers through which voluntary stimulation of the motor fibers is effected.

The fibers pass upward in the lateral column of the spinal cord and leave the latter in several bundles between the anterior and posterior cervical nerve-roots. Then the root-fibers that ascend through the great occipital foramen come together without uniting in the neighborhood of the jugular foramen and form the two branches of the nerve. Of the latter the inner enters wholly into the gangliform plexus (Fig. 246) and supplies the vagus with most of its motor fibers and also its cardiac inhibitory fibers. In man, accordingly, total paralysis of the accessory nerve is attended with immobility of the corresponding half of the larynx and soft palate.

According to Kreidl the inhibitory fibers for the heart are situated in the most anterior root-bundles of the inner branch of the accessory nerve. If these roots are divided, the cardiac inhibitory fibers undergo degeneration. If the trunk of the vagus in the neck is irritated four or five days after the operation the cardiac inhibitory action is no longer exhibited.

The external branch of the accessory nerve is derived from the spinal portion of the nucleus. This anastomoses with sensory filaments from the posterior root of the first, less commonly also of the second cervical nerve, which supply muscle-sense fibers to it. It then passes backward over the transverse process of the atlas and terminates as a motor nerve in the sternocleidomastoid and trapezius muscles (Fig. 246). The latter large muscle receives, however, motor filaments for its acromial portion from the cervical plexus.

The external branch anastomoses also with several cervical nerves. Either these fibers take part in the innervation of the muscles named, or the accessory nerve returns to them, in part, the sensory filaments received from the posterior roots of the two uppermost cervical nerves, which then constitute the cutaneous branches of these cervical nerves.

Pathological.—*Irritation of the external branch* causes clonic and tonic spasm of the muscles named, usually upon one side. If the branch for the sternocleidomastoid is alone affected, the head responds to the traction of this muscle in the presence of clonic spasm. If the disorder is bilateral, the traction is usually alternating; much less commonly the action is bilateral; so that the head executes a nodding movement. In the presence of clonic spasm of the trapezius, the head is drawn backward and to the side; the scapula generally follows the traction of the bundle of this great muscle that is most severely involved. *Tonic spasm* of the sternomastoid causes the characteristic position of caput obstipum (spasticum) —*spasmodic wry-neck.* Similar spasm of the trapezius usually involves only individual portions of the muscle, which then naturally cause special positions of the head or of the scapula. Irritation of the root causes at the same time spasmodic movements of the muscles of the larynx and of the uvula.

Paralysis of one sternomastoid causes the head to be turned toward the opposite side by the preponderant action of the muscle of that side (*paralytic torticollis*). *Paralysis of the trapezius* is usually confined to individual portions of the muscle. Paralysis of the *entire accessory* trunk, principally in consequence of central processes, gives rise, in addition to paralysis of the sternocleidomastoid and the trapezius, also to paralysis of the motor branches of the vagus previously mentioned. *Bilateral paralysis* is extremely rare and is said to be attended with acceleration of the heart-beat.

XII. HYPOGLOSSAL NERVE.

The elongated nucleus of origin of the hypoglossal nerve consists of large cells, and is a continuation of the anterior horn of the spinal cord. It is situated in the depth of the lowermost portion of the floor of the fourth ventricle. It receives anastomotic fibers from the cerebral cortex of the opposite side. The nuclei of both sides are connected by a commissure.

The nerve arises as a bundle of neurites of from ten to fifteen filaments, and makes its exit in a direction parallel with that of the anterior roots of the spinal nerves (Fig. 242). In its development the hypoglossal shows itself to be in part a spinal nerve.

Purely motor at its root, the hypoglossal is the motor nerve of all of the muscles of the tongue, including the geniohyoid and thyrohyoid. The trunk of the hypoglossal nerve anastomoses with: 1. The superior cervical ganglion of the sympathetic, through which it receives vasomotor fibers, for division of the hypoglossal (together with that of the lingual) is followed by redness of the corresponding half of the tongue. 2. Muscle-sense fibers enter the hypoglossal from the gangliform plexus and from the small lingual branch of the vagus, also from anastomoses with the cervical nerves and through those with the lingual beneath the tongue. After division of the lingual nerve the tongue still possesses dull sensibility. 3. The loop of the hypoglossal nerves anastomoses with the two upper cervical nerves. These anastomoses pass further through the descending branch (through which also muscle-sense fibers from the lingual descend) as motor branches for the sternohyoid, omohyoid and sternothyroid. Irritation of the roots of the hypoglossal affects the muscles named only rarely and in slight degree.

Division of both hypoglossal nerves paralyzes the tongue. Dogs are no longer able to drink and they bite the flabby, pendulous tongue. Frogs, which catch their prey with the tongue, must starve; in hanging out of the mouth the tongue prevents closure of this cavity and as a result the animals die from asphyxia, because they are able to pump air into the lungs only when the mouth is closed.

Pathological.—*Paralysis* of the hypoglossal nerve (*glossoplegia*) is generally of central origin, and it causes derangement of speech. The deviation of the tongue in case of unilateral paralysis is described on p. 276. Paralysis of the tongue renders chewing difficult, prevents formation of the bolus and swallowing in the mouth. In consequence of deficiency in the friction-movement of the tongue, the sense of taste is impaired. The singing of high notes and of falsetto notes, in the production of which special positions of the tongue appear to be necessary, is interfered with.

Spasm of the tongue, causing *aphthongia,* is generally of reflex origin, and, in any event, is extremely rare. Cases of idiopathic spasm of the tongue have also been described, the tongue being moved with great violence. The seat of irritation has been either in the cerebral cortex or in the medulla oblongata.

THE SPINAL NERVES.

The thirty-one spinal nerves are connected with the spinal cord by means of two roots: The anterior roots arise as neurites of the ganglia of the anterior horns; the posterior roots have in reality grown into the spinal cord from without. They arise from the pear-shaped bipolar cells of the ganglia of the posterior roots, one fiber of which enters the gray matter of the spinal cord as a neurite and here enters into contact with ganglion-cells (Fig. 251); and the other fiber of which passes to the ganglion as a dendrite from peripheral areas endowed with sensibility. The posterior roots make their exit from the sulcus between the posterior and lateral columns of the spinal cord; the anterior roots, from the groove between the lateral and anterior columns. The posterior forms the spindle-shaped spinal ganglion. Then the two roots enter into intimate union and form, still within the vertebral canal, a mixed trunk. The two branches originating from the trunk always contain fibers of both roots.   Each spinal nerve is derived from two or three or even several spinal segments.

The roots growing from the spinal ganglion into the spinal cord not only enter the spinal segment corresponding to them, but grow into other segments of the spinal cord and thus are connected with many segments.   The motor roots are localized exclusively in their spinal segments.   Toward the periphery the spinal nerves (motor and sensory) likewise not only pass to the segments of the body corresponding to them, but extend beyond these limits into the areas of other segments.   This is observed particularly in the extremities, less commonly on the trunk, especially on the skin, and in more marked degree in the fibers that pass to the sympathetic ganglia.

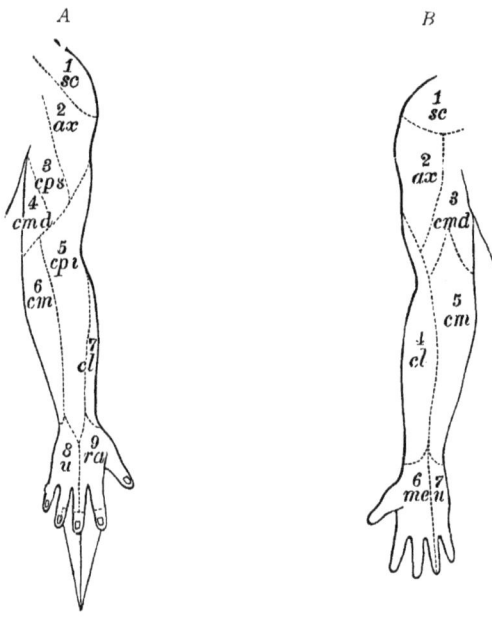

FIG. 247.—Distribution of the Cutaneous Nerves of the Upper Extremity (after Henle).

A. Dorsal aspect of the upper extremity: 1 *sc*, supraclavicular nerves; 2 *ax*, axillary nerve; 3 *cps*, posterior superior cutaneous nerve (radial); 4 *cmd*, middle or internal cutaneous nerve; 5 *cpi*, posterior inferior cutaneous nerve (radial); 6 *cm*, middle cutaneous or greater internal cutaneous nerve; 7 *cl*, lateral or external cutaneous nerve; 8 *u*, ulnar nerve; 9 *ra*, radial nerve; 10 *me*, median nerve.

B. Ventral aspect of the upper extremity: 1 *sc*, supraclavicular nerves; 2 *ax*, axillary nerve; 3 *cmd*, middle or internal cutaneous nerve; 4 *cl*, lateral or external cutaneous nerve; 5 *cm*, middle or greater internal cutaneous nerve; 6 *me*, median nerve; 7 *u*, ulnar nerve.

Charles Bell, in 1811, discovered the law named after him, namely, that the anterior roots contain the motor, and the posterior roots the sensory fibers.

Magendie, in 1822, noted the remarkable fact that the anterior roots of warm-blooded animals, but not of the frog, likewise contain sensory fibers, so that irritation of them causes pain.   This, however, is due to the fact that fibers from the sensory root pass in a centripetal direction in the anterior root after the junction of the two.   This phenomenon is designated *recurrent sensibility* (sensibilité récurrente). The sensibility of the anterior root, therefore, ceases at once as soon as the posterior root is divided. In conjunction with the loss of sensibility of the anterior roots thus brought about, that of the surface of the spinal cord in the vicinity of the root is also abolished. A considerable time after division of the anterior root (if degeneration has already taken place), a number of fibers that are not degenerated are found in its peripheral extremity, while a number of degenerated (sensory) fibers are present in its central stump. In cases in which

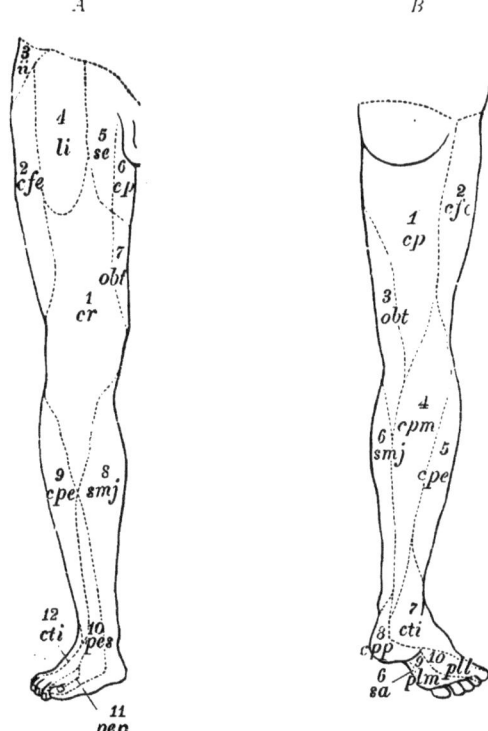

FIG. 248.—Distribution of the Cutaneous Nerve of the Lower Extremity (after Henle).

A. Anterior aspect: 1, crural nerve; 2, external or lateral cutaneous nerve of the femur, Henle; 3, ilioinguinal nerve; 4, lumboinguinal nerve; 5, external spermatic nerve; 6, posterior cutaneous nerve; 7, obturator nerve; 8 greater saphenous crural nerve; 9, communicating peroneal or fibular nerve; 10, superficial peroneal nerve; 11, deep peroneal nerve; 12, communicating tibial or sural nerve.

B. Posterior aspect: 1, posterior cutaneous nerve; 2, external or lateral cutaneous nerve of the femur, Henle; 3, obturator nerve; 4, posterior median cutaneous nerve of the femur (peroneal nerve); 5, communicating peroneal or fibular nerve; 6, greater saphenous nerve (crural nerve); 7, communicating tibial or sural nerve; 8, proper plantar cutaneous nerve (tibial nerve); 9, middle plantar nerve (tibial nerve); 10, lateral plantar nerve (tibial nerve).

the motor fibers were degenerated, Schiff found unaltered fibers in the anterior root, and these passed over to the spinal meninges. In rare cases the anterior root receives its sensibility, besides, from other sources than from its corresponding posterior root. The passage of sensory fibers into the motor root takes place either at the point of junction between the two roots, or in the plexus, or in the vicinity of the peripheral terminal distribution. Thus, sensory fibers passing in a centripetal direction also enter from the periphery into several motor branches of the cranial nerves. Even sensory branches of other sensory nerves may enter also into the trunks of sensory nerves. This fact explains the remarkable observation that, after division of a nerve-trunk, for example the median, its peripheral extremities remain sensitive. Landois offers the simple explanation for the conditions described that the tissue of the motor and sensory nerves contains (as do most of the tissues of the body) sensory fibers.

As a result of carefully observed experiments with division of the roots, as well as after discovery of the reflex relations of the sensory roots to irritation of the anterior root (reflex movement) by Johannes Müller and Marshall Hall, the following deductions may be readily made from the general law of Bell: (1) At the moment of division of the anterior root a contraction (mechanical irritation of the motor fibers) occurs in the muscles supplied from this root. (2) A sensation of pain, however, also results (recurrent sensibility). (3) After the section the related muscles are paralyzed. (4) Irritation of the peripheral stump of the anterior root causes (in the first period after the operation) contraction of the muscles, eventually also a sensation of pain in consequence of the recurrent sensibility. (5) Irritation of the central stump is entirely without effect. (6) Sensation is completely preserved in the paralyzed parts of the body. (7) Severe pain occurs at the moment of division of a posterior root. (8) A reflex movement occurs at the same time. (9) After the section, all regions supplied by the divided root are anesthetic. (10) Irritation of the peripheral stump of the divided root is without any effect. (11) Irritation of the central stump causes pain and reflex movement. (12) Motility is entirely preserved in the anesthetic parts, for example the extremities.

According to Waller, the peripheral portion always undergoes degeneration after division of the anterior root. Division of the posterior root in advance of or behind the ganglion leaves unaltered the peripheral fibers that have retained their connection with the ganglion. Those that are severed degenerate. Therefore, according to Waller, the spinal cord is the nutritional center for the anterior roots, and the spinal ganglion, on the other hand, for the posterior.

After division of the posterior roots, for example of the nerves for the posterior extremities, the muscles retain their motility, but, nevertheless, characteristic disturbances can be recognized in them. These consist in an apparent awkwardness with which the animal executes the movements (jumping about in an uncertain manner, holding the legs far apart in walking, etc.), which detracts from the normal harmony and elegance—*centripetal ataxia.* Landois observed that dogs in which the posterior roots for the posterior extremities were divided on both sides exhibited, after complete recovery in other respects, difficulty in balancing the posterior part of the body, which often fell over in running or in wagging the tail. The phenomena are due to the fact that in consequence of the anesthesia of the muscles and the skin, the animal is unconscious of the resistances opposed to its movements. Therefore, the measure of muscular force to be employed cannot be properly estimated. All aids excited through reflex influences are also naturally excluded. Animals with abolition of sensibility in individual extremities often hold these in abnormal positions, from which the animal with preserved sensibility would at once remove them. Analogous ataxic disorders of movement have been observed also in human beings with degenerated peripheral extremities of the cutaneous nerves.

Under some circumstances division of the sensory nerves in certain regions may indeed be attended with abolition of movement. In whole-hoofed animals, immobility of the upper lip was observed after resection of the infraorbital nerve, immobility of the corresponding side of the larynx after division of the superior laryngeal nerve, also loss of motility of the esophagus after paralysis of its sensory nerve. The motility is thus, in large measure, dependent upon preservation of the sensory nerves (*sensomobility*).

Harless, Ludwig and Cyon have made the observation, which, however, has been disputed by v. Bezold, Uspensky, Grünhagen, and G. Heidenhain, that the anterior roots possess a greater degree of irritability so long as the posterior remain intact and irritable, that, however, they exhibit signs of lessened irritability as soon as the posterior roots are divided. In explanation of this phenomenon it must be assumed that in the intact body a series of slight irritations pass successively through the posterior roots (from contact, position, the influence of temperature upon the parts of the body, and the like), and are transmitted reflexly through the spinal cord to the motor roots, so that, as a result, a slighter additional irritation is required in order to excite the anterior roots than if this reflex impulse from the posterior roots for the increase of the irritability were removed. Obviously, the irritation required for the excitation of an already slightly irritated nerve-fiber need be less than for a similar fiber that is not irritated, as, in the first instance, the existing irritation is added to that which is in constant action.

The *anterior roots* of the spinal nerve supply with centrifugal fibers:

1. All striated muscles of the trunk and of the extremities under the control of the will. Every muscle receives its motor fibers from several anterior roots, and not from a single root; while every root distributes fibers to a related group of muscles.

The experiments made by Ferrier and Yeo on the anterior roots in apes have shown, accordingly, that irritation of each root (in the brachial and lumbosacral plexuses) induces a synergistic coördinated movement. Division of one root failed also to cause complete paralysis of the muscles taking part in the combined movement, but these had suffered only loss in strength. These experiments confirm pathological observations made on man. The fibers for functionally related groups of muscles, for example flexors and extensors, arise from special circumscribed regions of the spinal cord. Thus the cervical and lumbar swellings of the cord represent centers for highly coördinated muscular movements.

2. The anterior roots supply, also, motor fibers to a number of organs provided with unstriated muscle-fibers, such as the urinary bladder, the vasa deferentia, the uterus, the skin.

3. Motor fibers for the unstriated muscles of the vessels, the vaso-motors.

4. Inhibitory fibers for the contraction of the vascular muscles (known only in part): vasodilators.

5. Secretory fibers for the sweat.

6. Trophic fibers for the tissues.

The *posterior roots* contain the sensory nerves for the skin and the internal tissues, with the exception of the anterior part of the head, the face and the inner portions of the head. They contain also the tactile nerves for the cutaneous surfaces indicated. Irritations, exciting reflex action, are also conveyed to the spinal cord through the posterior roots.

Every sensory root gives fibers to different peripheral nerves. Every posterior root corresponds to a circumscribed area of the skin, although adjacent cutaneous areas overlap in part, so that probably every portion of skin is innervated from at least two roots. Thus, for example, the nipple is supplied with sensory fibers from the fourth and from the third and fifth sensory thoracic roots. The areas even extend somewhat beyond the middle line of the abdomen and the back and into one another. The innervational areas of the sensory nerve descend lower than those of the nerve-fibers arising from the corresponding anterior roots.

Fig. 247 and Fig. 248 illustrate the areas of distribution of the sensory nerves of the extremities, Fig. 244 those of the sensory spinal branches on the head. In cases of neuralgia and anesthesia the nerves involved can be readily determined by comparison with these illustrations.

There receive sensory nerves as follows: Heart and lungs from the vagus and the upper thoracic nerves; stomach, small intestine, liver, spleen and pancreas from the vagus and the middle inferior thoracic and upper lumbar nerves; adrenals, kidneys, testicles, ovaries, uterus from the middle and lower thoracic and upper lumbar nerves; rectum, prostate, penis, uterus, vagina from the sacral nerves and the hypogastric plexus (from the lower dorsal and upper lumbar cord).

In the hen it is a remarkable fact that a few motor fibers pass out through the posterior roots; also in some fish; with extreme rarity also in the frog; further in the dog and the cat vasodilators (for the hind leg); in the frog motor nerves for the unstriated muscles of the digestive tract and the urinary bladder.

SYMPATHETIC NERVOUS SYSTEM.

Connected with the cerebrospinal system, the sympathetic occupies a special position in consequence of the peculiarity of arrangement of its tracts, as well as on account of the presence of non-medullated, gray fibers and characteristically constructed ganglion-cells. The anterior branch of each spinal nerve gives off a visceral branch (formerly designated communicating branch of the sympathetic), which is derived either from the anterior or the posterior root of the spinal nerve, and this naturally (in the sense of Bell's law) indicates the function of the nerve. All of the visceral branches collect on each side of the vertebral column to form the sympathetic chain, in the course of which ganglionic nodes are interpolated. From the first dorsal nerve downward there is a ganglion at the point where each visceral branch enters the sympathetic. In the cervical portion a contraction and partial coalescence of the ganglia has occurred, and the eighth and the seventh and also the sixth and the fifth nerves are represented by single ganglia, and the four upper cervical nerves together by the superior cervical ganglion. Sympathetic filaments also pass through the path of individual visceral branches from the sympathetic into the cerebrospinal nervous system.

From the sympathetic system fibers pass to the various viscera of the head, the chest and the abdomen, where again they form ganglionic plexuses, from which finally fibers endowed with varied functions pass to the different organs.

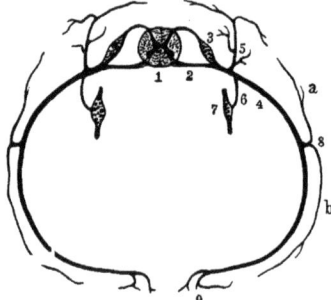

Visceral branches pass also from the cerebral nerves (although demonstrable with greater difficulty) and are connected with ganglia. The ciliary ganglion belongs to the third nerve as a part of the sympathetic system. The sphenopalatine nerves pass from the second division of the trigeminus as visceral branches into the sphenopalatine ganglion. The greater superficial petrosal nerve also is to be considered as a second visceral branch of this ganglion. The otic ganglion is to be looked upon as a sympathetic ganglion of the third division; likewise the submaxillary ganglion, the chorda tympani being the visceral branch. It appears that the glossopharyngeal, the vagus and the hypoglossal have their visceral branches in part in anastomotic filaments that they send to the superior cervical ganglion, which, therefore, gives off these cerebral nerves, together with the four upper cervical nerves, to the common ganglion.

FIG. 249.—Diagrammatic Representation of the Course of a Thoracic Branch of the Sympathetic: 1, spinal cord; 2, ventral root; 3, dorsal root with spinal ganglion; 4, intercostal nerve; 5, dorsal branch; 6, visceral branch; 7, ganglion of the sympathetic cord; 8, lateral cutaneous branch (pectoral and abdominal); a, posterior branch; b, anterior branch; 9, anterior cutaneous branch (pectoral and abdominal).

The sympathetic consists: (1) Of medullated fibers supplied to it as visceral branches by cerebral and spinal nerves, and (2) of fibers of Remak, which arise from sympathetic ganglia. The medullated fibers are (a) *sensory*, (b) *motor*, for vessels (vasomotors) and viscera, the latter entering into sympathetic ganglia, whence Remak's fibers, as well as medullated fibers, pass from the ganglion-cells to the innervated areas; (c) *inhibitory fibers* and *vasodilators*, in the course of which no sympathetic ganglia are intercalated. The fibers of Remak are all motor and they innervate directly or indirectly (that is entering again into ganglia) the unstriated musculature of the vessels, the viscera, the skin, and the muscles of the heart.

The conduction of the sympathetic nerve-fibers is in part direct and uninterrupted by means of sensory, inhibitory and vasodilator fibers. The medullated motor fibers from the visceral branches conduct indirectly, that is they pass at first in sympathetic ganglia, where they surround the cells, whose neurites then continue the conduction. The sympathetic contains further secretory fibers and fibers that control chemical processes, as in the thyroid gland and the adrenals. According to Langley all motor and sensory tracts derived from the spinal cord and situated in the sympathetic make their exit from the cord between the first dorsal and the second lumbar nerve.

Light is thrown upon the significance of the ganglia in the sympathetic system by poisoning with nicotin. In an animal thus poisoned the ganglion-cells are paralyzed, for irritation of the ganglia is without effect, as is also irritation of the nerves passing to the ganglion. On the other hand, irritation of the nerve-fibers that pass peripherally from the ganglion is still attended with results.

As to the *functions* of the sympathetic only a general summary will be given here.

I. *Independent functions* of the sympathetic are those of certain plexuses that persist after all the nervous connections with the cerebrospinal axis are severed. These include:

1. The automatic ganglia of the heart.

2. The myenteric plexus of the intestine.

3. The plexuses of the uterus, the oviducts, the vasa deferentia, and also of the blood-vessels and the lymphatics. The activity of these plexuses may be in part stimulated, in part inhibited, through centrifugal nerves from the cerebrospinal axis.

II. *Dependent Functions.*—The sympathetic contains also fibers that, like the peripheral nerves, functionate only in connection with the central nervous system, for example the sensory fibers in the splanchnic nerve. Other fibers convey to ganglia impulses received from the central nervous system, the ganglia in turn conducting the stimuli further on in the form of inhibition or motion to the respective organs.

A. CEREBRAL AND CERVICAL DIVISION OF THE SYMPATHETIC.

1. *Pupil-dilating Fibers.*—According to Budge, these arise from the spinal cord, and they pass, according to Langley, through the three or four uppermost dorsal nerves in the sympathetic cord and ascend to the head (in the cat). Division of the sympathetic cord or its communicating branches causes, therefore, contraction of the pupil. The central origin of these fibers is discussed on pp. 734 and 749.

2. The *motor fibers* for the unstriated muscles of H. Müller in the orbit and the lids and for the external rectus pass in part through the dorsal nerves from the first to the fifth (in the cat). According to Frl. Klumpke and Oppenheim the communicating branch of the first dorsal nerve in man is the path for 1 and 2.

3. *Vasomotor fibers* for the vessels of the external ear and the side of the face, the tympanic cavity, the conjunctiva, the iris, the choroid, the retina (only in part), the pharynx, the larynx, the thyroid gland, the brain and its membranes, derived from the thoracic nerves from the first to the fifth.

4. The cervical sympathetic contains centripetal fibers that stimulate the vasomotor center in the medulla oblongata.

5. *Secretory, trophic, and vasomotor fibers* for the salivary glands, appearing in the thoracic nerves between the first and the fifth.

6. The *sweat-fibers* are described on p. 537.

7. Also the lacrimal glands receive sympathetic *secretory fibers*.

B. THORACIC AND ABDOMINAL DIVISION OF THE SYMPATHETIC.

1. This division includes first the sympathetic portion of the *cardiac plexus*, which sends to the heart accelerator fibers from the inferior cervical and the superior thoracic ganglion arising (in the cat) from the upper cervical nerves between the first and the sixth.

2. The *vasomotors* for the extremities, the skin of the trunk, the lungs (in part from the vagus), passing through the sympathetic are described on p. 763, the *vasodilators* on p. 772.

3. The *pilomotor fibers* arise from the nerves between the fourth thoracic and the third lumbar. They pass to the sympathetic cord, where a ganglion-cell is

intercalated in each fiber. The sympathetic fibers have the same peripheral area of distribution as do the sensory fibers of the roots of the same spinal nerve.

4. The cervical sympathetic cord and the splanchnic nerves are believed to contain fibers irritation of which excites in a centripetal direction the *cardiac inhibitory system* in the medulla oblongata.

5. The function of the *splanchnic nerve* is described on pp. 288, 515 and 767. Its origin from the ganglia of the spinal cord extends from the sixth cervical to the fifth thoracic nerve. All, or almost all, of the filaments contain cells from the solar ganglion.

6. The significance of the *celiac* and *mesenteric plexuses* is discussed on pp. 328 and 359. After extirpation of the celiac ganglion Lamansky observed transitory derangement of digestion, in consequence of which undigested food was discharged from the anus.

7. *Sweat-fibers* are discussed on p. 536.

8. Finally, the abdominal division of the sympathetic contains *motor* and *vasomotor fibers* for the spleen, the large intestine (to which they pass with the arterial trunks), the bladder, the ureters (to which they pass in the hypogastric plexus), the vasa deferentia and the seminal vesicles. Irritation of any of these nerve-tracts causes increased movement of the organs in question, the diminished supply of blood also acting as an exciting factor. Section causes vascular dilatation, with secondary derangement of the circulation, and finally of the nutrition. The relations of the adrenal bodies to the sympathetic have been discussed on p. 107. The renal plexus is described on p. 515, and the cavernous plexus in connection with erection on p. 955.

From the lumbosacral portion of the spinal cord there issue as sympathetic filaments almost exclusively medullated fibers that pass in the sympathetic cord and thence partly to the inferior mesenteric ganglion and thence to the hypogastric and inferior mesenteric nerves, and partly through the sacral sympathetic ganglia to the sacral nerves to the skin.

The *lumbar branches* contain inhibitory fibers for the musculature of the descending colon and the rectum. Inhibitory and motor fibers pass to the internal sphincter ani. Irritation of the branches causes also contraction of the unstriated muscle-fibers of the skin surrounding the anus and pallor of the anal mucous membrane.

The *sacral branches* send motor fibers to the rectum and the colon, in addition inhibitory fibers to the internal sphincter ani, the adjacent musculature of the skin and vasodilators to the mucous membrane of the rectum and the external genitalia. The inferior mesenteric ganglion acting as a reflex center may transmit motor impulses to the bladder in response to irritation of sensory nerves of this viscus.

Pathological.—In accordance with the varied ramifications of the sympathetic it offers a wide field for pathological disturbances. It should be stated that affections of all of the fibers related to the vascular system are discussed elsewhere (p. 770).

The *cervical sympathetic* is most frequently paralyzed or irritated by direct traumatic influences. Gunshot-wounds or punctured wounds, tumors, enlarged lymphatic glands, aneurysms, inflammations of the apices of the lungs and the adjacent pleura, exostoses of the vertebral column may exert in part an irritant, in part a paralyzant effect. The resulting symptoms have been in part analyzed in the discussion of the ciliary ganglion (p. 684). *Irritation* of the cervical sympathetic causes in man dilatation of the pupil (spastic mydriasis), together with pallor of the face and occasionally hyperidrosis; disorders in near vision, the pupil being unable to contract, so that spherical aberration must have a disturbing effect; protrusion of the eyeball, with widening of the palpebral fissure. *Paralysis* causes an increased supply of blood to the affected side of the head, occasionally in association with anidrosis. The reddening may increase to a pathological degree. Later on, paralysis of the cervical sympathetic is attended with dilatation of the pupil (paralytic myosis), which in the act of accommodation undergoes change in diameter, but not on stimulation by light; it is slightly dilated by atropin. At the same time the palpebral fissure is narrowed, the eyeball retracted, the cornea somewhat flattened, and the tension of the eyeball diminished. Irritation of the sympathetic has been attended with increased secretion of saliva. Among the symptoms of irritation of the cervical sympathetic described, unilateral facial atrophy has been observed. Irritative phenomena in the distribution of the splanchnic nerve, especially as a result of lead-poisoning, are attended with severe pain (saturnine colic), inhibition of the movements of the intestine (and, therefore, obstinate constipation), reflex inhibition of the action of the heart (in the sense of the percussion-experiment of Goltz). Among the forms of irritation in the distribution of the sensory nerves of the sympathetic are the painful affection in the hypogastrium and sacral regions designated hypogastric neuralgia, hysteralgia, neuralgia of the testicle, which are localized in the respective plexuses of the sympathetic. In connection with affections of the abdominal sympathetic obstinate constipation is at times observed, and, in addition to irritation of the splanchnic, there may be deficient secretion on the part of the intestinal glands; at other times increased secretion from the intestinal mucous membrane. With respect to all of these subjects, however, there is as yet considerable obscurity.

COMPARATIVE. HISTORICAL.

Some of the cerebral nerves behave like the anterior, and others like the posterior roots of the spinal nerves. In selachians the nerve-branches arising as posterior roots supply upon the head the muscles of the visceral skeleton. In the vertebrates some of the cerebral nerves may be entirely wanting; others may be abortive or become branches of other nerves. Cetaceans possess no olfactory nerve. The facial nerve, which in man is the mimetic and the respiratory nerve of the face, grows smaller and smaller in the lower classes of vertebrates, in conjunction with reduction in the size of the facial muscles. In birds and reptiles it innervates the muscles attached to the hyoid bone or the superficial muscles of the neck and the nucha. In amphibia (frog), the facial is no longer present as a separate nerve. The branch equivalent to it is derived from the ganglion of the trigeminus. In fish the fifth and seventh nerves form a common complex. The portion corresponding to the facial (also designated opercular branch of the trigeminus) is especially the motor nerve of the muscles of the gill-cover and, therefore, proves itself to be a respiratory nerve. The cyclostomata (lamprey) possess an independent facial nerve. The vagus is present in all vertebrates. In fish and tadpoles the great lateral nerve of the abdomen is derived from it, passing in the middle line of the body along the lateral line. Its diminutive analogue in man is the auricular branch. In the frog, the ninth, tenth and eleventh, and also the seventh and eighth nerves arise from a common trunk. In fish and amphibia the hypoglossus is the first spinal nerve. In the amphioxus cerebral and spinal nerves are not to be distinguished from each other. In them also the posterior roots supply the muscles of the viscera. In other respects the spinal nerves in all classes of vertebrates exhibit marked uniformity. The sympathetic is wanting in cyclostomata, being replaced by the vagus. In the remaining fish its course is along the vertebral column, where it receives the communicating branches of the spinal nerves. In the region of the head its anastomoses with the fifth and tenth nerve, are especially conspicuous in fish. In frogs, and in still greater degree in birds, these anastomoses with the cerebral nerves are more extensive.

The vagus and the sympathetic were already known to the school of Hippocrates. Herophilus (307 B. C.) was the first to distinguish the nerves from the tendons, which Aristotle still confounded. He was aware of the decussation of the optic nerves. According to Erasistratus all nerves originate from the brain and the spinal cord. He distinguished motor and sensory nerves. Marinus (80 A. D.) was the first to describe seven pairs of cerebral nerves. Galen already possessed a comprehensive knowledge of the functions of the nerves. He, as well as Rufus of Ephesus (97 A. D.), was familiar with the embarrassed breathing following section of both vagi. He observed aphonia after ligation of the recurrent nerve. He was familiar with the accessory nerve and also with the ganglia connected with the abdominal nerves. He did not place the olfactory nerve in the same category as the other cerebral nerves. Achillini (died 1525) discovered the true olfactory filaments. Fallopius placed the glossopharyngeus in an independent position. The cauda equina is mentioned in the Talmud. Coiter (1573) described accurately the anterior and posterior roots of the spinal nerves. Van Helmont (died 1644) announced that the peripheral motor nerves are also sensitive to pain; and Cæsalpinus (1571) stated that interruption of the circulation in a part renders it insensitive. Thomas Willis described the portion of the accessory nerve derived from the spinal cord, as well as the principal ganglia (1664). The

46

first reference to reflex movements was made by Des Cartes (1670).   Stephen Hales and Robert Whytt showed that the spinal cord was necessary for their occurrence. Prochaska first demonstrated the reflex path.   Duverney (1761) discovered the ciliary ganglion; Varolius (1573) the chorda tympani.   Gall traced more accurately the third and the sixth nerves, as well as the spinal nerves, into the gray matter.   Up to this time only nine cerebral nerves were described.   Sömmering (1791) differentiated the facial and the auditory; Andersch (1797) the ninth, tenth and eleventh nerves.

PHYSIOLOGY OF THE NERVOUS CENTERS.

GENERAL CONSIDERATIONS.

The central nervous organs are in general characterized by the following properties:

1. They contain nerve-cells, which, arranged in groups, are situated either within the central organs of the nervous system or peripherally in the course of the nerves.

2. The nervous centers are capable of discharging reflexes, as, for example, reflex movements, reflex secretion, reflex inhibition.

3. The centers may be capable of automatic activity, that is, apparently without external stimulation, they may give rise to impulses that are conveyed to peripheral organs. This automatic stimulation may be either continuous, that is persisting without interruption (tonic automatism or tonus), or intermittent, pursuing a certain rhythm (rhythmic automatism).

[4] The central organs are the trophic centers for the nerves passing out from them. They may also act as centers for the nutrition of the tissues innervated by them. Psychic activity is dependent on an intact condition of the ganglionic central organs.

The foregoing functions are related to different centers, none of which is capable of representing several activities.

THE SPINAL CORD.

THE STRUCTURE OF THE SPINAL CORD.

The spinal cord (Fig. 250) contains within its structure the *gray matter*, which on section is H-shaped, and exhibits the anterior horns (*co.a*), the posterior horns (*co.p*) and the central connecting segment constituted of the anterior and the posterior gray commissures. In the middle of the last-named structure, from the calamus scriptorius downward, passes the central canal, which is the remains of the embryonal medullary tube and is lined by two or three layers of cylindrical epithelial cells.

The *white matter* surrounds the gray and it is divided into several *columns*. In the median line, anteriorly, a deep fissure (*s.a*) extends into the cord, but not quite to the gray matter, leaving at its bottom the *white commissure* (*c.a*) intact. The *anterior column* (*f.a*) is situated between the anterior longitudinal fissure and the groove for the exit of the anterior roots. The lateral portion of the white matter between the anterior and posterior roots is known as the *lateral column* (*f.l*). Finally, the area between the line of exit of the posterior roots and the posterior longitudinal fissure is designated the *posterior column* (*f.p*). The posterior longitudinal fissure (*s.p*) penetrates more deeply than the anterior into the cord, up to the gray matter.

The white substance consists of medullated nerve-fibers provided with horny sheaths, and arranged longitudinally in the columns. The incoming roots, as well as the longitudinal fibers entering the columns from the gray matter, have in part a transverse and in part an oblique course. In the anterior white commissure, fibers passing in a transverse direction decussate.

The gray matter exhibits in cross-section the anterior horns (*co.a*), from which, on each side, the anterior roots of the spinal nerves arise; also the posterior horns (*co.p*), with the incoming posterior roots (*r.p*); and in addition the smaller lateral horns (*co.l*). In the anterior horn the following groups of cells can be distinguished: (1) The large root-cells (*a*) situated anteriorly and laterally, from whose neurites the anterior roots arise directly. The dendrites of these cells pass into the anterior and lateral columns, and in part also into the anterior white commissure. (2) The commissural cells (*b*), principally anterior and median in situation, but in part also in the gray matter, which send their neurites through the anterior white commissure to the opposite side, where they divide into an ascending and a descending branch.

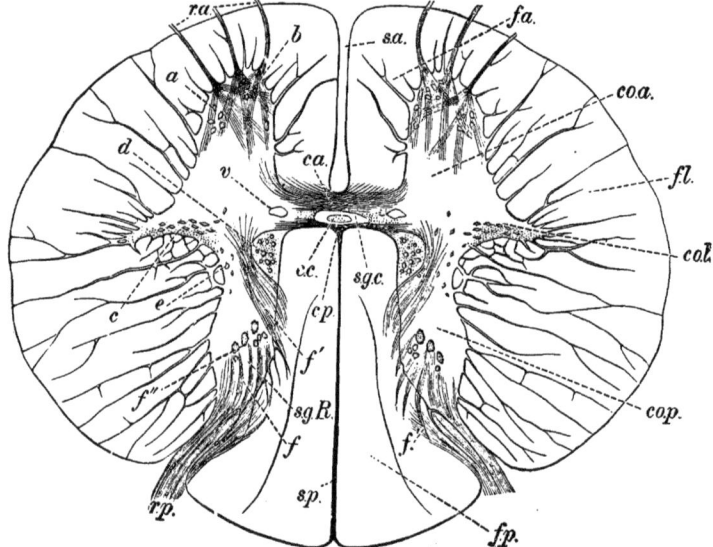

Fig. 250.—Transverse Section of the Spinal Cord at the Level of the Eighth Dorsal Nerve, × 10 (after Schwalbe). *s.a*, Anterior longitudinal fissure; *s.p*, posterior septum, occupying the posterior longitudinal fissure; *c.a*, anterior commissure; *s.g.c.* central gelatinous substance; *c.c*, central canal; *c.p*, posterior commissure; *v*, vein; *co.a*, anterior horn; *co.l*, lateral horn, and behind it the reticular process; *co.p*, posterior horn; *a*, antero-lateral and, *b*, anterior median group of ganglion-cells; *c*, cells of the lateral horn; *d*, cells of the columns of Stilling and Clarke; *e*, solitary cells of the posterior horn; *r.a*, anterior root; *r.p*, posterior root; *f*, its posterior-horn bundle; *f'*, posterior-column bundle; *f''*, longitudinal fibers of the posterior horn; *s.g.R*, gelatinous substance of Rolando; *f.a*, anterior column; *f.l*, lateral column; *f.p*, posterior column.

The *lateral horns* contain the so-called column-cells (*co.l*), that is small ganglia whose neurites form short connecting tracts between groups of cells at different levels of the cord. These connecting tracts are situated in the anterior, posterior and lateral white columns (Fig. 252, b, f, d), and they serve for the conduction of extensive coördinated reflexes. Internal to the origin of the posterior horns, adjacent to the posterior commissure, is situated a group of cells forming the column of Stilling and Clarke (*d*). This group of cells is plainly visible from the lower extremity of the cervical to the beginning of the lumbar enlargement, and its neurites pass partly in the direct cerebellar tract, and partly to the anterior white commissure.

In addition there are in the anterior portion of the gray matter ganglia with short neurites whose complex ramifications terminate in the immediate vicinity of the ganglia—further pluricordonal ganglia, whose neurites repeatedly send divided fibers into the white columns of the same side (or after passing through the commissure) of the opposite of the cord.

The *posterior horn* contains in addition to the gelatinous substance of Rolando (*s.g.R.*): (1) The exceedingly small, spindle-shaped ganglion-cells, whose neurites pass into the posterior column, and whose intricately branched dendrites penetrate the base of the posterior horn; (2) the rather superficial limiting cells situated near the apex of the posterior horn, whose neurites pass through the gelatinous substance into the lateral column; (3) star-shaped cells, whose dendrites in part enter the column of Burdach, in part the gelatinous substance.

The gray matter contains, in addition to the ganglion-cells, an exceedingly fine network of most delicate nerve-fibers. This is formed in part of intricately divided fine fibers, the dendrites of the ganglion-cells, but also of numerous fibrils given off by the longitudinal axis-cylinders present in all of the white columns of the spinal cord. These have been designated *collaterals* by Santiago Ramon y Cajal. The anterior columns send a rich network of collaterals to the anterior horns; the lateral columns to the region of the columns of Clarke and the central canal, and, in conjunction with the collaterals from all three columns, they form the posterior gray commissure. The fibers of the posterior columns send collaterals to the posterior horn, the anterior horn, and the columns of Clarke.

The motor fibers passing through the white columns of the spinal cord give off numerous collaterals to the gray matter throughout the entire length from above downward to the level at which the motor conducting tract reaches the motor cells of the anterior horn by contact (Fig. 251, m.c).

The neurites that form the *anterior roots* give off collaterals to the gray matter before they leave it.

If a posterior root-fiber be followed into the spinal cord, it will be found to divide into an ascending and a descending branch in the posterior column. From both of these branches, as well as from the root-fiber itself, collaterals are given off that enter the gray matter and terminate in arborescent ramifications (Fig. 251, s, c). The ex-

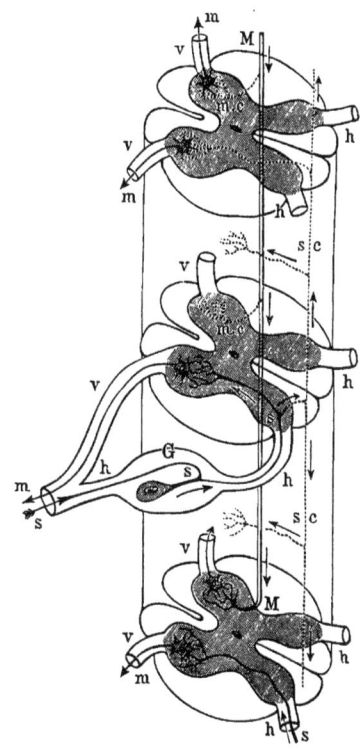

FIG. 251.

tremity of the descending branch forms a collateral in the gray matter of the cord, that of the ascending branch a collateral in the medulla oblongata.

The White Matter.—All of the longitudinal nerve-fibers composing the white matter of the columns of the spinal cord are arranged systematically, according to their function, into separate bundles.

Türck observed that after disease of certain portions of the brain the definite tracts of fibers in the spinal cord were secondarily degenerated. P. Schieferdecker confirmed this observation by animal experimentation. Finally, Flechsig demonstrated that the fiber-systems in the spinal cord receive their myelin-sheaths at different times in the process of development, and that those fibers whose course is the longest receive them latest.   In this way he established the following systems of longitudinal tracts (Fig. 252):

(1) The anterior column contains adjacent to the anterior median fissure (a) the direct pyramidal tract; externally to this (b) the anterior ground-bundle. (2) The posterior column contains (c) the column of Goll or slender column, and (d) the column of Burdach or wedge-shaped column.   (3) The lateral column contains (e) the antero-lateral tract of Gowers, (f) the lateral ground-bundle, (g) the crossed pyramidal tract, and (h) the direct cerebellar tract.

Of these, the pyramidal tracts, direct (a) and crossed (g), contain all the connections that pass from the central convolutions of the cerebral cortex as the path for voluntary motor impulses.   The direct cerebellar tract (k) connects in an ascending direction the superior vermis of the cerebellum on both sides through the restiform body with the columns of Stilling and Clarke.   As posterior roots of the same side enter the columns of Clarke, the direct cerebellar tract connects the cerebellum with the posterior roots of the trunk (not of the extremities). Gowers' tract (e) also terminates in the superior vermis almost entirely upon the same side (b, f), and a portion of d arises from the columnar cells of the gray matter and represents the short tracts connecting the reflex centers in the gray matter of the cord and in the medulla oblongata.   Sensory conduction-paths are present also in b, e, and f.   Finally, the columns of Goll (c) connect the posterior roots with the gray nuclei of the funiculi graciles of the medulla oblongata.   The column of Burdach (d) contains paths connecting the entering posterior roots with the nucleus funiculi cuneiformis and also tracts from the posterior roots through the restiform body to the vermis of the cerebellum.

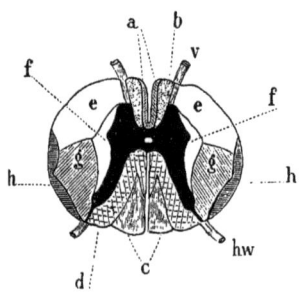

Fig. 252.—System of Conducting Tracts in the Spinal Cord, at the Level of the Third Dorsal Vertebra (after Flechsig): The black central portion of the figure is the gray matter; v, anterior root; hw, posterior root; a and g, pyramidal tracts; b, ground bundle of anterior column; c, column of Goll; d, column of Burdach; e and f, mixed tracts of the lateral column; h, direct cerebellar tract.

The direction of conduction in the posterior columns (the continuations of the posterior roots) is undoubtedly ascending, as they degenerate upward after destruction of the posterior roots.

The following additional points have been established with regard to these tracts: The pyramidal tracts (Fig. 253, 1 and 2), the direct cerebellar tracts (3), and the columns of Goll (5) exhibit progressive diminution in size in cross section from above downward.   They connect intracranial central parts with the groups of ganglia distributed throughout the gray matter of the spinal cord.   The columns of Burdach and the anterior ground-bundle, together with the bundle of Gowers and the ground-bundle of the lateral tract (6) show marked variations in the area of section at different levels of the spinal cord in proportion to the size of the incoming nerve-roots.   It can, therefore, be concluded that these tracts contain fibers that connect the gray matter at the different levels of the spinal cord and finally also in the medulla, without, however, penetrating into the higher parts of the brain itself.

The trophic center for the pyramidal tracts is situated in the cerebrum; that for the anterior roots of the spinal cord in the ganglia of the gray matter of the cord.   After division of the spinal cord the columns of Goll and the direct cerebellar tracts degenerate in an ascending direction.   The trophic center for the former is situated in the cells of the spinal ganglia of the posterior roots, that for the latter in those of the columns of Clarke.   Those fibers of the white substance, finally, that do not degenerate at all after section of the cord (and of which there are many in the anterior and lateral columns) are probably commissural fibers of the cord, which pass from ganglion to ganglion (columnar-cell fibers) and have their trophic centers in the ganglion at either extremity.

Flechsig makes the following statements with reference to the time of formation of the different systems: The first to form are the paths between the periphery and the central gray matter of the cord, especially, therefore, the nerve-roots.

Then there develop fibers that connect the different centers in the gray matter of the cord. Next there appear fibers that connect the gray matter of the cord with the cerebellum, and also the former with the tegmentum of the cerebral peduncle. Finally, there develop the fiber-systems that connect the ganglia of the pes of the cerebral peduncle and perhaps also the gray matter of the cerebral cortex with the gray matter of the spinal cord. The pyramidal tracts at the time of birth are still non-medullated. In cases of congenital absence of the cerebrum, neither the pyramids nor the pyramidal tracts develop. Even prior to birth myelinated fibers develop in the brain in the paracentral lobule, the central gyri, the occipital lobe, the island of Reil, and latest in the frontal lobes.

The *connective tissue* of the spinal cord is derived in part from the pia mater and penetrates with the vessels only into the white matter, to separate the nerve-fibers into separate bundles. From it the *neuroglia* must be distinguished — the true supporting tissue. This is not a connective tissue, being derived from the ectoderm. It consists of a homogeneous, structureless, semisolid ground-substance, together with spider-shaped, star-shaped, or tree-shaped glia-cells, intricately interwoven, and nucleated or non-nucleated fibers composed of keratin. The function of the neuroglia is to afford a supporting framework for the nerve-tissues, to protect them from pressure and to isolate them. In addition it forms channels for the fluids, or lymphatic passages, without endothelial lining, for the lymph so abundantly given off by the nervous elements, especially the ganglia, as a result of their activity, to eventually reach the perivascular spaces or the subpial space directly. This supporting tissue is much denser around the central spinal canal than the so-called central ependyma-fibers; further, it is more abundant at the apex and the margins of the posterior horns, where it is known as the gelatinous substance of Rolando. The neuroglia is present likewise in the cerebrum. The ganglion-cells are surrounded by cup-shaped lymph-spaces.

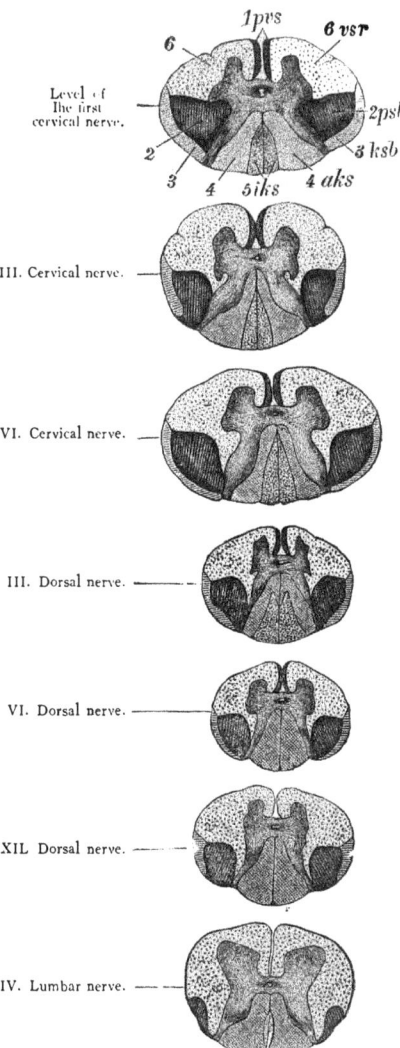

Level of
The first
cervical nerve.

III. Cervical nerve.

VI. Cervical nerve.

III. Dorsal nerve.

VI. Dorsal nerve.

XII. Dorsal nerve.

IV. Lumbar nerve.

Fig. 253.—Diagrammatic Representation of the Principal Tracts of the Spinal Cord: 1 *pvs*, Anterior pyramidal tracts (direct); (2) 2*psb*, lateral pyramidal tracts (crossed); (3) 3*ksb*, direct cerebellar tracts; (4) 4*aks*, external column of Burdach; (5) 5*iks*, internal column of Goll; (6) 6*vsr*, combined anterior ground-bundles, Gowers' column, and lateral ground-bundles.

THE SPINAL REFLEXES.

By reflex movements are understood such as are induced by irritation of a centripetal (sensory) nerve. The latter takes up the irritation, and conveys it to the spinal cord, the cellular gray matter of which acts as the reflex center. Here the impulse is transferred to the motor centrifugal path. Three factors are thus concerned in a reflex movement and together constitute the so-called *reflex arc*: the centripetal fiber, the transferring center in the gray matter, and the centrifugal fiber (Fig. 251, s h v m). The activity of the will is excluded in the occurrence of a reflex movement.

Three varieties of reflex movement are distinguished: 1. The *simple or partial reflex*, which is characterized by contraction of a single muscle or at most of a small group of muscles as a result of stimulation of a sensory nerve. Examples of this type of reflex movement are the contraction of the quadriceps femoris muscle following a tap on the knee; and the closure of the lids as a result of irritation of the cerebral nerves on the eye. 2. *The widespread incoördinated reflex*, or reflex spasm. This occurs in the form of tonic or clonic contractions involving entire groups of muscles, or even all of the muscles of the body. The reflex spasm is due to a double cause: (a) the gray matter of the spinal cord may be in a condition of excessive irritability, so that the conveyed stimulus can be readily transferred from its point of entrance to the readily irritated adjacent central areas. Such excessive irritability is caused by certain poisons, particularly strychnin, and also brucin, caffein, atropin, nicotin, carbolic acid, etc. The slightest touch of an individual poisoned by strychnin is sufficient at once to throw all of the muscles of the body into spasm. Reducing the temperature of the body to 23° C. likewise gives rise in the dog to marked reflex irritability. Also certain pathological and morbid conditions may bring about a similar result. An illustration is the excessive irritability in cases of hydrophobia and tetanus. Conversely, the central organs may be placed in a condition in which extensive reflex convulsions cannot occur. Thus, in the state of apnea the convulsions usually attending strychnin-poisoning do not occur, in consequence of the passive artificial respiratory movements, which cause stretching of the cutaneous nerves of the abdomen and chest. Also the practice of other periodic passive movements of portions of the body gives rise to a similar condition; and considerable reduction in the temperature of the spine inhibits reflex convulsions. (b) Extensive reflex convulsions may, however, occur when the reflex stimulation is severe. Examples of this kind are observed in man, as in the widespread convulsions attending intense neuralgias.

The general convulsion is extensor in type (involving the spinal column: opisthotonus), because the strength of the extensors is greater than that of the flexors. Nerves arising from the medulla oblongata may be excited reflexly also by stimulation of remotely situated central nerves, without the occurrence of general convulsions.

Strychnin, the most powerful of the poisons exciting reflex convulsions, acts directly upon the ganglia of the gray matter of the spinal cord. Therefore, the same reflex convulsions occur when the poison (in the frog, after ligation of the heart) is applied directly to the exposed spinal cord. The spasms occur after mechanical, thermal, or electrical stimulation, but not after chemical stimulation. During the spasm the heart stops in diastole from irritation of the vagus, and the pressure in the arteries undergoes a marked rise as a result of irritation of the vasomotor centers in the medulla and spinal cord. Mammals may die from asphyxia during the attack; although after large doses death results from spinal paralysis when the convulsions subside early. Fowl are as a rule immune to fairly large doses.

Elicitation of the Reflexes. Feeble stimuli that, applied once, are incapable of exciting a reflex may do so after repeated application. Under such circumstances a summation of the individual stimuli takes place in the spinal cord.

In order to obtain such a result three feeble stimuli in a second suffice; the best results are obtained from sixteen in a second, and beyond this no increase in the intensity of the effects is possible. Nevertheless, stimuli (induction-shocks) within other limits, namely an interval of from 0.05 to 0.04 second, have been found effective. W. Stirling has shown that the reflexes are probably due to repetition of the impulses sent to the nervous centers.

Diffusion of Reflexes. Pflüger has established the law according to which the diffusion of reflexes takes place: (1) The reflex movement takes place first on the same side as that on which the sensory nerve is stimulated, and only those muscles are thrown into action whose nerves arise from the same level of the cord. (2) If the reflex extends to the other side, it always occurs, as an associated movement only in the muscles that are already contracted on the primary side. (3) If the intensity of the spasm is different on the two sides, the movements are stronger on the primary side. (4) On diffusion of the reflex irritation to adjacent motor nerves those are involved always that are situated in the direction toward the medulla oblongata. Sherrington, however, has observed also diffusion of reflexes in a caudal direction. (5) Finally, all of the muscles become involved in the spasm.

In exceptional cases, however, deviations from these rules occur. If, for example, the region of the eye in a frog, after extirpation of the cerebrum, be stroked, a reflex in the hind leg of the opposite side often occurs. Tickling the foreleg of decerebrated tritons, lizards, turtles, and deeply narcotized dogs and cats, often causes a movement of the hind leg on the opposite side. These manifestations have been called crossed reflexes. If in animals a section be made throughout the length of the spinal cord in the median line the reflexes will naturally remain unilateral.

Every sensory root has, in its individual spinal segment, a motor reflex path, which offers the least resistance to the discharge (simple reflex). There are, however, also reflex paths into adjacent and remote segments; there is a functional relation between certain motor-cell groups and certain muscle-groups that act synergistically. The long association-paths in the spinal cord are primarily undecussated; crossed conduction appears, however, to exist between segments not widely separated. The crossed reflex path can be traversed with varying ease at different levels in the cord. The reflex readily passes in a crossed direction from the anterior to the posterior extremity; on the other hand, from the posterior extremity more readily to that of the opposite side.

The reflex can be conveyed within various levels of the cord. On applying feeble stimuli to the leg of a decerebrated frog, the reflex transference takes place at the junction of the cervical cord and the medulla; on applying stronger stimuli transference takes place at the lower portion of the spinal cord, which can be stimulated reflexly with greater difficulty. If alternating hemisections of the cord be made, the reflex irritation may nevertheless be propagated upward, passing through both sides of the cord in a serpentine manner. The greater the number of sections, the stronger must be the irritation of the sensory nerves.

3. *The widespread coördinated reflex* is characterized by the occurrence in entire and even different groups of muscles of complex movements having a purposive character or resembling voluntary movements and following irritation of a sensory nerve.

The observations are made either on cold-blooded animals (such as decapitated frogs, lizards, or eels) or on mammals, the four arteries passing to the brain being ligated (artificial respiration being maintained), so that the brain is rendered incapable of functionating. Reflexes involving the lower portion of the spinal cord may be studied also in animals (or man) after transverse section of the spinal cord in the upper dorsal region, but some time must have elapsed after the section so that the primary irritation of the lesion (so-called shock), which at first has a reflex-inhibiting effect, may subside.   Young mammals exhibit reflex activity for some time even after decapitation.

The coördinated reflexes include:

1. Protective movements and movements of escape are observed in decerebrated or decapitated frogs and turtles, as well as the removal of acids when applied to the skin, resistance to fixation-instruments, etc.   All of these movements are executed apparently with deliberation and with the employment of the most serviceable groups of muscle, so that Pflüger was led to attribute them to a spinal consciousness.   Even excised portions of eel turn away from an intense irritant such as a flame.   Also the tail of a decapitated triton, lizard, salamander, eel, or adder submits to gentle stroking, but turns away from intense irritation.

2. Goltz's croaking experiment, which consists in the croaking of a decerebrated frog when the skin of the back is stroked.

3. Goltz's embracing experiment: The portion of the trunk of a young male frog between the skull and the fourth vertebra, particularly during the breeding season, embraces every solid body that comes in contact with the skin of the chest and exerts a slightly stimulating effect.

In the intact animal the stimulating irritant consists in the degree of fulness of the male seminal organs.   The reflex immediately ceases after slight irritation of the optic thalamus.

4. In warm-blooded animals (dogs) the following are among the coördinated reflexes related to the posterior divided extremity of the cord: Scratching of tickled portions of the skin with the hind-paw, as in normal animals; the movements necessary for the evacuation of the bladder and the rectum, for erection and the act of parturition, the coördinated movements of the feet and the tail in decapitated ducks and pigeons.   Coördinated reflexes simultaneously in widely separated segments of the cord appear as a rule no longer to occur after removal of the medulla oblongata.   For this reason it is believed that the medulla may contain a complex organ of higher order connecting the different reflex areas in the spinal cord (by white fibers).

5. In man coördinated reflexes occur also during sleep, as well as in comatose states.

By far the majority of the movements executed unconsciously during the waking state, or when the mental activities are otherwise intently engaged, must be included among the coördinated reflexes.   Many complicated movements must first be learned before they can again unconsciously be executed harmoniously as coördinated reflexes, such as dancing, skating, and riding.   Coughing, sneezing, and vomiting are among the coördinated reflexes emanating from the spinal cord and the medulla oblongata.

With reference to the peculiarities of the reflexes the following points are noteworthy:

1. The reflexes can be elicited more readily and in more complete degree when the stimulus is applied to the specific end-organ of the centripetal nerve, rather than to the trunk of the nerve itself .

2. For the production of a reflex movement a stronger stimulus is required than for direct stimulation of the motor nerve.   The reflex movement induced by a stimulus of adequate intensity appears at once as a moderately strong contraction, which does not increase in intensity with increase in the intensity of the stimulus.

3. A reflex movement is of shorter duration than the same movement executed voluntarily.   Further, its occurrence after the moment of irritation is distinctly delayed, the interval until the appearance of the muscular contraction (in the frog) being twelve times as long as that required for conduction through the sensory and motor nerves.   The spinal cord thus interposes resistance to the rapid passage of the impulse.

The *reflex time* (that is the time required for the transference of the impulse within the ganglion-cells of the spinal cord) is in the case of closure of the eyelids in man 0.042 second, in the frog on an average in various muscles from 0.008

to 0.015 second. This time is increased by about a third if the impulse passes to the opposite side or through the length of the cord (from the sensory root of the anterior extremity to the motor root of the posterior extremity). Heat diminishes the reflex time and increases the reflex activity. Lowering of the temperature (winter-frogs), likewise the poisons previously mentioned that increase reflex activity, increase the reflex time, while at the same time increasing reflex irritability. Conversely, the reflex time diminishes with increase in the intensity of the stimulus and it may thus even be of minimal duration. It would appear as if the reflex resulting from the action of a strong stimulus traverses a shorter path (spinal cord in the frog) than that resulting from the action of a feeble stimulus, in which case the impulse must ascend to the portion of the cord below the calamus scriptorius, where the transfer takes place. There are, thus, two reflex paths, one for strong stimuli, the other for feeble stimuli, the latter being generally the normal.

The reflex time can be measured by noting the time of irritation of the sensory fiber and that of contraction. From the result thus obtained there must be deducted the time required for conduction through the two nerve-tracts, as well as the duration of the latent stimulation.

In accordance with their location Jendrássik divides the reflexes as follows:

I. *Spinal reflexes* (tendinous, muscular, periosteal, bony, articular, genital-muscle reflexes. The pathological spinal reflex occurs only in case of total or almost total transverse section of the spinal cord and is manifested as flexor, less commonly as extensor, movement of the lower extremities.

II. *Cerebral cortical reflexes*, elicitable especially by tickling the skin: scapular, abdominal, cremasteric, scrotal, gluteal, plantar, auricular, palpebral, palatal, conjunctival, anal reflexes.

III. *Complex reflexes*, for which a spinal and a cerebral reflex center are necessary: sneezing, vomiting, swallowing, coughing, evacuation of bladder and rectum, ejaculation, all of which belong to the vegetative functions.

INHIBITION OF REFLEXES.

There exist in the body mechanisms, by means of which the production of reflexes can be suppressed, and which accordingly have been designated reflex-inhibiting mechanisms:

1. Through the action of the will reflexes both in the cerebral and in the spinal distribution can be voluntarily inhibited; for example, keeping the eyes open when the bulb is touched; suppression of movement on tickling the skin. It should, however, be observed in this connection that the inhibition of the reflexes is possible only to a certain point. If the stimulus be strong and frequently repeated the reflex action finally predominates over the volitional inhibitory impulses. Moreover, such reflex movements as cannot under any circumstances be executed voluntarily, cannot be inhibited. Thus, erection, ejaculation, the act of parturition, and the movements of the iris can neither be executed voluntarily, nor if excited reflexly can they be inhibited by the will.

2. *Setschenow's inhibitory center* is the designation given to another cerebral apparatus, located in the frog on both sides in the optic thalamus and the quadrigeminate bodies. Separation of these parts by section increases reflex irritability, while irritation of the lower cut surface (by sodium chlorid or blood) suppresses reflex movements. This result can be observed also on one side. It is believed that analogous organs exist in the higher vertebrates in the quadrigeminate bodies and in the medulla oblongata. From what has been said, it is clear that reflexes occur more regularly and are more readily elicited after exclusion of the brain.

3. *Strong irritation of a sensory nerve* suppresses reflex movement. The reflex does not appear even when the centripetal nerve involved is strongly irritated; for example, inhibition of sneezing by friction of the

nose; inhibition of the movements usually elicited by tickling, by biting the tongue. Especially strong irritation may even suppress the reflexes coördinated for voluntary movement. Intense pain in the abdominal organs (intestine, uterus, kidneys, liver, bladder) renders impossible the act of walking or of standing. In the same category belongs the falling down which follows injury to viscera richly supplied with nerves, and which, either by involvement of motor nerves or by loss of blood, would of itself be insufficient to account for the inability to maintain the erect posture. Stimulation of the central organs through other centripetal paths (such as the organs of special sense, the sexual nerves, etc.) diminishes the reflexes in other paths.

4. During the discharge of an energetic reflex, such as ejaculation, the production of less active reflexes, such as coughing, is suspended.

5. Attention should also be called to the fact that inhibition of reflexes, whether through the will, or through the irritation of sensory nerves, therefore by reflex action, is often attended with the excitation of antagonistic movements. In some cases, further, it appears to be sufficient to inhibit a reflex to concentrate the attention on the execution of such a complicated reflex movement in order to prevent it. Some persons, for example, are unable to sneeze if they think intently of the motor processes concerned; as the will, in a measure prematurely, begins to control the reflex center by the thought, the normal course of the reflex excitation for the stimulus from the periphery is interfered with.

6. Certain poisons, such as chloroform, picrotoxin, morphin, quinin, potassium bromid, and others, diminish reflex irritability, probably after a transitory increase. A constant current passed longitudinally through the spinal cord enfeebles the reflexes, particularly a descending current. If a frog be paralyzed by asphyxia in air free from oxygen, the brain and the spinal cord are wholly unirritable and are, therefore, incapable of reflex excitation. The motor nerves and the muscles, however, suffer little impairment of irritability, even after the lapse of several days.

According to the method of Türck the degree of reflex irritability in decapitated frogs is tested by estimating the time that elapses between immersion of the foot in dilute sulphuric acid and withdrawal of the part. After application of blood to the optic lobes or after irritation of a sensory nerve, the time is increased.

Setschenow differentiates the reflexes into tactile, or those that are elicited by irritation of tactile nerves, and pathic, or those resulting from irritation of sensory (pain-transmitting) fibers. He believes, with Paschutin, that the tactile reflexes are inhibited by the will, and the pathic by the center described by him.

Theory of Reflexes.—The following theory has been proposed to explain the phenomena observed in connection with reflex movements. It is believed that the centripetal fiber, in the transmission of the impulse conveyed through it, encounters considerable resistance within the gray matter, with the ganglion-cells of which it is contact on all sides through the fibrous network of the gray matter. The least resistance is in the direction of those motor nerves that make their exit at the same spinal level on the same side. In this way the weakest irritation gives rise to the simple reflex, which in general can be recognized as a simple protective or defensive movement with respect to the seat of sensory stimulation. In the direction of other motor ganglion-cells the conduction of the impulse encounters still greater resistance. In order for the reflex to be transmitted also along these paths, either the stimulus must be considerably increased (for with increasing strength and duration of the stimulation the reflex movement may increase in extent), or the resistance in the course of the conduction between the cells of the gray matter must be diminished. The latter may be brought about by the action of the poisons mentioned, as well as under the influence of general increase in nervous irritability, such as is observed in cases of hysteria and neurasthenia. Thus, extensive reflex convulsions may occur as a result of increase in the intensity of the stimulus or of a diminution in the conduction-resistance in the spinal cord. Of those measures that have been shown by experience to diminish or prevent reflex manifestations, the conclusion is justified that they interpose increased resistance in the conducting paths of the reflex arc. The action of influences inhibiting reflexes must be interpreted in a similar manner. As, obviously, the fibers of the reflex arc must be connected with the reflex-inhibiting conducting tracts, it is believed that, through the reflex-inhibiting stimulation, resistance is at the same time interposed in the reflex arc. Difficulty is encountered in explaining the widespread, coördinated reflexes according to the view discussed. It has been assumed that from long use and also through heredity those groups of ganglion-cells that first receive the impulses are placed in communication, under conditions most favorable to conduction, with others that transmit the impulse to thôse groups of muscles whose activity best removes the body or the respective member from possible injurious effects of the stimulus by a coördinated, purposive movement. Thus, a stimulus always excites a group of ganglion-cells coördinated by practice and responding to the stimulus as an harmonious, coördinated motor mechanism.

Pathological.—Abnormalities in the reflex activity afford the physician a large and important field in the investigation of diseases of the nervous system. Enfeeblement or even abolition of reflex activity may occur (1) as a result of impairment or loss of irritability in the centripetal fibers; (2) as a result of analogous disorders in the central organs; (3) or, finally, in the centrifugal fibers. The reflexes are enfeebled or abolished when the entire nervous system is greatly depressed, as after concussion, compression, inflammation of the central organs, during asphyxia and deep coma, and in consequence of various intoxications. After division of the upper portion of the spinal cord in man, and in apes, abolition of the reflexes is often observed in the parts below the level of division. It appears that the upper portions of the cord contain the region where normally the reflexes are readily transferred, and after division of which the reflexes are in abeyance. Dogs and cats, like the frog, exhibit active reflexes after division of the spinal cord.

Under abnormal conditions special attention has been given to the behavior of certain reflexes, for example the so-called tendon-reflexes, which consist in reflex contraction of a muscle when a blow is struck on its tendon, for example the quadriceps extensor of the thigh, the tendo Achillis, etc. Thus, Westphal, Erb, and others have found that the tendon-reflexes, particularly the patellar-tendon reflex, also known as knee-phenomenon or knee-jerk, is almost constantly wanting in cases of ataxic tabes dorsalis, but is abnormally increased and extensive in cases of spastic spinal paralysis, which is characterized by a lesion of the pyramidal tracts. Division of the muscle-nerves abolishes the patellar phenomenon in rabbits, as does also division of the spinal cord between the fifth and sixth lumbar vertebræ. In Landois the contraction of the quadriceps occurred 0.040 second after the blow upon the patellar ligament; according to Jendrássik 0.039 second after the blow. A stronger blow has no effect upon the reflex time. According to Westphal these phenomena are not simple reflex processes, but complicated phenomena related to muscle-tone, so that, for example, diminution in the tone of the quadriceps femoris may abolish the phenomenon. The intact existence of the external segments of the posterior columns of the spinal cord is necessary for the preservation of the phenomenon. It is enfeebled by physical or mental fatigue and increased by transitory stimuli that involve the attention. Jendrássik found it especially marked when muscles of the body were contracted voluntarily, for example the muscles of the arm. It is enfeebled by strong and prolonged contraction and extreme tension.

Another reflex of diagnostic importance is the abdominal reflex, which consists in contraction of the abdominal muscles on stroking the skin of the abdomen with the handle of a percussion-hammer. Thus absence of this reflex on both sides is indicative of diffuse disease of the brain in the presence of a cerebral disorder. Absence on one side is indicative of a local disorder in the opposite half of the cerebrum. The hypochondriac, anal, cremásteric, conjunctival, mammillary, pupillary, nasal reflexes and others may also be made objects of investigation. Cerebral lesions attended with hemiplegia always exhibit diminution of the reflexes upon the paralyzed side, although not rarely the patellar reflex is increased.   In case of extensive cerebral disease, the reflexes are wanting on both sides if coma is present at the same time, naturally also those of the anus and the bladder.

In going to sleep there is transitory increase of the reflexes.   In early sleep the reflexes are enfeebled, the pupils small.   During sound sleep the abdominal, cremasteric, and patellar reflexes are wanting; tickling of the soles of the feet and of the nose is effective only when of a certain degree of intensity.   During narcosis, as, for example, that induced by chloroform or morphin, the abdominal reflex disappears first, then the conjunctival and the patellar reflex, and finally the pupils become contracted.

Abnormal increase in reflex activity is generally indicative of an increase in the irritability of the reflex center.   Abnormal irritability of the centripetal nerves may also be the cause, while injury is a cause of inhibition.   As the harmonious execution of voluntary movements is largely controlled and regulated by reflex activities, it will be readily understood that various derangements in those movements are observed in the presence of disease of the spinal cord, as for example the characteristic disorder of gait and of the movements of the hands in cases of tabes dorsalis.

CENTERS IN THE SPINAL CORD.

The spinal cord contains, in various situations, centers that on reflex stimulation permit the evolution of certain coördinated motor mechanisms.   These centers are capable of preserving their activity even when the spinal cord is separated from the medulla oblongata.   Further, the centers situated in the lower portion of the cord may remain active after division of the upper portion, but in the normal body these spinal centers are subordinate in their activity to other higher reflex centers in the medulla oblongata.   The centers may, therefore, be designated also subordinate spinal centers.   Further, the cerebrum may, partly through the formation of conceptions, partly as the organ of the will, exert an influence upon certain of the subordinate spinal centers by excitation or inhibition of the reflexes.   The following particulars are deserving of mention :

The *center for dilatation of the pupil* is situated in the lower cervical portion, extending downward to the level of the first, second, and third thoracic vertebræ—Budge's ciliospinal center.   It is stimulated by darkness.   In man both pupils react simultaneously if the retina on one side is darkened.   Extirpation of this portion of the spinal cord on one side is followed by contraction of the pupil on the same side.   The motor fibers, which have their trophic center in the same situation, pass through the anterior roots of the upper three thoracic nerves, in the cat, into the cervical sympathetic.

In goats and cats this center, separated from the medulla oblongata, may be stimulated directly by a state of the blood causing dyspnea, and likewise by reflex stimulation of sensory nerves, for example the median, especially if the irritability of the spinal cord has been increased by strychnin or atropin.   After total division of the upper portion of the cervical cord, subsequent section of the sympathetic is followed by contraction of the pupils.   The superior dilator center situated in the medulla oblongata is described on p. 749.

The *center for defecation*—Budge's anospinal center.   The centripetal nerves are contained in the hemorrhoidal and inferior mesenteric plexuses.   The center is situated at the level of the fifth (in the dog) or sixth and seventh (in the rabbit) lumbar vertebræ.   The centrifugal fibers are derived from the pudendal plexus and pass to the sphincter muscle. The excitation of this center and its domination by the cerebrum are discussed on p. 285.

After division of the spinal cord Goltz observed that the anal sphincter contracted rhythmically about a finger introduced into the rectum. The coördinated activity of the center is, therefore, possible only through connection with the cerebrum. After extirpation of the lumbar cord, the anal sphincter at first loses its tone, although this is partially restored later. The muscle does not degenerate: perhaps its trophic center is situated in the mesenteric ganglion.

The *center for micturition*. Budge's vesicospinal center, for the sphincter muscle is situated at the level of the fifth (dog) or the seventh (rabbit) lumbar vertebra, for the muscular wall of the bladder at a somewhat higher level. It functionates in a coördinated manner only when connected with the cerebrum (see p. 522).

The *center for erection* is situated in the lumbar portion of the cord. The centripetal fibers are the sensory nerves of the penis. The centrifugal fibers are, for the deep artery of the penis, the vasodilator nerves from the first, second and third sacral nerves (Eckhard's erector nerves), for the ischiocavernosus and the deep transverse perineal muscle the motor fibers from the third and fourth sacral nerves. The latter may be stimulated also voluntarily, the former also in part from the cerebrum by directing attention to sexual activity. Eckhard observed erection also after stimulation of higher portions of the spinal cord (Landois likewise in man), as well as of the pons and the cerebral peduncles.

In accordance with clinical observations the centers for the bladder and the rectum and for erection are situated at the point of exit of the first, second, third and fourth sacral nerves.

The *center for ejaculation*. The sensory (dorsal nerve of the penis) are the exciting nerves. The center (Budge's genitospinal center) is situated at the level of the fourth lumbar vertebra, in rabbits. The motor fibers of the vasa deferentia are derived from the fourth and fifth lumbar nerves, which enter the sympathetic and finally pass thence to the vasa deferentia. The motor fibers for the bulbocavernous muscle, the ejaculator of the seminal fluid from the bulb of the urethra, are contained in the third and fourth sacral nerves (perineal nerves). Ejaculation may be induced by mechanical stimulation of the lumbar cord, in guinea-pigs.

The *center for the act of parturition* is situated at the level of the first and second lumbar vertebræ. The centripetal fibers are derived from the uterine plexus, into which also the motor fibers from the spinal cord again enter. Goltz and Freusberg observed impregnation and delivery in a bitch with the spinal cord divided at the level of the first lumbar vertebra. Similar results have been observed in women with the spinal cord divided. Normal birth occurred with subsequent involution of the uterus and secretion of milk.

Centers for the vascular nerves, both vasomotor and vasodilator, are distributed throughout the entire spinal axis. Among these is to be included also the center for the spleen, which is situated between the first and fourth cervical vertebræ, in the dog. They can be excited reflexly, although they are subordinated to the dominating centers in the medulla oblongata. They may be influenced also by psychic stimulation, from the cerebrum.

Centers for the secretion of sweat have perhaps a distribution analogous to that of the centers for the vascular nerves.

The movements excited from the centers named are, in accordance with what has been stated, to be designated coördinated reflexes and fundamentally to be included with the coördinated reflexes of the musculature of the trunk and the extremities.

Muscular Tone.—Automatic functions also were formerly credited to the spinal cord, one of which was a certain moderate degree of active tension of the muscles that is designated tone. It was thought that the tone of the transversely striated fibers was demonstrated by the retraction of the extremities of a divided muscle, but this is due simply to the circumstance that all of the muscles are somewhat stretched beyond their normal length, and for this reason also paralyzed muscles, which must have lost their nervous tone, exhibit exactly the same phenomenon. Also the increased contraction of certain muscles after paralysis of their antagonists, and the distortion of the face toward the healthy side after unilateral facial paralysis, have been cited as examples of tone. These, however, are due to the fact that after activity of the intact muscle strength is wanting to restore the parts affected to the normal median position of rest. The following experiment of Auerbach and Heidenhain is opposed to the assumption of a tonic contraction. If the muscles of a decapitated frog's leg be made tense, they do not elongate after division of the sciatic nerve, or after paralysis of [this nerve by application of ammonia or carbolic acid. If, however, a decapitated frog is suspended in an abnormal position, it will be observed that if the sciatic nerve on one side or the posterior roots of the nerves of this extremity have been divided, then the member upon this side hangs in a relaxed manner, while the member upon the intact side is slightly retracted. The sensory nerves of the latter are by the weight of the member thrown permanently into a state of gentle stimulation, so that by this means a slight reflex retraction upward of the member is brought about, which fails to occur as soon as the sensory nerve-fibers of the member are paralyzed. If the slight retraction mentioned is to be considered as tone, then it is to be characterized as reflex tone. With this experiment that of Harless, C. Ludwig and Cyon should be compared (p. 716).

IRRITABILITY OF THE SPINAL CORD.

At the present time there is no unanimity of opinion as to whether the spinal cord, like a peripheral nerve, is irritable, or whether it is characterized by the remarkable peculiarity that most of its conducting paths and ganglia are without reaction to direct electrical and mechanical stimuli.

An outline of the views of the opposing investigators is as follows: If stimuli are carefully applied to the exposed white or gray matter, neither movement nor sensory perception results. In making this observation, however, the greatest care must be taken to avoid irritation of the roots of the spinal nerves, as these naturally react to stimuli and thus excite sensations, as well as reflex movements, on the one hand, and also directly excited movements on the other hand. As the spinal cord thus conveys to the brain the impulses brought to it through the stimulated posterior roots, but is incapable of reacting even to stimuli exciting sensory impressions Schiff has designated it as esthesodic, that is transmitting perceptions. Moreover, as the cord is capable, in like manner, of conducting both voluntary and reflex motor impulses, without, however, being itself receptive for motor impulses applied directly, it is kinesodic, that is movement-conducting.

According to Schiff, therefore, all of the results that follow stimulation of the uninjured spinal cord, such as spasm and contracture, are caused by simultaneous stimulation of anterior roots, or they are reflexes from the posterior columns alone or from the posterior columns and the posterior roots at the same time. Diseases involving only the anterior and lateral columns never cause irritative, but only paralytic symptoms. In the state of complete anesthesia and in that of apnea, all stimulation is without effect. According to Schiff's view all of the centers, both spinal and cerebral, cannot be stimulated by artificial means. The situation of a center can for this reason be determined only by the paralytic method.

Schiff concludes, therefore, that in the posterior columns the sensory root-fibers produce pain on stimulation, but not the actual paths for the posterior columns themselves. Schiff, however, observed as a sign that stimulation of the actual paths caused tactile sensations, dilatation of the pupils with each stimulation. Removal of the posterior columns causes anesthesia, loss of touch. Algesia, pain-sensation, is preserved, and at first there is even hyperalgesia.

The anterior columns cannot be stimulated so as to affect either transversely striated or unstriated muscles, if only the actual tracts are stimulated. Movements may occur, however, either if the motor root-fibers are stimulated or if the current reaches the posterior columns, in which it stimulates the sensory root-fibers and thus causes reflex movements.

Numerous investigators are opposed to these views and express themselves in favor of the possibility of direct stimulation of the spinal cord. Fick maintains that he is able to induce movements of the hind legs by irritating directly the anterior columns isolated for a considerable distance in order to eliminate diffusion of the current. Biedermann reaches the conclusion that the motor nerve is most irritable on its transverse section. Also on transverse section of the spinal cord, in the frog, feeble stimuli, such as descending opening shocks, are effective, but not further downward. This observation is in favor of analogous sensibility to stimuli on the part of both. According to Schiff's investigations in this connection, however, the anterior column of the spinal cord of the frog contains, in addition to the longitudinal fibers controlling movement, also sensory fibers, stimulation of which may cause reflexes. Therefore, all of the observations made on the anterior columns of the frogs are not available as evidence in favor of the direct irritability of the motor paths in the anterior columns. The sensory fibers in question are believed to rise from the gray matter and to pass within the spinal cord to the anterior columns, without first making their exit through the posterior roots—intracentral nerves.

The vasoconstrictors passing downward from the vasomotor center through the spinal cord can be irritated within the cord by all stimuli. Direct stimulation of any transverse section of the spinal cord causes contraction of all of the vessels innervated below that level. In a similar manner the fibers ascending in the spinal cord and exerting a pressor effect upon the vasomotor center can be irritated. Their stimulation causes no sensation. According to Schiff, the results obtained in these observations are, however, likewise not due to direct irritation

The spinal cord appears to be insensitive to chemical stimuli, such as moistening the cut surfaces with blood.

The motor centers can be irritated directly by blood at a temperature above 40° C. and by blood from an asphyxiated person or by sudden and total anemia in consequence of ligation of the aorta; likewise by certain poisons, such as picrotoxin, nicotin, barium-compounds.

In experiments of this character the spinal cord, for example at the level of the last thoracic vertebra, must have been divided some twenty hours previously, in order that it shall have recovered from shock. The posterior roots in the lower portion should also have been divided previously, in order to eliminate any possible reflex influences. If dyspnea be induced in cats thus prepared, or if their blood be overheated, extensor spasm, vascular contraction, secretion of sweat, evacuation of the bladder and the rectum, as well as contraction of the uterus or the vasa deferentia, occur in the distribution of the nerves from the lower portion of the cord. The administration of certain poisons, such as picrotoxin, has a similar effect. In animals with the medulla oblongata divided, rhythmic respiratory movements can even be induced in this way, if the spinal cord has been previously rendered highly irritable by the administration of strychnin or the action of heat.

Also mechanical stimuli are capable of irritating the ganglion-cells of the anterior horns, and, according to Biedermann, the gray matter responds to electrical stimulation.

47

The remarkable fact is worthy of mention that after unilateral division of the spinal cord, or, in the rabbit, of the posterior and the innermost portion of the lateral column, hyperesthesia below the level of the section appears upon the same side, so that rabbits cry aloud on slight pressure being made upon the toes. The phenomenon may persist for some three weeks, and it may be replaced by normal or subnormal sensibility. The healthy side exhibits permanent impairment of sensibility. Similar results have been observed in human beings with like lesions. An analogous phenomenon occurs after division of the anterior columns, that is, a marked tendency to contractions in the muscles below the level of the section—hyperkinesis.

In the intact body the normal irritability of the spinal cord is dependent upon the continuance of the normal circulation. Ligation of the abdominal aorta gives rise rapidly to paralysis and anesthesia in the lower portions of the body.

Sudden total anemia, as from occlusion of the aorta, in dogs, causes at first convulsions, lasting for twenty seconds; then paralysis, lasting for one minute; next sensory excitation, lasting for two minutes; and finally anesthesia, lasting for three minutes. Incipient degenerative processes appear in the ganglion-cells in the course of a few hours.

After prolonged ligation the anterior roots of the spinal cord and the entire gray matter of the lower portion of the cord that has been rendered anemic undergo degeneration. Motility and sensibility are permanently lost in the posterior extremities.

CONDUCTING PATHS IN THE SPINAL CORD.

Method.—The conducting paths in the spinal cord can be demonstrated by means of histological examination; of no less importance is a study of the functional disturbances exhibited by persons suffering from injury or degeneration in circumscribed areas. Animal experimentation is capable of affording confirmation in an analogous manner, although certain differences in the relations of the conditions are observed as compared with those present in human beings.

Flatau established the general law that the short paths in the spinal cords are situated nearer the gray matter, and the long paths nearer the surface.

Localized tactile impressions—pressure-sense, sensation of cold, muscle-sense—are conveyed upward through the posterior columns on the same side. The conduction of the sensation of heat is said to take place throughout the entire gray matter   Interruption of the posterior columns abolishes the sense of cold, the pressure-sense, and the muscular sense. The course in the brain is described on p. 745.

The columns of Goll in the cervical cord contain continuations of the posterior dorsolumbar and lower dorsal roots. The path for the muscle-sense is through the column of Burdach; where it approaches the medulla oblongata is situated the path for the muscle-sense in the arms.

In the rabbit the path for localized tactile sensation is situated in the lateral column in the lower portion of the dorsal cord. Division of certain portions of the lateral column, in the rabbit, abolishes this sensation for certain related cutaneous areas. Total division upon one side has the same effect for the entire half of the body below the level of the section. The condition of abolished tactile and muscular sense is designated anesthesia.

The impulses for *localized voluntary movements* are conducted in man through the anterior and lateral columns of the same side, through the pyramidal tracts. At the corresponding level of the spinal cord the fibers enter first into contact with the ganglion-cells of the anterior horns and thence the impulses pass into the appropriate anterior root (Fig. 255, M).

The exact division-experiments of C. Ludwig and Woroschiloff, Ott and Meade Smith demonstrated in the rabbit that in the lower dorsal portion of the cord the course of the fibers is exclusively in the lateral column. Partial division of the lateral column abolishes voluntary movement in individual related muscles below the level of the section. From what has already been stated it will be clear that the lateral columns increase progressively in size and in the number of fibers they contain from below upward. In the anterior horn every motor fiber enters into relation with one ganglion-cell, as has been demonstrated in the frog.

The fibers that intermediate the *tactile, widespread, coördinated reflexes* enter through the posterior roots and then pass to the columnar cells. At the various levels of the cord, the groups of ganglion-cells that control the coördinated reflexes are further connected by fibers that for the extremities pass within the anterior mixed tracts of the lateral column (the ground-bundle of the anterior column ?) and for the trunk in the posterior column. From the motor ganglion-cells the fibers for the stimulated muscles pass through the anterior roots.

Ataxic tabes dorsalis, in which, principally on a syphilitic basis, degeneration of the posterior columns is encountered, is noteworthy on account of its characteristic motor disturbances. Voluntary movements can, it is true, be executed with full strength, but they lack the fine harmonious gradation with reference to intensity and extent. This function is in part subserved by the normal existence of tactile sensations and of the muscular and articular sense, the paths for which are situated in the posterior columns. After degeneration of the latter, not only anesthesia, but also derangement in the execution of the tactile reflexes, occurs, as the centripetal segment of the arc is interrupted. Also the tone of the muscles, which depends essentially upon reflex stimulation, is considerably lowered, and in consequence the muscles exhibit an excessive degree of passive extensibility. An associated lesion of the simply sensory nerves, however, may in an analogous manner, as a result of anesthesia and loss of the pathic reflexes, give rise also to disturbances in coördination of movement. As the fibers of the posterior roots pass through the white matter of the posterior columns, it will thus be clear that disorders in the sensory sphere occur as a result of degeneration of these parts.

The view also is maintained by some that tabes represents a disease of the posterior roots extending to the spinal cord, for the roots also are found involved in the degenerative process, and their involvement may be responsible for the derangement in the sensory sphere. The latter consists in part in an abnormal increase of tactile or pain impressions, associated with lancinating pains, while in part it may be increased to the point of anesthesia or analgesia. At the same time tactile sensibility is altered, in consequence of irritation of the posterior columns, as indicated by sensations of numbness, softness, formication, or constriction. Often sensory conduction is delayed. The sensibility of the muscles, joints and internal parts is altered. If, finally, the posterior columns really contain fibers for the conduction of widespread coördinated reflexes, the ataxia can be explained in part by an interruption of these paths. The exceedingly rare cases of tabes without sensory derangement are to be interpreted only by assuming that either the conducting paths of the coördinated reflexes or the ganglion-cells are injured.

Inhibition of the tactile reflex takes place through the tracts of the anterior columns. At the proper level the conducting fibers pass from the anterior column into the gray matter, in order to enter into contact with the fibers of the reflex apparatus.

The transmission of *pain impressions* takes place through the posterior roots and thence through the entire gray matter. Loss of pain or of thermal sensibility occurs in man as a result of disease of the spinal cord in or near the posterior horn. In part decussation of the fibers that pass from one side to the other takes place in the cord. Their further course to the brain is described on p. 745.

If the gray matter is divided except for a small connecting band, this alone will suffice to convey pain impressions upward. According to Schiff, however, the transmission under such circumstances is delayed. Only after the gray matter has been wholly divided does the conduction of all pain sensation from the portions of the body below the level of the section cease. In this way the condition of analgesia is brought about, while tactile sensibility persists if the posterior columns are intact. A similar condition is not rarely observed in human beings during incomplete chloroform-narcosis, and particularly during the narcosis induced by the combined administration of chloroform and morphin. As these poisons benumb the nerves transmitting pain sensations earlier than the tactile nerves, those operated on maintain that they appreciate the operation as a tactile impression, as pressure, etc., but not as pain. As the conduction of pain takes place everywhere through the gray matter, and as, further, the excitation of pain extends the more widely throughout the gray matter the more intense the painful manipulation, the so-called irradiation of the pain impressions can be understood. In the presence of severe pain, the pain appears to radiate widely from its seat of origin. Thus, for example, in case of severe toothache beginning in a given tooth the pain soon radiates to the entire maxillary region and even to the entire half of the head. In contradiction of this statement Bechterew maintains that the path for pain impressions is situated in the lateral columns, in the rabbit, hen and dog.

The impulses for *spasmodic, involuntary, incoördinated movements* are transmitted through the gray matter and from the latter through the anterior roots.

Such transmission occurs, for example, in cases of epilepsy and as a result of certain intoxications, for instance with strychnin, in cases of uremic intoxication and of tetanus. Also the convulsions associated with anemia and dyspnea are transmitted downward from the medulla oblongata through the entire gray matter.

The impulses for *widespread reflex convulsions* are transmitted from the posterior roots to the ganglion-cells of the gray matter, further through the anterior horns and finally into the anterior roots, and under conditions that have been described in the discussion of this variety of reflex convulsions.

Inhibition of the pathic reflexes is effected through the anterior column downward and then in the gray matter to the connecting tracts of the reflex organs, into which resistance is introduced.

The *vasomotors* pass through the lateral columns and, after having entered the ganglion-cells of the gray matter at a given level, leave the spinal cord through the anterior roots. Later on, they approach vessels provided with a muscular coat, either simply through the path for the spinal nerves, or, more frequently, they pass through the communicating branches into the sympathetic and from this to the vascular plexuses.

Division of the spinal cord paralyzes all of the vasomotors below the level of the section. Stimulation of the peripheral stump of the cord conversely causes contraction of all of the vessels.

Fibers having a *pressor* action enter the cord through the posterior roots, then pass upward in the lateral column and undergo incomplete decussation.

Their final goal is the dominating vasomotor center in the medulla oblongata, which thus they stimulate reflexly. Analogous depressor fibers must pass through the spinal cord, although nothing concerning them is known.

From the respiratory center in the medulla oblongata there pass downward in the lateral column on the same side the *respiratory nerves*, which, after reaching the ganglion-cells of the gray matter, pass through the anterior roots into the motor nerves to the muscles of respiration.

Unilateral or total division of the spinal cord at progressively higher levels paralyzes, accordingly, respiratory nerves arising at successively higher levels on the same side or on both sides.

Pathological.—In case of degeneration or direct injury of the spinal cord or of individual portions of the cord, it should be especially noted that occasionally in recent cases irritative and paralytic phenomena occur side by side in closely adjacent portions of the cord, and as a result an analysis of the clinical picture is rendered difficult.

Degeneration of the posterior columns, without involvement of the afferent posterior roots, causes loss of tactile sensibility as the most conspicuous symptom, while thermal sensibility is preserved. Degeneration of the ganglion-cells in the anterior horns, for example in cases of spinal paralysis of infants, gives rise to paralysis in the distribution of the efferent motor nerves. At the same time the muscles supplied by these nerves undergo rapid atrophy. The ganglion-cells are the trophic-centers for the nerves and the muscles. The results, under such circumstances, are the same as those that follow permanent division of a peripheral motor nerve. As some fibers pass from above downward through the anterior horn to the opposite side, also some fibers on the contralateral side degenerate. Degeneration of the posterior gray horns gives rise to impairment of cutaneous sensibility and to trophic disorders in the skin. Degeneration of the central portion of the gray matter causes, in addition to trophic disorders in the skin, loss of thermal sensibility.

It is a highly interesting fact that temporary occlusion of the abdominal aorta, in rabbits, causes permanent sensory and motor paralysis in the entire area controlled by the portion of the spinal cord whose circulation is cut off. Ganglion-cells and nerve-fibers of the anterior horns undergo degeneration. Then secondary degeneration of the anterior roots follows (with the exception of the contained vasomotor fibers), and of the white matter adjacent to the anterior horns. Subsequently the posterior horns also undergo reduction in size. All of the tracts passing into the cord remain intact, namely the posterior roots, the spinal ganglion-cells, the posterior columns and the extreme periphery of the anterolateral tract.

Destruction of the lower portion of the spinal cord, in the dog, up to the cervical cord gives rise, in addition to loss of sensation and of motion, to reduction in the temperature of the lower portion of the body, but, if great care be taken, to no trophic disorder, except that the bones appear to be brittle. The anus is dilated only at first, but later the sphincter regains its normal tone and spontaneously contracts rhythmically (perhaps it contains its innervational center within itself), while all the other paralyzed muscles of the body undergo degeneration. The digestive processes, the intestinal movements, the act of parturition, the act of nursing are performed normally; but the temperature of the body can be regulated only within certain limits. The paralysis of the bladder present at first improves; the tone of the blood-vessels is restored in the course of a few days. A series of important vital processes are, therefore, not directly dependent upon the existence of the spinal cord, but they are rather decentralized.

THE BRAIN.

GENERAL OUTLINE OF THE STRUCTURE OF THE BRAIN.

COURSE OF THE MOTOR AND SENSORY TRACTS.

With respect to an organ exhibiting such a high degree of complexity of structure as the brain, it is of the greatest importance to be acquainted with its general arrangement, even if only from a brief description. It is to the credit of Meynert that he devised a practical system of this character based on extensive investigations. This will be used in the discussion of the subject that follows, although consideration will be given also to the results of more recent investigation.

The *weight of the brain* in man is on the average in the male 1372 grams, in the female 1231 grams. The uppermost and outermost layer of the cortex consists of a layer of glia containing nerve-fibers. Beneath this is the layer of small ganglion-cells, and next to this the layer of large pyramidal cells, which in turn covers the layer of irregularly formed ganglion-cells.   Each ganglion-cell possesses a neurite and numerous dendrites.   Between the cells everywhere lie large numbers of bundles of medullated nerve-fibers.

The cortex of the brain consists of peripheral gray matter with numerous convolutions and sulci (Fig. 254, C). This can be recognized as the central organ of the nervous system from the presence of large numbers of ganglion-cells.   From it pass all of the motor fibers that can be stimulated by the mind (will, conception), and to it pass all of the fibers derived from the organs of special sense and the sensory organs that intermediate the psychical perception of external impressions.

All of these tracts together, in part corticopetal, in part corticofugal, pursue in general a convergent course toward the central part of each cerebral hemisphere, where the large central ganglia of the brain are situated, corpus striatum (C. s.), lenticular nucleus (N. l.), optic thalamus (T. o.) and quadrigeminate bodies (v). Some fibers merely pass by these structures (5, 5), but many enter the central gray matter. The fiber-system mentioned, which has a radiating arrangement within the cerebral hemisphere, is known as the corona radiata, or the *projection-system of the first order*. In addition to this, the white matter contains two other groups of fibers, namely (a) the commissural fibers—corpus callosum and anterior commissure (c c), which connect the two hemispheres; and (b) the association-fibers, by means of which different areas of the cortex of the same side are connected (a a).

The large, cellular, gray masses of the central cerebral ganglia form the first stage in the course of a large number of fibers of the projection-system of the first order. On entering these central masses the fibers undergo an interruption in their course, while a reduction in the number of fibers of the corona radiata takes place. In detail the relation of the fibers of the corona radiata to the great central ganglia according to Meynert is as follows: The entire mass of fibers of the system of the corona radiata breaks up in general into as many bundles as there are ganglia on each side. There are thus systems for the striate body (1, 1), the lenticular nucleus (2, 2), the optic thalamus (3, 3) and the quadrigeminate bodies (4, 4).

According to Flechsig the convolutions of the brain can be divided into two groups: The first group contains only association and commissural fibers, by means of which the convolutions are connected with other cortical areas. The other group contains in addition bundles which pass to the optic thalamus in the corona-radiata. These Flechsig designates *sense-centers*, of which each possesses its own motor apparatus.

The central convolutions, the seat of the cutaneous and the muscular sense, serve as the principal source of origin for the association-tracts.   Then the auditory area, and in lesser degree the visual area, play an important rôle as the source of origin of the association-tracts.

From the large central ganglia there develops, later passing downward, the *projection-system of the second order*, whose longitudinal fibers reach their temporary termination in the so-called gray matter of the central cavity.   This is the cellular gray matter that extends from the third ventricle through the aqueduct of Sylvius, the floor of the fourth ventricle, to the lowermost portion of the gray matter of the spinal cord, occupying the interior of the medullary tube. It represents likewise the second stage in the course of the fibers. Accordingly, the projection-system of the second order extends from the large central ganglia of the cerebrum downward to the gray matter of the central cavity.   The fibers of this system must obviously be of widely varying length, some terminating in the gray matter of the central cavity above the medulla oblongata (oculomotor origin), while others extend to the level of the last spinal nerve.   The gray matter of the central cavity forms a mass for the interruption of the fibers, and an increase in the number of fibers takes place in it, for many more fibers pass out from the gray matter of the medulla oblongata and the spinal cord to the periphery than were sent to it from above from the central ganglia of the cerebrum.

With reference especially to the arrangement of the fibers of this projection-system of the second order, it is assumed that the fibers descending from the lenticular nucleus and the striate body (8, 8) unite to form a special tract passing through the upper portion of the crusta of the cerebral peduncle downward into the medulla oblongata, or only to the pons according to Flechsig. In a similar manner a bundle passes from the optic thalamus (7) and from the quadrigeminate bodies (6, 6) that descends through the tegmentum (H) of the cerebral peduncle. Both groups of fibers, those in the crusta as well as those in the tegmentum, unite below in the spinal cord.

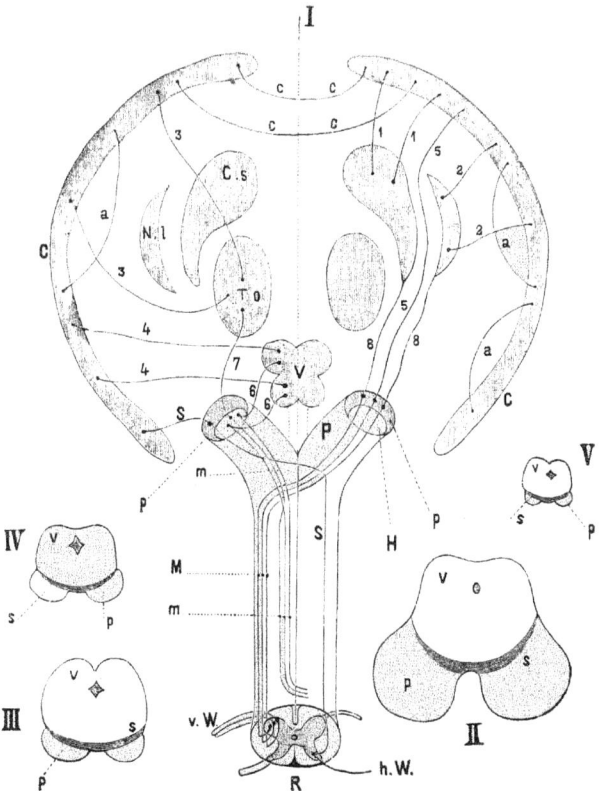

FIG. 254.—I. Diagrammatic Representation of the Structure of the Brain: C, C, cerebral cortex; C. s, striate body; N.l, lenticular nucleus; T.o, optic thalamus; V, quadrigeminate bodies; P, cerebral peduncle; H, tegmentum; p, crusta; 1, 1, fibers of the corona radiata to the striate body; 2, 2, to the lenticular nucleus; 3, 3, to the optic thalamus; 4, 4, to the quadrigeminate bodies; 5, direct fibers to the cerebral cortex; 6, 6, fibers from the quadrigeminate bodies to the tegmentum; 7, fibers from the optic thalamus to the tegmentum; m, their further course; 8, 8, fibers from the striate body and the lenticular nucleus to the crusta of the cerebral peduncle; M, their further course; S, S, course of the sensory fibers; R, transverse section of the spinal cord; v.W, anterior and, h.W, posterior root; a, a, association-fibers; c, c, commissural fibers. II. Transverse Section through the Posterior Pair of the Quadrigeminate Bodies and the Cerebral Peduncles of Man (after Meynert): p, Crusta of the peduncle; s, substantia nigra; v, the quadrigeminate bodies with the transverse section of the aqueduct. III. A Like Section from the Dog. IV. From the Ape. V. From the Guinea-pig.

According to Wernicke the lenticular and caudate nuclei are not parts of the cerebrum into which fibers of the corona radiata from the cortex enter, but they are independent structures analogous to the cortex, from which fibers originate. These fibers later on reach the tegmentum, where they lie side by side with the fibers derived from the optic thalami and the quadrigeminate bodies.

The fibers that pass from the thalamus and the quadrigeminate bodies through the tegmentum of the cerebral peduncle (6, 6, 7) represent, according to Meynert, reflex paths. Accordingly, the cerebral masses mentioned would be the centers for certain extensive coördinated reflexes. This is indicated by the fact that after destruction of the paths for the conduction of voluntary impulses in animals the technical perfection of the movements, in so far as these are induced by reflex activity, remains intact. The fibers named pass in the spinal cord at first downward upon the same side (m), but they probably cross below in the spinal cord itself.

Finally there passes out of the entire gray matter of the central cavity a group of fibers that constitute the *projection-system of the third order*. These are the peripheral nerves, sensory and motor. They exhibit in their totality an increase of fibers as compared with the number of fibers in the projection-system of the second order.

The cerebellum represents a separate central organ of special character containing gray matter in part as a cortical layer and in part as central accumulations. It is connected with the cerebrum (1) through the superior cerebellar peduncles, formed of fibers from the system of the corona radiata, passing then into the tegmentum, and reaching the cerebellum after total decussation, and (2) through the middle cerebellar peduncles to the pons and from the pons through the cerebral peduncles to the hemispheres. The cerebellum is connected also with the spinal cord, namely (1) with the posterior column (cuneate and gracile columns) and (2) with the anterior column (restiform body). The two hemispheres are connected by the transverse commissural fibers of the pons.

The *distribution of the cerebral vessels* is deserving of consideration from the practical standpoint. The artery of the fossa of Sylvius supplies the motor areas of the cortex in animals—in man the paracentral lobule is supplied by the anterior cerebral artery. The region of the third frontal convolution, which is of such importance for the function of speech, is supplied by a special branch of the Sylvian artery. Those portions of the frontal lobe, injury of which, according to Ferrier, causes derangement of intelligence, are supplied by the anterior cerebral artery. The middle cerebral artery supplies the internal capsule, with the exception of its most posterior portion, which, together with the uncinate gyrus, is supplied by the anterior choroid artery. Those portions of the cortex, lesions of which, according to Ferrier, give rise to hemianesthesia, are supplied by the posterior cerebral artery. Isolated anemia of this area is believed to have some connection with melancholic states in human beings.

Course of the Tracts for Voluntary Movements: Psychomotor or Corticomuscular Paths (Fig. 255).—From the motor regions of the cerebral cortex, from which impulses are sent for voluntary movements in the distribution of the cerebral and spinal motor nerves, the fibers that constitute the *pyramidal tracts* (Fig. 255, a, b, c) pass through the anterior two-thirds of the posterior limb of the internal capsule (Fig. 255, G.i; Figs. 263, 264), and then through the crusta of the cerebral peduncle (Fig. 254, middle portion of the lower free circumference of the crusta), through the pons on the same side (P) into the pyramid (Py) of the medulla oblongata. At this point the majority of the fibers cross through the pyramidal decussation to the opposite side and pass downward in the lateral column (lateral pyramidal tract, a) to the level of the spinal cord (Fig. 255), from which the anterior root for the transmission of the voluntary impulse in question makes its exit. Before entering the anterior root the fibers enter into communication with the ganglion-cells of the anterior horn (surrounding them by delicate arborizations). From each ganglion-cell there passes as a neurite a motor filament into the nerve-fiber of the anterior root. The largest number of decussated fibers in the pyramids pass to the motor nerves of the extremities. A smaller number of fibers (Fig. 255, b), however, do not decussate in the pyramid, but pass on the same side in the anterior column of the spinal cord (anterior pyramidal tract, b, z) and remain upon the same side. These, however, in their further course through the spinal cord likewise cross over to the opposite side through the anterior white commissure. A portion of these fibers, at first undecussated, appear, however, to remain upon the same side. It is perhaps their function to innervate those muscles of the trunk that, like the respiratory, the abdominal and the perineal muscles, are generally made to contract together on both sides.

With reference to the relations of the crossed and uncrossed fibers individual variations occur. In isolated cases the conditions are reversed, and in rare instances the pyramidal tracts remain upon the same side from the brain downward. In this way are to be explained the exceedingly rare cases in which paralysis of voluntary movement has been observed on the same side as the lesion of the cerebral cortex. Cases are uncommon also in which the muscles of the trunk and the lower extremities are moved bilaterally on voluntary efforts at unilateral movement. Finally, it should be mentioned that Unverricht occasionally observed in the dog a double decussation of the tracts: once in the pyramids and again in the spinal cord. Pyramidal tracts are present only in mammals.

The cerebral motor nerves naturally have their center for voluntary stimulation in the cortex of the cerebral hemisphere. From this point the fibers that transmit voluntary impulses pass through the internal capsule and the crusta of the cerebral peduncle, where they lie in front of and internal to the pyramidal tracts. Then their course is directed to their nuclei of origin. In Fig. 255, c represents the course of the facial nerve to its nucleus of origin. The hypoglossal nerve passes with the pyramidal tract and behaves like the anterior root of a spinal nerve.

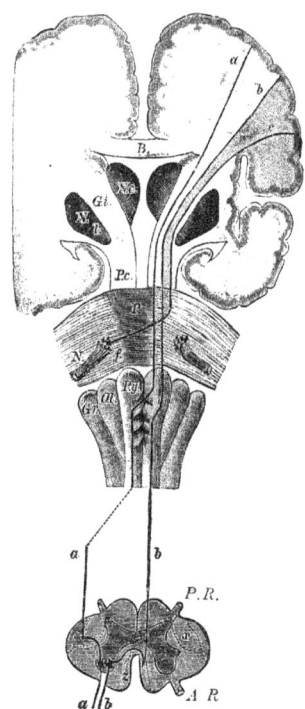

FIG. 255.—Course of the Paths for Voluntary Movement: a, b, Paths for the cerebral motor nerves; c, path for the facial nerve; B, corpus callosum; N.c, caudate nucleus; G.i, internal capsule; N.l, lenticular nucleus; P, pons; N. f, nucleus of origin of the facial nerve; Py, pyramid, with decussation; O.l, olivary body; G.r, restiform body; P.R, posterior root; A.R, anterior root; x, lateral pyramidal tract; z, anterior pyramidal tract.

Course of the Tracts for Conscious Sensation.—From the cortical area in which is situated the center for sensibility, which is designated the sensory sphere and is fully described on p. 785, the conducting tracts pass through the posterior third of the posterior limb of the internal capsule (Figs. 263, 264). The fibers for the transmission of the muscle-sense pass through the middle, those for the sensations of pressure, temperature, and pain through the inner half of the posterior third of the internal capsule. The tracts then pass through the tegmentum of the cerebral peduncle and their continuation through the pons and further to the medulla oblongata. The fibers for the transmission of cutaneous sensibility pass through the fillet and the ventral portion of the reticular formation.

The connections of the posterior roots of the nerves of the spinal cord, through which sensibility is transmitted, are, according to recent investigators, as follows:

The posterior columns transmit impulses from the afferent posterior roots upward. A lateral and a median bundle can be recognized in the posterior root. The median bundle of each afferent root in its course upward in the posterior column is generally situated to the outer side close to the posterior horn (Fig. 257, 2). Each root as it enters at a higher level (1) displaces further and further inward the fibers derived from the roots situated at a lower level. Therefore, the sensory fibers coming from the lower extremities are, in the cervical cord, situated principally in the columns of Goll, while the columns of Burdach still contain many fibers from the upper extremities. Ascending, the fibers of the posterior columns terminate above in the medulla oblongata, in the formations (nuclei of the gracile and cuneate columns) known as the nuclei of the posterior columns. From these nuclei many fibers pass into the fillet (L) of the opposite side. Other fibers pass to the cerebellum. The lateral fibers, coarse and fine, of the posterior root (3, 4) enter the delicate plexus of the posterior horn, in which the ganglion-cells of the posterior horn are lodged. From the plexus of the posterior horns there arise numerous

FIG. 256.—Course of the Motor and Sensory Paths through a Transverse Section of the Spinal Cord: 1, Anterior pyramidal tract; 3, lateral pyramidal tract; 4 and 5, decussating sensory paths in the spinal cord; 6, ascending sensory paths not decussating in the spinal cord; 7, sensory path to the columns of Stilling and Clarke, and thence undecussated upward through the lateral cerebellar tracts; 2, origin of a motor fiber as a neurite from a ganglion-cell of the anterior horn.

fibers that pass forward through the gray matter, undergo decussation, and then continue toward the cerebrum in the anterior and lateral columns. At a higher level these fibers, with their original accompaniments, reunite (at L), so that almost all of the posterior root-fibers (decussated) again lie together in the fillet or the intermediary layer of the olive. The further course of these fibers to the cerebral cortex is discussed on p. 801. A portion (5) of the fibers of the posterior roots, which are not connected with the cells of the spinal ganglia, terminate in the cells of the column of Stilling and Clarke, which are at the same time the trophic centers for those fibers. The fibers turn outward from the cells and ascend in the lateral cerebellar tract. Their further course is upward to the restiform body, and thence to the cerebellum. These fibers are concerned in the regulation of equilibrium, and their superior path of conduction is situated in the cerebellum. In cases of tabes these tracts and the columns of Stilling and Clarke are often degenerated.

In the human brain the sensory conducting path for the tactile and the muscular sense (continuation of the posterior roots and the direct cerebellar tract) develops first. This passes from the nuclei of the posterior columns (Fig. 257) through the thalamus and the lenticular nucleus to the central convolutions, especially the posterior.

The circumstance that a portion of the sensory cutaneous nerves cross in the spinal cord to the opposite side explains the fact that unilateral division of the spinal cord in man, and in apes, abolishes cutaneous sensation upon the opposite side of the body below the level of the lesion, while the muscular sense is preserved. Upon the side of the injury hyperesthesia is present below the level of the section.

From experiments on mammals Brown-Séquard concluded that the decussating sensory nerve-fibers cross in the spinal cord to the opposite side at various levels—the fibers that transmit tactile sensibility at the lowest level, then those that transmit sensations of tickling and pain, and at the highest level those that transmit thermal impressions.

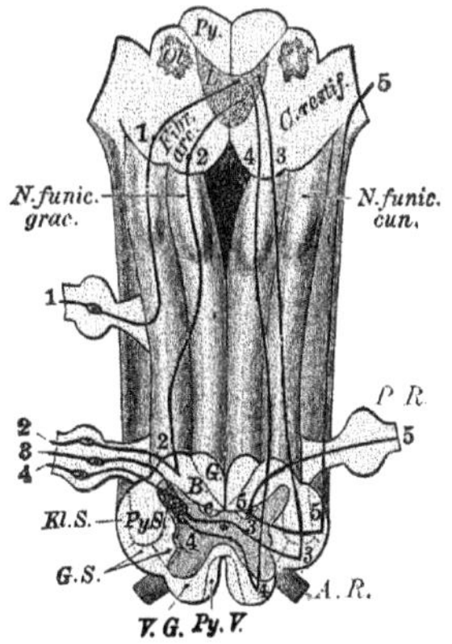

All of the fibers that, pursuing a longitudinal course, connect the spinal cord with the medullary mass of the cerebrum, thus, as a rule, undergo complete decussation in their course. Therefore, in man the result of a destructive lesion of one cerebral hemisphere is generally complete paralysis and abolition of sensation upon the opposite side of the body. Also the fibers arising from the nuclei of origin of the cerebral nerves decussate within the brain.

Only in those cases, not rare it is true, in which the lesion, as from pressure, inflammation, etc., involves the cerebral nerves situated at the base, are paralysis and anesthesia observed on the same side of the head.

Particulars as to the site of decussation have already been stated. Decussation takes place (1) in the spinal cord, (2) in the medulla oblongata, and finally (3) in the pons. Decussation is already complete in the peduncles.

FIG. 257.—Course of the Sensory Fibers from the Posterior Roots through the Spinal Cord upward to the Cerebrum. The explanation of the fibers will be clear from the description in the text, and with it also Fig. 256 should be compared: A.R. anterior root; P.R. posterior root; V.G, ground-bundle of the anterior column; Py.V, anterior pyramidal tract; Py.S, lateral pyramidal tract; G.S, ground-bundle of the lateral column; Kl.S, cerebellar tract of the lateral column; G, column of Goll; B, column of Burdach; Py, pyramid; Ol, olive; L, fillet or intermediary layer of the olive; fibræ arcuatæ; restiform body; nucleus of the slender column and nucleus of the cuneate column of the medulla oblongata.

Gubler observed, in cases of unilateral injury of the pons, paralysis of the facial nerve on the same side, but paralysis of the body on the opposite side. From this he concluded that the nerves from the trunk must undergo decussation below the pons, the facial fibers within the pons. Such rare instances are designated alternating hemiplegia. The conditions are made clear in Fig. 255.

Exceptions to decussation are formed by the olfactory nerves, which do not decussate at all (?), and the optic nerves, which decussate only partially in the chiasm.

THE MEDULLA OBLONGATA.

The medulla oblongata, which connects the spinal cord with the brain, resembles the cord in many respects, particularly from the fact that it contains centers, which, like those in the spinal cord, convey simple reflexes (for example that of closure of the eyelids). Further, it contains centers, however, that occupy a controlling relation to centers of analogous function in the spinal cord; for example the centers controlling the vasomotor nerves, the secretion of sweat, dilatation of the pupils, the reflex movements of the body. With reference to the irritation, some of the centers are reflex, others automatic.

The normal function of the centers is related to the gaseous interchange maintained by the normal circulation in the medulla. If this is interrupted by asphyxia or sudden anemia or venous stasis, the centers are, at first, thrown into a state of increased irritation, being later paralyzed by overstimulation. Overheating also acts as an irritant to the centers. Not all of the centers are active at the same time and they do not all exhibit the same degree of irritability. In the normal body the respiratory and vasomotor centers are in constant rhythmic activity. The cardiac inhibitory center is in some animals not continuously irritated; in others slight stimulation occurs normally only with inspiration (in conjunction with stimulation of the respiratory center). The spasm-center is not irritated at all under normal conditions, and the respiratory center not during intrauterine life. The medulla oblongata is, as the seat of many centers of importance with reference to the maintenance of life, as well as for the transmission of various nerve-paths, of the greatest significance. The details are considered in what follows. The reflex centers will be considered first and then the automatic centers.

REFLEX CENTERS IN THE MEDULLA OBLONGATA.

The medulla oblongata contains a number of reflex centers that permit the execution of coördinated movements.

The *center for closure of the eyelids*. The sensory fibers of the trigeminus to the cornea and the conjunctiva, as well as the skin in the vicinity of the eye, conduct centripetally the impressions received to the medulla oblongata, where they are transferred to the motor path of the facial branch that innervates the orbicular muscle of the lids. The center extends from about the middle of the ala cinerea upward to the posterior margin of the pons.

Intense illumination of the eye also causes closure of the eyelids through the intermediation of the optic nerve. The stimulation passes through the quadrigeminate bodies to the center.

Reflex closure of the eyelids takes place in man always on both sides, but voluntarily the eyelids can be closed on one side. On intense irritation the corrugator and the muscle-group that elevates the nose and the cheek toward the lower margin of the orbit also contract in order to secure more perfect protection and closure of the eye. The duration of voluntary and reflex closure of the eyelids is from 0.3 to 0.45 second.

The *center for sneezing*. The centripetal path is through the inner nasal branches of the trigeminus and probably also through the olfactory nerve (for strong odors). The motor path leads to the expiratory muscles. Sneezing cannot be practised voluntarily.

The *center for coughing*, according to Kohts, is situated above the inspiratory center and is excited centripetally through the sensory branches of the vagus. The centrifugal fibers are the expiratory nerves, including the constrictors of the glottis.

The *phonation-center*. The center for the control of the voice-mechanism is situated from the origin of the vagus upward to the quadrigeminate bodies. New-born animals from which the brain was removed, with preservation of this portion of the brain, are still able to cry.

The *center for the movements of sucking*, as well as of *chewing*. The centripetal nerves are the sensory fibers of the mouth, including those of the lips (second and third divisions of the trigeminus and the glosso-pharyngeal). The motor nerves for the movements of sucking are the facial (the lips), the hypoglossal (the tongue), the third division of the trigeminus (elevator of the lower jaw and the branches of the depressor of the lower jaw). After transitory (by cocain) or permanent paralysis of the trigeminus the sucking-reflex ceases. The same motor nerves take part in the act of chewing, but the action especially of the hypoglossal for the movement of the tongue and of the facial for that of the buccinator is necessary to keep the food between the teeth.

The *center for the secretion of saliva* is situated on the floor of the fourth ventricle, and can be stimulated reflexly. Irritation of the medulla oblongata, when the chorda tympani and the glossopharyngeal nerve are preserved, causes active secretion of saliva; a lesser amount of secretion when these nerves are divided; and, finally, none at all when the cervical sympathetic also is destroyed.

The *center for the act of deglutition* is situated on the floor of the fourth ventricle above the respiratory center and is stimulated through the sensory nerves of the palate and the pharynx (second and third divisions of the trigeminus and the vagus). The centrifugal path is through the motor branches of the pharyngeal plexus. Irritation of the glosso-pharyngeal does not cause swallowing, but rather inhibition of the act of deglutition. On the other hand, every act of swallowing excited by irritation of the palatine nerves or the superior laryngeal nerve causes rapid, abortive contraction of the diaphragm (deglutitional breathing).

The *center for vomiting* is discussed on p. 282. The relations between certain branches of the vagus and the act of vomiting have been pointed out on p. 710. The center may be set into activity by direct application of apomorphin or emetin.

The *upper center for the dilator muscle of the pupil* and the unstriated *muscles of the orbit and the eyelids* is situated in the medulla oblongata. The pupillary fibers passing through the trigeminus arise from the origin of this nerve and downward as far as the second cervical nerve (in the rabbit). Anastomotic fibers pass from this point downward through the lateral columns of the spinal cord to the ciliospinal region, and thence through the three or four uppermost thoracic nerves into the cervical sympathetic. The center is normally stimulated reflexly by cutting off the light from the retina. It is irritated directly by states of the blood causing dyspnea or by occlusion of the carotids.

The center may be irritated reflexly also by stimulation of sensory nerves of the trunk (sciatic). These fibers pass (from the sciatic) through the two lateral columns up to the center.

Finally, there is situated in the medulla a higher center, which establishes a connection among the various centers for the reflexes in the spinal cord. When Owsjannikow divided the medulla 6 mm. above the calamus scriptorius (in rabbits), the general reflexes of the body, in which the anterior and the posterior extremities took part, persisted. When the section was made 1 mm. lower, only partial, local reflexes usually appeared. The center extends upward to a little above the lower third of the medulla.

In the frog the medulla contains the sole center for movement from place to place. Division of this abolishes such movement in response to external irritation and only simple reflexes remain, but no reflex movement, such as jumping, crawling, swimming.

Pathological.—The medulla oblongata may be the seat of a typical disease designated bulbar paralysis, or glossopharyngolabial paralysis, attended with progressive paralysis of the bulbar (bulbus rhachiticus, medulla oblongata) nuclei of various cerebral nerves, which often represent the motor segments of important reflex mechanisms. From the latter point of view the clinical picture deserves consideration. Generally, the disorder begins with paralysis of the tongue, attended with fibrillary twitching, in consequence of which speech, the formation of the bolus, and swallowing in the mouth are rendered difficult. The secretion of an extremely viscous saliva indicates an inability to secrete a watery saliva by reason of paralysis of the facial nerve. Further, swallowing is rendered difficult or even impossible in consequence of paralysis of the pharynx and the palate. As a result of the latter the formation of consonants between the tongue and the soft palate is interfered with; speech, further, becomes nasal; and often especially fluid articles of food enter the nares on efforts at swallowing. Then, the facial branches for the lips become paralyzed. The mimetic expression of the mouth is extremely characteristic, "as if frozen stiff," and, at the same time, in consequence of horizontal enlargement of the opening of the mouth (as the orbicularis oris especially is paralyzed), marked by a lacrimose appearance. Later on, speech becomes more greatly interfered with. When the disorder is marked, all of the muscles of the face are paralyzed. Under such circumstances, the laryngeal muscles are not rarely paralyzed, so that phonation is abolished, and the ready entrance of fluids into the larynx is favored. The enormous retardation of the pulse-beat often present indicates irritation of the cardiac inhibitory fibers, derived from the accessory nerve. If, later on, attacks of dyspnea occur, such as have been observed after paralysis of the recurrent nerve, or such as are constant after section of the pulmonary branches of the vagi, death may take place suddenly amid signs of asphyxia if the attacks become more severe and more frequent. Rarely, the clinical picture is complicated by paralysis of the muscles of mastication (in consequence of paralysis of the motor portion of the fifth nerve), contraction of the pupils (in consequence of paralysis of the dilator-center) and paralysis of the abducens nerve.

THE RESPIRATORY CENTER AND THE INNERVATION OF THE RESPIRATORY APPARATUS.

Flourens determined the position of the respiratory center in the medulla oblongata, behind the point of exit of the vagi, on either side of the posterior extremity of the floor of the fourth ventricle, between the nuclei of the vagus and accessory nerves. He designated this the vital point or nœud vital, because its destruction is followed at once by arrest of respiration and therefore by death. The center occupies exactly the same situation in man, as Kehrer has demonstrated in a perforated new-born child. The center is bilateral, and it can be divided by a median section, the respiratory movements continuing symmetrically upon both sides. If one vagus is divided the respiration becomes slowed upon the corresponding side. If, however, both vagi are divided, the breathing is unequal in frequency and vigor on the two sides of the body. Irritation of the central stump of one of the two divided vagi causes arrest of respiration only upon the corresponding side, while respiration continues upon the other side. The same result is brought about if the trigeminal nerve upon one side is irritated. On unilateral transverse division of the center the respiratory movement ceases upon the side of the injury.

According to Schiff the respiratory center is situated near the lateral margin of the gray matter forming the floor of the fourth ventricle, extending posteriorly not so far as the ala cinerea. According to Gierke and Heidenhain and others that portion of the medulla, destruction of which is followed by cessation of respiratory movement is a single or double nerve-like strand, passing downward in the substance of the medulla, within which, however, gray matter with small ganglia is found. This is said to be constituted in part of the roots of the vagus, trigeminus, accessory and glossopharyngeal, connected with those of the opposite side by means of fibers and extending downward into the cervical enlargement of the spinal cord. The strand thus connects as an intercentral bundle the spinal cord (the seat of origin for the motor respiratory nerves) with the nuclei of origin of the cerebral nerves named, the relations of which to the respiratory movements are in part demonstrated.

It is most probable that the dominating center that controls the rhythm and the symmetry of the respiratory movements is situated in the medulla oblongata, but that in addition other centers of subordinate importance are situated in the spinal cord and are controlled by the center in the medulla, receiving their impulse to activity from that center. If in new-born animals the cord is divided below the medulla by means of an exceedingly sharp instrument respiratory movements of the chest will occasionally persist, from stimulation of the spinal centers, an observation that Landois was able to confirm in young dogs and cats.

The spinal respiratory centers are, moreover, susceptible even to reflex influences (excitation or inhibition). Nitschmann divided the spinal center situated in the upper cervical cord by means of a longitudinal section into two equal parts, both of which then had an excito-respiratory influence upon the diaphragm upon each side, even though the medulla just below the calamus scriptorius had been divided upon one side. Accordingly, the spinal centers of both sides must be connected in the spinal cord. Irritation originating in one-half of the center in the medulla may thus affect the spinal centers on both sides, for example the origins of both phrenic nerves. The spinal center for the phrenic nerve is situated between the third and seventh segments.

In addition to the spinal cord, the brain also contains subordinate *cerebral respiratory centers*. In the tissue between the striate body and the optic thalamus I. Ott found a center, irritation of which markedly increased the number of respirations. On destruction of this center the dyspneic respiratory acceleration (*heat-dyspnea*) induced by heat ceases.

In the optic thalamus, on the floor of the third ventricle, Christiani found further a special *inspiratory center*, which through stimulation of the optic and auditory nerves, also after previous extirpation of the cerebrum and the striate bodies, or also through direct irritation, causes deepening of inspiration and acceleration of respiration, and even arrest in inspiration. This inspiratory center can be extirpated and it can then be demonstrated that a *center controlling expiration* is situated in the substance of the anterior quadrigeminate body not far from the aqueduct of Sylvius. Finally, the posterior quadrigeminate body contains a second cerebral *inspiratory center* and also an inspiratory inhibiting center. Obviously all of these centers are connected with the center in the medulla.

According to Marckwald, the regular rhythm of the respiratory movements is maintained, in addition to the posterior quadrigeminate bodies, also by the sensory nucleus of the trigeminus.

The respiratory center consists of two central areas engaged in alternate activity; the *inspiratory* and *expiratory* centers, of which each forms the motor central point for the well-known group of inspiratory and expiratory muscles. The center is an automatic one, for, even after division of all of the sensory nerves that may exert a reflex influence upon it, it preserves its activity. The irritability and the stimulation of the center are dependent upon the state of the blood, particularly the amount of oxygen and carbon dioxid present.   In this connection the following distinctions are to be recognized:

1. *Apnea*, that is the cessation of the respiratory movements in consequence of deficient need therefor.   It occurs when the blood is saturated with oxygen and is deficient in carbon dioxid.   Blood in such a state fails to stimulate the center, and, therefore, the muscles controlled by it remain at rest.   The fetus is in this condition; likewise some animals in the state of hibernation.   If, further, an abundance of air is made to pass into the lungs of animals by means of artificial respiratory apparatus, they cease to breathe because the marked arterialization of their blood does not permit of stimulation of the respiratory center.   If, further, a similar state of the blood is induced by rapid and deep respirations *apneic pauses* of considerable duration occur.

A. Ewald found the blood in the arteries of apneic animals almost completely saturated with oxygen, while the amount of carbon dioxid was diminished.   The venous blood contained less oxygen than under normal conditions.   The latter fact is probably due to the circumstance that the apneic state of the blood greatly reduces the blood-pressure, and in consequence the circulation is slowed.   Therefore, the oxygen can be taken from the capillary blood in much larger amount. In general, however, the consumption of oxygen during the state of apnea is not increased.   Gad calls attention to the fact that on forced artificial respiration the pulmonary alveoli are greatly filled with atmospheric air, and in consequence they are capable of arterializing for a considerable time the blood entering the lungs, so that the necessity for respiration must be diminished.   According to Gad and Knoll the respiratory center during apnea is in a state of diminished irritability, which can be induced reflexly through the forcible stretching of the terminal pulmonary branches of the vagi in connection with the artificial respiratory movements.   This is shown also by the fact that the respiratory movements commence again only after the blood has already become dark, when, in consequence of this venous state of the blood, signs of irritation through the venosity of the blood appear in the heart, the vascular system and the intestine.   Apnea cannot be induced in young mammals.

2. The normal stimulus for the respiratory centers for quiet breathing, *eupnea*, is furnished by a state of the blood in which the amount of oxygen and carbon dioxid does not exceed the normal limits.

3. All factors that diminish the normal amount of oxygen and increase the amount of carbon dioxid in the blood circulating through the centers cause acceleration and deepening of the respiration, which finally may increase to the point of strained and laborious activity of all of the respiratory muscles.   This condition is designated *dyspnea*.

In case of normal breathing and beginning air-hunger the gases in the blood, according to Gad, irritate only the inspiratory center.   Expiration takes place reflexly through irritation of the pulmonary branches of the vagus stimulated by the distention of the lungs.   Gad believes that the normal respiratory movements are excited by the carbon dioxid.   If dyspneic blood be passed through the vessels of the brain of a normal animal, the latter becomes dyspneic.   In case of dyspnea in consequence of excessive physical activity, an as yet unknown body, formed as a result of the muscular activity, acts, in addition to the change in the gases of the blood mentioned, as an irritant for the center, perhaps as an acid.   The alterations in the dyspneic respiratory rhythm are described on p. 211.

4. If the abnormal conditions of the blood mentioned continue to exert an irritant effect, or if they are further intensified, there finally results a state of exhaustion in consequence of overstimulation of the respiratory centers; the respiration is again diminished with respect to the number and the depth of the movements, and later on only a few

gasping respirations occur. Then the exhausted muscles cease contracting entirely, and soon the movement of the heart also ceases. This condition is designated *asphyxia* and it may terminate fatally in *suffocation*. If, however, the causative factors can be removed, the asphyxia can be dissipated under favorable conditions by means of artificial stimulation of the respiratory muscles and of the cardiac activity, so that following the state of dyspnea that of eupnea may be again obtained. If the state of the blood becomes only gradually more and more venous, asphyxia may result without the signs of previous dyspnea, death taking place quietly and gradually. The condition here is in a certain measure one of insidiousness of the irritation.

Convulsions are associated with the dyspnea of acute onset. Extirpation of the cerebral hemispheres, likewise deep narcosis by means of chloroform, renders these slight or abolishes them. After removal of the optic thalami general convulsions appear not to take place.

Among the *causes of dyspnea* there should be mentioned:

1. Direct limitation of the activity of the respiratory organs: Diminution in the respiratory surface in consequence of inflammation, acute edema or collapse of the alveoli, occlusion of the alveolar capillaries, compression or collapse of the lungs in consequence of the entrance of air into the pleural cavities and stenosis of the air-passages. 2. Exclusion of the normal respiratory air through strangulation, enclosure in narrow spaces, drowning. 3. Failure of the circulation, in consequence of which a sufficient amount of blood is not sent to the medulla and as a result the necessary ventilation does not take place, as in connection with degenerations of the heart, valvular lesions, artificially through ligature of the carotid arteries, also obstruction to the escape of the venous blood from the cranial cavity, finally through the injection of large quantities of air or of indifferent bodies into the right heart. 4. Direct loss of blood, which may be effective likewise through interference with the gaseous interchange in the medulla. In this category belongs also the dyspneic gasping for air of the decapitated head, particularly of young animals.

If the rapidity with which these factors influence the respiratory activity be observed it will be noted that there occurs first accelerated and deepened breathing; there then follows, after the general convulsions and the associated expiratory spasm, a stage of complete respiratory rest, in relaxation, *asphyctic respiratory pause*. Finally, there occur only a few gasping inspirations before death takes place.

Generally the deficiency of oxygen and the excess of carbon dioxid tend at the same time to excite the dyspnea, although in the variations of the inspired air from the normal, the increase of carbon dioxid has an irritating effect earlier and in more marked degree than the diminution of oxygen. 1. Dyspnea from deficiency of oxygen occurs on breathing in closed space of moderate size, in a space the air of which is rarefied, as well as on breathing indifferent gases free from oxygen. On intense ventilation of the blood with nitrogen or hydrogen, the amount of carbon dioxid contained may even be diminished, and death takes place, nevertheless, amid the signs of asphyxia. 2. Dyspnea from excess of carbon dioxid occurs on breathing mixtures of gas rich in carbon dioxid (which form also on breathing for a long time in a closed space of considerable size or in an atmosphere of pure oxygen). Gaseous mixtures rich in carbon dioxid cause dyspnea even when the amount of oxygen contained is greater than that of the atmosphere. Even the blood itself may be found to contain more oxygen than normal.

Elevation of temperature also may stimulate the respiratory center to increased activity. This takes place even when the brain alone receives warmer blood, as was observed by A. Fick and Goldstein when they imbedded the exposed carotid arteries in heated tubes. In this experiment the heated blood obviously affects directly the medulla and the cerebral respiratory centers. Direct lowering of the temperature diminishes the irritability. When the temperature is elevated, apnea cannot be induced by means of forced artificial respiration and the resulting arterialization of the blood. Emetics act in the same manner.

Kronecker and Marckwald found *electrical irritation* of the center also effective. Irritation of the medulla oblongata separated from the brain excited respiratory movements or increased these if already present.   Langendorf observed in consequence of electrical, mechanical or chemical irritation (with salt) generally an expiratory effect; on the other hand, after irritation of the cervical cord (subordinate center), an inspiratory effect.   According to Laborde, a superficial lesion in the neighborhood of the apex of the calamus scriptorius causes arrest of the respiratory movements of a few minutes' duration.

If arrest of the heart is caused by irritation of the peripheral stump of the divided vagus, arrest of respiration for a few seconds ensues at the same time. In consequence of the arrest of the heart, transitory anemia occurs, as a result of which the irritability of the respiratory center is diminished, so that respiration ceases for a time.   The observations of Ahlfeld are most remarkable, in which in spite of abolition of the action of the heart in the new-born the respiratory movements still persisted, as in dogs after poisoning with antiar.

Reference has already been made to the great correspondence in the regulation of the respiratory and the intestinal nervous systems (p. 288).

In addition to direct irritation of the respiratory center locally, it may be influenced also by the will and reflexly through a number of centripetal nerves.

Through the *will* it is possible to suppress the breathing for only a short time, that is until the increased venosity of the blood excites the respiratory center to renewed activity.   The number and the depth of the movements can be increased for a considerable time; in addition the will has an influence upon the rhythm.   The influence of the cerebral cortex upon the respiration is considered on p. 790.

The respiratory center can be influenced *reflexly*, and there are both excitor and inhibitory nerves.   The nerves through which the respiratory center is stimulated reflexly are contained in the pulmonary branches of the vagus, also in the sensory nerves of the eye, the ear, and the skin.   Under normal conditions their action preponderates over that of the inhibitory nerves.   Thus, for example, a cold bath makes the respirations deeper and thus causes moderate acceleration of pulmonary ventilation.

Influence of the Vagus.—Division of the vagus on both sides causes, in consequence of removal of the influence of these stimulating fibers, slowing of the respiratory movements.   Under such circumstances, the entire amount of air exchanged remains for a time unchanged, but respiration is deepened and takes place with excessive and inadequate inspiratory effort.   In agreement with experimental section, subsequent feeble tetanizing irritation of the central stump of the vagus is again followed by acceleration of the respiratory movements.   More marked irritation causes arrest of breathing in inspiration or (particularly when the nerve is exhausted) in expiration.   Irritation of the sensory nerves of the thoracic and abdominal wall causes retardation of breathing when both vagi are divided.

According to Lewandowsky slight irritation of the central stump of the vagus by means of induced currents causes diminished depth of inspiration.   Then, as the strength of the current is increased, respiration becomes accelerated.   A still stronger irritation causes the thorax to assume the inspiratory position, then arrest in inspiration and finally irregular restlessness of the respiratory movements.   If the *constant current* be employed, closure of an ascending current or (in chloral-narcosis) an uninterrupted ascending current applied to the central stump of the vagus causes arrest of breathing in expiration or slowing of the respiratory rhythm (inhibitory effect); while an interrupted descending current, as well as closure of descending current, cause arrest of breathing in inspiration, or acceleration of breathing (stimulating effect).

Chemical irritation of the central stump of the vagus by means of sodium iodid or potassium chlorid causes expiratory arrest of breathing.   Momentary irritant influences cause inspiratory, continued irritation, expiratory effects.   The active fibers are situated in the uppermost root-bundles of the centers of the ninth, tenth and eleventh nerves in the rabbit.

If one lung is atelectatic (devoid of air), the pulmonary fibers of the vagus of the same side are inirritable. Section of the vagus upon the side of the healthy lung acts, therefore, in the same way as section of both vagi.

The *inhibitory fibers* which act upon the center for the respiratory movements pass in the superior and inferior laryngeal nerves to the respiratory center. The recurrent fibers are inactive in deep narcosis.

Even direct electrical, mechanical or chemical irritation of the center itself may inhibit respiration, perhaps because the irritation affects the central extremities of the inhibitory fibers at their point of entrance into the ganglia.

Irritation of the inhibitory fibers or their central stumps causes, therefore, slowing and even cessation of breathing in expiration. Irritation also of the nasal branches of the trigeminus, and the orbital branches, as well as the olfactory and the glossopharyngeal, causes arrest of breathing in expiration, as does also irritation of the pulmonary fibers of the vagus by the introduction of certain irritant gases into the lungs. Chemical irritation of the trunk of the vagus, by means of weak solutions of sodium carbonate, induces especially expiratory inhibition of breathing; mechanical irritation (rubbing with a glass rod), inspiratory inhibition. Also the irritation of sensory cutaneous nerves, particularly of the chest and the abdomen, for example by a sudden cold douche, and also of the splanchnic nerve, causes arrest in expiration, the former often after preceding clonic spasm of the respiratory muscles. The influence of irritation of sensory nerves upon respiration is in general widely distributed, thus, for example, that of the sensory fibers of the phrenic nerve, the heart, the aorta, the abdominal viscera. The slowing of the respiration in conjunction with pressure upon the brain is particularly noteworthy, the breathing not rarely becoming labored and stertorous.

In man irritation of the nasal mucous membrane in inspiration causes at first inhibition of breathing in the phase present at the time; then follows inspiration. During the period of reflex slowing of respiration the amount of work performed by the respiratory muscles is altered, and particularly the work in the slow respirations is increased through fruitless efforts at inspiration. On the other hand, it has been found that the volume of gases interchanged in the lungs remains the same during corresponding periods, and that, also, the respiratory interchange of gases is at first not altered directly.

Under normal conditions the pulmonary branches of the vagus appear to affect the respiratory centers through a mechanism of self-regulation in such a manner that the inspiratory dilatation of the lungs, and the associated rarefaction of the contained air, exerts a mechanical irritation upon the nerve-fibers stimulating the expiratory center reflexly. Conversely, the expiratory diminution in the size of the lungs, and the resulting increase in intrapulmonary air-pressure, causes irritation of the nerve-fibers passing to the inspiratory center. These fibers are situated in the posterior portion of the true trunk of the vagus.

According to Lewandowsky there is a special expiratory center, although in normal breathing the inspiratory center alone is active, rhythmically stimulating the inspiratory muscles and then permitting them to relax.

Deglutitional breathing, that is, a slight contraction of the diaphragm after each act of swallowing, occurs as an irradiation from the irritation of the swallowing-center to the respiratory center.

The Excitation of the First Respiratory Movements.—The fetus immediately after birth is in a state of apnea, as oxygen is abundantly supplied to it through the placenta. All factors that interfere with this supply, therefore especially compression of the umbilical vessels and persistent uterine contractions, cause reduction of oxygen and increase of carbon dioxid in the blood, and in consequence a state of the blood results that stimulates the respiratory center, and with this the impulse for the respiratory movement itself. Thus the fetus within the unopened membranes may be stimulated to respiratory movements. If the factors interrupting the gaseous interchange persist, the stimulated respiration becomes dyspneic, and finally death occurs from asphyxia. If the venosity of the fetal blood develops gradually, as, for example, in case of slow, quiet death of the mother, the medulla oblongata of the fetus may die gradually without the development of respiratory movement, without, therefore, the interruption of fetal apnea. This is a paralysis due to slowly insidious irritation.

Accordingly, the respiratory movement is excited in the medulla directly by the dyspneic state of the blood. Asphyxia of, the mother may have the same effect as compression of the umbilical vessels. In such an event the maternal blood rapidly becomes venous and abstracts the oxygen from the blood of the fetus. in consequence of which the death of the latter is accelerated. If the mother has been rapidly asphyxiated by carbon monoxid the life of the fetus may be prolonged, as the carbon-monoxid hemoglobin of the maternal blood naturally can remove no oxygen from the fetal blood. When the poisoning takes place slowly carbon monoxid also passes over into the fetal blood.

In many instances, especially when, after persistent uterine contractions, the irritability of the respiratory center is already greatly enfeebled, the dyspneic state of the blood, which becomes even more marked after birth, is not in itself sufficient to stimulate the respiratory movements in rhythmic and typical form. For this purpose there is required in addition irritation of the external integument, for example through lowering of the temperature on evaporation of the amnial liquor in the air. If, also, in consequence of the first movements that follow, air has entered the respiratory passages, the air may exert a stimulating influence upon the pulmonary branches of the vagus.

According to the observations of v. Preuschen the stimulation of the respiratory center through the nerves of the external integument is more effective than that through the branches of the vagus to the respiratory organ. Also in animals that have been made apneic by means of vigorous artificial respiration, this observer noted active respiratory movements setting in after application of cutaneous irritants, such as a douche of cold water. Mechanical cutaneous irritants, such as friction or slapping, support advantageously the stimulation of the respiratory center, as does also douching with cold water or irritation with the electric brush. If the placental circulation is completely intact, cutaneous irritation, however, alone induces no respiratory movements.

Artificial Respiratory Movements in the Asphyxiated.—In man, it is customary, for purposes of resuscitation in the presence of asphyxia, to practise artificial respiratory movements. The subjects under such circumstances have usually been suffocated, strangulated or drowned, or are children born in a state of asphyxia (intrauterine suffocation). The first duty in the presence of such a condition is the removal from the air-passages of foreign matters, such as mucus or edematous fluid in the newborn or the asphyxiated, water in the case of the drowned, by lowering the head; in desperate cases, even after tracheotomy, by suction through an elastic catheter introduced into the opening. Next, artificial respiration must be undertaken at once. Various devices and methods have been described for this purpose, but these cannot be considered in detail here. Alternate dilatation and contraction of the chest, and thereby gaseous interchange, can be effected by rhythmic compression of the thorax by application of the flat hand. The asphyxiated individual is placed in the dorsal decubitus, the vertebral column being flexed backward (with the aid of suitable support) as far as possible. The mouth is held open and the tongue (which would depress the epiglottis by falling backward) is drawn forward. Artificial dilatation of the thorax can be effected by stimulating the phrenic nerves at suitable intervals by means of sponge-electrodes connected with an induction-apparatus. The electrodes are placed in the situation of the anterior surface of the scalene muscle, irritation of which will augment the inspiration. In desperate cases air may be blown directly into the opened trachea through an elastic tube by means of a bellows or with the mouth. Care, however, is required in this connection, in order to avoid injury to the lungs. Artificial respiration has a vivifying effect through both the supply of oxygen to, and the removal of carbon dioxid from, the blood ; therefore particularly favoring the movement of the blood in the heart and in the large vessels of the thorax, and thus stimulating the circulation. If the action of the heart has already ceased, resuscitation cannot be hoped for. In the case of asphyxiated newborn children efforts at resuscitation should not be abandoned too early, that is before cessation of the heart-beat, even if at first they appear hopeless, as the medulla retains for a long time some measure of irritability. Pflüger and Zuntz observed the reflex irritability and the heart-beat persist in the fetus for several hours after death of the mother. In the case of resuscitated newborn children the resuscitating measures should be suspended only after loud crying has taken place.

Reference should be made here to the remarkable experiments of Böhm, who succeeded by means of rhythmic compression of the heart in conjunction with artificial respiration in resuscitating animals (cats) whose respiration and heart-beat had ceased entirely for forty minutes in consequence of asphyxia or poisoning with potassium-salts or chloroform and in which the carotid pressure had fallen. The compression of the heart causes a slight movement of blood (much like a feeble systole); at the same time the compression acts as a rhythmic stimulus for the heart. The heart-beat returns first, then also the respiration. The resuscitated heart-beat itself causes interchange of air. After restoration of breathing reflex irritability also returns and gradually likewise voluntary movements. The animals are blind for a few days, the brain torpid in function, the urine rich in sugar. The experiments show the great importance, in the resuscitation of asphyxiated individuals, of simultaneous action upon the heart.

For *physiological purposes* artificial respiration is practised by blowing air by means of a bellows into a tracheal cannula provided with a small lateral opening for the escape of the expired air. If the animal is at the same time paralyzed by curare it cannot be thrown into a state of disturbing restlessness in consequence of independent and reflex movements of the musculature of the body.

Pathological.—If the lung is distended with air, it cannot be deprived of this by direct compression, probably because in consequence of the direct pressure affecting the lung the small bronchi are compressed before air can escape from the pulmonary alveoli. If, however, a lung be filled with carbon dioxid instead of air and if it be suspended under water, the carbon dioxid will be absorbed by the water and the lung may thus become entirely airless (atelectatic). The occurrence of atelectasis in certain portions of the lung in connection with disease of this organ can be explained in this manner. If bronchi are occluded by mucus or exudate, marked accumulation of carbon dioxid takes place in the related pulmonary vesicles. This becomes the greater the more richly the blood in the lungs (in consequence of the existing disease of the lung) is itself impregnated with carbon dioxid. If, finally, the carbon dioxid is absorbed from the capillary blood of the alveoli, or from the lymph, the affected pulmonary area may become atelectatic.

Among the pathological phenomena that are caused by abnormal (direct or usually reflex) irritation of the respiratory center are spasm of the respiratory muscles, inspiratory, expiratory or complex spasm; also attacks of diminished respiratory frequency (spanipnea) or increased respiratory frequency (pyknopnea) observed in neurotic individuals, together with dyspnea and a sense of fear.

As the brain has relations to the respiratory movements the modification in these movements in connection with cerebral disorders are readily explained. The paralytic affections are, as a rule, upon the same side as the paralysis. Also Cheyne-Stokes breathing is observed.

THE CENTER FOR THE INHIBITORY NERVES (DIMINISHING THE FREQUENCY AND THE STRENGTH) OF THE HEART AND THE FIBERS PASSING TO THE VAGUS.

The fibers of the vagus nerve, moderate irritation of which diminishes the activity of the heart, strong irritation causing arrest of the heart, and which are conveyed to the vagus through the accessory nerve, have their center in the medulla oblongata far to the side of the floor of the fourth ventricle near the restiform body. The center sends to all portions of the heart, including the muscles of the superior vena cava, fibers that diminish the number of beats and others that diminish the vigor of the contractions. Slight irritation of the vagus occasionally exerts an inhibitory effect only upon the auricles. The force-diminishing fibers at the same time also prolong the diastole. If the diastole is rendered difficult by increased pressure within the pericardium, irritation of the vagus is believed to cause a prolongation of the diastolic distention.

This center can be stimulated both directly and reflexly from centripetal nerves.

Many investigators assume that this center is in a state of tonic innervation, that is that impulses pass out from it uninterruptedly through the vagus 'exerting a regulatory and inhibitory influence upon the heart-beat. According to Bernstein this tonic irritation is induced reflexly through the abdominal and cervical cords of the sympathetic. Landois did not accept this view, but maintained that under normal conditions of the respiration and the state of the blood, the center is not irritated, but that it is placed in a state of irritation only under special conditions.

Direct Irritation of the Center.—The center can be irritated locally by the same influences that affect the respiratory center. 1. *Sudden anemia* of the medulla oblongata (through ligation of both carotid and both subclavian arteries, or through decapitation of a rabbit, with preservation of the vagi alone) causes slowing and even temporary arrest of the heart-beat. 2. Sudden *venous hyperemia*, which can be brought about by ligation of the veins passing from the head, has a similar effect. 3. Also increased *venosity of the blood*, either through direct interruption of breathing (in the rabbit), or through insufflation into the lungs of a gaseous mixture containing much carbon dioxid, acts in a similar way. As, with marked uterine contractions, the circulation in the placenta, the actual lung of the fetus, is interfered with, the constant enfeeblement of the heart's action in association with severe uterine contractions is to be looked upon as a dyspneic, central irritation of the vagus. 4. The moment when *inspiration* takes place as a result of irritation of the respiratory center there is a fluctuation in the irritation of the cardiac inhibitory center. 5. Also *increased blood-pressure* in the cerebral arteries stimulates the cardiac inhibitory center.

That the center (in rabbits) is under normal conditions not in a state of tonic innervation was demonstrated in 1863 by Landois by the fact that when, after exposure of the vagi, care was taken, by means of artificial respiration, that the number of heart-beats remained exactly the same as in the intact rabbit, section of both vagi failed to cause increase in pulse-frequency. These observations were confirmed by Schiff. It is true that in dogs after division of' the vagi (in adult dogs and never in the newborn) sudden increase in pulse-frequency and in blood-pressure has been observed occasionally, but by no means constantly. The frequency of the pulse of the previously resting animal under observation should, however, be carefully determined first; and it should also be noted whether the preparations for the experiment did not cause' slowing of the pulse. Then, the section itself may cause irritation the accelerator fibers in the vagi or the pressor fibers, which likewise accelerate the heart-beat. In the dog whose vagi are paralyzed after injection of curare into the veins, with maintenance of artificial respiration, the heart-beat is not accelerated, and in the frog section of both vagi is invariably unattended with acceleration of pulse. Also, the increase in blood-pressure after division of both vagi is not solely dependent upon the associated increase in the pulse-rate that occurs.

The cardiac inhibitory center can be stimulated *reflexly:* 1. Through the *irritation of sensory nerves.* 2. Through *irritation of the vagus* itself, as by irritation of the central stump of one vagus, with preservation of the other. 3. Irritation of the sensory nerves of the abdominal viscera, by percussion of the abdomen of the frog (Goltz's percussion-experiment), has a cardiac inhibitory effect; as does also irritation of the splanchnic directly or of the abdominal and cervical cords of the sympathetic. Severe irritation of sensory nerves, however, inhibits the reflexes affecting the vagus described, and has a reflex inhibiting effect generally.

The experiment of Goltz succeeds at once if the irritation is permitted to act upon the exposed intestines (of the frog), which become inflamed on protracted exposure to the air. Also in dogs irritation of the stomach causes slowing of the pulse.

The irritation of the cardiac inhibitory center can be diminished reflexly, according to Hering, by vigorous distention of the lungs with atmospheric air. Under such circumstances there is marked reduction in blood-pressure.

In man forcible expiratory effort causes acceleration of the heart-beat in consequence of the increased intrapulmonary pressure, and this has been attributed by Sommerbrodt to reduction in the activity of the cardiac branches of the vagi, which are in a state of tonic innervation. At the same time a depressant effect is exerted upon the vasomotor center.

In the entire course from the center downward through the trunk of the vagus and further on through its cardiac branches, irritation causes slowing and enfeeblement and finally cessation of the activity of the heart. In the frog this result can be brought about even by irritation of the fibers of the vagus at the venous sinus of the heart. Feeble irritants slow the heart-beat, while stronger irritants cause diastolic arrest. If irritants of considerable intensity affect either the center or the course of the nerve for a considerable period of time, the irritated area becomes exhausted and the heart again pulsates more rapidly in spite of the persistent irritation. If, however, the site of irritation is displaced nearer to the heart, renewed inhibition takes place, as the irritation now affects a new nerve segment.

With reference to the irritation of the inhibitory fibers the following points are worthy of note: 1. It is probable that the fibers diminish the number of heart-beats and those diminishing the strength of the heart are distinct, both with reference to their anatomic arrangement, as well as with respect to their susceptibility to various poisons. The experiments of Heidenhain on frogs, confirmed by Löwit, have shown that electrical and chemical stimulation of the vagus have varying results with reference to the size and the number of the heart-beats. Either the contractions become only smaller, or they become only less frequent, or they become smaller and at the same time less frequent. Those branches of the vagus that in the frog are situated in the nerves of the septum exert an influence alone upon the strength and the tone. Those fibers, however, that enter the frog's heart outside of the nerves of the septum have an influence alone upon the number of heart-beats. In the turtle also, both sets of fibers are anatomically distinct. 2. To obtain the inhibitory effect persistent irritation is not necessary, but a moderately rapid rhythmic, interrupted irritation will suffice: from eighteen to twenty irritations in a second in warm-blooded and two or three in cold-blooded animals. 3. Donders observed in association with Prahl and Nuel that the inhibition manifested itself not immediately at the moment of irritation, but that from one-sixth to two-fifths of a second elapsed before the onset of the action.

After removal of the irritation, the heart still remains for a short time in a state of rest. Irritation of the vagus has thus an *inhibitory after-effect*. 4. Also chemical irritation of the center is effective: a crystal of sodium chlorid placed upon the medulla of the frog inhibits the heart-beat. 5. If the heart has been arrested by irritation of the vagus, it makes a single coördinated contraction on direct irritation (for example by a needle-prick), although the contractions of the heart in vagus-arrest, both after irritation, as well as those that arise secondarily in a portion of the heart in consequence of irritation of another portion, take place with greater difficulty, especially in the auricles 6. In the water-turtle the inhibitory fibers are, according to A. B. Mayer, contained only in the right vagus. Landois found this by no means constant in rabbits. 7. The nerve can occasionally be successfully stimulated mechanically also in human beings by digital compression against the cervical portion of the vertebral column; although alarming attacks of syncope have been observed to follow this procedure and for this reason its practice is to be cautioned against. 8. The behavior of the vagus nerve in the electrotonic state has been considered on p. 663 and the contraction-law for the same nerve on p. 665. 9. Schiff found that irritation of the vagus in the frog caused acceleration of the pulse (through an action upon the accelerator fibers contained in the vagus), after he had displaced the blood in the heart by a solution of sodium chlorid. If, subsequently, blood-serum is again introduced into the heart, the inhibitory action of the vagus is restored. 10. Many sodium-salts, naturally in suitable concentration, are capable of abolishing the inhibitory action of the vagi; while, conversely, potassium-salts possess the property of restoring the inhibitory function of the vagi suspended by the action of sodium-salts. Both sodium-salts and potassium-salts, however, can, after protracted action, induce a state in which the restitution of the suspended inhibitory function of the vagi is no longer possible. Under such circumstances the heart-beat is generally arrhythmic. 11. If the pulsations of the heart are greatly accelerated in consequence of high intracardiac pressure the activity of the cardiac branches of the vagus is correspondingly diminished. This is the case also in connection with the simultaneous action of direct cardiac irritants. The lessened activity of the vagus in conjunction with high internal pressure within the heart occurs only if the auricles and the venous sinus (in the frog) are at the same time greatly distended. 12. In the frog reduction in temperature impairs the inhibitory influence of the vagus, while elevation of temperature increases it. In the newborn and in the state of hibernation the irritation is ineffective.

Among *poisons* muscarin irritates the terminations of the vagus in the heart, and it may even cause diastolic arrest, which may then be neutralized by atropin. Digitalin reduces the frequency of the heart-beat by irritation of the vagus-center. Doses of considerable size diminish the irritability of the vagus-center and at the same time increase that of the accelerator ganglia of the heart, so that the frequency of the heart-beat is increased. In small doses digitalin also increases the blood-pressure by irritation of the vasomotor center and the structures of the vessel-walls. Nicotin first stimulates the vagus (and the resulting arrest can be neutralized by curare or atropin) and then paralyzes it; as does also hydrocyanic acid. Atropin and curare paralyze the vagi, as do marked reduction in temperature and high fever.

THE CENTER FOR THE ACCELERATOR AND AUGMENTING CARDIAC NERVES AND THE FIBERS TO WHICH IT GIVES RISE.

It is more than probable that the medulla oblongata contains a center that, on the one hand, sends to the heart accelerating fibers, increasing the number of heart-beats, and, on the other hand, fibers that increase its systolic force. These pass from the medulla, in which the exact situation of their origin has not yet been determined, downward in the spinal cord and enter through the communicating branches of the inferior cervical and the six upper thoracic nerves into the sympathetic. Thence, a main branch of these fibers passes principally through the first thoracic ganglion of the sympathetic and the loop of Vieussens, and hence to the cardiac plexus. This nerve is designated the accelerator nerve of the heart. Accordingly, irritation of the medulla, of the lower extremity of the divided cervical cord, of the inferior cervical ganglion (stellate ganglion), or of the superior dorsal node, is attended with acceleration of the heart-beat and increase of its strength (in the dog and the rabbit); or, if the heart's action has already ceased, with renewal of its beat, without change in blood-pressure.

It is probable that the accelerator nerves and those increasing the strength of the heart are distinct, both with respect to their anatomical arrangement, and with regard to their susceptibility to various poisons.

When the medulla oblongata or the cervical cord is irritated, the vasomotor nerves situated in them are also irritated. In consequence, the vessels that derive their motor fibers from the irritated area contract, and the blood-pressure is markedly increased. As, however, the increase in blood-pressure alone causes acceleration of the heart-beat, the irritation described does not directly demonstrate the existence of accelerator fibers in these central structures. The experiment would be convincing only if before the irritation is applied the blood-pressure were enormously lowered by destruction of the splanchnic nerves, so that the former could no longer exert an accelerating influence. Indirectly it can be demonstrated also that, if all of the nerves of the cardiac plexus, therefore also the accelerator fibers, are extirpated, after irritation of the medulla or the cervical cord the pulse-frequency does not rise (in consequence of increase in blood-pressure) in the same degree as before the extirpation.

The center is in any event not in a state of tonic irritation, for section of the nerve does not cause slowing of the heart. Destruction of the medulla or of the cervical cord itself likewise has a negative effect. Nevertheless, in this instance also, the splanchnic nerve must be previously destroyed, to bring about marked lowering of the blood-pressure, in order that the reduction in the number of heart-beats that occurs in consequence of the lowered blood-pressure after destruction of the cord shall not be incorrectly interpreted as being due to destruction of the accelerator center.

Cardiac accelerator fibers pass, according to the statements of earlier investigators and of v. Bezold, in part into the cervical sympathetic, in part through the vagus to the heart, and irritation accelerates the heart-beat or strengthens the cardiac contractions, or both. The inhibitory fibers of the vagus lose their irritability more readily than the accelerator fibers, but they are more irritable than the latter.

The fibers of the vagus that influence the force of the contractions are situated in the frog in the nerves of the septum. The acceleration of pulse attending increased muscular activity is attributable to irritation of the accelerator fibers, occurring in conjunction with stimulation of the motor nerves, while the irritation of the inhibitory nerves is diminished. The acceleration appears especially in debilitated convalescents. The heart, after a period of increased activity, later resumes its normal action. Practice in the form of activity favors such resumption. The cases described by Tarchanoff and van de Velde are most striking. In these, human beings were able, solely through the influence of the will (at rest and without alteration of respiration), to increase the number of pulse-beats even to twice the normal.

Direct irritation of the accelerator nerve gives rise to slowly developing effects, which disappear gradually after cessation of the irritation. If the vagus and the accelerator are irritated simultaneously, only the inhibitory action of the vagus makes its appearance. According to Hunt, in the case of this simultaneous irritation, the action of that nerve appears which is most strongly irritated. If during the activity of the accelerator nerve, the vagus is suddenly irritated, prompt reduction in the number of heart-beats occurs, and if the irritation of the vagus ceases the acceleration soon begins again. The activity of the accelerator nerves (in the frog) is enfeebled by cold and increased by heat.

According to experiments of Stricker and Wagner section of both vagi in the dog causes reduction in the number of heart-beats when the accelerator fibers on both sides are divided. This circumstance would indicate a state of tonic innervation of the latter.

The center can be stimulated reflexly through irritation of the central stumps of many sensory nerves.

THE VASOMOTOR CENTER AND NERVES.

The *dominating center*, which supplies all of the muscles of the arterial system with motor fibers (vasomotors, vasoconstrictors), is situated in the medulla oblongata at a point in part rich in large ganglia. It is 3 mm. long and 1½ mm. wide in the rabbit and extends from the region of the upper portion of the floor of the fourth ventricle to about 4 or 5 mm. above the calamus scriptorius. Each half of the body has as its own center, which is situated 2½ mm. from the middle line in that portion of the medulla on each side that represents the prolongation of the lateral columns of the spinal cord (lower portion of the superior olive). Irritation of this central point causes contraction of all of the arteries and in consequence increase in arterial blood-pressure, the veins and the heart becoming distended. Paralysis of the center causes relaxation and dilatation of all of the arteries, with enormous reduction in blood-pressure. Under normal conditions the vasomotor center is in a state of moderate tonic excitation. Like the cardiac inhibitory and the respiratory center, the vasomotor center can be stimulated directly and reflexly.

Direct Stimulation of the Center.—In this connection the amount of gases contained in the blood circulating in the medulla oblongata is of paramount importance. In the state of apnea the center appears to be in a condition of slightest excitation, as the blood-pressure is exceedingly low. With the state of the blood present under normal conditions the center is in a condition of moderate excitation. Fluctuations in the irritation of the center accompany the respiratory movements (Traube-Hering fluctuations), as can be seen from the simultaneous increase in blood-pressure. When the blood presents marked venosity, in consequence of asphyxia or insufflation of air rich in carbon dioxid, the center is more actively stimulated, so that all of the arteries contract, with marked increase in blood-pressure, and the venous system and the heart are greatly distended with blood. Under such circumstances the velocity of the blood-current is increased. The same effect is produced by sudden anemia of the medulla through ligation of both carotid and subclavian arteries, and likewise by sudden stagnation of the blood in the presence of venous hyperemia.

The venosity of the blood that always develops after death causes quite constantly active stimulation of the vasomotor center, as a result of which the arteries are strongly contracted. As, in consequence of this, the blood is driven to the capillaries and the veins, the state of emptiness of the arteries after death, which was familiar to the ancients, is explained.

Upon this circumstance is dependent also the fact, as Landois has found, that hemorrhage from large wounds takes place much more freely when the vasomotor center is preserved than when it has previously been destroyed (in the frog). As emotional disturbances have a corresponding influence upon the vasomotor center, their influence upon the control of hemorrhage is obvious. Thus, it has been observed in hysterical individuals that a wound yields only one-third as much blood as the same condition in a normal individual. If the hemorrhage is considerable, the anemic irritation of the medulla may finally exert a constricting influence upon the bleeding artery. In this way is to be explained the phenomenon familiar to surgeons that dangerous hemorrhage often ceases as soon as anemic syncope occurs. In the frog, after ligation of the heart, all of the blood is finally driven into the veins, likewise as a result of anemic irritation of the medulla. In mammals the equalization of the blood-pressure in the arterial and the venous system that follows exclusion of the heart takes place more slowly after destruction than after preservation of the medulla.

Among *poisons*, strychnin stimulates the center directly, even in curarized dogs; and nicotin and calabar bean have the same effect.

In animals in which the center is irritated electrically, it has been found that single induction-shocks of moderate strength are effective only when two or three shocks occur in a second. There is thus a summation of the effects of the individual stimuli. The maximum vasoconstrictor effect, which can be recognized from the maximum blood-pressure, is observed as a result of from ten to twelve strong or from twenty to twenty-five moderately strong shocks in one second.

The *course of the vasomotor nerves* is such that in part medullated and in part non-medullated nerve-fibers, partly mixed with ganglion-cells, pass to the muscular coats of the vessels. They pass from their center in part directly through the tract of some of the cerebral nerves to their distribution; through the trigeminus in part to the interior of the eye, through the hypoglossus to the tongue, through fibers of the vagus to the heart and in limited number to the lungs and to the intestines. All other vasomotor nerves descend in the lateral column of the spinal cord (so that irritation of the lower extremity of the divided cord causes constriction of the vessels supplied from a lower level), and are connected within the gray matter with centers of subordinate significance by means of contact. They make their exit through the anterior roots of the spinal nerves, then pass through the visceral branches into the ganglia of the sympathetic cord, where the ganglion-cells are intercalated in the course of the individual fibers. In the sympathetic cord they pass upward or downward and finally hence either to the vascular plexuses or through other visceral branches again into the trunks of spinal or cerebral nerves and from these to the respective vessels.

In detail, the distribution in the cerebral region is as follows: The cervical division of the sympathetic supplies in largest measure the head. In its area of innervation the great auricular nerve in some animals also supplies a number of vasomotors, which, in the rabbit, however, are derived from the inferior cervical ganglion of the sympathetic. The cerebral vessels are supplied principally by the sympathetic, irritation of which slows the blood-current in the small cerebral arteries and increases the resistance in them; on the other hand dyspnea, as well as administration of chloroform and amyl nitrite, causes acceleration of the blood-current. The nerves reach the cerebral vessels not only through the cervical sympathetic, but also through other tracts. The superior ganglion of the cervical sympathetic supplies the thyroid gland.

The upper extremities receive their vasomotor nerves through the anterior roots of the dorsal nerves from the fourth to the tenth and thence through the sympathetic cord to the first thoracic ganglion and from this through visceral branches to the brachial plexus. The vasomotors for the skin of the trunk are derived from the dorsal and lumbar nerves. The last three dorsal and the three uppermost lumbar nerves contain the fibers for the lower extremity (in the dog), which first pass through the sixth and seventh lumbar and the first and second sacral ganglia and then enter the trunks of the lumbar and sacral plexuses.

The lungs are supplied (in addition to a number of fibers in the vagus) by the first thoracic ganglion. According to Fr. Franck the sympathetic supplies vasoconstrictors to the lesser circulation, arising from the second and third dorsal nerves. They are stimulated reflexly through irritation of sensory nerves. The activity of the vasomotors of the lesser circulation is relatively slight. In the frog the vagus supplies the vasomotors of the lungs.

The splanchnic is the most important of all of the vasomotor nerves, supplying the abdominal viscera. Its vasoconstrictor fibers arise from the fifth dorsal nerve and below. Irritation of the communicating branches between the eleventh dorsal and the second lumbar nerve causes marked dilatation after primary contraction of the vessels. Dilatation is caused also by irritation of the vagus. Asphyxia causes contraction of all of the vessels of the entire intestine, the liver, and the pancreas. Irritation of sensory nerves, for example the crural, causes reflex contraction of the vessels of the small intestine, the kidney, the spleen, the pancreas, and dilatation of the vessels of the large intestine. Irritation of the centripetal fibers of the vagus causes dilatation of the vessels of the intestine, the pancreas and the kidney. The vasomotors of the liver are discussed on p. 313, those of the kidneys on p. 514, and those of the spleen on p. 195. The vasomotors are all medullated from their origin to the sympathetic cord.

In general the vessels of the skin of the trunk and the extremities are innervated by those nerves that supply the same parts also with other, for example sensory, fibers.

The various vascular areas respond differently with respect to the intensity of the action of the vasomotors. These affect in greatest degree the vessels of the peripheral portions of the body, for example the toes, the fingers, the ears, less markedly the central areas, for example the lesser circulation.

Reflex Stimulation of the Center.—The most varied centripetal nerves contain fibers, irritation of which has an influence on the vasomotor center. Irritation of some of these nerves causes stimulation of the center and, thus, increased contraction of the arteries, together with increased blood-pressure. These are designated *pressor fibers.* On the other hand, there are fibers irritation of which causes reflex diminution in the irritability of the vasomotor center. The result is, therefore, the opposite of the former. The nerves act really as inhibitory nerves of the center and are designated *depressor fibers.*

Pressor fibers have already been pointed out as present in the superior and inferior laryngeal nerves, also in the trigeminus, the direct irritation of which has a pressor effect, which occurs also as a result of insufflation of irritating vapors into the nose. Aubert and Roever discovered pressor fibers in the cervical sympathetic. S. Mayer and Pribram observed that mechanical irritation of the stomach, particularly of the serosa, has a pressor effect. Irritation of any sensory nerve is said to cause first a pressor effect.

Thus O. Naumann observed after feeble electrical irritation of the skin at first a pressor effect, namely contraction of the mesenteric vessels, the lungs and the web, with simultaneous excitation of the heart's action and acceleration of the circulation (in the frog). Strong irritation, however, had the opposite or depressor effect, with simultaneous reduction in the heart's action. Grützner and Heidenhain observed a pressor effect from contact with the skin, while manipulations causing severe pain were ineffective. Reflex alteration in the lumen of the vessels·and in the activity of the heart can be induced also through the cutaneous application of heat and cold. Schüller observed contractions of the vessels of the pia (in rabbits) after pinching of the skin, likewise after warm baths or the application of warm compresses, while cold baths or compresses caused dilatation of the vessels. Schüller interprets these phenomena in part as pressor and depressor effects, although he considers the principal cause to consist in the contraction of the cutaneous vessels in consequence of the cold, resulting in increase of the blood-pressure and therefore in dilatation of the vessels of the pia. Heat naturally has the opposite effect.

In man most forms of irritation of the sensory nerves, such as slight cutaneous irritation, tickling (also disagreeable odors, a bitter or a sour taste, optical or auditory irritation), cause reduction of temperature at the point of application, and diminution in the volume of the affected extremity, at times also increase in the general blood-pressure and alteration in the action of the heart. The opposite effects are induced by painful impressions, as well as by the action of heat (also agreeable odors and a sweet taste). The forms of irritation first mentioned cause at the same time dilatation of the cerebral vessels and increase in the amount of blood in the skull, while the latter bring about the opposite results. The time for the reflex is between three and five seconds.

Depressor fibers, irritation of which lowers the activity of the vasomotor center, are present in many nerves. The depressor nerve of the vagus has already received special mention. The trunk of the vagus below the latter contains depressor fibers, and so also do its pulmonary branches (in the dog). The latter have a depressant effect also in connection with marked expiratory pressure. In accordance with this fact Hering showed that marked distention of the lungs (at a pressure of 50 mm. of mercury) caused lowering of the blood-pressure and acceleration of the heart-beat. Irritation of sensory nerves, particularly if intense and long continued, causes dilatation of the vessels in the areas innervated by them. Also irritation of the muscle-nerves by pressure has a depressant effect. According to Latschenberger and Deahna all sensory nerves contain both pressor and depressor fibers.

Schiff observed after irritation of sensory nerves the periodic contractions in the ear of the rabbit, normally occurring from three to five times in a minute, give way to dilatation, after a preceding contraction of short duration. Direct pressure upon an artery within its distribution has a depressor effect, as can be seen, for example, from the fact that after long-continued pressure of the sphygmograph the pulse-tracing becomes larger and exhibits signs of lessened arterial tension.

In the intact body slowly alternating contraction and dilatation, without uniform rhythm, are observed in the arterial branches (arteries of the rabbit's ear, in the membrane of the bat's wing, the web of the frog's foot). This movement, discovered by Schiff, is for the purpose of supplying the organ in question at times with a larger, at other times with a smaller, supply of blood, accordingly as its nutrition or external influences demand. This phenomenon can appropriately be designated *periodic-regulatory vascular movement*. It is probably responsible, in case of occlusion of the vessels, for example after ligature, for the prompt establishment of the collateral circulation. This occurs with distinctly greater difficulty after division of the nerve.

According to Bier the final cause resides in the cellular activity in the tissues that have become anemic. After transitory anemia the small vessels of the skin open widely for the reception of the arterial blood-current, while they close to the venous current, and, moreover, independently of the central nervous system. Nothnagel agrees with v. Recklinghausen that the increased velocity with which the blood flows through the collateral branches of the obstructed vessel is the factor that causes hypertrophy of the walls of the vessel and dilatation of the lumen of the collateral branches.

Perhaps the arteries are capable of another form of movement, namely the *pulsatory*, which consists in active contraction after each pulsatory dilatation of the vessel. It would, therefore, correspond with the registration of the descending limb in the tracing. From what has been said as to the propagation-velocity of the pulse-wave, this contraction must be propagated centrifugally in a peristaltic manner with the same velocity as the pulse-waves. It should, however, be stated that, as yet, this form of movement has not been demonstrated with certainty.

The lumen of the vessel can be influenced directly by local application, cold and moderate electrical stimulation causing contraction, and, conversely, heat and strong mechanical or electrical stimulation, causing dilatation, the latter two probably after transitory preceding contraction.

Elevation of the temperature of the arm to 43° C. causes relaxation of the vessels, *reduction* to between 10° and 20° C. contraction. Abrupt alterations in temperature always cause transitory contraction of the vessels (also those of the opposite arm). Irritation by heat and cold is capable, in addition to its influence upon the vessels themselves locally at the seat of irritation, also of affecting the lumen of the vessels through reflex stimulation. Thus cold, for example, may cause dilatation of the vessels. *Electrical stimuli* likewise exert their effects principally in a reflex way. Among *poisons*, almost all of the members of the digitalis-group cause constriction. Quinin and salicin cause constriction of the vessels of the spleen. The remaining febrifuges cause dilatation of the vessels, as does also Witte's peptone.

The influence of the vasomotor nerves upon the *temperature*, both of limited portions of the body as well as of the entire body, is of great significance.

Local effects. Division of a peripheral vasomotor nerve, for ex-ample the cervical sympathetic, causes dilatation of the vascular area supplied by it, as the paralyzed vessels are readily distended by the intra-arterial pressure. In consequence, a larger amount of arterial blood at once enters this area, and as a result *injection-redness* develops, and, at the same time, also in parts that readily become cool, such as the ear and the skin of the face—*elevation of temperature*. Increased transudation takes place through the walls of the relaxed capillaries. Within the dilated vessels the velocity of the blood-current is, naturally, diminished, while the blood-pressure is increased. Further, the pulse is more readily palpated in such situations, because the lumen of the vessel is increased. With the increased size of the blood-current the blood may be bright red as it enters the veins and the pulse may even be followed into the veins. Every irritation of a peripheral vasomotor nerve gives rise to the opposite phenomena, namely pallor, diminished trans-udation and reduction in temperature in the external integument. Smaller arteries become contracted to the point of complete disappear-ance of their lumen. Long-continued irritation causes finally exhaus-tion of the nerve and gives rise at the same time to symptoms of paralysis of the vessel-wall.

The phenomena described as following paralysis of vasomotor nerves do not, however, remain unchanged. The paralysis of the muscular coat of the vessels must obviously give rise to stagnation in the circulation of the blood, as the muscular coat constitutes an important factor in the normal distribution of the blood in the vessels. The slower blood-movement is responsible for the fact that parts exposed to the air become more readily cooled. Thus, the primary stage of elevation of temperature after division of the vasomotor nerves may be followed by a second stage of reduction in temperature. As a result of numerous experi-ments Landois was able to confirm the observation of Schiff that in rabbits from which the cervical sympathetic had been removed some weeks previously, the ear upon the intact side was always warmer, and particularly if the animals were actively stimulated, in consequence of which the circulation in the intact vessels was accelerated. If, as, for example, in the paralyzed extremities of human beings, the muscle-nerves are paralyzed in addition to the vasomotors, the extremity will become cooler in the course of time, because the paralyzed muscles are no longer capable of generating heat by their contraction, and, further, because the dilatation of the muscle-vessels, which occurs with each con-traction of the muscles, is lost. If, finally, the paralyzed muscles undergo atrophy the vessels contained in them also become reduced in size. There is thus afforded an explanation for the fact that paralyzed extremities in human beings are as a rule cold in the further course of the case, although primarily the temperature is elevated.

If in consequence of the same procedure the vasomotor nerves of extensive areas of the skin are paralyzed, as, for example, in the lower half of the body after section of the dorsal cord, so much heat is given off by the dilated vessels that either an elevation of the temperature of the skin is observed for only a short time and in slight degree or reduction in temperature takes place at once. Thus, some observers have noted elevation of temperature after division of the cervical cord, although Riegel did not.

Pflüger found that a rabbit with division of the cervical cord produced more carbon dioxid when the surrounding temperature was elevated and less when the temperature was lowered. The human being injured in a similar manner exhibits analogous conditions, being more readily cooled when the surrounding temperature is low and more readily overheated when the surrounding temperature is high.

Influence upon the Temperature of the Entire Body.—Irritation or paralysis of vasomotor nerves within small areas has practically no influence upon the temperature of the entire body. If, however, the vessels in an extensive area of the skin are suddenly dilated by paralysis of their vasomotor nerves, the temperature of the entire body falls, because much more heat is given off from the dilated vessels than under normal conditions. This is the case, for example, after high division of the spinal cord. Inhalation of two or three drops of amyl nitrite is also attended with reduction in the bodily temperature in man in consequence of the resulting dilatation of the vessels of the skin. Under opposite conditions of irritation of extensive areas the temperature of the body is elevated because the constricted vessels give off less heat. This fact explains in part febrile elevations in temperature.

Also the cardiac activity, that is the number and the energy of the contractions of the heart, is greatly influenced by the state of irritability of the vasomotor nerves. If these nerves are paralyzed throughout considerable areas, the vessels whose walls contain muscle-fibers dilate, and the blood itself does not reach the heart with its usual rapidity and abundance, as the pressure under which it flows has become considerably lessened. The consequence is that the heart makes extremely small, slow and labored contractions, somewhat like a damaged pump, to which sufficient material is not sent for propulsion onward. Stricker even observed arrest of the heart in the dog on extirpation of the spinal cord between the first cervical and the eighth dorsal vertebra. Conversely, it is known that on irritation of the vasomotor nerves, in consequence of the resulting contraction of the vessels with a muscular coat, the blood-pressure rises considerably. As the arterial pressure is effective up to the left ventricle, it causes, as a mechanical irritant to the wall of the heart, an increase in the activity of the heart, with respect both to the number of beats and to their vigor, in the course of a short while. As a result, the circulation, already accelerated by the increase in pressure in the arterial system in consequence of the arterial contraction, is further accelerated.

By far the most extensive area of the circulation is controlled by the splanchnic nerve, as it innervates the large branches of all of the arteries of the abdomen. Irritation of this nerve is, therefore, followed by marked increase in the blood-pressure. Conversely, paralysis of the nerve is attended with such marked stagnation of blood in the dilated abdominal vessels that all the remaining portions of the body become anemic, and death may even result, in a measure in consequence of intravascular hemorrhage. For the same reason animals die of anemia after ligation of the portal vein.

The capacity of the interior of the vascular system, by reason of its dependence upon the vasomotor nerves, has obviously also an influence upon the *bodily weight*, especially in consequence of variations in the amount of fluid taken up into or given off from the blood. Strong irritation of the vasomotor apparatus may cause a fall in bodily weight through rapid loss of water. In this category belongs probably the loss of weight observed by some after epileptic convulsions, in consequence of polyuria, increased sweating, secretion of tears or of saliva. Conversely, paralysis or paresis of the vasomotor nerves causes dilatation of the blood-stream, with increase in the weight of the body.   Such a result is brought about by a number of poisons, for example alcohol in large doses. After disappearance of the intoxication the normal weight is restored after copious urination.

The *trophic disorders* that accompany affections of the vasomotor nerves are deserving of especial consideration. *Paralysis* of the vasomotors gives rise, in addition to vascular dilatation and local increase in the blood-pressure, also to increased transudation from the capillaries.   In consequence of the loss of the muscular activity in the vessels the blood-stream becomes slowed, and stagnates; as a result, the capillaries are dilated and the slowly moving blood in them becomes markedly venous, so that the skin acquires a livid color.   Further, normal transpiration is interfered with, so that dryness of the epidermis results, and often also desquamation and fissuration.   Passive hyperemia, a tendency to occlusion of the capillaries and to the formation of thrombi in the veins, together with passive transudates and edematous swelling, are not rare. Also the normal growth of the hair and the nails is readily interfered with, the skin exhibits increased vulnerability and the nutrition of all of the remaining tissues may suffer.   In consequence of long-continued irritation of vasomotor nerves the amount of blood passing through the affected vessels becomes diminished, and it may be conceived that, as a result, nutritive disturbances occur in the parts to be supplied.   Tangl found on long-continued faradic stimulation of the spinal cord a reduction in oxidation-processes in the tissues, as a result of which gaseous interchange, and finally also the bodily temperature, fall markedly.

In addition to the dominating vasomotor center in the medulla oblongata, the vessels are under the control of *subordinate centers in the gray matter of the spinal cord*.   This can be recognized from the following observation: If the spinal cord be divided in an animal, all of the vessels supplied by nerves arising below this level soon undergo paralytic dilatation, in consequence of section of the vasomotors from the medulla. If the animal survive, the vessels regain their previous caliber in the course of a few days, and the rhythmic movements of their muscular coat are now controlled by the subordinate vasomotor centers in the lower extremity of the spinal cord.

The subordinate spinal centers can be stimulated directly through a dyspneic state of the blood. Reflex stimulation also is possible; after destruction of the medulla oblongata the arteries of the web of the frog's foot contract on irritation of the sensory nerves of the opposite hind leg.   In the dog a spinal vasomotor center susceptible of reflex irritation is situated between the third and sixth dorsal nerves (origin of the splanchnic), and a similar center is present in the lower portion of the spinal cord.   According to Spina the cerebral vessels have such a center extending to the third cervical vertebra.

If, after the section, the lower extremity of the spinal cord is crushed, the vessels again undergo paralytic dilatation, in consequence of destruction of the subordinate centers. Even now, however, in surviving animals the dilatation is gradually replaced by normal contraction and rhythmic movement, and henceforth this movement of the vessel-wall is controlled by the *ganglia* everywhere distributed throughout it.   The latter are thus capable of acting independently and of maintaining the movement of the vessel-wall.   Increased tension in the vessel causes contraction of the muscular coat.   Even the vessels of excised surviving kidneys, through which blood is passed, exhibit these periodic fluctuations in caliber.   The observation is further worthy of mention that the vessel-walls contract as soon as the state of the blood becomes in marked degree venous.   The vessels oppose a greater resistance to the flow of venous blood than to that of arterial blood.   Perhaps the general disturbance of nutrition exhibited by individuals suffering from dyspneic states of long standing is to be explained in this way.   In any event the vessel-walls, however, appear after the series of procedures described not to attain again the complete mobility and reactivity that they possess under normal conditions.

Through the intermediation of these peripheral vessel-ganglia the movements of the vessels also appear to take place that are observed to occur on application of direct mechanical, chemical or electrical irritation to the vessels. The arteries contract often to the point of obliteration of their lumen. Amyl nitrite and digitalis have an effect upon the lumen. The pulsating veins in the bat's wing continue their movement after division of all of the nerves, and this is indicative of the local innervation through peripheral nerve-centers.

Finally, the cerebrum undoubtedly has an influence upon the vasomotor center, as is shown by the sudden pallor of the external integument in conjunction with emotional disturbances, such as fright, fear. This observation has received a satisfactory explanation in the discovery made by Eulenburg and Landois that the gray cortex of the cerebrum contains a circumscribed area (in the cruciate sulcus in the dog), irritation of which gives rise to reduction of temperature and destruction to elevation of temperature in the contralateral extremities. From this area fibers therefore probably pass to the center in the medulla, which they stimulate to either increased or diminished activity. In this way is to be explained the fact, as observed by Landois together with Budge, that irritation of both cerebral peduncles causes contraction of all of the vessels. Heidenhain noted, accordingly, that irritation in the further course, at the junction between the pons and the medulla oblongata, caused rapid rise in the bodily temperature.

Emotional influences generally increase the tone of the vessels (as observed in the arm), while fatigue and joy diminish it.

Although the medulla contains a dominating vasomotor center for all of the vessels in common, it is to be assumed that this is divisible into a number of central points lying close together and controlling definite vascular areas. In this connection there have been isolated the centers for the vessels of the liver and for those of the kidneys.

Finally, it should be mentioned that certain *poisons* especially stimulate the vasomotor apparatus, such as ergotin, tannic acid, balsam of copaiba and cubebs; while others at first stimulate and then paralyze, such as chloral hydrate, morphin, laudanosin, digitalin, veratrin, physostigma, alcohol; while still others rapidly paralyze, such as amyl nitrite, carbon monoxid, atropin and muscarin. The paralyzant action of poisons is recognized from the fact that after division or paralysis of the cardiac fibers of the vagus and of the accelerator nerve irritation of either the pressor or the depressor nerves is unattended with any effect. Certain agents having a pathological effect have an influence upon the vasomotor nerves. In surviving organs, narcotics and antipyretics cause dilatation and members of the digitalis-group contraction.

The *veins* are controlled by vasomotor nerves, for example, the ear-veins of the rabbit through the cervical sympathetic, the portal vein through the splanchnic, the veins of the hind leg through the sciatic. On the whole, the venomotors pursue the same course as the arteriomotors and the sweat-fibers.

Little is known with regard to the dependence of the *lymphatics* upon the nerves. Camus and Gley observed on irritation of the peripheral extremity of the splanchnic nerve that the receptacle for the chyle generally dilated. Irritation of other sympathetic fibers caused contraction

49

of the thoracic duct and the receptacle, as did also suspension of breathing. Irritation of the thoracic cord of the sympathetic is followed by dilatation or contraction of the thoracic duct. The constrictor fibers, however, are more readily exhausted.

Pathological.—Disorders in the function of the vasomotor nerves (angioneuroses) may occur in different forms. The points of attack for the abnormal irritations of the vasomotor nerves may be the ganglia in the vessels themselves or the spinal centers, together with the dominating center in the medulla, or, finally, the cortical vasomotor centers in the cerebrum. . The action may be either direct or reflex. In conformity with the phenomena of physiological experimentation, irritation of the vasomotor nerves will give rise to contraction of the blood-stream, pallor and reduction of temperature in the external integument and diminished diffusion in the tissues. Conversely, paralysis must give rise, in addition to dilatation of the vessels, to elevation of temperature and redness of the integument, as well as increased transudation in the tissues.

In the skin, affections of the vasomotor nerves give rise, first of all, to diffuse redness or pallor, which may be unilateral. There may, however, also be circumscribed disorders, such as the local cutaneous arterial spasm induced by irritation of individual vasomotor nerves. Later on, various forms of paralytic phenomena involving .the cutaneous vasomotor nerves appear upon the skin, in the sequence of a number of acute febrile diseases, after preceding initial severe irritation of the vasomotors, especially in the stage of chill in the course of various fevers. These may appear as simple circumscribed areas of redness or as increased transudation from the paralyzed vessels, with the formation of wheals, or even escape of red and white blood-corpuscles from the paralyzed, greatly dilated vascular areas, or edema, eruptions or even partial gangrene. In individuals suffering from epilepsy or other severe nervous affections, peculiar, red angioparalytic areas of geographical outline have ·occasionally been observed (Trousseau's tâches cérébrales). Weir Mitchell, in 1872, designated as *erythromelalgia* an angioneurosis in which paroxysmal redness and swelling of the skin appear at the periphery of the extremities usually in association with pain. As occasionally trophic and secretory disorders also appear, the condition is not exclusively a vasomotor but a combined neurosis. Long-continued, strong irritation of the vasomotor nerves may cause interruption of the circulation, in consequence of which the affected parts may undergo gangrene, which may involve deeper parts, as well as the skin. Inflammations in cutaneous areas whose vasomotors are paralyzed are aggravated in the course of time.

Among the angioneuroses of circumscribed distribution is the unilateral spasm of the branches of the carotid on the head, which is attended with severe headache, so-called *sympatheticotonic hemicrania*. Under such circumstances the cervical sympathetic is greatly irritated; and pallor, relaxation and coolness of one-half of the face, cord-like contraction of the temporal artery, dilatation of the pupil and discharge of viscid saliva are unequivocal signs of this affection. Eulenburg has described as the converse of this disorder a *sympatheticoparalytic hemicrania*. in which at the height of the attack the opposite symptoms appear, in conjunction with paralysis of the sympathetic. This form may succeed immediately upon the first, as paralysis in the sequence of intense irritation. Berger even observed both forms in alternation.

Exophthalmic goiter is a remarkable affection of the sympathetic, in which the vasomotor nerves are involved. It occurs in individuals of a neurotic disposition, and there develop consecutively palpitation of the heart (from 90 to 120 or 200 beats in a minute), enlargement of the thyroid gland (struma) and protrusion of the eyeballs (exophthalmos), with defective associated movement of the upper eyelid in elevating and depressing the plane of vision. It is believed that perverted function of the thyroid gland results in the formation of materials that act like a poison on the nerves affected. As a matter of fact exophthalmos is wanting in one-half and goiter in one-fifth of the cases. It is possible that this obscure disease depends upon simultaneous irritation of the accelerator nerve of the heart, of the motor filaments for the muscles of Müller in the orbits and the eyelids, and perhaps also of the filaments for the unstriated muscles discovered by Sappey in the orbital aponeurosis, as well as of the dilators of the thyroid vessels. The disorder might arise as result of direct irritation of the sympathetic paths named or of their final areas of origin, or finally it might be the result of reflex irritation. On the other hand, the clinical picture has been explained by assuming that the exophthalmos and the goiter are results of paralysis of the vasomotors, which gives rise to distention of the vessels. The increased action of the heart is looked upon as a sign of diminished or abolished action of the cardiac inhibitory fibers of the vagi. All of these phenomena, it is said, can be induced by injury of the upper portion of the restiform body on each side in rabbits, and according to Durduñ below the auditory tubercle.

Landois was the first, in 1866, to describe and designate as *vasomotor angina pectoris* an affection of paroxysmal occurrence involving either all or at least a large number of the vasomotor nerves. In consequence of intense irritation the vessels undergo contraction; the arteries are hard and small, the skin, especially of the hands and the feet, pallid and cold, and at the same time the seat of formication and prickling at the tips of the fingers. The increase in blood-pressure brought about by the vascular contraction causes enormous acceleration of pulse, together with a feeling of oppression, of vertigo, of fear, of abolition of the vital functions and even painful palpitation of the heart.

The appearance of sudden hyperemia, with transudation and ecchymoses in individual thoracic or abdominal viscera must likewise be referred to an angioneurotic origin. In this connection it should be recalled that Schiff, Brown-Séquard and others observed hyperemia and extravasations of blood in the lungs, the pleuræ, the intestines and the kidneys, after injury of the pons, the striate body and the optic thalamus. Crushing or section of one-half of the pons causes, according to Brown-Séquard, especially extravasations of blood into the lung of the opposed side. The same observer noted also extravasations of blood into the capsules of the kidneys after injury to the lumbar cord. The pulmonary vessels may be relaxed through the intermediation of the nerves, and attacks of asthma may thus be induced. Rarely, the vasomotor distribution upon an entire side of the body has been observed to be irritated or paretic.

The dependence of *glycosuria* upon vasomotor influences has been pointed out on p. 313, the influence of the vasomotors upon the *urinary secretion* on p. 514. The effect of *fever* upon the vasomotor nerves in the form of irritation is shown by the pale skin in the stage of chill attending some fevers, followed by redness in consequence of consecutive paralysis. Sudden elevation of temperature of paroxysmal occurrence has been considered as a sign of irritation of the vasomotor center in the medulla.

Little is known concerning affections in the distribution in the *veins* dependent upon the nerves. Moltschanoff observed that in the sequence of inflammation of the ulnar and median nerves, with anesthesia, venous dilatation occurred in the distribution of the basilic vein.

It should, finally, be pointed out that sensory nerves in the form of delicate networks have been found on the blood-vessels. Pathological manifestations of pain in the course of the vessels, in association with arterial spasm, aneurysm, arteriosclerosis, and thrombosis, are probably indicative of morbid states of irritation of the nerves.

THE VASODILATOR CENTER AND NERVES.

Although a center for vasodilator or vessel-relaxing nerves has not yet been demonstrated, the existence of such a center in the medulla may nevertheless be suspected. It would, thus, be the antagonist of the vasomotor center. The center is, in any event, not in a state of permanent (tonic) irritation. The vasodilator nerves are analogous in function to the cardiac branches of the vagus, as irritation of both causes relaxation in the state of rest. The nerves may, therefore, be designated *vaso inhibitory nerves*. A dyspneic state of the blood irritates the center (as it does also that for the vasomotors), and, as a result, especially the cutaneous vessels are dilated, while at the same time the vessels of the internal organs become anemic in consequence of simultaneous irritation of the vasoconstrictors. Chloral hydrate in small doses is a stimulant to the vasodilators. Irritation of the depressor nerve also excites them reflexly.

The Course of the Vasodilator Nerves.—The vasodilators pass to some organs as special nerves, while to other parts of the body they are distributed, in association with vasoconstrictor and other nerves. The buccofacial region receives dilators in part from the medulla oblongata directly through the trigeminus, and in part from the spinal cord. The latter, according to Dastre and Morat, make their exit with the first, second and third dorsal nerves (in the dog) and pass through the visceral branches (limb of the loop of Vieussens) into the sympathetic cord, then to the superior cervical ganglion and, finally, thence through the carotid plexus to the Gasserian ganglion of the trigeminus. The retina receives vasodilator nerves through the sympathetic and the trigeminus; the ear from the first dorsal and the inferior cervical ganglion; the brain from the sympathetic; the heart through the sympathetic and less through the vagus; the upper extremity from the thoracic sympathetic; the lower extremity from the posterior roots of the origin of the sciatic. The vasodilators for the submaxillary and sublingual glands pass in the chorda tympani, as do also those for the anterior portion of the tongue. Those for the posterior portion of the tongue are contained in the glossopharyngeus, those for the thyroid gland in the laryngeal branches of the vagus, those for the liver in the splanchnic, those for the pancreas in the vagus, those for the small intestine in the splanchnic, those for the kidney in the vagus. The lungs (in the rabbit) receive dilators from the cervical sympathetic; according to Henriques (in dogs and rabbits) from the vagus. Irritation of the nervi erigentes arising from the sacral plexus causes erection through dilatation of the arteries of the penis. The muscles receive the dilator fibers for their vessels through the trunks of the motor nerves. If the muscle-nerves or the spinal cord be irritated, the lumen of the vessels undergoes dilatation during the contraction of the muscle-fibers. The latter phenomenon appears even when the contraction of the muscles is prevented. The vasodilators remain medullated out to the terminal ganglia.

The vasodilators have *subordinate centers* in the spinal cord, as do the vasomotors; for example the fibers of the buccolabial region at the level of the first, second and third dorsal vertebræ. These can be influenced reflexly through the pulmonary fibers of the vagus, but also through the sciatic nerve. According to Goltz, a similar center is situated in the lower portion of the spinal cord. reflex irritation of which can be induced through the visceral nerves. The portion of the cerebral cortex having vasodilator functions is described on p. 789.

Goltz showed, in 1874, that vasomotors and vasodilators are situated side by side in the trunks of the extremities, for example in the sciatic (through the intermediation of the sympathetic). If the peripheral stump of this nerve is irritated immediately after division the action of the vasomotors predominates. If, however, the peripheral stump is irritated in the course of from four to six days, during which time the vasoconstrictors will have lost their irritability, the vessels become dilated in consequence of the action of the vasodilators. Irritants affecting the nerves at long intervals stimulate especially the vasodilators. Tetanizing irritants, however, stimulate the vasoconstrictors. The latent period of the vasodilators is longer and they are also more readily exhausted than the vasoconstrictors. Reduction in temperature lowers the irritability of the vasodilators in lesser degree than that of the vasomotors. Exposure of the nerves directly to high degrees of temperature (up to 50° C.) causes irritation of the vasodilators for a long time, as do also closing and opening, as well as permanent continued passage, of the constant current. The phenomena described (which have been observed by Goltz, Heidenhain and Ostroumoff, Putzeys and Tarchanoff and others) can be explained by assuming that the ganglia situated in the vessels, in analogy with the automatic ganglia of the heart, are influenced through both sorts of vascular nerves, the vasoconstrictors causing excitation, the vasodilators inhibition of the activity of these ganglia.

Certain nerve-trunks contain fibers through which reflex dilatation of vessels can be induced, and, in addition, others through which reflex vasoconstriction can be brought about. The former are less sensitive to cold, are more irritable and regenerate more quickly after injury.

Irritation of the loop of Vieussens gives rise to pseudomotor contractions in the muscles of the face paralyzed in consequence of destruction of the facial nerve. in the same way as does irritation of the chorda tympani in the tongue paralyzed in consequence of section of the hypoglossus.

In an analysis of the phenomena related to the vessels, inquiry should be directed especially to determine whether such dilatations as may be present and due to nervous influences are a result of irritation of the vasodilators or of paralysis of the vasoconstrictors. This is of great significance with respect to the interpretation of pathological phenomena. Emotional influences may also affect the vasodilator center. Thus, the blush of shame, which may not only involve the face, but may also extend to the entire skin, is probably due to irritation of the vasodilator center.

The vasodilator nerves obviously have a marked influence upon the bodily temperature and upon that of individual portions of the body, as may be inferred from what has been said with reference to the influence of the vasoconstrictors.

It cannot be denied that both vascular nerve-centers represent important regulators for the dissipation of heat through the vessels of the skin. Probably they are maintained in activity by reflex influences through sensory nerves. Derangement in the function of these centers may result in abnormal accumulation of heat (as in the presence of fever) or in abnormal reduction of temperature.

THE SPASM-CENTER. THE SWEATING CENTER.

The medulla oblongata at its junction with the pons contains a center irritation of which causes general convulsions. This can be excited through sudden venosity of the blood (*asphyxial convulsions*), also through sudden anemia of the medulla oblongata either in consequence of rapid hemorrhage or after momentary ligation of both carotid and subclavian arteries (*hemorrhagic or anemic convulsions*); finally, also through the action of sudden venous stasis as a result of constriction of the veins passing from the head. Under all these conditions the irritation of the center must be looked for in the sudden interruption of the normal gaseous interchange. If these influences operate gradually, death may take place without the occurrence of convulsions, as the uninterrupted gaseous interchange always associated with the onset of quiet death shows. Also direct irritation by means of the application of chemical substances, such as ammonium carbonate, potassium-salts, sodium-salts, and others, is capable of rapidly exciting severe general convulsions. Finally, it has long been known that intense direct mechanical irritation of the medulla oblongata, as, for example, by sudden crushing, causes general convulsions.

As the convulsions occur only when the cerebrum is preserved, Bechterew adopts the view that this portion of the nervous system contains only a motor center, but no spasm-center, as with its destruction the power of locomotion ceases. Convulsions occur on irritation of the area after the irritation is conveyed first to the cerebral cortex.

According to Nothnagel the spasm-center in the rabbit extends above the ala cinerea upward to the quadrigeminate bodies. It is bounded laterally by the locus cœruleus, together with the auditory tubercle, and internally by the eminentia teres. The center is generally irritated in connection with extensive reflex spasm.

Numerous *poisons*, most heart-poisons, nicotin, picrotoxin, the salts of ammonia and the compounds of barium cause death preceded by convulsions, as they irritate the spasm-center.

Pathological.—Schröder van der Kolk pointed out that in cases of general convulsions in epileptics the seat of irritation is situated within the medulla oblongata, the vessels of which he found repeatedly dilated and increased in number. Under such conditions the medulla would be in a state of increased irritability. Now, it has been shown, in the discussion of the vasomotor centers, that irritation of sensory nerves may cause both sudden contraction as well as dilatation of the cerebral vessels. If this takes place in the vessels of the medulla, sudden anemia or transitory hyperemia will develop in that structure. Both conditions are capable, however, of irritating the medulla in such a manner that epileptiform convulsions result. It is often the case in connection with general epileptic convulsions that the nerve can be distinctly demonstrated, irritation of which gives rise to the vascular change. The peculiar sensation, aura, that occurs in the course of such a nerve before the outbreak of the convulsions has long been known. Naturally, the convulsions may be induced by direct irritation of the medulla of different character.

Brown-Séquard observed that guinea-pigs became epileptic after injuries of the central and peripheral nervous system—spinal cord, medulla, cerebral peduncle, quadrigeminate bodies, sciatic nerve; and the disease thus induced was even inherited. Irritation of the cheek and the anterior aspect of the neck—epileptogenous zone—excites the attack, and in the presence of unilateral injuries of the spinal cord and the sciatic, if the same side be irritated; in the presence of injuries of the peduncle, if the contralateral region is irritated. Westphal made guinea-pigs epileptic by repeated gentle blows upon the skull. There developed a perfect epileptic condition, which likewise was inheritable. As the cause he found extravasation of blood into the medulla oblongata and the upper portion of the cervical cord.

Epileptic convulsions after intense irritation of the motor cortical region of the cerebrum are discussed on p. 783.

A *dominating center for the secretion of sweat* for the entire surface of the body, to which the local centers in the spinal cord are subordinate, is situated in the medulla oblongata. It is bilateral and unequally irritable in the rare cases of unilateral sweating.

Physostigma, nicotin, picrotoxin, camphor, ammonium acetate, stimulate the secretion of sweat by a direct action upon the sweating center. Muscarin causes local irritation of the peripheral sweat-fibers; it therefore causes sweating of the hind paw even after section of the sciatic nerve. Atropin neutralizes the effects of muscarin.

PSYCHIC FUNCTIONS OF THE CEREBRUM.

The cerebral hemispheres in man are the seat of all psychic activities. Only when the former are intact are the processes of thinking, feeling and volition possible. After destruction of the hemisphere the organism is reduced to the level of a complex machine, the entire activity of which can be considered only as the expression of internal and external stimuli acting upon it. The psychic activities appear to be localized in both hemispheres and in such a manner that after extensive injury of one hemisphere the other, or after injury upon both sides, the remaining cerebral tissue, is capable of assuming vicariously the functions of that which has been destroyed.

Cases in which after extensive unilateral destruction of one hemisphere the psychic activities apparently had not suffered are not rare. Even when both hemispheres are destroyed in moderate degree the intelligence may be apparently intact. The statement that the psychic faculties have remained intact under such circumstances should, however, be received with caution, as it is obviously infinitely difficult to determine to what extent these had been developed in various directions prior to the accident. Exceedingly rare cases are known in which alternately one or the other, and then both hemispheres together, took part in the mental processes. Recently a certain violence of manner, maliciousness and indifference have been observed repeatedly after injuries of the frontal portion of the brain in man—changes that Ferrier attributes to loss of the conceptions acquired by education and association with others, and which agree with the analogous changes in animals described by Goltz.

Developmental Defects of the Brain.—Microcephalus and hydrocephalus cause loss or impairment of mental functions to the degree of most profound idiocy. Extensive inflammation, degeneration, pressure, anemia of the cerebral vessels, as well as the influence of narcotic drugs, abolish these functions altogether. The extent to which the hemispheres are concerned in these processes is, as yet, a matter of doubt. Flourens believed that the hemispheres take part in every psychic act throughout their entire extent. Therefore, even a small remnant of healthy hemisphere is sufficient for the maintenance of all its functions. To the degree in which the hemispheres are removed, all of the functions of the brain are impaired. If the latter is entirely eliminated, all of the faculties are lost. Therefore, neither the different functions nor the different impressions are localized in special situations. Goltz agrees with Flourens in the view that an uninjured remnant of the same kind of cerebral tissue is capable to a certain degree of assuming the functions of a destroyed portion. This capability on the part of portions of the brain to act vicariously for portions that have been lost is designated by Vulpian as "loi de suppléance"—law of functional substitution.

As opposed to the opinion of Flourens the phrenological teachings of Goll (died 1828) may be recalled. According to this observer the various mental functions are localized in definite situations in the brain. A conspicuous faculty always corresponds with a voluminous development of the respective portion of the cerebral cortex, which may even be recognized externally from the configuration of the skull—*cranioscopy*. Thus, the different mental functions are to be referred to certain portions of the cerebral cortex. Spurzheim, who elaborated the system of his friend, set up the following categories: The first class comprises the sensations, including the instincts and the feelings. The second comprises the faculty of comprehension, including the power of recognition and that of thought. Although the detailed application of this system exhibits a certain inflexibility, obvious deficiencies and undeniable error, nevertheless the question is worthy of serious consideration whether the fundamental thought of the system is entirely to be rejected. The discovery of the localization of movements under the control of the will and of conscious impressions and their association in the cerebrum justify renewed examination of the phrenological system, although in quite another manner than that pursued by the originator.

After removal of both cerebral hemispheres in animals all voluntary and consciously performed movement ceases, as well as every conscious sensation and sensory impression. On the other hand the entire mechanism, the harmony and the equilibrium of the movements persist, as well as those functions that, independent of the memory, have been designated as lower or instinctive. The latter functions are localized in the midbrain and are controlled through important reflex paths.

Sudden arrest of the circulation in the brain, for example through decapitation, is attended with immediate cessation of the mental processes. On permitting arterial blood from a living horse to pass immediately through the carotids of the decapitated head of a dog, Hayem and Barrier observed signs of maintained consciousness and of volition in the head for more than ten seconds, but not later.

The *midbrain* is connected not only with the gray matter of the spinal cord and the medulla oblongata, the seat of the most extensive coördinated reflexes, but it contains also sensory elements, as well as fibers, derived from the higher sense-organs, which may likewise have a reflex effect upon motility. Finally, the midbrain contains inhibitory apparatus for reflexes. The associated action of all of these parts makes the midbrain a controlling organ for the harmonious execution of movements, and in a higher degree than the medulla oblongata. This is seen especially from the fact that animals with the midbrain preserved are capable under varied conditions of maintaining the equilibrium of their body, which is lost at once if the midbrain is destroyed. Christiani determined the situation of the coördinating center for locomotion and the maintenance of equilibrium in mammals to be in front of the inspiratory center of the third ventricle.

The significance of the coöperation between cutaneous sensibility and sense-impressions for the maintenance of equilibrium will be made clear from the following considerations: The frog deprived of its brain at once loses its power of equilibration as soon as the skin is removed from the hind legs. The influence of visual impressions is recognized from the inability to maintain the equilibrium that is observed in connection with nystagmus, and from the vertigo that often accompanies paralysis of the external ocular muscles. In human beings with impaired cutaneous sensibility, the eyes constitute the main dependence for the maintenance of the equilibrium. Such individuals fall on closing the eyes.

The frog with its cerebrum extirpated maintains the harmonious equilibrium of its body. Placed upon its back, it at once rolls over; irritated, it makes one or two jumps; thrown into the water, it swims to the margin of the reservoir, climbs upon this and remains quietly seated. Under the most complex inciting conditions it exhibits complete control, harmony and uniformity of its movements. Without external irritation, however, it makes, at least at first, no independent voluntary, purposeful movement. On the contrary, it sits constantly in the same place as if asleep, it takes no food, it has no conscious sense of hunger or thirst, it exhibits no fear and, finally, it dries into a mummy.

The pigeon behaves in the same way when its cerebral hemispheres are removed. Unirritated, it remains seated as if in sleep, although if stimulated it exhibits complete coördination in all movements in walking, flying, perching, and balancing of the body. In the course of several days it changes its position apparently without external excitation. The sensory nerves and those of special sense, it is true, still conduct impulses to the brain, but these are capable of exciting only reflex movements, and are no longer capable of exciting conscious sensations. Therefore, the bird starts when a firearm is discharged in its vicinity, its eyes blink when a flame is brought near them and the pupils contract, it turns its head when the vapor of ammonia is applied to its nose. All of these stimuli, however, are not appreciated consciously as such. Conception, will, memory are lost, and the animal spontaneously takes neither food nor drink. If these are placed in the pharynx the animal swallows, and in such a manner its life may be preserved for months.

Fish behave somewhat differently. A carp whose cerebrum has been extirpated is capable of seeing and even of selecting its food and of moving voluntarily. Under these circumstances the psychic function must be located also in the optic thalamus. According to Schrader the frog is said in the further course of observation to behave in a similar manner. Reptiles also are able later to move spontaneously, although they exhibit neither fear nor anger. Birds also are said later on to exhibit spontaneous movement. Their organs of special sense functionate, but they are mind-blind, mind-deaf, etc.

Mammals. Goltz was able to remove the cerebrum from dogs and to keep the animals alive for a long time. Subsequently they exhibited good powers of locomotion and the ability to take food, as well as taste, tactile sensibility, hearing and muscular sensibility. They were sensitive to bright light, without, however, being actually able to see. In other respects the dogs were in a state of most profound dementia. Feeding alone affected them agreeably and they showed also a sense of satiety. In other respects the loss was evident of all those manifestations from which conclusions are formed as to the existence of intelligence, memory and judgment.

Observations on somnambulists show that also in man complete coördination of all movements may be present without the aid of conscious volition or conscious sensation and perception. Most ordinary movements in the waking state, however, take place without the aid of consciousness, being controlled from the midbrain.

The degree of development of the mental activities in the animal kingdom varies in accordance with the size of the cerebral hemispheres in proportion to the remaining portions of the central nervous system. If, however, the brain alone is taken into consideration it will be found that those animals possess the higher grade of intelligence in which the cerebral hemispheres greatly preponderate over the midbrain. The latter is represented in the lower vertebrates by the optic lobes, in the higher by the quadrigeminate bodies. In Fig. 258, VI represents the brain of the carp, V that of the frog, IV that of the pigeon. In all of these figures the hemispheres are indicated by the numeral 1, the optic lobes by the numeral 2, the cerebellum by the numeral 3 and the medulla oblongata by the numeral 4. In carps the cerebrum is even smaller than the optic thalami, while in frogs it is already larger than the latter. In pigeons the cerebrum extends downward to the cerebellum. In correspondence with these variations in size is the degree of intelligence present in the animals named. In the dog's brain (Fig. 258, II) the hemispheres cover the quadrigeminate bodies entirely, but the cerebellum still lies behind the cerebrum. Only in man do the occipital lobes of the cerebrum entirely cover the cerebellum.

According to Meynert these relations can be made clear in another manner. It is well known that fibers pass downward from the cerebral hemispheres through the cerebral peduncles, particularly their lower portion, which is known as the crusta of the peduncle. This is separated by the substantia nigra from the upper portion, which is designated the tegmentum and is connected with the quadrigeminate bodies and the optic thalami. The larger, therefore, the cerebral hemispheres the more numerous are the fibers passing to the crusta. In Fig. 254 at II is shown a vertical section through the posterior quadrigeminate bodies, including the aqueduct of Sylvius, and the two cerebral peduncles from an adult man: p p is the crusta of each peduncle, over which is the substantia nigra (s). Fig. IV exhibits the same relations in the ape, Fig. III in the dog and finally Fig. V in the guinea-pig. It will at once be seen that the size of the crusta diminishes in the order named. In correspondence with this there is an analogous diminution in the size of the cerebral hemispheres and at the same time of the intelligence of the respective animals.

Finally, the degree of intelligence is dependent upon the complexity of the fissures in the hemispheres. While the fissures are yet wholly wanting in the lower animals (fish, frog, bird, Fig. 258, IV, V, VI) two shallow fissures are present on each side in the rabbit (III). The brain of the dog already exhibits numerous convolutions (I, II). The complexity of the convolutions in the elephant, the most intelligent of animals, is striking. Even in evertebrates, as, for example, a number of insects endowed with delicate instincts, convolutions have been observed in the cerebrum. Naturally, it cannot be denied that even some animals of low intelligence, such as the cow, possess hemispheres with complex convolutions. A similar condition has often been found in man in association with marked mental development, although brains rich in convolutions have been observed also in incompetent persons. In the male sex the average absolute weight of the brain of the first two decades is greater than in females. The absolute weight of the brain cannot be taken as an index of the degree of intelligence. The elephant has the absolutely heaviest, man the relatively heaviest brain.

The cerebrum consists in all vertebrates of three divisions: the *olfactory lobe*, the *striate body* and the *cortex*. The olfactory apparatus is situated at the base and is well developed in fish, although it varies greatly in size in vertebrates. It is large in reptiles and small in birds. The *striate body* is pretty uniformly developed and serves as a means of connection between the optic thalamus and the forebrain; from birds and mammals onward connections between the thalamus and the cerebral cortex appear. The *cerebral cortex* is the most important portion of the brain with respect to mental development. In the bony fish and the ganoids it is represented by merely a thin epithelial plate. In reptiles a cerebral cortex related to psychic activity first appears, but at the beginning there is only an olfactory sphere. These animals are, therefore, the earliest that are able to retain olfactory impressions in memory and to utilize them psychically. In birds the visual sphere and the optic radiation appear first and these animals are therefore the lowest that appreciate visual impressions psychically. In mammals the other spheres are added. In mammals the relations between the sensory and the sensorial nerve-paths to the cerebral cortex increase progressively. The brain of the mammal, however, is characterized particularly by the remarkable development of association-paths.

Time-relations of Mental Processes.—For the occurrence of mental processes it is necessary for a certain time to elapse between the application of the stimulus and the conscious reaction. This *reaction-time*, which is much longer than the simple reflex time, can be measured by noting the moment of irritation, and then having the individual under examination make a signal indicating the resulting correct conception. The reaction-time will then consist: (1) Of the duration of perception (entrance into consciousness); (2) of the duration of apperception (consciousness of the special qualities of the sensation, such as form, pitch, color, etc.), (3) of the duration of the voluntary impulse (for the making of the signal). In addition there is to be taken into account (4) the time required in the propagation through the centripetal nervous apparatus and (5) through the motor nerve. If the signal is, as usual, given with the hand, the reaction-time for impressions of sound is from 0.136 to 0.167 second, of light from 0.15 to 0.224 second, of taste from 0.15 to 0.23 second, of touch from 0.133 to 0.201 second.

Heat is appreciated later than cold, pressure earlier than heat. The reaction-time for the perception of odor, which naturally is dependent upon many circumstances, such as respiratory phases, draft, is from 0.2 to 0.5 second.

Irritation of considerable intensity, increased attentiveness, practice, anticipation of familiar impressions shorten the time. According to Lange in the case of the sensorial reaction after prepared attentiveness, the apperception coincides with the perception. The muscular reaction, on giving the signal, may finally, however, be converted also into a simple reflex. In the case of tactile impressions those are most rapidly perceived that affect situations endowed with the greatest acuity of the spatial sense. The time may be prolonged in the case of strong irritants and in that of complex objects to be distinguished. The duration of apperception for a number of from one to three figures was in observations of Tigerstedt and Bergquist from 0.015 to 0.035 second. Alcohol and anesthetics alter the time, occasionally shortening it, or they prolong it, in accordance with the intensity of their effects. If two different impressions are to be recognized psychically in rapid succession a certain interval of time is necessary, which for the ear is 0.002 to 0.007 second, for the eye from 0.044 to 0.047 second, for the tactile organ of the finger 0.0277 second (for two electrical cutaneous stimuli from 0.022 to 0.056 second).

During sleeping and waking the periodicity of the active and resting state of the mind can be recognized. During sleep there is diminished irritability of the entire nervous system, which is explicable only in part through fatigue of the centripetal nerves, but is especially attributable in a peculiar manner to the central nervous system. During sleep stronger irritation is required in order to excite reflexes. During deepest sleep the psychic activities appear to be wholly at rest, so that the sleeping person may be compared to a being with extirpated cerebral hemispheres. Toward the time for awaking, however, psychic activities may appear in the form of dreams, but in a manner differing from normal psychic processes. They comprise either sensations of which the objective cause is wanting (therefore hallucinations), or volitional impulses or conceptions that usually are not executed and that are for the most part absent in the healthy logic of the thinking process during the waking state. Often, especially toward the time of awaking, actual stimuli are interwoven with the dream-images and they may affect various organs of special sense. Reduction in the activity of the heart, of the blood-pressure in the arteries, of the amount of blood in the brain, of the irritability of the motor cortical centers, of the activity of respiration, of gastric and intestinal movement, in the generation of heat, in the secretions indicates a lessening in the activities of the respective nerve-centers, and the diminished reflex activity a lessening in the activities of the spinal cord. The pupils during sleep are the smaller the deeper the sleep, so that in deepest sleep they cannot be made to contract on exposure to light. They dilate in response to sensory or auditory stimulation and in greater degree the less deep the sleep. They attain their greatest size at the moment of awaking. It appears that during sleep a state of irritation of the central organ exists through which increased activity of certain sphincter-muscles, such as the sphincter of the iris and that for closure of the eyelids is brought about. The soundness of sleep can be determined from the intensity of the sound that is required to cause awaking. Thus, Kohlschütter found that sleep at first rapidly, then more slowly deepens, and after an hour, according to Mönninghoff and Priesbergen after 1¾ hours, is most profound; then, at first rapidly and later more slowly it becomes again shallow and finally several hours before awaking continues in almost uniform shallow depth. External or internal irritation is capable suddenly of diminishing the depth, although renewed deepening follows. The deeper the sleep the longer it lasts.

The *cause of sleep* is the consumption of potential energy in the nerves, principally in the central organs, which renders restitution necessary. Perhaps accumulations of decomposition-products in the body (? lactates) induce sleep. The advent of sleep is favored by the removal as far as possible of all sense-irritations. Sleep cannot voluntarily be postponed indefinitely, or be interrupted. The hypnotic effect of many narcotics is remarkable. Absolute sleeplessness causes death (in the dog after one hundred and twenty hours), with reduction of temperature, diminished reflex activity and changes in the brain.

Hypnotism.—In connection with the subject of sleep reference should be made here to the most important results of investigations into *hypnotism* or *animal magnetism* as disclosed by the studies of Weinhold, Heidenhain, Grützner, Berger and others. As the cause of this condition Heidenhain considers an inhibition of the activity of the ganglion-cells of the cerebral cortex, induced by slight persistent irritation of the face (by means of gentle stroking, feeble electrical currents), or of the optic nerve (staring at a bright button), or of the auditory nerves (uniform sounds). Intense and sudden irritation of the same nerves rapidly abolishes the state, particularly blowing on the face. Berger attaches especial significance to the psychological influence of the artificially excited conception and attention and their concentration upon certain portions of the body. Schneider believes that the abnormal exclusive concentration of consciousness upon the act of hypnotization furnishes the cause for the phenomenon. The first hypnotization of an individual is effected with greatest difficulty, and long fixation of a brilliant object, which Braid recommended as early as 1841 for the development of an anesthetic state, appears in this connection to be of special significance; although the ability to be hypnotized varies greatly in different individuals. On repeated hypnotization the condition can often be induced with extreme ease, for example by means of simple pressure upon the brow or by placing the subject passively in a definite position or by stroking. In some individuals the mere conception of the approach of the condition is sufficient to induce it, as Cardanus observed in himself in 1553.

The hypnotized individual is first incapable of opening the closed eyelids. There is then spasm of the accommodative apparatus of the eye, the range of accommodation being diminished, and abnormal positions of the eye are observed. Next there appear irritative phenomena in the distribution of sympathetic nerves arising from the medulla oblongata, such as widening of the palpebral fissure, dilatation of the pupils, exophthalmos, acceleration of respiration and of pulse. At a certain stage a marked increase in the acuity of the special senses can at times be demonstrated, and also of muscular sensibility. Later on, analgesia may appear, with preservation of tactile sensibility and loss of the sense of taste. The temperature-sense disappears with greater difficulty, and still later the senses of sight, smell and hearing become affected. The stimuli affecting the organs of special sense cause no conscious sensory impressions on account of the suspension of consciousness. At the same time, however, the irritation of the organs of special sense may induce movements on the part of the hypnotized individual; such as unconscious acts that appear to be voluntarily executed in imitation of others. In this way is to be explained the fact that the hypnotized individual appears to perform even foolish acts on command, while he imitates movements first made by the experimenter, without consciousness of the significance of his acts. In individuals with greatly increased reflex irritability voluntary movements may excite reflex spasm, for example inability to make coördinated speech-movements.

According to Grützner there are several fundamental *types* of hypnotism: (1) *Quiet sleep*, words being still understood, and occurring especially in girls. (2) In consequence of increased reflex irritability of the transversely striated muscles, which may persist for days, groups of muscles become *contracted*, especially in strong persons. At the same time, there may be ataxia, and the muscles may fail to perform their function. Hypnotized individuals can be placed in positions of varied kind—artificial catalepsy. In the stage of hysterical lethargy the tendon-reflexes are at times increased. At the same time the muscles become firmly contracted as soon as they or their nerves are pressed upon. Nerve and muscle in the cataleptic state exhibit increased irritability to the constant current and diminished irritability to the faradic current. In the condition of hysterical catalepsy the tendon-reflexes are often entirely absent. (3) *Command-autonomy*, in which the hypnotized individuals are obedient in shallow sleep, at first with still preserved consciousness. When grasped by the hand or stroked upon the head they perform involuntary movements, such as running about, dancing, riding upon a chair and the like. The effects of so-called *suggestion* are peculiar, that is conceptions can be aroused in the hypnotized subject by suggestion and these may dominate the impulses and sensations of the individual for a considerable time. (4) *Hallucinations* occur, and only in certain individuals, on gradual awakening from deep sleep. The hallucinations, generally of phenomena related to fire and olfactory impressions, are usually quite profound, both the agreeable as well as the frightful ones, and they often recur in dreams. (5) *Imitation* is rare. Gross movements, such as walking, are readily imitated; more delicate or even the most delicate, principally in the uneducated, occur less commonly. *Echospeech* can be induced by pressure upon the neck and speaking into the pharynx, against the epigastrium and against the nape of the neck. Pressure upon the right eyebrow often inhibits speech. Color-perception is abolished or disturbed by applying the warm hands upon the eye or by stroking the opposite side of the head. Stroking in a direction opposite to that in which stroking had previously been practised gradually abolishes the rigidity of the members in sleep; blowing does this immediately. Insane persons are susceptible to hypnotism equally with healthy persons. Disagreeable complications arise only if the practice be overdone; if, for example, it be repeated daily for one or two weeks with the same person, who then readily falls spontaneously into a state of hypnotism and catalepsy. .

Hypnotic states can be induced also in animals. Hens (also after removal of the cerebrum) assume a rigid position if an object be suddenly placed in front of the eye, or a straw be placed over the beak, or a chalk line be drawn in front of the head pressed upon the ground (Kircher's miraculous experiment, 1644). Birds, rabbits, frogs remain irresponsive when held for a time by gentle pressure in a fixed position upon the back; crabs stand upon the top of the head, as well as the tips of their claws.

Hypnotism may be employed *therapeutically* in cases of color-blindness, insomnia, hysterical convulsions and emotional disturbances. Also the influence of suggestion may be important, but great care is necessary in its employment.

THE MOTOR CORTICAL CENTERS OF THE CEREBRUM.

Fritsch and Hitzig, in 1870, discovered upon the surface of the convolutions of the cerebrum a number of circumscribed areas, electrical stimulation of which causes movement in definite groups of muscles on the opposite side of the body (Fig. 258, I, II).

Method.—To the exposed gyri of the cerebrum (dog, ape) two blunt unpolarizable electrodes are applied close together, and stimulation is practised by means of closure, opening or alternation of a constant current, the strength of which causes a distinct sensation at the tip of the tongue; or the induced current is employed, the strength of which causes a readily tolerated irritation at the tip of the tongue. Luciani observed movements appear in consequence of mechanical stimulation by scraping. The cerebrum is wholly insensitive to painful manipulations.

The regions of the cerebral cortex, stimulation of which causes characteristic movements, must be considered as true centers, as is evident from the fact that the latent period after irritation of the centers and the duration of the muscular contraction are longer than if the subcortical fibers passing from the centers into the depth are irritated. In favor of this view, further, is the circumstance that the irritability of the areas in question can be modified by stimulation of centripetal nerves. Probably it is these centers upon which the will operates in the execution of intended movements, and for this reason they are designated *psychomotor* centers by Landois. The motor zone of the brain is shown to be a center also from the presence of special, large pyramidal cells.

There are animals that come into the world with completely developed motor and sensory functions. In these the motor cortical centers of the new-born are already irritable. In such animals, however, that are born with incomplete motor and sensory functions either the irritability of the cortex is still wanting entirely, so that only the deeper fibers of the corona radiata are irritable, or movements cannot yet be induced separately, and they are at the same time slower and more sluggish, with a longer latent period. Man may exhibit an analogous condition.

Deep narcosis, as well as apnea and asphyxia, abolish the irritability of the centers, while the subcortical conducting fibers retain their irritability. Interference with the blood-supply to the head gives rise to loss of irritability of the cortical centers and of the conducting fibers passing from them. After restoration of the circulation in the brain the irritability returns. Small doses of narcotic poisons, of atropin, moderate loss of blood, increased blood-pressure in the brain and slight inflammation increase the irritability, while more profound influences of the same kind abolish it, as does also direct application of cold or of cocain.

If the cerebral cortex is removed from an animal, the irritability of the fibers of the corona radiata disappears completely at about the fourth day, exactly as does that of a peripheral nerve separated from its center.

As the fibers (of the corona radiata or projection-system of the first order) pass from the cerebral cortex toward the center of the hemispheres, it can be understood that after removal of the cortex, inasmuch as the course of the nerve-fibers in the depth of the hemispheres is followed, the same motor effect can be obtained by irritation of those fibers. If the stimulation be thus continued progressively to the internal capsule, where the conducting fibers lie close together, general contractions of the contralateral muscles will be observed. The motor fibers are irritable also within the crus cerebri.

Time-relations of the Irritation.—According to Franck and Pitres 0.045 second elapses between the moment of stimulation of the cerebral cortex and the movement, after subtraction of the muscular latent period and the time for conduction through the spinal cord and the nerve of the extremity. Bubnoff and Heidenhain found that in morphin-narcosis of moderate degree the contraction became greater and the reaction-time shorter with increasing strength of the stimulating current. After removal of the cortex the total delay in the onset of contraction, after beginning stimulation of the white medullary tissue, is diminished from one-quarter to one-third. The form of the muscular contraction (contraction-curve) is longer and more extended if the cortex is stimulated than if the subcortical conducting path is stimulated. If the animal (dog) is in a state of marked reflex irritability, these differences do not appear. In either event the contraction takes place rapidly. In case of strong irritation, the muscles of the same side also contract, though somewhat later than those of the opposite side. If the motor point for the foreleg and that for the hind leg are stimulated at the same time, the latter contracts the later. If the stimulus is applied to a motor point forty times in a second the muscles in question make forty individual contractions. With forty-six separate stimuli in a second persistent contraction results. In the same animal the same number of stimuli are necessary for the production of sustained contraction, whether the cortical center or the motor nerve or even the muscle is irritated.

In the case of exceedingly feeble stimulation the phenomenon of *summation of stimuli* is observed, the muscular contractions commencing only after several at first ineffective stimuli. The time required for the voluntary inhibition of a movement already present is about equal to the time for the voluntarily induced movement.

The **situation of the motor centers** in the brain of the dog can be seen in Fig. 258, I and II. For purposes of orientation it should be stated that the surface of the brain in the dog exhibits two *primary fissures*, the *cruciate sulcus* (S) which intersects the longitudinal sulcus, dividing the hemisphere in its anterior third, almost at a right angle. The second primary fissure is the *fossa of Sylvius* (F). Four *primordial convolutions* are arranged in a definite relation to these primary fissures. The first primitive convolution (I) surrounds with marked flexion the sharply defined fossa of Sylvius (F). The second primitive convolution (II) passes almost parallel to the first. The fourth primitive convolution is bounded in the middle line by the convolution of the opposite side. It surrounds the cruciate sulcus (S) anteriorly, so that the portion lying in front of this can be readily differentiated as the precruciate gyrus from the postcruciate gyrus lying behind it. The third primitive convolution (III) is in general parallel with the fourth.

In Fig. 258, I and II, the situations of the motor centers are indicated by dots, although their position varies somewhat and may even be different upon the two sides of the brain. It should, however, be stated that the individual centers do not have merely a punctate extent, but that in accordance with the size of the animal they represent areas the size of a pea and larger, whose central points are indicated by the dots in the illustration.

Fritsch and Hitzig isolated the following motor centers: (1)′ For the muscles of the nape of the neck: a second center was found by Werner below 7. (2) For the extensors and abductors of the foreleg. (3) For flexion and rotation of the foreleg. (4) For the movements of the hind leg, which Luciana and Tamburini were able to separate into two centers with antagonistic effects. (5) For the muscles of the face, or the center for the facial nerve (according to these investigators often more than 0.5 cm. in diameter). Ferrier has discovered the following additional centers: (6) For the lateral wagging movements of the tail.

(7) For retraction and abduction of the foreleg.   (8) For elevation of the shoulder and extension of the foreleg (walking movement).   The area 9 9 9 controls the movements of the orbicular muscle of the eyelids, the zygomatic (closure of the

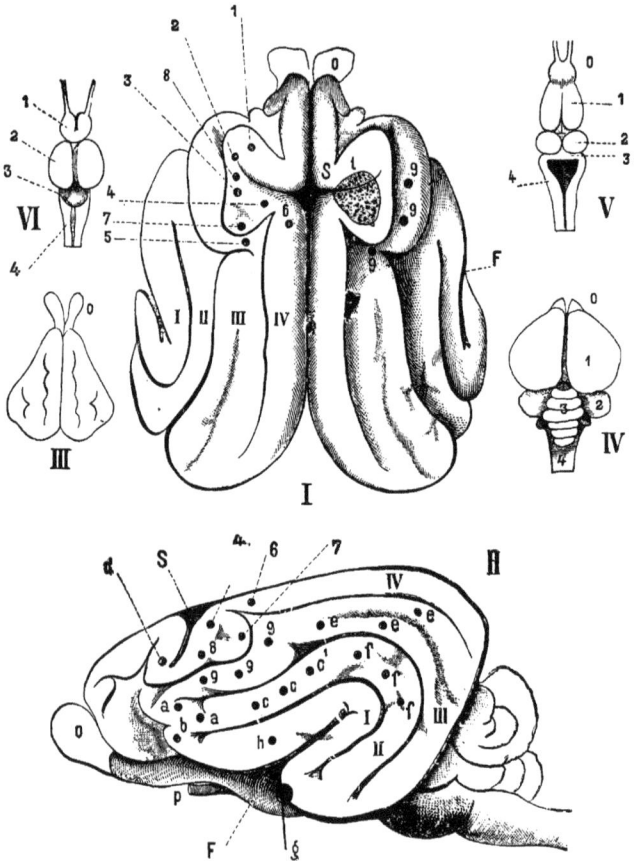

FIG. 258.—I, Cerebrum of the dog, viewed from above; II, from the side. I, II, III, IV, the four primitive convolutions; S, the cruciate sulcus; F, the fossa of Sylvius; o, olfactory bulb; p, optic nerve; 1, motor point for the muscles of the nape of the neck; 2, for the extensors and abductors of the foreleg; 3, for the flexors and rotators of the foreleg; 4, for the muscles of the hind leg; 5, for the facial nerve; 6, for lateral wagging movements of the tail; 7, for retraction and abduction of the foreleg; 8, for elevation of the shoulder and extension of the foreleg (walking movement); 9, 9, for the orbicular muscle of the eyelids, the zygomatic, closure of the eyelids.   II, a, a, for retraction and elevation of the angle of the mouth; b, for opening the mouth and for the movements of the tongue (mouth-center); c, c, for the platysma; d, for opening the eye.   I t, The thermic center, according to Eulenburg and Landois.   III, The cerebrum of the rabbit, viewed from above.   IV, The brain of the pigeon, viewed from above.   V, The brain of the frog, viewed from above.   VI, The brain of the carp, viewed from above.   (In all of these illustrations o is the olfactory bulb, 1 the cerebrum, 2 the optic lobe, 3 the cerebellum, 4 the medulla oblongata.)

eyelids, together with upward rotation of the eyeball and contraction of the pupil). In the anterior 9 is the point for the movements of the tongue, between the anterior and the middle 9 that for closure of the jaw.   Stimulation of the points a a (II) caused retraction and elevation of the angle of the mouth, with partial opening of the mouth. On stimulation of b Ferrier observed opening of the mouth, with protrusion and retraction of the tongue (bilateral action!), the dog not rarely making barking sounds. He designates this area the mouth-center. Stimulation of c c causes retraction of the angle of the mouth by the platysma, stimulation of c′ elevation of the angle of the mouth and of the side of the face to the point of closing the eye (the same as at 9). On stimulation of the middle e opening of the eye and dilatation of the pupil result, the eyes and the head being rotated toward the opposite side. Stimulation of the postcruciate gyrus causes contraction of the perineal muscles. Stimulation of the anterior declivous surface of the precruciate gyrus causes movements of the pharynx and the larynx. Stimulation of a definite point in the anterior half of the foot of the ascending frontal convolution (in the ape) gives rise to contraction of the glottis, as in phonation. Stimulation anteriorly and exteriorly to the center for the extremities (in the rabbit) causes movements of mastication and deglutition.

Observations on the ape showed likewise a strict localization of the centers. The movement caused by stimulation with induction-currents proved similar to those executed voluntarily. Rarely a single muscle contracted; generally a co-ordinated group. The antagonists are at times thrown into activity, in so far as the primary movement may be followed by that of the antagonists. The contraction exhibits an oscillatory rhythm of from ten to fifteen movements in one second.

On *more marked irritation*, in addition to the muscles of the opposite side, those of the same side may be made to contract, the irritation extending to the other side. Muscles such as those of the eyes, the perineum, the larynx, the pharynx, the muscles of mastication, that are moved on both sides at the same time, appear to have a center not only in the opposite hemisphere, but also in that on the same side. The question is of great practical significance from a diagnostic standpoint whether movements cannot be excited by irritation due to local disease, such as inflammation, tumor, degenerative processes, and the like, involving the motor areas in the brain of man. Hughlings-Jackson answers this question in the affirmative, and explains in this manner the occurrence of unilateral, localized epileptiform convulsions, which Ferrier and Landois observed as a result of inflammatory irritation. By means of marked irritation of the motor areas a complete general convulsive epileptic attack can be induced in dogs.

This begins with twitchings in the specially related group of muscles, passes then to the corresponding member of the opposite side and involves the entire musculature of the body, at first in clonic, then in tonic, and finally again in clonic spasms. Above the internal capsule feeble irritation is often sufficient to excite this form of epilepsy. The opposite side of the body has also been observed to be involved in convulsions, and from below upward, after the movements had been present in all parts on the side first affected. The spasmodic irritation passes from center to center, and intervening motor areas are never skipped. After a primary attack of such character, the slightest irritation is often sufficient for the excitation of other epileptic attacks. During the attack the circulation in the brain is accelerated and the vessels of the pia are dilated. As, at the same time, also the intracranial pressure is greatly increased, Kocher has suggested the making of a trephine-opening in the skull in epileptics and covering it only with soft parts, in order in this way to provide to a certain degree a safety-valve. With the object of preventing the variations in the fulness of the blood-vessels that are observed in epileptic attacks the suggestion to extirpate the cervical sympathetic in epileptics would appear justifiable. Perhaps it would be advisable to ligate the cerebral carotid at the same time.

The irritation of the centers appears to be followed by a brief state of lessened irritability, the refractory period. Also irritation of the subcortical white matter causes general convulsions, which, however, begin in the muscles of the same side.

If certain motor points are extirpated, the convulsions may be wanting in the epileptic attack in the muscles controlled from these points. Severance of the motor cortical points by means of a horizontal incision during an attack causes cessation of the attack. If an epileptic attack is of short duration, it is not rarely possible, by extirpation of the cortical center for one extremity, to exclude this alone, while the remainder of the body continues to be agitated by the convulsions.

Long-continued administration of potassium bromid prevents the possibility of causing epilepsy by irritation of the cortex.

Chemical irritation is further of particular interest. When in 1887 Landois applied to the motor regions a number of substances that occur in urine, for example kreatin, kreatinin, acid potassium phosphate, uratic sediment from human urine, and others, he observed the occurrence of marked eclamptic (clonic-tonic) convulsions, which were repeated spontaneously for a considerable time and were followed by profound coma (in the dog). Landois does not insist that the uremic convulsions in human beings, as well as epileptic convulsions induced through autointoxication, are to be compared with the phenomena observed in his experiments. The sensorial centers are also affected in the same way, the sense of vision suffering especially.

Certain *poisons* are capable of exciting convulsions by irritation of the cortical centers. Among these are santonin, physostigmin, carbolic acid, acetone (in cases of diabetes), also tannic acid on direct application. Under such circumstances convulsions upon both sides of the body may be excited by irritation of one hemisphere. The convulsions no longer occurred after the cortical centers were removed on both sides. Birds and lower vertebrates exhibit no convulsions.

Extirpation of the motor centers gives rise to characteristic derangement of movement in the affected contralateral muscles. Landois, together with other investigators, observed in the dog after destruction of the motor points for the extremities feeble and awkward movements of the latter, such as improper placing of the foot, slipping, yielding, dragging. While some investigators consider these phenomena as transitory only, Landois was able to observe them for months. In dogs, particularly the paws remain paralyzed with respect to all of those movements in which the paws are employed to a certain degree as hands, and which thus are acquired through education. In the course of time the pyramidal tracts degenerate downward and the related muscles undergo atrophy.

The higher the development of the intelligence in the animals and the more they are required to learn their movements and gradually to subordinate them to the control of the will, the more profound and persistent are the disturbances of movement after destruction of the cortical psychomotor centers. While, in the lower vertebrates, including birds, extirpation of the entire hemispheres does not appreciably affect the movements, the coördinated reflexes sufficing completely for the latter, in the dog extirpation of individual motor centers is attended at times with appreciable permanent derangement of motility, which in apes and human beings becomes intense and long continued.

Hitzig attributes the disturbances of movement following removal of the motor centers to the loss of *muscular consciousness*. According to Schiff, tactile sensation alone is lost in consequence of destruction of the motor cortical centers, and it never returns.

In a dog in which the motor centers for the extremities were destroyed on both sides Landois, in 1876, observed derangement of voluntary movement, which he was first to designate *cerebral ataxia;* that is the animal was unable to execute coördinated movements for the purpose of walking, standing, etc. He therefore believed, even at that time, that the cortical centers are the direct motor points for the operation of the will and also that conscious sensation of muscular contractions is localized in them.

The *irritability of the motor centers* may be considerably influenced in various ways. Thus, it is impaired by stimulation of sensory nerves, as this lowers the contraction-curve of the muscles and extends and prolongs the reaction-time. The irritability of the cortical centers appears to be increased only when active reflex muscular contractions occur in connection with severe sensory irritation. It is an especially remarkable fact that in a certain stage of morphin-narcosis a stimulus that is too feeble to induce a contraction becomes at once active if, shortly before its application to the cortical centers, the skin in certain portions of the body is exposed to even slight tactile irritation. The contractions acquire a tonic character on strong pressure upon the paw, so that all stimuli that in the normal state cause in the centers only transitory excitation now exert a permanent stimulating effect. If, during the tonic contraction, the skin on the dorsum of the paw is gently stroked, or the face is blown upon, or the nose is gently struck, or the animal is called, or the sciatic is irritated, relaxation of the muscles suddenly takes place. These phenomena are suggestive of the analogous observations on hypnotized individuals.

It is a further remarkable fact that if contracture of the muscles in question is induced by reflex irritation or strong electrical stimulation of the cortical center, feeble stimulation of the same center, and also of any other cortical region, suppresses the movement. There is thus afforded the peculiar phenomenon that irritation of the same cortical region, in accordance with the intensity of the current employed, excites irritation of the motor apparatus or inhibits an irritation already present. H. E. Hering and Sherrington observed on irritation of the motor centers of the ape, relaxation of the antagonistic muscles, which occurred even when the stimulus for the excitation of the movement in the muscles related to the center was still too weak. With currents of a certain strength they obtained such simultaneous contraction and relaxation not from the same cortical area, but from widely separated areas. Further, in addition to this reciprocal innervation of the true antagonists, there was a complicated relation between various groups of muscles. Thus, for example, on closure of the fist, dorsal flexion occurs at the wrist-joint. The relaxation of the antagonists takes place somewhat in advance of the contraction of the irritated muscle.

Sherrington stimulated the central stump of the flexor nerves of the leg containing the muscle-sense nerves, and observed at once loss of tone in the extensor muscles stimulated from the cerebrum.

According to the investigations of Fano and Libertini and others there is in the prefrontal region of the dog an inhibitory center for movements, therefore a *psycho inhibitory center* for the opposite side of the body. Irritation of the contralateral cerebral hemisphere, for example by application of a crystal of sodium chlorid (in the frog), causes inhibition of the irritability of the motor nerves. Also irritation of the contralateral basal portion of the midbrain or a transverse section of the spinal cord may have a similar effect.

THE SENSORIAL CORTICAL CENTERS.

The investigations of Ferrier and H. Munk have shown that areas are present in definite portions of the cerebral cortex in which the act of conscious sense-perception takes place. These areas are connected by means of fibers with the nerves of special sense. They are designated also *sensorial cortical centers*, *sense-centers*, or, according to the suggestion of Landois, *psychosensorial centers.* Total destruction of such a center abolishes conscious perception on the part of the organ of special sense in question. On partial destruction of such a center, the mechanism of sense-activity may remain intact, but the mental connection is wanting. A dog with such injured centers, it is true, sees, hears, and smells, but he no longer recognizes what he sees, hears, and smells. The centers are to a certain degree the repositories for experiences gained through the special senses. Irritation of these areas may give rise to movements such as occur when sudden, intense sense-impressions are produced; these movements are, therefore, reflex. In this group belongs also dilatation of the pupil and of the palpebral fissure, as well as lateral movement of the eyeball. It appears, however, that, in addition, each center possesses its own motor apparatus, by means of which the movements of the related organ of special sense are executed.

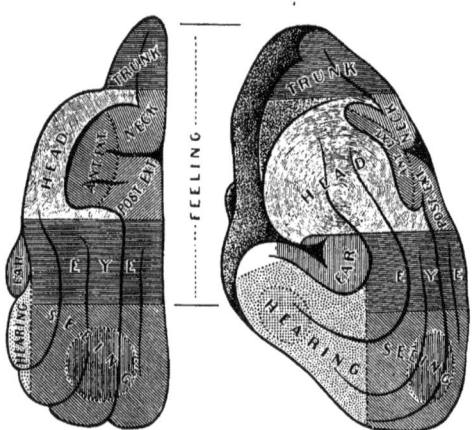

FIG. 259.—The Psycho-optic and Psycho-auditory Centers and the Sensory Sphere of the Dog's Brain (after H. Munk).

The *psycho-optic center* or the *visual sphere* comprises, according to Munk, the portion of the occipital lobe in the dog marked "Seeing" (Fig. 259). If this region is completely destroyed, the dog becomes permanently almost totally blind in the opposite eye—*cortical blindness.* If, however, only the more centrally situated portion (with circular outline) is destroyed, there will be loss of conscious sensation of vision on the opposite side, and this may be designated *mind-blindness* or *optic amnesia.* It is a remarkable fact that destruction of this area on one side is soon followed by compensation. It appears that other adjacent cortical areas of the visual sphere are capable of assuming the function of the injured portion. Under these circumstances it will be found that the animals with the affected eye must to a certain extent again learn to see as in earliest youth. Destruction of the entire center on each side causes total blindness on both sides, while that of the central (shaded) portions alone in the dog causes mind-blindness on both sides.

A psycho-optic center is observed first in birds, the optic nerve terminating in the midbrain in the lower vertebrates. In accordance with the extent of the decussation of the optic nerves in different animals the psycho-optic center is related to the retinas.

Munk determined in the dog, further, that both retinas are connected with each psycho-optic cortical center, and in such a manner that each retina receives the largest number of fibers from the opposite cortical center, and fibers only for the outermost lateral marginal portion from the center of the same side. If the surface of one retina be conceived as projected upon the centers, the outermost margin of the former will be connected with the center of the same side, the inner margin of the retina with the inner portion of the opposite center, the upper marginal portion with the anterior portion, and the inferior marginal portion of the retina with the posterior portion of the opposite center. The (shaded) middle of the center corresponds to the point of direct vision of the retina of the opposite side.

Irritation of the visual center causes in the dog movements of both eyes toward the opposite side, at times with movements of the head of like character and contraction of the pupils. If an eye is excised from a new-born dog the contralateral psycho-optic center will after an interval of months be found to be less well developed. After extirpation of the visual sphere in young animals the external geniculate body, the pulvinar (Fig. 263), the anterior quadrigeminate body (of the same side and in part also of the opposite side), undergo atrophy, together with degeneration of the sensory sphere for the eye (Fig. 259), and at a later period also the optic tract and nerve undergo atrophy. A similar condition has been observed after degeneration of the visual sphere in man.

The situation of the visual center has been outlined in a different manner by different investigators. According to Ferrier it is located in the dog in the region of the occipital portion of the third primitive convolution indicated by e e e (Fig. 258), and according to more recent statements in the occipital lobe and the angular gyrus.

According to Luciani, the visual field includes, in addition to the occipital, also the parietal lobe (in the dog and the ape). He also dissents from the precise projection of the retinas upon the cerebral cortex. He believes that both optic nerves are connected with all portions of the occipitoparietal region. Moreover, he is of the opinion that the visual images are only transformed in the cortex psychically, but that they arise in the quadrigeminate body, as he admits, even after most extensive destruction of the occipitoparietal region on both sides, the development only of mind-blindness, but not permanent actual blindness. Even previously Christiani had maintained that rabbits deprived of their cerebrum still avoided obstructions in running, because they were able to appreciate them with their eyes. In such animals, therefore, optic impressions must be advantageously utilized, and in such a manner that the optical impressions so affect the chief reflex and the coördination-center in the optic thalamus that the animals make appropriate reflex movements. Conscious vision is thus lost, while the coördinated reflex activities excited from the visual apparatus are still preserved.

In apes the center is situated at the apex of the occipital lobe. Destruction of the center on one side causes blindness for the halves of both retinas upon the side of the injury. In birds the visual sphere is situated in the portion of the cerebral cortex extending from the peduncle upward and forward and covering the ventricle. The retina of the opposite eye is supplied from one hemisphere, with the exception of its most posterior portion, which is supplied from the hemisphere of the same side. In the frog the visual center is situated in the optic lobe; frogs and fish thus see without a cerebrum.

The *psycho-auditory center* or the *auditory sphere* is situated in the dog in the region of the second primitive convolution indicated by the letters f f f (Fig. 258, II), according to Munk in that portion of the temporal lobe marked "Hearing" (Fig. 259). Destruction of the entire region causes deafness in the contralateral ear; destruction of the middle shaded portion alone causes *mind-deafness* or *auditory amnesia*, that is the animal has lost the memory-images of auditory impressions. Irritation of the center is followed by a reaction that corresponds to the abrupt start induced by a sudden and unexpected loud noise. Irritation of the center on one side causes movement of the ear on the opposite side. Under these circumstances also, the disturbances attending injury of the middle portion on one side disappear in the course of a few weeks (as in the case of the psycho-optic center), so that the animal must again learn to hear. Destruction of the middle portion on both sides causes mind-deafness on both sides. Dogs thus injured no longer prick their ears in response to auditory impressions and they gradually lose the faculty of barking. The anterior portions of the auditory sphere appear to subserve the perception of high notes and the posterior portions the perception of deeper notes. Munk observed after destruction of one ear in the new-born dog that the contralateral center was less well developed. Destruction of the entire region on both sides causes deafness (with mutism). Ferrier demonstrated the center in apes, rabbits, jackals, and cats.

According to Luciani the auditory center extends from the temporal lobe to the parietal and frontal lobes, the hippocampal gyrus and the cornu Ammonis. Each ear is in connection with both centers, although most intimately with that of the opposite side. After total extirpation of the auditory center on both sides mind-deafness alone develops.

Munk and Ferrier locate the *olfactory center* in the dog in the hippocampal gyrus. After destruction of the center on each side in the ape the sense of smell and that of taste were abolished. The psycho-osmic and psychogeusic centers located in this situation have as yet not been differentiated. According to Luciani the hippocampal gyrus and the cornu Ammonis constitute the olfactory center. Partial decussation is to be assumed also in this case, but the non-decussating bundle is the larger.

On irritation of this area Luciani observed in apes, dogs, cats, and rabbits distortion of the lips and partial closure of the nasal orifice on the same side. According to Zuckerkandl, who bases his conclusions upon comparative anatomical observations, the cortical portion of the olfactory center is constituted of the central extremity and the frontal extremity of the lobe of the corpus callosum, of the hippocampal lobe together with the uncus, of the cornu Ammonis including the marginal convolution (particularly the dentate fascia), of the cortex of the olfactory peduncle, of the cortex of the anterior perforated lamina and of the olfactory bulb.

A cortical olfactory center is present among vertebrates first in reptiles, and this is at the same time the earliest psychosensorial organ that appears. It may be concluded from this fact that, phylogenetically, the first psychic activity in the animal kingdom is concerned with the perception of odors.

Munk believes that the surface of the brain in the region of the motor centers is at the same time the *sensory sphere*, that is that it serves also for the reception of tactile, muscular, and innervational impressions from the opposite side. The boundaries of the areas for the individual portions of the body in the dog are indicated in Fig. 259. After injury of this region the function mentioned is lost. The sensory sphere in apes is situated in the parietal lobe and each individual area is related to a definite portion of the body. After total extirpation of the arm-area and the leg-area, tactile sensibility is lost permanently, while after partial extirpation return of sensibility takes place later.

Luciani, however, rejects such precise limitation for the individual regions of the body. According to Bechterew the centers in the dog for the perception of tactile impressions, the muscle-sense and sensations of pain are situated in the neighborhood of the motor zone, the first immediately behind and external to the motor area, the others in the region just above the beginning of the fossa of Sylvius. According to Schäfer extirpation of the gyrus fornicatus is followed by permanent impairment of sensibility in apes.

THE CORTICAL THERMIC CENTER.

DIVERGENT VIEWS AS TO THE LOCALIZATION IN THE CORTEX. OTHER CORTICAL FUNCTIONS.

A. Eulenburg and Landois succeeded in discovering on the surface of the cerebrum of the dog an area from which an undoubted influence is exerted upon the temperature and the size of the vessels in the contralateral extremities. This area (Fig. 258 I, t) comprises in general the region in which at the same time the motor centers for the flexors and the rotators of the foreleg (3) and for the muscles of the hind extremity (4) are situated. The effective areas for the anterior and posterior extremities are widely separated from each other. That for the foreleg is situated somewhat further forward, close to the lateral extremity of the cruciate sulcus. Destruction of this region is followed by elevation of the temperature in the contralateral extremities, and this may be variable in degree—from 1.5° to 2° or even as much as 13° C.

This observation has been confirmed by Hitzig, Bechterew, Wood, and others. The elevation of temperature bears no relation to muscular disturbances that may be present in the affected extremities. It is in almost all cases marked for a considerable time after the injury, although attended with considerable fluctuations. It has been observed to persist for as long as three months, while in other cases gradual return to normal sets in on the second or third day. In marked cases there is a reduction in the resistance of the wall of the femoral artery to pressure and a lowering of the pulse-tracings.

Localized electrical irritation of the areas causes slight transitory reduction in the temperature of the contralateral extremities. In dogs the same result may be brought about even by percutaneous irritation. The center can be irritated also by application of sodium chlorid, although under such circumstances the phenomenon of destruction soon follows. Irritation of the cortical center causes also in curarized animals marked elevation of blood-pressure in consequence of vascular contraction. The demonstration of a thermic center for the half of the head has not as yet been made. In cerebral-epileptic attacks the bodily temperature rises, in part in consequence of increased production of heat by the muscles, in part in consequence of lessened heat-dissipation through the vessels of the skin as the result of irritation of the cortical thermic centers.

According to Wood, destruction of this central area in the dog causes at the same time an increase in heat-production demonstrable calorimetrically, while irritation causes lessened heat-production. In dogs in which the internal capsule was divided by means of a small knife (which was made to close suddenly in the depth of the wound by traction on a cord), Landois observed likewise elevation of temperature, and he concluded, therefore, that the fibers controlling thermic influences traverse the internal capsule. Furthermore, injury of the cerebral peduncle is followed by obvious elevation of temperature. In rabbits destruction of the anterior portion of the cortex has no obvious influence, although that of the posterior portion has.

The experiments described explain the fact that psychic stimulation of the cerebrum may have an influence upon the size of the vessels and upon the temperature, as indicated by momentary pallor and blushing. v. Bechterew and Mislawski locate a vasodilator center in the external and middle portions of the anterior segment of the cruciate gyrus and in portions of the parietal region.

In opposition to the outlined doctrine of localization in the cerebrum the views of Goltz must be discussed in an unprejudiced manner. Goltz has described in detail the phenomena that appear in dogs subjected to extensive destruction of the cerebral cortex. He distinguishes on the one hand inhibitory phenomena, which are transitory and are referable to temporary suppression of the functions of nervous structures that are not injured anatomically. These are to be explained in the same way as the inhibition of the reflexes by strong irritation of sensory nerves. Opposed to these are the permanent phenomena of deficiency, which are due to the loss of function of the nerve-structures destroyed by the operative procedure. Such a dog with extensive loss of cortex may be looked upon as an eating, complex reflex machine. It behaves like a profoundly demented animal, walks slowly in an awkward manner, with head depressed, and exhibits impairment of cutaneous sensibility in all of its qualities. It is less sensitive to pressure upon the skin, pays less attention to variations in temperature and does not know how to touch objects. It is scarcely able to adjust itself with relation to the outer world or to its own body. This is noted especially in looking for and taking up its food. On the other hand there is no paralysis of its muscles. It is true the dog still sees, but without conscious appreciation of what is seen. It sees like a somnambulist, who avoids obstructions, without being perfectly conscious of their character. It is true, the animal hears, for it can be awakened from sleep by loud shouting, but it hears much like a human being who has just been aroused from deep sleep by being called, without at once comprehending the call with clear consciousness. The disturbance of the other senses is analogous in character. The animal howls when hungry, then eats until its stomach is entirely filled. It is absolutely indifferent and without sexual instinct.

With reference to the localization in the cerebrum Goltz holds dissenting views. He believes that every portion of the cerebrum takes part in the functions upon which volition, sensation, imagination, and thought are based. Every portion, independently of the others, is connected by conducting paths with all of the voluntary muscles, and on the other hand, with all of the sensory nerves of the body.

After removal of an anterior lobe, including the motor zone, there develop first unilateral motor and sensory paralysis and unilateral visual disturbance. Of these only the loss of muscle-sense is left after the lapse of months. Removal of the anterior lobe on both sides gives rise to these phenomena in more marked degree, and in addition there occur innumerable involuntary associated movements and increased reflex irritability. Goltz observed repeatedly general hyperesthesia, a remarkable motor propensity and an irritable aggressive character in the dog. Marked permanent disorders in the utilization of the senses of sight, hearing, smell, and taste are not necessarily present, even in connection with profound and extensive destruction of the forebrain.

Removal of the occipital lobes disturbs the utilization of the sense-perceptions in consequence of the defect of intelligence present. The sense of sight is injured most. Removal of the occipital lobe on both sides causes marked disturbance of vision, but not total blindness. In character the dogs become good-natured and considerate. There are never disorders of movement and of the muscular sense.

According to Loeb, a pupil of Goltz, the disorders of vision, of sensibility, and of motility that develop after partial injury of the cortex can be summarized as follows: On the opposite side stimulation of the retina and the sensory nerves causes less marked effects that appear more slowly. Likewise the stimulation of the muscles in connection with intended movements of the body is less marked upon the opposite side of the body.

Injuries of the cerebrum give rise also to inhibitory phenomena, including motor disturbances; and Goltz considers the complete hemiplegia that is not rarely observed after coarse unilateral injuries of the cortex as an inhibitory phenomenon. The injury exerts an inhibitomotor influence upon other (infra-cortical) organs, which resume their movement as soon as the inhibitory influence is removed.

Other Cerebral Functions.—Some investigators have observed variations in blood-pressure and change in the heart-beat after irritation of the cerebral cortex; thus, for example, Bochefontaine after electrical irritation of the motor area for the extremities. After irritation of the cortical center for the facial nerve (Fig. 258, 5) R. Danilewsky observed increase in the blood-pressure, the pulse being at first accelerated and later slowed (and also upon irritation of the caudate nucleus and the adjacent white matter). At the same time, he observed, under such conditions, slowing and at times interruption of the respiration. Balogh observed acceleration of the pulse after irritation of various portions of the cortex in the dog and slowing of the pulse after irritation of one point. Eckhard irritated the surface of the cerebrum in rabbits and found, as a rule, that so long as only a few contralateral movements take place in the anterior extremities, no influence upon the heart is observed, but that cardiac symptoms appear only in association with the occurrence of other movements. They consist in slower, stronger pulse-beats, intermixed with feebler beats, together with slight increase in the blood-pressure. If the vagus on each side be first divided the influence upon the pulse-beat is not exhibited, but the elevation of the blood-pressure persists. All of these experiments fail to afford a satisfactory explanation of the relations of the cerebrum to the action of the heart. That such a relation exists is indicated indubitably by the effect of psychic influences upon the heart-beat, as Homer and Chrysippus knew.

Irritation of the cortex laterally from the base of the olfactory tract exerts a slowing or inhibitory influence upon respiration, while irritation in the motor regions has an accelerating influence, and irritation of the uncinate gyrus causes sniffing. Unverricht observed arrest of respiration in the dog on irritation of a point in the third primitive convolution external to the center for the orbicularis of the eyelids. Preobraschensky observed inspiratory spasm of the diaphragm in the cat after irritation of a point behind this center. The increased secretion of saliva observed by Landois has been referred to on page 262.

According to v. Bechterew and Ostankow a point irritation of which causes swallowing movements is situated externally to the cruciate sulcus. Bochefontaine and Lépine observed further, particularly after irritation in the neighborhood of the cruciate sulcus in dogs, slowing of the movements of the stomach, peristalsis of the intestines, contraction of the spleen, the uterus, and the bladder, and increased respiratory frequency. Bufalini observed the occurrence of secretion of the gastric juice with elevation of the temperature in the stomach after irritation of the same cortical area whose irritation in rabbits caused movements of the jaw. The relations of the region about the cruciate sulcus in the dog to the cardia are considered on page 281. According to Bechterew and Mislawski irritation of various points in this region causes in part movements at the pylorus, and in part inhibition of such movements; occasionally the cardia is set in motion. From the same point and from the third primitive convolution situated posteriorly and externally contraction and relaxation of the muscles of the perineum can be induced, and also from the optic thalami. The conducting paths are contained in part in the vagi, in part in the spinal cord. From the latter the fibers for the small intestine pass through the eight lower dorsal and the uppermost lumbar nerves in the dog to the sympathetic plexus, those for the large intestine through the last two lumbar and the upper three sacral nerves. The foregoing statements afford an explanation of the fact that in epileptic attacks induced by irritation of the cortex contraction of the stomach, the intestine, and the bladder have also been observed.

Electrical stimulation of the inner portion of the sigmoid gyrus in the dog causes dilatation of the pupils (as does also chemical irritation of the parietal region in the rabbit); secretion of tears, protrusion of the eyeballs, and retraction of the third eyelid in the dog. Increased contractions of the vagina in rabbits could be induced by irritation of the anterior portion of the hemisphere, in dogs by irritation of the sigmoid gyrus. By irritation in the neighborhood, or by increased irritation, an inhibitory effect could be induced. Irritation of the optic thalamus or of the central stump of the vagus likewise gave rise to increase in the movements, while irritation of the peripheral stump of the vagus caused relaxation of the vagina.

Attention should finally be directed to a number of observations of pathological significance made after injuries to the brain. Thus, Schiff, Ebstein, Klosterhalfen, and v. Preuschen observed, after injuries to the pons, the striate body, the optic thalamus, the cerebral peduncle, the quadrigeminate bodies, the middle cerebellar peduncle, and the medulla oblongata, often hyperemia and extravasations of blood into the lungs, the pleuræ, the stomach, the intestines, and the kidneys. In this manner is to be explained the occurrence of hemorrhage into the stomach in new-born infants with injury to brain (melæna neonatorum); perhaps also the occurrence of gastric ulcer in adults in association with cerebral disease. v. Preuschen observed gastric and intestinal hemorrhage in rabbits also after injury of the cornu Ammonis, the floor of the anterior horn, the frontal lobe, and the upper portion of the spinal cord. Analogous phenomena have been observed in man after cerebral hemorrhage or softening. Brown-Séquard and Nothnagel induced hemorrhage into the lungs by irritation of basal portions or of a point on the surface of the brain.

The cerebral unilateral acute decubitus described by Charcot is particularly noteworthy, occurring always upon the paralyzed side, therefore on that opposite to the cerebral focal lesion. It may begin as early as the second or the third day and rapidly progress to a fatal termination, with profound destruction (buttock, lower extremity). The decubitus that appears in connection with disease of the spinal cord generally begins in the middle line of the buttocks and spreads thence symmetrically toward either side. In cases of injury of one side of the spinal cord this destruction takes place on the corresponding side of the sacrum.

PHYSIOLOGICAL TOPOGRAPHY OF THE SURFACE OF THE CEREBRUM IN MAN.

In the brain of man the physiologically analogous systems of fibers receive medullary sheaths approximately at the same time, the olfactory tract among the first.   According to Flechsig the cortex is developmentally divisible into forty different fields.   These can be arranged in three groups: (a) The *primordial areas*, which are formed even before mature birth; (b) intermediary areas, which begin to be surrounded with medullary tissue up to one month after birth; (c) the terminal areas, which are formed later than one month (from four to four and a half months) after birth.   The primordial areas coincide with the sense-centers in their rudimentary form; the terminal areas comprise association-areas; and the intermediary areas, amplifications in part of the sense-centers and in part of the association-centers.   The human brain is distinguished from that of the anthropoids principally through the terminal areas.   They vary in size in different individuals and they contribute in large measure to the shape of the human skull, under the eminences of which they are situated.   Thus, for example, the anthropoids are unprovided with the area destruction of which causes pure alexia. For this reason alone apes cannot acquire the faculty of speech.

The **motor areas** comprise the anterior (**Fig.** 246, **A**) and posterior (**B**) central convolutions, and the paracentral lobule, and extend posteriorly into the precuneus (**Fig.** 262).   They contain large ganglion-cells, which, however, are not present before the age of one and one-half months.   Degeneration of the entire area causes paralysis of the opposite side of the body.   This at first is total, but gradually passes into a condition in which especially all of the delicate movements under the control of the will and acquired by education and exercise are abolished, while associated and bilateral movements (which, for example, are present in animals that after birth are at once capable of executing various complex movements) are preserved more or less intact.   Therefore, the hand is paralyzed in man in greater degree than the arm, and this in turn in greater degree than the leg, the lower branches of the facial nerve in greater degree than the upper, and the nerves of the trunk finally almost not at all.

In hemiplegic individuals the strength of the unparalyzed side of the body is also impaired.   This is not fully explained by the fact that some fibers of the pyramidal tracts remain upon the same side of the body.   Among the movements in human beings there are some that have to be learned with great effort, and therefore gradually become subordinated to the varying impulses of the will, such, for example, as the delicate movements of the hands.   These movements are restored but slowly and incompletely or not at all after lesions of the psychomotor centers.   Those movements, however, that are at once at the command of the body, such as the associated movements of the eyes, the face, in part also of the lower extremities, either recover rapidly after the lesions described, or they appear to suffer but little at all.   Thus, the facial muscles appear never to be so completely paralyzed after a cortical lesion as after a lesion of the trunk of the facial nerve; for example the eye can yet be closed fairly well.   Sucking movements have been observed even in hemicephalic new-born children.

From the motor cortical centers the path for the fibers of the facial and hypoglossal nerves passes through the genu, that for the muscles of the extremities through the middle third of the posterior limb of the internal capsule (**Fig.** 263).   Irritation of these paths causes movement in the muscles on the opposite side.   After destruction of the cortical areas degeneration takes place in these corticomotor paths, which pass downward and whose continuation is designated the pyramidal tracts. This degeneration has been found within the white matter below the cortex, in the genu, and in the anterior two-thirds of the posterior division of the internal capsule, in the cerebral peduncle (middle portion of the lower free circumference of the crusta, where the tracts for the extremities and the nerves of the trunk-muscles lie externally and those for the motor nerves of the head internally), in the pons, in the pyramids of the medulla (**Fig. 261**), and thence in the pyramidal tracts of the spinal cord. It is obvious that lesions of these tracts at any point in their course must have the same effect, namely hemiplegia. In the progress of the degenerative process the paralyzed muscles may exhibit a certain degree of spastic rigidity and an increase of irritability to mechanical stimulation (tendon-reflexes), which must be considered as an irritative degeneration phenomenon. Later on, degenerative changes are observed in the ganglion-cells of the anterior horn and in consequence of these, atrophy and disappearance of the related muscles.

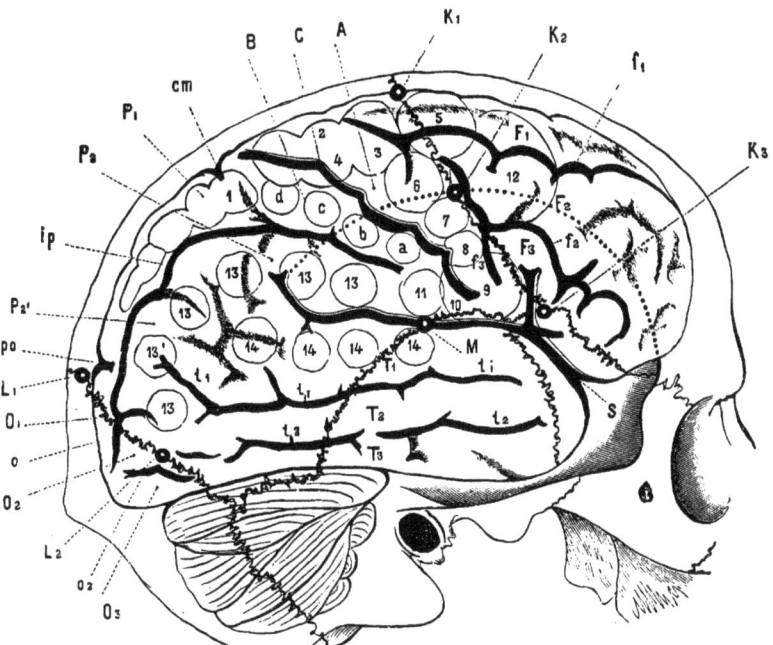

FIG. 260.—The Cerebrum with the Principal Convolutions and Sulci (after A. Ecker) in its Longitudinal Relation to the Skull: S, the Sylvian fissure, with its vertical ascending short anterior limb and its horizontal posterior longer limb; C, the central fissure or sulcus, or fissure of Rolando; A anterior, B posterior central convolution; F_1 superior, F_2 middle, F_3 inferior frontal convolution; f_1 superior, f_2 middle, f_3 vertical frontal fissure (precentral sulcus); P_1, superior parietal lobule; P_2, inferior parietal lobule, with P_2 indicating the supramarginal gyrus and P_2^1 angular gyrus; ip, interparietal sulcus; cm, extremity of the callosomarginal sulcus; O_1 first, O_2 second, O_3 third occipital convolution; po, parieto-occipital fissure; o, transverse occipital sulcus; o_2, inferior longitudinal sulcus; T_1 first, T_2 second, T_3 third temporal convolution; t_1 first, t_2 second temporal fissure; K_1, K_2, K_3, points in the sagittal suture; L_1, L_2, points in the lambdoid suture.

The muscular atrophy is not always proportionate to the intensity of the paralysis. Perhaps special trophic fibers, separated from the motor fibers, and situated nearer the sensory fibers, pass from the cerebrum downward through the internal capsule. The tract for the seventh, eleventh, and twelfth cerebral nerves is situated in the genu of the internal capsule. The tracts for the rotation of the eyes, the muscles of the trunk and the nape of the neck are situated on either side in the internal capsule for both sides of the body.

According to observations of Flechsig and Hoesel it appears that the motor area is at the same time the sensory center for the muscle-sense and innervational impressions. Therefore, the larger portion of the ascending fibers in the posterior columns of the medulla must pass to the central convolutions. By others the superior parietal lobule (P_1) is considered as the seat for impressions of position and movement. The conducting tracts are believed to be situated immediately behind the motor tracts in the internal capsule. It is a noteworthy fact that, in man, on the one hand exclusive loss of muscle-sense or of conception of position has been observed, and on the other hand also pure motor paralysis without a lesion of the former.

The psychomotor centers may also, at times, be stimulated to activity through psychic influences (grimace, pantomime, gesture), at times be inhibited through psychic shock ("paralyzed by fright," "spell-bound with fear," "speechless from grief," etc.). On stimulation of voluntary movements within certain muscles an inhibitory mechanism in the cortex at the same time becomes effective and renders the adjacent cortical centers inactive. If this inhibition is enfeebled, unintentional associated movements take place. Thus, in children, for instance, associated movements of the mouth are observed during writing exercises.

Pathological.—Irritation of the psychomotor areas from internal pathological causes may give rise to maniacal motor activity, for example in the state of so-called "possession." Involuntary twitchings in individual muscles due to irritation of the motor centers occur in the condition of paramyoclonus multiplex. Deficient activity of the previously referred to inhibition of the psychomotor centers is capable of causing cerebral chorea. In analogy with the ataxic motor states in animals first described by Landois there occurs also in human beings a condition of cerebral ataxia. The cerebral paralysis of childhood is due to degenerative inflammatory processes in the motor areas. In acephalous fetuses marked deficiency in the development of the pyramidal tracts has been observed.

In man the entire system of the pyramidal tracts may undergo degeneration also from internal causes. Paralysis, spastic contractures, and atrophy of the muscles of the body (on one side or upon both sides) are characteristic, as observed in the amyotrophic lateral sclerosis of the spinal cord of Charcot. In childhood the normal development of the pyramidal-tract system may fail to take place and cerebrospinal paralysis thus result.

Well-observed clinical cases aid in the localization of the individual motor subcenters. (1) The center for the movement of the leg is situated in the vicinity of the upper extremity of the fissure of Rolando (Fig. 260, C), and in the paracentral lobule (Fig. 262, A B). (2) The center for the upper extremity is situated in the middle third of the anterior central convolution or somewhat lower (Fig. 260). The center for the thumbs and the fingers is situated in the posterior central convolution below the center for the upper extremity. (3) The center for the facial nerve is situated at the lower extremity of the anterior central convolution (center for the mouth and the lower portion of the face). The lower third of the anterior central convolution on the left and the adjacent foot of the second and third frontal convolutions contains, on each side, the center for the trigeminus (movement of mastication). The anterior portion of the anterior central convolution is connected with the hypoglossal nerve. The most anterior and inferior portion of the anterior central gyrus appears to be the seat controlling the action of the tensors of the vocal bands. The island of Reil controls the movements of the vocal bands. (4) The portion of the frontal lobe lying in front of the middle third of the anterior central convolution controls the muscles of the nape of the neck. The centers for the muscles of the trunk are situated upon the surface of the anterior central convolution above the centers for the upper extremity. (5) The external ocular muscles appear to have their cortical center in the angular gyrus (Fig. 260, P_2^1). The centers for the lateral movement of the head and the eyes are situated in the posterior portion of the second frontal convolution.

The motor centers may be paralyzed either individually or collectively, and accordingly cortical oculomotor monoplegia, crural (rare), brachial, brachiocrural, linguofacial, and finally faciobrachial forms of monoplegia have been distinguished.

If the motor centers are irritated by morbid processes—particularly hyperemia and inflammation of syphilitic origin; rarely tubercle, tumors, cysts, cicatrices, splinters of bone—convulsive movements take place in the related groups of muscles. Those muscles that are usually moved upon both sides appear thus to be stimulated from one center.

In accordance with their seat these convulsive movements are designated facial, brachial, crural monospasm, and the like. Such movements naturally may also involve several centers at the same time. In men with the surface of the hemisphere freely exposed, the region of the motor centers has been successfully stimulated by electricity by Bartholow, Sciamanna, and others.

If intense irritation be applied upon one side bilateral convulsive movements with suspension of consciousness may occur (properly designated Jacksonian or cerebral epilepsy).

The following observations have been made bearing upon the center for voluntary combined movements of the eyes in the cortex in man: Both eyeballs are controlled from each hemisphere. In the presence of paralyzing lesions of the cortex or of the tracts that pass off from it both eyeballs are occasionally found in a state of lateral deviation. If the paralyzing lesion be situated in a cerebral hemisphere conjugate deviation of the eyeballs takes place toward the unaffected side. If, however, it be seated in the conducting tracts, where decussation has already taken place, namely in the pons, the deviation of the eyes takes place toward the paralyzed side. If the seat of the lesion is in a state of irritation causing contractions on one side of the body, the deviation of the eyes is naturally in a direction opposite to that in which it would be in association with paralysis. In cases of cerebral paralysis there is occasionally, instead of the marked lateral deviation of the eyeballs, only a paresis of the lateral rotators of the eyeballs, so that though during rest the eyes are not rotated toward the unaffected side, they cannot be adequately rotated toward the affected side. Also the elevator of the upper eyelid appears to have its center in the angular gyrus; but according to some observers this is situated in the posterior limb of the second frontal convolution, extending into the first frontal convolution.

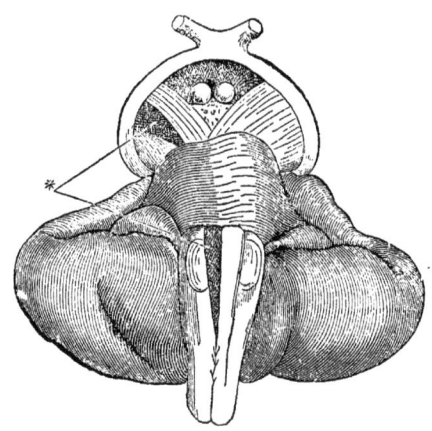

FIG. 261.—Secondary Degeneration of the Motor Tracts in the Cerebral Peduncle, the Pons and the Pyramid. The shaded areas (*) are degenerated (after Charcot).

The **motor speech-center,** which controls the voluntary movements of the tongue (hypoglossal nerve) and the mouth (facial nerve), including the lower jaw (third division of the fifth nerve), is situated in most persons in the left third frontal convolution (Fig. 260, F_3). The fact that most persons are right-handed also indicates a more refined development of the motor apparatus for the upper extremity in the left hemisphere. Human beings with well-developed right-handedness are obviously left-brained. Perhaps this arrangement is dependent upon an embryological basis. By far the majority of persons are thus left-brained speakers, although there are exceptions. As a matter of fact, left-handed persons have been observed to lose the faculty of speech after lesions of the right hemisphere.

Studies of the brains from distinguished men have shown that in them the third frontal convolution attains a greater size and a less simple form than in the brains from persons of lower grade of intelligence. In deaf-mutes this convolution is exceedingly simple and in microcephalic fetuses and in apes it is merely rudimentary.

Injuries of this speech-center, as well as transitory functional disorders, for example in consequence of copious hemorrhage, are followed either by loss or at least by more or less considerable derangement of the faculty of speech. The loss of the faculty of speech is designated *aphasia*. Stimulation of this region causes *sensations of speech-movement*, which occur rarely, the sensations often being referred by the patients, for example paralytics, to other parts of the body.

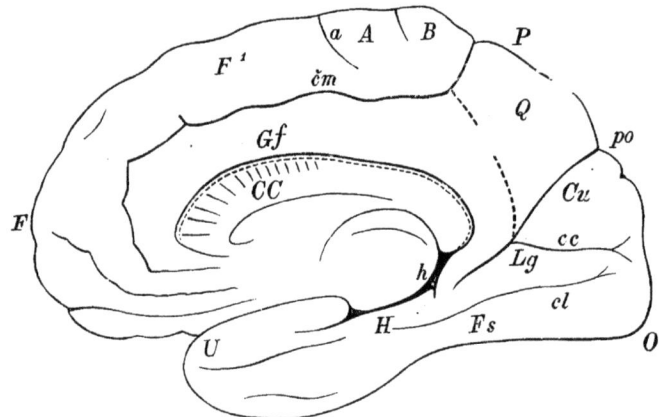

Fig. 262.—View of the Median Surface of the Human Brain: *CC*, the divided corpus callosum; F', first frontal convolution continuous at *a* with the anterior central convolution (*A*); *B*, posterior central convolution; between *A* and *B* is the median extremity of the fissure of Rolando (*AB* designated paracentral lobule); *Gf*, gyrus fornicatus, bounded by the callosomarginal fissure (*cm*) from the first frontal and the central convolutions. The callosomarginal fissure (cm in Fig. 260) passes upward between *B* and *P* (the superior parietal lobule); *po*, the parieto-occipito fissure (po in Fig. 260) separates the occipital lobe (*O*) from the parietal lobe (*P*); *Q*, quadrate lobe (precuneus); *Cu*, cuneus; *cc*, calcarine fissure; *Lg*, lingual lobe (median occipito-temporal gyrus); *Fs*, fusiform lobule (lateral occipito-temporal gyrus); *H*, hippocampal gyrus; *U*, uncinate gyrus; *h*, hippocampal sulcus; *F*, frontal, *P*, parietal, *O*, occipital lobe.

The motor tract for speech passes from the third frontal convolution first along the upper margin of the island of Reil, then in the depth of the hemisphere internally to the posterior margin of the lenticular nucleus, and then through the crusta of the left cerebral peduncle and the left half of the pons to the medulla oblongata, the seat of the nuclei of all the motor nerves concerned in the act of speaking—trigeminus, facial, hypoglossus, vagus, respiratory nerves. Total destruction of this motor tract causes, therefore, total aphasia. Partial lesions cause more or less coarse derangement of the mechanism of articulation, which has been designated *anarthria*.

Three types of activity are necessary for the function of speech: 1. The normal movement of the speech-apparatus—tongue, lips, mouth, respiratory apparatus. 2. A knowledge of the symbols for objects and ideas—speech, writing, and gesture. 3. The correct association of the two. Therefore, the following essentially different forms of aphasia must be distinguished:

1. *Ataxic aphasia*, or *psychomotor aphasia*, is loss of the power of speech in consequence of inability to execute in a coördinate manner the movements necessary for speech. Under such circumstances, the ability has been lost to form a conception of the movements for speech, as well as the power also of recognizing the position of the organs of speech. The intention to speak causes incoördinate grimaces and the utterance of inarticulate sounds. Therefore, the patients are unable to repeat what is spoken to them. At the same time, the mental processes necessary for the faculty of speech are wholly preserved, and all words are probably retained in memory, so that some are still able to express themselves in writing. If, however, the delicate movements acquired by education that are necessary for writing are lost in consequence of a lesion of possibly a special center at the extremity of the second frontal convolution there results at the same time *ataxic agraphia*, that is an inability to perform the movements necessary for writing. The intention to record thought on paper results only in an illegible scrawl. At times even pantomime-speech may be interfered with under such circumstances—*amimia*. There may be also a purely functional hysterical aphasia.

2. *Amnesic aphasia* or *psychosensorial aphasia*, a condition in which the memory of words is lost. Occasionally, only certain groups of words are lost, or even only portions of certain words, so that these may be produced in a deformed or partial manner. The movements necessary for speech are intact. Therefore, the patient is capable of repeating at once all that is spoken to him or of writing on dictation. In polyglot individuals all forms of language are lost and not only one. Amnesic aphasia has been observed in connection with destruction of the first temporal convolution on the left. There is also a combined form of *ataxo-amnesic aphasia*. In another variety of amnesic aphasia while the words are still retained in memory, they cannot be expressed fluently, that is the association of word and conception is inhibited. The failure to recall persons and the names of objects is, particularly in advanced age, a phenomenon observed within physiological limits, but which eventually may terminate in *senile amnesia*. Kussmaul has, further, included among the cerebral derangements of speech the following special varieties:

3. *Paraphasia*, or the inability to associate correctly the word-images with their conceptions, so that, instead of intelligent word-pictures, reversed or wholly incomprehensible word-pictures are aroused. There occurs to a certain degree permanent confusion of speech.

4. *Agrammatism* and *acataphasia*, or the inability correctly to form words grammatically and to arrange them syntaxically in sentences. In addition there may be:

5. *Abnormal slowness of speech, bradyphasia,* or *abnormal acceleration of speech* (*tumultus sermonis*), a lisping or abnormally slow speech, which likewise are dependent upon cortical disorders. Derangements of speech that are dependent upon affections of the peripheral nerve or the muscles of the organs of voice and speech have been described on pages 617, 697 and 713.

The *faculty of musical expression* may be preserved or lost in connection with aphasia—amusia, note-blindness, sound-deafness; it is perhaps represented by a special cortical center, possibly situated in the posterior portion of the first and second temporal convolutions.

The **cortical thermic center** for the extremities discovered by Eulenburg and Landois is at the same time related to the localization of the motor points. There are observations on record of injury or degeneration in these areas, with inequality in the temperature on the two sides of the body. After the existence of paralysis for a considerable length of time the temperature of the affected members, which at first is higher, may become lower than that of the unaffected side. Stimulation of this area gives rise also to increase in the blood-pressure, as, for example, in connection with epileptic convulsions. Wounds made for the exposure of the brain therefore usually bleed more freely during such an attack. According to Schüller the center for the entire contralateral half of the body is situated just in front of the precentral gyrus in the second temporal convolution.

In cases of progressive paralysis of the insane, attended with inflammation of the cerebral cortex, the temperature in the axilla is usually higher upon the side on which the paralytic phenomena are situated. Conversely, in the case of convulsions caused by inflammatory irritation of the cortical centers, the temperature upon the contralateral side is several tenths of a degree lower during their continuance. If extensive vascular areas are paralyzed, the temperature of the body may fall, for example in paralytics to as low as 25° C.

Degeneration of the internal capsule gives rise to vasomotor disorders and from this fact it is to be concluded that the tracts for the thermic fibers pass through this structure. The morbidly increased flushing from psychic influences, particularly from fear preceding the onset of the flushing (erythrophobia), has been attributed by v. Bechterew and Mislawski to irritation of the area discovered by them to have vasodilator effects as a result of experiments on dogs.

The **sensorial areas or the sense-centers** are the situations in which conscious perception of sense-impressions takes place. In addition, they constitute also the substratum of sensory conceptions and of sensory memory. The sense-centers are, according to Flechsig, developmentally *primordial*, that is in so far as they are indicated up to the time of birth, and, *secondary*, in so far as they attain complete development with their connections at a later period.

The *psycho-optic center*, visual center, visual sphere, comprises in its primordial rudiment up to the time of birth the lips of the calcarine fissure (**Fig. 262**) and the first occipital convolution. In its secondary development it comprises further the entire median surface of the occipital lobe, on the convexity only a small zone within the first occipital convolution and the occipital pole, but not the external occipital gyri and the angular gyrus.

According to clinical observation the first occipital convolution (Fig. 246, O[1]), including the cuneus, contains the *optical perception-field*. Accordingly, destruction of this region on one side cause homonymous hemiopia. To the patient the half of the visual field of the same side appears not as black, but only as if not present (deficiency of visual perception). In an analogous manner irritative conditions on one side give rise to photopsias in the homonymous halves of the visual fields. Hemiopia, occasionally associated with hallucinations within the blind halves, has been observed. Injury of the region named on both sides (also the effects of poisons, such as alcohol or lead) causes total blindness. Irritation of both centers gives rise to manifestations of light or color, or to visual hallucinations in the entire visual field. Cases, further, of cerebral lesions in which the spatial sense and the light-sense are wholly intact, while the color-sense alone is destroyed, indicate that the center for the color-sense must be especially located within the visual center, perhaps in the most posterior portion of the fusiform and lingual lobules (**Fig. 262**). Color-hemiopia has even been observed.

The clinical observations of hemiopia teach that the visual field of each eye can be divided into a larger external and a smaller internal portion, the two being separated by a vertical line passing through the yellow spot. The right or left halves of both visual fields are controlled from one hemisphere. The left halves must be projected upon the right occipital lobe and the right upon the left occipital lobe. Thus, every image, on binocular vision, if not too small, must be seen in two halves, the left half from the right and the right half from the left cerebral hemisphere. The yellow spot is in direct connection only with the external geniculate body, and the connecting fibers terminate in the wall of the calcarine fissure. It is a remarkable fact that in case of bilateral hemiopia a small central field of visual activity is preserved. In cases of hemiopia also the action of the pupils is impaired.

Exclusive irritation of the color-center gives rise to the appearance of color-hallucinations, as observed in the colored aura in cases of epilepsy. Colored vision occurs also in conjunction with other cerebral affections, for example as erythropia. Rarely, the subject has observed everything as yellow, or blue, or violet. Some poisons give rise to the same result through an influence upon the cerebral color-center: yellow vision due to santonin, red vision due to henbane, violet vision due to hashish. Lesions of the color-center have been found as the result of cerebral concussion and after the action of various poisons, permanent or transitory, total or partial color-blindness resulting.

The remaining portion of the center contains the *optical memory-field*, destruction of which gives rise to *mind-blindness*, in case of a lesion on one side especially upon the contralateral side. A special form of this condition is known as *word-blindness*, the individual no longer recognizing the symbols of writing—alexia. The area comprises, according to Flechsig, the supramarginal gyrus and the parietal lobule. Figures and letters appear to have special central memory-fields.

An interesting case of mind-blindness may be cited. After severe emotional disturbance loss of the memory of visual perceptions developed suddenly in an intelligent man. Everything with which he had been familiar—persons, streets, houses—appeared entirely strange to him and he even no longer recognized his image in the mirror. On attempting to read or figure he was compelled to speak the words and figures aloud. In his dreams visual images were entirely wanting.

In consequence of morbid irritation of the visual center, pronounced visual hallucination may develop in man, principally in the insane. Famous instances of visual hallucination are furnished by Jeanne d'Arc, Cardanus, Swedenborg, Nicolai, Justinus Kerner, Hölderlin. "The spirit and the demons of all time, the divine vision of the ascetics"—inanition-hallucinations in fasting persons—"the spiritual representation of the magician, the dream-object and the hallucination of the febrile and insane patient are one and the same phenomenon" (Johannes Müller). Cases have also been observed in which hallucinations were present only in one eye. Occasionally, these are seen, for example, in cases of delirium tremens, principally without color, therefore gray.

After degeneration of the cortical center, in the first and second occipital convolutions, cuneus and lingual lobe, the fibers degenerate that connect the occipital lobe with the external geniculate body, the anterior quadrigeminate body and the pulvinar of the optic thalamus; further, these structures themselves and later on the origin of the optic tract of the same side.

The lower in the animal kingdom one descends the less is the significance of the cortical center, and of the external geniculate body and the pulvinar, which together subserve the function of psychic vision in the higher vertebrates, with respect to the act of vision, while at the same time the anterior quadrigeminate body increases in size, until, finally, in fishes it constitutes the sole visual center.

In the new-born the optic radiation to the cortex is yet wanting, developing only in the course of weeks. The infant is also up to this time without psychic utilization of what is seen, that is it is for the time being still mind-blind. The deeper centers alone are at first active and excite only reflex action. With the development of the cortical center, the activity of the deeper centers later on diminishes to such a degree that, as soon as consciousness has developed, blindness occurs after destruction of the psycho-optic centers. In certain varieties of hysterical impairment of vision it appears that while the cortical center is still functionally active the mind of the patient does not appreciate what is seen.

The *psycho-auditory center* or *auditory sphere* is situated on each side (crossed) in the temporal convolutions, particularly in the root and the posterior portion of the first and concealed in the wall of the fossa of Sylvius. Total destruction of this center causes deafness; partial injury on the left side may give rise to *mind-deafness*. Among the phenomena of the latter is *verbal deafness*, which has been observed alone and also in association with verbal blindness. Wernicke found in cases of word-deafness softening in the posterior third of the first temporal convolution (T^1) on the left (!), and Naunyn designated the third and fourth fifths as the active areas. Complete deafness occurs only on destruction of the latter areas on both sides. Word-deafness is followed by secondary atrophy of the motor speech-center.

Verbal blindness and deafness may be included clinically in the group of aphasic disorders, in so far as they resemble the amnesic variety. The word-deaf or the word-blind patient resembles an individual who in early youth had learned a foreign language, which in later life he has completely forgotten. He hears, therefore, or he reads well the words and the symbols of writing and he is able also to repeat the words spoken to him and to write them on dictation, but he has entirely lost comprehension of the signs. While, therefore, the amnesic aphasic has lost only the key of the door to his speech-mechanism, the word-deaf or word-blind patient has lost this mechanism itself. From a case in which recovery took place it is known that the word sounds to the patient like a confused murmur. In left-handed persons destruction of the left temporal lobe is not followed by word-deafness, as in them the center is probably situated on the right side. The hallucinations of hearing induced by irritation of the psycho-auditory center appear usually in the right ear, although they may appear in both. Occasionally they are at the same time different in content and character in both ears. Huguenin observed atrophy of the temporal lobe after deafness of long standing.

Agraphia also may be due to word-blindness, the patient being unable to write from copy, although he can write spontaneously or on dictation; and likewise to word-deafness, the patient being unable to write on dictation, although he can write spontaneously or from copy.

According to Flechsig the *psycho-osmic center* or the *olfactory sphere* comprises the entire posterior margin of the base of the frontal lobe and the basal portion of the fornicate gyrus, the uncinate gyrus, and a portion of the adjacent inner pole of the temporal lobe. The *psychogeusic center* or the *gustatory sphere* is supposed by Flechsig to be situated within or at the margin of the center for bodily sensations or the olfactory sphere.

Subjective sensations of taste or smell in the insane and in epileptics are due to abnormal irritation in these regions, destruction of which will cause loss of the corresponding functions. In the new-born the olfactory center appears to be one of the earliest to enter upon functional activity. It degenerates after destruction of the olfactory tract.

The *sphere for bodily sensation, psycho-esthetic* and *psycho-algic center*, comprises the area between the fossa of Sylvius and the corpus callosum, including the central convolutions, the foot of all of the frontal convolutions, the paracentral lobule, and the gyrus fornicatus, especially in its middle third. The superficial tactile impressions and the sensations of movement are impaired after destruction of the central convolutions, while painful, thermic, and pressure sensations are preserved. After destruction of the fornicate gyrus and the hippocampal gyrus, tactile and thermic and common sensibility are partially lost. Destruction of certain regions (marginal gyrus) cause failure to recognize objects through the sense of touch.

On electric stimulation in a trephined human being sensory impressions (creeping) were observed in peripheral portions of the skin. All sensory impulses that rise from the posterior spinal roots pass through the lateral nucleus of the optic thalamus and from here they reach the central convolutions, which therefore are connected with the sensory nuclei of the posterior and lateral columns of the spinal cord.

Irritative disorders of sensibility occur in consequence of cortical irritation, including hallucinations of tactile, motor, and visceral sensations, the sensation of itching, prickling, and burning, which may reach a painful degree, as in epileptics and hysterics. Some cases of migraine, especially those associated with epilepsy, may be due to cortical irritation.

In cases of epilepsy marked excitation of the sensorial centers, manifested by excessive subjective impressions, often in association with psychic irritative disorders, for example the appearance of definite thoughts, has been observed as an irritative accompaniment of the convulsive seizure. Such excitation may appear even without accompanying convulsions as so-called *sensory epilepsy*, which may be partial, that is unilateral and confined to individual impressions, in the latter event without loss of consciousness. In cases of congenital inactivity of a psychosensorial center hallucinations in this region never develop.

Epileptoid hallucinations of the character described occur without convulsions but accompanied only by brief derangement of consciousness (absence). Under such circumstances amaurosis has also been observed, gradually disappearing later and being replaced by a concentric contraction of the visual field. Occasionally only the psychic cortical centers are affected, preëpileptic and postepileptic insanity, loss of memory for certain periods of time, derangement of consciousness resulting. Cases are extremely rare in which epilepsy occurs with loss of consciousness but without convulsions, and so also are cases in which the convulsions occur without derangement of consciousness.

The nerve-fibers passing from the sensorial and sensory organs to the psychosensorial cortical centers traverse the posterior third of the posterior limb of the internal capsule (Fig. 263). Destruction at this point causes, therefore, anesthesia on the contralateral half of the body. Only the viscera retain their sensibility. Also contralateral loss of hearing, of smell, and of taste, as well as hemiopia, appear. Whether the visceral sensations, sensations of internal processes associated with pleasure or displeasure, are localized in the cerebral cortex or in the midbrain is undetermined.

Pathological.—In human beings with more or less complete injury or degeneration of this tract, more or less well-marked loss of the pressure-sense and the temperature-sense, of cutaneous and of muscular sensibility, of taste, smell, and hearing is accordingly found. The eye is rarely entirely blind, but visual acuity is greatly impaired, the visual field is contracted and the color-sense may be partially or totally abolished. The eye upon the same side may suffer alone in lesser degree. In addition to material lesions of the brain, sensory anesthesia is observed also as a functional disorder in association with hysteria, neuroses, and psychoses. With reference to the mutual relations of the individual conducting paths within the internal capsule Redlich maintains that behind the pyramidal tracts there pass first the fibers for muscular sense, then those for cutaneous sensibility, and finally the visual fibers.

Cases of injury in the anterior frontal region without motor and sensory disorders have been collected in large number by Charcot, Pitres, Ferrier, and others. On the other hand, enfeeblement of intelligence and idiocy have been observed in connection with acquired or congenital deficiencies of the frontal region. According to Flechsig, there is no doubt, in accordance with clinical observation, that the frontal lobe and the temporo-occipital zone bear an intimate relation to mental processes, particularly those of a higher order, which disappear largely in the old and in epileptics.

The anterior portions of the first and second frontal convolutions, portions of the third and of the gyrus rectus in the frontal lobe, the island of Reil, the first and second parietal convolutions, the second and third temporal convolutions, the occipitotemporal gyrus, and the precuneus are association-centers. They connect the various sense-spheres and they have the function of associating irritative states of various sense-spheres.

Situation of the Cerebral Regions in the Skull.—In order to indicate the position of the principal fissures and convolutions in the uninjured head various

51

points suggested by Broca have been marked in Fig. 260, which shows the different parts of the brain according to **A**. Ecker. K_1 K_2 K_3 are points in the coronal suture that can be felt through the scalp. K_1 is placed, in order to avoid the longitudinal sinus, 15 mm. to one side of the median line. K_2 is the point of intersection of the coronal suture and the temporal line. At K_3 the coronal suture intersects the upper margin of the great wing of the sphenoid bone. L_1 and L_2 are situated in the lambdoid suture, the former 15 mm. to one side of the highest point, and the latter in the middle of the posterior border of the parietal bone. M corresponds to the highest point of the arch of the squamous suture. If horizontal lines be drawn backward from the points K_1 K_2 K_3, the central fissure C, which is so important in localization, is situated, at its upper extremity about 45 mm. and at its lower extremity about 30 mm. behind the coronal suture. According to Merkel the lower extremity is almost 5 cm. vertically above the inferior maxillary articulation. The bifurcation of the large fossa of Sylvius is 4 or 5 mm. behind K_3 or, according to Merkel, from 4 to 4.5 cm. above the middle of the malar arch. Its anterior branch is parallel with the coronal suture, and its posterior branch passes through the point M. The parieto-occipital fissure (po) is situated almost exactly in the lambdoid suture or, measured with compasses, 6 cm. above the external occipital protuberance. The frontal eminence forms the boundary between the first and second temporal convolutions. The parietal eminence covers the supramarginal gyrus.

The *corpus callosum* contains commissural fibers from both hemispheres (according to Mott and Schäfer between the two corticomotor centers), the angular gyri and the occipital and temporal lobes. Division of this structure in the dog causes no appreciable disturbance. In accordance with this fact almost total destruction has been observed in man without the development of noteworthy derangement of motility, coördination, sensibility, reflex activity, the special senses, speech, or considerable impairment of intelligence. The posterior portion of the anterior commissure serves for the connection of the two lingual gyri.

THE BASAL GANGLIA OF THE CEREBRUM.   THE MIDBRAIN. FORCED MOVEMENTS.   OTHER CEREBRAL FUNCTIONS.

The *striate body* and the *lenticular nucleus* (Figs. 263, 264) have no direct connection with the cerebral cortex, although fibers pass from their connections to the cerebral peduncle and the medulla oblongata. Their development in the animal kingdom keeps pace with that of the cerebral cortex. The general muscular contractions on the opposite side of the body observed on electrical stimulation are probably due to associated irritation of adjacent cortico-muscular tracts.

Gliky observed no movement on irritation of the striate body in the rabbit. It, therefore, appears that the motor tracts in this animal do not traverse the portion of the brain named, but pass by it.

Destruction of the lenticular nucleus or the striate body gives rise, according to earlier statements, to loss of voluntary movements on the opposite side of the body, with or without preservation of sensibility; although under such circumstances there is also associated injury of the cortico-muscular tracts. Recently, after injuries transitory weakness of the contralateral extremities (loss of muscular sense) has been observed, with increased general irritability and fear, as well as rapid (transitory) elevation of temperature. Irritation of the striate body is unattended with pain.

Pathological.—In man every lesion in the anterior portion of the striate body that is not too small causes contralateral paralysis, which is permanent if the internal capsule is affected, but which may gradually disappear if the lenticular and caudate nuclei are affected. Occasionally vascular dilatation occurs in consequence of vasomotor paralysis if the posterior portion is affected, and is attended with redness and slight elevation of temperature in the paralyzed extremities (at least for a time), swelling (edema), sweating, alterations in pulse demonstrable with the aid of the sphygmograph, acute decubitus on the paralyzed side, abnormalities of the nails, the hair, the skin, acute inflammation of the joints, particularly the shoulder-joint. Subsequently, contractures occur in the paralyzed muscles. In individual cases there may be, besides, cutaneous anesthesia, occa-

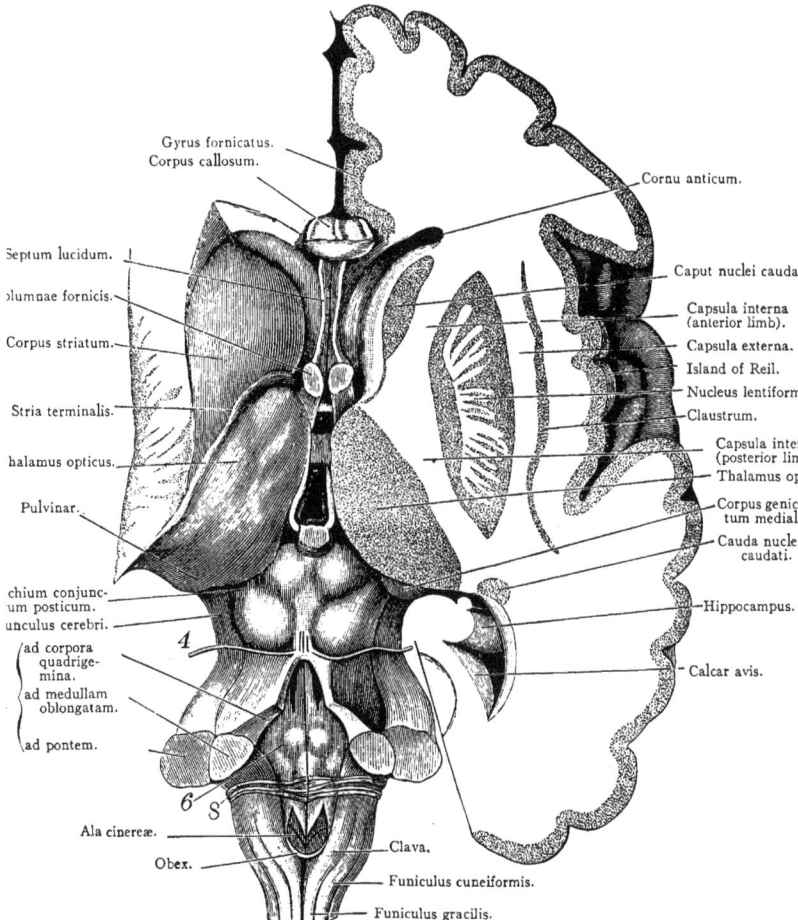

Gyrus fornicatus.
Corpus callosum.

Cornu anticum.

Septum lucidum.

olumnae fornicis.

Corpus striatum.

Stria terminalis.

halamus opticus.

Pulvinar.

chium conjunc-
um posticum.

unculus cerebri.

ad corpora
quadrige-
mina.

ad medullam
oblongatam.

ad pontem.

4

6 8

Ala cinereæ.

Obex.

Caput nuclei cauda

Capsula interna
(anterior limb).

Capsula externa.
Island of Reil.
Nucleus lentiform

Claustrum.

Capsula inte
(posterior lin

Thalamus op

Corpus genic
tum medial

Cauda nuclei
caudati.

Hippocampus.

Calcar avis.

Clava.

Funiculus cuneiformis.

Funiculus gracilis.

FIG. 263.—Cerebrum of Man. On the right the hemisphere is removed by a horizontal section. 4, Trochlear nerve; 8, auditory nerve; 6, origin of the abducens nerve.

sionally also impairment of sense-activity on the paralyzed side; both if the posterior segment of the internal capsule is affected. Generally hemiplegia and hemianesthesia exist together.

The *optic thalamus* is connected with all of the sense-centers. As it is connected with the cerebral cortex by fibers, principally as a partial origin of the optic nerve, it probably bears some relation to the sensation of vision. In man injury of the posterior third may give rise to visual disturbances. Removal of the optic thalamus or destruction of the parts in the neighborhood of the inspiratory center in the wall of the third ventricle impairs coördinated movement in rabbits. Also in man, contralateral disorders of coördination, choreiform twitchings or ataxia have been observed to follow degeneration of the optic thalamus. Destruction of the optic thalamus gives rise in man to loss of mimetic expression on the opposite side of the face in response to emotional influences, although the muscles can be moved voluntarily.

Bechterew concludes as a result of experiments and of pathological observations that the optic thalami play an important part with respect to the expression of varied perceptions, sensations, and emotional activities. They are motor centers through the intermediation of which principally the congenital movements of expression (such as laughing or crying) are executed, and which are excited under the influence of involuntary psychical impulses, such as emotions, or they can be stimulated reflexly through tactile stimulation and irritation of other sensory organs. The thalamus and the anterior quadrigeminate body contain the centers for complex reflexes. Both receive connections from the posterior nerve-roots of the spinal cord and from the sensory cerebral nerves, and the thalamus, in addition, from the olfactory and optic tracts. The anterior quadrigeminate body contains a common optico-auditory reflex path. The optic thalamus contains also the reflex center for the secretion of tears and from this situation the sensory irritation is conveyed to the path for the secretory branches of the trigeminus and the facial, as well as of the sympathetic.

After injury of one thalamus paresis or paralysis of the contralateral muscles, together with circular movements, have been reported in some cases, and contralateral hemianesthesia with or without involvement of the motor sphere in other cases. Fibers pass from the thalamus to the cortex of all of the cerebral lobes. also to the cornu Ammonis and the tegmentum of the cerebral peduncle, Extirpation of certain portions of the cerebral cortex in the rabbit is followed by atrophy of certain portions of the thalamus. The relations of the optic thalamus to reflex inhibition are discussed on page 731, to the movements of the stomach on page 288, to those of the intestines on page 791. The heat-center supposed to be situated in the thalamus is described on page 395.

Injury of the *cerebral peduncles* gives rise, first of all, to severe pains and spasms on the opposite side of the body, where the salivary glands secrete. These irritative phenomena are followed, as paralytic symptoms, in man by contralateral anesthesia and loss of voluntary control of the muscles as well as paralysis of the vasomotors. In case of lesions of the peduncles in man the oculomotor nerve should be observed, as it is often paralyzed on the same side.

The middle third of the cerebral peduncle comprises the well-known conducting path of the pyramidal tracts. The fibers of the inner third connect the frontal lobe through the superior cerebellar peduncle with the cerebellum. The outer third contains fibers that connect the pons with the temporal and occipital lobes of the cerebrum. The fibers passing from the tegmentum into the corona radiata serve for sensory conduction.

According to Goltz section of the cerebral peduncle in the dog is followed by a tendency to fall to the same side; the movements of the contralateral extremities appear larger and cutaneous sensibility is impaired on the entire contralateral side. The animal sees especially only objects that make an impression upon the right half of each retina. Therefore, each peduncle, according to Goltz, contains motor and sensory fibers for the entire body.

Irritation or section of the *pons* gives rise to pain and spasm. After section of the contained conducting fibers—sensory, motor, and vasomotor —paralyses appear, together with forced movements. An explanation is wanting as to the significance of the ganglia present in the pons. For diagnostic purposes in man attention should be directed to the presence of possible alternate hemiplegia.

The **quadrigeminate bodies** or the **midbrain.** Destruction of the quadrigeminate bodies on one side in mammals, or of the optic lobe in birds, amphibia, and fish, is followed by blindness, which may be situated upon the same or upon the opposite side in accordance with the conditions of decussation in the optic chiasm. Total destruction of the bodies on both sides causes blindness in both eyes. As a result the reflex between irritation of the retina and the oculomotor nerve is abolished, that is, the pupils no longer contract after illumination of the retina. If the cerebral hemispheres alone are removed, the pupils still contract on light-stimulation, as well as after mechanical irritation of the optic nerve.

Extirpation of the eyeball is followed by atrophy of the contralateral anterior quadrigeminate body.

According to Bechterew the fibers of one optic tract pass through the brachium conjunctivum anterius (Fig. 241) into the external periphery of the anterior quadrigeminate body. The fibers that decussate in the chiasm (Fig. 240) pass into the posterior quadrigeminate body. In accordance with this distribution are the symptoms of partial blindness after destruction of an anterior or posterior quadrigeminate body. Fibers pass onward to the cortex in the internal periphery of the anterior quadrigeminate body.

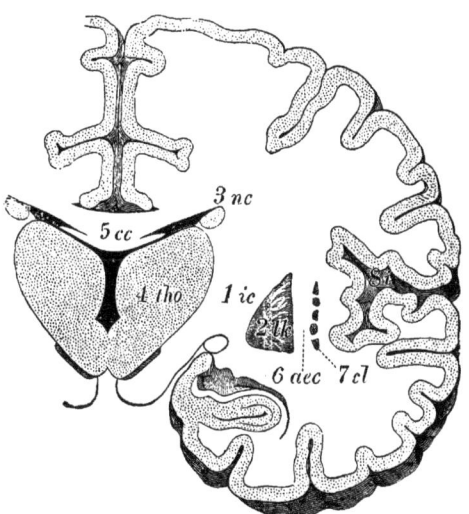

Fig. 264.—Frontal Section through the Cerebrum: 1 *ic*, internal capsule; 2 *lk*, nucleus lentiformis; 3 *nc*, caudate nucleus; 4 *tho*, optic thalamus; 5 *cc*, corpus callosum; 6 *aec*, external capsule; 7 *cl*, claustrum; 8 *i*, island.

In animals deafness has been observed to develop after destruction of the posterior quadrigeminate body. Animals exhibit under such conditions difficulty in phonation even to the point of loss. In man a paralyzing lesion of the tegmentum or of the internal capsule is present in all cases of midbrain deafness. Destruction of the quadrigeminate bodies is followed further by disturbance in the perfect harmony of movement; disorders of equilibration and incoördination of movement also occur.

The cochlear nerve undergoes partial decussation in the posterior quadrigeminate body and in the pons. The quadrigeminate bodies react to electrical, chemical, and mechanical stimulation. The reports are contradictory, however, as to the results of irritation. According to some observers dilatation of the pupil on the same side takes place; according to Ferrier the contralateral pupil dilates first and later also the pupil on the same side. The irritation extends from the quadrigeminate bodies to the medulla oblongata and further on to the origin of the sympathetic, for after section of the cervical sympathetic the dilatation no longer takes place. According to Knoll contraction of the pupil, such as was observed by earlier investigators, takes place only when the adjacent optic-nerve tract is irritated. In addition, irritation of the right anterior quadrigeminate · body causes rotation of both eyes to the left, and conversely. If the irritation be continued the head also is rotated toward the same side. Vertical section of the quadrigeminate bodies in the median line is followed, on unilateral irritation, by this result only upon the same side. Ferrier observed, further, signs of pain on irritation of the quadrigeminate bodies in mammals. Danilewsky, Ferrier, and Lauder Brunton observed, finally, increase in blood-pressure and slowing of the heart-beat, together with deep respirations.

Bechterew attributes all of the phenomena that occur after injury or irritation of the quadrigeminate bodies, except those referable to vision itself, to lesions of more deeply situated parts. Therefore, according to him, the quadrigeminate bodies themselves contain neither the center for the movements of the pupils nor that for the combined movements of the eyes, nor do they contain that for maintaining the equilibrium of the body. Irritation of the quadrigeminate body causes the animals to start back markedly as a reflex phenomenon. Nystagmus, forced movements, and uncertainty in walking occur also only in association with injuries of more deeply situated parts.

Pathological.—Lesions of the anterior quadrigeminate bodies in man give rise, in accordance with their extent, to visual disturbances, immobility of the pupils and even blindness. In addition profound injury may be attended with paralysis of the oculomotor nerves on both sides, in consequence of which the affected ocular muscles are not involved with entire symmetry and not in equal degree. An uncertain staggering gait, especially if it appears as the first symptom, is likewise characteristic.

Destruction of the posterior commissure in rabbits has the same effect as section of both oculomotor nerves; a lesion causes only diminution in the irritability of these nerves. An incomplete asymmetrical lesion causes asymmetrical diminution in the irritability of the two nerves, the nerve upon the side of the lesion being less irritable than that upon the opposite side.

Forced Movements.—The significance of the midbrain in relation to the harmonious execution of movements makes it clear that unilateral injuries of such parts as are connected with it by means of conducting fibers cause peculiar unilateral disturbances of equilibrium and deviations from the symmetrical movements of both sides of the body that have been designated *forced movements*. In this category belong the *circular movement* (mouvement de manége), in which the animal, with the intention of running onward, moves constantly in a circle; the *index-movement*, in which the fore part of the body is moved about the stationary posterior part, like an indicator about its axis; the *rolling movement*, by means of which the body is revolved about its longitudinal axis. All of these forms of movement may pass into one another, and they represent only gradual variations in the same disorder. The parts injury of which causes these forced movements are the striate body, the optic thalamus, the cerebral peduncle, the pons, the middle cerebellar peduncle, certain portions of the medulla; and even after injury of the surface of the cerebrum Eulenburg and Landois observed index-movements in rabbits, and Bechterew in dogs. Also in man forced movements have been observed, especially in association with lesions of the parietal convolutions. Forced movements, together with nystagmus and rotation of the eyes, are caused also by injury to the olive.

On pathological degeneration of one olive of the medulla oblongata pronounced rotatory movements toward the same side have been observed in man.

Statements differ as to the direction and the character of the movements after the individual injuries. The following observations have been made: Section of the anterior portion of the pons and the cerebellar peduncles causes index-movement and rolling movements toward the opposite (paretic?) side; section of the posterior portion of the same regions causes rolling movements toward the same (paretic?) side, as does also deeper puncture of the auditory tubercle or the restiform body. Incision of one cerebral peduncle causes circular movement, with the convexity directed toward the same side. The closer the incision is situated to the pons the narrower becomes the circle of movement. Finally, index-movement occurs. Injury of one optic thalamus causes much the same phenomena as puncture of the anterior portion of the cerebral peduncle, because the latter is injured at the same time. Injury of the anterior portion of one optic thalamus gives rise to forced movement in the opposite direction, that is with the concavity directed toward the side of the injury. Flexion of head and vertebral column, with the convexity toward the affected side, together with circular movement, is caused by injury of the spinal extremity of the medulla; the convexity is directed toward the unaffected side as a result of injury to the anterior extremity of the calamus and above.

Rotation (strabismus) and involuntary oscillation (nystagmus) of the eyes may be included among forced movements. Nystagmus occurs as a result of unilateral superficial lesions of the restiform body, as well as of the floor of the fourth ventricle, and as a result of irritation of the cerebellum. Unilateral, deep, transverse injuries from the apex of the calamus downward to the auditory tubercle cause strabismus of the eye of the same side downward and forward, and of the opposite eye backward and upward. Bilateral injuries cause this strabismus to disappear. It is, therefore, to be inferred that the medulla oblongata contains a mechanism controlling the ocular movements, which can be irritated as a result of sudden anemia (ligature of the cerebral arteries in the rabbit).

In explanation of the forced movements it has been in part assumed that they are due to unilateral incomplete paralysis, so that the animal, on attempting to move about, drags the paretic side somewhat (as, for example, in the circular movement on the side of the body directed toward the center of the circle), and therefore the symmetry of movement is lost. Others have attempted, in direct opposition to this view, to establish an irritation through the act of injury as the cause of an excessive activity upon one side of the body. Landois, as a result of his own observations, ranged himself on the side of those investigators who consider vertiginous sensations induced by the injury as the cause of the movements. He observed, at times, that immediately after the injury (stilet-puncture), the movement took place in a direction opposite to that appearing somewhat later. He considered this phenomenon as the effect of the irritation and paralysis induced in quick succession by the injury. The latter, by irritating or paralyzing the apparatus controlling locomotor sensations, may give rise to a false impression as if the body of the animal or also the objects of the external world moved in a definite direction. As a result of this motor deception the movements described are executed as a reaction, with the intention of correcting the abnormal fictitious movements by means of suitable counter-movements. The circular movements after injury of the optic thalamus may be induced by apparent movement in consequence of injury to the optic nerve.

In this connection it may be mentioned that injury of a point not far from the posterior extremity of the cerebral hemisphere causes after the lapse of some time marked forward or lateral movements, likewise probably as the result of a false motor impression. The unrestrained running movement after injury of an area in the middle of the striate body near the free margin directed toward the ventricle is probably to be explained in the same way. At first the animal remains quiet. If driven, however, it runs furiously until restrained by some obstruction. Landois has made the observation that every manipulation of the central organs that affects the equilibrium in considerable degree is attended with marked increase and deepening of the respirations.

FUNCTIONS OF THE CEREBELLUM.

Injuries of the cerebellum cause in marked degree disturbances in the harmony of the movements of the body. Probably the cerebellum represents a central organ for the more delicate gradation and the normal sequence of movements, inasmuch as it regulates especially continuous and tonic muscular contractions. Thomas designates it a reflex center for maintaining the equilibrium. Its connections with all of the ganglionic masses of the central organs render the cerebellum adapted to this purpose.

Through the lateral cerebellar tracts stimuli are conveyed to the cerebellum and these serve as guides to the position of the trunk. Connections of the vestibular nerve with the cerebellum have a similar effect with respect to the equilibrium. The cerebellum may influence the motor nerves of the spinal cord through fibers that pass downward through the restiform body into the lateral tract of the spinal cord. The cerebellum itself is insensitive to injuries.

The experiments of Luciani upon the functions of the cerebellum prove that each portion of this structure has the same function as the whole. The functions are threefold: 1. The cerebellum provides voluntary movements with sufficient strength. 2. It increases the tone of the muscles during rest. 3. It accelerates the rhythm of the individual motor impulses that constitute the movements and it fuses the impulses into a continuous act. 4. Russell has found, after extirpation of the cerebellum, incoördination of movement, rigidity of the muscles, and motor weakness as characteristic symptoms.

After almost complete removal of the cerebellum dogs exhibit paresis and deficient tone especially in the muscles of the vertebral column and the hind legs. The animal is able neither to stand nor to walk and the head oscillates to and fro. Immediately after the operation there appear as irritative phenomena: tonic spasm of the muscles of the nape of the neck, the back, and the forelegs, convergence of the eyes, occasionally falling forward of the body. Intelligence and sense-impressions, including the muscle-sense, remain intact.

Median division of the cerebellum without extirpation causes permanent enfeeblement of all voluntary movements, diminution of the muscular tone present during rest, as well as tremor, discontinuous muscular contractions, incoördination and uncertainty in voluntary movements. Extirpation of the vermis causes, as irritative phenomena, tonic contraction of the muscles of the nape of the neck and of the forelegs, which at times is followed by paresis especially of the hind legs. Complete removal of one-half of the cerebellum is followed, as irritative phenomena, by rolling movements about the longitudinal axis, as well as rotation of the eyes toward the unaffected side, curvature of the vertebral column toward the side operated on and tonic extensor spasm of the foreleg and less commonly of the hind leg upon the same side. These are followed, as paralytic phenomena, by relaxation of the muscles of the same side (atony), a somewhat less energetic contraction (asthenia) and a want of fusion of the composite movements, so that tremor, swaying and rhythmic oscillations (astasia) result. Superficial injury or partial removal of one hemisphere is neutralized by the assumption of its function by the intact portions of the cerebellum. Animals deprived of their equilibrium after extirpation of the cerebellum can regain it through the motor impulses gradually sent from the motor cortical centers of the cerebrum in standing, walking, and swimming.

Luciani observed, eventually, in animals after extirpation of the cerebellum, general marasmus, and he believed, therefore, that the organ exercises a trophic function. In accordance with this view, Friedeberg observed loss of weight after disease of the cerebellum.

Extirpation of the cerebellum is followed by secondary degeneration of the portion of the pons surrounding the pyramids, of the inferior olivary bodies, all of the cerebellar peduncles and the direct cerebellar bundle of Flechsig, principally on the same side, in lesser degree on the opposite side. Degeneration takes place, also, in some fibers within all of the cerebral nerves and the anterior roots of the spinal nerves.

In frogs an important organ for locomotion is situated at the junction of the medulla with the cerebellum. After its removal the animal is no longer able to hop about or to creep in a coördinate manner.

Pathological.—Asymmetrical or unilateral lesions of the cerebellum cause in man a tendency to fall toward the side of the injury, while bilateral injuries cause a tendency to fall backward. If the middle lobe is affected disorders of coördination occur, particularly a stumbling, staggering gait and marked vertigo, as well as atony, asthenia, and ataxia. Irritative disease of the middle cerebellar peduncle causes complete rotation of the body about its axis, with rotation of the eyes and the head in the same direction.

If an electrical current be passed through the head of a man, the electrodes being placed in the mastoid fossæ behind each ear, and in such a manner that the positive pole is applied upon the right and the negative pole upon the left, a marked feeling of vertigo occurs on closure, and the head and body fall toward the positive pole, while the objects of the outer world appear to move toward the left. If during the passage of the current the eyes are closed, the apparent movement is transferred to the individual himself, who then has a feeling of rotation toward the left. At the moment when the head falls toward the anode, the eyes also are rotated in the same direction and frequently exhibit nystagmus. The electrical current under such circumstances probably exerts an irritative effect upon the nerves of the ampullæ, disorders of which cause vertigo.

PROTECTIVE AND NUTRITIVE APPARATUS OF THE BRAIN.

The *cerebral dura mater* is intimately united with the periosteum of the cranial cavity. The *spinal dura mater* forms about the spinal cord a freely suspended long sac attached only on its anterior aspect. The dura mater is a fibrous membrane consisting of firm bands of connective tissue, interwoven with a large number of elastic fibers and provided with flat connective-tissue cells and Waldeyer's plasma-cells. The smooth inner surface is lined by squamous endothelium. Blood-vessels are present only in moderate number, though in somewhat greater abundance in the outer layers, while lymphatics are numerous. Nerves with unknown terminations (Pacinian bodies have been found on the petrous bone) endow the dura with great sensitiveness to painful impressions (also in the dog, but not in the rabbit).

Between the dura and the *arachnoid* is situated the lymphatic *subdural space*. The pia mater and the arachnoid united to it by means of a reticular network really form a common membrane, which cannot be separated. Between the two layers, as if enclosed in dropsical connective tissue, cerebrospinal lymph is present in a space, the *subarachnoid space*, which is lined by endothelium. The external limiting layer of this stratum, correctly designated also arachnoid in the strict sense, is thin, poor in vessels, without nerves and lined on both surfaces by squamous endothelium. It is, however, separated from the pia only over the spinal cord, so that between the two lies the lymphatic subarachnoid space. Over the brain the two are in large measure united, except where they form bridges over the sulci. Over these the arachnoid merely passes, while the pia penetrates into the depth. The cerebral ventricles communicate freely with the lymphatic subarachnoid space, but not with the subdural space. The subdural and subarachnoid spaces do not communicate with each other. The pia, made up of delicate bundles of connective tissue, without elastic fibers, exceedingly rich in blood-vessels and lymphatics, conveys nerves in association with the vessels into the structure of the central organs.

The lymphatics of the brain, apart from those accompanying the vessels, consist of spaces surrounding the ganglia and of the glia-cells of the cortex with their processes. They all empty finally into the subarachnoid space. The cerebrospinal fluid is described on p. 367. The Pacchionian granulations are connective-tissue villi that serve for the flow of lymph from the subdural and subarachnoid spaces into the sinuses of the dura mater, particularly the superior longitudinal sinus, into which they project. The subarachnoid space communicates also with the spongy cavities of the cranial bones and with the veins of the surface of the skull and of the face. The subdural space communicates, further, with lymphatic spaces of the dura, and the latter communicate directly with the veins of the dura. The two lymphatic intermeningeal spaces communicate also with the lymphatics of the nasal mucous membrane. The space external to the spinal dura (*epidural space*) may also be considered as a lymphatic space. From it the pleural and peritoneal cavities may be readily filled. It does not, however, communicate with the cranial cavity. The venous plexuses, which perhaps secrete the cerebrospinal fluid, consist of convolutions of vessels surrounded by undeveloped connective tissue. The telæ choroideæ in the newborn are still provided with ciliated epithelium.

The pulsations of the large blood-vessels at the base of the brain impart pulsatory movements to the latter. As a result of the physical conditions present in the calvarium, the large amount of blood thrown into the arteries with every systole causes the expulsion of an equal amount of blood from the veins. The act of breathing causes, besides, a respiratory movement of the brain, which is elevated on expiration and falls on inspiration. This movement is due in part to the respiratory fluctuation in pulse and in part to variations in the amount of blood in the veins of the cranial cavity. Finally, there is to be recognized a movement of vascular elevation and depression, occurring from twice to six times in a minute, corresponding to the periodic-regulatory dilatation and contraction of the vessels. This movement is influenced by emotional disturbances. It occurs most regularly during sleep.

The movements of the brain are apparent especially where its membranes offer slight resistance, therefore, for example, at the fontanels in children and in artificial trephine-openings. The presence of the cerebrospinal fluid is, however, exceedingly important with respect to this movement, probably because it propagates the pressure uniformly and, thus, concentrates every systolic and expiratory vascular dilatation upon the portion on the calvarium that does not offer resistance. If the fluid is drained away the movement becomes small to the point of disappearance.

As the arteries within the rigid calvarium undergo change in volume with the movement of the pulse a pulsatory variation in the volume of the veins (sinuses) is constantly observed, the opposite of that in the arteries. Emotional disturbances increase the pulsation of the brain. At the moment of awaking the amount of blood in the brain diminishes, while sensorial irritations during sleep, without awakening the subject, increase the amount of blood. In slight degree the brain may undergo passive movement within the cranial cavity on change in the position of the head.

The **vessels of the pia** are naturally in part under the influence of the vasomotor nerves accompanying them; in part their size may be influenced from remote parts of the body. If a trephine-opening be closed by means of a small glass window, the effects upon the lumen of the vessels can be observed with the aid of a microscope. Irritation of the sympathetic affects only the vessels of the same side, but does not alter the blood-pressure upon the other side (through the circle of Willis). Paralysis of the vasomotor nerves, also by means of narcotics, causes dilatation of the vessels. The vessels contract strongly in death. They are dilated in connection with cerebral activity, as well as during sleep. Transitory anemia of the cerebral arteries is followed by their secondary dilatation and hyperemia. Irritation of the vasomotor center, for example by asphyxia or strychnin or reflexly, causes the presence of an increased amount of blood in the arteries of the central nervous system in consequence of collateral hyperemia. These arteries, therefore, do not take part in the contraction of all of the remaining arteries. Excessive elevation of pressure in the cerebrospinal cavity in consequence of hyperemia is offset by the escape of cerebrospinal fluid into the lymph-sheaths of the cerebrospinal nerves. Cerebral irritation that excites epileptic attacks cause an increased supply of blood independently of the blood-pressure. Sudden ligation of all of the cerebral arteries causes immediate loss of the sensorium, and later on marked irritation of the medulla oblongata and its centers and finally rapid death with convulsions.

As a result of the free anastomoses at the base, the individual portions of the cerebrum are protected against anemia on compression or ligature of one or another vessel. Within the cerebrum the arteries are distributed as terminal arteries, that is in the area of their terminal distribution they do not form anastomoses with neighboring arterial branches. On the other hand, the peripheral arteries on the outer surface of the brain, the arteries of the corpus callosum, of the fossa of Sylvius, and the deep cerebral, form free anastomoses. The sudden assumption of the erect posture by persons who have occupied the recumbent position for a long time and are at the same time anemic is not rarely attended with cerebral anemia from hydrostatic causes, associated with loss of consciousness and obscuration of the senses. Alterations in the position of the body have otherwise no effect upon the pressure in the cerebral vessels. Death occurs in some animals after vertical elevation of the trephined skull and even more quickly if they are placed upon the centrifuge. Exceedingly severe muscular exertion as well as marked activity on the part of other organs greatly reduce the pressure in the carotid arteries.

Cerebral Pressure.—Enclosed within the unyielding calvarium there is on the one hand the brain together with the nutritive fluid (lymph) that permeates it, as well as the cerebrospinal fluid, and, on the other hand, the system of blood-vessels. If the volume of the latter is increased in consequence of the presence of an increased amount of blood in the skull, the brain becomes poorer in fluid, like an expressed sponge. Conversely, in case of excessive production of the fluids mentioned the blood must escape from the vascular system. That the latter, however, is possible must be concluded from the circumstance that the formation of the fluid is not, under all circumstances, dependent, as a simple transuding-filtrate, solely upon the blood-pressure, but that it may take place also independently of the latter as a result of the secretory activity of the vessels.

The brain and the fluid surrounding it are constantly under a certain mean pressure, which is influenced by the atmospheric pressure, so that the pressure within the skull is altered in correspondence with fluctuations in the atmospheric pressure. According to Grashey there prevails in the skull of an adult a negative pressure of —13 cm. of water; at the foramen magnum it is zero. . In the dural sac of the spinal cord there is a positive pressure, below (in the erect posture) greater than above, but on the average +60 cm. of water. The investigations of Naunyn and Schreiber upon pathological brain-pressure, or cerebrospinal pressure, have shown that this pressure must reach a level somewhat below the arterial pressure in the carotid artery before the distinctive symptoms of cerebral pressure appear. These consist of headache of paroxysmal occurrence, with marked vertigo to the point of unconsciousness, vomiting, slowing of the pulse, slow and shallow respiration, convulsions, injection of the conjunctiva, and increase of the pressure of the cerebrospinal fluid. The cause of these symptoms resides in anemia of the brain, so that bloodletting is to be avoided. In consequence of excessive tension of the cerebrospinal fluid the brain is expressed like a sponge. The blood escapes from it, and naturally from the capillaries most readily, as these can be most readily expressed on account of their lower internal pressure. Acute cerebral anemia is thus induced. If the degree of pressure attains only a moderate level the symptoms described may remain latent. Nevertheless, nutritive disorders develop in the brain, with consecutive phenomena, such as persistent slight headache, a feeling of vertigo, muscular weakness, visual disturbances (in consequence of neuroretinitis with papillitis). The symptoms may be relieved by elevation of the blood-pressure, while reduction of the pressure causes more marked symptoms of cerebral pressure.

At a pressure of from 70 to 80 mm. pain appears in dogs only in consequence of mechanical irritation of the dura; at a higher pressure, loss of consciousness; at a pressure of 100 mm., convulsions similar to those attending sudden occlusion of the arteries. A pressure of from 100 to 120 mm. gives rise to slowing of the pulse in consequence of central irritation of the vagus, while the respiratory frequency exhibits a transitory increase and later a reduction. Marked compression of long standing terminates fatally sooner or later. The blood-pressure is first increased in consequence of reflex stimulation of the vasomotor center, as a result of irritation of the sensory nerves by pressure. Then the blood-pressure falls, with marked slowing of the pulse. In addition, variations in blood-pressure of irregular occurrence are indicative of direct central irritation of the vasomotor center by pressure.

At the level of the cauda equina the pressure of the spinal fluid in the arachnoid sac is only between 7.5 and 12 mm. of mercury in the dog. After evacuation of the cerebrospinal fluid restoration takes place rapidly. Artificial increase is soon neutralized, the excess of fluid passing into the lymphatics and the veins.

COMPARATIVE. HISTORICAL.

Nerves are wanting in the protozoa. Among the celenterates the first indications of a nervous system are present in the neuromuscular cells of the hydroids and the medusæ. In the latter a closed nervous chain passes along the margin 'of the umbrella and, corresponding to the marginal bodies, exhibits cell-like thickenings from which filaments pass to the sense-organs. In worms a ring is often attached to the head and it surrounds the pharynx in those provided with intestines as a single or a double ring. From this there pass into the elongated body longitudinal trunks, frequently two, which are provided with ganglia corresponding to the body-segments, and here they anastomose. In the leech only one longitudinal trunk provided with ganglia, the so-called abdominal medulla, is present. In echinoderms the mouth is surrounded by a large nervous ring, from

which thick nerves pass off corresponding to the main trunks of the water-vascular system.   At the point of origin the nervous ring is provided with the
so-called ambulacral brains.   Arthropods possess above the pharynx a large cephalic
ganglion from which the sense-nerves arise.   Another ganglion below the pharynx
is connected on each side with the first by means of a commissure.   From this
point the abdominal chain of ganglia extends through the thorax and the abdomen.
At times several ganglia are fused into a nervous node of considerable size; at
other times they remain isolated for the majority of the segments of the body.
In molluscs also the pharyngeal ring is still present, although the ganglionic
masses occupy a varying position in it.   A number of remotely situated ganglia
connected with the pharyngeal ring by means of filaments represent the sympathetic.   In cephalopods a portion of the pharyngeal ring almost entirely devoid
of commissures is enclosed as the brain in a cartilaginous calvarium.   In addition
ganglia are found in the stomach and the heart.   In vertebrates the nervous
system is always situated on the dorsal aspect of the body.   In the amphioxus
it is not yet subdivided into brain and spinal cord.   The divisions of the brain
of vertebrates have been discussed on p. 776, the peripheral nerves on p. 721.

Historical.—Alkmaeon (580 B. C.) located consciousness in the brain, Galen
(131–203 A. D.) the impulse for voluntary movements.   Aristotle (384 B. C.)
described the brain of man as relatively the largest.   He designated it as unirritable to stimuli (insensitive).   He considered small persons as mentally superior.
He considered it a function of the brain to cool the heat arising from the heart.
Herophilus (300 B. C.) properly considered the region of the posterior horn as
the principal seat for sensation.   He described, further, the calamus scriptorius.
Probably as a result of experiment, he considered the fourth ventricle as the
most important for life.   Homer makes repeated references to the danger of
injury of the neck (the seat of the medulla oblongata).   Hippocrates, Galen,
Aretaeus, and Cassius Felix (97 A. D.) knew that a lesion of one-half of the brain
causes paralysis on the opposite side of the body.   Galen recognized the spinal
cord as containing the conducting tract for motion and sensation.   The ascetics
of the Middle Ages were familiar with visual hallucinations (visions) and the like,
and many important paintings are to be considered as representations of such
hallucinations, the eye of the expert now and again recognizing in them photoptic
secondary phenomena, for example scintillating scotoma.   Vesalius described
(1540) the five ventricles of the brain.   R. Columbo observed (1559) the movement of the brain synchronous with the action of the heart, while the respiratory
movement of the brain was first described accurately in 1811 by Ravinna.   Varolius (born 1543) described the pons.   Coiter noted (1573) the possibility for life
to continue after removal of the cerebrum.   Wepfer discovered in 1658 the
hemorrhagic nature of apoplexy ("sanguine extra vasa effuso ex rupto ramo"),
while Sylvius de le Boë described the fossa and the aqueduct named after him.
Schneider (1660) determined the weight of the brain of different animals.   Mis-
tichelli (1709) and Petit (1710) described the decussation of the fibers of the
medulla below the pons.   Haller and his pupil Zinn were familiar with the circular movements following injuries of the brain.   Lorry was the first to observe
disorders of coördination in a pigeon after puncture of the cerebellum (1760).
Gall demonstrated the partial origin of the optic nerve from the anterior quadrigeminate body and he gave the best descriptions of the fibers and of the convolutions of the brain from dissection of the brain from below.   Luigi Rolando (1809)
described the great central fissure of the brain.   He and Bellinger (1823) described
more fully the form of the gray matter of the spinal cord.   Carus described (1814)
the central canal of the spinal cord, which had already been observed in the
seventeenth century by J. Conrad Brunner.   A most extensive anatomical
work upon the brain was written by Burdach (1819–1826).

PHYSIOLOGY OF THE ORGANS OF SPECIAL SENSE.

INTRODUCTORY REMARKS.

The function of the organs of special sense is to transmit to the sensorium impressions of the various phenomena of the outer world; they act, therefore, as the intermediate apparatus of sensory perceptions. In order that these may be brought about, the following conditions must be fulfilled: (1) The sense-organ, with its specific end-apparatus, must be anatomically intact, and be capable of performing its physiological function. (2) A "*specific stimulus*" must be present and act upon the end-organ in a normal manner. (3) There must be an uninterrupted communication from the sense-organ through the course of the afferent nerve to the brain. (4) At the time of stimulation, psychic activity (attention) must be directed toward the process of stimulation; in this manner the sensation (as, for instance, of light or sound) first originates through the sense-organ. (5) If, finally, through a psychic act, the sensation is referred to its external cause (a process that takes place in the cerebral cortex of the psychosensorial centers), a conscious sense-percept is formed. Often, however, this reference is made unconsciously, inasmuch as it is deduced only from experiences previously made. (6) The sensory nerves are connected not only with the cerebral cortex, but also with more deeply situated central nuclei, whereby reflexes are produced, which (in the absence of a conscious sensory perception) appear as movements, for the purpose of guarding the sensory mechanisms against irritation, and protecting them. In the lower animals the instinctive movements for the material preservation of the animal that take place on irritation of the sense-organs are effectuated in this way.

Among the stimuli that affect the terminal apparatus of the sense-organs there are distinguished: (1) Adequate or homologous stimuli, that is, those for the activity of which the organ is especially constructed, for example the rods and cones of the retina for the undulations of the luminiferous ether. Thus, there is a specific stimulus for each sensory nerve-ending (Johannes Müller's law of specific energy). (2) Other stimuli of a different nature (mechanical, thermal, chemical, electrical, internal somatic) are also efficient, as, for instance, the seeing of stars in consequence of a blow upon the eye, or ringing in the ears as a result of cerebral hyperemia. These heterologous stimuli may affect the nervous elements of the sensory apparatus throughout their entire course from the terminal sense-organ to the cerebral cortex. On the other hand, the adequate stimuli act only upon the terminal apparatus; for example light thrown upon the trunk of the exposed optic nerve has no effect whatsoever.

The homologous stimuli are effective with respect to the sense-organs only within certain limits of intensity. Exceedingly feeble stimuli, to begin with, are without any effect. The degree of intensity of stimulation that originates the first trace of sensation is called the threshold of sensation, or the "threshold value." With increase in the intensity of the stimulus the sensations increase, and the sensations increase equally when the intensity of the stimulus increases in like proportions. For example, the same sensation of equal increase in brightness is produced by the light of 11 candles, instead of 10, or of 110 candles instead of 100 (the ratio of increase in each case being equal to one-tenth). As the logarithms of the numbers increase equally, the law has been expressed as follows: The sensations increase not as the absolute intensities of the stimuli, but approximately as the logarithms of the intensities. The universal applicability of this psychophysical law of **Fechner** has, however, been disputed recently by E. Hering. Specific stimuli of excessive activity give rise to peculiar painful sensations, as, for instance, the sense of blinding, of deafening of the ear, etc. The sense-organs react to adequate stimuli only within certain definite limits; as, for instance, the ear responds to vibrations of sonorous bodies only within the range of a definite number of vibrations, and the retina only to the undulations of the luminiferous ether between red and violet, although not to the heat-waves nor to the chemically active vibrations.

The designation *after-sensation* is applied to the phenomenon that the sensation, as a rule, lasts longer than the stimulus; to these belong the after-images, the persistent sensation after pressure upon the skin, etc. Subjective sensations, finally, are brought about by the irritation of the nervous part of the apparatus by internal, somatic causes. The highest order of these subjective sensations, which usually depend upon pathological irritation of the psychosensorial, cortical centers are known as hallucinations; as, for example, when a person in delirium sees forms or hears voices that are not present. In contradistinction to these, the designation *illusions* is applied to the modifications, by the mind, of a sensation actually present; as, for example, when the rolling of a wagon is thought to be thunder. Each of these subjects will be taken up in detail under the individual sense-organs.

In newborn infants the sense of touch is strongly developed, the pain-sense poorly; muscular sensations are doubtfully present; while smell and taste are frequently confused. Auditory stimuli are perceived from the second day on, visual stimuli immediately after birth, but a peripheral visual field does not yet exist. Toward the fourth or fifth week movements of convergence and accommodation are observed, while after four months colors are differentiated. Different stimuli are not perceived simultaneously—a reflex inhibitory center is not yet developed.

THE VISUAL APPARATUS.

PRELIMINARY ANATOMICAL AND HISTOLOGICAL OBSERVATIONS. THE INTRAOCULAR PRESSURE.

The following anatomical and histological sketch can refer only to the physiologically important points; it presupposes, naturally, a knowledge of the anatomical structure of the eye.

I. *External or fibrous tunic of the bulb, consisting of the cornea and the sclera.*

The cornea is, for the sake of simplicity, assumed to have a uniformly spherical curvature, although in reality it deviates from this form. It represents rather the vertical segment of a somewhat oblate ellipsoid, which must be conceived as produced by the rotation of an ellipse about its long axis; numerous deviations from such a regular figure occur, however. It is approximately of the same thickness throughout; except that in the newborn the central portion is somewhat thicker, while in adults it is rather thinner. The cornea is composed of the following layers: (1) The *anterior, stratified, nucleated epithelium,* 0.03 mm. thick (Fig. 266, *a*), consists of numerous layers of cells, all of which are connected by delicate processes of protoplasm. The deepest cells are rather cone-shaped, are arranged vertically side by side and are known as supporting cells. The

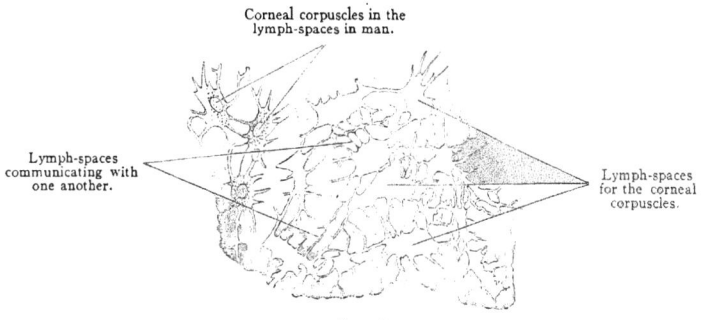

Corneal corpuscles in the lymph-spaces in man.

Lymph-spaces communicating with one another.

Lymph-spaces for the corneal corpuscles.

FIG. 265.

roundish cells of the middle layers are more arched, and possess tooth-like processes, which fit into corresponding depressions in the adjoining cells. The deeper cells contain a diplosoma. The superficial cells are flat, entirely smooth, and hard squamous epithelium, containing keratin. The conjunctiva contains scattered goblet-cells, which produce mucus. (2) The epithelial layer rests on the anterior surface of the rather uneven *anterior elastic membrane,* or Bowman's membrane, a structureless, hyaloid membrane (*b*) 0.01 mm. thick, under the action of reagents having a fibrillar appearance, and posteriorly passing gradually into: (3) The *true corneal tissue,* which consists of doubly refracting fibers, constituted of exceedingly delicate connective-tissue fibrils. These fibers are woven into about 60 mat-like lamellæ (*l*), which overlie one another and are cemented together in layers. Near the anterior elastic membrane these bundles bend forward as supporting fibers. Some fibers pass through this entire layer as "perforating" fibers. In the interstices of the meshwork there is a system of intercommunicating passages, which possess a sort of parietal layer. These anastomosing canals are lymphatic in nature and communicate with the lymph-vessels of the conjunctiva. (Fig. 266, *c*) lie in these spaces; they are provided with anastomosing processes, and possess the character of protoplasmic cells. Kühne saw these cells retract upon stimulation of the corneal nerves; the anatomical connection of the nerves with the cells has also been shown. According to v. Recklinghausen wandering cells may also penetrate these channels from without, and increase greatly in the presence of inflammation. (4) The transparent, structureless *posterior elastic membrane (d)*, 0.006 mm. thick, Descemet's or Demour's membrane, possesses in many animals a striated appearance indicative of a lamellar structure, and toward the corneal periphery, where it becomes thicker, it occasionally exhibits slight conical projections. This membrane is exceedingly tough and (in the presence of inflammation, etc.) resistant; when it is detached, it curls up toward its convex side. Its peripheral portion passes over into the fibro-elastic network of the pectinate ligament of the iris, the trabeculæ of which are lined with epithelial cells. (5) The *posterior corneal epithelium* is composed of a single layer of flat, delicate, hexagonal, nucleated

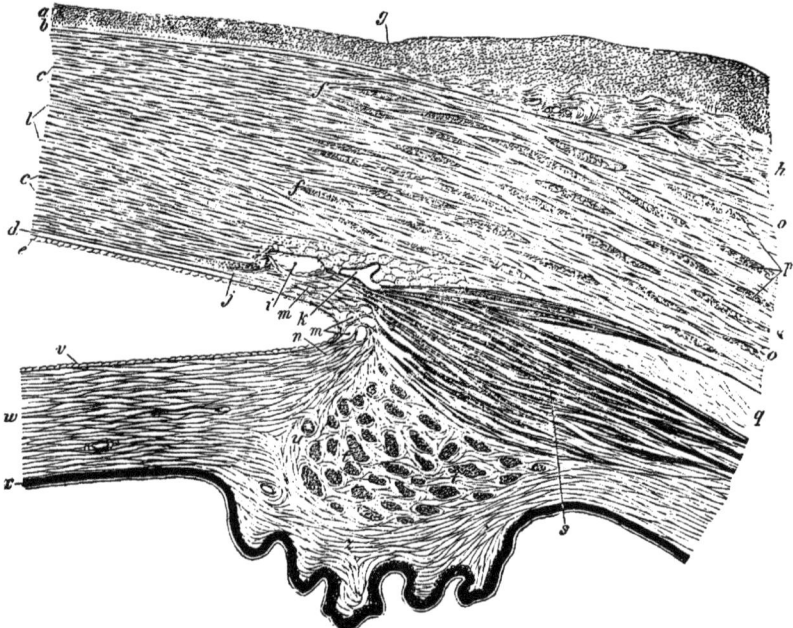

Fig. 266.—Meridional Section through the Corneoscleral Junction: *a*, Anterior corneal epithelium; *b*, Bowman's membrane; *c*, corneal corpuscles or lymph-spaces; *l*, corneal lamellæ; the layer between *b* and *d* is the true tissue of the cornea; *d*, Descemet's membrane; *e*, its epithelium; *f*, junction of the cornea and the sclera; *g*, limbus conjunctivæ; *h*, conjunctiva; *i*, Schlemm's canal; *k*, Leber's venous plexus, considered by Leber as belonging to Schlemm's canal; *m, m*, meshes in the tissue of the pectinate ligament of the iris; *n*, root of the iris; *o* longitudinal, *p* circular (transversely divided) fiber-bundles of the sclera; *q*, perichoroidal space; *s* meridional, *t* equatorial (circular) bundles of the ciliary muscle; *u*, section of a ciliary artery; *v*, epithelium of the iris (continuation of that on the posterior surface of the cornea); *w*, stroma of the iris; *x*, pigment of the iris; *z*, a ciliary process.

cells (*l*), bound together by fine processes, the attached portions of which have a fibrous appearance. These cells extend from the edge of the cornea onto the anterior surface of the iris (*v*). In the interspaces between the individual cells there are fine lymph-spaces that communicate with a delicate canal-system beneath the epithelial layer, and, further, through Descemet's membrane, with the corneal lacunæ.

The *nerves* of the cornea arise from the large and short ciliary nerves, and are partly sensory in function. They enter the margin of the cornea as trunks that possess medullary sheaths. Further inward the sheaths are lost, and the nerves form a network on the surface of the cornea. The branching, naked fibrillæ penetrate into the epithelial layer, again divide, ascending perpendicularly, and end finally between the epithelial cells as minute fibers with small knobs (visible on treatment with gold chlorid) (Fig. 334). The trophic fibers of the cornea are probably the deeper-lying twigs that are connected with the corneal corpuscles.

Blood-vessels are present only in the outer edge of the cornea (Fig 267, *v*), and extend inward from the limbus a distance of 2 mm. above, 1.5 mm. below, and 1 mm. laterally; the outermost capillary loops bend backward in an arched manner. The cornea is nourished from its outer margin. Opacities of the cornea produce corresponding disturbances of vision; pathologically, blood-vessels may be formed within it.

The cornea contains collagen and mucin (but no chondrin); the anterior epithelium two globulins. The "membranin" of Descemet's membrane stands between elastin and mucin, and is digested by trypsin.

The sclera is a dense, fibrous tunic composed of connective-tissue bundles, running in an equatorial (*p*) and a meridional (*o*) direction, with which are associated many elastic fibers. In its interstices, which communicate with those of the cornea, there are flat connective-tissue corpuscles, some of which are colorless, some pigmented, and also wandering lymph-cells. It is thickest posteriorly, and thinnest in the equatorial region; further forward it becomes thicker at the point of insertion of the tendons of the four recti muscles. It contains only a few blood-vessels, which form a wide-meshed capillary network immediately under its inner surface. Other vessels form an arterial circle around the optic-nerve entrance. In rare instances it is spherical, but usually it is more like an ellipsoid, which must be conceived as produced by the rotation of an ellipse either about its short axis (short eyes), or about its long axis (long eyes). Above and below, the sclera overlaps the transparent corneal margin, so that the cornea has an elliptical form if viewed from in front, and a circular form when viewed from behind. Following the margin of the cornea, but within the scleral substance, runs a circular canal, the canal of Schlemm (*i*), which anastomoses with other venous channels (Leber's venous plexus) (*k*); Schwalbe and Waldeyer regard Schlemm's canal as a lymph-channel. Posteriorly the sclera is continuous with the sheath of the optic nerve derived from the dura mater. The sclera also possesses nerves, which are said to terminate in the cellular elements within its substance.

II: *Median or vascular tunic of the bulb, consisting of the choroid, the ciliary processes, and the iris.*

The choroid is composed of the following layers: (1) On its inner surface there is a transparent boundary layer, only 0.7 μ thick, which becomes somewhat thicker anteriorly. (2) The extremely vascular capillary network of the choriocapillary layer or membrane of Ruysch, embedded in a homogeneous layer. Bounding this is: (3) A dense network of elastic fibers, which is lined on both surfaces by endothelium. Then follows (4) the choroid proper, a layer with pigmented connective-tissue corpuscles, which in the form of an elastic network contains numerous veins, with their accompanying lymph-sheaths, as well as arteries, which are provided with unstriated muscle-fibers in their connective-tissue sheaths. Finally, there is (5) the supra-choroid layer or lamina fusca, which bounds the large perichoroidal lymph-space (*q*); the latter is lined with endothelium and is crossed by branched and anastomosing trabeculæ covered with endothelial and connective-tissue cells. In newborn infants, who always have dark-blue irides, the uveal tissue contains no pigment; in bruns the pigment develops later, while in blonds the uvea remains unpigmented.

In the **ciliary portion of the uveal** tract, the pigmented connective-tissue cells are not so numerous. In this position lies the ciliary muscle (muscle of accommodation or tensor of the choroid), whose meridional fibers (*s*) arise by means of a branched, reticulated, connective-tissue insertion at the inner side of the corneo-scleral margin, near Schlemm's canal, and extend backward into the choroid; the radial fibers pass inward toward the interior of the eyeball; and the circular bundles (*t*) are situated more internally, just within the ciliary border (Heinr. Müller's muscle). The motor nerve of this unstriated muscle is the oculomotor. Within the ciliary processes ganglion-cells have been found, which probably belong to the trigeminus.

The iris consists of the following layers, from before backward: The anterior epithelium, a single layer of cells (*v*); a stroma with connective-tissue fibers and cells (vascular layer); and, finally, a posterior, structureless limiting membrane (membrane of Bruch), which is covered with the double layer of pigment-cells (x). This pigment-layer is lined by the exceedingly delicate limiting membrane of the iris, which is a continuation of the internal limiting membrane of the retina. Within the vascular layer (which contains pigmented connective-tissue cells in bruns) are the two unstriated muscles: the sphincter of the pupil (Fig. 281), which surrounds the pupil, and lies near the posterior surface of the iris (it is innervated by the oculomotor); and the dilator of the pupil. The latter consists of a thin layer of radially arranged fibers, some of which pass to the pupillary margin, while some bend around into the sphincter. At the outer extremity of the iris the radiating fibers are arranged in anastomosing arches and form a circular muscle-bundle. The chief nerve of the dilator of the pupil is the sympathetic. Ganglia are found on the ciliary nerves in the choroid. Gerlach has given the appropriate name of avenular ligament of the bulb to the prismatic bundle of fibrous tissue that bounds the periphery of the iris, and forms the point of union of the ciliary body, the iris, the ciliary muscle, the venous sinus of the iris, and the transition from the cornea to the sclera.

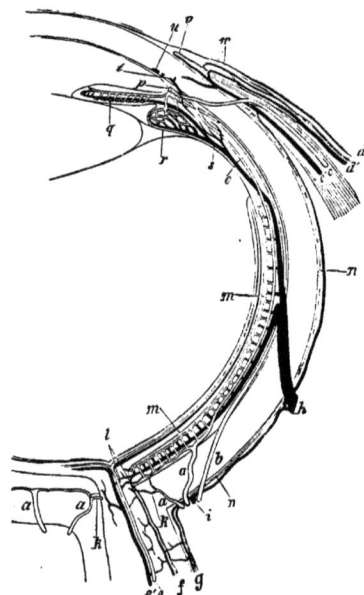

The course of the **choroidal vessels** is of great importance for the nutrition of the eye. This is described by Leber as follows: Among the arteries are: (1) The *short posterior ciliary* (Fig. 267, *a, a*), about 20 in number, which penetrate the sclera near the optic nerve. They terminate in the vascular network of the chorio-capillary layer (*m*), which reaches as far as the *ora serrata*. (2) The two *long posterior ciliary arteries*, one of which lies on the nasal, the other on the temporal side. They pass to the ciliary portion of the choroid (*b*), where they divide dichotomously and enter the iris, to help form the *circulus arteriosus iridis major* (*p*). (3) The *anterior ciliary arteries* (*c*), which arise from the muscular branches, perforate the sclera anteriorly, and give off branches to the ciliary portion of the choroid and to the iris. About 12 branches run backward (*o*) from them to the chorio-capillary layer. The veins carry off the blood as follows: (1) The *anterior ciliary veins* (*c¹*) receive the blood from the anterior part of the uvea; they pass outward and communicate with Schlemm's canal and Leber's venous plexus. They do not collect any blood from the iris, however. (2) The venous plexus

Fig. 267.—Diagrammatic Representation of the Blood-vessels of the Eye (after Th. Leber). Horizontal section—veins dark, arteries light, with a double contour: *a*, short posterior ciliary, *b*, long posterior ciliary arteries; *c c'*, anterior ciliary artery and vein; *d, d'*, conjunctival artery and vein; *e e'*, central artery and vein of the retina: *f*, vessels of the inner, *g* of the outer sheath of the optic nerve; *h*, vorticose veins; *i*, short posterior ciliary vein, running only to the sclera; *k*, branch of the short posterior ciliary artery to the optic nerve; *l*, anastomosis between the choroidal vessels and those of the nerve; *m*, chorio-capillary layer; *n*. episcleral branches; *o*, recurrent choroidal artery; *p*, circulus arteriosus iridis major (cross-section); *q*, vessels of the iris; *r*, ciliary process; *s*. branch of a vorticose vein from the ciliary muscle; *t*, branch of the anterior ciliary vein from the ciliary muscle; *u*, circulus venosus; *v*. marginal network of the corneal limbus; *w*, anterior conjunctival artery and vein.

of the ciliary body (*r*) collects the blood from the iris (*q*), and joins the choroidal veins further back. (3) The large *vasa vorticosa* of Stenon, the main trunks of which (*h*) perforate the sclera behind the equator of the globe. The inner edge of the iris glides over the anterior surface of the lens; the posterior chamber is quite narrow, even in adults, and in infants it is nearly obliterated. When Berlin blue is injected into the anterior chamber, it almost invariably enters the anterior ciliary veins, even in living animals; the same is true of carmine. It is, therefore, concluded that there must be a direct communication between the veins and the anterior chamber, as a diffusion of these substances through membranes is impossible.

Internally to the choroid lies the single layer of hexagonal epithelial cells, from 0.0135 to 0.02 mm. in diameter, which are filled with pigment. This layer belongs really to the retina. In front of the ora serrata it forms a double layer of cells, which extends to the posterior surface of the iris (Fig. 266, *x*). In albinos it is free from pigment; the outer cells on the ridges of the ciliary processes are also devoid of pigment.

III. *Internal tunic of the bulb, consisting of the retina (optic portion) and its continuations, the ciliary and iridic portions of the retina.*

The **retina** is bounded externally by the hexagonal pigment-epithelium (Fig. 268, *Pt*), which embryologically and functionally belongs to the retina. The cells are not flat, but send pigmented processes into the spaces between the rods. In some animals the cells contain drops of fat (rabbit) and other substances. At the *ora serrata* the cells are larger and darker. Of the true layers of the retina: (1) The visual cells, or the "rods" (*St*) and "cones," called also "neuroepithelium," lie most externally. They are absent at the optic-nerve entrance. The outer portions of the rods contain, during life, a red pigment, the "visual purple," which is preserved in the dark, but is bleached by daylight, and is continually reproduced in the eye. It may be extracted by 2.5 per cent. solution of the biliary acids, especially from retinas that have lain in 10 per cent. solution of sodium chlorid. The rods are from 0.04 to 0.06 mm. high, and from 0.0016 to 0.0018 mm. broad, and exhibit longitudinal striation, due to depressions; in the axis runs a fine fibril. The outer segment breaks up, occasionally, into numerous, exceedingly fine transverse discs. Krause found an ellipsoid body, the "rod-ellipsoid," at the junction of the outer and inner rod-segments. The flask-shaped cones are devoid of visual purple; the outer segment exhibits also longitudinal striation, and breaks up readily into transverse discs. In the macula lutea (the yellow pigment of which lies only in the outer retinal layers, and not in the cones) cones alone are present; near the macula each cone is surrounded by a garland of rods. The greater the distance from the macula, the fewer are the cones. Nocturnal animals (owl and bat) possess either no cones whatever, or only imperfect forms. In birds the retina has many cones, in the lizard cones alone. The rods and cones rest on the sieve-like, fenestrated external limiting membrane (*Le*); both send processes through the openings: the cones to the larger cone-granules, and those lying at a higher level, the rods to the transversely striated rod-granules. The granules belong to: (2) The outer nuclear layer (*äu K*); this and all the succeeding layers, are designated the cerebral layers. There then follows: (3) The narrow outer reticular (granular, plexiform) layer. (4) The inner nuclear layer (*inK*). The nuclei represent bipolar ganglion-cells (ganglion of the retina) and are called *rod-bipolars* or *cone-bipolars*, whose course is shown in Fig. 269, E. Each bipolar sends out, in addition, a fine fiber between the visual cells, and ends with a punctate knob near the limiting membrane. Ganglion-cells without demonstrable neurites, called amacrine cells, are of unknown nature. (5) The inner reticular (granular, plexiform) layer (*in.gr*). (6) Ganglion-cell layer (ganglion of the optic nerve) (*Ggl*). Finally: (7) The layer of optic-nerve fibers (*o*), which is next to the internal limiting membrane (*Li*). According to Salzer there are in all 438.000; according to W. Krause, however, 400,000 broad and an equal number of the finest optic-nerve fibers. For each fiber there are 7 or 8 cones, about 100 rods, and 7 pigmented cells (of the choroid). The fibers are naked axis-cylinders; they are absent in the macula lutea, where, however, the ganglion-cells are numerous.

The newer investigations have shown that there is no uninterrupted fiber-connection between the rods and cones and the optic-nerve fibers. According to Ramón y Cajal (Fig. 269) the fibers arising from the rods (a) end in the outer reticular layer (C) as tiny knobs, after passing through the outer nuclear layer (d); the cone-fibers below the cone-granules (c) like unraveled threads (z). The bipolar processes of the internal nuclear layer (e E) break up into fibrils in the outer reticular layer (C) and in the inner reticular layer (F), which are only approximately in contact, on the one hand with ganglion-cell processes (r), on the other hand, with the elements in the outer reticular layer (C). From each ganglion-cell (i, k) an axis-cylinder process is sent off centripetally, while one or more dendrites enter the inner reticular layer. The optic-nerve fibers contain also a number of centrifugal fibers (s, s). The further course of the optic nerve is discussed on p. 679. Between the two homogeneous limiting membranes (*Li* and *Le*) lies the supporting tissue of the retina (not true connective-tissue). It includes the supporting fibers of Müller (Fig. 268, *Rf*), which contain nuclei (*k*) and pass through all the cerebral layers, and end in expanded terminations at the internal limiting membrane (*Rk*). In addition, the supporting tissue forms a network throughout all the retinal layers, with openings for the penetrating nervous elements (*Sg*). In the outer reticular layer, there are also flattened, partly nucleated supporting cells, with long processes, and, at the optic-nerve entrance, glia-cells. The inner segments of the rods and cones are also surrounded by a basket-like supporting tissue. In the nerve-fiber layer there are flat, stellate cells.

From the ora serrata forward, the retina becomes suddenly thin, and, as the ciliary portion of the retina, consists only of a layer of cylindrical cells, which seems to have arisen from the coalescence of the two nuclear layers, and is

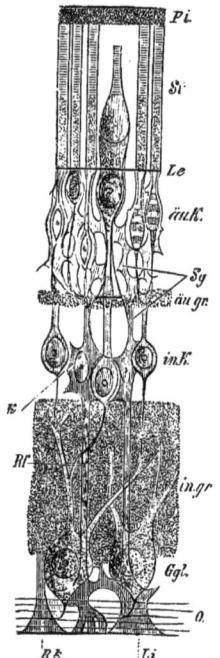

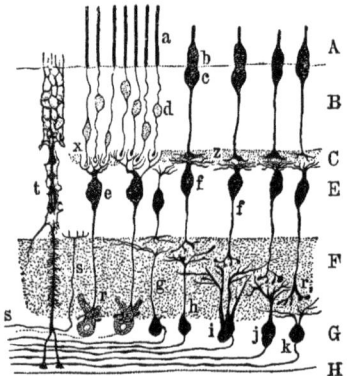

FIG. 269.—Transverse Section of a Mammalian Retina (after Ramón y Cajal): A, layer of rods and cones; B, visual cells (outer nuclear layer); C, outer reticular layer; E, bipolars (inner nuclear layer); F, inner reticular layer; G, ganglion-cells; H, nerve-fiber layer; a, rods; b, cones; e, a rod-bipolar; f, a cone-bipolar; r, lower ramification of the rod-bipolar; f, lower ramification of a cone-bipolar; g, h, i, j, k, ganglion-cells branching at various levels in F; x, z, contact of rods and cones with the bipolars; t, Müller's supporting fibers; s, centrifugal nerve-fiber.

FIG. 268.—Layers of the Retina.

covered on the inner side by the limiting membrane of the iris, a continuation of the internal limiting membrane of the retina. This layer extends to the posterior surface of the iris, as the iridic portion of the retina.

The *blood-vessels* of the retina lie in the inner layers, as far outward as the inner nuclei. They communicate with the choroidal vessels only by fine branches at the optic-nerve entrance; they are surrounded by perivascular lymph-channels. The greater number of the capillaries run internally to the inner nuclear layer. The fovea centralis has no vessels. Except in the mammalia, the eel, and several of the tortoise family, the retina contains no vessels at all. Destruction of the retina causes blindness.

The fresh retina has an acid reaction, but becomes alkaline if kept in the dark. The rods and cones contain albumin, neurokeratin, nuclein, and colored oil-globules (in the cones): so-called chromophanes. The other layers have the same constituents as the gray matter of the brain.

The lens is enclosed in a transparent, elastic capsule, which is thicker anteriorly than posteriorly, and is lined on the inner surface of its anterior portion by a layer of low, cuboidal cells. Toward the equator of the lens these cells become elongated into mononucleated fibers, all of which bend around the margin of the lens, and meet on both sides of the lens, their ends forming a stellate figure (lens-star), and being held together by a cement-substance. The lens-fibers contain globulin, enclosed in a sort of sheath. They are flattened mutually into hexagonal fibers, those of the central layers having their edges interlocked by means of toothed projections.

For the sake of simplicity, the lens may be considered as a biconvex body, with spherical surfaces, the posterior surface having the greater curvature. The anterior surface, however, really represents a part of an ellipsoid, produced by rotation about the short axis. The posterior surface resembles the vertical section of a paraboloid, that is it may be considered as produced by the rotation of a parabola about its axis. The outer layers of the lens have a lower index of refraction than the more internal layers. The central nucleus is of greater density than the lens as a whole, and it is, at the same time, more convex. The edge of the lens is always separated by an interspace from the ciliary process.

The **zonule of Zinn**, which arises from the ora serrata, is applied to the ciliary portion of the choroid in the form of a ruffled, folded membrane, in such a way that the ciliary processes occupy its folds, and are attached to them. It then passes to the edge of the lens, on the anterior portion of which it is inserted in a wavy manner. Behind the zonule of Zinn, reaching to the vitreous body, is the canal of Petit. The zonule is a fibrous fenestrated membrane; according to Merkel and H. Virchow, the canal of Petit also is occupied by exceedingly fine fibers: it is consequently not a true canal, but a complicated system of communicating spaces. The zonule is always stretched and keeps the lens in position, so that it may be considered as the suspensory ligament of the lens.

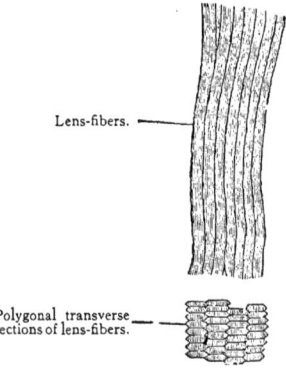

Lens-fibers.

Polygonal transverse sections of lens-fibers.

FIG. 270.

Opacities of the lens (gray cataract) hinder the entrance of rays of light into the eye. The absence of the lens (aphakia), following operations for cataract, may be compensated by the use of strong convex glasses. Such an eye, of course, possesses no power of accommodation. The lens contains albuminoid bodies, some of which are soluble in water and sodium chlorid (chiefly globulin and some albumin) and some insoluble.

The vitreous body is invested by the transparent hyaloid membrane, the outer surface of which, as far forward as the ora serrata is in contact with the internal limiting membrane of the retina. From this point forward the meridional fibers of the zonule of Zinn arise between the two, and are adherent to the surface of the vitreous and to the ciliary processes. A canal, 2 mm. in diameter, the hyaloid canal, runs from the optic papilla to the posterior surface of the lens; in fetal life it is occupied by blood-vessels. The peripheral portion of the vitreous body is laminated like an onion. The central portion is homogeneous. In the former, especially in newborn infants, there are spherical (leukocytes), spindle-shaped, or stellate, and also vacuolated cells of mucoid tissue; in the center there are only disintegrated remains of these cells. Running between them are transparent fibers and lamellæ. The vitreous body is gelatinous in character, and contains only 1.1 per cent. of solids, consisting of mucin, with albumin, and traces of globulin and glutin.

The **lymph-tracts** of the eye include an anterior and a posterior set. The anterior is composed of the anterior and posterior chambers, which communicate with the lymph-vessels of the iris, the ciliary processes, the sclera, the cornea and the conjunctiva. The posterior chamber communicates with the canal of Petit.

To the posterior lymph-system belongs, in the first place, the hyaloid canal, and secondly the large perichoroidal space situated between the sclera and the choroid. The latter communicates by means of lymph-vessels, which surround the emerging trunks of the vorticose vessels of Stenon, with the large lymph-space

of Tenon, which lies between the sclera and Tenon's capsule. Posteriorly
this is continuous with a lymph-space surrounding the surface of the optic nerve in
the form of a sheath; anteriorly it is in direct communication with the subcon-
junctival lymph-spaces of the eyeball. The optic nerve has three sheaths: (1)
The dural; (2) the arachnoid; and (3) the pial, arising from the corresponding
cerebral membranes. Between these three sheaths there are two lymph-spaces:
the subdural, between 1 and 2, and the subarachnoid, between 2 and 3 (Fig. 271).
Both are lined by endothelial cells: fine trabeculæ extending from one wall to
the other are similarly covered by cells. According to Axel Key and Retzius
these lymph-spaces communicate anteriorly with the perichoroidal space.

The **aqueous** humor resembles closely the cerebrospinal fluid, and contains
albumin, some sugar, urea and sarcolactic acid (which is present in the vitreous
body). The albumin increases when the difference between the blood-pressure
and the intraocular pressure is augmented. Such changes of pressure and likewise intense irritations applied to the eye cause the production of fibrin in the
anterior chamber.

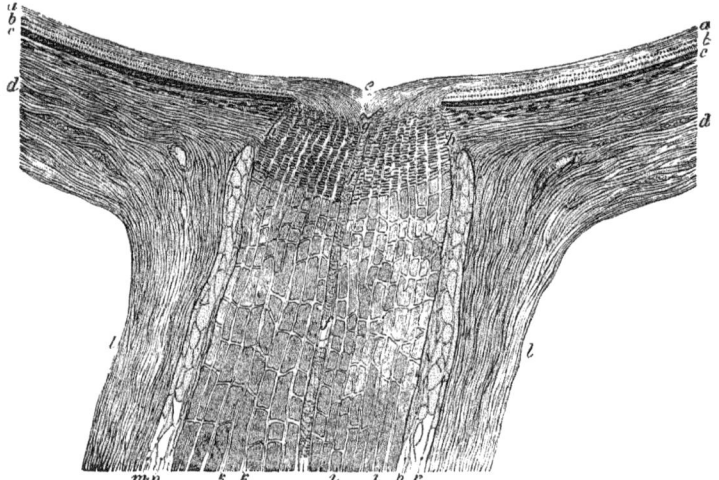

FIG. 271.—Horizontal Section through the Optic Nerve, at its Entrance into the Eyeball through the Coats of the
Eye: a inner, b outer layers of the retina; c, choroid; d, sclera; e, physiological cup; f, central artery of the
retina in the nerve; g, its point of bifurcation; h, lamina cribrosa; l, outer (dural) sheath; m, outer (subdural)
lymph-space; n, inner (subarachnoid) lymph-space; r, middle (arachnoid) sheath; p, inner (pial) sheath
t, nerve-fiber bundles; k, connective-tissue (longitudinal) septa.

The fluid within the eye is under a definite pressure, the intraocular pressure,
during life. This depends ultimately upon the pressure in the arteries in the
interior of the eye, and must rise and fall with the latter. It is determined by
testing the resistance or the yielding of the eyeball with the fingers. It may be
measured more accurately by an apparatus, the "ophthalmotonometer." Like
the arterial pressure, it is influenced by many circumstances. It is increased
with every pulse-beat and every expiration and it is decreased with every inspiration. The elasticity of the sclera and the cornea acts as a regulator with every
increase in the arterial pressure, causing, like the air-chamber of a fire engine,
more venous blood to be driven out when more arterial blood is pumped into the
eye. It is also important for the stability of the intraocular pressure that the
aqueous humor is secreted as rapidly as it is absorbed. Increase of the intraocular pressure makes the cornea flatter.

The secretion of the aqueous humor takes place with comparative rapidity, a fact
that Landois proved by the appearance of hemoglobin in the anterior chamber of a
dog half an hour after the introduction of free hemoglobin (transfusion of lamb's

blood) in the blood of the dog. It takes place more rapidly if the aqueous humor is previously removed from the chamber through a corneal wound. Ehrlich used fluorescein for the study of the movements of the fluids within the eye. This is an innocuous substance which, when introduced into the body, penetrates into the fluids of the eye, and may be recognized by its greenish fluorescence in reflected light, even in a solution of 1 part to two million parts of water. From observations on the entrance of this substance into the aqueous humor, it is now assumed that the ciliary body is the secreting organ for the aqueous humor, which passes through the pupil into the anterior chamber.

Section of the cervical symphathetic, and still more, that of the trigeminus, accelerates the secretion of the aqueous humor, but decreases its amount.

The cornea permits the entrance of fluids into the anterior chamber, for example atropin and fluorescein.

The excretion of the aqueous humor takes place by filtration in the angle of the anterior chamber; it passes through the clefts of the spaces of Fontana, which communicate with the anterior chamber, and enters the vessels of the circular canal of Schlemm, which lies directly on their outer side (*plexus ciliaris venosus* in animals). None passes through the cornea, although some is imbibed by its posterior layers, which are thus nourished, and there are no special lymph-vessels to remove it from the anterior chamber.

Under normal circumstances, the pressure is the same in the vitreous as in the aqueous, although atropin seems to increase the pressure in the former, and to decrease it in the latter, while physostigma has the opposite action. Arrest of the outflow of the venous blood often increases the pressure in the vitreous, and decreases that in the aqueous. Compression of the eyeball from without will cause more fluid to pass out of the eye temporarily than enters it. The decrease of the intraocular pressure after section of the trigeminus is striking; likewise its increase upon irritation of the same nerve—facts often observed by Landois. The statements with regard to a possible analogous action of the sympathetic vary. Interruption of the outflow of venous blood raises the pressure; an insufficient supply, associated with a normal outflow, decreases the pressure. The innervation of the vessels of the eyeball is discussed on p. 684.

PRELIMINARY DIOPTRIC CONSIDERATIONS.

The eye is comparable as an optical apparatus to the camera obscura. In both a diminished, inverted image of objects of the outer world is found on the background (the projection-surface). Instead of the simple lens of the camera, however, the eye possesses several refractive media, behind one another: cornea, aqueous humor, lens (the several parts of which: capsule, cortex and nucleus, also possess different refractive indices), and vitreous body. Each medium is separated from the one next to it by a refacting surface, which is assumed to be spherical. The projection-surface of the eye is the retina, which is colored by the visual purple. As this substance is bleached chemically by light, so that the images may even be fixed temporarily on the retina, the comparison of the eye to the camera is still more striking.

In order that the passage of the light-rays through the media of the eye may be accurately followed, the following factors must be known: (1) The refractive indices of the media; (2) the shape of the refracting surfaces; (3) the distance of the various media from each other and from the projection-surface.

The action of a convex lens will first be considered. There are to be distinguished in such a lens the centers of curvature, that is, the centers of the two spherical surfaces (Fig. 272 I, m m$_1$). The line connecting these points is called the principal axis: the center of this line is the optical center of the lens (O). All rays that pass through the optical center of the lens (and which may be countless) pass through unbent. They are called principal rays, or secondary axes (n$_1$). The following laws governing the refraction of rays by convex lenses must be remembered:

1. Rays falling on the lens parallel to the principal axis are so refracted that they meet on the opposite side of the lens at a point that is known as the focus or principal focus (f). The distance of this point from the optical center of the lens (O) is called the focal distance (f O) of the lens. The converse of this proposition is evident: Rays that diverge from the principal focus, and strike the lens, become parallel to the principal axis on the opposite side, and do not meet.

2. Rays proceeding from a point (IV, 1) in the prolonged principal axis beyond the principal focus (f), are united at a point on the opposite side of the lens (v) (conjugate focus).   The following conditions are possible: (*a*) If the distance of the point of light from the lens is double the focal distance, the conjugate focus is at an equal distance on the opposite side (double the focal distance).   (*b*) As the point of light moves nearer the lens, the conjugate focus moves further away. (*c*) If the point of light is more than double the focal distance from the lens, the conjugate focus is correspondingly closer to the lens.

3. Rays that proceed from a point in the principal axis (III, b) within the focal distance are rendered less divergent, but do not meet again; conversely, rays that converge on a convex lens are united at a point within the principal focal distance.

4. If the point of light (V, a) lies in a secondary axis (a b) the same laws hold good, provided that the angle made by the secondary axis with the principal axis is small.

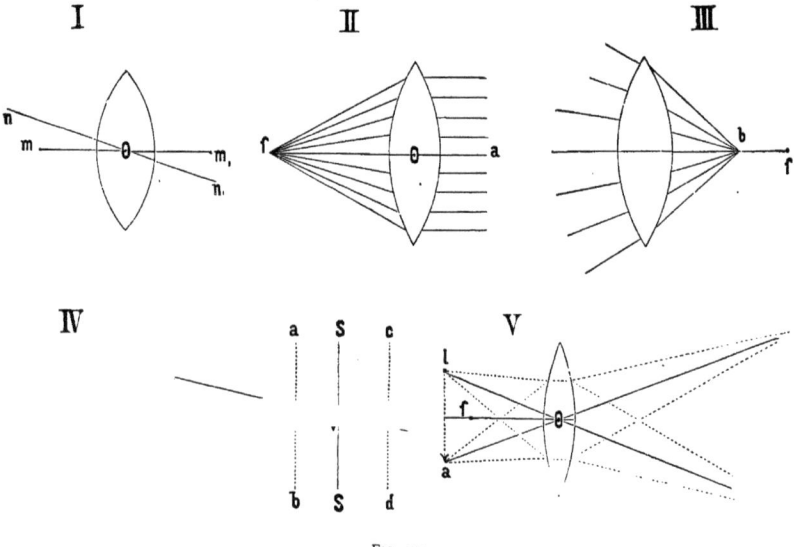

FIG. 272.

Formation of Images by Convex Lenses.—After what has been said about the conjugate foci of rays proceeding from a point of light, it is easy to construct the image of an object produced by a convex lens.   This is done simply by projecting the images of various points of the object.   For example (in V) b is obviously the image of the point a of the object, v the image of 1; the picture is therefore *inverted*.   Convex lenses form inverted and real images (upon a screen) only of such objects as are situated beyond the principal focus of the lens.

With regard to the size and distance of the image from the lens, the following conditions are to be noted: (*a*) If the object is double the focal distance from the lens, its image is of the same size as the object and at an equal distance from the lens; (*b*) As the object approaches the principal focus, the image recedes and at the same time becomes larger. (*c*) On the other hand, if the object is more than double the focal distance from the lens, the image approaches the lens, and at the same time becomes smaller.

The distance of the image from the lens may be readily calculated by the following formula, in which l represents the distance of the point of light, b the distance of the image, and f the focal distance of the lens:

$$\frac{1}{l} + \frac{1}{b} = \frac{1}{f}, \text{ or } \frac{1}{b} = \frac{1}{f} - \frac{1}{l}$$

Examples: Let $l = 24$ cm., $f = 6$ cm. Then $\frac{1}{b} = \frac{1}{6} - \frac{1}{24} = \frac{1}{8}$; therefore $b = 8$ cm., that is, the image is formed 8 cm. behind the lens. Further: Let $l = 10$ cm., $f = 5$ cm. (or $l = 2f$). Then $\frac{1}{b} = \frac{1}{5} - \frac{1}{10} = \frac{1}{10}$; therefore $b = 10$ cm., that is, the image is double the focal distance from the lens. Finally, let $l = \infty$. Then $\frac{1}{b} = \frac{1}{f} - \frac{1}{\infty}$, or $b = f$, that is, the focus for parallel rays, coming from an infinite distance, is in the principal focus of the lens.

Index of Refraction.—A ray of light passing from one medium to another medium of different density, in a direction perpendicular to the surface, passes through the latter without changing its direction. If, therefore (Fig. 273) G D. is perpendicular to A B, then D D is also perpendicular to A B. For a horizontal surface A B the axis of incidence is the vertical line G D, while for a spherical

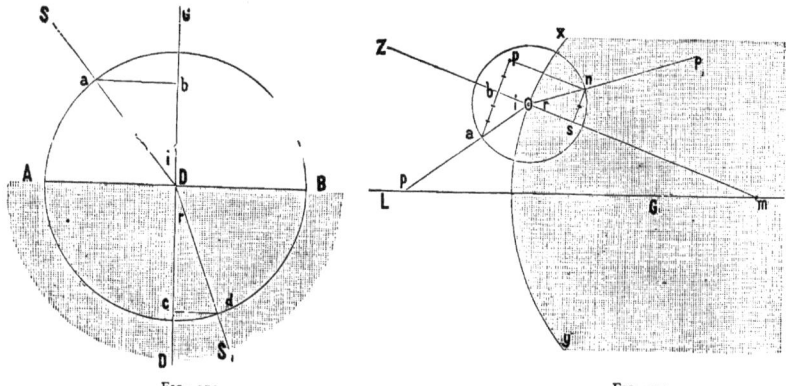

FIG. 273.   FIG. 274.

surface the axis of incidence is the prolonged radius. If the ray of light strikes the surface obliquely, it is refracted, that is deflected from its original direction. The incident and refracted rays lie, however, in the same plane. If the oblique, incident ray passes from a rarer medium to a denser one (for example, from air into water), the refracted ray is deflected toward the perpendicular. Conversely, if the ray passes from a denser medium into a rarer one, it is deflected away from the perpendicular. The angle that the incident ray (S D) forms with the perpendicular (G D), (angle i) is called the angle of incidence. The angle that the refracted ray (D.S₁) forms with the prolonged perpendicular (D D) is called the angle of refraction (angle r). The degree of the refraction is expressed by the index of refraction (or exponent of refraction); it is represented for each substance by the relation of the sine of the angle of incidence to the sine of the angle of refraction of a ray passing from air into that substance. Thus, n = sine i : sine r = a b : c d. In comparing the indices of refraction of two media, it is assumed that the ray of light passes from the air into the media. In passing from air into water, the ray of light is deflected to such a degree that the ratio of the sine of the angle of incidence to the sine of the angle of refraction is as 4 : 3; the index of refraction is therefore $\frac{4}{3}$ (more exactly = 1.336). With glass the ratio is 3 : 2

(more exactly the index of refraction is 1.535). The sines of the angles of incidence and refraction are in the same ratio as the velocities with which the rays of light pass through the two media.

The refracted ray is, therefore, easily constructed, if the indices of refraction are known. *Example:* Let L (Fig. 274) represent the air, G a denser medium (glass), with a spherical surface, x y, the center of which is at m; P O represents the oblique incident ray; m Z is then the axis of incidence, and the angle i the angle of incidence. Let the index of refraction be $\frac{3}{2}$; what is the direction of the refracted ray?

Construction.—Draw a circle of any radius, with its center at O; then from a draw a line a b perpendicular to the axis of incidence m Z; then a b is the sine of the angle of incidence, i. Divide the line a b into 3 equal parts, and prolong it a distance equal to two of these parts, that is to P. Now, draw from P the line P n, parallel to m Z. Then the line joining the two points O and n is the direction of the refracted ray. If the line n s is drawn perpendicular to m Z, n s = b P. Further, n s = the sine of the angle r. According to the construction, a b : s n (or : b P) = 3 : 2, or sine i : sine r = $\frac{3}{2}$.

Optical Cardinal Points of a Simple Collecting System.—Two refractive media (Fig. 275, L and G), which are separated from each other by a spherical surface (a b), form a simple collecting system. From a knowledge of certain properties of such a system, it is easy to construct an incident ray from the first medium (L) striking the separating surface obliquely, and also its direction in the second

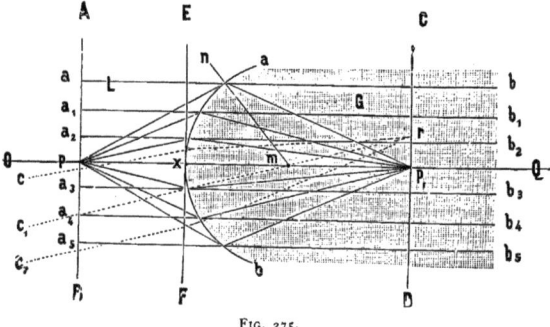

FIG. 275.

medium G, as well as to determine, from the position of a luminous point in the first medium, the position of its image in the second medium. The requisite properties and points of such a simple collecting system are as follows:

L (Fig. 275) is the first, and G the second medium; a b is the spherical surface separating them, and m its center of curvature. All radii drawn from m to a b (m x, m n) are, of course, normal to the surface, so that all rays of light coming in the direction of the radii, must pass through m without deviation. All such rays are called axial rays; and m, their point of intersection, is the nodal point. The line that connects m with the vertex of the spherical surface (x) and is prolonged in both directions is called the optical axis (O Q). A plane (E F) erected perpendicular to O Q at x is the principal plane, and x is its principal point. The following facts have been determined:

(1) All rays (from a to a_3) that fall upon a b and are parallel to each other and to the optic axis in the first medium will be so deflected by the second medium that they are united at one point (p_1) in the latter. This point is called the second principal focus. A plane erected at this point perpendicular to O Q is called the second focal plane (C D). (2) All rays (from c to c_2) that are parallel to each other in the first medium, but not parallel to O Q, are reunited at a point in the second focal plane (r) at the intersection of the undeflected axial ray (c_1 m r) with this plane (the angle, however, that the rays from c to c_2 make with O Q must be small). The converse of propositions 1 and 2 is also true: the rays diverging from p_1, and directed toward a b, continue through the first medium parallel to

each other and to the axis O Q (from a to a_5); and the rays coming from r run parallel to each other in the first medium, but not parallel to the axis O Q (from c to c_2). (3) All rays in the second medium that are parallel to each other (from b to b_3) and to the axis O Q are united at a point in the first medium (p), the first principal focus (the converse of this proposition is also true). A plane erected at this point perpendicular to O Q is known as the first focal plane (A B). The radius (m x) of the refracting surface is equal to the difference between the distances of the principal focal points (P and p_1) from the principal point (x); hence m x = p_1 x — p x.

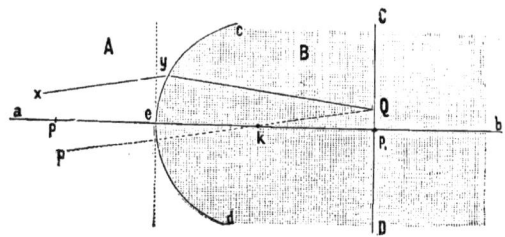

FIG. 276.

1. From a knowledge of these simple relations the **direction of the refracted ray** may be constructed. Let A (Fig. 276) be the first, B the second medium; c d the spherical surface between them; a b the optical axis, k the nodal point, p the first and p_1 the second principal focus; C D the second focal plane. If, now, x y is the direction of the incident ray, what is the direction of the refracted ray in the second medium?

Construction: Draw the undeviated axial ray P k Q parallel to x y. Then the line y Q must be the direction of the refracted ray (according to proposition 2).

2. **Construction of the Image of a Given Point in an Object.**—[The letters A, B, c d, a b, k, p and p_1, C D, in Fig. 277 have the same designations as before.] If now a point of light be given at o, where will be its image in the second medium?

Construction: Draw the undeflected axial ray o k P. Then draw the ray o x parallel to the axis a b. The parallel rays a e and o x are united at p_1 (according to

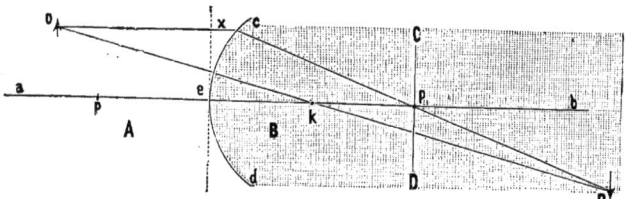

FIG. 277.

proposition 1). . If x p_1 is prolonged until it intersects the ray o P, P is the image of the point o, for the image will be situated at the intersection of the rays o x and o k in the second medium, consequently at P.

Construction of the Refracted Ray and of the Image when Several Refractive Media are Present.—If several refractive media are placed behind one another, the construction must be made from medium to medium in the manner already described. This, however, would be a troublesome procedure, especially in dealing with small objects. In 1840 Gauss calculated (by methods that cannot be explained in an elementary treatise) that in all such cases the method of construction may be greatly simplified. If the media are "centered," that is if all have the same optical axis, such a system may be represented by two surfaces having equal indices of refraction, separated by a definite distance. The rays that fall on the first surface are not refracted by it, but are only displaced laterally and run parallel to their original direction as far as the second surface. The refraction takes place at this point, and in the same way as previously constructed: that is as if only one refracting surface were present. For this calculation the indices of refraction of the media, the radii of the refracting surfaces, and finally the distance between these surfaces must be known: but this subject cannot be discussed in any further detail here. . The refracted ray is constructed in the following manner: Let a b (Fig. 278, I) represent the optic axis, H the first principal focus determined by calculation, h h the first principal plane, H_1 the second focal point, $h_1 h_1$ the second principal plane, k the first nodal point, k_1 the second nodal point, F the second principal focus, and $F_1 F_1$ the second focal plane. Let m n be the direction of the incident ray; what is the direction of the refracted ray?

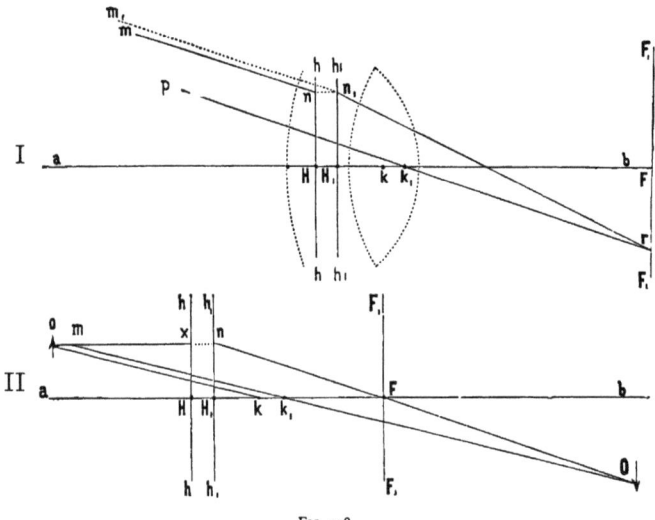

FIG 278.

Construction: Displace the ray m n parallel with itself as $m_1 n_1$, to the second principal plane. Now, draw the line p k_1, parallel to m_1 n. According to rule 2, p k_1 and m_1 n must intersect at a point of the plane F F_1. As p k_1 passes through unrefracted, the ray from n_1 must likewise pass through r; and n_1 r is therefore the direction of the refracted ray.

Construction of the Image of a Point.—Let o (Fig. 278, II) represent a point of light; where is the image of this point in the last medium? Draw from o the axial ray o k, and make o x parallel to a b. Displace both rays parallel to themselves to the second principal plane: then draw m k, parallel to o k, and prolong o x to u. The ray parallel to a b passes through F; m k_{11}, as the axial ray, is not refracted. The image of the point n will be found at the point of intersection of the prolonged rays n F and m k_1 (at O).

These constructions are not applicable to objects that are at some distance from the optical axis. For such a condition the eye is more advantageously constructed than a camera obscura (being periscopic), because its surface of projection is a hemisphere, and consequently the images are sharper in its lateral portions than would be possible on a flat surface.

APPLICATION OF DIOPTRIC LAWS TO THE EYE.   CONSTRUCTION OF THE RETINAL IMAGE.   THE OPHTHALMOMETER. ERECT IMAGES.

The eyeball represents a centered system, composed of several refracting media separated by spherical surfaces, the anterior surface of the cornea being in contact with the air.   In order to determine the course of the rays through these media, it is necessary to know the position of both principal foci.   Following the simplified solution of Gauss previously discussed, Listing and v. Helmholtz in particular have estimated the position of those points.   In order to make this calculation, a knowledge of the indices of refraction of the ocular media, the radii of the refracting surfaces, and the distance between them is necessary.   These will be taken up later.   According to this calculation:
(1) The first principal point lies 2.1746 mm., and (2) the second principal point 2.5724 mm., behind the anterior surface of the cornea; (3) the first nodal point is 0.7580 mm., and (4) the second nodal point 0.3602 mm. in front of the posterior surface of the lens; (5) the second

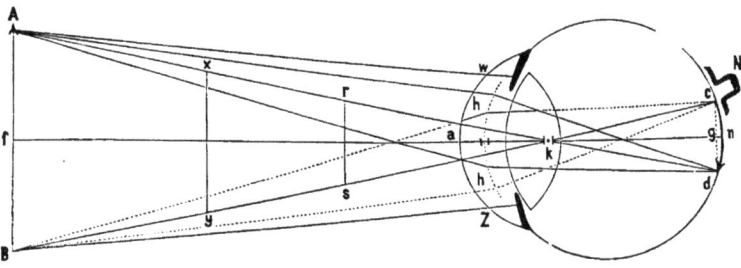

FIG. 279.

principal focus is 14.6470 mm. behind the posterior surface of the lens, and (6) the first principal focus is 12.8326 mm. in front of the anterior surface of the cornea.

As the distance between the two principal points and between the two nodal points is so small (only 0.4 mm.), a point in the middle of each pair may be adopted instead of the two separate points, without committing much error in the construction.   In this way there is only one refracting surface for all the media, and only one nodal point, through which all of the axial rays coming from without must pass unrefracted.   Such a simplified eye is known as the "reduced eye" of Listing.

The construction of the image on the back of the eye is now simple. The inverted image is formed on the retina with distinct vision.   Let A B (Fig. 279) represent an object standing perpendicularly before the eye.   From A a pencil of rays enters the eye; the axial ray A d passes through the nodal point k without being refracted.   As the image of A must lie on the retina, all the rays from A must come to a focus at d. The same is true for the rays from B, and of course for the rays coming from any point of the object **A B**. As all the axial rays must pass through the common nodal point k, this is also called the "point of intersection of the visual rays."

In the enucleated eye of an albino, or in any eye in which a piece of the sclera and choroid has been replaced by a piece of glass, the inverted image may be readily seen. In fact, the inverted image of a candle held in front of the eye may be seen through the sclera, if the eye is turned strongly to one side, and if it contains but little pigment.

By means of the construction of the retinal image the size of this image may be easily calculated, if the size of the object and its distance from the cornea are known. As the two triangles A B k, and c d k are similar, obviously A B : c d = f k : k g. Therefore, c d = (A B × k g) : f k. All of these values are known: namely k g = 15.16 mm.; further f k = a k ⊥ a f, of which a f may be measured directly and a k = 7.44 mm. The size of A B is obtained by measurement.

The angle **A** k **B** is designated the visual angle; the angle c k d is, of course, equal to it. It may be readily seen that the objects x y and r s, situated nearer the eye, must have the same visual angle. For this reason all three objects **A B**, x y, and r s have a retinal image of the same size. Objects whose peripheral points, when united with the nodal point, subtend a visual angle of the same size, and which consequently have retinal images of the same size are said to have the same "apparent size."

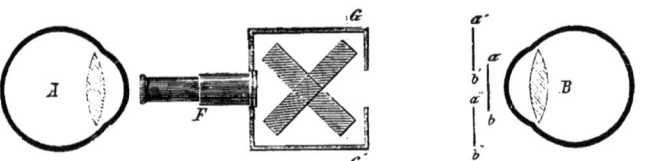

Fig. 280.—Ophthalmometer (after v. Helmholtz).

For the determination of the optical cardinal points by the method of Gauss a knowledge of the following relations is necessary:

1. The indices of refraction are: for the cornea 1.3739, for the aqueous humor and the vitreous humor 1.377, for the lens 1.4545 (the mean value of the different layers), air being taken as 1, and water as 1.33.

2. The radii of the spherical refracting surfaces are: for the cornea 7.7 mm.; for the anterior surface of the lens 10.3; for the posterior surface of the lens 6.1 mm.

3. The distances between the refracting surfaces are: from the anterior surface of the cornea to the anterior surface of the lens 3.4 mm.; from the latter to the posterior surface of the lens (axis of the lens) 4 mm.; the diameter of the vitreous body is 14.6 mm. The total length of the optical axis is therefore 22.0 mm.

As it is impossible to measure the normal curvatures of the eye after death accurately, on account of the rapid collapse of the eye, the calculation of the radii of the refracting surfaces is made according to Kohlrausch's method, from the size of the reflected images in the living eye. The size of a luminous object is to the size of the reflected image as the distance of each to half the radius of the convex mirror. The size of the reflected image must, therefore, be determined. This measurement is made by the ophthalmometer of v. Helmholtz. The apparatus is constructed on the following principle: If an object is seen through an obliquely set glass plate, it appears to be displaced laterally. This displacement is the greater the more obliquely the plate is set. Hence, if the observer A looks through the telescope F, in front of whose objective (in its upper half) the oblique plate G is placed, and sees the corneal image a b of the eye B, the latter appears to be displaced laterally, that is to a' b'. If a second plate G is placed before the lower half of the eye-piece of the telescope, and is inclined in an opposite direction (so that both plates meet at an angle, in the horizontal diameter of the objective) the observer sees the corneal image a b displaced laterally to a'' b''. As the two glass plates may be rotated on each other (at their points of intersection) they are so placed that the two reflected images have their inner edges exactly in contact (so that b' touches a''). From the size of the angle that the two plates make, the size of the image may be calculated (but the thickness of the glass plates, and the refractive index of the glass must be taken into consideration). In this way the size of the reflected image of the cornea and also of the lens can be determined, both in the state of rest, and in that of accommodation for near vision.

All of the ocular media, including the retina, have a certain amount of fluorescence, the lens most, the vitreous least.

As the retinal image is inverted, the perception of the object as an erect one is yet to be explained. By a psychical act the impulses from each point of the retina are projected outward: thus the stimulation of the point d (Fig. 279) to **A**, that of c to **B**. This projection outward is so accomplished that all the points appear to lie in a surface suspended before the eye, which is called the "field of vision." This field of vision is therefore the surface of the retina projected outward and inverted; consequently the field of vision appears erect, as the inverted retinal image is projected outward and inverted.

That the stimulation of each point is projected in an inverse direction through the nodal point is shown by the simple experiment that pressure on the outer side of the eyeball is referred to the inner aspect of the visual field. The entoptical phenomena of the retina are likewise projected outward and inverted; so that, for example, the point of entrance of the optic nerve is referred to the outer side of the yellow spot, and the like. All sensations of the retina are thus referred externally. „Wir sehen die Sonne, die Sterne an den Himmel, nicht an dem Himmel'' (v. Helmholtz).

ACCOMMODATION OF THE EYE.

The image of a point of light, as, for example, of a flame, formed by a convex lens, is always at a definite distance from the point (according to rule 2, page 824). If a projection-surface (a screen) is placed in this position, a real and inverted image is formed upon it. If, however, the screen is placed close to the lens (Fig. 272, IV, a b) or farther away from it (cd), the image is not distinct, and diffusion-circles are formed: in the first instance, because the rays have not yet united; in the second, because the rays have already crossed and are again diverging. When the point of light is alternately approached to and withdrawn from the lens, the screen must be correspondingly moved closer to the lens or farther away from it, if the sharpness of the image is to be preserved. If the screen is fixed, while the distance of the point of light from the lens varies, a sharp image could be formed only by increasing the curvature of the lens, and thus increasing its refractive power when the point of light is approached to the lens, or by diminishing its curvature, and thus making it less refractive, when the point of light is withdrawn.

As the eye has its surface of projection (retina) fixed at an unchanging position, and as the eye possesses the ability to form on the retina sharp images of both far and near objects, it must possess the power of altering its refractive strength (the form of the lens) to correspond in every case to the distance of the object.

By accommodation is meant the power of the eye to form sharp images of both distant and near objects upon the retina. This depends upon its ability to make the lens more or less convex (thicker or thinner), according to the distance of the object. If the lens is absent, accommodation is impossible.

When the eye is at rest, it is accommodated for the greatest distance; that is, sharp images are formed on the retina of objects at an infinite distance (as, for example, the moon). In other words, parallel rays (approximately) that enter the eye are united on the retina of the normal-sighted eye at rest; the principal focus is therefore in the retina. Distant vision is thus accomplished without the aid of any muscular action.

That no muscular activity is actually required for distant vision is proved by the following facts: (1) Normal-sighted persons see sharply and distinctly at a distance without the slightest feeling of exertion. On opening the eyelids after a considerable period of rest, distant objects are at once seen with sharp outlines.

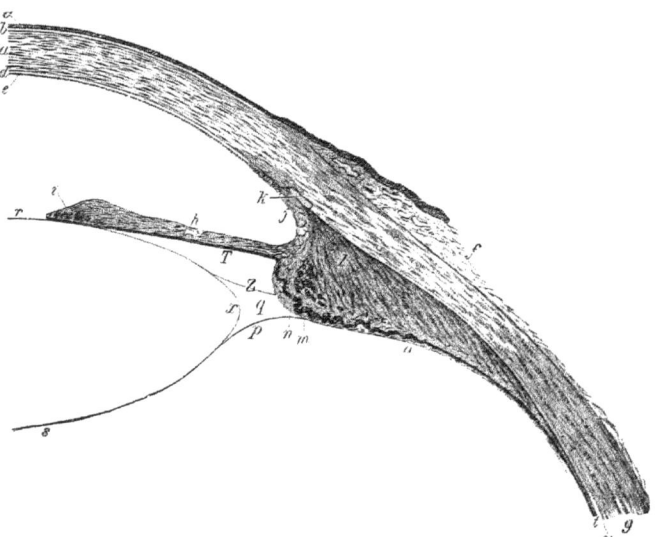

Fig. 281.—Anterior Quadrant of a Horizontal Section of the Eyeball. Cornea and lens cut in the middle of their vertical diameter: a, substantia propria of the cornea; b, Bowman's membrane; c, anterior corneal epithelium; d, Descemet's membrane; e, its epithelium; f, conjunctiva; g, sclera; h, iris; i, sphincter muscle of the iris; j, pectinate ligament of the iris with the adjacent fenestrated tissue; k, canal of Schlemm; l, longitudinal, m, circular fibers of the ciliary muscle; n, ciliary process; o, ciliary portion of the retina; q, canal of Petit, with the zonule of Zinn (Z) in front of it, the posterior leaflet of the hyaloid membrane (p) behind it; r, anterior, s, posterior, capsule of the lens; t, choroid; u, perichoroidal space; T, pigment epithelium of the iris; x, margin (equator) of the lens.

(2) If the eye has lost its power of accommodation in consequence of paralysis of the oculomotor nerve, sharp images of distant objects are still found on the retina. Paralysis of the accommodation-apparatus is always associated with disturbance of near vision, never of distant vision. Temporary paralysis, with the same result, is produced by the instillation of atropin or duboisin (and by taking toxic doses internally).

When the eye is accommodated for near objects, the lens becomes thicker and its anterior surface is more curved and protrudes further into the anterior chamber. The mechanism producing this change is as follows: While at rest the lens is kept flat against the vitreous by the traction of the stretched zonule of Zinn (Fig. 281, Z), which is attached to its margin. When the ciliary muscle (l, m) contracts to focus for near objects, it draws the margin of the choroid forward, and the zonule, which is in intimate relation with it, is relaxed. As a result, the lens assumes a more curved form, by virtue of its elasticity, so that it becomes more convex as soon as the flattening tension of the zonule relaxes. As the posterior surface of the lens rests in the saucer-shaped, unyielding depression of the vitreous, the anterior surface, in becoming more convex, must protrude further forward.

Hensen and Völckers discovered the origin of the nerve of accommodation in the most anterior portion of the oculomotor nucleus. Irritation of the posterior part of the floor of the third ventricle produces accommodation; if the irritation is applied a little further back, the pupil contracts. The fibers for the sphincter of the iris and for the ciliary muscle are derived from the upper oculomotor nucleus; close by this is the center for the elevator of the eyelid. If the boundary between the third ventricle and the aqueduct of Sylvius is irritated, the internal rectus muscle contracts; irritation of the posterior part of the aqueduct causes contraction of the superior rectus, the inferior rectus and the inferior oblique.

The movement during accommodation may be recognized by means of the following phenomena:

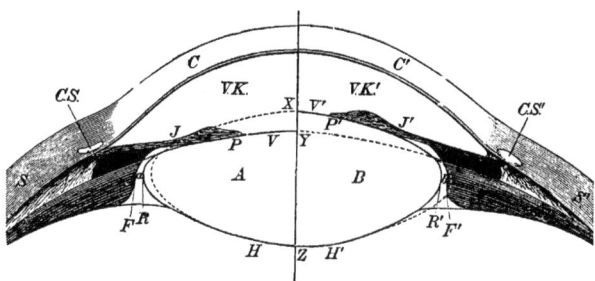

Fig. 282.—Diagrammatic Representation of Accommodation for Near and Far Objects. On the right the condition during accommodation is shown; on the left the condition during rest. On both sides one-half of the contour of the lens is drawn as a continuous line, the other half as a dotted line. The letters appearing twice—both on the right and left sides—have the same significance; those on the right side are primed. *A* left, *B* right half of the lens; *C*, cornea; *S*, sclera; *C.S.*, canal of Schlemm; *V.K.*, anterior chamber; *J*, iris; *P*, pupillary margin; *V* anterior, *H* posterior surface of the lens; *R*, equator of the lens; *F*, edge of the ciliary processes; *a* and *b*, interval between them. The line Z X shows the thickness of the lens in the act of accommodation for a near object; Z Y, the thickness of the lens when the eye is at rest.

1. The images of Purkinje-Sanson. If the light of a candle is allowed to fall on the human eye a little from the side, or, better still, if the light comes through two small triangular openings in a piece of cardboard, placed one above the other, the observer sees three pairs of reflected images in the eye. The brightest and most distinct pair (virtual) are formed by the anterior surface of the cornea (Fig. 283, a). The second pair (likewise virtual) are the largest, but at the same time the faintest; they are reflected from the anterior surface of the lens (b) and lie 8 mm. behind the plane of the pupil. (The images produced by convex mirrors are the larger the longer the radius of curvature.) The third pair are the smallest and stand midway in intensity; they are inverted, and lie about in the plane of the pupil (c). These images are also virtual, because they do not lie in the second medium, which is represented here by the air. The posterior capsule of the lens, which reflects these last images acts as a concave mirror. (If a luminous object is placed at a distance from a concave mirror, an inverted, real image, reduced in size, is formed near the focal point of the mirror, on the same side as the object.) While the observer watches these images, with the eye of the person at rest, the latter is requested to accommodate suddenly for a near object. Changes in the images are at once recognized. The middle pair (from the anterior surface of the lens) become smaller, and brighter, and approach each other (b), because the anterior surface of the lens becomes more convex. At the same time they approach the corneal images, because the anterior surface

53

of the lens is nearer to the cornea. Neither of the other pairs (*a,* and *c,*) change either in size or in position. By means of the ophthalmometer, the diminution of the radius of curvature of the anterior surface of the lens during accommodation for near vision can be determined.

2. As a result of the increased curvature of the lens during accommodation for near vision, the refractive conditions within the eye must be changed. According to v. Helmholtz the measurements for the resting eye, and for the eye accommodated for near vision are as follows (the first number is for the resting, the second for the accommodated eye): Radius of the cornea 8 mm., 8 mm. Radius of the anterior surface of the lens 10 mm., 6 mm. Radius of the posterior surface of the lens 6 mm., 5.5 mm. Distance of the vertex of the anterior surface of the lens from the vertex of the anterior surface oɪ the cornea 3.6 mm., 3.2 mm.; of the vertex of the posterior surface of the lens 7.2 mm., 7.2 mm.; of the anterior focal point 12.91 mm., 11.24 mm.; of the first principal point 1.94 mm., 2.03 mm.; of the second principal point 2.35 mm., 2.49 mm.; of the first nodal point 6.96 mm., 6.51 mm., of the posterior focal point 22.23 mm., 20.25 mm., behind the anterior corneal surface.

3. If the resting eye is viewed from the side, the pupil appears as a narrow black streak. This becomes broader as soon as the eye is accommodated for near   s on, as the entire pupil moves forward.

4ᵥᵢlf light is thrown obliquely into the anterior chamber, the focal line formed by the concave surface of the cornea falls upon the iris. If the experiment is made with an eye arranged for distant vision, so that the focal line lies near the pupillary margin of the iris, the line will recede immediately toward the scleral margin of the iris, as soon as the eye is accommodated for near vision, because the iris is placed more obliquely, as its pupillary margin moves forward.

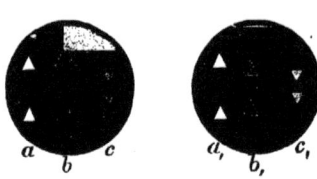

FIG. 283.—The Images of Purkinje-Sanson: *a b c,* in the eye at rest; $a_1 b_1 c_1$, in the eye accommodated for near vision.

5. In accommodation for near vision the pupil contracts; in distant vision it dilates. The contraction takes place, however, somewhat later than the accommodation. This phenomenon may be interpreted as an associated movement, as both the ciliary muscle and the sphincter of the pupil are innervated by the oculomotor nerve. Examination of Fig. 281 will show that the sphincter may assist the muscle of accommodation directly; for if the inner margin of the iris moves inward (toward r), this movement will be transmitted to the ciliary margin of the choroid, which must likewise follow inward to some extent. The movement of the choroid is, it is true, effected principally by the tensor of the choroid. Accommodation is, however, still possible even when the iris is absent or when its fibers are split.

6. On rotation of the eyeball inward, the eye is accommodated involuntarily for near vision. As both eyes rotate inward when the optic axes are directed toward near objects, it is evident that the eye must be involuntarily accommodated at the same time for near vision.

7. The accommodation from a near to a far object (simple relaxation of the tensor of the choroid) takes place more quickly than the reverse movement, from far to near. The process of accommodation requires a longer time the nearer the object is brought to the eye. The time required for the image formed by the anterior surface of the lens to complete its change of position is less than that required for subjective accommodation.

8. When the eye is accommodated for a certain distance, it obtains a sharp image not only of one point, but of a whole series of points behind one another. The line in which these points are situated is called the line of accommodation. The farther away the point is for which the eye is accommodated the longer this line becomes; beyond a distance of from 60 to 70 meters from the eye all objects, even the most remote, appear equally distinct. The shorter the distance, the shorter the line becomes; that is, during the highest degree of accommodation for near objects, a point situated a short distance behind the fixed point will appear indistinct.

9. The question whether it is possible for the lens to change its form partially,

that is in one or another meridian, and therefore for one section of the ciliary muscle to contract independently, is answered in the affirmative by Dobrowolsky, E. Fick, Michel and others, and in the negative by Hess.

10. In strong accommodative effort, the lens sinks downward from gravity, whatever the position of the head, on account of the relaxation of the zonula. Eserin causes strong contraction of the ciliary muscle.

The refractive action of the lens in accommodation for both distant and near vision is illustrated with especial clearness by Scheiner's experiment. A piece of cardboard (Fig. 284, K K,) containing two small openings (S, d) separated by a distance less than the diameter of the pupil is held before the eye, and the observer looks at two needles (p and r) placed behind each other; if the first needle (p) is fixed by the observer, the second one (r) appears double; and conversely. If the near needle (p) is fixed and the eye is accommodated for it, the rays passing from it are naturally focussed at the image on the retina (p,); the rays, however, from the distant needle (r) have already come to a focus within the vitreous, and, diverging, form two images (r, r,,) on the retina. If the right hole in the card (d) be closed, the left image of the distant needle (r,,) disappears. The result is analogous if the eye is accommodated for the distant needle (R). Then the near needle (P) forms a double image (P, P,,) because the rays passing from it have not yet come to a focus. Closure of the right hole (d,) makes the right image (P,) disappear. It must be noted especially (with regard to the projection of the retinal image outward into the field of vision) that when the observing eye is accommodated for the near needle, and one of the holes is closed, the homonymous double image of the distant needle disappears. If, however, the distant needle is fixed, and the opening is closed, the crossed image of the near needle disappears.

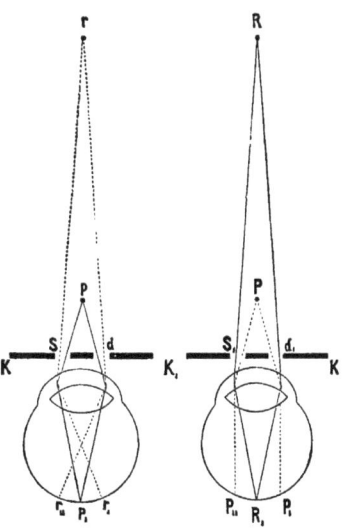

FIG. 284.—Scheiner's Experiment.

Statements have been made recently that in animals and even in man the sympathetic takes part in accommodation for distant vision. Irritation of the sympathetic is said to cause the lens to become flatter. This is shown by the fact that the pupillary sphincter cannot act as an auxiliary muscle of accommodation when the pupil is widely dilated.

Mammalia, birds, and reptiles exhibit the same mechanism for accommodation. In cephalopods and osseous fishes, whose eyes when at rest are accommodated for near vision, active accommodation for distant vision is effected by approximation of the lens to the retina, in fishes by the activity of a retractor muscle of the lens. In some amphibia and snakes active accommodation of the eye for near vision takes place through the separation of the lens from the retina in consequence of changes in the intraocular pressure. Some nocturnal animals and some sensitive to light have no power of accommodation whatever.

REFRACTIVE POWER OF THE NORMAL EYE. ANOMALIES OF REFRACTION.

The limits of distinct vision vary greatly for different eyes. A distinction is made between the far point (or resting point) and the near point. The far point is the greatest distance from the eye to which an object may be removed and still be seen distinctly; the near point is the shortest distance at which an object can still be seen distinctly. The distance between these two points is called the range of accommodation. Three types of eyes are distinguished:

1. The normal (emmetropic) eye is so constructed that, when it is at

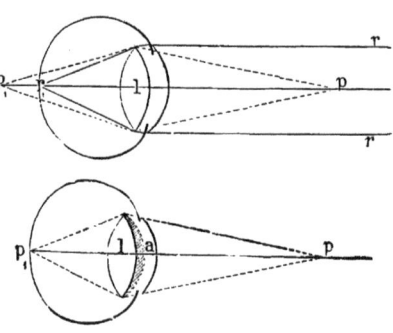

rest, parallel rays (**Fig.** 285, r r) from objects at an infinite distance come to a focus (r_1) on the retina. The far point therefore equals ∞. On the strongest effort of accommodation for near vision, during which the lens increases its convexity (**Fig.** 286, a), rays come to a focus upon the retina (p_1) that are emitted by a point of light (p) 5 inches from the eye, that is the near point is 5 inches (1 inch equals 27 mm.). The range of accommodation is, therefore, ∞.

FIG. 285 and FIG. 286.—Refractive Condition of the Normal Eye, at Rest and in Accommodation.

2. The short-sighted (myopic, hypometric, long) eye (**Fig.** 287) is unable, when at rest, to focus parallel rays on the retina. Such rays cross within the vitreous (at o), and then diverge and form a circle of diffusion on the retina. The objects must be at a distance of from 60 to 120 inches (at f) from the resting eye in order that the rays may be united on the retina. The resting short-sighted eye is, therefore, capable of bringing only divergent rays to a focus upon the retina. The far point, therefore, lies abnormally near. By the most powerful effort of accommodation, objects may be distinctly seen at distances of from 4 to 2 inches, or even less. The near point also is abnormally close; the range of accommodation is diminished.

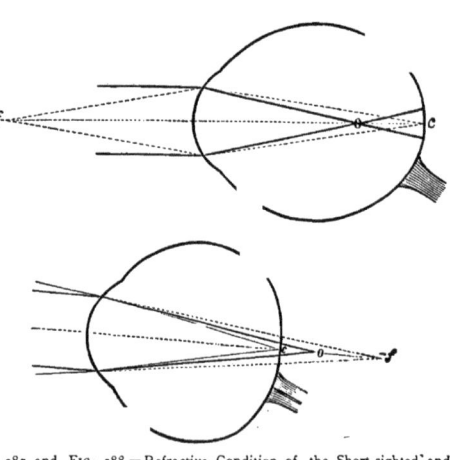

FIG. 287 and FIG. 288.—Refractive Condition of the Short-sighted and the Far-sighted Eye.

Myopia is usually dependent upon an elongation of the eyeball, which is congenital and often inherited. The correction of this anomaly of refraction is effected by the use of a concave glass, which causes parallel rays from a great distance to diverge, so that they can be brought to a focus upon the retina. It is remarkable that most infants are myopic at birth. This myopia, however, depends upon excessive curvature of the cornea and lens, and on excessive proximity of the lens to the cornea. As the eye grows this myopia disappears. Either the too constant activity of the tensor of the choroid (in reading, writing, etc.), or the continuous convergence of the eyeballs, whereby the external pressure on the eyeballs is increased, is considered as the cause of the myopia arising or increasing during school-life.

3. The far-sighted (hyperopic, hypermetropic, presbyopic, oversighted, flat) eye (Fig. 288) is capable, when at rest, of focusing only convergent rays on the retina (c). Distinct images can, therefore, be formed only when the rays from objects are made convergent by a convex lens, because parallel rays would come to a focus behind the retina (at j). All rays coming from natural objects are either divergent, or at most approximately parallel, never convergent. Therefore, no hyperope can see distinctly when the eye is at rest, without a convex glass. When the ciliary muscle contracts, slightly convergent, parallel and finally even somewhat divergent rays may be brought to a focus, by increasing efforts of accommodation. The far point is consequently negative, the near point abnormally remote (more than 8 or 10 inches), while the range of accommodation is infinitely great.

The cause of this defect is abnormal shortness of the eye, which is generally the result of imperfect development in all directions. In addition the lens becomes flattened in old age. The error is corrected by means of a convex lens.

The far point of an eye is determined by bringing toward it objects that subtend a visual angle of only 5 minutes (for example, Snellen's small letters, or the medium type—from 4 to 8—of Jaeger) and finding the point at which they first become distinctly visible. The distance from the eye indicates the far point: In obtaining the far point of a myope, the same objects (subtending a visual angle of 5 minutes) are placed at a distance of 20 inches from the eye, and the weakest concave glass is selected that will enable him to see the objects distinctly. The near point is found by bringing minute objects (for example fine print) closer and closer to the eye, until it becomes indistinct. The shortest distance at which distinct vision is possible is designated the near point.

The optometer may also be employed to determine the far and near points. A small object, such as a pin, is moved to and fro over a scale, along which the eye sights, as along a gun-barrel. The object is brought as close as possible and is then removed as far as possible, so as to permit of distinct vision. The scale shows directly the near and far points and also the range of accommodation.

Other optometers are based on Scheiner's experiment. By an arrangement similar to that described the object is viewed through two small openings in a card. When the object is within the near point, it appears double; and similarly when it is beyond the far point. This may be readily understood from a consideration of Scheiner's experiments. The instruments of Porterfield and Stampfer are constructed on this principle. In the latter a narrow, luminous slit, which can be moved in a dark tube, is used as the fixing object. The optometer of Th. Young and Lehot consists of a white thread stretched over a black scale. The thread is observed through two small openings, and appears single and distinct when within the range of accommodation; within the near point, and beyond the far point, however, it appears broken up into diverging lines.

MEASURE OF THE POWER OF ACCOMMODATION.

The range of accommodation, which is easily determined by investigation, does not of itself indicate the degree of force or the power of accommodation. The measure of this is the mechanical work done by the ciliary muscle. It cannot, of course, be determined directly in the eye itself. It is, therefore, necessary to take as its measure the optical effect produced by the change in the form of the lens that is brought about by the muscular activity.

These relations may first be considered in the emmetropic eye. In the condition of rest, those (dotted) rays that pass in a parallel direction from an infinite distance on the retina are united (Fig. 289, f). In order to focus rays that come from the near point at a distance of 5 inches (p), the muscle of accommodation must exercise its full strength, so that the lens may be made sufficiently convex. The power of accommodation, therefore, produces an optical effect by increasing the convexity of the previously passive flat lens (A) to an amount equal to B; or in other words, it is as though a new convex lens B, had been added to the original lens A. What, therefore, must be the focal distance of the lens B, in order that rays coming from the near point (5 inches) may be focused on the retina? Manifestly the lens B must render the divergent rays parallel; then A can focus them at f. Convex lenses render parallel rays that come from their principal focus. In the instance cited the lens consequently must have a focus of 5 inches. Therefore, the normal eye, with a far point of infinity, and a near point of

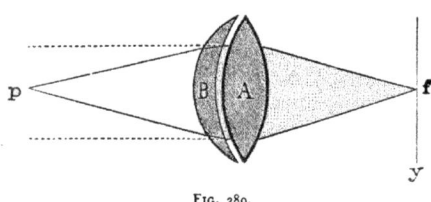

FIG. 289.

5 inches, has a power of accommodation equivalent to a lens of 5 inches focus. If, now, the lens is made more refractive by its power of accommodation, the increase may be readily eliminated by placing before the eye a concave lens that has an optical effect exactly the opposite of that due to the increase of accommodation (B). Hence, it is possible to use a lens of definite focus as the measure of the power of accommodation of the eye, that is for the optical effect produced by the latter. According to Donders the measure of the power of accommodation of an eye is the reciprocal value of the focal distance of a concave lens that, when placed before the accommodated eye, so refracts a bundle of rays coming from the near point (p) that it appears to come from the far point (resting point of the eye).

In accordance with the foregoing considerations, the power of accommodation, then, is calculated by the following formula: $\frac{1}{x} = \frac{1}{p} - \frac{1}{r}$; that is, the power of accommodation (expressed by the dioptric value of a lens of x inches focus) is equal to the difference between the reciprocals of the distances of the near (p) and far points (r) from the eye. *Examples:* In the emmetropic eye, as already mentioned, p = 5; r = ∞. Its power of accommodation is therefore $\frac{1}{x} = \frac{1}{5} - \frac{1}{\infty}$; therefore x = 5; that is, it is equal to a lens of 5 inches focus. In a myopic eye, p = 4, r = 12; so that $\frac{1}{x} = \frac{1}{4} - \frac{1}{12}$; and x = 6. Another myopic eye with p = 4, and r = 20, has x = 5, in other words, normal accommodative power. Two eyes with different ranges of accommodation may have the same power of accommodation. *Example:* One eye may have p = 4, r = ∞; the other p = 3, r = 4. For each eye $\frac{1}{x} = \frac{1}{4}$, or the power of accommodation of each is equal to the dioptric value of a lens of 4 inches focus. Conversely, two eyes may have the same range of accommodation, and yet have unequal power of accommodation. *Example:* One eye may have p = 3, r = 6, the other p = 6, r = 9 (both have a range of accommodation of 3 in.). The power of accommodation for the

first is $\frac{1}{x} = \frac{1}{3} - \frac{1}{6}$; or x = 6; for the second $\frac{1}{x} = \frac{1}{6} - \frac{1}{9}$; or x = 18. The general

law for these relations is as follows: If the ranges of accommodation of two eyes are equal, their powers of accommodation are equal, provided their near points are the same. If, however, the range of accommodation is the same for each eye, but the near points are unequal, then the powers of accommodation are unequal; and that eye has the greatest power of accommodation that has the shortest near point. The reason for this is that every difference in distance near a lens has a much greater influence on the image than the same difference in distance far from the lens. The normal eye can, in fact, see distinctly all objects at a distance between 60 or 70 meters and infinity without any accommodation.

While p and r can be directly determined for the emmetropic and the myopic eye, this is not possible for the far-sighted eye. The resting point (far point) in the latter is negative; in fact in cases of hyperopia of high grade even the near point may be negative. The far point, however, may be determined by means of the convex lens that renders the far-sighted eye emmetropic. The relative near point is then determined by means of the lens.

From the fifteenth year on the power of accommodation for near vision commences to diminish, perhaps because the lens gradually loses its elasticity.

SPECTACLES.

Older Measurement in Inches (1 Inch = 27 Mm.).—The focal length of both concave (diverging) and convex (converging) glasses depends, of course, upon the refractive index of the glass (usually 3 : 2), and upon the length of the radius of curvature. If the curvature of both sides of the lens is the same (biconcave or biconvex), then with the ordinary refractive index of glass the focal length is equal to the radius of curvature. If one side of the lens is plane, the focal length is twice the radius of curvature of the spherical surface. The glasses may be designated either in accordance with their focal lengths in inches, none shorter than 1 inch being usually taken; or in accordance with their refractive power. By this method the unit chosen is the refractive power of a lens with a focal distance of 1 inch. A lens with a focal distance of 2 inches, refracts the light only one-half as much as a lens with a focal distance of 1 inch; a lens of 3 inches focus has a refractive power only one-third as great, etc. This is true for both convex and concave lenses, the latter of course having negative focal distances. For example, the designation "convex $\frac{1}{5}$" would indicate a convex lens with a refractive power only one-fifth as great as a lens with a focal distance of 1 inch; or "concave $\frac{1}{8}$" would indicate a concave lens that caused the rays of light to diverge only one-eighth as much as the concave lens (negative) with a focal distance of 1 inch.

If the far point (always too close) of a myopic eye is determined, a concave lens of the same focal distance as the far point will be required in order to make the divergent rays coming from the far point parallel. The emmetropic eye has a far point of infinity. If, for example, a myopic eye has a far point of 6 inches, it needs a concave lens with a focus of 6 inches in order to see distinctly at an infinite distance. Therefore in a myopic eye, the readily determined distance of the far point from the eye is directly equal to the focus of the (weakest) concave lens that enables the eye to see distant objects distinctly; this lens is usually the number of the glass to be chosen. *Example:* A myopic eye with a far point of 8 inches needs, therefore, a concave lens with a focus of 8 inches, that is the concave glass No. 8. For the hyperopic eye the focal distance of the strongest convex lens that still allows the eye to see distant objects clearly is at the same time the distance of the far point from the eye. *Example:* A hyperopic eye that sees objects at a distance clearly through a convex lens with a focus of 12 inches has a far point of 12; the proper glass is likewise No. 12.

Newer Measurement in Diopters.—Instead of the older designation of the strength of lenses in inches, the meter has been adopted as a unit, following the suggestion of Donders, Nagel, Zehender, and others. By this system the lenses are designated according to their refractive power. The unit is a lens of small refractive power (large focus), that is one with a focus of 1 meter—40 inches. This unit is called a diopter (abbreviated, D). The refractive power of D is

$\frac{1}{4}$ meter. No. 2 is a lens of twice this strength, namely, 2 D; that is its refractive power = $\frac{2}{4}$ meter, and its focus = $\frac{1}{2}$ meter. No. 3 has three times the strength = 3 D, that is, its refractive power = $\frac{3}{4}$ meter, and its focus = $\frac{1}{3}$ meter. No. 4 is four times as strong = 4 D: its refractive power = $\frac{4}{4}$ meter, and its focus = $\frac{1}{4}$ meter. No. 5 is 5 times as strong = 5 D, etc. Weaker glasses than 1 D have been chosen: of 0.75 D, with a focus of 1.33 meter; further, of 0.50 D, with a focus of 2 meters; and 0.25 D, with a focus of 4 meters. Between the whole numbers of diopters $\frac{1}{4}$ and $\frac{1}{2}$ diopter may, of course, be introduced.

In cases of recognized myopia or hyperopia, glasses should by all means be worn for the preservation of the eye. If the far point in a case of myopia is beyond 5 inches, the glass may be worn constantly, but generally the distance for ordinary near work, such as reading, writing, and handwork, should always be about 12 inches. If the work is so fine (embroidering, dissection, drawing, etc.) that the object must be held closer to the eye in order to obtain a larger retinal image, the glass may be removed or a weaker one be substituted. The hyperope may use his glass for near vision, and especially in a poor light, because the diffusion-circles are then unusually large on account of the dilatation of the pupil. It is advisable to choose rather excessively strong convex glasses at first. Cylindrical glasses will be discussed under Astigmatism. Smoked or blue glasses are worn to protect the eye from unduly intense illumination when the retina is sensitive. Stenopaic glasses consist of narrow diaphragms placed in front of the eye, which compel the eye to look in a definite direction, namely through the opening in the diaphragm. Contact-glasses are discussed on p. 841.

CHROMATIC AND SPHERICAL ABERRATION.

DEFECTIVE CENTERING OF THE REFRACTING SURFACES. ASTIGMATISM.

Chromatic Aberration in the Eye.—All rays of white light that undergo refraction are at the same time decomposed into the prismatic colors of which white light is composed, because these colors possess different degrees of refrangibility. The violet rays are refracted the most, the red rays the least. A white point on a black ground does not form a sharp, simple image on the retina; many colored points are formed instead, one behind the other. If the eye is accommodated to focus the violet rays, the succeeding colors must yield concentric diffusion-circles, those near the red being the largest. In the center of the circles, where all the colors are superposed, a white point is formed by their union, while around it are the colored circles. The distance of the focus of the red rays from that of the violet rays in the eye is from 0.58 to 0.62 mm. v. Helmholtz calculated the focus for red in the reduced eye as 20.524 mm., that for violet as 20.140 mm. Therefore, both the near and far points for violet light are closer to the eye than those for red. Hence white objects beyond the far point seem to have a reddish tinge; those within the near point a violet shade. The eye must also accommodate more strongly for red rays than for violet; so that red objects are thought to be closer at hand than equally distant violet objects. This fact should be taken into account by artists.

Monochromatic or Spherical Aberration.—Apart from the decomposition of white light into its components, the rays from a point of simple light are prevented from coming to a single focus by the fact that the edges of refracting (even though only approximately) spherical surfaces refract the rays much more strongly than do the middle portions. Many images are, therefore, formed, instead of one. In the eye this defect is naturally corrected by the iris, which cuts off the marginal rays (Fig. 289), especially when the lens is strongly curved, and at the same time the pupil is most contracted. The marginal part of the lens, in addition, has less refractive power than the central nucleus. Finally, the marginal portions of the refracting surfaces in the eye are less curved than those lying nearer the optic axis, as will be seen by comparing the form of the cornea (p. 815) and that of the lens-surfaces (p. 821).

Defective Centering of the Refracting Surfaces.—The absence of exact centering of the refracting surfaces in the eye disturbs somewhat the sharp projection of the image. The vertex of the cornea does not lie exactly at the end of the optic axis. The same is true of the vertices of the lens and also of the various layers of the lens. The deviations and the visual disturbances produced by them are, it is true, usually but slight.

Regular Astigmatism.—When the curvature of the refracting surfaces of the eye is unequally great in its different meridians, rays of light cannot be united at a single point. Under such circumstances the cornea usually has the greatest curvature in the vertical meridian and the smallest in the horizontal meridian, as is shown by ophthalmometric measurement (p. 830). The rays that pass through the vertical meridian naturally come together first, and in a horizontal focal line; while the rays passing through the horizontal meridian are brought together further back in a vertical line. Such an eye, therefore, does not possess a common focus for light-rays: hence the name "astigmatism." The lens also exhibits some inequality of curvature in the various meridians, but just reversed. As a result, a part of the inequality of curvature of the cornea is thus compensated, and only part of it has any dioptric effect. The emmetropic eye possesses an exceedingly slight degree of this inequality (normal astigmatism). If two fine lines are drawn at right angles to each other on a piece of white paper, it will be found that the paper must be held closer to the eye, in order to see the horizontal line distinctly, than to see the vertical line; the normal eye is, thus, somewhat more short-sighted for horizontal than for vertical objects. If the inequality of curvature is more considerable, sharp vision naturally is altogether impossible. For the correction of this error, a glass is used that is ground in the form of a cylinder; that is in one direction it has no curvature, while in the other direction, perpendicular to the former, it is curved. The glass is so placed before the eye that the direction of its curvature corresponds to the direction of lesser curvature of the eye. For example, the section $\overset{+}{C}$ a b c d of the glass cylinder (Fig. 290) represents a plano-convex cylindrical glass; the section $C a \beta \gamma \delta$ a concavo-convex cylindrical glass.

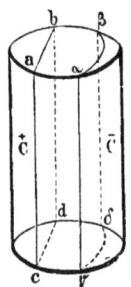

Fig. 290. — Cylindrical Glasses for Astigmatism.

Irregular Astigmatism.—As a result of the stellate arrangement of the fibers in the center of the crystalline lens, and of the unequal course of the fibers within different portions of one and the same lens-meridian, all of the rays passing through one meridian cannot be focused at the same point. For this reason sharp images of distant points of light (stars or lamps) are not obtained, but rather stellate, jagged figures, with projecting rays. The same thing may be seen by holding a card with a fine perforation toward the light, at a somewhat greater distance from the eye than the far point. Slight degrees of this irregular astigmatism are normal, but if developed to a high degree the condition disturbs the visual acuity greatly, by producing several images of each point of an object, instead of one image (monocular polyopia). This condition cannot, of course, be present in eyes deprived of their lens. Irregular curvatures of the cornea act in a similar way. A. E. Fick has eliminated these by the use of a lens in the form of a watch glass, placed in contact with the cornea (contact-spectacles); Lohnstein, by placing before the eye a chamber closed in front by a spherical glass, and filling the interspace (between the cornea and the spherical glass) with 0.85 per cent. solution of common salt (hydrodiascope).

THE IRIS.

1. The iris acts like a diaphragm in an optical apparatus by cutting off the marginal rays (Fig. 279), the entrance of which would produce a decided spherical aberration, and as a result, indistinct vision. 2. As the pupil contracts strongly in bright illumination, it regulates the amount of light that enters the eye; in this way fewer rays of light enter the eye when the light is strong than when it is feeble. 3. The iris acts, further, as an auxiliary to the muscle of accommodation.

As the retina can adapt itself to a comparatively wide range of illumination, the pupil (after the first reaction) can resume its size (from $3\frac{1}{4}$ to 4 mm.) if the limits of illumination are between 100 and 1100 meter-candles.

The size of the pupil increases from the first month of life up to from the third to the sixth year; and with it also the amplitude of reaction decreases, though more slowly.

With regard to the size of both pupils it may be remarked that when there is semidecussation of the optic nerves, the pupils are always of the same size, and they react symmetrically (man, cat). In animals in which the decussation is total (horse, owl), and in those that have only a few uncrossed fibers in the optic tract (rabbit), the pupillary reflex is confined to the eye tested.

The iris has two muscles: the sphincter, which surrounds the pupil and is supplied by the oculomotor nerve; and the dilator of the pupil, supplied chiefly by the cervical sympathetic and the trigeminus. The two muscles are antagonistic; the pupil dilates, therefore, after paralysis of the oculomotor, by the predominance of the sympathetic; conversely, it contracts after excision of the sympathetic. Simultaneous irritation of both nerves causes the pupil to contract; the excitability of the oculomotor is consequently the greater.

According to Arnstein and A. Mayer all the nerve-fibers lose their myelin-sheaths after a short course. Most of the motor fibers near the sphincter consist of naked bundles of fibers. Under the anterior epithelium there is a network of exceedingly fine sensory nerves. Numerous fibers pass to the capillaries and arteries as vasomotor nerves.

The **movements of the** iris take place under the following conditions:

1. Irritation of the retina by light causes a contraction of the pupil corresponding to the intensity and extent of the irritation. Irritation of the optic nerve itself has the same effect. This movement is a reflex action transferred to the path of the oculomotor nerve. The center is situated in the anterior pair of quadrigeminate bodies near the aqueduct of Sylvius. After section of the optic nerve the pupil dilates and subsequent section of the oculomotor causes no further dilatation. In the dark the pupil dilates, at first rapidly, later more slowly. Immediately after the darkening an illumination must have considerable strength to cause pupillary contraction. After the eye has become accustomed to the darkness, a weaker light is sufficient. A flash of lightning following a long period of darkness produces strong and prolonged contraction. A slow increase in illumination is almost without effect.

2. The center for the dilator fibers of the pupil is irritated by a state of the blood causing dyspnea. If the dyspnea passes into asphyxia, the dilatation of the pupil diminishes. Previous section of the peripheral dilator fibers makes these reactions impossible. Sudden anemia also has a stimulating action.

3. The center, as well as the ciliospinal region of the cord subordinated to it, is also susceptible to reflex irritation. Painful excitation of the sensory nerves produces dilatation of the pupils and protrusion of the eyeballs, as was demonstrated by the ancient acts of torture. Labor-pains, loud noises in the ear, irritations of the nerves of the sexual organs, and even slight tactile sensations have the same effect. According to Bechterew, these results are due to an inhibition of the light-reflex, in the sense expressed on p. 731.

4. The condition of the blood-vessels of the iris has an important influence on the size of the pupil. Everything that increases their injection contracts the pupil, while everything that diminishes the amount of blood dilates the pupil. The pupil is contracted, therefore, by forced expiration, which prevents the return of blood from the head; momentarily by each pulsation of the heart (by diastolic filling of the arteries); by decrease of intraocular pressure, for example after puncture of the anterior chamber, because more blood can enter the vessels of the iris, owing to the diminished intraocular pressure; further, by paralysis of the vasomotor fibers of the iris. Conversely, the pupil is dilated by conditions the reverse of those already mentioned, and also by strong muscular exertion, during which blood rushes into the dilated muscular branches, and further, when death takes place. The influence of the amount of blood accounts also for the fact that the pupil when dilated by atropin becomes narrower as soon as the superior cervical sympathetic ganglion, which supplies a part of the vasomotors

of the iris, is excised; and, further, that, after excision of this ganglion, atropin has less effect upon the pupil of the same side. The increased dilatation of the pupil by irritation of the sympathetic after instillation of atropin is probably also the result of diminished injection of the vessels of the iris. If an animal whose pupil is dilated by atropin be bled to death quickly, the pupil contracts on account of the irritation of the oculomotor center by the anemia. The dilatation of the pupil in cases of trigeminal neuralgia must be referred partly to irritation of the dilator fibers, and partly to irritation of the vasomotor fibers of the iris.

5. Contraction of the pupil occurs as an associated movement during accommodation for near vision, further, as a result of strong effort to close the lids, and in rotation of the eyeball inward, which is the case during sleep. Conversely, intense movement of the iris, caused by variations in the brightness of dazzling lights, for example of electric light, produces disturbing associated movements of the ciliary muscles. In connection with certain movements excited in the medulla oblongata (forced breathing, chewing, swallowing, vomiting) dilatation of the pupil occurs as a kind of associated movement.

Direct stimulation of the corneal limbus causes dilatation of the pupil. In fact partial dilatation may be produced by direct irritation of a circumscribed portion of the margin of the iris, by contraction of the dilator fibers, although also the sphincter contracts at the same time.

If a flame be placed in a dark room on one side of an eye directed straight ahead, and attention is suddenly directed to the flame, without changing the direction of vision, the pupil contracts. This movement is known as the "cortical reflex." Other things being equal, an analogous dilatation of the pupil also takes place; one may observe variations in the size of the pupil from the mere conception of light or darkness, even in the blind. Bechterew saw a case of unilateral voluntary dilatation of the pupil.

As to the **action of poisons** on the iris ignorance still prevails. The **mydriatics** cause dilatation: Atropin, homatropin, duboisin, scopolamin, daturin, hyoscyamin, hyoscin, probably through paralysis of the oculomotor chiefly. They must also stimulate the dilating fibers at the same time, for in the presence of complete oculomotor palsy the moderately dilated pupil is still further dilated by atropin. Minimal doses of atropin cause contraction of the pupil by stimulation of the pupil-contracting fibers. Excessive doses cause moderate dilatation as the result of paralysis of both the dilating and contracting fibers. Atropin acts even after destruction of the ciliary ganglion, in fact on the enucleated eye.

For the action of the constrictors, or **myotics:** physostigmin (or eserin, the alkaloid of physostigma), nicotin, pilocarpin, muscarin, and morphin, some investigators assume a stimulation of the oculomotor, others a paralysis of the sympathetic. As these drugs cause contraction of the ciliary muscle Grünhagen supposes an analogous action upon the sphincter. In all probability they paralyze the dilator fibers, and stimulate the oculomotor fibers at the same time.

Intravenous injection of suprarenal extract causes all signs of irritation of the cervical sympathetic in the eye.

If one pupil is contracted or dilated by these drugs, the other pupil is conversely dilated or contracted on account of the variation in the amount of light that enters the eye into which the drug has been introduced.

The Anesthetics.—Chloroform, in the excitation-stage of narcosis (beginning of unconsciousness), stimulates the center for the dilatation of the pupil. Later, this center is paralyzed (so that no dilatation occurs on the application of external stimuli). Then, the contracting center is stimulated (the pupil becoming reduced to the size of a pinhead), and finally (with danger of death) this center becomes paralyzed, and the pupil dilates.

The movements of the iris are always accompanied by variations in the intraocular pressure. Dilatation of the pupil increases, contraction of the pupil diminishes the intraocular pressure. Irritation of the sympathetic increases, section diminishes, the pressure. Instillation of atropin, after a short temporary lowering of the pressure, produces an increase. Eserin, after a primary increase, causes diminution of the pressure.

According to Hocker, atropin decreases the pressure; eserin increases it primarily and then decreases it on the appearance of myosis.

Reflex dilatation of the iris occurs slightly later than reflex contraction: respectively 0.5 and 0.3 second after the light-stimulus. A certain period of time always elapses before the size of the pupil adapts itself to the amount of illumination that stimulates the retina. In birds, irritation of the oculomotor produces exceedingly rapid contraction; in rabbits dilatation of the pupil does not appear until 0.89 second after irritation of the sympathetic.

In the enucleated eyes of amphibians and fishes, stimulation by light causes contraction of the pupil. In fact the iris of the eel, when removed from the eye and laid in salt-solution, contracts on stimulation by light, the green and blue rays being the most active. In these animals the cells of the sphincter muscle are pigmented; the contractile action of the light-rays seems to take place through the intermediation of the pigment.

Increase of temperature produces mydriasis of the enucleated eye of the frog or eel, while decrease of temperature causes myosis.

Grünhagen has disputed the existence of the dilator muscle. He explains the dilating action of the sympathetic by the contraction of the vessels of the iris, while Gaskell ascribes it to an inhibitory action on the sphincter. However, the dilatation of the pupil is not synchronous with the vascular contraction. Irritation near the center of the cornea causes contraction of the pupil.

Pathological.—Imperfect contraction of the pupil on illumination of the eyes may be caused: (1) By a lowered sensibility of the retina (loss of sensory reflex), or (2) by paralysis of the pupillary oculomotor fibers (loss of motor reflex), or (3) both may be combined. Such conditions have also been designated by the badly chosen term reflex immobility of the pupil. The remarkable cases of so-called paradoxical light-reaction exhibit dilatation of the pupil upon stimulation by light, perhaps as the result of profound exhaustion of the oculomotor, which is soon paralyzed by the light-stimulation.

ENTOPTIC PHENOMENA.

SUBJECTIVE OPTICAL MANIFESTATIONS.

The designation entoptic phenomena is applied to those that depend on the perception of objects that are present in the eye itself Subjective visual sensations are those that are not produced by the normal, homologous stimulation of the retina by light, but by internal, heterologous (mechanical, electrical, somatic) stimuli, which act upon the eye, the optic nerve, or parts of the central organs

Among entoptic phenomena are:

1. **Shadows** of various opaque bodies thrown on the retina. They may be recognized by the following method: the small image of a light is thrown upon a pasteboard screen by means of a convex lens; a fine hole is pricked through the image of the flame and the eye is placed on the other side of the screen, so that the illuminated point coincides with the anterior focal point of the eye (about 13 mm. from the cornea). As the rays from this point pass parallel through the ocular media, a diffusely illuminated field is formed, which is surrounded by the dark outlines of the pupillary margin. All dark objects that intercept these rays throw shadows on the retina and appear as spots (Fig. 291). Several varieties of these shadows can be distinguished: (a) The muco-lacrimal spectrum, especially on the lid-margins, arising from particles of mucus, fat-globules from the Meibomian glands, dust mixed with tears. These give rise to striated, nebulous or drop-like shadows, which are dissipated by winking. (b) Pressure on the cornea with the finger produces wrinkled shadows, due to temporary corneal pressure-folds. (c) Bead-like or dark specks, light and dark stellate figures, the former arising from deposits on and within the lens, the latter from the stellate structure of the lens. (d) The mouches volantes (muscæ volitantes), like strings of beads, circles, groups of tiny balls, or pale threads, are images of small, opaque particles in the vitreous cells, broken-down cells, granular fibers. They move about on sudden movement of the eye. Listing showed that it is possible to determine the approximate position of these objects. If the source of light (the illuminated opening) is raised and lowered, those shadows that retain their relative position in the bright field of vision are due to objects at the plane of the pupillary orifice (2); those that move apparently in the same direction as the light are due to bodies in front of the pupillary plane (1); those, however, that move in the opposite direc-

tion are due to bodies behind the pupillary plane (3). It should be noted in this connection that the impressions of the stimulated portions of the retina are projected outward in the opposite direction.

2. Purkinje's figure depends upon shadows thrown by the blood-vessels within the retina upon the posterior percipient layer, the layer of rods and cones. In ordinary vision they cannot be perceived. According to v. Helmholtz this is probably because the sensitiveness of these shaded portions of the retina is greater than that of the remainder of the retina, and their irritability is less exhausted. As soon as the position of the vessel-shadow is changed, so that it is thrown to one side, instead of directly backward, on places, therefore, that ordinarily do not receive shadows from the vessels, the Purkinje figure at once appears. The light must enter the eye as obliquely as possible. The experiment may be made in several ways: (1) A small bright image of a light may be thrown upon the sclera. As it moves up and down the vessel-figure moves with it. (2) Looking directly upward at the sky, the depressed upper lid is blinked, so that momentarily, in correspondence with the blinking movement, oblique rays of light enter the lowest part of the pupil from above downward. (3) One may look through a small opening toward the sky, and move the opening quickly to and fro, so that shadows fall rapidly from both sides of the vessels on the neighboring rods; or (4) one may look straight ahead in a dark room, and move a light to and fro below the eye. Occasionally in this experiment the macula lutea is seen —like a nonvascular shaded depression, appearing (on account of the inversion of objects) on the inner side of the optic-nerve entrance.

3. Recognition of the Movement of the Blood-corpuscles in the Retinal Capillaries.—On looking (without accommodation) at a large bright surface, or at the sun

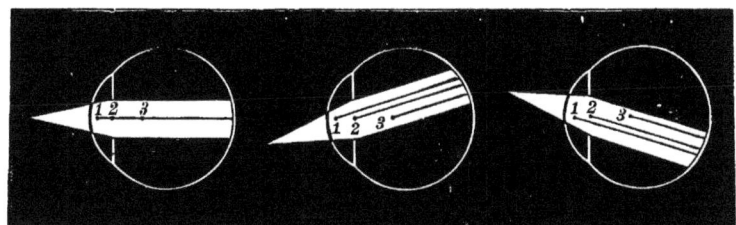

FIG. 291.—The Entoptic Shadows.

through a dark-blue glass, brilliant points like tiny sparks are seen to move in various tortuous paths over larger or smaller spaces. The movement appeared to Landois to resemble most that of a Gyrinus swarm (small water-beetles) on the surface of the water. The particles can often be seen to follow each other in definite, outlined paths. According to some observers, the phenomenon is due to the fact that the red blood-corpuscles—in the capillaries outside of the external nuclear layer—act as small concave discs, concentrating the light falling on them upon the rods of the retina. Each corpuscle must be in a suitable position; if it turn over, the light-phenomenon disappears. Vierordt, who projected the movement upon a screen, calculated, from its rapidity, that the velocity of the blood-current in the retinal capillaries is from 0.5 to 0.75 mm. per second, and this, in fact, corresponds to the direct observations of E. H. Weber and Volkmann on the blood-current in other capillaries. Compression of the carotid artery retards the movement, release of the vessel, as well as short expiratory pressure, accelerate it. As Landois occasionally observed the points as dark spots on a light ground, and as bright ones on a dark surface, the phenomenon is probably better explained as a pressure-phosphene (according to 5), from the friction of the blood-corpuscles in the capillaries against the rods.

4. The yellow spot appears also occasionally, when viewed with uniform blue illumination, as a dark circle. In stronger light the position of the yellow spot may be seen surrounded by a bright area, having a diameter about thrice as large (Löwe's ring).

5. Pressure-phosphenes, that is those phenomena that appear under the influence of pressure on the eyeball. (a) Partial pressure on the eyeball induces

the so-called luminous pressure-picture or phosphene, which was known to Aristotle. By the projection of this retinal stimulation outward, the phosphene is perceived on the side of the visual field opposite to that where the pressure was made upon the retina. For example, pressure on the outer side of the eyeball causes the light-phenomenon to appear on the inner side. If the retina is darkened, the phosphene appears bright; if the retina is illuminated, the phosphene appears as a dark spot within which the visual sensation is momentarily abolished. (b) If uniform pressure from before backward be made for some time on the eyeball, there appear in the field of vision after a time, as Purkinje pointed out, bright, changing figures which produce a strange phantastic play, often similar to the most brilliant kaleidoscopic pictures—probably comparable to the feeling of formication produced by pressure upon the sensory nerves (limbs "going to sleep"). (c) By applying equable and continued pressure, Steinbach and Purkinje saw appear a vascular network of a bluish-silvery color, with streaming contents, which seemed to correspond to the retinal veins. Vierordt and Laiblin recognized in addition the ramifications of the choroidal vessels, red on a dark background, as a network with the forms characteristic of these capillaries. (d) According to Houdin it is possible also to recognize the position of the yellow spot by pressure on the eyeball.

6. The **entoptic pulse-phenomenon** belongs to the pressure-phosphenes, and depends upon the mechanical stimulation of the optic-nerve fibers by the pulsating retinal vessels.

7. The **point of entrance of the optic nerve** may be perceived on rapid, jerking movements of the eye, especially inward, as a fiery circle or semicircle, slightly larger than a pea. Probably the retina around the nerve-entrance is irritated mechanically by the bending of the nerve. Landois saw this ring, as did Purkinje, remain persistent when the eye was turned strongly inward. If the retina is strongly illuminated, the ring appears dark, and when the visual field is colored, the ring has a different hue. If Purkinje's figure be produced at the same time, the blood-vessels appear to spring from this ring—a proof that the ring corresponds to the optic-nerve entrance.

8. **Accommodation-spot.**—If the eye is accommodated as strongly as possible for a white surface, a small bright, vibrating shimmer appears, in the middle of which a brownish spot the size of a pea may shortly be observed. If pressure be made on the eyeball at the same time, this spot becomes more distinct. When this phenomenon is once recognized, a brighter spot may be seen in the middle of the visual field when lateral pressure is made upon the open eye, another proof that the intraocular pressure rises during accommodation. By producing the preceding phenomenon (No. 7) it is demonstrated that the phenomenon takes place at the optic-nerve entrance.

9. The **accommodation-phosphene** consists in the appearance of a fiery ring at the periphery of the visual field when the eyes are suddenly allowed to come to rest after prolonged, intense accommodation for near vision in the dark. The sudden tension of the zonule of Zinn resulting from the relaxation produces a mechanical stretching of the edge of the retina, or more probably of the retina just beyond. Purkinje saw the phenomenon also after sudden cessation of pressure upon the eye.

10. **Mechanical Irritation of the Optic Nerve.**—When the optic nerve is severed in man, in the course of an operation, a bright flash appears at the moment of section. The incision through the optic-nerve fibers is painless; only the sheaths are sensitive.

11. **Electrical Phenomena.**—With variations in an electric current (one pole on the upper lid, the other on the neck) bright flashes of light shoot over the entire visual field. The closing flash is stronger with an ascending current, the opening flash stronger with a descending current. A uniform, constant, ascending current applied to the closed eye reveals the optic-nerve papilla as a dark disc in a whitish-violet field. At the same time sensibility for white is increased, that for black diminished. With a descending current, on the other hand, the visual field appears greenish-yellow, and darkened, and in its midst the position of the nerve appears light blue. If external colors are observed at the same time, these tones blend to form violet or yellow with the colors looked at. Under the influence of an ascending current external objects are said to be seen less distinctly and diminished in size when the eyes are open; while a descending current makes them more distinct and larger. During anelectrotonus of the retina (in conformity with the laws of electrotonus) the sensibility for the electrical light-

phenomena and also that for objective light are diminished. At times the macula lutea appears as a dark spot on a light ground, at other times as a light spot on a dark ground, according to the direction of the current. If the current is broken, the phenomena are reversed and the eye soon returns to rest.

When the eye is directed toward a source of polarized light, Haidinger's polarization brushes appear at the point of fixation. They may be seen if a bright cloud is looked at through a Nicol's prism. They appear as bright, bluish spots on a white ground, bounded by two similar hyperbolas; the dark bundle separating them is narrowest in the center, and has a yellowish color. Blue alone of the various colors of homogenous light exhibits the brushes. According to v. Helmholtz the seat of the phenomenon is the yellow spot, and it depends upon the fact that the yellow-colored elements of this spot are slightly birefringent, and they absorb more of the rays in one place than in others.

Finally, there should be mentioned the *sensations of light produced by internal causes*, by congestion of the retina (as from violent spells of coughing), increased intraocular pressure and the like, or by congestion of the central cerebral organs. Irritation of the psychooptical centers may induce distinct phantasms, which Cardanus, Goethe, Johannes Müller, Nägeli, and others could in fact excite in themselves voluntarily. "Video quae volo, nec omnino semper cum volo. Moventur autem perpetuo quae videntur. Itaque video lucos, animalia, orbes ac quaecunque cupio" (Cardanus). In men suffering from delirium tremens something similar at times takes place: they are able to call forth hallucinations even in daytime, as soon as they think of certain things—voluntary hallucinations.

ILLUMINATION OF THE EYE, AND THE OPHTHALMOSCOPE.

The light that enters the eye is partly absorbed by the black uveal pigment, and partly reflected again from the eye, and always in the same

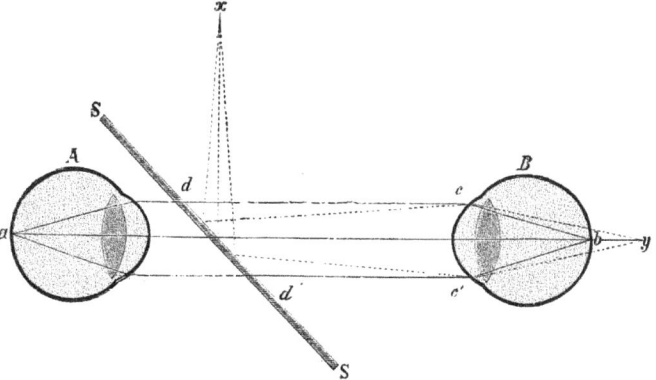

FIG. 292.—Apparatus for Illuminating the Back of the Eye *B*.

direction from which it has entered. If a person place himself directly before the eye of another, the head of the former, as an opaque body, naturally cuts off a considerable number of rays. As no rays can fall upon the eye from the direction of the head of the first person, none can be reflected from the eye of the other, which, therefore, appears black to the former, for the reason that he cuts off all those rays that could be reflected toward his eye. As soon, however, as it is possible to throw light into the eye of the second person in the same direction in which the first looks into the eye of the other, the eyeground at once appears brightly illuminated.

The simple apparatus shown in Fig. 292 is sufficient to corroborate what has been said: Let B represent the eye to be examined, and A the eye of the observer. If a flame be placed at x, its rays will be thrown upon the glass plate $S\,S$, which reflects them in the direction of the dotted lines into the eye B. The eyeground appears in this position brightly illuminated around b in diffusion-circles. As the observer A can see through the glass plate $S\,S$ without difficulty, and in the same direction as the reflected ray $x\;y$, he will see the retina brightly illuminated at b.

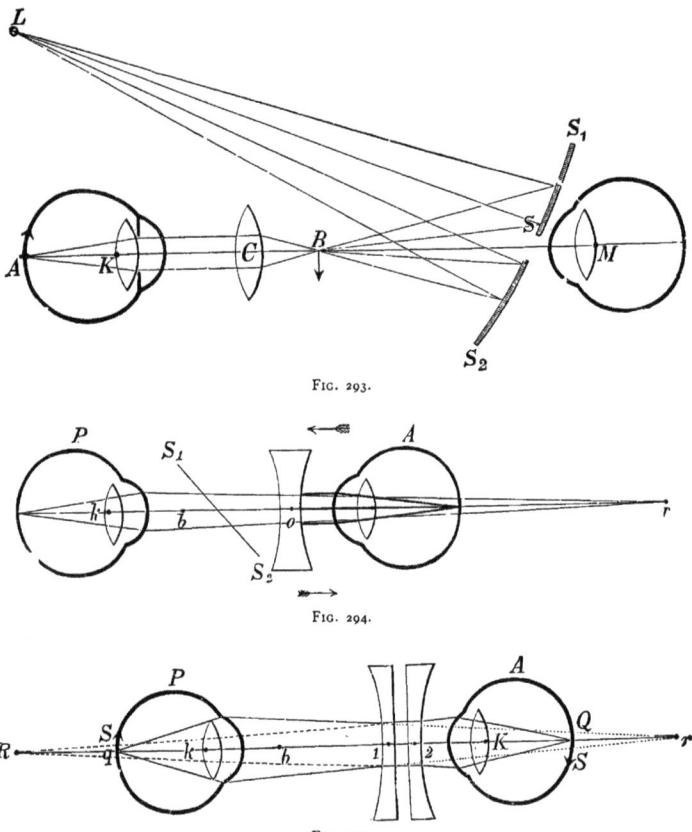

FIG. 293.

FIG. 294.

FIG. 295.

It is important for practical purposes to be able to recognize the details of the eye-ground: the blood-vessels, the macula lutea, the optic-nerve entrance, abnormalities of the retina, of the choroidal pigment, and the like. How this is to be done will be understood from the following considerations: As has been seen (and as **Fig.** 279, p. 829 shows), a small, inverted image is formed on the retina ($c\,d$) of an object ($A\,B$) for which the eye is accommodated. Conversely, according to the same dioptric law, an enlarged, inverted real image (*c d*) must be formed outside of the eye (at *A B*) of any illuminated, circumscribed portion of the retina (when the eye is accommodated for a certain distance). If the eyeground is sufficiently illuminated, this image formed in the air will possess a corresponding degree of brightness.

In order to examine more carefully the individual portions of this image of the retina, the observer must accommodate for the situation of the image. His eye is then separated from the retina of the eye under observation a distance equal to the sum of the focal distances of his own and the other eye. At this distance the finer details of the eyeground cannot be recognized. Moreover, as the pupil of the eye examined is narrow, only a small portion of the eyeground can be seen, and only under a low visual angle; aside from this, it is often impossible to accommodate for the image.

It is necessary, therefore, to bring the eye of the observer closer to the eye under examination. This may be done in two ways. (1) By placing before the eye under examination a strong convex lens with a focus of 1 inch (Fig. 293, *C*). As the image of the retina is thus brought closer to the eye (at *B*), as the result of the refraction of the rays by the lens, the observer *M* can approach much nearer and can still accommodate for the image. (2) Or by placing a concave lens (Fig. 294, *o*) before the eye examined. The rays emerging from this eye are either made parallel by the concave lens *o*, and are then focused on the retina of the emmetropic observer *A;* or, if the lens make the rays divergent (Fig. 295), an upright, virtual image of the eyeground is

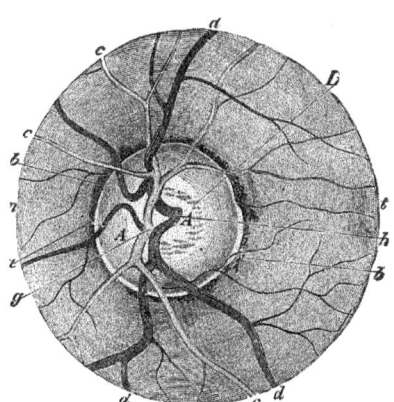

FIG. 296.—The Optic-nerve Entrance, with the Surrounding Structures, of a Normal Eyeground (after Ed. Jaeger): *A*, Optic disc (papilla); *a*, connective tissue ring ; *b*, choroidal ring; *c*, arteries; *d*, veins; *g*, point of division of the central artery; *h*, of the central vein; *L*, lamina cribrosa; *t*, temporal (outer) side; *n*, nasal (inner) side.

formed in the distance, behind the investigated eye (at *R*). Under such circumstances also the observer can approach much nearer,

The illuminating apparatus, together with one of these lenses forms the ophthalmoscope of v. Helmholtz, the basis of modern ophthalmoscopy, by means of which all the details of the eyeground can be examined.

For the illumination, v. Helmholtz used several plates of glass placed behind one another (for better reflection), in the same position as *S S* in Fig. 292. A plane mirror or a concave mirror, with a focus of 7 inches, through the center of which a hole is bored (Fig. 293, S_1, S_2) may also be employed. Fig. 296 shows the ophthalmoscopic appearance of the optic-nerve entrance, and its vicinity, in a normal eye. The letters indicate the details. In albinos the fundus appears light red, because light passes into the eye through the nonpigmented sclera and uvea. If a diaphragm be placed in front of the eye, so that only the pupil is free, the eyeground appears dark. In many animals the eyes have a bright-green luster. These possess a special layer, the tapetum, or the membrana versicolor of Fielding, which in the carnivora is composed of cells, in the herbivora of fibers; it lies between the chorio-capillary layer and the stroma of the uvea, yielding interference-colors, and reflecting a considerable amount of light, so that the eyes have a colored luster.

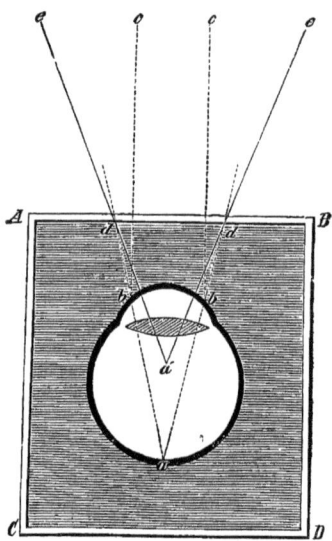

FIG. 297.—Mechanism of the Orthoscope.

For examination of the anterior chamber oblique illumination is employed. A bright beam of light, condensed by a convex lens, is thrown obliquely into the eye, upon the point to be examined, which appears clear and distinct. The point thus illuminated, for example a part of the iris, can then be magnified and examined with the help of a lens, or even of a microscope.

Czermak constructed the orthoscope (Fig. 297), by means of which the eye is placed under water. A small glass trough, one side of which is removed, is filled with water and pressed against the face, so that the eye and the face form the sixth side of the trough, and the cornea is covered with water. As the index of refraction of the water is the same as that of the media of the eye, the rays pass out of the eye unrefracted. Hence, objects in the anterior chamber can be seen directly, and appear as though outside of the eye. A further advantage lies in the fact that the objects are brought closer to the observer's eye. The rays from the point *a* of the eyeground would leave the eye parallel as *b c, b c,* if the eye were surrounded by air. Seen under water, however, these rays *a b, a b* continue in the same direction as far as *d, d,* where they are deflected from the perpendicular on emerging from the water, that is toward *d e, d e.* The observer's eye, looking in the direction *e d,* sees the point *a* closer, that is in the direction *e d a',* consequently situated at *a'.*

THE FUNCTION OF THE RETINA IN VISION.

The rods and cones are the only parts of the retina sensitive to light; they alone are stimulated by the vibrations of the luminiferous ether. This is confirmed by Mariotte's experiment, which shows that the optic-nerve entrance, where rods and cones are absent, has no light-perception. This is, therefore, called the blind spot.

If the letter f (Fig. 279, p. 829) of the two black letters B and f is fixated with one eye (the other being closed), so that its image falls on the fovea centralis (n), and the image of B falls on the optic-nerve entrance (N), the letter B disappears immediately. If three points, A f B, are drawn, and the eye fixes the middle point f, B disappears, but the points A and f are visible.

The optic-nerve entrance lies about 3.5 mm. to the inner side of the visual axis in the retina. It has a diameter of 1.8 mm. In the field of vision the horizontal diameter of the blind spot measures apparently 6° 56'; it lies from 12° 81' to 18° 55' external to the point of fixation. This diameter would cover 11 full moons, placed side by side, and would conceal a human face at a distance of more than 2 meters.

The proof that it is really the optic-nerve entrance that is not sensitive is furnished by the following observations: (1) Donders, by means of a mirror, threw a small image of a flame directly on the optic-nerve entrance of another person, who had no sensation of light. This appeared, however, as soon as the image was displaced to the neighboring portions of the retina. (2) If Mariotte's experiment be combined with the experiments that yield entoptic phenomena at the optic-nerve entrance, the latter coincide with the blind spot.

In order to determine the form and apparent size of the blind spot in one's own eye, the head should be placed at a distance of about 25 cm. from a piece of white paper. Then a small point should be fixed with the eye, and the position of the blind spot on the paper determined by moving a white feather about in various directions, making a mark wherever its point first becomes visible. In this way the blind spot may be mapped out, and it will be found to have an irregularly elliptical form, from which processes extend, representing the insensitive origins of the large central vessels of the retina. Mariotte concluded, from his experiment, that the choroid, which is perforated by the optic nerve, is the light-perceiving membrane, as the nerve-fibers are nowhere absent from the retina.

a   b   c

d   (e)   f

g   h   i

The blind spot in the eye causes no appreciable defect in the visual field. As the area is not excited by light, a black spot cannot appear in the visual field, for the sensation of black presupposes the presence of retinal elements, which, however, are absent at the blind spot. The circumstance that, despite the insensitive spot, no unoccupied spot in the visual field is perceived is due to psychical action. The unoccupied part of the field, corresponding to the blind spot, is probably filled out by a psychical process. Hence, when a white point on a black surface disappears, the entire surface appears black. A white surface of which a black point falls upon the blind spot appears entirely white, a printed page grayish throughout, etc. In the same way, parts of a circle, the middle parts of a long line, the middle portion of a cross are probably supplied. Such images, though, that cannot be reconstructed on a basis of probability are not completed, for example the end of a line, or a human countenance. Under other conditions a phenomenon contributes to the filling out of the empty space that has been designated the "contraction of the visual field." This becomes clear if the letter **e** is made to disappear from among the nine adjoining letters; the three letters of each side are no longer seen in a straight line, but **b, f, h, d** are drawn in toward **e.** The surrounding portions of the field seem to stretch out over the position of the blind spot and help to replace it.

The outer segments of the rods and cones possess rounded contours; they are placed close together, but there must be spaces between them (corresponding to the interspaces between circles placed in contact). These spaces are insensitive to light, so that the retinal image is constructed like a mosaic of small round stones. The diameter of a cone in the yellow spot measures from 2 to 2.5 μ. If two closely situated points form images on the retina, they will be perceived as isolated points, provided that the images fall upon two different cones. A distance of from 3 or 4 to 5.4 μ between the images on the retina is sufficient to enable them to be seen separately, as they will then fall on two neighboring cones. If the distance is so diminished that both images fall on one cone, or one on a cone, and the other on an interspace, only one point will be perceived. In the peripheral portions of the retina, the images must be still further apart in order to be perceived separately.

As the rounded ends of the cones are not placed in exactly straight lines, but so that a row of circles is adapted to the interstices of the succeeding row, exceedingly fine dark lines, drawn parallel, appear to have alternating twists, as their images must fall on the cones alternately to right and to left. In the same way every straight edge of an object appears wavy when its retinal image is moved across the retina with moderate rapidity.

The sharpest vision is obtained at the fovea centralis, where cones alone are found, placed closely together. In the peripheral portions of the retina the cones are placed less closely together, and here vision is less acute. From this it may be concluded that the cones are more important for vision than the rods. In order to see an object as distinctly as possible, the eyes are therefore turned involuntarily so that the retinal images fall on the fovea centralis. This adjustment is known

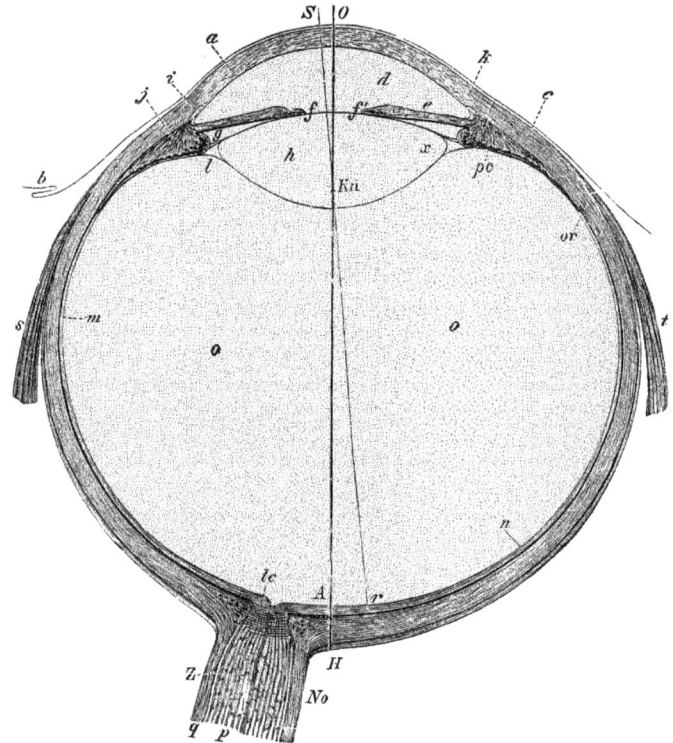

FIG. 298.—Horizontal Section of the Right Eye: *a*, Cornea; *b*, conjunctiva; *c*, sclera; *d*, anterior chamber, containing the aqueous humor; *e*, iris; *f*, *f'*, pupil; *g*, posterior chamber; *l*, canal of Petit; *j*, ciliary muscle; *k*, corneo-scleral junction; *i*, canal of Schlemm; *m*, choroid; *n*, retina; *o*, vitreous body; *No*, optic nerve; *q*, nerve-sheaths; *p*, nerve-fibers; *lc*, lamina cribrosa. The line *A O* is the optic axis, *S r* the visual axis, *r* the position of the fovea.

as *fixation;* the visual ray drawn from the fovea to a point in the object is called the *visual axis* (Fig. 298, *S r*). This forms with the *optic axis* of the eye (*O A*), which unites the centers of the spherical surfaces of the refracting media, an angle of only from 3.5° to 7°. The point of intersection lies, of course, in the nodal point (k n) of the lens (Figs. 279, 298). Vision with the optic axes directed upon the object is known as *direct vision*.

Faintly illuminated objects are not appreciated with the same degree of accuracy by the fovea centralis as by the surrounding retina.

If light be allowed to fall on the fovea centralis through a screen perforated like a sieve, it appears as a continuous bright surface, if one point of light falls on each cone. For this it is necessary that from 140 to 149 points of light fall on 0.01 sq. mm. of the fovea centralis. According to Salzer there are 138 cones in this area. If the points of light in the screen are to be appreciated separately, each illuminated cone must be surrounded by a circle of nonilluminated cones. In this case 72 points of light must fall on 0.01 sq. mm. of the fovea.

To test the visual acuity in direct vision, two fine parallel lines drawn close together, are gradually removed further from the eye, until they appear to fuse almost into one. From the distance between the two lines, and the separation of the drawing from the eye, the size of the retinal image is determined, and also that of the corresponding visual angle, which is ordinarily between 60 and 90 seconds—the lowest limit has been found to be between 50 and 27 seconds.

Indirect vision occurs when the rays of light from an object fall upon the peripheral portions of the retina. Indirect vision is much less sharp than the direct, but the periphery of the retina has a well-developed power of recognizing movements, changes or intermissions in visual impressions.

Perimetry.—For the determination of indirect vision the perimeter of Aubert and Förster is used. The eye is placed opposite a fixation-point, from which extends a semicircle, so that the eye lies at its center. As the semicircle can be revolved about the fixation-point, the surface of a hemisphere is formed by this rotation, in the center of which the eye is placed. An object is now pushed outward from the fixation-point along the semicircle toward the periphery of the visual field, until it becomes indistinct and finally disappears. This test is made along the various meridians by moving the arc into corresponding positions. The further away from the fixation-point two closely placed points are carried, the further they must be separated to prevent their being fused. The power of distinguishing colors diminishes more rapidly in the periphery than does that for distinguishing differences in brightness. The decrease is, moreover, more marked in the vertical meridian of the eye than in the horizontal, and decreases with the distance from the fixation-point. Aubert and Förster discovered the remarkable fact that in accommodation for a distant object the decrease of the differentiating power in the periphery occurs more rapidly than in accommodation for near vision. The sensibility of the retina for colors and for brightness is greater at points on the temporal side of the fovea than at equidistant points on the nasal side.

If the arc of the perimeter be divided into 90 degrees, commencing at the fixation-point (Fig. 299) and proceeding to L and M, and if a series of concentric circles be drawn about the fixation-point, a topographical chart of the visual power can be mapped out for the normal and the diseased eye. Fig. 299 will serve as an example. The thick lines refer to a diseased eye; the corresponding fine lines to a normal one. The continuous line represents the limit for the perception of white; the interrupted line, that for blue; the dotted and interrupted line that for red (m is the blind spot, according to Hirschberg). The limits for the normal eye are as follows:

	For White.	Blue.	Red.	Green.
Outward,	70–88°	65°	60°	40°
Inward,	50–60°	60°	50°	40°
Upward,	45–55°	45°	40°	30–35°
Downward,	65–70°	60°	50°	35°

The rods and cones alone possess the specific energy of being thrown by the vibrations of the luminiferous ether into the activity that is designated sight. Nevertheless, mechanical and electrical stimuli applied to any portion of the course of the nervous apparatus can also produce sensations of light. The mechanical stimulation is more intense than that produced by light-rays, as is shown by the fact that, when the dark figure is produced by pressure on the open eye in consequence of which the circulation of the retina is disturbed, external objects are not perceived by the retina.

When the eye is well rested, it has a diminished sensitiveness for colors in a poor light, and the green portion of the spectrum seems to possess the greatest degree of brightness. v. Kries believes that under such circumstances the rods are active, while with vision in strong illumination the cones functionate. As color-blind individuals perceive the brightness in the spectrum in the same way as does the well-rested eye in a poor light, perhaps in their case only the rods take part in vision.

The duration of the retinal stimulation need be but brief. The stimulation by light does not reach its full strength at once, but it increases gradually, so that a weak light that persists for some time may appear

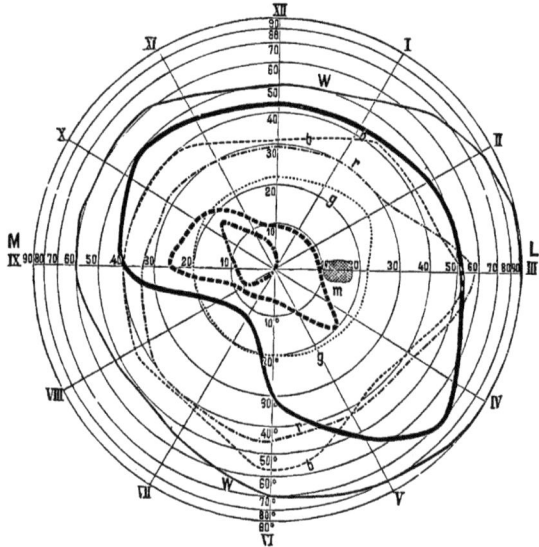

FIG. 299.—Perimetric Chart of a Healthy and of a Diseased Eye.

just as bright as a strong light that persists for but a short time. In general, the larger and the brighter the objects the less the time necessary for their perception. If light and darkness are quickly alternated, the same degree of illumination is produced as if the action of the light were divided uniformly over the entire time of the observation. A light-stimulus having 17 or 18 alternations in a second, produces the strongest sensation. In order that two flashes of light shall be perceived separately, 0.027 second must elapse between them. Further, an increase or decrease of 0.01 part of the light-intensity will be recognized. A shorter time is required for the perception of yellow than for that of red and violet. Long exposure to darkness, as during the night, makes the retina more sensitive to light. When the light-stimulus is of long duration and great intensity, retinal fatigue sets in, beginning earlier in the center than at the periphery. It progresses more quickly at first than later, and is most striking in the morning.

During direct vision, objects must have an angular velocity of from one to two minutes in a second in order to appear to be in motion.

The manner in which light acts upon the terminal apparatus of the retina has already been discussed in connection with the visual purple (p. 819). Kühne showed that by illuminating the retina, actual, permanent pictures could be produced on the retina, for instance the picture of a window, but these gradually disappear. The retina acts in some respects like the sensitive plate of a photographic apparatus, and the light is conceived as having a chemical action, especially as non-illuminated retinas have a more acid reaction than illuminated ones.

The visual purple is given off by the pigmented epithelium of the retina, as a sort of secretion, to the rods alone, and not to the cones. A bleached retina can take up the purple again if placed in contact with living pigmented epithelium. There are two modifications of the purple in the animal kingdom. The mammalian retina is bleached about sixty times more quickly than the frog's. In fixed rabbits' eyes, under atropin-mydriasis Ewald and Kühne obtained sharp optograms of bright objects at a distance of 24 cm., in from $1\frac{1}{4}$ to $1\frac{1}{2}$ minutes; the picture is fixed by 4 per cent. solution of alum. The visual purple is preserved when dissolved in bile and when saturated with sodium chlorid. It resists all oxidizing agents; zinc chlorid, acetic acid, and mercuric chlorid transform it into a yellow substance. It becomes white only through the action of light; the nonluminous heat rays have no effect; it is decomposed by temperatures above 52°.

Kühne found that in the illuminated frog's eye the pigment-granules of the pigmented epithelium extend further between the outer segments of the rods and cones, and in darkness withdraw again into the outer part of the pigmented epithelial cells. In the eye of the fish exposure to light produces also decrease of the chromatin in the granules and ganglia.

A further important fact should be mentioned, namely, that the inner segments of the cones become shorter under the influence of light, and elongate in the dark. The action is always bilateral, even when only one eye is exposed to light; but after destruction of the brain, the effect is confined to one side; strychnin-tetanus, thermal, chemical, or electrical irritations act in the same way as light. The optic nerve, therefore, must contain motor (retinomotor) fibers, in addition to the light-perceiving fibers. Motor phenomena are observed also in the ganglion-cells, and in the outer and inner segments of the rods, the cells of the external nuclear layer changing their form at the same time. In fact these movements cause electrical phenomena in the eye. Isolated inner cone-segments and granules likewise exhibit changes of form when exposed to light.

According to v. Kries the rods are entirely color-blind, and their chief function is related to vision in weak light; the cones are for the perception of colors.

In changing from a light to a dark room, or the reverse, the eye must adapt itself first to the action of the light. The eye adapted to light has been found superior in visual activity to the eye adapted to darkness.

Destruction of the rods or cones of the retina causes corresponding dark spots in the visual field.

Strong visual impressions render the retina insensitive to light, and permanent injury and blindness may result from necrosis of the retinal elements with edema.

PERCEPTION OF COLORS.

Physical Considerations.—The undulations of the luminiferous ether are perceived by the retina only within definite limits. If a beam of white light, for example from the sun, be allowed to pass through a prism, its rays are refracted, and are decomposed into the *prismatic spectrum* (Fig. 12). The white light contains rays of widely different wave-length, or number of vibrations. The dull heat-rays are the least refracted; their wave-length measures 0.00194 mm.; they do not affect the retina, and are, therefore, invisible, although as is well known they affect the sensory nerves. About 90 per cent. of these rays are absorbed by the ocular media. Commencing at Fraunhofer's line A (Fig. 15) the oscillations of the ether excite the retina, and the colors appear in the following order: red, with 481 billions of vibrations in a second; orange. with 532; yellow, with 563; green. with 607; blue, with 653; indigo, with 676; and violet, with 764 billions in a second. The perception of color depends, therefore, upon the number of vibrations of the ether, just as the pitch of a note depends upon the number of vibrations of the sounding body. The heat-rays that lie in the colored spectrum are transmitted by the ocular media in about the same way as by water. Beyond the violet rays lie the *chemically active* or actinic rays. The ultra-violet rays are largely absorbed by the ocular media, especially by the lens. On shutting off the entire spectrum, including the violet rays, the ultra-violet rays may yet be recognized from their pale, grayish-blue color. The ultra-violet rays can be most easily demonstrated by the phenomenon of fluorescence: on illuminating a solution of quinin sulphate with ultra-violet v. Helmholtz saw a bluish-white light arise from all parts of the solution that were reached by ultra-violet rays. As the ocular media themselves exhibit fluorescence, they must increase the power of the retina to distinguish these rays.

In order that color may be recognized, it is necessary for a certain amount of light to fall upon the retina. The lowest degree of brightness by which blue may be recognized as a color is 16 times less than that required by red. If, therefore, in a bright illumination, a red and a blue object appear equally bright, the blue will appear brighter as soon as the illumination is decreased: Purkinje's phenomenon. The retina is least readily stimulated by red, and the variations in intensity of red are recognized with the greatest difficulty. Therefore, according to Brücke, intermittent white light is perceived as greenish, because the red component in white light acts upon the retina with greater difficulty. Yellow, on the contrary, acts more powerfully, and then follows blue. In weak illumination green possesses the greatest brightness; then come yellow, blue, red. In strong illumination the analogous succession of colors is: yellow, red, green, blue.

While, therefore, light of varying rapidity of vibration produces in the eye the sensation of different colors, the amplitude of vibration (height of the waves) determines the intensity of the visual impression, just as the loudness of a note depends upon the amplitude of the vibrations of the sounding body. All of the colors are united in sunlight, and their simultaneous action on the retina produces the sensation that is designated as the sensation of white. If the colors of the spectrum obtained by means of a prism are again united, white light is once more produced. If the retina is not influenced by vibrations of the luminiferous ether, all sensation of light and of color is absent; but this cannot be designated black. It is rather the absence of sensation, as is the case also when a ray of light falls on the skin of the back. The skin perceives neither black nor any light-sensation whatever.

When a colored object is illuminated by a monochromatic light, it gives no impression of color. If a colored object is illuminated by two lights of different color, the color-impression appears best if one of the lights contains those rays that would be most strongly reflected by the color of the object, and the other light, on the contrary, contains such rays as stand closer to the color in the solar spectrum than does the complementary color.

The recognition of the impression of light requires lesser intensity of action than does that of a color. If the colored object is exceedingly small, if it is poorly illuminated, or if it is seen for only a short time, it appears colorless. The different colors show different gradations of activity in this respect, red furnishing the most unfavorable conditions.

Simple colors, for example those of the spectrum, are produced by the action of a definite number of oscillations upon the retina. *Mixed colors* are produced when the retina is stimulated by two or more simple colors, either simultaneously or in rapid alternation. The most complex mixed color is white, which is composed of all the simple colors of the spectrum. *Complementary colors* comprise any two whose admixture produces white. For the sake of completeness *contrast-colors* must be mentioned, as they are closely related to the complementary colors. They comprise any two colors that when mixed produce the tone of the general illumination that prevails at the time. In the blue light of the sky, the two contrast-colors must yield bluish white, in bright gaslight, yellowish white. In pure white illumination the contrast-colors are, naturally, the same as the complementary colors.

Methods of Mixing Colors.—(1) Two solar spectra are projected upon a screen, and the colors to be mixed are superposed. (2) The observer looks obliquely through a vertical glass plate at a color lying behind it. A second color is placed in front of the plate, so that its image is reflected by the glass into the eye of the observer. In this way transmitted light from one color and reflected light from the other enter the eye at the same time. (3) By means of the "color-top" small sectors of various colors are rotated rapidly on a disc. By rapid rotation the impressions produced by the individual colors are united to produce a mixed color. If the rotating disc which yields for example, a white color from the mixture of the prismatic colors, is viewed in a rapidly rotating mirror, the individual components of the white reappear. (4) Two different colored glasses are placed before the little holes in the cardboard used in Scheiner's experiment (p. 835, Fig. 284). The colored rays of light passing through the holes are united on the retina for the production of the mixed color.

Investigation has shown that the following colors of the spectrum are complementary, that is two together produce white: red + greenish blue; orange + cyan blue; yellow + indigo blue; greenish yellow + violet. Green has the compound complementary color purple. All of the mixed colors may be determined from the following table. At the head of the vertical and horizontal columns are placed the simple colors; the mixed color is found at the intersection of the respective horizontal and vertical columns:

	VIOLET.	INDIGO.	CYAN BLUE.	BLUISH GREEN.	GREEN.	GREENISH YELLOW.	YELLOW.
Red, ...	Purple.	Dark rose.	Whitish rose.	White.	Whitish yellow.	Golden yellow.	Orange.
Orange, ...	Dark rose.	Whitish rose.	White.	Whitish yellow.	Yellow.	Yellow.	
Yellow, ...	Whitish rose.	White.	Whitish green.	Whitish green.	Greenish yellow.		
Greenish yellow,...	White.	Whitish green.	Whitish green.	Green.			
Green	Whitish blue.	Watery blue.	Bluish green.				
Bluish green,......	·Watery blue.	Watery blue.					
Cyan blue,........	Indigo.						

Observations upon the mixing of colors have yielded the following results: (1) When two simple, but not complementary colors are mixed, they produce a color-sensation that may be represented by a color situated between them in the spectrum, to which a certain amount of white is added. Therefore, any color-mixture may be produced by a color of the spectrum plus white. (2) The less white the colors contain, the more saturated they are said to be; the more white they contain, the less saturated do they appear. The degree of saturation of a color diminishes with the intensity of the illumination. In a steadily increasing illumination, the colors become more nearly white, and at the same time they lose more and more their specific character; for example, in a bright light, yellow readily passes into white.

As the pigment of the macula lutea partly absorbs certain colored lights, an explanation is afforded for the fact that colors seen by the macula alone have a different appearance from those seen by other portions of the retina.

Since the time of Newton, attempts have been made to construct a so-called geometric color-chart, on which any mixed color can be found, according to the principle of construction of the center of gravity. The accompanying figure shows such a color-chart: white is placed in the middle, and different colors are represented at points in the curve surrounding it. From the center (white) to each of these points of the curve, lines may be drawn and each color may be conceived as applied to the line in such a manner that, commencing at white, there is the lightest tint, and then gradually more saturated ones, until, finally, at the point of the curve designated by the name of the color the latter appears

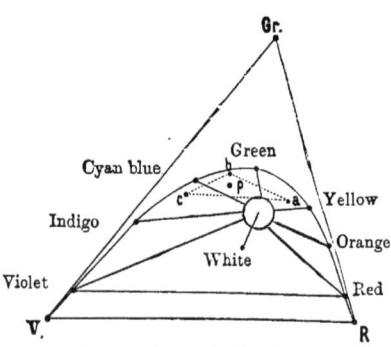

FIG. 300.—Geometric Color-chart.

in a pure saturated form. Between violet and red, their mixed color, purple, is indicated. In order to determine from this chart the mixed color produced by any two colors of the spectrum, the points of these colors should be connected by a straight line; in each point weights may be conceived as placed corresponding to the units of intensity of these colors. Then the position of the center of gravity of the two in the connecting line indicates the situation of the mixed color in the color-chart. The mixed color of two spectral colors lies in the straight line that connects the two color-points on the color-chart. It is easily seen, further, that the impression of the mixed color corresponds to an intermediate spectral color mixed with white. The complementary color of any spectral color is found by drawing a line from the point of this color through white, until it intersects the opposite edge of the color-chart. The point of intersection gives the complementary color. If pure white is to be made from a mixture of two complementary colors, the color lying nearest white on the connecting line must be especially strong, for then only would the center of gravity of the connecting line be situated in the point white.

The color-chart permits, further, of the determination of the mixed color produced by three or more colors. For example, the colors designated by the points a (pale yellow), b (comparatively saturated greenish blue), and c (comparatively saturated blue) are chosen for mixing. In the three points are placed weights that represent the intensities of the colors, and the center of gravity of the triangle a b c is determined; this will lie at p. It is obvious, however, that this mixed impression, whitish-greenish-blue can be produced also by the color greenish-blue + white (according to rule 1), for p can just as well be the center of gravity of two weights that lie on the line connecting greenish-blue and white.

Around the color-chart a triangle V Gr R may be drawn, enclosing the figure on all sides. The three primary or fundamental colors, red, green and violet, lie in the angles of the triangle. It is evident that every colored impression, that is any point of the color-chart, may be determined by placing at the angles of the triangle weights corresponding to the intensities of the primary colors, so that the point of the color-chart, consequently the mixed color sought, is the center of gravity of the triangle, with its angles thus weighted. In the production of the mixed color, the intensity of the three primary colors must be represented in the same proportion as the weights.

Various theories have been suggested to explain color-perception:

1. According to one theory, the elements of the retina, while uniform in type, are affected in different ways by the variously colored lights (vibrations of the ether of different wave-length, rapidity of vibration, and refractive exponent).

2. The theory of Thomas Young and Hermann v. Helmholtz assumes the existence of three different terminal elements in the retina, corresponding to the primary colors: Stimulation of the first kind produces the sensation of red, of the second green, and of the third violet. The red-perceiving elements are affected most strongly by the light of greatest wave-length (red rays), the green-perceiving by the light of medium wave-length (green rays), the violet-perceiving by the light of shortest wave-length (violet rays). For the explanation of many phenomena it must be assumed that each spectral color excites all forms of fibers, some slightly, and others strongly. If it be conceived that the spectral colors placed in their natural order in a horizontal direction in Fig. 301 (from red to violet), then the three curves shown may represent the amount of excitation of the three kinds of retinal elements: the continuous curved line that of the red-perceiving, the dotted line that of the green-perceiving, and the interrupted

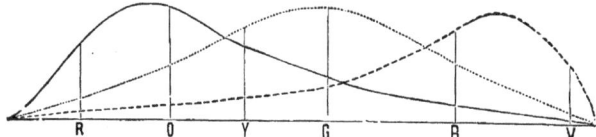

R   O   Y   G   B   V

FIG. 301.—Diagrammatic Representation of the Young-Helmholtz Color Theory.

line that of the violet-perceiving. Pure red excites the red-perceiving elements strongly, but the other two forms slightly (expressed by the heights of the ordinates erected at R), and the sensation of red results. Pure yellow excites the red-perceiving and the green-perceiving elements with moderate activity, the violet elements less actively, and the sensation of yellow results. Pure green excites the green-perceiving elements strongly, much less so the two other forms, and the sensation of green results. Pure blue excites the green and violet elements with moderate activity, the red element slightly, and the sensation of blue results. Violet excites the corresponding elements strongly, the others slightly, and the sensation of violet results. Stimulation of any two elements produces the impression of a mixed color; while an equal stimulation of all gives rise to the sensation of white. This hypothesis of the Young-Helmholtz theory, affords, in fact, a simple and clear survey and explanation of the phenomena of the physiological doctrine of color. The theory is a development of the doctrine of Joh. Müller as to the specific energy of the nerve-fibers. The findings in the structure of the retina, moreover, have been adapted to the theory. Accordingly, the cones alone are supposed to be the terminal apparatus for color-perception. In cases of congenital color-blindness the cones appear to be absent in the peripheral portions of the retina. The presence of longitudinal striations in their outer segments is regarded as proving that they represent multiple terminal end-organs. The degree of sensitiveness of any part of the retina to color is proportionate to the number of cones. It is most developed in the macula lutea, which has only cones; much less so further away from the macula, being lost, finally, in the periphery. The rods are supposed to be concerned only with the power of distinguishing between quantitative sensations of light. According to v. Kries, they are especially adapted for vision in poor illumination.

3. In explanation of visual perception, Ewald Hering proceeds upon the proposition that what reaches consciousness as a visual perception is the psychic expression of the metabolic change in the visual substance, that is in those nervous elements that are concerned in the act of vision. This substance, like every other organic substance, undergoes decomposition or dissimilation during the

process of metabolic change; while during rest it must be renewed, or undergo assimilation. For the perception of white (light) and black (dark) Hering assumes two different chemical processes in the visual substance, namely that the sensation of white or light corresponds with dissimilation (decomposition), that of black (dark) with assimilation (reconstruction) of the visual substance. Accordingly, the different degrees of distinctness or intensity with which these two sensations appear in the various transitional shades from pure white to deepest black, or, in other words, the proportions in which they appear to be mixed (gray), correspond to the relative intensity of these two psychophysical processes. Consequently the consumption and the restoration of the visual substance are the primary processes in the perception of white and black. The consumption of the visual substance in the perception of white is the result of the stimulation of the vibrating ether-waves, and the degree of the perception of brightness is proportional to the amount of material consumed. The restoration of the material produces the sensation of black; the more intensely this takes place the deeper is the sensation. The consumption of the visual substance in one place evokes greater reproduction in the neighborhood. Each process influences the other simultaneously and conjointly. In this way a physiological explanation is provided for the phenomenon of contrast (see p. 864), for which the older view could offer only a psychical interpretation.

In an entirely analogous manner, color-sensation is regarded as a sensation of decomposition (dissimilation) and of reconstruction (assimilation). In addition to white, red and yellow are the expression of decomposition; green and blue, on the other hand, represent the sensation of reconstruction. The visual substance is, thus, subject to three different forms of chemical change, or metabolism. The colored contrast-phenomena or after-images are thus explained. The black-white sensation may, further, be combined with all of the colors. It gives a dark or a light tone to each color-sensation, so that there are no absolutely pure colors. There are, thus, three different constituents of the visual substance: that which is sensitive to black-white (colorless), that sensitive to blue-yellow, and that sensitive to red-green. All rays of the visible spectrum decompose the black-white substance, but the different rays do so in different degrees. Only certain rays, on the contrary, decompose the blue-yellow, or the red-green substance; others reproduce them; and still others have no effect whatever. Mixed light appears colorless when it causes an equally strong dissimilation and assimilation in the blue-yellow and the red-green substances, so that the two processes neutralize each other, and the action upon the black-white substance alone appears. Two objective kinds of light, which together yield white, are consequently not to be considered as complementary, but as antagonistic, for they do not combine to form white, but, as antagonists, allow it to appear of itself because each neutralizes the effect of the other.

The weakness of the Young-Helmholtz color-theory lies in the fact that it assumes the existence of only one kind of irritability, stimulation and exhaustion (corresponding to Hering's dissimilation), and that it ignores the antagonistic relations of certain light-rays to the eye. Therefore, it does not recognize that white is produced from complementary colors by the neutralization of their action in the colored visual substances, but by their supplementing each other.

In applying Hering's theory to color-blindness, it must be assumed that the red-blind individual has no red-green visual substance. His solar spectrum contains only two partial spectra: the black-white and the yellow-blue. The green part appears colorless to him; the rays from the red portion are visible to the extent that the yellow and the white sensations that they arouse are strong enough to stimulate the retina sufficiently. He divides his spectrum into a yellow and a blue half. The violet-blind individual has no yellow-blue visual substance. His spectrum contains only two partial spectra: the black-white and the red-green. In cases of complete color-blindness both the yellow-blue and the red-green visual substances are absent, and the individual has only the sensation of light and dark. The sensibility to light and the length of the spectrum are preserved; the brightest area is in the yellow, just as in the normal eye.

COLOR-BLINDNESS: ITS PRACTICAL IMPORTANCE.

By *color-blindness* (*dyschromatopsia*) is understood a pathological condition. as the result of which the affected individuals are unable to recognize certain colors. It was recognized by Tuberville in 1684, and by Huddart in 1777, but it was first accurately described in 1794 by the physicist Dalton, who was himself red-blind. The designation color-blindness was given the condition by Brewster.

The adherents of the Young-Helmholtz theory assume the following varieties of color-blindness, corresponding to paralysis of the three color-perceiving elements of the retina: (1) *Red-blindness;* (2) *green-blindness;* (3) *violet-blindness.* In addition there is the most pronounced form—*total color-blindness.*

The adherents of E. Hering's color-theory distinguish the following varieties:

1. *Complete Color-blindness (Achromatopsia).*—The spectrum appears achromatic, the green-yellow portion is the brightest, and the adjacent parts on either side are darker. A colored painting appears like a photograph or an engraving. Occasionally the different degrees of light-intensity are recognized as one shade of color (for instance yellow), which cannot be compared with any other color. O. Becker and v. Hippel observed cases of unilateral, congenital complete color-blindness, in which the other eye had normal color-perception.

2. *Blue-yellow Blindness.*—The spectrum is dichromatic, consisting only of red and green; the blue-violet end of the spectrum usually is greatly shortened. In pure cases, only the spectral red and green are recognized correctly (Mauthner's erythrochloropia), not however the other colors. This has been observed also unilaterally.

3. *Red-green Blindness.*—The spectrum is dichromatic; yellow and blue are recognized correctly, while violet and blue are both seen as blue. The perception of red and green is absent. In this category the following types are further distinguished: (a) *Green-blindness* or red-green blindness without shortening of the spectrum (Mauthner's xanthocyanopia), in which light green and dark red are confounded. In the spectrum yellow passes directly into blue, or at most a band of gray lies between the two. The maximum of brightness lies in the yellow. This defect may be unilateral; it is often hereditary. (b) *Red-blindness* (or red-green blindness with shortening of the spectrum; also designated *Daltonism*), in which light red is confused with dark green. The spectrum consists of yellow and blue, but the yellow lies in the orange, and the red end of the spectrum is colorless or even dark. The greatest illumination, as well as the boundary between yellow and blue, lies more to the right. Between these two forms there are transitions. According to Hering the cause of the difference resides in a variation in the amount of absorption by the macula lutea of the rays of short wave-length.

4. *Incomplete color-blindness,* or diminished color-sense, is that condition in which the acuteness of color-perception is lowered, so that colors are recognized, for example, only in objects of considerable size or only at near range; also, on addition of white they are no longer perceived as such. A certain degree of this form is frequent, in so far as many are unable to distinguish between greenish blue and bluish green.

Acquired color-blindness occurs also in connection with diseases of the retina, and inflammation and atrophy of the optic nerve, with beginning tabes, with cerebral diseases and with intoxications (tobacco, alcohol, etc.). Green-blindness appears first, and is followed shortly by red-blindness. The peripheral zone of the retina suffers before the central portion. In cases of hysteria and of epilepsy there may be intermittent attacks of color-blindness; and also in hypnotized individuals.

The retina may be made temporarily color-blind for a given color by intense action of the color. Prolonged gazing into the dark-red, setting sun causes scarlet to appear black.

Finally, the remarkable observation of H. Cohn must be mentioned; he found that color-blindness in several individuals disappeared temporarily on heating the eyeball. Holmgren found that 2.7 per cent. of persons examined were color-blind, most of them being red-blind and green-blind; only a few were violet-blind.

The examination of the power of color-perception in the normal retina, best made with the Aubert-Förster perimeter, has revealed the surprising fact that complete color-perception is present only in the center of the visual field. Around this lies a middle zone, in which only blue and yellow are perceived, and in which there is, therefore, red-blindness. Beyond this zone there is, finally, a peripheral girdle, in which there is complete color-blindness. The red-blind individual is distinguished, therefore, from the normal by an absence of the central area of the normal visual field, which is included in the middle zone. The visual field of the green-blind individual is distinguished from that of the normal by the fact that its peripheral zone corresponds to the intermediate and peripheral zones of the normal eye. In the violet-blind, on the contrary, the normal peripheral zone is absent. Incomplete color-blindness of these two types is characterized by a uniformly contracted central field.

In the presence of hyperesthesia of the optic nerve resulting from cerebral conditions, there is, curiously, a widening of the normal color-limits toward the periphery.

When colored objects are exceedingly small, and illuminated for only a short time, the normal eye first fails to perceive red. It appears, therefore, that a comparatively strong stimulus is required for the perception of red. The observation of Brücke's, that rapidly intermittent white light appears green, is also in favor of this view, because the short duration of the stimulus is not capable of exciting the red-perceiving elements of the retina.

To Holmgren belongs the credit of having shown the necessity for examining all railroad-officials and all pilots as to the trustworthiness of their color-sense, as the correct recognition of red and green signal-lights is impossible for a color-blind person.

Method of Examination.—Holmgren, in conjunction with Seebeck, chooses embroidery-wool as the simplest material, in skeins of at least five shades each of red, orange, yellow, greenish-yellow, green, greenish-blue, blue, violet, purple, rose, brown, gray; it is best to have several different shades of the various colors. In the examination, one strand of this yarn (for example light green or rose) should be selected and put to one side. That color is chosen for which the individual is to be especially tested; and he is then requested to pick out the strands whose colors resemble most closely that of the sample, and place them by its side. According to the way in which he performs this task a judgment is reached as to his color-sense. A more accurate determination is made by means of the spectrum.

Macé and Nacati have measured the visual acuity for a small object when illuminated by different portions of the spectrum. They compared with the results of their investigation the observations on red-blind and green-blind individuals. It was shown that red-blind persons find green light much brighter than do normal persons. In the green-blind there is an excessive sensitiveness to red and violet. It seems, therefore, that what color-blind individuals lack in perception-power for one color, they gain for other colors. They possess also a greater power of distinguishing variations in brightness.

TIME-RELATIONS OF RETINAL STIMULATION.

POSITIVE AND NEGATIVE AFTER-IMAGES. IRRADIATION. CONTRAST.

As with irritation of every other nervous apparatus, a definite, though short time elapses after the entrance of the rays into the eye, before the visual effect is manifest, whether in the form of conscious perception, or a reflex effect on the iris. The intensity of the impression, here also, will depend essentially upon the irritability of the retina and the other nervous structures. If the visual impression continues for some time with the same intensity, the stimulation, after having reached its culminating point, soon diminishes, at first rapidly, then more slowly. If the light-stimulation of the retina is suddenly removed after it has continued for some time, the retina remains for a time in an excited condition, which is the more intense and persistent the stronger and longer the light-stimulation, and the more sensitive the retina. After every visual perception, therefore, especially when it has been quite bright and sharp, a so-called *after-image* persists. There is recognized, in the first place, the *positive after-image*, which persists with similar brilliancy and color.

A light-stimulus of short duration excites first a sensation of light, which lasts longer than the stimulus; thereupon a negative after-image appears (in the complementary color) and lasts $\frac{1}{4}$ of a second, being the more distinct the shorter the light-stimulation. Then follows the positive after-image, whose duration increases with the intensity of the illumination.

"That the impression of any image in the eye persists for some time we know as a physiological phenomenon; excessive duration of such an impression, however, may be considered pathological. The weaker the eye the longer does the image remain in it. The retina does not recover so quickly, and the effect may be looked upon as a sort of paralysis. This is not to be wondered at in the case of dazzling images. If one looks directly at the sun, he may carry the image about for several days. The same is also relatively true of images that are not dazzling. Büsch relates of himself that an engraving remained before his eye, completely with all its details, for seventeen minutes" (*Goethe*).

Experiments and Apparatus for Demonstrating After-images.—(1) The appearance of a ring of fire on rapid rotation of a coal. (2) The thaumatrope of Paris: a pasteboard card contains, for example, on one side the picture of a torso-statue, on the other side the picture of the remaining portions drawn in appropriate positions. If the card be rotated so that the two surfaces in alternation are rapidly turned toward the observer, the statue appears complete. (3) The phanakistoscope or the stroboscopic discs. Objects are drawn on a disc or cylinder in succession, so that the drawings represent successive details of a continuous movement. On rapid rotation of the disc, the observer, looking through a small opening, sees the moving images pass before the eye, each phase rapidly replacing the preceding. As the impression of each image remains until the following one takes its place, one and the same figure seems to go through the successive movements continuously. The apparatus, at present popularized by Anschütz in the form of the zoetrope (perfected by Edison as the kinematograph or kineto-scope), was not discovered by the two investigators mentioned, in 1832, as is generally supposed; but it was described by Cardanus as early as 1550. It may be used also scientifically for the representation of certain movements: as, for example, of spermatozoids and ciliated epithelial cells; likewise the movements of the heart and of walking may be instructively shown and analyzed. (4) The color-top contains in the sectors on its surface the colors that are to be mixed. As the color of each sector causes a stimulation of the retina, lasting throughout the revolution of the top, all of the colors must be seen simultaneously, and be perceived as a mixed color.

Occasionally, especially if the retinal excitation is of considerable duration and intensity, a *negative after-image* appears, instead of the positive. The former is characterized by the fact that bright portions of the object appear dark, and the colored portions in corresponding contrast-colors.

Examples of Negative After-images.—After gazing for a long time at a brightly illuminated white window, and then closing the eyes, there results the impression of a window with bright cross lines, and dark panes. Colored negative after-images are shown beautifully by Nörrenberg's apparatus: the eye is fixed for some time on a colored surface, such as a yellow card, in the center of which is pasted a small blue square. Suddenly a white screen is dropped in front of the card, and the white surface then appears bluish, with a yellow square in the center.

The usual explanation of the dark negative after-images is that the retinal elements are so fatigued by the light that they become less irritable for a time, so that in these portions of the retina the light can be only faintly perceived, and darkness, therefore, must prevail. Hering explains the dark after-images as resulting from the process of assimilation of the black-white visual substance.

For the explanation of colored after-images the Young-Helmholtz theory assumes that by the action of the color, for example red, the retinal elements for this particular color are paralyzed. If, now, the eye looks at a white surface, this mixture of all colors appears as white minus red, that is green (the contrast-color, which in bright daylight lies close to the complementary color). According to Hering, this contrast-colored after-image is the result of the assimilation of the corresponding colored visual substance, in the case cited, of the "red-green." From the beginning of a momentary illumination until the appearance of an after-image 0.344 second elapses.

Not infrequently, after intense stimulation of the retina, positive and negative after-images alternate, until they gradually fuse. Thus, after gazing at the dark-red, setting sun, one sees alternately discs of red and green. In the peripheral portions of the retina, the contrast-phenomena undergo some modification on account of the partial color-blindness that exists in these areas.

Irradiation is the term applied to certain phenomena that are the result of false estimates of visual sensations due to inexact accommodation. If, for example, the edges of objects are thrown upon the retina in diffusion-circles, the mind has a tendency to add the blurred edge to that part of the image which is the most prominent. Bright things appear larger and more prominent than dark ones, and an object itself, without reference to brightness or color, appears more prominent than the background. In the exercise of sharp accommodation the phenomenon of irradiation is not present.

"A dark object appears smaller than a bright one of the same size. If one look at the same time at a white circle on a black background, and a black circle on a white background, both of the same diameter, at some distance from the eye, the latter will appear to be one-fifth smaller than the former. If the black circle be made that much larger, both will appear of the same size. Tycho de Brahe observed that the moon appeared one-fifth smaller when in conjunction (dark) than when in opposition (full moon). The first quarter of the moon appears to belong to a larger disc than the dark part adjoining it which can often be distinguished at the time of the new moon. Dark clothing makes persons appear much thinner than light clothing. Lights seen behind an edge make an apparent indentation therein. A ruler held before a candle-light seems to be notched. The rising and setting sun appears to make a depression in the horizon" (*Goethe*).

By **simultaneous contrast** is understood, in the first place, the phenomenon that when light and dark parts are present in an image at the same time the light (white) parts always appear the more intense the greater the absence of light from the immediate vicinity, consequently the darker the latter; and conversely the light parts appear the less bright the greater the degree in which white tones are present around them. The analogous phenomenon in connection with colored pictures belongs in the same category: a color in a picture appears the more intense the more completely this color is absent from the immediate neighborhood, that is the more the neighborhood contains the tones of the contrast-color. The simultaneous contrast arises thus from two impressions simultaneously existing side by side and affecting two different but adjoining portions of the retina.

Examples of the contrast for light and dark: (1) If a white grating be viewed upon a black background, the points of intersection of the white lines appear darker, because the least amount of black is present in their vicinity. (2) If a point in a narrow strip of dark-gray paper be viewed against a dark background and a large piece of white paper is then inserted between the two, the strip appears much darker than before; if the white paper be removed, the strip immediately appears brighter. (3) The following is also a most instructive experiment: If, first, a grayish-white surface—for example, the ceiling of a room—be looked at with both eyes, and then, after a time, a paper tube blackened on the inside, about as long as the hand, and a finger's breadth in diameter, be brought before one eye: the part of the ceiling seen through the tube appears as a round, bright spot.

Examples of contrast for colors: (1) If a piece of gray paper be placed on a red, yellow or blue background, it appears immediately in the contrast-color, respectively green, blue or yellow. The appearance is still more distinct, if the whole is covered quickly with transparent tracing paper. Under similar conditions printed characters on a colored background appear in the complementary color. (2) An air-bubble in the deeply stained field of a thick microscopical preparation appears in an intense contrast-color. (3) On a rotating white disc are pasted four green sectors, each of which is interrupted in its center by a narrow band of black, concentric with the disc. On rotation of the disc, this ring appears red, and not gray. (4) If a grayish-white surface be looked at with both eyes and a tube about the length and diameter of a finger, made of transparent, colored oiled paper, through the walls of which light can pass, be placed in front of one eye, the part of the white surface seen through the tube appears in the contrast-color. The experiment also shows beautifully the contrast in the intensity of illumination. (5) A piece of white paper, with a round, black spot in the middle, when seen through a blue glass, appears blue with a black spot. If a white spot of the same size on a black background be placed in front of the glass, so that its reflected image covers exactly the black spot, it appears in the contrast-color, yellow. (6) The *colored shadows* also belong to the simultaneous contrasts. "Two conditions are necessary for the production of colored shadows—first, that the light casting the shadow shall in some manner color the white surface; and second that a second light shall illuminate the shadow to a certain degree. A short burning candle is placed on a white paper, in the twilight; between them and the diminishing daylight a lead-pencil is placed vertically so that the shadow thrown by the candle is illuminated but not extinguished by the feeble daylight; the shadow will appear of a beautiful blue. That this shadow is blue will be observed at once: but it is only by close observation that one can convince himself that the white paper acts as a reddish-yellow surface, through the luster of which the blue color is conveyed to the eye. One of the prettiest instances of colored shadows may be seen during the full moon. The light of a candle and that of the moon can be exactly equalized. Both shadows can be made of equal strength and distinctness, so that the two colors completely balance. A board is exposed to the light of the full moon, with the candle a little to one side, and an opaque body is held at a suitable distance in front of the board. A double shadow results, that thrown by the moon and illuminated by the candle-light appearing of an intense reddish-yellow color, while that thrown by the candle and illuminated by the moon appearing of a beautiful blue. Where the two shadows coincide and unite to form one a black shadow results (*Goethe*). (7) *The colored reflections* are the reverse of the colored shadows. If a piece of silverware be placed near a window, in the twilight, and the light from a candle be allowed to fall on it at the same time, the reflected image of the flame appears yellowish, that of the lessening daylight decidedly blue. (8) A piece of white paper is placed on the table and above it, separated by a horizontal line, a piece of black paper. Now a vertical, black strip is pasted on the white paper and on the black paper a white strip. If these strips are seen through a birefringent spar-prism, each will be doubled, and possess a gray color, because the strip is composed of white and black mixed. The strips on the dark background, however, appear brighter, and those on the white ground darker. Likewise, in an analogous way, with colored strips on a differently colored background the experiment shows the contrast-colors beautifully. Landois has found this excellent experiment especially convincing if the objects are covered with translucent tracing paper.

Some have tried to explain these phenomena as errors of judgment; in other words, when different impressions act simultaneously, the judgment is so deceived that if the action takes place in one position, it will have an extremely slight effect in the neighborhood. Thus, if light affects one portion of the retina, judgment wrongly assumes a slight illumination of the neighboring parts of the retina. The same would be true of colors. The phenomena are, however, much more correctly explained by Hering as true, physiological processes. Partial stimulation by light affects not only the parts so acted upon but also the surrounding retina; the directly stimulated portion by increased dissimilation, the (indirectly stimulated) vicinity by increased assimilation, in such a way, that the latter increase is most pronounced in the immediate vicinity of the illuminated spot, and decreases rapidly with the distance from it. As a result of the increase in assimilation at the part not occupied by the image of the object, the diffused light is ordinarily not perceived. As the increase in assimilation is greatest in the immediate neighborhood of the illuminated spot, the perception of this relatively strong diffused light is largely made impossible.

On looking for a long time at a dark or a bright object, or at a colored one (for example red), and then allowing the contrast-effects to appear on the retina, respectively bright or dark, or the contrast-color (green), these appear especially intense. This phenomenon has been designated *successive contrast*. In this connection the negative after-images obviously take part at the same time.

55

OCULAR MOVEMENTS AND OCULAR MUSCLES.

The spherical eyeball is capable of extensive and free movement in the correspondingly excavated cushion of fat in the orbit, like the head of a bone in the corresponding socket of a freely movable arthrodial joint. The motion is limited, in the first place, by the attachment of the muscles and in such a manner that in the action of one muscle its antagonist, acting like a rein, serves to limit the movement; and secondly by the insertion of the optic nerve. The soft elastic orbital pad on which the eyeball ₋ests may itself be moved backward and forward, so that the eyeball must follow these movements.

Protrusion of the eyeball takes place: (1) As a result of marked distention of the blood-vessels, especially of the orbital veins, when there is an obstruction to the outflow of venous blood (for example in the head after execution by hanging). (2) As a result of contraction of the unstriated muscle-fibers in Tenon's capsule, in the sphenomaxillary fissure, and in the eyelids, which are innervated by the cervical sympathetic. (3) As a result of voluntary, forcible opening of the palpebral fissure, because the pressure of the lids from before backward is diminished. (4) As a result of the action of the oblique muscles, which pull the eye inward and forward. If the superior oblique is made to contract, while the palpebral fissure is forcibly widened the eyeball may protrude about 1 mm. *Pathological* protrusion of the eyeball (especially caused by 2 and 1) is called *exophthalmos*.

Conversely, *retraction of the eyeball* is caused: (1) By forcible closure of the palpebral fissure. (2) By an empty condition of the retrobulbar blood-vessels, diminished succulence or disappearance of the orbital tissue. (3) In dogs, section of the cervical sympathetic causes recession of the eyeball. The smooth musculature of Tenon's capsule probably prevents the four rectus muscles from pulling the eye backward unduly. Many animals possess a special retractor muscle of the eyeball, for example amphibians, reptiles, and many mammals; the ruminants have, in fact, four of them.

The ocular movements are almost always accompanied by similar movements of the head, especially in looking upward, less in looking laterally and least in looking downward.

Difficult investigations into the ocular movements have been carried out especially by Listing, Meissner, v. Helmholtz, Donders, A. Fick, E. Hering and others.

Orchansky placed a closely fitting hemisphere against the eyeball, inside of the conjunctival sac (with an opening cut for the pupil), and in this way could observe the simple and combined movements, and also register them graphically by means of a writing lever.

All movements of the eyeball take place about its center of rotation (Fig 302, O), which lies 1.77 mm. behind the center of the visual axis, or 10.957 mm. from the vertex of the cornea. In order to study the movements more exactly, certain fixed data must be determined. Three axes, intersecting at right angles in the center of rotation, are conceived to be erected, namely: (1) The *visual axis* (S S_1) or sagittal axis of the eyeball, which connects the center of rotation with the fovea centralis, and is prolonged forward in a straight line to the vertex of the cornea. (2) The *transverse* or *horizontal axis* (Q Q_1), being the direct prolongation outward of the line connecting the centers of rotation of the two eyes (naturally at a right angle with 1). (3) The *vertical axis*, erected perpendicularly to 1 and 2 at the center of rotation. These three axes form a physical system of coördinates. Further, a similar, but always fixed system of axes may be conceived to be erected in the orbital cavity, the center of which coincides with the center of rotation of the eyeball. In the position of rest (primary position) of the eye, the three axes of the eye coincide exactly with the axes of the orbital system. If, however, the eyeball is moved, two or three of the axes cease to coincide, and must form angles with the fixed orbital system of axes.

For further study, partly also for further determinations, three planes may be imagined as passing through the eye, each of whose positions is determined by two of the axes. (1) The *horizontal plane* of division cuts the eyeball into an upper and a lower half; it is determined by the visual axis and the transverse axis. In its passage through the retina, it forms the horizontal line of division of this membrane, and it cuts the tunics of the eye in the horizontal meridian. (2) The *vertical plane* of division cuts the eye into an inner and an outer half; it is determined by the visual and vertical axes. It intersects the retina its vertical line of division, and the periphery of the eyeball in the vertical meridian of the eyeball. (3) The *equatorial plane* divides the eye into an anterior and a posterior half. Its position is determined by the vertical and transverse axes. It cuts the sclera in the equator of the eyeball. The horizontal and vertical lines of division of the retina intersect in the fovea centralis, and divide the retina into four quadrants.

v. Helmholtz has further, introduced the following terms for designating the positions of the eyes: the *line of fixation* is the straight line connecting the center of rotation with the fixed point in the external world. A plane passed through the lines of fixation of both eyes is called the *plane of fixation*. The *base line* of this plane is the line joining the two centers of rotation (consequently the transverse axis). A sagittal plane may further be imagined as passing through the head and divides it into a right and a left half. This plane will bisect the base line of the plane of fixation, and if prolonged anteriorly will intersect the plane of vision in its median line. The fixation-point of the eye may (1) be raised or lowered. The field that it traverses is called the *field of fixation*. It is part of a hemisphere, the center of which is the center of rotation of the eye. Starting from the primary position of both eyes, which is characterized by the fact that the two lines of fixation are parallel and horizontal, the elevation of the plane of fixation may be determined by the angle that it makes with the plane of the primary position. This angle is called the angle of elevation of fixation. It is termed *positive* when the plane of fixation is raised (toward the forehead), and *negative* when the plane is lowered (toward the chin). (2) The line of fixation may be turned also from the primary position in a lateral direction in the plane of fixation, that is toward the median line, or away from it. The amount of this lateral movement is measured by the *angle of lateral rotation*, that is the angle the line of fixation makes with the median line of the plane of fixation. The angle is called positive when the posterior extremity of the line of fixation moves toward the right, and negative when the extremity moves toward the left.

In accordance with the foregoing preliminary considerations the eyes may assume the following positions as the result of their movements:

1. *Primary position*, in which both lines of fixation are parallel, and the plane of fixation is horizontal. In this case the three axes of the eye coincide with the three fixed axes erected in the orbital cavity. 2. *Secondary positions* result from simple movements of the eyes from the primary position. There are two different kinds of secondary positions, namely: (a) The lines of fixation are parallel, but are directed upward or downward. The transverse axis of each eye remains the same as in the primary position. The deviation of the other two axes is expressed on the line of fixation by the size of the angle of elevation (as has already been mentioned). (b) The second kind of secondary position is produced by convergence or divergence of the lines of fixation. Here the vertical axes about which the lateral rotation is effected remain the same as in the primary position. The other axes form angles. The amount of the deviation (as already noted) is expressed by the angle of lateral rotation. The eye can be turned from the primary position outward 42°, inward 45°, upward 54°, and downward 57°. 3. A *tertiary position* is one assumed by the eye when the lines of fixation are convergent, and at the same time are inclined upward or downward. None of the three axes coincides now with its situation in the primary position. The exact direction of the lines of fixation is determined by the size of the angles of elevation and lateral rotation. In connection with tertiary

positions another important point comes into consideration, namely that the eyeball rotates at the same time about its line of fixation and its axis. As the iris rotates ·about the line of fixation as a wheel around its axle, these movements of rotation, which are always associated with the tertiary positions, are termed *wheel-movements*. Every oblique movement can be considered as composed of a rotation (1) about the vertical axis, and (2) about the transverse axis; or it may be conceived as a rotation about a single, constant axis, lying between these two axes, and passing through the center of rotation of the eye, perpendicular to the primary and the secondary direction of the visual axis (line of fixation). The amount of the wheel-movement (circular rotation) is measured by the angle that the horizontal line of division of the retina forms with the horizontal line of division of the retina when the eyes are in the primary position. This angle is positive when the eye has turned in the same direction as the hand of a clock that it observes, that is, when the upper end of the vertical line of division of the retina deviates toward the right.

According to Donders the angle of rotation increases with the angles of elevation and lateral rotation; it may increase to more than 10°. With equally great elevation or depression of the plane of fixation, the rotation is stronger the greater the elevation or depression of the line of fixation.

In looking upward in the tertiary position, the upper ends of the vertical lines of division of the retina diverge; in looking downward they converge. If the plane of fixation is raised, the circular rotation is toward the left when the eye turns laterally toward the right; and, conversely, the circular rotation is toward the right when the eye turns laterally toward the left. If the plane of fixation is lowered, however, the eye rotates in the same direction, to the right or the left, when it turns laterally respectively toward the right or the left. Expressed otherwise: if the angles of elevation and of lateral movement have the same signs (+ or —), the rotation of the eyeball is negative, but if they have dissimilar signs, the rotation is positive. In order to make the wheel-movement visible in one's own eye, a surface divided by vertical and horizontal lines is fixed with one eye, a positive after-image is excited, and then the eye is rapidly placed in a tertiary position. The lines of the after-image then form angles with the lines of the background. As the position of the vertical meridian of the eye is important from a medical point of view, it may be again particularly pointed out that in the primary and secondary positions of the eyes the vertical meridian retains its vertical position. In looking upward and to the left, and downward and to the right, the vertical meridians of both eyes are inclined to the left; conversely they are inclined to the right in looking downward and to the left, or upward and to the right.

In the secondary positions of the eye rotations never take place. Exceedingly slight rotation of the eyes occurs, however, when the head is inclined toward the shoulder, and in the opposite direction from the inclination. It amounts to 1° for every 10° of inclination of the head.

Ocular Muscles.—The movements of the eyeball are effected by the four straight and the two oblique ocular muscles. In order to determine the action of each of these muscles, a knowledge of the plane of traction of the muscle and of the axis about which it rotates the eye is necessary. The plane of traction is found by constructing a plane through the middle of the points of origin and insertion of the muscle and through the center of rotation of the eye. The axis of rotation is always perpendicular to the plane of traction, and passes through the center of rotation.

Measurement has yielded the following data: 1. The internal rectus (Fig. 302, I) turns the eye almost exactly inward, and the external rectus (E) outward. The plane of traction lies, therefore, in the plane of the paper: Q E is the direction in which the external rectus acts, Q_1 I that in which the internal rectus acts. The axis of rotation is perpendicular to the plane of the paper at the center of rotation O, and thus coincides with the vertical axis of the eyeball. 2. The axis of rotation of the superior and inferior recti (the dotted line R. sup.— R. inf.) lies in the horizontal plane of division of the eye, but it forms an angle of about 20° with the transverse axis (Q Q_1). The line of action for both muscles is indicated by the line s i. It will be seen at once that by the action of the superior rectus the cornea must move upward and somewhat inward; or downward and inward by the action

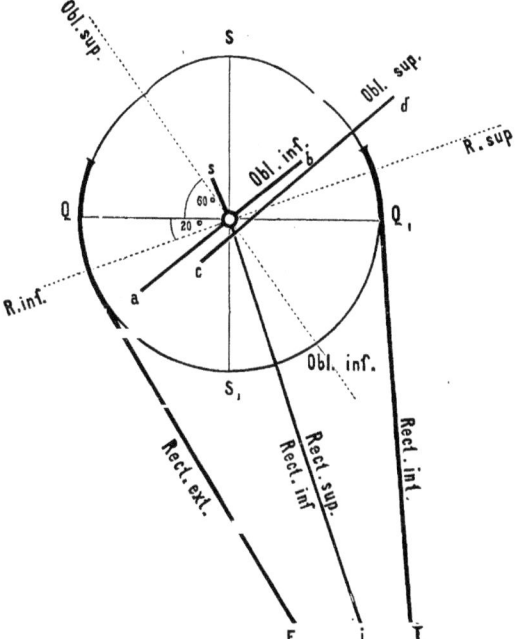

FIG. 302.—Lines of Traction and Axes of Rotation of the Ocular Muscles.

of the inferior rectus. 3. The axis of rotation of the two oblique muscles (the dotted line Obl. sup.—Obl. inf.) lies also in the horizontal plane of division of the eyeball, but it forms an angle of 60° with the transverse axis. The line of action of the inferior oblique is shown by the line a b, that of the superior oblique by the line c d. The action of these muscles causes the cornea to move respectively outward and upward, or outward and downward. These actions of the muscles are effective only when the eye is in the primary position; in every other position the axis of rotation of each muscle changes.

When the eye is at rest, the muscles are in equilibrium. Because of the greater strength of the internal recti, the visual axes converge somewhat, and if prolonged, would intersect 40 cm. in front of the eye. In the movements of the eye, only one, or two or even three muscles may be involved. One muscle acts alone in rotation of the eye directly outward and directly inward, namely the external rectus and the internal rectus respectively. Two muscles act in rotating the eye directly upward (superior rectus and inferior oblique) or directly downward (inferior rectus and superior oblique). Three muscles are employed in the diagonal directions, namely for inward and upward movement the internal rectus, the superior rectus, and the inferior oblique; for inward and downward movement, the internal rectus, the inferior rectus, and the superior oblique; for outward and downward movement, the external rectus, the inferior rectus, and the superior oblique; for outward and upward movement, the external rectus, the superior rectus, and the inferior oblique.

Ruete has imitated the movements of the eyes by means of a special model of the eyeballs and their muscles, and which he called the ophthalmotrope.

The extent of movement of the eyeball decreases with age, likewise the length of the eye. The mobility is less in the vertical direction than in the lateral; and less upward than downward. The emmetrope and the myope can move the eye further outward, the hyperope further inward. The external and internal recti act most vigorously in rotation of the eye outward; the oblique in rotation inward. One eye can be turned more strongly inward, if at the same time the other is turned outward, than if this eye also is turned inward. In near vision the right eye can be turned less toward the right, and the left eye less toward the left than in distant vision.

Both eyes are always moved simultaneously, even when one is totally blind; indeed, the ocular muscles move even after the eyeball has been extirpated. When the head is held erect, the movements proceed in such a manner that both lines of fixation (visual axes) lie in the same plane. The visual axes can diverge anteriorly to only a slight degree, but they can converge to a considerable extent. When single muscles are paralyzed, the position of the visual axes in the same plane is often disturbed (squinting). The individual can no longer direct both visual axes to one point at the same time, though each eye can be so directed singly in succession. Nystagmus also occurs in both eyes simultaneously, and in the same manner. The congenital, simultaneous movement of both eyes is termed an *associated movement*. E. Hering showed that all ocular movements are attended with a uniformity of innervation. Even with such movements in which one is apparently at rest, a movement takes place, nevertheless, of two antagonists, as may be recognized from slight to-and-fro movements.

The *nerves* of the ocular muscles are the oculomotor, the trochlear, and the abducens. The center is situated in the corpora quadrigemina, the cortical center in the angular gyrus.

BINOCULAR VISION.

The conjoint action of both eyes in the visual act has the following advantages: (1) The visual field of the two eyes is much larger than that of either one. (2) The conception of depth is facilitated, as the retinal images are obtained from two different standpoints. (3) A more accurate estimation of the distance and the size of objects is rendered possible, as a result of the estimation of the degree of convergence of the two eyes. (4) Certain errors in one eye may be corrected by the other.

If the head is fixed, an idea of the form of the common field of vision can be obtained by alternately closing one eye, and turning the other inward. It will then be seen that the field is pear-shaped, broad above, narrow below, and that the profile of the nose cuts out a portion corresponding to its size, between the upper broader portion and the lower narrow part. If a pasteboard card be held upright close to the face, the outline of the common field may be traced upon it with a pen.

SINGLE VISION. IDENTICAL RETINAL POINTS.

HOROPTER. SUPPRESSION OF DOUBLE IMAGES.

If the retinas of both eyes be considered as a pair of concave saucers placed one within the other, so that the two yellow spots, and the corresponding quadrants of the retinas coincide, all those points that correspond are called identical or corresponding points. The two meridians that separate the corresponding quadrants are known as lines of separation. The identical points are characterized physiologically by the fact that when light acts upon them at the same time, the stimulation is referred by a psychical act to one and the same place in the visual field (in a direction through the nodal point of each eye). The stimulation of the two identical points of the retina produces, therefore, only one image in the field. Hence, all of those objects of the external world, the rays from which pass through the nodal points to identical points of the retina, are seen singly, because their images are referred to the same part of the visual field, so that they coincide. *Double images* are produced by all other objects, whose images do not fall on identical portions of the retina.

The proof for what has been said is readily provided. If a linear object with the points 1, 2, 3 (Fig. 303) be looked at with both eyes, the corresponding points of the retinal images will be 1, 2, 3 and 3, 2, 1, which are obviously identical (corresponding) points on the two retinas. If there is, at the same time, a point (A) closer to the eye, or another point (B) farther from the eye, and the eyes are directed toward the object 1, 2, 3, the visual rays from neither A (A a, A a) nor B (B b, B b) will fall upon identical retinal points: therefore double images of A and B appear.

The following simple experiment also is instructive. If a point of ink (for example 2) upon white paper be looked at, the image will fall in both foveas (2, 2), which are of course identical points. By pressing laterally on one eye, so that it is somewhat displaced, two points immediately appear, because the image of the point no longer falls on the fovea of the displaced eye, but on an adjoining, not identical, point. Similarly in intentional squinting all objects appear to be double.

The vertical lines of division of the retinas do not coincide exactly with the vertical meridians; but exhibit a slight divergence above (from 0.5° to 3°), which varies in amount in different individuals, and even in the same individual at different times, while the horizontal lines of division coincide. Images that fall upon the vertical lines of division appear to be perpendicular to those on the horizontal, although they are in reality not so. Therefore, the vertical lines of division are the pseudovertical meridians.

Some investigators consider the identical points of the retina a congenital arrangement, while others think that they are acquired by ordinary use. Individuals who squint from birth have, however, single vision; under such circumstances the identical points must be arranged differently.

Horopter is the term used to indicate the aggregate of all those points in space, from which rays, entering both eyes, held in any given position, meet at identical points of the retinas. It varies for the different positions of the eyes.

1. In the *primary position* of both eyes, when the visual axes are parallel, rays drawn from two identical points of the two retinas are also parallel and intersect only at an infinite distance. The horopter for the primary position is, therefore, a vertical plane at an infinite distance.

2. In the *secondary position* of the eyes, with convergent visual axes, the horopter for the transverse lines of division is a circle passing through the nodal points of both eyes (Fig. 304, KK_1) and through the point fixed (I, II, III). The horopter of the vertical lines of division is in this position perpendicular to the plane of fixation.

3. In the (symmetrical) *tertiary positions*, in which horizontal and vertical lines of division form angles, the horopter for the vertical lines of division is a straight line inclined toward the horizon. For identical points of the horizontal lines of division there is no horopter for these positions, as the rays from these points do not meet at a distance.

4. In the unsymmetrical tertiary positions (with rotation), in which the point fixed is at an unequal distance from the two nodal points, the horopter is a curve of complicated form.

It will not be possible to enter into a more complete description of the difficult details of the horopter. For the determination of the horopter, v. Helmholtz constructs, in the primary position, similar meridians and parallel circles over both retinas; the identical points lie then as if on two globes of equal length and breadth. Hering draws two systems of planes through the eyeballs in the primary position: those of one system (the transverse) intersect in the transverse axis

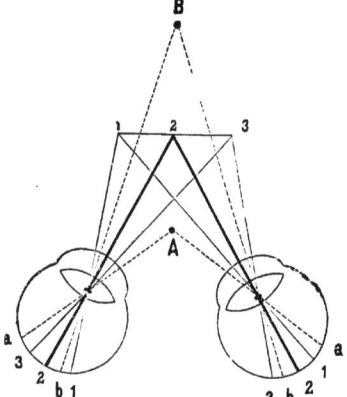

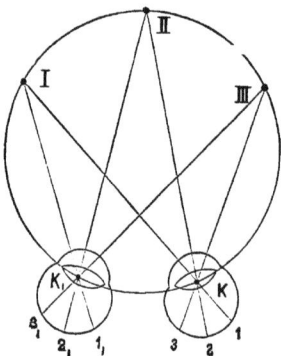

FIG. 303.—Diagrammatic Representation of Identical and Nonidentical Retinal Points.

FIG. 304.—Horopter for the Secondary Position, with Convergence of the Visual Axes.

connecting the nodal points of the eyeballs. Those of the second system intersect in a perpendicular drawn through the nodal point of each eye. The identical points lie at the intersections of the similar perpendicular and transverse planes of the retinas.

All objects whose rays fall on *nonidentical* (disparate) points of the retinas appear in *double images*. A distinction is made between *homonymous* and *crossed* (*heteronymous*) double images, according as the rays from the disparate retinal points intersect in front of or behind the fixed point.

In illustration, two fingers are held, one behind the other, in front of the eyes. If the far finger is looked at, the other appears double, while if the near one is fixed, the far one appears double. If, while looking at the far finger, the right eye is closed, the left (crossed) image of the first finger disappears. If the near finger is fixed and the right eye is closed, the right (homonymous) image of the second finger disappears.

The double images, like the single images, are referred to the proper distance by the eyes.

In spite of the large number of double images that are constantly formed, they are not a source of disturbance. They are usually *suppressed*, and to such an extent, in fact, that the attention must be directed to them, in order that they may be seen. The suppression of the double images is favored by the following factors: (1) Attention is always directed to that point of the visual field which for the time being is fixed. This throws its image on the two yellow spots, which are identical points. (2) Form and color are less sharply seen by the lateral portions of the retina. (3) The eyes are always accommodated for those points that are fixed. Therefore, only indistinct images arise from the objects that produce double images (in diffusion-circles), and these can be more easily suppressed. (4) Many double images lie so close together that when they are large, the greater portions of them overlap. (5) Images that, strictly speaking, do not coincide are often united by a psychical habit.

STEREOSCOPIC VISION. JUDGMENT OF SOLIDITY.

The images formed by the two eyes in looking at solid objects are not exactly alike, but differ somewhat, because the eyes look at the object from two different points of view. The right eye can see more of the side opposite to it, and the same is true of the left eye. Despite this dissimilarity, the two images are united.

The question as to how the impression of solidity is obtained by the combination of such different images may be best solved by analyzing two corresponding·stereoscopic pictures.

In Fig. 305, III, L and R are two such pictures that, when seen with a stereoscope, form a truncated pyramid, projecting toward the eye of the observer, and the similarly designated points coincide. If the distances between the corresponding points in the two figures be measured, it will be found that the distances A a, B b, C c, D d are equal, and are at the same time the greatest between any of the points of the two figures; further, the distances E e, F f, G g, H h are equal, but are·smaller than the first set. Considering, finally, the lines A E, a e, and B F, b f, which coincide, it may easily be seen that all points of these lines that lie nearer A a and B b are further apart than those that lie nearer E e and F f.

From a consideration of these relations in comparison with the stereoscopic images the following principles for stereoscopic vision appear: (1) All those points of two stereoscopic images (and naturally of two retinal images of solid objects) that are at equal distances from each other in the two images appear in the same plane. (2) All points that are closer together (than the others) project toward the observer. (3) Conversely, the points that are further apart recede perspectively into the background.

The reason for this phenomenon resides simply in the following principle: "In binocular vision we constantly refer the position of the individual points of the image in the direction of the visual axes where they intersect."

The following stereoscopic experiment proves this. In Fig. 305 I, two pairs of points are taken as the two images (a b and α β) that are at unequal distances from each other on the surface of the paper. If they are made to coincide, by means of a stereoscope, the point (A), formed by the union of a and α, appears in the plane of the paper, while the other (B) formed by the points b·and β, which are closer, seems to float in the air in front of A. Fig. 305, I, shows the construction clearly. The following experiment also illustrates the same condition. Two sets of lines, similar to B A, A E and b a, a e in Fig. 305, III, are drawn as the figures to be superposed. In the lines B A and b a all the points to be superposed lie equally distant from each other; on the contrary, all points in A E and a e, which lie nearer to E and e, are successively closer to each other. Looked at with a stereoscope, the superposed perpendicular line A B and a b lies in the plane of the paper, while the oblique line formed by A E and a·e projects obliquely outward from the plane of the paper toward the observer. From these two fundamental experiments all stereoscopic pairs of pictures may be easily analyzed; particularly, it will be seen also that if in Fig. 305, III, the two pictures be exchanged, so that R lies in the place of L, the impression of a truncated pyramidal hollow vessel must result.

Two stereoscopic pictures, which are so constructed that one contains an object taken from in front and above, the other the object taken from in front and below (for example, if the figures in Fig. 305, III, had the lines A B and a b as base line), are never united by means of the stereoscope. If these two figures be turned so that they have an oblique position (so that the corners C, c

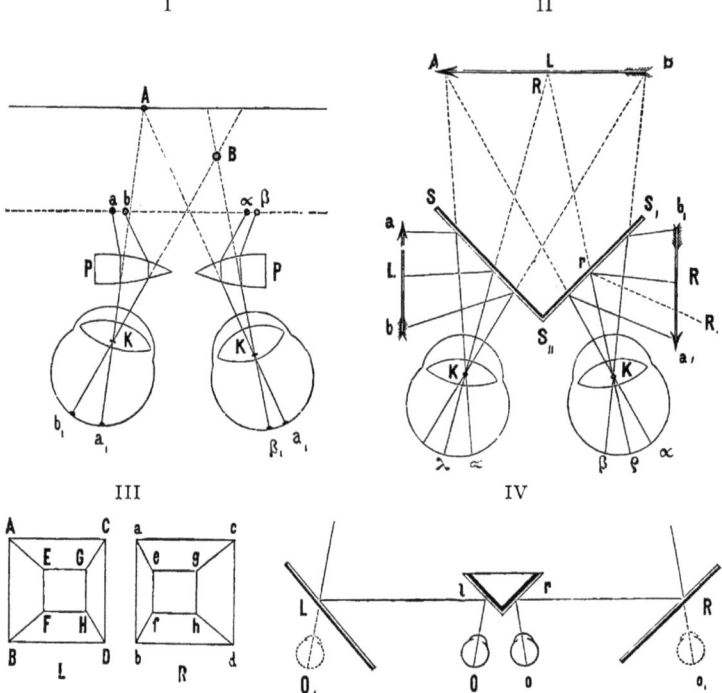

FIG. 305.—I, Diagrammatic Representation of Brewster's stereoscope; II, that of Wheatstone; III, two stereoscopic drawings; IV, v. Helmholtz's telestereoscope.

deviate toward the right and downward), the impression of solidity will be more and more interfered with; but Landois was able to preserve the stereoscopic appearance up to a similar rotation of 30°.

The process of stereoscopic vision has been explained also in another way. In the figure R and L (Fig. 305, III) only A B C D and a b c d fall on identical points of the retinas, and, therefore, these alone can coincide (or in another convergence of the visual axes only E F G H and e f g h coincide, for the same reason). If it be supposed that the quadrate bases of the figures coincided first, it has been further assumed that both eyes then make a quick "groping" motion toward the apex of the pyramid; and as the axes of the eyes would have to converge more and more in doing this, the apex of the pyramid appears to project; for all points appear closer when the eyes must be converged in order to see them. In this way, all corresponding parts of both figures would be brought successively on identical points by the movements of the eyes toward each other.

To this the objection has been urged that the duration of the electric spark is sufficient for stereoscopic vision; a length of time that is entirely too short for the ocular movements. Although this is true for many figures, this movement of the visual axes is not precluded for the correct combination of complex or unusual figures, and it is in fact of considerable advantage, especially for certain individuals.

It appears that not merely the movements that actually take place, but rather the sense of innervation of the muscles necessary to the movement is sufficient to produce the impression of solidity. Consequently,

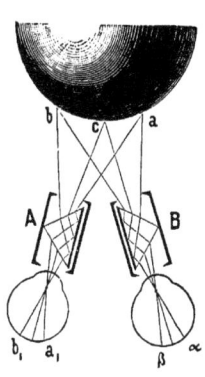

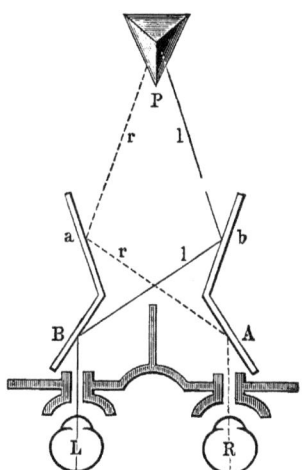

FIG. 306.—Wheatstone's Prism-pseudoscope.   FIG. 307.—Ewald's Mirror Pseudoscope.

stereoscopic vision may depend in part on a muscle-sensation: the feeling that a greater convergence of the visual axes is necessary for the superposition of two points in the stereoscopic figures produces the impression that the points are nearer; conversely, the feeling that to secure the congruence of two points a greater divergence of the visual axes is necessary produces the impression of greater distance.

If, now, in the momentary superposition of two figures making a solid image, a movement of the eyes does not take place, many points in the stereoscopic figures are apparently united that, strictly speaking, do not fall on identical retinal points. The latter therefore cannot be designated with mathematical accuracy as corresponding points of the retinas, but, from a physiological viewpoint, all points must be designated as identical, the simultaneous stimulation of which produces as a rule a single image. In this coalescence the mind plays a part: there is a certain psychical tendency to fuse the double impressions of both retinas into one, as experience has taught that they belong to one single image. If, however, the differences between the two stereoscopic figures are excessive, so that too widely separated retinal points are affected, or if new lines are introduced into a figure that do not harmonize with the solid figure, or would disturb the coalescence, the stereoscopic fusion ceases.

The **stereoscope** is an instrument by means of which two similar pictures, drawn in perspective, may be superposed, so that they appear single, and give the impression of solidity. Wheatstone accomplished this by means of two mirrors placed at an angle (Fig. 305, II), Brewster by means of two prisms (Fig. 305, I). The construction and mode of action are shown by the figures.

Even without a stereoscope some persons are able to unite two such pictures by directing the visual axis of each eye to the picture opposite to it.

Two exactly similar pictures, that is, those in which all corresponding points are at an equal distance from each other (for example the identical pages of two copies of a book), appear exactly on the same level under the stereoscope; but just as soon as one point in one or the other is closer to, or farther from, the corresponding point than the others, it appears immediately to project in front of or behind the plane. In this way Dove taught how to distinguish false banknotes from good ones, by their failure to yield perfectly flat images.

Solid objects seen from a great distance, such as the remote parts of a landscape, appear flat, as in a picture, because the difference in the position of the eyes is too small relatively to be taken into consideration. In order to obtain a stereoscopic view of such objects, v. Helmholtz constructed the telestereoscope (Fig. 305, IV), an instrument that, by means of parallel mirrors, moves the points of view of both eyes to some extent farther apart. The mirrors L and R throw their images respectively on the mirrors l and r, toward which the eyes O o are directed. According to the distance of L and R, both eyes may be apparently separated several feet (to $O_1 o_1$). The distant landscape thus has a distinct appearance of solidity. In order to see the distant objects more distinctly and at closer range, a telescope (field-glass) can be placed before each eye. Instruments of this sort, relief-telescopes, have been constructed in great perfection recently by Zeiss; instead of mirrors they contain similarly acting prisms.

If in two stereoscopic pictures the corresponding surfaces are made black in the one and white in the other (for example, if two truncated pyramids are drawn, as in Fig. 305, III, and one figure is drawn exactly like I, that is with white surfaces and black lines, while the other is drawn with black surfaces and white lines), the body appears to shine in the stereoscope. The explanation of luster is that the shining object, in a certain position, reflects bright light into one eye, and not into the other; because at a given angle the reflected ray cannot enter both eyes at the same time.

An interesting experiment for illustrating stereoscopic vision is furnished by Wheatstone's pseudoscope. This consists of two right-angled prisms enclosed in tubes (Fig. 306, A and B), through which the observer looks in a direction parallel to their oblique surfaces. If a spherical surface is seen through this instrument, the images falling in the eyes will be reversed laterally. The right eye thus obtains a view usually received by the left eye, and conversely; the shadow projected is also reversed. The result of this is that the ball appears hollow. R. Ewald constructed the apparatus with four mirrors, and its action will be readily understood from Fig. 307.

The stereoscope can also be used to explain the "rivalry of the visual fields." In other words, both eyes are almost never simultaneously and equally active in binocular vision, but they rather relieve each other more or less completely, so that at one time the image of one retina, at another time that of the other prevails. For example, if two differently colored surfaces are placed in the stereoscope, they will alternate in the common field of vision, especially if they are brightly illuminated, accordingly as one or the other eye is especially active. If two surfaces are used, on which lines are so drawn that they would cross if the surfaces were superposed, the lines first of one system and then of the other appear more prominently. The rivalry of the visual fields is similarly shown in looking through differently colored glasses at a landscape.

ESTIMATION OF SIZE AND OF DISTANCE.

FALSE ESTIMATES OF SIZE AND DIRECTION.

The judgment as to the size of an object depends, apart from all other factors, upon the size of the retinal image; thus the moon is estimated to be larger than a star. If, while looking at a distant landscape, a fly suddenly crosses the field of vision, close to the eye, its image may give the impression of a large bird, because of the relatively great size of the retinal image. If the image is seen in diffusion-circles, on account of a lack of accommodation, it may appear even larger. As, however, objects widely unequal in size yield equally large retinal images, especially when their distance is such that they subtend the same visual angle (Fig. 279), the estimate of the distance is of the greatest importance in estimating the actual size of an object, as opposed to the apparent size, which is determined by the visual angle alone.

An estimate of the distance of an object is formed from the feeling of accommodation, as a greater effort of accommodation is required for accurate vision of a near object than for seeing distant objects. As, however, of two unequally distant objects forming retinal images of the same size the one that is nearer is found from experience to be the smaller, the object for which greater accommodation is made is estimated to be the smaller.

This fact explains the following observation: beginners in microscopy usually make a strong accommodative effort, while trained observers work without exercising their accommodation. Hence, beginners estimate all microscopical images as too small, and in drawing them, make them much too small. The following experiment is a further proof. An after-image appears smaller if the eye accommodates for near vision, and much larger when the eye comes to rest. If a small object is held as close as possible to the eye, an object behind it, which is seen indirectly, appears to be smaller.

A much more important means of estimating the size of an object, by judging the distance, is given by the degree of convergence of the visual axes. The position of an object that is seen binocularly is referred to the point where these axes cross. The angle that is formed by the visual axes at their point of intersection is called the *angle of convergence of the visual axes*. The larger, therefore, this angle of convergence (with equal retinal images), the nearer the object is judged to be. The nearer, however, the object, the smaller it may be in order to subtend the same visual angle as a larger, more distant object. From this it may be concluded that when objects have the same apparent size (equal visual angles, or equal retinal images) that object is judged to be the smallest that requires the greatest convergence of the visual axes during binocular vision. The *muscular sense* of the ocular muscles gives the information as to the amount of muscular effort that is necessary.

The following experiments afford the proofs for this statement: 1. The *tapestry-phenomenon* described by Herm. Meyer. If a background with a regular, chessboard-pattern (tapestry or wickerwork) be looked at the squares appear of a certain size when the visual axes are parallel. If now the eyes are converged on an object closer to the eyes, so that the visual axes cross, the pattern appears to move into the same plane as the point fixed, the crossed double images, displaced laterally, coincide, and the pattern appears smaller. 2. Rollett looked at an object through two thick glass plates; in one case the plates were so placed (Fig. 308, II) that the apex of the angle between the two plates was turned toward the observer; in the other case (I) the opening of the angle is toward the observer. In order that both eyes f and i (in I) may see the object a, the eyes must converge more than if they were turned directly toward a, because the glass plates displace the rays a c and a g parallel with each other (e f and h i). Therefore the object appears nearer and smaller at a. In II the rays b_1 k and b_1 o from the smaller, nearer object b_1 fall on the glass plates. In order to see the object b_1, the eyes (n and q) must diverge more, and the object appears at f, enlarged and more distant. 3. By examination of Wheatstone's stereoscope (Fig. 305, II), it will readily be seen that the nearer the two pictures are brought to the observer, the more the latter must converge (because the angles of incidence and reflection become larger). Therefore, the fused image seems smaller. If the middle of the picture R is moved to R_1, the angle S_{11} r $\mathfrak{s}$ must be made equal to S_1 r R_1 (likewise naturally on the left). 4. As, in using the telestereoscope, the eyes are, in a manner, moved far apart, they must be converged more strongly, in looking at objects at a certain distance, than in normal vision. Objects in the landscape, therefore, appear as in a small model. As, however, it is customary to consider the distance great when the objects are so small, the latter appear to be moved at the same time to a remarkable distance.

With respect to the judgment of *distance* the following rules may be noted: With retinal images of equal size, the distance is estimated to be the greater the less the effort of accommodation (and conversely). In binocular vision, with retinal images of equal size that object is judged to be furthest away that requires the least convergence of the visual axes (and conversely).

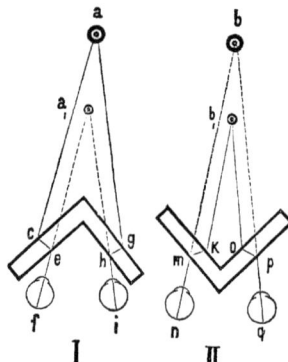

FIG. 308.—Rollett's Glass Plate Apparatus.

The estimation of size and distance, therefore, go largely hand in hand, and the correct judgment of distance affords also a correct estimate of the size of objects. A further aid to the estimation of distance is furnished by the observation of the *apparent displacement* of objects on movement of the head or body. During such movements, objects are apparently displaced laterally more rapidly the nearer they are. For this reason, in traveling in an express train, the objects change their position with great rapidity, and they appear nearer and consequently smaller than they are.

Finally, those objects appear nearest that are most distinct in the field of vision.

Examples: A light in a dark landscape, likewise a dazzling mountain-top covered with snow, appear exceedingly near. Viewed from a high mountain the silvery, gleaming, winding streams not rarely appear to be lifted above the level of the landscape. On looking at the railway embankment from a train, the ground passes indistinctly before the eyes. If suddenly a certain point is fixed for distinct vision, it appears momentarily to project from the general level toward the eye.

False Estimates of Size and Direction.—1. A given distance filled out by intermediate points appears larger than the same distance without them. Therefore, the sky appears elliptical, instead of hemispherical; and for the same reason the disc of the setting sun appears larger than when it is high in the heavens. 2. If a circle is moved slowly to and fro behind a slit, it appears as a horizontal ellipse. If it is moved rapidly, it appears as a vertical ellipse. 3. If a fine line be drawn obliquely across a heavy, black, vertical line, the former appears to deviate from its original direction on either side of the heavy line. 4. Three horizontal parallel lines, 1 cm. apart, are drawn, and through the upper and lower ones are drawn short parallel strokes obliquely from above and to the left downward and to the right, and through the middle line similar oblique lines from the right above downward and to the left. The three horizontal lines no longer appear parallel. 5. On looking at a bright, vertical line in a dark room, and inclining the head toward the shoulder, the line appears to be rotated in the opposite direction.

ORGANS FOR THE PROTECTION OF THE EYE.

The Eyelids.—The structure of the eyelids and the arrangement of their component parts are shown in Fig. 309 and the accompanying description. The tarsus is composed not of cartilage, but of a firm connective-tissue plate, in which the Meibomian glands are embedded. They are acinous sebaceous glands that anoint the lid-margin. At the basal margin of the tarsus, especially the upper, the acinotubular glands of Krause have their opening close to the transition-fold of the conjunctiva. The conjunctiva covers the anterior surface of the eyeball as far as the corneal margin, the cornea being covered only with the epithelium. The conjunctiva on the posterior surface of the lids has a papillary structure in places, the furrows of which have been considered small mucous glands in man and in several mammalia; a sharp distinction between furrows and glands, however, cannot be made. The epithelium is composed of layers of columnar cells, with intervening goblet-cells. Ruminants possess sweat-glands surrounding the cornea; external to the cornea, toward the outer canthus, the pig has simple, glandular, blind sacs. Waldeyer discovered modified sweat-glands at the margin of the tarsus in man. Small lymphatic follicles of the conjunctiva are called trachoma-glands. The lymph-vessels of the conjunctiva are connected with the lymph-spaces of the cornea and sclera. Stöhr saw leukocytes wander upon the free surface of the conjunctiva. Krause found end-bulbs in the conjunctiva of the globe, Dogiel at the lid-margin. The *secretion* of the conjunctiva, aside from some mucus, consists of the lacrimal fluid, which can be produced in as large a quantity by the numerous conjunctival vessels as by the lacrimal glands themselves.

Closure of the eyelids is effected by the orbicularis palpebrarum muscle (facial nerve), the upper lid falling by its own weight. The muscle contracts: (1) voluntarily, (2) involuntarily in individual contractions (winking), (3) as a reflex act from irritation of any of the sensory fibers of the trigeminus distributed to the eyeball and the surrounding tissues, likewise from intense stimulation of the retina by light; (4) persistent, involuntary closure occurs during sleep.

Opening of the lids is brought about by passive dropping of the lower and active elevation of the upper lid by the levator. The unstriated muscular fibers of the lids, which are in a state of tonic contraction, act in the same way by shortening the lid. In looking downward the lower lid is drawn down by bands of connective-tissue fibers running from the fascia of the inferior rectus muscle to the lower tarsus.

The Lacrimal Apparatus.—The straight and freely branching tubules of the lacrimal gland have secreting cells, which are tall when "loaded," and contain a reservoir for the secretion in the fine-meshed protoplasm; and smaller cells entirely filled with secretion in the form of large drops. A dumbbell-shaped pair of rods represents the nucleus in each cell. The secretion is discharged by contraction of the protoplasm. Intercellular secretory passages penetrate between the cells to the level of the nucleus. A terminal intercellular network is formed by exceedingly fine nerve-fibers. The secretory nerves have been described on p. 684. Four or five large, and from eight to ten small, excretory ducts convey the tears into the fornix conjunctivæ, just above the outer canthus. The lacrimal canaliculi, with their open extremities, the lacrimal puncta, dip into the lacrimal lake. The ducts are composed of connective tissue and elastic fibers, and are lined by stratified epithelium. Striated muscle-fibers accompany the ducts, and, by their contraction, keep them open. A sphincter surrounding the punctum was overlooked by Toldt; Gerlach found an incomplete sphincter-muscle. The canaliculi empty at separate points into a dilatation of the lacrimal sac. The connective-tissue covering of the sac and the canal is united to the neighboring periosteum. The thin mucous membrane, rich in lymphoid cells, has a double layer of (ciliated?) cylindrical epithelium, which becomes stratified and squamous below. The opening of the canal is often provided with a valve-like fold (Hasner's valve).

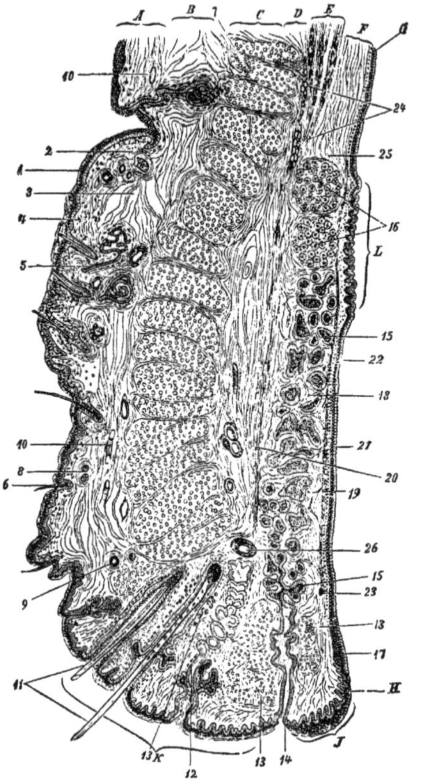

The tears are conducted between the lids and the eyeball by capillarity, being distributed evenly by the movements of winking. The Meihomian secretion prevents the tears from overflowing the lid-margins. The passage of the tears through the puncta, the canaliculus and the canal is effected largely by siphonage, but this is assisted materially by Horner's muscle (known also to Duvernoy in 1678), which, with every act of winking, draws the posterior wall of the sac backward, dilating the sac, and thus exerting an aspiratory influence on the tears. Scimemi has demonstrated this experimentally by introducing a fine tube through the wall into the lumen of the lacrimal sac (in human beings with lacrimal fistula): fluid is aspirated in this tube with every closure of the lids.

E. H. Weber and v. Hasner believe that the tears are aspirated by rarefaction of the air in the nasal cavities in the act of inspiration and in that of insufflation. Arlt thinks the sac is compressed by the contraction of the orbicularis, so that the tears must escape toward the nose. Finally, Stellwag believes that the tears are simply pressed into the puncta by closure of the lids; while according to Gad no such apparatus for pumping the tears into the lacrimo-nasal canal exists. Landois calls attention, however, to one point, namely that the tissue surrounding

FIG. 309.—Vertical Section through the Upper Lid (after Waldeyer): *A*, cutis; 1, epidermis; 2, corium; *B* and 3, subcutaneous connective tissue; *C* and 7, obicularis muscle, with its bundles; *D*, loose, submuscular connective tissue; *E*, insertion of Heinrich Müller's muscle; *F*, tarsus; *G*, conjunctiva; *J*, inner lid-margin; *k*, outer lid-margin; 4, pigment-cells in the cutis; 5, sweat-glands; 6, hair-follicles with hairs; 8 and 23, cross-section of nerves; 9, arteries; 10, veins; 11, cilia; 12, modified sweat-glands; 13, Riolan's ciliary muscle; 14, orifice of a Meibomian gland; 15, cross-section of its acini; 16, posterior tarsal glands; 18 and 19, tissue of the tarsus; 20, pretarsal or submuscular connective tissue; 21 and 22, conjunctiva with its epithelium; 24, adipose tissue; loose-meshed posterior extremity of the tarsus; 26, section of a palpebral artery.

the sac and the canal contains numerous large venous radicles. In expiration, especially in forced expiration, these radicles swell, and press the walls of the ducts together. For this reason air cannot be driven into the lacrimonasal canal, even by forced pressure. If strong inspiratory efforts are made, as in the act of deep frequent insufflation, the veins are emptied, and as the walls again retract they exert an aspiratory influence on the tears.

The *secretion of tears* results from direct irritation of the lacrimal nerve, the subcutaneus malæ, the facial, and the cervical sympathetic, which have been designated the secretory nerves. Reflex secretion of tears may be brought about by irritation of the nasal mucous membrane on the same side. The ordinary secretion in the waking hours is probably a reflex result of irritation of the anterior surface of the eyeball (by the air, by evaporation of the tears); the cornea and the conjunctiva possess sensibility to pain and touch, to cold and heat. Intense irritation by light also produces a reflex flow of tears through the intermediation of the optic nerve. In the rabbit the center does not extend further forward than the origin of the trigeminus, but it extends downward to the fifth vertebra. During sleep the factors mentioned are absent, and the tears dry up. Reichel under the direction of Heidenbain found that the active gland, after injection of pilocarpin, contains cloudy, granular diminutive cells, with obscure outlines, and spherical nuclei, whereas in the resting gland the cells are light and slightly granular, with irregularly formed nuclei. The overflow of tears produced by emotion is still unexplained, as is also that caused by hearty laughing. In coughing or vomiting, the secretion of tears is increased by reflex influences, and the drainage of the tears is impeded by the expiratory pressure.

The tears moisten the eyeball, protect it from desiccation, and carry off small particles, with the assistance of winking. Atropin diminishes their quantity.

The tears are alkaline in reaction and have a salty taste; they represent a "serous" secretion, containing from 98.1 to 99 per cent. of water, 1.46 of organic substances (0.1 of albumin and mucin, 0.1 of epithelial cells), from 0.4 to 0.8 of salts (principally sodium chlorid).

Pathologically, bacteria are present: in the secretion within the lacrimal canal, streptothrix.

COMPARATIVE. HISTORICAL.

Comparative.—The simplest form of visual apparatus consists of deposits of pigment in the external covering of the body connected with the terminations of afferent nerves. The pigment absorbs the light-rays and undergoes a chemical change as "visual substance," and, as a result of the action of the luminiferous ether, it discharges kinetic energy, which stimulates the terminations of the nervous end-apparatus. Deposits of pigment with efferent nerves, and in addition a bright, refractive body, are found in the margins of the swimming-bells of the higher medusæ, while the lower forms have spots of pigment only at the base of the tentacles. In many of the lower worms there are spots of pigment near the brain. In the earthworm the head is sensitive to light because of the presence of light-cells, the caudal end less so. In others the pigment surrounds the nerve-endings, which are represented by the so-called crystalline rods, or crystalline balls (for example the rotifera, or wheel-animalcules). In leeches the eyes, which are usually situated in the head, are not typically developed. Many of the lower worms, and especially the parasites, possess no visual apparatus whatever. In starfishes the eyes are in the ends of the arms, and consist of spherical crystalline organs, surrounded by pigment, and supplied with nerves. In all the other echinoderms, only deposits of pigment are present. Among the articulates eyes in different stages of development are met with: 1. Eyes without cornea that may consist of single crystalline cones, surrounded by pigment (nervous end-organ), are found in the neighborhood of the brain (the larvæ of some crabs), or there may be several crystalline rods in the compound eye (lower crabs). 2. Eyes with cornea, which consists of a lenticular shaped, chitinous formation of the outer external integument, are found to be either simple, consisting of a single crystalline rod, or compound. The latter have either only one large lens-shaped cornea, common to all the crystalline rods, as in spiders (Fig. 310); or each crystalline rod possesses for itself a special lens-shaped cornea. The numerous rods, surrounded by pigment, are placed close together, and form a curved sur-face. The chitinous covering of the head is faceted, and forms a cornea-lens on the surface of each rod (Fig. 311). There are two theories as to the way in which the image is produced by this compound eye of the arthropods. According

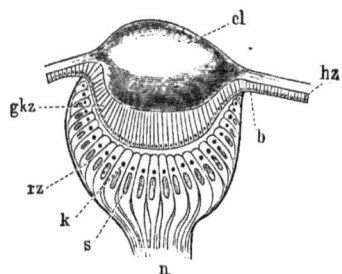

FIG. 310.—Eye of the Cross-spider, according to Grenacher; decolorized: cl, cornea-lens; hz, hypodermal cells; b, basal membrane; gkz, vitreous cells; rz, retinal cells; k, nuclei of the retinal cells; s, rods; n, nerve.

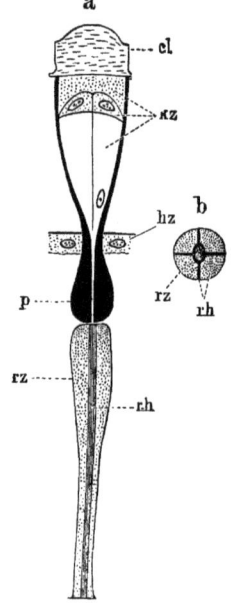

FIG. 311.—Individual Eye of a Libel-lula Larva (Dragon-fly), diagram-matic and simplified, according, Carrière: a, longitudinal section; b, cross-section; cl, cornea-lens; kz, crystal sphere (cells); hz, hy-podermis cells; p, pigment-cells; rz, retinal cells, surrounding rh , the retinal rod.

to one, each facet, with the lens and crystal sphere, is a separate eye: while man has two eyes, the in-sect is supposed to have many hundreds of eyes. Each eye sees the image of the outer world as a whole. The following experiment of Ant. Leeuwen-hoeck seems to indicate this: If the cornea is cut off, each of its facets forms a separate image of objects. If, for example, a cross is placed on the mirror of a microscope, while a piece of faceted cornea is placed as an object on the stage, an image of the cross can be seen in each facet. Consequently a separate image would be formed for each rod (crystal-sphere). This takes place, however, only when the crystal-sphere is removed. In combina-tion with the latter, each corneal facet forms only a part of the (upright) image of the external world, so that the image must be conceived to be composed like a mosaic (mosaic vision). The Röntgen rays appear to be visible to insects (flies).

Among molluscs the fixed brachiopods have two pigment-spots near the brain, but only in their free larval condition. Similarly, the larvæ of mussels have pigment-spots with refractive bodies. Adult mussels, however, have pigment-spots only at the margin of the mantle, but some of them have pedunculated, emerald-lustrous highly developed eyes. Among the snails several of the lower forms possess no eyes at all, others have a pair of pigment-spots on the head, and a number have eyes in various stages of development (Figs. 312, 313). The garden-snail has its eyes on a special pedicle, and they are provided with a cornea, optic nerve, retina, and finally, even lens and vitreous. Of the cephalopods the nautilus has no cornea or lens, and the seawater flows freely into the ocular cavity. Others possess a lens, but the cornea is absent, while still others have an opening in the cornea (sepia, octopus, loligo); all other parts of the eyes are well developed. The eye of the vertebrates needs no detailed description. The amphioxus is without eyes, which are undeveloped in proteus, and in the mammal spalax, whose life in the dark has caused the visual organ to atrophy. In many fishes, amphibians and reptiles the eye is covered by skin, which has become transparent. Several varieties of sharks, crocodiles, and birds have eyelids, and, in addition, a nictitating membrane in the inner angle of the eye. Connected with it is the Harderian gland. In mammals the nictitating membrane is reduced to the plica semilunaris. There is no lacrimal apparatus in fishes. The tears of reptiles remain under the watchglass-shaped cuticular covering that extends over the eye. The sclera of the osseous fishes has two bands of cartilage, which are often ossified; from the middle of the choroid, a muscular organ (falciform process) proceeds forward, and its anterior enlarged extremity, which is called the *campanula Halleri*, is inserted into the outer margin of the lens. The campanula, called by Beer the retractor muscle of the lens, pulls the lens nearer to the retina, and in this way produces an accommodation for distance (the eye being accommodated for near vision, when at rest). In birds the similar muscular structure, the pecten, often reaches nearly to the lens-capsule. The cornea of birds is surrounded by a

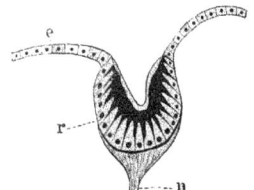

FIG. 312.—Eye of a Sea-snail (Patella coerulea), diagrammatic and simplified, according to Fraisse; the Nerve according to Hilger: e, body-epithelium; r, retinal cells; n, nerve.

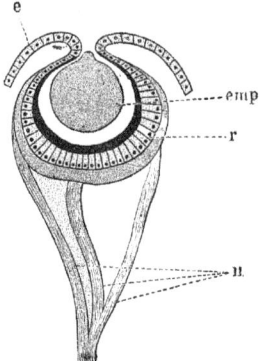

FIG. 313.—Eye of a Sea-snail (Haliotis tuberculata), diagrammatic and simplified, according to Fraisse: e, body-epithelium; emp, refractive jelly-like body inside; r, retina; n, branched nerve.

bony ring. In the birds of prey the cornea changes with the lens. The whale has a tremendously thick sclera. The lens in the aquatic animals is strongly convex. The muscles of the iris and the choroid are striated transversely in reptiles and in birds. It should yet be especially mentioned that the retinal rods of vertebrates (most reptiles have no rods in the retina and no visual purple) are directed from behind backward, while the analogous elements in invertebrates (crystalline rods, and spheres) are directed from behind forward. In the prehistoric salamanders, the existence of a third eye is assumed in the parietal region (parietal eye). The pineal gland of vertebrates appears to be the atrophic remnant of the parietal eye. In lizards the parietal eye is present beneath the skin, which is transparent in the iguana, so that it serves here probably in small measure as a visual apparatus.

The investigations of Loeb have shown that (as in plants) the direction of the visual rays has an influence on the direction of movement of many animals— *heliotropism*. In fact many animals without eyes exhibit heliotropism. Some turn toward the light, others away from it. By increasing the temperature or the concentration of the surrounding sea-water, Loeb was able to reverse this action.

Historical.—The Platonics and Stoics considered the visual act as material. Rays of light were supposed to proceed from the eye and from the objects, and to meet, and the rays from the eye to return to it with the feeling of the object The Epicureans believed that small corporeal images proceeded directly from the objects; the Peripatetics that the images were noncorporeal. According to Aristotle the eye does not take from the object any of its substance, but only its semblance, as the wax takes the impression of the seal. The Greeks were familiar with the ideas of fixation-point, field of vision, binocular single and double vision. Descartes originated the hypothesis of the vibrations of the ether, which were supposed to exist also in the eye, and to stimulate the nerve. The following may be mentioned with regard to the different parts of the eye, and their functions: The school of Hippocrates knew of the optic nerve and the lens. Aristotle (384 B. C.) records the fact that division of the optic nerve as a result of injury causes blindness. He was familiar with after-images, mentions hyperopia and myopia, states that blue eyes exhibit more vigorous iris-reactions on exposure to light than dark eyes, and that man alone has cilia on both eyelids. He mentions a man who was able to see visions, as Quinctilian relates of the painter Theon von Lamos. Herophilus (307 B. C.) discovered the retina; the ciliary body was first recognized in his school. Galen (131–203 A. D.) described the six ocular muscles, the lacrimal puncta, and the tear-ducts. According to him, the retina receives the impressions of light: he refers the origin of the optic nerve to the thalamus. Berengar (1521) was aware of the oily condition of the lid-margins; Stephanus (1545) and Casseri (1609) mentioned the Meibomian glands, which were named after Meibomius (1666). Aranzi described (1586) the muscles of the lid. Fallopia designated the hyaloid membrane and the ciliary ligament. Plater emphasized the greater curvature of the posterior surface of the lens (1583). Aldrovandi saw vestiges of the pupillary membrane (1599).

Even in the time of Vesalius (1540) the refractive power of the lens was discussed: Porta (1560) compared the eye to the camera obscura, and Maurolykos the action of the lens to that of a lens of glass, but Kepler (1611) was the first to show the true refractive indices of the eye, and the formation of the retinal image; he believed, however, that accommodation was effected by the movement of the retina backward and forward. The Jesuit father Scheiner (1619) proved, however, that the lens was made more convex by the ciliary processes, and he assumed the existence of muscle-fibers in the uvea. At the same time he recognized the simultaneous contraction of the pupil in accommodation for near vision. He believed myopia and hyperopia to be due to the curvature of the lens, and he first showed the inverted image on the retina of an enucleated eye. Briggs' remark (1676), "Ligamentum ciliare e fibris motricibus constans," likewise the analogous one of Ruysch (1743), led Morgagni to the correct interpretation of the process of accommodation. Edm. Mariotte recognized that the reflex from the pupil arose from reflected light (1668). As to the use of glasses, there is a note as early as Pliny. At the beginning of the fourteenth century, the Florentine, Salvino d'Armato degli Armati di Fir (died 1317), is said to have invented them; likewise the Pisan monk Alessandro de Spina (died 1313). Kepler in 1611, and Descartes in 1637 were the first to explain their action correctly. Huyghens made an apparatus in imitation of the eye, and showed upon it the action of glasses (1695). The struggle of the visual fields is ascribed to Gassendus (1658). Agulonius (1613) occupied himself with the horopter. Briggs (1676) surmised that single vision occurred when the object formed an image on homologous fibers of the retina; de Peiresc described positive and negative after-images (1634); v. Muschenbroeck knew of the color-top (1762). Leonardo da Vinci (died 1519) was well acquainted with contrast-phenomena, Otto v. Gericke (1672) with the colored shadows, Kepler (1611) with irradiation. The last named explained correctly upright vision, the perception of depth, and the estimation of distance. Nuck analyzed the aqueous humor (1688), Chrouet the lens (1688). De la Hire (the younger) ascribed to the aqueous and the vitreous the same refractive power, and tested that of the lens and the cornea (1707). Maitre-Jean referred the movement of the iris to its circular and radial fibers (1707). Knowledge of the eye was greatly advanced by Zinn (1755). Ruysch described the muscular fibers of the iris, Monro (1794) later the sphincter of the pupil more fully. Berzelius demonstrated chemically the presence of muscle-tissue in the iris. Jacob discovered the layer of rods of the retina. Sömmering (1791) first described the yellow spot. Ant. Leeuwenhoeck knew of the lens-fibers. Reil noted the star-shaped fissility of the lens. Berzelius examined chemically the lens, the aqueous, the vitreous, the pigment, and the tears. Young first observed astigmatism (1801). Brewster and Chossat (1819) tested the refractive power of the ocular media. Purkinje studied subjective vision thoroughly (1819). Helmholtz' "Physiological Optics" summed up the entire science in a classical work (1856–66).

THE AUDITORY APPARATUS.

PLAN OF THE STRUCTURE OF THE EAR.

The auditory nerve is excited normally by waves of sound, which are supposed to set in vibration the end-organs of the auditory nerve. These lie in the endolymph of the labyrinth of the inner ear, on membranous expansions of the cochlea, the saccule and utricle, and the semicircular canals. The waves of sound are first communicated to the labyrinthine fluid, producing wave-motions that set up similar vibrations in the nerve-endings. The stimulation of the auditory nerve is brought about, therefore, by the mechanical irritation produced by the undulations of the labyrinthine fluid.

The labyrinthine fluid is enclosed in the extraordinarily dense and hard mass constituting the petrous portion of the temporal bone (Fig. 314). At one situation in the shape of a small, rounded triangle (fenestra rotunda), the boundary is formed of a delicate, yielding membrane, the opposite side of which is in contact with the air in the tympanum (P). Not far from the fenestra rotunda is the fenestra ovalis (o), into which the basal plate of the stapes (s) is fixed by means of a yielding membranous ring. The outer surface of this also is in contact with the air in the tympanum. As the labyrinthine fluid is enclosed at these two places by flexible boundaries, it is evident that it is capable of an undulatory movement, as yielding limiting membranes are able to follow these undulations.

If it be asked further, in what ways the waves of sound can set the labyrinthine fluid in movement, three different methods suggest themselves:

1. Conduction through the bones of the skull. This takes place especially when solid, sounding bodies are placed directly on the head (for example, a tuning-fork; the sound is then propagated most strongly in the direction of the prolonged handle of the tuning-fork), also when the sound is transmitted to the head through fluids (for example, water under which the head is submerged). If the external auditory canal is stopped up, the vibrations of the tuning-fork are more strongly heard. From this it has been concluded that the vibrations in the bone set the air in the middle ear and the auditory canal in vibration, and that this is communicated to the tympanic membrane, so that the stimulation arises from this, as under normal circumstances—*craniotympanic stimulation*. Waves of sound in the air are practically not transmitted to the bones of the skull, as is shown by the inability to hear when the ears are closed.

Of the soft parts belonging to the head, only those that are directly in contact with the bones conduct sound well; of the detached portions, the cartilaginous part of the external ear is the best conductor. Even under the most favorable circumstances, conduction through the bones of the skull affords less favorable conditions for excitation of the auditory nerves than conduction of the sound through the auditory canal. For example, if a vibrating tuning-fork be held between the teeth until its sound is no longer heard, its tone may be still distinctly heard if it is brought quickly in front of the ear. Sounds are also better conducted through the bones of the skull if the oscillations are not freely transmitted by the bones to the tympanic membrane, and by this to the air of the auditory canal. Therefore, sounds are heard better if the ears are closed at the same time, as the transmission is thus restricted. If in persons hard of hearing conduction and hearing through the bones of the skull are still normal, the cause of the deafness is not in the nervous parts of the ear, but in the external sound-conducting part of the apparatus.

2. Normal conduction, in ordinary hearing through the external auditory canal, takes place as follows: the vibrations of the air first set the tympanic membrane (Fig. 314, T) into vibration; this in turn moves the malleus (h), the incus (a), and the stapes (s), the last of which transmits the vibrations of its base to the fluid of the labyrinth.

3. In individuals, in whom as a result of destructive disease of the middle ear, the tympanic membrane and the ossicles are destroyed, stimulation of the auditory apparatus can take place (to be sure, only in an impaired degree) also by a transmission of the atmospheric vibrations directly to the membrane covering the fenestra rotunda (r) and the structure closing the fenestra ovalis (o). The membrane of the fenestra rotunda can, in fact, be set in vibration alone, even if the parts closing the fenestra ovalis have become unyielding.

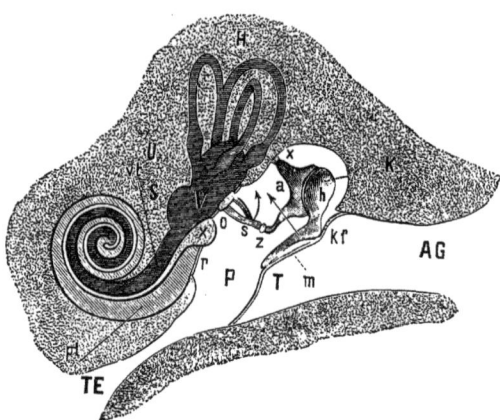

FIG. 314.—Diagrammatic Representation of the Auditory Apparatus: AG, external auditory canal; T, tympanic membrane, K, malleus, with its head (h), short process (kf) and manubrium (m); a, incus, with its short process (x) and long process, which is united with the stapes (s) by the os orbiculare (ossicle of Sylvius); P, tympanic cavity; o, oval window; r, round window; X, beginning of the lamina spiralis of the cochlea; pt, the scala tympani, and vt, the scala vestibuli; V, vestibule; S, saccule; U, utricle; H, semicircular canals; TE, Eustachian tube. The long arrow indicates the direction of action of the tensor tympani muscle, the short curved arrow that of the stapedius muscle.

PRELIMINARY PHYSICAL CONSIDERATIONS.

Sound is produced by the oscillations of elastic bodies capable of vibration. These oscillations cause alternate condensations and rarefactions of the surrounding air; or in other words waves, in which the particles vibrate longitudinally, that is in the direction of transmission of the sound. These condensations and rarefactions form concentric hollow spheres around the point of origin of the sound, which propagate the sound-vibrations to the ear. The vibrations of sonorous bodies are called stationary vibrations, that is all of their particles are always in the same phase of movement, as they begin to move simultaneously, reach the maximum of vibration at the same time, and begin the return motion at the same time as, for example, the particles of a sounding, vibrating metallic rod. Sound is produced, therefore, by the stationary vibrations of elastic bodies, and it is propagated by advancing wave-motions of elastic media (ordinarily of the air). The wave-length of a tone, that is the distance between one maximum of condensation and the succeeding one in the air (or between two condensation-spheres of the air) is proportional to the time of oscillation of the body whose vibrations produce the sound-waves.

If λ is the wave-length of a tone, t the time in seconds of an oscillation of the body producing the wave, then $\lambda = nt$, in which $n = 340.88$ meters the velocity in each second of sound-transmission through the air. The velocity of sound-transmission in water has been found to be 1435 meters in each second—or about four times as great as in air. In solid sonorous bodies, it is from 7 to 18 times greater than in air. Sound is conducted best when it remains in one medium; if it passes through different media, it is always weakened.

Reflection of sound-waves occurs when they strike a solid obstacle, in which case the angle of reflection is always equal to the angle of incidence.

At this place some additional facts relating to wave-movements may be stated. Two varieties of wave-movements are distinguished:

I. Progressive **Wave-movements.**—These can appear in two different forms: As *longitudinal waves*, in which the individual particles of the oscillating body vibrate about their center of gravity in the direction of the propagation of the wave. To these belong the waves in water and in air. In this form of motion, the particles are, of necessity, heaped up in certain places, for example, on the crests of water-waves, while in other places they are diminished in number. This form of wave is, therefore, called a *wave of condensation and rarefaction.* If, however, each particle in the advancing wave moves only up and down vertically, that is transversely to the direction of propagation of the wave, then there result simple *transverse waves* or progressive flexion-waves, in which there is no condensation or rarefaction in the direction of propagation, as the particles are merely displaced laterally. An example of this wave-motion is afforded by the progressive waves in a rope.

II. Stationary Flexion-waves.—If all the particles of an elastic vibrating body oscillate in such a manner that they are always in the same phase of movement, like the two prongs of a sounding tuning-fork, or a twanged cord, the resulting movements are designated stationary flexion-waves. As bodies of little extent in the direction of oscillation vibrate to and fro in stationary flexion-waves, it is evident that the small parts of the auditory apparatus also (tympanic membrane, auditory ossicles, endolymph) oscillate in stationary flexion-waves. Stretched strings, interrupted by nodal points, can also execute stationary flexion-waves in individual segments.

AURICLE. EXTERNAL AUDITORY CANAL.

When the cartilaginous (elastic) auricle is absent, the acuteness of hearing is but little altered. Consequently the auricle is physiologically of minor importance. It has been supposed that the elevations and depressions of the auricle have a favorable action in reflecting the sound-waves. Many of the latter are manifestly reflected outward again, and those that reach the deep part of the concha are supposed to be thrown against the tragus, to be reflected from this into the external auditory canal. It has also been suggested that the auricle intensifies the sound by oscillating in unison with it. By filling the depressions of the auricle with wax, up to the meatus, Schneider claims to have reduced the acuteness of hearing, but Harless and Esser found it unchanged. Against the assumption that there is an effective reflection of the sound-waves both from the parts of the auricle and from the walls of the canal, Mach with justice raises the objection that the dimensions of these parts are too small in comparison to the wave-lengths of sounds. Finally, it has been assumed that the auricle, as an independent, elastic plate, takes up the sound-waves, and conducts them to the cranial bones; so that, in this way, the stimulation of the auditory nerves is strengthened. As, however, the conduction of sound through the bones of the skull, from the air, is exceedingly slight, no serious consideration can be given to such a theory.

According to Kessel there are in the auricle five situations from which the sound is conducted to the ear, in varying degree, when the head is held still; or if the head is moved, variations in intensity will occur. If the posterior surface of the auricle is covered with rubber, the acuity of hearing and the ability to localize sound-impressions coming from behind are decreased.

Muscles of the External Ear.—(1) The entire auricle is moved by the retrahens, attrahens, and attollens. (2) The form of the auricle may be altered by the tragicus, antitragicus, helicis major and minor internally; and by the transversus and obliquus auriculæ externally. Individuals who can move their ears observe no alteration in hearing during the movement. The helicis major and minor are elevators of the helix; the transverse and oblique muscles of the auricle are dilators of the depressions in the auricle; the tragicus and antitragicus are constrictors of the canal; they correspond to analogous muscles in animals. In animals, however, the auricle and its muscular activity have a decided influence upon hearing. The muscles, in the first place, direct the openings of the auricles toward or away from the source of the sound (pricking up the ears). The internal muscles, moreover, contract or dilate the cavity of the auricle. In many diving animals, valvelike appendages close the canal. The human auricle may be most appropriately considered as a perfectly formed but functionally degenerate organ.

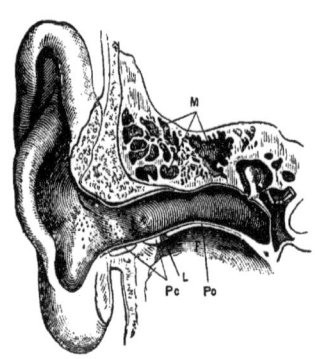

FIG. 315.—The External Auditory Canal, and the Tympanum: M, cavities in the temporal bone; Pc, cartilaginous portion of the canal; Po, osseous portion of the canal; L, membranous portion between them; F, glenoid cavity for the condyle of the lower jaw (Urbantschitsch).

The *external auditory canal* measures from 3 to 3.25 cm. in length, from 8 to 9 mm. in height, and from 6 to 8 mm. in breadth at the meatus; it is the conductor of the sound-waves to the tympanic membrane. As it has a slightly spiral curve (in order to look into the canal, the auricle should be drawn upward), almost all of the sound-waves strike first against its wall, and are reflected thence to the tympanic membrane. Occlusion of the auditory canal, especially by masses of inspissated cerumen (secreted by the ceruminous glands, which are similar to sweatglands), interferes, naturally, with the hearing.

THE TYMPANIC MEMBRANE.

The *tympanic membrane* (Fig. 316) is an unyielding, and almost inexpansible, elastic membrane, with a thickened border, set in a special bony groove, and stretched rather loosely. It is about 0.1 mm. thick, 50 sq. mm. in area (in small animals not much smaller), elliptical in shape (its larger diameter is from 9.5 to 10 mm., its smaller 8 mm.), and it is placed obliquely at the inner extremity of the external auditory canal at an angle of 40° from above downward and inward. Both membranes converge anteriorly so that, if prolonged, they would meet at an angle of from 130° to 135°. The oblique position allows the membrane to present a greater surface than if it were placed vertically, and thus many more waves of sound can fall vertically upon it. The membrane is not evenly stretched, but is drawn inward just below the center (umbo) by the handle of the malleus, which is attached to it; while the short process of the malleus projects forward somewhat at the upper edge of the membrane (Figs. 314 and 315).

The tympanic membrane consists of three layers: (1) The membrana propria is a fibrous membrane, composed of radial fibers on its outer surface, and of circular fibers on its inner surface. (2) The surface facing the external auditory canal has a thin covering of cuticle. (3) The side facing the tympanic cavity has a delicate mucosa with a single layer of squamous epithelium. Numerous nerves and lymph-vessels, as well as internal and external blood-vessels, are found in the membrane.

The tympanic membrane takes up the sound-waves that enter the auditory canal, and is set into vibration by them, in correspondence with the number and amplitude of the movements of the sound-waves in the air. Politzer connected the ossicles attached to the tympanic membrane of a duck with a recording apparatus, and was able to register the vibrations of the membrane produced by sounding any tone. On account of its small dimensions the tympanic membrane moves to and fro as a whole in accordance with the condensations and rarefactions of the undulating air (in the direction of the sound-waves). The membrane, therefore, makes transverse vibrations, for which it is especially adapted, owing to the relatively slight resistance.

Stretched cords and membranes are set into decided sympathetic vibration only when they are affected by tones that correspond with their own fundamental tone, or whose rate of vibration is some multiple of their own rate (octave, duodecime, etc.). If they are affected by other tones, their associated movement will be only inconsiderable. This may be illustrated by a simple example: if a membrane be stretched over a cylinder or a funnel, and a piece of sealing-wax be suspended by a silkworm-thread, so that it just touches the middle of the membrane, it will remain comparatively quiet when musical tones are struck in its vicinity. As soon, however, as the fundamental tone of the apparatus is sounded, the piece of wax will be greatly agitated by the marked vibrations of the membrane.

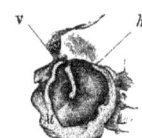

FIG. 317.—Tympanic Membrane of a Newborn Infant, viewed from the Outside, with the Handle of the Malleus shining through: At, tympanic ring, with its anterior (v) and posterior (h) end.

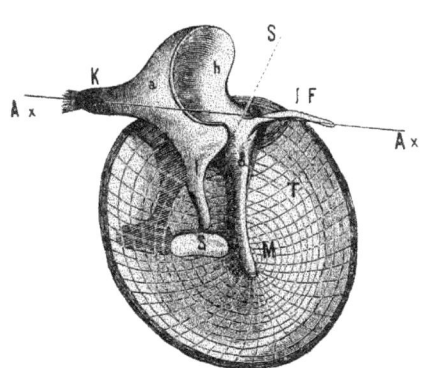

FIG. 316.—Tympanic Membrane and Auditory Ossicles (left) viewed from within (from the tympanic cavity): M. manubrium of the malleus; T, insertion of the tensor tympani; b, head of the malleus; I F, long process of the malleus; a, incus, with its short (K) and its long (I) process; S, plate of the stapes: A x, A x is the common axis of rotation of the ossicles. S the rachet-like arrangement between the malleus and the Incus.

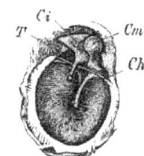

FIG. 318.—Tympanic Membrane and Ossicles (left) viewed from within: Ci, Cm. Ch, chorda tympani: T, pocketlike depression (Urbantschitsch).

If these conditions be transferred to the tympanic membrane, this would also be set into marked vibration when its fundamental tone is sounded, but only into slight vibration when other tones are produced. Such a state of affairs would be attended with an enormous inequality in the act of hearing, and provision is, therefore, made in the tympanic membrane for the neutralization of this inequality. This end is attained: (1) Through the great resistance to the vibrations of the membrane that the chain of attached ossicles offers. They act as a damping apparatus, which (as in the case of any damped membrane) prevents the tympanic membrane from vibrating excessively when its fundamental tone is struck. The damping reduces the amplitude of vibration of the membrane for all other tones also. In this way all of the vibrations of the tympanic membrane are moderated, but especially the excessive

vibration on the sounding of its fundamental tone is diminished. Therefore, the membrane is better adapted to respond to the vibrations of
different wave-lengths, although to a lessened degree. The damping
also prevents effectually disturbing after-vibrations. (2) The sympathetic vibrations of the membrane must be small, in accordance with
its diminutive size. Further, these slight elongations are quite sufficient
to transfer the sound-movement to the extremely delicate end-organs of
the auditory nerves. In fact in the description of the auditory ossicles
it will be seen that there exist other arrangements that still further
diminish the oscillations of the tympanic membrane.

As v. Helmholtz has pointed out, the increased associated vibration of the
tympanic membrane when its own note is sounded is not completely equalized
by the damping arrangement described. He calls attention to the fact that to most
men the tones of the sixth octave e and g are especially piercing and shrill (for
example the shrill tones of the cricket), and he supposes, therefore, that the
individual note of the auditory apparatus, including the tympanic membrane,
lies in this region, so that the membrane vibrates strongly in unison when these
tones are sounded. In general the sounds that are designated as' piercing seem
especially to cause the fundamental vibrations of the auditory apparatus.

According to Kessel, the individual portions of the tympanic membrane
have an independent relation to sounds: the shortest radial fibers in the upper
portion of the anterior segment and in the upper segment vibrate with the highest
tones, while the longest fibers on the posterior segment vibrate with the deepest
tones. Noises are supposed to be transmitted by the upper portion of the posterior
segment; therefore, deep tones are readily disturbed and extinguished by noises.

According to Fick the tympanic membrane possesses, in addition to the property of taking up all vibrations almost equally well, also that of a resonance-
apparatus, that is, it admits of an accumulation of the energy of successive vibrations. It owes this property to its funnel-shaped retracted form, as well as to
the radially placed, rigid handle of the malleus, as artificially constructed models
have shown.

Pathological.—Thickening and inflexibility of the tympanic membrane
diminish the acuity of hearing, in consequence of the lessened vibrating ability
of the membrane. Perforations and loss of substance have the same effect. In
cases of extensive destruction, artificial eardrums have been inserted into the
canal, the vibrations of which replace to a certain extent those of the lost membrane.

THE AUDITORY OSSICLES AND THEIR MUSCLES.

The auditory ossicles have a double function: (1) They transmit
the vibrations of the tympanic membrane to the endolymph of the
labyrinth by means of the chain that they form. (2) They afford
points of attachment for the muscles of the middle ear, which through
the bones alter the tension of the tympanic membrane and the pressure
on the fluid of the labyrinth.

Figs. 319 and 320 show the form and position of the ossicles, which constitute
an articulated chain connecting the tympanic membrane (M) with the labyrinthine fluid through the malleus (h), the incus (a), and the stapes (S). The manner
in which the ossicles move deserves especial attention. The handle of the malleus
(Fig. 320, n) is firmly attached to the fibers of the tympanic membrane. In
addition, the malleus is fixed by ligaments that regulate the direction of its movements. Two of them, the anterior ligament of the malleus, arising from the
processus Folianus, and the posterior ligament, arising from a small crest on the
neck of the malleus, form together a common axial band, which crosses the tympanic cavity from behind forward, consequently parallel to the surface of the
tympanic membrane. The neck of the malleus lies between the insertions of
the two ligaments. The united ligament determines the axis of rotation for the
movement of the malleus. When the handle of the malleus is drawn inward,
its head must naturally make the opposite outward movement. The incus

(a) is only partially fixed in its position by a ligament that secures its short process to the wall of the tympanic cavity, in front of the entrance to the mastoid cells (K). It is materially supported by the rather loose articulation with the head of the malleus (h), the saddle-shaped articulating surface of which is inserted into a depression in the incus. Special attention must be directed to the ratchet-like lower border of the incus (Fig. 316, S). When the handle of the malleus moves inward, this arrangement causes the long process of the incus (l), which is parallel to the manubrium of the malleus, and is attached to the stapes (S) almost at right angles through the sesamoid bone of Sylvius (s), to move inward at the same time. If, however, the tympanic membrane, together with the handle of the malleus, is moved outward, as by condensation of the air in the tympanic cavity, the long process of the incus does not make the same movement, as the malleus alone moves away from the ratchetlike edge of the incus. There is, consequently, no p   on the stapes, and, therefore, no disturbing agitation of the endolymph. As Edll Weber has well shown, the malleus and the incus represent a rectangular lever, whose movement occurs about a common axis (Fig. 316, and Fig. 320 Ax, Ax). In the movement inward, the incus follows the malleus as if the two were one piece. The common axis (Fig. 316) is not, however, the axial ligament of the malleus, but it is formed anteriorly by the processus Folianus (lF), which is directed forward, and posteriorly by the short process of the incus (K), which is directed backward. The rotation of the two ossicles about this axis takes place in a plane perpendicular to the plane of the tympanic membrane. During the rotation, the parts above this axis (head of the malleus and upper part of the incus) move in a direction opposite to that in which move those lying beneath it (manubrium of the malleus and long process of the incus), as is indicated in Fig. 320 by the direction of the arrow. The movement of the manubrium must always follow that of the tympanic membrane, and the reverse, while the excursion of the stapes is necessarily the same as that of the long process of the incus. Attention must be called to one more important point. As the long process of the incus is only two-thirds as long as the manubrium (Figs. 316, 317, 320) the excursion of the apex of the former,

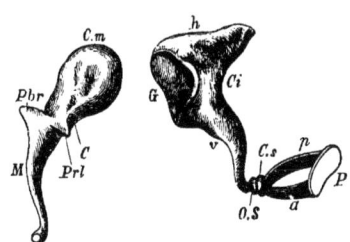

Fig. 319.—The Auditory Ossicles (right): *C.m*, head; *C*, neck; *Pbr*, short process; *Prl*, long process; *M*, manubrium of the malleus; *Ci*, body; *G*, articulating surface; *h* short and *v* long process of the incus; *O.S*, lenticular bone; *C.s*, head; *a* anterior and *p* posterior limb; *P*, base of the stapes.

and with it that of the stapes, must be correspondingly less than that of the apex of the manubrium. On the other hand, the force of the movement, corresponding to the diminution of the excursion, will be increased.

Movements of the tympanic membrane inward thus cause less extensive, but more powerful, movements of the base of the stapes against the fluid of the labyrinth, which v. Helmholtz and Politzer estimated to be about 0.07 mm. in amplitude.

The way in which the vibrations of the tympanic membrane are transmitted to the endolymph through the chain of ossicles is exactly analogous to the method of movement of these parts, as already explained. For the study of this movement, long delicate glass threads have been attached to the various portions of the ossicles, and the movements, when sounds were conveyed to the auditory apparatus, have been thus recorded on smoked paper. Bright particles have also been pasted on the individual parts, whose oscillating movements appear as lines of light, which have been followed and measured with the aid of the microscope. All experiments have proved that the transmission of the sound-vibrations takes place through the mechanism of the rectangular lever formed by the ossicles, as has been described.

Although the vibrations of the tympanic membrane are transmitted through the malleus to the incus, there is, however, a loss of about one-fourth of their original amplitude.

As the excursions of the ossicles caused by the sound-vibrations are extremely small, the articulations do not change position with every vibration. Such a change occurs probably only when larger movements are produced by the muscles, as will now be explained.

The *muscles of the auditory ossicles* affect the position of the latter and also the tension of the tympanic membrane, as well as the pressure in the endolymph. The tensor tympani muscle is situated in an osseous groove above the Eustachian tube; its tendon is deflected over a bony process of this prolonged groove in a direction outward almost at right angles, and is inserted on the malleus just below its axis of rotation (Fig. 321, M). When the muscle contracts (in the direction of the arrow t, Fig. 320) the handle of the malleus (n) pulls the tympanic membrane (M) inward and makes it tense. The incus and stapes are moved at the same time, and the stapes (S) is pressed more deeply into the fenestra ovalis, as has been already fully described. When the muscle relaxes, the original position is again assumed as a result of the elasticity of the rotated axial ligament and the tense tympanic membrane. The motor nerve of the muscle comes from the trigeminus and passes through the otic ganglion. C. Ludwig and Politzer observed the motion described follow irritation of the fifth nerve in the cranial cavity.

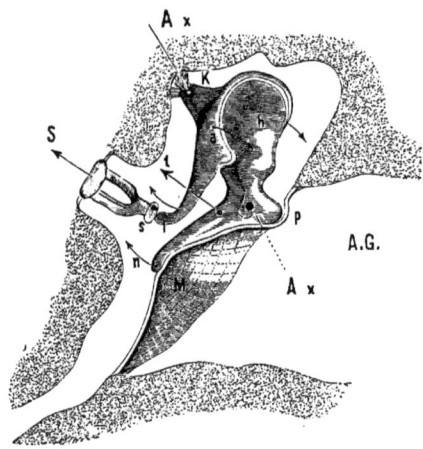

FIG. 320.—Tympanic Membrane and Auditory Ossicles (left), enlarged: A.G. External auditory canal; M, tympanic membrane with which the handle of the malleus (n) and its short process (p) are in contact; h. head of the malleus; a, incus; K, its short process with its fixation-ligament; l, its long process; s, ossicle of Sylvius; S, stapes. A x, A x. the axis of rotation of the ossicles (it is drawn in perspective and must be conceived as stuck through the surface of the paper); t, direction of action of the tensor tympani muscle. The other arrows indicate the movement of the ossicles when the tensor tympani contracts.

The stretching of the tympanic membrane effected by the tensor has a double purpose: (1) The tense membrane offers greater resistance to sympathetic vibration when the sounds are loud, as tense membranes are always the more difficult to set in sympathetic vibration the more they are stretched. In this connection the tensor acts as a protection for the ear, by preventing the transmission of excessively strong impulses through the tympanic membrane to the nerve-endings. (2) The tension of the tympanic membrane must vary according to the degree of contraction. In this way the membrane has a different fundamental tone according to the tension, and is therefore enabled always to vibrate more strongly in sympathy with the especial tone for which it is, as it were, adjusted.   By this means the perception of feeble tones is facilitated.

In this respect the tympanic membrane has been well compared with the iris. Both membranes, by contracting—narrowing of the pupil and stretching of the tympanic membrane—prevent the excessive action of the specific stimulus from causing excessive irritation, and both adapt the sensory apparatus to the action of moderate or weak stimuli.   The movement in both membranes is the result of a reflex action: for the ear through the auditory nerve, which constitutes the path for reflex stimulation of the motor fibers of the tensor.

That increased tension of the tympanic membrane makes it less sensitive to sound-vibrations can be readily shown by closing the mouth and nostrils, and either making a forcible expiration, so that air is forced through the Eustachian tube into the tympanic cavity, and the tympanic membrane is bulged outward, or by making a strong inspiration, so that the tympanic membrane is drawn inward as a result of rarefaction of the air in the tympanic cavity.   In both cases hearing is interfered with as long as the increased tension persists, as may be distinctly observed on listening for a note to die out.

If air is blown into the external auditory canal of a normal individual, by means of a rubber bag, both tensors of the tympanum contract; and in consequence the ear not blown into becomes momentarily hard of hearing.   Johannes Müller made the same action clear by means of the following experiment: If a funnel, with a small lateral opening, be placed in the auditory canal, and the wide end be covered with a tense membrane, hearing is less acute as soon as the membrane is made more tense by means of a traction-apparatus. In other words, the membrane of the funnel represents a second tympanic membrane, which is placed in front of the ear.

The normal mode of stimulation of the tensor tympani is, as has been said, reflex.   The muscle is not directly and solely under the influence of the will. L. Fick explains the following phenomenon as an associated movement of the tensor: When he pressed his jaws firmly together, he heard in his ear a high peeping-singing tone, and in a capillary tube, placed airtight in the auditory canal, he saw a drop move quickly inward.   During this experiment an individual with normal acuteness of hearing perceives a reinforcement of all musical tones,

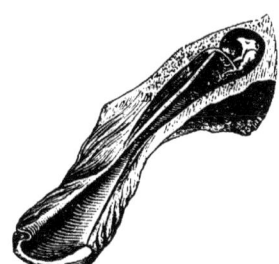

FIG. 321.—Tensor Tympani Muscle; the Eustachian Tube (Left).

but a weakening of all high, nonmusical tones.   In yawning, with great stretching of the muscles of the face and jaws, v. Helmholtz and Politzer found an impairment of hearing for certain tones, which Landois also was able distinctly to perceive in himself, and which he was more inclined to ascribe to an increased activity of the stapedius.

Hensen found that the tensor tympani muscle takes part in the act of hearing by sudden movements, and not by tonic contraction.   At the commencement of the act a contraction occurs that facilitates the perception because the membrane, when set in motion by the muscle, vibrates more readily in sympathy with the higher tones than when at rest.   On exposing the tympanum in dogs and cats, he showed that the contraction takes place only at the commencement of the sound, and that it then quickly ceases, although the sound may continue.

The stapedius muscle, which is situated within the pyramidal eminence, and is inserted from behind forward on the head of the stapes and the sesamoid bone of Sylvius, has the following action: by pulling on the head of the stapes (indicated in Fig. 314 by the small curved arrow) it places the bone in an oblique position, so that the posterior extremity of the base of the stapes is pressed more deeply into the fenestra ovalis, and the anterior extremity is displaced outward.   The stapes is, in this way, more firmly fixed, as through the oblique position mentioned the ligamentous mass inserted around the edge of the base must be more strongly stretched.   Hence, the action of the muscle prevents unduly strong impulses communicated by the incus from being transmitted in their full strength to the endolymph.   The nerve is derived from the facial.

In some persons the stapedius nerve is innervated through associated movement by forcible closure of the eyelids, a rumbling noise being heard at the same time.   Landois was able to excite such innervation through reflex influences by scratching with the fingernail directly in front of the auditory meatus; Henle accomplished the same thing by gently stroking the outer margin of the orbit. The nerve seems to be susceptible to reflex irritation also in many ear-patients by syringing the tympanic cavity.   Under these circumstances Voltolini and Politzer observed contractions of the auricle as associated movements and Ziem, blepharospasm.

Opinions are still much divided as to the action of the stapedius.   In the oblique position of the stapes, the head of the bone forces the long process of the incus, and with it the malleus and tympanic membrane, outward.   Consequently the stapedius has been designated also the antagonist of the tensor tympani.   Politzer observed a decrease of the pressure in the labyrinth on irritation of the muscle.   According to Toynbee the stapedius is supposed to raise the stapes out of the fenestra ovalis, and make it more movable, so that it will vibrate more readily.   The stapedius would, therefore, be the true listening muscle of the ear.   Henle believed that the stapedius is more concerned in fixing the stapes than in making it movable, and that it acts only when there is danger of a violent motion being transmitted to the stapes from the malleus through the incus.   Landois agreed with this view, and considered the orbicularis palpebrarum and the stapedius as muscles for the protection of important sense-apparatus.   Both are innervated by the facial nerve, and both can be stimulated reflexly by irritation of the sensory nerves in the vicinity of the sense-organ.   Strong contraction of the orbicularis induces associated movement of the stapedius.   Lucae, who demonstrated an associated movement of the stapedius with powerful movements of the facial muscles, for example with closure of the lids (in association with which a deep entotic sound is heard), believes that the muscle effects an accommodation of the tympanic membrane for the highest, nonmusical tones (just as the tensor does for musical tones).   These highest tones sound louder, therefore, in this experiment.

FIG. 322.—Stapedius
Muscle (Right).

Pathological.—Immobility of the ossicles in consequence of cicatricial adhesions of their joints or of ankyloses causes impairment of hearing in accordance with the degree of immobility.   Firm adhesions of the stapes within the fenestra ovalis have the same effect.   In the presence of contractures of the tensor tympani, the tendon of this muscle has occasionally been cut.   Paralysis of the tensor is discussed in connection with the otic ganglion (p. 691) that of the stapedius on p. 698.

EUSTACHIAN TUBE. TYMPANIC CAVITY.

The Eustachian tube, which is 4 cm. long, is the ventilating tube for the tympanic cavity.   It keeps the air in the interior of the cavity of the same density as the outer air by means of the communication that is established between the two in the pharynx (Figs. 314, 321).   The normal vibration of the tympanic membrane is possible only under this condition.   The tube is ordinarily closed.   In swallowing, however, the canal is dilated by the traction of the fibers of the tensor of the veil of the palate (sphenosalpingostaphylinus or abductor tubæ, or dilatator tubæ) upon its membranocartilaginous portion, into which they are inserted (Fig. 323).   As the tube is closed, the vibrations of the tympanic membrane can be transmitted with less impairment to the ossicles than if the tube were open and the air allowed to escape through it during the vibrations. If, however, the tympanum were permanently closed, the air within it would soon be so rarefied, that the tympanic membrane would be drawn inward, under abnormal tension, and hardness of hearing would result. The tube serves, moreover, as a drainage-canal for the secretion of the tympanic cavity by means of the ciliated epithelium.

If, after destruction of the tympanic cavity, gas is allowed to stream into the ear of a narcotized dog, through the external canal, it passes through the tube into the throat only when the tensor tympani contracts.

The tube opens its valvelike mechanism more easily in the direction of the pharynx than in the opposite direction. The valve is placed behind the orifice of the tube; after each opening of the mouth of the tube the valve is again closed through the elasticity of the tube-walls.

A tuning-fork held before the nostrils is heard more strongly during a swallowing movement, because the tube is opened. One's own voice seems deafening at the moment that the tube is opened by an influx of air, and the voice seems to sound as if within the ear. Patulousness of the tube as a result of pathological conditions may produce a similar result—*autophony*. The pulsation of the vessels and the respiratory sounds are then also abnormally distinct.

If the act of swallowing is performed slowly in the pharynx, while the tensors of the palate are stretched, a sharp hissing, or loud crackling noise. is heard distinctly. This sounds much like the noise produced by forcing saliva between the incisor teeth by pushing the tongue forward when the mouth is closed, and it results from the separation of the moistened walls of the tube from each other. Another person can hear this noise by applying his ear, or by using a stethoscope. It was formerly thought to be a cracking of the joints of the ossicles through the action of the tensor tympani.

In Valsalva's experiment air enters the tube as soon as the air-pressure equals between 10 and 40 mm. of mercury. Under such conditions Landois heard first the same noise, and then he felt suddenly the increased tension of the tympanic membrane due to the entrance of air into the tympanic cavity. During forced inspiration, while the mouth and nostrils are held closed, air is sucked out, and finally the tympanic membrane is drawn inward.

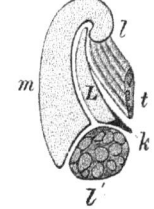

FIG. 323.—S e c t i o n through Eustachian Tube (Diagrammatic): *m*, median plate; *l*, lateral plate; *k*, margin of tube; *l*, elevator; *t*, tensor of the palate; *L*, lumen.

The elevator of the veil of the palate forms in this situation the levator-cushion as it passes under the floor of the pharyngeal orifice of the tube (Fig. 330). Consequently, when this muscle contracts, and its belly thickens (in the commencement of the act of swallowing), and also with every elevation of the soft palate during inspiration, the lower wall of the pharyngeal opening is forced upward, and the opening is narrowed. The subsequent contraction of the tensor of the veil of the palate, in the further course of the act of swallowing, then dilates the tube. (The subject is further discussed with the act of swallowing.) The result is that by this action of the levator the tension of the air in the tympanum is at first increased; it is then diminished by the action of the tensor, as may be recognized under favorable conditions from corresponding movements of the tympanic membrane.

The *tympanic cavity* forms a protective chamber for the auditory ossicles and their muscles. Its air-capacity, amplified by the communications with the mastoid cells, permits free oscillations of the tympanic membrane.

The assumption that the tympanum strengthens by resonance the sound-vibrations that strike the ear, for the purpose of delicate hearing, must be considered erroneous. That, further, the air of the tympanum can transmit vibrations to the membrane of the fenestra rotunda must be admitted. but with normal hearing this slight conduction is of but little significance in comparison with that of the ossicles.

The Eustachian tube and the tympanum have a common mucous membrane; the parts within the tympanum are lined by the mucosa. The epithelium is composed of ciliated columnar cells; it is not ciliated on the surface of the ossicles and the promontory. Tröltsch and Wendt found racemose mucous glands in the mucous membrane.

Pathological.—Among the diseases of the Eustachian tube, obstruction attending chronic catarrhal conditions, and narrowing from scars, hypertrophy of the mucous membrane or pressure by tumors may be mentioned. The impairment of hearing thus produced can often be corrected by catheterizing the tube through the nares. Effusions and collections of pus in the tympanum must, of course, disturb the normal function of all the sound-conducting parts within the tympanum. Inflammatory processes often have also injurious effects upon the tympanic plexus. Moreover, progressive destruction of the temporal bone by caries, commencing in the tympanic cavity may finally cause fatal inflammation of the neighboring portions of the brain.

SOUND-CONDUCTION IN THE LABYRINTH.

The oscillations of the basal plate of the stapes in the fenestra ovalis of the vestibule produce waves in the fluid of the labyrinth, so-called flexion-waves, that is the labyrinthine fluid moves as a whole before each impulse of the stapes. This yielding of the fluid is possible only because in one place a flexible membrane, the membrane of the fenestra rotunda (of the cochlea) or secondary tympanum, which, when at rest, projects into the scala tympani, can be forced outward toward the tympanic cavity by the movement of the stapes (Fig. 314, r). These waves must correspond in number and intensity to the movements of the auditory ossicles, and must also excite the terminations of the auditory nerve, which float free in the fluid of the labyrinth.

Fig. 324.—External Conformation of the Labyrinth: The oval window leading into the vestibule, the cochlea, the superior (f), posterior (s) and horizontal (h) semicircular canal (left).

As both the cochlea anteriorly and the semicircular canals posteriorly communicate with the saccules of the vestibule, the fluid of which first receives the impulse of the vibrations, the movement of the fluid must be propagated through these canals. In the cochlea the movement passes upward from the sacculus (hemisphæricus) through the scala vestibuli to the apex of the cochlea, here through the helicotrema into the scala tympani, at the extremity of which the membrane of the fenestra rotunda moves outward. In a similiar manner the wave-motion commencing in the utricle (sacculus hemiellipticus) passes along through the semicircular canals. Thus, Politzer saw the fluid of the labyrinth mount upward into the superior canal (which was exposed) when he caused contraction of the tensor tympani by stimulating the trigeminus, which must force the base of the stapes against the labyrinthine fluid, with each sound-vibration of the tympanic membrane.

Hensen showed that a membrane that is set in motion by water exercises a strong attraction. This attraction may be noticed also on the oval membrane of the labyrinth, which is weighted by the base of the stapes; the fluid must, consequently, move toward the stapes, and then away from it. Otoliths are probably also under the influence of this attraction, and their mechanical action upon the endings of the auditory nerve is thus clearly explained.

STRUCTURE OF THE LABYRINTH AND THE TERMINATIONS OF THE AUDITORY NERVE.

The vestibule of the labyrinth (Fig. 325, III) possesses two separate sacs: one, the *saccule* (*sacculus*, or *S. hemisphæricus*, S) communicates with the cochlear duct (Cc) of the cochlea; the other, the *utricle* (*utriculus* or *sacculus hemiellipticus*, U) communicates with the semicircular canals (Cs, Cs). The interior of the cochlea, which consists of 2½ spiral turns, is divided into two compartments by a horizontal septum (lamina spiralis ossea et membranacea), which is bony internally and membranous externally (Fig. 325, I). The lower compartment is the scala tympani; it is separated from the tympanic cavity by the membrane of the fenestra rotunda. The upper compartment is the scala vestibuli, which leads into the vestibule of the labyrinth (Fig. 314). These two passages are in direct communication with each other through a small opening (helicotrema) at the apex of the cochlea. A smaller space is separated from the upper passage by the obliquely placed membrane of Reissner. (Fig. 325, I), which bridges over the lower outer angle. This space (ductus or canalis cochlearis, Cc) is bounded below chiefly by the lamina spiralis membranacea, on which the organ of Corti, the end-organ of the cochlear nerve, is placed. The lower end of the cochlear canal (III) is blind and faces the saccule, with which it is united by the fine canalis reuniens (Cr). The three semicircular canals (Cs, Cs) communicate with the

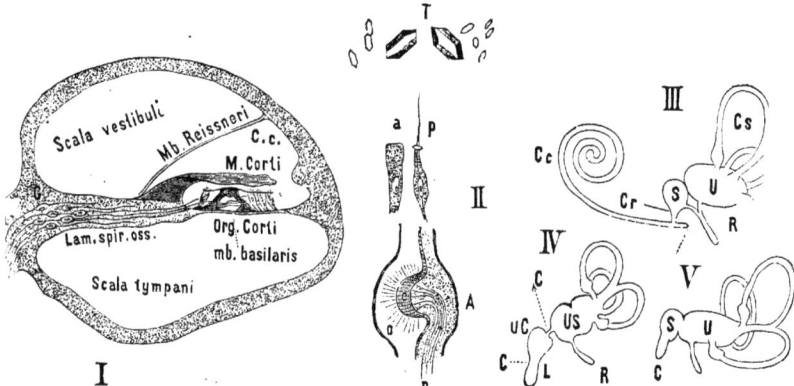

Fig. 325.—I. Cross-section of the cochlea; II, A, ampulla with the crista acustica; a p, a hair-cell and its bristle; T, otoliths; III, diagrammatic representation of the human labyrinth; IV, diagrammatic representation of the labyrinth of a bird; V, diagrammatic representation of the labyrinth of a fish.

utricle (Fig. 325, III, U). Each begins in an ampulla, within which lie the terminations of the ampullar nerves; while at the other extremity they have only two openings, as the posterior and superior canals unite so as to form one common canal. A membranous lining continues from the utricle through the semicircular canals. The limpid perilymph, which is present also in both cochlear passages, and the viscid endolymph fill the entire cavity of the labyrinth. All of these compartments are lined by low, cylindrical epithelium.

Only those portions of the system of cavities that are filled with endolymph contain the nervous end-organs. The cochlea, the semicircular canals, and the ampullæ belong to the organs of hearing. After extirpation of the cochlea on each side there is still a distinct reaction to coarse sounds. The cavities of the labyrinth are all in communication with each other, the semicircular canals directly with the utricle, the cochlear duct with the saccule through the canalis reuniens. Finally, the saccule and the utricle communicate through the endolymphatic duct, which arises as an isolated branch from each sac. The canal thus formed passes through the osseous aqueduct of the vestibule, to end beneath the dura in an endolymphatic sac on the posterior aspect of the petrous portion of the temporal

57

bone (Fig. 325, III, R). Another small canal, the aqueduct of the cochlea, is a narrow passage that begins in the scala tympani, just in front of the round window and emerges near the jugular fossa. It forms a communication between the perilymph of the cochlea and the subarachnoid space.

Semicircular Canals and Saccules.—The membranous semicircular canals do not fill the corresponding osseous cavities completely, but are separated from the walls by a rather wide space, which is filled with the perilymph. On the concave margin alone they are more closely attached to the bone by connective tissue. The ampullæ, however, fill the bony cavities completely. Semicircular canals and saccules consist of an outer vascular connective-tissue layer, upon which lies a hyaloid membrane, bearing a single layer of squamous epithelium. The vestibular branch of the auditory nerve sends a twig to each ampulla and to the saccule and the utricle. In the ampullæ (Fig. 325, II, A) the nerve-ending (c) lies on a yellowish, equatorial ledge, which projects into the interior (crista acustica). The medullated nerve-fibers (n), passing through ganglia, form a plexus in the connective-tissue layer, then lose their sheaths near the basement membrane and end by means of telodendrites by contact in the characteristic cells, each of which is provided with an immovable, rigid bristle (o, p), 90 *u*

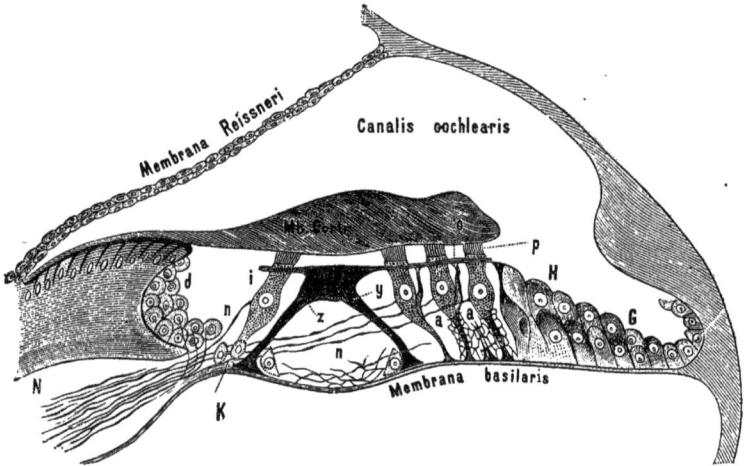

Fig. 326.—Organ of Corti.

long, and which are situated on the crista; between them are indifferent cylindrical cells (hair-cells, a), which often contain yellowish pigment-granules. The bristles or *auditory hairs* are composed of many fine fibers. An exceedingly delicate membrane (membrana tectoria) covers the hairs. The nerve-endings in the maculæ acusticæ of the saccule and utricle are exactly the same as in the ampullæ, except that the free surface of the membrana tectoria is covered with small chalky-white otoliths (II, T) composed of calcium carbonate. These are partly amorphous, and partly in the form of arragonite, with a minute central nucleus, and they lie fixed in the homogeneous membrane of the otoliths. Here also the nonmedullated axis-cylinders of the saccular nerves come into contact with the hair-cells, through the medium of telodendrites.

Cochlea.—Only that portion of the cochlear canal or duct (Fig. 325, I, C.c, and III, Cc, and Fig. 326) that is covered by the membrane of Reissner, and whose endolymph surrounds the organ of Corti, contains in the latter the end-organs of the cochlear nerve. The organ of Corti lies on the fibrous lamina spiralis membranacea (membrana basilaris) and consists of a supporting structure composed of the so-called arches of Corti, each of which consists of two rods of Corti (z y), which are inclined toward each other and meet above like the beams in the roof of a house; but every two rods do not form an arch, as there are always three inner to two outer rods. There are about 4500 outer rods.

The cochlear duct becomes larger toward the apex of the cochlea, and the rods also become longer. The inner ones are 30 μ long in the first, and 34 μ in the upper turns; the outer rods respectively 47 μ and 69 μ. Likewise, the width of the arches increases. The cylindrical *hair-cells* (cells of Corti), observed by Corti, of which there are from 16,400 to 20,000, serve as the actual end-organs of the cochlear nerve. There is one row of inner cells (i) which rest on a layer of small granular cells (k); the outer cells (a a) number 12,000 in man, and rest upon the basement membrane, in three or even in four rows. The cells are directly connected by fibrous processes with the fibers of the basilar membrane, so that each cell is connected with two or three fibers, and must, therefore, vibrate in unison with the latter. Between the outer hair-cells there are other cellular structures, which are regarded either as special cells (Deiter's cells), or merely as processes of the hair-cells. Following the outer cells of Deiter come the cylindrical cells of Henle, which gradually pass into the ordinary epithelium of the cochlear duct.

The fibers of the cochlear nerve (N) emerge from the bony spiral lamina, and, after passing through the intercalated ganglion-cells, (Fig. 325, I, G) end by fine varicose fibrils on the hair-cells, with which their telodendrites are in contact (Fig. 326). The bristles of the hair-cells consist in vertebrates of closely massed fine fibrils.

The arches of Corti and the hair-cells are covered by a special membrane (o, reticular membrane), through openings in which project the upper extremities of the hair-cells with the hairs. This membrane consists of cement-substance holding these parts together. Mention should be made finally of the soft membrane of Corti, which is comparatively thick, and extends from above outward over the organ of Corti. Waldeyer regards this as a damping apparatus for the organ of Corti.

The fluid within the labyrinth also is under a constant pressure—the intralabyrinthine pressure. Every diminution in the pressure of the air in the middle ear is accompanied by a temporary diminution in the intralabyrinthine pressure, while, conversely, every increase in air-pressure is accompanied by an increase in the intralabyrinthine pressure.

The perilymph of the internal ear flows chiefly through the aqueduct of the cochlea within the jugular foramen into the peripheral lymphatic system, which also takes up the cerebrospinal fluid of the subarachnoid space, while a small portion passes through the internal auditory meatus to the subdural space.

QUALITY OF AUDITORY PERCEPTIONS.

PERCEPTION OF THE PITCH AND INTENSITY OF TONES.

Every normal ear is able to recognize musical tones and noises as such, and to distinguish between them. Physical experiments have proved that musical tones are produced when a vibrating, elastic body executes a periodic movement, that is a movement that is exactly reproduced at equal intervals of time, as in the vibration of a twanged cord. A noise is produced when the vibrating object executes nonperiodic movements, that is, when unequal movements occur at equal time-intervals. This is readily proved by means of the siren. If there be on the circular disc of this instrument a number of holes, for example forty, arranged in a circle and placed exactly the same distance from each other, and if, on rotating the disc, a current of air is blown against it, the air will be alternately rarefied and compressed exactly 40 times with every revolution, and every two rarefactions and condensations will be separated from each other by an equal interval of time. This arrangement produces a characteristic musical tone. If, however, holes are made in another circle of the same disc perforated at unequal distances apart, the current of air directed against the disc gives rise to a whirring, rushing nonmusical noise, because the movements of the sounding body, the condensations and rarefactions of the air, are non-periodic.

Every sound must last a certain length of time in order to be heard by the ear (the feeblest sound at least two seconds); on the other hand, after a sound is once heard, the stimulation of the ear persists for some time. Hence, when sounds recur at short intervals, no intermission can be detected.

The normal ear distinguishes in every tone three distinct qualities:

1. *The Intensity of the Tone.*—This depends upon the amplitude of the vibrations of the sounding body. It is well known that a gradually weaker and weaker sounding string exhibits correspondingly smaller amplitude of vibration. The intensity of a sound corresponds to the degree of illumination or brightness in vision.

2. *The Pitch of the Tone.*—This depends upon the number of vibrations that occur in a given unit of time. This also is demonstrated by means of the siren. If the rotating disc have a series of 40 holes, and another of 80 holes, at equal intervals, on blowing a current of air against the rotating disc, two sounds of unequal pitch will be heard, one being an octave above the other. The perception of pitch corresponds to the sensation of color in vision.

3. *The quality* or *timbre of the tone*, which is peculiar to different sonorous bodies. As will appear later, this depends upon the peculiar form of the vibration of the sonorous body. There is no analogous sensation in the case of light.

Perception of Pitch.—Through the sense of hearing, it is learned that different tones have a different pitch. In this connection the established difference in the pitch of the notes of the so-called musical scale or gamut is characteristically distinct to the normal ear. In addition, there are four tones in the scale that, when sounded together, cause, in the normal ear, the sensation of pleasing sound; and that, when once recognized, may be easily reproduced always with characteristic difference in pitch. These are the tones of the so-called *accord* or *major chord*, consisting of the first, third, and fifth tones of the scale, to which the eighth tone or octave is added. It is necessary to determine first the pitch of the tones of the accord, and then that of the other tones of the scale. The siren serves for the determination of the first, and from this the others can easily be calculated. Four concentric circles are drawn upon the disc of the siren, the inner one containing 40 holes, the second 50, the third 60, and the outer one 80, all of the holes being at an equal distance from one another. If the disc be rotated, and a current of air be forced against each series of holes in turn, there will be heard successively the four tones of the accord (major chord). When the entire four series are blown upon simultaneously, the major chord is produced in complete purity. The relative number of the holes in the four series indicates, in the simplest manner, the relative pitch of the tones of the major chord. While 40 condensations and rarefactions of the air in each revolution are necessary to produce the fundamental tone, double this number in the same time (one revolution) are required to produce the octave. Hence, the relation of the number of vibrations of the fundamental tone or keynote to the octave next above it is $1 : 2$. In the second series there are 50 holes, which produce the pitch of the third. Therefore, the relation of the fundamental tone to the third in this case is $40 : 50$ or $1 : 1\frac{1}{4} = \frac{5}{4}$; that is for every vibration of the fundamental tone there are $\frac{5}{4}$ vibration in the third. In the third series there are 60 holes, which, when blown upon produce the fifth. Hence, the ratio of the fundamental tone to the fifth in the disc is $40 : 60$, or $1 : 1\frac{1}{2} = \frac{3}{2}$. In this way the pitch of the four tones of the major chord is determined experimentally; it is found that the number of vibrations of the first, third, fifth, and octave are to each other as $1 : \frac{5}{4}; \frac{3}{2} : 2$.

The *minor chord* is just as agreeable to the normal ear as the major, from which it differs in the fact that its third is a half-tone lower. It may be readily shown by the siren that the *minor third* is produced by a number of vibrations that have the relation of 6 : 5 to the fundamental tone; that is if 5 vibrations occur in a given time in the fundamental tone, then 6 occur in the minor third; and its vibration number is, therefore, $\frac{6}{5}$.

From these relations of the major and minor chords, the relations of other agreeable tones in the scale may readily be calculated, and it must be remembered that the octave of a tone always yields the fullest and most complete harmony. It is evident that if the major third, the minor third, and the fifth harmonize with the fundamental tone, or keynote, they must also harmonize with its octave. Hence, from the major third, with the vibration-ratio $\frac{5}{4}$, there is obtained the minor sixth $= \frac{8}{5}$; from the minor third, with $\frac{6}{5}$, the major sixth $= (\frac{12}{10} =) \frac{5}{3}$; from the fifth, with $\frac{3}{2}$, the fourth $= \frac{4}{3}$. This process is known as *the inversion of the intervals.* These tone-relations are, collectively, the consonant intervals of the scale.

The dissonant intervals of the scale may be estimated from these consonant relations as follows: There are known the fundamental tone C, with the vibration-number 1, the third E $= \frac{5}{4}$, the fifth G $= \frac{3}{2}$, the octave $C^1 = 2$. From the fifth, or dominant, G there is constructed a major chord; this is G, B, D^1. The vibration-ratio of these three tones is evidently the same as in the major chord C, E, G. Hence, the number of vibrations of G : B, is as that of C : E. Substituting the values in this equation, we have $\frac{3}{2}$: B $= 1 : \frac{5}{4}$; so that B $= \frac{15}{8}$. Further, $D^1 : B = G : E$; therefore, D : $\frac{15}{8} = \frac{3}{2} : \frac{5}{4}$, or $D^1 = \frac{18}{8}$, or D an octave lower $= \frac{9}{8}$. If a major chord is constructed upon F (subdominant), that is F, A, C^1, the relation of A : $C^1 = E : G$; or A : $2 = \frac{5}{4} : \frac{3}{2}$, and A $= \frac{5}{3}$. Finally, F : A $= C : E$; or F : $\frac{5}{3} = 1 : \frac{5}{4}$, and F $= \frac{4}{3}$. Consequently, all the tones of the scale have the following vibration-ratios: I. C = 1; II. D $= \frac{9}{8}$; III. E $= \frac{5}{4}$; IV. F $= \frac{4}{3}$; V. G $= \frac{3}{2}$; VI. A $= \frac{5}{3}$; VII. B $= \frac{15}{8}$; VIII. $C^1 = 2$.

Since 1885 it has been agreed to call a tone of 435 vibrations per second a. The previous agreement was 440 vibrations for a. From this the absolute number of vibrations for the tones of the scale is estimated, using the foregoing vibration-ratios: C = 33 vibrations; D = 37.125; E = 41.25; F = 44; G = 49.5; A = 55; B = 61.875. The number of vibrations of the tones of the octave above are found by multiplying these figures by 2.

The lowest notes used in music are: double-bass E, with 41.25 vibrations; piano C with 33; grand piano A^1 with 27.5, and organ C^1 with 16.5. The highest notes in music are the piano c^V, with 4224 vibrations, and d^V on the piccolo-flute, with 4752 vibrations in the second.

The limits of audible sounds lie between 16 and 23 vibrations per second, on the one hand, and 20,480—e^{VII}—(at the most a^{VII}) on the other; they embrace about $10\frac{1}{2}$ octaves. These boundaries, however, depend a good deal upon the intensity of the tone. Fewer vibrations than 16 in the second (organ-tones) are not heard as tones, but as separate, rumbling impulses. Beyond the highest tones, produced by stroking small tuning-forks, or by metallic rods, harmonica-tongues, or small whistles, the ear likewise no longer appreciates the vibrations as tones, but these cause instead a piercing, painful impression upon the ear. The highest tones, which the ear is no longer capable of appreciating, still affect the sensitive flame.

The power of hearing high notes decreases with advancing age about $\frac{1}{4}$ an octave. In rare cases tones of 35,000 vibrations can be perceived. During contraction of the tensor tympani, tones of from 3000 to 5000 vibrations higher may be heard, but rarely more. According to Lucae there are among normal individuals, and especially among those hard of hearing, some whose ears are better adapted for hearing deep tones, others for hearing high tones. He calls them deep-hearing or high-hearing persons respectively. Both conditions are disadvantageous for the normal perception of speech. The deep-hearing individuals hear the high consonants imperfectly, for example ch in "Kirche"; while the high-hearing individuals hear the deep consonants indistinctly, for example ch in "Auch." Diminished tension of the sound-conducting apparatus decreases the perception for high tones. Abnormal power of hearing low tones is present also in cases of rheumatic facial paralysis, that for hearing high tones in cases of absence of the tympanic membrane, the malleus and the incus. The stapedius is said to possess the power of making the highest high tones (even up to 80,000 vibrations) perceptible at the expense of the low ones. Pathologically, an increased perception for high tones is found in conjunction with any condition producing increased tension of the sound-conducting apparatus.

If the eye be compared with the ear, it is evident that the ear greatly exceeds the eye in its range of perception. As the red of the spectrum makes about 456 billions of vibrations in the second, and the visible violet only 667 billions, the eye can take cognizance only of. vibrations of the other that are less than 1 octave from each other (double number of vibrations).

How many vibrations must follow successively for the ear to receive the impression of a tone? Two are sufficient in the case of low tones up to 3168 vibrations, 5 for a tone of 6000, 10 for one of 7040 per second, 20 for all tones. When tones follow one another in rapid succession, they are heard as separate tones if there is an interval of at least 0.1 second between them; if the interval is less, tones become fused, although for many musical tones a shorter interval is sufficient.

A person is said to have an "*accurate ear*" who is able to distinguish a difference in the pitch of two tones of nearly the same number of vibrations. This power can be greatly increased by practice, so that musicians can distinguish tones that have only a difference of pitch of $\frac{1}{500}$ or even $\frac{1}{1200}$ of their number of vibrations. It is easier to determine differences in pitch from the purity of musical intervals than when tones are almost in unison.

With reference to the *time-sense* of the ear, it should be remarked that time is appreciated with greater precision by the ear than by any other sense-organ.

Pathological.—Many normal persons are said to hear the same tone higher with one ear than with the other; v. Wittich found that, during an attack of inflammation of the ear, he heard a tone a half-note higher with one ear than with the other, Spalding even a minor third higher. In a case seen by Moos, the deep tones were heard one-third of a tone too high, the high ones too low. Perhaps the cause of the unilateral heightening of tone-perception associated with this condition, which has been designated binaural diplacusis, consists in an abnormal change in those portions of the labyrinth that are set in sympathetic vibration. The condition designated monaural diplacusis, in which a note sounded in one ear is perceived as two notes, is rare. It is due to the irritation of the elements producing the second tone in addition to those producing the first tone. In rare cases, sudden loss of the perception of certain tones has been observed, for example *bass-deafness;* in a case described by Magnus, the tones from d¹ to h¹ were not heard.

Perception of Intensity.—With respect to the intensity of the tone it has been established that it is dependent upon the amplitude of the vibrations of the sounding body. The intensity of the tone is proportional to the square of the amplitude of the vibrations; consequently, with an amplitude multiplied 2, 3, or 4 times, the intensity of the tone is 4, 9, 16 times as great. As the sound-vibrations are transmitted to the ear by the wave-movements of the air, it is evident that, just as the waves in water become progressively smaller and smaller with the distance from their point of origin, until they finally disappear, so also the intensity of the sound diminishes with the distance of the sounding body from the ear, and finally it must become zero. The sound-intensities, however, are not exactly as the inverse ratios of the squares of the distances from the ear to the source of the sound, but the intensity diminishes slowly near the source of the sound, and more rapidly as the distance increases. The ear is little sensitive to differences in intensity; differences can be distinguished, if the intensities are in the proportion of 72 : 100.

For the determination of the sound-intensity that is necessary to stimulate the ear the following methods may be pursued: (1) A feeble source of sound, such as a ticking watch, is placed horizontally at a distance from the ear, and, by bringing it closer and removing it further away, the most remote point is determined at which the ticking can be heard. The distance is determined by measurement. (2) Itard uses a small hammer, suspended like a pendulum, which strikes on a hard surface when allowed to fall. The sound is increased 4-fold, 9-fold, and 16-fold when the angle of elevation is 2, 3, or 4 times as great, although this is true only when the elevation does not exceed 60. (3) In a similar manner, balls of different weight may be dropped from different heights on a sounding-plate. In this case the sound-intensities are proportional to the product of the weight of the ball by the height of the fall. (4) If a tuning-fork is permitted to sound before the ear, always with the same amplitude of vibration, a normal ear hears the note longer than a diseased ear.

It has been determined with respect to the limits of barely appreciable tone-intensities that a cork sphere, weighing 1 milligram, falling from a height of 1 mm. on a glass plate, may be heard at a distance of 5 cm. There are, however, individual variations, and also differences in acuity between the two ears of the same person. Töpler and Boltzmann estimate the amplitude of vibration of the air-particles that are capable of setting the tympanic membrane in vibration so that an auditory sensation results as equal to only 0.00004 mm.; Rayleigh estimates it as only 0.000001 mm. Direct observation of movements so minute would exceed the limits of the best microscope, through which it is possible to recognize objects not smaller than 0.000217 mm. in diameter. The author's brother made the discovery that animals make sounds that, on account of their weakness, cannot be heard by human ears. Thus, some capricorn beetles (Cerambyx) produce shrill tones by rubbing a grooved plate on the neck against the sharp edge of the chest. For example, Gracilia pygmacea produces the tone f^{III}, with 1413 vibrations, which cannot be heard because of its weakness. [The number of vibrations (s) of the tone is estimated from the length (l) of the rubbing plate of the insect in mm., the number of grooves (a) to each mm., and the time of the rubbing motion; s = (l . n) : t.] Larger capricorn beetles produce sounds that can be heard.

PERCEPTION OF TIMBRE. ANALYSIS OF VOWELS.

By *tone-quality or timbre* is meant a special property of tones, by means of which they may be distinguished independently of their pitch and intensity. For example, a flute, a horn, a violin, and a human voice may produce the same note with equal intensity, and yet each is immediately recognized by its quality or timbre. What constitutes the quality? Investigations, especially those of v. Helmholtz, have shown that of all the sound-producing instruments, only the metal rod fastened at one end and swinging to and fro like a pendulum, and the tuning-fork, produce simple oscillatory and continuous vibrations. This may be shown by fastening a fine point to one branch of a tuning-fork, and registering its movements on a moving strip of smoked paper, on which there will appear then perfectly uniform wave-lines, with equal elevations and depressions. Only those sounds that are produced by such simple oscillatory vibrations are called "tones."

Further investigations have shown that the tones of all musical instruments and of the human voice, all of which have a characteristic·quality, are composed of many individual simple tones. Of these, there is one that is especially conspicuous by reason of its intensity, and which at the same time determines the pitch of the whole composite "tone-picture." This is known as the *fundamental tone* or keynote. The other, weaker tones, which are added to this fundamental tone or keynote vary in number and intensity for the different instruments. They are called *overtones*. Their rate of vibration is always 2, 3, 4, or 5 times that of the fundamental tone. In general, it may be said that all those musical tones that possess numerous strong overtones, especially high ones, have a sharp, cutting, rough quality (for example trumpet, clarionet), and, on the contrary, tones with few and weak, and especially deep overtones are peculiarly soft and mild (for example flute). Only a trained, musical ear is able, without assistance, to detect the overtones present in a given note, in addition to the fundamental tone. This is easily done, however, with the aid of so-called resonators. These are funnel-shaped hollow receivers connected with the external auditory canal by means of a short tube. They are so attuned that each succeeding resonator possesses as its fundamental tone that of the next following multiple of the first.

If, for example, the first resonator has B as its fundamental tone (which is easily heard by blowing upon it), then the tone of the second resonator is b (of the following octave), that of the third is f^1 (three times the rate), that of the fourth b^1 (the second higher octave), that of the fifth d^{11} (five times the rate). Then come f^{11}, as^{11}, b^{11}, etc.

If such a resonator be applied to the ear, it is easy to distinguish the weakest overtone of the same rate of vibration in the sound of a musical instrument and v. Helmholtz found that each instrument possesses a definite number of overtones, differing in pitch and intensity. The tuning-fork, and the simple swinging metal bar, however, have no overtones, but yield only the single fundamental tone. Following v. Helmholtz, only the simple oscillatory sound-vibrations are designated simple tones, while sound-vibrations consisting of fundamental tone and overtones are designated musical tones (Klänge).

If it be borne in mind that each musical tone possesses a fundamental tone, and a number of overtones of definite intensity, which determine its quality, it becomes possible to construct geometrically the vibration-curve of the tone by a combination of the vibrations of the fundamental tone and those of the overtones.

In Fig. 327 the continuous curved line **A** represents the vibration-curve of the keynote, and B that of the first, moderately weak overtone. The combination of these two curves is made by putting together the heights of the ordinates, whereby the ordinates lying above the horizontal are added to those of the keynote, and those below the horizontal are subtracted. In this way the curve C is obtained, which does not correspond to a simple oscillation, but to an unsteady movement. To the curve C a new curve of the second overtone, with three times the rate of vibration, can be added, etc. The final result of all such combinations is that the vibration-curves corresponding to compound musical tones are irregular, periodic curves. All of these curves must, naturally, differ according to the number and the height of the combined overtone-curves. Hence, if the number and the intensity of the overtones in the sound of an instrument have been analyzed by the resonators, the geometrical vibration-curve of the sound can be constructed therefrom.

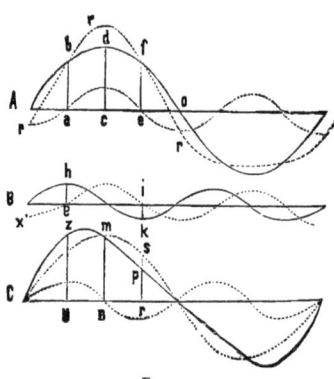

FIG. 327.

The form of vibration of the same tone may vary considerably, if in combining the curves **A** and B the curve B is displaced laterally. If B is displaced to such an extent that the depression r falls under **A**, the addition of the two curves yields the curve r r r with narrow summits and broad valleys. If B is displaced still further, until the summit h coincides with A, still another form is produced. Hence, by displacing the phases of the wave-movements of the simple oscillatory vibrations that are to be combined, there arise numerous different forms of the same musical tone, but this displacement of the phases has no influence whatever upon the ear.

The simple tones, produced by simple oscillations, have a uniform increase and decrease in the oscillations, while the musical tones, according to the number and the strength of their overtones, have a characteristic form of elevation and depression of the vibration-curve.

Just as the irregular curve of vibration of a musical tone may be constructed from several simple oscillating tones, so every such curve may be analyzed. In fact, Fourier has shown that each complicated, irregular curve of vibration may be resolved into a sum of simple oscillatory vibrations, whose number has a ratio of $1:2:3:4 \ldots \ldots$ There can be only one set of simple tones in such an analysis. On the other hand, every complicated irregular movement may be resolved in many ways into movements that are likewise irregular. The result of this deduction is that the quality of a musical tone depends upon the characteristic form of the vibratory movement.

Analysis of the Vowels.—The human larynx represents a wind-instrument, with vibrating, elastic reeds (vocal bands). In producing the various vowels, the mouth assumes a characteristic form, so that its cavity sustains a definite fundamental tone, which is produced when the air passes into it from the larynx. In this manner certain overtones are added to the fundamental tone produced by the larynx, and they give to the voice the vocal quality. The vowel-sound, therefore, is the timbre of a musical sound produced by the larynx. The timbre depends upon the number, strength and pitch of the overtones, and the latter depend upon the configuration of the vocal-cavity in producing the various vowels.

If the different vowels are sung one after the other in a definite pitch, for example b, it can be determined with the aid of the resonators what overtones are added to the fundamental tone, and in what strength. According to v. Helmholtz, if the note b is sung, there is one characteristic overtone of definite, absolute pitch for three vowels, namely b^{II} for A; b^{I} for O, and f for U. The other vowels (and, in German, the modified vowels) have each two especially characteristic overtones, because the oral cavity is so shaped, while producing them, that there is a fundamental tone both for the posterior, more capacious portion, and for the anterior, narrow portion (I and E, p. 612). According to v. Helmholtz these two overtones are for E, b^{III} and f^{I}; for I d^{IV} and f; for Ä g^{III} and d^{II}: for Ö, cis^{III}, and f^{I}; for Ü g^{III} and f. These are, however, only the especially characteristic overtones. Fundamentally there exist for the vowels almost generally many others, which, however, are considerably less conspicuous.

Thus the partial tones present in the same absolute pitch are always characteristic of the vowels; according to v. Helmholtz, Hensen, Pipping and others, they are harmonic overtones of the note of the vocal bands strengthened by resonance. According to Hermann the overtone is an independent note produced in the oral cavity, and it need have no harmonic relation with the sound produced by the larynx.

Just as it is possible to resolve a vowel into its fundamental tone and overtones by means of resonators, so the vowel can be reproduced by sounding together the strong fundamental tone and the weaker overtones. This may be done in the following ways: (1) Most simply by singing loudly a vowel, for example A, at a certain pitch, into an open piano against the free strings, while the damper is at the same time raised by the pedal. If the voice suddenly ceases, the vowel is sounded by the strings of the piano. In other words, all those strings are set into sympathetic vibration whose overtones (apart from the fundamental tone) occur in the vowel-sound. They continue to sound, therefore, for some time after the voice has been interrupted. This experiment may be modified by raising the damper from those notes only that occur as overtones (by holding down the keys). In this way it is possible to combine the vowel-sound, note for note. (2) The vowel-apparatus of v. Helmholtz consists of a number of tuning-forks, which are kept in constant vibration by electromagnets. The lowest fork yields the fundamental tone B, the others in succession the overtones. In front of each fork there is placed a resonance-tube, which can be opened and closed by a lid. When the tube is closed, the tone of the corresponding tuning-fork cannot be heard, but when one or more of the tubes are opened their notes are heard distinctly with an intensity proportional to the size of the opening. In this way different combinations of the fundamental tone with one or more harmonic overtones, in various degrees of intensity, can be made, and musical tones of varying quality (the vowels) produced. v. Helmholtz made the following vowel-combinations: U = B, together with faint b and f^{I}; O = subdued B, and strong b^{I} and weaker b, f^{I}, d^{II}; A = b (as fundamental tone), with moderately loud b^{I} and f^{II}, and strong b^{II} and d^{III}; Ae = b as fundamental tone, with b^{I} and f^{II}, somewhat stronger than for A, d^{II} strong, b^{II} weaker, d^{III} and f^{III} as strong as possible; E = b as fundamental tone, rather strong, with b^{I} moderate, f^{I} likewise, and f^{III}, as III flat, and b^{III} as strong as possible; I cannot be produced in this way. (3) G. Appunn has constructed a vowel-apparatus of organ-pipes. There are 20 open, loud-sounding pipes from the fundamental tone to the 19 succeeding overtones, and 20 stopped, weakly sounding pipes, placed in two rows on a special air-chest. Each pipe can be opened and closed by a valve. A large valve, at the entrance of the air-chest allows all the opened pipes to sound together.   The two rows of pipes make possible three degrees of tone-intensity, namely loud tones, when both rows sound; moderately loud, when the open pipes sound; and weak, when the stopped pipes alone sound.   The formation of the vowels by this apparatus is not so satisfactory as that by the tuning-forks, because the pipes do not yield simple tones, but contain several weak (especially the uneven) overtones. Moreover, the graduation of tone-intensity cannot be made as fine as with the resonators of the tuning-forks. However, several of the vowels can be beautifully reproduced. They always sound the best when they are quite short. Thus, a good A is produced by b and b^I weak, f^{II} moderately strong, b^{II} strong, d^{III} weak and f^{III} moderately strong. U is produced by B strong, and b moderately strong. Deep O = B and b moderately strong, f^I and b^I strong, with f^{II} weak. A high O is produced by b^I weak, d^{II} moderately strong, f^{II} and b^{II} strong, d^{III} and f^{III} weak. The other vowel-sounds are produced imperfectly: E = d^{II} weak, with b^{III}, d^{III}, a^{III} strong. Ä = b^I, f^{II}, b^{II} weak, d^{III}, f^{III} moderately strong, a^{III} flat strong and a^{III} moderately strong. Ö = b^I weak, f^{II}, b^{II} strong, f^{III} weak, b^{III}, c^{IV}, d^{IV} moderately strong. Ü = f^I, f^{II} weak, f^{III}, c^{IV} strong. I cannot be produced. The highest pipe d^{IV} yields approximately the character of I. Similarly the stopped pipe B yields an obscure U and the open B a rather clearer U.

According to the foregoing considerations, the vowels, being composed of a fundamental tone and overtones, must have definite vibration-curves. These may be demonstrated in various ways. If a vowel be spoken against a delicate glass membrane closing the extremity of a hollow cylinder, at whose center is a fine curved style, applied to a cylinder covered with a layer of paraffin-wax and capable of revolving uniformly and of being displaced laterally, the style will trace the vowel-curve on the layer of wax. If, now, a small point connected with the membrane is allowed to run in the groove traced by the style, the resulting vibrations of the membrane will reproduce the sound (Edison's phonograph). Enlarged curves of the sound-impressions may be obtained by transmitting

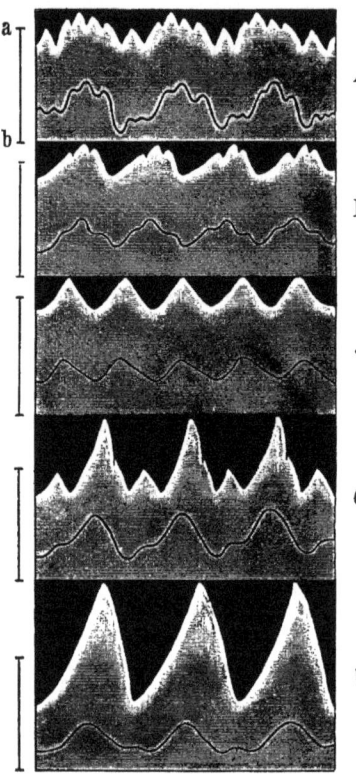

FIG. 328.—Flame Pictures and Phonautographic Tracings of the Vowels.   The vowels were sung in the key of C' (= 256 vibrations in the second). The measure a b shows the height of the flame at rest. The curves traced below are registered by the phonautograph.

the impressions on the cylinder to a suitable apparatus.   The vowels yield the same sound only when the rapidity of revolution of the cylinder remains the same.

If on the other side of such a membrane, there is a small, closed gas-chamber, from which a gas-burner passes, a characteristic tracing of the vibrating flame can be obtained in a rotating mirror when the vowel is produced (Fig. 328). Nagel and Sawojloff made use of the tympanic cavity and the tympanic membrane for this purpose—gas being introduced into the tympanic cavity of a fresh animal's head, and this being connected with a gas-burner—and they were thus enabled to see characteristic vowel-curves in a rotating mirror.

If one limb of a Y-shaped tube be fitted into the nostrils, while the second is connected with a gas-fixture, and the third with a burner, every time a vowel is uttered the flame is set into sonorous vibrations, which reproduce exactly the sound of the vowel. If the vowel is given a nasal sound, the flame shoots up high, because the air is forced into the nasal cavity. Such a flame also may be analyzed in the rotating mirror.

The movements of the membrane may be drawn or photographed by means of a writing lever placed in contact with it. In this way characteristic curves are obtained for each vowel: phonautograph of Hensen, A. Fick, and others. Fig. 328 shows the flame-pictures of the vowels, and under each the corresponding tracing as registered by the phonautograph.

FUNCTION OF THE LABYRINTH IN THE ACT OF HEARING.

With respect to the part played by the ear in the appreciation of timbre it may be said that, just as a musical tone can be resolved into its fundamental tone and overtones by means of resonators, so the ear is able to make such an analysis. The ear resolves the complicated wave-motions into their components, which it perceives as separate tones harmonizing with one another. As a result of adequately trained observation the ear can bring these components separately to the notice of consciousness, and it distinguishes as different qualities of sound only different combinations of these simple tone-sensations. This resolution of the complicated vibrations into simple pendulum-like vibrations is a most striking property of the ear. What are the mechanisms in the ear through which this resolution is effected? If with the dampers raised the vowel-sound A be sung loudly in a certain note (for example b) against the strings of an open piano, all of those strings, and only those strings, are set into vibration that are contained in the vowel-sound. It must, therefore, be assumed that a similarly acting apparatus is present in the ear, which is tuned for certain pitches, and is set into sympathetic vibration when a note is sounded, like the strings of a piano. "If we could connect each string of a piano with a nerve-fiber in such a way that the nerve-fiber would be stimulated and receive an impression every time the string was set in motion, each musical tone that strikes the instrument would, in fact, as is actually the case in the ear, excite a series of sensations, corresponding exactly to the oscillatory vibrations into which the original movement of air could be resolved; and thus the existence of every individual overtone would likewise be perceived exactly as it is by the ear. Under these circumstances the perceptions of the various high tones would devolve upon different nerve-fibers, and, therefore, would occur separately and independently of one another. Now, in fact, the recent discoveries of the microscopists as to the intimate structure of the ear permit the assumption that similar arrangements exist in the ear, such as we have just considered. Thus, the end of each fiber of the auditory nerves is connected with small elastic particles, of which we must assume that they are set in vibration in sympathy with the sound-waves" (v. Helmholtz).

v. Helmholtz believed formerly that the arches of Corti are the apparatus attuned to the individual tones, stimulating the nerve by sympathetic vibration; in other words that they represent a sort of keyboard. As, however, amphibians and birds have no arches of Corti, although they are certainly able to hear musical tones, the stretched radial fibers of the basilar membrane (on which the organ of Corti rests), which are shortest in the first turn of the cochlea, and become longer near the apex, must be considered as the organ that takes up the vibrations. Hence, there would be a fiber of the basal membrane vibrating in sympathy with each possible simple tone. According to Hensen the hairs of varying length in the labyrinth may also subserve the same purpose.

The foregoing assumption is sufficient also to explain the perception of noises. Many of these may be resolved into a confused mass of simple pure tones. True noise in the physical sense must, like separate explosions, be perceived by the saccules and ampullæ.

R. Ewald has proposed a new theory, the so-called *sound-picture theory*. He believes that the impulses produced by the sound on the basilar membrane give rise to a wave-picture (sound-picture), the special form of which enables the basal membrane to form a link in the chain of transmitting mechanisms that are interposed between the sound and its perception.

If the parts played by the cochlea and the saccules together with the ampullæ be compared, it may be said that only the fundamental sensation, the general perception of hearing from concussion of the auditory nerves, as through blows and noises, is excited by the saccules and ampullæ; whereas, on the other hand, the pitch and the depth of the vibrations and their musical character are appreciated by the cochlea.

According to another view each nerve-cell of the cochlea hears every tone; therefore separate cells are not attuned for different tones. The sharpness of hearing is supposed to result from the sum of the sensitive auditory cells, all of which hear the same thing. According to Held, several hair-cells are connected with one nerve-fiber; hence tones of different pitch can excite one and the same fiber.

The relations between the semicircular canals and the bodily equilibrium are treated in the consideration of the auditory nerve (p. 699).

Pathological.—In the presence of varying degrees of deafness, loss either of all or of only certain tones in greater or lesser amount has been found. Labyrinthine affections and those of the auditory nerve both cause disturbances of hearing, but with the following differences: In the presence of affections of the labyrinth tones having from 12 to 64 vibrations are heard poorly with air-conduction; while, in the presence of so-called torpor of the auditory nerve, such tones are well heard. Bone-conduction is good in both cases for the lowest tones. In the presence of torpor of the nerve high tones are well perceived, but in that of affections of the labyrinth, poorly. The hearing of spoken sounds and bone-conduction are much reduced in both cases. Double hearing is rarely produced by affections both of the middle and of the internal ear.

SIMULTANEOUS ACTION OF TWO TONES.

HARMONY. BEAT. DISCORD. DIFFERENTIAL TONES AND SUMMA-TION-TONES.

If two tones of different pitch are heard at the same time, they produce different sensations in accordance with the difference in pitch.

If the number of vibrations of the two tones is in the ratio of simple multiples, or as 1 : 2 : 3 : 4, so that, while the lower tone makes one vibration, the higher one completes 2, or 3, or 4, the ear obtains an impression of complete harmony or concord.

If the number of vibrations of the tones is not in the ratio of simple multiples interference must result if the two are sounded together. The summits and valleys of one wave can no longer always coincide with the corresponding summits and valleys of the other, but in accordance with the difference between the number of vibrations there must be places where the summits and valleys come together. Consequently, if two summits fall together there must be an increase in the strength of the tone, but if the summit of one wave coincides with the valley of another, there must be a diminution in the strength of the tones. In this way there is obtained an impression of variation in tone-intensity that is designated *beat* or *tremor* (battements).

The number of beats is, naturally, equal to the difference between the number of vibrations in the two tones. The beats are most clearly perceived when two deep tones of the same pitch, for example, of organ-pipes, are slightly out of tune. If of two organ-pipes, each of which produces C with 33 vibrations in the second, one is made to yield 34 vibrations, one distinct beat will be heard every second. It is evident, further, that the beats are fewer the less the difference between the two vibration-numbers, and that they are more frequent the greater this difference. Further, with equal relative difference in pitch of two tones, the beats are fewer the deeper the tones, and they are the more frequent the higher the tones. If, for example, the tone c with 66 vibrations is sounded with a second tone with 68 vibrations in the second, two beats must occur in every second (while in the preceding example, with equal relative differences in pitch, only one beat is heard).

The beats produce widely different impressions upon the ear, according to the rapidity with which they follow one another.

When they occur at long intervals, they may be perceived as completely isolated reinforcements of the tone, with subsequent enfeeblements; they thus produce the sensation of completely isolated beats.

If the beats follow one another more rapidly, the inequality produced causes a continuous, disagreeable, whirring impression that is designated a discordant sensation. The highest degree of disagreeable, painful discord is felt when there are 33 beats in the second.

The intense unpleasantness of this sensation may be well likened to the disagreeable impression produced by a flickering light before the eye. It is evident that in order to produce this intense discord, two low tones must have a much greater difference of pitch than two high tones.

If, by an increase in the difference in the number of vibrations of the tones, the beats follow oftener than 33 in the second, the sensation of harsh discord gradually disappears, as the beats become more frequent. Hence the sensation progresses from moderately inharmonious tone-ratios (which in music demand a resolution in the succeeding chords) to more and more consonant, and finally to completely harmonious ratios. These tone-ratios are successively the second, seventh, minor third, minor sixth, major third, major sixth, fourth, and fifth.

As 33 beats in the second produce the greatest discord, it is evident that for the production of discord in tones of low pitch, the tones of the scale must lie further apart than when they are of high pitch. In deep tones the major third may easily be discordant; in high tones, on the contrary, even those lying close together sound much less discordant, because the number of beats quickly exceeds 33 in the second, on account of the high number of vibrations. In general, therefore, musical passages that possess but little harmony are much less inharmonious in high notes than in low ones.

The conditions are exactly the same for two musical tones that are heard at the same time by the ear as for two simple tones. Under such circumstances, however, the overtones come into consideration, as well as the fundamental tones that determine the pitch. The degree of discord of two musical tones is, therefore, all the more prominent the more the two fundamental tones and the overtones (and finally the differential tones, which will be considered presently) produce beats that number about 33 in the second.

Finally, two simple tones or musical tones sounded together may give rise to new tones if they sound simultaneously and in suitable intensity. In addition to these two primary tones or musical sounds, a third new tone is heard on listening intently, the number of whose vibrations is equal to the difference between the two primary tones. These tones are called differential tones, or Andreas Sorge's or Tatini's tones.

If, for example, two tones in the relation of the fifth (2 : 3) or of the fourth (3 : 4) or of the third (4 : 5), are sounded, the fundamental tone = 1 is heard as a differential tone. Musical tones that are rich in overtones yield differential tones of higher order. Thus, if the third (produced by two metal bars) in a high register, namely 16 : 20 (= 4 : 5) is sounded, the tone = 4 (fundamental tone) is readily heard as the first differential tone. This tone 4 forms, however, with 16 another differential tone of second order, that is 16 — 4 = 12. In fact, with the aid of resonators the differential tone of third order may be heard, namely 12 — 4 = 8.

Helmholtz showed, further, that new tones may also result through addition of their vibration-numbers (so-called summation-tones). These are difficult to hear, though best when the two primary tones belong to the middle and lower register, and are rich in overtones.

When musical tones are sounded together, the harmony of the differential tones must also be taken into account. In the major chord these are consonant; in the minor chord there is dissonance of the differential tones. Therefore, the first have a finished, complete, satisfying character, while the latter produce a feeling of unsatisfactoriness, melancholy, contention, which requires a resolution into more finished consonant harmonies.

AUDITORY PERCEPTION. FATIGUE OF THE EAR. OBJECTIVE AND SUBJECTIVE HEARING. ASSOCIATED SENSATIONS. AUDITORY AFTER-SENSATIONS.

When the stimulations of the nerve-endings in the labyrinth are referred, by a psychical act, to the source of the sound in the outer world, there result objective auditory perceptions. Only such stimulations, however, are referred outward as are transmitted to the tympanic membrane by vibrations of the air. This is proved by the fact that, when the head is held under water, so that the external auditory canals are filled, all sound-vibrations will be heard as if originating in the head; the same is true of one's own voice, if the auditory canals are held closed, and also of sound-waves conducted through the bones of the skull.

As to the direction from which a sound comes a judgment is formed from the relation of the auditory canals to the source of the sound, especially if this direction is estimated from time to time by turning the head. The direction from which musical tones combined with noises come is more easily recognized than that from which simple tones come. With equally strong stimulation of both ears, the source of sound is referred to the median plane in front as a single sound; but if one ear is more strongly affected, the sound is referred to that side. The position of the auricles, which act as collecting funnels for the sound-waves, is naturally important in judging the direction from which these come. According to Eduard Weber it is much more difficult to determine the direction when the auricles are held firmly pressed against the head. This observer states that if the hands are placed over the ears in such a manner as to form cavities opening backward, a sound coming from in front will be heard as though coming from behind. The semicircular canals probably also possess the function of determining the direction of sound, as a sound coming from a certain direction must always strike one canal (or the same one of both sides) more strongly than the others. For example, the left horizontal canal is most strongly excited by a horizontal sound-impulse coming from the left side. Other investigators ascribe to the tympanic membrane the function of localizing the sound, inasmuch as certain portions of the membrane are often affected alone.

As to the distance of the sound, the strength of the sound-vibrations serves as a guide, an estimate of this having been formed as a result of experience in the case of familiar sounds, but error in this connection is not rare.

A certain time always elapses before a tone is heard by the ear, especially if the tone is faint (from 1 to 2 seconds). Likewise, the auditory sensation persists for some time after the sound has ceased.

Among *subjective* auditory sensations there may be distinguished:

After-vibrations, especially of loud and persistent musical sounds. *Roaring in the ears*, which often is caused by abnormalities in the circulation of the blood (hyperemia or anemia) in the ear, depends upon mechanical irritation of the auditory nerve-fibers (by the blood-current). Abnormal pressure in the labyrinth may also cause subjective noises. There are also undoubted subjective sensations of a purely nervous character in the entire nervous apparatus. *Ringing in the ears* is ascribed partly to tetanic contraction of the tensor tympani muscle, and partly to circulatory abnormalities. Also, many poisons, such as quinin, and others, cause subjective noises. *Entotic perceptions*, which are due to processes within the ear itself, consist in hearing the pulse-beat in the neighboring arteries, and rushing noises in the blood-current, which are especially loud when there is increased resonance in the ear, as from occlusion of the external canal or of the tympanic cavity, or a collection of fluid in the latter; further, when the action of the heart is increased, or in association with hyperesthesia of the auditory nerve. Entotic sounds are produced also by crunching and crackling noises in the articulations of the lower jaw, by muscular traction on the Eustachian tube, and by the entrance of air into the tube or when the drum is moved inward or outward. Other instances of subjective auditory sensations are referred to on p. 701: Pathological.

The ear exhibits the phenomena of *fatigue*; and this confines itself to that tone or group of tones to which the ear is exposed, while its sensitiveness to other tones is not demonstrably diminished. In the course of a few seconds, however, complete recovery takes place.

The auditory phenomena resulting from applications of the galvanic current are discussed on p. 701.

The following *auditory after-sensations* can be distinguished: (1) Those that correspond to positive after-images, and may be designated *echoes* or *resonances*, that is the after-sensation is so intimately related to the original sound that they appear to be continuous. (2) There are also auditory after-sensations attended with a pause between the end of the objective and the beginning of the subjective tone. A splashing sound has been heard as a peculiar after-sensation for a minute after a tone has been listened to for some time. (3) A third variety of after-sensation may be compared with negative after-images. As such may be designated the sense of striking stillness noted by Landois after interruption of a long-continued, loud sound.

Some persons associate the perception of tones with the appearance of subjective sensations of color or of light (colored hearing), for example, the tone of the trumpet with the sensation of yellow. Photisms of this kind are more rarely observed when the nerves of taste, smell, and sensation are stimulated. There are persons in whom every form of sensory impression necessarily calls forth another subjective one. It is more frequent to find a sympathetic irritation of sensory nerves in connection with loud, sharp sounds. In this category belongs the cold chill that many feel when they hear the squeaking of a slate-pencil, or any similar shrill tone.

According to Urbantschitsch analogous relations exist between all of the sensory organs: Shading the eyes usually weakens the hearing; subjective auditory sensations are usually increased by light; gustatory sensations are frequently strengthened by red and green, etc.   Color-blind individuals exhibit also typical defects of musical sense; those that are green-blind confuse different tones that they hear or repeat in a way that is different from those that are red-blind.

It is often observed that the auditory impulse conveyed to one ear strengthens the function of the other ear, as a result of stimulation of the auditory centers of both sides.

The auditory apparatus may be excited not only by sound-vibrations, but also by other heterologous stimuli.   It is mechanically excited by a sudden blow or shock to the ear.   If the fingers are placed tightly in the canal, and a trembling motion is made, a singing, ringing sound is caused by the condensation and rare-faction of the air in the canal.   Stimulation of the auditory nerve by electricity is discussed on p. 699 and pathological conditions of irritation on p. 700.

COMPARATIVE. HISTORICAL.

The lowest forms of fishes, the cyclostomata (lampreys) possess only a saccule, provided with auditory hairs and otoliths, communicating with two semicircular canals; the myxinoids have only one semicircular canal.   Most of the other fishes have a utricle, with three semicircular canals typically developed.   The osseous fishes have in the cysticula of Brechet (Fig. 325, V, C) the first indication of the cochlear canal leading from the saccule.   In the carp and the shad posterior prolongations and diverticula of the labyrinth are connected with the air-bladder by means of a chain of three auditory ossicles.   In several of the herrings and perches, bladder-like processes of the air-bladder are either in immediate contact with the labyrinth, or in close proximity to it.   According to Kreidl the carps, and according to Beer the crustaceans, do not react at all through the auditory apparatus to auditory stimuli, and the fishes only through their highly developed cutaneous sense, which is set into activity by the sound-waves.   The organs of the "side line" in fishes are intended for the preservation of the equilibrium. The amphibia are in general closely related to the fishes with respect to the con-struction of the labyrinth, but the cochlea is not typically developed.   Most of them, except the frog, have no tympanum.   The fenestra ovalis alone exists, and not the fenestra rotunda, the former being connected in frogs with the exposed tympanic membrane by means of three ossicles   In reptiles the saccule, appended to the cochlear canal, is quite prominent; in tortoises it is still a simple sac, but in crocodiles it is longer and somewhat curved and dilated at its extremity.   In all reptiles the round window is found for the first time; through it the cochlea communicates with the vestibule.   The cochlea is divided into a scala tym-pani and a scala vestibuli in crocodiles and birds.   Snakes have no tym-panic cavity.   In birds the saccule and the utricle are fused (Fig. 325, IV, ÜS). The cochlear canal (ÜC), which is connected with the saccule by means of a fine tube (C), is already longer.   It exhibits indications of a spiral arrangement, and it possesses a flask-like, blind end, the lagena (L), which is present likewise in crocodiles.   The auditory ossicles in reptiles and birds are reduced to one, which is columnar in shape, and corresponds to the stapes; it is known as the columella.   The lowest mammals (echidna and duck-bill) are still more like the birds in structure; the higher mammals, however, exhibit the same type of auditory apparatus as man (Fig. 325, III).   In whales the Eustachian tube is always open.   According to G. Retzius all vertebrates possess so-called hair-cells as end-organs of the auditory nerves.

Among *invertebrates* the ear is found in a simple form in several of the medusæ, annelids, and molluscs.   It is a round vesicle, filled with fluid, on the wall of which are the auditory nerves with ganglionic enlargements.   The inner wall of the vesicle bears cells provided with cilia (auditory cells), which contain either only one otolith composed of concentric layers, or numerous crystalline movable otoliths.   The otoliths consist of an organic base, which is impregnated with lime-salts.   In the medusæ the auditory vesicles lie in the margin of the bell (*marginal bodies*).   According to more recent views, however, the otoliths regulate the equilibrium of the animal, by pressing harder in one direction on the surface beneath them, with every change of position.   Verworn proposes, therefore, to call them statoliths.   Extirpation of the saccules containing the otoliths disturbs the equilibrium of the animals.

In molluscs the ears are situated on the side of the gullet, and in several they are connected with the surface of the body by a fine tube (helix). In the crustacea there are otolith-saccules, partly closed and partly open. The auditory bristles are supplied with nerves, and are of various lengths; they support the otoliths. Other auditory bristles, supplied by the same nerve-trunk, are found on the surface of the body, on the antennæ, and on the tail. When a sound was conducted into the water Hensen observed several bristles to be set in vibration, which were attuned to various pitches. The lining membrane of the auditory vesicle is lost with every shedding, and the animals voluntarily replace their otoliths by grains of sand. In insects the ear is represented by a tympanic membrane, to which a tracheal vesicle is attached, and between which there is a ganglionic nervous expansion. In the acridia (cricket) the ear lies over the base of the third foot, in grasshoppers in the forefeet, in beetles at the root of the hind wings, and in flies at the bases of the poisers. There are, however, also in the antennæ, bristles connected with ganglionic fibers, and still other formations that are considered as auditory organs, as, for instance, the "*auditory pencils*" of arthropods. In cephalopods the ear is connected with the head-cartilage, and the first indications of a membranous and cartilaginous labyrinth are found. The nerve passes to a plate or ledge of horn, on which ciliated epithelial cells represent the end-organs.

Historical. Empedocles (473 B. C.) referred auditory impressions to the cochlea. The school of Hippocrates was familiar with the tympanic membrane; Aristotle (384 B. C.) knew of the Eustachian tube. According to Cassius Felix (97 A. D.) hearing is dulled during the act of yawning. Vesalius (1571) described the tensor tympani muscle, Ingrassias the stapes; the latter connected the function of the tensor with accurate hearing. Cardanus (1560) first mentioned sound-conduction through the cranial bones. More exact descriptions of the finer parts of the ear were made by Fallopius (1561), who described the vestibule, the semicircular canals, the chorda tympani, the two windows, the cochlea, and the aqueduct; by Eustachius (died 1570), who described the modiolus, and the bony staircase of the cochlea, the Eustachian tube, and the muscles of the auricle; by Plater, who described the ampullæ (1583); by Casseri (1600), who described the spiral lamina of the membranous cochlea. Sylvius de le Boë discovered (1667) the ossicle named after him, Vesling the stapedius muscle (1641). Mersenne (1618) knew of overtones. Gassendius determined the velocity of sound (1658). Follius described accurately the membranous labyrinth and the process of the malleus named after him (1645). Tulpius (1641) considered the possibility of air passing through the ears (when the drum is perforated), a condition that, curiously, was spoken of by Alkmäon (580 B. C.) as normal in goats. Subsequently, there was much discussion as to the possible existence of a normal opening in the tympanic membrane (foramen Rivini). Scarpa made a masterly dissection of the ear. Perrault (1666) suggested a theory similar to that of v. Helmholtz as to the perception of pitch by the cochlea. Berzelius investigated the cerumen chemically, Krimer the labyrinthine fluid. According to Authenrieth, the three differently placed, semicircular canals are supposed to aid in hearing sounds from the respective directions. The study of acoustics was greatly advanced by Chladni (1802). A most complete work on the ear of the vertebrates was written by G. Retzius (1881–84).

THE ORGAN OF SMELL.

STRUCTURE OF THE OLFACTORY APPARATUS.

The entrance to the nasal cavity is formed by the *vestibular region* or the *vestibule*. Its mucous membrane is covered with papillæ and it is lined with squamous epithelium, which reaches to the anterior extremity of the inferior meatus and the inferior turbinate bone. Near the opening of the nostril there are hairs (vibrissæ) with greatly developed sebaceous glands. Mucous glands are found toward the cartilages. The area of the terminal expansions of the olfactory nerve, the *olfactory region*, measures about 500 sq. mm., and in man it includes only the upper part of the septum, and the islands of the superior turbinate (Fig. 330, *Cs*); detached islands or peninsulas are found in the vicinity of this chief olfactory region. The remainder of the nasal cavity is designated the respiratory region. The differences between the olfactory and

58

respiratory regions are as follows: (1) The olfactory region possesses a thicker mucous membrane; (2) it is covered with a single layer of cylindrical epithelium, 0.06 mm. thick (Fig. 329, E), the branched basal portions of which often contain a yellow or brownish-red pigment (denser in animals), while the respiratory region has a double layer of ciliated epithelium, mixed with goblet-cells; (3) the olfactory region is, therefore, distinguished by the coloration mentioned; (4) it contains peculiar club-shaped tubular glands (the glands of Bowman), which are considered mucous glands. According to A. Heidenhain. the latter are serous, according to Stöhr (in man) mixed glands. Lymph-follicles are found in the mucosa beneath the epithelium. and from them numerous leukocytes make their way on to the free surface. (5) Finally, the olfactory region contains the end-organs of the olfactory nerve. The olfactory cells (N) lie scattered between the long, cylindrical epithelial cells (E) of the surface. A spindle-shaped cell-body, with a nucleus and large nucleolus sends upward, between the cylindrical cells a smooth

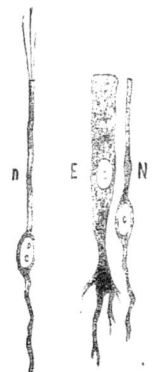

FIG. 329.—N, Olfactory cell from man (the hairs have fallen off); n, from the frog; E, epithelial cell from the olfactory region.

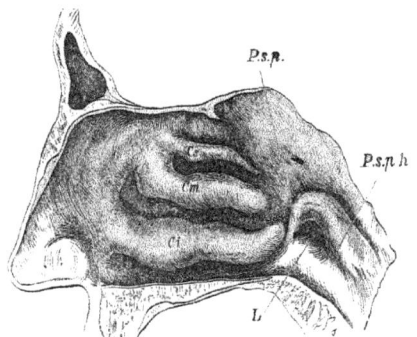

FIG. 330.—Nasal Cavity and Nasopharynx: L, levator pad; P.s.p., salpingopalatine fold; P.s.ph., salpingopharyngeal fold; Cs, Cm, Ci, the three turbinates (Urbantschitsch).

rod, from 0.9 to 1.8 μ thick, from the extremity of which from 6 to 8 fine olfactory hairs project through the pores of a delicate, structureless limiting membrane covering the surface of the epithelium. The olfactory cells become continuous with fine varicose nerve-fibrils in the depths of the mucosa. and these pass into the olfactory nerve. According to C. K. Hoffmann and Exner, after section of the olfactory nerves in frogs the specific end-organs are converted into a nonciliated cylindrical epithelium, while in warm-blooded animals they undergo fatty degeneration; but, at the same time. the epithelial cells between them exhibit signs of degeneration.

The olfactory cell of the olfactory region is a ganglion-cell whose neuron is represented by a nerve-fiber, which enters the olfactory bulb. The telodendrites of the neurons come in contact within the spherical glomeruli with dendrites from ganglion-cells of the bulb. Passing through further layers of the bulb (gelatinous layer, layer of pyramidal cells, granular layer) the origin of the fibers in the olfactory tract is reached, the course of which is described on p. 678.

SENSATION OF SMELL.

The sensation of smell is brought about by the action of odorous substances in a gaseous state, which come in direct contact with the olfactory cells, especially in their passage through the nares during inspiration. In the act of inhalation, the air passes along the septum, upward beneath the bridge of the nose, and under the roof of the nasal cavity, and it then curves backward and downward. But little air passes through the meatuses, especially through the superior; most passes through the middle meatus. Odorous substances received through the mouth and then expired through the choanæ may also be smelled, although not so well.

The first moment of contact of the odorous substance with the olfactory cells seems to be the most effectual for the sensation; consequently it is customary to repeat these inspiratory acts with closed mouth when it is desired to smell accurately: sniffing. By this means the air in the accessory cavities is rarefied, and as the air-pressure gradually becomes equalized, the odorous fumes are capable of diffusing over the entire region. Nothing is practically known as to the nature of the action of odorous substances, but many odorous vapors have a decided power of absorbing heat.

There is as yet no criterion for a special classification of odorous substances. The observation that certain categories of olfactory sensations can be abolished, while others remain intact, would seem to indicate that there are qualitatively different forms of olfactory nerves or end-organs.

The strength of the sensation depends: (1) Upon the extent of the surface affected; hence animals with great acuteness of smell (for example the seal) are found often to have exceedingly complex turbinates, which are covered with the olfactory membrane. (2) Upon the frequency with which the fumes are conducted to the olfactory cells (sniffing). (3) Upon the concentration of the odorous air-mixture; many substances, however, can be detected even in remarkable dilution. (4) There are many connections between smell and taste; chloroform has an ethereal odor and a sweet taste at the same time. Moreover, it excites the pain-producing and cold-perceiving nerves. Ether has a similar action, but it has a bitter taste.

Bromin may be detected by its odor in a dilution of $\frac{1}{30000}$; hydrogen sulphid in a dilution of $\frac{1}{30000}$ mgm. when contained in 1 cu. cm. of air. The odor of $\frac{1}{1500000}$ mgm. of chlorphenol, and of $\frac{1}{460000000}$ mgm. of mercaptan can be detected.

Odorous substances dissolved in indifferent solutions (for example 0.73 per cent. sodium chlorid solution) and introduced into the nose excite a feeble smell. The olfactory nerve is exhausted by olfactory sensations that persist for more than a few minutes; the exhausted nerve may recover, however in the course of a minute. The sensation is impaired by fever, and also by cocain. Mechanical and thermal stimuli do not excite olfactory sensations.

Variations of the olfactory sensation are described on p. 679. If both nostrils are filled with substances of different odors, some individuals do not appreciate a mixture of the odors, but at times one and at other times the other prevails; in some, however, there is a mixture of odors. Many odors cause others to disappear, when they act upon the nose at the same time, for example bitter almonds and musk, caoutchouc and wax. Under such circumstances both odors may be taken either into both nostrils, or into one and the same nostril.

The extremely sensitive sensory nerves of the nasal cavity are painfully irritated by some pungent fumes, for example of ammonia and of acetic acid; the latter act upon the olfactory nerves even in great dilution. The nose is important as a sentinel to guard against the introduction of bad air and food. The sense of smell frequently assists the sensations of taste, and conversely. Earlier and recent investigators speak of a connection between the nose and sexual activity.

To test the olfactory acuity Zwaardemaker makes use of the *olfactometer*, that is a hollow cylinder of an odorous substance (for example vulcanized caoutchouc), through which air is drawn into the nostril. A nonodorous tube can be introduced into this, so that any desired length of the odorous surface may be covered. The intensity of the smell is proportional to the length of the cylinder used.

The galvanic current—one electrode being placed in or on the nose, the other (indifferent) being held in the hand—upon kathodal closure and persistence of the current, likewise upon anodal opening, excites a sensation of smell that Kieselbach compares to the smell produced upon striking flint. Induced currents have no effect.

Comparative.—In the lowest vertebrates the olfactory apparatus is represented by depressions to which the olfactory nerve passes. Amphioxus and the cyclostomes have only one olfactory depression, while all other vertebrates have two. In many selacians the olfactory depression communicates with the mouth by means of a canal. In frogs the olfactory organs open into the mouth through short passages. In the higher vertebrates the nose develops together with the palate, and becomes more and more independent. In the gymnophions, a group of amphibians, the olfactory apparatus is extraordinarily developed from the presence of four nerves, while, on the other hand, the ears and eyes are stunted. The cetaceans have no olfactory nerve. In many mammals there is in the anterior part of the septum a hollow cavity, lined with cells similar to the olfactory cells, opening either into the nasal cavity or into the canalis incisivus, and to which a branch of the olfactory nerve runs; it is known as Jacobson's organ, and is undeveloped in man. Cephalopods have olfactory depressions, lined with ciliated olfactory cells, back of the eyes; the olfactory nerve arises near the optic nerve. In *molluscs* also, there are ciliated places that are considered olfactory organs. In arthropods the olfactory organs lie in the feelers and antennæ, in the first as cilia in connection with a ganglion and nerve. In crabs they are situated in the outer arms of the antennula. Ciliated, shallow or flask-shaped depressions supplied with nerves represent the olfactory apparatus in the higher worms. All other animals appear to possess no especial organ.

Historical.—Theophrastus (born 311 B. C.) mentions the short nose of man; that animals enjoy their food only from its odor; that strong perfumes cause headache; that many fragrant salves impart an odor to the urine; that there are many connections between smell and taste. Rufus Ephesius described the passage of the olfactory nerves through the cribriform plate of the ethmoid (97 A. D.). According to Galen the sense of smell has its seat in the cerebral ventricles. The monk Theophilus Protospatharius (end of the eighth century) described the olfactory nerve as the nerve of smell. Rudius (1600) dissected a man with congenital anosmia, in whom the olfactory nerves were absent. Sömmering wrote a masterly description of the olfactory apparatus, Cloquet (1815) of its physiological and pathological phenomena.

THE ORGAN OF TASTE.

SITUATION AND STRUCTURE OF THE ORGANS OF TASTE.

There are still many contradictory views as to the extent of the region in which the sensation of taste is developed, and accordingly as to whether the various nerves in question are to be considered as possessing taste-fibers or not. (1) The root of the tongue in the region of the circumvallate papillæ, the area of distribution of the glossopharyngeal nerve, is undoubtedly endowed with taste. (2) So also is the tip of the tongue and its margins, through the intermediation of most of the fungiform papillæ (the filiform papillæ and about 20 per cent. of the fungiform papillæ are insensitive to taste), but with many individual variations; so that often not all varieties of taste are appreciated. The relations of the nerves to these situations are pointed out in the descriptions of the lingual nerve and the chorda tympani. (3) The lateral portion of the soft palate, its posterior surface, the glossopalatine arch, and the inner surface of the epiglottis are endowed with taste through the glossopharyngeal nerve. (4) It is uncertain whether the hard palate also possesses the sense of taste; it is usually said not to be present in the middle of the tongue.

The end-organs of the gustatory nerves are the taste-buds or taste-goblets discovered by Schwalbe and Lovén. These are found on the lateral surfaces of the circumvallate papillæ (Fig. 331, I) facing the capillary cleft R R of the surrounding furrow, more rarely on the surface of the papillæ, and on the opposite side of the furrow. They occur also on the fungiform papillæ, on the papillæ of the soft palate and on the uvula, but also on the under surface of the epiglottis, the upper portions of the posterior surface of the larynx and the inner aspect of the arytenoid cartilage, and on the vocal bands. Many of these buds are said to disappear with age. The gustatory goblets, 81 *μ* high and 33 *μ* thick, are bud-shaped or barrel-shaped cellular structures, embedded in the thick squamous epithelium of the tongue. The outer portions are made up of curved, fusiform,

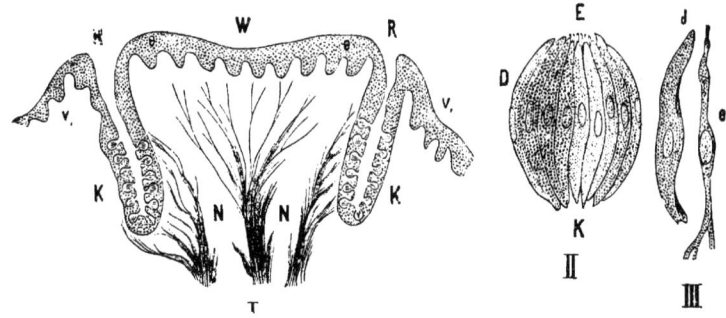

FIG. 331.—I. Transverse Section through a Circumvallate Papilla: W, the papilla; v₁ v₁, the wall in section; R R, the ring-shaped cleft; K K, the taste-bud in position; N N, nerves. II, Isolated taste-bud; D, cortical portion; K, lower extremity; E, free open extremity, with projecting tips of the taste-cells. III, Isolated cortical cells (d) and taste-cells (e).

nucleated investing or supporting cells, like the staves of a barrel (Fig. 331, II, D; isolated III, d). Toward the free surface they surround an opening, the *porus*, and beneath this a small depression. Surrounded by these cells, in the axis of the bud, there are from one to ten taste-cells (II, E), some of which possess a delicate process at their upper extremity (pin-cells—III, e), while others do not (rod-cells). The gustatory nerves lose their myelin-sheaths, and form plexuses, always ending free in the taste-buds, either by surrounding the buds with delicate fibrils on the outer side like a basket or by penetrating their interior. They terminate, ultimately, free, on a level with the opening of the taste-bud. After section of the glossopharyngeal nerve, the taste-buds degenerate within thirty hours, and the protecting cells are converted into ordinary epithelial cells in the course of twelve days. Leydig found in the skin of fresh-water fishes goblet-shaped organs similar to the taste-buds.

The glands of the tongue, to which the ninth cranial nerve sends secretory fibers, are discussed on p. 256; the follicles likewise.

GUSTATORY SENSATIONS.

There are four different qualities of taste: the sensations of sweet, bitter, sour and salty. Sour and salty substances irritate also the sensory nerves of the tongue. In greatest dilution, however, they stimulate only the endings of the specific nerves of taste. In all probability a special perceptive fiber exists for each quality of taste (in accordance with the doctrine of the specific energies).

The proof of this is as follows: Oehrwall, and after him Goldscheider and H. Schmidt, found that among the fungiform papillæ some reacted to sugar, but not to tartaric acid; some to quinin, but not to tartaric acid; and some to quinin, but not to sugar. Electrical stimulation of some papillæ excited a bitter taste, of others a salty taste, and of still others a sweet taste. With the constant current, the purest sensation was at the anode.

The leaves of Gymnema sylvestre, when applied to the tongue, destroy the sensation of sweetness and bitterness.

Continued stimulation of taste is followed by phenomena of fatigue for the various taste-sensations. This fact may be readily explained by the assumption of specific end-apparatus for the different categories of taste, which are present in relatively different number on the various papillæ.

With regard to the character of the stimulation of the gustatory nerves, no real advance has been made since the time of Democritus (469 B. C.), who attributed the taste-impression to the form of the tasting atoms. In order that a gustatory impression may be made, a solution of the substance in the fluids of the mouth is necessary, especially if it is in a solid or gaseous state. The intensity of the gustatory sensation depends: (1) Upon the extent of the surface affected, as Camerer especially showed by placing the substance upon 1, 2, 3 or 4 circumvallate papillæ. By rubbing the substance into the furrows and between the papillæ (rubbing movements of the tongue in the act of tasting) the perception is facilitated. (2) The concentration of the sapid substance is of great importance. Valentin found that the following series of bodies ceased to be tasted in the order stated, as they were progressively diluted: sirup, sugar, salt, aloes, quinin, sulphuric acid. Quinin can be diluted 20 times more than salt before it becomes tasteless. (3) The time that elapses between the application of the substance and the appearance of the sensation varies with different substances. Salt is most quickly tasted (after 0.17 second), then sweet, sour and bitter (quinin after 0.258 second). This is true also of mixtures of these substances. The last-named substances produce the longest after-taste. (4) The delicacy of taste is in the first place congenital (the newborn infant is said to be able to distinguish qualities of taste), but it can be greatly improved. Prolonged tasting of the same substance, or of similar or of strongly tasting substances, quickly impairs correct gustatory judgment. (5) The sense of taste is greatly assisted by the sense of smell, and the one is often confounded with the other. Thus, musk and asafetida affect only the organ of smell without stimulating the sense of taste. Even the eye is capable of assisting the sense of taste by the excitation of conceptions of familiar tastes. Thus, alternate testing of red and white wine, with the eyes bandaged, soon results in uncertainty. (6) The most suitable temperature for taste lies between 10° and 35° C.; hot and cold water abolish taste temporarily.

Ice placed on the tongue suppresses temporarily all power of taste, cocain only the bitter taste, chewing the leaves of Gymnema sylvestre the bitter and the sweet tastes. Two per cent. sulphuric acid makes water taken subsequently taste sweet. Sugar dissolved in a tasteless solution of salt or quinin, tastes sweeter then when dissolved in water. Children and the insane who refuse food may sometimes be induced to partake of substances repugnant to them by the smell of an agreeable perfume.

Electrical Taste-sensations.—The constant current excites an acid sensation at the positive pole, both on closing and on opening, as well as during the passage of the current; and an alkaline, or more correctly an astringent-burning, sensation at the negative pole. This cannot be the result of electrolysis of the saliva, for even when the tongue is moistened with an acid solution, the alkaline taste persists at the negative pole. The most probable explanation is that electrolytes are formed in the interior of the taste-bud that irritate the end-organ during the passage of the current. This is proved by the fact that the taste-sensation changes on the use of currents of different tension, so that it is dependent upon the ions liberated by the current. The constant current as such irritates the end-organs of the gustatory nerves directly only at the moment of closure and of opening. The sensation thus produced is added to the preceding excitation. If one electrode is placed on the tongue and the other (indifferent) in the hand the following phenomena appear: Kathodal closure and the passage of the current excite no taste-sensation on the root of the tongue; and the same is true of anodal opening. If, however, the anode is placed on the tongue, a sour taste is excited, both on closure and during the passage of the current, and also on kathodal opening. On the tip of the tongue, and on its middle portion, a salty or a bitter taste is excited when the kathode is placed on the tongue, upon closure and while the current is passing; likewise on opening, when the anode is placed on the tongue. No sensation results on anodal closure, or while the current is passing, or on kathodal opening. Rapidly interrupted currents cause no taste-sensation. Applications of cocain to the tongue abolish the electrical taste temporarily. The experiments of v. Vintschgau, whose taste was imperfect at the tip of the tongue, showed that the electrical current never excited a taste-sensation when applied there (although a distinct tactile sensation).

In experiments on Hönigschmied, who had normal taste-sensation at the tip of the tongue, the positive pole often excited a metallic taste at the tip but not rarely also an acid taste; while at the negative pole, taste was often absent, and when present, it was almost always alkaline, exceptionally acid. It is important to note that after interruption of the current a metallic after-taste could be recognized with both directions of the current.

Pathological.—Diseases of the tongue, coating of the tongue, and dryness disturb or destroy the sensation. Subjective tastes are common among insane or nervous patients, probably from irritation of the psychogeusic center. A bitter taste has been noted after poisoning with santonin, bitter and acid tastes after subcutaneous injections of morphin. Gymnemic acid is capable of destroying subjective tastes and parageusis.

The designations *hypergeusis*, *hypogeusis* and *ageusis* are applied respectively to increase, decrease and abolition of taste-sensations. Many forms of tactile sensation on the tongue are confused with gustatory sensations, for example so-called biting, cooling, pricking, sandy, mealy, pasty, astringent, bitter tastes.

Comparative.—In cattle there are as many as 1760 taste-buds to a circumvallate papilla. A large taste-organ, with numerous folds is described as the foliate papilla in the lateral posterior portion of the tongue in rabbits. This has an analog in man in the form of parallel furrows on the posterolateral edge of the tongue, the fimbriæ linguæ. Reptiles and birds have no taste-buds, which are numerous in the gill-slits of the tadpole, although the tongue of the adult frog is lined only with an epithelium suggestive of taste-cells. The goblet-shaped organs in the epidermis of fishes and tadpoles are similar in structure to the taste-buds, and probably have the same function. Taste-buds are present on the palate of the carp, and in the mouth of the shark and ray. In aquatic amphibians and in fish, the end-organ of the olfactory nerve is probably stimulated like the taste-buds, that is the stimulation takes place through the action of substances dissolved in the water.

The tongue of the cyclostomes serves as a suction-apparatus, while in other fish it has no muscular tissue. Salamanders and most of the batrachians can extrude the tongue from the mouth and again withdraw it. In many of the lower vertebrates the entoglossal bone serves as a support for the tongue, while in the higher forms it is replaced by the cartilage or the septum of the tongue. The nerve-endings in the proboscis (flies), jaw and tongue (ants), palate and epipharynx are the seat of the taste-organs in insects. Taste-organs have been found also in snails.

Historical.—Bellini considered the papillæ of the root of the tongue as the gustatory organs (1665). Sulzer reported in 1760 as to electrical taste-sensations. Baur was the first to describe accurately the course and the division of the muscles in the tongue; and Rudolphi the course of the nerves. Elsässer (1834) showed that the sensation of taste was most intense for all substances on the vallate papillæ, and on the posterior portion of the lateral margin of the tongue.

Richerand, Foderà, and Mayo considered the lingual nerve alone to be the gustatory nerve. Magendie showed, however, that after section of this nerve, the posterior part of the tongue retained its taste-sensation. Panizza (1834) designated the glossopharyngeal as the gustatory, the lingual as the tactile, and the hypoglossal as the motor nerve of the tongue.

TOUCH.

TERMINATIONS OF THE SENSORY NERVES.

The tactile corpuscles, discovered by Meissner in 1852, are ellipsoidal in form, from 40 to 200 μ long, and from 60 to 70 μ wide, and they lie in the papillæ of the corium. They are abundant in the palm of the hand, and on the sole of the foot, likewise on the fingers and toes (21 to each sq. mm. of skin, or 108 for each 400 vascular papillæ). They are less abundant on the back of the hand and foot, on the mammilla, the lips, and the tip of the tongue; rare on the glans clitoridis, isolated on the volar aspect of the forearm (also in anthropoid apes and the raccoon).

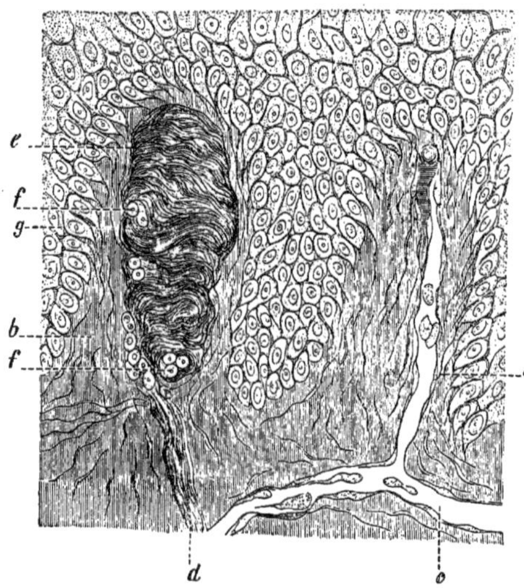

The tactile corpuscles have in their interior an ellipsoidal inner bulb composed of nucleated epithelial cells. The supplying nerve-fiber loses the sheaths of Henle and of Schwann at the base of the inner bulb, and these in turn surround and enclose the bulb. The nerve-fiber, at first medullated, then nonmedullated, makes spiral turns around the inner bulb, after which it breaks up into fibrils and penetrates the bulb. Here the isolated nerve-fibrils terminate with nodular enlargements between the cells of the bulb.

FIG. 332.—*a*, Vascular papilla; *b*, touch-papilla; *c*, blood-vessel; *d*, nerve-fiber passing to the tactile corpuscle; *e*, tactile corpuscle; *f*, nerve-fibers divided transversely; *g*, cells of the Malpighian layer (Biesia-decki).

Arth. Kollmann distinguishes especially on the hand three principal tactile areas: (1) The finger-tips, where there are 24 tactile corpuscles for each 10 mm. of length; (2) the three eminences on the palm behind the interdigital spaces, where there are from 5.4 to 2.7 tactile corpuscles for every 10 mm. of length; (3) the thenar and hypothenar eminences, where there are from 3.1 to 3.5 tactile corpuscles for each mm. The first two areas contain also numerous corpuscles of Vater, the third only a few. On the remaining surfaces of the hand the nervous end-organs are much less numerous.

The Corpuscles of Vater and Pacini (Fig. 333) are from 1 to 2 mm. long, and lie in the subcutaneous tissue, especially on the nerves of the flexor aspect of the fingers and toes (from 600 to 1400), in the mammillary region, in the neighborhood of joints and muscles, on the interosseous membrane, on the perimysium, on the tendons, on the plexuses of the abdominal sympathetic, at the side of the abdominal aorta and of the coccygeal gland, in the pancreas, on the pericardium, at the side of the knee of the facial nerve, on the back of the penis and of the clitoris, as well as in the mesocolon of the cat. Numerous connective-tissue capsules, separated from one another by fluid, like the layers of an onion, surround the homogeneous central bulb, which is filled with neuroplasm and is lined by flat epithelial cells. The lamellæ of the corpuscle are formed by hypertrophy

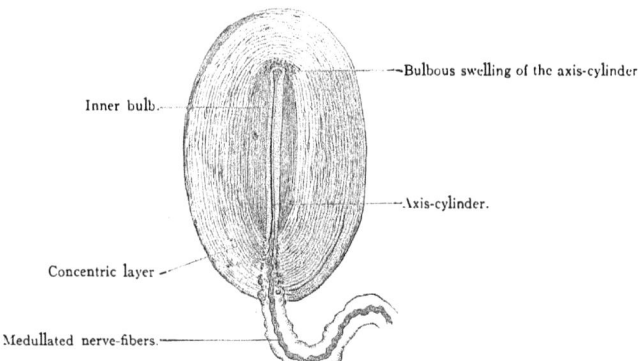

Inner bulb.

Bulbous swelling of the axis-cylinder.

Axis-cylinder.

Concentric layer

Medullated nerve-fibers.

Fig. 333.—Vater-Pacinian Corpuscle.

of Henle's layer of the nerve-fiber, and are composed of nucleated, flat cells. The medullated nerve-fiber, which enters the pedicle, loses its myelin-sheath, and its sheath of Schwann, and terminates as an axis-cylinder either in a single or in a bifurcated extremity, with a slight terminal enlargement, the end-bulb. within which each nerve-fibril ends in a most delicate terminal nodule.

Krause's longitudinal end-bulbs (Fig. 335) are found in the conjunctiva of

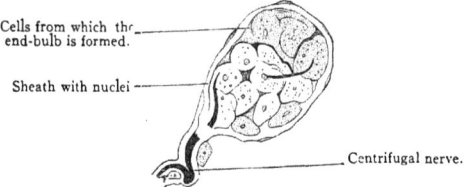

Cells from which the end-bulb is formed.

Sheath with nuclei

Centrifugal nerve.

Fig. 334.—Spherical End-bulb in the Human Conjunctiva (Longworth).

Fig. 335.—Longitudinal End-bulb: a, the nucleated sheath.

the eyeball, on the floor of the mouth, at the margin of the lips, in the nasal mucous membrane, on the epiglottis, on the fungiform and circumvallate papillæ, on the glans penis and clitoridis, in the tendilemma, in the tendons, on the sole of the foot in man, on the plantar surfaces of the toes (porpoise), on the ear and trunk (mouse) and in the wing of the bat. They are from 0.075 to 0.14 mm. long and are probably present in all mammals in the cutis and the mucous membranes as the regular form of nerve-ending. The adventitia of the double-contoured fiber passes over into the connective-tissue covering of the bulb, the sheath of Schwann becomes thickened and it develops into the inner bulb, consisting of cells of the longitudinal bulb. The spheroidal end-bulbs in man (nasal mucous membrane, conjunctiva, mouth, epiglottis, folds of the rectal mucous membrane) consist, according to Longworth and Waldeyer, in the interior of a spherical connective-tissue sheath of numerous closely grouped cells, between which the terminal fibrils of the nerves end (Fig. 334). Waldeyer compares these cells with those of the Grandry-Merkel corpuscles. These structures evidently are closely allied to the genital and articular corpuscles. The first appear to be end-bulbs fused together in varying degree in the skin of the glans penis and clitoridis. The joint-corpuscles are found in the synovial membrane of the finger-joints; they are larger than the end-bulbs, and exhibit on the outer surface numerous oval nuclei; as many as four nerves penetrate their interior.

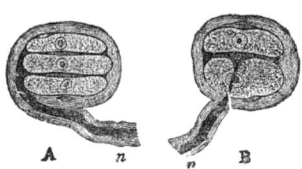

FIG. 336.—Grandry-Merkel Corpuscles: A consisting of 3 cells; B of 2 cells; *n*, nerve (tongue of the duck).

The **Grandry-Merkel corpuscles** occur in the so-called waxy covering of the bill, and in the tongue of ducks and geese. They are large cells with spherical nuclei and nucleoli, surrounded by a fibrous sheath, and between them a naked nerve-fiber is interposed by means of a protoplasmic disc—*tactile disc*. Two or more cells are often found on top of one another, with a nerve-end disc between them. When a number of such cells are placed upon one another and side by side larger structures are produced that appear to be transitional forms to the tactile corpuscles. In animals there are many other kinds of terminal corpuscles on the sensory nerves: The *corpuscles* of Herbst in birds, resembling small corpuscles of Vater, with longitudinal striation in the periphery and transverse striation within, but without a distinct capsule; the *tactile cones* in the snout of the mole and allied animals; the *end-capsules* on the penis of the hedgehog, and on the tongue of the elephant; the *tactile bulbs* on the beak and the tongue of several birds; the *nerve-rings* in the auricles of the mouse. Terminal ganglion-cells, connected with cilia, form the tactile organ in the rotifera, crustaceans, and insects.

The termination of the nerves by means of most delicate fibrils with knob-like ends (terminal nodules) between the epithelial cells of the cornea has already been described (p. 816). A similar arrangement exists also between the cells of the epidermis and between the epithelial cells of the genital organs.

In sensitive situations the peripheral ends of the nerve-fibers form distinct patelliform *tactile discs (tactile menisci)* within the epidermis, and upon

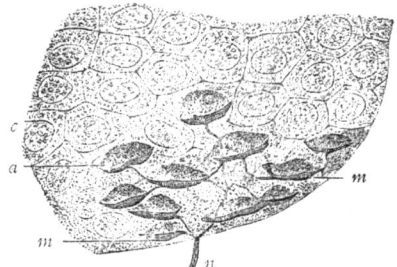

FIG. 337.—Tactile Discs with Nerves from the Epidermis (snout of the pig): *c*, epidermal cells; *a*, tactile cells; *m*, tactile discs; *n*, nerve.

them the lower cells of the Malpighian layer are placed. These structures are found in man and in animals, for example in the snout of the pig (Fig. 337).

On the hairs, which are in many places connected with the tactile apparatus, there is below the opening of the sebaceous gland a nervous end-organ in the external root-sheath, consisting of longitudinal and circular fibers, forming a network. Tactile discs are present in the cells of the outer root-sheaths of the tactile cilia in mammals.

SENSORY AND TACTILE SENSATIONS.

The sensory nerve-trunks contain two functionally different sets of nerve-fibers, namely: (1) Those that convey *painful* sensations, and are *sensory* nerves in the narrow sense of the word, and (2) those that receive *tactile impressions*, and are consequently designated *nerves of touch* or *tactile fibers*. Tactile sensations include the perceptions of *temperature* and of *pressure*. Some observers assume that the sensory and tactile nerves possess separate end-organs and nerve-fibers, and that they likewise have special perception-centers in the brain, although little of a definite nature is known in this connection. This view is supported: (1) By the fact that both sensory and tactile sensations are not excited at the same time in all of the areas endowed with feeling. Tactile sensations (including pressure and temperature) are transmitted only by the coverings of the external integument, the oral cavity, the entrance and floor of the nasal cavity, the pharynx, the end of the rectum, and the urogenital openings; feeble, indistinct sensations of temperature are appreciated also in the esophagus. On the other hand, tactile sensations are wanting in all the viscera, as experiments on men with fistulas of the stomach, intestine, and bladder teach; in these situations pain alone can be excited. (2) The paths for the tactile and sensory nerves are far apart in the spinal cord; this renders probable the assumption that also their central and peripheral extremities are distinct. (3) The reflexes (tactile and painful) excited by the two kinds of nerves are probably controlled or inhibited respectively by special central organs. (4) Under pathological conditions and under the influence of narcotics, the one kind of sensation may be abolished, while the other is preserved.

Bier and Hildebrandt found in themselves that after injection of cocain into the dural cavity of the spinal cord, the sensation of pain was abolished, while the sensation of touch persisted; sensations of cold and heat were preserved, but intense heat caused no pain. According to another view, the sensation of pain belongs to the nerves both of pressure-sense and of common sensation, and represents only an increase in the irritation of these nerves.

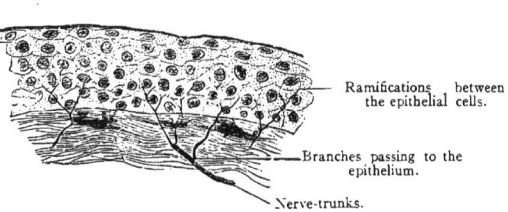

Ramifications between the epithelial cells.

Branches passing to the epithelium.

Nerve-trunks.

FIG. 338.—Nerve-endings in the Corneal Epithelium.

The nerves of sensation must be subjected to relatively strong irritation in order that pain may be excited. The irritant may be mechanical, electrical, thermal, chemical, or somatic, the last in connection with inflammatory processes, nutritive disorders, etc. The nerves are sensitive to irritations, not only at the peripheral extremity, but also throughout their entire course; and the central extremity is sensitive to irritation by pain The pain, however, is, according to the *law of peripheral perception*, always referred to the periphery.

The tactile nerves can convey pressure-sensations only as a result of moderately strong, mechanical irritations causing differences in pressure, and temperature-sensations as a result of thermal stimuli; and in both instances only when the peripheral end-organs are irritated. If pressure or cold be applied in the course of a nerve-trunk, for example to the ulnar in the depression in the inner condyle, sensations of pain—never of touch—are excited in the peripheral distribution. All intense irritants disturb normal tactile sensations by over-stimulation, and, therefore, excite only pain.

v. Vintschgau discovered that if two electrical tactile stimuli are applied to the middle of the forehead in succession, a shorter interval of time is usually required to perceive them as separate (0.022 sec.) than if they are applied to the dorsal surface of the lower arm (0.033 sec.).

If rapidly interrupted electrical or mechanical stimuli which can still be perceived as separate irritations, are permitted to act on the skin, and if the stimuli are then suddenly withdrawn, a new sensation arises after a short interval of rest. This *secondary sensation* appears as a short stinging sensation. It is supposed to result as a summation within the cells of the sensory paths in the spinal cord, and to be identical with the phenomenon of delayed sensation of pain.

The law of specific energies presupposes the existence among the cutaneous nerves of different fibers with different end-organs, which conduct the various forms of sensation (pressure, temperature, pain). In fact Blix and Goldscheider have found such fibers. Electrical stimulation causes different sensations in different minute punctate areas of the skin: in one place pain alone is perceived, in another cold, in a third heat, and in a fourth the sensation of pressure. At each *temperature-point* there is insensitiveness to pain or pressure. The *pressure-points* are much closer together and usually more numerous than the temperature-points. There are also special *pain-points* and *ticklish points*. These sensory points are arranged in linear chains, which usually radiate from the hair-papillæ. *Ticklish points* coincide with the pressure-points and pain-points. The sensations of tickling and itching correspond to the feeblest irritation of the nerve-fiber, that of pain to the strongest irritation. The pain-points may be shown by the needle and by electricity, especially in the wrinkles of the skin, in which the pressure-sense is absent.

Goldscheider removed small pieces of his own skin, in which he had previously determined the various points, and examined the tissues microscopically. At every sensory point be found an extraordinary number of nerves; at the pressure-points there were no tactile corpuscles.

The best way of testing the tactile sense in general, according to E. Hering, is by means of numerous rods, wrapped with wire of different size. The coarse wire is, naturally, the easiest to distinguish on account of its unevenness, while fine wire appears, on the contrary, almost smooth when the tactile sense is not acute. The different portions of skin exhibit varying degrees of tactile delicacy, and they may be arranged in the following order, from the most delicate to the least delicate: finger-tips, palm of the hand, inferior surface of the toes, back of the hand, flexor surface of the forearm, buttocks, extensor surface of the forearm, leg, upper arm, thigh, scapular region.

SENSE OF SPACE.

Man is able not only to distinguish differences in pressure or in temperature and also pain as such, by means of his nerves, but also to locate the point where the impression is made; this faculty is designated the spatial sense.

Method of Testing.—(1) Two blunt *compass-points* are placed at different distances upon the portion of skin to be examined, and the greatest distance is determined at which the two points are still perceived as one. Instead of the compass, Sieveking's esthesiometer may be used. This consists of two points, one fixed, and the other movable on a scale like a cobbler's measure. (2) With the points fixed at a distance at which they can be perceived as separate they are moved over other parts of the skin, and the subject is asked whether the points seem to move closer or further apart. (3) Two compasses with their points separated unequally are placed on two different portions of the skin, and the subject is asked to state when they seem to be equally separated: *Fechner's method of equivalents.* Thus, a separation of four lines on the forehead seems to be equal to a separation of 2.4 lines on the upper lip. Camerer found, in general, that the separation of the points applied to a portion of the skin endowed with delicate tactile sensitiveness is equivalent to a much greater separation in a less sensitive area. (4) A portion of the skin can be touched with a blunt rod, and the subject with his eyes closed be asked to indicate exactly where he was touched.

Investigation has yielded the following results: The spatial sense in a given portion of skin is the more highly developed:

1. The more numerous the tactile nerves that terminate in the area in question.

2. The greater the mobility of the part; hence it is most delicate in the extremities, toward the fingers and toes; also in parts of the body that are moved with great rapidity.

3. In the extremities the sensitiveness is greater in the transverse than in the long axis. It is one-eighth greater on the flexor surface of the upper arm, and one-fourth greater on the extensor surface. Likewise, the flexor surface is more sensitive than the extensor—one-sixth more in the upper extremity.

4. The method of application of the compass-points has an influence: (*a*) if they are applied in succession, instead of together, or if they are considerably warmer or colder than the skin, or if they are unequally warm, it is possible to distinguish a separation of shorter distances; (*b*) if the examination is begun with the

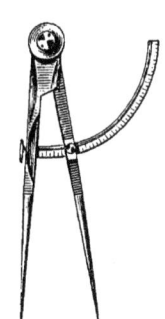

FIG. 339.—Compasses for Testing Sensation.

points far apart and the distance is gradually lessened, it is possible to recognize shorter distances than when the examination is begun with the points separated by an indistinguishable distance and the distance is gradually increased; (*c*) if one point is cold, and the other hot, two impressions are felt when the minimum distance is exceeded, but it is impossible to determine their relative position.

5. The spatial sense can be sharpened by practice; hence its delicacy in the blind, and the improvement is always bilateral.

6. Moistening the skin with indifferent fluids increases the delicacy of the spatial sense. If, however, the skin between two points that are still recognized as separate be gently tickled or be traversed by imperceptible electrical currents, the impressions become fused.

FIG. 340.—Sieveking's Esthesiometer.

The spatial sense is sharpened at the kathode on applying the constant current, likewise by congestion of the skin in consequence of irritation, and also by slight stretching of the skin; further after carbonated baths or warm sodium chlorid baths, and temporarily by the use of caffein.

7. Anemia (induced by elevation of the extremities) and venous hyperemia (induced by compression of the veins) impair the spatial sense; likewise too frequent repetition of the tests (as a result of fatigue).

The same influence is exerted by the application of cold to the skin, by the action of the anode, strong stretching of the skin, for example of the abdominal walls during pregnancy, likewise previous exertion of the muscles situated beneath the cutaneous area; as well as certain poisons: atropin, daturin, morphin, strychnin, alcohol, potassium bromid, cannabin, and chloral hydrate.

The shortest distances in millimeters at which two compass points were recognized as separate by an adult are as follows (the analogous figures for a boy twelve years old are enclosed in parenthesis): Tip of the tongue 1.1 mm. (1.1); palmar aspect of the third phalanx of the finger 2–2.3 (1.7); red part of the lip 4.5 (3.9); palmar aspect of the second phalanx of the finger 4–4.5 (3.9); palmar aspect of the first phalanx of the finger 5–5.5; dorsal aspect of the third phalanx of the finger 6.8 (4.5); tip of the nose 6.8 (4.5); palmar aspect of the head of a metacarpal bone 5–5.5–6.8 (4.5); thenar eminence 6.5–7; hypothenar eminence 5.5–6; middle of the palm of the hand 8–9; middle and border of the back of the tongue, white part of the lips, metacarpus of the thumb 9 (6.8); plantar aspect of the third phalanx of the great toe 11.3 (6.8); dorsal aspect of the second phalanx of a finger 11.3 (9); cheek 11.3 (9); eyelid 11.3 (9); middle of the hard palate 13.5 (11.3); palmar aspect of the lower third of the forearm 15; skin over the front part of the zygoma 15.8 (11.3); plantar aspect of the metatarsal bone of the toe 15.8 (9); dorsal aspect of the first phalanx of a finger 15.8 (9); dorsal aspect of a metacarpal bone 18 (13.5); inner aspect of lip 20.3 (13.5); skin over the posterior part of the zygoma 22.6 (15.8); lower portion of the forehead 22.6 (18); posterior portion of the heel, 22.6 (20.3); lower portion of the occiput 27.1 (22.6); back of the hand 31.6 (22.6); submental region 33.8 (22.6); top of the head 33.8 (22.6); patella 36.1 (31.6); sacrum and gluteal region 40.6 (33.8); forearm and leg 40.6 (36.1); back of the foot near the toes 40.6 (36.1); sternum 45.1 (33.8); upper portion of the neck 54.1 (36.1); spine (fifth dorsal vertebra), lower dorsal and lumbar 54.1; middle portion of the neck 67.7: arm, thigh, and middle of the back 67.7 (31.6–40.6).

By experimenting according to method 4 (p. 925), it is found that the spatial sense is best developed in the face and in the furrows of the finger-joints; then follow: the palm of the hand, the back of the hand (error as high as 1½ cm.), the neck, the arm (error up to 2 cm.), the clavicular region, the upper arm, the abdomen (error up to 3 cm.), the chest, the back of the foot, the leg (error up to 4 cm.), thigh (error up to 7 cm.). The touching of one toe is often confused. Pregnant women localize poorly upon the abdomen.

Illusions of the spatial sense are quite common. The most striking are: (1) A uniform movement over the surface of the skin seems to be more rapid on those parts that possess the most delicate spatial sense. (2) If the skin be merely touched by two compass-points, they seem further apart than when they are stroked over the skin. (3) A sphere provided with short rods appears larger than one with long rods. (4) When two fingers are crossed, small objects placed between them seem to be doubled (Aristotle's experiment). (5) If, however, the terminal phalanges of two fingers are first touched in the normal position of the fingers, and then the same places when the fingers are crossed, the two points touched seem to lie in the same relative position. (6) If flaps of skin are transplanted, for example a flap with a pedicle from the forehead to the nose, the patient will often for months have the feeling in the new portion of the nose as though it were the forehead, providing the nerves of the forehead remain intact.

Many attempts have been made to explain the phenomena of the spatial sense. E. H. Weber started from the assumption that one and the same nerve-fiber, passing from the brain to the skin could receive and transmit only one kind of impression within the area supplied by it. He gave the name of *sensory circle* to each region of the skin to which a single fiber was distributed. If two impressions act simultaneously upon the tactile apparatus, they are recognized as double if one or more sensory circles lie between the two points. This interpretation, based on anatomical considerations, cannot be reconciled with the fact that the circles of sensation may be reduced in size by practice, and, further, that only one sensation arises when the two points are so applied that, although further apart than the diameter of such a circle, they at times lie upon two adjoining circles, and at other times upon two other circles separated by a third.

Following Lotze, Wundt assumes from a psychophysiological standpoint that each area of the skin sends to the brain, together with tactile impressions, information as to the localization of the sensation. Hence, each area is able to give to the tactile sensation a *local coloring*, which is made use of as a *local sign*. Wundt assumes that this local coloring varies from point to point of the skin. The gradation is abrupt on those parts of the skin in which the spatial sense is highly developed, but gradual in those where it is comparatively poor. Separate impressions become fused wherever the gradation of this local coloring is imperceptible. As it is possible by exercise and attention to distinguish differences of sensation that cannot ordinarily be appreciated, the reduction in the size of the circles of sensation by practice can be thus explained. The circle of sensation is an area of skin within which the local coloring of the sensation is so little changed that two separate impressions are fused into one.

Loeb has made experiments upon the *tactile area* of the hand, that is the total number of points that an individual can reach with the tip of the forefinger, without change in the position of the body. If, with closed eyes, the hands are moved along a cord stretched transversely to the right and the left respectively, an inequality within the distance traversed will be apparent: in right-handed persons the distance to the right is generally smaller, and in left-handed the distance to the left. Nervous patients often exhibit marked deviations. In the attempt to make movements of the same extent, the movement executed will be the smaller the more the muscles are already contracted. The perception of the size and the direction of voluntary movements depends upon the voluntary impulse sent to the muscles.

THE PRESSURE-SENSE.

Through the pressure-sense information is obtained as to the amount of weight placed upon the skin; v. Frey considers the hair-nerves and Meissner's corpuscles as the organs of the pressure-sense. The pressure-sense is subserved by specific nervous end-organs having a punctate arrangement. These *pressure-points* possess different degrees of sensibility; in many places (such as the back and the thigh) they are characterized by an especially marked after-sensation. The distribution of the points corresponds to that of the temperature-points. The chains of pressure-points usually take a different direction from that of the hot and cold points; in general their density is greater; but this varies in different regions. In places provided with hair the number of pres-

FIG. 341.—Pressure Points: *a*, from the middle of the sole of the foot; *b*, from the skin of the zygoma; *c*, from the back (after Goldscheider).

sure-points does not correspond exactly with the number of hairs, but the points always lie in circles around the hairs. Hence, the hair when touched can also transmit the pressure-sensation, as it presses like a lever on the nerves of the root-sheaths. The smallest distances at which two pressure-points applied simultaneously can be felt as double have been found to be as follows: on the back, from 4 to 6 mm.; on the chest, 0.8; on the abdomen, from 1.5 to 2; on the cheek, from 0.4 to 0.6; on the arm, from 0.6 to 0.8; on the forearm, from 0.5 to 1; on the back of the hand, from 0.3 to 0.6; on the palm of the hand, from 0.1 to 0.5; on the palmar aspect of the distal phalanx of a finger, 0.1; on the dorsal aspect, from 0.3 to 0.5; on the leg, from 0.8 to 2; on the back of the foot, from 0.8 to 1; on the sole of the foot, from 0.8 to 1 mm. In the parts of the body devoid of hair, the tactile corpuscles are supposed to transmit the pressure-sensation.

Method of Examination.—(1) Weights of different amount are placed successively on the parts of the skin to be tested, and the subject is asked to form an estimate of the differences in pressure. In order to exclude, so far as possible, the influence of temperature, displacement and inequality in application, the area of skin should be previously covered by a plate, which is allowed to remain throughout the experiment. The influence of the muscular sense must also be eliminated. (2) A projecting arm from a scale-beam is placed on the skin, and by the addition or removal of weights the differences in pressure are learned that the subject is capable of estimating. (3) In order to avoid the troublesome changing of weights A. Eulenburg constructed his *baresthesiometer*, an apparatus constructed upon the principle of the spiral-spring balance. It is provided with a small button directed downward, which is depressed by the force of the spring. An indicator marks directly the pressure in grams, and this can at once be readily varied. (4) Goltz employed a pulsating, elastic tube, in which waves of different height could be produced. He determined how high they had to be before they were perceived as pulse-waves on the different areas of skin on which the tube was placed. (5) The *mercurial pressure-balance* constructed by the author satisfies all requirements completely (Fig. 342).

A scale-beam (W) resting on knife-blades (O O) is supported on the horizontal arm (b) of a heavy stand (T). One arm of the scale possesses a thread (m) on which a balancing weight (S) can be moved to and fro. The other arm (d), which passes vertically upward, terminates in a graduated tube (R). From the latter there projects downward a pressure-button (P), to which weights (G) can be added at will, and which rests upon the area of skin to be tested (H). From a buret (B) near by, which is supported on an upright (A), mercury can pass through the hollow arm of the scale, in the direction of the arrow, and mount upward in the tube (R). A thin, readily movable piece of rubber tubing connects the arm (O) with a fixed glass tube, and the latter passes subsequently to the rubber tube (D D) of the buret. If the cock (h) is closed, the mercury moves onward in d and rises in R, thus increasing the pressure of the button (P) whenever pressure is made on the tube (D D). The weight of the mercury filling one division of the tube (R) is known. The apparatus permits, without any agitation whatever, of rapid or slow variation in pressure, with any selected initial weight (through G). In the figure a indicates a screw for varying the position of the supporting arm (b): t is an arrangement with two screws to prevent the scale-arm from tipping over. The more extensively pressure is made upon the tube (D D), the greater, naturally, will be each increase of pressure. By raising the buret (B), if h is open, the pressure can also be increased.

If P is at first supported the mercury can be allowed to rise in R to different heights (in order to produce different amounts of pressure), and after closing the cock (h), the pressure of the button can be permitted to act suddenly by quickly releasing its support. In general, those methods are to be preferred in which the different pressures act at distinct intervals of time, instead of allowing the original pressure to increase or decrease gradually, because in the latter method the cutaneous nerves are gradually fatigued. Both the pressure-sense and the temperature-sense (to be discussed presently) may be most reliably tested by the principle of the least perceptible difference, that is by permitting different pressures (or temperatures) to act in graduated order, commencing either with great differences or with the smallest ones, and seeking the limit at which or within which a positive recognition of the difference takes place.

The results of the investigations of the pressure-sense are as follows:

1. The minimal pressure that can just be perceived on different parts of the body varies greatly in accordance with the locality. The most delicate areas are the forehead, the temple, the back of the hand, and the forearm, which perceive a pressure of 0.002 gm. The fingers do not recognize a pressure of less than from 0.005 to 0.015 gm.; the chin, the abdomen, the nose a pressure of less than from 0.04 to 0.05 gm.; nor the finger-nails of less than 1 gm. In order to test the pressure-sense of individual small points, they may be pressed upon with a flexible, elastic hair. The most delicate pressure-sensations in these situations were: the face, as low as 0.0007 gm., the arm and the leg, as low as 0.012 gm. Pressure is more readily perceived when applied

suddenly than when gradually increased. Decrease of pressure is less readily perceived than increase.

2. Intermittent variations of pressure are more readily recognized than light pressure, rapid variations with more difficulty than those occurring at longer intervals.

The greater the sensitiveness of a portion of the skin the more rapidly may individual impulses or blows follow one another and yet be perceived as separate: on the posterior aspect of the thigh 52, on the back of the hand 61, on the finger-tips 70 impulses in a second.

3. Differences between two weights are perceived by the finger-tips when they are in the ratio of 29 : 30 (by the forearm when the ratio is 18.2 : 20), provided that the weights are not too light or too heavy.

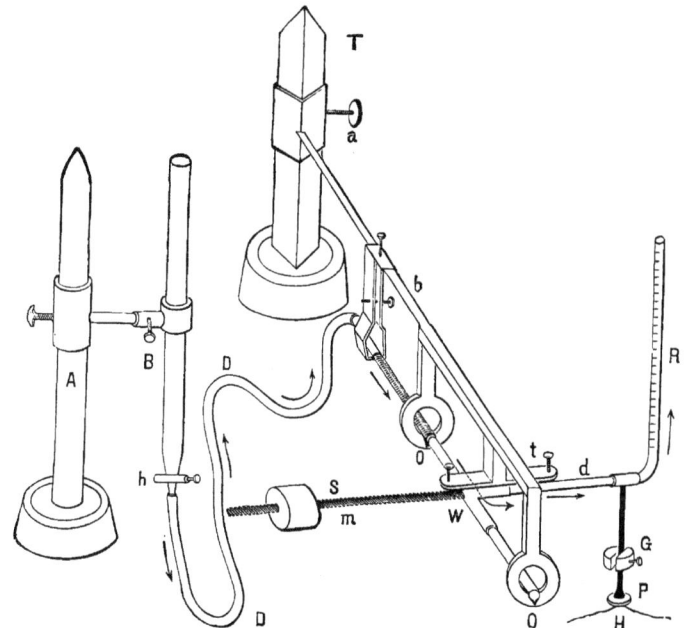

Fig. 342.—Landois' Mercurial Pressure-balance.

Ascending from light to heavier weights, the accuracy in distinguishing between two weights increases at first, and then decreases rapidly for heavier weights. This observation is contradictory of the psycho-physical law of Fechner.

4. A. Eulenburg found the following gradations in the accuracy of the pressure-sense: the forehead, the lips, the back of the tongue, the cheek, and the temple showed differences of from $\frac{1}{40}$ to $\frac{1}{30}$ (from 200 : 205 to 300 : 310 gm.). The dorsal aspect of the last phalanx of the fingers, of the forearm, of the hand, of the first and second phalanges, the palmar aspect of the hand, and of the forearm and the arm perceived

59

differences of from $\frac{1}{10}$ to $\frac{1}{20}$ (from 200 : 220 to 200 : 210 gm.). The anterior surface of the leg and thigh resembled the forearm. Then followed the back of the foot and of the toes; the sensitiveness was much less on the plantar aspect of the toes, the sole of the foot, and on the posterior aspect of the thigh and leg. Dohrn tried to determine the smallest increase of weight that in the presence of a weight of 1 gm. could be appreciated by different portions of the skin. This was for the third phalanx of the finger 0.499 gm., the back of the foot 0.5 gm., the second phalanx of the finger 0.771 gm., the first phalanx 0.82 gm., the leg 1 gm., the back of the hand 1.156 gm., the palm of the hand 1.018 gm, the patella 1.5 gm., the forearm 1.99 gm., the sternum 3 gm., the umbilical region 3.5 gm., the back 3.8 gm. The delicate lanugo-hairs of the skin are especially sensitive to pressure.

5 Too long a time must not elapse between the application of two weights, but as much as 100 seconds may elapse if the difference in weight is in the ratio of 4 : 5.

6. The after-effect in connection with the pressure-sense is especially pronounced when pressure of considerable amount is applied for some time. Even slight pressures, however, when applied repeatedly and successively, must be separated by intervals of at least from $\frac{1}{610}$ to $\frac{1}{480}$ second in order to be perceived individually. More rapid succession causes confusion of the impressions. Valentin found that, when he held the finger-tip against a wheel set with blunt teeth, he had the impression of a smooth edge if the teeth struck the skin at the intervals mentioned; when the revolution was slower, each tooth excited a separate pressure-sensation. Vibrations of strings are recognized as such when they make from 1506 to 1552 vibrations in the second.

7. It is remarkable that pressure effected by thoroughly uniform compression of a part of the body, for example by immersing an arm in mercury, is not perceived as such; the finger dipped in mercury perceives the pressure only at the limit of the fluid on the palmar surface of the finger.

8. The points that are sensitive to pressure are also sensitive to traction. If pressure and traction are applied alternately to the same area of the skin, the distinction is possible only when the irritation has a certain extent, duration and intensity. Traction (acting somewhat as a negative pressure) is tested by the application of small pieces of plaster, which can be drawn upon by means of a thread. The forehead and the temple are capable of recognizing 0.05 gm. of traction-force, the finger-tips and the lower lip 0.5 gm., the forearm 9 gm., the leg 20 gm.

THE TEMPERATURE-SENSE.

Through the temperature-sense information is obtained as to the variations of the temperature of the surface of the body. The temperature-sense is subserved by specific nerve-endings having a punctate arrangement. These *temperature-points* are arranged in chains or lines, which usually are slightly curved. They radiate from certain points of the skin (chiefly from the roots of the hairs). The chains of the *cold-points* do not coincide, as a rule, with those of the *heat-points*, although both have the same areas of radiation. These lines of points are frequently not complete, but are indicated only by isolated points, between which points of another sensation are frequently interposed. In this way mixed punctate chains arise. Near hairs there are almost always temperature-points; in areas of the skin with feeble temperature-sensibility, the temperature-points are present only near the hairs. According to Abrutz the cold-points are situated more superficially in the skin than the heat-points; according to v. Frey, Krause's bulbs are the organs for the perception of cold, the nerve-plexus that for the perception of heat.

The heat-points are larger than the cold-points; the slightest mechanical irritation of the latter excites a sensation of cold. The cold-points react to feebler electrical stimuli than do the heat-points; chemical irritants also are capable of exciting the heat-points. The feeling of cold occurs at once, that of heat appears gradually.

A gentle touch is not perceived as such at the temperature-points, which seem to be anesthetic for pressure and pain. In general, the cold-points preponderate upon the entire body, and they are denser, while in many places the heat-points are entirely wanting. With regard to the degree of sensibility the points may be divided into those that are

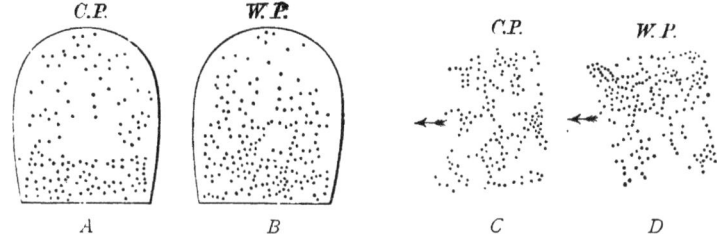

C.P.   W.P.   C.P.   W.P.

A   B   C   D

FIG. 343.—A Cold-points; B heat-points on the palmar aspect of the distal phalanx of the index-finger to the margin of the nail (Goldscheider). C, Cold-points and D heat-points on the radial half of the dorsal aspect of the wrist (the arrow indicates the direction in which the hair points (Goldscheider).

extremely sensitive, those that are moderately so, those that are slightly so, and those not at all sensitive. It is possible to indicate the intensity of the temperature-stimulus and the point where it is applied. The heat-points are, on an average, perceived as double at greater distances than the cold-points. The minimal distances upon the forehead are for the cold-points 0.8 mm., for the heat-points from 4 to 5 mm., on the chest the respective values are 2 and from 4 to 5 mm., on the back from 1.5 to 2, and from 4 to 6, on the back of the hand from 2 to 3 and from 3 to 5, on the palm of the hand 0.8 and 2, on the thigh and leg from 2 to 3, and from 3 to 4 mm.

For testing the heat-points and the cold-points a pencil-shaped metallic rod heated to from 45° to 49°, or cooled to 15°, is employed. When the cold-points are lightly touched only the cold rod is felt and as cold; and only heat is appreciated by the heat-points. Both kinds of points are insensitive to lightly applied objects of the same temperature as the skin.

The determining factor with respect to temperature-sensibility is, according to E. Hering, the temperature of the thermal end-organ itself. Whenever the temperature of the latter in any part of the cutaneous surface is above its own zero-temperature, that is, its normal temperature, there will be the sensation of warmth; under opposite conditions, the sensation of cold. The greater the deviation of the thermal apparatus from its zero-temperature, the more distinct or the more intense will be the sensation of heat or cold. The zero-point may, however, be displaced quite rapidly within certain limits as a result of external conditions.

Method of Examination.—Areas of the skin are successively touched with objects of different temperature, of equal size and possessing equal heat-conducting powers. (1) Nothnagel employs for this purpose small wooden cups, with metal bottoms, which are filled with cold or warm water, and are placed on the skin; the temperature of the water is indicated by a thermometer. (2) Two thermometers, which are heated unequally (if necessary by electrical means), can be applied directly to the skin for comparison.

The following facts have been ascertained with regard to the temperature-sense:

1. In general, the feeling of cold arises when a body applied to the skin withdraws heat, and, conversely, that of warmth, when heat is communicated to the skin.

2 The greater the heat-conducting power of the body touching the skin the more intense is the feeling of heat or of cold.

3. Between the limits of 15.5° and 35° C. the finger-tips are able to distinguish differences of temperature of from 0.20° to 0.25° C. The temperatures most exactly determined are those that lie close to the temperature of the blood (from 33° to 27° C.), differences as small as 0.05° C. being recognized (in the most sensitive situations). Temperatures between 33 and 39° C. and also those between 14° and 27° C. are less accurately determined. Temperatures of 52.6° C. and +2.8° C. and lower cause in addition to the temperature-sensation, marked pain, but in this respect there are variations in different individuals, and in different places between —11.4° and +36.3° C.

4. The sensitiveness for cold is in general greater than that for heat, and greater on the left hand than on the right. The different portions of the skin differ in their acuteness of heat-perception, and in the following order: tip of the tongue, eyelids, cheeks, lips, neck, trunk. Nothnagel found the minimum of differentiation on the chest 0.4°, on the back 0.9°, on the back of the hand 0.3°, on the palm 0.4°, on the arm 0.2°, on the back of foot 0.4°, on the thigh 0.5°, on the leg 0.6°, on the cheek from 0.4 to 0.2°, on the temple from 0.4 to 0.3° C. Curiously, the skin in the median line (for example of the nose) has a less acute heat-perception than the lateral portions (alæ of the nose). According to Dessoir, the glans penis has no perception of heat.

Fig. 343 shows the difference in the topographical arrangement of the heat-sense and the cold-sense on the same portion of the skin.

Goldscheider assumes 12 empirically determined grades of sensitiveness for the perception of cold, and 8 for that of heat. Each part of the skin has a comparatively constant grade of sensitiveness. For example, the skin of the mammilla has the grade 11 for cold-perception, the grade 8 for heat-perception; the middle of the sole of the foot respectively 7 and 2.

Application of a 10 per cent. solution of cocain to the tongue and the mucous membrane of the mouth abolishes completely the sensibility for heat and cold. The cooling sensation due to menthol depends upon stimulation of the nerves for cold; carbon dioxid irritates the heat-nerves of the external skin.

5. The differences in temperature are best recognized when different degrees of temperature are applied to one portion of the skin in rapid

succession. Gradual variations of a temperature continuously in operation are the more imperfectly recognized the more slowly they take place. If two different temperatures are allowed to act at the same time, near each other, the impressions are readily confused, especially if the two positions are close together.

6. Practice sharpens the temperature-sense; venous congestion of the skin blunts it; ischemia increases its delicacy. The power of differentiation is greater when the application is made to a large cutaneous surface than when made to a small surface. Rapid variations, further, cause more pronounced sensations than gradual changes of temperature. Fatigue occurs readily.

7. According to Abrutz strong thermal irritants excite sensations of cold as well as of heat: the *feeling oj heat* is supposed to result from a combination of these two sensations. Direct application of carbon dioxid excites a sensation of warmth; menthol gives rise to a sensation of coldness and burning.

Fig. 344.—Topography of the Cold-sense and the Heat-sense on the Same Part of the Anterior Surface of the Thigh: *a*, cold-sense; *b*, heat-sense. (The dark areas are the highly sensitive, the shaded areas the moderately sensitive, the dotted areas the slightly sensitive, and the clear areas the nonsensitive points.)

Various illusions may occur also in connection with the temperature-sense: (1) Occasionally the sensation of heat and cold will alternate in a paradoxical manner: for example if the skin be immersed in water at a temperature of 10° C., a sensation of cold results; if it then be immediately transferred to water at a temperature of 16° C., there is first a sensation of heat, but soon again that of cold. (2) The same temperature will be estimated as higher when applied to a large surface of the skin than when applied to a small surface. Thus, the entire hand immersed in water at a temperature of 29.5° feels warmer than when the finger is dipped into water at a temperature of 32° C. Cold weights feel heavier than warm ones.

Pathological.—Sharpening of the tactile sense (hyperpselaphesia) occurs but rarely, although greater sensitiveness for differences in temperature has been observed in places where the epidermis is thinned after the use of vesicants and after vesicular eruptions (zoster); likewise in tabetic patients. Sharpening of the spatial sense has been noted also under the two conditions first named and in cases of erysipelas. Brown-Séquard describes as an abnormality of the tactile sense the sensation of three points when only two are in contact with the skin, or of two when only one is applied. The author observed in himself as a peculiar paradoxical localization of sensation that pressure with the edge of the finger-nail over the junction of the manubrium with the body of the sternum always caused a sense of pricking in the chin. In this case irritation of one of the terminal branches of the subcutaneous nerves of the neck is referred to the periphery of another branch of the same nerve. A similar phenomenon is observed in other nervous territories, especially where exact localization is rarely or never attempted. The latter is the case with the sensory nerves of the intestines. Hence, it is not astonishing that in the presence of painful affections of the intestines, pains appear in distant parts of the skin supplied by nerves from the same level of the spinal cord into which also the sensory visceral nerve enters. In the same category is the familiar occurrence of pains in the left arm in connection with heart-disease. Analogous conditions prevail in the head and the neck. A remarkable alteration of the spatial sense consists in the circumstance that, when the eyes are closed, the individual feels his body to be abnormally large, or greatly reduced in size, or at times has a sensation of the trunk being doubled. The author has observed the first condition also in connection with moderate morphin-intoxication. In cases of degeneration of the posterior columns of the spinal cord Obersteiner observed that the patient was uncertain whether he was touched on the right or the left side (*allochiria*). Brown-Séquard observed after section of a lateral half of the cord that irritants applied to the left side were felt on the right side, and conversely. Rarely in cases of brain-disease irritation applied to both sides of the body has been felt only on one side.

Impairment of tactile sensibility to the point of abolition (*hypopselaphesia* and *apselaphesia*) may be present in association with corresponding disorders of the sensory nerves or occur independently. Less commonly individual qualities of the tactile sensations are lost, for example the pressure-sense, or the tempera-ture-sense, or only sensibility for heat and cold, conditions that have been desig-nated *partial paralysis of the tactile sense*. Limbs that are *sound asleep*, and are insensitive to slight pressure-irritations, do not appreciate cold; the function of the pressure-fibers and the heat-fibers is not disturbed until much later.

COMMON SENSATION. PAIN.

By common sensations are understood pleasant or unpleasant sen-sations in parts of the body endowed with feeling, which are not referable to external objects, and which in their peculiarity can be neither described nor compared with other sensations. These include pain, hunger, thirst, disgust, fatigue, horror, vertigo, tickling, sensuality, comfort and discomfort, and the respiratory sensations of free or em-barrassed breathing.

Pain can appear wherever sensory nerves are present. The organs subserving this sensation appear to be the free interepithelial nerve-endings. The pain-points do not coincide, in general, with the pressure-points, and they are about 1000 times less sensitive than the latter. The cause of the pain is always an irritation of the sensory nerves exceeding the normal. All kinds of irritation: mechanical, thermal, chemical, electrical, and somatic (inflammatory processes, disturbances of nutrition and the like) may excite pain. Especially the last named appear to be particularly effective, as some tissues are exceedingly painful when inflamed (for example muscles and bones), while they are comparatively insensitive when incised. Pain may be excited throughout the entire course of a sensory nerve, from its center to the periphery, but the sensation is invariably referred to the peripheral extremity, in accordance with the law of peripheral reference. Thus, it may happen that irritation of the nerves, as in the scar of an amputation-stump, may cause pain that is referred to parts that have been long since removed. As a result of frequent irritation in the course of a sensory nerve the latter may lose its function at the site of the affection, so that peripheral impressions can no longer be perceived. If the painful irritation affects the central extremity of the nerve-tract, it will still be referred to the peripheral extremity of the nerve. In this way there arises the apparently paradoxical condition of *painful anesthesia.* It is noteworthy that pain-sensations cannot be localized with precision. The localization succeeds best when the irritation is applied peripherally to a small area (for example a pin-prick). When, however, the stimulation is applied in the course of the nerve, or in the center, or to nerves whose peripheral extremities are inaccessible (such as the intestines), pain results that cannot be localized (for example belly-ache). When the pains are severe the phenomenon of *irradiation of pain* is added, in consequence of which localization is impossible The pain rarely continues in uniform degree, but there occur, as a rule, exacerbations and remissions in the intensity and also paroxysmal exacerbations. This probably is due to the fact that pain often results from a summation of irritations, each of which causes no pain of itself.

The intensity of the pain depends first upon the irritability of the sensory nerves. In this respect there are important individual variations, some nerves, for example the trigeminus and the splanchnic, being extremely sensitive as compared with others. The greater the number of nerve-fibers affected the more severe is the pain. Finally, the duration is of importance, as the same irritation, long continued, may cause an intensification of the pain beyond the point of endurance.

According to the character of the sensation the pain is described, as stinging, cutting, boring, burning, shooting, throbbing, pressing, gnawing, tearing, twitching, and dull, the causes for the differences being, however, entirely unexplained. Painful sensations are abolished by anesthetics and narcotics, such as ether, chloroform, morphin, etc.

The best means for testing cutaneous sensibility consists in the employment of constant or induced electrical currents. The *minimum of sensibility* is determined as that strength of current that excites the first trace of sensation; and also the *minimum of pain,* that is, the weakest current that first excites distinct pain. The electrodes are metallic, about the size of a knitting-needle, and they are placed from 1 to 2 cm. apart. According to Bernhardt the following distances of the cylinders of the induction-apparatus represent the minima of sensation, and the figures in parenthesis the minima of pain in a healthy person: tip of the tongue 17.5 (14.1); palate 16.7 (13.9); tip of the nose, eyelids, gums, back of the tongue, red lips 15.7–15.1 (13–12.5); cheek, lips, forehead, 14.8–14.4 (13–12.5); acromion, sternum, nape of the neck 13.7–13 (11.5–12.2); back of the arm, buttocks, occiput, loin, neck, forearm, vertex, coccyx, thigh, back of the first phalanx, back of the foot 12.8–12 (12–9.2); back of the second phalanx, back of the metacarpal bone, back of the hand, leg, distal phalanx, knee 11.7–11.3 (10.2–8.7); palmar aspect of the head of the metacarpal bone, tip of the toe, palm of the toe, palm, palmar aspect of the second phalanx, hypothenar eminence, plantar aspect of the first metatarsal bone, 10.9–10.2 (8–4). Motczutkowski found the pelvic region the least sensitive to painful impressions, the sensitiveness increasing from this situation in all directions. The ventral aspect of the body is less sensitive than the lateral, and the latter less so than the dorsal. Those regions exhibit less sensitiveness that have a thick epidermis; those that are less exposed to external injuries and areas over joints and interosseous sutures exhibit increased sensitiveness.

Pathological.—When there is increased sensibility of the nerves transmitting painful impressions even slight contact with the skin, or a mere breath of air upon it, may cause the most violent pain (*cutaneous hyperalgia*), especially in the presence of inflammatory or exanthematous conditions of the skin. The designation *cutaneous paralgia* may be applied to certain unpleasant or painful abnormalities of sensation that are frequently localized in the skin, namely itching, formication, burning, and cold.   In cases of cerebrospinal meningitis a prick on the sole of the foot has occasionally been observed to cause a double sensation of pain and a double reflex contraction.   Perhaps this phenomenon may be explained by supposing that the conduction is delayed in a part of the irritated nerve.   *Neuralgia* occurs in the form of characteristic paroxysms of pain of great violence, with radiation elsewhere (for example neuralgia of the fifth nerve, p. 692).   It is due to pathological processes in the nervous apparatus.   Frequently during the attacks excessive pain is produced by pressure on the points where the nerve-trunks emerge from the bony canals or openings in the fasciæ, or grooves (Valleix' points douloureux).   The skin itself to which the sensory nerve passes may, especially at first, be the seat of great sensitiveness, but if the neuralgia be of long duration, the sensibility may be much impaired, up to the point of analgesia. In the latter event there may be pronounced painful anesthesia.

Diminution or abolition of the sense of pain (hypalgia and analgia) may be due to affections of the nerve-terminations, or of the nerve-trunks, or of the central insertions of the nerves.

In hysterical subjects, suffering from hemianesthesia, the remarkable observation has been made that the feeling of the affected side is restored when small metallic plates or compresses are applied to the skin (metalloscopy).   At the same time that the affected part recovers its sensibility, the corresponding part of the opposite, healthy side or limb becomes anesthetic.   It has been thought that a transference of sensibility takes place from the healthy to the affected side of the body.   The application of the metallic plates gives rise to galvanic currents whose intensity varies with the character of the metal, but the resulting phenomena cannot be attributed to these currents.   The explanation of the fact is found in the circumstance that a similar result occurs under entirely normal, physiological conditions.   In a healthy person every increase in sensibility on one side of the body, produced by the application of warm metallic plates or compresses, is followed by a diminution in sensibility on the opposite side.   Conversely, it is found that when one side of the body is made less sensitive by the application of cold metallic plates, the homologous part of the other side becomes more sensitive.

THE MUSCULAR SENSE.   POWER-SENSE.

The sensory nerves of the muscles constantly convey impressions as to the inactivity or activity of the muscles, and in the latter event, as to the degree of contraction (power of distinguishing the weights of various objects).   They furnish information also as to the amount of the contraction to be employed to overcome resistance (power-sense). The power of differentiating the weight of objects lifted is based upon a comparison of the degree of innervation with the duration of the latent period, that is of the time that elapses between the willing of the movement to raise the object and the actual commencement of the movement.   In a wider sense the muscular sense includes also the appreciation of active and passive movements, the recognition of position, and finally that of resistance and weight.   Obviously, the muscular sense must be largely aided by the pressure-sense, and conversely; although E. H. Weber showed that the muscular sense exceeds the pressure-sense in delicacy, as by its aid weights in the ratio of 39 : 40 can be distinguished, while with the aid of the pressure-sense only those in the ratio of 29 : 30 can be distinguished.   In some cases in man perfectly retained muscular sense has been observed in conjunction with total cutaneous insensibility.   A parallel condition is the ability of a frog deprived of the skin on its legs to jump without material disturbance.   The muscular sense is also greatly aided by the sensibility of the joints, the bones, the fasciæ and the tendons.   By the associated action of several sensations, especially in the muscles and tendons, there results the recognition of the temporary position of the extremities. Some muscles, for example the respiratory muscles, possess only slight muscular sensibility, which seems to be absent normally from the heart and unstriated muscles.

Method of Testing.—Weights are wrapped in a cloth and are suspended from the part to be tested (for example the leg) by a sling. The subject estimates the amount of the weight by lowering and raising it; and also the *difference in resistance* (of the weights), as well as the *minimum of resistance* (appreciation of the smallest weight). The *electromuscular sensibility* also may be tested by causing the muscles to contract by means of induction-currents and having the subject report as to the sensation thereby produced. In this way also the minimum of sensibility and of pain can be determined.

A healthy person recognizes a weight of 1 gram applied to his upper extremity; likewise an addition of one gram when the original weight was 15 grams, an addition of two grams when the original weight was 50 grams; and an addition of 3 grams when the original weight was 100 grams. The power-sense differs in different fingers. The lower-extremity (with the weight suspended from the knee), recognizes from 30 to 40 grams; but often only a heavier weight. Often, a difference of from 10 to 20 grams can be detected, or from 30 to 70 grams. In general, the same differences are detected whether the original weights are light or heavy. In blind persons the muscular sense is often heightened.

Section of the sensory nerves causes derangement of the fine gradations of movements. Meynert supposed that the motor cortical centers represented the cerebral center for the muscular sense, the muscles being connected by motor and sensory paths with the ganglion-cells in these centers. Support is given this view by the occurrence of complete ataxia as a result of destruction of those areas in which the psychomotor cortical centers of the extremities are situated.

Illusions occur in the range of the muscular sense. A weight held by one limb appears to become lighter as soon as other muscles of the limb are contracted, although these do not themselves aid in supporting the weight. Under converse conditions the weight appears to be heavier. If equally heavy objects of different size are lifted, the larger appear to be the lighter. A weight raised with both hands seems lighter than when raised with one hand. The following illusion has been observed with respect to the muscular sense of the tongue. If the tip of the tongue is pressed against a narrow interval between the teeth, and is moved to and fro, there results a feeling as if the teeth yielded in movement.

Excessive muscular activity causes the sensation of fatigue, of oppression and weight in the limbs, which is referable to the muscular sense.

Pathological.—Abnormal heightening of the muscular sense is rare (muscular hyperalgia and hyperesthesia). It occurs in the distressing condition of unrest designated *anxietas tibiarum* (fidgets), which is attended with continual change in the position of the limbs, and not rarely may be a source of annoyance even to healthy persons at night. Cramp is a condition attended with intense pain as a result of irritation of the sensory nerves of the muscles; it occurs also in association with inflammatory processes. Impairment of the irritability of the nerves of muscular sensibility appears also in part to be responsible for certain choreic and ataxic movements. In tabetic patients the muscular sense in the upper extremities may be normal or diminished; in the lower extremities it is usually considerably diminished. Occasionally, the electromuscular sensibility is impaired or even lost; in other cases the subjective sense of muscular activity is lost (paralysis of muscular consciousness). Adequate doses of cocain or alcohol are capable of heightening the muscular sense, while amyl nitrite blunts it.

PHYSIOLOGY OF REPRODUCTION AND DEVELOPMENT.

VARIETIES OF GENERATION.

Abiogenesis (Spontaneous or Equivocal Generation).—Even until modern times it was believed that inanimate substances, derived from the decomposition of organized matter, could under certain conditions again be transformed spontaneously into living matter. While Aristotle believed that spontaneous generation could be extended to include the insects (vermin), the few modern adherents of this theory applied it only to the lowest forms of life. As a result of much research along this line it has been demonstrated conclusively that when organized matter is subjected to a high temperature (200° C.) within hermetically sealed tubes and all bacteria therein are actually destroyed, spontaneous generation cannot take

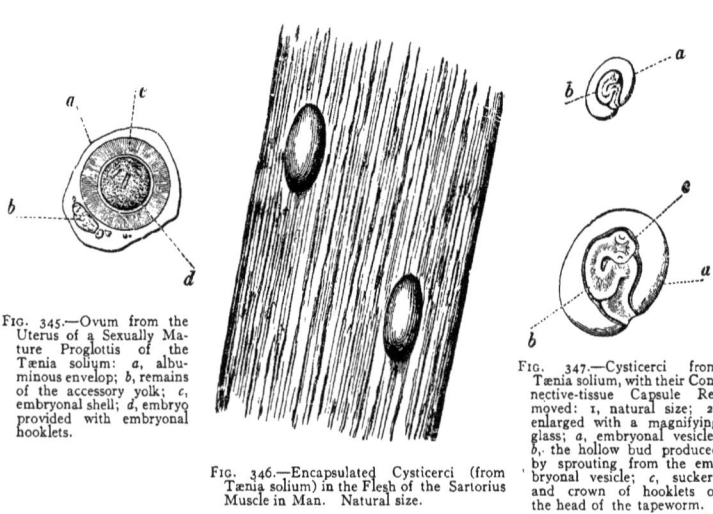

FIG. 345.—Ovum from the Uterus of a Sexually Mature Proglottis of the Tænia solium: *a*, albuminous envelop; *b*, remains of the accessory yolk; *c*, embryonal shell; *d*, embryo provided with embryonal hooklets.

FIG. 346.—Encapsulated Cysticerci (from Tænia solium) in the Flesh of the Sartorius Muscle in Man. Natural size.

FIG. 347.—Cysticerci from Tænia solium, with their Connective-tissue Capsule Removed: 1, natural size; 2, enlarged with a magnifying glass; *a*, embryonal vesicle; *b*, the hollow bud produced by sprouting from the embryonal vesicle; *c*, suckers and crown of hooklets of the head of the tapeworm.

place. This fact sustains the doctrine that all life is derived from previous life ("omne vivum ex ovo" or "ex vivo"); or as Harvey says: ut omnibus viventibus primordium insit, ex quo et a quo proveniant.

It is a noteworthy fact that even some of the higher invertebrates (gordius, anguilula, tardigrada, rotatoria) may, when kept dry for some time, apparently die and lie dormant for a considerable time, but when supplied with moisture may be resuscitated—*anabiosis*. Rotatoria (wheel-animalcules) recovered after having been kept in a dry vacuum for eighty-two days, and immediately after exposure for thirty minutes to dry heat at a temperature of 100° C. Rotatoria dried gradually in their natural habitation proliferated when again moistened after the lapse of eleven years, and the same is true of anguilulæ after twenty-eight years. Spores and seeds can be placed in a state in which metabolic activity is no longer demonstrable, and from which, as has long been known, they may under suitable conditions, again germinate. According to Decandolle, vegetable seeds may exhibit this property after from sixty to one hundred and fifty years.

Division takes place in many protozoa (amoebæ, infusoria), and in such a manner that, in accordance with the character of the cellular division, the organism, including its inner nuclear structure, and the cell-body, divides by an active process into two organisms. Starfish (ophidiaster) divide spontaneously, or they

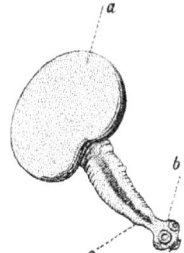

FIG. 348.—Cysticercus from Tænia solium with Everted Hollow Bud (Cephalic Segment): a, caudal vesicle (embryonal vesicle); b, the head of the tapeworm with suckers and ring of hooklets (scolex); c, cervical portion. Enlarged with magnifying glass.

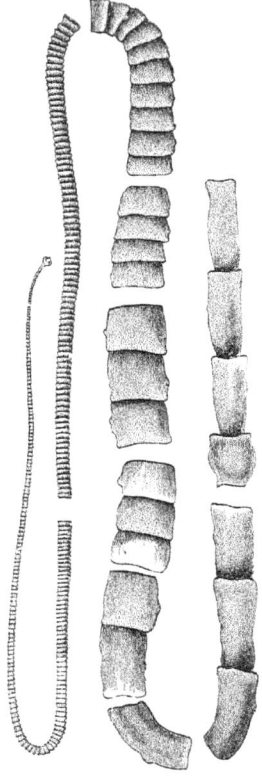

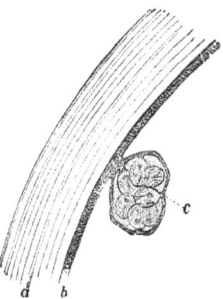

FIG. 349.—Portion of an Echinococcus-cyst with Brood-capsule: a, capsule; b, parenchymatous layer; c, brood-capsule filled with scolices (Figs. 345–349 after Sommer).

FIG. 350.—Tænia mediocanellata. Natural size.

eliminate an arm, which may develop into a complete animal. The artificial division of lower forms of animal life and the development of the fragments into entire beings were first demonstrated by Trembley in the hydra.

Budding or sprouting takes place in most marked degree in polyps; but also in infusoria (vorticellidæ) and others. It consists in the sprouting of a bud-like structure from the maternal body, which it gradually comes to resemble. The bud-like formations either remain attached permanently to the maternal organism, so that gradually a complete animal of considerable size is formed (polypariæ), the bodies of the individual remaining directly connected with one another (they may even possess a common "colonial" nervous system, like the bryozoa); or they may detach themselves and individually enter upon an independent existence. In some animal forms (siphonophores), the individual beings at times exhibit a definite differentiation of function, so that, for example, digestive, motor and reproductive activities may be distinguished—physiological division of labor. The formation of buds within the organism, which subsequently are detached, has been observed in rhizopods. Among animals that multiply by division or by budding, there has been observed in part also the formation of spermatozoa and ova (polyps, infusoria), so that here, together with asexual reproduction, there is at the same time sexual reproduction.

Conjugation or concrescence is the name applied to a variety of generation that is already suggestive of sexual reproduction, for example that of unicellular gregarines. In such a being the anterior extremity unites with the posterior extremity of another. Both become encysted into one round, resting body. The two nuclei coalesce and, after previous spindle-formation, a polar body

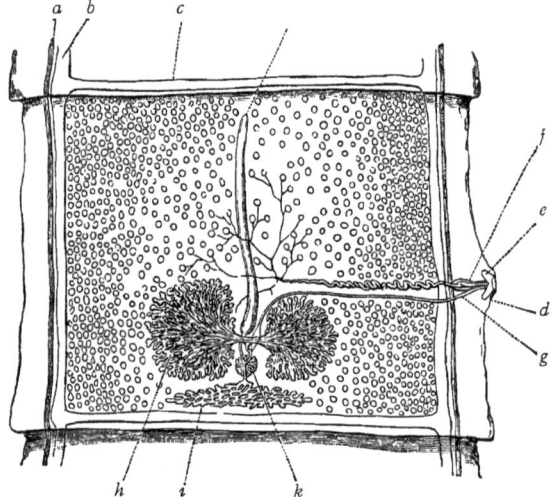

Fig. 351.—Sexually Active (Middle) the Proglottis of Tænia mediocanellata (after Sommer): *d*, sexual eminence with the genital pore *e*,—into the latter the penis (cirrus, *f*) projects from above, advancing in the segment into the tortuous vas deferens, which exhibits extensive ramification and leads to numerous testicular vesicles (most of the testicular vesicles are not yet connected by excretory ducts with the vas deferens); *g*, vagina; *h*, ovary; *i*, albumin-gland; *k*, group of shell-glands; *l*, uterus; *b*, excretory longitudinal trunk with transverse anastomosis *c*; *a*, lateral nerve.

is expelled. The united body-mass is resolved into a shapeless structure, from which numerous vesicles arise. In each vesicle many boatlike figures appear (pseudonavicellæ). These give rise to ameboid organisms, which by the formation of a nucleus and a protecting envelop are in turn transformed into gregarines. Concrescence has been observed also among some infusoria.

Sexual reproduction requires the formation of the offspring from the union of the male and the female generative elements (semen and ovum). These elements may be derived from two different individuals, male and female, or they may belong to the same individual (hermaphroditism, for example in tapeworms, snails, etc.). Sexual reproduction embraces also the following forms of generation:

Metamorphosis is the name applied to that form of sexual reproduction in which, from the fertilized ovum on, the organism appears in a succession of outwardly different forms (for example caterpillar, chrysalis), which possess no power of propagation. Then, finally, there develops the sexually mature form (imago, for example butterfly), which yields, through the union of sperm and ovum, the fecundated initial form in the developmental process. Metamorphosis occurs extensively among insects, either with many (holometabola) or with few intermediate stages (hemimetabola); likewise in other arthropods and in some worms (for example trichina). The sexually mature male and female descendants of the trichina are set free in the intestine, where they enter into sexual union and live for but a short time. They are known as *intestinal trichinæ*. They produce many eggs, which penetrate the muscular tissues of the host and constitute the *larvæ*. The encapsulated sexually immature muscle-trichinæ, which may remain quiescent for more than thirty years, are the *pupæ*, and these, when ingested in the living state by another suitable organism, develop in the intestine of the latter into sexually mature and active individuals. Among vertebrates, metamorphosis occurs also in the amphibia (frogs), and among fish in lampreys (petromyzon).

The **alternation of generations** (metagenesis) exhibits in common with metamorphosis the series of externally different forms in the process of development. It differs materially from metamorphosis, however, in the circumstance that the animal can multiply asexually in one or the other of the stages. The final stage alone then exhibits the sexual reproduction. Medically, the most important

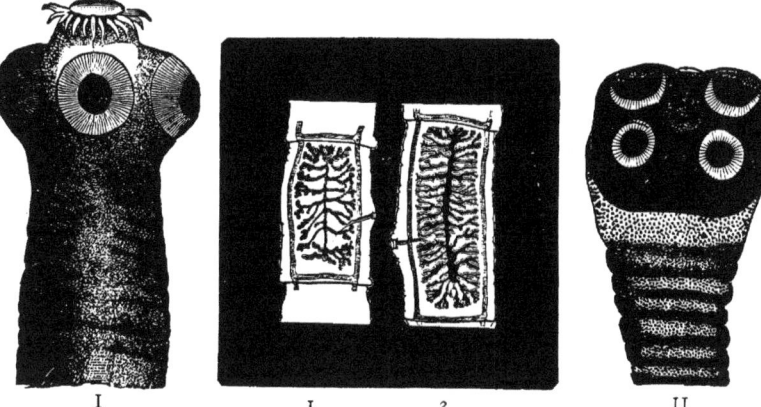

FIG. 352.—Heads of Tænia solium (I) and Tænia mediocanellata (II) and Mature Proglottids of each (1, 2).

example is furnished by the tapeworms (teniæ). The sexually mature, hermaphroditic individual, with hundreds of testicles, vasa deferentia, penises, ovaries, yolks, shell-glands, vaginas and uteruses (Fig. 351), is the protoglottis (tapewormsegment), which becomes detached and evacuated with the feces; it is motile and occasionally continues to grow. From an ovum (Fig. 345), rendered capable of reproduction by self-impregnation, there results an elliptical embryo, provided with six hooklets. This gains entrance with food into the intestine of another animal, whence it penetrates into the tissues and there develops into the third stage, the bladder-worm—cysticercus, cenurus, echinococcus. Within this vesicle there develops only one (cysticercus, Fig. 347) or several (cenurus) short-pedicled tapeworm-heads; or within the vesicle there develop at first numerous daughter-vesicles, and within those many heads (echinococcus, Fig. 349). For further development the bladder-worm must be consumed alive by another being. Then, the tapeworm-heads (scolex) attach themselves to the wall of the intestines by means of their hooklets or suckers and by budding form a chain of numerous segments (Fig. 350), each developed link of which constitutes a sexually mature offspring of the tenia. The most important tapeworms are as follows: Tænia solium is found in the intestine of man. Its bladder-worm, cysticercus cellulosæ (Fig. 352), occurs in swine, seldom in man. Tænia mediocanellata is found in the intestine of man (Fig. 352); its bladder-worm in the cow. Tænia cœnurus is found in the intestine of the dog, its cysticercus in the brain of the sheep (cœnurus cerebralis, the cause of staggers). Tænia echinococcus possesses but two or three segments, a few millimeters long, which are present in innumerable quantity in the intestine of the dog. Its encysted form (acephalocyst, with daughter-cysts) often attains the size of a child's head in man. It occurs in the liver, but also, though less frequently, in all other tissues. It is often dangerous to life and it is found also in slaughter-animals. Bothriocephalus latus is found in the intestine of man, its bladder-worm in the meat of the pike.

Among lower animals the medusæ also exhibit alternation of generation: among insects gall-gnats (cecidomyia, with endogenous larval multiplication) and plant-lice. The latter develop in the spring from impregnated, hibernated ova as asexual organisms. These produce successively in numerous generations unfertilized, living, likewise asexual offspring. In the late autumn the last of the young males and females are thus produced, and the latter, impregnated, deposit the fertilized ova.

Parthenogenesis or virgin reproduction is characterized by the circumstance that, in addition to sexual procreation, generation without sexual connection may also take place at the same time. The asexually produced brood is always of but one sex. The beehive serves as an example. It contains the queen-bee (sexually mature procreative female) the workers (imperfect females) and the drones (males). In swarming (nuptial flight) the queen is impregnated by a drone. The semen stored for three or four years in the reproductive life in the receptaculum seminis can apparently be added by the queen to the ova to be deposited for purposes of fertilization or be withheld from them. It is also possible that the matter of impregnation depends upon mechanical conditions related to the size of the comb in which the ova are lodged. Impregnated ova give rise to females only, unimpregnated ova to males only. If the queen is

FIG. 353.—Seminal Crystals.

incapacitated for flying and if she cannot be impregnated, she deposits ova that produce drones only. Generous feeding of the larvæ of the impregnated ovum, perhaps also the size of its comb (queen-bee cradle), favors the development of a perfect female (queen-bee), while if the nourishment be insufficient, sexually deficient working females result. This description has recently been challenged, it being maintained that the normally impregnated queen always deposits only impregnated ova, whose sexual development depends upon nutritive influences on the part of the workers.

In many of the higher animals the ova may pass through the first stages of development without impregnation, for example the hen, swine, rabbits, salpians, to the stage of division. Unfecundated ova of starfish even develop to the larval form.

Sexual reproduction without intermediate forms occurs in mammals, birds, reptiles and most fish.

THE SEMINAL FLUID.

The ejaculated seminal fluid is intermixed with the secretion of the alveolar glands of the tubules of the epididymis, the racemose glands of the vas deferens, the glands of Cowper and the prostate gland, and with the fluid of the seminal vesicles. It has a neutral or alkaline reaction and it contains 82 per cent. of water, serum-albumin, alkali-albuminate (propeptone from the accessory glands), nuclein (nucleinic acid + protamin), lecithin, cholesterin, fats, also fat containing phosphorus, and of salts (somewhat more than 2 per cent.) particularly alkaline and earthy phosphates, together with sulphates, carbonates and chlorids.

The testicle contains besides a hyaline-like albuminous body, leucin, tyrosin, kreatin, xanthin-bodies, inosite and glycogen (starch-like granules in birds).

The tenacious, whitish-yellow seminal fluid, for the greater part a mixture of the secretions from the organs previously mentioned, is, on exposure to air, at first coagulated into a gelatinous mass, then becomes again diffluent, on addition of water gelatinous and separating in the form of whitish, translucent flocculi. It forms, on standing for some time, elongated tapering, rhombohedral crystals that consist of the phosphate of an organic base, spermin ($C_5H_{14}N_2$), which results on the decomposition of nuclein.

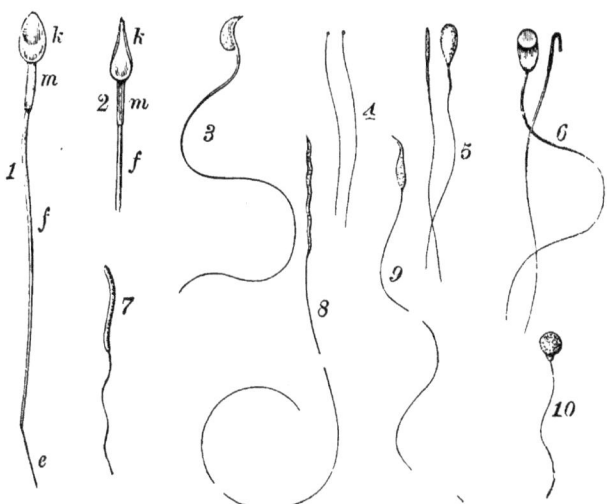

FIG. 354.—Spermatozoa: 1, human (× 600), the head viewed from the surface; 2, the head viewed from the side; k, head; m, middle piece; f, tail; e, terminal filament (after Retzius); 3, spermatozoon from the mouse; 4, from bothriocephalus latus; 5, from the deer; 6, from the mole; 7, from the green woodpecker; 8, from the black swan; 9, from the bastard of a goldfinch (male) and a canary bird (female); 10, from the cobitis (weather-fish) (after A. Ecker).

These crystals (Fig. 353) are derived in part also from the prostatic fluid (and they resemble the so-called Charcot's crystals observed in sputum). A small amount of seminal fluid (also stains dissolved in water), heated with a concentrated watery solution of iodin and potassium iodid, yields a crystalline formation (not unlike hemin-crystals). The formation is caused by a step in the disintegration of lecithin. Also other bodies containing lecithin form, through putrefaction, neurin, cholin and other decomposition-products and then yield the same reaction.

The *prostatic fluid* is a thin, milky fluid, amphoteric or of a feebly acid reaction, and it possesses the odor of the seminal fluid, which is given off by the base of Schreiner in solution. The phosphoric acid necessary to the formation of the crystals is supplied by the semen. Perhaps the prostatic secretion furnishes to the spermatozoa the motor stimulation essential for their power of impregnation. The ovary. the thyroid gland, the spleen, the pancreas and the leukocytes likewise contain spermin, though in less amount. An odor similar to that of the seminal fluid is possessed by Brieger's cadaverin (pentamethyldiamin), a nontoxic cadaveric alkaloid, as well as by the free diamins. It is these substances that give the odor to the sawdust of macerated bones, and occasionally to stale eggs or pike, and probably also to some plants, such as rhubarb, rhus, berberis. The secretion of the seminal vesicles (of the guinea-pig) contains a considerable amount of fibrinogen.

The seminal thread (spermatozoon, spermatosoma), 50 μ long, consists of a flattened pear-shaped head (**Fig.** 354, 1 and 2 k), a bodkin-shaped middle segment (m) attached to the broader pole of the head and the thread-like elongated cilium (flagellum or tail) (j), through whose to-and-fro movements the sperm, often rotating on its axis, traverses 400 times its own length in one minute, or from 0.05 to 0.15 mm. in one second. This activity is most pronounced directly after ejaculation.

The head, containing chromatin (mammals), consists of an anterior and a posterior segment. From the posterior segment a process projects like a sphere into the interior of the anterior segment. A delicate membrane covers the anterior segment of the head like a hood. The spermatozoa of some vertebrates possess at the anterior extremity of the head a projection furnished with barbs, which corresponds to the hood. The middle segment of the spermatozoon sometimes presents transverse striations, due to a spiral structure.

G. Retzius describes the spermatozoon as possessing a special, detached terminal segment of the tail, which represents the extremity of the latter. An axial fiber (Fig. 354, 1, e), surrounded by a protoplasmic sheath, passes through the middle segment and the tail. The sheath is lacking only at the extremity of the tail. The axial fiber consists of two filaments, each of which in turn is made up of numerous primitive fibrils. Also the terminal segment may be resolved into four fibrils. Some vertebrates possess a marginal filament arising from the middle segment, also an accessory filament parallel to the axial fiber and a steering membrane in advance of the terminal segment. In insects and amphibia the nonfibrillated axial fiber forms the supporting structure. In some organisms the spermatozoon is still more complicated. Only axial fibers having a fibrillar structure exhibit motility; those not so constructed are motionless.

The number of spermatozoa in man reaches 60,900 in 1 cu. mm.; it is increased after sexual excitement. To each mature human ovule, there are approximately 850,000,000 spermatozoa.

The **movement of the spermatozoa** is due to the circular, whip-like oscillation of the tail, which at the same time causes rotation about the long axis and is brought about by the protoplasm of the middle segment and the tail. Both of these, even if detached, are capable of movement. Ciliated cells, whose individual cilia consist of numerous filaments lying side by side, swarm-spores in plants and ameboid cells show analogous motility, as transitions between ciliated and ameboid movement have been observed, as in monera.

Human spermatozoa preserved without heat have exhibited motility after the lapse of nine days; spermatozoa from the guinea-pig after the lapse of eleven days. Permitted to rest passively in the testicle, in the absence of any diluting fluid, the spermatozoa possess no motility. They remain especially active in the normal secretions of the female genitalia. They retain their motility for a considerable length of time also in all normal animal secretions, except saliva. On addition of water they become rolled into rings and cease their movement. Also alcohol, ether, chloroform, creosote, gum, dextrin and vegetable mucus, concentrated solution of grape-sugar, as well as excessively alkaline uterine, and exceedingly acid vaginal mucus, acids and metallic salts and excessively high and excessively low temperature inhibit the activity of the spermatozoa. Their motility is unaffected by narcotics, insofar as they are chemically indifferent, and by solutions of urea, sugar, albumin, common salt, glycerin, amygdalin and other substances of moderate strength; although these inhibit motility like water if greatly diluted and by abstraction of water if unduly concentrated.

It is noteworthy that the rest occurring after the action of water, as well as that on gradual cessation of movement, may be terminated by the action of weak alkalies, as may be observed also in ciliated epithelium. Perhaps the alkalies neutralize an acidity of the protoplasm induced by fatigue; although Engelmann attributes restorative power to small quantities of acid, alcohol, and ether.

The spermatozoa of the frog may be frozen four times successively without injury; they endure a heat of 43.75° C. and continue to live for seventy days in the testicle transplanted to the abdominal cavity of another frog.

On account of the large proportion of earthy salts they contain, spermatozoa can be fused on a glass slide and nevertheless retain their form, like the cells of some plants rich in ash, for example the equisetaceæ. Nitric, sulphuric, hydrochloric and boiling acetic acid, and caustic alkalies do not destroy the form of the spermatozoa. Solutions of sodium chlorid and potassium nitrate, of from 10 to 15 per cent., transform the spermatozoa into amorphous clumps. The organic substance resembles the semisolid albumin of epithelial cells.

In addition to spermatozoa the seminal fluid contains seminal cells, a few epithelial cells from the vasa deferentia (isolated examples of which are in a state of colloid degeneration), numbers of lecithin-granules and occasionally laminated amyloid bodies, granular or scaly yellow pigment, especially in later life, leukocytes and sperm-crystals.

The development of the spermatozoon (Fig. 355) has been made clear only in recent times after considerable research, especially by v. Ebner, whose results were obtained simultaneously and independently by the author. From the nuclear protoplasmic layer (Fig. 355, I, b and IV, h) lining the inner surface of the wall of the seminal tubules (I, a and IV, n), which is composed of several layers of interlacing elastic fibers (interspersed with flat cells), large columnar processes,

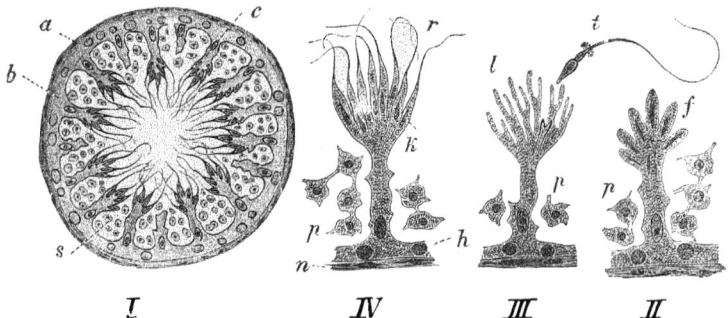

FIG. 355.—Spermatogenesis (Semidiagrammatic): I, Transverse section of a seminal tubule; a, its sheath; b, its protoplasmic internal layer; c, spermatoblast; s, seminal cells. II, Immature spermatoblast; f, its rounded upper lobules; p, seminal cells. III, Spermatoblast with released spermatozoon; t, spermatozoon; p, seminal cell. IV, Spermatoblast with mature heads (k) and cilia (r); n, wall of the seminal tubule; h, protoplasmic layer of the tubule; p, seminal cell.

0.053 mm. long (I, c and II, III, IV), project into the lumen. These break up at their free extremities into several oval lobules (II) like ears of corn and are known as spermatoblasts or seminal ears. These formations were first discovered by Sertoli, who considered and designated them "supporting cells." They consist of soft, finely granular protoplasm and contain an oval nucleus usually in their lower portion. In the course of development each lobule of the spermatoblast is prolonged into a long cilium, like the grains of an ear of corn (IV, r), and in the depth of the lobule the head and the middle segment of the spermatozoon (IV, k) are developed from a condensation of the protoplasm. In this stage the spermatoblast resembles a very large, irregularly formed ciliated cylindrical cell. When development is complete the head and the middle segment become separated from the mother-cell (III, t), and the remainder of the spermatoblast, with its resulting goblet-shaped depressions, resembles a threshed ear of corn (III, l). Later it undergoes fatty degeneration. The spermatozoon itself often exhibits for a long time an adherent mass of protoplasm at the junction of the head and the middle segment, representing a part of the spermatoblast (III, t). In accordance with its development, the spermatozoon may be regarded as a detached, independently motile cilium of a huge ciliated epithelial cell. Between the spermatoblasts lie numerous, round, ameboid, unencapsulated cells,

60

undergoing division and toward the termination of this process still connected with filaments. These are designated seminal cells (*I, s* and *II, III, IV, p*).

Contrary to the description just given the formation of spermatozoa has recently been described as follows: The spermatozoa originate from seminal cells, spermatogonia or primitive seminal cells. These multiply by indirect division, move toward the center of the seminal tubule, increase in size and are known as spermatocytes or seminal mother-cells. Each one now further subdivides into four cells—the spermatids or seminal cells, which move further toward the lumen. Each of these develops into a seminal filament or spermatocoma, the nucleus of the cell forming the head. a small portion of the cell-protoplasm the cilium of the spermatozoon, while the axial fiber of the latter develops from the central body of the cell. For the complete development of the spermatozoa, it is now necessary that the spermatids (seminal cells) unite with the free extremity of Sertoli's "supporting cells" through a form of copulation; and here mature as upon a nourishing stem and finally become detached. The united formations are the spermatoblasts of Ebner.

In most animals, the spermatozoa are capillary in form, with larger or smaller heads. The latter are elliptical or pear-shaped (mammals) or cylindrical (birds. amphibia, fish) or spiral (singing birds, sharks, viviparidæ) or simply capillary (insects and other animals) (Fig. 354). Nonmotile seminal cells differing entirely from the capillary form are found in myriapods and oysters.

Considerable interest has been aroused by the subcutaneous injection of orchitic extract recently made by Brown-Séquard. This greatly increases the ability to indulge in muscular exercise; coincidently with the diminution in fatigue there is a diminution in the subjective sense of fatigue. There is greater endurance and recuperation has an increased influence.

The significance of the secretion of the accessory sexual glands (prostate, seminal vesicles, Cowper's gland) has not been fully made clear. That these play some part in the act of procreation is evident from the fact that after their extirpation the impregnating power of the semen often ceases. In rats castrated before puberty, the accessory sexual glands do not develop. Hammar found secretion also in the epididymis of dogs. Castration or division of the seminal tubules causes atrophy of the prostate. The secretory nerves of the prostate are the hypogastric. The seminal vesicles preserve their function in sexually mature individuals even after castration.

THE OVUM.

The human ovum (from o.18 to o.2 mm. in diameter) is a globular, cell-like structure, presenting a thick, firm, elastic capsule, with delicate radiating striations (oolemma or zona pellucida), protoplasmic, granular contractile contents (yolk, vitellus), including a clear vesicular nucleus from 40 to 50 μ in diameter and possessing a nuclear framework (germinal vesicle) and a nucleolus from 5 to 7 μ in diameter endowed with ameboid movement (germinal spot). The chemical composition of the ovum is described on p. 423.

The zona pellucida (o.o2 thick) (Figs. 356, 357). to whose surface cells of the cumulus oophorus often adhere, may be regarded as a cuticular membrane developed secondarily from the follicle. Internally to it in many mammals and directly upon the yolk lies a delicate membrane, which is probably the original cell-membrane of the ovum. Between the zona pellucida and the yolk lies a small perivitelline space (Fig. 356). The finely radiate striations of the zona pellucida are due to the presence of numerous pore-canals. through which the adjacent cells of the granulosa send processes for purposes of nutrition.

In the ova of many animals—holothurians. many fish. for example sticklebacks, mussels, etc.—a special micropyle is observed. In addition. some ova possess a number of pore-canals collected together at a special area of the ovular membrane (many insects, for example the flea), and these serve partly as a means of ingress to the spermatozoa and partly for the respiratory interchange of gases.

The yolk has a peripheral clear layer, which encloses a finely granular layer and the latter finally a central mass containing numerous granules—yolk-granules or van Beneden's deutoplasm (Fig. 356).

The development of the ovum takes place in the following manner: The surface of the ovary is covered with cylindrical epithelium, the so-called *germinal epithelium*, between which here and there lie round *primordial ova* (Fig. 358, *I, a a*). In places the epithelial layer dips down to form tubular depressions in the surface of the ovary (*II*). These tubules, which, according to Waldeyer, are derived from the germinal layer of the ovary, become deeper and deeper, and within them are observed isolated, large globular cells, with nuclei and nucleoli, and also a larger number of smaller parietal cells. These tubes are the ovarian or ovular tubes; the larger round cells are the ova (primitive ova), the smaller, the epithelial cells of the tubes (*I*). At the bottom of the tubes the ovular cells, which may undergo mitotic division, predominate. Later on the orifices of the tubes close and the

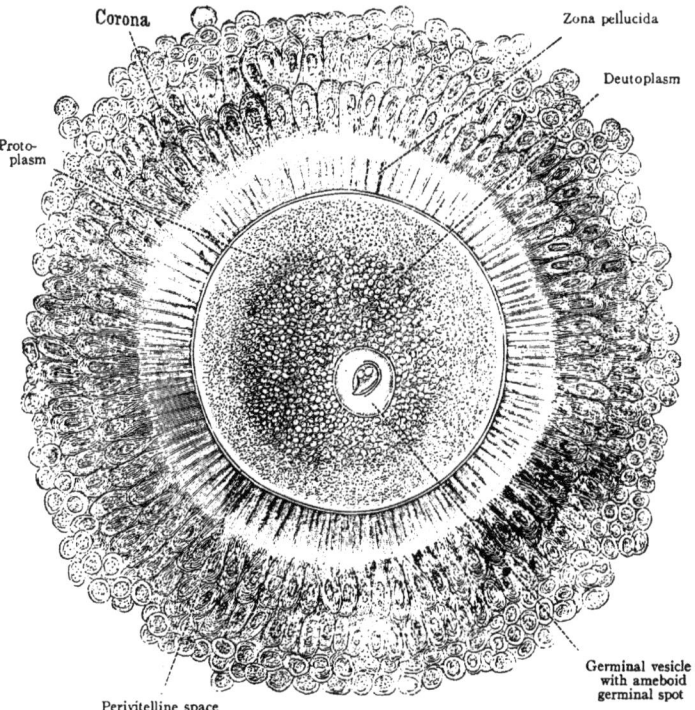

Corona   Zona pellucida

Deutoplasm

Proto-plasm

Germinal vesicle with ameboid germinal spot

Perivitelline space

Fig. 356.—A Fresh Ovum from the Ovary of a Woman Thirty Years Old. The side of the vitellus where the germinal vesicle is situated is directed toward the observer, who thus looks directly upon the germinal vesicle which lies upon the deutoplasm.

tubes are constricted off by the growth of the ovarian stroma into isolated rounded compartments (*I, c*). Each constricted compartment, which contains usually one, occasionally two ova (*IV, o o*), becomes a Graafian follicle. The follicles become distended with fluid; their parietal cells become the epithelium of the follicle or the cells of the granulosa, which at particular points surround the ova (*IV*). Such areas, designated cumuli oöphori, are spindle-shaped or cylindrical and consist of several layers; they produce the zona pellucida. According to some observers the yolk also is in part secreted by these cells into the ovule, and a number of the cells are believed to penetrate into the ovum. The follicles, at first only 0.03 mm. in diameter, attain complete development only at the time of puberty. The maturing follicles (*IV*) at first sink more deeply into the stroma of the ovary, become distended by taking up water (liquor folliculi), acquire a vascular, independent well-differentiated capsule (theca folliculi), and their epithelium (*IV, g*) (membrana granulosa) increases through mitosis in a similar manner, to form a layer of several rows of small cells. In the last stages of ripening the follicle leaves the depths of the stroma, again to reach the surface: it now attains a diameter of from 1 to 1.5 mm. and is ready to rupture. Only a small number of Graafian follicles attain normal final development; the majority previously undergo atrophy. In some animals (rabbits) the occurrence of furrowing has been observed as a noteworthy phenomenon.

The medullary substance, which extends from the hilus into the interior of the ovary, consists of vascular, fibrous connective and elastic tissue, with bundles of unstriated muscle-fibers; in contradistinction to the cortical substance, which contains principally cellular connective tissue, with the epithelial constituents in various stages of development. The ovary possesses numerous nonmedullated nerves (connected with sympathetic ganglia), of which the majority terminate in the walls of the vessels (also the capillaries), and others between the follicles and upon their surface.

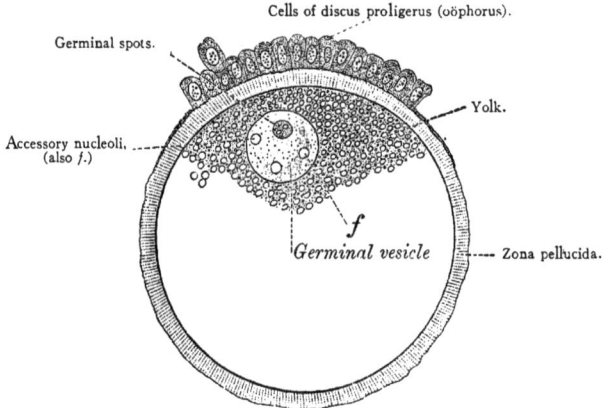

FIG. 357.—Mature Rabbit Ovum (after Waldeyer).

According to Paladino the ovary of woman is in a state of continuous involution and true new-formation through invagination of the germinal epithelium.

According to Waldeyer the mammalian ovum is not a simple cell, but a more complex structure. The original ovular cell, he believes, is formed only from the germinal vesicle and germinal spot, and the surrounding clear unencapsulated portion of the yolk (Fig. 358, *III*). The remaining portion of the yolk is derived from transformed granulosa-cells, which also constitute the zona pellucida.

In animals the following peculiarities may be observed in the formation of the ovum itself. The first ovular cells are known as primitive ova or ovogonia. They divide several times by mitosis at first into small, then into larger ovamother cells or ovocytes. These mature and after undergoing division by mitosis once or twice give rise to the polar bodies and thus form the true fully developed ovules.

Holoblastic and Meroblastic Ova.—The ova of batrachians and cyclostomata are formed according to the same type as those of mammals. They are designated holoblastic ova, because their contents are entirely transformed into the formative cells that serve for the development of the embryo. In contrast with these, birds, monotremata among mammals, reptiles, and the remaining fish have so-called meroblastic ova. These contain, in addition to the (white) formative yolk, which corresponds to the yolk of holoblastic ova, and yields the embryonal cells, also the so-called nutritive yolk (yellow in birds), which serves as a source of nutrition for the embryo during the period of development. This nutritive material penetrates into the originally small and simple ovular cell and causes it to swell considerably. The embryology of the bird's egg has shown that only the small, round, white protoplasmic germinal layer at the center of the surface of the yolk (cock's treadle, cicatricula), from 2.5 to 3.5 mm. wide and from 0.28 to 0.37 mm. thick, corresponds to the contents of the mammalian ovum and is therefore the formative yolk. It contains the germinal vesicle and the germinal spot (Fig. 359). From this, which contains also the characteristic white yolk-elements (Fig. 360, a) processes extend into the yellow yolk (Fig. 359). In addition, a flask-shaped mass of white yolk extends into the center of the yellow yolk (Purkinje's latebra); the yolk is surrounded by an extremely thin membrane (white yolk-membrane or the cortical protoplasm) (Fig. 359 and Fig. 360).

The *yellow yolk* (nutritive yolk) consists of soft, yellow non-nucleated cellular structures from 23 μ to 100 μ in diameter, and somewhat polyhedral in shape from

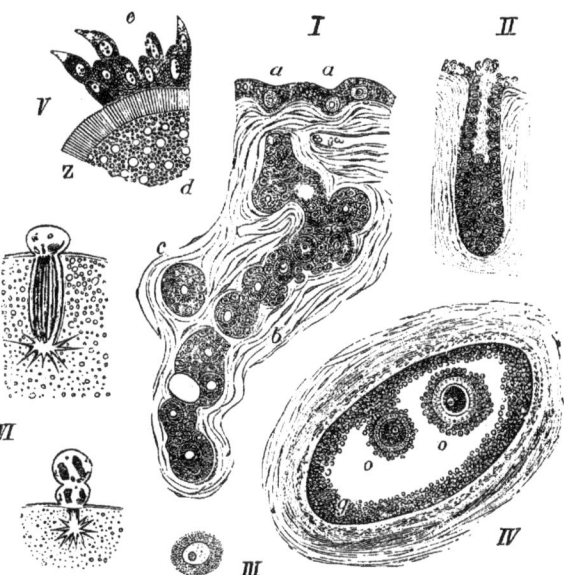

FIG. 358.—*I*, Ovarian Tube (from a Newborn Child) in Process of Follicle-formation: *a a*, Ova in the midst of the epithelial cells of the surface of the ovary; *b*, ovarian tube with ova and epithelial cells; *c*, a constricted-off small follicle, with ovum. *II*, Open ovarian tube of a bitch six months old. *III*, Isolated human primordial ovum. *IV*, Older follicle with two ova (*o o*) and the cells of the granulosa (*g*) (dog). *V*, Portion of the surface of a mature rabbit's ovum; *z*, zona pellucida; *d*, yolk; *e*, adherent cells of the granulosa (after Waldeyer). *VI*, Expulsion of the first polar body. *VII*, Expulsion of two polar bodies (after Fol).

mutual pressure (Fig. 360, b). These result from proliferating hyperplasia of the granulosa-cells of the Graafian follicle, which finally give rise also to the granulo-fibrous two-layered yolk-membrane (Fig. 359). The entire yolk of the bird's egg has been considered equivalent to the mammalian ovum, together with its corpus luteum.

When the yolk-globule in the bird's ovary is fully developed, the capsule of the Graafian follicle is ruptured and the yolk-globule passes in a rotatory fashion through the oviduct, the folds of whose mucous membrane, like the riflings of a gun-barrel, always cause the rotation to take place in a definite manner. Numerous glands in the oviduct secrete the albumin in which the yolk is enveloped in layers, the chalazæ being formed at either pole. As the tenacious layers of albumin tend to unroll again, the albuminous layer is rotated about the yolk in the bird's egg, and if freshly laid eggs are permitted to float in concentrated solution of sodium chlorid, all will rotate in the same direction.

The albumin in the eggs of nesting birds is vitreous and translucent when boiled, but it is transformed in the process of hatching into a mass like the albumin in the eggs of nonnesting birds (hen). On the other hand, the albumin of hen's eggs coagulates on addition of dilute sodium hydroxid into a vitreous transparent mass.

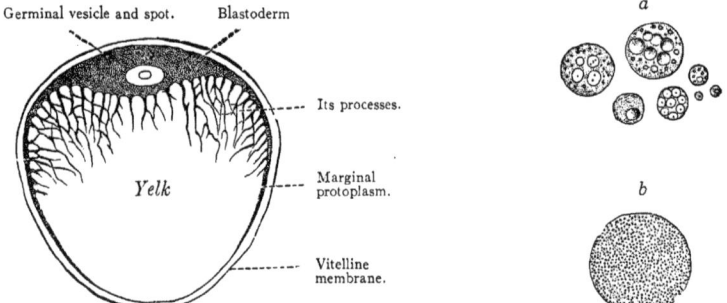

Germinal vesicle and spot.   Blastoderm

Its processes.

Marginal protoplasm.

Yelk

Vitelline membrane.

FIG. 359.—Diagrammatic Representation of a Mesoblastic Ovum (after Waldeyer).

FIG. 360.—*a* White, *b* yellow yolk-globules.

The fibers of the membrana testacea are secreted, spontaneously coagulated keratin-like filaments, wound spirally about the albumin, upon which a porous cement (testa) consisting of a mixture of albumin and lime is deposited in the lower portion of the oviduct. A structureless, porous, mucinous, occasionally fatty cuticula represents the outermost shell-layer in some birds. The limeshell of the bird's egg is utilized in part for the formation of the bones of the chick. The coloring-matters of the surface of the egg, which are often present in several superposed layers, appear to be derivatives of hemoglobin (hematoporphyrin) and biliverdin. Between the albumin and the shell-membrane there is detached epithelium of the oviduct (Fig. 361).

The *white yolk* (hen) contains albumin, nuclein, lecithin, potassium, glycogen (?); the yolk-membrane keratin. The yellow yolk contains a nuclein containing iron, a vitellin resembling globulin, lecithin, cholesterin, fat, coloring-matters (iron-lutein), containing neuridin, glucose, mineral matters,

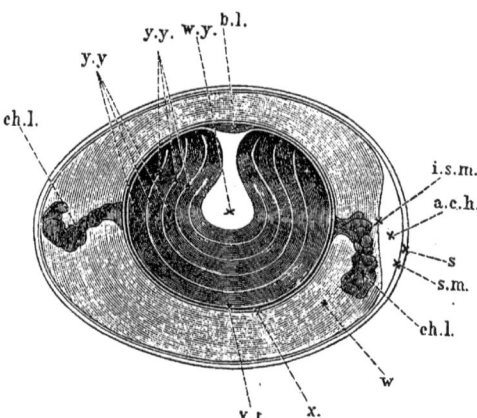

FIG. 361.—Diagrammatic Longitudinal Section of a Hen's Egg: b.l., germinal layer (cicatricula); w.y., latebra, filled with white yolk; y.y., a number of layers of yellow yolk, surrounding the latebra concentrically; x, yolk-membrane; v.t., white yolk-cortex (cortical protoplasm); w, mass of surrounding albumin in layers; ch.l., chalazæ; a.c.h., air-chamber at the blunt pole of the egg; i.s.m. internal and s.m. external lamella of the shell-membrane (membrana testacea); s, lime-shell (testa).

cerebrin (?), amyloid granules (?), sodium, potassium, calcium, magnesium, iron, phosphoric acid, silicic acid. The white of egg contains crystallizable ovalbumin (a mixture of several albumins), together with globulin, a body resembling mucin, sugar, and keratin. The ash contains more chlorin and alkalies, but less calcium, phosphoric acid and iron than the yolk.

PUBERTY.

The time at which man begins to be sexually mature is designated the age of puberty. In females this occurs between the thirteenth and the fifteenth year, in males between the fourteenth and the sixteenth year. In hot climates girls are often sexually mature as early as the eighth year. Between the forty-fifth and the fiftieth year, with the cessation of menstruation, the reproductive period terminates in the female (climacteric, involution); while in the male the production of spermatozoa is observed even to most advanced age. From the time of puberty sexual desire is awakened and the matured germinal material is expelled. All of the internal and external sexual organs, together with their accessory structures, undergo increase in size and become more vascular; the pelvis of the female acquires a characteristic shape. The evolution of the breasts is described on p. 418. The pubic and axillary hairs, in the male the beard, make their appearance in conjunction with increased sebaceous secretion.

The period of puberty is attended with alterations in many other organs: the larynx of the boy increases in size considerably in a sagittal direction, and the vocal bands become longer and thicker, so that the voice becomes at least one octave deeper (and therefore "breaks"). In the female the larynx becomes longer in its entirety, and the range of the voice is also increased. The vital capacity increases considerably in correspondence with the enlargement of the thorax. The entire figure and face acquire the contour characteristic of the sex, and the mental tendencies also receive a characteristic stamp at puberty. The vegetative development with relation to the individual is ended and the stream of growth in organic strength now passes in the direction of new production or procreation.

MENSTRUATION.

At regular intervals of from $27\frac{1}{3}$ to 28 days (solar month) there occurs in the sexually mature woman rupture of one or several mature Graafian follicles, with the coincident appearance of a bloody discharge from the external genitalia. This phenomenon is designated *menstruation* (menses, catamenia, courses, periods, monthly purification). Most women menstruate during the first quarter of the moon, only a few at the time of the new or full moon. In mammals the analogous process is termed *heat*; especially in carnivora, horses, and cows there is a bloody discharge from the genitalia, and the apes of the old world have a well marked menstrual bleeding.

The onset of menstruation is usually preceded by signs indicative of increased flow of blood to the internal genitalia, such as drawing pains in the sacral regions and the loins, as well as in the region of the uterus and the ovaries, which are sensitive to pressure, fatigue in the legs, flushes, alternate heat and cold, and even slight elevation of temperature in the external integument. In addition there may be sluggishness of gastric digestion, abnormalities in the evacuation of feces and of urine and secretion of sweat. It is noteworthy that during menstruation the decomposition of the nitrogenous elements of the body in the metabolic process is diminished.

The menstrual discharge is at first mucoid, then bloody, and it lasts three or four days (rarely from one day to two weeks). The blood has the characteristics of venous blood and, if admixed with a copious alkaline genital secretion. it exhibits a lessened tendency to coagulation, which may, however, take place in clumps if the bleeding be active. The amount of blood discharged approximates between 100 and 200 grams. After cessation of the bleeding itself there is a moderate discharge of mucus. Subsequently sexual desire is generally increased.

The essential characteristic internal phenomena of menstruation concern: (1) The alterations in the uterine mucosa and (2) the rupture of the ovarian follicle.

The uterine mucosa is the actual source of the hemorrhage. The ciliated epithelium of the reddened, greatly swollen, spongy and soft endometrium, from 3 to 6 mm. thick, is exfoliated. The openings of the numerous convoluted uterine glands are distinct, but their cells are in a state of fatty degeneration, as is also the interglandular tissue of the cells and the blood-vessels. This fatty degeneration and the desquamation of the degenerated tissue after disintegration take place only in the superficial layers of the mucosa, whose lacerated vessels give rise to the hemorrhage. The deeper layers of the mucosa remain intact and from them reconstruction of the entire mucosa takes place at the close of menstruation.

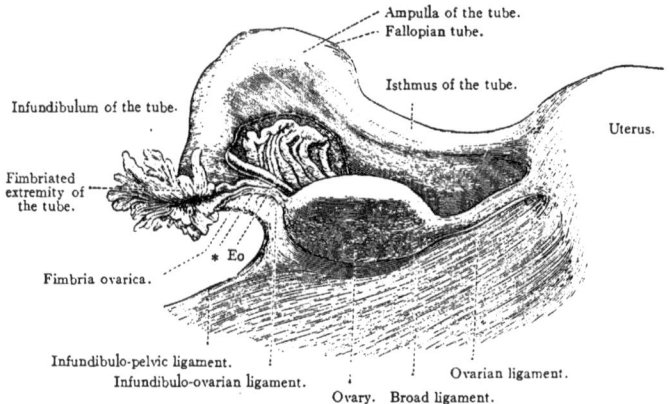

FIG. 362.—The Ovary and the Fallopian Tube (after Henle).
* Blood-vessel following the margin of the ovary. Eo. Epoöphoron exposed by removal of a portion of the broad ligament.

The second important internal process, *ovulation*, takes place in the ovary. The latter receives a greatly increased supply of blood, and the most mature follicle becomes more fully distended, projects above the surface, and finally ruptures its wall and the ovarian capsule, with hemorrhage from the laceration. At the same time the fimbriated extremity of the tube, in a state of erection from the engorgement of the vessels, lies in close apposition to the ovary in such a manner that the ovum, carried out with the liquor of the follicle and the surrounding granulosa-cells, passes along the ovarian fimbriæ and falls into the tube. The ciliated cells of the tube and the fimbriæ, moving toward the uterus, cause a movement of the fluid moistening the ovary that carries the ovum into the funnel of the tube. Ducalliez and Küss were able by tense injection of the vessels to bring about artificial erection and application of the abdominal orifice of the tube to the ovary. Rouget calls attention to the unstriated muscle-fibers of the broad ligament, which it is thought may by constriction cause the necessary injection of the tubal vessels.

. With regard to the connection between ovulation and the discharge of blood from the endometrium, there are at the present time two views. Pflüger considers the bloody exfoliation of the superficial layer of the endometrium as a preparatory freshening of the tissues (in the surgical sense) occurring physiologically, as a result of which it is rendered capable of uniting firmly by adhesion (as in case of thrombosis or cicatrization) with the ovum that finds its way into the uterus, so that the ovum is further nourished from the new lining membrane of the uterus like a developed or adherent part. This view is entirely at variance with another, according to which there develop within the uterus marked engorgement, sponginess, and swelling of the mucosa, under normal conditions, even

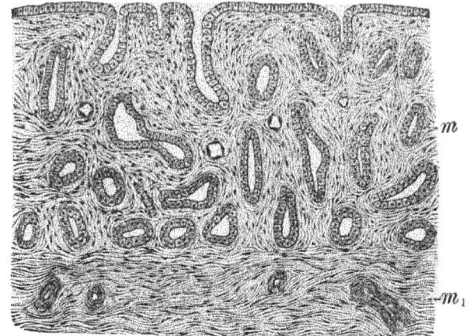

FIG. 363.—Sagittal Section through the Normal Endometrium, *m*, together with a Portion of the Contiguous Muscular Layer, *m*₁.

before the discharge of the ovum from the follicle, in consequence of a sympathetic formative process. The endometrium thus prepared is designated the menstrual decidual membrane. From this point of view it is capable, as a suitable place of incubation, of receiving an impregnated ovum. If, however, the ovule has not been impregnated and if, therefore, it is lost after its passage through the genital canal, destruction of the uterine mucosa takes place with hemorrhage, as already described. Accordingly, the hemorrhage from the uterine mucosa would be a sign of

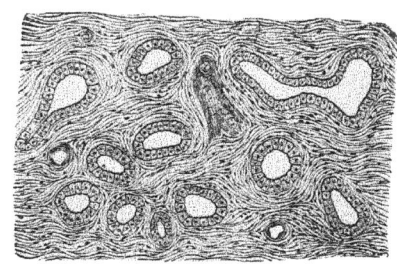

FIG. 364.—Horizontal Section of the Normal Endometrium (after Orthmann).

the nonoccurrence of pregnancy. The mucosa undergoes destruction because it could not be utilized for the time being, and the menstrual hemorrhage is, therefore, an external sign that the discharged ovum has not been impregnated. Accordingly, pregnancy, that is the development of the fetus in the uterus, must be reckoned not from the last menstruation that occurred, but from the first menstruation that was absent.

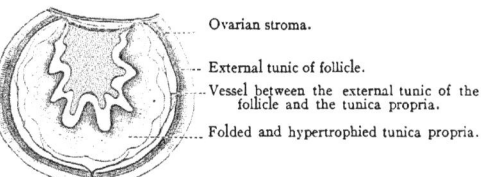

Ovarian stroma.

External tunic of follicle.

Vessel between the external tunic of the follicle and the tunica propria.

Folded and hypertrophied tunica propria.

FIG. 365.—Fresh Corpus luteum (after Balbiani).

With the rupture of the follicle, the cumulus oöphorus first is detached from the wall of the latter, the most superficial portion of the follicle, designated the stigma, becomes thin, its vessels are obliterated at this point and the tissue undergoes atrophy, so that with increasing pressure rupture must take place here.

After menstruation, the epithelium of the uterine mucosa is regenerated by indirect division, particularly from the sixteenth to the eighteenth day after the beginning of menstruation; the premenstrual swelling of the mucosa begins again between the eighteenth and the nineteenth day.

In individual cases ovulation and the formation of the menstrual decidua may take place independently; so that menstruation may occur without ovulation (more frequent) or ovulation without menstruation (seldom). Menstrual bleeding occurs only in the presence of ovarian tissue and a sufficient development of the uterine mucosa. Although many facts tend to support this new conception, there still remains the difficulty that in animals that have several placental sites (for example the cow), bleeding takes place from these situations at the time of heat.

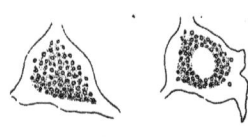

FIG. 366.—Lutein-cells from the Corpus luteum of the Cow (after His).

Formation of the Corpus Luteum.—The follicle whose contents have been discharged collapses. In its interior there remains the lining of granulosa-cells and a small amount of blood, which quickly coagulates. The small wound of rupture undergoes cicatrization after the serum has been absorbed. The wall of the follicle, which has become vascular, now swells as a result of mitotic division of the cells of the inner thecal wall and forces inward villous granulations of young connective tissue (Fig. 367), rich in capillaries and cells. Leukocytes wander into the cavity. Lutein-cells are formed anew through proliferation of the internal connective tissue layer of the wall of the follicle (Fig. 366). The corpus luteum is not an epithelial, but a connective tissue structure. Internal to the lutein-cells a layer of connective tissue develops later on. The lutein-cells subsequently undergo degeneration and there remains a cicatricially contracted "corpus albicans." The capsule becomes gradually more and more fused with the ovarian stroma.

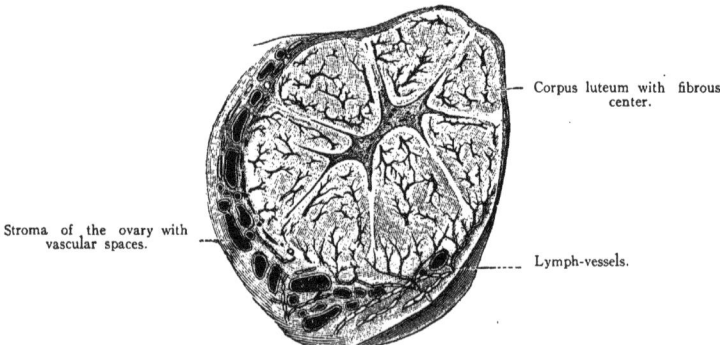

Corpus luteum with fibrous center.

Stroma of the ovary with vascular spaces.

Lymph-vessels.

FIG. 367.—Corpus luteum of the Cow, enlarged one and one-half times (after His).

Should pregnancy not occur after the menstruation, absorption of the fat formed takes place, with the formation of a crystalline body formerly supposed to be hematoidin, but shown to be lutein or lipochrome, and of other pigment-derivatives, while the yellow body undergoes uniform contraction within four weeks, down to a small remnant. Such yellow bodies, when pregnancy does not subsequently take place, are designated spurious corpora lutea. If, however, pregnancy results, the size of the body, in accordance with the greatly increased formative processes, is quite considerable (especially in the third or fourth month).

The wall is thicker and the color is deeper, so that the body at the time of labor still measures from 6 to 10 mm. in diameter and its remains may be recognizable even after the lapse of years. The yellow body after pregnancy is designated the true corpus luteum (Fig. 367).

ERECTION.

The knowledge of the distribution of the blood in the penis is due to the in- vestigations of C. Langer. The albuginea of the cavernous bodies consists of tendinous connective tissue, closely reticulated elastic tissue and unstriated muscle-fibers, which form a firm fibrous envelop, from which innumerable trabec- ulæ of similar structure pass inward, so that the cavernous bodies acquire the con- figuration of a sponge. The anastomosing spaces thus produced form a labyrinth of venous sinuses, which are lined by endothelium. The largest of these spaces are situated in the lower, outer portion of the cavernous body; in the upper portion the spaces diminish in number and size. The smaller arteries of the cavernous bodies arise from a branch of the arteria profunda of the penis running along the septum and they reach the trabeculæ in a tortuous course. Some of the small arterial branches in the cortical areas pass directly over into the larger venous sinuses; but similar direct transition from arteries to venous spaces takes place also in the interior of the cavernous bodies. A capillary network occurs also in the cortex and in the interior of the cavernous bodies, opening into the venous spaces. The helicine arteries of the penis described by Johannes Müller are only more or less incompletely injected arterial loops bent upon themselves, whose occurrence is due to the cord-like course of the trabeculæ. From the in- terior of the cavernous bodies, the venæ profundæ of the penis arise by means of fine branches. In addition venous branches pass from the cavernous spaces to the dorsum of the penis, uniting to form the dorsal vein of the penis. As these branches pass through the meshes of the vascular network in the cortex of the cavernous bodies, it is obvious that constriction of the meshes resulting from congestion of the network must cause compression of the efferent branches.

The spongy body of the urethra consists for the greater part of an outer layer of anastomosing veins lying close together, surrounding the longitudinal vessels of the urethra.

In the dog all of the arteries of the penis pass toward the surface, where they divide in tuft-like fashion. The veins arise from the capillary loops of the papillæ and they convey their blood into the cavernous bodies. Only a small amount of blood reaches the cavernous spaces through internal capillaries and veins; and arterial blood never flows directly into them.

The mechanism of erection consists in a marked distention of the blood-vessels of the penis, with fourfold or fivefold increase in volume, elevation of temperature, increase of blood-pressure within its vessels to one-sixth of the carotid pressure and initial pulsatory movement, increased consistency and erection, with a direction of the organ in conformity with the curvature of the vagina. The preliminary process consists in a marked increase in the arterial supply of blood, the arteries becoming dilated and pulsate strongly. This process is controlled by the erector nerves, which arise principally from the second (less commonly from the third) sacral nerve (in the dog) and possess ganglion-cells in their course. These nerves, belonging to the vaso- dilators, may be in part stimulated reflexly by irritation of the sensory nerves of the penis, the transference of the irritation taking place in the erection-center in the spinal cord. Thus, also, sensory irritation induced by voluntary movements of the genitalia, may excite this reflex through the ischiocavernosus and bulbocavernosus and the cremaster muscles, even the conception of sensory irritation of the penis may be attended with the same results.

Some vasodilators pass (in the dog) also through the lumbar sympathetic and the internal pudendal nerve. The last-named nerve usually contains vaso- constrictor fibers for the penis, although the erector nerves contain some.

The *erection-center* in the spinal cord is, however, naturally subordinate to the dominating vasodilator center in the medulla oblongata, from which connecting fibers pass downward through the cord to the erection-center. Therefore, stimulation of the spinal cord above causes erection, as, for example, by mechanical irritation, asphyxiation, or the use ot muscarin (pathologically also in cases of spinal irritation). Finally, the psychical activity of the brain has a distinct influence upon the genital vasodilators. In the same way as the psychical emotions of anger or shame cause dilatation of the vessels of the head by stimulation of the dilators, the direction of the attention to the sexual sphere has an effect upon the erector nerves. This influence of the brain has been explicable since the dependence of the local lumen of the vessels upon the cerebral cortex has been known. From the cerebral cortex the fibers whose irritation Eckhard observed to cause erection probably pass through the cerebral peduncles and the pons.

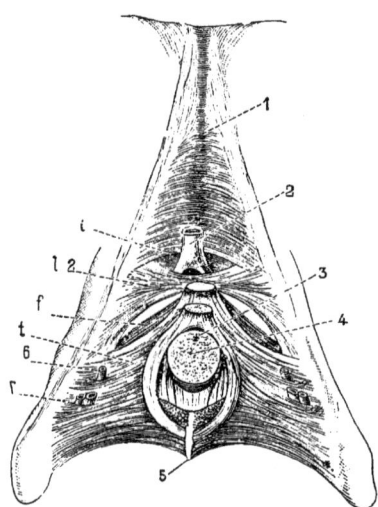

FIG. 368.—Anterior Pelvic Wall with the Urogenital Diaphragm, viewed from in front (externally), after Henle. The corpus cavernosum of the urethra, 4, with the urethra, 3, is divided beneath the point of exit from the pelvis. 1, pubic symphysis; 2, dorsal vein of the penis; 5, portion of the bulbocavernosus muscle, arising fsom the perineal septum; t, deep transverse perineal muscle, together with its fascia, f; 6, deep vein of the penis; 7, bulbocavernosus artery and vein.

If, thus, the impulse to erection is given by the arterial fluxion, the complete development of the process may take place through the activity of the following transversely striated muscles:

1. The ischiocavernosus muscle (Fig. 108) arises from the ischium, and surrounds the root of the penis by its tendinous union like a sling. In its contraction it compresses the root of the penis from above and the sides, so that the escape of the venous blood is prevented. It has no effect upon the dorsal vein of the penis, as this vessel is protected in the dorsal groove of the penis from the pressure of the tendon.

2. The deep transverse perineal muscle is penetrated by the deep veins of the penis coming from the cavernous bodies and later uniting with the common pudenal vein and the plexus of Santorini in such a manner that its contraction must compress them between the highly contracted horizontal fibers (Fig. 368, 6). 3. Finally the bulbocavernosus muscle also aids in stiffening the corpus spongiosum, by compressing the bulb of the urethra (Fig. 368, 5 and Fig. 108). All of these muscles can in part be moved voluntarily, and as a result the erection becomes more marked. Under ordinary conditions, however, their contraction follows reflex excitation from the sensory nerves of the penis.

The stagnation of blood in the penis is not complete; otherwise, long-continued erection (priapism, satyriasis) would under pathological conditions give rise to gangrene.

Blood-stasis in the penis is favored by the fact that the veins of the penis originate in the cavernous bodies, by whose hardening the veins must be compressed. Furthermore, there are on the walls of the large veins of the plexus of Santorini trabeculæ of unstriated muscle, which in contracting act as columns penetrating into the lumen of the veins and in part hinder the flow of blood.

The dependence of erection, as a complex motor mechanism, upon the nervous system was demonstrated by the experiments of Hausmann, who observed that erection failed to appear after section of the nerves of the penis in stallions. The erection that occurs in women is less complete and extends to the cavernous bodies of the clitoris and the bulb of the vestibule. During erection the communication between the urethra and the bladder is closed partly by swelling of the caput gallinaginis, a portion of the spongy body of the urethra, and partly through the action of the urethral sphincter, which is connected with the deep transverse perineal muscle.

EJACULATION.   RECEPTION OF THE SEMINAL FLUID.

In the expulsion of the seminal fluid two distinct factors are to be distinguished, namely: (1) The passage of the seminal fluid from the testicle to the seminal vesicles, and (2) the act of ejaculation itself. The first takes place continuously in consequence of the advance of newly formed seminal fluid through the activity of the ciliated epithelium (from the epididymis to the beginning of the vas deferens) and as a result of the gradual peristalsis of the vasa deferentia, which are provided with a well-developed muscular layer.

For the initiation of ejaculation, however, a strong peristalsis of the vasa deferentia and of the muscular walls of the seminal vessels is necessary. This is brought about through reflex excitation of the ejaculatory center in the spinal cord. As soon as seminal fluid enters the urethra by this means, rhythmic contraction of the bulbocavernosus muscle takes place as a result of distention of the urethra acting as a mechanical irritant, and the seminal fluid is vigorously expelled from the urethra. Both seminal vesicles and both vasa deferentia do not always discharge their contents into the urethra at the same time. With moderate stimulation only one of these may empty itself at a time. Coincidently with the contraction of the bulbocavernosus, the ischiocavernosus and the transversus perinei profundus also contract, but these have no influence upon ejaculation itself.

Also in the female there occurs under normal conditions, at the height of sexual excitement, a reflex motor process corresponding to ejaculation in the male. This consists of movements analogous to those observed in the male. There occurs, first, a peristaltic movement of the tubes and the uterus from its cornua to the vaginal portion, induced by reflex irritation of the genital nerves. Dembo observed in animals general uterine contractions after irritation of the anterior upper wall of the vagina. As a result of the movement of the tubes and the uterus (which corresponds to the peristalsis of the vasa deferentia in the male), a certain amount of mucoid fluid normally moistening the uterine wall is expressed into the vagina. This is followed by rhythmic contraction of the sphincter cunni (analogous to the bulbocavernosus), the insignificant ischiocavernosi and the deep transversus perinei being at the same time active. As a result of the vigorous contraction of the muscular fibers of the uterus and its muscular round ligaments, the organ becomes erect and descends toward the vagina, its cavity becoming more and more reduced in size, while its mucus contents are expressed. If the uterus later on, after the excitation has ceased, gradually returns to its relaxed state of rest, it aspirates into its cavity the seminal fluid deposited at its orifice (conception).

Such aspiration of the seminal fluid by the uterus irritated to maximum degree is by no means necessary to fecundation. The spermatozoa are capable by their own movement of entering the uterus from the vaginal portion through the clear mucus that normally occupies the cervical canal. Indeed, observations as to impregnation without entrance of the penis, in consequence of pathological obstructions, such as partial atresia of the vulva or vagina, show that spermatozoa may traverse the entire vagina into the uterus.

IMPREGNATION OF THE OVUM.

The ovum is fecundated by the penetration of one spermatozoon.

Since the time of Swammerdam (died 1685) it has been known that for fecundation to take place contact of the ovum with the seminal fluid is necessary and indeed with the spermatozoa, which according to Hartsoecker penetrate into the ovum. Barry saw spermatozoa enter the interior of the rabbit's ovum. This takes place with considerable rapidity by a boring-movement through the capsule of the ovum. The invasion takes place eventually through pores that are present or through the micropyle.

In the mouse and some other mammals the ovary is surrounded by a space filled with fluid (periovarial space), to which both the ovum and the spermatozoon gain access; both are conveyed by aspirating movements of the tube into the uterus.

The viscid surface of the ovum affords a means for the attachment of the spermatozoon. In the case of meroblastic ova the spermatozoon penetrates in the situation of the nucleus; in that of holoblastic ova at the animal pole, when this is present. At the spot where the head of the spermatozoon meets the yolk, the latter throws out toward it a humplike elevation. As soon as a spermatozoon has penetrated into the yolk, the entrance of other spermatozoa seems to be opposed by the appearance of a firm membrane—the yolk-membrane—upon its surface, which, acting as a protecting wall, prevents the penetration of other spermatozoa. Nevertheless in the case of meroblastic ova (selachians, reptiles, insects and others) the penetration of several spermatozoa takes place normally for the purpose of fecundation—polyspermism.

The place where fecundation (impregnation) takes place is either the ovary (as indicated by the occurrence of abdominal pregnancy) or the tube, whose numerous mucous folds constitute a suitable place of lodgment for the spermatozoa. That fecundation may take place also in the tube is shown by the occurrence of tubal gestation. Spermatozoa must, accordingly, pass from the uterus through the tube to the ovary and they do this probably by their own movement. Whether the peristaltic movements of the uterus and the tube assist in this transportation is uncertain. The ciliary movement of the tubal epithelium can, however, have nothing to do with the phenomenon, as the movement is directed outward. If the ovum enters the uterus unimpregnated, it does not undergo fecundation here, as it perishes. It is believed that the extruded ovum reaches the uterus within two or three weeks (in dogs from eight to fourteen days).

Double impregnation (twins) occurs once in 87 times (in tropical regions more commonly); triplets once in 7600 times; quadruplets once in 330,000 times; sextuplets are extremely rare; septuplets (?) were born by **Anna Breyers** of Hamlin in 1600. The average number of conceptions in women is 4½. The largest number of children observed is from 32 to 38.

By **superfecundation** is understood the occurrence of the impregnation of two ova discharged at the same menstrual period, as a result of different copulations. For example, a mare may throw a foal and a mule, after having been covered first by a stallion and then by an ass. Thus, also, a woman has been observed to give birth to a negro and a white twin. If, however, the second fecundation occurs at a later time during pregnancy, as for example, at the second or the third month (as in a case cited in the Talmud), then the rare phenomenon of *superfetation* occurs. This, however, is possible only in the presence of a double uterus and the persistence of menstruation until the time of the second impregnation. Hippocrates explained superfetation as due to independent pregnancies in the horns of the uterus, a condition that according to Aristotle occurs with especial frequency in hares. Superfetation cannot occur in the normal uterus, as a plug of mucus occludes the cervical canal during pregnancy, as Herophilus knew, and in addition from the fact that menstruation usually ceases.

Hybrids.—Impregnation is possible also between related species (horse, ass, zebra; dog, jackal, wolf; goat, ibex; goat, sheep; varieties of the llama; camel, dromedary; tiger, lion; varieties of pheasants; varieties of finch; goose, swan; carp, crucian; varieties of the butterfly). Most of the hybrids thus produced are sterile, chiefly because of a deficiency of developed spermatozoa in the male. The female hybrid, however, may be impregnated by males of the species of either of her parents; for example the mule. The progeny, however, tends to revert in type to the species of the parents. Only a few hybrids are capable of procreation among themselves, as hybrids in dogs. In different species of frogs, the cause of the frequent failure of hybridization is to be found in mechanical obstructions to the penetration of the spermatozoa into the ovum. Only such spermatozoa as are more slender and more vigorous in movement than those of

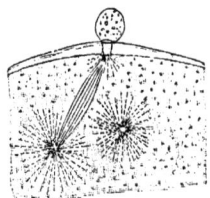

FIG. 369.—Ovum of Scorpæna scrofa. The germinal vesicle has extruded a polar body and has withdrawn to the center of the ovum as the nucleus; it is being approached by the male pronucleus.

the other species are capable of impregnating ova of the latter. Therefore, the possibility of hybridization between two species is almost always one-sided. In some amphibia hybrid fertilization is possible, but development does not take place beyond the first stages. This appears to be due to the circumstance that only a portion of a spermatozoon that has incompletely entered the ovum becomes active. According to O. and R. Hertwig, hybridization can be more readily effected in echinodermata the more virile the spermatozoa and the feebler the ova.

In breeding among close blood-relationships (rats) increase of sterile pairings, diminution in the number of offspring, greater mortality among the young and marked inability on the part of the mother to nourish them occur. Certain bodily defects and weaknesses appear to be increased.

Exceptionally the ovum from the ruptured follicle of one ovary may enter the tube of the opposite side, as indicated by the cases of tubal pregnancy and of pregnancy within a rudimentary uterine horn abnormally present, in which the true corpus luteum has been found in the ovary of the opposite side (external transmigration). In accordance with this observation is the fact that fine granules suspended in water (India ink, etc.), and injected into the peritoneal cavity, penetrate into both tubes, as a result of the action of the cilia, and reach the uterus. In animals ova may also wander through

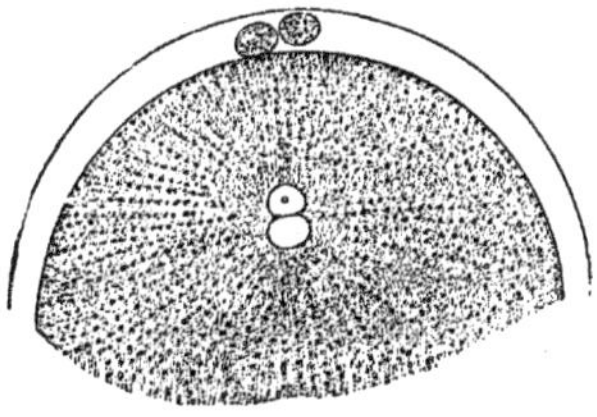

FIG. 370.—Ovum of a Starfish (asteracanthion) with two Extruded Polar Bodies: male and female pronuclei in apposition.

the double uterine mouth: out of the one and through the other into the opposite uterine horn (internal transmigration).

In the maturing ovum the first characteristic change affects the germinal vesicle, which divides by mitosis. At the same time it moves toward the surface of the ovum, and loses its capsule; its chromatin-fibrils begin to form convolutions and become converted into a longitudinal structure known as the *nuclear spindle.* At both poles of the spindle, the granular elements of the protoplasmic yolk become collected each into a radiate form (diaster). When this has taken place, the peripheral pole of the nucleus of the ovum thus altered appears above the surface of the ovum, becomes constricted off and expelled from the ovum like a waste product in the form of a small body (Fig. 358, *VI* and *VII*). The formation of a second polar body takes place again by mitosis in the same way during the penetration of the spermatozoon. Both of the bodies thus eliminated, which are of no further use in the growth and development of the ovum, are designated *directing bodies* or *polar cells*. (Figs. 369 and 370.) The remaining portion of the germinal vesicle lying near the center remains within the yolk, wanders back toward the center of the ovum, increases in size, and thus forms the egg-nucleus or female pronucleus, which has no centrosome.

The spermatozoon that has entered the ovum moves toward the female pronucleus, its head becoming surrounded by a radiating crown; then

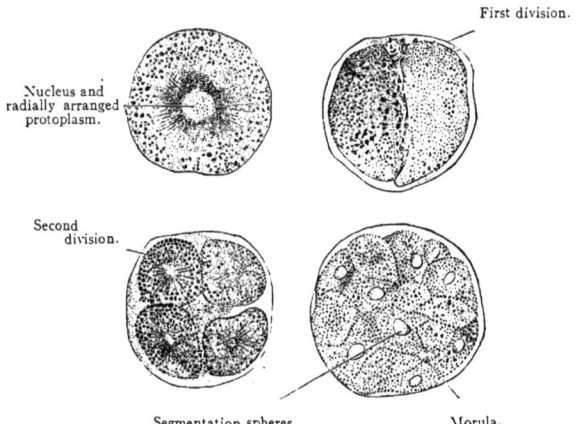

Fig. 371.—Four Stages of Division of an Impregnated Ovum of Echinus saxatilis.

its cilium is dissolved and its head, alone remaining, forms a chromatic mass and swells into a second new nucleus, the *sperm-nucleus* or the *male pronucleus*. From the connecting segment a centrosome (surrounded by rays) develops and this soon becomes directed toward the interior of the ovum. The centrosome of the male pronucleus also divides. The male and female pronuclei now unite to form the new nucleus of the impregnated ovum, with which both of the sperma-spheres resulting from the division are in contact; and the yolk assumes a radiating appearance (Figs. 370 and 371).

The entrance of several spermatozoa into the ovum (polyspermism) takes place normally in large ova rich in yolk. From these accessory sperm-nuclei develop, all of which probably disappear later. O. Hertwig and Fol made the remarkable observation (in echinoderms) that several embryos form from one ovum, when several spermatozoa enter the ovum abnormally. The male pronuclei resulting from the individual spermatosomes then each unite with a fragment of the disintegrated female pronucleus.

CLEAVAGE, MORULA, BLASTULA, GASTRULA, FORMATION OF THE GERMINAL LAYERS. FIRST RUDIMENTS OF THE EMBRYO.

In the fecundated ovum the yolk-mass contracts more closely about the newly formed nucleus, becoming somewhat separated from the yolk-membrane, and there now follows division first of the nucleus and then of the yolk into two nucleated globules or blastomeres. This process, designated total cleavage, is repeated in accordance with the method of cell-division in the two globules formed, so that first 4, then 8, 16, 32, etc., globules result. The division ceases only after the entire yolk has been subdivided into numerous small globules, the nucleated *cleavage-spheres*, or the unencapsulated protoplasmic *primitive cells* (from 20 to 25 μ in diameter). The yolk now consists of a collection of primitive cells and is designated the *morula* or the *mulberry mass*.

The division of the nucleus of the ovum takes place by mitosis after the previous formation of a spindle-form. The centrosomes of this first cleavage-spindle are derived from the centrosome of the spermatozoon. According to the observations of van Beneden the constituents of both the male and the female pronucleus pass over into the cleavage-spheres, so that all cells of the body are formed from a combination of the male and female procreative elements. This fact explains the process of inheritance from the paternal and the maternal organism. Deficiency of oxygen gives rise in the ova of some fish to an involution in the process of cleavage. The globules become dissolved and coalesce. A renewal of the supply of oxygen stimulates the process of cleavage anew.

The uniform mode of cleavage described, such as occurs in mammals and amphioxus, is designated equal or adequal. A second variety of cleavage is the total unequal, in which, for example in the frog's egg, one-half of the yolk, designated the animal pole, which often is pigmented, yields much smaller cleavage-cells than the other or vegetative pole. The embryo forms in the animal pole. When, finally, the yolk-mass has become so large that cleavage remains confined to the animal pole, then partial cleavage (described later) occurs.

Numerous attempts have recently been made to trace the development from a single isolated blastomere. Some investigators found that at first only a right or a left half of an individual (echinodermata) is formed from one blastomere, but that this in the further course is capable through post-generation to develop into an entire being. Other observers, on the other hand, from the outset obtained from a single blastomere (for example from an ascidian ovum) an entire individual (even to the sixteenth division), but of smaller size. Under special experimental conditions, finally, it was possible to produce double malformation, from partially isolated, but partially connected blastomeres.

Under normal conditions the first line of cleavage (frogs) passes, according to Roux, in the same direction as the central nervous system. The second fissure intersects the first at right angles and divides the ovum into two unequal parts, of which the larger serves to form the cephalic portion of the embryo.

Meanwhile the ovum has increased in size through the absorption of fluid. All of the cells are polyhedral in shape from mutual pressure, and form a cellular vesicle, the germinal vesicle, which is applied throughout its periphery to the zona pellucida.

The human ovum has reached this stage of development during the first week; that of rabbits in 4 days, of guinea-pigs in 3½, of cats in 7, of dogs in 11, of foxes in 14, of ruminants and pachydermata in from 10 to 12, of deer in 60 days. In some animals (for example rabbits) the zona pellucida is further surrounded by a layer of albumin. A small collection of blastomeres do not participate in the formation. They are apparently not utilized, and apply themselves at one point on the interior of the blastula, and here, later, the embryo develops.

The *germinal vesicle*, which as a typical stage in the development of numerous animal species is also designated the blastula, thus consists of a vesicle with walls made up of a single layer of cells, as represented diagrammatically in Fig. 372, 1, in sagittal section (from amphioxus). The subsequent important formative process consists in the development from the blastula of a hollow structure, whose walls consist of a double layer of cells. This process of transformation can be best followed in the ovum of the lancet-fish (amphioxus). Here the blastula is invaginated at one point into the interior of the vesicle (Fig. 372, 2), and progressively until the invaginated layer of cells comes in contact with the opposite layer (3). At the same time the invagination-opening becomes smaller and smaller. The stage in development thus attained is designated *gastrula*. The external layer of cells is the *ectoblast* (or epiblast), the internal layer the *entoblast* (or hypoblast). The opening is designated the *blastopore* (or primitive mouth), and the central space the primitive gut (archenteron). In vertebrates the primitive mouth closes fully in the course of further development.

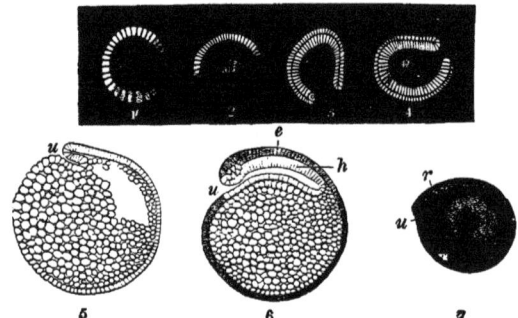

FIG. 372.—1–4. Formation of the hypoblast by invagination of the blastula, and the resulting gastrula from amphi oxus (lancet-fish); 5, early and, 6, later development of the hypoblast by invagination in petromyzon; *u*, blastopore (primitive mouth); *e*, epiblast; *h*, hypoblast in vertical section; 7, the ovum at this stage viewed from the side; *u*, primitive mouth; *r*, spinal furrow (after Kupffer).

Wholly identical gastrula-larvæ are found in some radiates and worms that move about independently in water and nourish themselves like the celenterates through the primitive mouth.

The formation of the hypoblast (*h*) by invagination in the situation of the primitive mouth is exhibited distinctly in a similar manner by the species of fish, the lampreys (petromyzon). In Fig. 372, 5 and 6 illustrate these processes of formation in diagrammatic section, after Kupffer. It will be observed that invagination takes place from the primitive mouth (*u*) and thus the epiblast (*e*) and the hypoblast (*h*) are formed in layers one upon the other, the primitive intestinal cavity being situated beneath the hypoblast. These formations develop in much the same manner also in batrachia.

It appears justifiable to interpret the analogous early developmental processes in mammals in a similar manner. According to van Beneden the ovule after segmentation is completed likewise exhibits two layers of cells, the epiblast (Fig. 373, *I*, *e*), which lies next the zona pellucida (Z), and the hypoblast (h). The primitive mouth (u) here also leads to the central cavity of the ovum. When the rabbit's ovum has reached a diameter of 2 mm. there appears at one point the oval embryonal spot or embryonic shield (germinative or embryonal area). The cells of the ectoderm multiply so as to form several layers in the region of the shield. Careful examination leads to the detection further at the border of the latter of a small longitudinal area (II, u), from which the duplication of the cell-layer of the blastula takes place, and which, therefore, must be looked upon as the blastopore. From the blastopore the lower layer of cells (hypoblast) extends in the region of the embryonal spot, although its growth continues uninterruptedly, until finally the entire blastula consists of two layers. The site of the primitive mouth (II, u) becomes the so-called primitive streak (III, pr), which at first appears as an oval elevation, and later as a longitudinal furrow.

The primitive streak (like the primitive mouth in general in vertebrates) is a temporary structure. It is, however, still present when the medullary groove is formed in the epiblast (IV, rf) in front of it; then it gradually atrophies. This subject will later on be discussed at greater length. The primitive streak presents a nodular swelling (Hensen's nodule) anteriorly, and posteriorly a terminal enlargement. The furrow of the primitive streak is also designated the primitive groove, its borders the primitive folds.

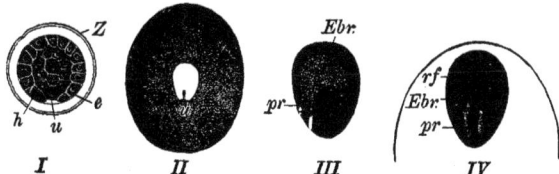

FIG. 373.—I, Ovum of the rabbit, after van Beneden; Z, zona pellucida; e, epiblast; h, hypoblast; u primitive mouth. II, Ovum of the rabbit with the (clear) rudimentary embryo; at u the earliest formation of the primitive streak (or primitive mouth) can be recognized. III, Ebr, Rudimentary embryo from a somewhat older rabbit-ovum; pr, the primitive streak, with groove. IV, Still further developed embryo (seventh day); the rudimentary embryo (Ebr) exhibits above the primitive streak the first indication of the spinal furrow (after Kölliker).

The embryonal area later on loses its pear-shaped form and becomes dumbbell-shaped. The portions of the germinal vesicle adjacent to the rudimentary embryo become more transparent, so that the latter is surrounded by an area pellucida, about which the dark embryonal spot, or opaque area, is situated. The zona pellucida now acquires a villous appearance, becomes covered with a gelatinous layer and is designated the *primitive chorion* or *prochorion*.

In the dog the zona becomes covered in the uterus with a coating of mucoid secretion. Bonnet was able to demonstrate that this tenacious secretion penetrates into the lumina of the glandular ducts and thus forms gelatinous filaments, which formerly were erroneously looked upon as villi springing from the zona. They serve as a means of attachment and of nourishment for the ovum.

Later on a new layer of cells extends from the primitive streak between the epiblast and the hypoblast, namely the *mesoblast* (Fig. 376, I), which soon advances over the region of the embryonal spot and continues to grow into the germinal vesicle. Blood-vessels form, further, within the mesoblast, and their area of distribution upon the germinal vesicle is known as the *vascular area*. The discovery of an analogous formation in the meroblastic ovum, as, for example, that of birds, has been attended with no little difficulty. In such ova only *partial cleavage* takes place, that is only the white yolk in the region of the cock's treadle undergoes division into many blastomeres in the process of segmentation as a result of processes in other respects analogous to those in the ova of mammals. The cells thus resulting form on the surface of the yolk the *germinal membrane;* and they later on become arranged into two superposed, thin, circular layers or *germinal plates*. The upper layer (*ectoblast*) is the larger and contains smaller, paler cells. The lower layer (*hypoblast*), which at first is not arranged continuously, is smaller, and its cells are larger and darkly granular.

Observation of germinal plates during the first hours of hatching permits the recognition of conditions indicative of a formative process in the development of the hypoblast analogous to that occurring in

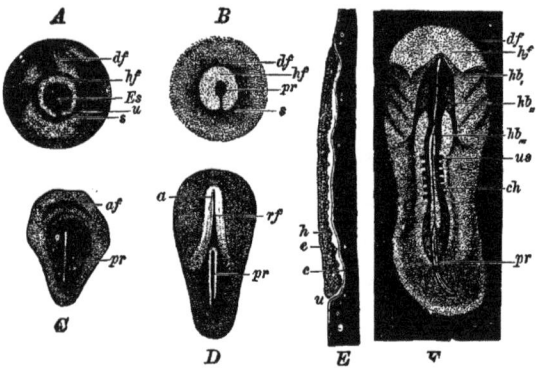

Fig. 374.—*A*, Germinal plate of hen's egg in the first hours of incubation (after Koller); *df*, dark germinal area; *hf*, clear germinal area; *Es*, rudimentary embryo; *u*, point from which the hypoblast is formed by invagination, or the blastopore (primitive mouth) becomes sickle-shaped below (*s*). *B*, Somewhat older preparation; *pr*, primitive streak. *E*, Longitudinal section through the germinal plate of a nightingale at this stage (after Duval); *u*, primitive mouth; *e*, epiblast; *h*, hypoblast, beneath which is the primitive intestinal cavity, *c*. *C*, Distinctly developed primitive streak with the primitive groove (*pr*), in front of which is the earliest indication of the amniotic folds, *af*. *D*, The spinal furrow (*rf*) is formed in front of the primitive streak (*pr*) (18th hour, after Balfour). *F*, Hen of 33 hours (after Duval), the primitive streak (*pr*) in process of involution; *df* dark, *hf* clear germinal area; *hb₁ hb₁₁ hb₁₁₁*, the three anterior cerebral vesicles; *us*, primitive vertebra; *ch* chorda dorsalis.

holoblastic ova. A formation corresponding to the primitive mouth has been encountered also on the germinal plate of the bird (Fig. 374, *A*, *u*). This at first is short and is expanded in its lower area into the shape of a sickle. This blastopore, gradually becoming longer, develops into the primitive streak (*B*, *C*, *pr*), which undoubtedly is comparable with that of mammals. That in the ova of birds the hypoblast must likewise be considered to have resulted from invagination of the blastopore is rendered probable by the study of a longitudinal section of the two germinal plates at this first period. Fig. 374 *E* represents such a sagittal section of the germinal plate from a nightingale-ovum. The lower germinal plate (*h*) appears to be pushed out from the blastopore (*u*) under the ectoblast. Both plates rest upon the cavity of the archenteron (*c*) filled with fluid.

Between the ectoblast and the hypoblast there now develops, from the primitive streak, as a product of the cellular hyperplasia of the ectoblast, the *mesoblast*, which, growing peripherally, insinuates itself between the two former. The three germinal layers (in birds) in their growth arrange themselves according to size in such a manner that the uppermost is the largest, the middle the next in size, and the under-most the smallest. All three grow at the periphery. As the middle

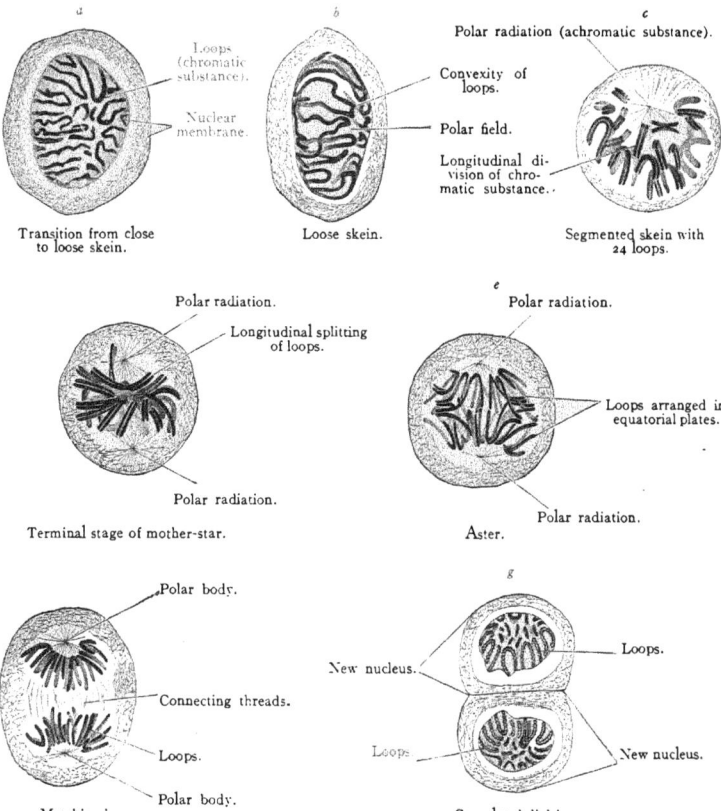

Loops (chromatic substance).

Nuclear membrane.

Transition from close to loose skein.

Polar radiation (achromatic substance).

Convexity of loops.

Polar field.

Longitudinal division of chromatic substance.

Loose skein.

Segmented skein with 24 loops.

Polar radiation.

Longitudinal splitting of loops.

Polar radiation.

Terminal stage of mother-star.

Polar radiation.

Polar radiation.

Loops arranged in equatorial plates.

Polar radiation.

Aster.

Polar body.

Connecting threads.

New nucleus.

Loops.

Loops.

Loops.

New nucleus.

Polar body.

Metakinesis.

Completed division.

FIG. 375.—Stages of Nuclear Division (after Rabl).

layer develops into vessels, its border is always easily recognizable from the sinus—the future terminal vein. The border of the upper layer encloses the yellowish-white, wavy vitelline area; the border of the middle layer, the vascular area; the embryo lies in a portion of the pellucid area that is dumbbell-shaped and clear as glass. As all three plates, finally, surround the entire yolk, their borders come in contact with the pole of the yolk lying opposite to the embryo.

Thus, there are developed in all vertebrates three germinal layers. From the ectoblast there results the central nervous system, the epidermal formations, and also the epithelium of the organs of special sense. From the hypoblast is formed the epithelium of the intestinal tract, including the cells of the glands that result through evagination from the intestinal tube. All of the other tissues of the body, with the exception of the parts forming the vascular system and the connective-tissue substances, develop from the mesoblast.

The cells of the ectoblast, but particularly those of the hypoblast, take up during development, in the bird, the constituents of the yolk through direct active incorporation, and in this process the ameboid movement of the cells plays a role. The parts taken up are transformed (digested) within the cells, and employed in the process of building up.

The division of the cells of the growing tissues takes place in the following manner: 1. By *direct cell-division*, in which first the nucleus and then the cell-body breaks up into two halves, for example in the division of the embryonal erythrocytes (page 41); 2, by *indirect cell-division* (mitotic division), in which the following processes are observed in the cell: (a) The nucleus becomes enlarged and its chromatin-network increases and takes on a definite grouping. There form loops, which at one pole of the nucleus (polar field) exhibit especially bendings, and at the other (opposite polar field) the extremities of the limbs of the loops. This is the stage of the close skein or the spireme. (b) The close skein is transformed into the looser, the threads becoming separated. In this process part of the loops turn toward one pole, and part to the other (segmented skein, c). (c) All the loops move with their bendings toward the center; and there is thus formed the star (first mother-star, d). (d) Meanwhile, there is formed the achromatic nuclear spindle, which bears at each extremity a polar body, from which the nuclear protoplasm passes in a radiate manner—polar rays (d, e, f). (e) The loops divide lengthwise; each half moves away from its fellow (e, f). (f) The loops undergo a rearrangement—metakinesis (e, f), and form two equatorial plates. (g) After atrophy of the connecting threads, the protoplasm of the nucleus, and later also that of the cells, undergoes division, and the network of both nuclear halves (dispireme, g) appears as in the original form of the undivided nucleus (a).

The cell consists of body, nucleus, and nucleolus. The cell-body forms a movable protoplasm, which appears as a threadwork, or network, or honey-combwork in the midst of which, in a softer substance, lie small granules. The nucleus possesses a capsule, and consists of a nuclear ground-substance, in which (colorable) chromatin and achromatin lie as enclosures. The nucleolus is a dense mass of chromatin. Occasionally, accessory nucleoli are present (Flemming's reticular nodes). Finally, the cell-body contains also the centrosome, surrounded by an area, the actual center of motion of the cell, which also breaks up in the process of cell-division. All cells are derived from parent-cells; Omnis cellula ex cellula (Virchow). From cells at first apparently similar the elements of the different organs and tissues develop through transformation. If, therefore, all tissues are referable morphologically to a single form of primitive cell, it follows that the physiological activity of the different organs and tissues must be referred to a single primitive form of function, to an "identity of physiological activity," present from the beginning. The proof of the development of the special activity of the tissues from this as yet undifferentiated primitive form of vital manifestation will, with certainty, at a later period constitute an important chapter of the subject of physiological development.

FORMATIONS FROM THE EPIBLAST.

Upon the ectoblast there is formed, in mammals, as in birds, in front of the primitive streak, and at a later period, a longitudinal furrow (Fig. 373, *IV*, and Fig. 374, *D*), whose margins, curved anteriorly, pass over into each other; while posteriorly they pass side by side, though in a somewhat divergent manner. This is the *medullary* or *spinal groove*. Later on the adjacent margins, the *medullary* or *spinal folds* approach each other at their free edges, and finally join in the median line, to

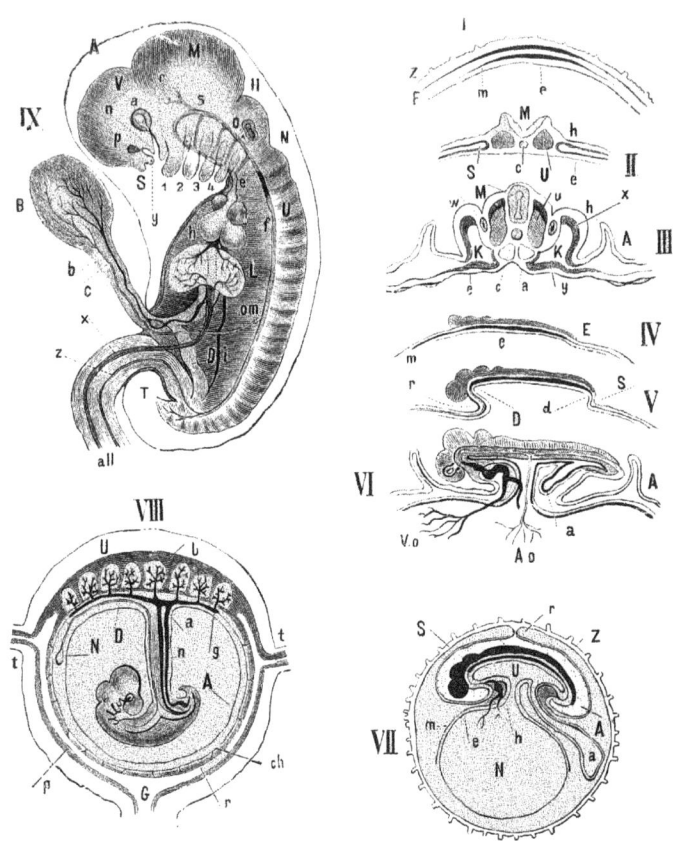

FIG. 376.—I, The three germinal layers of the ovum of mammals: Z, Zona pellucida; E, epiblast; m, mesoblast, e, hypoblast. II, Cross-section of chick (with six primitive vertebræ) on the first day: M, spinal furrow; h, epidermis; U, primitive vertebra; c, chorda dorsalis; S, the lateral plates divided into two lamellæ; e, hypoblast. III, Cross-section of chick on the second day, in the region behind the heart: M, medullary canal; h, epidermis; u, primitive vertebra; c, chorda; w, Wolffian duct; K, cœlom; x, cutaneous plate; y, splanchnopleura; A, amniotic fold; a, aorta; e, hypoblast. IV, Diagrammatic representation of the first embryonal rudiment in longitudinal section. V, Diagrammatic representation of the beginning of the process of constriction: r, headfold; D, cavity of the foregut; S, caudal fold; d, hind-gut cavity in an early stage of formation. VI, Diagrammatic longitudinal section of the embryo after constriction: A o, omphalomesaraic artery; V o, omphalomesaraic vein; a, rudimentary allantois; A, amniotic fold. VII, Diagrammatic longitudinal section through a human ovum: Z, zona pellucida; S, serous capsule; r, union of amniotic folds; A, amniotic cavity; a, allantois; N, umbilical vesicle; m, mesoblast; h, heart; U, primitive gut. VIII, Diagrammatic longitudinal section through the pregnant uterus at the time of the formation of the placenta: U. muscular wall of the uterus; p, mucous membrane of the same or true decidua; b, maternal placenta or serotine decidua; r, reflex decidua; ch, chorion; A, amnion; n, umbilical cord; a, allantois with urachus; N, umbilical vesicle with D, the omphalomesaraic duct; t, t, tubal openings; G, cervical canal. IX, Human embryo at the time of the visceral arches (diagrammatic): A, amnion; V, forebrain; M, midbrain; H, hindbrain; N, afterbrain; U, primitive vertebra; a, eye; p, nasal depression; S, Frontal process; y, internal nasal process; n, external nasal process; r, superior maxillary process of the first visceral arch; 1, 2, 3, 4, the four visceral arches with the intervening clefts; o, auditory vesicle; h, heart with, e, the primitive aorta. which divides into the five aortic arches; f, descending aorta; om, omphalomesaraic artery; b, the same artery upon the umbilical vesicle, B; c, omphalomesaraic vein; L, liver with the venæ advehentes and revehentes; D, gut; i, inferior cava; T, coccyx; all, allantois with, z, an umbilical artery, and, x, an umbilical vein.

form a linear union. A tube is thus formed from the furrow, the *medullary canal* (**Fig.** 376, II, III). The cells lying next the lumen of the canal become the ciliated cylindrical epithelium of the central canal of the spinal cord; the remaining cells produce the ganglia of the central nervous system and their processes. At the cephalic portion, the medullary canal widens out into the following dilatations, of progressively diminishing size: the *forebrain, prosencephalon* (rudiment of the cerebrum; the *midbrain, mesencephalon* (quadrigeminate bodies); the *hindbrain, metencephalon* (cerebellum); and the *afterbrain, myelencephalon* (oblongata) (**Fig.** 376. IV and V; **Fig.** 374, *F*; **Fig.** 377), which gradually passes over into the spinal cord. Below the hindbrain, in the vicinity of the afterbrain, the spinal furrow does not close, and there remains here an open passage-way to the contiguous lower portion

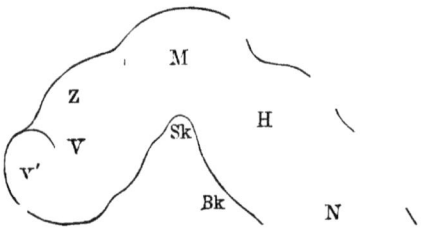

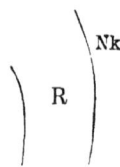

FIG. 377.—Lateral View of the Brain of a Human Embryo (after His). V, Primary forebrain vesicle. v′, secondary forebrain or hemisphere vesicle; Z, interbrain vesicle; M, midbrain vesicle: H, hindbrain vesicle; N, afterbrain vesicle; R, spinal portion of the medullary canal; Nk, nuchal flexure; Bk, pontal flexure; Sk, cranial flexure (anterior).

of the fourth ventricle (calamus scriptorius). At the caudal extremity there appears also a dilatation of the medullary canal, the lumbar enlargement. In birds the spinal furrow remains permanently open in this situation, and forms the rhomboid sinus.

While the medullary canal develops in this way, the primitive streak gradually atrophies, and finally disappears entirely (**Fig.** 374, *F*). The medullary canal does not continue in a straight course, but it bends in several places; namely at the junction of the spinal cord and oblongata (nuchal flexure); at the juncture of the afterbrain and the hindbrain (pontal flexure); finally almost at a right angle between the midbrain and the forebrain (parietal flexure). At first all of the brain-vesicles are without sulci or gyri. From the forebrain vesicle there grows on each side a pedunculated hollow vesicle (**Fig.** 376, VI, IX), the primary optic vesicle. The entire remaining portion of the epiblast furnishes the epidermal layer of the body, and is known as the horny layer. The stratum corneum can be differentiated early from the Malpighian network: from the first arise hairs, nails, feathers, etc.

FORMATIONS FROM THE HYPOBLAST AND THE MESOBLAST.

From the hypoblast there forms from above a cordlike arrangement of cells, which is placed lengthwise under the spinal furrow—the chorda dorsalis (Fig. 376, II, III, e). In man it is relatively thin. It forms the foundation of the spinal column, around which the substance of the vertebræ subsequently becomes so arranged that it pierces them like a thread through a string of pearls. After its formation, the chorda is soon surrounded by a double sheathlike covering. Further formations from the hypoblast do not occur at this time; it lies as a thin layer of single cells directly on the splanchnopleure.

While, formerly, the chorda dorsalis was in general believed to originate from the mesoblast, most observers at present, incline to the view that its development takes place from the hypoblast. The chorda begins to form at the anterior nodular swelling of the primitive streak, and grows toward the head. At first it represents a tube (Kupffer's canal, chordal canal) which opens posteriorly in the primitive groove, later breaking into the yolk-cavity. The chorda occurs in ascidia as well as in all vertebrates, although during their development it soon undergoes retrogressive changes.

On both sides of the chorda, the cells of the mesoblast group themselves into cubical structures, always arranged in pairs one after the other: *primitive vertebræ* (primitive segments or somites, Fig. 376, II, u; III, u; and Fig. 374, F, us). The first pair of these represent the atlas. At a later period a cellular cortical and a nuclear region can be distinguished in each primitive vertebra. The body of the primitive vertebra is used only in part for the formation of the later vertebra.

The portion of the mesoblast that lies peripherally from the primitive vertebræ, the *lateral plates* (Fig. 376, II, S), produces through the dehiscence of its cell-layers two lamellæ, which, however, remain united opposite the primitive vertebræ through the *middle plates*. The space thus resulting within the lateral plates is designated the pleuroperitoneal cavity or the cœlom (III, K). The upper lamella of the divided lateral plate is closely applied to the ectoblast and is known as the *musculocutaneous plate, somatopleure* (Fig. 376, III, x); the inner layer unites with the hypoblast, and is designated the *gut-fiber plate* or *splanchnopleure* (III, y). On the opposed surfaces of these two plates there develops the flat epithelium of the large pleuroperitoneal cavity. On the surface of the middle plate turned toward the cœlom there remain cylindrical cells, the germinal epithelium of Waldeyer, from which the oviducts and the ova are developed.

From the somatopleure, according to Remak, originate the skin and the musculature of the trunk, as well as the vessels; according to His, only the musculature of the trunk. According to both observers the smooth muscle of the digestive tract is derived from the splanchnopleure.

Especial emphasis should be placed on the views of His, who believes that the vessels, together with the blood and connective-tissue structures, do not arise autochthonously from the mesoblast, but that certain cells wander from the margins of the germinal layers, between the epiblast and the hypoblast, to form the structures named. They are not formed through the process of cleavage, but are derived from the elements of the white yolk lying external to the situation of the embryo, and they are thought originally to have wandered into the ovum as derivatives of the epithelium of the Graafian follicle. His designates these formations parablastic, in contradistinction to the archiblastic, which belong to the three germinal layers of the embryonal rudiment. Waldeyer also believes in the parablastic formation of blood, vascular endothelium, and connective tissue, although he considers the material from which the latter are derived as cohesive, and as living protoplasm of the same significance as the elements of the germ. The doctrine of archiblast and parablast has recently experienced many modifications.

The development of the middle germinal layer and the formation of the organs derived from it constitute one of the most difficult problems for investigation. The work of recent investigators, particularly that of the brothers Hertwig, has shown that in the lower vertebrates (amphioxus, triton), the chorda dorsalis and both walls of the cœlom-cavity result from evaginations of the hypoblast, as

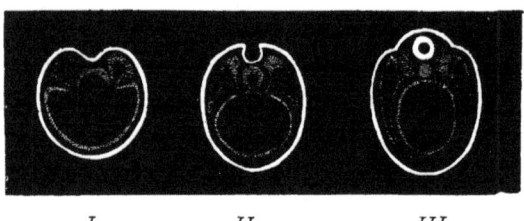

I   *II*   *III* .

FIG. 378.—Scheme of the Formation of the Chorda and the Cœlom through Evagination of the Hypoblast after the Theory of the Brothers Hertwig.

Fig. 378 illustrates in a diagrammatic way. In *I* is the beginning of the central evagination (for the chorda); the two lateral evaginations (for the walls of the cœlom) are still in free communication with the hypoblast; in *II* the points of the evagination are narrowed; and in *III* the chorda (which now lies below the medullary canal likewise constricted off) is fully detached and appears in cross-section as a round body. In the same way the walls of the cœlom-cavity have become detached, and they exhibit their two plates, the somatopleure and the splanchnopleure, and between the two the large body-cavity has expanded. The intestinal tube and the body-cavity have thus each obtained an independent wall. According to many new investigations both ectoderm and entoderm participate in the formation of the mesoderm, which, in its turn, is capable of producing the most varied tissues, with the exception of the nerves.

FOLDING OFF OF THE EMBRYO. FORMATION OF THE HEART AND THE FIRST CIRCULATION.

Up to this time the embryo with its three germinal layers has occupied the level of the layers themselves. Now (**Fig.** 376, **V**) the cephalic portion raises itself above this level and, becoming free, it grows more and more forward. There is thus formed in front of and under the head an invagination of the germinal layers known as the head-fold (**V, r**). The prominent cephalic portion is hollow within and an entrance may be gained from the interior of the germinal vesicle into the cephalic cavity. The latter is designated the fore-gut cavity (**V, D**), and the entrance to it the anterior intestinal portal. The formation of the fore-gut through the elevation of the head from the level of the germinal layers occurs in the chick as early as the second day (in dogs on the twenty-second day). In an entirely similar manner, although somewhat later (in the chick on the third day, in dogs on the twenty-fourth day), the analogous formation of the caudal portion takes place, and in consequence of which also this projects free, with the formation of the tail-fold (S) and the hind-gut (d), to which the posterior intestinal portal leads. The embryonal body thus communicates with the germinal vesicle by means of a pedicle that is as first wide open. This pedicle is known as the omphalomesenteric or vitellointestinal duct. The saccular vesicle attached to it is designated in mammals the umbilical vesicle (VII, X), while the analogous much larger sac in birds, which contains nourishment from the yellow yolk, is known as the yolk-sac. Toward the end of the third month of pregnancy the entodermal lining of the human umbilical vesicle develops genuine liverlike glandular tissue. The omphalomesenteric duct becomes in its further course narrower and finally is obliterated in the chick on the fifth day. Where the duct is inserted into the abdominal wall there results the abdominal umbilicus; where it is inserted into the primitive gut there results the intestinal navel.

On the ventral surface of the fore-gut and the hind-gut there are points where the mesoderm is wanting, and where, therefore, the epiblast and the entoblast come in contact. These are known as the pharyngeal and the cloacal membrane. The openings for the formation of the oral and the anal orifices are later found in these situations.

Even before this process of constriction takes place the primitive heart develops from that portion of the splanchnopleure that is in contact below with the fore-gut, in the chick at the conclusion of the first day as a rhythmically moving point ($\sigma\tau i\gamma\mu\eta$ $\chi\iota\nu\nu\nu\mu\dot{\epsilon}\nu\eta$ of Aristotle, punctum saliens); in mammals however, much later. The heart (Fig. 376, VI) develops as a cellular, hollow, bladderlike bud of the splanchnopleure (originally as a paired structure). Its cavity soon dilates and it grows into the cœlom suspended from a mesentery-like duplicature (mesocardium): that part of the cœlom situated in the vicinity of the heart is now designated the cardiac fossa (fovea cardiaca). The heart acquires a longitudinal tubular form, with its aortic portion directed anteriorly and its venous portion directed posteriorly. It then undergoes a moderate S-shaped curvature (Fig. 384, 1). From the middle of the second day the heart in the chick beats regularly, about 40 times per minute. At the anterior (aortic) extremity of the heart, the aorta originates from the bulbus aortæ; it bends forward, and, dividing into two arches (primitive aortas), it curves beneath the brain-vesicles and descends posteriorly in front of the primitive vertebræ. Both primitive aortas originally terminate blind at the caudal extremity of the embryo. Opposite the omphalomesenteric duct each primitive aorta in chicks gives off one, in mammals several (in the dog 4 or 5) omphalomesenteric arteries (Fig. 376, VI, Ao) which divide within the mesoblast upon the yolk-sac, or the umbilical vesicle, into a rich network of vessels. These unite and, passing backward (in birds arising from the terminal sinus of the subsequent terminal vein of the area vasculosa), form omphalomesenteric veins (Vo), which ascend on the duct and empty into the two venous trunks of the heart by means of two branches.

Thus the first or primitive circulation is completed. Its purpose is to convey nutritive material for growth and oxygen to the embryo. The latter, in birds, passes through the porous shell of the egg from the air; the first is supplied by the yolk-sac until the end of the incubation. In mammals both are supplied to the ovum from the vessels of the uterine mucosa. In birds, on account of the consumption of the contents of the yolk-sac the vascular area becomes steadily diminished. Finally, toward the end of the period of hatching, the yolk-sac, which has become smaller, slips into the abdominal cavity. Upon the umbilical vesicle of mammals the circulation usually disappears at an early date and the vesicle becomes transformed into a tiny appendage, while the second circulation develops to supplant the omphalomesenteric circulation. The first vessels in birds are formed outside the embryonal body in the area vasculosa as early as the last quarter of the first day, even before the heart can be seen. The vessels develop from vasoformative cells of the blood-islands, which at first appear isolated and then become confluent, and whose origin, whether from mesoblast or entoblast, has not yet been determined. At first solid, they later become hollowed out. In mammals (sheep) the first vessels also appear outside the embryo; the first blood-corpuscles are formed in the region of the vascular area as a product of the endothelium

Within the area vasculosa of the chick, there develops a closer-meshed lymphatic canal-system, which communicates with the amniotic cavity.

FURTHER DEVELOPMENT OF THE BODY.

The formative processes still wanting and necessary for the typical development of the body are as follows:

1. The cœlom gradually increases in extent, and in consequence the differentiation between the body-wall and the intestinal canal becomes the more distinct. The latter moves away from the primitive vertebræ, the middle plate becoming elongated to form the beginning mesentery. The body-wall, which, at first, still consists of the epidermis and the external lamella of the lateral plate (cutaneous plate), undergoes thickening, the primitive muscle growing from the muscle-plate, and the primitive bone, together with the spinal nerves, from the primitive vertebræ beneath the epidermis into the body-wall.

2. From the primitive vertebræ there is detached a portion situated dorsally, which is designated the muscle-plate. The remaining portion of the primitive vertebra (true primitive vertebra) now unites with its fellow of the opposite side, both growing completely around the chorda (membrana reuniens inferior; in dogs on the third, in rabbits on the tenth day), and also enclosing the medullary canal (membrana reuniens superior; in chicks on the fourth day). Thus, there has taken place in front of the medullary canal a union of the primitive vertebral masses that enclose the chorda and therefore form the basis of all of the vertebral bodies, while the membrana reuniens superior, interposed between muscle-plates, and epidermis on the one side and the medullary canal on the other side, represents the rudiment of the entire system of vertebral arches, together with the intervertebral ligaments between them. The spinal column is in this membranous stage an exact reproduction of the spinal column of the cyclostomes (lamprey). From the membrana reuniens superior there are formed, besides, the membranes of the spinal cord and the spinal ganglia and nerves.

In rare cases the formation of the membrana reuniens superior does not occur. Under such circumstances the medullary canal is covered posteriorly by the horny layer (epidermis) alone, either throughout its entire extent, or only in limited areas. This defect in development is known as spina bifida (at the head, hemicephalus). Failure in the development of the membrana reuniens inferior is exceedingly rare. This arrest of development gives rise to permanent separation of the bodies of the vertebræ into two lateral halves.

The cutaneous plates finally grow also toward the middle line of the back and insinuate themselves between the muscle-plates and the epidermis; in this manner the dorsal skin is formed. In the membranous spinal column the individual cartilaginous vertebræ are formed successively (in man between the sixth and seventh weeks), but these do not at first possess closed vertebral arches; the latter close in man during the fourth month. Each cartilaginous vertebra, however, does not develop from a pair of primitive vertebræ (thus, the sixth cervical does not develop from the sixth pair of primitive vertebræ); but a new articulation of the spinal column takes place, and in such a manner that the lower half of the preceding and the upper half of the following primitive vertebra unite to form the definitive vertebra. In the process of chondrification of the vertebral bodies the chorda suffers a reduction, remaining larger, however, in the intervertebral discs. The body of the first vertebra unites with that of the second to form its odontoid process; in addition, it forms the anterior arch of the atlas and the transverse ligament. The chorda can be followed upward through the ligamentum suspensorium dentis to the posterior portion of the sphenoid bone.

The histogenetic formation of cartilage from the indifferent formative cells takes place through multiplication and enlargement of the cells that finally become clear nucleated vesicles. The cement-substance probably originates from the union of the cells at the periphery and their outer portion (parietal substance) giving off the intercellular substance. Whether the latter possesses fine canals that connect the interstices of the cartilage is asserted by some and denied by others. According to the statements of some investigators, the ground-substance after special treatment appears to be made up of fine fibrils.

3. In the cervical portion, on each side, there develop four cleft-like openings: the *visceral clefts* or *branchial openings*. Above the clefts are recesses in the lateral wall, the *visceral arches* (in the chick formed at the end of the third day). The clefts result from rupture of the fore-gut from within (although, perhaps, this does not always take place in the chick, in mammals, and in man), and they are surrounded by endoblastic cells. Upon the visceral arches, above and below each cleft, there pass on each side the *aortic arches*, of which there may be as many as five (Fig. 376, IX). These formations are permanent only in fish. In man, all of the clefts are obliterated except the uppermost, which forms the auditory canal, the tympanum, and the Eustachian tube. The four visceral arches are for the greater part later transformed into other formations.

In the middle line beneath the forebrain is a thin point where (in the region of the pharyngeal membrane) an invagination with an embankmentlike or craterlike border first takes place, followed by rupture, and forming the *primitive oral orifice* (which still comprises the mouth and the nose together). Later, a depression at the caudal extremity (in the situation of the cloacal membrane) ruptures into the hind-gut, forming the *anus*. Should this fail to take place, *atresia ani* results. The lungs, the liver, the pancreas, the cecum (in birds), and the allantois (to be described later) develop from the entoblast and the adjacent splanchnopleure as diverticula from the primary intestinal tube. The extremities appear as short stumps upon the trunk at first devoid of members.

FORMATION OF THE AMNION AND THE ALLANTOIS.

During the process of folding-off of the embryo there results, first (at the end of the second day in the chick) in front of the head, a foldlike elevation, consisting of epiblast and the outer layer of mesoblast. This is reflected like a cowl to form the head-fold for the cephalic portion of the embryo (Fig. 376, VI, A). Later and more slowly there develops the caudal fold from behind, and, finally, also between these two the lateral folds are formed (Fig. 376, III, A). As all of these folds tend toward the back of the embryo they finally grow together and form the amniotic sac (in the chick on the third day). There is thus formed about the embryo a cavity that becomes filled with amniotic fluid. Also in mammals the amnion develops early and in the same way as in ·birds (Fig. 376, VII, A). From the middle of pregnancy the amnion lies in immediate contact with the chorion, with which it is united by a layer of gelatinous tissue (tunica media).

Both the amnion and the allantois develop only in mammals, birds, and reptiles, which therefore are designated also amniota, while the lower vertebrates, the anamnia, are without these structures. The *amniotic liquor* is a clear, serous alkaline fluid, having a specific gravity of from 1002 to 1028. It contains. in addition to epithelium, lanugo-hairs and from $\frac{1}{2}$ to 2 per cent. of fixed solids. The latter comprise albumin (from $\frac{1}{15}$ to $\frac{1}{3}$ per cent.), mucus, globulin, a body resembling vitellin, some grape-sugar (cow), allantoin, urea, ammonium carbonate (probably transformed from urea), sometimes lactic acid and kreatinin, calcium sulphate and phosphates, and sodium chlorid. The total amount of fluid at the middle of pregnancy is from 1 to 1.5 kilos; at the end of pregnancy 0.5 kilo.

The amniotic liquor is of fetal origin, as its presence in birds indicates, and it may be a transudate from the ovular membranes. In mammals the urine of the fetus probably contributes to the accumulation of the fluid in the second half of pregnancy. In cattle, in which the allantoic and the amniotic fluids remain permanently separate, the first may be regarded as fetal urine, the latter as transudate. In the presence of the pathological condition of hydramnios, also the vessels of the uterine mucosa may secrete serum, especially when there is stasis in the distribution of the umbilical vein in the placenta. The amniotic fluid protects the fetus and the vessels of the fetal membranes from external injuries; it affords free movement to the limbs, and thus prevents them from forming adhesions; finally, it is important during the act of parturition for the dilatation of the mouth of the uterus. The amnion is contractile (in the chick from the seventh day on), from the presence of smooth muscle-fibers that develop in the cutaneous plate (mesodermal portion). Nerves have not been found.·

From the anterior extremity of the hind-gut there grows a vesicular sac, which appears at first as a small double tubercle and then becoming hollow (Fig. 376, VII, a); it projects into the cœlom-cavity. This is the *allantois* or urinary sac (in the chick before the fifth day; in man during the second week). As a true evagination from the hind-gut, the allantois has two layers: one from the entoblast, and the other from the splanchnopleure. From each side there passes upon the sac the allantoic or umbilical artery, arising from the hypogastric artery, and ramifying upon the surface of the sac. The allantois grows (like a steadily filling urinary bladder) in front of the hind-gut in the abdominal cavity toward the· umbilicus, and finally out of this (at the side of the omphalomesenteric duct), together with its vessels (VII, a), and it exhibits different relations in birds and in mammals.

In birds, the allantois, after passing out at the umbilicus, undergoes excessive growth, in a short time lining the entire inside of the shell as a vascular sac. Its arteries, at first branches of the primitive aorta, appear with the development of the posterior extremities, as branches of the hypogastric artery. From the rich capillary network of the allantois there originate two allantoic or umbilical veins. These enter the navel and pass, at first in association with the omphalomes-enteric veins, into the venous portion of the heart. In birds this allantoic circu-lation (or second circulation) subserves the purpose of respiration, as its vessels maintain an interchange of gases through the porous egg-shell. This circulation gradually assumes the respiratory function of the yolk-circulation; this is neces-sary, because the yolk-sac, steadily decreasing in size, no longer presents a suffi-ciently large respiratory surface. Toward the end of the period of hatching the bird can breathe and cry within the shell, a sign that the respiratory function of the allantois is taken up, at least in part, by the lungs. The allantois is further-more the excretory organ for the urinary constituents. Especially in mammals the excretory ducts of the primitive kidneys, the Wolffian or Oken ducts, empty into the cavity of the allantois (in birds and snakes, which possess a cloaca, they empty into the posterior wall of the cloaca). The primitive kidney, consisting of many glomeruli, discharges its secretion through the Wolffian duct into the allantois (in birds into the cloaca); and the secretion passes by way of the allan-tois through the navel into the peripheral portion of the urinary sac. Remak found in the allantoic contents, ammonium and sodium urates, urea, allantoin, grape-sugar, and salts. From the eighth day on, the allantois of the chick is contractile from the presence of fibrillar cells that are derived from the splanch-nopleure. Lymphatic vessels accompany the arterial branches.

In mammals and in man, the relation of the allantois is somewhat different. From the first part, the urinary bladder is formed; from the vertex of this the urachus, at first still open, passes as a tube out through the navel (Fig. 376, VIII, a).

The blind sac of the allantois, which is situated outside the abdomen, is in some animals filled with a fluid resembling urine. In man, how-ever, this sac atrophies in the course of the second month. The vessels alone remain and these apparently lie in the splanchnopleural portion of the allantois. In some animals the allantoic sac continues to grow, without undergoing atrophy, and then conveys from the bladder through the urachus an alkaline, cloudy fluid that contains some albumin, sugar, urea, and allantoin. The relations of the allantoic vessels will be described in connection with the fetal membranes.

HUMAN FETAL MEMBRANES. PLACENTA. FETAL CIRCULATION.

When the fecundated ovum gains entrance into the uterus, it be-comes surrounded by a particular membrane, which William Hunter described as the *deciduous membrane*, because it is expelled in the act of parturition. A distinction is made with regard to the *basilar or true decidua* (Fig. 376, VIII, p), which is nothing else than the thickened, hy-peremic, spongy endometrium, loosely attached to the uterine wall. From this there develops an especial formation around the ovum, which receives the latter as in the pocket of a swallow's nest; this thinner membrane is known as the *capsular decidua* or *decidua reflexa* (VIII, r). Between the second and the third month there is still a space in the uterus outside of the decidua reflexa, but in the fourth month the entire cavity is occupied by the ovum and the decidua. At one point the ovum is thus applied directly to the endometrium (*basal or true de-cidua*); but in its greatest extent, however, it is in contact with the decidua reflexa. The first layer forms later the *placenta*.

The decidual swelling and softening of the endometrium begins in the mucosa of the tubes of the cervical canal; in the third month the membrane is from 4 to 7 mm. thick, in the fourth month only from 1 to 3 mm., and it is devoid of epithelium, rich in blood-vessels, and has lymph-spaces around the glands and vessels; its spongy tissue contains large round cells (decidual cells), which in the depth are often transformed into fibrillar and spindle-shaped cells; in addition, leukocytes are present. The uterine glands, which, at the beginning of pregnanty, are enormously developed, undergo a transformation between the third and the fourth month to large, noncellular dilated tubes. In the last months these become indistinct and their epithelium (which, according to Friedländer, Lott, and Hennig was originally ciliated), disappears progressively toward the depth.

The capsular decidua. much thinner than the true decidua, is devoid of epithelium and also of vessels and glands from the middle of pregnancy. Toward the end of pregnancy both deciduas unite completely with each other.

The basilar decidua and likewise the uterine placenta consist of a compact layer (pars caduca), which is detached during labor, and of a deeper spongy layer, in which the process of detachment takes place and of which a portion remains upon the surface of the muscularis (pars fixa). From the latter the regeneration of the new mucosa after labor takes place. Also the tubes exhibit during pregnancy hyperplasia of the mucosa and of the muscularis.

The ovum reaches the endometrium as a vesicle without villi. The mucosa has become softened and hyperemic, and the ovum sinks into its tissue, quickly to become completely encapsulated. In the basal decidua there form lacunar maternal blood-passages, which undergo progressive enlargement With the formation of the amnion, there occurs, after its closure, the production from the epiblast of a special entirely closed vesicle, which passes over the embryo, the amnion and the umbilical vesicle, and thus lies next to the primitive chorion; this is the *serous capsule* (Fig. 376, VII, S), which applies itself closely to the chorion. The allantois, rich in vessels, passes out of the navel, and lies directly upon the ovular membrane; its vesicle atrophies about the second month in man, but its vascular layer grows rapidly and lines the entire interior of the ovular cavity, where it can be found on the eighteenth day. From the fourth week the vessels, together with a connective-tissue framework, form many intricately branching *villi*, while the original ovular membrane (prochorion or primitive chorion) disappears (in dogs it is absorbed and serves for nourishment). There is thus reached a stage of general vascularization of the chorion; the derivative of the zona pellucida is now replaced as the ovular membrane by the villous vascular layer of the allantois, which is covered by the cells of the serous capsule (derived from the epiblast). The chorionic villi grow downward in the direction toward the decidual vascular spaces. The villi are separated from the vascular space by two layers of specialized cells: the chorionic epithelium, or the layer of Langhans (derived from the fetal ectoderm), and a second layer designated syncytium, whose cells with large nuclei and indefinite outline are, according to most recent investigators, derived from transformed uterine epithelium. In the protoplasm of the latter vacuoles appear and the cilia atrophy when the existing spaces are filled with blood. The ovum adheres to the syncytium after the disappearance of the zona pellucida. The stage of general vascularization continues, however, until the third month; at that time the vegetation of the vascular villi ceases upon the entire ovular membrane that is in relation with the decidua reflexa. On the other hand, those villi of the chorion that are in direct contact with the decidua vera become larger and more branched. There thus results the distinction between the chorion læve and the chorion frondosum.

The chorion læve. which has a connective-tissue stroma and is covered by a double layer of epithelium, possesses besides at great intervals diminutive villi that pass to the decidua reflexa. Between the chorion and the amnion there is a gelatinous layer (membrana intermedia) of immature connective tissue.

The large villi of the chorion frondosum (**Fig.** 379) penetrate more deeply into the uterine mucosa and first of all into the ducts of the glands, like roots into loose soil. According to Selenka this penetration takes place from the first week. From the syncytial covering of the villi there arise in especially large number between the second and the third month cell-buds, which are probably related to the nourishment of the embryo. In the further penetration of the villi through the ducts of the glands they make their way through the walls of the large contiguous blood-vessels, which in structure are similar to the capillaries, so that the villi, bathed in the maternal blood (uterine vessels), float in these enormous decidual capillaries—the so-called *intervillous spaces* (Fig. 376, VIII, b). The villi within the blood-spaces are covered by the epithelium of the latter.

Some villi that have no epithelium unite by means of bulbous extremities with the tissue of the uterine placenta and thus form adherent villi; a means of firm union. In this way the *placenta* is formed; a distinction is made between the fetal placenta, which includes the entire mass of villi, and the uterine or maternal placenta, the portion of the endometrium in relation with the ovum, which is especially rich in vessels at this point. The two parts are not separable even at the time of birth. Venous maternal vessels of considerable size course around the border of the placenta, constituting the *marginal sinus*. The placenta is the nutritive and respiratory organ of the fetus, which obtains the necessary material through endosmosis from the maternal blood-spaces through the coverings and vessel-walls of the villi, in which the fetal blood circulates.

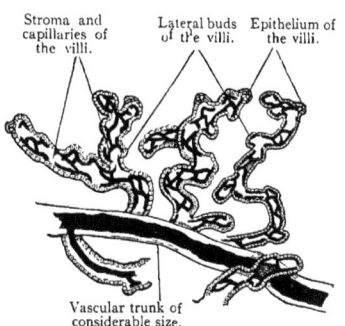

FIG. 379.—Isolated Portion of Villi from a Human Placenta.

Between the placental villi there is a clear fluid that contains numerous small, albuminoid globules and is designated *uterine milk* (abundant in the cow); it is believed to originate from degeneration of decidual cells. It is thought, together with the blood, to take part in the process of nutrition.

The investigations of Walter have shown that when pregnant animals are poisoned with strychnin, morphin, veratrin, curare, and ergotin, these substances cannot be demonstrated in the fetus. Certain other chemical substances, for example phosphorus, potassium chlorate, potassium bromid, potassium iodid, arsenic, mercury, alcohol, phenol, morphin, and methylene-blue, do, however, pass over to the fetus. Some substances pass also from the fetus to the maternal body. An examination of the placenta shows that its villi are grouped in individual sections of considerable size, between which are furrow-like indentations. These individual complexes may be compared with the cotyledons of lower animals.

The position of the placenta is, as a rule, upon the anterior or posterior uterine wall, less commonly at the fundus uteri, or laterally in front of or beneath a tubal opening (lateral placenta) or in front of the internal orifice of the uterus

62

(placenta prævia).   The last position is a dangerous one, because rupture of the vessels at birth may cause death of the mother from hemorrhage.   Implantation of the ovum in the cervical canal is extremely rare.

The *umbilical cord* may be inserted either into the center of the placental disc (central insertion) or more toward the border (marginal insertion); or the cord may be attached to the chorion læve, so that the vessels must pass to the placenta through the thin chorion læve (velamentous insertion).   Rarely there is an accessory placenta separated from the placenta proper (placenta suc-centuriata).   Kölliker designates as marginate placenta one that has villi only at its center.   If the placenta consists of two halves it is known as duplex or bipartite (constant in the apes of the old world).

The umbilical cord (mature, from 48 to 60 cm. long and from 11 to 18 mm. thick) is covered by the amniotic sheath.   The vessels make about 40 spiral turns (beginning after the middle of the second month) passing from the embryo from left to right toward the placenta: they consist of two arteries with a well-developed muscular coat and one

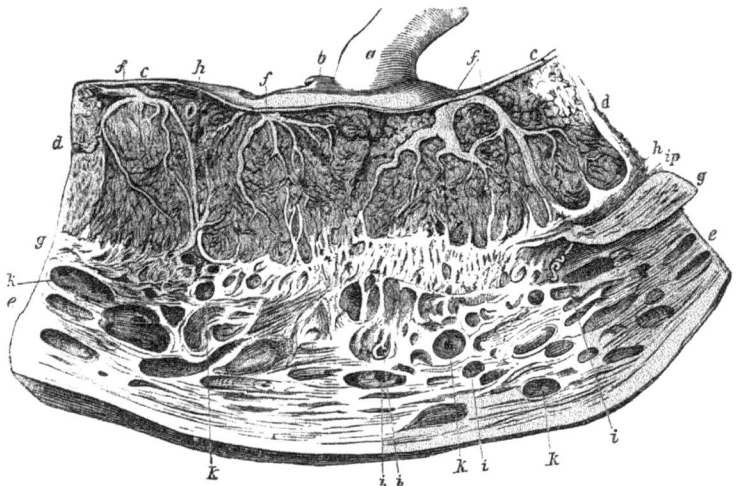

Fig. 380.—Section through the Uterus and the Attached Placenta at the Thirtieth Week (after Ecker): *a*, root and insertion of the umbilical cord; *b*, amniotic covering of the umbilical cord; *c*, chorion; *d, d*, fetal portion of the placenta; *e, e*, uterine wall; *f, f*, villous radiation forming the framework of the fetal placenta; *g, g*, decidua; *h, h*, processes of the decidua penetrating into the fetal placenta; *i, i*, branches of the uterine artery *ip*, an artery entering the placenta: *k, k, k, k*, uterine veins.

(left) umbilical vein.   Both arteries anastomose in the placenta.   In addition, the cord contains the continuation of the urachus, the ento-dermal portion of the allantois (Fig. 376, VIII, a), which, persisting until the second month, is often atrophied later.   The omphalomesen-teric duct can still be dissected at the time of birth near the umbilical vesicle as a filamentous pedicle (VIII, D) of the umbilical vesicle (N) that persists and as a rule is situated beyond the placental border.   The ves-icle contains in its interior small villi, squamous epithelium, and the obliterated vessels of the first circulation.   Persistent, though diminu-tive omphalomesenteric vessels are rare.   Wharton's jelly, a gelatinous connective tissue, surrounds all of these parts; it contains connective-tissue fibrils, connective-tissue corpuscles and lymphoid cells, even elastic fibers. The gelatinous substance contains mucin. Numerous lymph-channels, lined with endothelium, traverse the jelly; lymph-vessels and blood-vessels are absent. Nerves are found from 3 to 8 or 11 cm. from the navel.

The jelly contains two forms of mucin, like that of the tendons, also globulin (myosin?) and albumin.

The **fetal circulation** that exists after the development of the allantois pursues the following course: the blood of the fetus passes by way of the two umbilical arteries (from the hypogastric) through the umbilical cord to the placenta, where the arteries break up into the capillaries of the placental villi. Returning from these the blood collects in the umbilical vein (its color is scarcely a little brighter as compared with that of the venous blood in the umbilical arteries). The umbilical vein (Fig. 387, 3, u_1) turns upward from the navel and, passing under the border of the liver, it anastomoses with the portal vein (a) and continues as the ductus venosus of Arantius to the inferior vena cava, which then conveys the blood to the right auricle. From here the Eustachian valve and the tubercle of Lower (Fig. 384, 6, tL) deflect most of the blood through the foramen ovale into the left auricle, from which, on account of the presence of the valve of the foramen ovale, it cannot flow back into the right auricle. From the left auricle the blood passes through the left ventricle, the aorta and the hypogastric artery back into the umbilical arteries. The blood of the superior vena cava in the fetus, by reason of its peculiar entrance, passes from the right auricle into the right ventricle (Fig. 384, 6, Cs). From here it enters the pulmonary artery (Fig. 384, 7, p), which conveys it into the aorta through its prolongation, the ductus arteriosus Botalli (B), which empties into the aortic arch. Only a little blood passes by way of the small branches of the pulmonary artery (1, 2) through the lungs. The course of the blood makes it clear that the head and the upper extremities are supplied with purer blood than is the remainder of the trunk, which also receives an admixture of the blood from the superior vena cava. After birth the umbilical arteries are obliterated, and become the lateral ligaments of the bladder; their lower portion, however, persists as the superior arteries of the bladder. The umbilical vein also is obliterated and becomes the round ligament; and likewise the ductus venosus of Arantius. Finally the oval foramen closes, and the ductus arteriosus Botalli becomes obliterated to form the ligamentum arteriosum.

The **relation of the fetal membranes** in multiple pregnancies is as follows: (1) In the presence of twins there are two entirely separate ova, with two placentas and two reflex deciduas. (2) Two entirely separate ova have but one reflex decidua, the placentas becoming adherent, while their vessels are separate. The chorion is double, but not separable into two lamellæ at its surface of contact. (3) When there are one reflex decidua, one chorion, one placenta, two umbilical cords, and two amnia, the vessels anastomose in the placenta, and, therefore, the central stump of the umbilical cord of the first born of twins should always be tied. Under such circumstances there has been either one ovum, with a double yolk, or two germinal vesicles in one yolk; or it must be assumed that two separate ova have subsequently united, with absorption of the contiguous parts of the chorion. (4) When the conditions just described are present, except that there is but one amnion, they are due to the formation of two embryos in the same germinal area of the same germinal vesicle.

Brief mention should be made here of the formation of the fetal membranes in animals, which has been followed since the time of Home, Blainville, H. Milne-Edwards, Owen, and others, in the classification of mammals.

1. The oldest mammals have no placenta or allantoic vessels at all: these are the *mammalia implacentalia*, namely marsupials and monotremata (duck-bills and echidna). In addition to a serous capsule and an amnion devoid of villi, these animals have only a large yolk-sac which contains vessels, but which never undergoes placental formation. The allantois remains rudimentary (in the kangaroo-bear it becomes larger, and together with the yolk-sac, serves as a respiratory organ). (In the oviparous monotremata the ovum develops outside the maternal body.)

2. The second group includes the *mammalia placentalia*. Among these: (a) the *mammalia nondeciduata* possess only chorionic villi (supplied by the allantoic vessels), which project into depressions in the uterine mucosa, from which they retract during parturition (placenta diffusa, for example pachyder-mata, cetacea, solidungula, camelida). The umbilical vesicle, which, at the earliest period, contains vessels, subsequently undergoes a marked involution, in the different groups of animals, into manifold modifications. In the ruminants the large villi are arranged in groups, and they grow into the greatly hypertro-phied rolls of mucosa (cotyledons) corresponding to the uterine glands, from which they retract at birth. The ovum is for a long time spindle-shaped. (b) The *mammalia deciduata* form such an intimate union between the chorionic villi and the endometrium that the corresponding portion of the latter must be thrown off at birth. Here, the placenta is either girdle-shaped (placenta zonaria), as in carnivora, pinnipedia, elephant, hyrax; or it is disc-shaped (placenta discoidea), for example, in apes, insectivora, rodents, alipeds, and edentates.

The same placental formation as occurs in man is found in the anthropoid apes, but in none of the others.

Certain variations occur in different animals in detail with reference to the formation of the fetal membranes. In rabbits the umbilical vesicle also is greatly expanded, and the large omphalomesenteric vessels participate in the formation of the placenta through the development of a yolk-sac placenta. Also in guinea-pigs (which, remarkably, have the three germinal layers in a reverse order, the epiblast within, so that in the folding-off of the embryo, the latter sinks into the interior of the umbilical vesicle) the omphalomesenteric vessels play a prom-inent part in the formation of the placenta. In some carnivora (cats), the um-bilical vesicle is provided with vessels until the time of birth. It is to be noted, finally, that, in the uterus of the smooth shark (mustela lævis) a yolk-sac placenta is formed.

CHRONOLOGY OF HUMAN DEVELOPMENT. FETAL MOVE-MENTS.

Development in the First Month.—The youngest ovum is described by Hub. Peters. It had a diameter of 1.6 × 0.8 × 0.9 mm., and consisted of a vesicle about three days old. The small villi had already a covering of two layers of cells. It is the only observed ovum around which the capsular decidua had not yet closed. At the point of attachment of the ovum to the uterine mucosa, there were large blood-lacunæ, in which the ovum appeared embedded.

Ova of from six to eight days have been described by Merttens and Siegen-beck van Heukelom. They possess small short villi covered by a double layer of cells; an outer layer of large cells (syncytium), derived from the uterine epi-thelium, and a subjacent layer, formed from the ectoderm—cellular layer of Lang-hans. An ovum seven or eight days old was 3.7 × 4 mm. in size; the villi already had small ramifications. From the twelfth to the thirteenth day: (5.5 mm. and 3.3 mm. in diameter) there exists a simple germinal vesicle, which, at one point, contains the germinal area consisting of two layers of cells. Ova from the fifteenth to the sixteenth day have a diameter of from 5 to 6 mm., with simple cylindrical villi, or are provided with bulbous processes from the base to the apex. The youngest ovum of Allen Thomson he estimated to be about fifteen days old. It was 13.2 mm. long, oval, and provided with villi; the germinal vesicle was 2.2 mm. (abnormally small), the rudimentary embryo 2.2 mm. with a spinal furrow and spinal ridges, projecting beyond the vesicle at each extremity; the rudimen-tary heart was present (and the amnion?). A somewhat older ovum studied by the same investigator was 6.6 mm. long, with short, thin villi, and a large germinal vesicle, from which the embryo (2.2 mm.), with a closed medullary canal, had begun to be constricted off.

There now follows the stage in which the formation of the allantois appears. It is at present a much-disputed point whether in man a free allantoic vesicle, growing from the navel, exists or not. The youngest embryo that bears upon this point has been examined by v. Preuschen and the author. In the fresh state, this measured 3.78 mm. in length; it was divided into sections and thoroughly studied. The brain-vesicles were indicated; the organs of special sense were wanting; the ganglia in the cephalic region were visible. The visceral arches were visible as thickenings in cross-section, but not yet isolated; the visceral clefts, the mouth, and the anus were absent. The sella turcica was in process of formation. Heart, lungs and liver were in their earliest form. The umbilical vesicle (torn) was apparently still provided with a wide opening. The allantois was distinct as a free vessel outside of the abdomen; its lamella from the mesoderm was yet without vessels. The extremities were entirely absent. The chorda dorsalis was indicated, and on either side the primitive vertebral masses. A free projecting allantoic vesicle has also been described in embryos by W. Krause and Bruch, but these were older.

An ovum of from the fifteenth to the eighteenth day has been described by Coste; it was 13.2 mm. long; the villi were small, and slightly branched; the embryo was 4.4 mm. long, of curved form, with a moderately thickened cephalic portion. Amnion, umbilical vesicle (with a large omphalomesenteric duct), and allantois were fully developed, the last already adherent to the serous covering. The S-shaped heart, lying in the cardiac cavity, exhibits a cavity and the bulb of the aorta, but no ventricles or auricles. The visceral arches and clefts are indicated, but the latter are not yet broken through. Upon the umbilical vesicle the first circulation of the two omphalomesenteric arteries is developed; the folding-off is only moderately advanced; the duct is still widely open; two primitive aortas pass in front of the primitive vertebræ. The allantois, adherent to the fetal membranes, possesses its vessels. The two omphalomesenteric veins, united with the two umbilical veins, empty into the lowermost, venous portion of the heart. The mouth is in process of formation. The extremities and organs of special sense are wanting; the Wolffian bodies are probably present. Similar descriptions have recently been made by His, although the length of the embryo was somewhat less.

There now follows a stage wherein all of the visceral arches are indicated, and the clefts are broken through. The midbrain forms the highest point of the brain; the two auricles appear in the heart. The communication with the umbilical vesicle is still pretty free. The embryo is from 2.6 mm. to 3.3 or 4 mm. long. The head undergoes a deflection to the side. At a still later period there appear on the brain the parietal and the nuchal curvature; the hemispheres appear more distinct; the entrance to the umbilical vesicle becomes constricted, the rudimentary liver is discernible; and the extremities are still absent. In addition to the embryo of His one of the twentieth day described by Johannes Müller belongs here. The ovum was between 15.2 by 17.6 mm. in size; the embryo from 5 to 6 mm. long; the umbilical cord 1.3 mm. thick. The umbilical vesicle was in free communication with the bowel. The amnion surrounded the embryo and formed a sheath for the umbilical cord. The visceral arches and clefts were present; and behind them the projecting heart-tube; the extremities were wanting.

Third week (R. Wagner): The ovum measured 13 mm., the embryo from 4 to 4.5 mm., the umbilical vesicle 2.2 mm.; the bowel was almost entirely closed. Three visceral clefts, the Wolffian bodies, the first rudiments of the extremities, three brain-vesicles, and the auditory vesicles were present. A similar embryo described by Hensen should be included here. Twenty-first day (Coste): The nasal pits, the buccal orifice, the eyes, the auditory vesicles, four visceral arches and the mouth (toward which the frontal and the superior maxillary process grow) were especially marked; the heart with two ventricles and two auricles and the vessels of the umbilical vesicle were present.

End of the First Month.—The embryos of from twenty-five to twenty-eight days are characterized by the distinct pedunculation of the umbilical vesicle and by the definite appearance of the extremities. The length of the ovum is 17.6 mm.; of the embryo from 8 to 11 mm., of umbilical cord 4.5 mm. with its vessels.

Second Month.—Embryos of from twenty-eight to thirty-five days begin to extend in greater degree; the visceral clefts are closed with the exception of the first; the allantois has but three vessels, as the right umbilical vein is obliterated. In the fifth week the trunk is from 0.85 to 1.28 cm. long; the olfactory pits are connected with the angles of the mouth by furrows, which close in the sixth week and form canals. In embryos from thirty-five to forty-two days old the trunk has a length of between 1.1 and 1.3 cm.; the buccal and nasal openings are separate; the face is smooth; the extremities have three divisions; on the feet the toes are not so well developed as the fingers. The auricle of the ear forms first a low projection in the seventh week. The Wolffian body is considerably reduced. The length of the trunk at between the seventh and the eighth week is from 1.6 to 2.1 cm.

End of the Second Month.—The ovum is 6½ cm., the villi 1.3 mm. long; the circulation in the umbilical vesicle is obliterated; the embryo is 26 mm. long and weighs up to 4 gm.; the eyelids and the nose are present. The umbilical cord is 8 cm. long; the abdominal cavity closed; ossification has begun in the lower jaw, the clavicles, the ribs, the vertebræ; sex is undeterminable; the kidneys are indicated.

Third Month.—The ovum is as large as a goose-egg; the formation of the placenta has begun; the embryo (trunk-length between 2.1 and 6.8, total length from 6 to 11 cm.; weight 11 gm.) is from now on known as the fetus. The auricles of the ear are developed; the umbilical cord is 7 cm. long; the external sexual differentiation has begun; the navel is in the lower fourth of the linea alba.

Fourth Month.—The length of the fetal trunk is between 6.9 and 9 cm.; the total length from 10 to 17 cm.; the weight 57 gm.; the sex is distinct; hair and nails have begun to form; the placenta weighs 80 gm.; the umbilical cord is 19 cm. long; the navel is in the lower third of the linea alba; jerking movements of the extremities occur; the bowels contain meconium; vessels are visible through the translucent skin; the eyelids are closed.

Fifth Month.—The length of the fetal trunk is from 9.7 to 14.7, the total length from 18 to 28 cm.; the weight is up to 284 gm. The hair of the head and lanugo-hairs are distinct; the skin is still somewhat light pink and thin, covered with vernix caseosa, and not perfectly transparent. The weight of the placenta is 178 gm.; the umbilical cord is 31 cm. long.

Sixth Month.—The length of the fetal trunk is from 15 to 18.7, the total length from 26 to 37 cm.; the weight is 634 gm. The face acquires more fat and has a less aged appearance; the lanugo is thick and fluffy; the amount of vernix is increased; the testicles are in the abdomen; the pupillary membrane and the eye-lashes are present; meconium is present down to the large intestine.

Seventh Month.—The length of the fetal trunk is from 18 to 22.8, the entire length from 35 to 38 cm.; the weight is 1218 gm.; the large intestine has the same length as the body (at an earlier date it is shorter, at a later date longer); the descent of the testicles has begun, one testicle finding its way into the inguinal canal; the eyes are open; the pupillary membrane often has disappeared at its center in the twenty-eighth week; in addition to the primitive fissures, other fissures begin to form. The fetus is viable. At the beginning of this month there is a center in the os calcis.

Eighth Month.—The length of the fetal trunk is from 24 to 27.5, the total length from 41 to 42 cm.; the weight is 1569 gm.; the hair of the head is thick, 13 cm. long; the nails have narrow margins; the navel is below the middle of the linea alba; one testicle is in the scrotum.

Ninth Month.—The length of the fetal trunk is from 27 to 30, the total length from 42 to 65 cm.; the weight is 1971 gm.; the fetus at this age is not distinguishable from the mature fetus.

Tenth Month.—The length of the trunk is from 30 to 37, the total length from 45 to 67 cm.; the weight is 2334 gm.

The Mature Fetus.—The length of the body is 51 cm.; the weight 3 kilos (from 2.5 to 5 kilos); lanugo-hair is still present only on the shoulders; the skin is white; the cartilages of the nose and the ears are hard to the touch. The nails project beyond the finger-tips. The navel is somewhat below the middle of the linea alba. The center of ossification in the lower epiphysis of the femur, from 4 to 8 mm. in transverse diameter (it begins at the commencement or the middle of the ninth month, and is from 2 to 5 mm. wide at the end of the ninth month), is a characteristic of the mature fetus. There is often a center of ossification in the upper epiphysis of the tibia at the end of the tenth month.

In conclusion, the duration of development in the following animals will be given: Colibri, twelve days; hen, duck, twenty-one days; goose, twenty-nine days; stork, forty-two days; cassowary, sixty-five days; mouse, three weeks; rabbit, hare, four weeks; rat, five weeks; hedgehog, seven weeks; cat, marten, eight weeks; dog, fox, polecat, nine weeks; badger, wolf, ten weeks; lion, fourteen weeks; pig, seventeen weeks; sheep, twenty-one weeks; goat, twenty-two weeks; roe, twenty-four weeks; bear, small apes, thirty weeks; deer, from thirty-six to forty weeks; man, forty weeks; horse, camel, thirteen months; rhinoceros, eighteen months; elephant, twenty-one months. Limitation of the supply of oxygen to the incubating egg of birds is followed by dwarf-formation.

Various intrauterine movements of the fetus are discernible through the abdominal wall of the mother: extension-movements of the trunk, movements of the extremities, and in the later period of pregnancy (and during labor) a regular rhythmical movement of the respiratory muscles, recurring at intervals and usually continuing for some time. In addition the fetus makes sucking and swallowing movements.

DEVELOPMENT OF THE OSSEOUS SYSTEM.

Spinal Column.—The ossification of the vertebræ begins between the eighth and the ninth week, a center appearing first in each half-arch, then a center, probably consisting of two lying close together, in the body behind the chorda. In the fifth month the bony tissue has reached the surfaces, the chorda in the body is displaced. The three pieces unite in the first year. The atlas has a center in the anterior arch and two in the posterior arch; union occurs in the first year. A center appears in the epistropheus (axis) during the first year. The three points of the sacral vertebræ unite between the second and the sixth year; and all the vertebræ (sacral) join together between the eighteenth and the twenty-fifth year. The four coccygeal vertebræ receive each a center of ossification between the first and the tenth year. The vertebræ produce further in later life one or two centers in each spinous process, one or two in each transverse process, one in the mammillary process of the lumbar vertebræ, and one center in some of the articular processes between the eighth and the fifteenth year. Each surface of a vertebra develops further an epiphysis-like, thin plate of bone, which may still be visible at the age of twenty years. Groups of chorda-cells persist in the adult in the intervertebral discs. So long as the coccygeal vertebræ, the odontoid process of the axis and the base of the skull are cartilaginous, they contain remains of chorda. The coccygeal vertebræ form the tail, as the continuation of which an invertebrate caudal filament is prolonged. The coccyx consists originally in man of a free projecting tail, *vertebral tail* (Fig. 376, IX, T), which later becomes covered and enclosed by the growth of the soft parts over it. Rarely, a free, projecting tail persists; if the caudal filament alone remains free, there is formed the so-called *soft tail*.

The number of rudimentary vertebræ is at first small, then larger than even in adults; and, finally, again smaller. Eventually the embryo has 25 true vertebræ, the ilium fusing with the twenty-sixth. Later, the ilium moves so far forward that the twenty-fifth vertebra becomes the first sacral. The persistence of 25 true vertebræ is to be regarded as a developmental defect.

The ribs bud out from the primitive vertebræ; their first rudiments reside in each vertebra. The thoracic ribs become catilaginous in the second month and grow forward into the chest-wall, the upper seven being joined together by a median strip of cartilage. The latter represents the rudimentary half of the sternum; by the union of the rudiments of the two sides in the median line the sternum is formed. The developmental defect of fissure of the sternum occurs in some howling apes in which the manubrium is permanently divided. The lower, false ribs normally exhibit, to a certain extent, a fissure of the sternum; openings in the sternum as the remains of a fissure are frequent.

In the sixth month a center of ossification appears in the manubrium; then from 4 to 13, in pairs, in the gladiolus, and one in the ensiform process. Each rib acquires a center of ossification in the body in the second month; between the eighth and the fourteenth year one each in the tubercle and the head of the bone; fusion takes place between the fourteenth and the twenty-fifth year. The rudimentary ribs in front of the transverse processes in the neck become the anterior portions of these processes. Rarely isolated, short, true cervical ribs persist in conjunction with the sixth and seventh cervical vertebræ (in birds the cervical ribs are better developed). In the lumbar region the cartilaginous rudiments of the ribs become later the processus costarii (transversi of earlier writers). Occasionally a thirteenth rib is formed. The accessory process of the lumbar vertebræ is the true transverse process, as is easily demonstrable in the skeleton of the ape. The sacral vertebræ have likewise 3 or 4 rudimentary ribs which after the sixth year unite with the superficies auricularis. The rib-piece has not yet been found on the coccygeal vertebræ.

The cranium, the closed extremity of the vertebral canal, contains the chorda in the axial part of its base up to the anterior sphenoid body. It is at first entirely membranous (membranous primordial cranium); then the basal portions become cartilaginous in the second month, being held together as if cast from a mold: the occipital bone, with the exception of the upper half, the anterior and posterior sphenoids with the wings, the petrous and mastoid portions of the temporal bone, the ethmoid with the nasal septum, and the imperfectly developed external cartilaginous portion of the nose. The other portions of the cranium remain membranous. Accordingly, a membranous and a cartilaginous primoidal cranium have been distinguished. In animals (pigs) the entire occipital and a portion of the parietal region may become cartilaginous.

Ossification of the individual bones of the skull is completed as follows:

I. The *occipital bone* receives in the third month a center of ossification in the basilar portion, one each in the condyloid portion and in the fossa for the cerebellum. In addition two centers occur in the membranous fossæ for the cerebrum. The four centers of the bone unite during intrauterine life, although a cleft can be seen on each side from the border between the upper and the lower portion of the squamous portion. Between the first and the second year all the other points unite. Rarely the upper half of the squamous portion persists, as the analogue of the interparietal bone, which is constant in many animals; this is an independent semilunar-shaped bone (of which the author possessed a beautiful example); occasionally one-half of this portion. It should be pointed out as particularly important (also with reference to the development of the brain) that in man the upper portion of the occipital bone enlarges in the process of development, while in apes, on the contrary, it diminishes in size. In some skulls the upper and lower halves of the occipital bone exhibit differences in development. According to Albrecht, the anterior part of the basilar portion forms a special piece of bone, the basioticum.

II. The *postsphenoid* has the following centers of ossification from the third month on: two in the sella turcica; two in the carotid groove, and two in both great wings, which form also the external plate of the pterygoid process (while the previously formed noncartilaginous internal plate is derived from the superior maxillary process of the first visceral arch). In the second half of fetal life, these centers unite to form the great wings. The dorsum sellæ and the clivus remain cartilaginous up to the spheno-occipital synchondrosis, which ossifies from the thirteenth year on.

III. The *presphenoid* has from the eighth month two centers in the small wings; then two in the body. In the sixth month these unite, although cartilage is still found within them, remains of which persist until the thirteenth year.

IV. The *ethmoid* contains at the fifth month a center in the labyrinth, together with the os planum, the spongy bones and the cribriform plate; then in the first year there is a center in the perpendicular plate and the crista galli. Fusion takes place between the fifth and sixth years.

V. Among the bones developed from membrane are the inner lamina of the pterygoid process (one center); the upper half of the occipital bone (two centers); the parietal bone (one center in the parietal eminence); the frontal bone (a double center in the frontal eminence); in addition three small centers in the nasal spine, the trochelear spine, and the zygomatic process; the nasal bone (one center); the squamous portion of the temporal bone (one center); the tympanic ring (one center); the lacrimal bone, the vomer and the intermaxillary bones. All of these bones are designated covering or protecting bones; they are formed in a special membranous deposit, which is applied externally to the primordial cranium. O. Hertwig considers them as due to ossification of skin and mucous membrane.

Goethe appreciated that the cranium of mammals was "derived from vertebral bones. The three first vertebræ are admitted: the occipital bone, the postsphenoid, and the presphenoid." The arch of the middle cranial vertebra is closed by the great wings and the parietal bones; the anterior by the frontal bones. The condition is, however, much more complicated. Gegenbaur and Stöhr, after careful investigation as to the distance the chorda extends anteriorly, the number of cranial nerves that correspond to spinal nerves, and the number of visceral arches, concluded that there are at least nine cranial vertebræ through which the chorda passes. To this vertebral portion of the head is subjoined a prevertebral or an evertebral portion, comprising the ethmoid and anterior orbital region. Froriep showed that the hypoglossal nerve represents the fusion of several spinal nerves. In mammals the cranium up to the vagus results from the fusion of four rudimentary vertebræ, while the anterior portion of the skull, through the cranial nerves, situated further forward, permits the recognition of a systematic arrangement of the vertebræ.

The **development of the bones of the face** is intimately related to the transformation of the visceral arches and clefts. Toward the large buccal opening there projects from each side the medial extremity of the first visceral arch, with two processes: the superior maxillary process (Fig. 382, *A*, 3), which grows more toward the side of the buccal opening: and the inferior maxillary process (*u*), which extends along the lower border of the mouth. From above downward there now grows the frontal process (*f*) as a prolongation of the base of the skull, a thick process, provided at its lower outer angle with a spine (1, the internal nasal process). The frontal process and the superior maxillary process (3) unite in such a manner that the former (*f*) insinuates itself between the latter on each side. At the same time a small external nasal process (2), a continuation of the lateral portion of the cranium, situated above the superior maxillary process, unites with the latter. Between the superior maxillary process and the external nasal process there is a cleft leading to the eye (*a*), which grows together to form the lacrimal duct (*B, O*). The buccal opening is thus separated from the nasal opening above it. The division, however, extends also in the depth of the mouth; the superior maxillary process produces the hard palate, the frontal process the intermaxillary bone (Fig. 382, *B, Z*), which occurs also in man, and later unites with the upper jaw. In many animals the intermaxillary bone persists as a separate bone (os incisivum), and bears the incisor teeth. The hard palate is closed in the ninth week; and upon it the nasal septum, which is derived from the frontal process, is supported at right angles. From the inferior maxillary process there develops the lower jaw (*B, U*). At the margins of the buccal cavity the lips and alveolar border are formed. The tongue (*z*) develops behind the junction of the second and third pairs of visceral arches, according to Born from an intercalated piece between the inferior maxillary processes; its root from the second visceral arch.

These formations may suffer interruption.

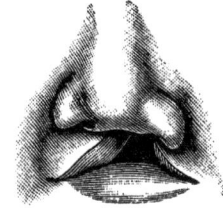

FIG. 381.—Left-sided Hare-lip.

1. *Hare-lip* (oronasal fissure, Fig. 382, *C*) results from nonunion of the internal nasal process on the one hand and of the superior maxillary and external nasal processes on the other hand. The cleft runs into the nasal orifice. As a rule, it passes between the incisor teeth, although rarely also in front of the canine tooth. In the presence of maxillary fissure there are often supernumerary incisor teeth. The intermaxillary bone has two centers of ossification, one in the internal nasal process, the other in the region of the superior maxillary process. From the external nasal process, which does not extend all the way down, no especial bone results. The nose and the mouth may be united either only through the soft parts (hare-lip, Fig. 381), or entirely, also through the hard palate (wolf's throat); both malformations may be unilateral or bilateral. The formation of cleft palate may be due to the circumstance either that the superior maxillary and the frontal process wholly or in part remain too short, so that they do not come in contact; or the frontal process grows too far forward like a snout, and often also is diminished in size; so that the superior maxillary process cannot reach it.

2. A failure of union between the internal and external nasal process on the one side and the superior maxillary process of the other side results in the *oblique facial cleft* (oro-orbital cleft, Fig. 382, *D*); the nasal orifice is not slit.

3. The *oral cleft* (*macrostomia*) is an abnormally large lateral cleft between the superior and inferior maxillary processes, which may extend as far as the ear (Fig. 382, *B, m*).

4. The occurrence of a *fistula of the lower lip* is extremely rare; it is regarded

as the remains of a fetal cleft between the middle and lateral portions of the forming lower lip.

From the posterior portion of the first visceral arch there develop the incus, the malleus (which undergo ossification in the fourth month), and the long cartilaginous process of Meckel, which arises from the malleus behind the tympanic ring, and passes forward, and which extends on the inner side of the lower jaw almost to its median union. This process begins to atrophy at the sixth month. Nevertheless, its posterior portion forms the internal lateral ligament of the temporomaxillary joint. Close to it, at its origin from the malleus, the processus Folli is formed. A portion of its median extremity in ossifying unites with the inferior maxilla. The lower jaw originates in membrane as a protecting bone upon the first visceral arch; the angle and the condyle develop from a cartilaginous deposit. The symphysis of the lower jaws unites in the first year. From the superior maxillary process there develops in addition to the upper jaw also the internal plate of the pterygoid process, as well as the palatine process of the upper jaw and the palatine bone at the end of the second month; and, finally, the zygomatic bone.

The second visceral arch, originating from the temporal bone, and running parallel with the first visceral arch, forms successively the stapes (according to Salensky, however, this originates from a cartilaginous mass connected with the first arch), the pyramidal eminence with the stapedius muscle, the styloid process; the (previously cartilaginous) stylohyoid ligament, the lesser cornu of

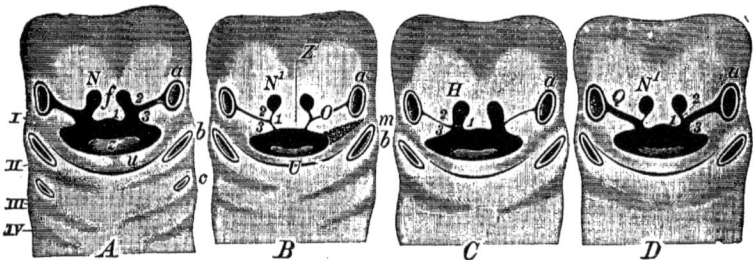

FIG. 382.—Formation of the Face and Developmental Defects of the face: *A*, First fetal rudiment; *I, II, III, IV*, the four visceral arches; *J*, the frontal process; 1, internal and, 2, external nasal process; 3, superior maxillary process; *u*, inferior maxillary process; *b, c*, first and second visceral clefts; *a*, eye; *z*, tongue. *B*, Normal union of the embryonal parts; *Z*, intermaxillary bone; *N¹*, nasal orifice; *O*, lacrimal canal ; *U*, lower jaw; *m*, abnormal enlargement of the oral cleft, macrostomia. *C*, Arrested development of the oronasal cleft (harelip or wolf's throat). *D*, Arrested development causing oblique facial cleft, Q.

the hyoid bone (Landois saw the styloid process transformed into bone down to the lesser cornu on both sides), and finally the glossopalatine arch.

From the third visceral arch there develop the greater cornu and the body of the hyoid bone and finally the pharyngopalatine arch.

The fourth visceral arch contains the rudimentary thyroid cartilage.

The branchial or visceral arches may in general be regarded as the analogues of the ribs.

Of the visceral clefts only the first remains—as the auditory canal, which is transformed into the tympanic cavity and the Eustachian tube; all of the others fuse. If one or another remains open (arrest of development, occasionally hereditary in certain families) there then results the congenital complete cervical fistula (mostly arising from the second cleft alone). The passages may persist with either an inner or an outer opening only; there then result blind passages or diverticula, which are designated incomplete cervical fistula. Also branchiogenous tumors and cysts take their origin from the visceral formations. Partial duplication of the lower jaw, which is exceedingly rare, is to be attributed to an increase in the number of visceral arches.

The thymus and thyroid glands are formed as paired diverticula or thickenings of the epithelium covering the visceral arches. The thyroid gland results (in swine) from a middle and two lateral rudiments, which subsequently fuse. The epithelium of the last two pharyngeal clefts does not atrophy (swine); it proliferates, drives the cylindrical processes inward and develops into two epithelial vesicles (the paired rudiment of the thyroid gland). These vesicles have a central cleft, which at first still communicates with the pharynx. The paired rudiment of the thyroid gives off from the original cavity buds that are at first solid, but subsequently become hollow; later the paired rudiments fuse. According to His, the thyroid (man, fourth week) in the region of the second pair of visceral arches lies in front of the tongue as an epithelial vesicle. Of the epithelial portion of the thymus gland (which, according to Born and Fischeles, originates from the third visceral cleft) there persist only the so-called concentric bodies. His describes in man (fourth week) epithelial diverticula from the sides of the fourth and fifth aortic arches as the rudimentary thymus. The carotid gland also is of epithelial origin, a variety of the thyroid.

The Extremities.—The course and the origin of the nerves of the brachial plexus indicate that the upper extremity has had a position more toward the cranium on the vertebral column (last cervical and first dorsal vertebræ). The rudiment of the lower extremity corresponds to the region between the last lumbar and the third or fourth sacral vertebra.

The clavicle, preformed not in connective tissue, but in cartilage, like the furcula of birds, exhibits marked growth, so that in the second month it is four times as large as the thigh. It ossifies, the first of all of the bones, in the seventh week. At the time of puberty a sternal epiphysis is added. Episternal formations must be referred to the clavicle. Ruge regards cartilaginous pieces between the clavicle and the sternum as the analogues of the episternum of animals. The clavicle is wanting in many mammals (hoofed animals, beasts of prey); in alipeds it is exceedingly large, in rabbits half membranous. The furcula of birds represents the united clavicles.

The scapula when first indicated is united with the clavicle; at the end of the second month it exhibits a central nucleus, which soon enlarges. Of the accessory centers of ossification, that in the coracoid process is of interest morphologically; the latter forms at the same time the uppermost portion of the articular surface. In birds this rudiment grows as the coracoid bone up to the sternum, while in man only a membranous band passes from the apex of the coracoid process to the sternum. The long basal separate strip of bone corresponds to the suprascapular bone of some animals. Other centers of ossification are as follows: One in the lower angle; two or three in the acromion; one in the articular surface; an inconstant one in the spine. Complete consolidation occurs at the time of puberty.

The humerus undergoes ossification between the eighth and the ninth week in the diaphysis. Other centers of ossification are as follows: One in the upper epiphysis and one in the eminentia capitata (first year); one in the greater tuberosity, and one in the lesser tuberosity (second year); two in the condyles (between the fifth and the tenth year); one in the trochlea (twelfth year). The diaphysis unites with the epiphysis between the sixteenth and the twentieth year.

The radius ossifies in the third month in the diaphysis. In addition, a center occurs in the lower epiphysis (fifth year); one in the upper epiphysis (sixth year); the centers in the tuberosity and in the styloid process are inconstant. Union occurs at the time of puberty.

The ulna ossifies in its middle portion in the third month. In addition there is a center in the lower extremity (sixth year) and two in the olecranon (between the eleventh and fourteenth years); a center in the coronoid process and one in the styloid process are inconstant. Consolidation of the bone takes place at puberty. In the flying dog, pteropus, the olecranon persists as a special bone, cubital patella.

The carpal bones in vertebrates are arranged in two rows. The first row contains three bones side by side, the radial, the intermediate, and the ulnar. These are represented in man by the scaphoid, the semilunar, and the cuneiform (the pisiform is only a sesamoid bone in the tendon of the flexor carpi ulnaris). The second row contains actually (for example, in the salamanders) as many bones as there are digits; in man the common rudiment of the fourth and fifth fingers corresponds to the unciform bone.

Morphologically it is noteworthy that between both rows there is at first formed a central bone (corresponding to the central carpal bone of reptiles, amphibia, and some mammals), but this atrophies in the third month or fuses with the scaphoid. Only in rare instances does it persist. It persists constantly in the orang. All of the carpal bones are still cartilaginous at birth. They ossify as follows: Os magnum unciform (first year), cuneiform (third year), trapezium, semilunar (fifth year), scaphoid (sixth year), trapezoid (seventh year), pisiform (twelfth year).

The **metacarpal bones** exhibit a center in the diaphysis at the end of the third month; the same is true of the phalanges. All of the phalanges and the first bone of the thumb have cartilaginous epiphyses at the proximal extremity; the remaining metacarpal bones at the distal extremity. Accordingly, the first bone of the thumb is to be regarded as a phalanx. The epiphyses of the metacarpal bones ossify in the second year; those of the phalanges in the third year; union takes place at the time of puberty. The assertion of Schenk is noteworthy, that in the first formation a greater number of fingers (up to nine) are indicated, which later diminish to five. Accordingly, polydactyly may be regarded as a malformation due to arrest of development. Moreover, rudimentary indications of a sixth finger (radial aspect) and of a sixth toe (tibial aspect) are present in many mammals, for example in moles. v. Bardeleben designates them respectively prepollex and prehallux, They are rare in man as an animal analogue.

The **innominate bone,** in the cartilaginous stage, consists of two parts, the pubis and the ischium. Ossification begins at three centers: One in the ilium (between the third and fourth months), one in the descending ramus of the ischium (between the fourth and fifth months), one in the horizontal ramus of the pubis (between the fifth and seventh months). Between the sixth and the fourteenth year three centers appear where the bodies of the three bones unite to form the acetabulum; another in the superficies auricularis and one in the symphysis. Further accessory centers are as follows: One each in the anterior inferior spine, the crest of the ilium, the tuberosity and spine of the ischium, the spine of the pubis, the iliopectineal eminence and the floor of the acetabulum. The descending ramus of the pubis and the ascending ramus of the ischium are the first to unite between the seventh and eighth years. The Y-shaped suture in the acetabulum persists until puberty. A special center in the margin of the acetabulum appears as the os acetabuli (twelfth year), which fuses with the adjacent bones in the eighteenth year.

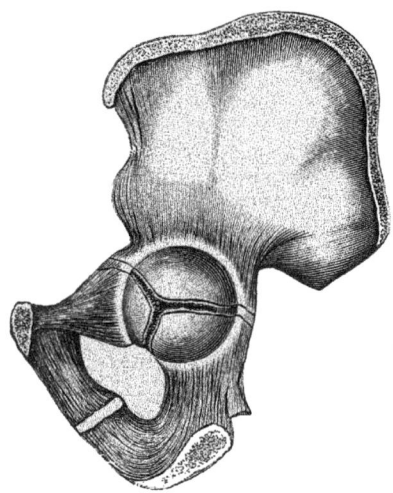

FIG. 383.—Ossification of the Innominate.

The **femur** acquires its middle center of ossification at the end of the second month. At birth there is a center in the lower epiphysis, and somewhat later one in the head.

There are in addition: One in the greater trochanter (between the third and eleventh years), one in the lesser trochanter (between the thirteenth and fourteenth years), two in the condyles (between the fourth and eighth years). Union of all occurs at about the time of puberty. The patella is a sesamoid bone in the tendon of the quadriceps femoris. In some marsupials it unites with the fibula as the fibular olecranon. The patella is cartilaginous in the second month; it ossifies between the first and third years.

The tibia and the fibula undergo ossification in their diaphyses at the beginning of the third month. The upper epiphysis first contains a center (between the first and third years), then the lower. Accessory centers are present in the tuberosity of the tibia and in the malleoli. Consolidation of all takes place at puberty.

The **tarsal bones** are formed in a manner analogous to those of the carpus:

In the first row the astragalus corresponds to the tibia, the calcaneum to the fibula. As a third bone of the first row there is particularly worthy of note a small piece of bone adherent to the astragalus at the insertion of the posterior fibular liga·ment of the tarsus. This corresponds to the semilunar bone of the carpus, is indicated in the second month as an independent cartilage, and appears in urodela and marsupials as a typical intermediate tarsal bone, although it does not undergo development in man.

In the second row (as in the carpus) the rudiments of the fourth and fifth bones are united as the cuboid. The tarsal bones ossify in the following order: Calcaneum (beginning of the seventh month): astragalus (beginning of the eighth month); cuboid (end of the tenth month); scaphoid or central (between the first and fifth years); first and second cuneiform (third year); third cuneiform (fourth year). In the heel of the calcaneum an accessory center develops between the fifth and the tenth year, and unites after puberty.

The **metatarsal bones** are formed in the same way as the metacarpal bones, but later.

After numerous measurements of the diaphyses of long bones in embryos and fetuses, the writer has been able to establish the following general principles: 1. Up to the ninth or tenth week the ossified middle portions of the long bones in the upper part of the body are the largest, and in the following order: Inferior maxilla, clavicle, humerus, radius, ulna, femur, tibia, fibula. 2. From the sixth month they range in size as they do in the adult. 3. The diaphyses of the tubular bones of the upper extremity are at all periods of fetal life relatively larger than those of the lower extremity. 4. In the first half of fetal life, the diaphyseal bones grow in the same ·length of time much more than they do later; even twice as much and more. Of the epiphyses, ossification takes place earliest in those that have the greatest relation in weight as compared with their diaphyses.

In the **formation of bone from cartilage,** the cartilage-cells multiply in the dilating spaces in which they are contained. The latter unite to form large cavities, upon whose walls the new bone-mass is deposited in layers. Whether, under such circumstances, the descendants of the cartilage-cells, greatly increased in number by division, become transformed into bone-cells, or whether the cells utilized for this purpose grow, together with the blood-vessels, into the ossifying cartilage (while the cartilage-cells degenerate) is still an open question.

Dried bone consists one-third of organic matrix (bone-cartilage, by boiling reduced to gelatin), further of neutral calcium phosphate (57 per cent.), calcium carbonate (7 per cent.), magnesium phosphate (from 1 to 2 per cent.), calcium fluorid (1 per cent.) and a trace of chlorin. Fresh bone contains about 23 per cent. of water; the marrow, fluid bone-fat, albumin, hypoxanthin, cholesterin, and extractives. Red marrow contains more iron in conformity with the hemoglobin present.

The bones (for example, the tubular bones) grow in thickness by deposition from the periosteum, the cells of the latter being transformed as osteoblasts into bone-corpuscles.

In part the peripheral regions (parietal layer) of the osteoblasts, resembling epithelium lying close together, are transformed into the hardened matrix of the bone, the cells becoming star-shaped bone-corpuscles. In part, however, isolated star-shaped periosteal cells also are transformed into bone-cells, a hardening blastema being poured out between them and taking up the fibers of the periosteum as Sharpey's fibers into the substance of the bone. Coincident with the growth of the bone at its periphery, the medullary cavity becomes larger through absorption. Rings placed around the shaft of long bones in young animals subsequently come to be within the medullary cavity.

The growth of the bones in length takes place in such a manner that the strip of epiphyseal cartilage adjacent to the diaphysis undergoes constant ossification, while new cartilage is being ·constantly produced at the peripheral extremity. When the growth of the bone is completed, the epiphyseal cartilage finally ossifies as a whole. Whether in addition to this growth of the bone through apposition, there is growth also through intussusception or interstitial expansion, investigation (whether two pegs driven into the shaft of a growing bone become further removed or not) has not yet been determined conclusively.

The form into which the bones develop depends also upon external influences. They develop the more rapidly the greater the activity of the muscles attached to them. In overcoming pressure normally applied to a bone it yields in the direction of least resistance, and becomes thicker in this direction. The bone grows more slowly, further, upon the side of the greater external pressure; and it bends in the presence of one-sided pressure.

DEVELOPMENT OF THE VASCULAR SYSTEM.

Heart.—The simple pouchlike rudimentary heart acquires the shape of the letter S (Fig. 384, 1) and soon exhibits a differentiation into the upper aortic portion (a) with the bulb (b), the middle ventricular portion and the lower venous portion (v). The ventricular portion now bends upon itself in the shape of the stomach (2), the venous portion in consequence taking a higher position (A) and, later on, one somewhat back of the arterial portion. From the venous portion there grows to right and to left a blind sac, the forerunner of the large auricle ($3, o, o_1$). The bend of the heart body corresponding to the greater curvature ($2, V$) is divided externally by a shallow groove into two large parts (3).

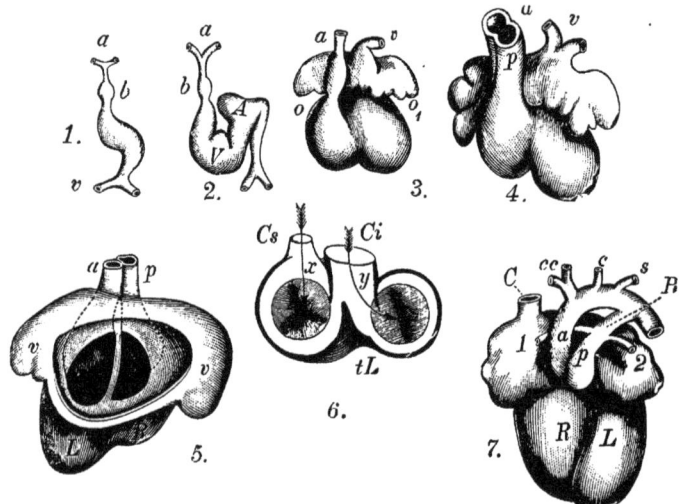

FIG. 384.—Development of the Heart (in Part Diagrammatic): 1, First indication of the heart; a, aortic portion with the bulb (b); v, venous portion. 2, Stomach-shaped flexure of the heart; a, aortic portion, with the bulb (b); V, ventricle; A, auricular portion. 3, Development of the auricular appendages o, o_1 and the external groove on the ventricle. 4, Beginning division of the aorta (p) into two longitudinal tubes (a). 5, View from behind through the freely opened auricle (v, v) into the left (L) and the right (R) ventricle, between which the septum projects, and in which respectively the two large arterial vessels (a) aorta and (p) pulmonary artery empty. 6, Relations of the entrance of the superior (Cs) and inferior (Ci) cavæ in the auricles (diagrammatic view from above); x, direction of the blood-stream from the superior cava into the right ventricle; y, that of the blood-stream from the inferior cava into the left ventricle; tL tubercle of Lower. 7, Heart of the mature fetus; R, right, L, left ventricle; a, aorta with the innominate artery (cc); carotid (c) and left subclavian (s) artery; B, duct of Botal; p, pulmonary artery with the still diminutive pulmonary branches 1 and 2.

The large truncus venosus (4, v), which becomes embedded in the middle of the posterior wall of the auricular portion, is formed by the union of the superior and inferior cavæ. Later, this common trunk is drawn into the wall of the enlarging auricle, and in this way there result the separate openings for the two cavæ. In man the development of a special cavity in which the heart lies occurs early; a portion of the rudimentary diaphragm bounds this cavity.

Between the fourth and the fifth week the division of the heart into a right and a left begins. In correspondence with the perpendicular groove in the ventricle there first grows vertically upward into the ventricle a septum (5) that divides the ventricular portion into a right and a left half (5, R, L). Between the ventricular portion and the auricular portion the heart undergoes constriction, forming the auricular canal. This contains a communication between the auricle and both ventricles, lying between an anterior and a posterior lip of projecting endothelium, from which the auriculoventricular valves are formed. The ventricular septum grows upward toward the auricular canal and fuses at this point with two endothelial proliferations (endothelial cushions), which in the lumen of the auricular canal traverse the mouth of the canal from before and behind. In the eighth week the ventricular septum is fully developed. A free communication thus exists between the large undivided auricle and the corresponding ventricle (5) by way of a right and a left auriculo-ventricular orifice. There then grow into the large truncus arteriosus (4, p) two wing-like septa (4, p a), which finally meet in union and divide the tube into two tubes (5, a p). The latter lie parallel to one another like the barrels of a double-barreled gun (aorta and pulmonary artery). The septum between the two assumes a downward direction in such a way as to meet the interventricular septum (5). In this way the right ventricle communicates with the pulmonary artery and the left with the aorta. The division of the truncus arteriosus, however, takes place only in its first part. Further on, it is not complete, that is the pulmonary artery and the aorta unite above to form a common trunk—the ductus arteriosus Botalli (7, B). In the auricle there grows from before and behind a portion of a septum that ends within the cavity in a concave border. The superior cava (6, Cs) enters to the right of this fold, so that its blood would have a tendency to pass into the right ventricle (in the direction of the arrow 6, x). The inferior cava (6, Ci), on the contrary, opens directly opposite the margin of the fold. There is thus formed to the left of its

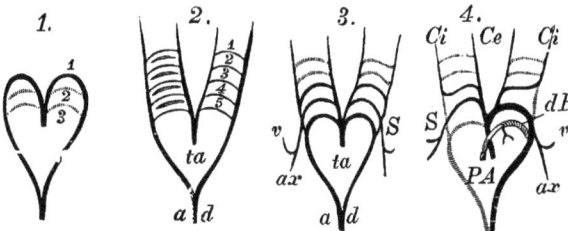

Fig. 385.—Development from the Aortic Arches: 1, The rudimentary 1st, 2d, and 3d aortic arches. 2, Five aortic arches: ta, common aortic trunk; a d, descending aorta. 3, Atrophy of the two uppermost arches on each side; S, subclavian artery; v, vertebral artery; ax, axillary artery. 4, Transition to the definitive stage of formation; P, pulmonary artery; A, aorta; dB, ductus Botalli; S, right subclavian artery, joined with the right common carotid, which divides into the internal (Ci) and the external (Ce) carotid; ax, axillary artery; v, vertebral artery. (Diagrammatic.)

entrance, opposite the auricular fold, the valve of the oval foramen, which permits the blood to flow only to the left in the direction of the arrow y. To the right of the mouth of the cava, opposite the fold, there is formed the Eustachian valve, which together with the tubercle of Lower (tL) guides the stream of the inferior cava to the left into the left auricle. After birth the opening is closed by the valve of the oval foramen. In addition, the ductus Botalli is obliterated (by increase of the pressure in the aorta, which closes the lumen of the mouth), so that the blood of the pulmonary artery is now forced to pass through the distending pulmonary branches.

The persistence of the oval foramen is a developmental defect that causes severe circulatory disturbances. Kergeradek discovered the fetal heart-sounds. Many anomalous peculiarities in the development of the heart, which have been studied especially by Born and Rose, cannot be described here.

Arteries.—With the development of the visceral arches and clefts, the number of aortic arches on each side becomes increased from one up to five (Fig. 385), and these pass above and below each visceral cleft, but subsequently unite to form a common trunk (2, a d). The vessels persist only in gill-breathers, Fig. 74. In man, the two uppermost aortic arches on each side (3) atrophy first. In the division of the truncus arteriosus into the pulmonary artery and the aorta (4, PA) the lowermost arch on each side, together with its origin, becomes the pulmonary artery (4) and therefore arises from the right heart. Of these, the left lowermost arch forms the ductus Botalli (dB), and at its origin the pulmonary branches of the pulmonary artery arise. Of the arches joined to the aorta, the left middle one (into which the ductus Botalli empties) becomes the permanent aorta, the right, the right subclavian (S). The uppermost arch on each side becomes the origin of the carotids (Ci, Ce). According to Zimmermann there occur in man and in rabbits the rudiments of a hitherto unknown transitory arch between the lowermost and the next higher aortic arch on each side.

The arteries of the first and the second circulation have already been considered. With the disappearance of the omphalomesenteric circulation, there is but one omphalomesenteric artery present, and this soon gives off a branch to the intestine. Later, the umbilical artery atrophies, so that the trunk of the intestinal artery (the superior mesenteric artery, the largest of all arteries) is originally an omphalomesenteric artery.

The Veins of the Body.—The veins that first develop in the body of the embryo itself are the two cardinal veins: on each side an anterior (Fig. 386, I, cs) and a posterior (ci), which, passing toward the heart, unite on each side to form a large trunk, the duct of Cuvier ($D\,C$). The latter joins the venous portion of the heart. The anterior cardinal veins give off the subclavian veins ($b\,b$) and the common jugular veins, which divide into the internal (Ji) and external (Ie) jugular veins. In addition, there is a transverse anastomosing branch passing obliquely from the left (where it divides) toward the right, and emptying into its trunk at a somewhat lower level. In the definitive development (II) this anastomosing branch (As) becomes large (forming the left innominate vein); besides the subclavian veins ($b\,b$) increase in size with the growth of the extremities; and, finally, the caliber of the two jugular veins is reciprocally altered, so that the rudimentary internal jugular vein becomes large (Ji), while the external jugular vein becomes smaller (Ie); in many animals, for example dogs and rabbits, the embryonal proportions persist. The portion of the left superior cardinal vein from the point of anastomosis down to the left duct of Cuvier atrophies. The posterior cardinal veins divide in the pelvis into the hypogastric (I, h) and the exter-

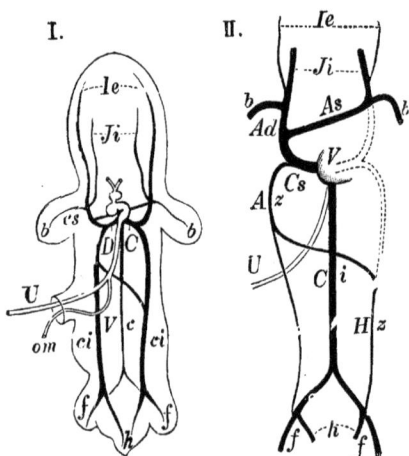

FIG. 386.—I, Rudimentary veins of the body of the embryo. II, Transformation of the same rudimentary into the definitive veins. (Diagrammatic.)

nal iliac ($f\,f$). The inferior cava is at first small (I, $V\,c$); it divides at the pelvic inlet, and passes over on each side to the point of division of the cardinal veins. In addition, there exists a transverse ascending anastomosing branch between the right and left cardinal veins. For the establishment of the definitive condition, the inferior cava dilates (II, $C\,i$), and with it, downward, the hypogastric and external iliac on each side. The right cardinal vein is replaced by the small vena azygos (Az), and on the left side up to the transverse anastomosing branch in an analogous manner the vena hemiazygos (Hz). On the other hand, the upper portion above the anastomosing branch up the left duct of Cuvier atrophies. The site of entrance of the vena magna cordis is the remains of the left duct of Cuvier. Finally, the combined venous trunk is so withdrawn into the wall of the auricle (V) that the two cavæ acquire independent orifices. All vertebrates possess the same rudimentary venous system in the embryonal state; it persists, however, only in fishes (Fig. 74, I).

Veins of the First and the Second Circulation, and the Development of the Portal System.—Originally the two omphalomesenteric veins (om, om_1) empty into the truncus venosus of the first pouch-shaped rudimentary heart (Fig. 387, 1, H).

The right of these, however, soon atrophies. As soon as the allantois is formed, both allantoic or umbilical veins unite to form the truncus venosus (1, u u_1). At first the omphalomesenteric veins are larger than the umbilical veins: later, the conditions are reversed, and also the right umbilical vein atrophies. As soon as the veins of the trunk itself are formed, the inferior cava also empties into the truncus venosus (2, $C i$). Gradually the umbilical vein becomes the chief path (2, u_1), to which the small omphalomesenteric vein (2, om_1) sends but little blood.

The umbilical and omphalomesenteric veins pass in part directly beneath the liver to the heart; in part, however, they send also branches, carrying arterial blood, into the liver, which grows around the vessels from above—the venæ advehentes (2 and 3, a). The blood from the latter passes back through other

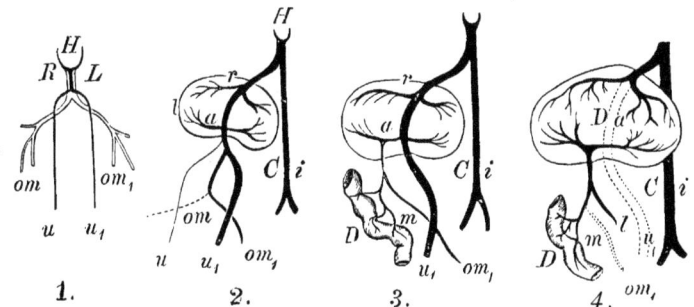

FIG. 387.—Development of the Veins of the First and the Second Circulation, and of the Portal System: H, heart; R, right, L left side of the body; om, right omphalomesenteric vein, om_1, left omphalomesenteric vein; u, right umbilical vein; u_1, left umbilical vein; Ci, inferior vena cava; a, venæ advehentes; r, venæ revehentes; D, intestine; m, mesenteric vein; $4 l$, splenic vein; $2 l$, liver. (Diagrammatic.)

veins (venæ revehentes, 2 and 3, r), which again unite with the main trunk of the umbilical veins at the blunt border of the liver. In the liver, the umbilical vein (3, u_1) anastomoses with the omphalomesenteric vein (3, om_1).

With the development of the bowel (3, D) the mesenteric vein (m) empties into the omphalomesenteric; as does also the splenic vein (4, l) with the development of the spleen. When, later, the omphalomesenteric vein atrophies (4, om_1), the mesenteric vein is the sole trunk of these previously united vessels. It is, therefore, the vein that unites with the umbilical vein in the liver, and it thus represents the trunk of the portal vein. When, finally, at birth the umbilical vein atrophies (4, u_1), the mesenteric vein alone remains as the portal vein. This must, however, send all its blood through the liver, as the ductus venosus of Arantius (4, $D a$) becomes obliterated. In this way the portal circulation is completed.

DEVELOPMENT OF THE ALIMENTARY CANAL.

The primitive intestine is originally a straight tube, passing from the cranial to the caudal extremity. The omphalomesenteric duct is inserted at a point that corresponds later to the lower part of the ileum. Here the tube in the fourth week makes a slight bend toward the navel (Fig. 388, I). It has already been pointed out that the duct later is obliterated (intestinal navel) and is finally detached as a thread from the intestinal tube; it is still discernible in the third month. In rare cases a short blind tube joined to the bowel persists as a vestige of the incompletely obliterated duct. This is the so-called true intestinal diverticulum: occasionally a cord (the obliterated omphalomesenteric vessels) passes from it to the navel; in rare cases the duct may remain open through the navel even after birth, so that a congenital fistula of the ileum results; or finally the diverticulum may be the seat of cyst-formation. In human embryos four weeks old, His was able to differentiate the mouth, the pharynx, the esophagus, the stomach, the duodenum, the mesenteric intestine, the end-gut and the cloaca.

63

At a later period the bowel forms the first loop (Fig. 388, *II*), being rotated on itself at the situation of the intestinal umbilicus, so that the lower portion of the bowel lying next to the knee-shaped flexure is turned upward, and the upper portion is turned downward. From the lower limb of this loop there grow the

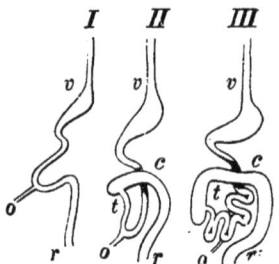

FIG. 388.—Development of the Intestine: *v*. Stomach; *o*. insertion of the omphalomes-enteric duct; *t*, small intestine; *c*, colon; *r*, rectum. (Diagrammatic.)

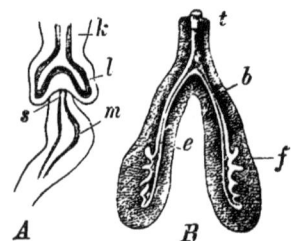

FIG. 389.—Development of the Lungs: *A*, Evagination of the lungs as double sacs; *k*, mesoblastic layer; l, entoblastic layer; *m*, stomach; *s*, esophagus. *B*. Further ramification of the lungs: *t*, trachea; *b*, *e*, bronchi; *f*, budding glandular vesicles.

coils of small intestine, constantly increasing in length (*III*, *t*). From the upper limb, which increases in length, the large intestine is formed in such a manner that first the descending colon, then by elongation the transverse colon, and finally also the ascending colon results.

The intestinal canal gives rise through evagination to various glands. This process is participated in by the cells of the hypoblast, which become the secretory cells of the glands, as well as the splanchno-pleure, which supplies the limiting membrane of the glands. These diverticula are in order: (1) The *salivary glands*, which at first are solid, but soon develop from the oral cavity as intricately ramifying glandular bodies. (2) The *lungs*, which develop as two separate hollow vesicles (Fig. 389, *A*, *l*), which subsequently take their origin from a common tubular evagination of the esophagus. The upper portion of the united tracheal tube becomes the larynx. The epiglottis and the thyroid cartilage are derived from the rudimentary tongue. The two vesicles grow according to the type of a branching, tubular gland, with hollow buds (*B*, *f*). In the earliest stages of development there exists no particular difference between the epithelium of the bronchi and that of the primitive air-vesicles. The spleen and the suprarenal bodies, however, do not develop in this manner. The first is formed as a fold of

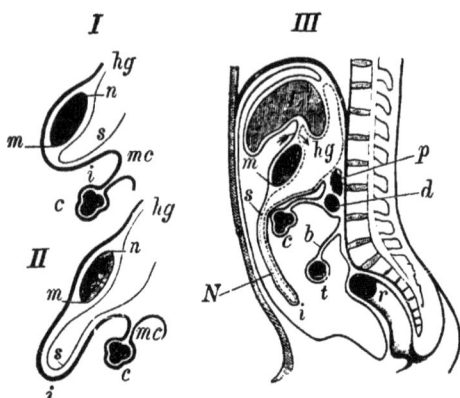

FIG. 390.—Development of the Great Omentum. *I* and *II*: *hg*, hepatogastric ligament; *m*, greater and *n* lesser curvature of the stomach; *s* posterior and *i* anterior layer of the omentum; *mc*, mesocolon; *c*, colon. *III* (in addition to the letters in *I* and *II*): *L*. liver; *t*, small intestine; *b*, mesentery; *p*, pancreas; *d*, duodenum; *r*, rectum; *N*, great omentum. (Diagrammatic.)

the mesogastrium in the second month; the latter are at first larger than the kidneys. (3) The *pancreas* develops in the same way as the salivary glands, and originally in two rudimentary parts, a dorsal and a ventral. It is, however, not yet indicated in the fourth week. (4) The *liver* appears early, beginning as an evagination by means of two hollow, primitive ducts, which break up to form the biliary passages. At their periphery they exhibit solid cellular masses, the liver-cells, which thus also are derived from the hypoblast. As early as the second month the liver is large; it secretes as early as the third month. According to Kupffer a single large gland, which in lower animals extends throughout the length of the mid-bowel, corresponds to the liver, spleen and pancreas. From this the three organs are subsequently differentiated. (5) In birds two small sacs are formed from the hind-gut. (6) The fetal respiratory organ, the allantois, has already been considered (p. 974). The inner surface of the cœlom, the surface of the bowel and of the mesentery become covered with a serous membrane, the peritoneum. This encloses the bowel, which for a time remains simple, in a duplicature or fold. On the stomach, which, at first, occupies a vertical position as a spindle-shaped dilatation of the digestive tract, this fold is known as the mesogastrium. Subsequently the stomach comes to lie upon its side, and in such a way that the left aspect becomes the anterior, the right the posterior. In this way, the insertion of the mesogastrium, which at first was directed posteriorly, toward the vertebral column, becomes directed toward the left. The line of insertion is formed by the region of the greater curvature, which subsequently becomes more markedly curved. From the greater curvature, the mesogastrium is prolonged as a pouch-shaped appendage (Fig. 390, *I* and *II*, *s i*), the omental bursa, so far downward as to extend over the transverse colon and the coils of small intestine *III*, *N*). As the mesogastrium consisted originally of two layers, the duplicature formed from it, the omental bursa, must consist of four layers. In the fourth month the posterior surface of the omental bursa becomes adherent to the surface of the transverse colon.

DEVELOPMENT OF THE URINARY AND SEXUAL ORGANS.

Urinary Organs.—The urine-forming gland originates developmentally from three organs, which succeed one another in function: (1) The rudimentary kidney (pronephros). (2) The primitive kidney (mesonephros). (3) Definitive kidney (metanephros).

1. The rudimentary kidney is in the amniota (and salachians) only a rudimentary embryonal organ; in the remaining vertebrates, it still exhibits some functional activity in the embryonal or larval period; it is here the provisional embryonal kidney (as the Wolffian body is for the amniota). In bony fish canals can be differentiated that begin anteriorly within the abdominal cavity by means of funnel-shaped openings and unite to form a common excretory duct that empties into the cloaca. In front of the funnels lies the glomerulus, whose secretion is carried outward in the canals.

2. In the amniota the primitive kidney (mesonephros) is the fetal uriniferous organ. From this there arises as the first formation, in the chick on the second day, in rabbits on the ninth day, the duct of the primitive kidney or Wolffian duct (Fig. 392, *I*, *W*), which is formed from cells of the ectoderm and at first is solid, to the side of and somewhat behind the primitive vertebræ and extending from the fifth to the last primitive vertebræ. Seated within this duct, there arise in the mesoblast from the level of the liver downward a series of tubules, which in the chick are believed at first to open free at their other extremities into the peritoneal cavity, and which become transformed into structures similar to the glomerulus of the kidney by the ingrowth of vascular convolutions into their extremities. The tubules elongate, become twisted into convolutions; and increase by the addition of newly formed communicating accessory tubes. The caudal extremity of the Wolffian duct is at first closed; its lower extremity, which lies in a fold projecting into the abdominal cavity (plica urogenitalis), opens (in the rabbit on the eleventh day) into the urogenital sinus. In the anamnia the primitive kidney is the permanent uriniferous gland.

3. Just above the outlet of the Wolffian duct the definitive kidney (metanephros) arises in an upward direction as the renal duct. The elongated duct divides at its upper extremity like a shrub. These accessory branches finally form convolutions. Each canal at its extremity assumes the form of a pedunculated hollow rubber-ball, with a cup-shaped depression, into which the vascular

convolution formed independently penetrates, and here it becomes closely surrounded. The duct of the kidney later empties separately into the urogenital sinus, and becomes the ureter. The point where the branching begins becomes the site of the pelvis of the kidney; the branches themselves become the urinary tubules. Toldt found as early as the second month complete Malpighian bodies in the human kidney; in the fourth month Henle's loops. The urinary bladder arises in its first indication as early as the fourth week, becoming more distinct in the second month from the first portion of the allantois (Fig. 392, 4, a). The upper portion passes over into the middle ligament of the bladder as the obliterated urachus, which often remains permeable for a short distance beyond the bladder; although even in adults there often persist, in the lower third of the urachus, unobliterated portions, which may give rise to cyst-formation.

According to Keibel, the development of the bladder takes place in such a manner that the common cloacal space is divided by two lateral folds into an anterior (the rudimentary urinary bladder) and a posterior space (rectum). Congenital communications between the bladder and the rectum are thus easily explained as developmental defects. Congenital fissure of the abdominal wall and the bladder results from persistent patulousness of the blastopore.

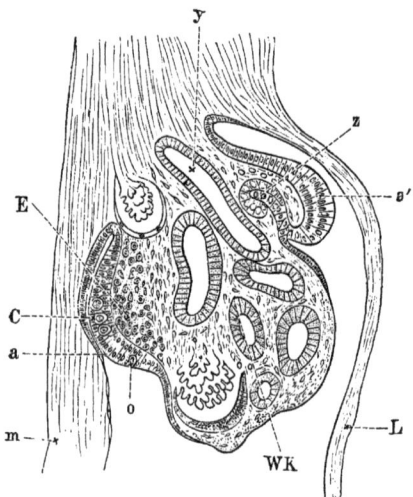

Internal Organs of Generation.—In front, and internally to, the Wolffian bodies there develops in the mesoblast the longitudinal, projecting sexual gland (Fig. 392, *I, D*), which originally is the same in both sexes (hermaphroditic stage). In addition, there forms parallel to the Wolffian duct (*W*) a canal, which empties downward likewise into the urogenital sinus, the duct of Müller or the sexual duct (*M*). The sexual gland appears first as a longitudinal protuberance, and is covered with the high epithelium of the mesoblast, the germinal epithelium of Waldeyer. The duct of Müller (which is not yet present at the fourth week) appears at first as a linear furrow in the germinal epithelium, which then becomes deeper and constricts itself off to a cord that is at first solid, but later becomes hollow. The upper outlet of the duct opens free into

FIG. 391.—Transverse Section through the Primitive Kidney, the Rudimentary Duct of Müller, and the Sexual Gland in a Chick at the Fourth Day (after Waldeyer); enlarged 160 times: m, mesentery; L, abdominal wall; a', the region of the germinal epithelium from which the anterior extremity of the duct of Müller (z) has invaginated itself; a, thickened layer of the germinal epithelium, in which the primary germ-cells (G and o) lie; E, mesenchyma, from which the stroma of the sexual gland is formed; WK, primitive kidney; y, duct of the primitive kidney.

the abdominal cavity; the lower extremities of both ducts fuse for a short distance. In the evolution of the female sex, ova develop in the germinal epithelium, and sink into open tubular formations of the ovary (in man up to the time of birth). In the female the duct of Müller becomes the oviduct (*II, T*) and the lower fused extremities of both, the uterus (*U*).

In the male sex the germinal epithelium is lower (although at first it still exhibits rudimentary ova). According to Waldeyer, of the two kinds of tubules that can be distinguished in the Wolffian body, the smaller penetrate the rudimentary sexual gland (sexual portion of the Wolffian body). These tubules, which communicate with the Wolffian duct, become the seminiferous tubules, and the Wolffian duct in man becomes the vas deferens (*III, V*), together with the seminal vesicle. According to Sernoff, Bornhaupt, Egli and Biegelow, autochthonous strands of cells develop within the sexual gland of man and these are transformed into the seminal ducts, and later communicate with the Wolffian ducts.

The Müllerian ducts (the true excretory ducts of the sexual glands) undergo atrophy in man, with the exception of the lowermost portion, which becomes the masculine utricle or the prostatic vesicle (*III, u*); this is the analogue of the uterus. In carnivora and ruminants the Müllerian ducts attain a greater size, to form a rudimentary vagina and a uterus bicornis; in rare cases a true, small uterus has also been found in man. The upper tubules of the Wolffian body unite in the third month with the sexual gland, and become the coni vasculosi of the epididymis, which is furnished with ciliated epithelium (*E*); the remaining portion of the primitive kidney undergoes atrophy. A number of detached tubules become the vasa aberrantia (*a*) of the testicle. The pedunuculated hydatid of Morgagni (*h*) at the head of the epidermis is, according to v. Luschka, Becker, and M. Roth, a constricted-off vesicle of the epididymis, occasionally containing semen and lined by ciliated epithelium; according to Waldeyer it is the homologue of the infundibuliform portion of the oviduct, while according to Toldt it is derived from the abdominal extremity of the duct of Müller. The

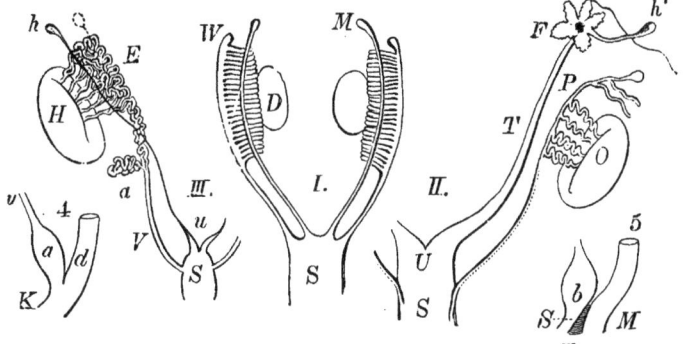

Fig. 392.—Development of the Internal Organs of Generation. *I*, Undifferentiated stage: *D*, sexual gland lying upon the tubules of the Wolffian body; *W*, Wolffian duct; *M*, duct of Müller; *S*, urogenital sinus. *II*, Transformation into the Female type: *F*, fimbria with the hydatid (*h'*); *T*, oviduct; *U*, uterus; *S*, urogenital sinus; *O*, ovary; *P*, parovarium. *III*, Transformation into the male type: *H*, testicle; *E*, epididymis, with the hydatid (*h*); *a*, vas aberrans; *V*, vas deferens; *S*, urogenital sinus; *u*, utriculus masculinus; *4 d*, end-gut; *a*, allantois; *u*, urachus; *K*, cloaca; *5 M*, rectum; *m*, perineum; *b*, rudimentary bladder; *S*, urogenital sinus. (Diagrammatic.)

organ of Giraldès (convoluted tubules with ciliated epithelium) at the upper extremity of the testicle is probably also a vestige of the Wolffian body. The Wolffian duct itself becomes the vas deferens (*V*), together with the seminal vesicle (as an outgrowth). The two Wolffian and the two Müllerian ducts lie close together at the pelvic inlet in a cord (genital cord). Later, when the Müllerian ducts have undergone atrophy, the seminal ducts formed from the Wolffian ducts become more widely separated.

In the female sex, the tubules of the primitive kidney, with the exception of a vestige within ciliated tubes (parovarium or Rosenmüller's organ) and a portion in the broad ligament resembling the organ of Giraldès, undergo atrophy (*II, P*); as do also the Wolffian ducts; although they are still visible in fetuses of five months, but downward only as far as the region of the vaginal vault; below this and toward the urethral orifice they disappear completely. Diminutive vestiges of the ducts are often found anteriorly and laterally embedded in the uterine and vaginal muscularis, chiefly on the right. They persist permanently in ruminants, the horse, the pig, the cat, the fox, as Gärtner's ducts; in man they may give rise to pathological cyst-formation. The Müllerian ducts become fringes at their upper opening to form the fimbriæ (F), upon which often a hydatid is situated (h¹).   According to Thiersch and Leuckart the two Wolffian and the two Müllerian ducts lie together below in the genital strand.   The two Müllerian ducts now unite at their lower extremities (end of the second month) and form in their combined lumen the vagina and the uterus (U), while the upper, free portion of each becomes the oviduct (T).   It is thus clear that the condition of double uterus and vagina is due to a developmental defect, resulting from a failure in union.   The vagina is originally closed by epithelium; arrest of development may result in atresia of the vagina.   The Müllerian ducts empty originally into the lowermost posterior portion of the urinary bladder, below the ureters—urogenital sinus (S); later this portion of the bladder becomes elongated posteriorly in such a manner that the vagina (the united Müllerian ducts) and the urethra are united only at a point deep down in the vestibule of the vagina.

The vagina and the uterus are first distinctly separated from each other in the fourth month; between the fifth and sixth months the uterus becomes characteristically differentiated.   The hymen is formed in the fifth month.

The testicle lies originally in the inguinal region of the abdomen (Fig. 393, V, t), supported by a fold of peritoneum (mesorchium, m).   From the hilus of the testicle there passes through the inguinal canal to the base of the scrotum (according to C. Weil, only to the root of the penis) a cord, the gubernaculum of Hunter. At the same time there is formed independently from the peritoneum, a sheath-

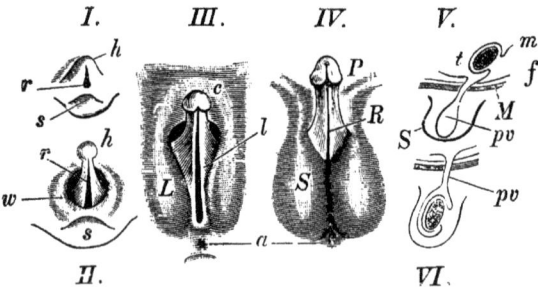

FIG. 393.—Development of the External Genitalia. I and II: Genital eminence; r, genital groove; s, coccyx; w, cutaneous elevations. IV: P, penis; R, raphé of the penis; S, scrotum. III: c, clitoris; l, labia minora; L, labia majora; a, anus. V and VI: Descent of the testicle; t, testicle; m, mesorchium; pv, vaginal process of the peritoneum; M, abdominal wall; S, scrotum. (Diagrammatic.)

like process extending down to the base of the scrotum (pv).   Arrested development or atrophy of the gubernaculum of Hunter causes the testicle to be drawn down through the inguinal canal into the scrotum.   In its passage the testicle takes with it from the superficial or transverse abdominal fascia the tunica vaginalis communis as a covering; and with it the muscular fibers carried down from the ascending and transverse oblique form the cremaster muscle.   The peritoneal covering of the testicle becomes the double sac of the tunica vaginalis propria; the vaginal process of the peritoneum is obliterated as a rule, and leaves irregular vestiges as the vaginal ligament.   If this vaginal process, communicating with the peritoneal cavity, remains patulous, a passage is afforded for the development of a congenital external inguinal hernia.

The ovaries also pass somewhat downward.   A strand similar to the gubernaculum of Hunter, passing through the inguinal canal, later becomes the muscular round ligament of the uterus.   Also in women the peritoneum sends a vaginal process through the inguinal canal (canal of Nuck).   Rarely even the ovaries descend into the labia majora, while, conversely, a retention of the testicles in the abdominal cavity (cryptorchism) must be looked upon as an arrest of development.

The **external genitalia** are at first not to be differentiated in the two sexes (Fig. 393, I).   In the fourth week there is a single orifice at the caudal extremity, constituting at the same time the anus and the opening of the urachus, thus a cloaca (Fig. 392, 4, K).   In the sixth week an elevation appears in front of the opening (Fig. 393, I, h), the genital eminence, then laterally from the opening on either side a large cutaneous elevation (*II*, *w*).   At the end of the third month there passes on the under surface of the genital eminence to the cloaca a groove, on whose two sides distinct folds appear (*II*, *r*).   In the middle of the third month the cloacal orifice becomes divided, prolongations from above and from each side insinuating themselves as the perineum (*m*) between the urachus, which has now become the bladder (Fig. 392, 5, *b*), and the rectum (*M*).

In the male (*II*) the genital eminence now becomes large, and its groove closes from the orifice of the bladder (the urachal orifice of the former cloaca) to the apex of the eminence in the tenth week.   The entrance to the bladder is thus displaced to the apex of the genital eminence.   Should this closure fail to take place, either wholly or in part, there occurs the arrest of development known as hypospadias.   In the fourth month the glans is formed, in the sixth month the prepuce; both are at first adherent.   The cutaneous folds that unite in the raphe form the scrotum.

In the female (*III*) the undifferentiated condition of the original rudimentary sexual organs remains, to a certain degree, permanent; the small genital eminence becomes the clitoris, the genital folds the nymphæ; the cutaneous folds remain separate as the labia majora.   The urogenital sinus remains short, as it was, and it becomes the vestibule of the vagina; while in the male, through closure of the genital groove, a long additional canal is formed.

Hermaprodism.—In rare cases the external genitalia persist in their original undifferentiated rudimentary stage (somewhat as is shown in Fig. 393, *II*), constituting an arrest of development.   Under such circumstances an external determination of sex is impossible (pseudohermaphrodism).   In isolated cases there occurs the development on one side of male, and on the other side of female internal organs of generation: the external genitalia are then not typically developed.   Such cases are designated true lateral hermaphrodism.   The condition is not rare in swine, goats and beeves, but it has probably never been established in man beyond all doubt.

The **cause of sexual development** in one or the other direction has not, as yet, been determined with certainty.   From statistical data (80,000 cases) the influence of the age of the parents has been established.   If the husband is younger than the wife, boys and girls will be produced in equal number.   If both parents are of the same age, there will 1029 boys and 1000 girls; if the husband is older, as many as 1057 boys to 1000 girls.   The general application of this law is contested by some.   Nutrition, further, appears to have some influence.   Fetuses with adherent placentas that communicate through the fetal vessels are always of the same sex.   Acardiac twins that receive blood that has already nourished the normal twin are always of the same sex as the well-developed twin.   These facts find explanation in the remarkable observations upon armadillos.   In these mammals, the numerous young of the same brood, all of which develop normally within the same chorion, are always of the same sex.   In insects the nutrition plays an important rôle, the best-nourished brood forming females in preponderant degree.   In man, impaired nutrition of the mother leads to the expectation of male children.   It has, further, been maintained that more male progeny result when greater demands are made upon the father, also when impregnation of the wife occurs late, and finally when the father is very young or very old (when the father has reached middle age, more girls are born).   According to Düsing, in general the impregnation of a young ovum with an old spermatozoid, when the mother is well nourished, more frequently results in female progeny, and conversely the impregnation of an old ovum with a young spermatozoid, especially when the nutrition of the mother is somewhat impaired, more frequently results in male children.   Thury believed that animals (cows) that are covered shortly after heat more frequently bear female offspring.   Fürst believes the opposite is the rule for man.   Fiquet maintained that female calves can be produced if the cow is poorly nourished while the bull is well nourished for weeks before intercourse.   Other investigators have come to the conclusion that the sex is unalterably established already at the time of conception.   Also Pflüger's investigations have shown that all external influences (in frogs) during development are without effect upon the development of the sex, that the latter, therefore, is definitely established before impregnation.   Hermaphrodites are common among tadpoles, later becoming males or females.

DEVELOPMENT OF THE CENTRAL NERVOUS SYSTEM.

Upon each side of the fore-brain vesicle, prosencephalon, which is covered externally by epiblast and internally by ependyma, there grows a large pedunculated hollow vesicle, the rudimentary cerebral hemisphere, telencephalon (the forerunner of the rhinencephalon, pallium and corpus striatum). The relatively narrow opening in the pedicle is the rudimentary foramen of Monro. The small middle portion behind the two hemispheres is the interbrain, diencephalon (the forerunner of the thalamus, together with the metathalamus and the epithalamus). The interior of this contains the third ventricle, which about the second month becomes elongated toward the base in the shape of a funnel as the tuber cinereum, with the infundibulum. The thalami, developing on each side from the floor of the interbrain, reduce the foramen of Monro to a semilunar cleft.

In the second month the corpora albicantia also develop at the base, in the third month the chiasm; the commissures are formed within the third ventricle, in the third month. The hypophysis, belonging to the mid-brain, is a diverticulum of the pharyngeal mucosa through the base of the cranium toward the hollow infundibulum, which grows to meet it, and which subsequently become constricted off. There is thus an effort at union between the cavity of the fore-gut and the medullary canal. It should, further, be mentioned here that in the amphioxus, the goose, some parrots and the lizard, the medullary canal originally communicates with the rudimentary hind-gut by means of a passage-way (myeloenteric canal). The choroid plexus, which grows into the cavity of the hemispheres by way of the foramen of Monro, is a vascular hyperplasia of the ependyma. In the fourth month the conarium develops, and at this time the quadrigeminal bodies are already covered by the hemispheres. Within the cavity of the hemisphere there develops in the second month the striate body, in the fourth month the cornu Ammonis. In the third month there develops the fossa of Sylvius, at the bottom of which the insula is formed as a part of the original trunk of the fore-brain and over which, at the end of fetal life, the operculum projects. From the seventh month the remaining convolutions of the brain are formed. Medullated fibers are already present in the cortex of the newborn in the central convolutions, as well as the paracentral lobule. Finally, in the third month such fibers appear in some regions of the frontal and parieto-temporal lobes.

The mid-brain vesicle, mesencephalon (rudiment of the quadrigeminal bodies and cerebral peduncles) becomes gradually covered by the growth backward of the hemispheres; its cavity becomes reduced to the aqueduct of Sylvius. The surface of the vesicle becomes divided into four parts, the quadrigeminal bodies; a longitudinal groove appearing in the third month, and a transverse groove in the seventh month. The cerebral peduncles are formed on the floor as thickenings.

From the hind-brain, metencephalon (rudiment of the pons and cerebellum), there develop separately the hemispheres of the cerebellum, which, growing backward, unite in the middle line. In the sixth month the hemispheres are more fully developed, and the vermiform process is formed. The cerebellum covers the subjacent unclosed portion of the medullary tube down to the calamus. The opening of the medullary tube at the calamus further the tendency of the third ventricle to communicate with the pharynx, facilitates an understanding of the structure of articulates, in which the mouth traverses the central nervous system, and the latter passes down on the ventral aspect. The pons is formed on the floor of the hind-brain in the third month.

The spindle-shaped after-brain, myelencephalon, grows narrower in its course downward, and becomes the oblongata, the upper portion of which exhibits the open medullary cavity.

From the medullary canal, downward from the after-brain, the spinal cord develops, the gray substance nearest the cavity; later, this becomes surrounded by the newly formed white matter. The ganglion-cells (amphibia) increase by division. At first the spinal cord extends to the coccyx. As in adults the extremity of the spinal cord extends only to the first or second lumbar vertebra, the spinal cord does not keep pace in growth with the spinal column; and in consequence the lower spinal nerves must undergo increase in length. The extent to which disparity in the growth of the spinal column and the spinal cord respectively, so that, for instance, the former grows too rapidly, or the latter too slowly, may produce sensory derangement or paralysis in the lower extremities in children should be kept in mind. The tactile nerves of the fetus are capable of executing reflex movement, for example on pressure upon the palpable fetal parts. The first indications of the muscles appear upon the back in the second month; in the fourth month, they become reddish. The first appreciable fetal movements occur about the middle of pregnancy; these are reflex, as they are observed also in acephalous fetuses. It is noteworthy that, in the early periods of development, the central nervous system has no functional influence upon the vital processes, having no sensory, or motor, or trophic (morphogenetic) function, as has been demonstrated by the extirpation-experiments of Alf. Schapee.

The spinal ganglia develop from a special band, situated on each side of the medullary canal, and forming the direct connection between this and the epidermis. The spinal ganglia are the nuclei of origin of the sensory nerves, whence a communication with the spinal cord is established and the peripheral nerve-trunks grow in a centrifugal direction. The nerves of special sense also

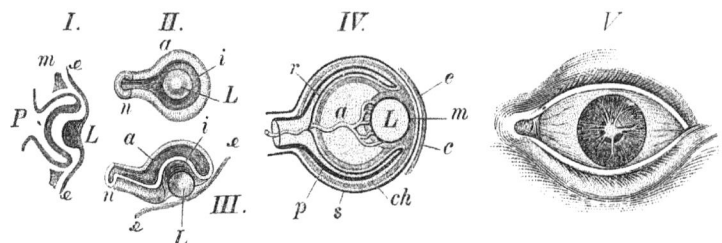

Fig. 394.—Development of the Eye: *I*, Invagination of the lenticular sac (*L*) into the primary optic vesicle (*P*); *e*, epiderm; *m*, mesoblast; *II*, the invaginated primary optic vesicle viewed from below; *n*, optic nerve; *a* the outer, *i* the inner layer of the invaginated vesicle; *L*, lens; *III*, the same formation in longitudinal section; *IV*, further development: *e*, corneal epithelium; *c*, cornea; *m*, capsulopupillary membrane; *L*, lens; *a*, central artery of the retina; *s*, sclera; *ch*, choroid; *p*, pigment-epithelium of the retina; *r*, retina; *V*, persistent vestige of the pupillary membrane. (Diagrammatic.)

grow from the periphery into the central organ. The motor nerve-roots grow from the rudimentary ganglia in the central organ (neuroblasts) into the periphery. At first the nerves are non-medullated. Human embryos four weeks old possess spinal ganglia, anterior roots and in part the trunks of the spinal nerves, whereas the posterior roots are absent. The ganglia of the fifth, seventh, eighth, ninth, and tenth cranial nerves, and in part their origins, are present; on the other hand, His failed to find the first, second, third, and twelfth cranial nerves, as well as the sympathetics. Fetuses with absence of the spinal cord show that the posterior roots present and the sensory nerves originate from the spinal ganglia. In the new-born the cranial motor nerves and the auditory are already provided with medullary sheaths; the others are not. Their envelopment progresses peripherally. In the peripheral spinal nerves the formation of the medullary sheath does not take place before the second and third years.

The sympathetic ganglia of the viscera make their way from the sympathetic cord into the organs.

DEVELOPMENT OF THE ORGANS OF SPECIAL SENSE.

Eye.—The primary optic vesicle grows out to the external covering of the head (epiblast) and then becomes invaginated into itself from before backward (as has been seen to take place in human embryos four weeks old), so that the pedunculated vesicle has acquired the shape of an egg-cup (Fig. 394, *I*). The interior of this cup, the subsequent cavity of the eye, is now called the secondary optic vesicle. The portion of the original vesicle that has undergone invagination, namely the anterior convex portion, now made concave, becomes the retina (*IV*, *r*); the posterior portion of the vesicle becomes the pigmented choroidal (retinal) epithelium (*IV*, *p*). The pedicle is the subsequent optic nerve. The invagination of the primary optic vesicle takes place, however, not exactly according to this simple plan; but there is formed below on the egg-cup-shaped structure a cleft, which permits certain portions of the mesoblast to enter the ocular cavity. This cleft, which extends from the pedicle of the optic vesicle to the border of the invaginated cup (*II*), is known as the coloboma. It is delimited anteriorly as an unpigmented cleft. At the pedicle of the optic vesicle it continues as a furrow to the base of the cerebral vesicle; and in this furrow lies the central artery of the retina. The margins of the coloboma subsequently unite completely; if, however, in rare cases, this union fails to take place, a strip will be wanting in the retina and in the choroidal pigment. There then results a congenital malformation, or arrest of development, or coloboma of the choroid and retina. In birds, the embryonal coloboma-cleft does not close at all, but through it a vascular process of the mesoderm penetrates into the interior of the eye; this is the subsequent pecten. A similar condition occurs in fish, in which the especially large invaginated process, consisting of portions of mesoblast and epiblast, persists as the falciform process.

Why does the primary, pedunculated optic vesicle become invaginated into itself in the form of an egg-cup? Because a sac, derived from the ectoderm, in the fourth week still pedunculated, becomes lodged in the primary optic vesicle (*I, L*). From this the crystalline lens is formed, whose epithelial origin (from epiblast) is indicated even in later life by its peculiarities of growth. The capsule of the lens is a cuticular formation of the ectodermal cells. The portion of the ectoderm that covers the optic vesicle in front of the lens subsequently becomes the laminated anterior corneal epithelium. The cornea exists as early as the sixth week. The pigmentary layer of the invaginated optic vesicle passes from the margin of the egg-cup over the ciliary body and over the posterior surface of the subsequently formed iris. It is clear that a persistent coloboma must thus give rise to the formation of an unpigmented strip in the iris, or even a cleft, the coloboma of the iris. The substance of the choroid, the sclera, and the cornea is formed from the mesoblast surrounding the rudimentary eye (*m*). The capsule of the lens is at first wholly surrounded by a vascular membrane, the capsulo-pupillary membrane. Subsequently, the lens moves further backward into the ocular cavity, but the anterior portion of the capsulopupillary membrane remains in the anterior portion of the eye, and toward it the margin of the iris grows (seventh week), so that the pupil is closed by this portion of the vascular capsule (pupillary membrane). The vessels of the iris are continuous with those of the pupillary membrane; those of the posterior capsule of the lens are given off by the hyloid artery, a continuation of the central artery of the retina; the veins empty into those of the iris and the choroid. The vitreous body is first represented as early as the fourth week by a large collection of cells between the lens and the retina. In the seventh month the pupillary membrane disappears. It may, however, persist throughout life as an arrest of development (*V*).

The Organ of Smell.—On the inferior, lateral border of the fore-brain, the epiblast forms a small pit lined with thickened epithelium, which becomes depressed toward the brain, but always remains a pit—the olfactory depression. The olfactory nerves arise in the epithelium of the pit, and growing centripetally unite with the olfactory lobe. The nasal cavity appears at first as a blind sac; the choanæ develop only as secondary formations.

The Organ of Hearing.—On either side of the after-brain an invaginated pit develops from the epiblast, and becomes depressed toward the brain from without—the labyrinthine depression. This subsequently becomes entirely closed off from the ectoderm (as in the case of the lens), and is known as the vesicle of the labyrinth. It obviously represents the vestibular vesicle, from which, in the second month, the semicircular canals and the cochlea are formed by budding. In the same way, the union of the brain with the labyrinth takes place subsequently through the intermediation of the auditory nerve. The first visceral cleft becomes an irregularly shaped, relatively small passage; in the sixth week the auditory bones are present. Externally, the auricle develops in the seventh week; at the bottom of the auditory canal the tympanic membrane is formed; the innermost portion becomes the Eustachian tube.

The Organ of Taste.—The gustatory papillæ develop in the last period of uterine life; the taste-buds appear only a few days before birth.

PARTURITION.

With the growth of the ovum the uterus becomes more distended and its walls richer in muscle-fibers and in vessels. In the last period, the neck of the uterus also becomes obliterated, and after ten periods of ovulation, therefore about the two hundred and eightieth day of pregnancy, labor-pains set in for the expulsion of the contents. The pains are separated by intervals of freedom; each pain, further, begins gradually, then reaches its height, and diminishes slowly. With each pain the temperature of the uterus increases. The activity of the fetal heart is, further, somewhat slowed and enfeebled with each pain, as a result of irritation of the vagus in the oblongata of the fetus.

The uterine contraction passes in a peristaltic manner from the tubes to the os in from twenty to thirty seconds. The curve traced by this movement has usually a much more steep ascending than descending limb; rarely the reverse; occasionally, both limbs are alike. The curve of contraction increases slowly, persists on the average about eight seconds at its height, and then falls in from five to twenty-five seconds. The frequency of the pains increases to the conclusion of labor. The pains are shortest in the first half of the period of dilatation, while the elevation of the curve is lowest, and the intervals are long; in the second half the pains become longer and stronger with the dilatation of the os; and combined pains appear (like superposed contractions). In the first half of the period of expulsion the curves are higher, in the second half more frequent and higher, but of shorter duration and with shorter intervals.

The pressure within the uterine cavity during a maximal contraction increases from $1\frac{1}{2}$ to 6 fold in the course of labor in consequence of the progressive expulsion. The increase in pressure depends upon the increased thickness of the uterine walls, somewhat also upon their increased curvature. Both factors would of themselves tend to increase the degree of pressure, were it not that the strength of the muscular fibers is considerably reduced by the shortening that occurs in the process of evacuation of the uterus.

Polaillon estimates the pressure that the uterus exerts upon the ovum with each pain at 154 kilos; and that the uterus with each pain performs work equal to 8820 kilogram-meters. The intra-uterine pressure is greatest up to the rupture of the membranes, after which it diminishes, to regain its maximum toward the end of labor (on making bearing-down efforts it may reach 400 mm. of mercury).

After expulsion of the fetus the placenta remains behind for a time, and about it, with further pains, the uterus contracts tightly. In consequence a not inconsiderable amount of placental blood flows to the child. Therefore, it may be advisable not to tie the umbilical cord immediately after the birth of the child. After some time placenta, fetal membranes, and decidua are expelled as the after-birth.

With respect to the dependence of the movements of the uterus upon the nervous system, the following is known: (1) Irritation of the hypogastric plexus causes contraction of the uterus. The fibers arise from the spinal cord (the last dorsal and the 3d and 4th lumbar vertebræ), enter the abdominal sympathetic, and pass from here into the plexus named. (2) Also irritation of the nervi erigentes, arising from the sacral plexus, has a motor effect. (3) Irritation of the lumbar and sacral portions of the spinal cord causes strong movements. A center for the act of parturition is situated in the spinal cord. (4) The uterus probably possesses, like the intestine, parenchymatous centers of its own, which can be stimulated to movement by suspension of respiration and anemia (through compression of the aorta or rapid hemorrhage). Reduction in the bodily temperature diminishes, while increase augments the contractions, which cease in the presence of high fever. The experiments made by Rein on pregnant bitches, in which he divided all of the nerves passing to the uterus, have yielded the remarkable result that, in the uterus freed from all connection with the cerebro-spinal centers, all of the principal phenomena are possible that are connected with impregnation, pregnancy, and parturition. The uterus must, therefore, possess its own automatic ganglia, under whose control the processes named take place. According to Dembo, a center is situated in the upper portion of the anterior vaginal wall (rabbits). According to Jastreboff the vagina of the rabbit undergoes independent rhythmical contractions. Sclerotic acid excites the movements energetically, as does likewise anemia. (5) v. Basch and Hoffmann observed reflex contractions after irritation of the sciatic; Schlesinger after central irritation of the brachial plexus; Scanzoni after irritation of the nipples in man. (6) The uterus contains for its vessels both vasoconstrictors (by way of the hypogastric plexus), derived from the splanchnic, and vasodilators (by way of the nervi erigentes). The vasomotor nerves may be excited reflexly; also through irritation of the sciatic.   The internal os is especially rich in nerves.

After birth the entire uterus is deprived of its mucosa (decidua); its inner surface, therefore, is like a wound-surface, upon which a new membrane is formed, with a secretion at first resembling an infusion of meat, later containing a larger number of cells, and finally becoming mucoid (lochia).   The thick muscular layer of the uterus undergoes gradual reduction through partial fatty degeneration of its fibers.   Within the lumen of the large vessels an obliterating connective-tissue hyperplasia begins from the intima, and in the course of several months diminishes the lumen of the vessels or occludes them.   The unstriated muscle-fibers of the media undergo fatty degeneration.   The relatively large blood-spaces at the placental site are plugged by thrombi, and the latter are invaded by connective-tissue from the walls of the vessel.

After birth secretion of milk sets in, with a peculiar effect upon the vas-cular nervous system (milk-fever?), an increased amount of blood being sent to the mammary glands on the second or third day.   The institution of the first respiratory movements in the new-born is discussed on p. 755.

COMPARATIVE.   HISTORICAL.

Embryology must not omit to take into consideration the general develop-ment of the entire animal kingdom.   The question "How have the innumerable animal forms at present living originated?" has in part been answered by the statement that all species have been created as such from the beginning, "every form is an embodied idea of creation"; all species, further, remain as such without alteration; the "constancy of species prevails."   In opposition to this view, held by Linnæus, Cuvier, Agassiz, and others, Jean Lamarck in 1809 developed the doctrine of "the unity of the animal kingdom," embodying the old idea of Empedocles, namely that all species have developed by variation from a few funda-mental species, that originally only a few fundamental species of lower formation existed, from which the new, numerous species have evolved—a view supported also by Geoffroy St. Hilaire and Goethe.   After a long interval this thought was de-veloped in a particularly fruitful way by Charles Darwin (1859).   He supported his "monistic conception" of the animal kingdom by a description of the manner in which gradual evolution of species can be explained.   Among the creatures of the earth there takes place, for the preservation of life, a struggle of all against all, and from this "struggle for existence" only those will go forth victorious that are char-acterized by particularly striking qualities.   Such qualities: strength, speed, size, color, fruitfulness, are, however, hereditary, and thus it is evident that, in this manner, to a certain degree through natural selection, an uninterrupted process of improvement and thereby a gradual variation in species takes place.   In addition, the creatures are capable, within certain limits, of adapting themselves to their surroundings and the prevailing necessities of external influences.   In this way, certain organs may undergo a useful transformation, while inactive parts can gradually undergo involution to rudimentary organs.   The gradual alteration of animal forms thus resulting through "natural selection" finds its prototype in "artificial selection" among animals and plants.   It is known, for instance, that breeders of animals are able, in a relatively short time, to produce variations in form that are much more considerable than those between two well-characterized species of animals.   Thus, the skull of a mastiff and that of an Italian grayhound exhibit a much greater difference than that of a fox as com-pared with that of a similar species of dog.   As in the case of artificial selection, however, there is observed a sudden reversion to an ancestral type, so also in the development of natural species atavism may occur.   Obviously the ease of varia-tion is increased by the widespread distribution of a given species in different climates, as in this way different influences become operative.   Thus, the migra-tion of organisms may gradually contribute to variations in species (M. Wagner's law of migration).   Inheritance of mutilations does not occur.

Without entering upon the details in the development of the different varieties of animals, the *biogenetic fundamental law* may be briefly discussed.   According to this, "the history of the individual (ontogeny) is a brief repetition of the history of the family (phylogeny)."   Applied especially to man, this law implies that the separate stages in the development of the human embryo, for example its exist-ence as a unicellular ovum, as a collection of cells after cleavage has been completed, as a cellular vesicle (germinal vesicle), as a two-layered vesicle, as a being without a cœlom, etc., indicate an equal number of animal varieties, from which the human race, in the course of inconceivable time, has gradually evolved. The separate steps through which the human race has passed in the process of transformation have been briefly repeated in its embryonal development. This exposition has, naturally, not escaped criticism. In any event, the comparison of human development with relation to the individual organs with the corresponding fully developed organs of the vertebrates is important. Thus also mammals possess in the development of their organs, originally, the simple heart, the visceral clefts, the undeveloped rudimentary brain, the cartilaginous chorda dorsalis, various arrangements of the vascular system, etc., that are peculiar to the lowest forms of vertebrates throughout life. In the higher classes, these incomplete rudimentary structures gradually approach perfection. The morphological differences between man and the gorilla or the chimpanzee are slighter than those between the anthropoids mentioned and other apes. The fossil Pithecanthropus erectus was at first regarded as an extinct link between anthropoids and man, but recently more correctly as a powerful long-armed ape (hylobates, gibbon); the Palæopithecus sivalensis may occupy an analogous position, with its cranial cavity two-thirds the size of the human cranial cavity and occupying an intermediate position between the cranial cavity of the anthropoids and the lower races of man. In detail, however, there are still many difficulties in the way of establishing the Darwinian theory and the fundamental biogenetic law.

Historical.—Although the discoveries in embryology, more than those in any other branch of biological science, belong especially to modern times, it is, nevertheless, interesting to consider the views of the ancients upon different points. Pythagoras (550 B. C.) rejected the theory of spontaneous generation: All life results from seed. According to Alkmaeon (580 B. C.) both sexes furnish the fecundating material; the sex of the offspring corresponds to the sex supplying the most seed. In development the head is formed first. Anaxagoras (500 B. C.) believed that boys came from the right and girls from the left sexual gland. Empedocles (473 B. C.) recognized the nutrition of the embryo through the umbilicus; he was the first to designate the chorion and the amnion, and the segmentation of an embryo as complete on the thirty-sixth day. He taught that the first animals of creation were the most incomplete. Hippocrates considered the seventieth day the earliest time for movement and the two hundred and tenth day as term. He taught, with Democritus, that the sexual material came together from all parts of the body (Darwin's pangenesis), thus accounting for the resemblance of the offspring. He observed incubating eggs from day to day, and saw in them the allantois emerge from the umbilicus, and the chick escape on the twentieth day. He taught that seven-month children are viable, explained the possibility of superfetation from the horns of the uterus and described the lithopedion. According to Plato (430 B. C.) the spinal cord is formed first, as the appendix of which, the brain appears anteriorly. The writings of Aristotle (born 384 B. C.) are rich in observations of which many have already been cited in the text. He taught that the embryo received its blood-supply through the vessels of the umbilical cord, and that the placenta absorbs blood from the vascular uterus, as a tree absorbs moisture through its roots. He differentiated the polycotyledonary and the diffuse placenta; he attributed the former to animals that do not have complete rows of teeth in both jaws. In the incubated bird's egg he recognized the vessels of the yolk-sac, which convey nourishment from the latter to the embryo, and the vessels of the allantois. The statement is correct also that the chick rests with its head on the right leg, and that the yolk-sac finally enters the body. In the birth of mammals when the head alone is born it does not breathe. The formation of double monsters is ascribed to the junction of two germs or two embryos lying in close proximity. In the process of conception, the female supplies the material, the male the principle that is responsible for form and movement. With regard to reproduction in the lower animals, reference may be made to the generative arm of the cephalopods, the yolk-sac of the cuttle-fish, the yolk-sac placenta of the smooth shark, the conjugation of snakes and the absence of the amnion and the allantois in fish and amphibia. Diocles (a contemporary of Theophrastus, born 371 B. C.) appears to have seen the ovum as early as the second week as a cutaneous vesicle, marked by bloody points (villi?); he describes also the cotyledons of the uterus. Erasistratus (304 B. C.) taught the development of the embryo by a neoplastic process in the ovum (epigenesis); he considered scar-formation in the uterus as a cause for sterility. His contemporary Herophilus found that the pregnant uterus is closed.

He noted the glandular character of the prostate and named the seminal vesicles and the epididymis. Aretæus (81 A. D.) recognized the decidua; Galen (131-203 A. D.) the oval foramen and the passage of the blood in the fetus through it and through the ductus arteriosus. He was familiar with the physiological relations between the vessels of the breasts and the uterus, and he knew that the uterus contracted upon pressure. The Talmud contains the statement that an animal with an extirpated uterus can live; that the pubic bones separate during labor, and an account of a successful Cesarean section, with a living mother and child, performed at the request of Cleopatra. Sylvius (1555) described the valve of the oval foramen, Vesalius (1546) the follicles of the ovary, Eustachius (died 1570) the ductus arteriosus (Botalli) and the branches of the umbilical vein to the liver. Arantius examined the duct named after him, and stated that the umbilical arteries do not anastomose with the maternal vessels in the placenta. Libavius (1597) makes the statement that a child had cried aloud in the uterus. Riolan (1618) recognized the corpus Highmori. Pavius (1657) examined the position of the testicles in the inguinal region of the fetus. Harvey (1633) laid down the fundamental principle: Omne vivum ex ovo. Fabricius ab Aquapendente (1600) described the embryological development of birds. Regner de Graaf (1668) described the ovarian follicle named after him; he found the ovum of mammals in the oviduct. He produced erection in the cadaver through tense injection of the cavernous body. Mayon (1679) observed in the placenta the respiratory activity of the lung. Schwammerdam (died 1685) discovered metamorphosis; he developed the butterfly from the caterpillar before the Grand Duke of Tuscany. He described the cleavage of the frog's egg. Malpighi (died 1694) described the embryology of the chick, with illustrations. The first half of the eighteenth century was given up to a discussion as to whether the ovum or the semen was the more important in development (ovists and animalculists); further, whether the progeny was newly formed in the ovum (epigenesis), or whether it merely evolved and grew, thus lodged in the ovum as a being already formed (evolution). The ancients attributed the fructifying power to the odor of semen (aura seminalis). The question of spontaneous generation has been studied exhaustively particularly since the time of Needham (1745), and it has, until recent times, been made the subject of numerous investigations, until it was finally overthrown chiefly through the efforts of Pasteur and of Robt. Koch and his pupils.

A new epoch began with Caspar Fried. Wolff (1759), who first taught the formation of the embryo from germinal layers, and who, besides, first described the tissues as composed of minute "globules" (cells)—an idea that was first thoroughly investigated by Schleiden (1838) with respect to plants, and by Schwann (1839) with respect to animals. Wolff published, as a model of investigation in special embryology, a monograph upon the development of the gut. Will. Hunter described (1775) the fetal membranes and the pregnant uterus, Sömmering (1799) the development of the external bodily form of man, Oken and Kieser that of the intestine. The intermaxillary bone in man was viewed by Goethe (1786) in its correct significance; he also suggested the correct morphological conception of the development of cleft palate. Even prior to 1791 Goethe recognized the construction of the cranium from vertebræ. Tiedemann (1816) described the development of the brain, Meckel that of monstrosities. The work of Pander (1817), Carl Ernst v. Baer (1828-1834), Rathke, Th. Bischoff, Robert Remak and many other living investigators, laid the foundation for studies of the development of individual organs from the three germinal layers. Theodore Schwann first (1839) traced the development of all of the tissues from the primordial germinal cells to the stage of complete evolution.

INDEX.

65

COLUMBIA UNIVERSITY LIBRARIES

This book is due on the date indicated below, or at the expiration of a definite period after the date of borrowing, as provided by the rules of the Library or by special arrangement with the Librarian in charge.

DATE BORROWED	DATE DUE	DATE BORROWED	DATE DUE
	OCT 19		

C28(1141)M100

Lightning Source UK Ltd.
Milton Keynes UK
UKHW011330020119
334852UK00011B/1006/P

9 781527 722354